DIGITAL MICROWAVE COMMUNICATION

IEEE Press
445 Hoes Lane
Piscataway, NJ 08854

DIGITAL MICROWAVE COMMUNICATION

Engineering Point-to-Point Microwave Systems

GEORGE KIZER

For general information on our other products and services or for technical support, please contact our Customer Care Department within the United States at (800) 762-2974, outside the United States at (317) 572-3993 or fax (317) 572-4002.

Wiley also publishes its books in a variety of electronic formats. Some content that appears in print may not be available in electronic formats. For more information about Wiley products, visit our web site at www.wiley.com.

Library of Congress Cataloging-in-Publication Data:

Kizer, George M. (George Maurice), 1945-
Digital microwave communication : engineering point-to-point microwave systems / George Kizer.
pages cm
ISBN 978-0-470-12534-2 (hardback)
1. Microwave communication systems. 2. Digital communication. I. Title.
TK7876.K548 2013
621.382–dc23

2012048284

Printed in the United States of America

ISBN: 9780470125342

10 9 8 7 6 5 4

CONTENTS

PREFACE

As a young engineer, with only one previous significant project as experience, I was tasked with an overwhelming project: expand the existing South Korean intercity microwave network by 140%. I had a copy of Bob White's *Engineering Considerations for Microwave Communications Systems*, a couple of volumes of the *Lenkurt Demodulator*, some *Collins Engineering Letters*, and a couple of Dick Lane's propagation papers. While these were excellent resources, I was totally unprepared for the job ahead of me. As the old cowboy said, "There were a lot of things they didn't tell me when I signed on for my first cattle drive." The Korean project was, as you might imagine, rather exciting for a young, enthusiastic engineer. I was introduced to problems I could never have imagined. With the help of many others, I was successful and learned from the experience. However, my technical preparation could have been better.

It has been several years since that first big project. I have done many others and been involved in numerous technical areas related to microwave transmission. However, I continue to be disappointed in the technical information available for the practicing microwave transmission engineer. If I were a new engineer starting on a project, I don't know where I would go to get in-depth technical knowledge on designing fixed point-to-point microwave communication systems. This book is my attempt to remedy the situation.

When I approach a complicated subject for the first time, I like to grasp the overall concepts before diving into the details. I have always admired Dumas and Sands's little blue book, *Microwave System Planning*. It covers most of the important considerations of microwave path design in less than 140 pages. To provide similar coverage I have organized this book so that the first six chapters address general topics of universal interest. Equations have been kept to a minimum. Figures and tables have been used extensively. The other chapters go into detail on a wide range of topics. The depth of coverage varies. If the topic has been covered adequately in the literature, I attempt to summarize. If the topic has not been covered adequately (e.g., path diversity, dispersive fading, or antenna near field), I go into considerably more detail. Appendix A summarizes the important formulas, and Appendix B covers safety, a critical topic ignored in all other books to date.

This book covers universal design principles. While the agencies performing frequency planning and path design are quite different in North America from those in Europe, the methodologies are similar. I address both North American and European (ITU-R) methods. Several other authors have covered the European (ITU-R) methods; for the first time, this book also covers the North American approach.

To augment the text, Internet resources are also available. Understanding multipath (Chapter 9) is critical to path engineering. After you grasp the concept of a spectrum analyzer (a device that displays received power (on the *Y*-axis) in a narrow bandwidth around a specific frequency (on the *X*-axis), take a look at the following videos on YouTube: *Digital Radio Multipath Experiment (authored by Eddie*

Allen) http://youtu.be/AR8Nee-GmTI and *Digital Radio Dispersive Fading (authored by Ron Hutchinson)* http://youtu.be/ugaz4R3babU. These videos graphically illustrate the received signal distortion caused by multipath propagation.

Wiley has graciously provided a Website for additional data associated with the book: http://booksupport.wiley.com. Enter the ISBN, title, or author's name to access the files. The following folders of information are provided:

Site Index and Book Updates or Corrections. A detailed index of the site folder contents is provided in one document. The other document describes any updates or corrections that may be discovered.

Computer Code. This folder contains actual working code for several of the important algorithms described in the book. The code is Microsoft QuickBasic but can be easily converted to other languages.

Data. This folder contains critical data required to implement many of the algorithms discussed in the book.

Figures. This folder contains detailed color pictures from Chapters 2 and 8. Like the book, they are copyrighted by Wiley.

Public Domain References. This folder contains resources related to the book's topics. Most of the publications are from the US government. A few are from the National Spectrum Manager's Association. While NSMA documents are not public domain, NSMA has granted the right to distribute their documents freely as long as they are attributed to NSMA.

Rain fading is a complex, difficult subject. Defining high frequency microwave path performance in a rain environment is subject to considerable variability between short-term estimates and actual performance in all cases. Spatial and temporal variations of an order of magnitude or more are common. Rain-related documentation (and climatic data in general) is just too extensive to be easily described or provided. To gain an appreciation for the problem, a good start would be to go to the NOAA Website http://www.nws.noaa.gov/oh/hdsc/currentpf.htm#PF_documents_by_state and download the basic documents found there. For more detailed study, you may need to contact NOAA directly for archival support. Be prepared to be surprised by the challenge of this topic.

My goal is to provide you with the technical background to understand and perform the significant tasks in microwave path design. While no book can make you an expert, I believe this book can significantly enhance your knowledge. As you probably know, success is a combination of ability, preparation, and opportunity. I can't help you with the first and last requirements, but I am confident this book can help you with the preparation.

ACKNOWLEDGMENTS

First, I would like to thank Mike and Cathy Newman. Mike suggested this project and was a great supporter and facilitator. Cathy connected me with Wiley. I would also like to thank all my reviewers: Michael Newman (Editorial Coordinator and general whip wielder), Prof. Donald Dudley, Thomas Eckels, Ted Hicks, William Ruck, and last (alphabetically but not technically), Dr. William Rummler. They have given me many great suggestions and corrections. I am in their debt. I especially want to thank the late Dr. Dudley who convinced Wiley this book needed to be published. Also, Dr. Rummler's many technical and ITU-R-related comments and corrections are very much appreciated.

I would also like to thank my editor, Mary Hatcher, production editor, Stephanie Loh, and project manager, Jayashree Saishankar. Moving a concept from text to book is a daunting task; this book was especially demanding. Without their tireless efforts and creative ideas, the project could not have been completed successfully.

I don't want to forget all my associates at Collins Radio, Rockwell International, and Alcatel-Lucent who have contributed do my day to day experiences in microwave radio. I appreciate the friends I have made in many industry associations and government offices I have frequented over the years. I fondly remember the many trips Bob Miller and I made to Washington, D.C. in support of industry regulatory matters. The many customers I have worked with have helped me improve as an engineer; I have enjoyed our mutual experiences. I have many friends throughout the industry but I would like to single out four: Dick Lane has been a longtime associate. I appreciate his knowledge and advice. Eddie Allen is always helpful with path design advice. He is a world-class microwave propagation expert. Of course, it is hard to say too much good about Bill Rummler. He and I have worked together in FCC, FWCC, ITU-R, and TIA matters, and his political and technical capabilities cannot be overstated. Mike Newman has been a longtime associate. He and I started working together 20 years ago when the industry created the FCC Part 101 rules and regulations. This pleasant association has continued ever since.

Although it took me a couple of years to assemble this book, it is based on decades of projects, courses, and presentations. I would like to thank my wife Anne and our children, Amy and Mark, who over the years have put up with the seemingly endless trips and other interruptions that were a constant part of my professional life—and a source of the material for this book.

ABOUT THE AUTHOR

George Kizer has been a microwave engineer for the US Air Force, Collins Radio, Collins Microwave Radio Division of Rockwell International, and Alcatel (now Alcatel-Lucent). He has been a systems engineer, project manager, and product manager for microwave products. From 1991 to 1996, George served as Chairman of the Fixed Point-to-Point Communications Section of TIA in Washington, D.C. During this time, the Section, in coordination with the National Spectrum Management Association, assisted the FCC in the creation of Part 101, the rules that govern licensed microwave communication in the United States. George retired from Alcatel in 2001 and has been a private consultant since then. He lives in Plano, Texas, with his wife Anne and two dogs, Jax and Zoey.

1

A BRIEF HISTORY OF MICROWAVE RADIO FIXED POINT-TO-POINT (RELAY) COMMUNICATION SYSTEMS

1.1 IN THE BEGINNING

Message relaying and digital transmission seem like recent inventions. Not true—these go way back. The first known message relay system was created by the Egyptian king Sesostris I about 2000 BCE. The earliest recorded digital relay transmission by electromagnetic means was around the same time during the Trojan War. King Agamemnon and his troops used signal fires located on mountaintop repeater stations to communicate with each other. The king even used that method to send a message to his wife Clytemnestra. The binary message was either the war was continuing (no fire) or the war was over and he was returning home (fire). The Greek general Polybius, in 300 BCE, developed a more complex message set to allow greater information transfer per transmitted symbol. One to five torches were placed on top of each of two walls. Since each wall had five independent states, this allowed 24 Greek characters plus a space to be transmitted with each symbol. This basic concept of using two orthogonal channels (walls then and in-phase and quadrature channels today), with each channel transmitting independent multiple digital states, is the basis of the most modern digital microwave radio systems of today (Bennett and Davey, 1965).

Digital transmission systems continued to advance using the basic concept developed by Polybius. Systems used in the eighteenth and early nineteenth centuries were direct descendents of this approach. In 1794, the French government installed a two-arm optical system, developed by the Chappe brothers 2 years earlier, which could signal 196 characters per transmitted symbol. This system used several intermediate repeater sites to cover the 150 miles between Paris and Lille. In 1795, the British Admiralty began using a 64-character dual multiple shutter optical system. Versions of this semaphore system are in use in the military today (Bennett and Davey, 1965).

Synchronous digital transmission began in 1816 when Ronalds installed an 8-mile system invented by the Chappe brothers. Each end of the system had synchronized clocks and a synchronized spinning wheel that exposed each of the letters of the alphabet as it spun. At the transmitting end, the operator signaled when he or she saw the letter of interest. At the receiving end, a sound (caused by an electric spark) signaled when to record the exposed letter (Bennett and Davey, 1965).

Sömmering proposed a telegraphic system in 1809. Wire (cable)-based electromagnetic telegraphic systems began in the early 1800s with the discovery of the relationship between electricity and magnetism by Aepinus, Oersted, Ampère, Arago, Faraday, Henry, Ohm, Pouillet, and Sturgeon and chemical

Digital Microwave Communication: Engineering Point-to-Point Microwave Systems, First Edition. George Kizer.

batteries by Volta, Becquerel, Daniell, Bunsen, and Grove (although a chemical battery from 250 BCE was discovered in Baghdad, Iraq, by Konig in 1938). In 1886, Heaviside introduced the concept of impedance as the ratio of voltage divided by current. In 1892, he reported that an electrical circuit had four fundamental properties: resistance, inductance, capacity, and leakage. In 1830, Joseph Henry used an electromagnet to strike a bell over 1 mile of wire. In 1834, Gauss and Weber constructed an electromagnetic telegraph in Gottingen, Germany, connecting the Astronomical Observatory, the Physical Cabinet, and the Magnetic Observatory. In 1838 in England, Edward Davy patented an electrical telegraph system. In 1837, Wheatstone and Cooke patented a telegraph and in 1839 constructed the first commercial electrical telegraph. Samuel Morse, following Henry's approach, teamed with Alfred Vail to improve Morse's original impractical electromagnetic system. The Morse system, unlike earlier visual systems, printed a binary signal (up or down ink traces). Vail devised a sequence of dots and dashes that has become known as *Morse code*. Morse demonstrated this system in 1838 and patented it in 1840. This design was successfully demonstrated over a 40-mile connection between Baltimore and Washington, DC in 1844. About 1850, Vail invented the mechanical sounder replacing the Morse ink recorder with a device allowing an experienced telegraph operator to receive Morse code by ear of up to 30 words per minute. Morse and Vail formed the Western Union to provide telegram service using their telegraphic system (Carl, 1966; IEEE Communications Society, 2002; Kotel'nikov, 1959; O'Neill, 1985; Salazar-Palma et al., 2011; Sobol, 1984; AT&T Bell Laboratories, 1983).

While the early systems were simple optical or sound systems, printing telegraphs followed in 1846 with a low speed asynchronous system by Royal House. In 1846, David Hughes introduced a high speed (30 words per minute) synchronous system between New York and Philadelphia. Gintl, in 1853, and Stearns, in 1871, invented telegraphic systems able to send messages in opposite directions at the same time. In 1867, Edward Calahan of the American Telegraph Company invented the first stock telegraph printing system. In 1900, the Creed Telegraph System was used for converting Morse code to text.

Soon systems were developed to provide multiple channels (multiplexing) over the same transmission medium. The first practical system was Thomas Edison's 1874 quadruplex system that allowed full duplex (simultaneous transmission and reception) operation of two channels (using separate communications paths). In 1874, Baudot invented a time division multiplex (TDM) system allowing up to six simultaneous channels over the same transmission path. In 1936, Varioplex was using 36 full duplex channels over the same wire line. Pulse code modulation (PCM), the method of sampling, quantizing, and coding analog signals for digital transmission, was patented in 1939 by Sir Alec Reeves, an engineer of International Telephone and Telegraph (ITT) laboratories in France. In the 1960s, PCM telephone signals were time division multiplexed (TDMed) to form digital systems capable of transmitting 24 or 30 telephone channels simultaneously. These PCM/TDM (time division multiplex) signals could be further TDMed to form composite digital signals capable of transmitting hundreds or thousands of simultaneous telephone signals using cable, microwave radio, or optical communications systems (Bryant, 1988; Carl, 1966; Fagen, 1975; Welch, 1984).

From 1847, wire-based terrestrial systems were used on oversea cables beginning in 1847. These long systems could not use repeaters and were quite slow (about one to two words per minute). The use of Lord Kelvin's mirror galvanometer significantly increased transmission speed to about eight words a minute. Basic transmission limitations were analyzed using the methods of Fourier and Kelvin. In 1887, Oliver Heaviside, by analyzing the long cable as a series of in-line inductances and parallel resistances, developed a method of compensating the cable to permit transmission rates limited only by loss and noise. Distributed inductors (loading coils) were patented by Pupin in 1899 and further developed by Krarup in 1902. By 1924, distributed inductance allowed the New York to Azores submarine cable to operate at 400 words per minute (Bryant, 1988; Carl, 1966; Fagen, 1975).

About 585 BCE, Thales of Miletus discovered both static electricity (attraction of dry light material to a rubbed amber rod) and magnetism (attraction of iron to a loadstone). In 1819, Hans Orsted demonstrated that a wire carrying electric current could deflect a magnetized compass needle. Wireless transmission, utilizing orthogonal electric and magnetic fields, began in 1840 when Joseph Henry observed high frequency electrical oscillations at a distance from their source. James Maxwell, besides making many contributions to optics and developing the first permanent color photograph, predicted electromagnetic radiation mathematically. He first expressed his theory in an 1861 letter to Faraday. He later presented his theory at the Royal Society of London in December 1864 and published the results in 1873. His theory can be expressed as four differential or integral equations expressing how electric charges produces electric fields (Gauss' law of electric fields), the absence of magnetic monopoles (Gauss' law of magnetism), how changing magnetic fields produce electric fields (Faraday's law of induction), and how currents and changing electric fields produce magnetic fields (Ampere's law). The modern mathematical formulation of Maxwell's equations is a result of the reformulation and simplification by Oliver Heaviside and Willard Gibbs. Heinrich Hertz (an outstanding university student and an associate of Helmholtz) demonstrated the electromagnetic radiation phenomenon in 1887. In 1889, Heinrich Huber, an electric power station employee, questioned Hertz if radio power transmission between two facing parabolic mirrors was possible. Hertz said that radio transmission between parabolic antennas was impractical. In 1892, Tesla delivered a speech before the Institution of Electrical Engineers of London in which he noted, among other things, that intelligence would be transmitted without wires. In 1893, he demonstrated wireless telegraphy (Bryant, 1988; Carl, 1966; Fagen, 1975; Maxwell, 1865; Salazar-Palma et al., 2011; Tarrant, 2001).

In the early 1860s, several people, including Bell, Gray, La Cour, Meucci, Reis, and Varley demonstrated telephones. In 1876, Alexander Bell patented the telephone (Fagen, 1975) in the United States. In 1880, Bell patented speech over a beam of light, calling this the *photophone*. This device was improved by the Boston laboratory of the American Bell Telephone Company and patented in 1897. E. J. P. Mercadier renamed the device the "radiophone," the first use of the term *radio* in the modern sense (Bryant, 1988; Carl, 1966; Fagen, 1975).

It is not often appreciated that before Marconi, several "wireless" approaches were attempted that did not involve radio waves. In the 1840s, Morse developed a method of sending messages across water channels or rivers without wires. He placed a pair of electrodes on opposite sides of a channel of water. As long as the electrodes on the same side of the water were spaced at least three times the distance across the water, practical telegraphic communication was possible. He demonstrated communication over a river a mile wide. In 1894, Rathenau extended Morse's concept to communicate with ships. Using a sensitive earpiece and a 150-Hz carrier current, he was able to communicate with ships 5 km from shore using electrodes 500 m apart. In 1896, Strecker extended the distance to 17 km (Sarkar et al., 2006).

In 1866, Loomis demonstrated the transmission of telegraph signals over a distance of 14 miles between two Blue Ridge Mountains using two kites with 590-ft lines. The two kite lines were conductors. He transmitted a small current through the atmosphere but used the Earth as the return path. This was somewhat like Morse's transmission through water (Sarkar et al., 2006).

In 1886, Edison devised an induction telegraph for communicating with moving trains. He induced the telegraphic signals onto the metal roof of the train by wires parallel to the train tracks. The grounded train wheels completed the circuit. While this system worked, it was not a commercial success (Sarkar et al., 2006).

In the 1880s, Hertz experimented with radio waves in the range 50–430 MHz. In 1894, Sir Oliver Lodge demonstrated a wireless transmitter and receiver to the Royal Society. In the early 1890s, Augusto Righi performed experiments at 1.5, 3, and 15 GHz. Soon several investigators including Marconi, Popov, Lebedew, and Pampa were performing wireless experiments at very high frequencies. In 1895, Bose used 10- to 60-GHz electromagnetic waves to ring a bell. About the same time, he made the first quantitative measurements above 30 GHz. In the 1920s, Czerny, Nichols, Tear, and Glagolewa-Arkadiewa were producing radio signals up to 3.7 THz. Very high frequency research of up to 300 GHz is currently underway. Commercial applications are currently being deployed as high as 90 GHz. High frequency microwaves in the 11–40 GHz range are finding applications in wide area networks and backhaul networks in urban areas. Higher frequency systems are being used for high density industrial campus and building applications (Bryant, 1988; Meinel, 1995; Wiltse, 1984).

In 1825, Munk discovered that a glass tube with metal plugs and containing loose zinc and silver filings tended to decrease electrical resistance when small electrical signals were applied. Using this principle to create a "coherer" detector, in 1890, Edouard Branly demonstrated the detection of radio waves at a distance. The coherer detector was improved by Lodge and others. Braun invented the galena crystal ("cat's whisker") diode in 1874 (Bryant, 1988). Crystal detectors were applied to radio receivers by Bose, Pickard, and others between 1894 and 1906 and were a big improvement over the coherer.

In 1894, Lodge detected "Hertzian" waves using Branly's coherer. In 1897, Tesla sensed electrical signals 30 miles away and received his basic radio patent. In 1898, Tesla demonstrated a radio controlled boat. In 1894, Marconi became interested in Hertzian waves after reading an article by Righi. After visiting the classes and laboratory of Professor Righi, Marconi began radio experiments in 1895. His radio receiver detector was the newly improved coherer. (Later he transitioned to Braun's crystal diode.) He created a wireless communications system that could ring a bell. Perhaps, he and Bose were the first to use radio for remote control. In 1896, he demonstrated a 1.75-mile 1.2-GHz radio telegraph system to the British Post Office. This was probably the first microwave radio link. In 1897, Marconi installed the first permanent wireless station on the Isle of Wight and communicated with ships. The next year he added a second station at Bournemouth. This was the first permanent point to point wireless link. In 1899, the link was used to send the first paid wireless digital transmission, a telegram. In 1899, Marconi sent messages across the English Channel, and in 1901, he sent signals across the Atlantic between St. Johns, Newfoundland and Poldhu, England. In 1897, Lodge patented a means of tuning wireless transmissions. In 1898, Braun introduced coupling circuits to obtain accurate frequency tuning and reduce interference between radio stations. Marconi and Braun were cowinners of the Nobel Physics Prize for their work (one of the few times the Nobel Prize was awarded to engineers rather than scientists) (Bryant, 1988; Tarrant, 2001).

The first audio transmission using radio was by the Canadian Reginald Fessenden in 1900. He also performed the first two-way transatlantic radio transmission in 1906 and the first radio broadcast of entertainment and music in the same year. However, commercial applications awaited the de Forest Audion. Early transmitters were broadband Hertzian types (spark gaps exciting tuned linear radiators). In 1906, the quenched spark transmitter was introduced by Wien. Continuous wave oscillations were introduced by Poulsen in 1906 (using an arc). Alexanderson, Goldschmidt, and von Arco quickly demonstrated continuous waves by other methods (Bryant, 1988).

On the basis of the comments by Crookes, in 1892, Hammond Hayes, head of the Boston laboratories of the American Bell Telephone Company, had John Stone and later G. Pickard investigate the possibility of radiotelephony using Hertzian waves. These investigations did not result in practical devices. Further investigation was delayed until 1914. By 1915, one-way transmissions of 250 and 900 miles had been achieved. Later that year speech was successfully transmitted from Arlington, Virginia, to Mare Island, California, Darien, Panama, Pearl Harbor, Hawaii, and Paris, France. In the 1920s, radio research in the Bell System was divided among the American Telephone and Telegraph (AT&T) Development and Research Departments and the Bell Laboratories, all in New York. The Bell Laboratories moved to New Jersey to be less troubled by radio noise and became the primary radio investigation arm of the Bell System. In the 1920s, terrestrial radio propagation was an art, not a science. In 1920, Englund and Friis of the Bell Labs began developing radio field strength measuring equipment. This was followed by field measurements to, as Friis stated, "demystify radio." By the early 1930s, an interest began to develop for long-distance relaying of telephone service by radio. It was clear that wide-frequency bandwidth was needed. The only spectrum available was above 30 MHz, the ultrashortwave frequencies (later termed *very high frequencies*, *ultrahigh frequencies*, and *superhigh frequencies*). Radio theory was developed and propagation experiments were carried out to validate it. By 1933, surface reflection, diffraction, refraction, and K factor (equivalent earth radius) were understood (Bullington, 1950; Burrows et al., 1935; England et al., 1933, 1938; Schelleng et al., 1933). By 1948, the theory and technology (Friis, 1948) had advanced to the point that fixed point to point microwave radio relay systems were practical (Bryant, 1988; Fagen, 1975; Friis, 1948).

At least three major technologies have been used for microwave antennas. In 1875, Soret introduced the optical Fresnel zone plate antenna. It was adapted to microwave frequencies in 1936 by Clavier and Darbord of Bell Labs (Wiltse, 1958). Dielectric and metal plate lenses were also tested (Silver, 1949, Chapter 11). However, by far, the most practical antenna was the reflector antenna. The first use of optical parabolic reflectors was by Archimedes during the siege of Syracuse (212–215 BCE). This reflector was used by Gregory (1663), Cassegrain (1672), and Newton (1672) to invent reflector telescopes. Hertz (1888) was the first to use a parabolic reflector at microwave radio frequencies. The World War II saw the widespread use of this type of antenna for radio detection and ranging (radar) systems. They remain the most important type of microwave antenna today (Rahmat-Samii and Densmore, 2009).

Radio transmission and reception antennas need to be above path obstructions. Convenient locations for the transmitter and receiver equipment are usually somewhere else. A transmission line that was free from reflecting or absorbing objects was needed to connect antennas and radio equipment. Coaxial cable was patented in Germany by Ernst Werner von Siemensin in 1884 and in the United States by Nikola Tesla in 1894. Hertz demonstrated the use of coaxial lines in 1887. Transmission by two parallel wire lines was demonstrated by Ernst Lecher in 1890. While this method had significantly less loss than coaxial cable, extraneous radiation made it impractical at microwave frequencies. Until the late 1930s, all radio transmission lines were two-conductor lines: two wire balanced line, one conductor (with implied ground plane mirror conductor), and coaxial cable. Two-conductor lines were popular for transmission at radio frequencies below 30 MHz (Bryant, 1984, 1988; Fagen, 1975; Millman, 1984).

Stripline- and microstrip-printed circuit technologies developed in the 1950s were used extensively in high frequency radio products. V. H. Rumsey, H. W. Jamieson, J. Ruze, and R. Barrett have been credited with the invention of the stripline. Microstrip was developed at the Federal Telecommunications Laboratories of ITT. Coaxial cable, while the most complex, tended to be the choice for most long-distance applications because of its low radiation and cross-talk characteristics. However, its relatively high loss was an issue for transmission of high frequency radio signals for long distances.

Coaxial cable was patented in England in 1880 by Oliver Heaviside and in Germany in 1884 by Siemens and Halske. The first modern coaxial cable was patented by Espenschied and Affel of Bell Telephone Laboratories in 1929. The first general-use coaxial connector was the UHF (ultrahigh frequency) connector created in the early 1940s. It was suitable for applications up to several hundred megahertz. The N connector, a connector for high frequency applications, was developed by Paul Neill at Bell Labs in 1944. This was followed by several derivative connectors such as the BNC (baby-type N connector) and the TNC (twist-type N connector). The N connector was limited in frequency to about 12 GHz, although precision versions were used up to 18 GHz. The SMA (SubMiniature connector version A) connector was adopted by the military in 1968 and became the industry standard for radio signals up to 18 GHz (precision versions are rated to 26 GHz). By extending the SMA design, the connectors 3.5 mm (rated to 34 GHz), 2.9 mm (or K) (rated to 46 GHz), 2.4 mm (rated to 50 GHz), 1.85 mm (rated to 60 GHz), and 1 mm (rated to 110 GHz) have been developed (Barrett, 1984; Bryant, 1984).

In 1887, Boys described the concept of guiding light through glass fibers. In 1897, Lord Rayleigh published solutions for Maxwell's equations, showing that transmission of electromagnetic waves through hollow conducting tubes or dielectric cylinders was feasible. R. H. Weber, in 1902, observed that the wave velocity of a radio signal in a tube was less than that in free space. He suggested that the wave was equivalent to a plane wave traveling in a zigzag path as it is reflected from the tube walls. DeBye, in 1910, developed the theory of optical waveguides. The first experimental evidence of radio frequency (RF) waveguides was demonstrated by George Southworth at the Yale University in 1920. The hollow waveguide was independently researched by W. Barrow of Massachusetts Institute of Technology (MIT) and George Southworth of Bell Laboratories in the mid-1930s. Southworth discovered the primary modes and characteristics of rectangular and circular waveguides. Southworth (1962) wrote a highly readable history of waveguide, waveguide filters, and related developments at Bell Labs. At nearly the same time, characteristics of various shapes of waveguides were also being developed by Brillouin, Schelkunoff, and Chu and also William Hansen began working on high Q microwave frequency resonant cavity circuits. Waveguide flanges of various types were invented to provide a cost-effective yet accurate way to attach waveguide components (Bryant, 1988; Fagen, 1975; Millman, 1984; O'Neill, 1985; Packard, 1984; Southworth, 1950).

During World War II, most of the critical waveguide and coax elements had been developed. Waveguide flanges (and flange adapters) provided cost-effective coupling of waveguide, directional couplers provided signal monitoring and sampling, filters provided frequency selectivity, and isolators and circulators provided two- and three-port directional routing of signals (Fig. 1.1) (Marcuvitz, 1951; Montgomery et al., 1948; Ragan, 1948; Southworth, 1950).

In 1897, Braun invented the Cathode Ray Tube with magnetic deflection. Fleming invented the two-electrode "thermionic valve" vacuum tube rectifier in 1904. de Forest improved on Fleming's rectifier by inventing the three-electrode "Audion" vacuum tube in 1906. In 1912, Colpitts invented the push–pull amplifier using the Audion. Meissner used the three-electrode (triode) tube to generate RF waves in 1913. About the same time, other oscillators were developed by Armstrong, de Forest, Meissner, Franklin, Round, Colpitts, and Hartley. Colpitts developed a modulator circuit in 1914. In 1918, Armstrong invented the superheterodyne receiver that is commonly used today. In 1919, Barkhausen and Kurz used a triode to generate radio frequencies as high as 10 GHz and Transradio, a subsidiary of Telefunken, introduced duplex radio transmission. Oscillators using coaxial line and waveguide (hollow cavities) were introduced in 1932 and 1935, respectively. Armstrong invented frequency modulation (FM) (Armstrong, 1936) in 1935. With the exception of a short period in the early 1970s [when single-sideband amplitude modulation

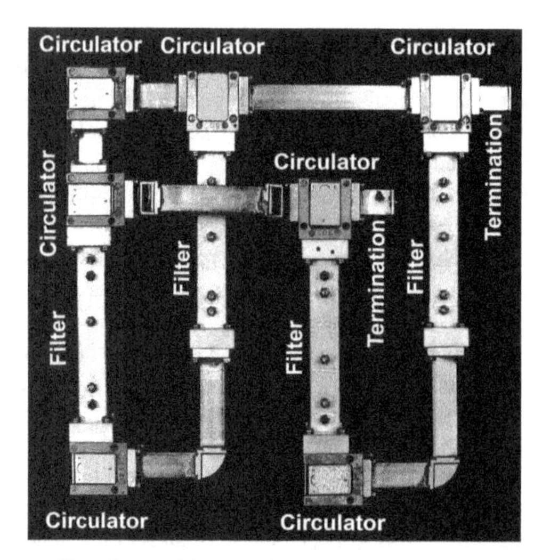

Figure 1.1 Waveguide coupling for multiple radios. *Source*: Reprinted with permission of Alcatel-Lucent USA, Inc.

(AM) was used briefly], FM was the primary modulation used in wideband microwave radios from the 1940s until the beginning of the digital radio era in the mid-1970s (Bryant, 1988; O'Neill, 1985).

Radar was the beginning of widespread applications of microwave radio frequencies. Radar was developed independently in the 1930s by Great Britain, Germany, Canada, Italy, Russia, Japan, the Netherlands, and the United States. In 1940, the United States formed the National Defense Research Committee (NDRC). Shortly thereafter, the NDRC's Microwave Committee began meeting at the private laboratories of Alfred Loomis. Microwave radar and navigation were the primary interests. At this time, the primary microwave research centers were at MIT and the Stanford University and the laboratories of Bell Laboratories, General Electric, Radio Corporation of America (RCA), and Westinghouse. In coordination with Sir Henry Tizard of the British Scientific Mission, the US government selected MIT as the contractor to carry out the radio research needed for the US and British military. This was organized as the Radiation Laboratory (Rad Lab). The 28-volume Radiation Laboratory Series of books detailing the results of the laboratory from 1940 to 1945 are beyond a doubt the most impressive single group of research reports on radio. Volumes 8, 9, 10, 12, and 13 (Kerr, 1951; Marcuvitz, 1951; Montgomery et al., 1948; Ragan, 1948; Silver, 1949) are still useful reading for microwave radio engineers. In roughly the same time period, low noise concepts such as noise figure and noise factor and low noise design concepts were discovered (Bryant, 1988; Fagen, 1975, 1978; Okwit, 1984; Sobol, 1984).

In 1937, Sigurd and Russell Varian demonstrated the first klystron oscillator. It was further developed by General Electric, the Stanford University, Sperry Gyroscope, and Varian Associates. Eventually, 3-GHz klystrons were manufactured by Bell Telephone Laboratories (Bell Labs), MIT Radiation Laboratories, Federal Telephone and Radio, General Electric, Westinghouse, Varian Associates, CSF in France, and the Alfred Loomis Laboratory in England. Klystrons have had a long use for low and medium power applications. However, their relatively low power conversion efficiency (30% conversion of DC power input to microwave power output) limited high power applications (Bryant, 1988; Fagen, 1975, 1978; Sobol, 1984).

A 200-MHz two-pole magnetron was first demonstrated by Albert Hull at General Electric in the 1920s. By 1930, both the Americans and the Japanese were using magnetrons to generate microwave signals. In 1935, a 3-GHz multiple cavity magnetron was developed by Hans Hollmann. In 1940, John Randall and Harry Boot produced a high power water-cooled magnetron and a 6-kW version was produced for the US government by GECRL of Wembley, England. During World War II, Percy Spencer, a Raytheon engineer, significantly improved magnetron efficiency and manufacturability. After the war, he invented the first microwave oven (Bryant, 1988; Fagen, 1975, 1978).

In 1942, Rudolf Kompfner invented the traveling wave tube, a medium power microwave amplifier. In 1947, Brattain, Bardeen, and Shockley at Bell Laboratories invented the point-contact transistor. In 1948, they invented the junction transistor. The first n–p–n transistor was demonstrated in 1950. Townes published the principle of the MASER (microwave amplification by stimulated emission of radiation) in 1951. In 1957, Esaki developed the germanium tunnel diode and Soulde described the LASER (light amplification by stimulated emission of radiation). In 1959, Jack Kilby of Texas Instruments and Robert Noyce of Fairchild independently developed the integrated circuit. The two shared the 2000 Nobel Prize in Physics for this achievement. In 1960, Khang and Atalla developed silicon–silicon dioxide field-induced surface devices [which led to metal–oxide–semiconductor field-effect transistors (MOS FETs)]. In 1961, Biard and Pittman of Texas Instruments invented gallium arsenide (GaAs) diodes. Today most microwave low and medium power applications use solid-state devices such as galium arsenide field-effect transistors (GaAs FETs). In 1962, Holonyak invented the first practical visible light-emitting diode (LED). In 1975, Ray Pengelly and James Turner invented the Monolithic Microwave Integrated Circuit (MMIC), although the concept had been mentioned in the early 1960s by Kilby. These devices were later further developed through support from the Defense Advanced Research Projects Agency (DARPA) (Bryant, 1988; Fagen, 1975, 1978; Millman, 1983).

In 1963, the Institute of Electrical and Electronic Engineers was formed by the merger of the Institute of Radio Engineers (IRE) and the American Institute of Electrical Engineers (AIEE) (Tarrant, 2001).

In 1909, Sommerfeld (1909) published his theoretical integral equation solution to free space radio wave propagation. This was the beginning of theoretical analysis of radio waves (Oliner, 1984). In the 1920s, Nyquist (1924, 1928) and Hartley (1928) published the first significant papers addressing information theory. In the 1920s, 1930s, and 1940s Kotel'nikov (1959), Nyquist, and Shannon developed the theoretical concepts of sampled signals and the relationship between the time and frequency domains. In 1939, Philip Smith, at Bell Telephone's Radio Research Lab in New Jersey, developed what is known as the *Smith chart* (Smits, 1985), a circular chart that shows the entire universe of complex impedances in one convenient circle. From the mid-1930s to the mid-1940s, considerable research was applied to (Norton, 1962). In 1944, 1945, and 1948, Rice (July 1944, 1948) published his mathematical analysis of random noise with and without a sine wave. In 1943, North (1963) defined what became called *Matched Filters* and Friis (1944) discovered the concept of noise figure. This work has been used extensively in the analysis of microwave fading statistics. In the late 1940s, Shannon (1948, 1949, 1950) and Tuller (1949) published significant papers on information theory of communications. In 1949, Weiner (1949) published his theory of linear filtering of signals in the presence of noise. About this same time he reported what has become known as the *Weiner–Hopf equation* that defines the relationship between signals in the time and frequency domains. In the 1950s, Cooley and Tukey developed the fast fourier transform (FFT) algorithm. Blackman and Tukey (1958) introduced the concept of signal power spectrum.

The Hamming (1950) and Reed–Muller (Muller, 1954; Reed, 1954) convolutional (Elias, 1955) and cyclic (Prange, 1957) codes were invented in the 1950s. Friis (1946) and Norton (1953) developed the modern radio wave transmission loss formula. In the 1950s, Bullington (1947, 1950, 1957, 1977) was developing the fundamental characteristics of practical microwave propagation. About the same time, Norton et al. (1955) were expanding Rice's work on the combination of constant and Rayleigh-distributed signals. In the 1950s and 1960s, Medhurst, Middleton, and Rice were developing the theory of analog FM microwave transmission. In 1960, Kalman (1960) published his theory of linear filtering of signal in the presence of noise and the Bose, Hocquenghem, Chaudhuri (BCH) (Bose and Ray-Chaudhuri, 1960; Hocquenghem, 1959), and Reed and Solomon (1960) coding were invented. In 1965, Wozencraft and Jacobs introduced the concept of geometric representation of signals. This is the basis of "constellations" now popular in modulation theory. In 1967, Viterbi (1967) published the algorithm currently used for most digital radio demodulators. Ungerboeck (1982) invented trellis coded modulation in 1982. In 1993, Berrou et al. (1993a) invented Turbo Coding. In 1996, Gallagher's (Gallagher, 1962) low density parity codes (LDPCs) were rediscovered by MacKay (Kizer, 1990; MacKay and Neal, 1996).

1.2 MICROWAVE TELECOMMUNICATIONS COMPANIES

Ericsson was started in 1876 as a telephone repair workshop in downtown Stockholm. It eventually became the primary supplier of telephones and switchboards to Sweden's first telecommunications operating company, Allmänna Telefonaktiebolag.

In 1897, Guglielmo Marconi formed the Wireless Telegraph and Signal Company (also known as the *Marconi Company Limited*, as well as the Wireless Telegraph Trading Signal Company). Marconi and his company created the first commercial radio transmission equipment and services. English Electric acquired the Marconi Company in 1946. The company was sold to the General Electric Corporation in 1987 and renamed the Marconi Electronic Systems. In 1999, most of the Marconi Electronic Systems assets was sold to British Aerospace (BAE) and it became part of BAE Systems. However, General Electric retained the Marconi name, Marconi Corporation, which it sold to Ericsson in 2006 (Bryant, 1988; Sobol, 1984).

Alcatel-Lucent was created in 2006 when Alcatel acquired Lucent Technologies. Alcatel was started in 1898 as Compagnie Générale d'Electricité. In 1991, it became Alcatel Alsthom. In 1998, it shortened its name to Alcatel. ALCATEL stands for "ALsacienne de Constructions Atomiques, de TELecommunications et d'Electronique" (Alsacian Company for Atomic, Telecommunication, and Electronic Construction). Over several years ITT, SEL, Thomson-CSF, Teletra, Network Transmission Systems Division (NTSD) of Rockwell International (including the former Collins Microwave Radio Division), Newbridge Networks, DSC Communications, Spatial Wireless, Xylan, Packet Engines, Assured Access, iMagicTV, TiMetra, and eDial. Lucent Technologies was formed in 1996 by AT&T when it spun off its manufacturing and research organizations (primarily Western Electric and Bell Labs). Lucent acquired Ascend Communications in 1999.

Alcatel-Lucent has three centers of microwave radio development and marketing: Velizy (southwest Paris), France; Vimercate (northeastern Milan), Italy (the former Teletra); and Plano (north Dallas), Texas (the former Collins Microwave Division of Rockwell International). The Alcatel-Lucent North American microwave radio facilities in Plano, Texas traces its roots to the Collins Radio Company, which was founded in 1933 by Arthur A. Collins in Cedar Rapids, Iowa. The Collins Radio Microwave Radio Division was founded in Richardson (north Dallas), Texas, in 1951. The first prototype of Collins commercial microwave equipment was placed in service between Dallas and Irving, Texas, in the spring of 1954. Later that year, the first Collins microwave radio system was sold to the California Interstate Telephone Company. By 1958, Collins was mass-producing microwave equipment and was providing the FAA (Federal Aviation Administration) with microwave systems providing communications and radar signal remoting networks. In 1973, Collins radio merged into Rockwell International. The Texas based Collins Microwave Radio Division ultimately became Rockwell's NTSD. During the 1970s, this division was the sole supplier of microwave radio equipment to the MCI (Microwave Communications, Inc.), with most of its other sales to the Bell operating companies. In 1976, NTSD introduced its first digital microwave radio, the MDR-11, an 11-GHz multiline system delivered to Wisconsin Bell (Madison to Eau Claire). In a parallel evolution, the Alcatel Network Systems' Raleigh, North Carolina, facility was originally operated by the ITT Corporation, which opened its first plant in 1958. From this facility, ITT first established its T1 spanline business in 1971, T3 fiber-optic transmission systems in 1979, and the first commercial single-mode fiber-optic transmission system in 1983. In 1987, ITT and Compagnie Generale d'Electricitie (CGE) of France agreed to a joint venture, creating Alcatel N.V.—the largest manufacturer of communications equipment in the world. Alcatel completed a buyout of ITT's 30% interest in the spring of 1992. The company was incorporated in the Netherlands, operated from Paris, and had its technical center in Belgium. This organization also created the Alcatel Network Systems Company, which was headquartered in Raleigh, North Carolina. In 1991, Alcatel purchased NTSD from Rockwell International and combined it with Alcatel Network Systems Company to form the Alcatel Network Systems, Inc., headquartered in Richardson, Texas. After the acquisition of DSC, the headquarters was moved to the former DCS facilities in Plano, Texas. In addition to microwave radios, this facility develops and markets fiber optics and digital cross-connect systems.

Founded in 1899 in Japan as the first US/Japanese joint venture with Western Electric Company, Nippon Electric (now NEC), headquartered in Tokyo, Japan, established itself as a technological leader early in its history by developing Japan's telephone communications system. Recognizing the impact information processing would eventually have on the world community, NEC was one of the earliest entrants into the computer and semiconductor markets in the early 1950s. NEC also supported much microwave research. NEC has also manufactured microwave radios since the early 1950s and introduced its first microwave radio product to the US market in the early 1970s. Later, NEC delivered its first digital microwave radio to the United States in the mid-1970s. The NEC Corporation of America's Radio Communications Systems Division (RCSD) is headquartered in Irving, Texas (Morita, 1960).

In 1903, the German wireless company Telefunken was formed. It was the first significant commercial wireless telegraph competitor to Marconi's company.

In 1878, Alexander Graham Bell and his financiers, Gardiner Hubbard and Thomas Sanders, created the Bell Telephone Company. The company name was changed to the National Bell Telephone Company in 1879 and to the American Bell Telephone Company in 1880. By 1881, the company bought a controlling interest in the Western Electric Company from Western Union. In 1880, the AT&T Long Lines was formed. This group became a separate company named the *American Telephone and Telegraph Company* in 1885. In 1899, the AT&T Company bought the assets of American Bell and became the Bell System. In 1918, the federal government nationalized the entire telecommunications industry, with national security as the stated intent. In 1925, AT&T created Bell Telephone Laboratories ("Bell Labs"). In 1956, the Hush-A-Phone v. United States ruling allowed a third-party device to be attached to rented telephones owned by AT&T. This was followed by the 1968 Carterfone Decision that allowed third-party equipment to be connected to the AT&T telephone network. On January 8, 1982, the 1974 United States Department of Justice antitrust suit against AT&T was settled. Under the settlement AT&T ("Ma Bell") agreed to divest its local exchange service operating companies in return for a chance to go into the computer business. Effective January 1, 1984, AT&T's local operations were split into seven independent Regional Bell Operating Companies (RBOCs), or "Baby Bells." Western Electric was fully absorbed into AT&T as AT&T Technologies. After its own attempt to penetrate the computer marketplace failed, in 1991, AT&T absorbed the NCR (National Cash Register) Corporation. After deregulation of the US telecommunications industry via the Telecommunications Act of 1996, NCR was divested again. At the same time, the majority of AT&T Technologies and Bell Labs was spun off as Lucent Technologies. In 1994, AT&T purchased the largest cellular carrier, McCaw Cellular. In 1999, AT&T purchased IBM's Global Network business, which became AT&T Global Network Services. In 2001, AT&T spun off AT&T Wireless Services, AT&T Broadband, and Liberty Media. AT&T Broadband was acquired by Comcast in 2002. AT&T Wireless merged with Cingular Wireless in 2004 to become Cingular; in 2007, it became AT&T Mobility. In 2005, SBC Communications acquired AT&T Corp. and became AT&T Inc.

General Telephone & Electronics (GTE), founded in Wisconsin in 1918, was started as the Richland Center Telephone Company. It changed names many times: Commonwealth Telephone Company (1920), Associated Telephone Company (1926), General Telephone Corporation (1935), and finally, GTE Corporation (1959, when it merged with Sylvania Electric Products). In 1964, the Western Utilities Corporation merged with GTE. In 1955, GTE acquired Automatic Electric, the largest independent manufacturer of automatic telephone switches. In 1959, it acquired Lenkurt Electric Company, Inc., a manufacturer of microwave radio and analog multiplex equipment. Lenkurt Electric was established in 1933 as a wire-line telephone multiplex manufacturer. It moved from San Francisco to San Carlos in 1947. Its radio product line was terminated in 1982 and a number of employees migrated to Harris Farinon. At the same time, the company adopted the name GTE Corporation and formed GTE Mobilnet Incorporated to handle the company's entrance into the new cellular telephone business. In 1983, Automatic Electric and Lenkurt were combined as GTE Network Systems. GTE became the third largest long-distance telephone company in 1983 through the acquisition of Southern Pacific Communications Company. At the same time, Southern Pacific Satellite Company was also acquired, and the two firms were renamed GTE Sprint Communications Corporation and GTE Spacenet Corporation, respectively. Through an agreement with the Department of Justice, GTE conceded to keep Sprint Communications separate from its other telephone companies and limit other GTE telephone subsidiaries in certain markets. In 1997, Bell Atlantic merged with NYNEX but retained the Bell Atlantic name. In 2000, Bell Atlantic merged with GTE and adopted the name Verizon. In 2005, Verizon acquired MCI (formerly WorldCom).

The company that eventually became IBM was incorporated in the state of New York on June 16, 1911, as the Computing-Tabulating-Recording (C-T-R) Company. On February 14, 1924, C-T-R's name was formally changed to International Business Machines Corporation. In 1944, IBM and Harvard introduced the Mark 1 Automatic Sequence Controlled Calculator based on electromechanical switches. In 1952, IBM introduced the 701, a computer based on the vacuum tube. The 701 executed 17,000 instructions per second and was used primarily for government and research work. The IBM 7090, one of the first fully transistorized mainframes, could perform 229,000 calculations per second. In 1964, IBM introduced the System/360, the first large "family" of computers to use interchangeable software and peripheral equipment.

Farinon Electric was established in San Carlos, California, in 1958. Seeing a need for a high quality, light-route radio for telecommunications, Bill Farinon left his job with Lenkurt Electric to begin business in a Redwood City cabinet shop. Farinon Electric was started as a limited partnership with $140,000, of which $90,000 was cash (Farinon and his wife came up with $30,000, Farinon's father invested $20,000, and a friend invested $40,000.). With two employees, he started designing and building the first Farinon Electric PT radio offering 36 channels at 450 MHz for the telephone and industrial market. In 1980, Farinon Corporation was sold to Harris Corporation and became known as *Harris Farinon*. In 1998, the division was renamed the Harris Microwave Communications Division. Meanwhile, elsewhere in California, in 1984, Michael Friedenbach, Robert Friess, and William Gibson formed the Digital Microwave Corporation (DMC) to serve the short-haul microwave market. In 1998, DMC acquired MAS Technology and Innova Corporation. In 1999, the company name was changed to Stratex Networks. Plessey Broadband was acquired in 2002. In 2007, Stratex Networks and the Microwave Communications Division of Harris Corporation were merged to create Harris Stratex Networks. This independent company was majority-owned by Harris Corporation. In 2010, Harris spun off the company as Aviat Networks.

In 1899, Cleyson Brown formed the Brown Telephone Company in Abilene, Kansas. The company's name was changed to United Utilities in 1938 and to United Telecommunications ("United Telecom") in 1972. In 1980, United Telecom introduced a nationwide X.25 data service, Uninet.

The Southern Pacific Railroad operated its telephone system as an independent company, called the *Southern Pacific Communications Corporation* (*SPCC*). In the late 1950s, this primarily wire-based communications system began transitioning into a microwave radio relay network. In 1983, the GTE Corporation, parent company of General Telephone, purchased the network and renamed it GTE Sprint Communications. In 1986, Sprint was merged with US Telecom, the long-distance arm of United Telecom, to form the US Sprint. This partnership was jointly owned by GTE and United Telecom. Between 1989 and 1991, United Telecom purchased controlling interest in US Sprint. In 1991, United Telecom changed its name to Sprint. In the mid-1980s, Sprint began building a nationwide fiber-optics network. In 1986, a highly successful "dropping pin" ad was used to describe the superiority (clear sound depicted by the pin's "ting" when it hit a hard surface) of the "all-fiber" digital network. In 1988, Sprint ran an ad showing a microwave radio tower being blown up, thereby emphasizing their claim of an "all-fiber" network (in an effort to differentiate the Sprint network from the largely microwave based AT&T and MCI networks). While this marketing approach was very successful, it did not win friends within the microwave community (who were quick to point out that radio was more reliable than multiline fiber—a problem later solved with ring architecture). Some microwave "old timers" are happy to point out that Sprint rebuilt the microwave tower and now has many fixed point to point microwave links in their all-digital network.

1.3 PRACTICAL APPLICATIONS

In the early 1860s, several people, including Bell, Gray, La Cour, Meucci, Reis, and Varley demonstrated telephones (transmitters and receivers). In 1876, Alexander Bell and Thomas Watson, as well as Elisha Gray, demonstrated the first practical telephones. Both Bell and Gray filed patent documents the same day in 1886. Meucci sued Bell for patent infringement but died in 1889 before the suit could be completed. Western Union, in conjunction with Edison and Gray's company, Gray and Barton, began offering telephone service and entered into litigation with Bell concerning patent rights. Edison had invented a superior carbon telephone transmitter but Bell had the better receiver. About the same time, Western Union formed the Western Electric Manufacturing Company. In 1878, the New England Telephone Company was formed. The same year the Bell Telephone Company was formed by Bell and his financiers. The next year the two companies merged to form the National Bell Telephone Company. In 1880, Bell's company became the American Bell Telephone Company. In 1882, the company bought controlling interest in Western Electric Company from Western Union. About the same time, Western Union and Bell settled their long-standing patent infringement conflict. The same year AT&T was formed to create a nationwide long-distance telephone network. This organization eventually became AT&T Long Lines. In 1899, AT&T purchased the assets of the American Bell Telephone Company and added local service to its long-distance services. In 1925, AT&T created Bell Telephone Laboratories. In the United

States, much of the radio development was conducted by Bell Labs. These laboratories were funded from federally regulated telephone service income. After much legal maneuvering, the US Department of Justice imposed the Consent Decree of 1956 on AT&T. Bell Labs was treated as a national resource. The results of its employees were viewed as national property. Until this decree was changed in 1984, AT&T could not receive royalties for any of its inventions (such as the transistor and laser) (Bryant, 1988; IEEE Communications Society, 2002; O'Neill, 1985; Schindler, 1982; AT&T Bell Laboratories, 1983; Thompson, 2000).

The first commercial telephone exchange was opened in New Haven, Connecticut, in January 1878. In 1889, the first coin-operated telephone was installed in a bank in Hartford, Connecticut. In 1892, the Strowger Automatic Telephone Exchange Company (later Automatic Electric Company) installed the first automatic telephone exchange, the Stroger Step by Step. Later other automatic mechanical switches were invented and installed: the Panel in 1930 and the Crossbar in 1938. The cost of mechanical switch upgrades was increasing with each generation. In the 1950s, the concept of a software-defined switch was envisioned as a way to reduce the cost of upgrades and changes to telephone switches. Electronic switches were trialed in the 1960s. The first digital switching toll office system in North America was the Western Electric 4ESS introduced in 1972. Initially, these switches had analog telephone interfaces. Later they evolved to strictly digital DS1 interfaces. The digitization of the national telecommunications transportation network followed the evolution of the 4ESS. Northern Telecom introduced the first local office switch, the DMS10, in the late 1970s followed by Western Electric's 5ESS in early 1989 (Schindler, 1982; Thompson, 2000).

With the proliferation of telephone service, the need for telephone lines increased dramatically. A solution to this problem was the development of frequency division multiplex (FDM) systems that "stacked" multiple telephone channels into one composite wide bandwidth analog signal. At first they connected cities via coaxial cable. Later analog FM microwave radios were used to transport the FDM signals (Fig. 1.2).

The use of these telecommunication systems was universal. Compared to long-distance cable systems, the microwave radio systems were relatively inexpensive and could be placed practically anywhere (Fig. 1.3). Many companies evolved worldwide to supply this telecommunication equipment.

In 1916, ship to shore two-way radio communication was demonstrated between the USS New Hampshire and Virginia based transmitter and receiver locations. Bell Labs developed the CW-936 500-kHz to 1.5-MHz radio telephones for the Navy. About 2000 of these radiotelephones were installed on US and British ships during the World War I. Western Electric produced the 600-kHz to 1.5-MHz SCR-68 ground to air radiotelephones for the US Navy (Bryant, 1988; Fagen, 1978).

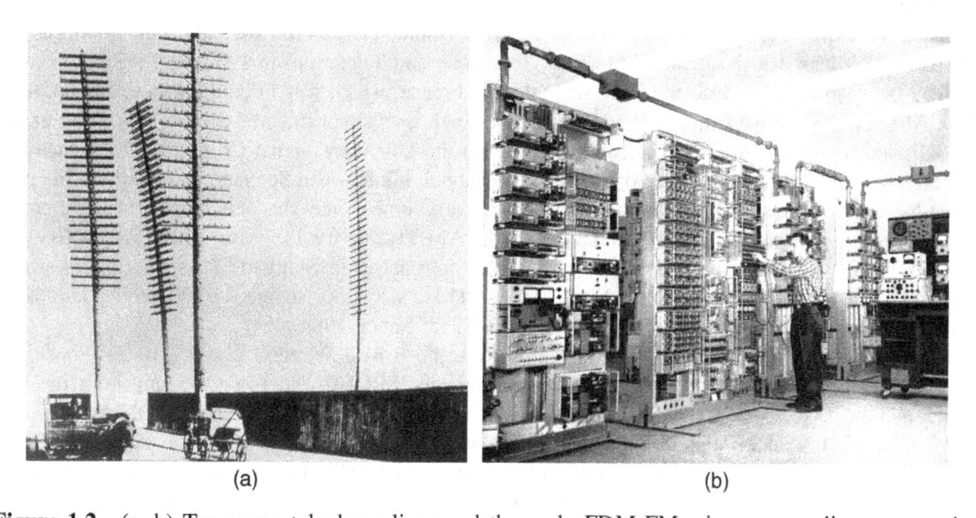

(a) (b)

Figure 1.2 (a, b) Too many telephone lines and the early FDM-FM microwave radio system solution. *Source*: Photos from Collins Microwave Radio Company archives. Reprinted with permission of Alcatel-Lucent USA, Inc.

(a) (b)

Figure 1.3 (a, b) Microwave radio locations. *Source*: Photos from archives of Collins Microwave Radio Division of Rockwell International. Reprinted with permission of Alcatel-Lucent USA, Inc.

About the same time, commercial applications began. Long-distance telephone service was needed on the Catalina Island off the coast of California. However, due to wartime shortages, a cable system could not be provided. Radio was a logical choice. The project was started in April 20, 1920, and engaged in service on July 16, 77 days later. Rapid deployment has been a standard radio system feature ever since. Transatlantic telephony experiments were conducted in the 1920s. Radio telephone service between the United States and England began on January 7, 1927. The cost for a 3-min call was $75. By 1939, shortwave radio telephone service was available between most major cities in the world (Fagen, 1978; Sobol, 1984).

Although Guarini had suggested radio relay communications in 1899, this was not attempted for several years. In 1925, the RCA installed an experimental radio link across the English Channel. The first commercial radio telephone service was initiated in 1927 between Great Britain and the United States. In 1931, French and English engineers of Les Laboratoires Standard (later Laboratoire Central de Telecommunication) and Standard Telephone and Cables (later International Telephone and Telegraph), under the direction of Andre Clavier, experimented with a 40-km microwave radio link across the English Channel (one telephone/telegraph channel between Calais and St. Margarets Bay). It operated with a 1-W 1.7-GHz transmitter with 10-ft (3-m) parabolic antennas. Reports on this project (Armstrong, 1936) first used the terms *micro waves* (two words). The first commercial microwave radio link was installed between the Vatican and the Italian PTT (Post, Telephone and Telegraph) in 1932. In 1933, a link was installed between Lympne, England, and St. Inglevert, France, which was in continuous operation until 1940. Also, in 1933, at the Chicago World's Fair, Westinghouse demonstrated 3.3-GHz radio links using parabolic antennas. Two of these systems were sold to the US Army Signal Corps for $2500 each. In 1936, the British General Post Office opened a multichannel link between Scotland and Northern Ireland. This 65-km link operated at 65 MHz using AM to carry nine voice channels. In Germany, Lorenz and Telefunken produced a single-channel 500-MHz AM system for the Army in 1937. In 1939, a 1.3-GHz FM 10-channel magnetron-based system was introduced in Stuttgart. These German systems were deployed widely in Europe and North Africa. These networks covered 50,000 route km, with terminals as far apart as 5000 km (Carl, 1966; Fagen, 1978; Sobol, 1984).

The first microwave radio relay system radios were the British Wireless Set No. 10 developed by the UK Signals Research and Development Establishment (SRDE). The Pye Company built the RF section and the TMC Company built the multiplex. It was an eight-telephone-channel TDM pulse width modulation 5-GHz radio system designed to operate in tandem as a radio relay. It was demonstrated to the US Signal Corps Labs and Bell Laboratories on September 1942. This spurred the development of similar systems in the United States: the RCA AN/TRC-5 and the Bell Labs AN/TRC-6 (Carl, 1966; Fagen, 1978; Sobol, 1984).

In 1941, Bell Laboratories tested a 12 voice channel AM system between Cape Charles and Norfolk. In 1943, Western Union installed the first intercity commercial microwave radio system using the RCA microwave equipment. In 1945, AT&T Corporation was operating a multichannel FM system between New York and Philadelphia. In 1948, Western Union had a 1000-mile 24-hop microwave radio system connecting New York, Washington, DC, and Pittsburgh. This was the first system to use unattended radio repeater locations and was the first use of loop topology to increase system reliability. A 7-hop (tandem radio path) 100 voice channel 4-GHz system between New York and Boston (300 km) was introduced in 1947. This experimental system, named *TD-X*, was the basis of the widely deployed improved system, named *TD-2*, which provided 489 telephone channels or one television channel per radio channel. In 1951, AT&T completed the TD-2 107-hop, 12 RF channel per hop, system between New York and San Francisco. This system spanned 4800 km and reached 12,000 km of total hop length. The short-haul 11-GHz TJ system was announced in 1957. In 1955, AT&T began the development of the 1800 voice channel 6-GHz TH system. By 1960, it had been deployed in parallel with the TD-2 system (Fagen, 1978; Friis, 1948; Sobol, 1984; Thayer et al., 1949).

In 1954, the US Air Force SAGE system employed the first (1200-baud analog telephone channel) modems to communicate between computer systems. In the 1960s, digital cable systems began to be deployed worldwide to interconnect telephone operating company switches. In France, TDM using PCM had been studied beginning in 1932. These techniques were used extensively in the United States beginning in 1962 with T1 digital cable spans with 24 voice channel banks connecting the 4ESS tandem switches. The transport was digital but the connections to the 4ESS were analog. Later direct DS1 interfaces were added to the switch. The introduction of the electronic 4ESS switch in 1976 spurred the development of terrestrial digital systems. In Europe, 30 voice channel TDM/PCM E1 links were being deployed (Fagen, 1978; Sobol, 1984; Welch, 1984).

In 1971, Hoff at Intel invented the silicon microprocessor that is used in personal computers. Jobs and Wozniak created the Apple I computer in 1976. The 16-bit microprocessor computer was introduced in 1980 by IBM using the Microsoft disk operating system (MS-DOS) developed by Gates and Allen. Jobs and Wozniak introduced the Macintosh computer in 1984.

Before 1949, the Federal Communications Commission (FCC) assigned microwave spectrum only to telecommunications common carriers. After that date, it began to license private microwave systems on a case by case basis if no common carrier service was available.

In 1959, the FCC, in its "Above 890" ruling, decided to allow licensing private intercity microwave systems for voice or data service at frequencies above 890 MHz. After that ruling, in the United States, microwave frequencies were defined as starting at 890 MHz. This definition is still in common use in the United States today. In 1962, the first telecommunication satellite, Telstar, was placed into orbit. In 1963, the American Standard Code for Information Exchange (ASCII) was defined. In 1968, DARPA began deployment of ARPANET (Advanced Research Projects Agency Network) and placed it in service in 1971. This was the first step in creating the Internet. This was to have a profound impact on telecommunications worldwide. In the United States, the 1968 Carterfone Decision created the opportunity to interconnection of customer-owned telephone equipment. In 1969, the first digital radio relay system went into operation in Japan. It operated at 2 GHz with a transmission capacity of 17 Mb/s. The FCC's Specialized Common Carrier Decision of 1969 decreed that new microwave companies could compete with the existing regulated telephone companies to sell private network transmission services. This brought a flood of Specialized Common Carriers utilizing microwave radio. In 1968, Western Microwave merged with Community Television cable system and became American Tele-Communications, with Western Tele-Communications (WTCI) and Community Tele-Communications (CTCI) subsidiaries. The same year the parent company's name was changed to Tele-Communications Inc. (TCI) and the headquarters was moved to Denver, Colorado. WTCI used the 1969 ruling to begin building an extensive microwave network used primarily for video distribution. By 1974, WTCI had become a large US microwave common carrier, second only to AT&T. MCI was founded as Microwave Communications, Inc. on October 3, 1963. Initially it built microwave relay stations between Chicago and St. Louis. In 1969, the umbrella company Microwave Communications of America, Inc. (MICOM) was incorporated and MCI began building its national private long-distance microwave radio network. MCI was purchased by Verizon in 2006. In 1979, the Times Mirror Company entered the cable network business. Not long after that the company formed Times-Mirror Microwave Communications Company of Austin, Texas. This company operated a large microwave network that provided, on a long-term contractual basis, transmission capacity to Telcos (long-distance resellers, independent telephone companies, and regional Bell companies) to carry their voice calls.

In 1970, Data Transmission Co. (DATRAN) filed for FCC approval of a nationwide system exclusively for data transmission over digital microwave radios. The DATRAN system was a nationwide all-digital-switched microwave radio network, which linked subscriber terminals in 35 metropolitan areas. The same year Norman Abramson and Franklin Kuo at the University of Hawaii introduced ALOHANET, the first large-scale deployment of data packets over radio. In 1986, this concept was refined by Robert Metcalfe at Xerox PARC into Ethernet, the technology that led to the IEEE 802.3 Local Area Network data interface standard. The 1972 FCC "Open Skies" ruling created domestic satellite communications carriers. These systems shared the terrestrial microwave radio 4- and 6-GHz common carrier bands. The same year Southern Pacific Communication (the forerunner of Sprint) got the FCC approval for an 11-state common carrier microwave radio network. In 1977, Bell Labs installed the first Advanced Mobile Phone System (AMPS). This was the first cellular radio system. The need for transmission circuits between cell sites would eventually expand the use of microwave radio systems. In the late 1970s, AT&T digitized its network enabling it to carry data traffic. In 1983, Judge Greene approved divestiture of AT&T. The AT&T divestiture (the US Department of Justice's Modified Final Judgment of the 1956 Consent Decree), effective January 1, 1984, separated AT&T from seven new RBOCs (Ginsberg, 1981).

1.4 THE BEAT GOES ON

In the late 1940s, all fixed point to point microwave relay systems used analog FM transmission. It carried video and telephony exclusively. FDM was used to aggregate the 4-kHz-wide analog telephone channels for transmission over FM radios.

In the 1970s, single-sideband analog radios were used to increase the analog transmission capacity. However, by the late 1970s, digital transmission began to be deployed worldwide. TDM was used to aggregate the PCM analog telephone channels. Various standards were developed in different countries to multiplex various levels of TDMed digital signals ("asynchronous" systems in North America and Plesiochronous Digital Hierarchy in Europe) (Gallagher, 1962; AT&T Bell Laboratories, 1983).

In 1988, Bellcore's Synchronous Optical Network (SONET) and in 1989, the ITU-T's Synchronous Digital Hierarchy (SDH) were finalized, setting new standards for worldwide digital transport interconnectivity. The SONET and SDH systems were widely deployed in radio networks in the 1990s.

Beginning in the 1970s, while data equipment was being developed, data network architectures were beginning to become standardized. The IBM Systems Network Architecture (SNA) introduced the concept of layered hierarchical peer processes. Its six-layer architecture was very popular. The Digital Equipment Corporation (DEC) also provided a five-layer Digital Network Architecture (DNA). The International Standards Organization (ISO) defined a seven-layer Reference Model [Open Systems Interconnection (OSI) "seven-layer stack"]. ARPANET developed its four-layer architecture that has become the standard for the Internet. There were many other architectures that achieved various levels of popularity. However, today the Internet ("Ethernet and IP") architecture is by far the most popular (Green, 1984; IEEE Communications Society, 2002; Konangi and Dhas, 1983).

J. C. R. Licklider, in his January 1960 paper, Man-Computer Symbiosis, proposed "a network of such [computers], connected to one another by wide-band communication lines [which provide] the functions of present-day libraries together with anticipated advances in information storage and retrieval and symbiotic functions." During the 1960s, Paul Baran and Donald Davies independently proposed data networks based on the principle of breaking down all digital messages into message blocks called *packets*. AT&T engineers and management discounted the concept as unworkable. Unlike the AT&T approach of circuit-switching networks, the proposed packet networks would store and forward message blocks over different routes based on various criteria. With adequate path redundancy, these networks were inherently highly reliable in the face of localized network outages. Leonard Kleinrock, in 1961, was the first to develop a mathematical theory of this technology (Hafner and Lyon, 1996).

In 1962, Licklider was appointed head of the US Department of DARPA. Licklider created a computer science community associated with DARPA. In 1964, Ivan Sutherland took over as head of DARPA. Sutherland recruited Robert Taylor, from Dallas, Texas, to manage the DARPA computer networks. Taylor's office had three different communications terminals to three different computers at three different locations. The complexity of interacting with each computer and the inability to transfer information from one computer to another prompted Taylor to propose a data network to connect all facilities performing research for DARPA using a common interface. Taylor proposed this network and the project was approved for implementation (Hafner and Lyon, 1996).

Taylor's network began as a network of four nodes connecting the University of California, Los Angeles (UCLA), Stanford Research Institute, University of Utah, and University of California, Santa Barbara. The nodes were controlled by Interface Message Processors (IMPs), the forerunner of the modern router. The IMPs and the network concept were specified and managed by Bolt Beranek and Newman (BBN), and the IMPs were designed and manufactured by Honeywell. The data connections among the nodes were data modems connected to audio circuits leased from AT&T. The IMP packet switches and their connections were called the *ARPANET*. The UCLA Network Measurement Center would deliberately stress the network to highlight bugs and degradations. The IMPs reported various quality metrics and statistics to a central Network Control Center (NCC) to facilitate effective management of network transmission quality. The NCC was also the focal point for coordinated software upgrade of all IMPs via remote download; the concept of a data Network Operations Center (NOC) was introduced. Request for Comments Number 1 (RFC 1), entitled "Host Software," was written by Steve Crocker in 1969. About the same time, an informal group, which was eventually called the *Network Working Group* (*NWG*), was formed to oversee the evolution of the network. This group eventually became the Internet Engineering Task Force (Fial, R., private communication, 2010, Hafner and Lyon, 1996).

The first electronic mail (e-mail) between two machines was sent in 1971 by Ray Tomlinson at BBN. Tomlinson chose the @ symbol as the separator between the user name and the user's computer. In 1972, Robert Metcalfe and others at Xerox PARC adapted the packet techniques from ALOHANET (Norman Abramson and others) to create a coaxial cable network connecting Alto computers. Metcalfe first called the new network *Alto Aloha* and later *Ethernet*. In the early 1970s, AT&T was asked if it wanted to take over ARPANET. AT&T and Bell Labs studied the proposal but declined. About the same time, ITU developed a packet network standard X.25. In 1974, Vint Cerf and Robert Kahn described the end to end routing of packets called *datagrams*, which encapsulated digital messages. The paper also introduced the concept of gateways. In 1975, Yogen Dalal, using the Cerf and Kahn concepts, developed a specification for transmission control protocol (TCP). The original concept of TCP included both packet protocol and packet routing. In a TCP review meeting in 1978, Vint Cerf, Jon Postel, and Dan Cohen decided to split the packet protocol and routing functions of TCP into two separate functions: Internet protocol (IP) and TCP. All ARPANET host computers were converted to Transmission Control Program and Internetwork Protocol (TCP/IP) operation in 1983 (Fial, R., private communication, 2010, Hafner and Lyon, 1996).

Also in 1993, Mosaic, the first graphical Internet browser, was released by Marc Andreessen and Eric Bina at the National Center for Supercomputing Applications (NCSA) at the University of Illinois Urbana-Champaign. The next year the Netscape Navigator appeared, which quickly expanded the Web's presence and made it a viable commercial medium. In 1988, the ISO produced the OSI protocol standard. It was intended to replace TCP/IP. Its complexity and insistence to replace (rather than supplement) existing standards made its adoption and implementation very difficult. In 1991, Tim Berners-Lee of CERN published a summary of the World Wide Web. For the first time the Internet was introduced to the concepts of HyperText Transfer Protocol (HTTP), HyperText Markup Language (HTML), the Web browser, and the Web server. HTML is the markup language used for documents served up by a Web server. HTTP is the transfer protocol developed for easy transmission of these hypertext documents by the Web server. A Web browser consumed these documents and drew them on a page. While these sat upon the already existing infrastructure of the Internet, they were one of the several formats at the time used for sharing information. Some of the other formats popular at the time were Gopher and FTP. The HTML format was a little more user friendly, embedding navigation and display together, but it was not until the graphical browser was created that it became the de facto standard. In 1995, the US government formally turned over operation of the Internet to private Internet service providers (ISPs). The world would never be the same (Kizer, M., private communication, 2010, Hafner and Lyon, 1996).

The new millennium has shown a significant increase in adoption of IP technology for interconnecting all forms of digital transmission. For now, the SONET and SDH systems are maintaining a hold on the long-distance transmission market. However, the user community drop and edge connections are rapidly transitioning to IP. IP, with its evolving Quality of Service features, is the new wave of digital transmission. Fixed point to point microwave network evolution mirrors that transition.

REFERENCES

Armstrong, E., "A Method of Reducing Disturbances in Radio Signaling by a System of Frequency Modulation," *Proceedings of the IRE*, pp. 689–740, May 1936.

AT&T Bell Laboratories, *Engineering and Operations in the Bell System, Second Edition*. Murray Hill: AT&T Bell Laboratories, 1983.

Barrett, R. M., "Microwave Printed Circuits—The Early Years," *IEEE Transactions on Microwave Theory and Techniques*, Vol. 32, pp. 983–990, September 1984.

Bennett, W. R. and Davey, J. R., *Data Transmission*. New York: McGraw-Hill, 1965.

Berrou, C., Glavieux, A. and Thitimajshima, P., "Near Shannon Limit Error-Correcting Coding and Decoding: Turbo Codes," *IEEE International Conference on Communications Proceedings*, pp. 1064–1070, 1993.

Blackman, R. B. and Tukey, J. W., *The Measurement of Power Spectra from the Point of View of Communications Engineering*. New York: Dover Publications, 1958.

Bose, R. and Ray-Chaudhuri, D., "On a Class of Error-Correcting Codes," *Information and Control*, Vol. 3, pp. 68–79, 1960.

Bryant, J. H., "Coaxial Transmission Lines, related Two-Conductor Transmission Lines, Connectors and Components: A U. S. Historical Perspective," *IEEE Transactions on Microwave Theory and Techniques*, Vol. 32, pp. 970–983, September 1984.

Bryant, J. H., "The First Century of Microwaves—1886 to 1986," *IEEE Transactions on Microwave Theory and Techniques*, Vol. 36, pp. 830–858, May 1988.

Bullington, K., "Radio Propagation at Frequencies Above 30 Megacycles," *Proceedings of the IRE—Waves and Electrons Section*, pp. 1122–1136, October 1947.

Bullington, K., "Radio Propagation Variations at VHF and UHF," *Proceedings of the IRE*, pp. 27–32, January 1950.

Bullington, K., "Radio Propagation Fundamentals," *Bell System Technical Journal*, Vol. 36, pp. 593–626, May 1957.

Bullington, K., "Radio Propagation for Vehicular Communications," *IEEE Transactions on Vehicular Technology*, Vol. 26, pp. 295–308, November 1977.

Burrows, C. R., "Propagation Over Spherical Earth," *Bell System Technical Journal*, Vol. 14, pp. 477–488, July 1935.

Burrows, C. R., Hunt, L. E. and Decino, A., "Ultra-Short Wave Propagation: Mobile Urban Transmission Characteristics," *Bell System Technical Journal*, Vol. 14, pp. 253–272, April 1935.

Carl, J., *Radio Relay Systems*. London: MacDonald, 1966.

Elias, P., "Coding for Noisy Channels," *IRE Convention Record*, Vol. 3, pp. 37–47, 1955.

England, C. R., Crawford, A. B. and Mumford, W. W., "Some Results of a Study of Ultra-Short-Wave Transmission Phenomena," *Bell System Technical Journal*, Vol. 12, pp. 197–227, April 1933.

England, C. R., Crawford, A. B. and Mumford, W. W., "Ultra-Short-Wave Transmission and Atmospheric Irregularities," *Bell System Technical Journal*, Vol. 17, pp. 489–519, October 1938.

Fagen, M. D., *A History of Engineering & Science in the Bell System, The Early Years (1875-1925)*. Murray Hill: Bell Telephone Laboratories, 1975.

Fagen, M. D., Editor, *A History of Engineering and Science in the Bell System, National Service in War and Peace (1925–1975)*. Murray Hill: Bell Telephone Laboratories, 1978.

Friis, H. T., "Noise Figures of Radio Receivers," *Proceedings of the IRE*, pp. 419–422, July 1944.

Friis, H. T., "A Note on a Simple Transmission Formula," *Proceedings of the IRE—Waves and Electrons Section*, pp. 254–256, May 1946.

Friis, H. T., "Microwave Repeater Research," *Bell System Technical Journal*, Vol. 27, pp. 183–246, 1948.

Gallagher, R. G., "Low Density Parity Check Codes," *IRE Transactions on Information Theory*, Vol. 8, pp. 21–28, January 1962.

Ginsberg, W., "Communications in the 80's: The Regulatory Context," *IEEE Communications Magazine*, Vol. 19, pp. 56–59, September 1981.

Green, P. E., Jr., "Computer Communications: Milestones and Prophecies," *IEEE Communications Magazine*, Vol. 22, pp. 49–63, May 1984.

Hafner, K. and Lyon, M., *Where Wizards Stay Up Late, the Origins of the Internet*. New York: Simon & Schuster, 1996.

Hamming, R., "Error Detecting and Error Correcting Codes," *Bell System Technical Journal*, Vol. 29, pp. 41–56, January 1950.

Hartley, R., "Transmission of Information," *Bell System Technical Journal*, Vol. 7, pp. 535–563, July 1928.

Hocquenghem, A., "Codes Correcteurs d'Erreurs," *Chiffres*, Vol. 2, pp. 147–156, 1959.

IEEE Communications Society, *A Brief History of Communications*. Piscataway: IEEE, 2002.

Kalman, R. E., "A New Approach to Linear Filtering and Prediction Problems," *Transactions of the ASME*, Vol. 82, pp. 35–45, January 1960.

Kerr, D., *Propagation of Short Radio Waves*, *Radiation Laboratory Series*, Volume 13. New York: McGraw-Hill, 1951.

Kizer, G. M., *Microwave Communication*. Ames: Iowa State University Press, 1990.

Konangi, V. and Dhas, C. R., "An Introduction to Network Architectures," *IEEE Communications Magazine*, Vol. 21, pp. 44–50, October 1983.

Kotel'nikov, V. A., *The Theory of Optimum Noise Immunity*. New York: McGraw-Hill, 1959.

MacKay, D. J. C. and Neal, R. M., "Near Shannon Limit Performance of Low Density Parity Check Codes," *Electronics Letters*, Vol. 32, pp. 1645–1655, August 1996.

Marcuvitz, N., *Waveguide Handbook*, *Radiation Laboratory Series*, Volume 10. New York: McGraw-Hill, 1951.

Maxwell, J. C., "A Dynamical Theory of the Electromagnetic Field," *Philosophical Transactions of the Royal Society of London*, Vol. 155, pp. 459–512, 1865.

Meinel, H. H., "Commercial Applications of Millimeterwaves History, Present Status and Future Trends," *IEEE Transactions on Microwave Theory and Techniques*, Vol. 43, pp. 1639–1653, July 1995.

Millman, S., Editor, *A History of Engineering and Science in the Bell System, Physical Sciences (1925–1980)*. Murray Hill: AT&T Bell Laboratories, 1983.

Millman, S., Editor, *A History of Engineering and Science in the Bell System, Communications Sciences (1925–1980)*. Indianapolis: AT&T Technologies, 1984.

Montgomery, C. G., Dicke, R. H. and Purcell, E. M., Editors, *Principles of Microwave Circuits*, *Radiation Laboratory Series*, Volume 8. New York: McGraw-Hill, 1948.

Morita, K., "Report of Advanced in Microwave Theory and Techniques in Japan - 1959," *IRE Transactions on Microwave Theory and Techniques*, Vol. 8, pp. 395–397, July 1960.

Muller, D., "Application of Boolean Switching Algebra to Switching Circuit Design," *IEEE Transactions on Computers*, Vol. 3, pp. 6–12, September 1954.

North, D. O., "An Analysis of the Factors which Determine Signal/Noise Discrimination in Pulse-Carrier Systems," *Proceedings of the IEEE*, pp. 1016–1027, July 1963.

Norton, K. A., "Transmission Loss in Radio Propagation," *Proceedings of the IRE*, pp. 146–152, January 1953.

Norton, K. A., "Radio-Wave Propagation During World War II," *Proceedings of the IRE*, pp. 698–704, May 1962.

Norton, K. A., Vogler, L. E., Mansfield, W. V. and Short, P. J., "The Probability Distribution of the Amplitude of a Constant Vector Plus a Rayleigh-Distributed Vector," *Proceedings of the IRE*, pp. 1354–1361, October 1955.

Nyquist, H., "Certain Factors Affecting Telegraph Speed," *Bell System Technical Journal*, Vol. 3, pp. 324–346, April 1924.

Nyquist, H., "Certain Topics in Telegraph Transmission Theory," *AIEE Transactions*, Vol. 47, pp. 617–644, April 1928.

O'Neill, E. F., *A History of Engineering & Science in the Bell System, Transmission Technology (1925–1975)*. Murray Hill: AT&T Bell Laboratories, 1985.

Okwit, S., "An Historical View of the Evolution of Low-Noise Concepts and Techniques," *IEEE Transactions on Microwave Theory and Techniques*, Vol. 32, pp. 1068–1082, September 1984.

Oliner, A. A., "Historical Perspectives on Microwave Field Theory," *IEEE Transactions on Microwave Theory and Techniques*, Vol. 32, pp. 1022–1045, September 1984.

Packard, K. S., "The Origin of Waveguides: A Case of Multiple Rediscovery," *IEEE Transactions on Microwave Theory and Techniques*, Vol. 32, pp. 961–969, September 1984.

Prange, E., "Cyclic Error-Correcting Codes in Two Symbols," *Air Force Cambridge Research Center Technical Report TN-57-103*. Cambridge: United States Air Force, 1957.

Ragan, G., Editor, *Microwave Transmission Circuits*, *Radiation Laboratory Series*, Volume 9. New York: McGraw-Hill, 1948.

Rahmat-Samii, Y. and Densmore, A., "A History of Reflector Antenna Development: Past, Present and Future," *SBMO/IEEE MTT-S International Microwave & Optoelectronics Conference*, pp. 17–23, November 2009.

Reed, I., "A Class of Multiple-Error-Correcting Codes and a Decoding Scheme," *IEEE Transactions on Information Theory*, Vol. 4, pp. 38–49, September 1954.

Reed, I. and Solomon, G., "Polynomial Codes Over Certain Finite Fields," *Journal of the Society of Industrial Applied Mathematics*, Vol. 8, pp. 300–304 1960.

Rice, S. O., "Mathematical Analysis of Random Noise," *Bell System Technical Journal*, Vol. 23, pp. 282–332, July 1944, and Vol. 24, pp. 46–156, January 1945.

Rice, S. O., "Statistical Properties of a Sine Wave Plus Random Noise," *Bell System Technical Journal*, Vol. 27, pp. 109–157, January 1948.

Salazar-Palma, M., Garcia-Lamperez, A., Sarkar, T. K. and Sengupta, D. L., "The Father of Radio: A Brief Chronology of the Origin and Development of Wireless Communications," *IEEE Antennas and Propagation Magazine*, Vol. 53, pp. 83–114, December 2011.

Sarkar, T. K., Mailloux, R. J., Oliner, A. A., Salazar-Palma, M. and Sengupta, D. L., *History of Wireless*. Hoboken: John Wiley & Sons, Inc., 2006.

Schelleng, J. C., Burrows, C. R. and Ferrell, E. B., "Ultra-Short Wave Propagation," *Bell System Technical Journal*, Vol. 12, pp. 125–161, April 1933.

Schindler, G. E., Jr., Editor, *A History of Engineering and Science in the Bell System, Switching Technology (1925–1975)*. Murray Hill: AT&T Bell Laboratories, 1982.

Shannon, C. E., "A mathematical theory of communication," *Bell System Technical Journal*, Vol. 27, pp. 379–423, 623–656, July and October 1948.

Shannon, C. E., "Communication in the Presence of Noise," *Proceedings of the IRE*, pp. 10–21, January 1949.

Shannon, C. E., "Recent Development in Communication Theory," *Electronics*, Vol. 21, pp. 80–83, April 1950.

Silver, S., Editor, *Microwave Antenna Theory and Design*, *Radiation Laboratory Series*, Volume 12. New York: McGraw-Hill, 1949.

Smits, F. M., Editor, *A History of Engineering and Science in the Bell System, Electronics Technology (1925–1975)*. Indianapolis: AT&T Technologies, 1985.

Sobol, H., "Microwave Communications—An Historical Perspective," *IEEE Transactions on Microwave Theory and Techniques*, Vol. 32, pp. 1170–1181, September 1984.

Sommerfeld, A., "Über die Ausbreitung der Wellen in der drahtlosen Telegraphie," *Annals of Physics*, Vol. 28, pp. 665–736, 1909.

Southworth, G. C., *Principles and Applications of Waveguide Transmission*. New York: Van Nostrand, 1950.

Southworth, G. C., *Forty Years of Radio Research*. New York: Gordon and Breach, 1962.

Tarrant, D. R. *Marconi's Miracle*. St. John's: Flanker Press, 2001.

Thayer, G. N., Roetken, A. A., Friis, R. W. and Durkee, A. L., "A Broad-Band Microwave Relay system Between New York and Boston," *Proceedings of the IRE—Waves and Electrons Section*, pp. 183–188, February 1949.

Thompson, R. A., *Telephone Switching Systems*. Boston: Artech House, 2000.

Tuller, W. G., "Theoretical Limits of the Rate of Transmission of Information," *Proceedings of the IRE*, pp. 468–478, May 1949.

Ungerboeck, G., "Channel Coding with Multilevel/Phase Signals," *IEEE Transactions on Information Theory*, Vol. 28, pp. 55–67, January 1982.

Viterbi, A. J., "Error Bounds for Convolutional Codes and an Asymptotically Optimum Decoding Algorithm," *IEEE Transactions on Information Theory*, Vol. 13, pp. 260–269, April 1967.

Weiner, N., *Extrapolation, Interpolation and Smoothing of Stationary Time Series with Engineering Applications*. Cambridge: The MIT Press, 1949.

Welch, H., "Applications of Digital Modulation Techniques to Microwave Radio Systems," *Proceedings of the IEEE International Conference on Communications*, Vol. 1, June 1978, pp. 1170–1181, September 1984.

Wiltse, J. C., "History of Millimeter and Submillimeter Waves," *IEEE Transactions on Microwave Theory and Techniques*, Vol. 32, pp. 1118–1127, September 1984.

Wiltse, J. C., "History and Evolution of Fresnel Zone Plate Antennas for Microwaves and Millimeter Waves," *Antennas and Propagation Society International Symposium*, Proceedings, Vol. 2, pp. 722–725, July 1999.

2

REGULATION OF MICROWAVE RADIO TRANSMISSIONS

The International Telecommunication Union—Radiocommunication Sector (ITU-R) is the branch of the United Nations that regulates radio transmissions internationally. The ITU began in 1865 as the International Telegraph Union. In 1903, the first International Radiotelegraph Convention was held in Berlin. The result of that convention was the formation of the International Radiotelegraph Union in 1906. It produced the first international regulations governing wireless telegraphy. A Table of Frequency Allocations was introduced in 1912. In 1927, the International Radio Telegraph Convention established the International Technical Consulting Committee on Radio (CCIR, Comite Consultatif International des Radiocommunications) to study issues pertaining to radio communication. In 1932, the International Telegraph Convention and the International Radiotelegraph Convention were merged to form the International Telecommunication Convention. The name of the merged group was changed to International Telecommunication Union (ITU) in 1944. In 1947, the ITU became an agency of the United Nations. The same year the ITU entered the UN, the International Frequency Registration Board (IFRB) was established, and conformance to the Table of Frequency Allocations became mandatory for all ITU signatory nations.

In 1993, the name of the Comite Consultatif International des Telegraphes et Telephones (CCITT), which dealt with international telecommunications standards, was changed to International Telecommunication Union—Telecommunication Standardization Sector (ITU-T). In 1992, the name of CCIR, the group that dealt with international radio matters, was changed to International Telecommunication Union—Radio Communication Sector.

International frequency allocations and technical rules are listed in the ITU-R Radio Regulations ("Red Books"). Allocations and rules are defined by region:

Region I	Africa, Europe, Middle East, and Russian Asia
Region II	North and South America and Greenland
Region III	South Pacific, Australia, and South and East Asia

ITU-R radio regulations (http://www.itu.int/publ/R-REG-RR/en) are updated and modified during World Radiocommunication Conferences (WRCs) that are held at 2- to 5-year intervals, most often at the ITU headquarters in Geneva. The result of the WRC is considered a treaty obligation by participating countries. By this international treaty, all subscribing nations agree to abide by the ITU-R worldwide regulations contained in the ITU-R Radio Regulations. Those regulations apply to all radio transmissions

Digital Microwave Communication: Engineering Point-to-Point Microwave Systems, First Edition. George Kizer.

that extend beyond a single country. For transmissions completely contained within a single country, the transmissions are considered "internal matters." Country administrations are not bound by the ITU-R regulations for the management of these "internal" transmissions (with the exception of radio transmission near an international border where such transmission could interfere with another nation's transmission systems and with international satellite systems that terminate within that nation).

For frequency bands that share dissimilar services, the difference in governing regulations between "internal matter" services and international services can be significant. In the United States, the fixed point to point terrestrial commercial services are governed by Part 101 of CFR 47, Chapter 1 (http://www.fcc.gov/encyclopedia/rules-regulations-title-47). Fixed satellite services (FSSs) are governed by Part 25. Part 101 uses coordination standards developed by the Telecommunications Industry Association (TIA) (Committee TR14-11, 1994). Part 25 uses coordination standards developed by the ITU. Coordination of potential interference into the satellite systems from the fixed radios and protection to the fixed radios by the satellite systems, as defined in Part 25, is different than coordination within the fixed community, as defined within Part 101. Differences in coordination and licensing methodologies complicate frequency sharing between fixed point to point microwave and satellite services.

Within the United States, national telecommunication law is codified within the Code of Federal Regulations (CFR), Title 47—Telecommunications (Mosley, published yearly). These laws provide for three agencies: Chapter I defines the Federal Communications Commission (FCC), Chapter II defines the Office of Science and Technology Policy and National Security Council, and Chapter III defines the National Telecommunications and Information Administration (NTIA), Department of Commerce.

The FCC has regulatory authority over all non-Federal-Government radio spectrum, as well as all international communications that originate or terminate within the United States, the District of Columbia, and the US possessions. It is the result of a long history of regulatory attempts to manage radio transmission in the United States (Linthicum, 1981).

The Wireless Ship Act passed by the US Congress in 1910 required all ships of the United States traveling over 200 miles off the coast and carrying over 50 passengers to be equipped with wireless radio equipment with a range of 100 miles. The Radio Act of 1912 (enacted at least in part because of the Titanic disaster) gave regulatory powers over radio communication to the Secretary of Commerce and Labor and required all seafaring vessels to maintain 24-h radio watch and keep in contact with nearby ships and coastal radio stations. It did not mention broadcasting and limited all private radio communications to what is now the AM band. The Radio Act of 1927, which superseded the Radio Act of 1912, created the Federal Radio Commission (FRC). In 1934, Congress passed the Communications Act, which abolished the FRC and transferred jurisdiction over radio licensing to the new FCC. The Commission was created to regulate radio use "as the public convenience, interest, or necessity requires."

The Office of Science and Technology Policy and National Security Council is responsible for procedure for the use and coordination of the radio spectrum during a wartime emergency. It establishes emergency restoration priority procedures for telecommunications services. Within an official disaster area, it may preempt frequency allocations and procedures temporarily to provide short-term telecommunication restoration.

The NTIA, an agency of the US Department of Commerce, serves the president in an advisory role regarding telecommunications policy. Its director is appointed by the president. The Office of Spectrum Management (OSM) within the NTIA has regulatory authority over all Federal Government radio spectrum within the United States, the District of Columbia, and the US possessions. The NTIA was created in 1978 as a result of Executive Branch reorganization. This reorganization transferred and combined various functions of the White House's Office of Telecommunications Policy (OTP) and the Commerce Department's Office of Telecommunications (OT).

2.1 RADIO FREQUENCY MANAGEMENT

Nationally and internationally, frequency management begins with a decision regarding how to manage radio paths. One of three approaches is usually chosen:

Individual Licensing. This is conventional link-by-link coordination. It is usually managed under a national administration although the technical tasks may be assigned to private entities or, in the so-called case of "light licensing," this responsibility may be assigned to the users. It is usually implemented in such a way that multiple users have access to the spectrum. It does limit the utilization of the frequencies to specified technologies. This is generally regarded as the most efficient method of spectrum usage.

Block Assignment. A block of spectrum in a defined geographic area is licensed to an individual user (typically by auction). The user defines the usage within that block. However, the user is responsible for establishing appropriate guard bands or spectrum powers and/or masks to protect other users in other spectrum and/or geographic blocks. Since only one user controls the spectrum, this method does limit user access to the spectrum. This method is generally regarded as a compromise between spectrum usage and user flexibility.

License Exempt. In this methodology, a block assignment is made but access is open to any eligible user whose equipment meets defined standards. Frequency assignments are ad hoc and no guarantee of interference protection is provided. This is the most flexible and cost-effective method of radio usage, but quality and availability of service are unpredictable.

We limit our discussion to conventional individual licensing. We start with a segregation of compatible radio services into similar contiguous blocks of frequencies ("frequency bands"). Different services are assigned to different frequency bands (Withers, 1999). Rules are then adopted for the implementation of radio service. If the band is licensed, rules for licensing are established in such a way that the introduction of a new service or user has minimal negative impact on the service quality of existing services or users. Owing to the limited number of frequency allocations, most new services attempt to "share" bands of established services. Convincing the regulatory agencies and the incumbent services that successful sharing is possible is an interesting process.

A database of existing users and their equipment's significant characteristics is developed and maintained. New users use this database to design their systems to be compatible with the existing users. The licensing process causes the new user's equipment to be entered into the database for future use. In the United States, Canada, and Australia, the licensing process is controlled by the national government. In many other countries, the spectrum is controlled by private organizations.

Before a commercial satellite system is deployed, it must be coordinated with the ITU, as well as with the countries that could be affected by the deployment of such a satellite. Both the affected countries and the ITU-R maintain records of the satellite systems. In the United States, commercial fixed point to point microwave radio systems are licensed by the FCC of the US government. The FCC maintains its own database. Commercial firms and some private companies also maintain private databases that are used to prepare license applications and provide related services. The commercial databases are the ones usually used to prepare license applications and resolve frequency disputes.

A frequency allocation ("frequency band") for a particular type of radio is typically subdivided into equally spaced subdivisions ("channels") for use by individual transmitters. The bandwidth of the channels is sized on the basis of the anticipated data transmission requirements. For most radio applications, the communication between two sites is duplex (simultaneous transmission in both directions along the radio path). Therefore, each radio path requires a transmit and a receive radio channel.

The earliest frequency plan, developed for the 4-GHz Bell System TD-2 multiple channel microwave system, interleaved transmit and receive frequencies consecutively. Transmit and receive frequencies at a station were on opposite polarizations to take advantage of antenna cross-polarization discrimination. Having the transmit and receive frequencies so close together complicated equipment design. All subsequent microwave frequency channel plans arranged the radio frequency (RF) transmission channels into two groups (a high sub-band and a low sub-band) with the intent that one group would be used for transmission and the other for reception. The concept of using opposite polarizations on consecutive channels was retained. A small portion of the spectrum, called a *guard band*, separates the two sub-bands to reduce the cost of radio filtering. When the transmit and receive channels are grouped into two sub-bands, the frequency plan is called a *two-frequency plan*. When the channels are further subdivided into four different sub-bands, the plan is called a *four-frequency plan*. Four-frequency plans may be required to solve a bucking station issue (see following paragraphs). Since they increase the guard bands required, they are less efficient than the two-frequency plans using the same spectrum (the guard bands block usage of those frequencies by other users in the area).

When analog radios employing FM modulation were common, using two plans, one a conventional two-frequency plan and another plan (called an *offset or interstitial plan*) offset by half a channel bandwidth, was common. Offsetting the analog received signals by half a channel bandwidth significantly reduced the interference due to the interfering signal as the analog signal had most of its energy concentrated at the carrier frequency. However, modern radios are digital. These radios spread the transmitted energy evenly through the radio channel. The interstitial frequency plans have little advantage in this situation and are no longer used.

Microwave frequency bands for the United States are defined by the FCC in the CFR Title 47 (Telecommunication), Chapter I, Part 101.147 (http://wireless.fcc.gov/index.htm?job=rules_and_regulations) (http://www.access.gpo.gov/nara/cfr/waisidx_10/47cfr101_10.html). In Canada, microwave frequency bands are defined by Industry Canada (http://www.ic.gc.ca/eic/site/smt-gst.nsf/eng/h_sf06130.html). Figure 2.1 graphically depicts the frequency channel allocation of the US lower 6-GHz frequency band.

The most popular channels are 30 MHz wide with a smaller number of smaller bandwidth channels. The smaller bandwidth channels are grouped in such a way as to minimize their impact on the higher bandwidth channels. It is expected that, to the extent possible, channels will be used in numbered order with transmit and receive channels being of the same number (for complicated, dense systems, this is not always feasible). The odd-numbered channels typically operate on one polarization, and the even-numbered channels on the other. This is to take advantage of antenna cross-polarization discrimination to reduce receiver filtering requirements.

Grouping transmit ("go") and receive ("return") channels into two separate groups is the most frequency-efficient method of using the most channels in a geographic area. However, this does place some constrains on the use of the channels. As noted in Figure 2.2, the use of a two-frequency plan assumes that you will transmit using one group of channels and receive using the other group.

At one site, the radio transmitters transmit in the high sub-band and receive in the low sub-band. This is termed a *High site*. The next site receives in the high sub-band and transmit in the low sub-band. This second site is called a *Low site*. For any given site, the intent is to transmit using the same sub-band. For junction stations (as illustrated in Figure 2.3) with many radio paths converging at a site, frequencies may be exhausted. While this methodology optimizes the reuse of a frequency spectrum, it is challenging to implement in practice. In congested areas, frequency coordination is especially difficult where many different paths of varying lengths and capacities must share the same area.

Since the sites cycle between high and low sub-bands down a line, closed loops or rings need to include an even number of sites. An odd number is difficult to accommodate, as noted in Figure 2.4. Linear networks, as depicted in Figure 2.5, also have constraints. Once a system is installed, adding or eliminating sites may be desirable for any of a number of reasons (e.g., capacity upgrades or path propagation performance improvement). This can cause significant frequency planning problems (it may be difficult to re-coordinate the system with existing users or may require frequency retuning of several sites). Also, changing routes to cause sites to appear at locations with other radio systems in the same frequency band can also prove challenging (especially if the existing transmitters are transmitting in one sub-band and the proposed new transmitters need to transmit in the other sub-band). If two consecutive stations are both "high" or both "low," one of the stations is considered a "bucking" or "bumping" station. It needs to transmit high in one direction and receive low in the other. Transmitter to receiver isolation for operation at the same frequency needs to be at least 120 dB. Achieving this in practice is very difficult (because of antenna spillover and unanticipated foreground reflections, as well as inadequate feeder isolation within the station). The only practical solution for this situation is to subdivide the band into a four-frequency plan (use frequencies in each of the two normal sub-bands to both transmit and receive), use another frequency band, or connect the sites by another transmission medium (such as fiber-optic cable).

Maintaining the high–low pattern for the many different systems within a geographic area can be challenging. Typically, all sites within approximately one-half mile of each other need to be using the same high or low transmit frequencies to avoid creating a "buck." Having to use a four-frequency plan to solve a "bucking station" can block channels for all users in the area.

In general, frequency planning requires careful consideration of potential interference from and to other radio systems within a coordination area (Fig. 2.6). One of the significant challenges of frequency planners is in maintaining a high/low frequency pattern for new systems once such a plan has been established in an area.

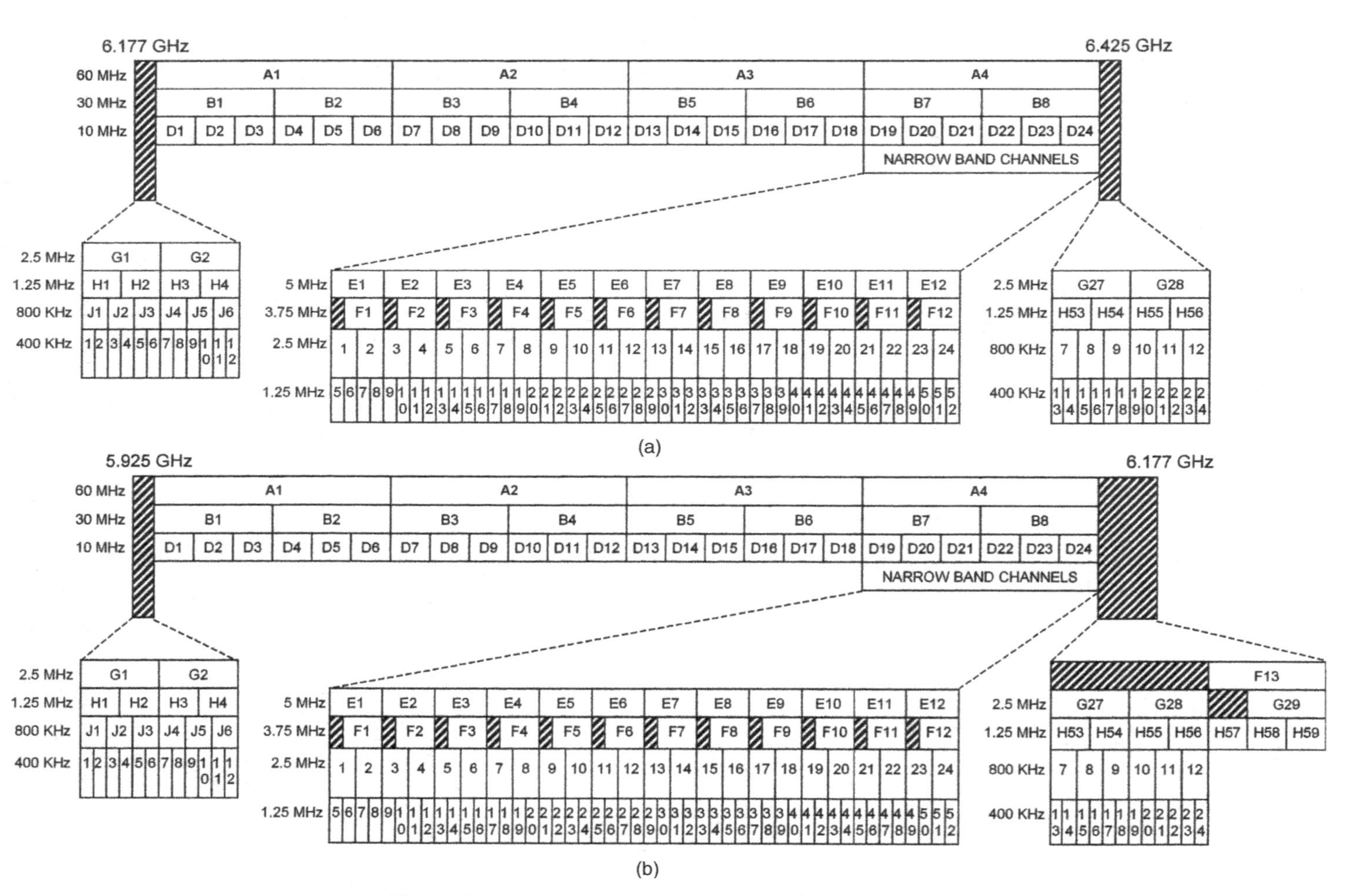

Figure 2.1 The US lower 6-GHz band channel allocations.

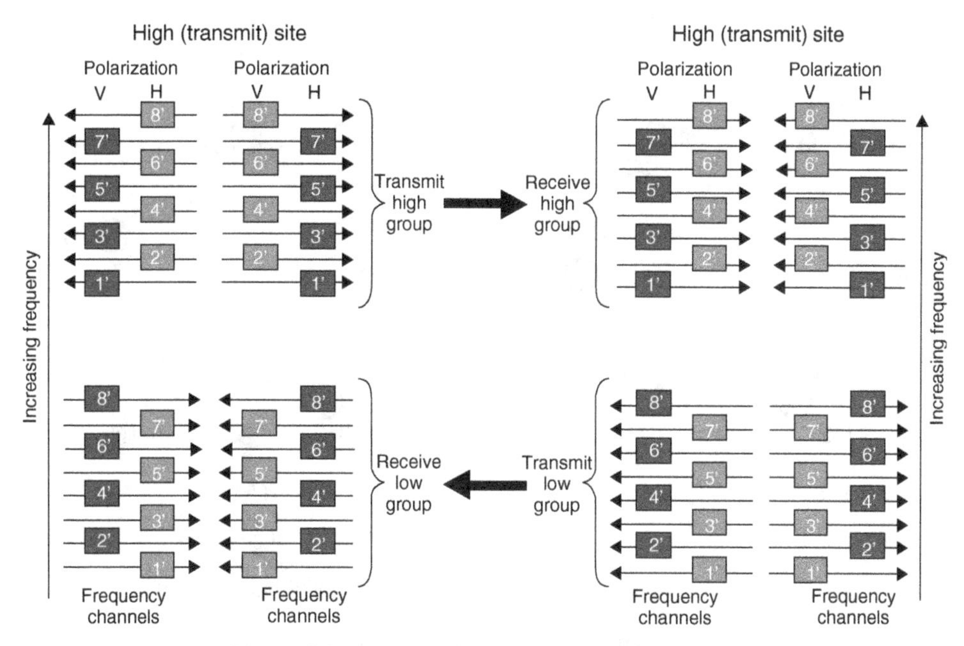

Figure 2.2 Typical two-frequency plan utilization.

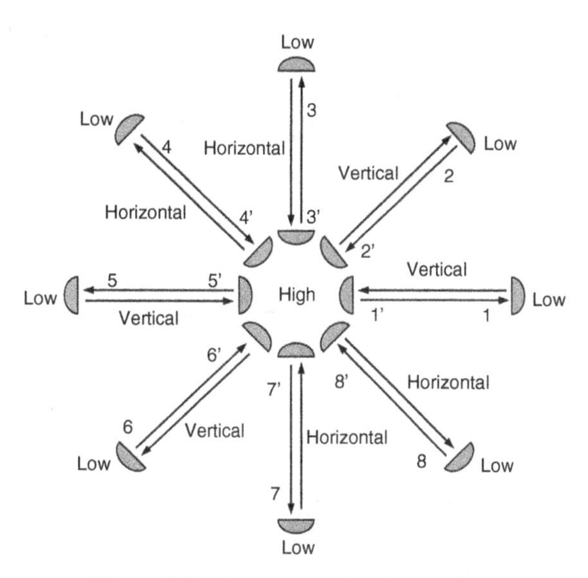

Figure 2.3 An example junction station.

In North America, coordination is accomplished on the basis of carrier-to-interference (*C/I*) ratio. This is different than the international *C/I* methodology described in Chapter 14.

The coordination area defines the area within which the effect of transmitters must be evaluated into existing receivers. In the United States, that is defined by the National Spectrum Managers Association guidelines (Fig. 2.7) (Working Group 3, 1992). A similar but more loosely defined international coordination area is defined by ITU-R Recs. F.1095 and SF.1006 (ITU-R Recommendations).

For United States commercial digital systems, coordination is accomplished on the basis of threshold to interference (*T/I*) objectives (degradation to the receiver threshold) (Committee TR14-11, 1994; Working

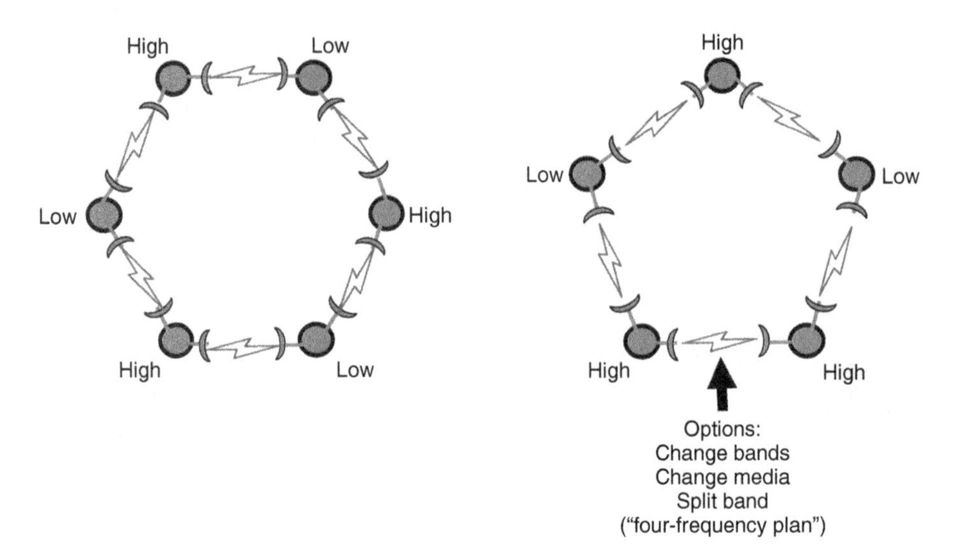

Figure 2.4 Closed loop networks.

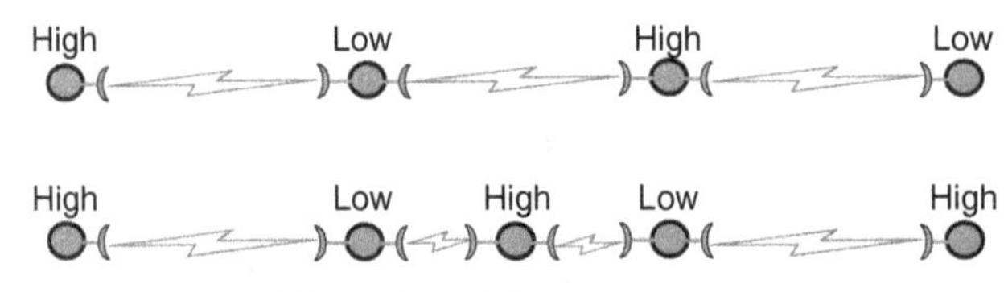

Figure 2.5 Linear networks.

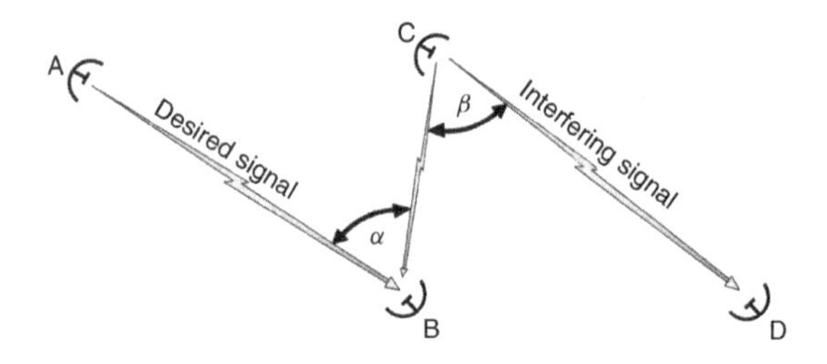

Figure 2.6 Generalized interference situation.

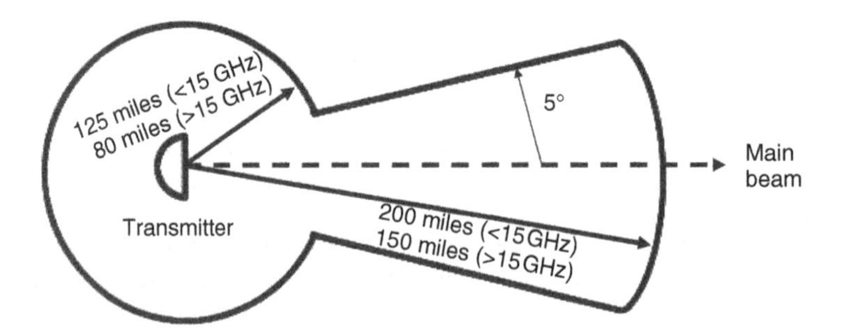

Figure 2.7 The US FCC band coordination area.

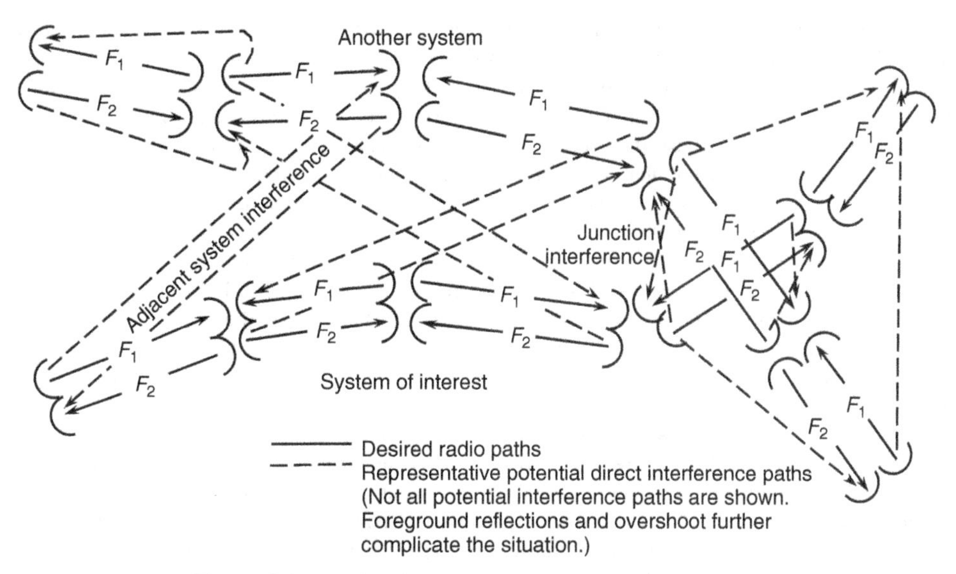

Figure 2.8 Typical interference cases to be investigated.

Group 3, 1986; Working Group 3, 1995; Working Group 3, 1987; Working Group 18, 1992). In the United States, commercial analog systems (and most international administrations) use *C/I* objectives (Working Group 5, 1992) (See Chapter 14).

As illustrated in Figure 2.8, the number of frequency interference cases usually requires computer analysis for practical optimization. Even with a computer, this can be a complicated, time-consuming, and iterative process.

Interference is analyzed as both long-term interference, which represents interference that is present most of the time, and short-term interference, which represents high power levels that may occur for short periods. Long-term interference may affect radio performance by degrading the fade margin of the receiver. Short-term interference could cause errors in a receiver even if the received signal is unfaded. Internationally, long-term interference is analyzed in terms of the interference power level that is exceeded no more than 20% of the time. This level is called the *80% interference level.* Domestically, long-term interference is analyzed in terms of the median value of the interference power. Short-term interference is a term used in analyses of the effects on a receiver of interference power levels that are exceeded less than 1% of the time. Interference criteria are usually specified as interference power levels that can only be exceeded no more than specified percentages of the time (Rummler, W. D., private communication with George Kizer in 2009).

In the United States, in bands where frequency coordination is carried out between fixed service (FS) systems, frequency coordination is usually based only on the long-term interference criteria. Short-term interference criteria are only invoked to clear exceptional cases as needed. For coordination purposes in the United States, short-term interference is defined (Working Group 9, 1985) as a level 10 dB worse than the long-term (median) interference power level. In bands shared with FSS earth stations, both long-term and short-term interference criteria are used in frequency coordination because of the high power used by transmitting earth stations and the extreme sensitivity of earth station receivers (Rummler, W. D., private communication, 2009).

Internationally, frequency coordination within the FS is implemented under the rules specified by each administration. The most important international application of short-term interference criteria is in studies of the use or potential use of spectrum shared between services. The short-term interference criteria developed for this purpose are specific to the frequency band and the applications in each of the two services. Because of the widely differing characteristics of some of the other proposed services, the short-term interference criteria vary widely (Rummler, W. D., private communication, 2009).

Several principles are employed in developing interference criteria to protect the FS from interference from other services. Recommendation ITU-R F.1094 specifies that shared services are allowed to take 10%

of the (international) performance and availability budgets. Where there is more than one other service in a frequency band, it may be necessary to further subdivide the allowance for the service under consideration. The most vulnerable FS application in the band must be identified, and the interference objective for this application be allocated to long-term and short-term interference. Then, appropriate interference criteria must be developed. This process may require iterations in redefining the characteristics of the other service and the FS criteria (Rummler, W. D., private communication, 2009).

In some cases it may not be necessary to develop a long-term interference criterion because of the intermittent presence of the interference sources (see, for example, Report ITU-R M.2119). In other cases, it may be necessary to develop multiple short-term interference criteria to ensure the protection of the FS. Guidance for the development of interference criteria can be found in ITU-R Recommendations F.758, F.1108, and F.1094. The results of some of the sharing studies carried out over recent years may be found in ITU-R Recommendations F.1494, F1495, F1606, F1669, F1706, SF1006, SF1482, SF1483, and SF1650. As might be expected, the specific results vary widely depending on the sharing scenario (Rummler, W. D., private communication, 2009) (ITU-R Recommendations).

Several frequency bands are shared between the fixed point to point terrestrial microwave services and the FSSs. The coordination requirements and procedures are different. Terrestrial satellite transmitters are usually of much higher power than terrestrial point to point transmitters. For frequency bands shared with synchronous (stationary) satellite uplinks (e.g., lower 6 GHz), this imposes a couple of locations on the horizon, which must be excluded from transmission to protect satellite receivers. Terrestrial stations must usually stay away from satellite earth station transmitter sites. Satellite earth station transmitters are usually limited to urban areas. This can be an issue in major cities, but is generally not an issue elsewhere.

For terrestrial services, the primary issue with sharing frequency bands with satellite service has been the satellite receivers. Satellite earth station antennas do not use shrouds and their sites usually have no shielding (earth hills or RF fences). Without additional shielding mechanisms, satellite earth station receivers and antennas are much more sensitive to interference than fixed terrestrial microwave services (Curtis, 1962). Terrestrial services coordinate specific frequencies actually to be used. FSSs coordinate all frequencies at all azimuths regardless of anticipated need. Satellite earth station receivers can enter an area if they avoid the existing terrestrial users' frequencies. However, owing to their receiver sensitivity, new terrestrial users are often excluded. The impact of this on fixed microwave deployment can be seen by comparing Figure 2.10 with Figure 2.11 and Figure 2.12.

The FSS uses geostationary and nongeostationary orbit satellites. The earth synchronous satellites are located 22,500 miles above the Earth. The synchronous satellite locations are nearing saturation over many areas of the Earth. New satellite services are often required to use lower orbits. Medium Earth Orbit (MEO) and Low Earth Orbit (LEO) systems with satellites approximately 6000 and 480–1000 miles above the Earth, respectively, are being contemplated. Since these systems are always moving relative to the Earth, their potential interference into terrestrial services (and vice versa) is always changing. Traditional FSs are coordinated on the basis of median estimated ("long-term") interference limits (as both satellite and terrestrial equipment are stationary). Nonstationary satellites are always moving relative to the Earth. Nonstationary satellite services sharing the spectrum with the FSs impose significant short-term interference limits (higher than long-term limits and often representing a short outage of terrestrial service) for frequency coordination.

2.2 TESTING FOR INTERFERENCE

Fixed point to point microwave systems are designed for performance objectives that place minimum fade margin (difference between typical receiver received signal level and receiver out of service received signal level) requirements on path design. Fade margin is directly limited by radio system characteristics and external radio system interference. Frequency coordination is intended to protect the designed path fade margin. In areas with dense deployment of radio systems, unexpected interference can happen. If it occurs, path performance can be significantly impaired. A common test used during the commissioning phase of a microwave radio link is a fade margin test (Fig. 2.9).

After the radio link is turned up and optimized, an attenuator is placed between the transmitter and the antenna. The attenuator is increased until the far end receiver threshold is reached. The amount of

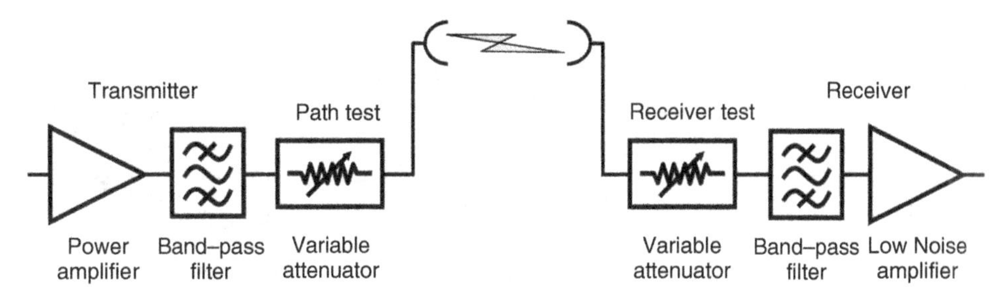

Figure 2.9 The fade margin test.

attenuation is a measure of path fade margin. Of course, this must be done when path fading is at a minimum. If the measured fade margin is significantly different from the anticipated value, the test is repeated at the receive end. If the measured fade margin at the receiver is the expected value, path interference is probable. If the measured fade margin is similar to that measured using the attenuator at the transmitter, a defective receiver is probable.

Placing an attenuator between a transmitter or receiver and the waveguide can be challenging. For high frequency split package radios, this may be impossible (although this test is important for initial testing of urban radios where unanticipated interference is common). Some radios allow Automatic Transmit Power Control (ATPC) to reduce transmitter power by 30 or 40 dB below the nominal level. This feature, if available, can be used to fade the test to validate the flat fade margin.

The above interference test is quite powerful for constant interference. However, it is an "out of service" test that is most suitable for use before commissioning the path. Interference can occur intermittently or after the path is installed. Identifying this condition is much more difficult after the path is in service. The impact of interference is to reduce path fade margin. If a path experiences significantly more fading outages in one direction than the other, one should suspect interference (or the receiver front end could have become defective).

If interference is probable, one must find the source. The direct approach is to take a standard gain horn antenna, low noise amplifier and spectrum analyzer around the area and try to find the interference. Discussions with a coordinating agency can be helpful to pick probable transmitters. Errors in transmitter polarization and transmitter location (errors in tower location, errors in antenna type and placement, reversal of hop transmitters) do happen. Finding and eliminating this interference can be challenging.

2.3 RADIO PATHS BY FCC FREQUENCY BAND IN THE UNITED STATES

Figures 2.10–2.17 graph the paths in the FCC license database by frequency band. Notice that the 4 (3.7–4.2)-GHz band is being significantly underutilized from a fixed point to point microwave radio perspective. Every year a couple hundred satellite earth stations are coordinated. No new fixed point to point paths are coordinated although in the comparable lower 6 (5.925–6.425)-GHz band, thousands of fixed MW paths are coordinated each year. The reason for this disparity is the FCC's curious policy of licensing earth stations for all possible frequencies and all possible path angles regardless of need. In addition, earth station antenna and location standards are not strict from a MW compatibility perspective. For these reasons, the 4-GHz band is essentially dead for new MW paths except for very isolated locations.

Clearly, frequency coordination has been successful in the 6- and 11-GHz bands. Those bands are highly utilized. Notice that 10.5 and 11 GHz are long-distance bands in the western United States where the rain rates are moderate (Fig. 2.13 and Fig. 2.14). All the other higher frequency bands are only applicable for short distance paths and typically are used in large metropolitan areas (Fig. 2.15, Fig. 2.16, and Fig. 2.17).

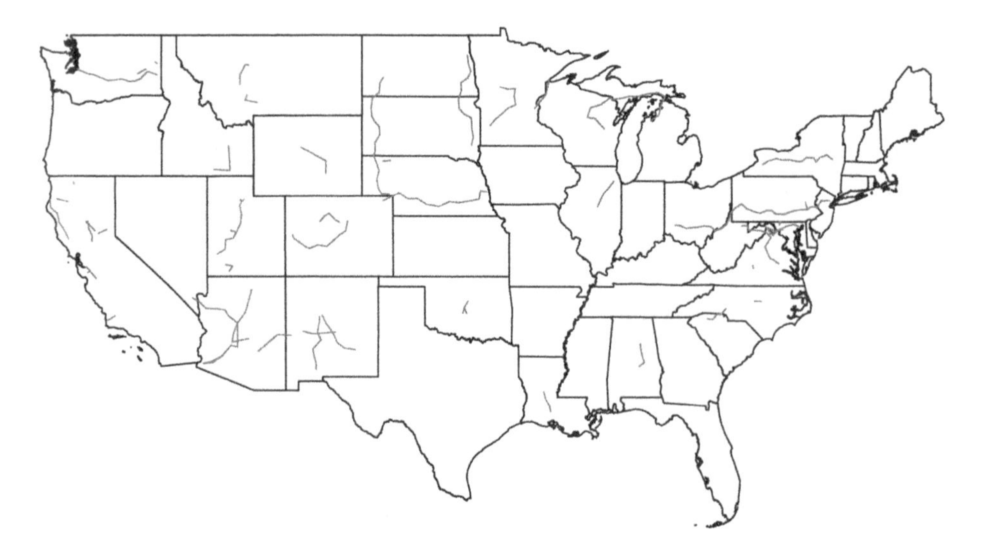

Figure 2.10 The 4 (3.7–4.2)-GHz band (shared between fixed point to point microwave and satellite earth stations).

Figure 2.11 Lower 6 (5.925–6.425)-GHz band.

2.4 INFLUENCES IN FREQUENCY ALLOCATION AND UTILIZATION POLICY WITHIN THE WESTERN HEMISPHERE

2.4.1 United States of America (USA)

2.4.1.1 Governmental

FCC This governmental agency has regulatory authority over all US non-federal-government wireline and radio communications. Frequency coordination within FCC frequency bands is governed by the FCC Rules and Regulations (Title 47 of the CFR, Parts 0 through 101, http://wireless.fcc.gov/index.htm?job=rules_and_regulations).

The FCC is composed of five commissioners appointed by the president and confirmed by the Senate. One of the commissioners is designated the chairman. The commission sets broad frequency policies. Its

Figure 2.12 Upper 6 (6.525–6.875)-GHz band.

Figure 2.13 The 10.5 (10.55–10.68)-GHz band.

members are strongly influenced by high-ranking industry officials and the US Congress. Serious industry policy issues may be discussed personally with the commissioners. However, most commercial activity is coordinated with the various FCC Bureaus.

The Office of Engineering and Technology (OET) advises the commission on technical matters. However, the individual bureaus have considerable influence on technical matters also. The OET Laboratory Division Equipment Authorization Branch sets standards for radio transmitters. The OET Policy and Plans Division includes the Technical Rules Branch, the Spectrum Policy Branch, and the Spectrum Coordination Branch. The Common Carrier Bureau, typically, is involved in wire-line regulatory issues and does not get involved directly in frequency issues. The Wireless Telecommunications Bureau is the most significant bureau for most FCC radio issues. However, it seldom gets involved in satellite-related issues. The International Bureau is the group that is involved in WRC, ITU-R Study Groups, various industry

Figure 2.14 The 11 (10.7–11.7)-GHz band.

Figure 2.15 The 18 (17.7–19.7)-GHz band.

ad hoc committees, and satellite frequency allocation proceedings. This agency is quite influential at the working level in the setting of FCC international and domestic satellite spectrum policy.

The FCC attempts to operate through consensus. The recommendations of industry associations are important factors in FCC decision making. A positive working relationship with the various governmental agencies, industry groups, unofficial coalitions and groups of common interest is crucial to success in influencing fixed terrestrial and satellite policy.

NTIA This organization has regulatory authority over all US federal government radio communications. The most important parts of this organization are the Institute for Telecommunication Sciences (ITS) and the OSM. ITS is the federal laboratory that addresses the technical telecommunications issues. OSM develops and implements policies and procedures for the use of the spectrum controlled by the federal government in the United States. The most important group in OSM is the Frequency Assignment & IRAC

Figure 2.16 The 23 (21.2–23.6)-GHz band.

Figure 2.17 The 38 (38.6–40.0)-GHz band.

(Interdepartmental Radio Advisory Committee) Administrative Support Division. Its IRAC is comprised of members from the 20 most active federal users and the FCC. IRAC is responsible for developing and executing policies and procedures pertaining to frequency management of the federal government spectrum. IRAC is composed of the Frequency Assignment Subcommittee (FAS), the International Notification group, the Radio Conference Subcommittee, Spectrum Planning Subcommittee, Technical Subcommittee, and approximately 20 ad hoc subcommittees. The FAS is the group that coordinates with the FCC to manage FCC/NTIA shared spectrum.

Frequency coordination within NTIA frequency bands is governed by the Manual of Regulations and Procedures for Federal Radio Frequency Management (NTIA "Red Book" manual, http://www.ntia.doc.gov/osmhome/redbook/redbook.html). It has little in common with FCC rules and regulations. This complicates coordination with commercial FCC-governed services sharing NTIA frequency bands.

ITU-R National Committee This group, sponsored by the US Department of State, is the official interface of the United States with the ITU. This organization is made up of industry and government members. The major suborganizations are the following:

Radiocommunication Advisory Group (RAG) The Radiocommunication Advisory Group (RAG) is the overall steering committee of the ITU. It sets the overall direction regarding strategic planning, work programs, and ongoing activities. The members of the United States from this group provide high level guidance.

International Telecommunications Advisory Committee (ITAC) This organization is similar to the RAG. It provides strategic planning recommendations. It is a feeder organization to the ITU Working Groups (WGs).

The telecommunications industry participates in WRC preparation through participation in the FCC advisory committees. The WRC Advisory Committee (WAC) takes the consensus views developed by the industry participants in the Informal Working Groups (IWGs). This process is managed by the FCC. Details of structure, activities, and all documents are available on the FCC Web site (Rummler, W. D., private communication, 2009).

WRC Preparation Committee This group takes input from the NTIA and the FCC and prepares the US WRC position documents. This is where the pre-WRC Conference Preparatory Meeting (CPM) United States position is set. This organization significantly influences US international frequency policy.

The US WRC Delegation This group is the official US delegation at the WRC. The United States' positions change through the entire WRC process. Contact with this delegation is crucial to monitoring the process of US international frequency policy setting and allocation.

2.4.1.2 Industrial Organizations Many professional organizations lobby Congress and the FCC for policy and rules favorable to their interests. Congress and the FCC attempt to provide rules favorable to the most significant users. The following Washington, DC organizations are among those who have considerable influence on FCC policy and rules.

Fixed Wireless Communications Coalition (FWCC) Fixed Wireless Communications Coalition (FWCC) is a coalition of companies, associations, and individuals interested in the terrestrial fixed microwave communications. It is the single most significant organization representing the interests of both, the fixed point to point microwave radio users and manufacturers.

National Spectrum Managers Association (NSMA) This organization represents the microwave (both fixed terrestrial and satellite) coordination organizations within the United States. Its primary focus is to define the implementation methodology to support the reduction of interference among all users. Since it represents all radio interests, it is usually policy neutral. It does not establish rules or policy. It establishes procedures to implement them.

Telecommunications Industry Association (TIA) This organization represents manufacturers of telecommunications equipment in the United States. Satellite and fixed terrestrial interests are represented by different divisions. This organization can be very influential regarding FCC plans and policy for telecommunications users and manufactures.

Utilities Telecommunications Council (UTC) This organization represents US utilities.

Association of American Railroads (AAR) This organization represents the railroads.

American Petroleum Institute (API) The telecommunications subcommittee of this organization represents the oil companies in telecommunications matters.

Association of Public-Safety Communications Officials (APCO)—International This organization represents the domestic and international police, fire, and local government organizations in telecommunications matters.

Cellular Telecommunications Industry Association (CTIA) This organization represents the cellular and some licensed PCS (personal communications services) users.

National Association of Broadcasters (NAB) This organization is the most influential private telecommunication group in Washington. It wields enormous influence in all FCC and congressional frequency policy matters.

Various Manufacturers Various manufacturers lobby Congress and the FCC for policy and rules favorable to their interests. They typically do this individually and as part of industry groups.

2.4.1.3 Intergovernmental

ITU This specialized agency of the United Nations has several support groups within the United States. The most significant are the ITU-R USA Study Groups.

The ITU-R USA Study Groups, aligned with the ITU-R groups, are sponsored by the FCC. They are quite influential in developing ITU recommendations.

Study Group 1 (SG 1)	Spectrum management
Study Group 3 (SG 3)	Radiowave propagation
Study Group 4 (SG 4)	Satellite services
Study Group 5 (SG 5)	Terrestrial services
Study Group 6 (SG 6)	Broadcasting service
Study Group 7 (SG 7)	Science services

The cost of participation in these groups at the national and international level is significant. New radio services take an active interest. They develop methodologies of "spectrum sharing," which they then feed through the ITU study group and WRC Preparation Committee process. If these positions are successfully adopted at the WRC, the new services then lobby the FCC to adopt the new rules within the rules for domestic FSs in the interest of "world telecommunications harmonization." Mature FSs fail to participate in this process at their own peril.

North American Free Trade Agreement (NAFTA) The Telecommunications Standards Subcommittee (TSSC), established pursuant to the North American Free Trade Agreement (NAFTA) and comprised of governmental representatives from the United States, Mexico, and Canada, is charged with facilitating the implementation of NAFTAS's telecommunications-related provisions . The Consultative Committee on Telecommunications (CCT) is comprised of private sector representatives and assists the TSSC.

NAFTA is aimed at facilitating telecommunications equipment deployment. Thus, much of the TSSC's work deals with standards and conformity assessment procedures that are often used as a means to limit market access. NAFTA limits the types of standards that can be imposed on telecommunications terminal equipment to those that can be justified under certain criteria. One such criterion is to prevent electromagnetic interference, and ensure compatibility with other uses of the electromagnetic spectrum. Thus, the TSSC and the CCT indirectly may be involved in frequency policy. The country most affected by these NAFTA criteria is Mexico, as Canadian and US regulations already meet these criteria.

Inter-American Telecommunications Union (CITEL) The Inter-American Telecommunications Commission (CITEL) is the advising entity to the Organization of American States (OAS) in telecommunications matters as a specialized commission for the OAS Inter-American Economic and Social Counsel. CITEL is formed by an assembly, a permanent executive committee, and three permanent consultative committees (PCCs).

Most concrete work of CITEL is carried out in the PCCs. PCC 1 deals with public telecommunication. PCC 2 deals with broadcasting issues. PCC 3 deals with radiocommunications issues. The main goals of PCC 3 are the harmonization of services, the reduction of harmful interference, and the promotion of ITU regulations and standards. Specific issues are studied in detail by WGs chaired by member nations.

PCCs meet once or twice a year in plenary sessions. Each CITEL member (i.e., government of an OAS member country) has one vote during PCC plenary sessions. Over 60 nongovernmental organizations, such as private sector companies and associations, pay annual membership dues and have the status of associate members of CITEL, with a voice but no vote. Nongovernmental groups from the United States participate extensively in CITEL PCC meetings. CITEL PCCs report their findings to the member state telecommunications regulators, and thereby influence the ITU standardization process. CITEL is especially active in WRC matters. It provides the forum for developing WRC inputs that represent the consensus of ITU-R Region 2.

2.4.2 Canada

2.4.2.1 Governmental

Industry Canada Industry Canada is similar to a combination of the US FCC and NTIA. This governmental organization (http://www.ic.gc.ca/eic/site/ic1.nsf/eng/h_00006.html) is responsible for managing both private and government spectrum in Canada. Its spectrum policy branch develops frequency policies through the use of *gazette notices* (the Canadian equivalent of an FCC notice of proposed rulemaking). Its spectrum engineering branch implements these policies and relies heavily on the advice and recommendations of the Radio Advisory Board of Canada (RABC). Preparation for WRCs is the ongoing responsibility of the Canadian preparatory committee. The Industry Canada ad hoc group that handles ITU-R matters is the Canadian National Organization (CNO/ITU-R).

2.4.2.2 Industrial

Radio Advisory Board of Canada (RABC) The RABC is an industry advisory group comprised mainly of associations of users and manufactures of radio equipment with Industry Canada sitting as an observer. The terrestrial microwave manufactures sit at the RABC radio relay committee under the umbrella of Electro-Federation Canada (EFC). Industry Canada pays considerable attention to this group. The RABC is the single most powerful body for influencing frequency policy in Canada and implementing it. It operates through consensus and meets three to four times a year.

Frequency Coordination System Association (FCSA) This association is similar to the National Spectrum Managers Association in the United States. It is the umbrella organization for the various RF coordination groups. This group prepares recommendations for coordination methods and procedures but is less politically active than its US equivalent, the National Spectrum Managers Association.

2.5 FCC FIXED RADIO SERVICES

Chapter I of the CFR Title 47—Telecommunications establishes the FCC and the following Fixed Radio Services:

Experimental Radio	Part 5
Unlicensed Radio	Part 15
Domestic Public Fixed Radio	Part 21
International Fixed Public Radiocommunication	Part 23
(Public Fixed) Satellite Communications	Part 25
TV Studio-Transmitter Links (STLs)	Part 74 Subpart F
Fixed Point-to-Point Microwave Services	Part 101

The frequency bands used by these services are defined in Part 2. However, deployment within these allocations is prohibited until rules are included within the CFR. Regulations are always changing. Frequency bands are reallocated (services are moved, eliminated, or created) and sharing among services may be allowed. Use of the licensed bands is based on the class of service. A user licensed with a primary status is accorded protection from harmful interference from any other user (whether primary, secondary, or unlicensed). A user licensed with secondary status is allowed to use the band but has no legal recourse if interfered with. Harmful interference is defined as "Any emission, radiation or induction that endangers the functioning of a radio navigation service or of other safety services or seriously degrades, obstructs or repeatedly interrupts a radiocommunications service." [15.3(m)] *For the above and following sentences the citation within brackets [ ] indicates the paragraph (Part) and subparagraph within the CFR 47, Chapter 1, containing the reference.*

Part 101 defines three FSs:

Private Operational Fixed Point-to-Point Microwave Service—Part H.
Common Carrier Fixed Point-to-Point Microwave Service—Part I.
Local Multipoint Distribution Service (LMDS)—Part L.

The following is a summary of the rules that apply to these services:

Microwave radio is defined as radio operation above 890 MHz [101.3].
No foreign government can hold a radio license [101.7].
No foreign corporation may operate a common carrier radio service [101.7].
Common carrier services may be concurrently licensed for noncommon carrier communications purposes [101.133(a)].
Private carrier and common carrier transmission facilities may be interconnected [101.135].
Private carriers may offer for-profit private carrier service [101.135].
More than one private carrier may use the same transmission facilities [101.133].
License applications are typically one of the following [1.929]:
- Application for initial authorization
- Application for renewal of authorization (typically once every 10 years)
- Application to change ownership or control (including partitioning and disaggregation)
- Application requesting authorization for a facility that would have a significant environmental effect
- Application for an amendment that requires frequency coordination, including adding new frequency or frequencies
- Application for special temporary authority [1.931], or temporary or conditional authorization [101.31].

Emergency operations are allowed in some cases [101.205].
Licenses normally authorize operation between or among individual stations. Operation at 38.6–40.0 GHz is based on a Partitioned Service Area (PSA) [101.56] and [101.64].
License applications for new authorization must contain the following [101.21(e)]:
- Applicant's name and address
- Transmitting and receiving station name
- Transmitting and receiving station coordinates (within 1 s)
- Frequencies and polarizations to be added, deleted, or changed
- Transmitting equipment, its stability, effective isotropic radiated power (EIRP), emission designator, and type of modulation
- Transmitting antenna(s), model, gain, and radiation pattern (if required)
- Transmitting and receiving antenna center line height(s) above ground level and ground elevation above mean sea level [within 1 m (3.3 ft)]
- Path azimuth and distance.

Licensee must file a modification application if major changes are made [1.947(a)].

Major changes are defined as any of the following [1.929(d)(1)]:

Any change of transmitter antenna location by more than 5 s in latitude or longitude

Any increase in frequency tolerance

Any increase in bandwidth

Any change in emission type

Any increase in EIRP of more than 3 dB

Any increase in transmit antenna height of more than 3 m (9.8 ft)

Any increase in transmit antenna beamwidth

Any change in transmit antenna polarization

Any change in transmit antenna azimuth greater than 1°

Any change since the last major modification that may produce a cumulative effect exceeding any of the above criteria.

License applications for any major change to an existing authorization must contain the following [101.21 (e)]:

Applicant's name and address

Transmitting and receiving station name

Transmitting and receiving station coordinates (within 1 s)

Frequencies and polarizations to be added, deleted, or changed

Transmitting equipment, its stability, EIRP, emission designator, and type of modulation

Transmitting antenna(s), model, gain and radiation pattern (if required)

Transmitting and receiving antenna center line height(s) above ground level and ground elevation above mean sea level [within 1 m (3.3 ft)]

Path azimuth and distance.

To operate channels with a bandwidth of at least 10 MHz and with channel frequency between 3.7 and 11.7 GHz, transmitters must meet the minimum payload capacity requirements [101.141(a)(3)].

This capacity must be loaded (utilized) within 30 months of licensing [101.141(a)(3)].

Attachment of appropriate multiplex equipment meets minimum loading requirements [101.141(a)].

Minimum transmit and receive antenna standards are imposed (these standards do not apply to diversity antennas) [101.115].

Antenna structures (towers, buildings, etc.) higher than 200 ft must be registered with the FCC [17.7] (Fig. 2.18).

Antenna structures within the specified glide slope of enumerated airports and heliports must be registered with the FCC [17.4] (Fig. 2.19).

See the subsequent section for a detailed discussion regarding path clearance near airports and heliports.

Exceptions are provided for minor additions to existing structures [17.7 (a) and 17.14 (b)].

FAA must be notified of proposed antenna structure construction [17.7].

Rules are imposed for painting and lighting these structures [17.21].

Transmitter frequency tolerance [74.661][101.107] and power (EIRP) limitations [74.636][101.113] apply.

For transmitters using ATPC, this power limitation applies to the maximum transmit power, not to maximum coordinated power [101.143(b)].

Paths shorter than the following [74.644(a)][101.143(a)] require transmit power reduction:

1.850–7.125 GHz: 17 km (10.6 miles)

10.550–13.250 GHz: 5 km (3.1 miles)

The transmitter power reduction formula is

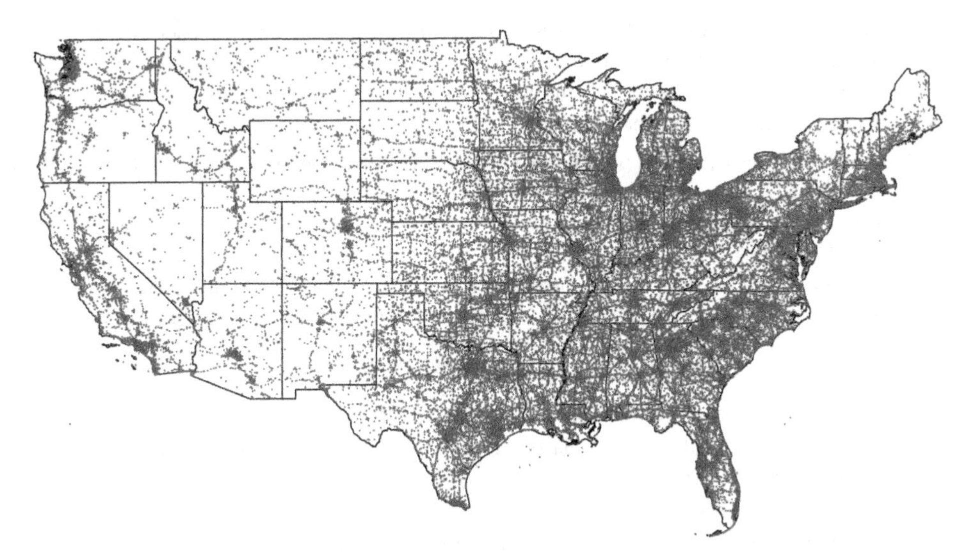

Figure 2.18 Locations of FCC-registered antenna structures.

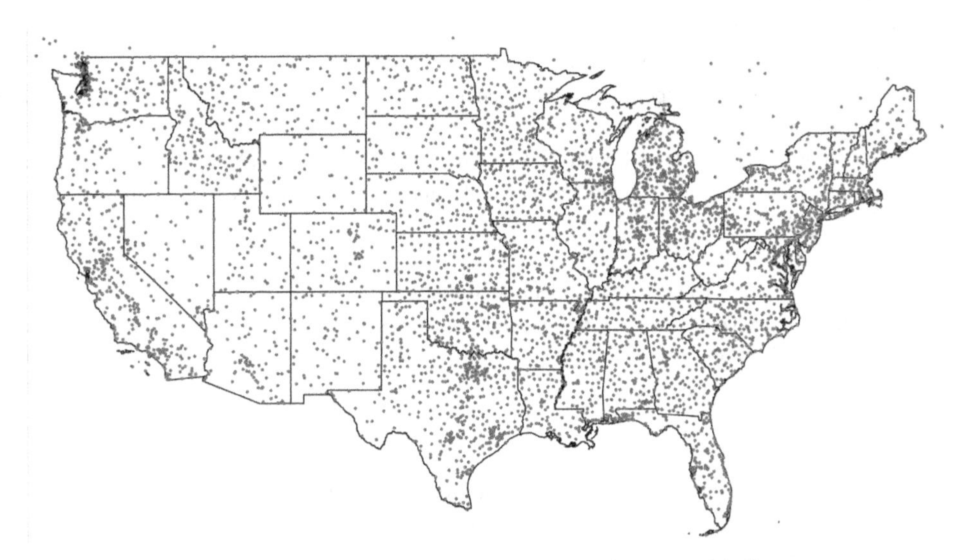

Figure 2.19 Locations of FCC-registered airports and heliports.

EIRP = Max EIRP − 20 $\log_{10}$ (Limit/Actual) [74.644(b)]

EIRP = Max EIRP − 40 $\log_{10}$ (Limit/Actual) [101.143(b)]

EIRP = usually allowable transmitter power

limit = above minimum distance limit

actual = the actual path length.

Frequency diversity requires at least 1 : 3 channels within 3 years [101.103(c)]. Note that there are at least three exceptions to this rule. First, collapsed rings operation is allowed (Wireless Telecommunications Bureau, 2000). Second, if a frequency diversity radio system provides for the different traffic to be placed on different frequencies (the so-called *protect channel access* with high priority traffic on one frequency channel and low priority traffic on the other frequency channel) and the higher priority channel preempts operation of the lower priority channel as needed, this type of operation would be allowed (Knerr, 1998). Third, if the multiple radio channels are connected to an IP router, different traffic will be applied to the different radio channels. If a channel fails, the router will not transmit low priority traffic.

Waivers of technical rules may be requested [1.925][101.23].

Common carriers must provide special showing for renewal of systems using frequency diversity [101.705].

Coordination applies for new applications, major amendments, or major modifications to existing licenses [101.103].

License applications must contain evidence of prior coordination of proposed frequency use with existing licensees, permittees, and applicants in the applicable coordination area [101.21 (f)][101.103(d)(1)].

Coordination must use procedures in [25.251][101.103 (d)].

Coordination must include geostationary satellite users [101.145].

Coordination within 35 miles of Canada or Mexico has special requirements and involves the US government [101.31(b)(v)].

"Quiet zones" must be respected [0.121][1.924].

License with prior coordination notice must contain the following [101.103(d)(2)(ii)]:

Applicant's name and address

Transmitting and receiving station name

Transmitting and receiving station coordinates (within 1 s)

Frequencies and polarizations to be added, deleted, or changed

Transmitting equipment type, its stability, actual output power, emission designator, and type of modulation

Transmitting antenna(s), model, gain, and radiation pattern (if required)

Transmitting and receiving antenna center line height(s) above ground level and ground elevation above mean sea level [within 1 m (3.3 ft)]

Path azimuth and distance

Estimated transmitter and receiver transmission line loss in dB

For systems employing ATPC, maximum transmit power, maximum coordinated transmit power, and nominal transmit power.

Coordination is a two-part process:

Notification

Response.

The maximum coordination period is 30 days after receipt by the entity being notified.

The notifying party must secure positive responses from all notified entities.

No response within 30 days is assumed a positive response (most coordinators allow 35 days to provide time for the recipient to receive the coordination notice).

Expedited prior coordination may be requested by the notifying party.

New applicants are required to technically resolve any potential interference. Both parties are encouraged resolve interference disputes. The Commission can be contacted as a last resort [101.106(e)].

New applicants must make reasonable effort to avoid blocking existing coordinated systems. Any frequency reserved by the current licensee for future use must be released for use by pending applicant on showing that the use of any other frequency cannot be coordinated. Prior coordination is canceled if license application is not filed in 6 months or within 10 days of the end of the 6-month period. When co-pending applicants file, the earliest file date has priority for license.

The following is a summary of licensed station operator requirements:

All point-to-point services may begin station construction before station authorization at applicant's risk [101.5].

On filing a properly completed application, successful completion of prior coordination, and tower clearance by the FAA, conditional authorization to operate is granted [101.31(b)] with the following exceptions:

Operation of paths within 35 miles (56.3 km) of the Canadian or Mexican border must not begin until the license is granted.

Operation of paths in bands shared with the federal government (NTIA) must not begin until the license is granted.

Operation should not be within a "quiet zone" or other similarly designated area.

Although not stated, if a station license application is modified after submission, operation may not begin until the final license is granted.

Operation must begin within 18 months from the granting of the license (except LMDS & 38.6–40.0 GHz service) [1.946] and [101.63 (a)].

Licensee must file a notification of compliance within 15 days of expiration of the 18-month construction period [1.946(d)].

License for station authorization is issued for 10 years [101.67].

Station operator must retain the license application certification form [101.31(b)(viii)(3)].

Station identification (from over the air) is not required [101.212].

Tower lights must be inspected (ideally automatically) at least once every 24 h [17.47].

Light monitoring equipment must be inspected every 3 months [17.47].

Record of light inspections must be maintained [17.49].

Record must be maintained of any known malfunction of the tower lighting system [17.49].

Repairs or modifications must be recorded.

Date and time of FAA notification of malfunction must be recorded.

Towers ("antenna structures") must be cleaned or painted as often as necessary to maintain good visibility [17.50].

Transmitters must be installed so as to limit operation to those authorized by the licensee [101.131].

Station operator must post station authorization and transmitter identification [101.215].

Name, address, and telephone number of custodian must be posted at each station [101.215].

Station operator must maintain station records of the following actions and the name and address of the individual performing them [101.217]:

Results and dates of transmitter measurements

Pertinent details of all transmitter adjustments (Note that monitoring transmitter frequency and power is no longer required. However, it is still a good idea that they be monitored.).

Records must be kept in an orderly manner and readily available [101.217].

All records must be retained for at least 1 year [101.217].

Licensee must make the radio station available for inspection by the Commission [101.201].

Operation of an intentional, unintentional, or incidental radiator is subject to the conditions that no harmful interference is caused ... [15.5(b)]. The operator of an RF device shall be required to cease operating the device on notification by a Commission representative that the device is causing harmful interference. Operation shall not resume until the condition causing the harmful interference has been corrected [15.5(c)].

Harmful interference is defined as "Any emission, radiation or induction that endangers the functioning of a radio navigation service or of other safety services or seriously degrades, obstructs or repeatedly interrupts a radiocommunications service" [15.3(m)].

2.6 SITE DATA ACCURACY REQUIREMENTS

FCC rule 101.21 (e) requires the horizontal accuracy of coordinates to be no less than 1 s and the vertical accuracy to be no less than 1 m. One second of latitude represents approximately 101 ft. One second of longitude represents approximately 92 ft in the southern most parts of the United States and approximately 66 ft in the northern most parts of the United States. USGS 7.5-min maps have an average elevation accuracy of 5 ft (1.52 m). USGS digital seamless NED data has an average elevation accuracy of 8 ft (2.44 m). Neither of these sources is suitable for meeting the 1 m site elevation accuracy requirement.

During FAA studies of building structures, the FAA may impose measurement accuracy standards. The following FAA Obstacle Accuracy Codes are defined in FAA Order 8260.19E, Appendix 3:

Horizontal Measurements		Vertical Measurements	
Code	Tolerance	Code	Tolerance, ft
1	±20 ft	A	±3
2	±50 ft	B	±10
3	±100 ft	C	±20
4	±250 ft	D	±50
5	±500 ft	E	±125
6	±1000 ft	F	±250
7	±½ mile (nautical)	G	±500
8	±1 mile (nautical)	H	±1000
9	Unknown	I	Unknown

Current FCC license requirements (101.21 (e)) dictate at least 2A accuracy.

2.7 FCC ANTENNA REGISTRATION SYSTEM (ASR) REGISTRATION REQUIREMENTS

The *Antenna Structure Registration Program* (Part 17) is the process under which each antenna structure that requires FAA notification—including new and existing structures—must be registered with the FCC by its owner. The owner is the single point of contact to resolve antenna-related problems and is responsible for the maintenance of those structures that require painting and/or lighting. Note that because the ASR requirements apply only to those antenna structures that may create a hazard to air navigation (either by their height or by their proximity to an airport), the registration files do not contain a comprehensive record of all antenna structures. The ASR does not replace the FAA notification requirement. When the antenna structure is registered, an ASR number is assigned. This number is seven digits long, with the first (leftmost) digit being one. ASR number assignments may be determined at the Web site (http://wireless2.fcc.gov/UlsApp/AsrSearch/asrRegistrationSearch.jsp).

The FCC rules specifically define the term *antenna structures* as "[T]he radiating and/or receive system, its supporting structures and any appurtenances mounted thereon." In practical terms, an antenna structure could be a free-standing structure, built specifically to support or act as an antenna, or it could be a structure mounted on some other man-made object (such as a building or bridge). Note that in the latter case, the structure must be registered with the FCC, and not the building or the bridge. Objects such as buildings, observation towers, bridges, windmills, and water towers, the *primary* function of which is *not* to mount antenna, *are not* antenna structures and should not be registered. Keep in mind that the FCC only has jurisdiction over antenna structures and, thus, other objects that do not normally house antennas are not required to be registered with the FCC—regardless of their location or height (but the FAA will have an interest in them).

Per CFR, Title 47, Telecommunication, Chapter 1, FCC, Part 17, Construction, Marking and Lighting of Antenna Structures, Subpart B, FAA Notification Criteria, some antenna structures are required to be registered with the FCC.

Sec. 17.7—antenna structures requiring notification to the FAA are discussed here.

A notification to the FAA (CFR), Title 14: Aeronautics and Space, Part 77, Safe, Efficient Use and Preservation of the Navigable Airspace) is required, except as set forth in Sec. 17.14, for any of the following constructions or alterations:

(a) Any construction or alteration of height of more than 60.96 m (200 ft) above ground level at its site; antenna structure heights are recorded without and with appurtenances (structures added to the antenna structure). The FCC considers the "with appurtenances" height when reviewing the 200-ft limit.

(b) Any construction or alteration of height greater than an imaginary surface extending outward and upward at one of the following slopes (*Glide Slope Rules*):

  (1) 100 to 1 for a horizontal distance of 6.10 km (20,000 ft) from the nearest point of the nearest runway of each airport specified in paragraph (d) of this section with at least one runway of more than 0.98 km (3200 ft) in actual length, excluding heliports

  (2) 50 to 1 for a horizontal distance of 3.05 km (10,000 ft) from the nearest point of the nearest runway of each airport specified in paragraph (d) of this section with its longest runway no more than 0.98 km (3200 ft) in actual length, excluding heliports

  (3) 25 to 1 for a horizontal distance of 1.52 km (5000 ft) from the nearest point of the nearest landing and takeoff area of each heliport specified in paragraph (d) of this section.

(c) When requested by the FAA, any construction or alteration that would be in an instrument approach area (defined in the FAA standards governing instrument approach procedures) and when available information indicates it might exceed an obstruction standard of the FAA.

(d) Any construction or alteration in any of the following airports (including heliports):

  (1) An airport that is available for public use and is listed in the Airport Directory of the current Airman's Information Manual or in either the Alaska or Pacific Airman's Guide and Chart Supplement

  (2) An airport under construction that is the subject of a notice or proposal on file with the FAA, and except for military airports, it is clearly indicated that the airport will be available for public use

  (3) An airport that is operated by an armed force of the United States.

Sec. 17.14 lists certain antenna structures exempt from notification to the FAA.
A notification to the FAA is not required for any of the following constructions or alterations:

(a) Any object that would be shielded by existing structures of a permanent and substantial character or by natural terrain or topographic features of equal or greater height, and that would be located in the congested area of a city, town, or settlement where it is evident beyond all reasonable doubt that the structure so shielded will not adversely affect safety in air navigation. The applicant claiming such exemption under Sec. 17.14(a) shall submit a statement with their application to the FCC explaining the basis in detail for their finding.

(b) Any antenna structure of height 6.10 m (20 ft) or less, except one that would increase the height of another antenna structure (*20 Foot Rule*).

(c) Any air navigation facility, airport visual approach or landing aid, aircraft arresting device, or meteorological device, of a type approved by the administrator of the FAA, the location and height of which is fixed by its functional purpose [currently Navigation Aids (i.e., a glideslope, VOR (VHF Omnidirectional Range), or nondirectional beacon) are the main facility of concern].

FCC Form 601, Schedule I lists the following antenna structure codes as appropriate for filing for a MW path license:

Code	Description
B*	Building (with a side mounted antenna)
BANT	Building with antenna on top
BMAST	Building with mast (and antenna) on top
BPIPE	Building with pipe (and antenna) on top
BPOLE	Building with pole (and antenna) on top
BRIDG*	Bridge

BTWR	Building with tower
GTOWER	Guyed structure used for communication purposes
LTOWER	Lattice tower
MAST	Mast (self-support structure used to mount an antenna)
MTOWER	Monopole
NNGTAMM#	Guyed tower array (grouping of guyed towers)
NNLTAMM#	Lattice tower array (grouping of lattice towers)
NNMTAMM#	Monopole tower array (grouping of monopoles)
PIPE	Any type of pipe
POLE	Any type of pole (used only to mount an antenna)
RIG*	Rig used for oil or water extraction or other purpose
SIGN*	Any type of sign or billboard
SILO*	Any type of silo
STACK*	Smoke stack
TANK*	Any type of tank (e.g., water or gas)
TREE*	Tree when used as a support for an antenna
UPOLE*	Utility pole (or tower) used to provide utility service (e.g., electric or telephone service)

The following codes have been used for years, but are obsolete as of June 2012:

NNTAMM#	Antenna tower array
NTOWER	Multiple antenna structures
TOWER	Free-standing or guyed structure used for communications purposes

*This structure, as its primary function is not related to antenna support, is not considered an antenna structure (although it may perform that function in addition to its primary function) and, therefore, is exempt from ASR requirements. However, it still must conform to FAA glide slope requirements.

#The NN indicates the number of towers in the array. The MM is optional and indicates the position of that tower in the array. The value of MM would be between 1 and NN (inclusive).

The above abbreviations are also used in the FCC ASR database (ULS Downloads/Databases/Database Downloads/Antenna Structure Registration at http://wireless.fcc.gov/uls).

If an antenna structure is attached to any of these exempted structures, then only that attached structure is considered an antenna structure (e.g., BANT, BMAST, BPIPE, BPOLE, and BTWR above). If that attached structure exceeds 20 ft above the nonantenna structure, then it is subject to FAA glide slope rules and possibly ASR requirements. Nonexempted structures are considered antenna structures and additions are treated as extensions of that structure. Therefore, if a 19 ft structure is attached to the top of a 190-ft smoke stack, no ASR is required (the smoke stack is not an antenna structure and the antenna structure is less than 20 ft high). However, if a 19 ft structure is added to the top of a 190-ft self-supporting tower, an ASR is required (because the antenna structure height is now 209 ft).

The FCC Web site (http://wireless2.fcc.gov/UlsApp/AsrSearch/towairSearch.jsp) has a tool (TOWAIR Determination) that can be used to test whether or not a site passes the FAA glide slope requirements. The FAA Web site (https://oeaaa.faa.gov/oeaaa/external/portal.jsp) has a tool (Notice Criteria Tool on the left border of Web site) that tests whether or not a site passes both the glide slope and navigational facilities requirements.

2.8 ENGINEERING MICROWAVE PATHS NEAR AIRPORTS AND HELIPORTS

When designing microwave networks, airport or heliports occasionally appear under or near paths. The following guidelines are proposed for microwave paths that might be adversely affected by aircraft:

Regarding *heliports*, as helicopters can move up or down vertically without limitation, no microwave path should pass directly over a helipad. Movement laterally around a helipad is unrestricted and unpredictable, so microwave paths within one-quarter mile of the pad should be avoided.

Airports should be investigated for potential helicopter use. Helicopter landing pads are often located in airports. Helicopters can take off or land from any airport ramp. They may take off and land on runways if they have to make an emergency landing. Microwave paths over a runway used by helicopters can be problematic.

Radar transmitters can interfere with fixed point to point microwave paths. The National Weather Service operates the NEXRAD weather radar on hills near airports. It operates between 2.70 and 3.00 GHz. Sometimes, its second harmonic interferes with nearby 6-GHz receivers. However, this is rare. Of more concern is the High Resolution Terminal Doppler Weather Radar (TDWR) that operates between 5.60 and 5.65 GHz and is located near most large airports. The transmitters are often operated without filtering to avoid filter losses (and increase detection range). Microwave paths operating at lower than 6 GHz should avoid passing over or near commercial airports that use this radar. Airports using TDWR may be found at http://www.wunderground.com/radar/map.asp.

Airport runways in the United States are 800 to 18,000 ft in length (Fig. 2.20). For commercial airports, the touchdown area (area within which the aircraft must land on the airport runway) is a maximum of 3000 ft long (from threshold to last mark). For airports that have runways shorter than 7000 ft, the touchdown area may be shorter. The end of the touchdown area is marked with a large number of hash marks. This group of hash marks is the threshold marker.

The part of the runway between the threshold markings is used for *landing*. Smaller airports may not have a marked touchdown area. However, all pilots are expected to land in the first half of the runway regardless of airport size. Therefore, for unmarked airports, the touchdown area is the first half of the runway that the plane approaches. The airplanes are allowed to use the second half of the runway (including the displaced threshold) to stop.

The short area beyond the landing threshold is the displaced threshold. This is the lineup area for takeoffs and for taxiway entrance and exit. Airplanes can take off from this area but are not permitted to land on it (but can use it to stop). This area is usually marked with large white arrows pointing in the direction of the runway.

Large commercial airports have a short portion of the runway extending beyond the displaced threshold called the *blast pad*. It keeps the jet exhaust from eroding the ground and provides a safety area for troubled aircraft that require additional landing distance. This area, if it exists, is typically marked with yellow chevrons (V shapes). Planes should never be on this area except in an emergency.

Commercial airliners flying the precision glide slope appear 50 ft above the runway threshold marker and, with 3° glide slope, touch the runway 1000 ft from the beginning of the touchdown area threshold (on the touchdown marker). Private planes are allowed to use glide slopes between 2.7° and 4° and may land anywhere in the touchdown area. When approaching the runway, private pilots at small airports are expected to maintain a minimum of 500 ft above ground level while approaching the runway until they begin the glide slope. Pilots of large commercial airliners are expected to maintain 1200 ft of minimum elevation until they begin to land.

During *takeoff*, the commercial airplane climb angle varies from 10° to 20°. Private plane climb angles are more in the 2°–5° range. Climb angles vary greatly depending on plane type, its weight loading and runway altitude, and temperature and barometric pressure.

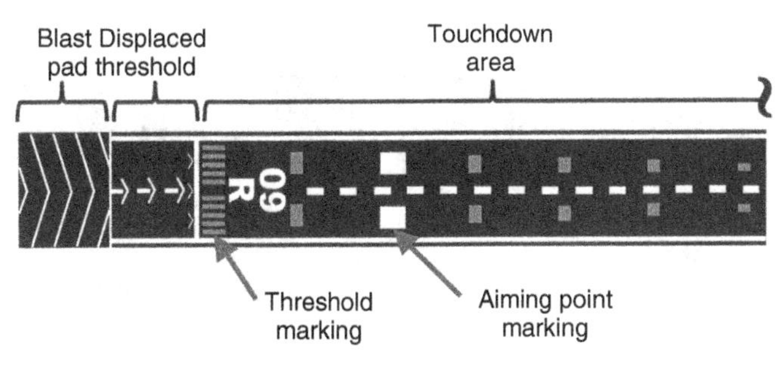

Figure 2.20 Airport runway.

The launch location (takeoff rotation point) is controlled by speed rather than location on the runway. However, it usually occurs in the final 40% stretch of the runway for medium and large commercial airplanes. Private planes are unpredictable. Some can takeoff in as short a distance as 200 ft. Their takeoff point can be anywhere on the runway. Private jets with light loads often takeoff using much shorter distances than heavily loaded commercial planes. Military planes can take off at very steep angles using very little of runway. Microwave links over private or military airport runways are highly speculative.

If the airplane is *not landing* or *taking off*, it can be anywhere on the airport tarmac. Therefore, a basic limitation applied to all runways, taxiways, and parking areas is the maximum vertical height of the tail of the largest aircraft in use at that runway. For commercial airports, that would be the Boeing 747 with a 64 ft tail height. For military airports, the largest plane is the Lockheed C-5 with a tail height of 65 ft. The microwave height minus first Fresnel zone distance (below) should exceed these limits

2.8.1 Airport Guidelines

For microwave paths traversing commercial airports, all microwave path heights *minus* first Fresnel zone clearance should *exceed* 65 ft or the expected height of the tail of planes expected to land at the airport.

For private airports, microwave paths should not cross between the landing threshold markers because the planes may takeoff from anywhere on the runway. For heavily loaded scheduled commercial airplanes, the possible takeoff height Y_{TO}: is estimated by the following:

$$Y_{TO} = (\tan 20^\circ \times d) + h_T = (0.364 d_{TO}) + h_T$$

d_{TO} = distance out (away from the center of the runway) from point D;
h_T = worst case airplane tail height $\approx$65 ft;
D = a point 50–75% of the way from one threshold marker to the other threshold marker. Of course, both directions of takeoff must be considered. For large commercial runways, the value is near 50%. For smaller commercial airports the value is closer to 60–75%.

For microwave paths traversing airports limited to scheduled commercial airplanes, all microwave path heights *minus* first Fresnel zone clearance should *exceed* Y_{TO} (Fig. 2.21).

For microwave paths traversing private or commercial airports, all microwave path heights *plus* first Fresnel zone clearance should *be less than* the possible landing height Y_L:

$$Y_L = \tan 2.7^\circ \times d = 0.0472 d_L$$

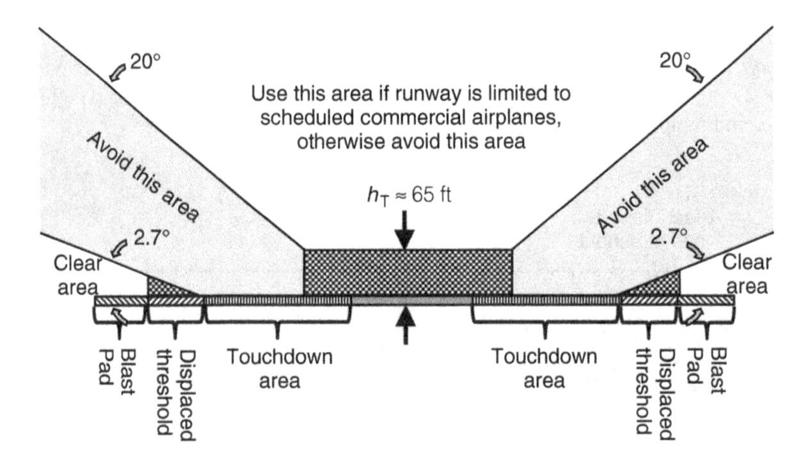

Figure 2.21 Airport exclusion area.

d_L = distance out (away from the center of the runway) from the runway threshold marker.

First Fresnel Zone Radius F_1

$$F_1\,(\text{ft}) = 72.1 \text{ sqrt}\left[\frac{d_1\,(\text{miles})\, d_2\,(\text{miles})}{F(\text{GHz})D(\text{miles})}\right]$$

$$F_1\,(m) = 17.3 \text{ sqrt}\left[\frac{d_1\,(\text{km})\, d_2\,(\text{km})}{F(\text{GHz})D(\text{km})}\right]$$

$$F_n = F_1 \text{sqrt}(n)$$

sqrt(x) = square root of x;
n = Fresnel zone number (an integer);
d_1 = direct distance from one end of the path to the reflection;
d_2 = direct distance from other end of the path to the reflection;
D = total path distance = distance = $d_1 + d_2$;
F = frequency of radio wave.

REFERENCES

Committee TR14-11, *TIA/EIA Telecommunications Systems Bulletin 10-F, Interference Criteria for Microwave Systems.* Washington, DC: Telecommunications Industry Association, June 1994.

Curtis, H. E., "Interference between Satellite Communications Systems and Common Carrier Surface Systems," *Bell System Technical Journal*, Vol. 41, pp. 921–943, May 1962.

ITU-R Recommendations. Geneva: International Telecommunications Union—Radiocommunication Sector, available online for a fee or by subscription to biannual DVD, 2013.

Knerr, A., Chief, Technical Analysis Section, Public Safety and Private Wireless Division, private letter to M. Blomstrom, State of Nevada, Reference: PS&PWD-LTAB-647, September 3, 1998.

Linthicum, J. M., "A Guide to the FCC's Rulemaking Procedures," *IEEE Communications Magazine*, Vol. 19, pp. 34–37, July 1981.

Mosley, R. A., Director, *Code of Federal Regulations (CFR), Title 47—Telecommunication, Chapters 1, 2 and 3.* Washington: Office of the Federal Register, published yearly.

Wireless Telecommunications Bureau Order, Alcatel USA, Inc. Request for Ruling that Part 101 Frequency Diversity Restrictions are not Applicable to Collapsed Ring Architecture for Microwave Systems, Adopted January 19, 2000.

Withers, D., *Radio Spectrum Management*. London: Institution of Electrical Engineers, 1999.

Working Group 18, *Automatic Transmit Power Control (ATPC), Recommendation WG 18.91.032*. Washington: National Spectrum Managers Association, 1992.

Working Group 3, *The Contents of Prior Coordination Notifications, Recommendation WG 3.86.002*. Washington: National Spectrum Managers Association, 1986.

Working Group 3, *Primer on Frequency Coordination Procedures, Recommendation WG 3.87.001*. Washington: National Spectrum Managers Association, 1987.

Working Group 3, *Coordination Contours for Terrestrial Microwave Systems, Recommendation WG 3.90.026*. Washington: National Spectrum Managers Association, 1992.

Working Group 3, *Coordination Procedures for Automatic Transmit Power Control (ATPC), Recommendation WG 3.94.041*. Washington: National Spectrum Managers Association, 1995.

Working Group 5, *Report & Tutorial, Carrier-to-Interference Objectives, Recommendation WG 5.92.008*. Washington: National Spectrum Managers Association, 1992.

Working Group 9, *Long Term/Short Term Objectives for Terrestrial Microwave Coordination, Recommendation WG 9.85.001*. Washington: National Spectrum Managers Association, 1985.

3

MICROWAVE RADIO OVERVIEW

3.1 INTRODUCTION

The purpose of a communication system is to transport information from one location (the transmitter or source) to another (the receiver, destination, or sink). Information in its simplest form is knowledge previously unknown to the receiver. The signal conveying information will have the characteristics of a random process (similar to noise) if it is to convey information most efficiently. However, for the information to be interpreted, the signal must have some predefined nonrandom components to define when the information is being received. By definition, this "framing" information must be known previously and must therefore contain less information than that being conveyed. As Hartley (1928) noted, "The capacity of a system to transmit a particular sequence of [information conveying] symbols depends upon the possibility of distinguishing at the receiving end between the results of the various selections made at the sending end."

The transmitted information is contained in a time-varying electrical signal called the *payload* or *baseband*. The fundamental quality of analog payloads, which may have an arbitrary ("infinite") number of states at any given instant of time, is characterized by average signal-to-noise (power) ratio. The fundamental quality of digital payloads, which may have a predefined number of states at any instant of time, is characterized by the probability of message error. Digital radio transmission systems process payload signals in a digital environment but transport those signals between two locations in an analog environment. Depending on where the payload signal is observed, the signal may be regarded as analog or digital.

A payload may be described as being sampled at repetitive time intervals T (samples per second) and encompassing a frequency range (bandwidth) F (Hz). If the payload is statistically time invariant ("stationary"), the information content of the signal may be loosely associated with the product FT. For a digital payload signal, information is delivered in discrete periodic times using a symbol from a predefined symbol set ("alphabet"). For a statistically stationary digital signal from a source with no memory (each symbol is statistically independent from the others), the average information content of a symbol is the entropy of that symbol: the sum of (the negative logarithm of the probability of each possible symbol multiplied by the probability of that symbol). If the logarithm is base 2, the information content unit is a bit. If the base is 10, the content unit is a digit (Hartley, 1928; Shannon, 1948). Information content is directly related to symbol uncertainty and entropy. For the systems we will consider, all baseband symbols will be equally likely.

Digital Microwave Communication: Engineering Point-to-Point Microwave Systems, First Edition. George Kizer.

In 1623, Francis Bacon was the first to notice that information could be completely described using binary symbols, "... a man may expresse and signifie the intentions of his minde, at any distance ... by objects which ... be capable of a twofold difference onely ..." (Aschoff, 1983). *Bit*, a term attributed to J. W. Tukey (Shannon, 1948), represents the state of a single binary symbol. (The term *bit* was first used in an article by Claude Shannon in 1948.) This state is usually represented as "0" or "1". Digital signals (of uniform probability) are usually described in units of bits. The average transmission capacity of a system is generally expressed as bits per second (b/s). While the *bit* is the most common measure of symbol information content, other units are also used, but rarely. One bit equals approximately 0.301 decits or 0.693 nepits.

A symbol (single signaling element) represents a single transmission event. The term *baud*, the unit of signaling speed, is equal to the number of symbols per second. If a symbol of a memory-less system (trellis and partial response transmission systems use memory in their coding and are exceptions) may assume any of the m states with equal probability, then the number of bits n that may be transmitted by the symbol is defined by the relationship $m = 2^n$ or $n = \log_2(m) \cong 3.32193 \log_{10}(m)$. For example, 64 quadrature amplitude modulation (QAM) may assume any of the 64 (constellation) states ($m = 64$). Each symbol represents 6 bits ($n = 6$). If the Nyquist bandwidth signaling limitation (discussed below) is considered, 64 QAM is often described as having a spectral efficiency of 6 bits/s/Hz.

If digital communication signals are distorted by noise or channel distortions, the receiver may make mistakes ("errors") in determining the transmitted digital signal. For radio systems, the most common measure of error performance is termed *bit error ratio* (*BER*) (Kizer, 1995). This is the ratio of binary errors to total number of transmitted bits. BER is measured in radios using many different methodologies (Newcombe Pasupathy, 1982). While BER is a common criterion for radio performance, it is well known that many practical systems have errors in bursts (Johannes, 1984). Measuring error performance using errored-second ratio (ESR, ratio of seconds with at least one received error to the total number of seconds of data transmission) and background block error ratio (BBER, ratio of transmitted blocks with errors to the total number of transmitted blocks) is common method of quantifying error performance. BER is popular in North America and BBER in Europe (ITU-T and ITU-R).

Communication of information requires a previously agreed signal format to encode the information for transmission and the capacity to transmit that signal. Shannon (1950) observed, "The type of communications system that has been most extensively investigated ... consists of an information source which produces the raw information or message to be transmitted, a transmitter which encodes or modulates this information into a form suitable for the [transmission] channel, and the channel on which the encoded information or signal is transmitted to the receiving point. During transmission the signal may be perturbed by noise ... . The received signal goes to the receiver, which decodes or demodulates to recover the original message, and then to the final destination of the information."

Shannon (1950) described this process using the lumber mill analogy: "A basic idea in communication theory is that information can be treated very much like a physical quantity such as mass or energy. ... The system ... is roughly analogous to a transportation system; for example, we can imagine a lumber mill producing lumber at a certain point and a conveyor system for transporting the lumber to a second point. In such a situation there are two important quantities, the rate R (in cubic feet per second) at which lumber is produced at the mill and the capacity C (cubic feet per second) of the [lumber] conveyor. If R is greater than C it will certainly be impossible to transport the full output of the lumber mill. If R is less than or equal to C, it may or may not be possible [to transport the full output of the lumber mill], depending on whether the lumber can be packed efficiently in the conveyor. Suppose, however, that we allow ourselves a saw-mill at the source. Then the lumber can be cut up into small pieces in such a way as to fill out the available capacity of the conveyor with 100% efficiency. Naturally in this case we should provide a carpenter shop at the receiving point to glue the pieces back together in their original form before passing them on to the consumer. If this analogy is sound, we should be able to set up a measure R in suitable units telling how much information is produced per second by a given information source, and a second measure C which determines the capability of a channel for transmitting information. Furthermore, it should be possible, by using a suitable coding or modulation system, to transmit the information over the channel if and only if the rate of production R is not greater than the capacity C."

The rest of this chapter discusses digital sawmills and the methods of reliably gluing our digital lumber back together after transportation from the source to the destination.

3.2 DIGITAL SIGNALING

For transmission channels constrained by limited bandwidth and transmission power and corrupted by noise, Shannon (1948, 1949) developed the theoretical limit C for digital transmission channel information capacity:

$$
\begin{aligned}
C &\leq W\log_2\left[1+\left(\frac{\mathrm{s}}{\mathrm{n}}\right)\right] \\
&\leq 3.322\ \ W\log_{10}\left[1+\left(\frac{\mathrm{s}}{\mathrm{n}}\right)\right] \\
&\leq\approx 3.322\ \ W\log_{10}\left(\frac{\mathrm{s}}{\mathrm{n}}\right) \\
&\leq 0.3322\ \ W\left[\frac{\mathrm{S}}{\mathrm{N}}\ (\mathrm{dB})\right]
\end{aligned}
\tag{3.1}
$$

where

C = channel capacity (Mb/s);
W = channel bandwidth (MHz);
$\mathrm{s/n}$ = channel signal to noise (power ratio);
$\mathrm{S/N}$ = channel signal to noise (dB) = $10\log_{10}(\mathrm{s/n})$.

In Equation 3.1, replacing the power ratio $(\mathrm{S/N}+1)$ with S/N introduces less than 1% dB error for all S/Ns $\geq$ 10 dB. (All practical systems require S/N $>$ 10 dB for acceptable operation.)

Shannon's limit may be rewritten to define the minimum S/N required to achieve a given spectral efficiency:

$$
\begin{aligned}
&\mathrm{S/N\ (dB)} \geq 3\ \frac{C}{W} \\
&\mathrm{S/N\ (dB)} \geq 3\ [\text{spectral efficiency (bits/s/Hz)}]
\end{aligned}
\tag{3.2}
$$

The primary Shannon assumptions are filtering is rectangular ("brick wall") and noise is Gaussian. While this limit may be approached with appropriate (but undefined) signal processing, Shannon noted, " ... any system which attempts to use the capacities of a wider band to the full extent possible will suffer from an threshold effect ..." (Shannon, 1949) and "... as one attempts to approach the ideal, the transmitter and receiver required become more complicated and the delays increase."(Shannon, 1950)

Shannon set limits on communication channel performance but offered no guidance for approaching those limits. Our first challenge is to find a practical method of signaling through a frequency-band-limited transmission channel with noise and distortion using a power-limited transmitter. We will then improve our transmission capacity by intelligent choice of data coding. However, first let us consider the effect of receiver noise.

3.3 NOISE FIGURE, NOISE FACTOR, NOISE TEMPERATURE, AND FRONT END NOISE

In the absence of external interference, the limiting factor in radio system gain (or transmission distance or transmission speed) is the noise introduced by the first amplifier in the receiver ("front end noise") (Kerr and Randa, 2010). This noise has the effect of degrading the signal-to-noise ratio of the incoming received signal.

The minimum noise of an ideal amplifier that is perfectly impedance-matched to its receive antenna would be the noise introduced by a (hypothetical) resistor of the interface impedance (typically 50 Ω) operating at temperature T (usually assumed to be 290 K = 17 °C = 63 °F). In general, the noise P delivered to a matched device by the noise source resistor at temperature T may be shown (Kizer, 1990) to be the following:

$$
n = KTb(\mathrm{W}) \tag{3.3}
$$

n = noise produced by a matched resistor operating at temperature T;
K = Boltzmann's constant = 1.38×10^{-23} (J/K);
T = noise temperature of the resistor (kelvin = °C + 273);
b = noise bandwidth of the device (Hz).

If the amplifier adds noise to the received signal, that noise is characterized by adding another noise temperature to characterize the added noise. The relationship of amplifier signal-to-noise ratio is

$$\text{nf} = 1 + \left(\frac{T_e}{T_o}\right) = \frac{\frac{s}{n_I}}{\frac{s}{n_O}} \tag{3.4}$$

nf = noise factor;
T_o = amplifier operating ("room") temperature (nominally 290 K);
T_e = amplifier additional ("excess") noise temperature (K);
= device "noise temperature";
= $T_o(\text{nf}-1)$;
s/n_I = signal-to-noise power ratio at input to amplifier;
s/n_O = signal-to-noise power ratio at output of amplifier.

$$\text{NF(dB)} = 10\log(\text{nf}) = \left(\frac{S}{N_I}\right) - \left(\frac{S}{N_O}\right) \tag{3.5}$$

NF(dB) = noise figure;
nf = $10^{\text{NF}/10}$;
S/N_I = signal-to-noise ratio at input to amplifier (dB);
= $10\log(s/n_I)$;
S/N_O = signal-to-noise ratio at output of amplifier (dB);
= $10\log(s/n_O)$.

Friis (1944) derived the noise figure for cascaded (series) active amplifiers:

$$\text{nf} = \text{nf}_1 + \frac{(\text{nf}_2 - 1)}{g_1} + \frac{(\text{nf}_3 - 1)}{g_1\, g_2} + \cdots \tag{3.6}$$

nf = overall noise figure of the cascaded amplifiers;
nf_1 = noise factor of the first device;
nf_2 = noise factor of the second device;
nf_3 = noise factor of the third device;
g_1 = gain (power ratio) of the first device;
g_2 = gain (power ratio) of the second device.

The implied assumption is all devices are matched impedances and bandwidth shrinkage of cascaded devices is insignificant.

The noise factor of an attenuator is simply the attenuation (1/gain) of the device since the output signal and noise are the input signal and noise multiplied by the gain ($1/g$) of the attenuator:

$$\text{nf} = \frac{\frac{s}{n_I}}{\frac{s}{n_O}} = \frac{\frac{s}{n_I}}{\frac{g_1 s}{n_I}} = \frac{1}{g_1} \tag{3.7}$$

Since the device is an attenuator, attenuator gain g_1 is between 0 and 1. If an attenuator (device 1) and an amplifier (device 2) are cascaded, the overall noise figure of the pair is the sum of the attenuator loss (dB) and the noise figure (dB) of the device:

$$\mathrm{nf} = \frac{1}{g_1} + \frac{\mathrm{nf}_2 - 1}{g_1} = \left(\frac{1}{g_1}\right)(\mathrm{nf}_2) \tag{3.8}$$

nf = overall noise factor of the cascaded amplifiers.

$$\mathrm{NF(dB)} = 10\log\left[\left(\frac{1}{g_1}\right)(\mathrm{nf}_2)\right] = 10\log\left(\frac{1}{g_1}\right) + 10\log(\mathrm{nf}_2) \tag{3.9}$$

NF(dB) = overall noise figure of the cascaded attenuator and amplifier;
= attenuation (dB) + noise figure of amplifier (dB).

The above equations show that the noise figure of an attenuator is simply the attenuation (dB, > 0) of the attenuator. The noise figure of a cascaded attenuator and an amplifier is the sum of the two (dB values).

The front end noise produced by an amplifier may be calculated as follows:

$$\begin{aligned} n &= K(T_o + T_e)B10^6(W) \\ &= 1.38 \times 10^{-17} T_o \left[1 + \left(\frac{T_e}{T_o}\right)\right] B \\ &= 4.00 \times 10^{-15} \mathrm{nf}\, B \end{aligned} \tag{3.10}$$

n = noise produced by a matched "internal" resistor;
B = noise bandwidth of the device (MHz);
N = front end noise = $10\log(n)$.

$$\begin{aligned} N(\mathrm{dBW}) &= 10\log(n) \\ &= -144 + \mathrm{NF(dB)} + 10\log(B) \\ N(\mathrm{dBM}) &= N(\mathrm{dBM}) + 30 \\ &= -114 + \mathrm{NF(dB)} + 10\log(B) \\ N(\mathrm{dBW/MHz}) &= -144 + \mathrm{NF} \\ N(\mathrm{dBW/4\,kHz}) &= -168 + \mathrm{NF} \end{aligned}$$

A common problem is to determine the signal associated with a known radio threshold signal-to-noise ratio. Let us assume that the radio receiver is limited by front end noise.

$$\begin{aligned} S(\mathrm{dBM}) &= \mathrm{S/N(dB)} + N(\mathrm{dBM}) \\ &= \mathrm{S/N(dB)} - 114 + \mathrm{NF(dB)} + 10\log(B) \end{aligned} \tag{3.11}$$

S = received signal power level (dBm) at threshold;
S/N = receiver threshold signal-to-noise ratio (dB);
NF = receiver noise figure (dB);
B = receiver bandwidth (MHz).

It should be remembered that the receiver noise figure is the noise figure of the front end amplifier plus the loss (dB) between the amplifier and the measurement location. The typical amplifier noise figure for low frequency microwave radios is about 2 dB. The typical waveguide and receiver filter loss in front of a receiver is about 2 dB. Therefore, the typical receiver noise figure is 4 dB.

Receiver front end noise, along with channel bandwidth, will be a primary limitation of microwave radio receivers.

3.4 DIGITAL PULSE AMPLITUDE MODULATION (PAM)

The simplest digital symbol is the single bit. It represents a "0" or "1" by one of two voltage levels at a defined repetitive sampling time. This is termed *pulse amplitude modulation* (*PAM*) of two or 2-level PAM. A more complex symbol may be constructed to signal multiple bits per symbol. In this case, various voltage levels are used at a defined repetitive sampling time (*N*-level PAM) (Fig. 3.1).

The advantage of multilevel coding is that the baud (signaling) rate can be slower than that in the binary case. The disadvantage is that multilevel coding is increased susceptibility to noise and sampling time error.

If an oscilloscope is used to view the voltage amplitude versus time of a PAM pulse train, the display looks like Figure 3.2 (QAM is a form of modulation discussed later):

At the sampling time, the PAM pattern is relatively clear or has an open "eye." A subjective measure (Breed, 2005) of pulse impairment may be made by measuring the eye. The width is a function of timing accuracy and stability, and the height is a function of noise and channel distortion (primarily noise and intersymbol interference).

If a transmission channel had unlimited bandwidth, digital pulses could be signaled as rectangular pulses (see Chapter 14 for various PAM formats currently in use). However, all radio transmission

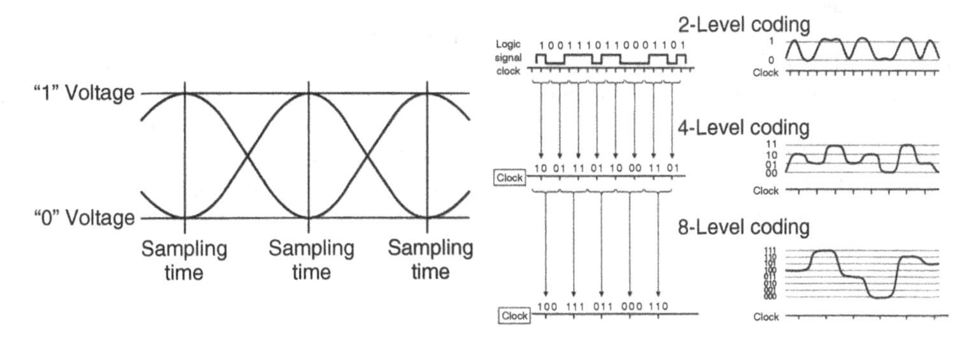

Figure 3.1 Binary and multilevel digital coding.

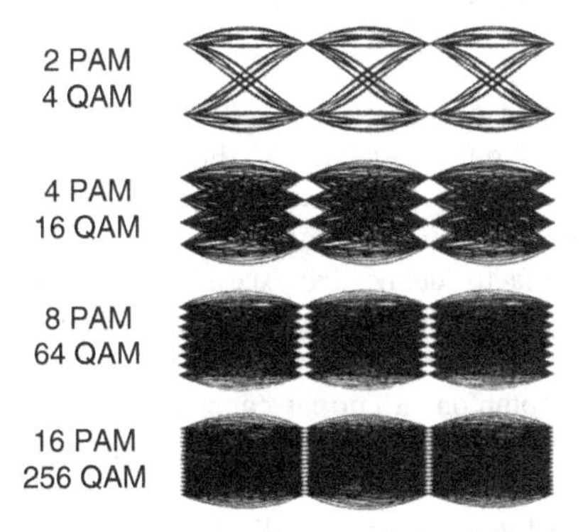

Figure 3.2 Binary and multilevel digital coding.

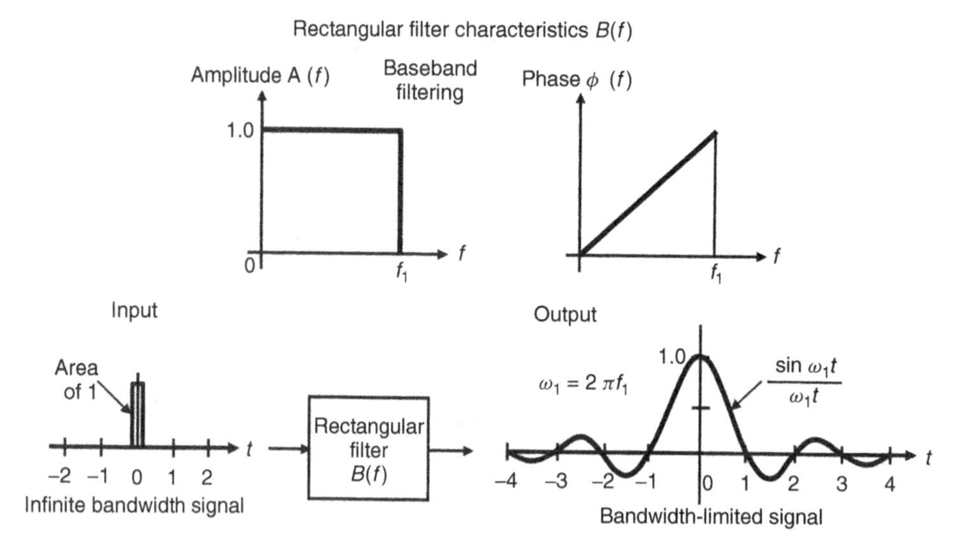

Figure 3.3 "Brick wall" filtering.

channels have significantly limited bandwidth and that limitation must be taken into account. Nyquist (1928) determined that (for baseband digital systems) the minimum frequency bandwidth (often called the *Nyquist bandwidth*) required to pass a PAM signal without distortion was half the PAM sampling rate. The sampling rate is the signaling (baud) rate. For double sideband orthogonal modulation (e.g., QAM), the Nyquist bandwidth and the baud rate are the same (Forney and Ungerboeck, 1998) (Fig. 3.3).

If an impulse signal with symbol period T is passed through a transmission channel that has a low pass filter (LPF) strictly limited to the Nyquist bandwidth $W = 1/(2s)$ (a "brick wall" filter), the filter output would be a $\sin \omega t/\omega t$ signal (where t is time) for a single digital pulse input. While this signal could in theory be used for signaling (since it has zero value at all sampling instants except one), it is of no practical significance. A "brick wall" filter would have infinite time delay and is physically unrealizable, and the output pulse becomes unbounded when time sampling is not perfect. Nyquist defined several criteria (Bennett and Davey, 1965; Nyquist, 1924, 1928) that must be met if the "brick wall" filtering requirement were to be relaxed to allow practical filter implementation. For our application, they may be summarized as the following:

1. The filter impulse response must have zero voltage axis crossings equally spaced in time.
2. The area under the filter impulse response around the signaling time must be proportional to the area of the signal entering the filter and zero for all other signaling times.

Nyquist demonstrated that these two criteria are satisfied if the frequency response of the filter is relaxed symmetrically about the Nyquist frequency W (Fig 3.4).

Nyquist proposed a specific curve for the relaxed filter frequency response now called *raised cosine* filtering. This filtering requires more bandwidth than the brick wall filter but is physically realizable (Bayless et al., 1979). The alpha factor defines the excess bandwidth of the filter. The smaller the alpha (α), the narrower the filter. However, the impact of reducing alpha is to increase overshoot between sampling instants (Fig. 3.5).

Alphas as small as 0.2 are common in current commercial radios. Since all power amplifiers are peak-power-limited from a distortion perspective, this requires the power amplifier average output power to be reduced ("backed off") as alpha is reduced. The impact of complex multilevel signaling can be several decibels. The relative peak-to-average ratio (PARR) for various alphas (Noguchi et al., 1986) is listed in Table 3.1.

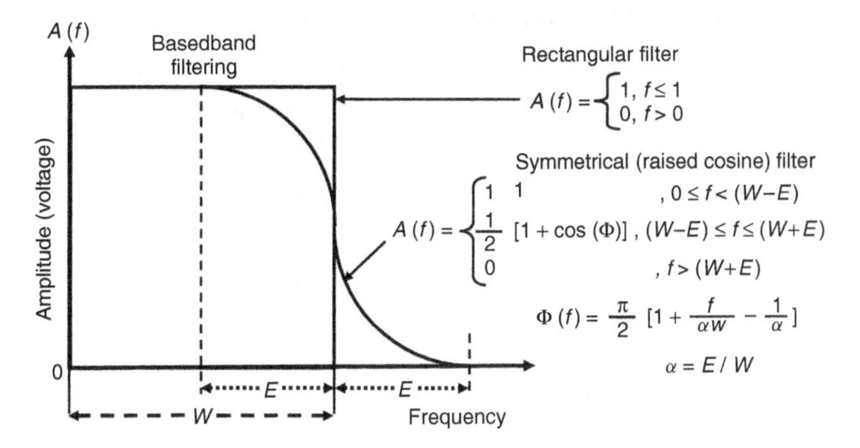

Figure 3.4 "Raised cosine" filtering.

Figure 3.5 Impact of alpha.

These values are independent of the constellation PARRs noted below. For the range of $0.1 \leq \alpha \leq 1.0$, the following equation estimates the above values:

$$Y(\text{PARR, in dB}) = e^F$$

$$e \approx 2.7182818285$$

$$F = A + B\alpha + C\alpha^2 + D\alpha^3 + E\alpha^4$$

$$A = 1.701169502177906$$

$$B = -2.651499059441027$$

$$C = -6.772188272114639$$

$$D = 19.65951075385066$$

$$E = -16.93991626467497 \tag{3.12}$$

The relaxed filtering curve is not unique. Gibby and Smith (1965) more precisely defined Nyquist's conditions for signaling without intersymbol amplitude distortion. This and their later work would suggest

TABLE 3.1 Alpha Peak-to-Average Ratios

α	Peak-to-Average Ratio, dB
1.0	0.00 (reference)
0.9	0.05
0.8	0.20
0.7	0.45
0.6	0.75
0.5	1.10
0.4	1.45
0.3	2.00
0.2	2.80
0.1	4.00

that Nyquist's two criteria for signaling without intersymbol interference through a filter could be restated as the following two criteria:

1. $f(nT) = 1$ for $n = 0$; $f(nT) = 0$ for $n \neq 0$'
2. The frequency domain transform of $f(t)$ is symmetrical about frequency $T/2$.

For the above, $f(t)$ is a time domain impulse response signal out of the filter, t is the time, T is the data sampling rate, and n is a positive or negative integer.

While Nyquist's raised cosine filter is quite popular, several other curves are "better" (have more open eyes that are less sensitive to timing error). The optimum curves (Franks, 1968; Lee and Beaulieu, 2008) use a two-step ("double jump") discontinuous frequency response curve that must be approximated. Continuous frequency response curves with eyes more open than those of Nyquist have been demonstrated (Assalini and Tonello, 2004; Beaulieu et al., 2001; Scanlan, 1992).

Some work (Liveris and Georghiades, 2003; Mazo, 1975; Rusek and Anderson, 2009) has been done using Gaussian shaped sync pulses signaled slightly faster than the Nyquist rate. However, these pulses have intersymbol interference and have not found practical application.

To this point, our discussion has assumed impulses (very narrow pulses) driving the transmission channel filter. We would usually prefer to use rectangular pulses. While the frequency spectrum of an impulse is flat ("white"), the spectrum of a rectangular pulse is sin x/x, where x is the normalized frequency. If we drive the Nyquist filter with rectangular pulses, we must multiply the filter frequency response by the inverse of the rectangular filter ($x/\sin x$) to "whiten" the frequency spectrum of the rectangular pulse (Bennett and Davey, 1965). Also, we usually would like to transfer the digital signal to a limited frequency band in an appropriate portion of the radio spectrum using double sideband quadrature modulation (Fig. 3.6).

The preceding process is the total radio filtering between the transmitter and the receiver. Part of the filtering occurs in the transmitter and the other part in the receiver. Filtering is required in both locations for different reasons. See discussion below regarding square root raised cosine filtering.

For a transmitter, the bandwidth the digital signal may be specified in many ways (Amoroso, 1980). For modern QAM-derived radios with relatively small α factors, the radio transmitter spectrum 3-dB bandwidth is approximately 80% of the channel bandwidth. The 99% bandwidth is about 90% and the 20-dB bandwidth is essentially the same as the channel bandwidth. The radio baud rate (symbols per second) is typically about 85% of the channel bandwidth. One difficulty is how to quantify the small intermodulation energy produced by the transmitter power amplifier. It has little impact on some definitions of bandwidth but can create excessive adjacent channel interference (Fig. 3.7).

Regulatory requirements (Mosley, published yearly) usually sidestep a formal bandwidth definition and address the spectrum issue by requiring the transmitter-modulated spectrum to fit within a "spectrum mask" (Fig. 3.8).

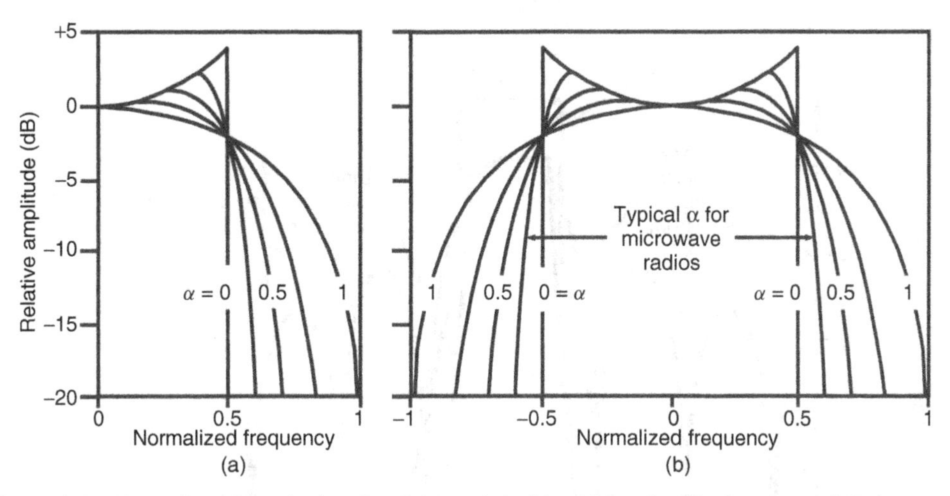

Figure 3.6 Normalized Nyquist baseband (a) and double sideband radio frequency (RF, b) spectra for rectangular pulse signaling.

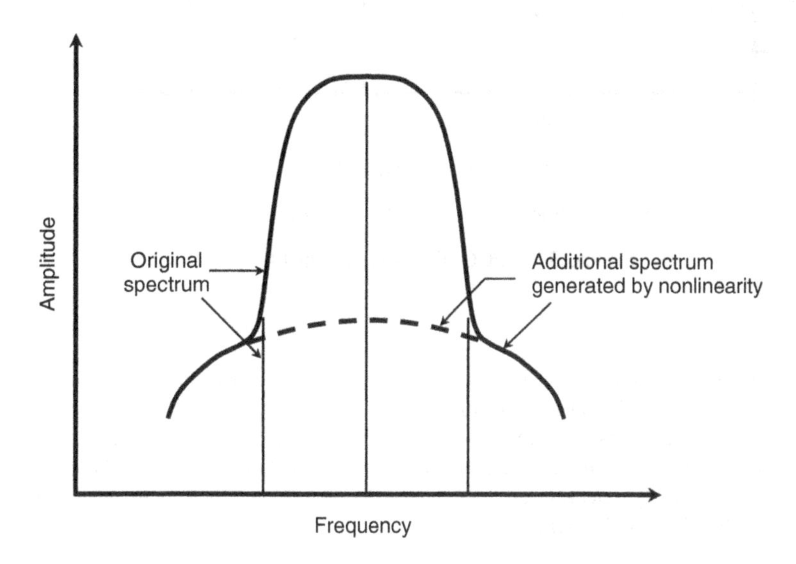

Figure 3.7 Spectrum "skirt" produced by transmitter power amplifier nonlinearity.

The receiver needs to reject adjacent potentially interfering signals. The optimum way to split the filtering requirement has been studied extensively (Lucky et al., 1968). The optimum detection performance (Noguchi et al., 1986) for Gaussian noise interference is to make the transmit and the receive filters the same (ignoring the prewhitening ($x/\sin x$) term that occurs at the modulator). Each filter voltage amplitude response is the square root of the raised cosine filter frequency response. Each filter is termed a *square root raised cosine filter*.

A receiver-matched filter is a filter whose frequency domain response matches the frequency domain spectrum of the transmitted signal (actually the filter is the complex conjugate of the transmitted spectrum but assuming the spectrum has no complex component, this is an accurate statement). It has the desirable property of being the filter that optimizes the receiver signal-to-noise ratio in the presence of white Gaussian noise. While North (1943) originated this concept, Van Vleck and Middleton independently derived this result and defined the name "matched filter." The matched filter optimizes the detection of pulse amplitude signals being sampled at the center of the sampling window. Using square root raised

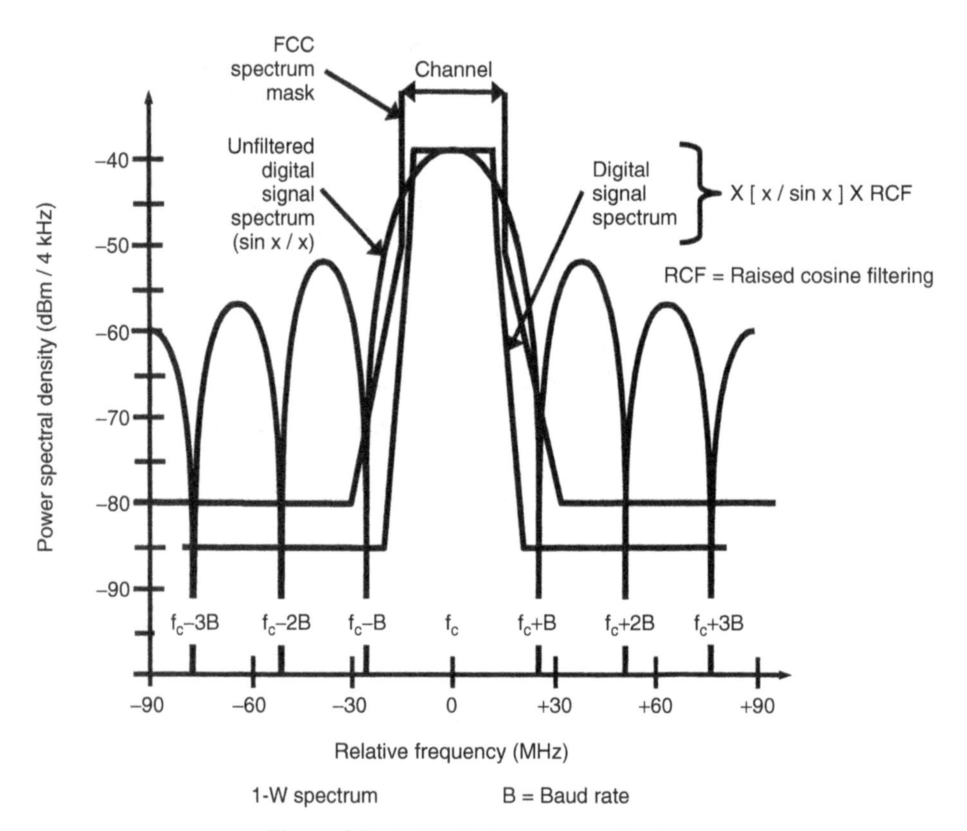

Figure 3.8 FCC digital spectrum filtering.

cosine filtering (assuming a prewhitened transmit spectrum followed by a square root transmit filter) creates a matched filter at the digital radio receiver.

These pulse shaping filters reside in the modulator and the demodulator of the radio and are usually digital. They have limited rejection and overload capabilities so additional physical flat pass-band filters with deep adjacent frequency rejection are also added for additional filtering to meet regulatory and adjacent frequency filtering requirements.

3.5 RADIO TRANSMITTERS AND RECEIVERS

Microwave radio transmitters and receivers are paired to convey information from one location to another. They are subjected to the many potential impairments of this process (Fig. 3.9).

The degree of impairment is a function of the external environment and internal design choices (Borgne, 1985; Johnson, 2002a, 2002b; Yin, 2002). Overall performance will be fundamentally limited by the radio BER for very low received signal (power) levels (RSLs) (due to receiver noise) and receiver overload distortion (due to excessive received signal) (Fig. 3.10).

The nominal RSL will lie somewhere within the low BER range of the microwave radio. The decibel difference between the nominal RSL and the radio threshold for larger RSL is termed *head room*. The difference between the nominal RSL and the small RSL radio threshold is termed *flat fade margin*. These concepts are shown in Figure 5.38. Dispersive fade margin is a concept unrelated to "flat" power fading (see Chapter 9 for a discussion of this concept).

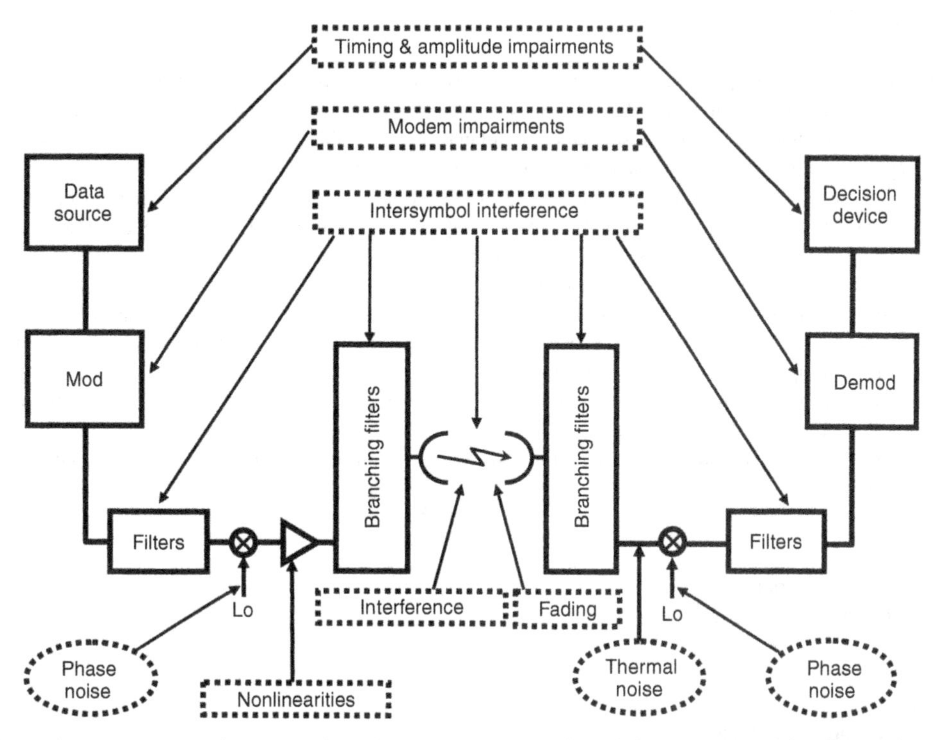

Figure 3.9 Typical radio system impairments. *Source*: Adapted from Borgne, M. "Comparison of High-Level Modulation Schemes for High-Capacity Digital Radio systems," *IEEE Transactions on Communications*, Vol. 33, p. 442, May 1985. Adapted with permission of IEEE.

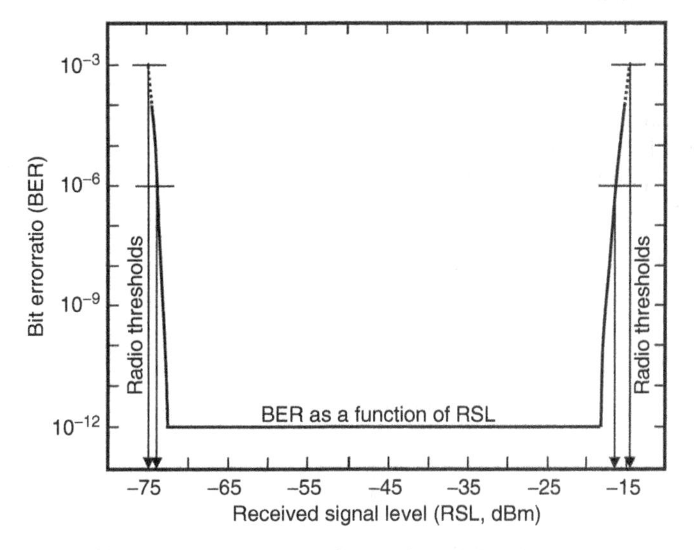

Figure 3.10 Typical radio receiver dynamic range.

3.6 MODULATION FORMAT

As noted previously, Nyquist's criteria for distortionless data transmission was that digital transmission could not occur faster than twice the frequency bandwidth at baseband (unmodulated) signals or the transmission bandwidth for double sideband modulated signals. This limits how fast we may produce digital symbols without intersymbol interference. (Some early-generation digital radios used oversampling techniques called *partial response signaling* that used predictable intersymbol interference, but the techniques were limited in their spectral efficiency. They are no longer produced.) Of course, the primary purpose of modulation is to impress a digital data stream on an RF sine wave ("carrier") for transmission over the air to another location. The other purpose of modulation is to increase spectral efficiency by creating symbols that represent multiple bits. This is accomplished by changing (modulating) a carrier sine wave (or waves) amplitude and/or phase in such a way as to map a set of bits into a unique symbol.

PAM is a simple one-dimensional (voltage amplitude as a function of time) signaling method. Microwave radio modulation signaling methods typically operate in at least two dimensions (amplitude and phase as a function of time).

The modulated signal can be represented at a unity length phasor multiplied by a time-varying multiplicative scalar M:

$$C = Me^{j(\omega+\phi)} \tag{3.13}$$

C = modulated carrier;
M = carrier amplitude modulation;
$\omega = 2\pi ft$;
F = frequency;
t = time;
ϕ = carrier phase modulation.

For QAM radios, the amplitude and phase modulation is created by modulating two orthogonal signals: a cosine wave and a sine wave. We will express the orthogonal signals as a complex number with cosine amplitude on the real axis and the sine amplitude on the imaginary (j) axis. Radio carrier phase will be referenced to the cosine wave. The amplitude modulation of the cosine wave will be termed the *in-phase modulation*. The amplitude modulation of the sine wave will be termed the *quadrature modulation*.

$$C = V_I \cos(\omega) + jV_Q \sin(\omega) \tag{3.14}$$

C = modulated carrier;
V_I = amplitude of in-phase modulation (in phase with carrier);
V_Q = amplitude of quadrature modulation (orthogonal to carrier);
j = imaginary number (i) used to create a complex number.

The modulated carrier can be represented as two orthogonal carrier signals (V_I and V_Q) that are multiplied by modulation signals.

$$C = Me^{j(\omega+\phi)} = Me^{j(\phi)}e^{j(\omega)}$$

$$V = \text{sqrt}\left[(V_I)^2 + (V_Q)^2\right]$$

$$\phi = \arctan\left(\frac{V_I}{V_Q}\right) \tag{3.15}$$

We wish to focus on the modulation. Euler's formula allows us to represent the modulated carrier as a set of quadrature signals.

$$V_I \cos(\phi) \;+\; jV_Q \sin(\phi) = Ve^{-j\phi} = \text{carrier signal modulation} \tag{3.16}$$

Our modulation is a unity amplitude phasor (vector) $e^{-j\phi}$ multiplied by a time-varying amplitude V. We would like to display the modulation symbol states in two-dimensional Cartesian coordinates (referenced to the carrier with constantly changing phase ω). This display is called a *phasor diagram* (Fig. 3.11).

This phasor representation of all possible signaling states (symbols) is called a *constellation*. For the typical system with no memory, to transport n bits of information per signaling interval, a constellation of $M = 2^n$ symbols is required. The transmission system is often described as having a spectral efficiency of n bits/s/Hz. Many different constellations are possible (Cahn, 1960; Campopiano and Glazer, 1962; Dong et al., 1999; Foschini et al., 1974; Hancock and Lucky, 1960; Simon and Smith, 1973; Thomas et al., 1974) (Fig. 3.12).

The above constellations are referenced to normalized average signal-to-noise ratio (S/N) values for a 10^{-6} BER relative to the best case constellation. Most radios are limited by the peak S/N values, not average values. As the number of points increases, the optimum constellation converges toward a grouping of equilateral triangles within a circle (Foschini et al., 1974). The most popular constellations are the QAM and the phase shift keying (PSK). Although they are not optimal, the performance penalty is nominal and they are much easier to implement than the other constellations.

In general, the constellation PARR will impact the transmitter amplifier design (Table 3.2).

Since PSK uses a circular constellation, the PARR never changes. QAM constellations vary between a square (4, 16, 64, 256, 1024, and 4096 QAM) and a cross pattern (32, 128, 512, and 2048 QAM). The lower PARR for the cross pattern is significant.

Xiong (2006) showed that for a square QAM constellation, the PARR (dB) may be calculated using the following formula:

$$\text{PARR(dB)} = 10\log\left\{\frac{3\left[\text{sqrt}\,(N) - 1\right]^2}{N-1}\right\} \tag{3.17}$$

where N is the number of states in the QAM constellation (e.g., 64 for 64 QAM).

The relationship between constellation pattern and required signal-to-noise ratio for a given BER has been established by many sources [including the work by Craig 1991, with the π in Equation 3.13 replaced by 2π (Khabbazian et al., 2009), (Szczecinski et al., 2006)]. Proakis and Salehi (2002, Table 7.1) noted the relationship between QAM and PSK (Table 3.3).

Proakis and Salehi (2002, Eq. 7.6.69) derived the relationship between signal-to-noise ratio (S/N) and probability of error (BER):

$$\text{BER} = 2\left\{1-\left[\frac{1}{\text{sqrt}\,(M)}\right]\right\} Q\left\{\text{sqrt}\left[\left(\frac{3}{M-1}\right)\left(\frac{\text{s}}{\text{n}}\right)\right]\right\} \tag{3.18}$$

M = QAM level = 2^{η};
η = spectral efficiency (bits/s/Hz) = an integer > 0;
s/n = signal-to-noise ratio (power ratio) = $10^{(\text{S/N})/10}$.

$$\text{S/N(dB)} \approx \frac{E_\text{b}}{N_\text{o}} + 10\log_{10}(\eta)$$

S/N(dB) = average signal-to-noise power ratio (dB).

$$\begin{aligned}\frac{E_\text{b}}{N_\text{o}} &= \text{energy per bit to noise power spectral density ratio (dB)}\\ &= \text{S/N (dB)} - 10\log_{10}\left[\text{bits per symbol} \times \frac{\text{symbols per second}}{B}\,(\text{Hz})\right]\\ &\approx \text{S/N (dB)} - 10\log_{10}[\text{spectral efficiency (bits/s/Hz)}]\\ &= 10\log[(\text{s/n})/\eta](\text{assuming the modulated signal essentially fills the transmission channel})\end{aligned} \tag{3.19}$$

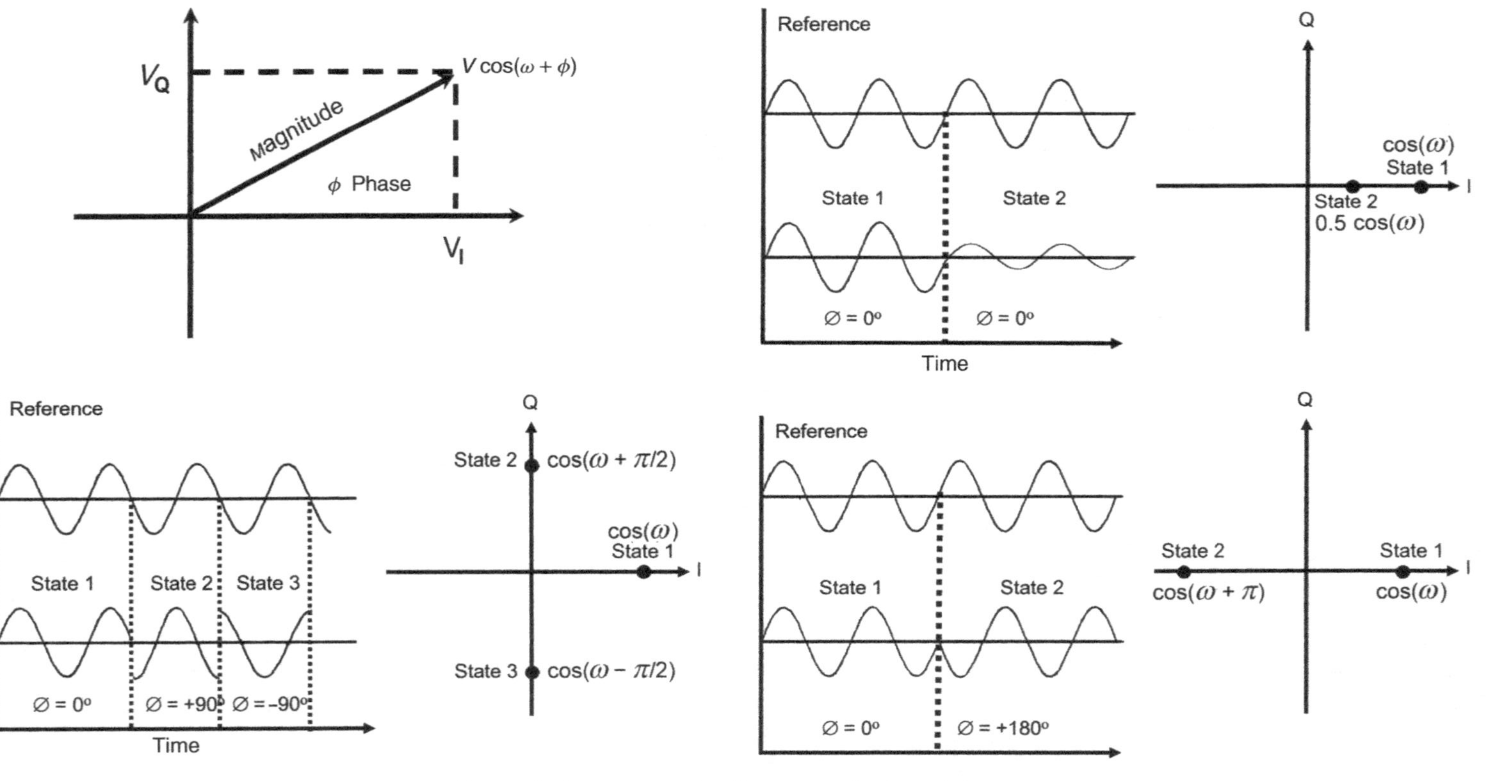

Figure 3.11 Phasors of different amplitude and phase.

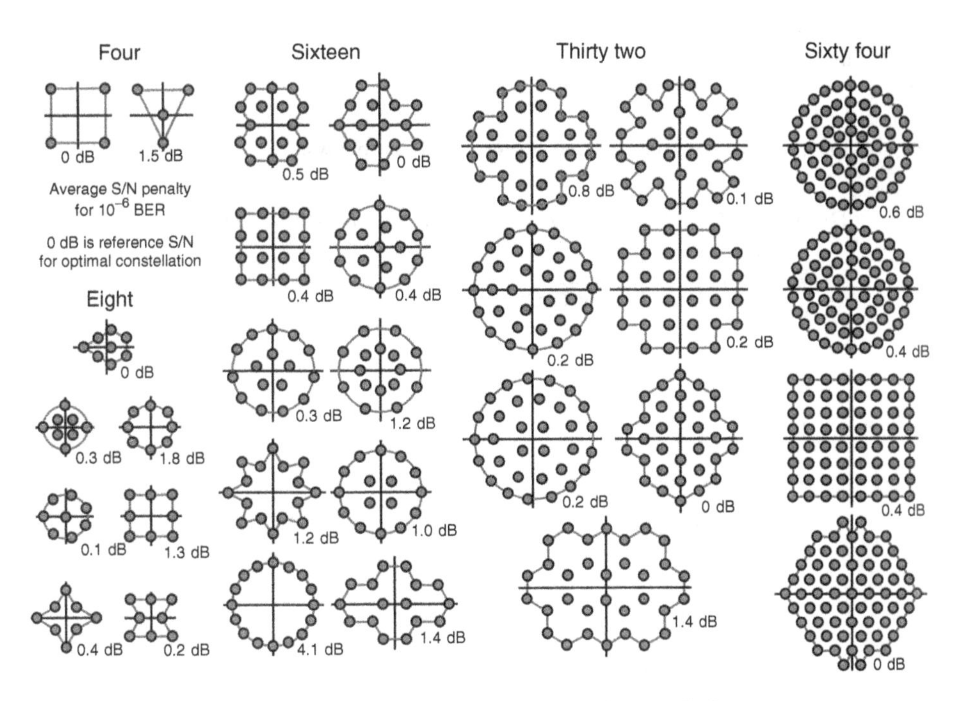

Figure 3.12 Typical modulated signal constellations.

TABLE 3.2 M-ary Constellation Peak-to-Average Ratio (dB)

M	QAM	PSK
4	0.0	0.0 (reference)
16	2.6	0.0
32	2.3	0.0
64	3.7	0.0
128	3.2	0.0
256	4.2	0.0
512	3.4	0.0
1024	4.5	0.0
2048	3.6	0.0
4096	4.6	0.0

Abbreviations: M, signaling states; QAM, Quadrature Amplitude Modulation; PSK, phase shift keying.

TABLE 3.3 Average S/N Advantage of M-ary QAM over M-ary PSK

M	S/N Advantage, dB
4	0.0
8	1.7
16	4.2
32	7.0
64	10.0

Abbreviations: M, signaling states; QAM, Quadrature Amplitude Modulation; PSK, phase shift keying.

B = channel bandwidth;
sqrt (x) = square root of x;
Q = tail probability of a Gaussian random variable.

Q may be approximated (Abramowitz and Stegun, 1968) as follows:

$$
\begin{aligned}
Q(X) &\approx Z[(B_1 \times T) + (B_2 \times T^2) + (B_3 \times T^3) + (B_4 \times T^4) + (B_5 \times T^5)] \\
Z &= \left[\frac{e^{-(X \times X)/2}}{\text{sqrt}(2\pi)}\right] = \frac{1}{\text{sqrt}(2\pi)[e^{(X \times X)/2}]} \\
X &= \text{sqrt}\left(\frac{3}{M-1}\right)\left(\frac{\text{s}}{\text{n}}\right) \\
T &= \frac{1}{1 + (R \times X)} \\
\pi &\approx 3.1415926536 \\
e &\approx 2.7182818285 \\
R &= 0.2316419 \\
B_1 &= 0.319381530 \\
B_2 &= -0.356563782 \\
B_3 &= 1.781477937 \\
B_4 &= -1.821255978 \\
B_5 &= 1.330274429
\end{aligned}
\tag{3.20}
$$

Abreu's tight upper bound (12) (Abreu, 2012) on the Q-function also is an excellent approximation for Q. Using the above relationships, we may calculate the signal-to-noise ratio (S/N) relationship for QAM to BER (Table 3.4). These values are for coherent demodulation. Most practical receivers use differential demodulation since the original carrier phase is difficult to determine. Differential demodulation is a couple of decibels worse than coherent demodulation but this is usually made up by adding forward error correction to the modulation/demodulation process (Fig. 3.13).

Figure 3.13 graphically displays the carrier-to-noise requirements for a 10^{-6} BER. For commercial products, the term *carrier to noise* (C/N) is typically used in lieu of signal-to-noise ratio (S/N). They represent the same quantity. See Table A.22 for signal-to-noise threshold requirements, spectral efficiency, and transmitter PARRs for popular modulation formats.

TABLE 3.4 Average S/N of M-ary QAM for 10^{-6} BER

M	S/N, dB	η, bits/s/Hz
4	13.5	2
8	17.1	3
16	20.2	4
32	23.2	5
64	26.2	6
128	29.1	7
256	32.0	8
512	34.9	9
1024	37.7	10

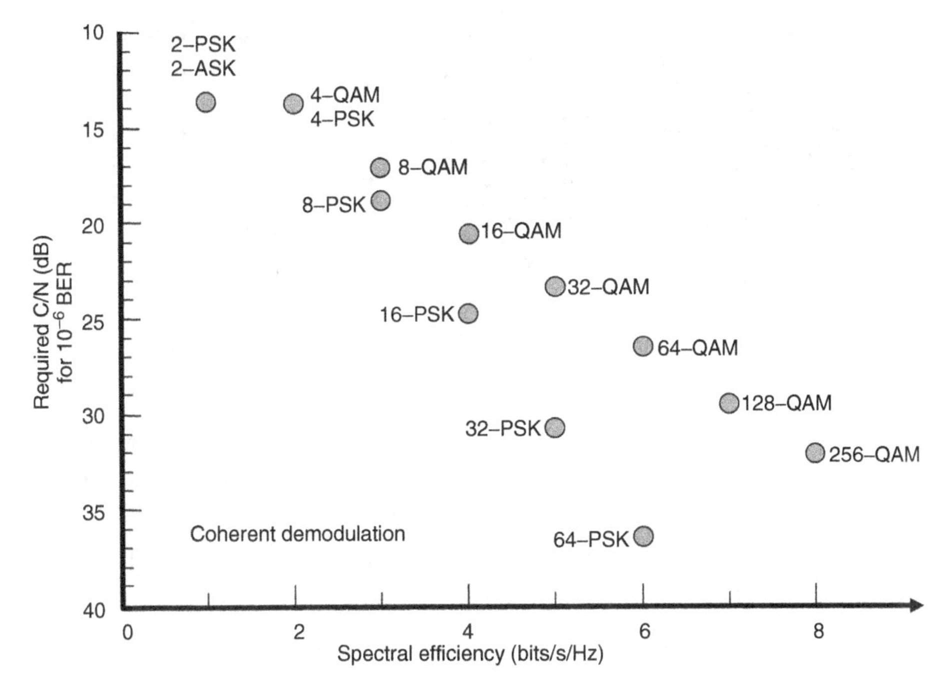

Figure 3.13 Spectral efficiency for QAM and PSK.

Constant amplitude envelope PSK is popular for satellite systems where maximum transmit power is the primary interest. For most fixed point to point radios, spectral efficiency is most important so variable amplitude envelope QAM constellations are chosen.

In Figure 3.14, 8 QAM is not shown. Although it is theoretically possible, there is no industry agreement as to the appropriate constellation and it is not used commercially. (Several logical choices have been proposed but when eight states are required, 8 PSK is typically used.) As systems transition from lower to higher order QAM, the constellations transition naturally between square and cross patterns. Cross patterns are desirable for peak-power-limited systems because their peak to average power ratio is lower than that of square constellations. Cross patterns for 16, 64, and 256 are possible, but they, like 8 QAM, require complicated multiple levels for different constellation symbols and are not used commercially.

Figure 3.15 shows the constellation pattern and the digits associated with each symbol for 16 and 64 QAM. The digits are gray coded in such a way that the closest symbols to any symbol only change by one digit. This minimizes the impact on BER of moderate noise symbol errors. (For moderate noise, one symbol error is one bit error.)

Figure 3.16 is a simplified diagram of 16 and 64 QAM modulators. QAM is usually formed by summing the output of two PAM modulators operating with carriers in phase quadrature (90° relative phase). Each modulator varies the amplitude of the carrier as well as the phase between the two states in phase or 180° out of phase. The two modulated carriers are usually termed the *in-phase* (I) and *quadrature* (Q) signals and the modulator is called an I–Q *modulator*. For square patterns, the modulators are independent. For cross constellations, there is dependency.

3.7 QAM DIGITAL RADIOS

The power amplifier for a QAM transmitter is by far the largest consumer of power in the radio. For an I–Q modulator, it must operate in linear mode with some output power reduction due to the constellation

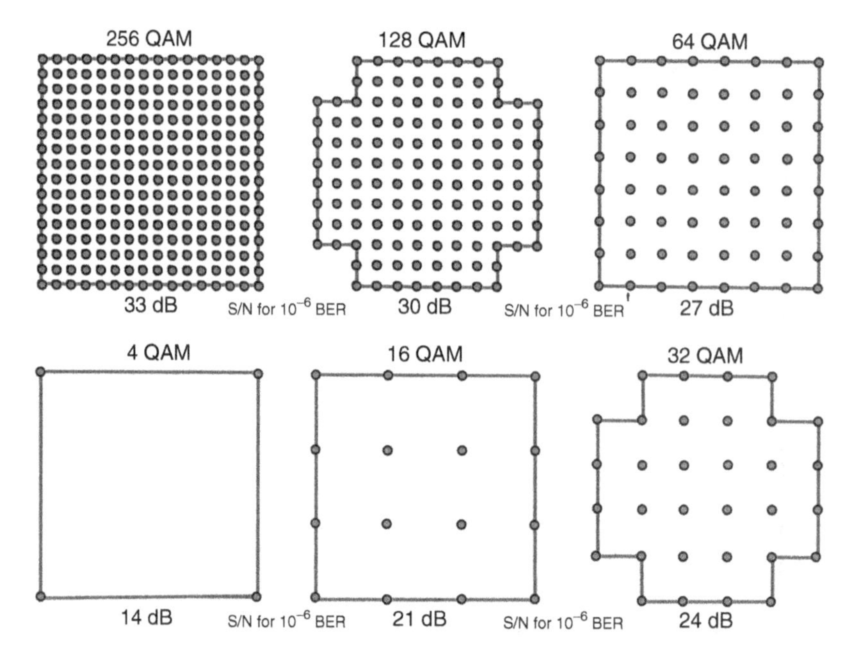

Figure 3.14 Typical QAM constellations.

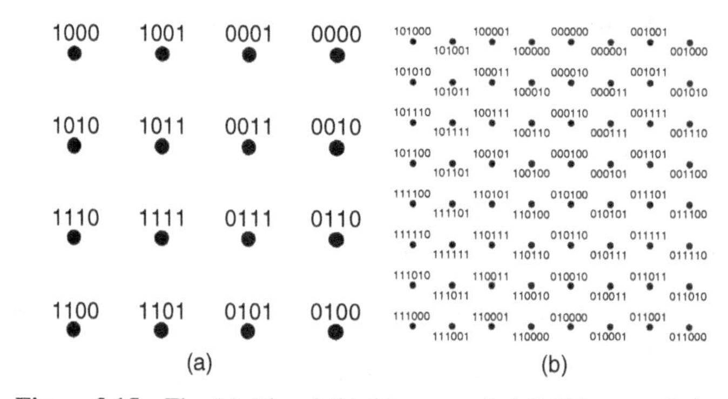

Figure 3.15 The (a) 16 and (b) 64 gray-coded QAM constellations.

and filter alpha PARRs. Relatively low speed radios are now starting to implement novel approaches to improve transmitter amplifier efficiency (Birafane et al., 2010; Groe, 2007; Kim et al., 2010a, 2010b; Lavrador et al., 2010). These modulators offer the opportunity to reduce amplifier power consumption while maintaining amplitude and phase linearity. These approaches have yet to be implemented in high speed digital radios.

QAM radio architecture is fairly standardized (Dinn, 1980; Noguchi et al., 1986) (Fig. 3.17). Serial binary data enters the transmitter. It is scrambled by a self-synchronizing scrambler to remove periodic patterns (such as tributary data stream framing, which would create undesired coherent spectral lines in the modulated signal) from the incoming data. The data is then converted from serial to parallel data (S to P). The parallel data is used by the modulator to create I and Q amplitude states. These I and Q signals are up-converted to the desired transmit frequency and summed to form a constellation state. "Direct" modulation radios typically modulate a low frequency (e.g., 2-GHz) signal and up-convert it to the appropriate transmission frequency.

The received signal is amplified to an appropriate level and applied to two down converters being driven by a voltage controlled oscillator (VCO) with quadrature outputs. The VCO has been phase and

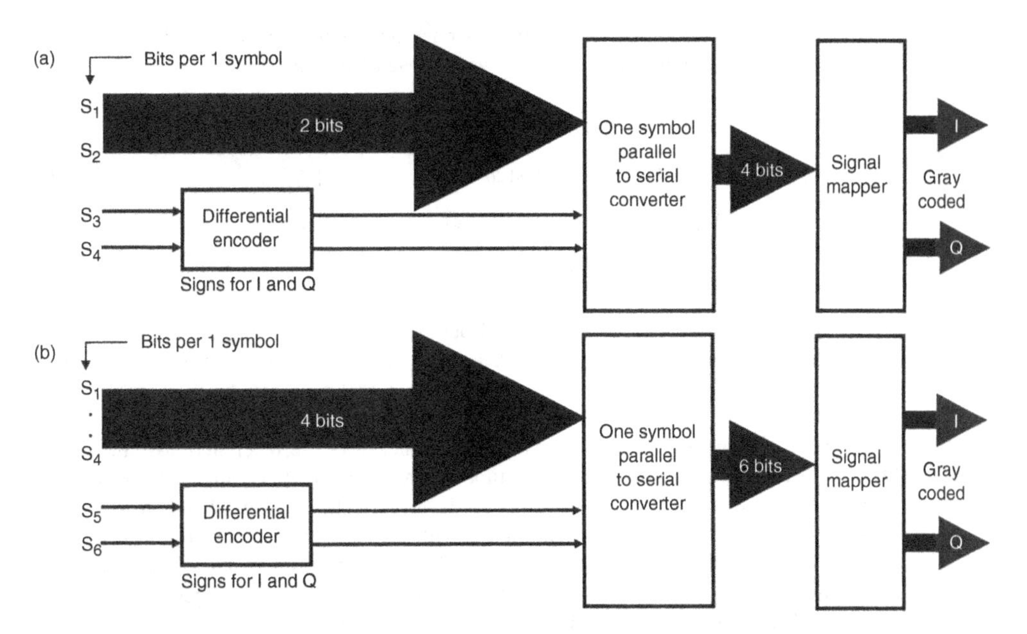

Figure 3.16 The (a) 16 and (b) 64 QAM modulators.

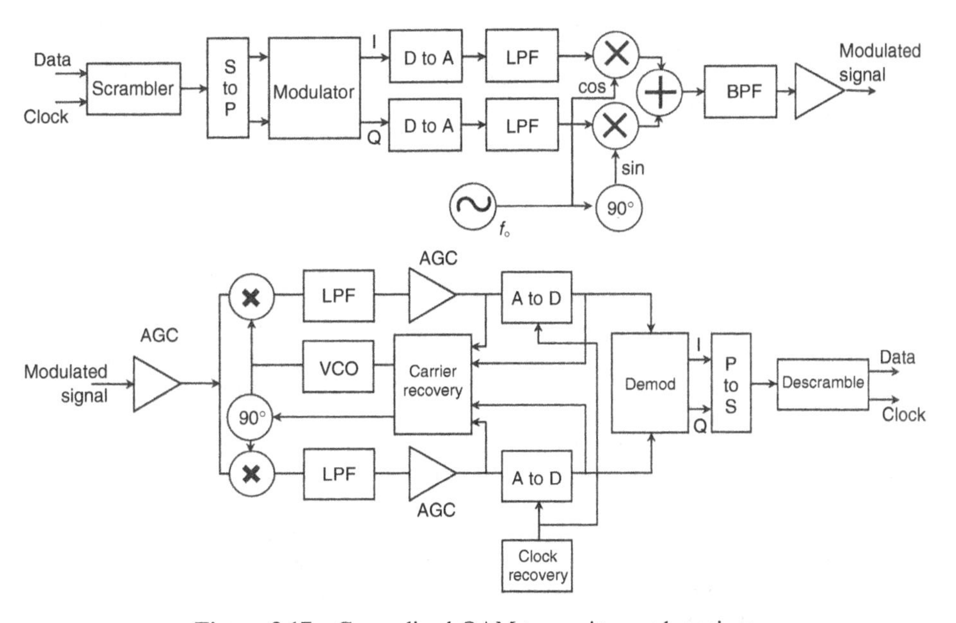

Figure 3.17 Generalized QAM transmitter and receiver.

frequency locked to the incoming received signal. The quadrature VCO signals extract the incoming V_I and V_Q modulation signals.

The following illustrates this process. The incoming signal is C.

$$C = V_I \cos(\omega) + jV_Q \sin(\omega) \tag{3.21}$$

C = modulated carrier.

The variables are as noted previously. One receive channel multiplies the incoming signal by a cosine wave and the other channel multiplies it by a sine wave and each LPFs the result.

$$V_\mathrm{I} = \mathrm{LPF}\{2\cos(\omega)[V_\mathrm{I}\cos(\omega) + \mathrm{j}V_\mathrm{Q}\sin(\omega)]\}$$
$$= \mathrm{LPF}\{V_\mathrm{I}[1 + \cos(2\omega)] + \mathrm{j}V_\mathrm{Q}\sin(2\omega)\} \tag{3.22}$$

$$V_\mathrm{Q} = \mathrm{LPF}\{2\sin(\omega)[V_\mathrm{I}\cos(\omega) + \mathrm{j}V_\mathrm{Q}\sin(\omega)]\}$$
$$= \mathrm{LPF}\{V_\mathrm{I}\sin(2\omega) + \mathrm{j}V_\mathrm{Q}[1 - \cos(2\omega)]\} \tag{3.23}$$

The recovered V_I and V_Q signals are digitized and demodulated into a sequence of parallel groups of bits. These are converted to a sequential binary data stream and descrambled before being delivered to the data user. The descrambler feedback loops multiply the output error rate by the number of loops (which is kept small).

During heavy received signal fading (see Chapter 9), a microwave receiver may lose the received signal repetitively. To minimize transmission outage time, the receiver should regain operation as quickly as possible after the received signal returns to acceptable levels. The receiver should recover carrier frequency, lock to its phase, and then start synchronizing with the data stream. This process is a function of many factors (Franks, 1980; Mueller and Muller, 1976; Noguchi et al., 1986). Regaining carrier frequency and phase lock usually happens quickly for short-duration receiver outages. It can be much longer if the receiver has been without a received signal for some time. Lockup on the demodulated data is a function of the radio symbol signaling (baud) rate (the slower the baud, the longer the lockup time). This can have a significant effect on system availability.

Use of differential demodulation means the receiver is insensitive to 180° phase constellation orientation. However, the receiver must still determine whether the receiver has been locked up in the left/right (in-phase I) or up/down (quadrature Q) orientation. This is usually done using unique frame sequences on the I and Q channels ("rails").

3.8 CHANNEL EQUALIZATION

Microwave paths are subject to dispersive fading (see Chapter 9). This is caused by multiple transmission paths in the atmosphere between the transmitter and the receiver. The multiple paths produce a signal at the receiver that is the original digital signal corrupted by time-shifted "echos" of this signal. These additional signals produce a broadened ("dispersed") digital signal, and this linear distortion is called *dispersion*. The digital radio receiver must compensate for this distortion before demodulation can be performed reliably.

Linear distortion can, in theory, be compensated for by frequency domain equalizers. However, in this case, that is not always possible. The dominant signal may occur from any of the multiple paths at any given time. If the strongest signal is from a relatively long time-delayed path, the echo (e.g., signal from the normal main path) may precede the dominant signal in time. This produces a nonminimum phase linear distortion that cannot be compensated for by any realizable frequency domain equalizer. This type of distortion is known to occur approximately half the time when multipath fading occurs.

The Wiener–Khinchin theorem implies that for linear (approximately) time-invariant transmission channels, equalization may be performed in either the frequency domain (as observed on a spectrum analyzer) or in the time domain (as observed on an oscilloscope), or both. It has long been known (Aaron and Tufts, 1966) that the optimum linear distortion equalizer is a frequency-domain-matched filter followed by a (infinite tap) time domain transversal equalizer. The radio transmission channel changes over time and the ability to provide a frequency domain filter to match the channel distortion at any instant is limited. Most microwave radio receivers use a slope equalizer to compensate for out-of-band dispersive notches and leave all other compensation to a time domain transverse equalizer.

While linear feedback (infinite impulse response) adaptive filters have been considered, they are not popular because they are difficult to stabilize and lack inputs from the opposite quadrature transmission channel (Qureshi, 1985). Today, use of automatically adaptive (finite impulse response) linear transversal

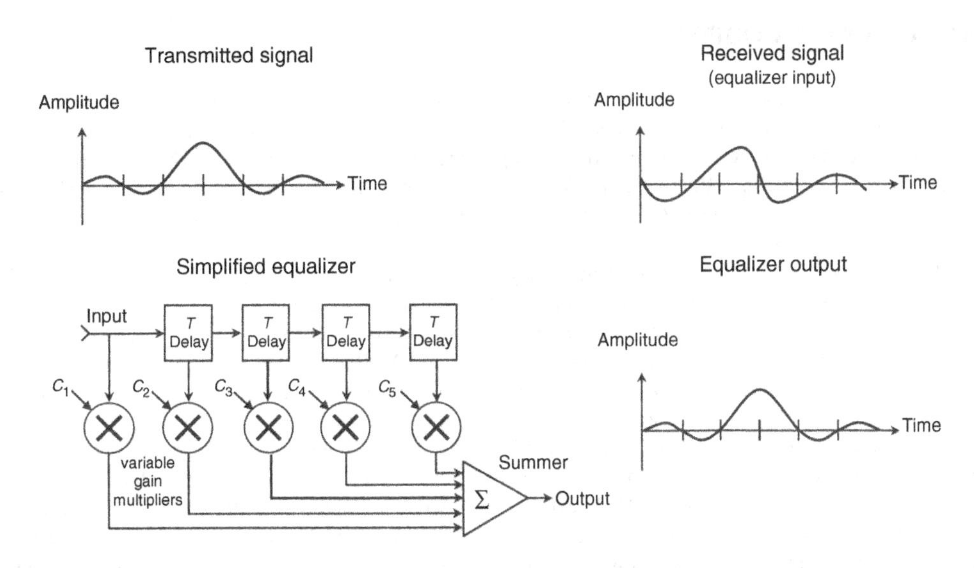

Figure 3.18 Simplified transversal equalizer.

equalizers (Lucky, 1966) is the standard method of compensating for path intersymbol interference (dispersive fading). The transversal equalizer is a digital tapped delay line, with the output signal composed of weighted delayed samples of the original received signal (Lucky, 1965; Qureshi, 1982, 1985). The equalizer sums the weighted portion samples that bracket the sample of interest. The value of interest is delayed so that samples ahead of it and behind it in time may be used. Typically, there are an equal number of samples ("taps") before and after the decision circuit to enhance convergence (Brunner Weaver, 1988) (Fig. 3.18).

The weighting coefficients are automatically varied to satisfy a criterion of goodness. The speed at which coefficients are changed is a trade-off between stability and dynamic performance (how fast the equalizer produces a useful signal). The early equalizers took signal samples at the symbol (baud) rate synchronized to the incoming signal (synchronous equalizers) (Lucky, 1965, 1966). They used a zero forcing (ZF) criterion that varied the weighting to achieve zero composite signal for all samples except the sample of interest. This had disadvantages in that it enhanced high frequency (band edge) noise and could not equalize a signal with a fully collapsed eye pattern. Equalizers using the least mean square (LMS) criterion (Widrow, 1966) were an improvement. They minimized the mean square error of all the taps. Using this criterion, the equalizer could both suppress noise and equalize a severely distorted signal.

Later equalizers were designed to take samples faster than the signaling rate (Gitlin and Weinstein, 1981). These were called *fractional equalizers* because they sampled the received signal between the digital sampling instants. The advantage of high rate (fractional equalizer) sampling was better control of distortion near the frequency edge of the transmission channel (no sampling fold-over distortion) and insensitivity to the sampling time (better dynamic performance since precise signal sampling timing is not required). The fractional equalizer could synthesize the best combination of the characteristics of an adaptive matched filter and a synchronous equalizer with the constraints of its number of taps (Qureshi, 1982). For the same number of taps, fractionally spaced equalizers outperform synchronous ones (Baccetti Raheli Salerno, 1987; Niger Vandamme, 1988).

Figure 3.18 shows a simplified analog equalizer for only one channel. Today all equalizers are digital. For I and Q demodulators, both channels are sampled and fed back to the channels to produce the output for each channel. The weighting coefficients must be complex (both real and imaginary numbers since the received signals are in quadrature).

The transverse equalizer is a critical element to a modern digital receiver. It (and the receiver filtering) will determine the shape of the dispersive fading W or M curve (see Chapter 9) and thereby determine the dispersive fade margin of the receiver.

3.9 CHANNEL CODING

We know how to shape digital signals for transmission. We know how to map these signals into efficient constellations. We know how to compensate for transmission channel distortion. The final step is to improve the dynamic range of the receiver by improving error performance. This takes us to the subject of error-correcting coding (Bhargava, 1983; Costello Forney, 2007; Forney, 1991; Forney et al., 1984; Forney and Ungerboeck, 1998; Goldberg, 1981; Kassam and Poor, 1983; Lucky, 1973; Sklar, 1983a, 1983b; Whalen, 1984).

The first codes were block codes because they encoded data in fixed blocks of bits. Hamming (1950) invented the first such code. He took a block of 4 bits and created three check sums to create a 7-bit word. Using that word, a single error could be corrected. Golay (1949) produced two codes that had the ability to correct two or three block errors. Reed 1954 and Muller 1954 produced even more powerful codes. The Hamming, Golay, and Reed–Muller codes were linear, meaning if they encoded n blocks, the modulo-n sum of any two code words produced a new code word.

Cyclic codes were invented by Prange (1957). These block codes had the property that any cyclic shift of a code word produced a code word. Shortly thereafter, Bose and Ray-Chaudhuri (1960) and Hocquenghem (1959) produced the BCH cyclic code. About the same time, Reed and Solomon (1960) created the powerful Reed–Solomon code. This code could correct continuous groups of errors besides individual block errors. This code was more complex than any previous code and initially this limited its use. However, today, it is the most popular block code. It is used in CDs, DVDs, cell phones, and NASA deep space missions (Berlekamp et al., 1987; Liu and Lee, 1984).

Forney (1966) introduced concatenated codes by cascading two block encoders (Fig. 3.19). This two-stage coding scheme was capable of correcting a wide variety of error patterns not correctable by individual block codes.

While quite useful, block codes have some limitations. Since they are frame oriented, they require an entire frame before an output signal is available and this introduces considerable delay (latency). They obviously require frame synchronization, which causes start-up delay and framing complexity. These limitations were overcome by convolutional coding (Forney, 1970, 1971, 1974; Viterbi, 1971) first introduced by Foschini et al. (1974) and Elias (1955). Viterbi (1971) noted that for the same order of complexity, convolutional codes considerably outperform block codes. Rather than segregating data into distinct blocks, convolutional encoders add redundancy to a continuous stream of input data using a linear shift register. Each set of n output bits is a linear combination of the current set of k input bits and the m bits stored in the shift register. The total number of bits that each output depends on is called the *constraint length*. The rate of the encoder is the number of data bits k taken in by the encoder in one coding interval divided by the number of code bits n output during the same time. While various decoding algorithms for convolutional codes were invented, the optimal solution was not available until the invention of the Viterbi decoding algorithm (Forney, 1973; Viterbi, 1967). This algorithm allowed soft decisions to be modified by the history of the data stream and the constraints of the encoding architecture.

While convolutional coding with the Viterbi decoder is very powerful, convolutional codes suffer from a serious problem—when the Viterbi decoder fails, it creates long, continuous bursts of errors. This limitation was moderated by Odenwalder [as reported by Liu and Lee (1984)]. He used Forney's concatenated coding architecture with the Reed–Solomon outer coding but replaced the inner block encoder/decoder pair with a convolutional encoder and Viterbi decoder pair. The powerful Reed–Solomon

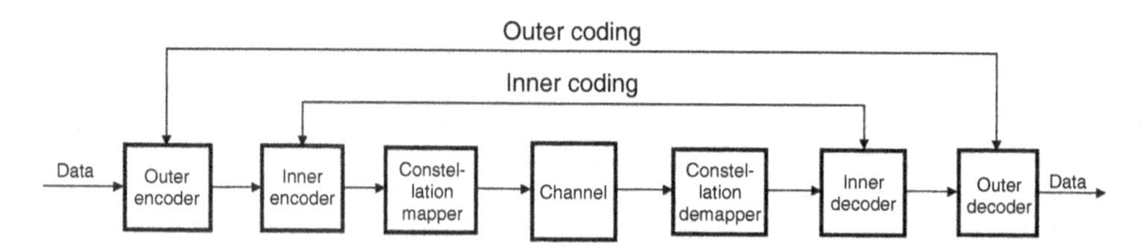

Figure 3.19 Concatenated coding.

coding was used to "tame the Viterbi." Sometimes a block interleaver and deinterleaver (Forney, 1971b) are placed between the inner and outer coders to break up the decoder burst errors so that a shorter Reed–Solomon code can be used. Bhargava (1983) evaluated several coding techniques (for binary phase shift keying (BPSK) and quadrature phase shift keying (QPSK)) and concluded that concatenating the Reed–Solomon block coding outside a system using a convolutional encoder with Viterbi inner decoder resulted in the optimum error performance (if the error source was Gaussian noise). This architecture is commonly used in modern QAM and trellis coded modulation (TCM) radios.

The next step in coding was to merge constellation mapping and coding. Imai and Hirakawa (1977) realized that if subsets of constellations could be used at any signaling instant, the larger space between symbols would improve error performance (Forney, 1988a, 1988b, 1989; Forney and Wei, 1989). Their approach, called *multilevel signaling*, used independent binary codes to pick the PSK or QAM constellation subsets. While they mentioned convolutional coding and Viterbi decoding, most actual commercial products used block coding and hard decision decoding. Shortly thereafter, Ungerboeck (1982; 1987a, 1987b) disclosed two-dimensional (2D) TCM for QAM and PSK constellations, which incorporated convolutional encoders and the Viterbi decoder [Wolf and Ungerboeck (1986) even developed trellis coding for partial response signaling]. Since commercial products using multiple levels used hard decisions at each level of the decoding process, multilevel performance was slightly worse than products using 2D TCM. Later, Yamaguchi and Imai (1987) improved multilevel performance by explicitly including convolutional encoding and Viterbi decoding to achieve slightly better performance than Ungerboeck's 2D trellis coding.

Wei (1984a, 1984b; 1987) discovered rotational invariant multidimensional trellis modulation. Wei took Ungerboeck's two-dimensional trellis constellation decomposition and added multiple consecutive time dependencies to define 4, 8, 16, and even higher dimension trellis coding. As the number of dimensions increased, the complexity of decoding (and latency) increased dramatically, with only incremental coding gain improvement. Four-dimensional trellis coding is the typical choice for current commercial products. Multilevel is used only if low latency or limited computational complexity is important.

The current state-of-the-art codes are turbo codes (Berrou Glavieux Thitimasjshima, 1993; Lodge Young Hoeher Hagenauer, 1993) and the recently rediscovered LDPCs (Gallagher, 1962; MacKay and Neal, 1996). An LDPC has been demonstrated (Chung et al., 2001) to come within 0.04 dB of the Shannon capacity limit. However, this code requires a block length of 10^7 and typically requires about 1000 processing iterations to achieve 10^{-6} BER. While these codes are popular in low speed wireless equipment (Brink, 2006), their computational complexity and recursive processing induce latency and the computational difficulties have delayed their recent introduction into high speed radio systems. Currently, for QAM, LDPCs achieve up to 7-dB threshold coding gain when compared to uncoded QAM. While these codes dramatically improve 10^{-3} and 10^{-6} BER thresholds, they exhibit a relatively high BBER. For critical requirements, they are paired with a powerful background error-correcting scheme (such as the Reed–Solomon coding) to suppress residual errors.

The primary limitation to improving radio S/N performance is processing latency. The 150-Mb/s QAM radios typically have latency less than 100 μs (transmitter plus receiver). TCM radio may push that to a few hundred microseconds. The primary difference is the degree of the Reed–Solomon coding used in the radios. LDPCs can significantly improve the radio S/N performance.

For current digital microwave systems, the modulation methods of choice are concatenated LDPCs with QAM (IP baseband) or the Reed–Solomon coding with 4D TCM (SONET baseband) (Fig. 3.20).

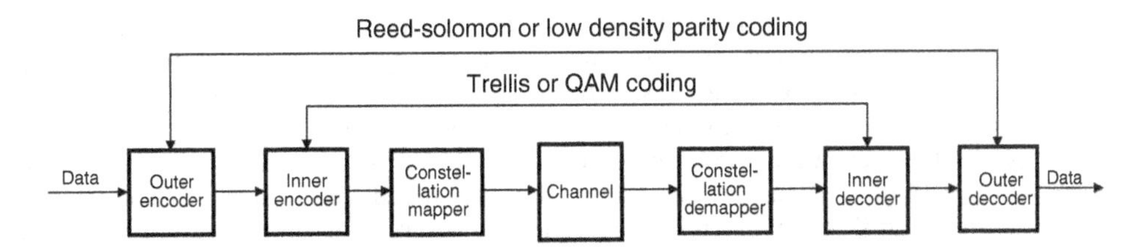

Figure 3.20 Typical digital microwave radio coding.

Currently, typical 10^{-6} BER threshold improvements of 4 dB with the Reed–Solomon coding and 7 dB with LDPCs are commercially available. For 3 DS3 and 4 E3 transmissions, 64 QAM is typically used. It has the spectral efficiency (6 bits/s/Hz) to transport those signals in the typical nominally 30-MHz radio channel. For the SONET/SDH equivalents (STS-3/OC-3 or STM-1), 4D trellis provides the additional transmission bandwidth (6.5 bits/s/Hz) required without sacrificing system gain or increasing channel bandwidth. In the United States, for radio channels narrower than 10 MHz (transporting multiple DS1s or E1s), 16 QAM or 32 4D trellis are commonly used. Modern adaptive modulation IP radios typically use QAM to facilitate rapid switching between different transmission speed modulation schemes.

When compared to the Shannon theoretical S/N limit, QAM is a little more than 9 dB away (requires slightly more than 9 dB S/N than Shannon's limit).

Compared to the 2^n constellation QAM, the TCM 4D 2^{n+1} constellation achieves slightly better system gain and spectral efficiency. Its spectral efficiency (increased transmission bandwidth for a given radio channel bandwidth) is the primary reason it is used with most SONET radio system.

3.10 TRELLIS CODED MODULATION (TCM)

To transport n bits of information per signaling interval, a two-dimensional (I and Q) constellation of 2^n symbols is required. Trellis coding expands the two-dimensional constellation to 2^{n+1} symbols. The constellation is partitioned into 2^{m+1} subsets (called *cosets* or *subfamilies*), which have greater voltage separation between signaling states than the 2^{n+1} constellation. Of the n bits that arrive into the trellis modulator as a group (from the serial to parallel converter), m bits enter an $m/(m+1)$ convolutional encoder. The $m + 1$ bits out of the encoder specify which of the constellation subsets will be used for signaling at the next signaling intervals.

If the number of consecutive symbol pairs constrained by the convolutional encoder is p, the dimensionality of the TCM is $D = 2 \times 2^p$. If each signaled symbol (using one of the cosets) is independent of the other symbols (no pairs, just a single set of cosets), then $p = 0$ and the TCM is two-dimensional (it only has the two dimensionality of an individual I and Q symbol). If the set of coset symbols is consecutively chosen in single pairs, then $p = 1$. This defines 4D TCM, the most common form. Higher order TCM is possible (e.g., 8D with two consecutive pairs, $p = 2$, and 16D with three consecutive pairs, $p = 3$). The higher order TCMs are much more complex to decode, have much greater latency, and have only marginal error threshold improvement. They are not popular commercially.

Each of the subset constellations has significantly improved noise immunity relative to the original 2^{n+1} constellation. However, the receiver must determine which of the 2^{m+1} subsets was used for the signaling interval. The Viterbi decoder is used to make that determination.

In North America, the most popular trellis formats are Wei's original (Wei, 1987) 32 4D or 128 4D implementations (Fig. 3.21 and Fig. 3.22). Trellis 32 and 128 have a 0.3- and 0.5-dB PARR advantage over 16 and 64 QAM, respectively. The trellis 32 4D and 128 4D have a 1-dB average signal-to-noise advantage over 16 and 64 QAM, respectively, while achieving improved spectral efficiency.

Comparing QAM and TCM is not straightforward. TCM expands the constellation by 1 bit relative to QAM but spreads out signaling over multiple intervals. Composite transport efficiency is slightly greater than QAM since more composite bits are absorbed by the transmitter per multiple signaling intervals. When compared to QAM, the error threshold improvement is relatively small while the spectral efficiency is slightly greater. For SONET transport radios, this increased spectral efficiency was the main reason TCM was the popular choice (Table 3.5).

The challenge for trellis coding demodulation is to determine the subset constellation being used for signaling at any given time. Since the particular subset used at any given signaling instant is not known initially, it must be determined based on the signaling history and a knowledge of the possible signaling states. The Viterbi decoding algorithm is an optimal method for determining the subset.

For a typical 4D coder, only four subsets are used (Fig. 3.21). Since they are paired by the convolutional encoder, there are 16 (4×4) possible pairs. This can be diagramed using a trellis diagram. A trellis diagram is a graph that illustrates the transitions between modulation states for a modulation method with memory. For a modulation system with memory, current states are constrained by previous states (Fig. 3.23).

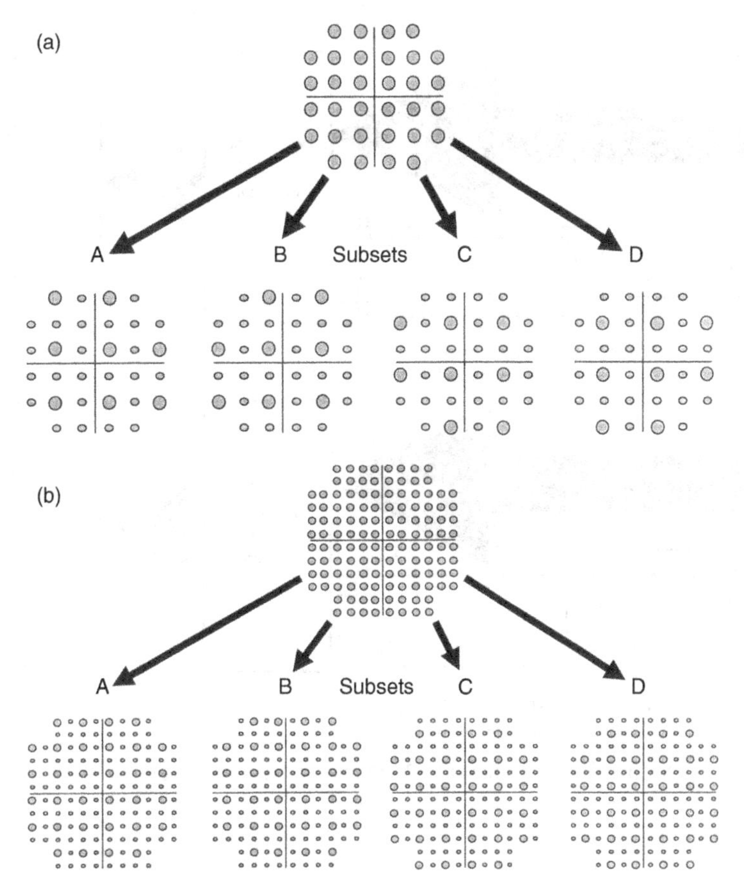

Figure 3.21 Constellation decomposition for trellis (a) 32 4D and (b) 128 4D.

TABLE 3.5 Comparing Typical QAM and TCM

Relative System Gain		
16 QAM = Ref	32 TCM 2D = 0.0 dB	32 TCM 4D = 1.0 dB
64 QAM = Ref	128 TCM 2D = 0.0 dB	128 TCM 4D = 1.0 dB
128 QAM = Ref	256 TCM 2D = 0.0 dB	256 TCM 4D = 1.0 dB
256 QAM = Ref	512 TCM 2D = 0.0 dB	512 TCM 4D = 1.0 dB
Spectral Efficiency, bits/s/Hz		
16 QAM = 4	32 TCM 2D = 4	32 TCM 4D = 4.5
64 QAM = 6	128 TCM 2D = 6	128 TCM 4D = 6.5
128 QAM = 7	256 TCM 2D = 7	256 TCM 4D = 7.5
256 QAM = 8	512 TCM 2D = 8	512 TCM 4D = 8.5

For any set of two subsets the next possible set of subsets is constrained by the convolutional coder to four (rather than 16) possible pairs. Each of the sets of four pairs is chosen to have the largest space between symbols in consecutive space (I and Q) and time subsets. The constraint of consecutive subsets allows the Viterbi decoder to estimate the subset used (Fig. 3.24).

The Viterbi decoder stores the analog received signal I and Q coordinates at each sampling instant. If the Viterbi could store all states, it would be the optimum decoder. Since the storage must be terminated

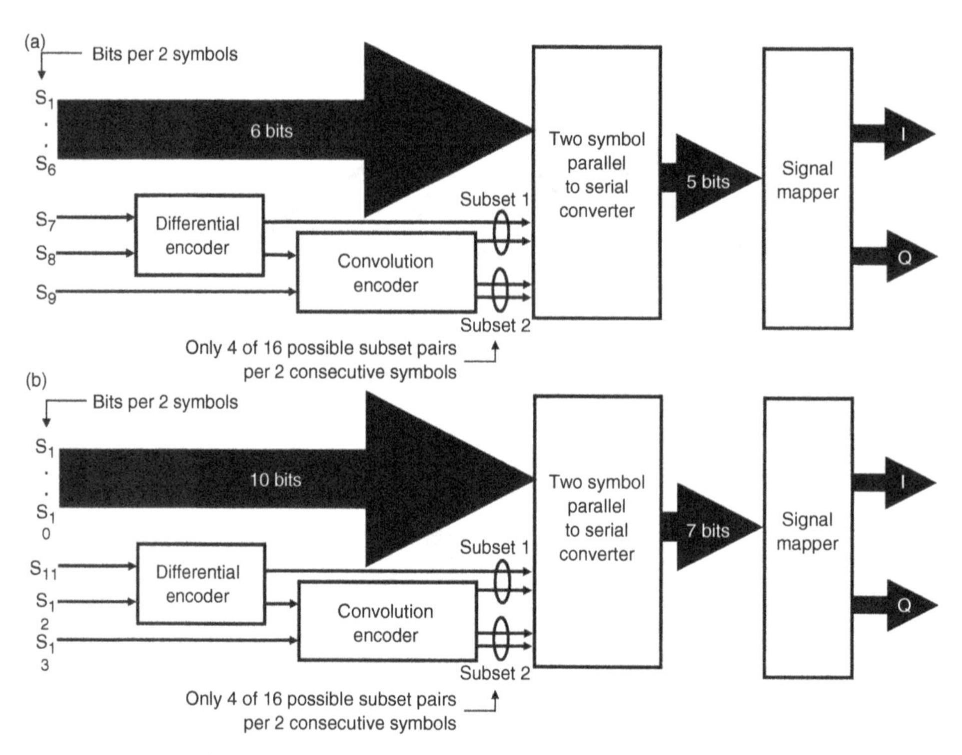

Figure 3.22 Trellis (a) 32 4D and (b) 128 4D modulators.

Current demodulator coset pairs	Current demodulator state	Next demodulator state	Next demodulator coset pairs
AA	1	1	AA
AB	2	2	AB
AC	3	3	AC
AD	4	4	AD
BA	5	5	BA
BB	6	6	BB
BC	7	7	BC
BD	8	8	BD
CA	9	9	CA
CB	10	10	CB
CC	11	11	CC
CD	12	12	CD
DA	13	13	DA
DB	14	14	DB
DC	15	15	DC
DD	16	16	DD

Figure 3.23 Trellis 4D subset combinations and allowable states.

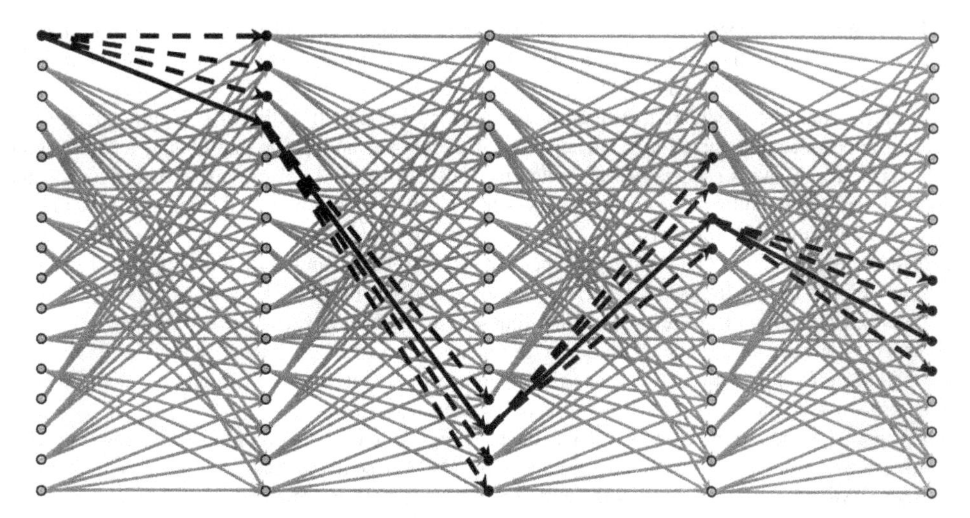

Figure 3.24 Trellis (state diagram) of five pairs of consecutive subset choices.

at some defined length, actual implementations are slightly suboptimal. Let us assume that the Viterbi memory length is five times the constraint length of a 4-bit convolutional encoder. The Viterbi must store 40 (5 coset pairs × 4 bits/encoder state × 2 subsets per set of encoder bits) consecutive signal samples. The Viterbi calculates and stores the squared distance between each received signal and the closest symbol in each of the 16 subsets. This squared distance is stored for all 40 signal samples. The Viterbi calculates the cumulative squared error for all possible paths across the 40 sets of subsets and picks the path with the lowest squared error (the "survivor"). This determines the subset most likely to have been used for signaling 40 samples delayed from the current sample. After signaling 40 samples, the digital decoding decision is made based on the minimum distance to a symbol in the just determined constellation subset. Since the Viterbi must wait for many samples (40 in this example) before a signal decision is made, signal latency is significant. Also, since many samples are used, if some of the samples are so significantly flawed that an error is made in the estimated constellation subset, the error is propagated for several signal estimates (the Viterbi exhibits burst errors). At start-up after radio loss of signal, many (40 in this case) samples are required before starting to provide valid outputs. There is a design trade-off between the Viterbi depth (for improved stable-state error performance) and dynamic performance and latency.

3.11 ORTHOGONAL FREQUENCY DIVISION MULTIPLEXING (OFDM)

If a digital transmission channel is subjected to noise and intersymbol interference (usually caused by multipath propagation), the conventional approach is to use an appropriately modulated single carrier operating at the Nyquist symbol rate at the transmitter followed by an adaptive time domain equalizer at the receiver. While this approach is effective for fast changing multipath of low order, its use in very complex multipath environments (as in urban radio propagation in a highly reflective environment or for long-distance cable modems subjected to multiple signal reflections) is more challenging. An alternative approach is to subdivide the available channel bandwidth into a number of equal bandwidth subchannels operating at the same relatively slow symbol rate. The bandwidth of the sub-channels is sufficiently narrow that the frequency response characteristics of the sub-channels require only simple amplitude compensation but not complicated phase compensation. While this approach has been around for some time (Doeltz Heald Martin, 1957), it has become popular fairly recently (Bingham, 1990; Chow et al., 1995). For example, it is the technology of choice for asymmetric digital subscriber line (ADSL) communication.

Unlike previous generation analog frequency division multiplex systems that required complex analog filters, these systems avoid that complexity by separating the individual carriers by the inverse of the symbol rate. When this separation is used, each carrier is orthogonal to all the others (the time integral of any two carriers over the time duration of a single symbol is zero). This allows each carrier to be extracted by digital means (typically FFT techniques).

Each carrier may be modulated differently (or may be turned off entirely) if significant noise or attenuation appears at that carrier's frequency. All of the above modulation and coding techniques are available. This technique is especially well suited to deal with spectral noise (narrow band interference) and multiple reflection multipath distortions. However, since it uses very narrow ("sluggish") transmission carriers, it is best suited for transmission channels whose noise and multipath distortions change slowly. It has no advantage over any other approach when subjected to broadband noise (such as spread spectrum interference or flat fading front end noise). It is well suited for obstructed urban paths which attempt to benefit from reflections from terrain and buildings to achieve a transmission path. However, system performance is difficult to predict for these environments. It can be designed to minimize the effect of narrow band interference. Otherwise, it has no advantage over conventional modulation methods when operated on conventional unobstructed point to point microwave paths.

When applied to radio transmission systems, the most significant limitation of this technology is PARR. For conventional systems the PARR for high capacity QAM systems is roughly 8 dB. For multicarrier systems, the PARR can be tens of decibels more. Since transmitter performance is peak power limited but system gain is average power limited, this issue can dramatically impact reliable transmission distance for multicarrier systems. This issue is the focus of much current research on radio orthogonal frequency division multiplexing (OFDM) systems.

3.12 RADIO CONFIGURATIONS

All the preceding examples describe communication in one direction (simplex transmission). Most actual systems communicate two directions simultaneously (full duplex). By their very nature, they transfer data to and from other devices, which have a transmit and a receive function. However, the concepts of transmitter and receiver as well as the transmit and the receive functions are not standardized within the telecommunications community (Fig. 3.25).

A transmit signal from one device may be a receive signal on the device it is connected to. This can be quite confusing when discussing these functions with different people at different locations. The concepts of in and out are less confusing but not as commonly used. For the following discussion, the transmit signal enters the radio transmitter and the receive signal exits the radio receiver. This convention may or may not conform to the definitions of equipment (e.g., add/drop multiplexers or routers) to which the radio's baseband signals are connected.

Commercial fixed point to point microwave radios have two basic hardware configurations: integrated radio (all functions in one box) and split package radio (baseband functions in one box near the other telecommunication equipment and RF functions in another box usually collocated with the antenna). For radios that accept several different signal formats, the integrated radio is typically an all-indoor unit. If the radio supports only an IP interface, the radio may be a single package intended for all-outdoor installation (basically a split package radio with no indoor unit). The radio will have an IP interface that connects directly to a router (Fig. 3.26).

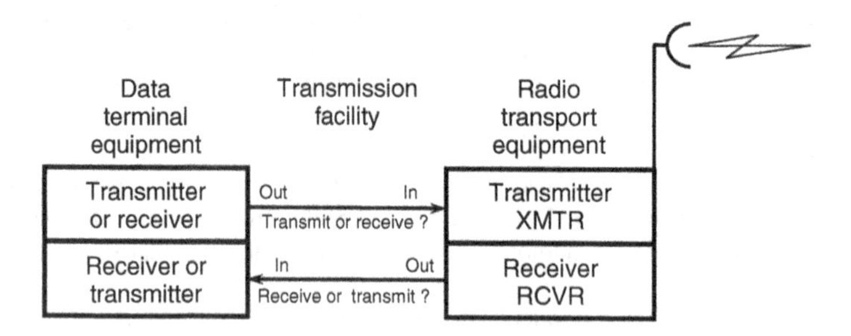

Figure 3.25 Interfacing the radio transmitter and receiver.

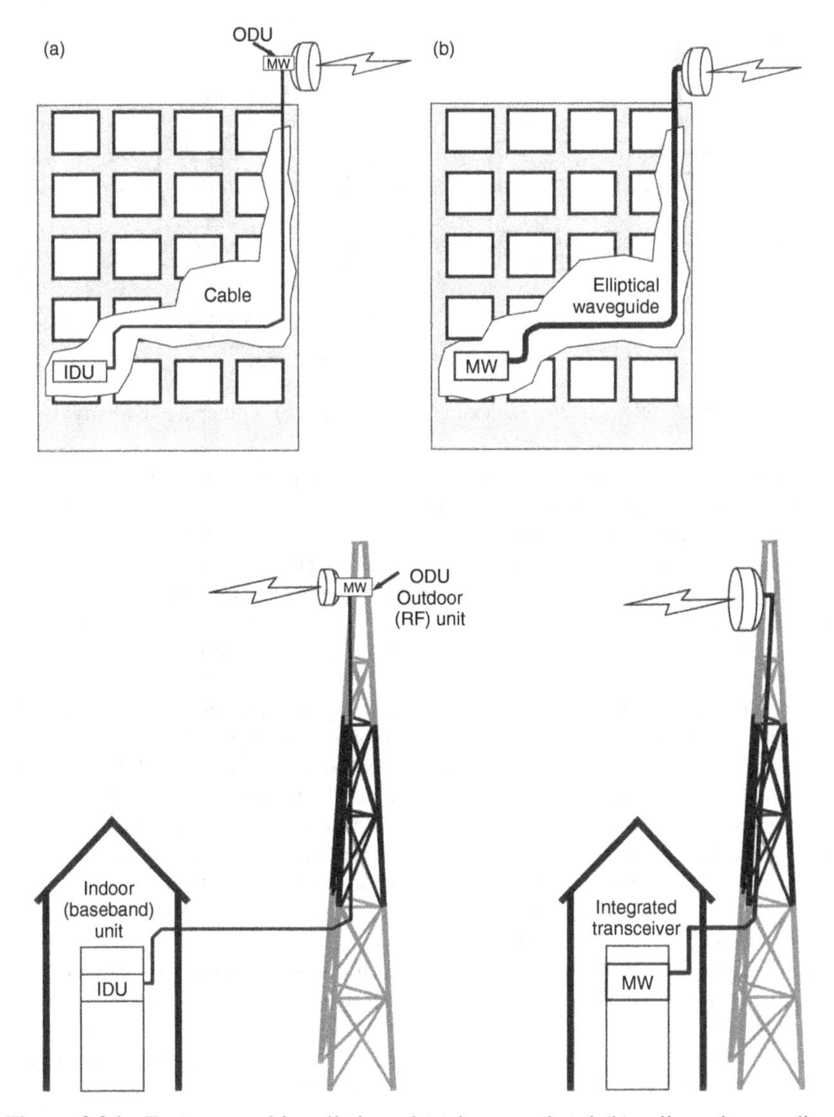

Figure 3.26 Two types of installation of (a) integrated and (b) split package radios.

The split package and all-outdoor configurations have advantages in an urban building environment. However, they pose operational constraints in suburban and rural tower installations.

Functionally, radios are either terminal or nodal. Terminal radios connect baseband signals between two ends of a single radio path (point to point). Nodal radios connect several end locations (drop and insert). The nodal radio connects to several antennas which communicate to multiple far end locations. The radio may allow some traffic to pass through (act as a repeater) and may drop/insert other traffic at their location. Segregation of traffic may be via IP routing or digital cross-connect. Often, both TDM and IP traffic are transported via radio. TDM radios transport IP using TDM data blocks, and IP radios transport TDM using IP packets (pseudowire or circuit emulation). Since IP transport is asynchronous, transport of TDM traffic over IP circuits requires some form of external or encapsulated synchronization. IP radios invariably have more transport latency than TDM radios. IP latency is a function of packet size and transport bandwidth.

The following paragraphs diagram the basic structures of microwave radio transmitters and receivers. These may be used in any of the above radio configurations.

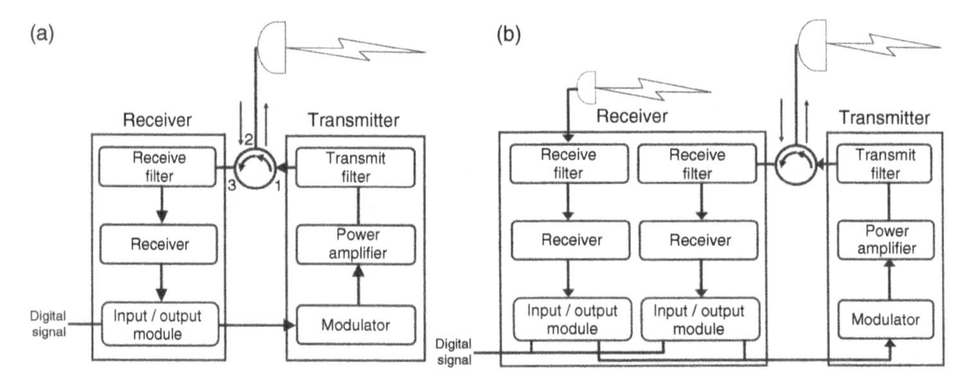

Figure 3.27 Nonstandby radios (a) without and (b) with space diversity.

The simplest duplex radio is nonstandby without or with space diversity (Fig. 3.27). A key component is the circulator (shown as a circle with three numbered ports). The circulator is a passive device that manages the transfer of microwave signals between ports. A transmit signal entering port 1 exits port 2 toward the antenna. A receive signal from the antenna entering port 2 is transferred to port 3.

Another popular configuration is monitored hot standby without or with space diversity (Fig. 3.28). The nondiversity radio receiver has an asymmetrical receive signal power splitter that typically reduces the main receiver input power by about 1 dB and the protection (offline) receiver by 10 dB. In integrated radios, the transmitter radio outputs are switched using a relay switch (as shown in figure). For split package radios, since each radio transmitter is in a separate box, the outputs are combined using a waveguide coupler and the transmit outputs are switched on and off electronically.

Figure 3.29 shows the simplified diagrams of the major radio configurations. The hot standby space diversity configuration that switches transmit antennas when the transmitters are switched is often used in locations where loss of an antenna (due to snow or wind loading) is a significant concern. This affects the antenna structure design. Most path clearance criteria (see Chapter 12) are between the main transmit and main receive antennas. In this case, the main transmit antenna may be the (typically lower) diversity antenna. This will cause the antennas to be placed higher on the antenna structure than would be the case with a conventional space diversity configuration.

Frequency diversity (Fig. 3.30) depends on all frequencies being adequately separated so that RF filters (on the receiver and the transmitter front ends) tuned to frequencies other than the receiver or transmitter will be reflected back into the circulator (and propagated on down the waveguide).

Frequency and space ("quad") diversity is sometimes used for difficult paths. The configuration in Figure 3.30b is commonly used. The configuration in Figure 3.30c is the preferred configuration. This is the most powerful radio diversity configuration that is commercially available but is costlier than the former configuration because it requires same-sized antennas for main and diversity and, due to clearance requirements for the lower antenna, it usually requires a taller tower. It is used for the most difficult (e.g., long overwater or mountaintop to mountaintop) paths subjected to multipath and reflection distortions.

Hybrid diversity is sometimes used when space diversity is not possible at one end of the radio link (Fig. 3.31). Angle diversity is also sometimes used if space diversity is not practical.

Multiline is a method of placing several radios on the same antenna to provide several radio channels on the same radio path (Fig. 3.32). Each radio is nonstandby. Equipment protection is achieved by switching traffic from a failed radio to a separate nonstandby radio reserved for circuit restoration.

3.12.1 Cross-Polarization Interference Cancellation (XPIC)

The flexibility to increase path transmission bandwidth has created a renewed interest in cross-polarization interference cancellation (XPIC). Historically, this methodology was applied to low frequency (lower than 12 GHz) radios to maximize the number of transmission channels on a path. Significant degradation of the radio signal due to multipath or rain usually limited this methodology to relatively short paths.

This configuration is a method of placing many radios on the same radio path (Fig. 3.33). Different radios using the same frequencies are operated on the antenna horizontal and vertical polarizations. The

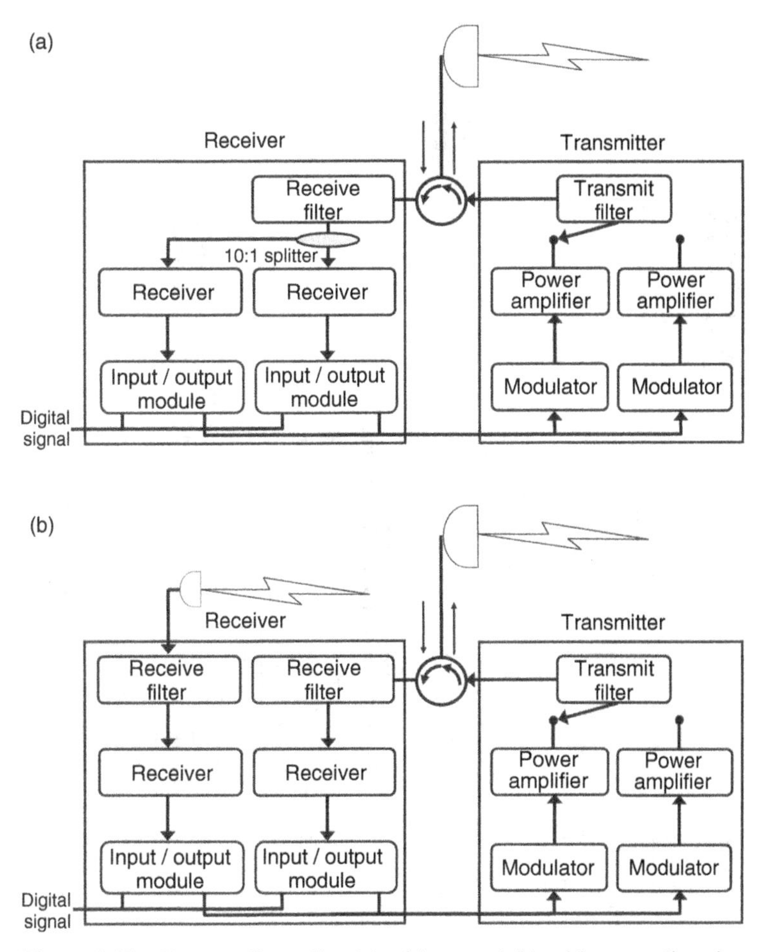

Figure 3.28 Hot standby radios (a) without and (b) with space diversity.

vertical signal receiver must sense and cancel the horizontal signal using its frequency and vice versa for the horizontal signal receiver. Every received frequency signal must appear at two receivers. This is done by using hybrids before the receiver lineup (a technique in older systems that reduced system gain) or by taking an output from one receiver and connecting it to another (a modern approach impacting reliability). XPIC can only reduce the cross-polarized signal by about 20 dB, and at least 40-dB channel isolation is required for typical high order QAM to avoid impacting performance (inducing "dribbling" errors). It means precisely oriented high performance antennas with high cross-polarization discrimination are required. Maintaining antenna alignment over time is difficult and some operators only assume 20-dB cross-polarization discrimination for design purposes even when very high polarization discrimination antennas are used.

XPIC radios are sometimes used in the multiline configuration for integrated package radios. They are also used for split package or all-outdoor radios to provide two nonstandby radio channels using one antenna. This minimizes tower loading and leasing costs and is popular in dense urban high frequency networks.

High frequency radio paths (higher than 10 GHz) are naturally short due to rain and multipath limitations. However, spectrum congestion is becoming greater in urban areas. Cross-polarization-cancelling radios increase the flexibility of radio channel selection while minimizing radio transceiver sparing requirements.

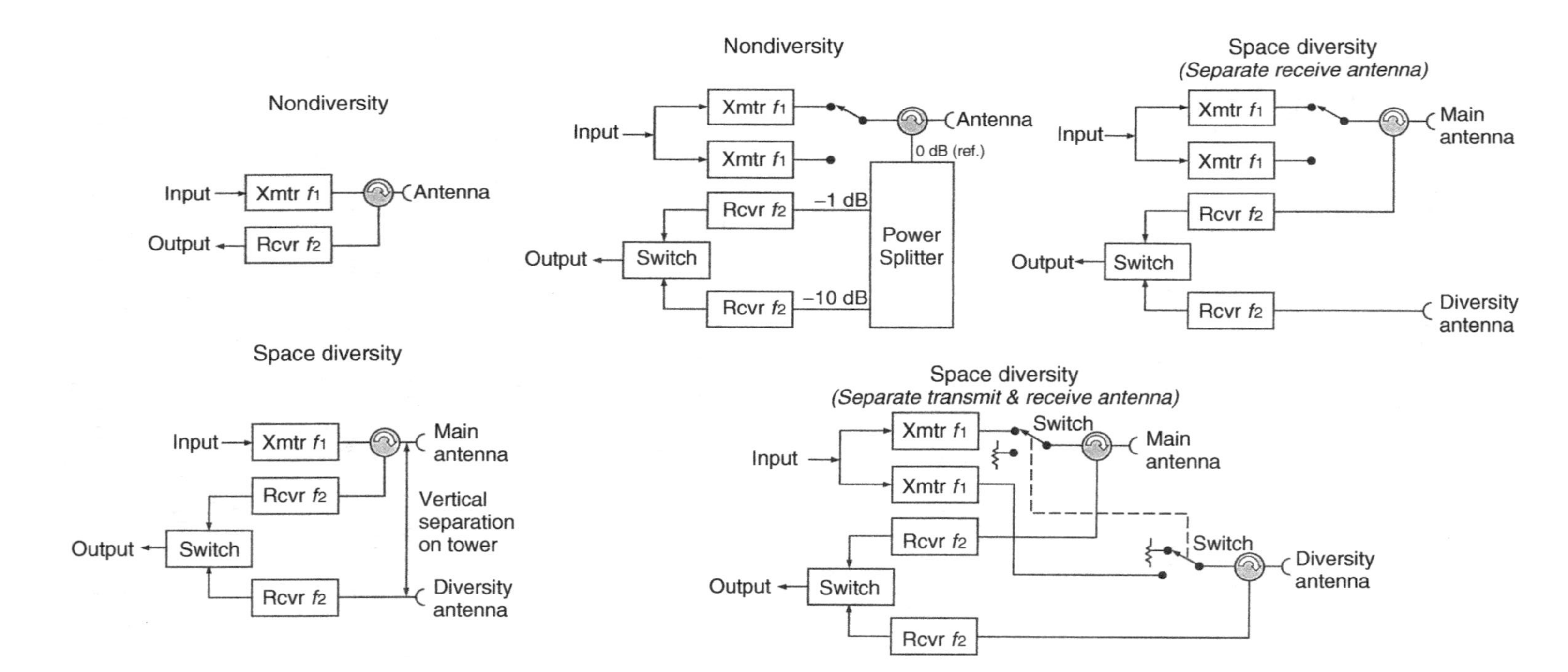

Figure 3.29 Nonstandby and hot standby radios without and with space diversity.

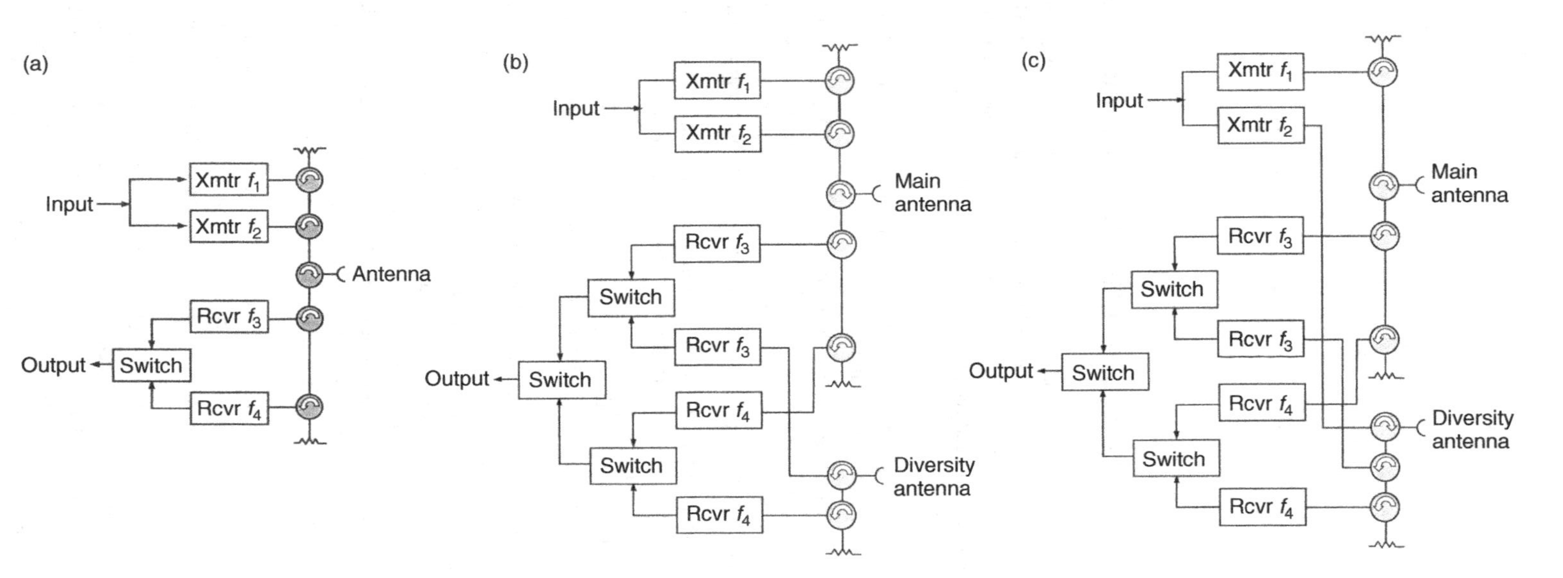

Figure 3.30 (a) Frequency diversity without space diversity and (b,c) quad diversity (frequency diversity with space diversity).

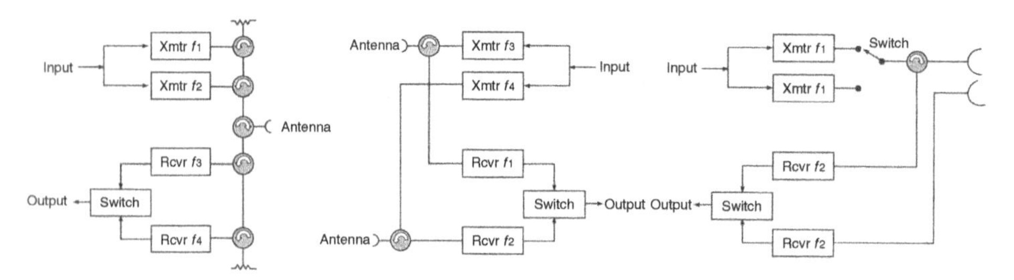

Figure 3.31 Hybrid diversity (frequency diversity with asymmetrical space diversity) and angle diversity.

3.13 FREQUENCY DIVERSITY AND MULTILINE CONSIDERATIONS

Historically, high capacity 6-GHz radio links used multiple transmitter/receiver pairs on the same path to increase overall path transmission capacity on "long-haul" circuits. Interest in increasing the capacity of short-haul high frequency radio paths has led to more designs with multiple transmitter/receivers on the same path. When multiple transmitters and receivers are placed on the same radio path, two degradations (of receiver threshold) are possible:

Out-of-band transmitter noise can affect receivers operating on nearby channels. This potential problem is resolved by filtering the transmitter output.

Very slight amplitude nonlinearity in waveguide flanges (multiple flanges in long waveguide runs common in low frequency systems), ferrite-combining circulators, or receiver front end amplifiers produces intermodulation products from the multiple transmitter signals. Those products may be predicted and eliminated using the procedures described in the following paragraphs.

When two or more transmitters share the same antenna as two or more receivers, intermodulation interference can occur (Fig. 3.34). This intermodulation can occur due to nonlinearities between the transmitters and receivers (Fig. 3.35).

If a composite signal made of several discrete signals of slightly different frequencies (e.g., *A*, *B*, *C*, ...) is passed through a nonlinear device, intermodulation products are produced at the output. If the nonlinearity is constant and relatively small compared to the normal linear output components, the input V_i to output V_o relationship may be characterized by a low order polynomial of V_i. The even-order components have frequencies much higher than the original signals. The odd-order products are similar in frequency to the original signals and can cause undesired interfering signals (Table 3.6).

If high order products are present, the low order products are also present. Typically, system nonlinearity is so small that only third-order intermodulation products need to be considered. The $(2A - B)$ product (where *A* and *B* are any two combinations of transmit center frequencies) applies to all frequency diversity systems. The $(A + B - C)$ product (where *A*, *B*, and *C* are any three combinations of transmit center frequency) only applies to multiline systems (systems with multiple transmitters on the same waveguide).

On the basis of the Monte Carlo simulation, the expected spectral density of the intermodulation components is shown in Figure 3.36. When performing intermodulation checks with channel center frequencies, products as far away as one channel frequency should be considered as potentials for interference.

If the intermodulation is likely to occur in the receivers (typical for high frequency radio designs with short or no waveguide runs), a filter can be placed in front of the receivers to eliminate the potentially interfering transmit signals (band-pass filters can be used to filter the receiver so that only the desired signal enters the receiver preamplifier). However, if the nonlinearities are in the waveguide joints or the circulators (common for complex low frequency systems), there are only two choices: The simplest

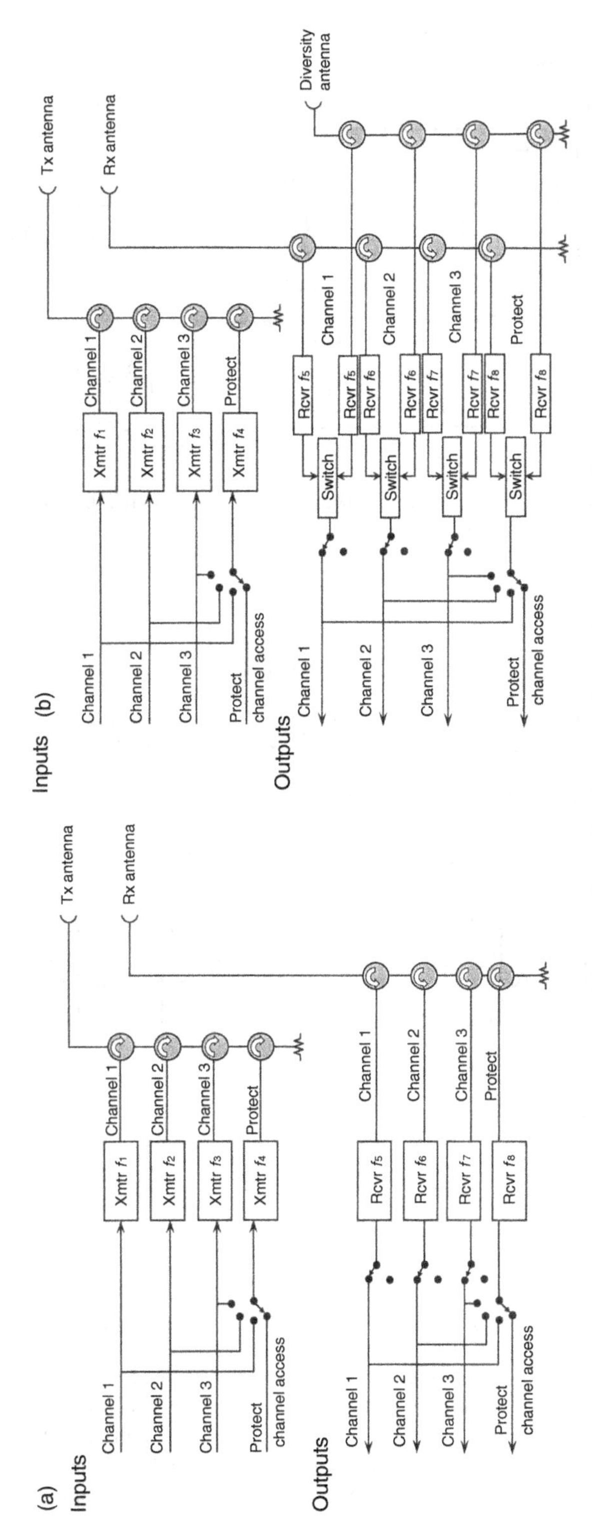

Figure 3.32 Multiline (a) without and (b) with space diversity.

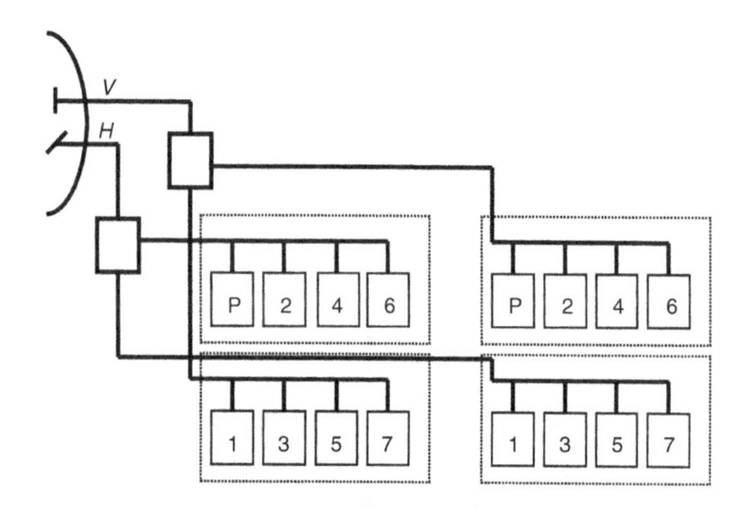

Figure 3.33 Multiline cross-polarization interference cancellation (XPIC).

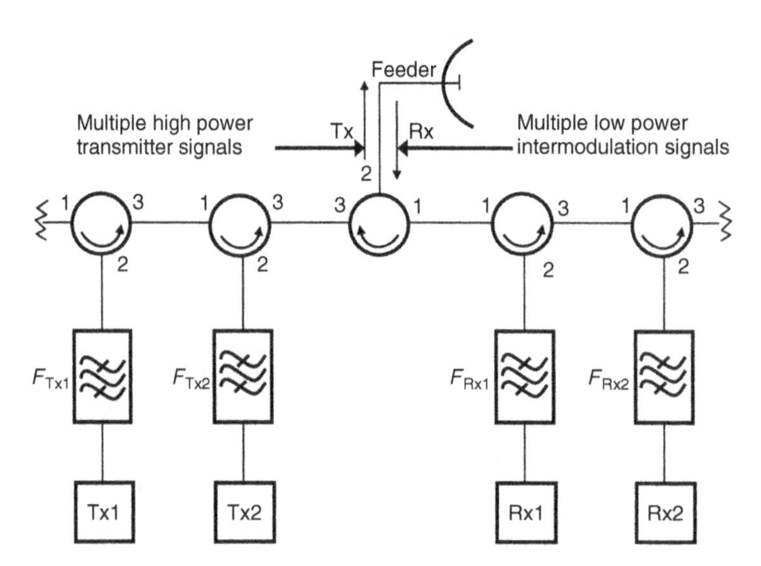

Figure 3.34 Frequency diversity "$2A - B$" intermodulation interference.

is to predict the interfering signal (i.e., perform a "$2A - B$" calculation under the assumption that the nonlinearity is only third and fifth order) using Table 3.6 and choose transmit or receive channel center frequencies that do not produce intermodulation products near the receiver frequencies (typically assumed to be a spurious carrier frequency within the victim receiver's receive channel). If appropriate frequencies are not available, the intermodulation interference can be eliminated by placing the transmitters on one antenna and the receivers on another. The antenna to antenna isolation is adequate to keep the transmitter products entering the receivers. For multiline systems with more than four duplex ("go/return") channels, separate antennas are usually required.

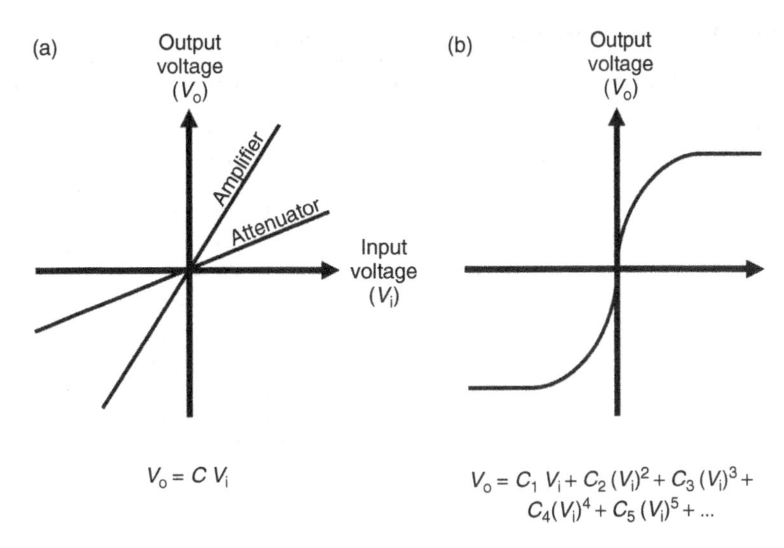

Figure 3.35 Transfer functions of (a) linear and (b) nonlinear devices.

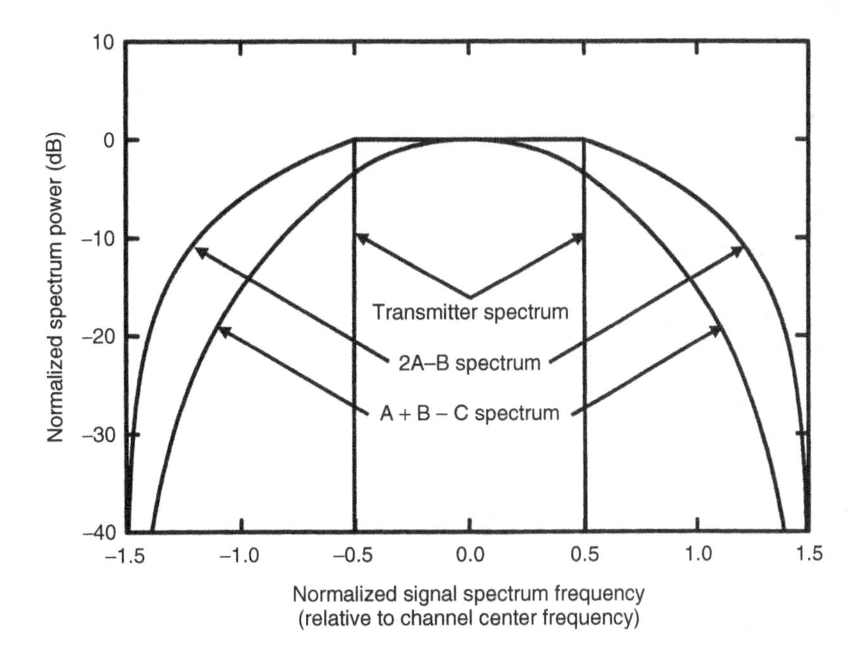

Figure 3.36 Low order intermodulation spectrum.

3.14 TRANSMISSION LATENCY

Some radio applications (such are simulcast and transfer trip circuits) are quite sensitive to transmission delay. Radio transmission delay is typically a function of forward error correction and block interleavers, as well as time slot interchange buffers used in cross-connects and first-in first-out (FIFO) buffers used in IP transport. In general, the higher the transmission speed of the radio, the lower the latency. The following are typical "ball park" estimates of single hop (transmitter to receiver) transmission latency:

TABLE 3.6 Low Order Intermodulation Products

Third-Order Products	Fifth-Order Products	Seventh-Order Products
$A + B - C$	$A + B + C - D - E$	$A + B + C + D - E - F - G$
$2A - B$	$A + B + C - 2D$	$A + B + C + D - 2E - F$
	$2A + B - C - D$	$A + B + C + D - 3E$
	$2A + B - 2C$	$2A + B + C - D - E - F$
	$3A - B - C$	$2A + B + C - 2D - E$
	$3A - 2B$	$2A + B + C - 3D$
		$2A + 2B - C - D - E$
		$2A + 2B - 2C - D$
		$2A + 2B - 3C$
		$3A + B - C - D - E$
		$3A + B - 2C - D$
		$3A + B - 3C$
		$4A - B - C - D$
		$4A - 2B - C$
		$4A - 3B$

Air (radio)
  5.4 μs/mile (1 ms/190 miles)

Fiber-optic cable (optical)
  8.3 μs/mile (1 ms/120 miles)

Voice frequency channel bank (FDM looped at group distribution bay)
  180–290 μs delay over the range 1000–2600 Hz (relative to 2000 Hz)

FDM master group/super group filters
  75–200 μs absolute delay

DS1 channel bank

No buffering, no cross-connect	250 μs
Buffering with cross-connect	35 ms

M13 multiplexer

DS1 to DS3 to DS1	50 μs

SONET add/drop multiplexer (OC-3 to OC-3 connection)

VT1.5 cross-connect	30–50 μs
STS-1 cross-connect	10–25 μs
DS1 to OC-3 to DS1	25–200 μs
DS3 to OC-3 to DS3	140–200 μs
STS-1 to OC-3 to STS-1	40–50 μs

Radio (one transceiver)

2 DS1 to 2 DS1	120–220 μs
4 DS1 to 4 DS1	60–120 μs
8 DS1 to 8 DS1	30–60 μs
12 DS1 to 12 DS1	20–40 μs
16 DS1 to 16 DS1	15–30 μs
1 DS3 to 1 DS3	75–200 μs
2 DS3 to 2 DS3	45–110 μs

3 DS3 to 3 DS3	35–85 μs
1 STS-1 to 1 STS-1	85–200 μs
1 STS-3 to 1 STS-3	55–85 μs
OC3 to OC3	65–95 μs
Ethernet to Ethernet	100–400 μs
Pseudowire to pseudowire	Multiple milliseconds

Ethernet test results are highly variable depending on the block size and the test methodology. Most vendors use the Internet Engineering Task Force (IETF) RFC 2544 as the Ethernet testing method. The emerging Ethernet and pseudowire (TDM over Ethernet) circuit's latency currently is highly variable from manufacturer to manufacturer.

3.15 AUTOMATIC TRANSMITTER POWER CONTROL (ATPC)

Unlike analog microwave radios, baseband performance does not change with RSL until the RSL is within a few decibels of radio BER threshold. Digital radio operators soon took advantage of this fact to facilitate frequency-coordinating microwave radios into areas already densely populated with other microwave radios (Vigants, 1975). Automatic transmit power control (ATPC) is a feature of a digital microwave radio link that adjusts transmitter output power based on the varying signal level at the receiver. ATPC allows the transmitter to be operated at a less than maximum power level (typically 10 dB) most of the time. This is the power level that is frequency-coordinated. When received signal fading occurs, the transmit power is increased as needed until the maximum power level is reached.

In the United States, there is a limit as to how much reduced power level can be frequency-coordinated (10 dB), the maximum time for which maximum power can be on (5 min), what is used to monitor receiver performance (RSL, not BER), and the maximum operating frequency at which ATPC can be used (not >11 GHz) (TIA/EIA, 1994; Working Group 18, 1992).

Although ATPC was originally conceived as a frequency coordination tool, some operators use it to reduce power consumption and to lengthen the life of transmitter power amplifiers (since they are running cooler during the reduced power output). It has even been used to reduce $2A - B$ frequency diversity interference (since this intermodulation interference is very power sensitive). It is a very popular microwave radio feature.

3.16 CURRENT TRENDS

Worldwide, digital networks are evolving toward a "converged" all-IP transport solution. SONET and very high speed "SONET-like" fiber systems will remain for the foreseeable future. However, all other legacy TDM systems (DS1, DS3, E1, and E3 interfaces) as well as ATM are going away in favor of IP connectivity [ATM has been reinvented in the IP world as MPLS (Multiprotocol Label Switching)]. Support for the legacy TDM traffic during this transition, as well as maximum transport capacity, is important. This has led to the emphasis on the evolving areas discussed in the following sections.

3.16.1 TDM (or ATM) over IP

Microwave radios are evolving toward all-IP transmission. The TDM/ATM traffic will be packetized and encapsulated for transmission over IP. TDM (DS1, DS3, E1, or E3) or ATM over IP is usually termed *pseudowire*, Circuit Emulation over Packet (CEP) or TDM over IP (TDMoIP). Several standards are evolving:

IETF pseudowire emulation edge to edge (PWE3) RFC 3985 (architecture) and RFCs 4xxx and 5xxx for specific implementations.

IP/MPLS Forum suite of specifications.

Metro Ethernet Forum (MEF) MEF 3 (service definitions) and MEF 8 (implementation).

ITU-T Y.1411 (ATM) and Y.1413 and Y.1453 (TDM).

TDM signals are inherently synchronous. IP packet transmission is inherently plesiochronous ("asynchronous"). This requires an IP transport product to provide a method of synchronizing the received TDM signal.

3.16.2 TDM Synchronization over IP

Another issue to be dealt with is received TDM signal frequency synchronization. The point to point TSM signal will be transported by a series of separate packets. These packets are routed and received individually. TDM signals have maximum allowable wander and jitter requirements. This requires that the TDM signal be buffered and then clocked out at the same rate as it was received at the packet source. Since the packet network is not synchronous, recovering the TDM signal clock requires additional consideration. Three methods of clock recovery are currently used.

Adaptive Clock Recovery. In this approach, the TDM clock is recovered by averaging the received TDM signal clock. When averaged over a long time, the received clock will have the same average frequency of the transmitted signal. The clock stream format may be proprietary (requiring similar source and sink devices) or a standard pseudowire flow (simplifying interoperability with third-party equipment). Typically, proprietary methods achieve better performance. Sometimes a multicast pseudowire signal is used for general clock distribution. Owing to packet arrival time variation, the short-term frequency will tend to wander. This wander is uncontrolled and may exceed the requirements of the terminating equipment. For small networks this may be satisfactory, whereas for large networks with considerable packet arrival time variation, this approach may not be satisfactory.

Differential Clock Recovery. If the network is closed, some equipment can provide for one transmission device to serve at the master clock and send clock reference to all appended equipment. The frequency of the incoming TDM signal is differentially referenced to this master clock at the transmit end and recovered at the receive end. The disadvantage of this approach is that all equipment must support the same clock synchronization mode. At present, Sync Messaging to validate synchronization quality is not standardized.

External Synchronization. For large networks composed of different vendor's equipment or for networks composed of different operators, the most practical method may be to provide external synchronization to each transmission device. Clock networks similar to those used in SONET and SDH networks have been created. At present, Sync Messaging to validate synchronization quality is not standardized in IP radio networks.

3.16.2.1 Timing over IP Connections and Transfer of Clock (TICTOC) An IETF Working Group has been developing standards for distribution of time and frequency over IP and MPLS networks. This working group appears to be transitioning to IEEE 1588 V2 and NTP V4.

3.16.2.2 IEEE Precision Time Protocol 1588 V2 (IEEE 1588-2008 and IEC61533 Ed. 2) This is a frequency and time-of-day distribution protocol that is based on time-stamp information exchange in a master–slave hierarchy. Timing information originated at a Grandmaster Clock function that is usually traceable to a Primary Reference Clock (PRC) or Coordinated Universal Time (UTC). Similar to NTP (network time protocol), it nonetheless offers better accuracy (fractional nanosecond precision). This standard defines the packet format for timing distribution but does not specify the actual clock recovery algorithm. Although it can be implemented end to end, use of intermediate network elements ("boundary clocks" and "transparent clocks") improves performance. This technology is primarily concerned with very accurate time stamps. Currently, the multiple queues in the process create excessive short time bias and jitter for use in synchronizing TDM traffic in an Ethernet network.

Synchronous Ethernet (ITU-T G.8261/Y.1361 and G.8262/Y.1362) is an evolving methodology whereby methods such as the aforementioned ones are used to synchronize selected nodes. Those nodes then synchronize slave nodes in a traditional hierarchal timing network. These approaches show promise for synchronizing TDM traffic in Ethernet networks.

Traditional SONET/SDH (GR-436-CORE, G.803) synchronous clock distribution networks have matured over the past 20 years and are suitable for synchronizing TDM traffic in Ethernet networks.

3.16.3 Adaptive Modulation

The need to increase fixed point to point microwave radio bandwidth is well known. For many radio paths, adequate fade margin exists for radio transmission using relatively high modulation formats for most of the time. Traditionally, radio transmission bandwidth was fixed and the priority of all baseband signals was considered equal. However, IP traffic priority is not equal. Encapsulated TDM signals require full time operation but many packet services can sustain short interruptions in service and still provide adequate performance. Radios are starting to utilize adaptive modulation. When the microwave path is distortion free, the highest capacity modulation format is used.

When the path experiences rain of multipath-induced distortion, the radio transmitter and receiver dynamically reduce the modulation complexity to limit the effect of path distortions. Modulation reduction to 4 QAM or QPSK for short periods is common. Most of the time, high bandwidth is available. For short periods, part of the transmission bandwidth is blocked. The signals that must be transmitted and those that can be blocked are differentiated using one of several possible quality-of-service (QoS) bandwidth identification methods.

Adaptive modulation radios are now available, which can provide large bandwidth transmission most of the time but revert to low bandwidth (as low as 4 QAM) when propagation anomalies occur. Some manufacturers use BER or BBER thresholds, and some use RSL thresholds for transmission rate changes. The error thresholds work for any degradation but require errors before a rate shift occurs. The RSL method does not sense all degradations but switches before any error degradation due to low RSL. Both methods work well and have their supporters. Since switching from high capacity to low capacity is attempted before significant errors (for a BER lower than normal threshold), the high capacity circuit fade margin (and path availability) for an adaptive modulation radio will be less than that for a fixed modulation radio of the same characteristics.

Given the embryonic nature of adaptive modulation, router considerations could be important. How will the routers in the network deal with changing transmission bandwidth? Will they disable the radio path because it is "flapping"? An open issue is whether or not OSPF (Open Shortest Path First) routing information (which is based on path distance and bandwidth) must be updated to account for bandwidth capacity changes. At present, the prevailing opinions are that the rate changes are so infrequent and transient that updates are not needed.

3.16.4 Quality of Service (QoS) [Grade of Service (GoS) in Europe]

QoS refers to resource reservation control, not the achieved service quality. QoS is the ability to provide different priority to different applications, users, or data flows, or to guarantee a certain level of performance to a data flow. QoS guarantees are important if the network capacity is insufficient. In the absence of network congestion, QoS mechanisms are not required.

Standards for quality and availability of IP circuits are still in the early stages of development. The ITU-R Report F2058 outlines design considerations for wireless IP transmission. The most significant factors mentioned are bandwidth, delay, and lost packets. The ITU-T Recommendation Y.1540 defines the QoS/CoS (class of service) parameters. Currently, the parameters of interest include successful packet transfer, errored packets, lost packets, spurious packets, average packet delay, and packet delay variation.

As noted in the ITU-R Report F2058 (Chung et al., 2001), "There are two schemes to achieve QoS and Cos. One is the prioritized scheme which offers a priority control among the service classes without specifying service specific parameters. The other is a parameterized scheme to assure required communication quality parameters. Only a parameterized scheme has the possibility to guarantee QoS. ... CoS control is often explained by an 'airplane model.' Service quality is classified into several service classes just like airplane seats which are classified into first, business and economy class. Higher service classes than usual best effort class are used to offer high-level services, e.g. ensuring minimum delay time or available bandwidth. High-quality service is provided if the request from the user is accepted. Admission control or policy control methods are used to determine which service classes are allowed for data transfer. According to the service class, each data transfer is transferred based on that quality. However, the amount of traffic carried in such higher service classes is limited because the available bandwidth is limited."

TABLE 3.7 IP Classes of Service

Network Performance Parameter	Nature of Network Objective	Quality-of-Service (QoS) Classes					
		Class 0	Class 1	Class 2	Class 3	Class 4	Class 5 Unspecified
IPTD	Upper bound on the mean IPTD	100 ms	400 ms	100 ms	400 ms	1 s	Undefined
IPDV, ms	Upper bound on the 1×10^{-3} (0.999) quantile of IPTD minus the minimum IPTD	50	50	Undefined	Undefined	Undefined	Undefined
IPLR	Upper bound on the packet loss probability	1×10^{-3}	1×10^{-3}	1×10^{-3}	1×10^{-3}	1×10^{-3}	Undefined
IPER	Upper bound	1×10^{-4}					Undefined

Abbreviations: IPTD, IP Packet Transfer Delay; IPDV, IP Packet Delay Variation; IPLR, IP Packet Loss Ratio; IPER, IP Packet Error Ratio.
The suggested measurement time is 1 min.

Currently, there are several internationally established methods of QoS/CoS:

IP Differentiated Services (DiffServ) field in the IP header;

IEEE 802.1d Annex H2 Tagging and 802.1p (MAC layer 2) (similar to DiffServ);

TCP/IP TOC (layer 3 or 4);

IP Integrated Services (IntServ);

Resource reSerVation Protocol (RSVP);

IETF MPLS.

Informal and proprietary methods are also sometimes used. A prospective user of an adaptive modulation radio should confirm whether his or her preferred method of QoS is supported by the radio.

The ITU-T Recommendation Y.1541 defines bounds on network performance between User Network Interfaces (UNIs). Six classes or services are defined: Class 0 is the strictest. It is for real time, jitter sensitive, high interaction traffic using constrained routing and distance. Class 5 is least strict (with no defined objectives). It is for traditional applications of IP networks using any route or path. The provisional values in Table 3.7 are defined.

Packet-switched networks (PSNs) operate on a contention basis. If bottlenecks occur in the network, packets are queued or dropped if transmission bandwidth is unavailable. QoS/CoS provides a means of prioritizing traffic when this transmission dilemma occurs. For legacy TDM radios, baseband transmission bandwidth is fixed. With the new generation of IP radios, baseband transmission bandwidth can be variable (although the RF signal bandwidth remains constant). Often, microwave radios have more fade margin than necessary to support spectrally efficient modulation formats. Error-free transmission at relatively high data rates is quite possible much of the time. Multipath and rain outages occur only for a small fraction of the total transmission time. With IP radios, baseband transmission bandwidth can be varied to suit the real-time transmission qualities of the radio path.

REFERENCES

Aaron, M. R. and Tufts, D. W., "Intersymbol Interference and Error Probability," *IEEE Transactions on Information Theory*, Vol. 12, pp. 26–34, December 1966.

Abramowitz, M. and Stegun, I., *Handbook of Mathematical Functions (NBS AMS 55, seventh printing with corrections)*. Washington, DC: US Government Printing Office, pp. 931–933, 1968.

Abreu, G., "Very Simple Tight Bounds on the Q-Function," *IEEE Transactions on Communications*, Vol. 60, pp. 2415–2420, September 2012.

Amoroso, F., "The Bandwidth of Digital Data Signals," *IEEE Communications Magazine*, Vol. 18, pp. 13–24, November 1980.

Aschoff, V., "The Early History of the Binary Code," *IEEE Communications Magazine*, Vol. 21, pp. 4–10, January 1983.

Assalini, A. and Tonello, A. M., "Improved Nyquist Pulses," *IEEE Communications Letters*, Vol. 8, pp. 87–89, February 2004.

Baccetti, B., Raheli, R. and Salerno, M., "Fractionally Spaced Versus T-Spaced Adaptive Equalization for High-Level QAM Radio Systems," *IEEE Global Telecommunications Conference (Globecom) Conference Record*, Vol. 2, pp. 31.1.1–31.1.5, November 1987.

Bayless, J. W., Collins, A. A. and Pedersen, R. D., "The Specification and Design of Bandlimited Digital Radio Systems," *IEEE Transactions on Communications*, Vol. 27, pp. 1763–1770, December 1979.

Beaulieu, N. C., Tan, C. C. and Damen, M. O., "A Better Than Nyquist Pulse," *IEEE Communications Letters*, Vol. 5, pp. 367–368, September 2001.

Bennett, W. R. and Davey, J. R., *Data Transmission*. New York: McGraw-Hill, 1965.

Berlekamp, E. R., Peile, R. E. and Pope, S. P., "The Application of Error Control to Communications," *IEEE Communications Magazine*, Vol. 25, pp. 44–57, April 1987.

Berrou, C., Glavieux, A. and Thitimasjshima, P., "Near Shannon Limit Error-Correcting Coding and Decoding: Turbo-Codes," *Proceedings, IEEE International Conference on Communication*, pp. 1064–1070, May 1993.

Bhargava, V. K., "Forward Error Correction Schemes for Digital Communications," *IEEE Communications Magazine*, Vol. 21, pp. 11–19, January 1983.

Bingham, J. A. C., "Multicarrier Modulation for Data Transmission: An Idea Whose Time Has Come," *IEEE Communications Magazine*, Vol. 28, pp. 5–14, May 1990.

Birafane, A., Mohamad, E., Kouki, A. B., Helaoui, M. and Ghannouchi, F. M., "Analyzing LINC Systems," *IEEE Communications Magazine*, Vol. 11, pp. 59–71, August 2010.

Borgne, M., "Comparison of High-Level Modulation Schemes for High-Capacity Digital Radio Systems," *IEEE Transactions on Communications*, Vol. 33, pp. 442–449, May 1985.

Bose, R. C. and Ray-Chaudhuri, D. K., "On a Class of Error-Correcting Binary Group Codes," *Information and Control*, Vol. 3, pp. 68–79, March 1960.

Breed, G., "Analyzing Signals Using the Eye Diagram," *High Frequency Electronics*, Vol. 4, pp. 50–53, November 2005.

Brink, S. T., "Coding over Space and Time for Wireless Systems," *IEEE Communications Magazine*, Vol. 13, pp. 18–30, August 2006.

Brunner, K. S. and Weaver, C. F., "A Comparison of Synchronous and Fractional-Spaced DFE's in a Multipath Fading Environment," *IEEE Global Telecommunications Conference (Globecom) Conference Record*, Vol. 1, pp. 44.4.1–44.4.5, November 1988.

Cahn, C. R., "Combined Digital Phase and Amplitude Modulation Communication System," *IRE Transactions on Communications*, Vol. 8, pp. 150–155, September 1960.

Campopiano, C. N. and Glazer, B. G., "A Coherent Digital Amplitude and Phase Modulation System," *IRE Transactions on Communications*, Vol. 10, pp. 90–95, March 1962.

Chow, J. S., Cioffi, J. M. and Bingham, J. A. C., "A Practical Discrete Multitone Transceiver Loading Algorithm for Data Transmission over Spectrally Shaped Channels," *IEEE Transactions on Communications*, Vol. 43, pp. 357–363, October 1995.

Chung, S-Y., Forney, G. D., Jr., Richardson, T. J. and Urbanke, R., "On the Design of Low-Density Parity-Check Codes within 0.0045 dB of the Shannon Limit," *IEEE Communications Letters*, Vol. 5, pp. 58–60, February 2001.

Costello, D. J. and Forney, G. D., Jr., "Channel Coding: The Road to Channel Capacity," *Proceedings of the IEEE*, pp. 1150–1177, June 2007.

Craig, J. W., "A New, Simple and Exact Result for Calculating the Probability of Error for Two-Dimensional Signal Constellations," *Military Communications Conference (MILCOM) Record*, Vol. 2, pp. 25.5.1–25.5.5, November 1991.

Dinn, N. F., "Digital Radio: Its Time Has Come," *IEEE Communications Magazine*, Vol. 18, pp. 6–12, November 1980.

Doeltz, M. L., Heald, E. T. and Martin, D. L., "Binary Data Transmission Techniques for Linear Systems," *Proceedings of the IRE*, pp. 656–661, May 1957.

Dong, X., Beaulieu, N. C. and Wittke, P. H., "Signaling Constellations for Fading Channels," *IEEE Transactions on Communications*, Vol. 47, pp. 703–714, May 1999.

Elias, P., "Coding for Noisy Channels," *IRE Convention Record*, Part 4, pp. 37–46, March 1955.

Forney, G. D., Jr., *Concatenated Codes*. Cambridge: MIT Press, 1966.

Forney, G. D., Jr., "Convolutional Codes I: Algebraic Structure," *IEEE Transactions on Information Theory*, Vol. 16, pp. 720–737, November 1970.

Forney, G. D., Jr., "Correction to "Convolutional Codes I: Algebraic Structure"," *IEEE Transactions on Information Theory*, Vol. 17, p. 360, May 1971a.

Forney, G. D., Jr., "Burst-Correction Codes for the Classic Bursty Channel," *IEEE Transactions on Communications Technology*, Vol. 19, pp. 772–781, October 1971b.

Forney, G. D., Jr., "The Viterbi Algorithm," *Proceedings of the IEEE*, pp. 268–278, March 1973.

Forney, G. D., Jr., "Convolutional Codes II. Maximum-Likelihood Decoding," *Information and Control*, Vol. 25, pp. 222-266, July 1974.

Forney, G. D., Jr., "Coset Codes—Part I: Introduction and Geometrical Classification," *IEEE Transactions on Information Theory*, Vol. 34, pp. 1123–1151, September 1988a.

Forney, G. D., Jr., "Coset Codes—Part II: Binary Lattices and Related Codes," *IEEE Transactions on Information Theory*, Vol. 34, pp. 1152–1187, September 1988b.

Forney, G. D., Jr., "Multidimensional Constellations—Part II: Voronoi Constellations," *IEEE Journal on Selected Areas in Communications*, Vol. 7, pp. 941–958, August 1989.

Forney, G. D., Jr., "Combined Equalization and Coding Using Precoding," *IEEE Communications Magazine*, Vol. 29, pp. 25–34, December 1991.

Forney, G. D., Jr. and Ungerboeck, G., "Modulation and Coding for Linear Gaussian Channels," *IEEE Transactions on Information Theory*, Vol. 44, pp. 2384–2415, October 1998.

Forney, G. D., Jr. and Wei, L.-F., "Multidimensional Constellations—Part I: Introduction, Figures of Merit, and Generalized Cross Constellations," *IEEE Journal on Selected Areas in Communications*, Vol. 7, pp. 877–892, August 1989.

Forney, G., Jr., Gallager, R., Lang, G., Longstaff, F. and Qureshi, S., "Efficient Modulation for Band-Limited Channels," *IEEE Journal on Selected Areas in Communications*, Vol. 2, pp. 632–647, September 1984.

Foschini, G. J., Gitlin, R. D. and Weinstein, S. B., "Optimization of Two-Dimensional Signal Constellations in the Presence of Gaussian Noise," *IEEE Transactions on Communications*, Vol. 22, pp. 28–38, January 1974.

Franks, L. E., "Further Results on Nyquist's Problem in Pulse Transmission," *IEEE Transactions on Communications Technology*, Vol. 16, pp. 337–340, April 1968.

Franks, L. E., "Carrier and Bit Synchronization in Data Communication—A Tutorial Review," *IEEE Transactions on Communications*, Vol. 28, pp. 1107–1121, August 1980.

Friis, H.T., "Noise Figures of Radio Receivers," *Proceedings of the IRE*, pp. 419–422, July 1944.

Gallagher, R. G., "Low Density Parity Check Codes," *IRE Transactions on Information Theory*, Vol. 8, pp. 21–28, January 1962.

Gibby, R. A. and Smith, J. W., "Some Extensions of Nyquist's Telegraph Transmission Theory," *Bell System Technical Journal*, Vol. 44, pp. 1487–1510, September 1965.

Gitlin, R. D. and Weinstein, S. B., "Fractionally-Spaced Equalization: An Improved Digital Transversal Equalizer," *Bell System Technical Journal*, Vol. 60, pp. 275–296, February 1981.

Golay, M. J. E., "Notes on Digital Coding," *Proceedings of the IRE*, p. 657, June 1949.

Goldberg, B., "Applications of Statistical Communications Theory," *IEEE Communications Magazine*, Vol. 19, pp. 26–33, July 1981.

Groe, J., "Polar Transmitters for Wireless Communications," *IEEE Communications Magazine*, Vol. 45, pp. 58–63, September 2007.

Hamming, R. W., "Error Detecting and Error Correcting Codes," *Bell System Technical Journal*, Vol. 29, pp. 147–160, April 1950.

Hancock, J. C. and Lucky, R. W., "Performance of Combined Amplitude and Phase Modulated Communications System," *IRE Transactions on Communications*, Vol. 8, pp. 232–237, December 1960.

Hartley, R., "Transmission of Information," *Bell System Technical Journal*, Vol. 7, pp. 535–563, July 1928.

Hocquenghem, A., "Codes Correcteurs d'Erreurs," *Chiffres*, Vol. 2, pp. 147–156, September 1959.

Imai, H. and Hirakawa, S., "A New Multilevel Coding Method Using Error-Correcting Codes," *IEEE Transactions on Information Theory*, Vol. 23, pp. 371–377, May 1977.

Johannes, V. I., "Improving on Bit Error Rate," *IEEE Communications Magazine*, Vol. 22, pp. 18–20, December 1984.

Johnson, K. K., "Optimizing Link Performance, Cost and Interchangeability by Predicting Residual BER: Part I—Residual BER Overview and Phase Noise," *Microwave Journal*, Vol. 45, pp. 20–30, July 2002a.

Johnson, K. K., "Optimizing Link Performance, Cost and Interchangeability by Predicting Residual BER: Part II—Nonlinearity and System Budgeting," *Microwave Journal*, Vol. 45, pp. 96–131, September 2002b.

Kassam, S. A. and Poor, H. V., "Robust Signal Processing for Communication Systems," *IEEE Communications Magazine*, Vol. 21, pp. 20–28, January 1983.

Kerr, A. R. and Randa, J., "Thermal Noise and Noise Measurements—A 2010 Update," *IEEE Microwave Magazine*, Vol. 11, pp. 40–52, October 2010.

Khabbazian, M., Hossain, M. J., Alouini, M. and Bhargava, V. K., "Exact Method for the Error Probability Calculation of Three-Dimensional Signal Constellations," *IEEE Transactions on Communications*, Vol. 57, pp. 922–925, April 2009.

Kim, B., Kim, I. and Moon, J., "Advanced Doherty Architecture," *IEEE Communications Magazine*, Vol. 11, pp. 72–86, August 2010a.

Kim, B., Moon, J. and Kim, I., "Efficiently Amplified," *IEEE Communications Magazine*, Vol. 11, pp. 87–100, August 2010b.

Kizer, G. M., *Microwave Communication*. Ames: Iowa State University Press, pp. 589-602, 1990.

Kizer, G. M., "Microwave Radio Communication," *Handbook of Microwave Technology*, *Volume* 2, *Applications*. Ishii, T. K., Editor. San Diego: Academic Press, pp 449–504, 1995.

Lavrador, P. M., Cunha, T. R., Cabral, P. M. and Pedro, J. C., "The Linearity-Efficiency Compromise," *IEEE Communications Magazine*, Vol. 11, pp. 44–58, August 2010.

Lee, J. S. and Beaulieu, N. C., "A Novel Pulse Designed to Jointly Optimize Symbol Timing Estimation Performance and the Mean Squared Error of Recovered Data," *IEEE Transactions on Wireless Communications*, Vol. 7, pp. 4064–4069, November 2008.

Liu, K. Y. and Lee, J., "Recent Results on the Use of Concatenated Reed-Solomon/Verterbi Channel Coding and Data Compression for Space Communications," *IEEE Transactions on Communications*, Vol. 32, pp. 518–523, May 1984.

Liveris, A. D. and Georghiades, C. N., "Exploiting Faster-Than-Nyquist Signaling," *IEEE Transactions on Communications*, Vol. 51, pp. 1502–1511, September 2003.

Lodge, J., Young, R., Hoeher, P. and Hagenauer, J., "Separable MAP 'Filters' for Decoding of Product and Concatenated Codes," *Proceedings, IEEE International Conference on Communication*, pp. 1740–1745, May 1993.

Lucky, R. W., "Automatic Equalization for Digital Communication," *Bell System Technical Journal*, Vol. 44, pp. 547–588, April 1965.

Lucky, R. W., "Techniques for Adaptive Equalization of Digital Communication," *Bell System Technical Journal*, Vol. 45, pp. 255–286, February 1966.

Lucky, R. W., "A Survey of the Communication Theory Literature: 1968-1973," *IEEE Transactions on Information Theory*, Vol. 19, pp. 725–739, November 1973.

Lucky, R. W., Salz, J. and Weldon, E. J., Jr., *Principles of Data Communications*. New York: McGraw-Hill, 1968.

MacKay, D. J. C. and Neal, R. M., "Near Shannon Limit Performance of Low Density Parity Check Codes," *Electronics Letters*, Vol. 32, pp. 1645–1655, August 1996.

Mazo, J. E., "Faster-Than-Nyquist Signaling," *Bell System Technical Journal*, Vol. 54, pp. 1451–1462, October 1975.

Mosley, R. A., Director, *Code of Federal Regulations (CFR), Title 47 - Telecommunication, Chapter I, Part 101.111*. Washington: Office of the Federal Register, published yearly.

Mueller, K. H. and Muller, M. S., "Timing Recovery in Digital Synchronous Data Receivers," *IEEE Transactions on Communications Technology*, Vol. 24, pp. 516–531, May 1976.

Muller, D. E., "Application of Boolean Algebra to Switching Circuit Design and to Error Detection," *IRE Transactions on Electronic Computers*, Vol. 3, pp. 6–12, September 1954.

Newcombe, E. A. and Pasupathy, S., "Error Rate Monitoring for Digital Communications," *Proceedings of the IEEE*, pp. 805–828, August 1982 and Correction to "Error Rate Monitoring for Digital Communications", *Proceedings of the IEEE*, p. 443, March 1983.

Niger, Ph. and Vandamme, P., "Outage Performance of High-Level QAM Radio Systems Equipped with Fractionally-Spaced Equalizers," *IEEE Global Telecommunications Conference (Globecom) Conference Record*, Vol. 1, pp. 8.3.1–8.3.5, November 1988.

Noguchi, T., Daido, Y. and Nossek, J. A., "Modulation Techniques for Microwave Digital Radio," *IEEE Communications Magazine*, Vol. 24, pp. 21–30, October 1986.

North, D. O., "An Analysis of the Factors Which Determine Signal/Noise Discrimination in Pulse-Carrier Systems," RCA Report PTR-6C, 1943; also *Proceedings of the IEEE*, pp. 1016–1027, July 1963.

Nyquist, H., "Certain Factors Affecting Telegraph Speed," *Bell System Technical Journal*, Vol. 3, pp. 324–346, April 1924.

Nyquist, H., "Certain Topics in Telegraph Transmission Theory," *AIEE Transactions*, Vol. 47, pp. 617–644, April 1928.

Prange, E., *Cyclic Error-Correcting Codes in Two Symbols, Technical Report TN-57-103*. Cambridge, MA: Air Force Cambridge Research Center, September 1957.

Proakis, J. G. and Salehi, M., *Communications Systems Engineering*, *Second Edition*. Upper Saddle River: Prentice Hall, 2002.

Qureshi, S., "Adaptive Equalization," *IEEE Communications Magazine*, Vol. 20, pp. 9–16, March 1982.

Qureshi, S., "Adaptive Equalization," *Proceedings of the IEEE*, pp. 1349–1387, September 1985.

Reed, I. S., "A Class of Multiple-Error Correcting Codes and the Decoding Scheme," *IRE Transactions on Information Theory*, Vol. 4, pp. 38–49, September 1954.

Reed, I. S. and Solomon, G., "Polynomial Codes over Certain Finite Fields," *Journal of SIAM*, Vol. 8, pp. 300–304, June 1960.

Rusek, F. and Anderson, J. B., "Multistream Faster Than Nyquist Signaling," *IEEE Transactions on Communications*, Vol. 57, pp. 1329–1339, May 2009.

Scanlan, J. G., "Pulses Satisfying the Nyquist Criterion," *Electronics Letters*, Vol. 28, pp. 50–52, January 1992.

Shannon, C. E., "A Mathematical Theory of Communication, Parts I, II & III," *Bell System Technical Journal*, Vol. 27, pp. 379–423, and 623–656, July and October 1948.

Shannon, C. E., "Communication in the Presence of Noise," *Proceedings of the IRE*, pp. 10–21, January 1949.

Shannon, C. E., "Recent Developments in Communication Theory," *Electronics*, Vol. 23, pp. 80–83, April 1950.

Simon, M. K. and Smith, J. G., "Hexagonal Multiple Phase-and-Amplitude-Shift-Keyed Signal Sets," *IEEE Transactions on Communications*, Vol. 21, pp. 1108–1115, October 1973.

Sklar, B., "A Structured Overview of Digital Communications—a Tutorial Review—Part I," *IEEE Communications Magazine*, Vol. 21, pp. 4–17, August 1983a.

Sklar, B., "A Structured Overview of Digital Communications—a Tutorial Review—Part II," *IEEE Communications Magazine*, Vol. 21, pp. 6–21, October 1983b.

Szczecinski, L., Gonzalez, C. and Aissa, S., "Exact Expression for the BER of Rectangular QAM with Arbitrary Constellation Mapping," *IEEE Transactions on Communications*, Vol. 54, pp. 389–392, March 2006.

Thomas, C. M., Weidner, M. Y. and Durrani, S. H., "Digital Amplitude-Phase Keying with M-ary Alphabets," *IEEE Transactions on Communications*, Vol. 22, pp. 168–180, February 1974.

TIA/EIA, Interference Criteria for Microwave Systems, Telecommunications Systems Bulletin TSB10-F, June 1994.

Ungerboeck, G., "Channel Coding with Multilevel/Phase Signals," *IEEE Transactions on Information Theory*, Vol. 28, pp. 55–67, January 1982.

Ungerboeck, G., "Trellis-Coded Modulation with Redundant Signal Sets, Part I: Introduction," *IEEE Communications Magazine*, Vol. 25, pp. 5–21, February 1987a.

Ungerboeck, G., "Trellis-Coded Modulation with Redundant Signal Sets, Part II: State of the Art," *IEEE Communications Magazine*, Vol. 25, pp. 12–21, February 1987b.

Vigants, A., "Space-Diversity Engineering," *Bell System Technical Journal*, Vol. 54, pp. 103–142, January 1975.

Viterbi, A. J., "Error Bounds for Convolutional Codes and an Asymptotically Optimum Decoding Algorithm," *IEEE Transactions on Information Theory*, Vol. 13, pp. 260–269, April 1967.

Viterbi, A. J., "Convolutional Codes and Their Performance in Communication Systems," *IEEE Transactions on Communications Technology*, Vol. 19, pp. 751–772, October 1971.

Wei, L-F., "Rotationally Invariant Convolutional Channel Coding with Expanded Signal Space—Part I: 180°," *IEEE Journal on Selected Areas in Communications*, Vol. 2, pp. 659–671, September 1984a.

Wei, L-F., "Rotationally Invariant Convolutional Channel Coding with Expanded Signal Space—Part II: Nonlinear Codes," *IEEE Journal on Selected Areas in Communications*, Vol. 2, pp. 672–686, September 1984b.

Wei, L-F., "Trellis-Coded Modulation with Multidimensional Constellations," *IEEE Transactions on Information Theory*, Vol. 33, pp. 483–501, July 1987.

Whalen, A. D., "Statistical Theory of Signal Detection and Parameter Estimation," *IEEE Communications Magazine*, Vol. 22, pp. 37–44, June 1984.

Widrow, B., *Adaptive Filters, I: Fundamentals, Technical Report No. 6764-6*. Stanford: Stanford Electronic Laboratories, Stanford University, December 1966.

Wolf, J. K. and Ungerboeck, G., "Trellis Coding for Partial-Response Channels," *IEEE Transactions on Communications*, Vol. 34, pp. 765–773, August 1986.

Working Group 18, *Automatic Transmit Power Control (ATPC), National Spectrum Managers Association (NSMA) Recommendation WG 18.91.032*, April 1992.

Xiong, F., *Digital Modulation Techniques, Second Edition*. Boston: Artech House, pp. 694–696, 2006.

Yamaguchi, K. and Imai, H., "Highly Reliable Multilevel Channel Coding System Using Binary Convolutional Codes," *Electronics Letters*, Vol. 23, pp. 939–941, August 1987.

Yin, P., "Specifications and Definitions for Quadrature Demodulators and Receiver Design Measurements," *Microwave Journal*, Vol. 45, pp. 22–42, October 2002.

4

RADIO NETWORK PERFORMANCE OBJECTIVES

Telecommunications providers may offer a telecommunications service with a specified end to end quality of service and service availability. This availability objective is expected to cover all causes of outage. Customer service objectives are established to support this offering. Design, commissioning, and maintenance objectives are defined to support overall customer service objectives. Design objectives are the most stringent. They are estimates based on idealized average performance of many individual systems. They ignore any affects of maintenance actions or unpredictable events. They define expected median or average performance. Actual performance will vary above or below these levels. Commissioning objectives are a little more lax in acknowledgment that their actual performance is less than ideal and subject to significant variation. Maintenance objectives are the least stringent. They are intended to insure end to end customer objectives are maintained. The ITU concepts of differentiation of objectives by function and by different networks and equipment have been accepted worldwide. This overall concept in network design is outlined in the ITU-T (a branch of the United Nations) Recommendation G.102, Transmission Performance Objectives and Recommendations, Transmission Systems and Media (ITU-T Recommendation G.102, 1993). This recommendation describes the design, commissioning, and maintenance objectives.

4.1 CUSTOMER SERVICE OBJECTIVES

Customer service objectives are often a service availability (often specified as 98% or 99.8%) end to end. Most of the availability objective is allocated to nontransmission impairments. The transmission circuits are specified in such a way as to make their impairment of the end to end circuit minor. The various components of the transmission system are then allocated a portion of this objective. Since most transmission systems support many customers, this is generally an economical approach. The customer service objective is a "not to exceed" objective.

4.2 MAINTENANCE OBJECTIVES

Maintenance objectives are more stringent than customer service objectives but less stringent than commissioning objectives (ITU-T Recommendation M.35, 1993). The difference in performance between

Digital Microwave Communication: Engineering Point-to-Point Microwave Systems, First Edition. George Kizer.

maintenance and commissioning objectives is usually termed *system margin* or *aging factor*. Maintenance objectives are often formalized as levels of alarm and performance with associated maintenance actions. These are "not to exceed" objectives.

ITU-T general maintenance quality objectives are contained in M.2100 [Plesiochronous Digital Hierarchy (PDH)] (ITU-T Recommendation M.2100, 2003) and M.2101 (SDH) (ITU-T Recommendation M.2101, 2003). ITU-R radio-specific quality objectives are contained in F.1566-1 (ITU-R Recommendation F.1566-1, 2007). As noted in F.1566-1 (ITU-R Recommendation F.1566-1, 2007), currently there are no ITU-R unavailability (outage) maintenance objectives. The ITU-R maintenance quality objectives are based on a maintenance performance limit (MPL), which is quite similar to the bringing into service performance objective (BISPO). In accordance with M.20 (ITU-T Recommendation M.20, 1992), these objectives must be modified to determine unacceptable, degraded, and acceptable performance limits. These limits are left to the network administrations to determine. Specific objectives vary widely among administrations.

Telcordia defines North American maintenance objectives by defining an alarm type and an alarm level (Telcordia (Bellcore) Staff, 2009a). Alarm types are service affecting (SA) and nonservice affecting (NSA). An SA alarm indicates an equipment failure that causes loss of the transported (baseband) signal. An NSA alarm indicates that an equipment failure has occurred but functionality was automatically restored by backup equipment.

Critical alarms are SA alarms indicating failures affecting many users or considerable bandwidth. Telcordia defines (Telcordia (Bellcore) Staff, 2000) critical alarms as failures requiring immediate corrective action independent of the time of day. Major alarms are SA alarms affecting fewer users or less bandwidth. Major alarms are failures requiring immediate attention. Minor alarms are NSA alarms or SA alarms indicating failure of few customers or little bandwidth. Action on minor alarms is usually deferred until normal operating hours. Nonalarmed is a nonalarm condition, status, or event. Typically it is not reported or recorded.

Telcordia defines specific alarms and performance levels for TDM circuits. However, operators of other services are expected to define appropriate similar maintenance objectives.

Maintenance performance is directly related to equipment quality [as measure by equipment two-way mean time between failure (MTBF)] and maintenance staff performance.

$$\begin{aligned}\text{System availability} &= \frac{\text{Operational time}}{\text{Total time}} \\ &= \frac{\text{Operational time}}{\text{Operational time} + \text{Outage time}} \\ &= \frac{\text{MTBF}}{\text{MTBF} + \text{MTTR}} = \frac{1}{1 + (\text{MTTR}/\text{MTBF})} \qquad (4.1)\end{aligned}$$

MTBF = composite mean time between failure of the cascaded components in the system;
MTTR = mean time to restore (mean down time).

MTTR includes the whole time from the outage occurrence until the network is restored. This includes the time to detect and diagnose the problem, acquire a working replacement, and travel to and enter the failed site. A fully qualified repair person with working spare unit is assumed available who performs maintenance without error. It is also assumed that each repair makes the system "as good as new." Sparing philosophy, maintenance personnel training and staffing, site access, and effective fault management systems will significantly influence the actual MTTR achieved. Some authors use the term *mean time to replace* (*MTR*) for the MTTR function. MTTR is then defined as only the actual time to perform the repair (after being notified, obtaining the spare unit, and reaching the site). This is not common usage and the use of MTTR to encompass all restoral action time is preferred. In evaluating equipment specifications, it is critical that all specifications use the same MTTR value. If not, the comparisons can be quite misleading. To simplify comparing various vendors, Telcordia has suggested that MTTR be standardized to 2 h for central office equipment with separable modules, 4 h for remote-site equipment with separable modules, and 48 h if the equipment must be replaced as a complete unit or the site has limited access (Telcordia (Bellcore) Staff, 2009b).

4.3 COMMISSIONING OBJECTIVES

Commissioning objectives are always less strict than design objectives (ITU-T Recommendation M.35, 1993). This is an acceptance that actual performance varies from designed performance due to variations in equipment and media performance, software errors in transmission and protection equipment, maintenance and operations personnel actions, and the effects of other equipment such as power and fault alarm systems. These are "not to exceed" objectives.

ITU-T general commissioning objectives (bringing into service performance limits) are contained in M.2100 (PDH) (ITU-T Recommendation M.2100, 2003) and M.2101 (SDH) (ITU-T Recommendation M.2101, 2003). ITU-R radio-specific objectives are contained in F.1330-1 (ITU-R Recommendation F.1330-1, 1999). These objectives only apply to quality. Currently there are no unavailability (outage) commissioning recommendations. Telcordia offers no commissioning objectives. These are determined by the individual operator.

4.4 DESIGN OBJECTIVES

Design objectives are always made more stringent than required for actual operation. This is necessary for practical reasons. Design objectives represent average (mean) performance. Individual path performance will vary above or below this value. Design objectives are "typical" or "average" objectives. It is seldom economical to design telecommunications systems to "not-to-exceed" or "worst-case" objectives. Typically, the telecommunications systems design objectives are about 1% of the customer service objectives. The worldwide telecommunication transmission design objectives (sometimes termed *engineering standards*) focus on two areas: quality and availability.

4.4.1 Quality

Quality is the performance of the end to end telecommunications circuit under normal conditions. This is usually defined as the residual BER, BBER, or a percentage of error-free seconds (EFS). *It is measured one-way* (each direction of an end to end circuit is evaluated separately) during the time the end to end circuit is considered available. Quality is usually specified over 1 month in the ITU and 1 year for North American systems.

This book uses the term *quality* for the performance of the end to end telecommunications circuit under normal conditions. This concept is also identified as error performance by some North American, some ITU-T, and all ITU-R documents. Worldwide, the use of the terms *quality* and *error performance* is inconsistent; usually, the preferred term is *quality*. Replace the word *quality* with *error performance* when appropriate.

4.4.2 Availability

Availability defines the percentage of time over which the end to end circuit achieves a minimum level of performance. *It is measured two-way* (each direction of an end to end circuit must meet the criterion simultaneously to be considered available). Generally, availability is specified over a 1-year period. All ITU-T and ITU-R sources use the term *availability*. Some North American sources such as Telcordia use the term *reliability* (Telcordia (Bellcore) Staff, 2002; Telcordia (Bellcore) Staff, 2005).

For the ITU (ITU-T Recommendation G.821, 2002; ITU-T Recommendation G.827, 2003) and most North American telecommunications systems (Telcordia (Bellcore) Staff, 2009b), service is said to become unavailable if the transmission system has experienced severely errored seconds (SESs), loss of frame (LOF), or loss of signal (LOS) for the last 10 consecutive seconds. The 10 s before declaring the circuit unavailable are considered unavailable time once the declaration has been established. Once the service becomes unavailable, it remains so until 10 consecutive non-SESs occur, after which it becomes available. The 10 s before declaring the circuit available are considered available time once the declaration has been established. Availability is defined in both directions simultaneously. The duplex path is considered unavailable when either (or both) simplex paths are unavailable.

North American radio objectives (AT&T, 1984; TIA/EIA, 1994) depart from this definition. North American availability objectives are based on "instantaneous" path BER performance. The ITU requirement for waiting 10 s before declaring a circuit available or unavailable is not used. Unlike the ITU SES criterion [originally taken as a 10^{-3} BER criterion but currently a variable BBER criterion (ITU-T Recommendation G.826, 2002; ITU-T Recommendation G.828, 2000)], North American objectives use a 10^{-6} BER availability criterion (TIA/EIA, 1994). As with ITU, the duplex path is considered unavailable if either simplex path is unavailable.

Using ITU definitions, quality (error performance) objectives include the effects of multipath fading, (short-term) upfading, residual error rate, error bursts, and (short-term) interference for radio systems. Error performance is measured one-way. Availability includes the effects of rain fading, obstruction (bulge) fading, (long-term) upfading, and (long-term) interference. These effects are measured two-way.

Since North American objectives use an "instantaneous" availability definition, all performance degradations except residual errors are included in measurements of availability. In North America, quality is used to define residual error performance objectives.

4.5 DIFFERENCES BETWEEN NORTH AMERICAN AND EUROPEAN RADIO SYSTEM OBJECTIVES

Everyone agrees on the following definitions.

Quality: The performance of end to end telecommunications circuits under normal conditions. Quality criteria define residual ("background") performance thresholds (typically BER, ES, SESs, or BBER) to be met over a defined measurement period when the system is available. This is a one-way (simplex) criterion.

Availability: The performance of end to end telecommunications circuits when they achieve a minimum level of performance. Availability criteria define limits of path (media) and equipment outages (usually defined as an unavailability objective). This is a two-way (duplex) criterion.

Differences in the concept of availability can lead to misunderstandings of performance expectations.

4.5.1 North American Radio Engineering Standards (Historical Bell System Oriented)

Availability: It is an "on/off" criterion. All equipment and media defects are included. Equipment and media defect apportionments are predefined and individually specified. Usually, the equipment will have one set of objectives (typically MTBF and MTTR oriented) and the media will have another (typical path performance oriented), and the equipment and media (radio path) each typically have half the overall system availability objective. Path availability objectives will include all predictable sources of path degradation (multipath, rain, and interference). The on/off threshold is a 10^{-6} BER. Often path availability objectives are specified unavailability objectives (when the system does not meet the minimum level of performance).

Quality: It is usually defined as residual ("background") performance thresholds (typically BER, ES, SESs, or BBER) during normal operation (when the system is available).

4.5.2 European Radio Engineering Standards (ITU Oriented)

Availability: It has a 10-s measurement window. It is usually specified as an unavailability objective. This objective only includes defects that last for at least 10 s. The on/off threshold is an SES. These long-term media defects include rain and (long-term) interference, as well as equipment and maintenance outages. The mixing of equipment, maintenance, and media objectives can add confusion. The division of the objective among these three defect groups is not specified in the ITU recommendations. These decisions are left to the system designer.

Quality: Quality defects are limited to short-term defects that individually last no longer than 10 s. The on/off threshold is an SES. For microwave radio systems, this objective is limited to multipath, short-term interference, and residual error performance. Quality objectives are usually named performance objectives in the ITU recommendations.

The above discussion leads to the following differences in understanding.

Availability (two-way objectives): In North America, these objectives include all path defects. These objectives are highly defined. In Europe, these objectives include only long-term media (radio path) defects (typically rain and long-term interference degradations) and maintenance and equipment outages. These objectives vary widely among recommendations because they depend on the class of service and whether or not the system was designed before or after 2002 (the adoption of ITU-T G.826). In Europe, the proration of the objective to the media is not defined. (Some administrations use the entire objective for media defects; others only take a portion.)

Quality (one-way objectives): In North America, these standards are relatively loosely defined. These objectives do not include media (radio path) defects. When quality measurements are being taken, it is assumed that the system is operating normally (radio paths are not fading). In Europe, these objectives (termed *performance objectives* by the ITU-R) include only short-duration system defects. For microwave systems, the only defects covered by this objective are multipath fading (including fade margin degradation by long-term interference), short-term interference, and residual errors. These European objectives typically are the primary system objectives and are highly defined.

An SES is defined by block errors in ITU but as a BER in most North American performance standards. The ITU-R standards for SESs are formulated for SONET/SDH TDM networks. However, their use in IP networks is not currently defined. The differences in definition lead to some confusion.

Given the above definitions, discussions between North American customers and international telecommunications systems suppliers can lead to misunderstandings. In North America, a two-way availability (or unavailability) objective sets the limits for radio path performance. It includes all sources of path degradation, including multipath, rain outage, and interference. The threshold is a 10^{-6} BER.

In Europe, radio path performance has two specifications. One specification is quality (termed *performance* in the ITU-R recommendations), which includes multipath and interference and is defined one-way. Two error performance objectives must be met: ESs and SESs.

Because of the existence of two error performance objectives, rain fading can cause error performance degradations. For fade levels between the fade margin for ESs and the level for SESs, the error performance is degraded without regard for 10-s constraints. The amount of time the received signal is expected to be between these two levels depends on the particular levels and the statistics of the rain fading. Some designers estimate the time between these two levels to be between 5% and 15% of the time that the SES threshold is exceeded, but there is no known documentation supporting these estimates.

The other European path performance specification is availability (or unavailability). It includes rain fading and interference, as well as maintenance and equipment outages. (Some path designers forget to assign a portion of the European unavailability specification to maintenance and equipment outages and erroneously give the entire objective to rain outages.) It is defined two-way. The threshold is SESs. Notice that the terms *quality* and *availability* define different defects in the North American and European radio systems.

4.6 NORTH AMERICAN TELECOMMUNICATIONS SYSTEM DESIGN OBJECTIVES

In North America, the customer service objective is typically a 98% overall system (two-way) availability. This outage objective is not more than 175 h per year. The transmission system is allocated 1% of that objective (99.98%). Half the objective is usually assigned to media and the half to hardware and personnel (typically 99.99% each). Individual link objectives are based on typical percentage objective per link or outage time per kilometer or mile. The objective allocations are summarized in Figure 4.1, Figure 4.2, and Figure 4.3. Details of the derivation of these summaries and the references are provided in Chapter 7.

4.7 INTERNATIONAL TELECOMMUNICATIONS SYSTEM DESIGN OBJECTIVES

The ITU-T and ITU-R have established various recommendations for the design of international telecommunications systems. These relate to all transmission media. The end to end customer service objective is

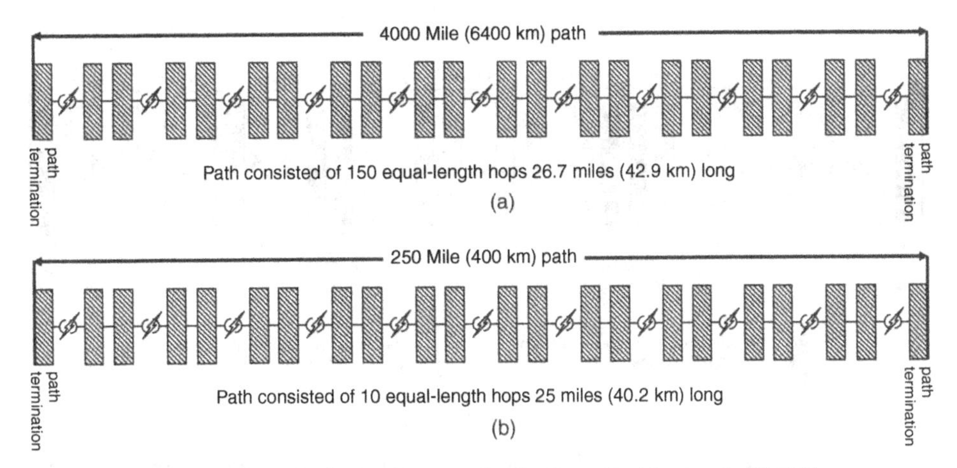

Figure 4.1 Bell System hypothetical reference circuit. Overall objective is 99.98% average annual two-way availability end to end. Path objectives include hop diversity but not multiplex/demultiplex or path-protective switching equipment. (a) Long haul and (b) short haul. *Source*: Bell System Technical Journal, pp. 2085–2116, September 1971 and pp. 1779–1796, October 1979.

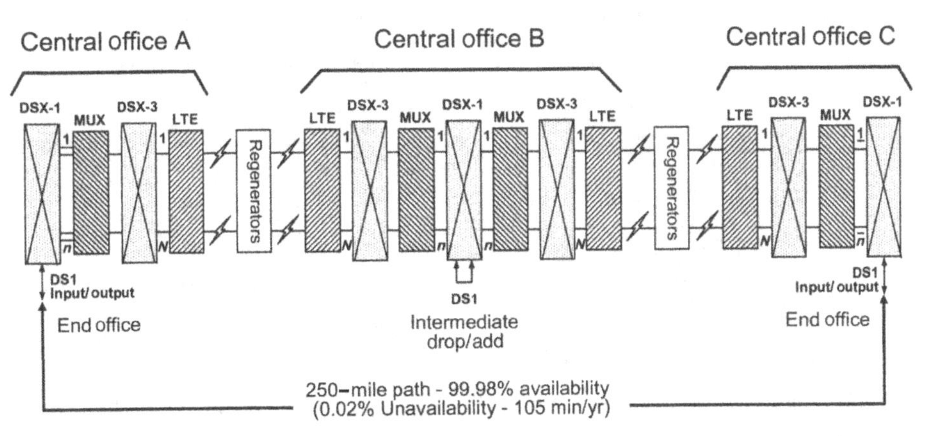

Figure 4.2 Telcordia hypothetical reference circuit, average 250-mile path. The conventional assumption is that each radio hop is 25 miles long. Path objectives include but not multiplex/demultiplex or path-protective switching equipment. Source: Telcordia GR-499.CORE.

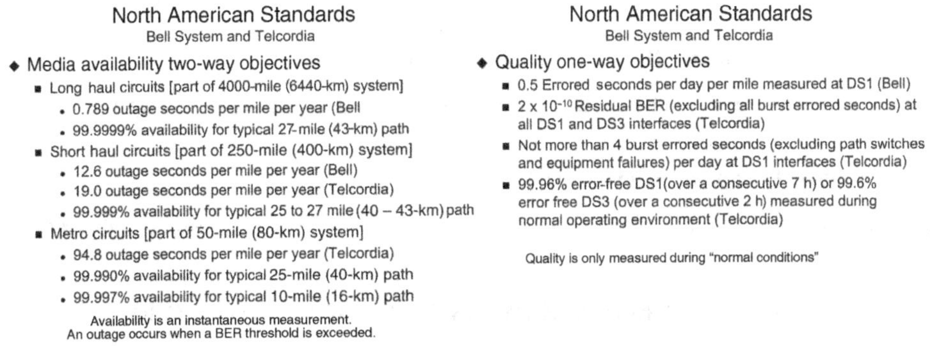

North American Standards
Bell System and Telcordia

- Media availability two-way objectives
  - Long haul circuits [part of 4000-mile (6440-km) system]
    - 0.789 outage seconds per mile per year (Bell
    - 99.9999% availability for typical 27-mile (43-km) path
  - Short haul circuits [part of 250-mile (400-km) system]
    - 12.6 outage seconds per mile per year (Bell)
    - 19.0 outage seconds per mile per year (Telcordia)
    - 99.999% availability for typical 25 to 27 mile (40 – 43-km) path
  - Metro circuits [part of 50-mile (80-km) system]
    - 94.8 outage seconds per mile per year (Telcordia)
    - 99.990% availability for typical 25-mile (40-km) path
    - 99.997% availability for typical 10-mile (16-km) path

Availability is an instantaneous measurement.
An outage occurs when a BER threshold is exceeded.

North American Standards
Bell System and Telcordia

- Quality one-way objectives
  - 0.5 Errored seconds per day per mile measured at DS1 (Bell)
  - 2×10^{-10} Residual BER (excluding all burst errored seconds) at all DS1 and DS3 interfaces (Telcordia)
  - Not more than 4 burst errored seconds (excluding path switches and equipment failures) per day at DS1 interfaces (Telcordia)
  - 99.96% error-free DS1(over a consecutive 7 h) or 99.6% error free DS3 (over a consecutive 2 h) measured during normal operating environment (Telcordia)

Quality is only measured during "normal conditions"

Figure 4.3 North American objectives. Availability is an instantaneous measurement. An outage occurs when a BER threshold is exceeded. Quality is only measured during "normal" conditions. Availability is per year. Quality is typically per short term measurement.

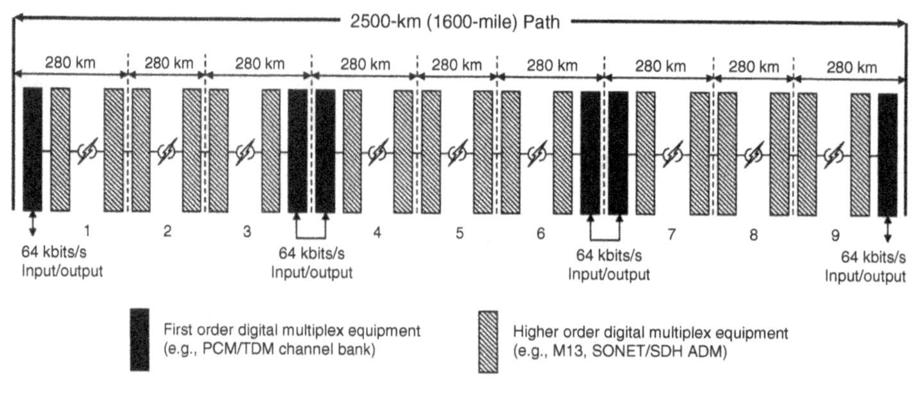

Figure 4.4 Legacy ITU-R hypothetical reference digital path for high grade performance. Reference path does not include multiplex/demultiplex or (path) protective switching equipment. Conventional assumptions are to assume each radio hop is 40.0 or 46.7 km long. Source: ITU-T Rec. G.821 and ITU-R Rec. F.1556-1.

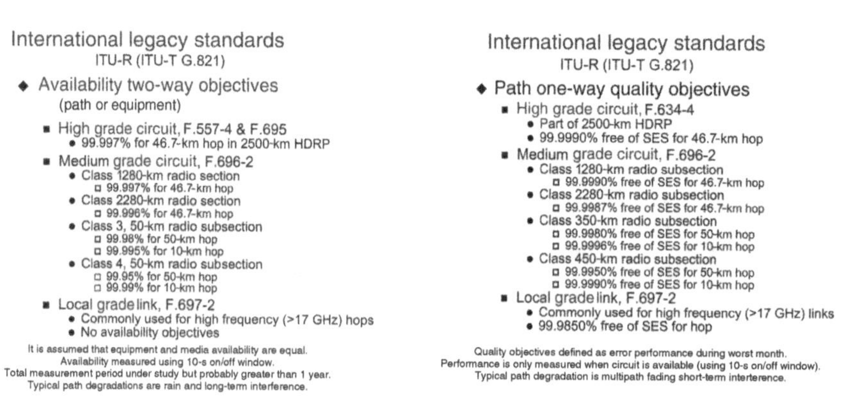

Figure 4.5 Legacy ITU-R objectives. Availability objectives are per year. Quality objectives are per worst month.

98% available for high priority systems and 91% for standard priority systems (ITU-T Recommendation G.826, 2002). The following sections overview these recommendations applicable to fixed point to point microwave paths. See Chapter 7 for the derivation of this overview as well as the applicable references.

4.7.1 Legacy European Microwave Radio Standards

These are for systems designed before December 2002, the adoption of G.826. See Fig. 4.4 and Fig. 4.5.

4.7.2 Modern European Microwave Radio Standards

These are for systems designed after December 2002, the adoption of G.826. See Fig. 4.6 and Fig. 4.7.

4.8 ENGINEERING MICROWAVE PATHS TO DESIGN OBJECTIVES

The process of designing a microwave path begins with the overall system design. After that has been determined, the appropriate end to end and path objectives are established. The transmission engineer's task is to design a system that meets a defined end to end objective.

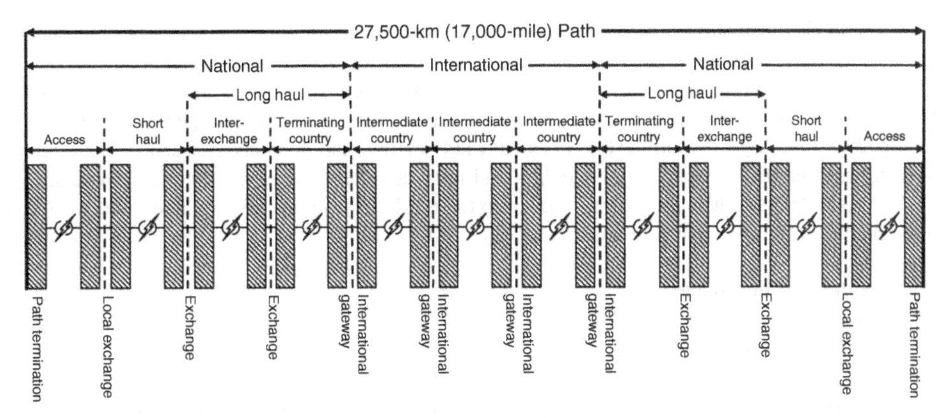

Figure 4.6 Modern ITU-T hypothetical reference path, the technology or media of each section is not explicitly defined. The actual number of spans/hops of equipment is not explicitly defined. Path objectives do not include multiplex/demultiplex or protective switching equipment. Source: ITU-T Recs. G.801, G.826, G.827 and G.828 as well as ITU-R Recs. F.1668 and F.1703.

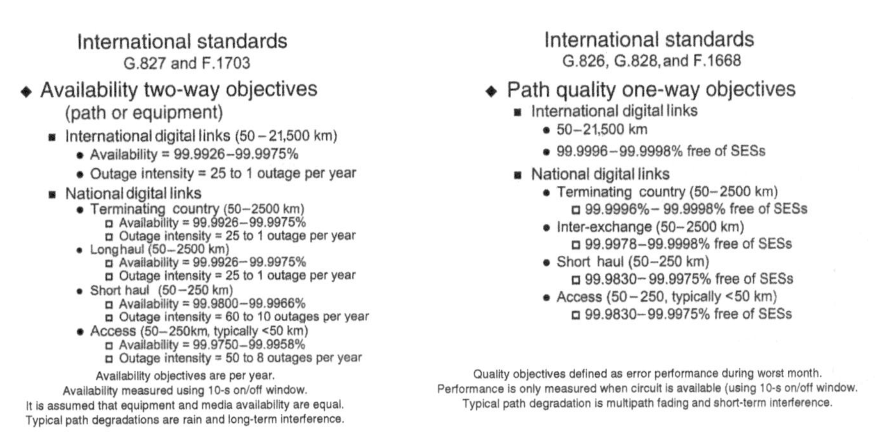

Figure 4.7 Modern ITU-R objectives. Availability objectives are per year. Quality objectives are per worst month.

The traditional approach is to divide the end to end performance objective of the hypothetical reference circuit by the total circuit distance to arrive at a per mile or per kilometer objective for an individual system. Each transmission system's path objective is taken as the per unit distance objective multiplied by the path length. While this is the historical methodology of designing paths, it often leads to uneconomical designs when applied to radio systems composed of paths of varying length. Most radio system degradations are not a linear function of path distance. (Multipath fading, both flat and dispersive, increase as distance cubed. Rain outage increases with distance up to the typical thunderstorm cell size. Obstruction fading tends to remain constant for paths exceeding 25 miles.) Slavish adherence to per mile objectives typically causes most of the system money to be spent on the long paths. The skillful transmission engineer can tailor each path's objective to respect the overall system objective while prorating the outage objective between paths in a nonlinear fashion. By increasing the objective for short paths (which are economical to improve) more performance degradation can be allowed for longer paths (which are much more costly to improve). Use of per mile or kilometer objectives with careful tailoring of objectives to balance short- and long-path requirements lead to more economical successful designs.

Another approach commonly used is to define a typical path, determine the objectives for that path, and use those as the requirements for all paths in the network. This is commonly expressed as "all paths must meet a 99.999% availability." As with the traditional approach, this can lead to uneconomical designs if path characteristics vary greatly. It is recommended that the designer always keep in mind the end to end objective to achieve the most cost-effective design consistent with the user objectives.

There are three main steps in designing a microwave path. The *first* step is to size the transmitter power and antenna sizes in such a way that the path performance objectives are met. This is usually done by making path availability calculations based on path fade margin and diversity (if needed). Consideration is given to flat (thermal), dispersive rain fading and inter- and intrasystem interference. There are many factors (often undefined) that significantly influence the results. The following are among the choices that must be made before objective estimation can begin.

- Path engineering methodology
  - North American
  - European (ITU-R)

Atmospheric attenuation models vary slightly. Use of receiver hysteresis is rare. Antenna gains (mid-band or frequency specific) are not used consistently.

- Antenna height (obstruction fading) criteria
  - ITU-R
  - Bell Labs/Alcatel-Lucent
  - Lenkurt/Aviant
- Terrain data source
  - USGS Seamless 10 m (one-third arc second)
  - USGS Seamless 30 m (arc second)
  - USGS Space Shuttle [Shuttle Radar Topography Mission (SRTM)]
  - Commercial Private GIS Data
- Upfading criteria
  - Bell Labs
  - ITU-R
  - Do not consider
- Multipath objectives
  - Two-way (duplex)
  - One-way (simplex)
  - Transmitter output power
    - Peak or average or guaranteed
    - Measured where (antenna, waveguide, amplifier)
  - Receiver thresholds
    - Typical or guaranteed
    - BER threshold (10^{-3} or 10^{-6})
    - Measured where (antenna, waveguide, receiver)
- Rain point rate data (>6 GHz)
  - Crane (1980, 1996, or 2003)
  - ITU-R (1978 [530-1], 2001 [530-10])
  - Other
- Rain point to path fading model (>6 GHz)
  - Crane (which version)
  - ITU-R (which version)
  - Other
- Wet radome loss (>6 GHz)
  - Determine value for wet radome

Apply at one or both ends

Not considered (most engineers ignore this)

Interference

Intrasystem

Intersystem due to similar services

Intersystem due to dissimilar services such as FSS

Interference allowances vary from 1 (single instance) to 5 dB (multiple instance) depending on the network operator.

Field margin (additional miscellaneous path loss)

Used (determine value)

Not used.

Choices on the above factors will significantly affect the results.

After choosing the appropriate system characteristics, the path designer must pick the appropriate calculation methodologies to estimate path quality and availability. The typical choices for path performance estimates are the following:

Multipath flat fading

Multipath dispersive fading

Rain fading.

Both Bell Labs and ITU-R have calculation methods to address these path performance limitations. The different methods provide different estimates. The following path performance limitations are well documented but typically ignored:

Wet radomes attenuation (Anderson, 1975; Blevis, 1965; Burgueno et al., 1987; Effenberger and Strickland, 1986; Lin, 1975; Lin, 1973; Rummler, 1987).

Ducting upfading (Anderson and Gossard, 1955; Day Trolese, 1950; Dougherty, 1968; Dougherty and Dutton, 1981; Dougherty Hart, 1976; Dougherty, 1979; Dutton, 1982; England et al., 1938; Fruchtenicht, 1974; Hubbard, 1979; Ikegami, 1959; Katzin Bauchman Binnian, 1947; Mahmoud Boghdady El-Sayed, 1987; Schiavone, 1982; Stephansen, 1981) (Bell Laboratories, Upfade Margin and Outage Due to Upfades, unpublished results of experiments on two Palmetto, Georgia, paths, 1981.)

Earth bulge (obstruction) fading (Dougherty, 1968; Dougherty Hart, 1976; Lee, 1986; Lee, 1985; McGavin et al., 1970; Schiavone, 1981; Vigants, 1981; Vigants, 1972; Wheeler, 1977).

Industry standards for interference mitigation are currently available. However, they seldom cover multiple exposure limits—only limits for a single interference case.

There is no industry standard defining which methodologies must be used and which path performance limitations must be estimated. These are determined by the designer and the user.

It should be remembered that there are unusual weather situations that give rise to "anomalous propagation." These events, as with other weather-dominated phenomena, cannot be predicted but can adversely affect microwave radio propagation.

The *second step* is to place the transmit and the receive antennas at appropriate locations on the antenna-supporting structure in such a way that adequate path terrain clearance is achieved. This is done using agreed path clearance guidelines (see Chapter 12). These guidelines are intended to keep obstruction fading at an insignificant level. Other than the Bell Labs obstruction fading estimation procedures discussed in Chapter 12, there is no performance expectation associated with meeting these guidelines.

If operating in a licensed frequency allocation, the *third step* is to perform frequency planning (see Chapter 2) if the radios are operating in a licensed band. In North America, frequency planning is based on the *T/I* concept (see Chapter 14). If *T/I* objectives are met, the performance objectives in the first step are not significantly degraded. If the *T/I* objective is not met, the difference between the *T/I* objective and the estimated interference is simply a decibel degradation of the radio path thermal fade margin. This simplifies the recalculation of system performance to evaluate the impact of the estimated interference.

4.9 ACCURACY OF PATH AVAILABILITY CALCULATIONS

For microwave paths, the largest variation is due to multipath flat and dispersive fading and rain outage. These phenomena can vary an order of magnitude (or more) from design estimates (Achariyapaopan, 1986; Babler, 1972; Crane, 1996; Giger, 1991; ITU-R Recommendation P.530-13, 2009; Osborne, 1977; Ranade Greenfield, 1983; Stephansen, 1981; Vigants, 1971).

4.9.1 Rain Fading

Rain fading will be quite different from path to path and from year to year. Lin (1975) reviewed the statistics of 96 rain gauges located in a grid with 1.3-km spacing. The incidence of 100 mm/h rain was higher by a factor of 5 for the upper 25% of rain rates as opposed to the lowest 25%. In another study of four rain gauges spaced in a square with 1-km sides, Lin noted that for rain rates greater than 80 mm/h (the rates of interest for high frequency path engineering), rain rates varied by a factor of 3. He observed "... on a short-term basis, the relationship between the path rain attenuation distribution and the [point] rain rate distribution measured by a single rain gauge is *not unique*." Different paths in the same area will experience difference average rain attenuation when averaged over the same period. The actual rain rate measured over any one specific year will be different than the average value. (Rain rates associated with time shorter than the gauge integration time, typically 1 or 5 min, clearly did not occur for each year in the observation period.) Osborne (1977) observed that the worst-case 1-year rain rates can exceed long-term averages by a multiplicative factor of 2–20. Worst-case month or hour rates can exceed long-term averages by extremely large factors. Data taken over a period of less than 10 years is generally unreliable for moderate rain rates. High rain rates are rarely observed.

4.9.2 Multipath Fading

The effect of *multipath flat fading* has been studied extensively (Babler, 1972; Barnett, 1974; Vigants, 1971) and is fairly well understood. There is general industry agreement as to how to calculate long-term outage associated with multipath flat fading (Vigants, 1975). What is not generally appreciated is the short-term variability of this data. Some engineers and managers assume that yearly outage calculations should be met at all times. This is simply not the case. All radio outage events are statistical and vary widely. First of all, flat fading only occurs during a few months of the year (typically the warm months). When it does occur, the fading usually occurs at night (with the exception of path reflective fading that is usually most intense during the day). Fading varies considerably even on any given path.

Over a 2-month period in the summer, Vigants (1971) observed a wide variation in multipath-influenced received signal power levels on different radio channels (frequencies) on the same 26-mile 6-GHz path. At the 40-dB fade point, on some channels, fades occurred at much as 25% more than expected. In another case, Babler (1972) observed relatively consistent performance of different radio channels (frequencies) on one antenna but significantly greater variation on another antenna. Amazingly both antennas were 20 ft apart on the same tower of the same 28-mile 6-GHz path. For the horn antenna, for 40-dB fades, an order of magnitude difference in radio channel performance was observed. One channel experienced four times the expected fading outage over the 2-month observation time.

Vigants (1975) of Bell Labs devised a multipath estimation methodology that is widely used in North America to estimate fixed point to point microwave radio multipath outage. Giger (1991), also of Bell Labs, observed, "It is our experience that a prediction of r [fading occurrence factor, directly related to fading outage time], based on the best available information, may still be off by an order of magnitude either above or below the measured worst month value."

Stephansen (1981) observed, "... it is well known that a considerable year-to-year variability exists in measured data; for example, for deep [multipath] fades, year-to-year variations of more than a factor of 10 in time percentage are often seen." ITU-R (ITU-R Recommendation F.1093-2, 2006) noted, "Propagation conditions vary from month to month and from year to year, and the probability of occurrence of these conditions may vary by as much as several orders of magnitude. It may therefore take some three to five years before drawing a proper conclusion on the results of a propagation experiment."

4.9.3 Dispersive Fading Outage

Dispersive fading outage estimates are made based on the dispersive fade margin (DFM) concept (Dupuis Joindot Leclert Rooryck, 1979; Rummler, 1982). The advantage of this approach is that DFM can be used in the path design calculations exactly like the flat fade margin. It also allows similar bandwidth radios to be compared on their ability to discriminate against dispersive fading. One disadvantage is that the current industry practice is to use 6.3-ns delay to characterize the reference path length (26.4 miles). Rummler (1979) showed that his typical 26.4-mile path was characterized by a median delay of 9.1 ns. The universal use of 6.3 ns (rather than 9.1 ns) as the typical path-dispersive echo delay leads to 2-dB optimistic DFM estimates.

The Bellcore (now Telcordia) method of calculating DFM was developed as a method to facilitate the comparison of different radio receivers. It fails to accommodate the characteristics of actual paths. Modifying W curves to accommodate different path delays is well understood. However, there is no general industry agreement for estimating path delay on actual paths. AT&T Bell Labs attempted to introduce the concept of dispersion ratio (Rummler, 1988) to account for differences in path-dispersive fading characteristics. This concept has not been fully developed.

The concept of DFM was developed using 6-GHz data. It is not clear how DFM should be modified to be applicable at other frequencies. Currently, a linear frequency dependency is assumed. Limited data does not substantiate this assumption.

An issue not currently addressed by industry methodologies is receiver hysteresis. All published W curves are based on "static" measurements (BER after the receiver has recovered from any anomalous performance). It is well known (Lundgren and Rummler, 1979) that dispersive events occur quickly and can cause momentary loss of synchronization for large BERs. "Dynamic" measurements more accurately represent this actual performance. Nevertheless, these "dynamic" measurements, described in Bellcore specifications (Bellcore (Telcordia) Staff, 1989) and typical DFM measurement equipment manuals, are generally not available. The use of static W curves leads to optimistic path outage estimates ranging from 1 dB to several decibels depending on the particular receiver.

Perhaps, the most significant issue with dispersive fading estimates is the variability from year to year of measured fading time. Rummler (1981) observed yearly outages as much as twice the long-term average (on different antennas on the same tower of a 26-mile 6-GHz path). Ranade and Greenfield (1983) observed 6-GHz yearly outages as much as three times the long-term average. This year to year variability limits the accuracy of total fading time estimates.

4.9.4 Diversity Improvement Factor

Diversity improvement factor varies throughout the industry. For space and frequency diversity, engineers typically use the Vigants (Vigants, 1975; Vigants, 1968) diversity improvement factors for flat fading. There is no complete agreement on dispersive fading. Some engineers use space and frequency diversity improvement factors developed for flat fading, whereas others use factors developed by Bell Labs (Lee Lin, 1986; Lee Lin, 1985; Lin et al., 1988). The Bell Labs factors are more optimistic than flat fading improvement factors. Some of the flat fading models include a factor for threshold hysteresis; however, these are typically ignored. None of the dispersive diversity improvement models includes a factor for practical considerations such as threshold hysteresis. Experience suggests actual systems do not always achieve calculated diversity improvement. There are very few published statistics of calculated versus achieved radio diversity improvement. Giger (1991) observed that " ... the [space diversity] improvement factor *I* [for dispersive fading] can vary by at least an order of magnitude ... over the [measurement] period of 1 year ... ."

Angle diversity can be implemented in different ways. There is no industry agreement on how to engineer or install these systems. Likewise there is no agreement on what angle diversity improvement to expect. AT&T Bell Labs developed an angle diversity improvement estimate model (Giger, 1991), but it has not gained acceptance. Everyone treats angle diversity differently. Rummler and Dhafi (1989) observed, "As yet there are no algebraic formulas which permit the estimation of improvement by using angle diversity. However, recent studies by Lin have shown that improvements in performance over conventional space diversity are possible. It is clear ... that the treatment of angle diversity lacks completeness, and further work is required. In particular, the dependence of the observed improvement factors on the nature of the path has yet to be determined."

Everyone agrees that diversity improves microwave radio performance. However, other than for flat multipath fading, the industry is not in agreement on diversity improvement estimation.

4.10 IMPACT OF FLAT MULTIPATH VARIABILITY

Paths are engineered to design objectives. Each path's performance will vary above and below the path's design objectives. Many propagation limitations (including multipath and rain) are believed to have a lognormal distribution. Path calculations attempt to estimate the performance mean (average) value. System or path performance can be expected to vary around the mean value M based on the standard deviation σ.

$$\text{System or path performance} = M \pm [m\sigma] \tag{4.1}$$

About 68% of the systems can be expected to perform within one standard deviation of the mean ($m = 1$), 95% within two standard deviations ($m = 2$), and 99% within three standard deviations ($m = 3$). Although variation from path to path can be significant, end to end variation is reduced by the following relationship:

$$\sigma_{\text{system}} = \frac{\sigma_{\text{individual path}}}{\text{sqrt}\ (n)} \tag{4.2}$$

Each path is assumed to be identical and n is the number of cascaded paths.

Bellcore (Achariyapaopan, 1986) studied several microwave radio paths influenced by flat multipath fading and determined that the standard deviation of actual paths from the Vigants model (typical result) was 10.2 dB. ITU-R estimates (International Telecommunication Union—Radiocommunication Sector (ITU-R), 2007) its multipath propagation model to have a standard deviation of 5.2–7.3 dB depending on the type of path. This means that if an engineer wanted a path to be designed so that the Vigants estimated value of outage would not be exceeded for more than 10% of the paths, the path fade margin would have to be increased 13 dB from the Vigants fade margin (Alternatively, it may be expected that annually only 10% of the paths will fade more than 20 times the estimated outage time when averaged over many years.). If the expected outage of only 1% is desired, the Vigants fade margin must be increased 24 dB. This is generally not practical (or necessary). Engineers use average results and expect the cumulative performance of several cascaded paths to average to the expected result.

4.11 IMPACT OF OUTAGE MEASUREMENT METHODOLOGY

Multipath outage time estimates are different than outage time measurements. Multipath outage estimates attempt to predict the total time a receiver's receive signal level is below its digital threshold (10^{-3} or 10^{-6} BER). Actual receiver outages are usually measured in cumulative threshold ESs. If a digital test set is used, usually the only threshold measurement available is SESs (currently, in North America, this is equivalent to a threshold of 10^{-3} BER). The difference between the digital test set threshold and the threshold of interest (typically 10^{-6} BER) will introduce some time measurement error. More importantly, the duration of a fade will not exactly match a second. Barnett (1974) and Vigants (1969; 1971) determined that for a typical 28.5-mile path, the average fade duration at 4, 6, or 11 GHz is given by the following:

$T = 410\ L$ for nondiversity receivers;
$= 205\ L$ for space or frequency diversity receivers;
= average duration of multipath fade in seconds;
$L =$ square root of P;
$P = 10^{-\text{FFM}/10}$
= inverse of flat fade margin expressed as a power ratio;
FFM = flat fade margin expressed in positive dB.

For a 40-dB fade margin, the average nondiversity fade lasts for about 4 s. The diversity fade lasts half that time. If the fade margin is less than 40 dB or the path is shorter, the outage will be longer; if

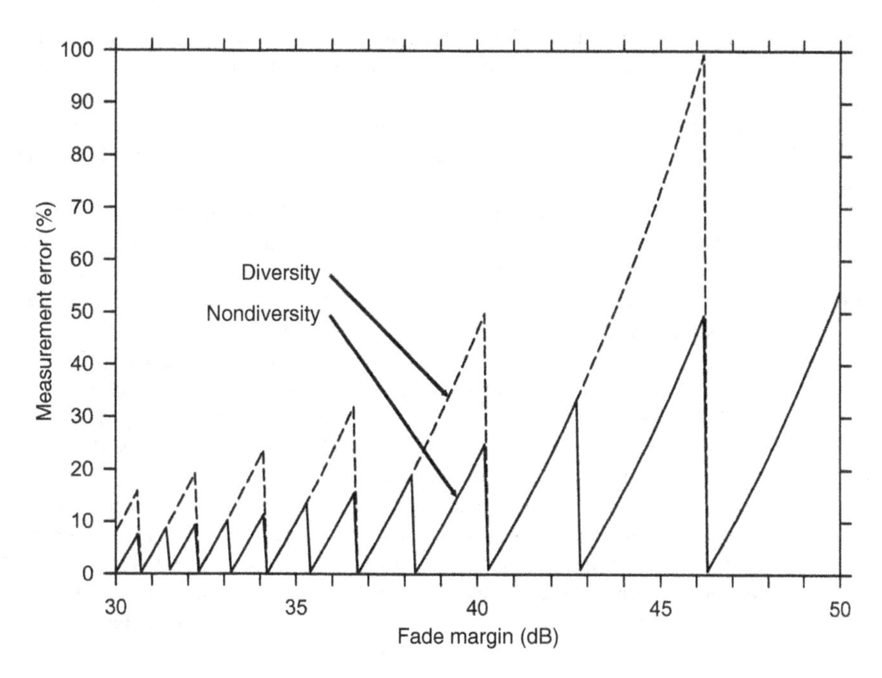

Figure 4.8 Errored-second measurement error—actual outage versus threshold errored-second count.

the fade margin is greater or the path longer, it will be less. Every time a fade occurs, several outage seconds are counted. If the outage is less than an integer number of seconds, an integer number of outage seconds is still counted. This creates measurement error that can be considerable for deep fades (Fig 4.8).

This "outage stretching" effect causes measurements to be longer than the estimates predict even if the actual outages exactly match the estimates.

The "outage stretching" effect can be made worse by poorly synchronizing receivers. However, measurements of well-functioning radios demonstrate that for fade durations of a few seconds, receivers resynchronize within a few tens of milliseconds. This is a relatively insignificant increase in outage time.

4.12 IMPACT OF EXTERNAL INTERFERENCE

In the frequency planning aspect of path engineering, the intent is to reduce external interference due to other radio systems to a nominal influence (typically not more than 1 dB threshold degradation). In many cities today, for a variety of reasons (e.g., installation errors causing reversed paths, undocumented changes in antennas or transmitters), interference can, in fact, be a significant degradation. For low frequency paths primarily influenced by flat multipath fading, a 10-dB loss in flat fade margin increases path outage time by a factor of 10. Performing a path fade margin test before commissioning the path is strongly recommended.

4.13 CONCLUSION

Network engineering has several universally accepted objectives. The most common are customer, maintenance, commissioning, and design. Of these the relative performance objective (Ivanek, 1989) is quite different. Differentiation of performance limits is universally recognized as critical to the successful and independent operation of the various network-engineering functions. The ITU establishes general international recommendations. Telcordia has further defined these objectives for North American systems. For international connections, ITU standards are universally applied. For connections within the US Public Switched Telephone Network, Telcordia standards are imposed. For all other operations, each operator must establish their own standards for internal use.

Microwave path design objectives are achieved through the use of path calculations of typical path degradations. The choice of the objectives and the methods of estimating them vary widely. The calculations are an attempt to estimate typical path performance. Actual path performance can be expected to vary from those estimates. Those estimates do not attempt to estimate performance in unusual, atypical situations. Since all radio path performance is ultimately limited by weather conditions, unusual propagation conditions occasionally happen.

As an independent microwave consultant, Thrower (1977) observed "On the philosophical side of the dB ledger, one doesn't like to see a path fade but, being practical, you can't make absolute predictions as to whether a path will fade or not. We live in a terrestrial environment rather than in theoretical free space. Even path testing is not an absolute way of establishing the reliability of a path. To do it properly, one would have to run a propagation test over the path, with the planned tower height and with the planned antenna sizes for a minimum of a year in order to obtain data under all environmental conditions. Even that will vary from year to year; witness, the drought stricken areas of the country for 1976–1977. Tests made, for example, in the Pacific Northwest during that winter would result in totally different results compared to "normal" wet years in that region. A planned 11-GHz radio path would be defective when weather conditions return to their more normal saturated state.

It is for these reasons that one cannot and should not guarantee a path. The potential system user should be wary of those who offer to guarantee the path because it just can't be done. The experienced systems manufacturer and the experienced consultant don't and won't and shouldn't. The user should be cautioned that there is always the possibility of [excessive outages due to] fading although the path was designed using the techniques that have been found to offer best protection against fading."

The microwave propagation researcher Millington (1959) noted "... where practical applications are concerned, we should not try to be too precise. For instance, some of the propagation curves that have been calculated with great accuracy from an idealized theory may give the impression that we can estimate field strengths with much greater precision than is actually feasible, and the engineer should always strive to appreciate the limits of accuracy set by the practical conditions. ... I wish to make [my] concern [known regarding] the use of statistical results when dealing with specific situations ... The chief difficulty in applying statistical methods arises when the basic material is very complex ... field strength [i.e., received signal level] at a given distance from the transmitter may vary greatly from place to place ... As a result of a measurement survey or of a prediction based on a study of ground profiles, a certain field strength will be obtained at 50% of locations ... This may be the best scientific way of assessing the problem in general ... But when it comes to the serving of a particular area where the terrain may be exceptionally rugged or of a particular town that is unfavorably placed, this general picture may be inadequate ... These [field strength] curves are drawn through a spread of points, each one of which represents a time average at a given location for a specific link. This means that, for a point that lies a long way off the curve, the performance over the circuit to which it corresponds will on the average differ considerably from the value given by the curve at the same distance. I wish, therefore, to plead that in applying statistical methods we should keep a sense of perspective."

REFERENCES

Achariyapaopan, T., "A Model of Geographic Variation of Multipath Fading Probability," *Bellcore National Radio Engineer's Conference Record*, pp. TA1–TA16, 1986.

Anderson, I., "Measurements of 20-GHz Transmission Through a Radome in Rain," *IEEE Transactions on Antennas and Propagation*, Vol. 23, pp. 619–622, September 1975.

Anderson, L. J. and Gossard, E. E., "Prediction of Oceanic Duct Propagation from Climatological Data," *IRE Transactions on Antennas and Propagation*, Vol. 3, pp. 163–167, October 1955.

AT&T, Microwave Radio, Radio Engineering Standard, Western Electric Practices, Section 940-300-130, Issue 2, March 1984.

Babler, G. M., "A Study of Frequency Selective Fading for a Microwave Line-of-Sight Narrowband Radio Channel," *Bell System Technical Journal*, Vol. 51, pp. 731–757, March 1972.

Barnett, W. T., "Multipath Propagation at 4, 6 and 11 GHz," *Bell System Technical Journal*, Vol. 51, pp. 321–361, June 1974.

Bellcore (Telcordia) Staff, *Bellcore (Telcordia) Technical Reference TR-TSY-000752, Microwave Digital Radio Systems Criteria*, pp. 7–13, October 1989.

Blevis, B. C., "Losses Due to Rain on Radomes and Antenna Reflecting Surfaces," *IEEE Transactions on Antennas and Propagation*, Vol. 13, pp. 175–176, January 1965.

Burgueno, A., Austin, J., Vilar, E. and Puigcerver, M., "Analysis of Moderate and Intense Rainfall Rates Continuously Recorded Over half a Century and Influence on Microwave Communications Planning and Rain-Rate Data Acquisition," *IEEE Transactions on Communications*, Vol. 35, pp. 382–395, April 1987.

Crane, R. K., *Electromagnetic Wave Propagation Through Rain*. New York: John Wiley & Sons, Inc., pp. 107–184, 1996.

Day, J. P. and Trolese, L. G., "Propagation of Short Radio Waves Over Desert Terrain," *Proceedings of the IRE*, pp. 165–175, February 1950.

Dougherty, H. T., *A Survey of Microwave Fading Mechanisms: Remedies and Applications, Environmental Science Services Administration Technical Report ERL 69-WPL4*. Washington, DC: US Department of Commerce, pp. 4–32, March 1968.

Dougherty, H. T., "Recent Progress in Duct Propagation Predictions," *IEEE Transactions on Antennas and Propagation*, Vol. 27, pp. 542–548, July 1979.

Dougherty, H. T. and Dutton, E. J., *The Role of Elevated Ducting for Radio Service and Interference Fields, NTIA Report 81-69*. Washington, DC: US Department of Commerce, March 1981.

Dougherty, H. T. and Hart, B. A., *Anomalous Propagation and Interference Fields, Office of Telecommunications Report 76-107*. Bolder: Institute of Telecommunications Sciences, US Department of Commerce, pp. 20–31, December 1976.

Dupuis, P., Joindot, M., Leclert, A. and Rooryck, M., "Fade Margin of High Capacity Digital Radio System," *IEEE International Conference on Communication*, Vol. 3, pp. 48.6.1–48.6.5, June 1979.

Dutton, E. J., "A Note on the Distribution of Atmospherically Ducted Signal Power Near the Earth's Surface," *IEEE Transactions on Communications*, Vol. 30, pp. 301–303, January 1982.

Effenberger, J. A. and Strickland, R. R., "The Effects of Rain on a Radome's Performance," *Microwave Journal*, Vol. 29, pp. 261–272, May 1986.

England, C. R., Crawford, A. B. and Mumford, W. W., "Ultra-Short-Wave Transmission and Atmospheric Irregularities," *Bell System Technical Journal*, Vol. 17, pp. 489–519, October 1938.

Fruchtenicht, H. W., "Notes on Duct Influences on Line-of-Sight Propagation," *IEEE Transactions on Antennas and Propagation*, Vol. 22, pp. 295–302, March 1974.

Giger, A. J., *Low-Angle Microwave Propagation: Physics and Modeling*. Boston: Artech House, pp. 214–218, 1991.

Hubbard, R. W., *Investigation of Digital Microwave Communications in a Strong Meteorological Ducting Environment, NTIA Report 79-24*. Washington, DC: US Department of Commerce, August 1979.

Ikegami, F., "Influence of an Atmospheric Duct on Microwave Fading," *IEEE Transactions on Antennas and Propagation*, Vol. 7, pp. 252–257, July 1959.

ITU-R Recommendation F.530-12, "Propagation Data And Prediction Methods Required for the Design of Terrestrial Line-of-Sight Systems," 2007.

ITU-R Recommendation F.1093-2, "Effects of Multipath Propagation on the Design and Operation of Line-of-Sight Digital Fixed Wireless Systems," 2006.

ITU-R Recommendation F.1330-1, "Performance Limits for Bringing Into Service of the Parts of International Plesiochronous Digital Hierarchy and Synchronous Digital Hierarchy Paths and Sections Implemented by Digital Radio-Relay Systems," 1999.

ITU-R Recommendation F.1566-1, "Performance Limits for Maintenance of Digital Fixed Wireless Systems Operating in Plesiochronous and Synchronous Digital Hierarchy-based Paths and Sections," 2007.

ITU-R Recommendation P.530-13, "Propagation Data and Prediction Methods for the Design of Terrestrial Line-of-Sight Systems," pp. 4–8, 13–14, 2009.

ITU-T Recommendation G.102, "Transmission Performance Objectives and Recommendations, Transmission Systems and Media," pp. 1–4, 1993.

ITU-T Recommendation G.821, "Error Performance of an International Digital Connection Operating at a Bit Rate Below the Primary Rate and Forming Part of an Integrated Services Digital Network," 2002.

ITU-T Recommendation G.826, "Error Performance of an International Digital Connection Operating at a Bit Rate Below the Primary Rate and Forming Part of an Integrated Services Digital Network," 2002.

ITU-T Recommendation G.827, "Error Performance of an International Digital Connection Operating at a Bit Rate Below the Primary Rate and Forming Part of an Integrated Services Digital Network," 2003.

ITU-T Recommendation G.828, "Error Performance of an International Digital Connection Operating at a Bit Rate Below the Primary Rate and Forming Part of an Integrated Services Digital Network," 2000.

ITU-T Recommendation M.20, "Maintenance Philosophy for Telecommunication Networks," 1992.

ITU-T Recommendation M.2100, "Performance Limits for Bringing-Into-Service and Maintenance of International Multi-Operator PDH Paths and Connections, International Transport Network," 2003.

ITU-T Recommendation M.2101, "Performance Limits for Bringing-Into-Service and Maintenance of International Multi-Operator SDH Paths and Multiplex Sections, International Transport Network," 2003.

ITU-T Recommendation M.35, "Principles Concerning Line-up and Maintenance Limits," 1993.

Ivanek, F., *Terrestrial Digital Microwave Communications*. Boston: Artech House, pp. 21–71, 1989.

Katzin, M., Bauchman, R. W. and Binnian, W., "3- and 9-Centimeter Propagation in Low Ocean Ducts," *Proceedings of the IRE*, pp. 891–905, September 1947.

Lee, J. L., "Refractivity Gradient and Microwave Fading Observations in Northern Indiana," *IEEE Global Telecommunications Conference (Globecom) Conference Record*, Vol. 3, pp. 36.8.1–36.8.5, December 1985.

Lee, J. L., "Observed Atmospheric Structure Causing Degraded Microwave Propagation in the Great Lakes Area," *IEEE Global Telecommunications Conference (Globecom) Conference Record*, Vol. 3, pp. 1548–1552, December 1986.

Lee, T. C. and Lin, S. H., "More on Frequency Diversity for Digital Radio," *IEEE Global Telecommunications Conference (Globecom) Conference Record*, Vol. 3, pp. 36.7.1–36.7.5, December 1985.

Lee, T. C. and Lin, S. H., "A Model of Space Diversity Improvement for Digital Radio," *International Union of Radio Science Symposium Proceedings*, pp. 7.3.1–7.3.4, July 1986.

Lin, S. H., "Statistical Behavior of Rain Attenuation," *Bell System Technical Journal*, Vol. 52, pp. 557–581, April 1973.

Lin, S. H., "A Method for Calculating Rain Attenuation Distributions on Microwave Paths," *Bell System Technical Journal*, Vol. 54, pp. 1051–1086, July-August 1975.

Lin, S. H., Lee, T. C. and Gardina, M. F., "Diversity Protections for Digital Radio—Summary of Ten-Year Experiments and Studies," *IEEE Communications Magazine*, Vol. 26, pp. 51–64, February 1988.

Lundgren, C. W. and Rummler, W. D., "Digital Radio Outage Due to Selective Fading - Observation vs Prediction From Laboratory Simulation," *Bell System Technical Journal*, Vol. 58, pp. 1073–1100, May-June 1979.

Mahmoud, S. F., Boghdady, H. N. and El-Sayed, O. L., "Analysis of Multipath Fading in the Presence of an Elevated Atmospheric Duct," *Proceedings of the IEE*, Vol. 134, pp. 71–76, February 1987.

McGavin, R. E., Dougherty, H. T. and Emmanuel, C. B., "Microwave Space and Frequency Diversity Performance Under Adverse Conditions," *IEEE Transactions on Communication Technology*, Vol. 18, pp. 261–263, June 1970.

Millington, G., "Random Thoughts of a Propagation Engineer," *Proceedings of the IEE*, Vol. 106, Part B, pp. 11–14, January 1959.

Osborne, T. L., "Applications of Rain Attenuation Data to 11-GHz Radio Path Engineering," *Bell System Technical Journal*, Vol. 56, pp. 1605–1627, November 1977.

Ranade, A. and Greenfield, P. E., "An Improved Method of Digital Radio Characterization from Field Measurements," *IEEE International Conference on Communications*, pp. C2.6.1–C2.6.5 (Vol. 2, 659–663), June 1983.

Rummler, W. D., "A New Selective Fading Model: Application to Propagation Data," *Bell System Technical Journal*, Vol. 58, pp. 1037–1071, May-June 1979.

Rummler, W. D., "More on the Multipath Fading Channel Model," *IEEE Transactions on Communications*, Vol. 29, pp. 346–352, March 1981.

Rummler, W. D., "A Comparison of Calculated and Observed Performance of Digital Radio in the Presence of Interference," *IEEE Transactions on Communications*, Vol. 30, pp. 1693–1700, July 1982.

Rummler, W. D.,"Advances in Microwave Radio Route Engineering for Rain," *IEEE Conference on Communications (ICC) Proceedings*, pp. 10.8.1–10.8.5, June 1987.

Rummler, W. D., "Characterizing the Effects of Multipath Dispersion on Digital Radios," *IEEE Global Telecommunications Conference (Globecom) Record*, Vol. III, pp. 52.5.1–52.5.7, November 1988.

Rummler, W. D. and Dhafi, M., "Route Design Methods," *Terrestrial Digital Microwave Communications*. Ivanek, F., Editor. Norwood: Artech House, pp. 326–329, 1989.

Schiavone, J. A., "Prediction of Positive Refractivity Gradients for Line-of-Sight Microwave Radio Paths," *Bell System Technical Journal*, Vol. 60, pp. 803–822, July-August 1981.

Schiavone, J. A., "Microwave Radio Meteorology: Fading by [Duct] Beam Focusing," *IEEE International Conference on Communications (ICC) Conference Record*, Vol. 3, pp. 7B.1.1–7B.1.5, June 1982.

Stephansen, E. T., "Clear-air Propagation on Line-of-Sight Radio Paths: A Review," *Radio Science*, Vol. 16, pp. 609–629, September and October 1981.

Telcordia (Bellcore) Staff, *Telcordia Special Report SR-2275, Telcordia Notes on the Networks*, pp. 8.49–8.52, October 2000.

Telcordia (Bellcore) Staff, *Telcordia Generic Requirements GR-929-CORE, Reliability and Quality Measurements for Telecommunications Systems (RQMS-Wireline), Issue 8*, December 2002.

Telcordia (Bellcore) Staff, *Telcordia Generic Requirements GR-1929-CORE, Reliability and Quality Measurements for Telecommunications Systems (RQMS-Wireless), Issue 2*, February 2005.

Telcordia (Bellcore) Staff, *Telcordia Generic Requirements GR-253-CORE, Synchronous Optical Network (SONET) Transport Systems: Common Generic Criteria, Issue 5*, pp. 6.86–6.87, September 2009a.

Telcordia (Bellcore) Staff, *Telcordia Generic Requirements GR-499-CORE, Transport Systems Generic Requirements (TSGR): Common Requirements, Issue 4*, pp. 2-1 to 4-3, November 2009b.

Thrower, R. D., "Curing the Fades, Part III," *Telephone Engineer and Management*, p. 82, 1 September, 1977.

TIA/EIA, Interference Criteria for Microwave Systems, Telecommunications Systems Bulletin TSB10-F, June 1994.

Vigants, A., "Space-Diversity Performance as a Function of Antenna Separation," *IEEE Transactions on Communication Technology*, Vol. 16, pp. 831–836, December 1968.

Vigants, A., "The Number of Fades and Their Durations on Microwave Line-of-Sight Links With and Without Space Diversity," *IEEE International Conference on Communications (ICC) Proceedings*, pp. 3.7–3.11, June 1969.

Vigants, A., "Number and Duration of Fades at 6 and 4 GHz," *Bell System Technical Journal*, Vol. 50, pp. 815–841, March 1971.

Vigants, A., "Observations of 4 GHz Obstruction Fading," *IEEE International Conference on Communications (ICC) Conference Record*, pp. 28.1–28.2, June 1972.

Vigants, A., "Space-Diversity Engineering," *Bell System Technical Journal*, Vol. 54, pp. 103–142, January 1975.

Vigants, A., "Microwave Radio Obstruction Fading," *Bell System Technical Journal*, Vol. 60, pp. 785–801, July-August 1981.

Wheeler, H. A., "Microwave Relay Fading Statistics as a Function of a Terrain Clearance Factor," *IEEE Transactions on Antennas and Propagation*, Vol. 25, pp. 269–273, March 1977.

5

RADIO SYSTEM COMPONENTS

Microwave radio signals are generally regarded as those that extend from about 1 GHz to roughly 1000 GHz (1 THz), just below the optical infrared frequency band. In optical terms, 1000 GHz is 300,000 nm (300 μm) in wavelength. The so-called "low frequency" microwave is generally from 1 to 7 GHz, which is the range of frequencies typically used for "long haul" applications that are relatively unaffected by rain attenuation, but highly influenced by multipath fading. The frequency range 7–10 GHz is a transition region where rain attenuation becomes a significant factor. "High frequency" microwave signals are generally those above 10 GHz. High frequency radio paths are limited in length by rain attenuation. Examples of path lengths by frequency band are given in Section 5.13.

The ITU-R has allocated frequency bands to various services up to 400 GHz. The FCC currently has rules and regulations that govern operation up to 100 GHz. Most licensed applications in the United States are in the 6–25 GHz range. Unlicensed applications occur above and below these frequencies. See Appendix A, Table A.18, for more details on these frequency allocations.

Fixed point-to-point microwave radio systems (Fig. 5.1) use transmitters and receivers deployed miles apart to transport high speed digital signals. Wireless transmission between a transmitter and a receiver is influenced by several factors. A transmission line [coaxial cable (coax) or waveguide] connects the transmitter or receiver to an antenna; an antenna structure (typically a building or a tower) holds the antenna at an appropriate height and orientation and the antenna launches or receives a signal that is propagated between the antennas along the radio wave path. Power loss between the transmit and receive antennas is variable in time or location because of many factors.

Path outage time is a function of external interference (Chapter 2, frequency planning section and Chapter 14), rain attenuation (Chapter 11) and received signal variations due to upfades (Chapter 12, ducting), earth bulge (Chapter 12, obstruction fading), flat multipath fading (frequency insensitive time variable path attenuation), and dispersive multipath fading (frequency selective time variable path distortion). Multipath fading is discussed in Chapter 9.

Microwave path engineering can be rather complicated. At least two tasks are always required. The first is to reduce multipath and rain fading to an acceptable level. This is done by sizing the transmit power and antenna sizes appropriately. This process is accomplished by performing path performance calculations using a spreadsheet or a computer program. The result is the estimated average path outage due to multipath and rain. The calculation process uses path performance equations discussed in Chapter 16. Performance objectives are reviewed in Chapter 4.

Digital Microwave Communication: Engineering Point-to-Point Microwave Systems, First Edition. George Kizer.

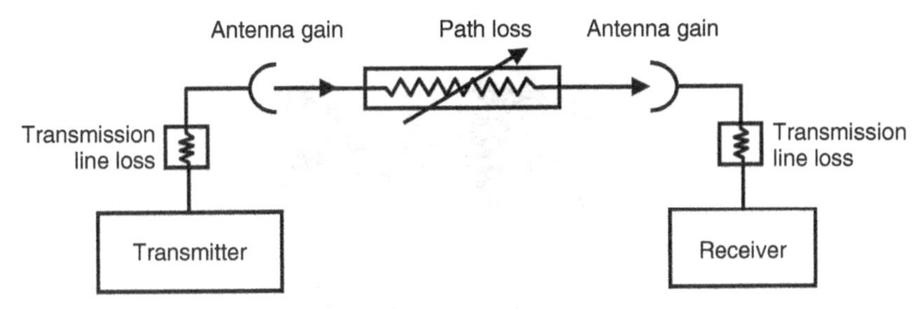

Figure 5.1 Simplified radio transmission model.

The second task is to place the antennas at an appropriate height to reduce obstruction and reflective fading. This is usually done by using a computer-generated path profile (vertical plot of terrain and antenna heights) to place the antenna appropriately. Antenna placement methodology is reviewed in Section 12.4.1.

After the path is designed, if the radio operates in a licensed frequency band, a third task must be accomplished: that path must be frequency coordinated with other users and then licensed (Chapter 2).

This chapter presents an overview of the building blocks of a microwave radio path between the transmitter and the receiver. The transmitter and receiver are discussed in Chapter 3.

5.1 MICROWAVE SIGNAL TRANSMISSION LINES

Microwave transmission lines are typically waveguide for frequencies above 3 GHz and coaxial cable ("coax") for lower frequencies. The reasons are primarily their physical size and transmission loss. See Appendix A for typical waveguide and coax cable attenuation.

In Figure 5.2, a 1-ft ruler and a sheet of paper the size of a US dollar bill were included for reference. Waveguides are usually connected to other devices using flanges that bolt together. Rectangular waveguides are usually used indoors and are unpressured. Elliptical waveguides are used outdoors and are usually pressured to prevent the entry of water. When they are connected together, a pressure window is used to maintain pressure on the elliptical waveguide.

Rectangular waveguides have lower loss than elliptical waveguides. However, they are difficult to adapt to complex installations. The primary advantages of elliptical waveguides are the simplicity of installation and resistance to water penetration. Circular waveguides have the lowest attenuation of all commercial waveguides. However, they require long, straight installation with provision to expand and

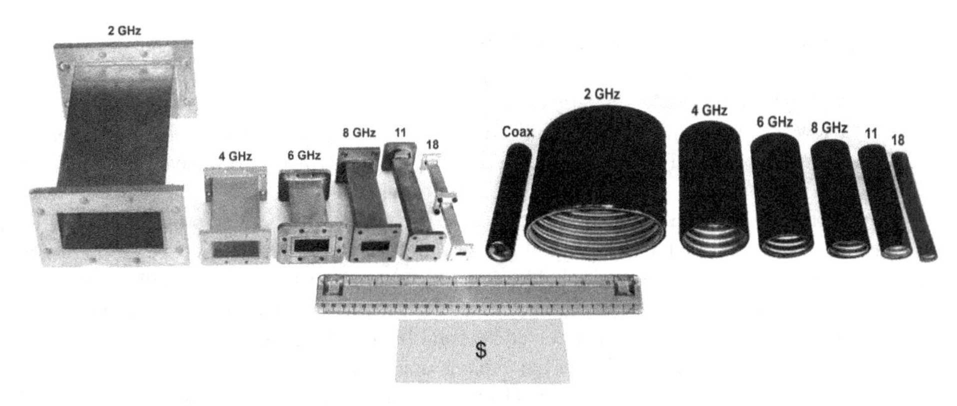

Figure 5.2 Rectangular and elliptical waveguide with coaxial cable. *Source*: Reprinted with permission of Alcatel-Lucent USA, Inc.

Figure 5.3 Typical coaxial connectors.

contract as a result of temperature variation. With proper couplers (containing mode filters), circular waveguides can be used over a very large frequency range. However, significant installation considerations limit their utilization to situations where very long transmission lines or multiple frequencies are important.

Waveguide flanges are quite distinctive. Occasionally, waveguides must be transitioned to coax. Coaxial connectors come in two impedances, 50 and 75 Ω. As shown in Figure 5.3, they appear very similar. Be careful not to mate connectors of different impedances (a common risk with coax adapters). The connection will be loose or one of the connector center pins will be deformed permanently.

A waveguide is a hollow metal tube that is rectangular, elliptical, or circular. In free space, Maxwell's equations (Schelkunoff, 1943) dictate that a traveling electromagnetic wave will have electric and magnetic fields orthogonal to the direction of wave transmission. Inside a metal tube, Maxwell's equations force either the electric or the magnetic field to be orthogonal to the side of the waveguide and the other field to be parallel to the direction of wave transmission down the guide. If the electric field (Schelkunoff, 1963) is always perpendicular to the direction of propagation, the mode of transmission is termed *transverse electric* (*TE*). If the magnetic field is always perpendicular to the direction of propagation, the mode of transmission is termed *transverse magnetic* (*TM*). Modes are usually described as TM or TE with a two-number subscript following the mode designation. The first subscript describes the number of half-cycle variations across the wide dimension of the waveguide (when viewed as cross-section). The second subscript describes the number of half-cycle variations across the narrow dimension of the waveguide.

A waveguide (Marcuvitz, 1986; Schelkunoff, 1963; Southworth, 1950a, 1950b) operates similarly to a high pass filter. Many modes of operation are possible. Each mode has a cutoff frequency. Below that frequency, the waveguide acts similarly to an attenuator. Above that frequency, the waveguide acts similarly to a low attenuation transmission line. Lower order modes (smaller number subscripts) have lower cutoff frequencies. Usually, a waveguide's operating frequency is between the cutoff frequency for the lowest order (fundamental) mode and that of the next higher mode (Fig. 5.4).

Operation with multiple modes is undesirable because each mode's velocity of propagation is different. If multiple modes are generated at the transmit end of a transmission line and reconverted into a usable signal at the receive end, significant undesirable signal dispersion (pulse widening) will occur.

Figure 5.5 illustrates the fundamental (lowest frequency) mode of propagation for typical waveguide shapes and coax. The electric field is represented by solid lines and the magnetic field is represented by dashed lines. Waveguide propagation in coaxial cable is undesirable.

Coax is constructed by surrounding one conductor (center conductor) with a second one (shield). If the shield is solid, radiation outside the transmission line is essentially eliminated. In coax, the normal mode of propagation is by transverse electric and magnetic (TEM) field. However, at high frequencies

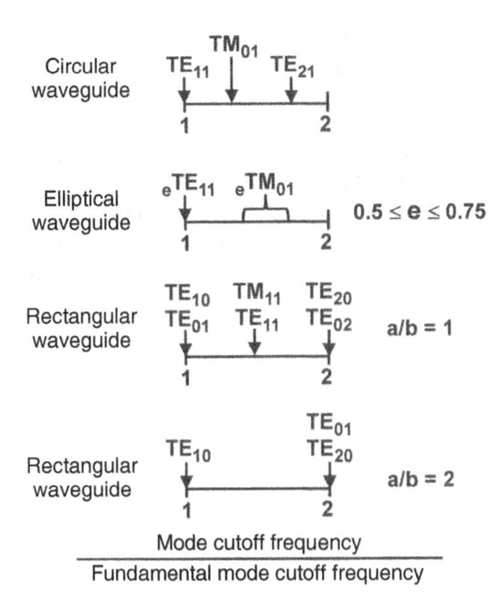

Figure 5.4 Waveguide modes cutoff frequencies.

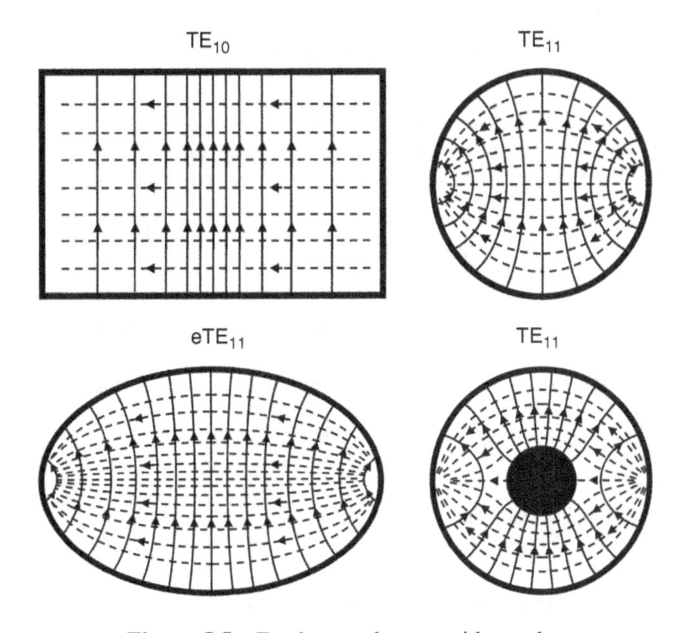

Figure 5.5 Fundamental waveguide modes.

the coax will become a waveguide that is able to support both TE- and TM-mode propagation. As with traditional waveguides, coax has a lowest waveguide mode, TE_{11}.

The cutoff frequency for coax TE_{11} mode is given by the following equation:

$$F_{CO} = \frac{(7.50 \times V_F)}{[D(\text{in}) + d(\text{in})]} = \frac{(190 \times V_F)}{[\,D(\text{mm}) + d(\text{mm})]} \tag{5.1}$$

$D =$ inside diameter of outer conductor;
$d =$ outside diameter of inner conductor;
$V_F =$ velocity factor = 1/sqrt[dielectric constant(relative permittivity)].

Bends and connectors will generate evanescent waveguide modes (highly attenuated modes operating below the cutoff frequency for that mode). As long as the coax is operated at frequencies no higher than 3/4 the lowest mode cutoff frequency, evanescent modes will not affect performance.

In waveguide or coax, the velocity of propagation of an electromagnetic wave is slower than in free space. The velocity of propagation in the transmission line is termed *group velocity*.

$$V_G = \text{Group velocity} = V_0 V_F \tag{5.2}$$

$V_0 =$ velocity of propagation in free space;
$= 0.9833$ ft/ns;
$= 0.2998$ m/ns;
$V_F =$ velocity factor.

The absolute delay, D of a transmission line is given by the following:

$$D = \frac{L}{V_G}$$

$L =$ physical length of the transmission line.

The effective length of the transmission line (when compared to a RF signal traveling in free space) is given by the following:

$$L_{EFF} = \text{Effective length} = \frac{L}{V_F} \tag{5.3}$$

For coax, the velocity of propagation is independent of frequency because the velocity of propagation is a function of the dielectric constant (relative permittivity). See Eq. 5.2.

For a waveguide, the velocity of propagation is a function of frequency (Ramo et al., 1965).

$$V_F = \text{Velocity factor} = \text{sqrt}\left[1-\left(\frac{f_c}{f}\right)^2\right] \tag{5.4}$$

$f_c =$ cutoff frequency for the waveguide mode of interest. $f =$ frequency of interest.

For a rectangular waveguide with $a = 2b$, the following applies for the fundamental mode, TE_{10}:

$$f_c(\text{MHz}) = \frac{5902}{a(\text{inches})} = \frac{14{,}990}{a(\text{cm})} \tag{5.5}$$

$a =$ larger cross-sectional dimension;
$b =$ smaller cross-sectional dimension.

The next higher modes (TE_{01} and TE_{20}) have cutoff frequencies that are twice the fundamental waveguide cutoff frequency.

A smooth, elliptical waveguide has not been studied as deeply as rectangular and circular waveguides. Currently, all commercial elliptical waveguides have corrugations. This type of waveguide has never been studied theoretically. A corrugated waveguide is designed with periodic corrugation spacing set so that reflections caused by the corrugations occur at frequencies below cutoff. Its loss is slightly higher than that of the theoretically smooth elliptical waveguide. Elliptical waveguide cutoff frequencies are generally a function of waveguide ellipticity (Chu, 1938; Stratton et al., 1941).

$$\text{Ellipticity}(e) = \text{sqrt}\left[1-\left(\frac{b}{a}\right)^2\right] \tag{5.6}$$

$a =$ larger cross-sectional dimension;
$b =$ smaller cross-sectional dimension.

The parameter is not well defined for corrugated waveguides. Fortunately, the fundamental cutoff frequency is relatively insensitive to ellipticity (e). For elliptical waveguides with a typical ellipticity of 0.5–0.75, the following applies for the fundamental mode, ${}_{e}TE_{11}$:

$$f_c(\text{MHz}) = \frac{7005}{a(\text{inches})} = \frac{17{,}790}{a(\text{cm})} \tag{5.7}$$

$a =$ larger cross-sectional dimension.

The next higher mode (${}_{e}TM_{01}$) has a cutoff frequency that is 1.4–1.7 times the fundamental waveguide cutoff frequency for ellipticity between 0.5 and 0.75, respectively.

For a circular waveguide, the following defines the cutoff frequency for the fundamental mode, TE_{11}:

$$f_c(\text{MHz}) = \frac{6917}{D(\text{inches})} = \frac{17{,}570}{D(\text{cm})} \tag{5.8}$$

$D =$ inside diameter of the circular waveguide.

The circular waveguide is usually specified as WCXX, where XX is the inside diameter in inches. For instance, WC281 is 2.812 in. in diameter, WC109 is 1.09 in., and WC75 is 0.75 in. The next higher mode (TM_{01}) has cutoff frequency 1.3 times the fundamental waveguide cutoff frequency. A circular waveguide is sometimes operated over a very wide frequency range. Several higher modes are possible. If this is done, input and output couplers with mode suppressors are necessary. It is also important to not attempt operation near cutoff frequencies for any of those higher order modes. Significant attenuation and phase distortions occur at those frequencies. The first 31 mode cutoff frequencies (Kizer, 1990) for circular waveguides are listed in Appendix A.

Waveguide attenuation is given by the following (Kizer, 1990):

$$\text{Attn}\left(\frac{\text{dB}}{100\ \text{m}}\right) = \frac{A\left(\frac{f}{fc}\right)^2 + B}{\sqrt{\left(\frac{f}{fc}\right)\left[\left(\frac{f}{fc}\right)^2 - 1\right]}} \tag{5.9}$$

$$\text{Attn}\left(\frac{\text{dB}}{100\ \text{ft}}\right) = 0.3048\ \text{Attn}\left(\frac{\text{dB}}{100\ \text{m}}\right) \tag{5.10}$$

$f =$ frequency of interest (GHz);
$f_c =$ cutoff frequency (GHz).

A and B are coefficients determined from measured waveguide attenuation versus frequency tables. See Appendix A for the coefficients for common rectangular and elliptical waveguides. A and B may be determined from the following:

$$D_N = \frac{\left(\frac{f_N}{fc}\right)^2}{\sqrt{\left(\frac{f_N}{fc}\right)\left[\left(\frac{f_N}{fc}\right)^2 - 1\right]}} \tag{5.11}$$

$$E_N = \frac{1}{\sqrt{\left(\frac{f_N}{fc}\right)\left[\left(\frac{f_N}{fc}\right)^2 - 1\right]}} \tag{5.12}$$

$$B = \frac{C_2 - \left(\frac{C_1 D_2}{D_1}\right)}{E_2 - \left(\frac{E_1 D_2}{D_1}\right)} \quad (5.13)$$

$$A = \frac{(C_1 - BE_1)}{D_1} \quad (5.14)$$

f_1 = lowest frequency of interest (GHz);
f_2 = highest frequency of interest (GHz);
N = 1 or 2;
C_1 = attenuation measured through 100 m of waveguide at frequency f_1;
C_2 = attenuation measured through 100 m of waveguide at frequency f_2.

Coax attenuation is determined from the following formulas:

$$\text{Attn}\left(\frac{\text{dB}}{100\ \text{m}}\right) = A\sqrt{f} + Bf \quad (5.15)$$

$$\text{Attn}\left(\frac{\text{dB}}{100\ \text{ft}}\right) = 0.3048\ \text{Attn}\left(\frac{\text{dB}}{100\ \text{m}}\right) \quad (5.16)$$

$$A = \frac{\text{Attn}_1\sqrt{f_2}}{f_1\left(\sqrt{\frac{f_2}{f_1}} - 1\right)} - \frac{\text{Attn}_2}{\sqrt{f_2}\left(\sqrt{\frac{f_2}{f_1}} - 1\right)} \quad (5.17)$$

$$B = \frac{\text{Attn}_2}{\sqrt{f_1 f_2}\left(\sqrt{\frac{f_2}{f_1}} - 1\right)} - \frac{\text{Attn}_1}{f_1\left(\sqrt{\frac{f_2}{f_1}} - 1\right)} \quad (5.18)$$

A = conductive ("skin effect") loss coefficient;
B = dielectric loss coefficient;
f = frequency of interest (MHz);
Attn_1 = cable loss (dB/100 m) at f_1;
Attn_2 = cable loss (dB/100 m) at f_2;
f_1 = lowest frequency of interest (GHz);
f_2 = highest frequency of interest (GHz);
= $f_1 < f_2$ and $f_1 \leq f \leq f_2$.

The microwave radio is connected to an antenna by a transmission line (coax for low frequencies and waveguide for higher frequencies). The power match between the transmission line and the radio and antenna can affect performance (Wu and Achariyapaopan, 1985). For low frequency transmission lines, the power match between a transmission line and a terminating element is measured as a voltage standing wave ratio (VSWR). In a waveguide, the concept of voltage and current do not exist; only electric and magnetic fields are relevant. For this situation, the concept of return loss is used. The return loss of a transmission line and terminating element interface describes the relative amount of incident energy returned toward the source. For example, if a 30-dBm pulse enters a transmission line interface from a radio and 10 dBm of energy is reflected back toward the radio (the pulse source), the transmission line has a return loss of 20 dB (when it is connected to the radio). Reflections from both ends of the transmission line cause a signal echo that can degrade radio receiver performance (Fig. 5.6).

This echo represents a secondary signal that, when it appears at the receiver, introduces dispersion (pulse widening) distortion. Radiowave pulses in free space travel about 1 ns/ft. In coax and waveguide, they are approximately half that speed (2 ns/ft).

If the transmission line is short enough that the echo delay is much shorter than the transmitted symbol, return loss at the transmission line interface does not matter (as long the interface does not cause a significant power loss). For example, consider a radio operating in a 30-MHz radio channel. Nyquist's signaling rate limits radio symbols to a maximum rate of 30×10^6 symbols per second (for

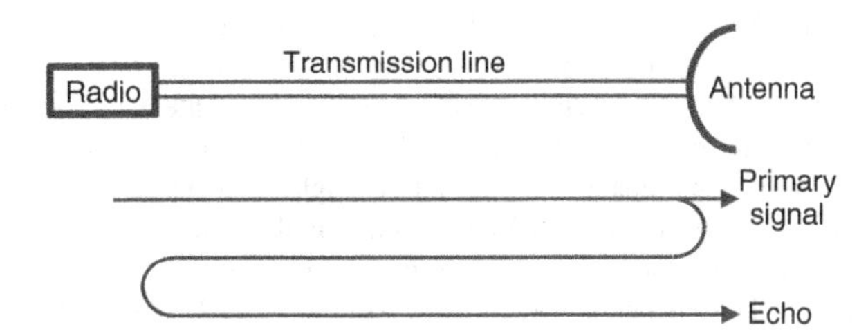

Figure 5.6 Transmission line echo.

distortion-free transmission). This signal would be $[1/(3 \times 10^7)]\ 10^9 \approx 33$ ns wide. If the echo is delayed by less than 4 ns (transmission line <1 ft long), transmission line return loss is unimportant. However, for longer transmission lines, return loss is a factor.

As the transmission line becomes longer, the effect of echoes becomes more significant. All modern radios use scramblers to decorrelate successive data pulses (to achieve uniform transmit spectrums). After the echo delay exceeds the delay of a single symbol (in this case, about 33 ns), the echo appears as a co-channel-like modulation interfering source. For this example, a waveguide longer than [(pulse width/2) = (33/2)/2] ≈ 8 ft would fit this criterion. For this "long echo" case, we can use the common receiver interference T/I specification (see Chapter 14) to estimate the transmission component return loss such that radio performance will not be impaired.

T/I defines an interference level that will result in a 10^{-6} receiver BER at receiver threshold. If we wish to define an interference level that will allow the receiver to perform unaffected by the interference, we could conservatively add 6 dB to the T/I value. Next, we assume that we know the return loss at each end of the transmission line (assume the worst case for the radio, transmission line, and antenna). For essentially error-free operation, we want the following condition to apply:

$$\left(\frac{T}{I}\right) + 6\ \text{dB} \leq (\text{Return loss at radio interface} + \text{Return loss at antenna interface}$$

$$+2 \text{ x Attenuation of the transmission line}) \qquad (5.19)$$

If we assume that the return loss at each end of the transmission line is the same, then the required return loss for error-free operation is the following:

$$\text{Return loss} \geq \frac{(T/I)}{2} + 3 - \text{Transmission line loss(dB)} \qquad (5.20)$$

As an example, consider a radio with co-channel-like modulation interference T/I of 33 dB. The minimum return loss for the transmission line (or radio or antenna) would be (ignoring waveguide loss) approximately 20 dB.

5.2 ANTENNA SUPPORT STRUCTURES

Antenna support structures are the towers, buildings, or other structures used to mount antennas. They are typically one of the following types: AM radio towers, billboards, bridges, buildings, chimneys, church steeples, clock towers, cell (site) on wheels (cow), flag poles, guyed towers, high voltage transmission towers, self-supporting towers, light poles, monopalms, monopines, monopoles, monotrees, pent houses, private homes, rooftops, (roadside) signs, (grain) silos, smoke stacks, stealth structures (of various types), pipes, (wood or steel) poles, (oil) rigs, (oil or water) tanks, trees, (high voltage) power transmission towers, or utility poles/towers. These may be generalized as one of seven major types: guyed lattice towers, self-supporting lattice towers, monopole towers, architecturally designed structures, building mounts, camouflaged structures, or temporary structures. Most of the others are exceptions typical of high frequency urban radio systems. Currently, the FCC only recognizes the following structure types in Form

601 Schedule I: antenna tower arrays, bridges, buildings, free standing/guyed towers, multiple structures, pipes, poles, (oil) rigs, self-supporting structures, signs/billboards, silos, smoke stacks, tanks, trees, or utility poles/towers. From the FCC's perspective, not all of these are antenna structures (see Chapter 2).

The US FCC and FAA require the antenna structure's height to be measured with and without appurtenances. The height without appurtenances is the tallest height to which a substantial device (such as an antenna) can be mounted. The height with appurtenances is the tallest height including the height of any item (such as whip antennas, small masts, flag poles, lightning rods, or light strobes) attached to the structure. See Chapter 2 for FCC Antenna Structure Registration (ASR) requirements.

Most antenna structures are towers. They are designed and constructed using industry standards (Institute of Electrical and Electronics Engineers, 2009; Telcordia Technologies, 2008, 2011; TIA Subcommittee TR-14.7, 2005). Antenna structures are usually one of the following types.

5.2.1 Lattice Towers

These towers are typically self-supporting or guyed. The tower is interlaced with bracing members to stabilize and strengthen the structure. They are typically manufactured in 20-ft sections although the first section of very large self-supporting towers may be 10 ft. For heights above approximately 100 ft, the guyed tower is less expensive than the self-supporting one.

Lattice towers are usually composed of solid rods, tubular pipes, or angular steel. Tubular rod antennas have relatively low wind drag and are easy to inspect and maintain. They are relatively heavy towers (impacting shipping and erecting costs), difficult to modify, and difficult to mount antennas on (mounting hardware relies on friction). Pipe towers are lighter than tubular rods but have the same disadvantages as rod towers. Rust and ice are significant concerns for hollow pipe towers. Rust typically starts inside the tower, is virtually impossible to detect, and can lead to unanticipated structural failure. Ice due to trapped water is also common. Hollow sections cannot be inspected adequately. Angular steel is significantly of lower weight than rod or tubular towers. They are easy to inspect and maintain. They are relatively easy to modify for increased loading. They are usually totally bolted construction which must be inspected routinely. They have relatively high wind loading (Communication Equipment Specialists and Grasis Towers LLC, 2002).

Drilling holes in tower members is strictly forbidden because this weakens the structure. Most lattice towers are galvanized steel. As welding removes the galvanization (and potentially weakens the structure), modifications, additions, or mounting structures must be mechanical (typically clamps or bolts in predesigned holes).

5.2.2 Self-Supporting Towers

Self-supporting lattice towers of over 600 ft have been built. However, typical designs are of 300 ft or less. They may have three or four legs. They require less land than guyed towers. They have excellent structural integrity, are relatively easy to modify, and have unrestricted antenna placement. They are relatively expensive to construct and limited in height (Fig. 5.7).

5.2.3 Guyed Towers

Guyed lattice towers as high as 2000 ft have been built. However, typical microwave radio towers are less than 400 ft high. For flat terrain, the guys are typically attached to the ground at 80% of the height away from the tower center. Guys can be attached as close as 40% of the tower height away from the tower center, but, as the attachment distance decreases, the cost of the guys and attachment points increases significantly. Most guyed towers have three sets of guys. Guyed towers are the most common type of towers for private operators in North America. The initial cost is about half the cost of a self-supporting antenna. They require the most land space compared to any other type of tower. They have relatively high maintenance costs, are expensive to modify, and are highly vulnerable to vandalism. Antenna placement is limited by the guys. They have low resistance to wind torque and typically require additional torque stabilizers (Fig. 5.8).

Figure 5.7 Self-supporting towers.

Figure 5.8 Guyed towers.

Figure 5.9 Monopole towers.

5.2.4 Monopoles

These are steel tubular structures composed of a single or multiple pipe sections. Monopole towers are typically either step tapered or continuously tapered. A step-tapered monopole is composed of sections of constant diameter with a round flange plate at each end. Each section is typically of a different diameter. The tower is constructed by bolting together sections of decreasing diameter. Continuously tapered towers may be a single section or several sections connected by slip joints. If several sections are used to create the tower, pairs of the continuously tapered sections are placed one inside the other and then pulled into place using hydraulic jacks.

Monopoles are typically chosen when aesthetics, minimal land use, or construction time are primary considerations. They are relatively expensive when compared to lattice towers and are expensive to modify for increased loading. While monopoles can be as high as 250 ft, owing to the cost of tall monopoles most are in the 50–100 ft range. Monopoles accommodate antennas at virtually any location on the tower. Compared to other towers, they have limited loading capability. Transmission cables typically go inside the tower making cable wind loading unimportant. The smooth, relatively slender tower has minimal wind loading. These towers have poor resistance to wind sway. Cellular antennas have relatively lax twist requirements. Using the towers to support microwave parabolic or square antennas usually requires a stiffer tower because of the relatively stringent twist sway requirements, as well as the increased asymmetrical wind torque loading (Fig. 5.9).

5.2.5 Architecturally Designed Towers

These towers have an attractive appearance, fit virtually any land space, and have low maintenance, outstanding structural integrity, and unrestricted antenna placement. They are relatively expensive and difficult to modify (Fig. 5.10).

Figure 5.10 Architecturally designed towers.

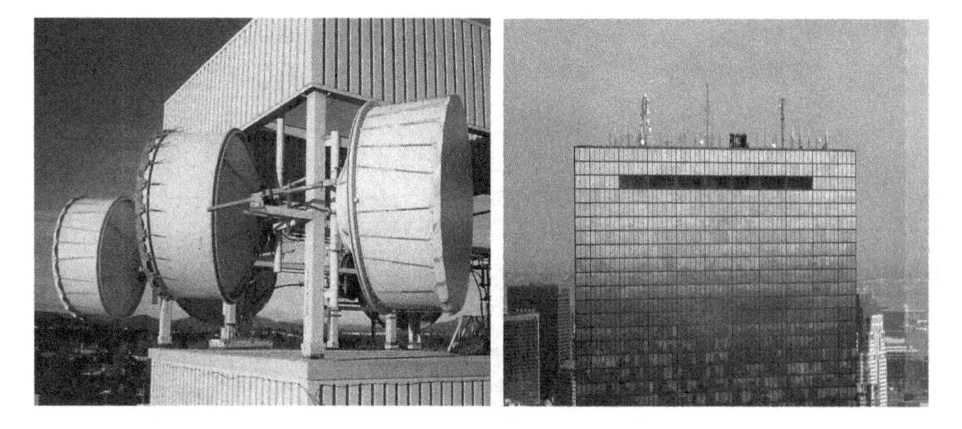

Figure 5.11 Building-mounted antennas.

5.2.6 Building-Mounted Antennas

Building-mounted antennas have the significant advantages that no additional land is required and they provide flexible antenna placement. They are relatively expensive, are limited by building integrity, typically require civil engineering studies, have limited maintenance access, and may have challenging cable placement restrictions (Fig. 5.11).

5.2.7 Camouflaged Structures

The primary motivation to use these structures is their ability to solve difficult zoning issues. They may take the form of a tree, a palm, a boulder, or another common structure. They have attractive appearances, fit virtually any land space, require low maintenance, and have outstanding structural integrity. They have very high relative cost, restricted antenna placement, and are difficult to modify (Fig. 5.12).

5.2.8 Temporary Structures

Temporary or nonpermanent structures are often used for path or site restoral. In cellular cases, cells on wheels (cows) utilize relatively small crank-up towers to create a site while it is being built, reworked, or restored. Common carriers often use temporary towers for site or path restoral. Some users place heavily weighted tripods or other simple structures on tops of buildings to hold relatively small antennas. These antennas do not permanently attach to the building and therefore are easier to gain approval from the landlord (Fig. 5.13).

Figure 5.12 Camouflaged structures.

Figure 5.13 Temporary structures.

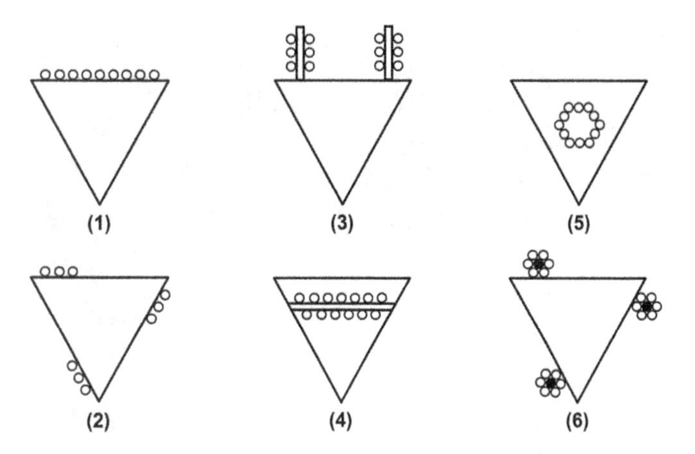

Figure 5.14 Typical transmission line distribution on towers.

5.3 TOWER RIGIDITY AND INTEGRITY

Antenna structures must maintain the antenna position within acceptable limits. In the United States, the current standard, ANSI/TIA/EIA-222-G (TIA Subcommittee TR-14.7, 2005), allows 10-dB loss of received signal level due to antenna structure twist or tilt under standardized wind and ice loading. See Section 5.A.7 for methods of estimating those twist limits.

When antennas and transmission lines are attached to antenna structures, the attachment must be done in such a way that the integrity of the structure is maintained (Telcordia Technologies, 2008).

5.4 TRANSMISSION LINE MANAGEMENT

The antenna transmission lines (waveguide and coax) extending from the ground to the antenna will add additional static (normal weight plus ice) and dynamic (wind) loading to the antenna structure (Andrews, J. A. 1996, Unpublished lecture notes). Placement of these lines can significantly affect the rigidity requirements of the structure. Figure 5.14 depicts typical transmission line distributions on a three-legged tower.

If the transmission lines are distributed across the tower (on the left side of Figure 5.14), they are relatively easy to maintain but significantly increase the tower's wind and ice loading. Bundled transmission lines (on the right side of Figure 5.14) minimize tower loading but are difficult to maintain (Fig. 5.15).

Note Fig. 5.15. At site A, the cables and elliptical waveguides are spread across a vertical cable ladder in the center of the tower. The transmission lines are placed on the climbing ladder side of the cable ladder. Maintenance will be simple. At site B, the transmission lines are bundled on top of each other on a vertical cable ladder. If the lines are damaged (with a rifle shot, for example), repair will be quite difficult. "If you can't fix it or maintain it, don't do it" (Andrews, J. A. 1996, Unpublished lecture notes).

Lastly, the transmission line will transition from the tower or other structure to the equipment room. The transmission line should be supported and protected by a horizontal ladder. If ice is a potential issue, the transmission line and the antennas on the tower should be protected by an ice shield.

Grounding of the tower, transmission lines (and tower lights), and the inside facility is very important but is a subject well beyond the scope of this chapter.

5.5 ANTENNAS

Antennas come in a wide variety of configurations for a wide range of purposes (Ramsay, 1981; Ramsdale, 1981; Tilston, 1981). Microwave radio antennas used in fixed point-to-point applications are devices that

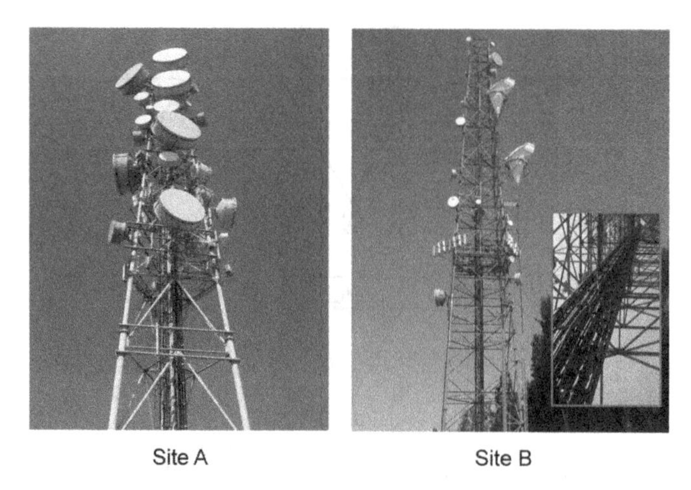

Site A Site B

Figure 5.15 Transmission line distribution on towers.

provide an efficient means of transmitting and receiving radio waves between two (or more) locations. They are a transition between the transmission line (waveguide or coax) and "free space."

At microwave frequencies, antennas are usually passive reflectors, standard gain horns, circular ("parabolics" or "horns"), or rectangular ("flat panel") antennas. Passive reflectors (Norton, 1962) are used to solve difficult path problems in mountainous terrain. Standard gain horns (Slayton, 1954) are relatively small antennas used extensively for interference studies both in laboratory facilities as well as in the field.

The rectangular antennas are an array of microstrip patches (Pozar and Schaubert, 1995) in front of a reflective sheet. A single antenna patch is essentially a dipole plated on a printed circuit. The concept was originally introduced by Munson (1974) and Howell (1975). A patch antenna is a metal patch or shape above a ground plane. It can be of any shape that possesses a well-defined resonant mode. Typical shapes include circles, ellipses, annular rings, squares, rectangles, and triangles. The patches can be fed directly with strip lines or indirectly using electromagnetic coupling. They are arranged and fed in such a way as to achieve the desired electromagnetic field amplitude and phase necessary to achieve the desired antenna characteristics. For fixed point-to-point microwave radio applications, the patches are grouped evenly and symmetrically about a common feed point. The structure is very similar to the "pine tree" antenna described by Schelkunoff and Friis (1952). It usually achieves a uniform power distribution across the antenna aperture. These antennas are small (0.5–2 ft) and have rather poor radiation characteristics (restricted bandwidth, high loss, and excessive side lobes) when compared to conventional feedhorn-fed parabolic antennas. They are intended for unlicensed applications where aesthetics is more important than technical performance. In general, circular antennas are the most popular microwave antennas.

Circular ("parabolic") antennas usable for licensing by the FCC are classified as Class A or Class B (currently, no square "panel" antennas meet Class A or B standards and are only used in unlicensed bands) (Fig. 5.16).

Most Class A antennas have a shroud ("barrel") around the antenna to block side lobes emissions. Most Class B antennas are just plain parabolic antennas (with a radome to protect the feedhorn). Properties of Class A and B antennas are discussed later.

From an engineering perspective, parabolic antennas are usually classified (International Telecommunication Union—Radiocommunication Sector (ITU-R), 2005) as follows:

Grid: Useful up to about 2.7 GHz. The primary advantage is low wind loading. The primary disadvantage is its relatively poor side lobe performance.

Standard: Unshielded solid antenna. Average side lobe performance and front-to-back ratio on the order of 40–55 dB.

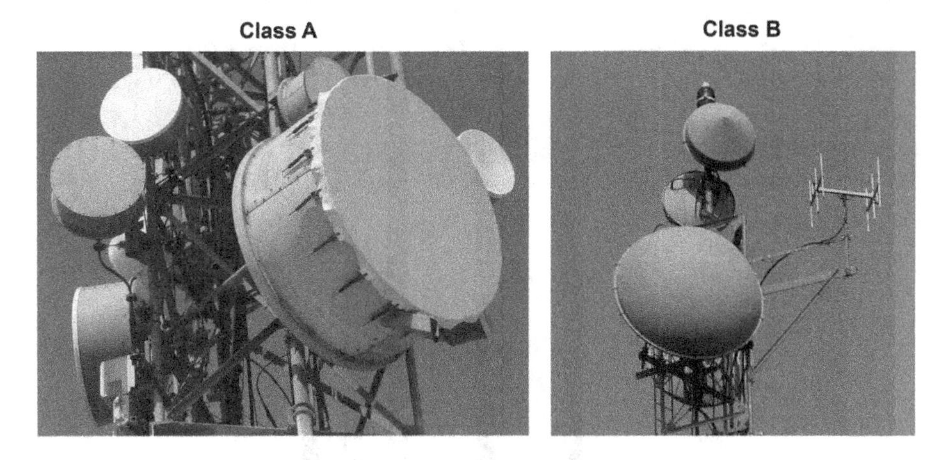

Figure 5.16 Class A and Class B antennas.

High performance: Shrouded dish that offers improved side lobe suppression and front-to-back ratio on the order of 70 dB.

Ultra high performance: Shrouded dish with optimized feedhorn, very high side lobe suppression, and front-to-back ratios in excess of 80 dB.

Fig. 5.17 graphs are the most popular circular antennas in the lower 6- and 11-GHz US frequency bands. For these charts, all transmit and receive antennas were counted separately. The charts are based on an unpublished report by Comsearch commissioned by the author.

From an installation engineering perspective, significant characteristics are aesthetics, radome type (and power loss under various conditions), wind loading, weight, and azimuth and elevation (tilt) adjustment. We will limit the discussions to transmission engineering characteristics.

Antennas are the devices that couple electromagnetic energy into and out of the air. The principle of antenna reciprocity, postulated by Lord Rayleigh and John Carson and formally proven by Sergei Schelkunoff (1943), states that the power transfer between an electromagnetic wave and an antenna is independent of the direction of power flow between them. Therefore, the significant electrical characteristics of an antenna are independent of whether it is transmitting or receiving.

The characteristics of an antenna are different at different distances from it. Very near the antenna, the dominant energy is nonradiating "reactive" fields. This is called the *reactive near field*. Farther away, but still near the antenna, is the radiating near field (sometimes called the *Fresnel region*). In this region, the energy radiates (it produces the antenna radiation pattern when viewed from the far field). Most of the energy is directly in front of the antenna. The energy in front of the antenna is roughly independent of the distance from the antenna. Much farther away from the antenna is the far-field (Fraunhofer) region. In this region, energy is distributed in a pattern that is dependent on the viewing angle relative to boresight. The pattern relative to boresight is independent of the distance from the antenna. At a given angle relative to boresight, energy is a function of the distance squared. For an aperture antenna, the transition point from reactive near field to radiating near field (Balanis, 2008) is nominally 0.62 $[(D^3/\lambda)]^{(1/2)}$ where D is the largest antenna dimension (diameter or width) and λ is the wavelength (both variables in the same distance units). The nominal transition point from radiating near field to far field (Friis, 1946) is $2D^2/\lambda$. These transition points are only rough guidelines. They are conservative or optimistic depending on the nature of the measurement of interest. This topic is discussed in more detail in Chapter 8.

Many antenna characteristics are important (Aubin, 2005): Boresight horizontal and vertical polarization gain, cross (opposite) polarization gain, input impendence or waveguide flange size and type, input return loss (or VSWR), directivity at different angles relative to boresight, efficiency, and overall radiation pattern. Probably the most important antenna parameter for the transmission engineer is far-field (boresight) gain. Gain is a measure of an antenna's ability to concentrate radiation in a given direction.

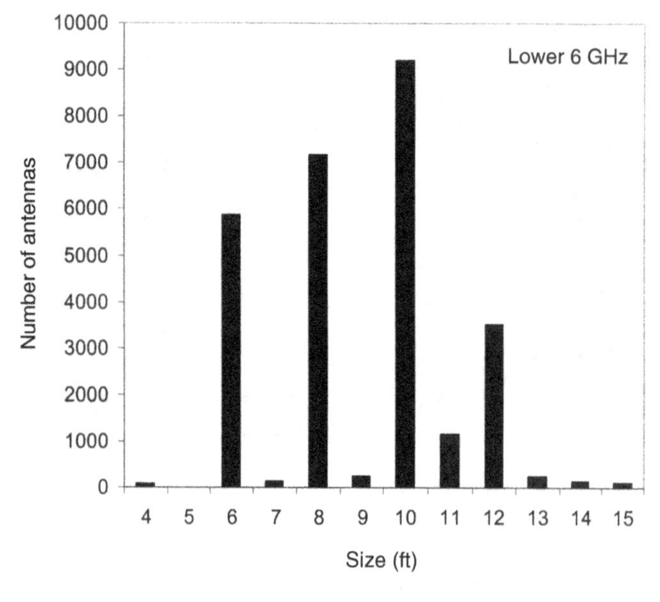

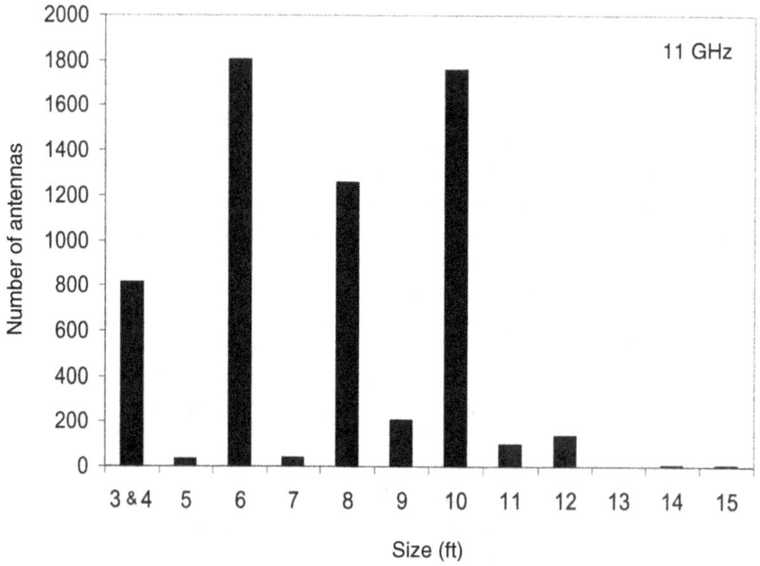

Figure 5.17 Antennas used on US Paths.

The gain of microwave antennas is usually compared to that of an isotropic radiator (a hypothetical radiator that transmits or receives energy uniformly in all directions). Isotropic gain is usually expressed as dBi (Fig. 5.18).

Occasionally, antenna gain is referenced to the maximum gain of a half wave dipole (dBd). Unlike an isotropic radiator, the half wave dipole radiation pattern is not uniform.

$$\mathrm{dBd} = \mathrm{dBi} - 2.15 \tag{5.21}$$

These two different references for antenna gain are used to define the effective radiated power (ERP) of transmitters connected to an antenna: ERP is the transmitter power in watts $\times$ antenna dipole gain as a power factor. EIRP is the transmitter power in watts $\times$ antenna isotropic gain as a power factor.

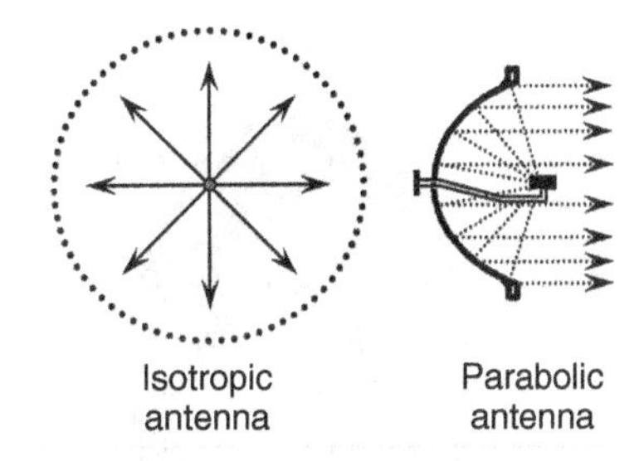

Figure 5.18 Isotropic and parabolic antennas.

$$\text{ERP(W)} = 10^{P(\text{dBW})/10} \times 10^{G(\text{dBd})/10} = 0.001[10^{P(\text{dBm})/10} \times 10^{G(\text{dBd})/10}] \tag{5.22}$$

$$\text{EIRP(W)} = 10^{P(\text{dBW})/10} \times 10^{G(\text{dBi})/10} = 0.001[10^{P(\text{dBm})/10} \times 10^{G(\text{dBi})/10}] \tag{5.23}$$

$$\text{ERP}(W) = \text{EIRP}(W) \times 10^{-2.15/10} = 0.61\text{EIRP}(W) \tag{5.24}$$

P = transmitter power (at the antenna input);
G = antenna gain.

Friis (1946) developed a simple antenna and transmission formula based on the concepts of isotropic radiation (radiation in all directions) and antenna effective area. Using his concepts, it can be shown (see Section 5.A.1) that the isotropic gain (dBi) of a parabolic antenna is a function of both frequency and area.

$$\text{Gain (dBi)} = 11.1 + 20\ \log F(\text{GHz}) + 10\ \log A(\text{ft}^2) + 10\ \log\left(\frac{E}{100}\right)$$

$$\text{Gain (dBi)} = 21.5 + 20\ \log F(\text{GHz}) + 10\ \log A(\text{m}^2) + 10\ \log\left(\frac{E}{100}\right) \tag{5.25}$$

F = frequency of radio wave;
A = antenna physical area;
E = antenna power transmission efficiency expressed as a percentage;
= antenna illumination efficiency.

If you double the frequency or the area, the antenna gain increases to 6 dB (and the 3-dB beamwidth is halved). Antenna catalog data typically lists antenna gain at only three frequencies (low, middle, and upper frequency). Gains at other frequencies may be estimated using the following formula:

$$G = G_{\text{ref}} + 20\log\left(\frac{F}{F_{\text{ref}}}\right) \tag{5.26}$$

G_{ref} = antenna gain (dBi) at F_{ref} (GHz);
F_{ref} = reference frequency (GHz);
G = antenna gain (dBi) of interest;
F = frequency of interest (GHz).

Practical parabolic antenna gains are listed in Table A.4. Notice that, in general, a change from a 1-ft antenna to a 2-ft antenna represents a 5-dB increase in gain. Changing from a 2- to 3 ft-, 3- to 4 ft-, or 4- to 6 ft-antenna increases gain approximately by 3 dB. Changing from a 6 ft- to 8 ft-, 8 ft- to 10 ft-, 10 ft- to 12 ft-, or 12 ft- to 15 ft-antenna increases gain approximately by 2 dB.

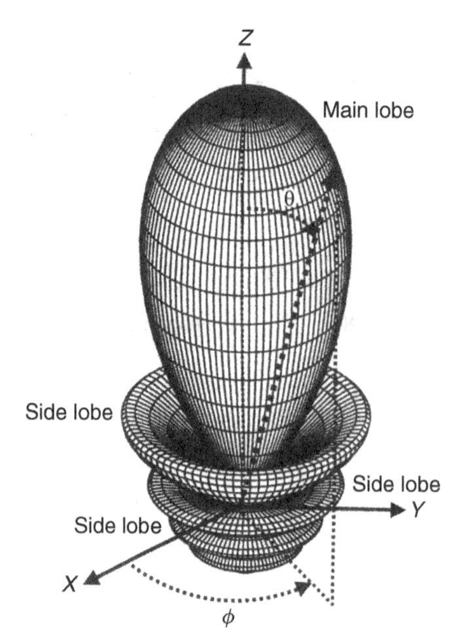

Figure 5.19 Circular antenna three-dimensional radiation pattern. *Source*: Reprinted with permission of Wiley.

Antenna illumination efficiency, E, defines antenna gain relative to an antenna which is fully ($E = 100\%$) illuminated. It does not represent a power loss. Antenna illumination efficiency, E, as well as most other antenna characteristics, are a function of the distribution of energy (illumination) across the face (aperture) as well as the shape of the antenna (Friis, 1948; Friis and Lewis, 1947; Hansen, 1981; Hollis et al., 1970; Silver, 1949; Western Electric Company, 1970). If two antennas have the same aperture shape, aperture illumination, and the same normalized dimension (width or diameter divided by wavelength), they are electrically identical.

The radiation pattern of an antenna is three-dimensional (Fig. 5.19). Sometimes, the antenna radiation patterns are displayed in three-dimensional Cartesian coordinates so that all the side lobes may be seen clearly. However, it should be remembered that the pattern is actually a function of the spherical coordinates as well as illumination (Fig. 5.20 and Fig. 5.21).

Most microwave radio paths are essentially parallel to the earth. All radio sites lie in a horizontal plane parallel to the earth. Therefore, when radios are coordinated (see Frequency Planning, Chapter 2) only the radiation pattern in the horizontal plane is needed. That is why an antenna radiation pattern is usually measured in a two-dimensional plane parallel to the earth. However, it should be remembered that in the absence of reflections from the feedhorn and terrain, the circular antenna pattern is the same regardless of the plane of measurement (as long as that plane includes the boresight axis). For the square antenna, as its radiation pattern is not symmetric relative to the boresight axis, two-dimensional radiation patterns will be different depending on the orientation of the antenna. As Silver (1949) observed, in a plane of symmetry, the radiation pattern is determined by the aperture dimension in that plane only.

The radiation pattern is a plot of antenna gain as a function of angle relative to the angle of maximum gain (Fig. 5.22). Usually, the radiation pattern is referenced to the maximum antenna gain (at boresight). Both main and cross-polarization responses are usually plotted. Button hook feedhorn high performance category A antennas often have different patterns on alternate sides of boresight. Installation of the correct feedhorn can be critical. Commercial antenna patterns display envelopes (straight lines that touch the envelope peaks and ignore the pattern nulls) rather than the actual antenna pattern.

Circular antennas

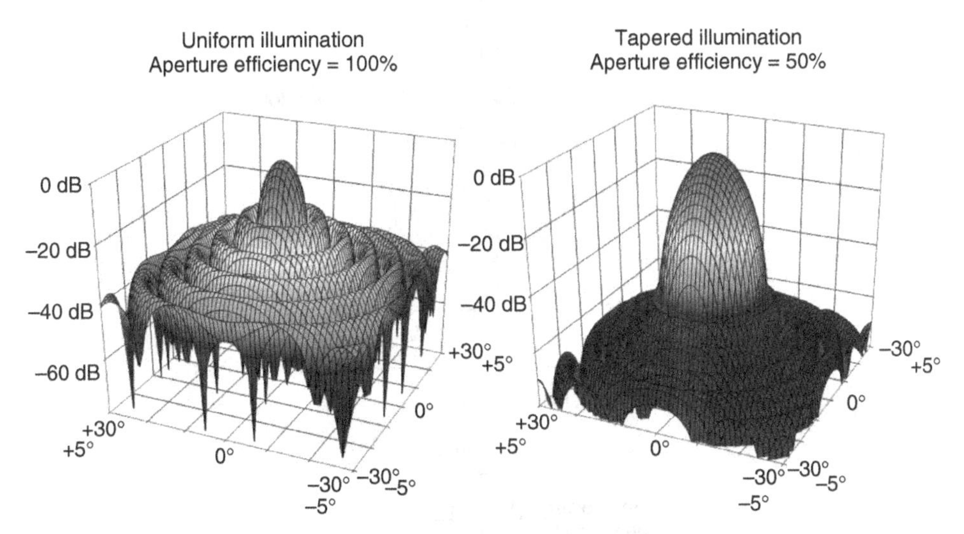

2 ft diameter, 5 GHz unlicensed band (5.5 GHz), D/λ = 11 (+30° to –30°)
10 ft diameter, lower 6 GHz licensed band (6.2 GHz), D/λ = 66 (+5° to –5°)

Figure 5.20 Circular antenna radiation as a function of efficiency.

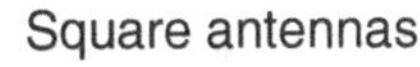

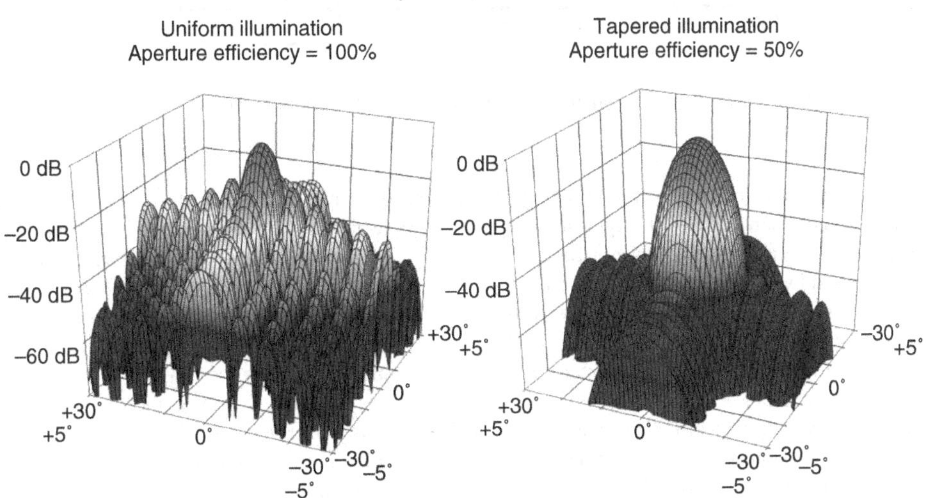

2 ft width, 5 GHz unlicensed band (5.5 GHz), D/λ = 11 (+30° – –30°)
10 ft width, lower 6 GHz licensed band (6.2 GHz), D/λ = 66 (+5° – –5°)

Figure 5.21 Square antenna radiation as a function of efficiency.

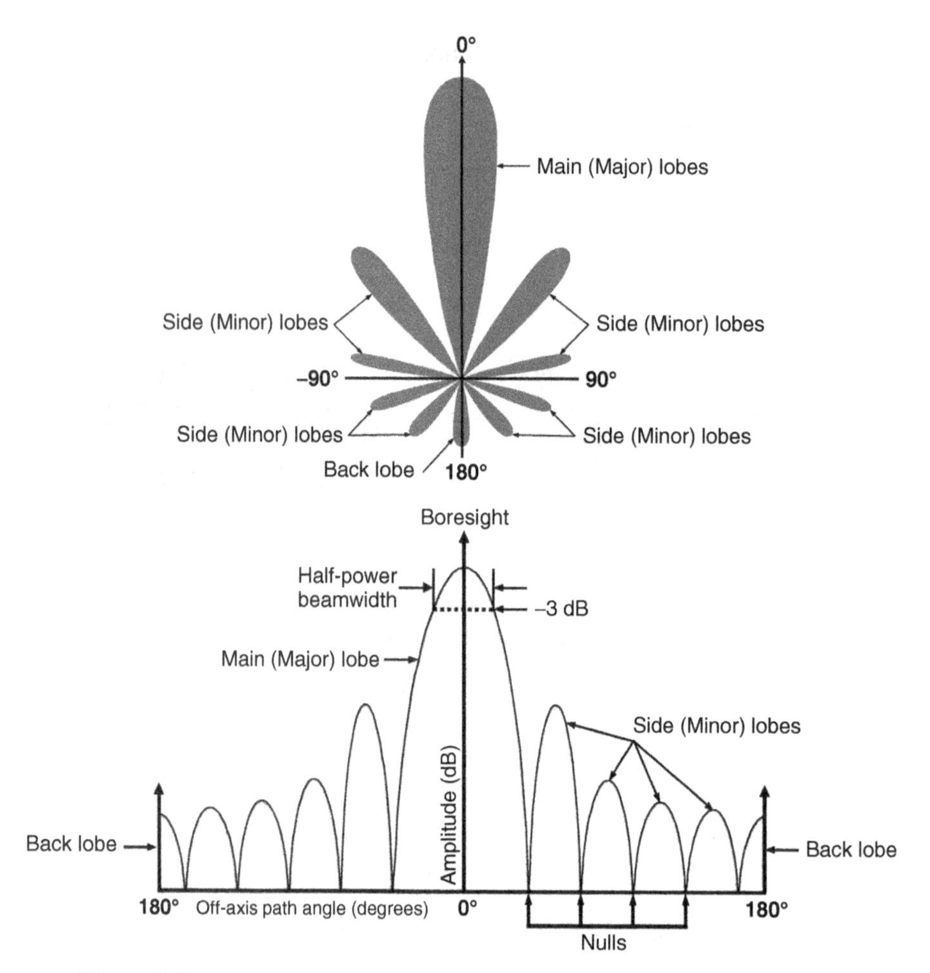

Figure 5.22 Two-dimensional polar and cartesian antenna radiation patterns.

Antennas are not located in free space; they are located near other objects that may be on or near the paths. A common engineering question is "Does an object near the path affect the antenna or path performance?" That question may not be easy to answer. As can be observed in Chapter 8, the signal power received by the receive antenna is the integral over the receive antenna aperture of energy from the transmit antenna. Unless a structure between the transmit and receive antennas significantly shadows the transmit or receive antenna, there will be little received signal power loss. (Power transfer occurs within the cylinder formed by the lines between the edges of the transmit and receive aperture antennas. See Chapter 8, Antenna to Antenna Near Field Coupling Loss, for details.) However, the on-path structure can cause reflections that affect received signal quality. These reflections can also significantly reduce side lobe suppression and cause unexpected frequency interference. A common example is guy wires near an antenna (Davies Hurren Copeland, 1985). They have virtually no effect on received signal level. However, if they extend in front of the transmit antenna, they can cause appreciable side lobe degradation. If the antennas are moved so that the guy wires are at the edge of the antenna or beyond, the side lobe reduction effect is negligible.

Since most microwave radio paths are essentially in a plane parallel to the earth, this two-dimensional pattern is adequate to describe the antenna characteristics as they influence and are influenced by other radiators. For circular antennas (in the absence of reflections), the antenna pattern is actually the same for any plane surface that intersects the line perpendicular to the plane of the antenna ("boresight"). For most long microwave paths, the horizontal plane antenna pattern is adequate. However, for short high

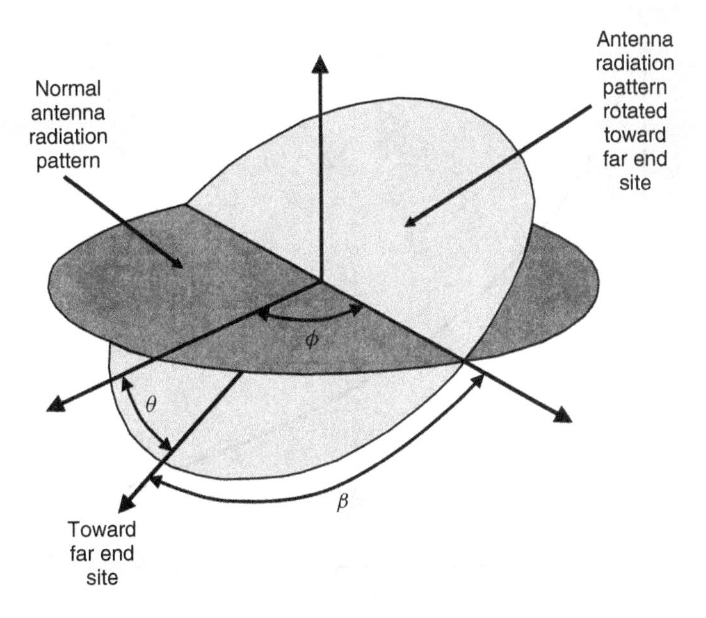

Figure 5.23 Angle associated with foreign site on short paths.

frequency paths, often a foreign site is at a significantly different elevation (at angle θ below the horizontal plane) than the desired transmit or receive antenna. If the antenna is a circular parabolic antenna, the three-dimensional antenna pattern is a three-dimensional torus of the two-dimensional pattern rotated about the boresight axis. (Reflections can have a significant effect on antenna patterns and are ignored in this analysis. While paths between desired transmitters and receivers are engineered to have adequate clearance, paths toward foreign antennas typically have minimal clearance and reflections are blocked.) If the two-dimensional antenna pattern is tilted toward the site of interest, the usual radiation pattern can be used as long as the actual angle toward the site in this tilted plane β is known (in general, it is different than the angle ϕ of the projection of the path to the foreign site in the horizontal plane) (Fig. 5.23).

$$\begin{aligned}\beta &= \text{angle toward far end site measured in the plane intersecting the site}\\ &= \text{arc}\ \cos(\cos\theta \quad \cos\phi) \qquad (5.27)\end{aligned}$$

$\phi =$ bearing of the far end site measured in the horizontal plane;
$\theta =$ vertical angle of the site relative to the horizontal plane.

Beamwidth, side lobe suppression, front-to-back ratio, and cross-polarization discrimination are common parameters of interest. Beamwidth and side lobe suppression are functions of illumination efficiency (see Chapter 8 for details) (Fig. 5.24).

Antenna radiation patterns are measured with other antennas in the far field. At radio junction stations, the coupling between antennas on the same tower is not generally known yet is of interest since intrasystem interference must be considered in frequency planning. Coupling between antennas for arbitrary relative orientation is a function of fringing effects not modeled by antenna theory. Experimental data is useful as a guideline (Fig. 5.25).

Cross-polarization discrimination is a function of relative angular orientation both in relative rotation and in the horizontal plane (Fig. 5.26).

Cross-polarization discrimination is based on $-20\log[(\text{upper limit})(1 - \sin\Delta\phi)]$ where $\Delta\phi$ is the angular misalignment of the two antennas (Kizer, 1990). The figure illustrates the importance of feedhorn rotational alignment to cross-polarized antenna systems. Feedhorn angular misalignment is based on

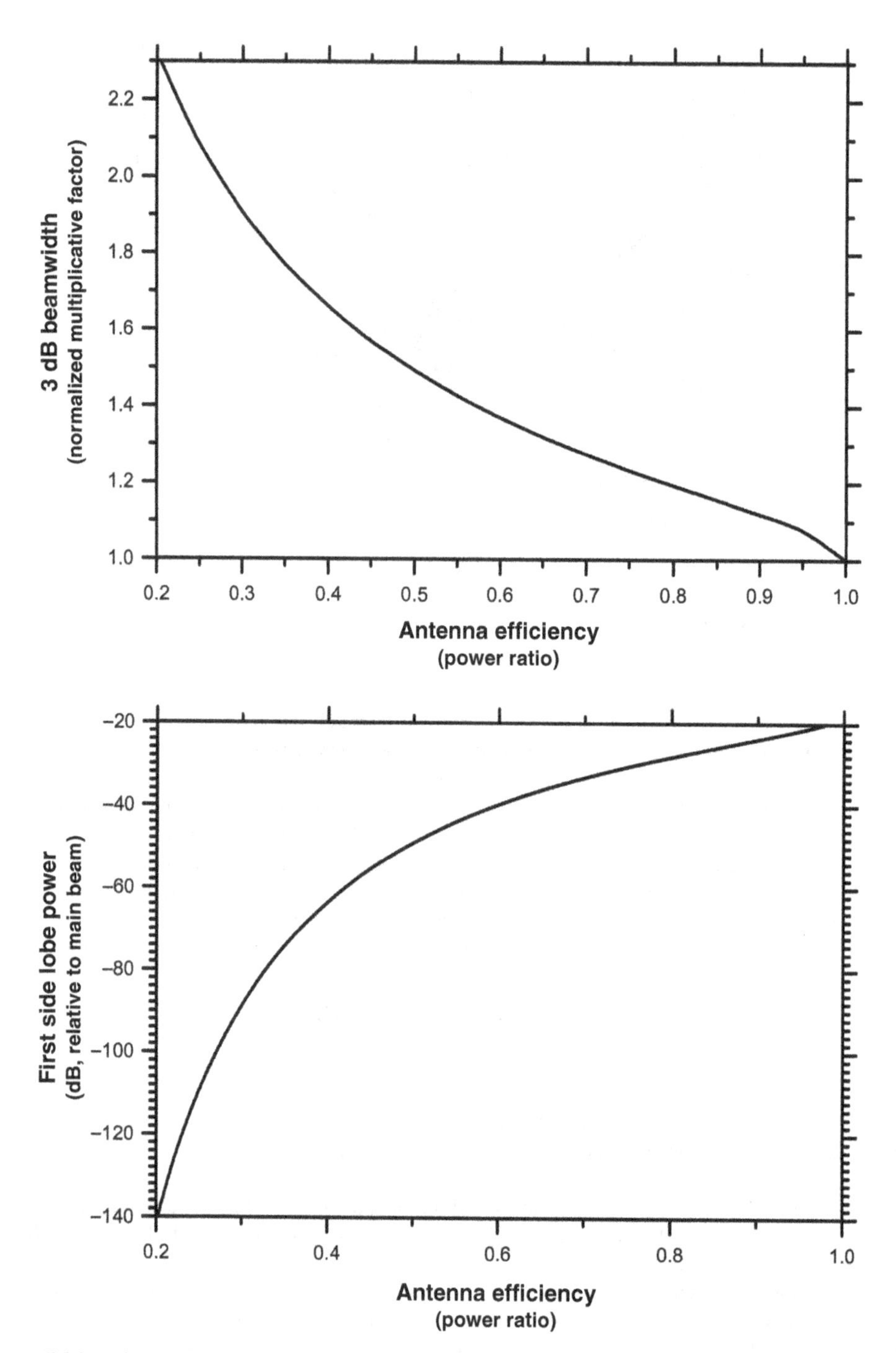

Figure 5.24 Circular antenna 3-dB beamwidth and first side lobe as a function of illumination efficiency.

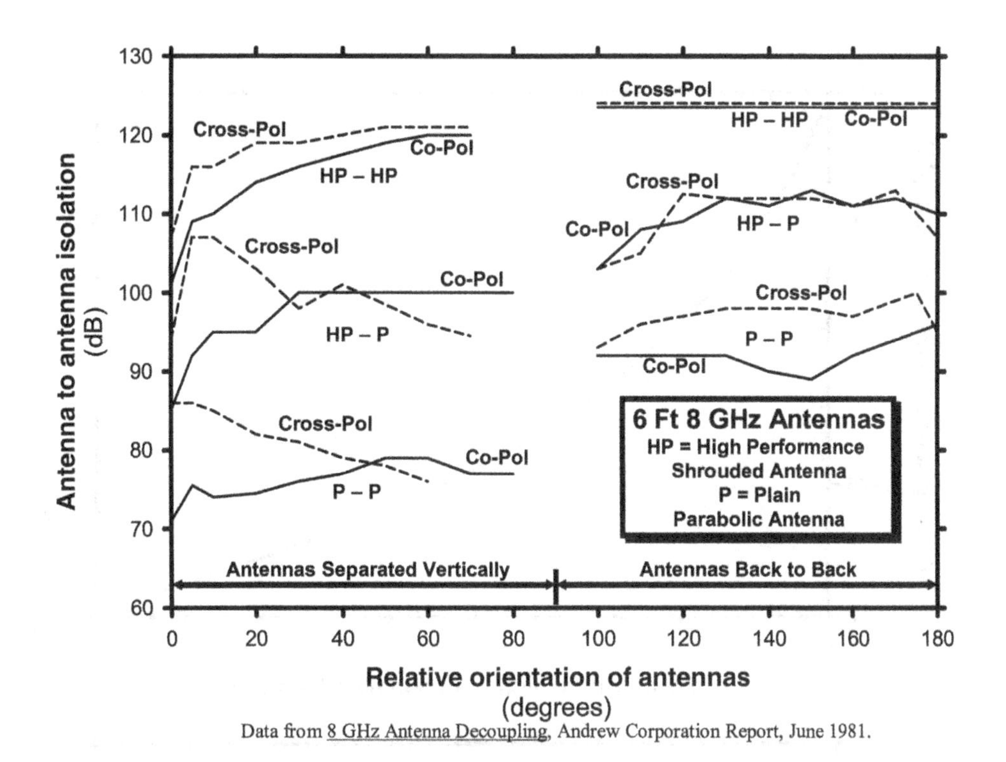

Figure 5.25 Antenna to antenna near field power coupling.

the averages of antenna data sheets. The figure illustrates the importance of antenna horizontal angular alignment to cross-polarized antenna systems.

In the United States, the FCC sets standards for radiation from antennas (Fig. 5.27) (Mosley, published yearly). The FCC classifies the antennas as Class A (high standard) or Class B (minimum quality). Most operators try to install Class A antennas as they can be required to upgrade to Class A antennas if they cause interference to or receive interference from an existing licensed station or a station for which a license application is in process (Mosley, published yearly). FCC (and NTIA) rules effectively limit the size of Class A antennas (Mosley, published yearly) (Table 5.1).

5.6 NEAR FIELD

To this point the assumption has been made that all antennas are a long way away from each other. In this situation the antennas are operating in the "far field" or Fraunhofer region. The region is roughly when the antennas are separated by a distance of at least $d = 2W^2/\lambda$ where W is the widest dimension (in the E field of interest) of the larger antenna and λ is the operating wavelength. The far-field radiation patterns are based on the assumption that all radiation components of the transmit antenna irradiated aperture arrive at the receiving antenna at the same angle.

As the antennas come closer together, the different components of radiation from the transmit antenna appear at the receive antenna at different angles. The effect of this is that the antenna gain is reduced relative to the far-field gain. In this region, called the *near field* or *Fresnel region*, most of the transmit antenna energy is concentrated directly in front of the antenna (Fig. 5.28, Fig. 5.29, and Fig. 5.30).

$$\Delta = \frac{d}{(2D^2/\lambda)} = \text{normalized distance parameter of circular antenna} \tag{5.28}$$

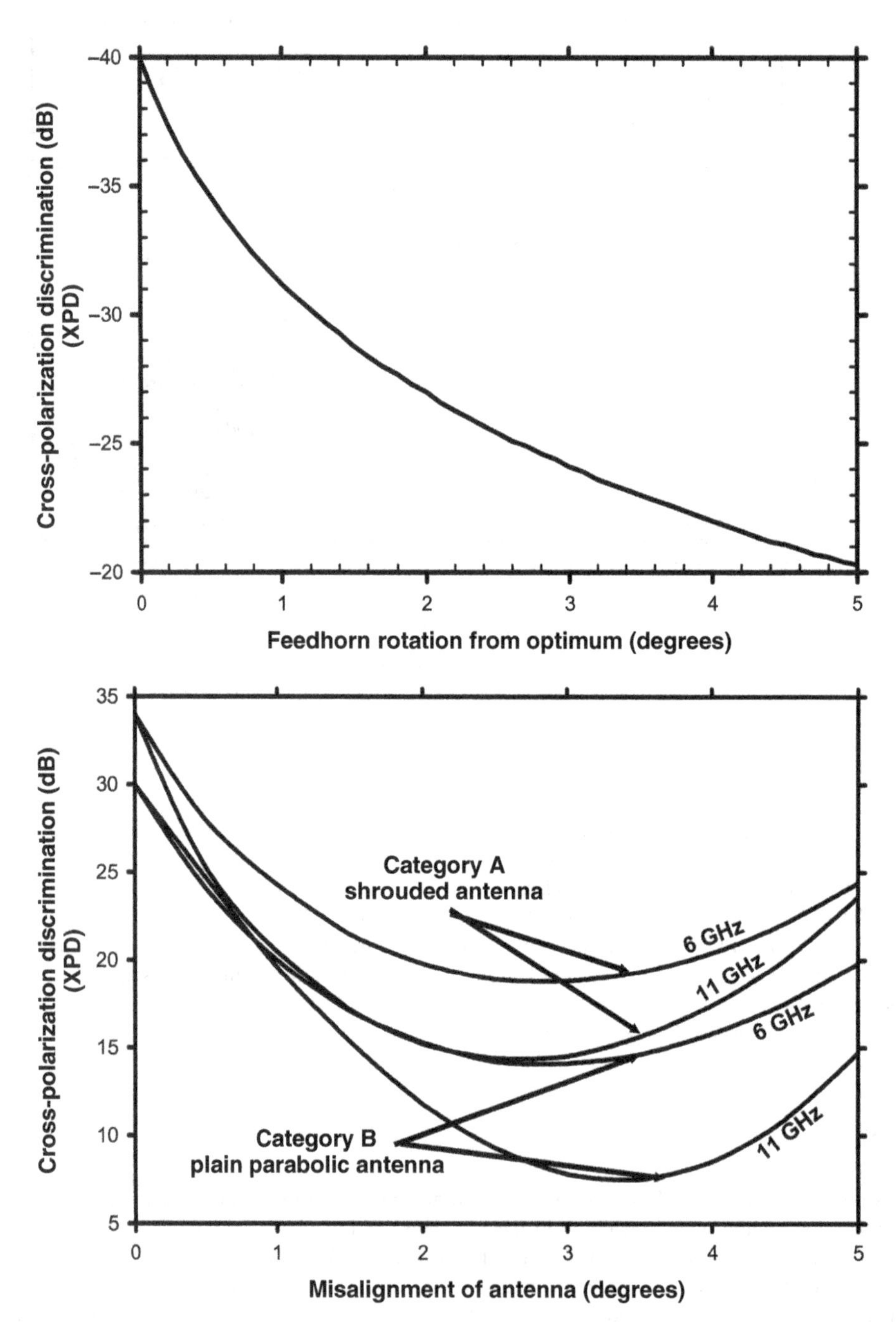

Figure 5.26 Cross-polarization discrimination versus relative feedhorn rotation or feedhorn angular misalignment.

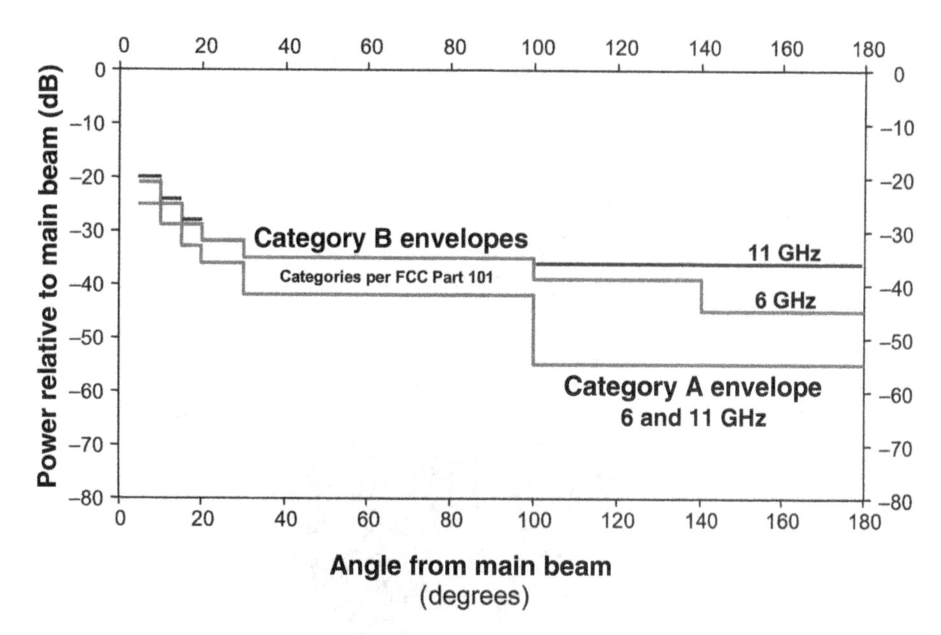

Figure 5.27 FCC Class A and Class B antenna requirements.

TABLE 5.1 Class A Antenna Size Limitations

Frequency Band, GHz	Smallest Antenna, ft	Regulatory Agency
2.200–2.300	4	NTIA Federal Government
4.400–4.940	4	NTIA Federal Government
5.725–5.850	1	FCC Part 15 Unlicensed Band
5.925–6.425	6	FCC Part 101 Licensed Band
6.525–6.875	6	FCC Part 101 Licensed Band
6.875–7.125	6	FCC Part 74 Licensed Band
7.125–8.500	4	NTIA Federal Government
10.55–10.68	2	FCC Part 101 Licensed Band
10.7–11.7	3[a]	FCC Part 101 Licensed Band
14.5–15.35	4	NTIA Federal Government
17.7–19.7	2	FCC Part 101 Licensed Band
21.2–23.6	1	FCC Part 101 Licensed Band

[a] A Class B 2 ft antenna can have most of the properties of a Class A antenna when using limited power. See Section 5.A.6.

$$\Delta = \frac{d}{(2W^2/\lambda)} = \text{normalized distance parameter of square antenna} \quad (5.29)$$

$d =$ distance from the center of the antenna to the point of interest;
$D =$ diameter of the circular antenna (aperture);
$W =$ width of the rectangular antenna (aperture with $\zeta = 0$);
$\zeta =$ angular rotation (usually 0 or 45 degrees) of a rectangular antenna (aperture);
$\zeta =$ 0 when one side of the rectangular aperture is parallel to the plane of the earth;
$\lambda =$ radio free space wavelength.

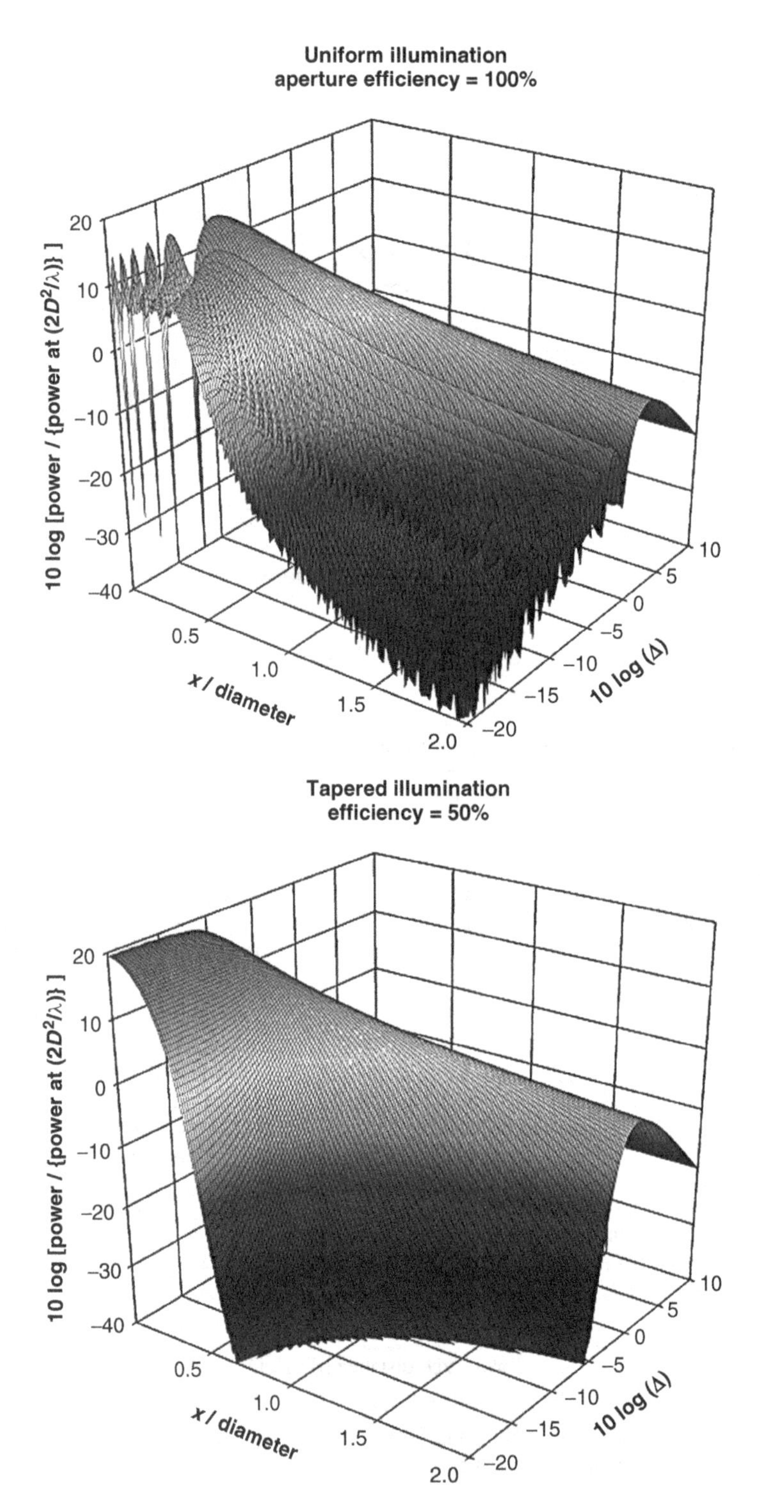

Figure 5.28 Circular antenna near-field power density ($D/\lambda = 66$).

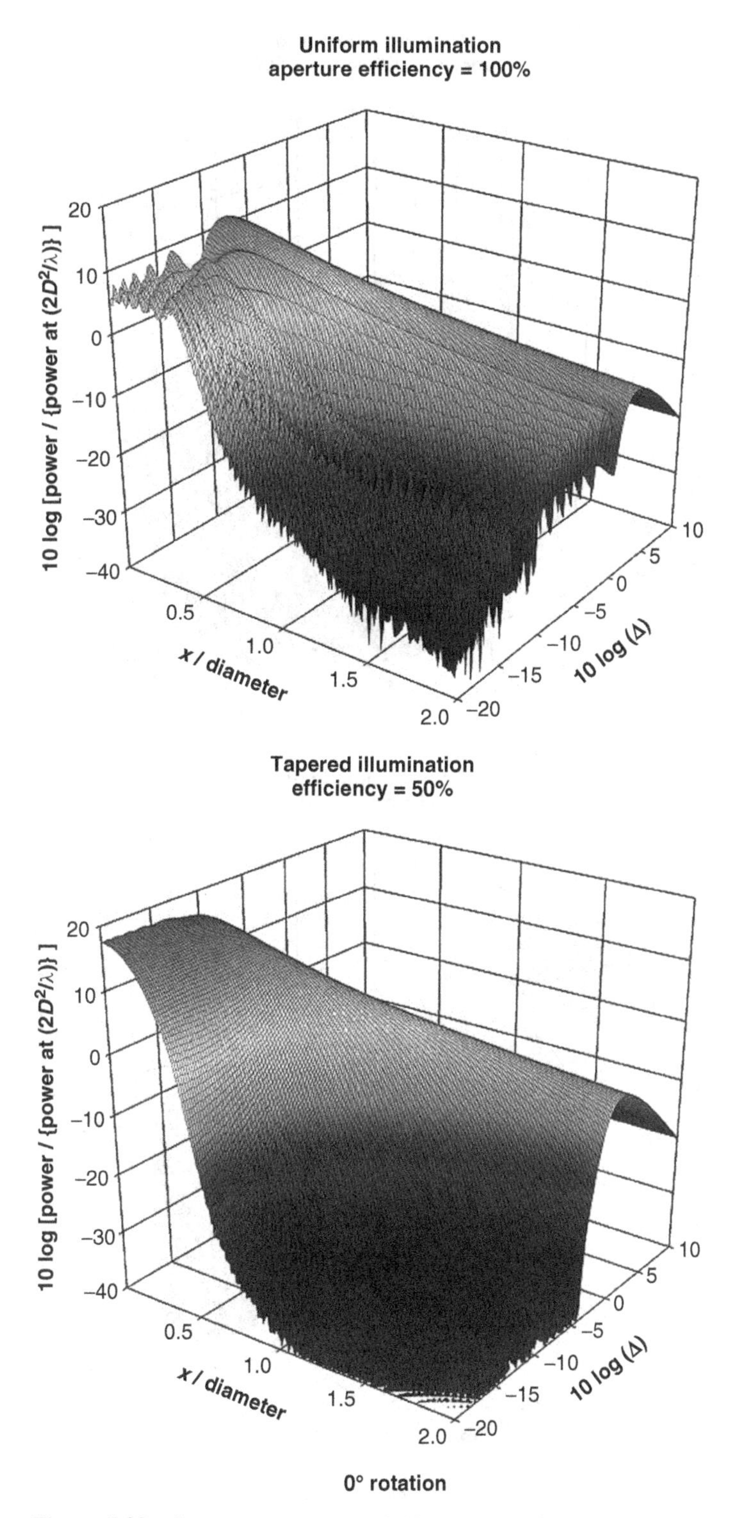

Figure 5.29 Square antenna near-field power density ($D/\lambda = 66$).

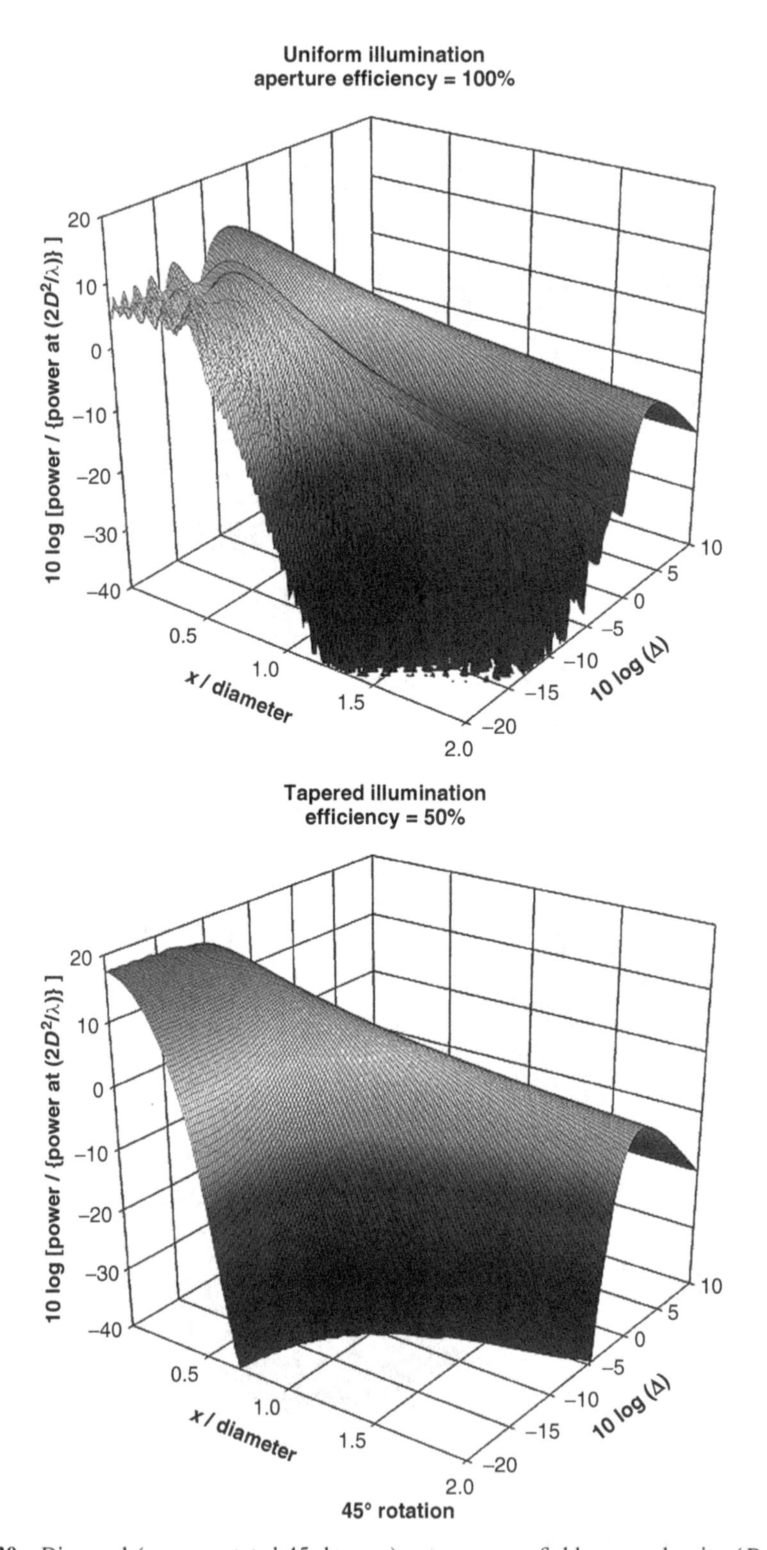

Figure 5.30 Diamond (square rotated 45 degrees) antenna near-field power density ($D/\lambda = 66$).

All distance units are in the same unit of measure. In these graphs, the logarithmic depth axis is the distance normalized to the far-field crossover point [$10 \log(\Delta = 1) = 0$]. The longitudinal axis is the distance parallel to the plane of the antenna normalized to the half width of the antenna ($D/2$ or $W/2$). Zero represents the center of the antenna and 0.5 is the edge. The vertical axis is the power (in decibel) relative to the power at the far-field crossover point ($10 \log(\Delta = 1) = 0$). In all cases in the near field, the maximum (peak) power is directly in front of the center of the antenna. Near-field power is of considerable interest from a public safety perspective. This region is covered in some detail in Chapter 8 and Appendix B.

5.7 FUNDAMENTAL ANTENNA LIMITATIONS

All fixed point-to-point antennas are circular parabolic dishes or square flat panels. They are aperture antennas whose major transmission characteristics are defined by their aperture shape and the distribution of radio energy across that aperture. These antennas have theoretical limitations as well as limitations not currently addressed by theory.

Limitations not addressed by theory include coupling between parallel or back to back near-field antennas. This coupling is dominated by fringing and physical dispersive effects. Cross-polarization characteristics require actual measurement. Current antenna theories do not address cross-polarization effects or near-field antenna to antenna coupling effects. In addition, the path can adversely affect cross-polarization isolation. Reflections from terrain, rain, and multipath fading, all cause polarization rotation that results in loss of cross-polarization discrimination in linear polarized antennas.

There are three primary theoretical limitations. The first is due to surface roughness and deformation (Ruze, 1966). With aperture antennas, greater gain is achieved with larger physical area. There is no theoretical limitation to the performance of an aperture antenna but there is a practical limit to their size (and surface accuracy). Boresight gain and side lobe reduction are a direct function of size and aperture illumination. If the surface has mechanical imperfections (primarily roughness and deformation), the side lobe and cross-polarization characteristics are degraded. With larger size comes challenges from manufacturing tolerances, gravity, wind, and thermal stress. The size limit depends on the application (and the user's available funding). Currently, the large single aperture microwave antennas used in radio telescope applications can achieve up to 80-dB isotropic gain (Cogdell et al., 1970). For conventional fixed point-to-point applications, the antenna isotropic gain is usually between 30 and 50 dBi.

The second limitation is size relative to operating signal wavelength (Hansen, 1981). Aperture antennas need to be large relative to operating wavelength. Small antennas suffer from lowered efficiency (boresight gain), degraded side lobe performance, and reduced operating bandwidth (when compared to larger antennas) (Wheeler, 1993). These limitations are especially noticeable for small microwave antennas.

Lastly, there are fundamental limitations on the relationship between side lobe performance and boresight gain (Hansen, 1981). Most commercial microwave antennas use power illumination that is spread with equal phase across the aperture. Side lobes are reduced by tapering the (equal phase) illumination to reduce illuminated power near the edge of the antenna. This produces the greatest boresight gain for a given illumination efficiency. Increased side lobe reduction can be achieved by varying the phase as well as the amplitude of the aperture illumination. (Currently, there is much theoretical work using antennas composed of many small antenna "patches" used to vary spot amplitude and phase.) This technique is termed *superdirectivity* (in the past this was referred to using the misleading term, *supergaining*). Theoretically, any side lobe pattern can be produced by an aperture antenna (by manipulating illumination phase) but the boresight gain and operating bandwidth will be reduced (Hansen, 1981) when compared to equal phase illumination.

5.8 PROPAGATION

The transfer of energy between transmit and receive antennas occurs by electromagnetic waves. Maxwell's equations require that in free space the wave be composed of electric and magnetic waves oriented perpendicular to each other and the direction of propagation (Kerr, 1951) (Fig. 5.31).

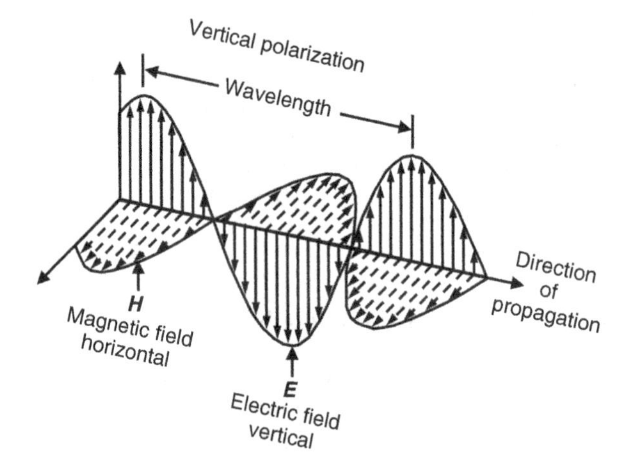

Figure 5.31 Electromagnetic radio wave.

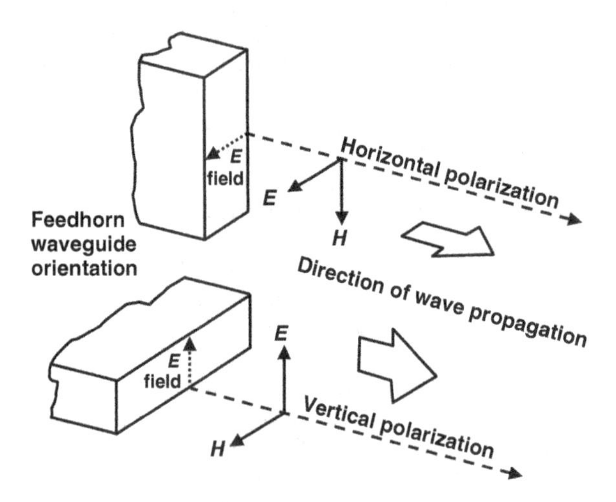

Figure 5.32 Radio wave polarization.

Radio waves are classified by polarization. If the electric field of the wave is horizontal, the wave is horizontally polarized. If it is vertical, it is vertically polarized. If the polarization is changing between vertical and horizontal periodically, the signal is circularly polarized if vertical and horizontal electrical fields are the same, and elliptical if one is larger than the other (Fig. 5.32).

Under unobstructed propagation conditions, the power loss between the transmit and the receive antennas is called the *free space loss*.

$$\text{Loss (dB)} = 96.6 + 20\log F(\text{GHz}) + 20\log\ D(\text{miles}) > 0$$

$$\text{Loss(dB)} = 92.4 + 20\ \log\ F(\text{GHz}) + 20\ \log\ D(\text{km}) > 0 \qquad (5.30)$$

$F =$ frequency of radio wave;
$D =$ path length.

As we will see in other chapters, atmospheric and terrain factors can significantly change the actual loss experienced on a radio path. Reflections from the terrain and nearby structures are addressed in Chapter 13. Abnormal atmospheric refractive index effects are covered in Chapter 12. Multipath fading caused by atmospheric layering is covered in Chapter 9. Rain fading is addressed in Chapter 11. However, other factors can also be significant.

The atmosphere contains pollutants described as aerosols and hydrometeors. Aerosols are particulate matter suspended in the atmosphere with diameter of 1 micron (1 micron = 1 μm = 1 micrometer = one-millionth of a meter) or less. Examples include smog, smoke, haze, clouds, fog, and soil. Hydrometeors have a diameter greater than 1 μm. They include mist, rain, freezing rain and ice pellets, snow, hail, ocean spray, clouds, fog, dust, and sand.

Fog and rain are addressed in Chapter 9. Attenuation by ice (and snow) is significantly less than by rain (Ishimaru, 1978). With the exception of sleet (typically treated as rain), ice is usually limited to an accumulation on antenna radomes and is not deep enough to significantly attenuate the radio signal (as contrasted with rain which can extend for several miles over a path).

For most areas of the world, the climate is relatively humid or arid. Humid areas subject to rain are addressed in Chapter 9. Arid areas subjected to soil, dust, or sand storms are the opposite extreme. Soil, dust, and sand particles are chemically similar and have similar highly irregular shape. Soil particles have diameters less than 1 μm. Dust is soil particles between 1 and 60 μm in diameter. Fine dust is between 1 and 10 μm. Course dust is between 10 and 60 μm. Sand is greater than 60 μm in diameter. All these particles are significantly smaller than the radio wavelength and have similar propagation characteristics.

Dust and sand storm attenuation is a function of particle density as well as the particle size distribution (Ahmed, 1987). This is specified as a visual distance inside the storm. Attenuation and cross-polarization performance of circularly polarized signals are much more sensitive to sand and dust storms than are linearly polarized signals. Except for very dense storms (visibility of less than 10 m), linearly polarized signal attenuation is negligible for frequencies below 30 GHz.

Figure 5.33 was created from the results of research by Chen and Ku (2012) for vertically polarized signals. Other researchers have used other assumptions and derived results which estimate greater (Ahmed et al., 1987) or less (Dong et al., 2011; Goldhirsh, 2001) attenuation for a given visibility. The results of other researchers vary from Chen and Ku's by ±40% to ±80%. The Chen and Ku results represent an average of current research. They may be estimated from the following equations.

$$A_1 = 9.286 + 0.2911F + 0.0001426F^3 - \frac{102.5}{F} \tag{5.31}$$

$$\log(A) = \log(A_1) - 1.25\log(V) \tag{5.32}$$

$$A = 10^{\log(A)} \tag{5.33}$$

$A =$ path attenuation (dB/km);
$F =$ frequency (GHz), $10 \leq F \leq 100$;
$V =$ visibility (m).

Small amounts of water increase dust propagation attenuation (5% moisture increases attenuation at 11 GHz by 75% relative to dry dust attenuation). Horizontal polarization has about twice the attenuation of vertically polarized signals. Linear signal cross-polarization discrimination can be significantly affected by dust and sand even for relatively low frequencies (Ahmed et al., 1987; Chen and Ku, 2012; Dong et al., 2011; Ghobrial and Sharief, 1987; Goldhirsh, 2001).

5.9 RADIO SYSTEM PERFORMANCE AS A FUNCTION OF RADIO PATH PROPAGATION

If the atmosphere is homogeneous (e.g., well stirred by wind or rain), propagation between the transmitter and the receiver is along one path. However, if the atmosphere is allowed to stratify (e.g., during quiet summer nights), invisible vertical microlayers of slightly different temperatures and humidity levels can form. These thin atmospheric layers parallel to the earth provide multiple propagation paths between the transmitter and the receiver. Signals traveling these paths take different time to propagate to the receiver. During times when the atmosphere is quiet, the receive antenna may receive two or three (or more) signals from the transmitter over slightly different paths. These signals are exact replicas of the originally

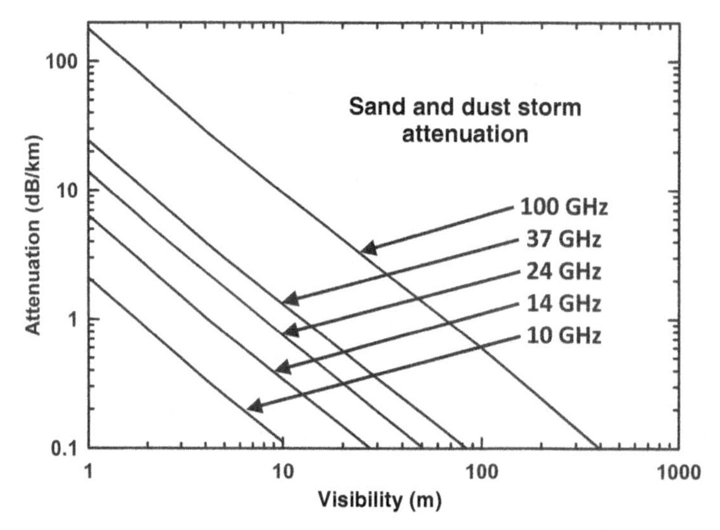

Figure 5.33 Dust storm path attenuation.

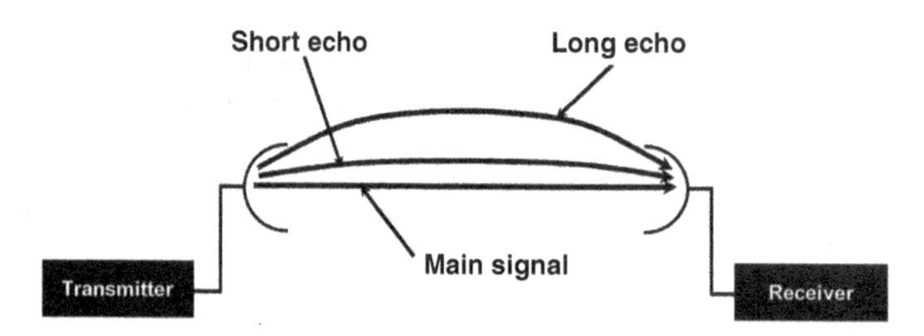

Figure 5.34 Desired and multipath signals.

transmitted signal but delayed in time. For this reason, they are often termed *echoes*. Typically, the main received signal occurs over a fairly direct path from the transmit antenna. The other signals typically occur over slightly longer paths that are physically slightly above the main path (Fig. 5.34).

Most atmospheric multipath occurs because of paths at elevations slightly above the main path. The delayed signal's angle of arrival is typically slightly greater (0.25°–2°) than the angle of arrival of the main signal. The difference in path length between the main signal and a short echo is less than a foot (less than a nanosecond in time) for a typical 26 mile path. The long echo path difference is usually 9 or 10 ft (9 or 10 ns) (Rummler, 1979, 1980) for a typical 26 mile path. When the different echo signals combine, the resultant signal is a distorted received signal (Fig. 5.35).

5.9.1 Flat Fading

If the primary echo is a short delayed echo, the result is an enhancement ("upfade") or reduction ("downfade") of the composite received signal power ("received signal level") that is essentially constant ("flat") in frequency across the radio transmission channel. This type of fading is termed *flat fading* because it represents an overall depression in received signal level but otherwise causes no distortion of the received signal (Fig. 5.36).

Flat fading is also termed *scintillation* fading. This multipath fading increases as the path length increases. It is the same mechanism that causes star light to "twinkle" at night (although in the case of light, the multipath is caused by air turbulence rather than layering). The visual effect can be seen by

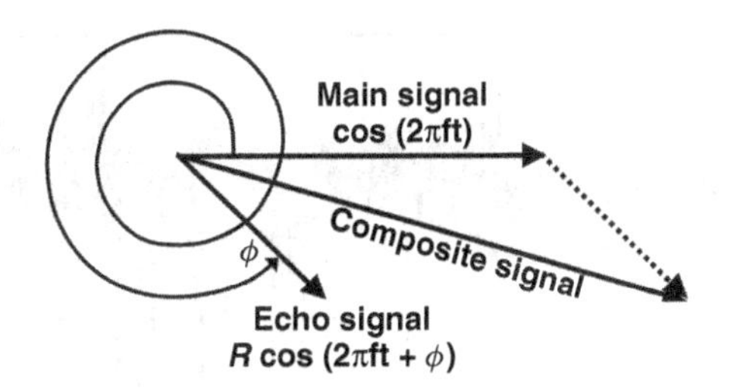

Maximum composite signal = 20 log (1 + R)
Minimun composite signal = 20 log (1 − R)
0 <= R <= 1

Figure 5.35 Result of multiple received signals.

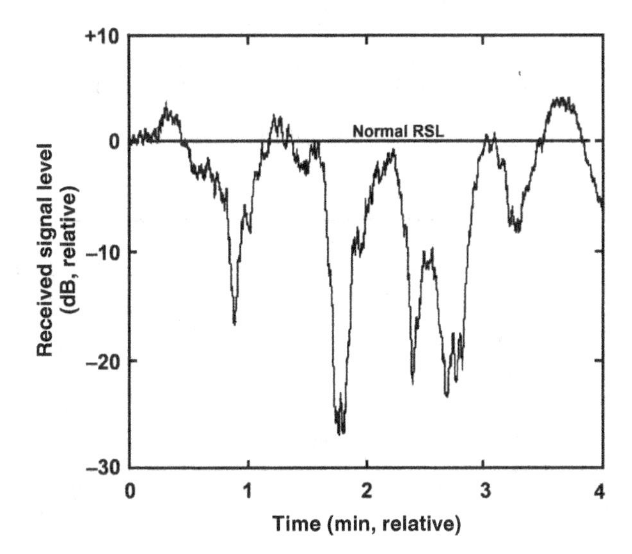

Figure 5.36 Example of flat fading.

an airplane traveler when landing in an urban airport on a quiet evening. The lights near the airport are constant but the lights farther and farther away from the airport flicker more and more.

Flat fading usually occurs at night during summer and fall (Fig. 5.37). However, if the fading is due to earth surface reflective fading (as may occur in the desert), the fading usually occurs during daylight hours (because of thermally induced reflective path changes).

Since up and downfades are relatively rare events, a radio path is typically designed so that upfades are mitigated by "headroom" and downfades are minimized by "flat fade margin" (Fig. 5.38).

For flat fading, the relationship between radio BER and flat fade margin is well defined. Flat fading is minimized by designing appropriate flat fade margins (and occasionally adding diversity techniques). Typical fade margins for lower 6- and 11-GHz paths in North America are summarized in Figure 5.39. In these graphs, each simplex path is counted separately. The graphs are based on an unpublished report by Comsearch commissioned by the author. Later in this chapter, frequency band statistics are graphed by pairs of (duplex) paths.

The number of paths for greater than 60 and 70 are the cumulative number of paths with fade margin greater than 60 or 70 dB, respectively.

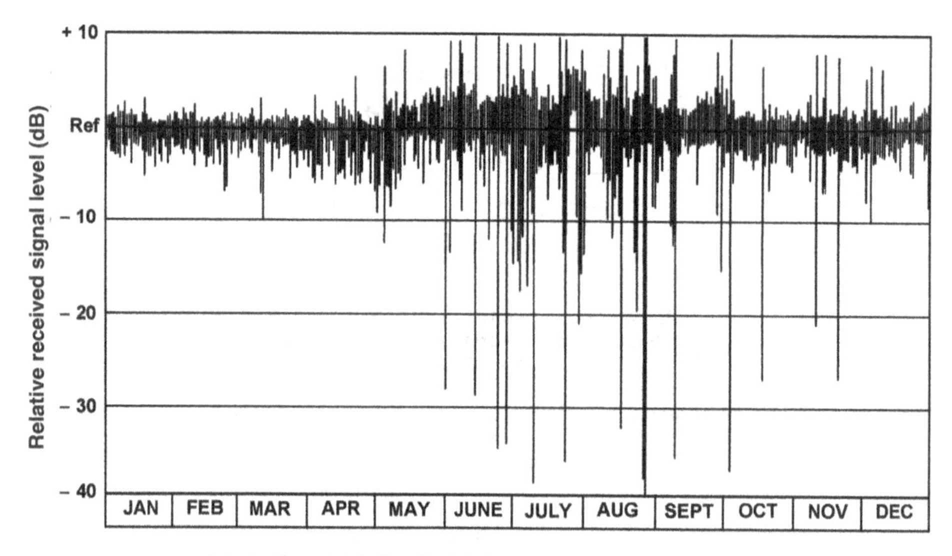

Figure 5.37 Seasonal flat fading.

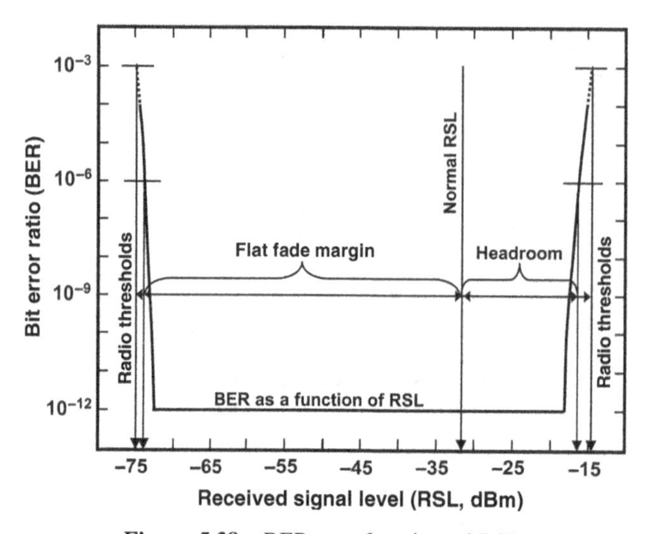

Figure 5.38 BER as a function of RSL.

5.9.2 Dispersive Fading

If the primary echo is a long delayed echo, the result is a composite received signal that has a very narrow frequency dependent enhancement or depression when viewed over the radio transmission channel. This type of fading is termed *dispersive fading*. Since the echo produces such a narrow distortion across the radio channel, the effect on received signal power is insignificant. However, the shape (amplitude vs time) distortion of the digital signal can be quite significant.

Viewed in the frequency domain, the distortion is a significant amplitude and group delay discontinuity. In the time domain, the distortion is more obvious. During dispersive fading, the typical echo delay is about 9 or 10 ns (Rummler, 1979, 1980) (no, not the often quoted 6.3 ns) behind (or ahead) of the primary signal (based on measurements on a 26.4 mile path in Atlanta, Georgia). Path delays of 5–20 ns were

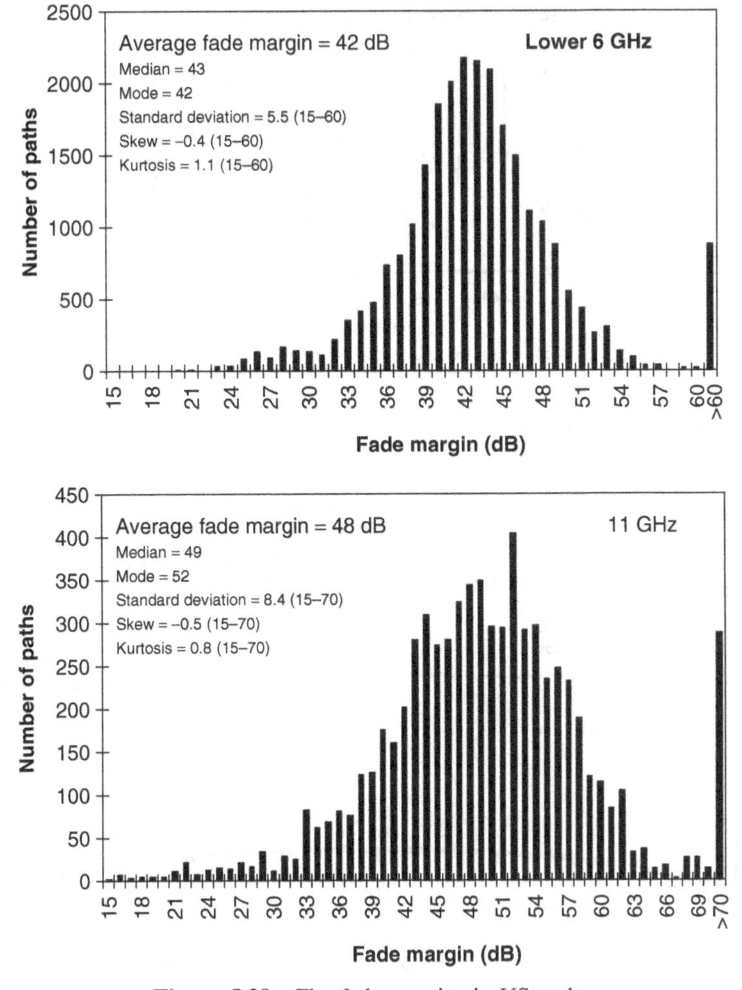

Figure 5.39 Flat fade margins in US paths.

measured on this path (Rummler, May-June 1979, Introduction and Table 1). When both signals are demodulated at the receiver, the digital pulse is significantly widened or "dispersed."

Dispersive fading is usually produced by relatively long atmospheric echoes. However, it can also be caused by reflections from the terrain for paths with excessive clearance and from off-path structures.

Dispersive fading has no relationship to received signal level (Giger and Barnett, 1981). Figure 5.40 shows the relationship between RSL and BER for a 26-mile path dominated by dispersive fading.

Since the effective fade margin for dispersive fading is not related to RSL, a statistical concept for fade margin, termed *dispersive fade margin*, has been developed (Giger and Barnett, 1981). It relates the statistical composite effects of dispersive fading to a particular radio receiver. The effects of dispersive fading can only be reduced by the use of transversal equalizers and diversity techniques.

5.10 RADIO SYSTEM PERFORMANCE AS A FUNCTION OF RADIO PATH TERRAIN

Another fundamental transmission limitation is the terrain near a radio path. A primary task of path design is to mitigate the effects of path terrain (which is interrelated with atmospheric refractivity) by vertical antenna placement.

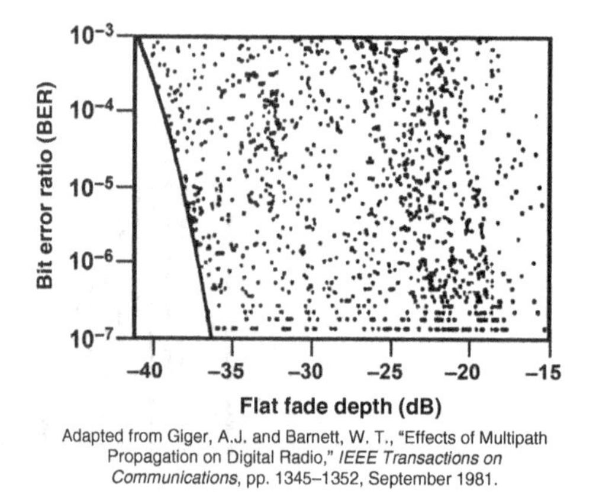

Figure 5.40 Example of dispersive fading. Reprinted with permission of IEEE.

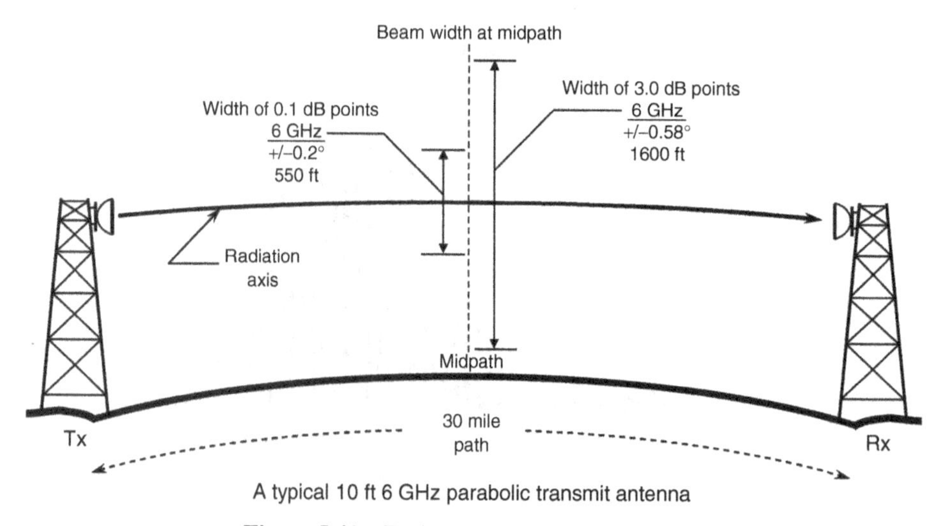

Figure 5.41 Typical path signal illumination.

Microwave radio transmit antennas do not just send a thin beam to the receive antenna. They actually illuminate a wide area of terrain along the microwave radio path (Fig. 5.41). The effect of the terrain in reflecting the transmitted energy toward the receive antenna can significantly influence the received signal (Fig. 5.42).

Analysis of terrain reflections is done on the basis of Fresnel zones (Fig. 5.43 and Fig. 5.44). A Fresnel zone is described as the locus of points above or below the direct path from the transmitter to the receiver where the distance from one end of the path to the point and then to the other end of the path is an integer number of 1/2 wavelengths longer than the direct path. The first Fresnel zone, F_1, has a total additional path length of 1/2 wavelength. The second Fresnel zone, F_2, has $2 \times 1/2$ wavelengths, the third Fresnel zone, F_3, has $3 \times {}^1\!/_2$, and so on.

A Fresnel zone radius, F_n, is the distance perpendicular to the path from a location of interest to a point on the Fresnel zone.

$$F_n\,(\text{ft}) = n\text{th Fresnel zone radius}$$

$$F_n\,(\text{ft}) = 72.1\ \text{sqrt}\left\{\frac{[n \times d_1\,(\text{miles}) \times d_2(\text{miles})]}{[F(\text{GHz})D(\text{miles})]}\right\}$$

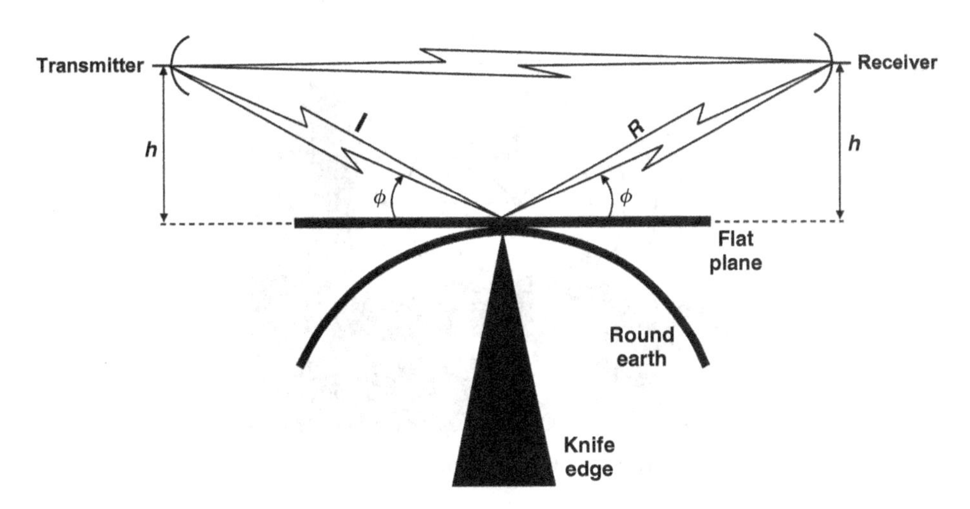

Figure 5.42 Major terrain reflection models.

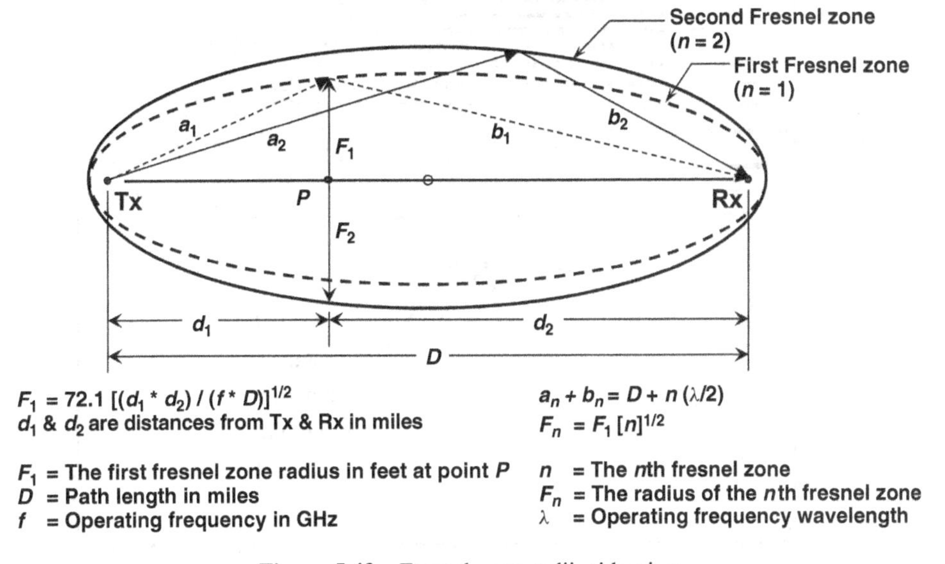

Figure 5.43 Fresnel zone radii, side view.

$$F_n(\text{m}) = 17.3 \text{ sqrt}\left\{\frac{[n \times d_1(\text{km}) \times d_2(\text{km})]}{[F(\text{GHz}) \times D(\text{km})]}\right\} \tag{5.34}$$

n = Fresnel zone number (an integer);
d_1 = distance from one end of the path to the reflection;
d_2 = distance from the other end of the path to the reflection;
D = total path distance = $d_1 + d_2$;
F = frequency of radio wave.

Microwave paths are essentially parallel to the earth. As noted in Chapter 13, virtually all surfaces (since they have a significantly different refractive index than the atmosphere) reflect high frequency radio

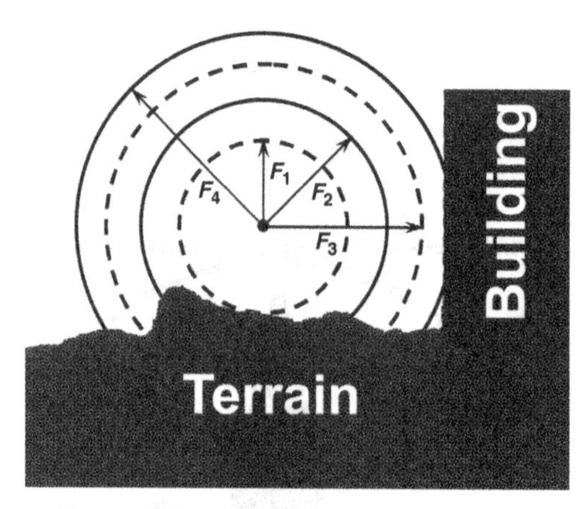

Figure 5.44 Fresnel zone radii, end view.

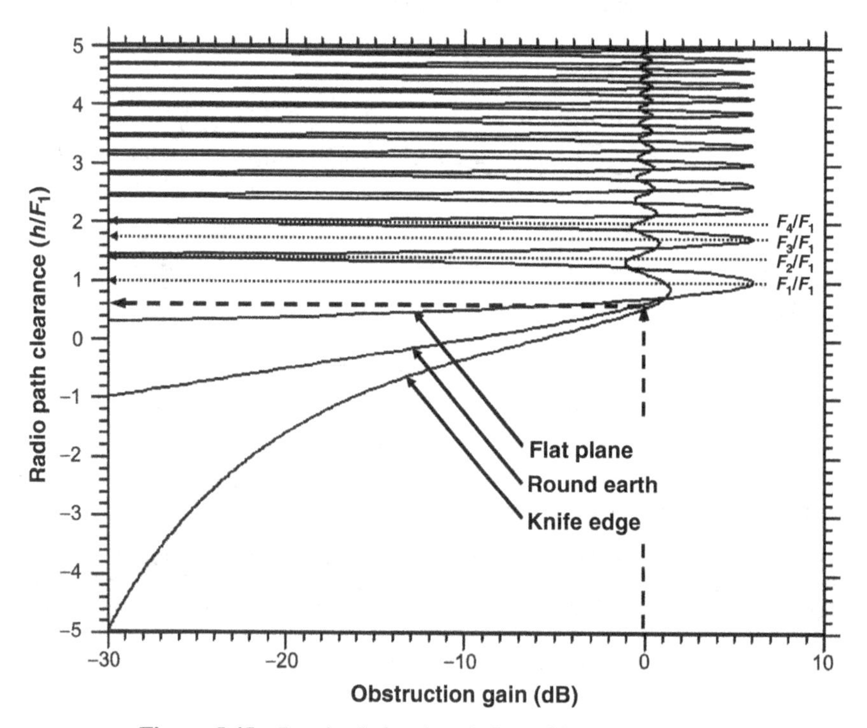

Figure 5.45 Received signal variation with antenna height.

signals (unless the signal is blocked by terrain or trees or dispersed by rough surfaces). Tangential reflections from the earth have a 180° phase reversal relative to the direct wave (a requirement of—you guessed it—Maxwell's equations). Therefore, all reflected signals with odd Fresnel zone clearance produce signal enhancement at the receive antenna. All reflected signals with even Fresnel zone clearance produce signal cancelation (of the received direct signal). These concepts are important in path engineering.

As the receive antenna is raised above the terrain, the composite received signal increases or decreases depending on the type of terrain and the height above the terrain (Fig. 5.45). For paths with significant surface reflections, two vertically spaced receive antennas ("space diversity") can be used to mitigate the effects of the reflected signal (Fig. 5.46).

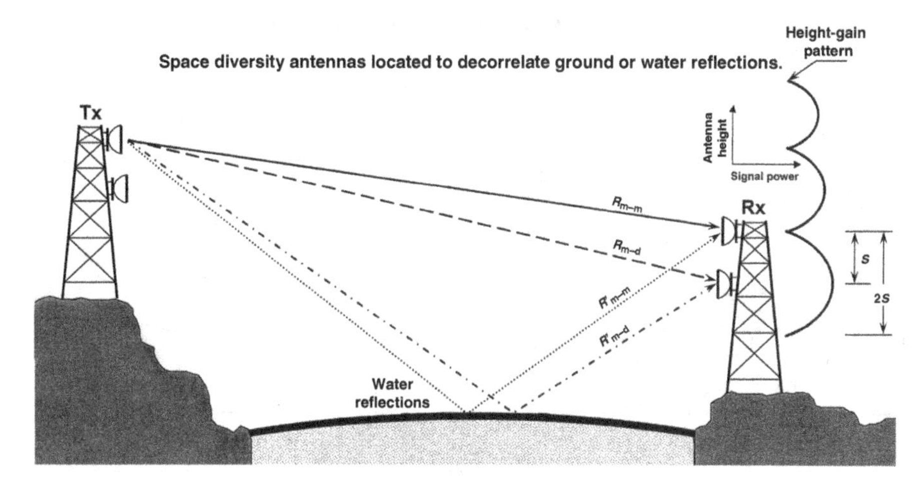

Figure 5.46 Space diversity.

The two antennas are placed so that the reflection effects on the antennas are complementary. Since path clearance for long paths varies with atmospheric refractivity (K factor), this requires an understanding of expected refractivity.

Dual antenna spacing will be a compromise for the expected range of refractivity. For long paths with very large path clearance (such as mountain top to mountain top), the even and odd Fresnel zones are so close together that exact placement is not possible. Experimentation or placement of the antennas to avoid reflection may be the only practical choices.

For ground-based reflections, another technique is to tilt the antenna up to place the reflected path in the first null of the receive antenna (Hartman and Smith, 1977). This has the disadvantage of degrading the side lobe performance and cross-polarization discrimination of the antenna. However, this is an effective technique for relatively short (i.e., stable propagation) remote area (i.e., few other transmitters) paths.

5.11 ANTENNA PLACEMENT

One of the main tasks of path engineering is the proper placement of the antennas to mitigate the effects of terrain reflection and atmospheric refractivity variations.

The refractivity of the atmosphere is a function of atmospheric pressure, temperature, and relative humidity. Under normal conditions, atmospheric humidity and pressure are relatively constant with height. However, atmospheric temperature decreases with increasing height so atmospheric refractivity decreases with height. This causes the radio wave to bend down toward the earth. (Diagrams showing the radio wave bending toward the earth are misleading. The radio wave is bowed slightly down but the earth curvature bows up more. This causes the path clearance between the radio wave and the earth surface to decrease near the center of the path.) If the vertical gradients of temperature and/or humidity change, the vertical direction of the radio wave can change significantly. This change in refractivity gradient is usually described as a change in K factor (see following equations).

Under normal atmospheric conditions (Bean and Dutton, 1966; Bullington, 1957; Schelleng et al., 1933), radio and light waves curve toward the earth. If the earth were replaced by a sphere with radius aK (where a is the physical radius of the earth and K is a function of refractivity), a radio or light wave launched parallel to the earth would remain parallel to this modified "earth." For radio waves nearly parallel to the earth, the following equation approximates K (Fig. 5.47):

$$K \approx \frac{1}{\left[1 + a\left(\frac{dn}{dh}\right)\right]}$$

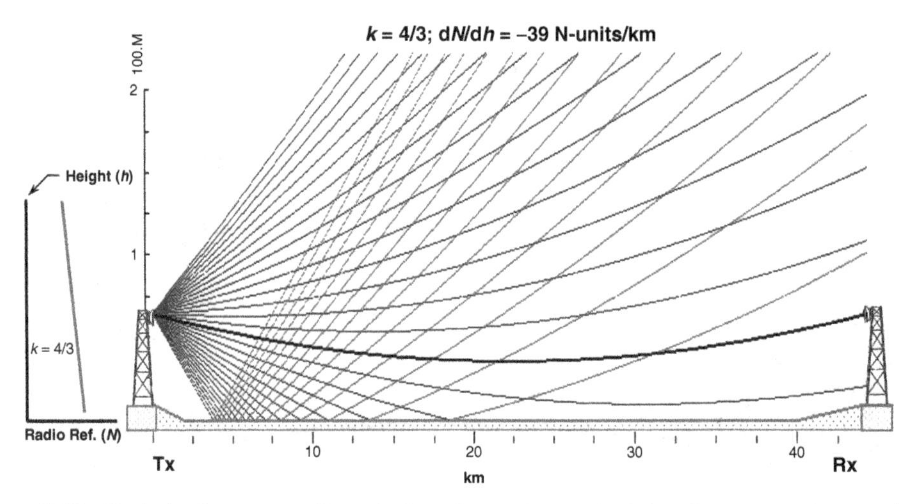

Figure 5.47 Typical microwave radio path. Drawing courtesy of Eddie Allen. Used with permission.

$$
\begin{aligned}
&= \frac{253}{\left(253 + \left[\frac{dN}{dh}\text{ (N units per mile)}\right]\right)} \\
&= \frac{157}{\left(157 + \left[\frac{dN}{dh}\text{ (N units per kilometer)}\right]\right)} \\
&= \text{typically } \frac{6}{5}\text{(average) to } \frac{7}{5}\text{(midday) for light} \\
&= \text{typically } \frac{4}{3}\text{(average) for radio waves} < 40\text{ GHz}
\end{aligned} \qquad (5.35)
$$

K = effective earth radius factor;
a = physical earth radius.

Microwave path antenna locations are typically designed and analyzed using a plot of the terrain between the transmitter and the receiver with a line representing the maximum power of the radio wave front as it moves from transmitter to receiver. The radio wave path will bend up or down depending on the K factor (atmospheric refractivity). The convention is to always plot the radio wave path as a straight line and to move the earth up or down as a function of K factor to preserve the vertical distance between the radio wave and the earth at any location on the path. In the past, different graphs were plotted with the earth surface predistorted to account for different K factors. Today, path profiles are created by computers, but the "earth bulge" convention remains (Fig. 5.48).

The physical height of the height measurements on the path profile are modified by adding the following values (Fig. 5.49):

$$
\begin{aligned}
h(\text{ft}) &= \frac{[d1(\text{miles})\, d2(\text{miles})]}{(1.500\ K)} \\
h(\text{m}) &= \frac{[d1(\text{km})\, d2(\text{km})]}{(12.74\ K)}
\end{aligned} \qquad (5.36)
$$

Path profiles are used to place the vertical location of antennas (see Chapter 10). The path profile is usually based on digitized path elevation data. However, this data is modified to account for actual path obstructions, such as trees and structures. "Driving the path" is important to make sure potential obstructions and reflections are identified.

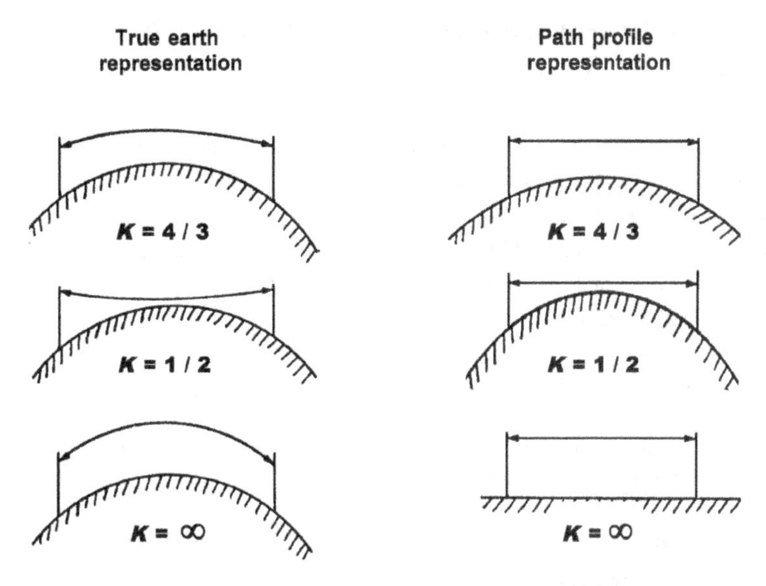

Figure 5.48 Typical microwave radio path profile convention.

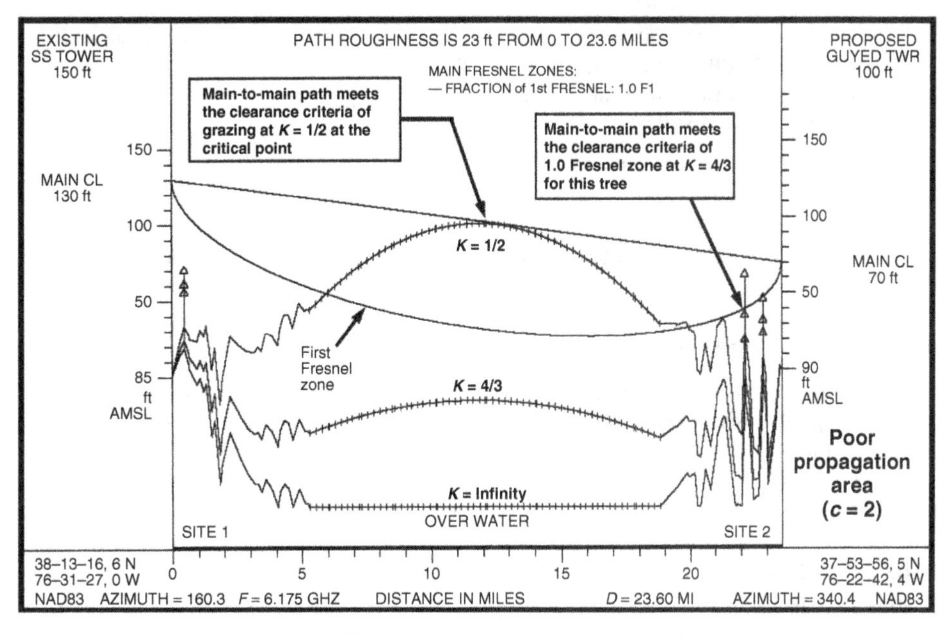

Figure 5.49 Computer-generated path profile.

5.12 FREQUENCY BAND CHARACTERISTICS

For commercial applications, two general types of bands are available: licensed and unlicensed.

Unlicensed bands have the advantage of rapid installation but have unpredictable performance because interference is not controlled. Nevertheless, these bands are quite popular. Several unlicensed bands are available:

Unlicensed National Information Infrastructure (UNII) Bands (FCC Part 15.407):

5.2-GHz Band (5150–5250 MHz)

- Transmit power: 50 mW maximum
- Transmit power must be reduced by 1 dB for every decibel of antenna gain that exceeds 6 dBi

5.3-GHz Band (5250–5350 MHz) and 5.6-GHz band (5470–5725 MHz)

- Transmit power: 250 mW maximum
- Transmit power must be reduced by 1 dB for every decibel of antenna gain that exceeds 6 dBi
- User must sense radar and cease operation if detected

5.8-GHz Band (5725–5825 MHz)

- Transmit power: 1 W
- Transmit power must be reduced by 1 dB for every decibel of antenna gain that exceeds 23 dBi

Because of the significant transmitted power and antenna gain limitations, the UNII bands are not popular for fixed point-to-point applications.

Unlicensed Industrial, Scientific, Medical (ISM) Bands (FCC Part 15.247)

900-MHz Band (902–928 MHz)

- Not useful for wideband fixed point-to-point applications

2.4-GHz Band (2400–2483.5 MHz)

- Transmit power: 1 W maximum
- Transmit power must be reduced by 1 dB for every 3 dB of antenna gain that exceeds 6 dBi
  - 4 ft parabolic = 27 dBi → transmit power = 200 mW
  - 8 ft parabolic = 33 dBi → transmit power = 125 mW

5.8-GHz Band (5725–5850 MHz)

- Transmit power: 1 W maximum
- No antenna limitations

The ISM 2.4 and 5.8 bands are quite popular. However, owing to transmit power and antenna limitations at 2.4 GHz, 5.8 GHz is the more popular band for fixed point-to-point applications.

60-GHz Band (57.0–64.0 GHz)

This band's primary feature is the very high atmospheric attenuation due to oxygen absorption. This band is typically used where frequency reuse within a geographic area and communication security are important.

The licensed bands are quite important as they offer the user a relatively predictable performance:

4 GHz (3.7–4.2 GHz)

- 20 MHz channels
- Very good propagation band
- Owing to coordination difficulties with existing satellite receivers, the band is effectively unavailable for new fixed point-to-point applications

Lower 6 GHz (5.9–6.4 GHz)

- 0.4, 0.8, 1.25, 2.5, 3.75, 5, 10, 30, and 60 MHz channels
- Very good propagation band
- Highly desirable but congested in many urban areas

Upper 6 GHz (6.5–6.9 GHz)

- 0.4, 0.8, 1.25, 2.5, 3.75, 5, 10, and 30 MHz channels

Very good propagation band
Highly desirable but congested in many urban areas

10 1/2-GHz Band (10.6–10.7 GHz)
0.4, 0.8, 1.25, 2.5, 3.75, and 5 MHz channels
Path length rain limited in most areas except the west coast

11-GHz Band (10.7–11.7 GHz)
1.25, 2.5, 3.75, 5, 10, 30, and 40 MHz channels
Path length rain limited in most areas except the west coast

18 GHz (17.7–18.7 GHz)
1.25, 2.5, 5, 10, 20, 30, 40, 50, and 80 MHz channels
Path length rain limited

23-GHz Band (21.3–23.6 GHz)
2.5, 5, 10, 20, 30, 40, and 50 MHz channels
Path length rain limited

28, 29, and 31 GHz (27.5–28.4, 29.1–29.3, and 31.0–31.3 GHz)
75–850 MHz channels
Spectrum auctioned and not available to the general public
Path length rain limited

38 GHz (38.6–40.0 GHz)
50 MHz channels
Spectrum auctioned and not available to the general public
Path length rain limited

70-GHz Band (71.0–76.0 GHz)
Spectrum unchannelized
Often paired with the 80-GHz band to achieve very wide duplex channels
Path length rain limited

80-GHz Band (81.0–86.0 GHz)
Spectrum unchannelized
Often paired with the 70-GHz band to achieve very wide duplex channels
Path length rain limited

90-GHz Band (92.0–94.0 and 94.1–95.0 GHz)
Band unchannelized and not currently popular
Path length rain limited

The above frequency ranges are rounded off. See Appendix A for more detailed frequency ranges. In the following section we will observe the typical path distances for the most popular licensed frequency bands.

5.13 PATH DISTANCES

The distance over which a fixed point to point microwave radio system transmission can operate reliability depends heavily on the frequency of operation and geographic and weather characteristics of the terrain near the path. The details of path characteristics are discussed in the later chapters. There are thousands of paths already licensed and in use in the United States today. Looking at their path lengths will give an understanding of the path distances that can be expected for each microwave frequency band.

To perform a statistical evaluation of all US licensed frequency bands, the entire FCC microwave point-to-point license data of simplex paths was utilized. The following statistics are based on that data. In these graphs, a duplex path is a pair of matched simplex paths (Fig. 5.50).

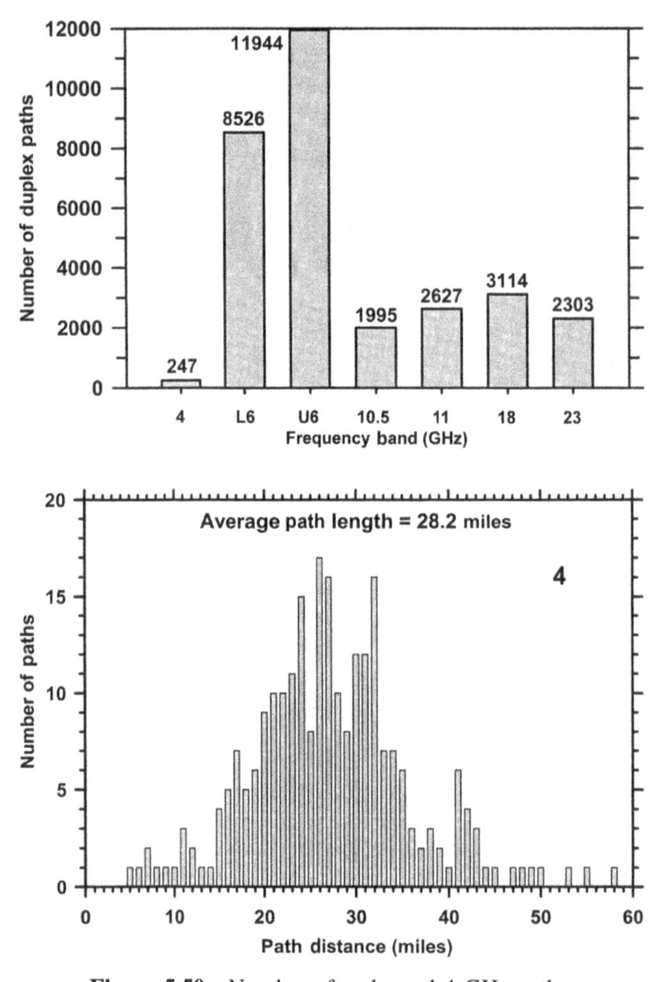

Figure 5.50 Number of paths and 4-GHz paths.

The national averages of path lengths only give a part of the picture. As we will learn in Chapter 11, high frequency paths are limited by intense rain. The southeastern United States has very intense rainfall but the western United States has very light rainfall. The southeastern United States is a poor propagation area for low frequency microwave paths but the upper western United States is a good propagation area. When we look at the statistics of paths in those areas, we will have more insight as to what is possible in those areas.

For low and high frequency paths, we will define a poor propagation area (SE) for latitudes south of 32 and longitudes east of −92. For low frequency paths, we will define a good propagation area (UW) as latitude north of 36 and longitude between −114 and −105. For high frequency paths, we will define a good propagation area (W) as anywhere west of −114 longitude. The following statistics are based on these filters (Fig. 5.51, Fig. 5.52, Fig. 5.53, Fig. 5.54, Fig. 5.55, Fig. 5.56, Fig. 5.57, Fig. 5.58, Fig. 5.59 and Table 5.2). The 4-GHz paths were not analyzed because there were not enough paths in the areas of interest to be statistically significant.

Obviously not all paths are created equal.

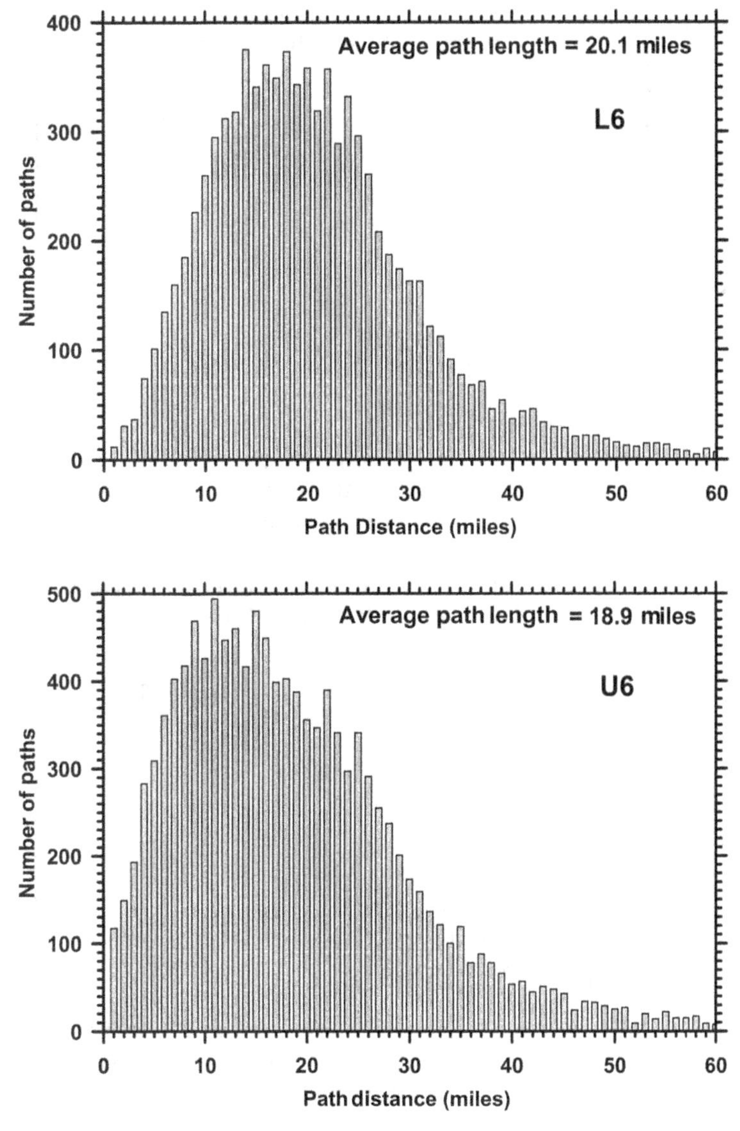

Figure 5.51 Lower 6-GHz and upper 6-GHz paths (entire United States).

5.A APPENDIX

5.A.1 Antenna Isotropic Gain and Free Space Loss

Consider an isotropic transmit antenna radiating a radio signal that is received at a remote location. The power received by the receiving antenna may be estimated as follows:

Pr = receiving antenna power;
Pt = transmitting antenna power;
Are = receive antenna effective area;
Ate = transmit antenna effective area;
D = distance between the transmitting and receiving antennas;
λ = radio signal free space wavelength.

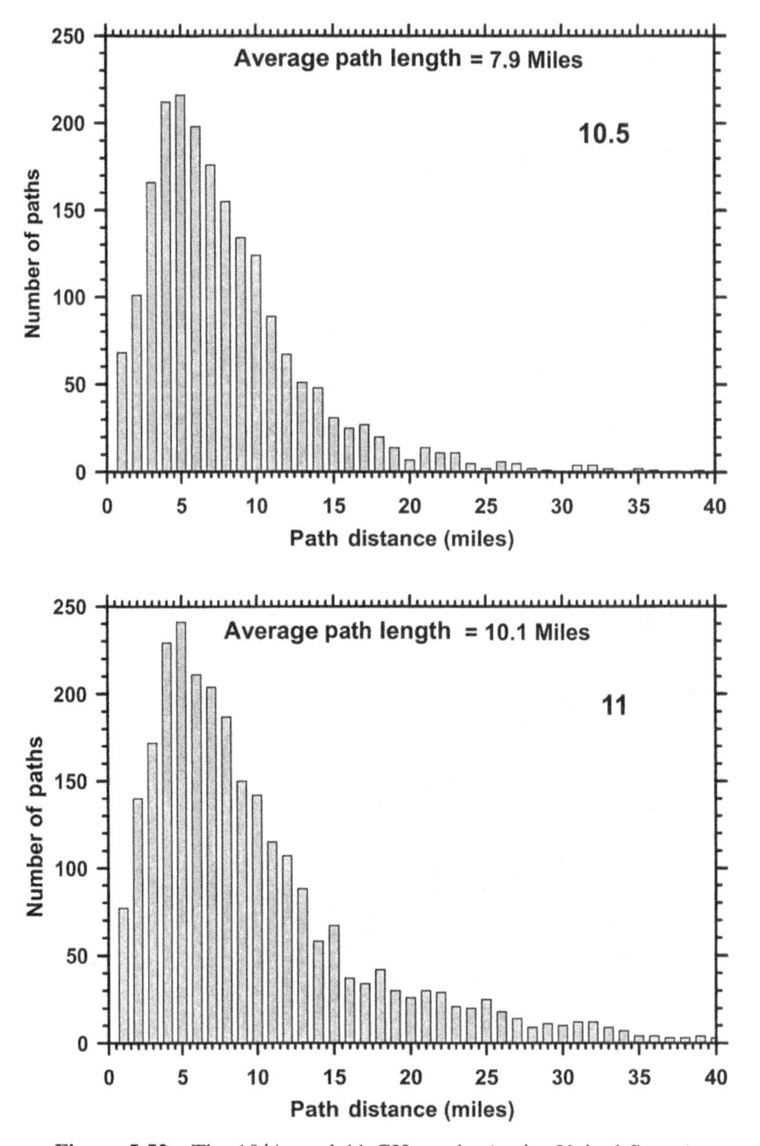

Figure 5.52 The $10\frac{1}{2}$- and 11-GHz paths (entire United States).

$$
\begin{aligned}
\text{Pr} &= \text{Transmitted power received by the receive antenna} \\
&= (\text{Received power density})(\text{Receiver effective area}) \\
&= \left[\frac{\text{Pt}}{(\text{Area of a sphere at distance } d)}\right] \text{Are} \\
&= \frac{\text{Pt Are}}{(4\pi d^2)} \qquad (5.\text{A}.1)
\end{aligned}
$$

$$
\begin{aligned}
\frac{\text{Pr}}{\text{Pt}} &= \text{Isotropic transmit antenna to receive antenna power loss} \\
&= \frac{\text{Are}}{(4\pi d^2)} \qquad (5.\text{A}.2)
\end{aligned}
$$

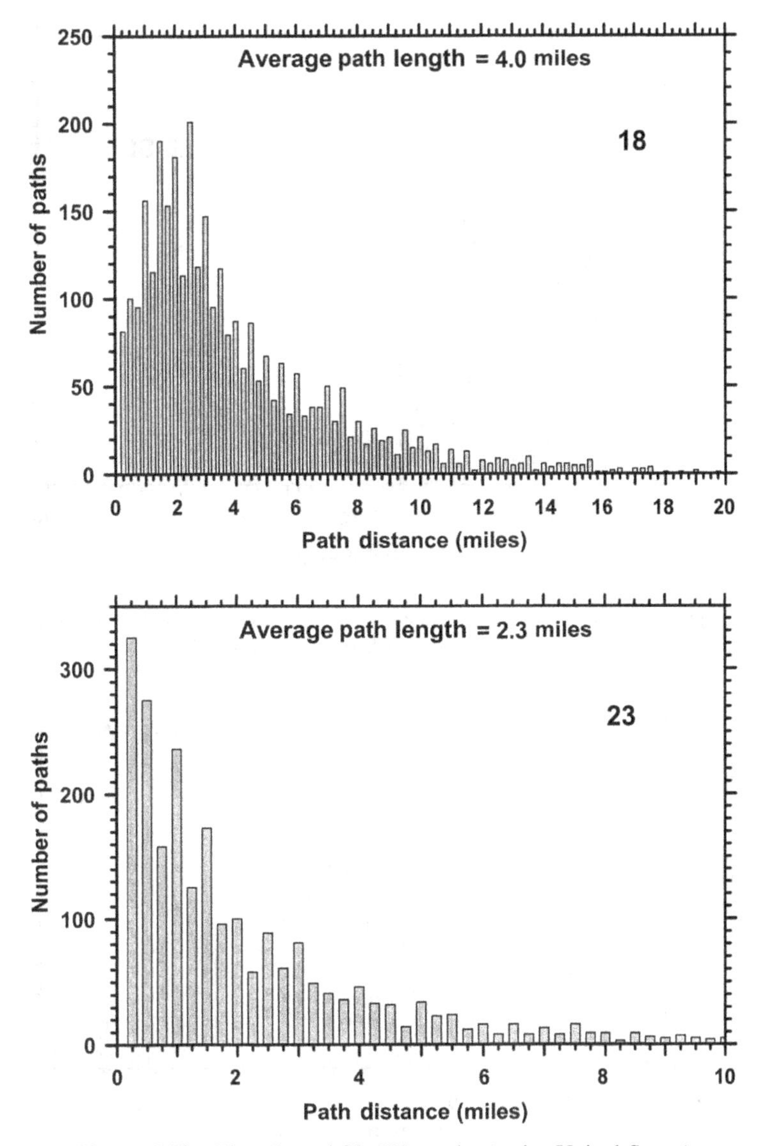

Figure 5.53 The 18- and 23-GHz paths (entire United States).

Now replace the isotropic transmit antenna with a directional antenna.

$$\frac{\text{Pr}}{\text{Pt}} = \text{Directional transmit antenna to receive antenna power loss}$$

$$= (\text{Transmit antenna gain relative to an isotropic radiator})\frac{\text{Are}}{(4\pi d^2)} \tag{5.A.3}$$

Silver (1949, Eq. 20, p. 177) showed that the gain of an antenna relative to an isotropic radiator is the following:

$$\text{Gi} = \text{Antenna gain relative to an isotropic radiator}$$

$$= \frac{4\pi Ae}{\lambda^2} \tag{5.A.4}$$

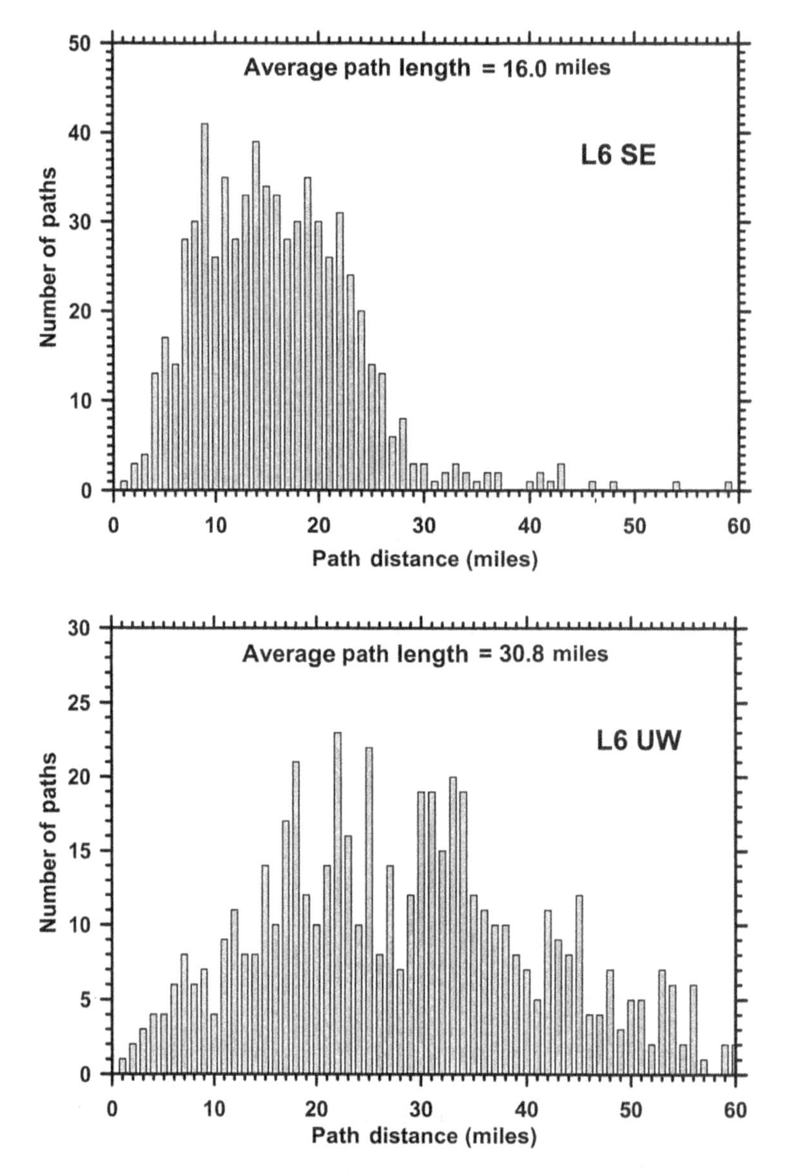

Figure 5.54 Lower 6-GHz poor propagation area and good propagation area path lengths.

$$\frac{\text{Pr}}{\text{Pt}} = \frac{\text{Ate Are}}{(\lambda^2 d^2)}$$

This is the well-known Friis transmission loss formula (Friis, 1946). Now, we will reformat the formula into the more familiar path loss form:

$$\frac{\text{Pr}}{\text{Pt}} = \frac{\text{Gti Gri}}{L_{\text{FS}}} = \frac{\left(\frac{4\pi\,\text{Ate}}{\lambda^2}\right)\left(\frac{4\pi\,\text{Are}}{\lambda^2}\right)}{\left[\frac{(16\pi^2 d^2)}{\lambda^2}\right]} \tag{5.A.5}$$

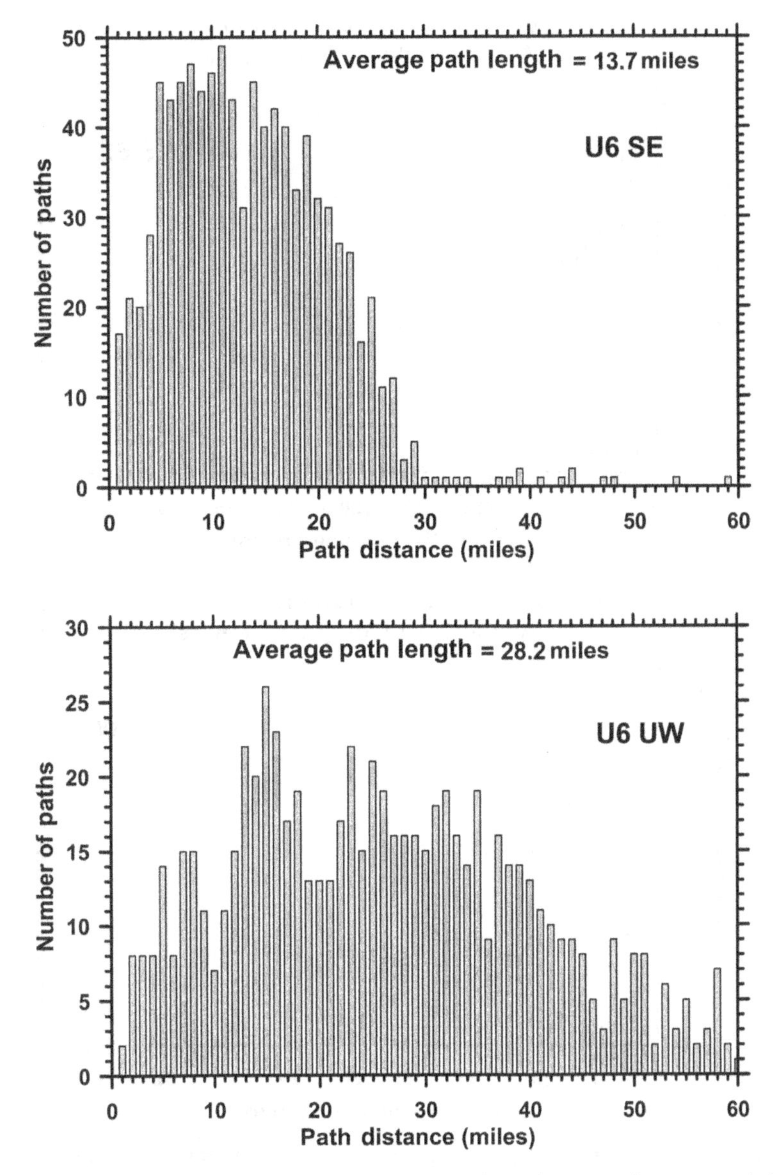

Figure 5.55 Upper 6-GHz poor propagation area and good propagation area path lengths.

Gti = transmit antenna gain relative to an isotropic radiator;
Gri = receive antenna gain relative to an isotropic radiator;
L_{FS} = free space loss.

Now, we convert these to the popular decibel format:

$$\mathrm{Pr(dBm) - Pt(dBm) = Gt(dBi) + Gr(dBi)} - L_{FS}\mathrm{(dB)} \tag{5.A.6}$$

5.A.2 Free Space Loss

$$L_{FS}\mathrm{(dB)} = 10 \log \left[\frac{(16\,\pi^2 d^2)}{\lambda^2} \right] \tag{5.A.7}$$

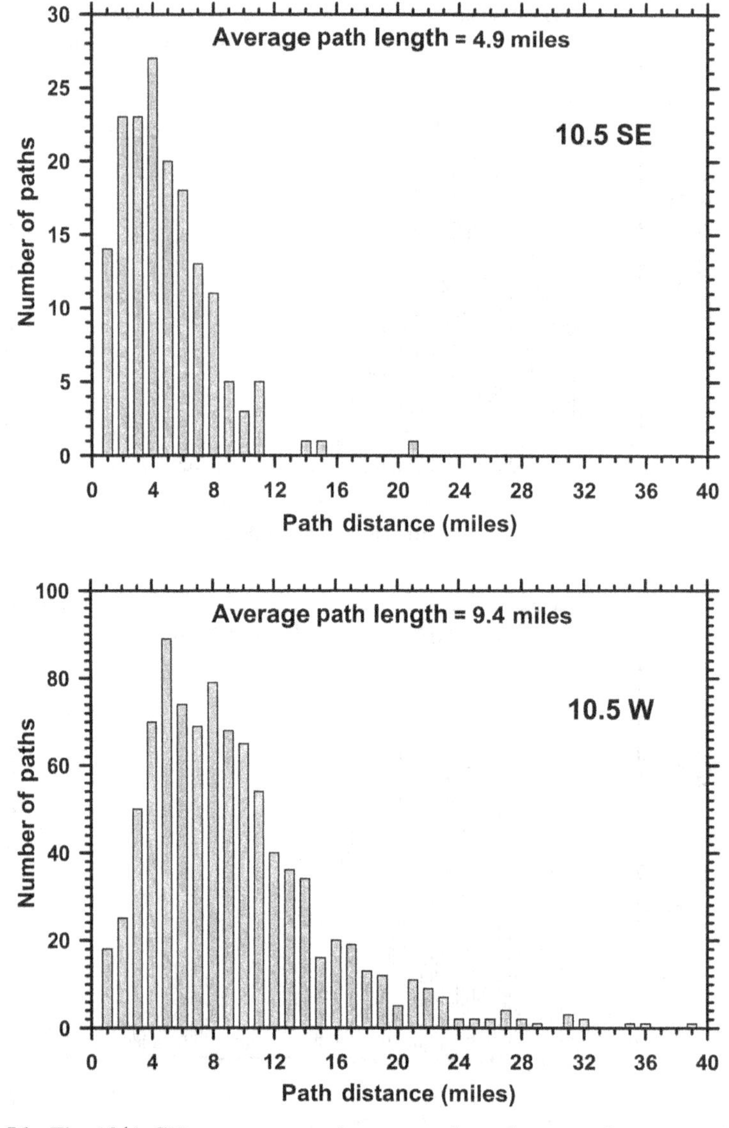

Figure 5.56 The 10 ½-GHz poor propagation area and good propagation area path lengths.

d = distance between the transmit and receive antennas (ft or m);
λ (ft) = 0.98357/F (GHz);
λ (m) = 0.29980/F (GHz).

$$L_{FS} = 96.58 + 20\ \log[d(\text{miles})] + 20\ \log[F(\text{GHz})]$$
$$= 92.45 + 20\ \log[d(\text{km})] + 20\ \log[F(\text{GHz})] \quad (5.A.8)$$

5.A.3 Antenna Isotropic Gain

$$\text{G(dBi)} = \text{antenna gain(relative to an isotropic radiator)}$$
$$= 10\ \log\left(\frac{4\pi\,\text{Ae}}{\lambda^2}\right)$$

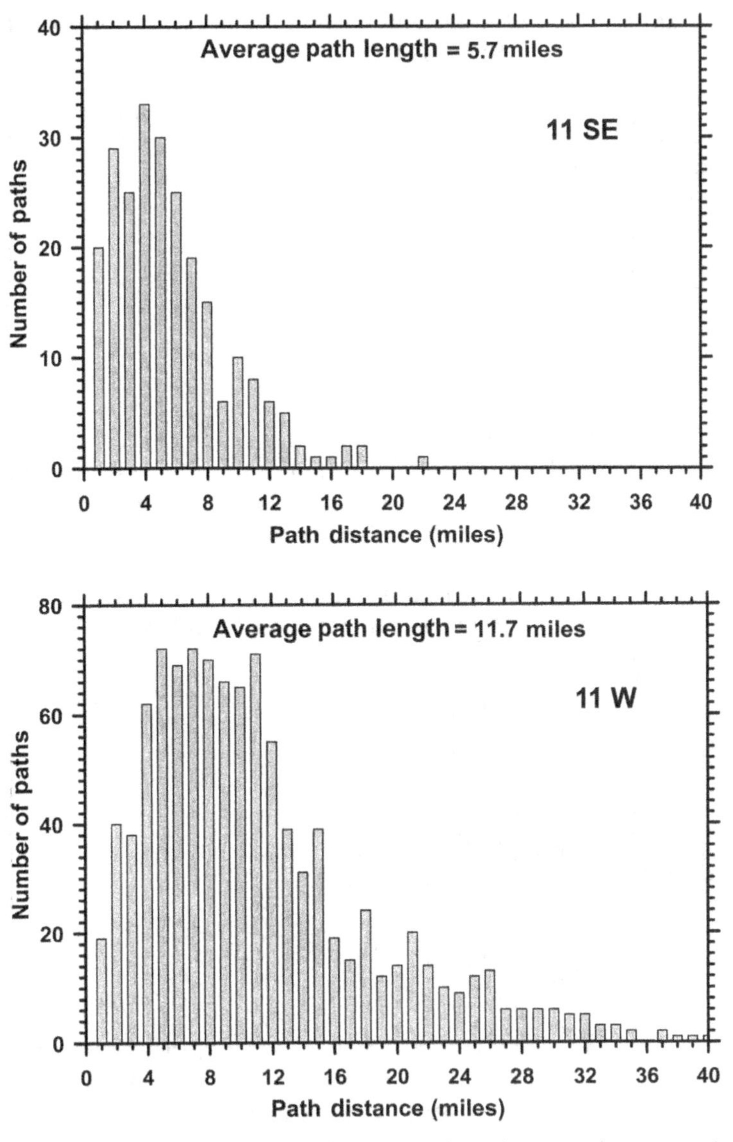

Figure 5.57 The 11-GHz poor propagation area and good propagation area path lengths.

$$= 10\ \log\left(\frac{4\pi\eta A}{\lambda^2}\right) \tag{5.A.9}$$

$A =$ antenna physical area;
$\eta =$ antenna illumination efficiency (power ratio ≤ 1) $= E/100$
$E =$ illumination efficiency (percentage) $= 100\ \eta$.

$$G(\text{dBi}) = 10\ \log(4\pi) + 10\ \log\left(\frac{E}{100}\right) + 10\ \log(A) - 20\ \log\lambda \tag{5.A.10}$$

$$\text{G(dBi)} = 11.14 + 10\ \log\left(\frac{E}{100}\right) + 10\ \log[A(\text{ft}^2)] + 20\ \log[F(\text{GHz})] \tag{5.A.11}$$

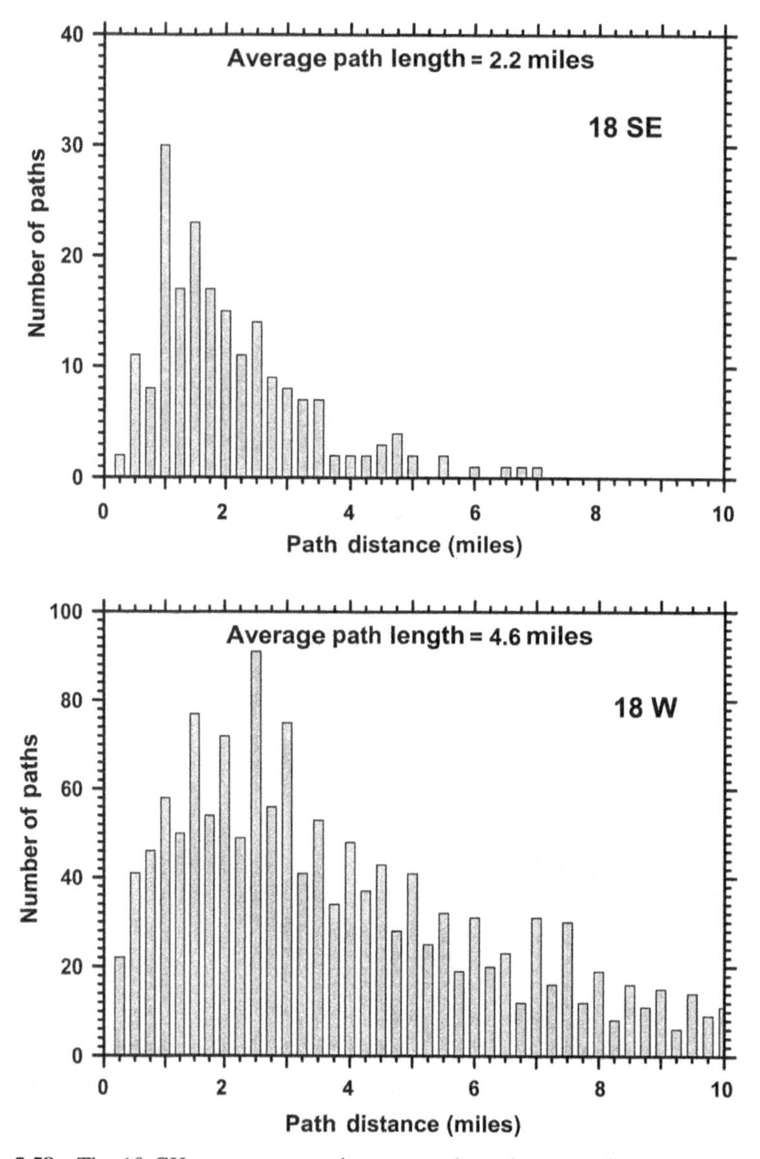

Figure 5.58 The 18-GHz poor propagation area and good propagation area path lengths.

$$\mathrm{G(dBi)} = 21.46 + 10 \log\left(\frac{E}{100}\right) + 10 \log[A(\mathrm{m}^2)] + 20 \log[F(\mathrm{GHz})]$$

For parabolic antennas, $E \approx 55\%$ (generally between 45% and 65%). For panel antennas and passive reflectors, $E \approx 100\%$. For passive reflectors, the area is the area projected onto the path (see Appendix A).

5.A.4 Circular (Parabolic) Antennas

$$G(\mathrm{dBi}) = 10.09 + 10 \log\left(\frac{E}{100}\right) + 20 \log[D(\mathrm{ft})] + 20 \log[F(\mathrm{GHz})]$$

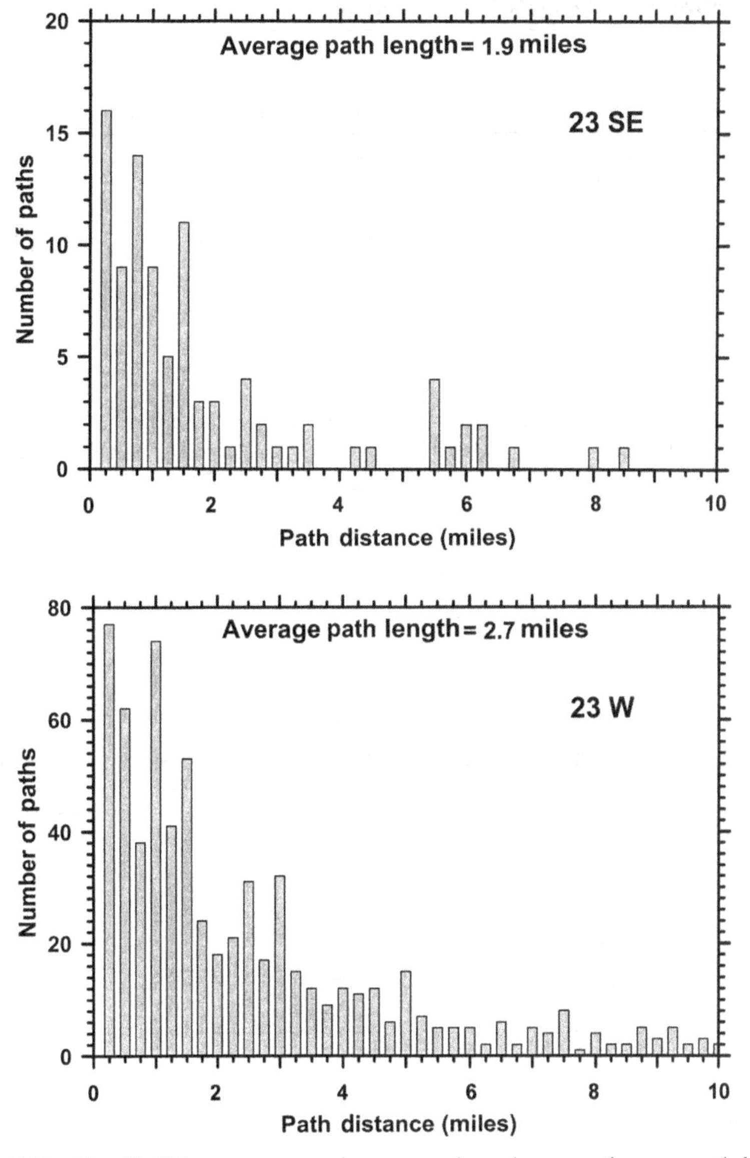

Figure 5.59 The 23-GHz poor propagation area and good propagation area path lengths.

$$G(\text{dBi}) = 20.41 + 10\log\left(\frac{E}{100}\right) + 20\log[D(\text{m})] + 20\log[F(\text{GHz})] \tag{5.A.12}$$

$D =$ diameter of the antenna.

5.A.5 Square (Panel) Antennas

$$G(\text{dBi}) = 11.14 + 10\log\left(\frac{E}{100}\right) + 20\log[W(\text{ft})] + 20\log[F(\text{GHz})]$$

$$G(\text{dBi}) = 21.46 + 10\log\left(\frac{E}{100}\right) + 20\log[W(\text{m})] + 20\log[F(\text{GHz})] \tag{5.A.13}$$

TABLE 5.2 Microwave Radio Path Length Statistics

	Entire United States					
	Arithmetic Mean	Median	Mode	Standard Deviation	Skew	Kurtosis
4 GHz	28.2	26.9	26	11.5	2.2	9.6
Lower 6 GHz	21.0	19.4	14	11.2	1.6	4.9
Upper 6GHz	18.9	16.7	11	12.2	1.5	3.6
10.5 GHz	7.9	6.6	5	5.4	1.9	6.3
11 GHz	10.1	7.7	5	8.7	2.7	12.4
18 GHz	4.0	3.0	2.50	3.5	2.8	22.6
23 GHz	2.3	1.4	0.25	2.5	2.7	12.5
	Poor Propagation Area (Southeast United States)					
	Arithmetic Mean	Median	Mode	Standard Deviation	Skew	Kurtosis
Lower 6 GHz	15.9	15.3	9	7.7	1.1	3.0
Upper 6GHz	13.6	12.8	11	7.6	0.93	2.3
10.5 GHz	4.7	4.2	4	2.9	1.2	2.9
11 GHz	5.6	4.7	4	3.7	1.3	1.9
18 GHz	2.1	1.8	1.00	1.5	3.7	27.5
23 GHz	1.7	1.0	0.25	1.8	1.7	2.1
	Good Propagation Area (Upper Western United States for 6 GHz, Western United States for Others)					
	Arithmetic Mean	Median	Mode	Standard Deviation	Skew	Kurtosis
Lower 6 GHz	30.7	29.3	22	16.5	1.0	1.3
Upper 6GHz	27.9	25.7	15	16.2	0.78	0.65
10.5 GHz	9.3	8.2	5	5.8	1.6	4.4
11 GHz	11.4	9.5	6	8.5	2.2	9.3
18 GHz	4.5	3.4	2.50	3.9	3.2	28.4
23 GHz	2.7	1.6	0.25	2.9	2.2	7.1

$W =$ width of the antenna.

For the aperture antennas typically used in microwave applications, their gain is in the 20 to 50 dBi range. For reference, a short (Hertzian) dipole has 1.76-dBi gain (ignoring resistive losses). A half wave dipole has 2.15-dBi gain.

5.A.6 11-GHz Two-foot Antennas

In metropolitan areas, cellular operators use many 11-, 18-, and 23-GHz radio paths. These paths are typically short and installed on leased towers. Lease costs are directly related to the size of the antenna. Most operators limit their antenna size to a maximum of 2 ft. Since essentially everyone uses Class A antennas, this has excluded the use of 11 GHz. However, many operators feel the need to use 11 GHz for long paths or in high rainfall regions. The FCC recently revised their rules (Mosley, 2011) to provide "almost Class A" properties for 2-ft antennas. FCC rule 101.115 (f) states, "In the 10,700-11,700 MHz band, a fixed station may employ transmitting and receiving antennas meeting performance standard B in any area. If a Fixed Service or Fixed Satellite Service licensee or applicant makes a showing that it is

TABLE 5.A.1 FCC Antenna Radiation Pattern Requirements

Antenna	Boresight Gain (dBi)	Gain Relative to Boresight						
		5°–10°	10°–15°	15°–20°	20°–30°	30°–100°	100°–140°	140°–180°
Class A	38	−25	−29	−33	−36	−42	−55	−55
Class B	33.5	−17	−24	−28	−32	−35	−40	−45
Diff. (dB)	4.5	8	5	5	4	7	15	10

likely to receive interference from such fixed station and that such interference would not exist if the fixed station used an antenna meeting performance standard A, the fixed station licensee must modify its use. Specifically, the fixed station licensee must either substitute an antenna meeting performance standard A or operate its system with an EIRP reduced so as not to radiate, in the direction of the other licensee, an EIRP in excess of that which would be radiated by a station using a Category A antenna and operating with the maximum EIRP allowed by the rules."

Although 2-ft antennas are still Class B, they may be used similarly to Class A antennas [i.e., once they are licensed, they do not have to be changed in the future unless impacted by a "major change" (see Chapter 2)]. The conditions of "almost Class A" operation may be inferred from the rules (Mosley, published yearly):

Per FCC 101.115 (b), 11-GHz Class A antennas must meet the standards listed in Table 5.A.1. Per FCC 101.113 (a), 11-GHz maximum allowable EIRP is +55 dBW (85 dBm). If the boresight gain of a Class A antenna is +38 dBi, then the maximum allowable transmit power into the antenna is +47 dBm.

The worst case difference between Class A and Class B antenna side lobes is 15 dB. Therefore, as long as transmitter power does not exceed +47 dBm − 15 dB = +32 dBm, the antenna may be treated as Class A. Very few 11-GHz transmitters exceed this transmit power.

Keep in mind that typical 11-GHz Class B 2-ft antennas meet or exceed Class A antenna side lobe standards between 100° and 180°. For these cases, the worst case difference in side lobe power is 8 dB. For these cases, the transmit power level limit becomes +47 dBm − 8 dB = +39 dBm. For all practical purposes, this gives all Class B 2 ft 11-GHz antennas the rights of Class A antennas.

5.A.7 Tower Rigidity Requirements

Antenna structures must maintain the antenna position within acceptable limits. In the United States, the current standard, ANSI/TIA/EIA-222-G (TIA Subcommittee TR-14.7, 2005), allows 10-dB loss of received signal level due to antenna structure twist or tilt under standardized wind and ice loading. Obviously, the direct approach is to use the antenna pattern of the proposed antenna. If the specific antenna decision has not been made when the tower is being specified, the standard offers the following formula for estimating that limit:

5.A.7.1 Parabolic (Circular) Antenna

$$\begin{aligned}\theta &= \text{Maximum allowable twist or tilt (degrees) relative to normal position}\\ &= 54\left(\frac{\lambda}{D}\right)\\ &= \frac{53.1}{[D(\text{ft})\ F(\text{GHz})]}\end{aligned}$$

$$= \frac{16.2}{[D(\text{m})\ F(\text{GHz})]} \tag{5.A.14}$$

$D =$ antenna diameter;
$F =$ radio operating frequency.

Reflectors and square antennas are not addressed in the current standard.

The earlier version of this standard (ANSI/TIA/EIA-222-F (TIA Subcommittee TR-14.7, 1996)) listed the following guidelines for θ:

5.A.7.2 *Parabolic (Circular) Antenna*

$$\theta = \text{Maximum allowable twist or tilt (degrees) relative to normal position}$$
$$= 60\left(\frac{\lambda}{D}\right)$$
$$= \frac{59.0}{[D(\text{ft})F(\text{GHz})]}$$
$$= \frac{18.0}{[D(\text{m})F(\text{GHz})]} \tag{5.A.15}$$

$D =$ antenna diameter;
F $=$ radio operating frequency.

5.A.7.3 *Rectangular Reflector*

$$\theta = \text{Maximum allowable twist or tilt(degrees)relative to normal position}$$
$$= 44\left(\frac{\lambda}{W}\right)$$
$$= \frac{43.3}{[\,W(\text{ft})\ F(\text{GHz})]}$$
$$= \frac{13.2}{[\,W(\text{m})\ F(\text{GHz})]} \tag{5.A.16}$$

$W =$ reflector width (as projected along the path);
$F =$ radio operating frequency.

Square antennas were not addressed but would be similar to rectangular reflectors.

Chapter 8 provides the following limits (for a 10-dB power loss):

5.A.7.4 *Circular (Projection) Reflector*

$$\theta = \text{Maximum allowable twist or tilt(degrees)relative to normal position}$$
$$= 49.8\left(\frac{\lambda}{D}\right)$$
$$= \frac{49.0}{[\,D(\text{ft})F(\text{GHz})]}$$
$$= \frac{14.9}{[\,D(\text{m})F(\text{GHz})]} \tag{5.A.17}$$

$D =$ reflector diameter (as projected along the path);
$F =$ radio operating frequency.

5.A.7.5 *Rectangular Reflector*

$$\theta = \text{Maximum allowable twist or tilt(degrees)relative to normal position}$$

$$= 42.3\left(\frac{\lambda}{W}\right)$$

$$= \frac{41.6}{[\,W(\text{ft})\ F(\text{GHz})]}$$

$$= \frac{12.7}{[\,W(\text{m})\ F(\text{GHz})]} \tag{5.A.18}$$

$W =$ reflector width (as projected along the path);
$F =$ radio operating frequency.

5.A.7.6 *Diamond (Projection) Reflector*

$$\theta = \text{Maximum allowable twist or tilt (degrees) relative to normal position}$$

$$= 45.2\left(\frac{\lambda}{W}\right)$$

$$= \frac{44.5}{[W(\text{ft})\ F(\text{GHz})]}$$

$$= \frac{13.6}{[W(\text{m})\ F(\text{GHz})]} \tag{5.A.19}$$

$W =$ reflector width (as projected along the path, measured along the edge of the reflector);
$F =$ radio operating frequency.

5.A.7.7 *Parabolic (Circular) Antenna*

η(illumination efficiency) $= 0.65$ (65%, worst case)

$$\theta = \text{Maximum allowable twist or tilt (degrees) relative to normal position}$$

$$= 68.1\left(\frac{\lambda}{D}\right)$$

$$= \frac{67.0}{[D(\text{ft})\ F(\text{GHz})]}$$

$$= \frac{20.4}{[D(\text{m})\ F(\text{GHz})]} \tag{5.A.20}$$

η(illumination efficiency) $= 0.55$ (55%, typical)

$$\theta = \text{Maximum allowable twist or tilt (degrees) relative to normal position}$$

$$= 74.2\left(\frac{\lambda}{D}\right)$$

$$= \frac{73.0}{[D(\text{ft})F(\text{GHz})]}$$

$$= \frac{22.2}{[D(\text{m})\ F(\text{GHz})]} \tag{5.A.21}$$

h(illumination efficiency) = 0.45(45%)

$$\begin{aligned}\theta &= \text{Maximum allowable twist or tilt (degrees) relative to normal position}\\ &= 82.1\left(\frac{\lambda}{D}\right)\\ &= \frac{80.8}{[D(\text{ft})F(\text{GHz})]}\\ &= \frac{24.6}{[D(\text{m})F(\text{GHz})]}\end{aligned} \tag{5.A.22}$$

D = antenna diameter;
F = radio operating frequency.

5.A.7.8 Square Antenna

$$\eta(\text{illumination efficiency}) = 1.00(100\%, \text{typical}) \tag{5.A.23}$$

For the typical square antenna, the twist limits are exactly the same as for the square or diamond reflector (as reflectors in the far field have illumination efficiency of 100%).

REFERENCES

Ahmed, A. S., "Role of Particle-Size Distributions on Millimetre-Wave Propagation in Sand/Duststorms," *IEE Proceedings*, pp. 55–59, February 1987.

Ahmed, A. S., Ali, A. A. and Alhaider, M. A., "Airborne Wave into Dust Storms," *IEEE Transactions on Geoscience and Remote Sensing*, Vol. 25, pp. 593–599, September 1987.

Aubin, J. F., "A Brief Tutorial on Antenna Measurements," *Microwave Journal*, Vol. 48, pp. 92–108, August 2005.

Balanis, C. A., *Modern Antenna Handbook*. New York: John Wiley & Sons, Inc., 2008.

Bean, B. R. and Dutton, E. J., *Radio Meteorology*. Washington, DC: U. S. Government Printing Office 1966.

Bullington, K., "Radio Propagation Fundamentals," *Bell System Technical Journal*, Vol. 36, pp. 593–626, May 1957.

Chen, H. and Ku, C., "Calculation of Wave Attenuation in Sand and Dust Storms by the FDTD and Turning Bands Methods at 10–100 GHz," *IEEE Transactions on Antennas and Propagation*, Vol. 60, pp. 2951–2960, June 2012.

Chu, L. J., "Electromagnetic Waves in Elliptic Hollow Pipes of Metal," *Journal of Applied Physics*, Vol. 9, pp. 583–591, September 1938.

Cogdell, J. R., McCue, J. J. G., Kalachev, P. D., Salomonovich, A. E., Moiseev, I. G., Stacey, J. M., Epstein, E. E., Altshuler, E. E., Feix, G., Day, J. W. B., Hvatum, H., Welch, W. J. and Barath, F. T., "High Resolution Millimeter Reflector Antennas," *IEEE Transactions on Antennas and Propagation*, Vol. 18, pp. 515–529, July 1970.

Communication Equipment Specialists and Grasis Towers LLC, *Tower 101*. Lee's Summit: Communication Equipment Specialists and Grasis Towers LLC, 2002.

Davies, W. S., Hurren, S. J. and Copeland, P. R., "Antenna Pattern Degradation Due to Tower Guy Wires on Microwave Radio Systems," *IEE Proceedings*, pp. 181–188, June 1985.

Dong, X., Chen, H. and Guo, D., "Microwave and Millimeter-Wave Attenuation in Sand and Dust Storms," *IEEE Antennas and Wireless Propagation Letters*, Vol. 10, pp. 469–471, May 2011.

Friis, H. T., "A Note on a Simple Transmission Formula," *Proceedings of the I. R. E and Waves and Electrons*, pp. 254–256, May 1946.

Friis, H. T., "Microwave Repeater Research," *Bell System Technical Journal*, Vol. 27, pp. 183–246, April 1948.

Friis, H. T. and Lewis, W. D., "Radar Antennas," *Bell System Technical Journal*, Vol. 26, pp. 219–317, April 1947.

Ghobrial, S. I. and Sharief, S. M., "Microwave Attenuation and Cross Polarization in Dust Storms," *IEEE Transactions on Antennas and Propagation*, Vol. 35, pp. 418–425, April 1987.

Giger, A. J. and Barnett, W. T., "Effects of Multipath Propagation on Digital Radio," *IEEE Transactions on Communications*, Vol. 29, pp. 1345–1352, September 1981.

Goldhirsh, J., "Attenuation and Backscatter from a Derived Two-dimensional Duststorm Model," *IEEE Transactions on Antennas and Propagation*, Vol. 49, pp. 1703–1711, December 2001.

Hansen, R. C., Fundamental Limitations of Antennas, *Proceedings of the IEEE*, pp. 170–182, February 1981.

Hartman, W. J. and Smith, D., "Tilting Antennas to Reduce Line-of-Sight Microwave Link Fading," *IEEE Transactions on Antennas and Propagation*, Vol. 25, pp. 642–645, September 1977.

Hollis, J. S., Lyon, T. J. and Clayton, Jr., L., *Microwave Antenna Measurements*. Atlanta: Scientific-Atlanta, 1970.

Howell, J. Q., "Microstrip Antennas," *IEEE Transactions on Antennas and Propagation*, Vol. 23, pp. 90–93, January 1975.

Institute of Electrical and Electronics Engineers, *IEEE Std 951–1996, IEEE Guide to the Assembly and Erection of Metal Transmission Structures*, Revised 2009.

International Telecommunication Union—Radiocommunication Sector (ITU-R), "Report F.2059, Antenna characteristics of point-to-point fixed wireless systems to facilitate coordination in high spectrum use areas", pp. 1–18, 2005.

Ishimaru, A., *Wave Propagation and Scattering in Random Media*, Volume 1. New York: Academic Press, pp. 41–68 (Fig. 3–5), 1978.

Kerr, D. E., *Propagation of Short Radio Waves*, Volume 13. New York: McGraw-Hill, 1951.

Kizer, G. M., *Microwave Communication*. Ames: Iowa State University Press, 1990.

Marcuvitz, N., *Waveguide Handbook*. London: Peter Peregrinus Ltd., 1986 (reprint of the 1951 McGraw-Hill Rad Lab Series, Volume 10, with errata).

Mosley, R. A., *Code of Federal Regulations (CFR), Title 47 - Telecommunication, Chapterl, Parts 101.113 and 101.115*. Washington, DC: Office of the Federal Register, published yearly.

Munson, R. E., "Conformal Microstrip Antennas and Microstrip Phased Arrays," *IEEE Transactions on Antennas and Propagation*, Vol. 22, pp. 74–78, January 1974.

Norton, M. L., "Microwave System Engineering Using Large Passive Reflectors," *IRE Transactions on Communications Systems*, pp. 304–311, September 1962.

Pozar, D. M. and Schaubert, D., *Microstrip Antennas*. New York: John Wiley & Sons, Inc., 1995.

Ramo, S., Whinnery, J. R. and Van Dozer, T., *Fields and Waves in Communication Electronics*, *First Edition*. New York: John Wiley & Sons, Inc., 1965.

Ramsay, J., "Highlights of Antenna History," *IEEE Communications Magazine*, Vol. 19, pp. 4–16, September 1981.

Ramsdale, P. A., "Antennas for Communications," *IEEE Communications Magazine*, pp. 28–36, September 1981.

Rummler, W. D., "A New Selective Fading Model: Application to Propagation Data," *Bell System Technical Journal*, Vol. 58, pp. 1037–1071, May–June 1979.

Rummler, W. D., "Time- and Frequency-Domain Representation of Multipath Fading on Line-of-Sight Microwave Paths," *Bell System Technical Journal*, Vol. 59, pp. 763–795, May–June 1980.

Ruze, J., "Antenna Tolerance Theory—A Review," *Proceedings of the IEEE*, pp. 633–640, April 1966.

Schelkunoff, S. A., *Electromagnetic Waves*. New York: Van Nostrand, pp. 476–479, 1943.

Schelkunoff, S. A., *Electromagnetic Fields*. New York: Blaisdell, pp. 224–238, 1963.

Schelkunoff, S. A. and Friis, H. T., *Antennas, Theory and Practice*. New York: John Wiley & Sons, Inc., p. 40, 1952.

Schelleng, J. C., Burrows, C. R. and Ferrell, E. B., "Ultra-Short Wave Propagation," *Bell System Technical Journal*, Vol. 21, pp. 125–161, April 1933.

Silver, S., *Microwave Antenna Theory and Design, Radiation Laboratory Series*, Volume 12. New York: McGraw-Hill, 1949.

Slayton, W. T., *Design and Calibration of Microwave Antennas Gain Standards, NRL (Final) Report 4433*. Washington, DC: Naval Research Laboratory, November 1954.

Southworth, G. C., *Principles and Applications of Waveguide Transmission*. New York: Van Nostrand, 1950a.

Southworth, G. C., "Principles and Applications of Waveguide Transmission," *Bell System Technical Journal*, pp. 295–342, July 1950b.

Stratton, J. A., Morse, P. M., Chu, L. J. and Hutner, R. A., *Elliptic Cylinder and Spheroidal Wavefunctions*. New York: John Wiley & Sons, Inc., 1941.

Telcordia Technologies, *Generic Requirements GR-180-Core, Generic Requirements for Hardware Attachments for Steel, Concrete and Fiberglass Poles*, May 2008.

Telcordia Technologies, *Special Report SR-1421, Blue Book—Manual of Construction Procedures*, October 2011.

TIA Subcommittee TR-14.7, *Structural Standards for Steel Antenna Towers and Antenna Supporting Structures, ANSI/TIA/EIA-222-F*, Arlington: Telecommunications Industry Association, 1996.

TIA Subcommittee TR-14.7, *Structural Standard for Antenna Supporting Structures and Antennas, ANSI/TIA/EIA-222-G*, Arlington: Telecommunications Industry Association, 2005 (Addendum 1, 2007).

Tilston, W. V., "On Evaluating the Performance of Communications Antennas," *IEEE Communications Magazine*, Vol. 19, pp. 18–27, September 1981.

Western Electric Company, "Significant Characteristics of Bell System Microwave Antennas," *Engineering Handbook, Systems Equipment & Standards*. New York: Western Electric Company, Engineering Division, 1970.

Wheeler, H. A., "Small Antennas," *Antenna Engineering Handbook*, *Third Edition*, Johnson, R. C., Editor. New York: McGraw-Hill, pp. 6-1–6-18, 1993.

Wu, K. T. and Achariyapaopan, T., "Effects of Waveguide Echoes on Digital Radio Performance", *IEEE Global Telecommunications Conference (Globecom)*, Vol. 3, pp. 47.5.1–47.5.5, December 1985.

6

DESIGNING AND OPERATING MICROWAVE SYSTEMS

6.1 WHY MICROWAVE RADIO?

Digital transmission is usually accomplished by cable or wireless methods. For long-distance transmission, fiber optics and microwave radios are the typical choices. Fiber optics has many obvious advantages, the primary advantage being transmission bandwidth. The disadvantages are also obvious: high cost of installation and maintenance, lead time for implementation, and inability to redeploy assets once installed. Fiber optics is also not practical in extreme terrain and climate locations. Microwave radio has a transmission bandwidth limitation due to channel bandwidth restrictions. However, it has several attractive features. The cost of installation and time for implementation can be much less than for fiber optics. Since only the terminal locations need to be maintained, maintenance cost is lower and transmission security is greater than for fiber optics. Radio is much easier to restore after natural disasters and it often survives them with little degradation. Fiber optics is economical for communication between major metropolitan areas and within an urban or suburban environment, but is seldom economical when smaller cities and rural areas are involved. Perhaps, the greatest advantage of radio is its ease of deployment. Microwave radio greatly simplifies system planning. As long as site-to-site path clearance is available, a radio path can be installed almost anywhere (Fig. 6.1). For short paths, even lack of line of sight (LOS) can sometimes be overcome.

6.2 RADIO SYSTEM DESIGN

There are many ways to design microwave networks—nearly as many as there are designers. Although the major tasks are usually the same, their order and importance is not. The following tasks are typical:

Create preliminary network design

Validate the site coordinates and tower information

Determine the system architecture and design objectives (transmission and network management)

Create candidate design based on terrain data and site and fiber hub candidate list

Perform path LOS surveys to confirm path availability and clearance

Digital Microwave Communication: Engineering Point-to-Point Microwave Systems, First Edition. George Kizer.

Figure 6.1 Microwave radio can be placed nearly anywhere. *Source*: Reprinted with permission of Alcatel-Lucent USA, Inc.

Perform site surveys

Perform site and tower mapping

Perform tower loading and attachment analysis

Perform inside and outside plant feasibility analysis.

Create final path design (based on actual cleared paths and site availability)

Perform regulatory requirements tasks: FAA antenna structure studies (if needed), frequency coordination, prior coordination notice and Form 601 data preparation

Perform final detail engineering

Create site integration plan

Create installation engineering plan

Generate bill of materials

Perform installation, test, and documentation.

The first order of business is to formalize the need. This is usually the defining of the end points requiring communication with each other. For a corporate data network or telephone system, this is typically a number of "core" sites and a number of "edge" user locations. For high capacity data networks, all sites may be "core" or all "edge." In any case, the first step is to establish termination points with capacity, reliability (robustness), and QoS (priority and timeliness) needs.

The next step is to establish a hierarchy (meta-architecture). A logical segregation of sites or functions may be necessary. A typical case is a large grouping of radio backhaul paths tied together or to a central office by a broadband fiber network. The actual hierarchy will be heavily influenced by the choice of transport technology [plesiochronous/"asynchronous" TDM, synchronous TDM (SONET/SDH) or ATM or IP packets]. Many, if not most, new microwave networks are converged IP technology.

Next is the consideration of architecture or topology. This will be influenced by transmission needs among users. Clusters may be needed or the network may be relatively uniform. Next, the physical span of the network must be considered. If a large number of nodes are all in a metropolitan area with relatively close spacing (e.g., a cellular network or a local business campus), a high frequency radio network (11 GHz or higher frequency) is usually most practical. If the network is a small number of nodes spaced over a large geographic network (e.g., a railroad or pipeline network), a low frequency (lower or upper 6 GHz) radio network is typically required. This is illustrated by the following maps of

Figure 6.2 Lower 6-GHz band utilization.

Figure 6.3 18-GHz band utilization.

the lower 6- and 18-GHz networks currently in place in the United States (based on the FCC licensed service database) (Fig. 6.2 and Fig. 6.3).

The basic architectures are as shown in Figure 6.4. The decision of architecture is based on perceived needs, such as network length (for coverage), diameter (for delay minimization), scalability (for flexibility to handle changing requirements), and connectedness (for robustness/reliability/survivability). Backbone and spur is the most common for long-distance designs where the transmission requirement is dispersed along a defined route.

Concatenated rings are often used to cover a metropolitan area (telecommunications providers use this to advantage in the Los Angeles and San Francisco areas where system redundancy and coverage are critical). Star, hub and spoke, and mesh are common high frequency metropolitan design approaches. Ring architectures (tied to a fiber aggregation hub), as they typically use nonredundant hardware configurations,

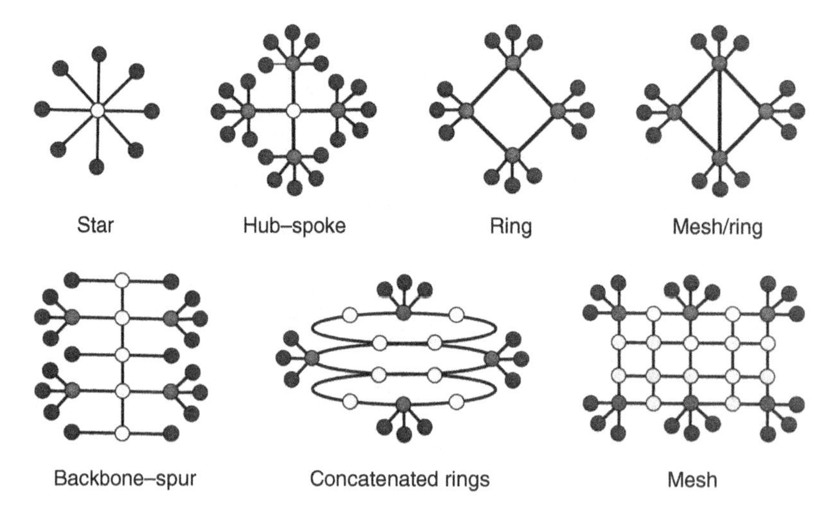

Figure 6.4 Typical architectures.

provide the highest coverage for the lowest cost. However, they are the least flexible. An entire ring is required regardless of the final coverage requirement. Equipment cannot be redeployed if requirements change and capacity expansion is awkward (adding a second ring or bifurcating the initial ring with additional hubs) as initial deployment of rings usually requires high bandwidth radios.

Rings usually home to a single-fiber aggregation hub (fiber point of presence or POP). An alternative is to use two-fiber aggregation hubs with a backbone and spur architecture connecting them ("necklace"). The main design characteristics of both approaches are similar.

The hub and spoke provides more flexibility than the ring but at a higher cost. The hub and spoke equipment can be redeployed as coverage requirements change without impacting existing network functionality. Capacity upgrades are usually easier because the initial implementation is relatively low bandwidth radios. With the migration of microwave networks to IP technology, hot standby configurations are being replaced with dual radios connected to routers. With dual IP radios, the path capacity is restricted on failure but under normal operation, capacity is double that of a hot standby radio. Replacing hot standby with IP-routed radios is simple. However, if both radios are to be used for active traffic, route link aggregation (point-to-point LAG) is required. Currently, not all routers can separate traffic for transmission on multiple physically parallel (collocated) paths and reaggregate it at the far end.

Small mesh networks (diamonds and bifurcated rings) require symmetric loads on all nodes to be effective. Large mesh networks are the most flexible and redundant, but are also the most expensive, owing to the large number of redundant paths. The performance of these networks (transmission capacity, reliability, and latency) is difficult to predict. Gupta and Kumar (2000) note that for two-dimensional meshes, per node traffic throughput capacity degradation with additional nodes is generally proportional to $[W/(n \log n)^{1/2}]$ for randomly located nodes and traffic patterns. W is a single-path (node to adjacent node) traffic capacity in bits per second and n is the total number of nodes in the network. If the nodes are optimally placed in a disk pattern and traffic is optimized, the capacity degradation becomes proportional to $(W/n^{1/2})$. Jun and Sichitiu (2003) noted that if gateways to other networks [such as to the Internet or to a high speed fiber optic metropolitan area network (MAN)] are added to a random two-dimensional mesh network, the extra congestion created by the gateways cause the per node traffic throughput capacity degradation to be proportional to (W/n). Sometimes, 60-GHz radios are spread throughout a building to create a three-dimensional mesh network. Gupta and Kumar (2000) note that for spherical three-dimensional meshes, per node traffic throughput capacity degradation is generally proportional to $[W/(n\log^2 n)^{1/3}]$. Since channels near the center of a mesh tend to experience higher traffic than near the periphery, this result would be effectively the same for square networks. Anticipated network growth should also be considered in metropolitan designs. Mesh networks are relatively easy to expand. However, consideration of throughput reduction when nodes are added should be kept in mind.

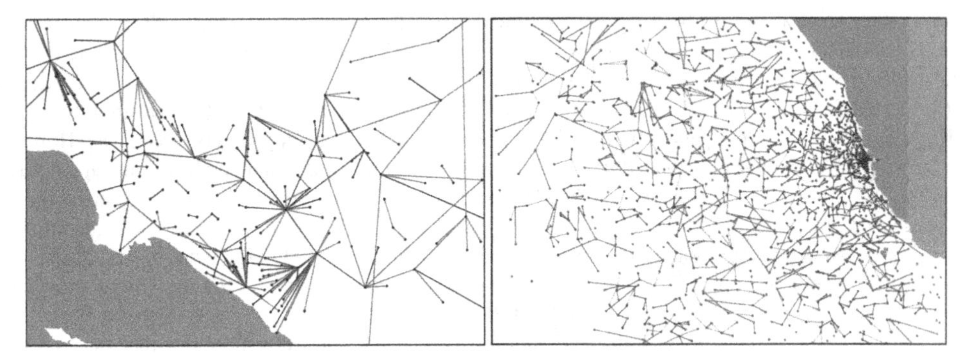

Figure 6.5 Metropolitan radio networks.

Integrated radio and fiber optics subnetworks are quite common. The examples in Figure 6.5 show two different approaches for cellular networks. Both network designs use extensive fiber-optic rings to pull the networks together and add survivability to the network.

Low frequency networks using all indoor radios for long-distance networks have been engineered since the late 1940s. With the emphasis on metropolitan networks, high frequency radios (often split package radios with RF on the tower behind the antenna and a baseband interface unit in the telecommunications facility or just an IP radio directly behind the antenna) have become more common. High frequency networks represent unique challenges for the network designer. Typically, they are very large (300 to over 1000 site networks are common). All microwave radio paths are limited by terrain clearance. However, high frequency radios are also limited in performance by rain attenuation ("fading"). Minimizing the impact of rain outages is a dominant concern for the network designer. While necklace (backbone connecting two-fiber aggregation hubs) and hub/spoke architectures are the most common, some designers use novel designs to minimize the rain effect. While heavy rain will always impact a node, the architecture is usually designed is such a way that the impact to the network is minimized.

6.3 DESIGNING LOW FREQUENCY RADIO NETWORKS

Low frequency radio paths (frequency $<10\,\text{GHz}$) have been designed for decades. These systems are usually composed of a relatively small number of sites covering a large (state or multiple state) geographic area. Usually, the design process revolves around defining good locations for terminal or repeater sites and site acquisition. In general, paths should be chosen to avoid flat terrains that will support surface reflections (as noted in Chapter 13, all flat surfaces reflect radio waves). If flat terrain is unavoidable, increasing path inclination can be used to move the reflection point to rough terrain or behind an obstruction to reflections. Sometimes, terrain features such as trees or hills can be used to block reflections. Space diversity is usually required to make reflective paths perform adequately.

Paths between two high locations (such as between two mountain sites) are especially challenging. Invisible atmospheric layers below the mountains cause unpredictable reflections resulting in unusually large amounts of dispersive fading (Chapter 9). The use of space diversity is usually required but excessive clearance over the reflective layers (and their unpredictable movement up and down) makes antenna spacing optimization difficult. Use of quad diversity on these paths is sometimes required. If possible, move the antennas as far away from the edge of the site as possible so that the antennas only see the far-end antenna but do not receive signals from below the path. This can significantly improve path performance.

Candidate sites are determined based on height and anticipated clearance considerations. A terrain database and anticipated tree or building heights are used to draw a path profile (vertical plot of terrain height vs. distance). The path is analyzed for path clearance on the basis of predefined path clearance criteria and minimum multipath (and rain) availability calculations. Currently a significant unsolved problem is in anticipating path obstructions. While Internet sites offer detailed vertical pictures of many geographical areas, tree and building heights are difficult to determine. Existing USGS land use and canopy data is too granular to offer much guidance. Radar and laser (light detection and ranging or lidar) land clutter ("canopy") mapping is expensive and not always adequate. Both radar and lidar mapping perform well for vegetation in the northwestern and northeastern United States (if measured during full foliage seasons when depth penetration is slight) where vegetation typically covers large areas. However, they usually miss the narrow isolated trees common in the southwestern and southeastern states (radar's relatively large aperture averages the tree height with surrounding terrain and the laser measurement grid is often too granular). Both methodologies have difficulty accurately measuring building heights in dense urban areas (reflections and multipath are a serious measurement issue). This issue is one of the most challenging issues for simplifying and reducing path design cost. Currently, the designs of most paths require some form of physical survey to confirm the path's viability.

If the path appears suitable, the sites and path are evaluated by actually surveying the sites and path. Usually, a site survey is performed to evaluate any existing site or potential site. If the site exists, the site and tower are usually mapped (Fig. 6.6).

The tower's vertical mappings should clearly define and locate existing antennas, whips, lights and beacons, ice shields, transmission line ladders, star mounts, and guy wires. This will affect the placement of new antennas. For FAA and FCC filings, the tower height with and without appurtenances will be required (an appurtenance is any item attached to an antenna structure). The site layout map and its accompanying building in the plant report will define the location of the tower and potential new equipment. This will help estimate the transmission line distances between the tower and the proposed equipment. The tower's longitude and latitude should be precisely determined. If the equipment is available, ground elevation should also be determined.

During the site survey, many practical factors should be considered:

- Inside plant
- Location of available equipment mounting locations
- Radio transmission line access between the indoor and outdoor equipment
- Power access and adequacy
- Interface to other transmission equipment or user interface equipment
- Interface to alarm and management systems
- Outside plant (in addition to the tower or antenna mounting structure)
- Anticipated building code issues
- Site security
- Relationship to nearby rivers, lakes, ocean, and mountains (sources of unanticipated water, high winds, and boulders)
- Anticipated weather factors (wind, sand, rain, ice, and snow)
- Location and height of nearby buildings and trees (potential path blockages)
- Road and power access
- Relationship to nearby airports or helicopter pads (commercial, private, or military)
- Feasibility of providing anticipated radio transmission lines, power, and alarm cables into and around the facilities (transmission, power, and alarm cables often have maximum practical run distances).

A surveyor walks, drives, or flies (by plane or helicopter) the path and notes all obstructions along the path. The path is reevaluated based on a path profile using actual surveyed data. GPS receivers are commonly used for path surveillance (Fig. 6.7).

If the path still passes appropriate design criteria (path has adequate clearance and adequately sized antennas can be attached at the end locations), site acquisition (buy, rent, or lease) is attempted. If this process fails at any point, the path and site selection process starts over again.

For low frequency system, because they typically cover a large geographical area, the network design often takes a relatively long time (on the order of weeks). Current software, which requires the user to manually enter all data, is adequate for this relatively slow process.

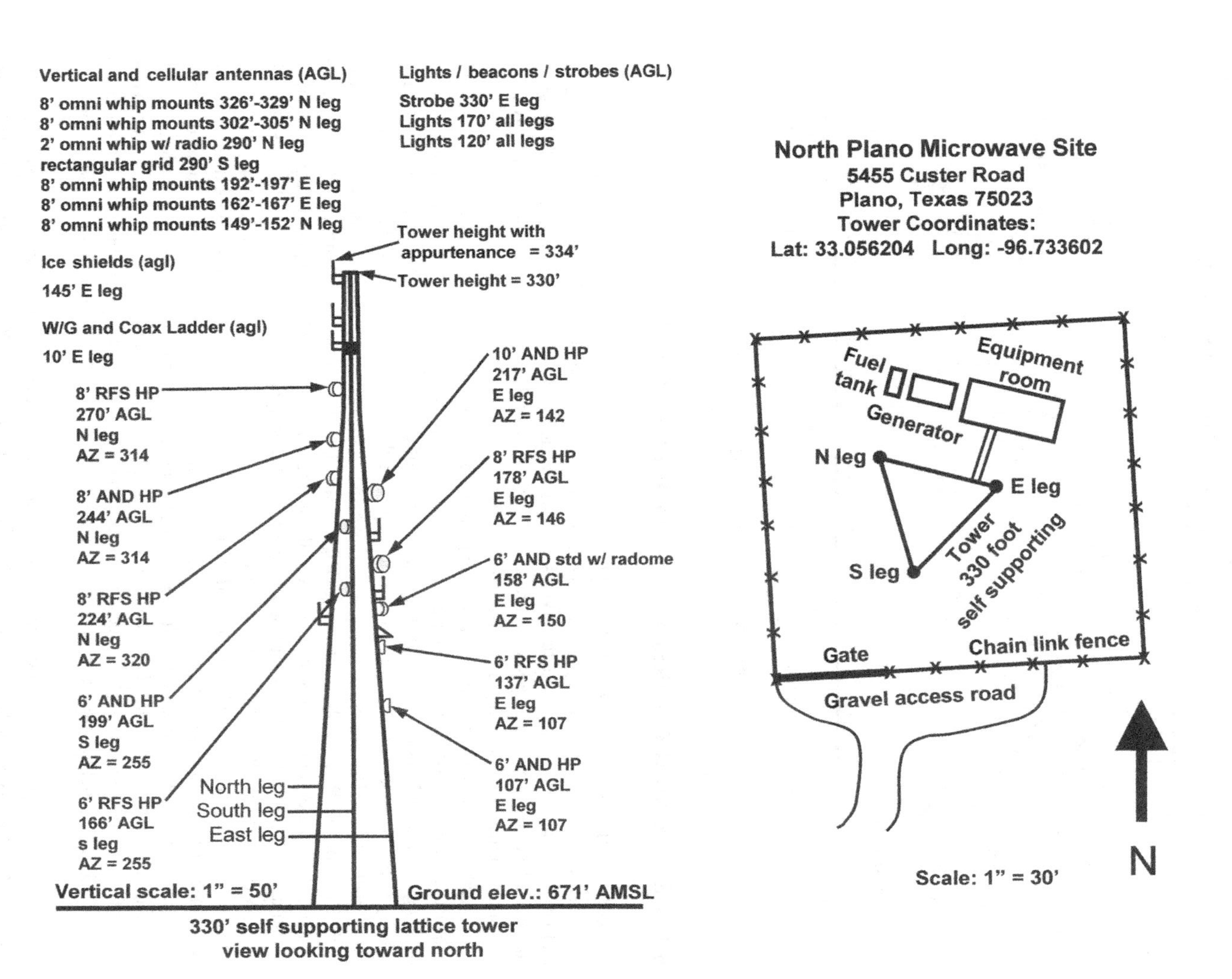

Figure 6.6 Example tower and site mappings.

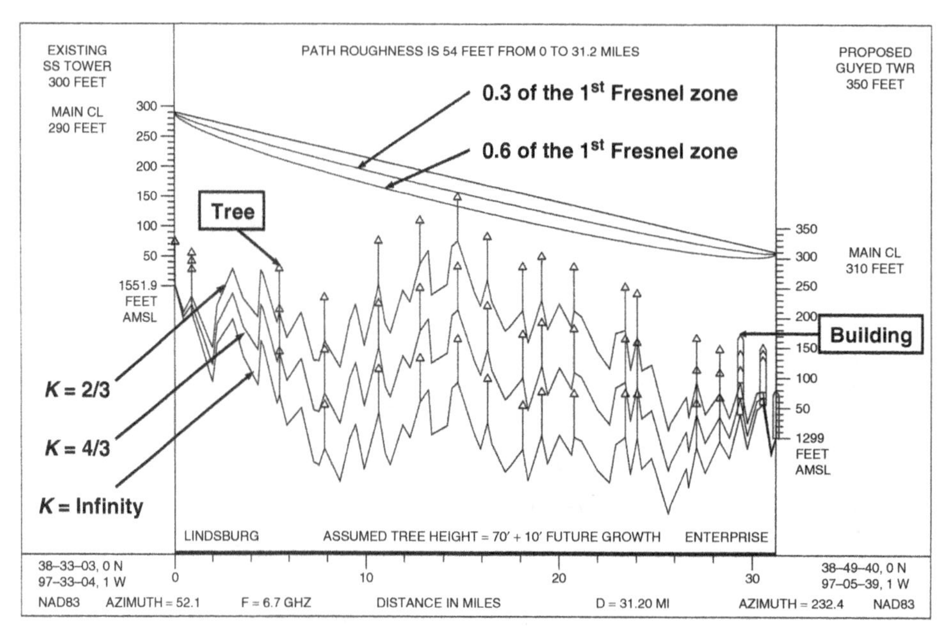

Figure 6.7 Example surveyed path profile.

6.4 DESIGNING HIGH FREQUENCY RADIO NETWORKS

Sometimes high frequency networks (frequency >10 GHz) are a few links connecting a business campus. However, more commonly, today these systems comprise of hundreds of sites covering an entire city and providing backhaul for new data services. Low frequency and high frequency radio path design has much in common. However, the site volume and churn of many high frequency network designs can be a daunting transmission engineering challenge.

Typically, the service provider is attempting to utilize known sites on a rental or lease basis. Site acquisition or evaluation may be going on simultaneously with the path design. Hundreds of paths must be designed (and redesigned) very quickly (on the order of days, not weeks). If time is important (and it usually is for large networks), automating at least part of the design process is highly desirable.

Virtually all modern metropolitan designs have the radio backhaul networks converging at a fiber POP. The fiber POP is usually part of a SONET ring connecting to a central office or Internet access location. The first design step is critical to overall coverage and success of the market design. That step is to estimate which potential POPs will be suitable for radio designs (Aoun et al., 2006; Dutta Kubat Liu, 2003; McGregor and Shen, 1977). If the network uses IP radios, end to end concatenated path latency must be evaluated. If the candidate network is hub–spoke, the criterion might be how many sites can be connected to the POP by paths meeting the maximum path length, minimum path clearance, and the maximum number of cascaded hops. If the candidate network is ring and spoke, the criterion might be the number of paths that meet the maximum path length and minimum path clearance. If fiber is not currently at the potential sites, the process is complicated by finding sites where fiber ring access can be provided at a reasonable price. Currently, this is a semiheuristic process of matching sites to expected coverage areas.

Many urban high frequency designs involve hundreds of sites (and thousands of potential paths). Manual path and system design is costly from a time and labor perspective. To speed up the path design process, software can yield significant labor and time savings. Candidate radios and antennas are chosen. Potential sites are recorded in a database with name(s), site coordinates, site address, and maximum practical height. Sites are also graded by capability to support microwave radios. An example would be high grade (buildings or self-supporting towers capable of supporting many radios), medium (guyed towers and monopoles capable of supporting some radios), and low (small or skinny structures capable of supporting only one radio—suitable only for end sites). Design rules are established for using those sites (Doshi and Harshavardhana, 1998; Dutta and Mitra, 1993; Gersht and Weihmayer, 1990; Ruston and Sen, 1989; Dijkstra, 1959). Some are common to all designs and some are architecture specific. Most current network design packages do not support sophisticated radio network designs (Bragg, 2000; Kasch et al., 2009). However, new versions of microwave radio path design programs are starting to include some of these features.

First a "spider web" of all possible radio paths is created. This will typically be thousands of candidate paths. These "cleared paths" are then used to synthesize an appropriate system design (Fig. 6.8). Perhaps the most challenging aspect of semiautomated high frequency design for cellular backhaul networks is determining accurate antenna structure locations, structure types, and potential antenna heights. At this time, another challenging part of this design process is in obtaining adequate terrain clutter (trees, buildings, and other structures) data at a reasonable price (as noted earlier for low frequency design). All current radar and laser terrain mapping methodologies have their limitations as noted earlier.

Adding to the difficultly of automating high frequency design is the lack of accurate site geographic coordinates (longitude, latitude, and elevation). For cellular networks, this data is notoriously unreliable. Location inaccuracies of the order of over 100 ft are common. This significantly increases the error in path clearance analysis (since vegetation and building clearance is highly location dependent).

The basic idea is to take a group of sites (typically predefined), look at a subset of all possible paths and discover a set of paths that provides reasonable coverage while meeting predefined design rules (Chattopadhyay et al., 1989; Ju Rubin, 2006; Ko et al., 1997; Kubat et al., 2000; Luss et al., 1998; Soni et al., 2004). The optimum design is not required (and often would take too long). Since time to market is generally the overriding consideration, an adequate design that meets minimum standards is usually the criterion.

Two architectures popular in metropolitan high frequency design are discussed in the following sections.

6.4.1 Hub and Spoke

A critical aspect of this type of design is the proper placement of hub sites. This is a challenging issue because coverage will be a factor of antenna structure height, but usually access to a fiber ring is also required for backhaul. For high capacity systems, two or three tiers (layers of paths in series with

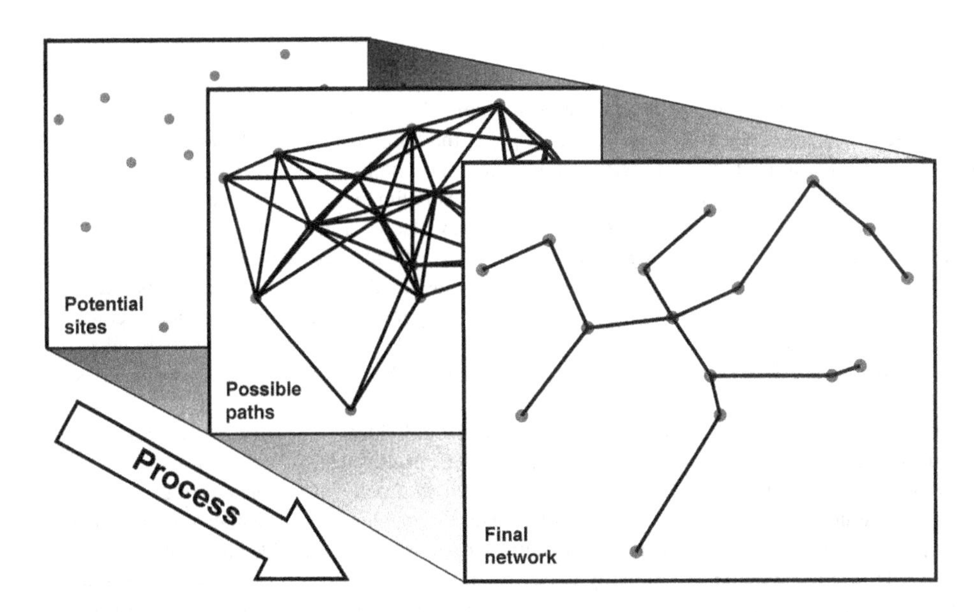

Figure 6.8 Semiautomated design process.

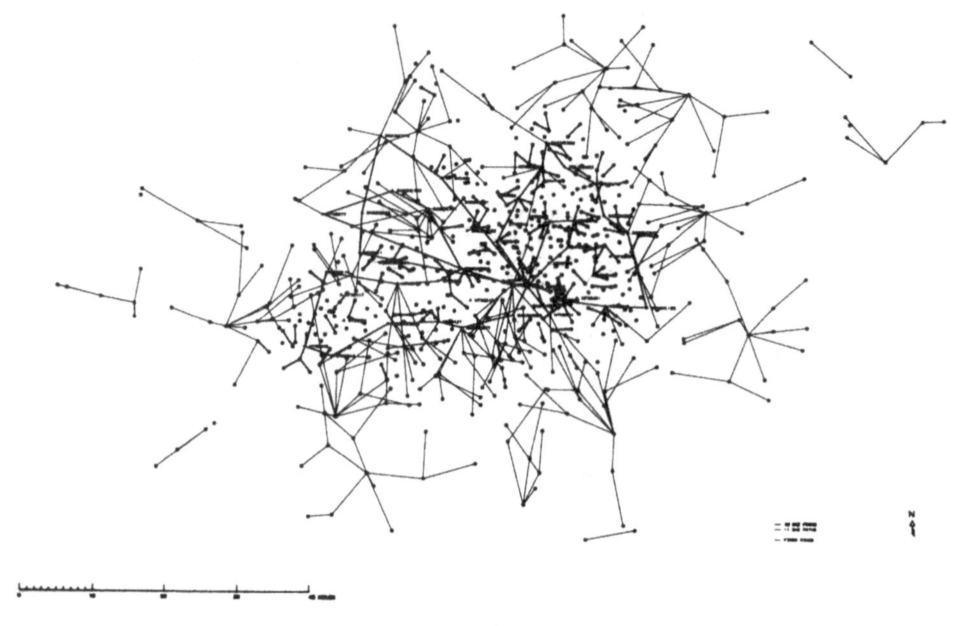

Figure 6.9 Automated hub–spoke design.

the hub) are usually acceptable. For low capacity systems, four or five may be acceptable. Eventually, the end to end rain outage path availability and/or concatenated path latency become the dominant limitations.

When an automated design process is used, it is usually based on a breadth-first (Knuth, 1997) search of paths meeting the site and path length maximums. This process is continued until adequate area coverage is obtained. Figure 6.9 is an example (Dallas/Fort Worth) of the results of this type of automated program (the long dark lines are fiber rings).

6.4.2 Nested Rings

Here, the task is to define methods of creating multiple rings from each of several fiber POPs that do not interfere with each other. As with hub and spoke, the hub location is important. However, it is not as challenging because only a few hubs are required. The limitation to the ring size is primarily the ring transport capacity (which limits per site capacity). This limitation must be defined before beginning network design. For IP networks, end to end latency may also be a criterion.

In the ring design, paths that begin and end at a node ("hub") are designed. Either spurs or additional ("secondary") rings are added to the primary ring if additional sites need to be reached. The limitation of additional rings and spurs is the primary ring transport capacity. Usually secondary rings are used if significant additional distance coverage is needed (using the loop to minimize the cascaded path unavailability). If only a site or two needs to be added, simple concatenated spurs are used.

Once the area is grouped and sectored around the hubs, individual rings must be determined. Typically, this uses a depth-first (Cormen et al., 2001) (to yield a result quickly) or breadth-first (Knuth, 1997) (to yield all possible paths) search method. The search is usually based on meeting a minimum and maximum number of nodes in the ring and, of course, the maximum path distance per frequency band. After primary rings are created, secondary and tertiary rings and spurs are added to complete the coverage. This is challenging to automate primarily because customers typically have many design requirements to be met. Generally, each set of rings (and finally spurs) must be selected from the set of all possible rings ordered by the appropriate criteria (such as overall ring distance). Each network layer was added until the design was complete. Totally automated algorithms have also been utilized but are challenging to implement if customer constraints are significant.

For networks where hub reliability is a concern, a variant of this design method is to use two hubs rather than one ("necklace" design). All paths that were a ring now terminate on two hubs rather than one. While this improves overall reliability, it complicates the design by requiring two hubs for each set of paths. Otherwise, the design methodology is similar to the basic nested ring.

6.5 FIELD MEASUREMENTS

After an initial tabletop design effort, the sites and paths must be finalized. Site surveys validate the antenna mounting structure and equipment locations. Actually locating the correct location and structure is a surprisingly difficult task. Many locations are hundreds of feet from any street so street names and numbers can be problematic. Site coordinates are notoriously wrong (for many reasons). Often, structure types have been modified significantly or new structures built since the last drawings were made. Sometimes, owing to faulty records, the site is at one location, the address of record at another, and the site coordinates of record at a third. Even when the correct location is found, there may be several different potential antenna structures at that location. Owing to its criticality to the path design, finalizing site locations and structures is not trivial.

The next significant task is verifying that the potential path (with anticipated antenna heights) meets the minimum path clearance. If the site coordinates and antenna heights are known (and that is a big "if"), in theory, satellite images could be used to validate a path. While this is helpful, lack of current data (and lack of height information) is a significant limitation to this approach. In Canada, some operators use helicopters equipped with terrain tracking radar to map terrain heights. In urban areas, stereo images that yield terrain heights with 1- to 2-m vertical and horizontal resolution are available. However, their cost is currently prohibitive for large urban networks. Although it is time consuming and difficult to meet typical urban high frequency development schedules, using path surveyors to verify path clearance is the current method of final path verification.

6.6 USER DATA INTERFACES

In North America, most national and regional networks are gravitating to a three-layer model (Fig. 6.10). At the highest, fastest layer, very high speed proprietary synchronous systems dominate. At the intermediate level, virtually all systems are standardized synchronous systems. At the lowest user interface layer, there is a wide range of interfaces. Most systems today use the TDM technology, but migration to IP interface and transport is clearly the future.

When specifying a physical data interface, at least three basic questions must be answered:

What defines a bit (signaling format)?

Where are the bits located (method of synchronization)?

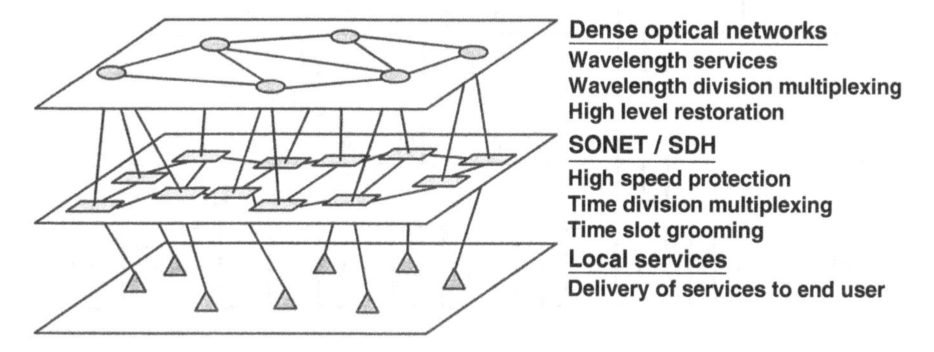

Figure 6.10 National networks.

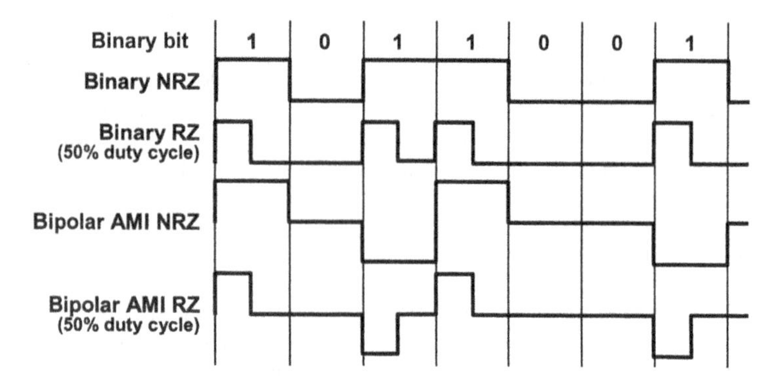

Figure 6.11 Basic binary coding formats.

Where are the sequenced continuous bits that represent a message (packet or frame message format and sequencing)?

In the early development of TDM digital transmission, several different signaling techniques were used (Stallings, 1984). Today, there are three basic TDM bit signaling formats (Fig. 6.11):

Nonreturn to Zero (NRZ). Signaling voltage maintains its level until the next signal.

Return to Zero (RZ). Signaling voltage returns to zero level before the next signal.

Bipolar Alternate Mark Inversion (AMI). Binary 0s are represented by zero amplitude. Successive binary 1s are represented by alternating positive and negative levels of the same amplitude.

The bit coding formats shown in Figure 6.12 are used in telecommunications signals (American National Standards Institute (ANSI), 1996). BPV means bipolar violation (explained later in this chapter). B means normal bipolar signal and V means bipolar violation. Bipolar signals and the above abbreviations will be explained subsequently.

Messages are subdivided into frames (or packets) (Fig. 6.13).

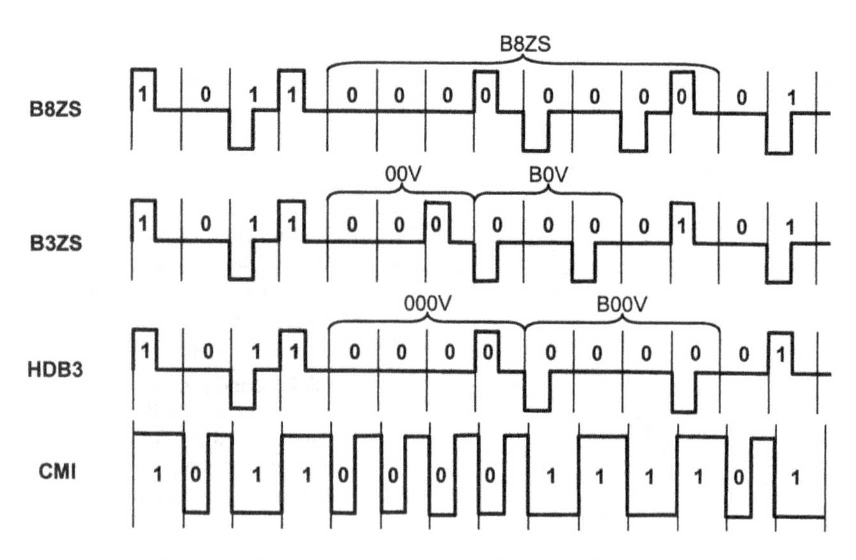

Figure 6.12 Telecommunications binary coding formats.

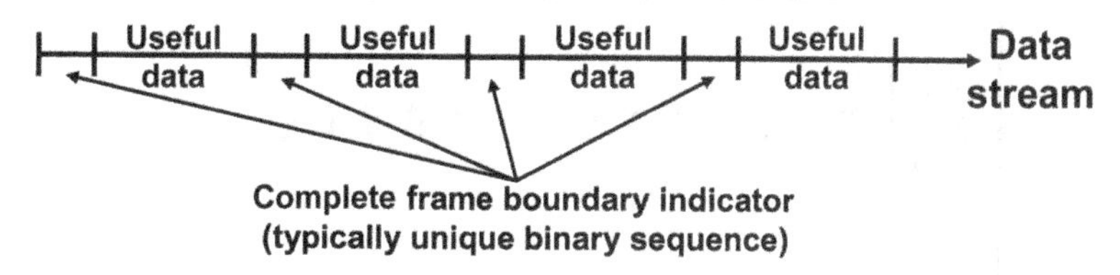

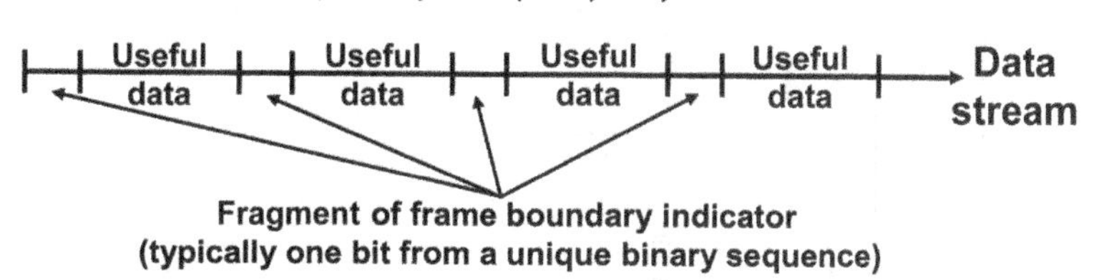

Figure 6.13 Message frames.

The choice of user data interface will impact the way data is managed and therefore, can influence architecture. Both North American and European TDM *plesiochronous systems* (American National Standards Institute (ANSI), 1993; American National Standards Institute (ANSI), 1995a; International Telecommunication Union—Telecommunication Standardization Sector (ITU-T), 1993; International Telecommunication Union—Telecommunication Standardization Sector (ITU-T), 2001; International Telecommunication Union—Telecommunication Standardization Sector (ITU-T), 1998) have been the primary user data interface since the late 1960s (Fig. 6.14 and Fig. 6.17).

The North American "asynchronous" hierarchy defines a common digital interface location termed a *DSX point*. This is a location where digital signals of common rate and signal shape can be interconnected (patched or "rolled") and tested. It is a concept only used in North American Digital Signal (DS-N) formats (it is included in the ANSI T1-102 (American National Standards Institute (ANSI), 1993) specification but is not in the ITU-T G.703 (International Telecommunication Union—Telecommunication Standardization Sector (ITU-T), 2001) specification for DS1 signals) (Fig. 6.15 and Fig. 6.16).

Equipment line build out (LBO) circuits are required in digital equipment to achieve cross-connect power and waveform requirements. The LBO circuits are typically specified for a reference length of a reference cable. If a different cable is used, the cable reference length changes.

All data sources and sinks must be synchronized (Bregni, 2002; Okimi and Fukinuki, 1981). The early data networks were composed of pairs of point to point channel banks with the master channel bank synchronizing the slave channel bank. All pairs of channel banks were synchronized but plesiochronous ("asynchronous") relative to other pairs. For all these channel banks, all traffic terminated at

Digital signal level	DS1 signals	Line rate (kb/s)	Line code	Voice channel equivalent
"DS0"	1/24	64	-	1
DS1	1	1544	AMI/B8ZS	24
DS2	4	6312	B6ZS	96
DS3	28	44736	B3ZS	672

Ref: ITU-T G.702/703/704; ANSI T1.102/107

Figure 6.14 North American "asynchronous" (plesiochronous) digital hierarchy.

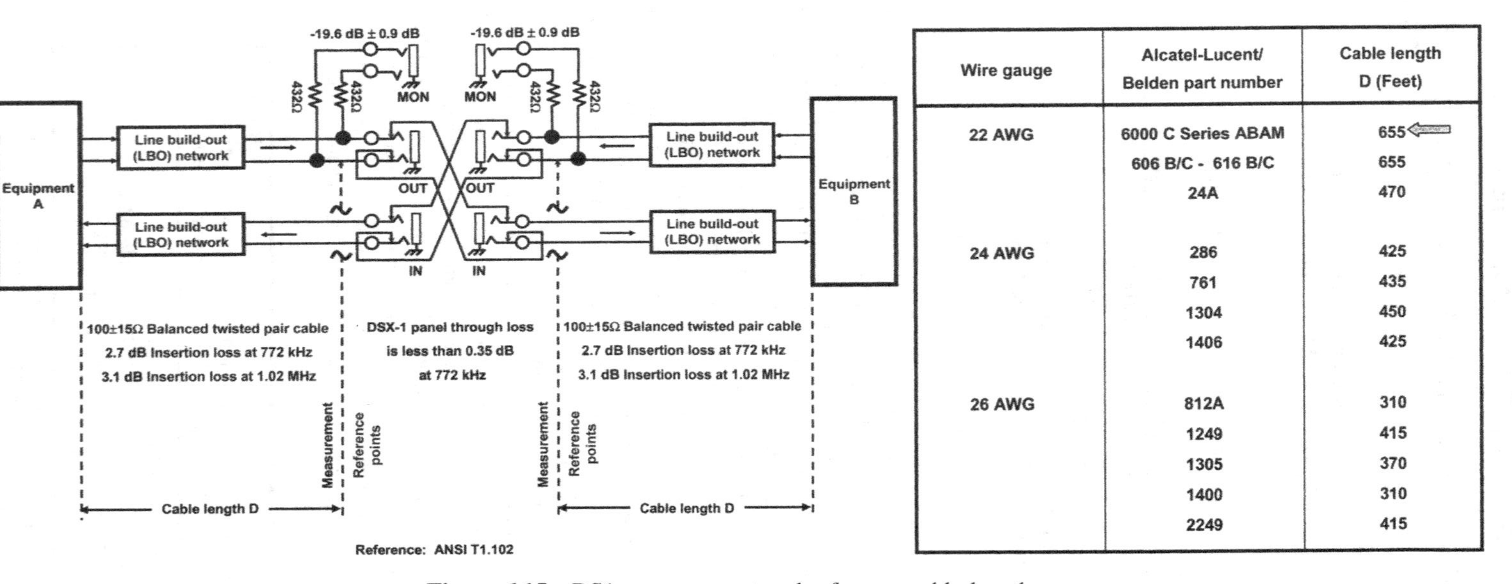

Wire gauge	Alcatel-Lucent/ Belden part number	Cable length D (Feet)
22 AWG	6000 C Series ABAM	655
	606 B/C - 616 B/C	655
	24A	470
24 AWG	286	425
	761	435
	1304	450
	1406	425
26 AWG	812A	310
	1249	415
	1305	370
	1400	310
	2249	415

Figure 6.15 DS1 cross-connect and reference cable lengths.

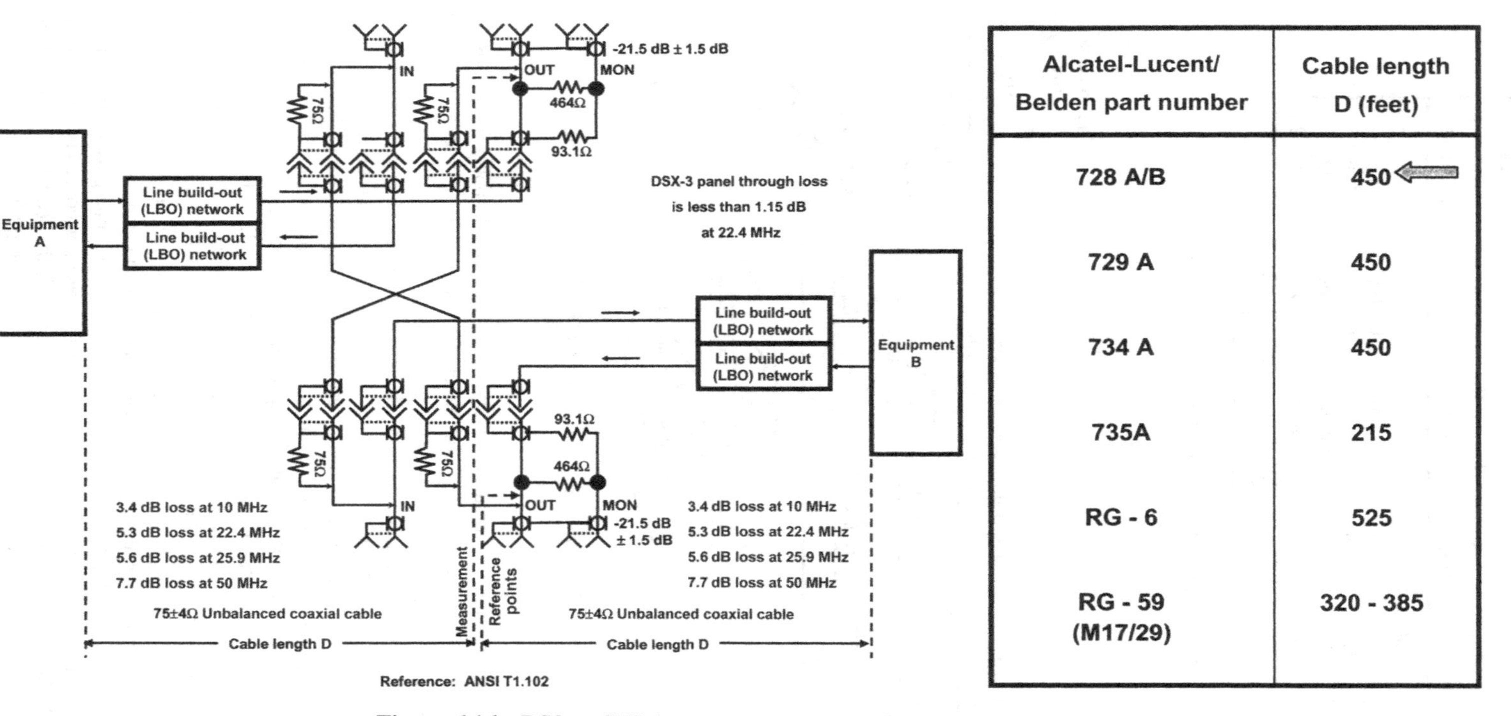

Alcatel-Lucent/ Belden part number	Cable length D (feet)
728 A/B	450
729 A	450
734 A	450
735A	215
RG - 6	525
RG - 59 (M17/29)	320 - 385

Figure 6.16 DS3 or STS-1 cross-connect and reference cable lengths.

Designation	E1 signals	Line rate (kb/s)	Line code	Voice channel equivalent
"E0"	1/30	64	-	1
E1	1	2,048	HDB3	30
E2	4	8,448	HDB3	120
E3	16	34,368	HDB3	480
E4	63 or 64	139,264	CMI	1890 or 1920

Ref: ITU-T G.702/703/704

Figure 6.17 European plesiochronous digital hierarchy.

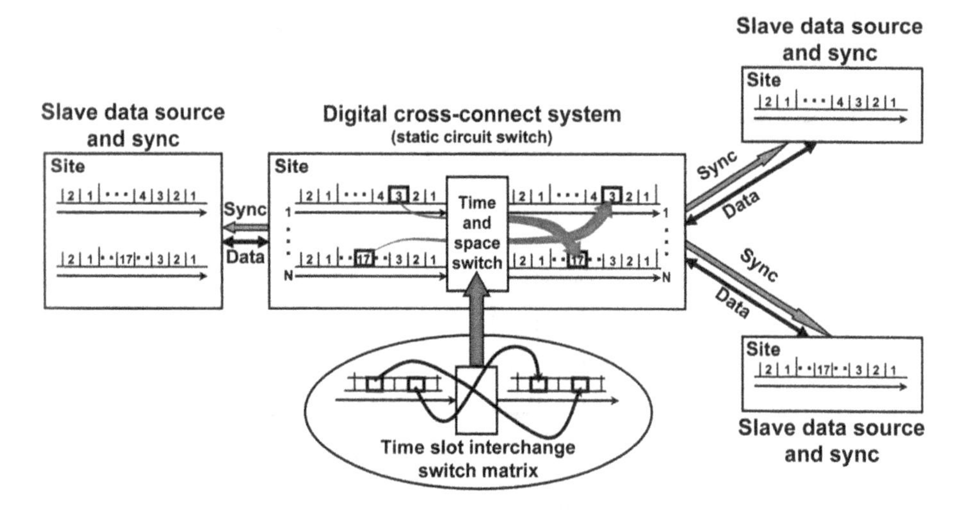

Figure 6.18 Mutually synchronized DS1/E1 data network.

each bank. With the introduction of subrate ($n \times$ "DS0" or $n \times$ "E0") drop and insert cross-connects, not all traffic dropped at a single digital terminal. In this environment, mutual synchronization was required to avoid frame slips (Fig. 6.18).

If a single cross-connect connects all low speed digital terminals, they all sync slave to the cross-connect. If multiple cross-connects are used, the cross-connects must each be mutually synchronized. If the cross-connects are distributed throughout the network in intelligent channel banks containing small cross-connects, each channel bank must be synchronized. Radios transporting the low speed data merely loop time to that signal and transport the data transparently and need no synchronization to the data.

Plesiochronous hierarchies support subrate (slower than DS0 or E0) signals. For these, the nominal signaling speed is predefined. The short data strings make absolute timing accuracy unnecessary. Standard speeds are 110, 150, 300, 600, 1200, 2400, 3600, 4800, 9600, and 19,200 bits per second (b/s). Typical electrical format is V.24/28 or RS-232. Data is organized into 8- to 11-bit strings. Zero (space) is high voltage and one (mark) is low voltage (inverted data). The first bit is a space start bit followed by 5-, 6-, 7-, or 8-character bits. Usually, the character is 8 bits (7-character bits plus a parity bit or 8-character bits and no parity bit). The parity bit can be even, odd, mark, space, or none. The 8-character bits usually represent an ASCII character. The 8-character bits are sent with the least significant bit first and most significant bit last. The bit string ends with 1, 2 (and rarely $1\frac{1}{2}$) stop bits.

Higher speed signaling (integer multiples of 64 kb/s) is organized by the data source and sink. For these signals, transmission is usually synchronous with the transmission clock being provided by the transmission equipment. The data source and sink loop time to this clock reference. Typical electrical format is V.35 or RS-422.

For plesiochronous DS1, DS3, E1, and E3 signals, signal timing is determined loosely by a predefined nominal frequency with specified absolute accuracy (ppm or b/s). The data source uses an external or internal reference frequency to transmit the digital signal. Received signal synchronization is achieved by frequency and phase locking to the incoming data stream of bits. Successful receiver synchronization depends on maintaining a minimum of data transition activity. Data activity is maintained through the use of an appropriate line code. The data source encapsulates the baseband signal to be transmitted into a predefined digital frame. The data sink locates the digital frame and retrieves the transported data.

For DS1, DS3, E1, and E3 signals, the basic signal is Bipolar AMI RZ with 50% duty cycle. Pulse shape must meet predefined shapes ("pulse mask") at the data receiver. The signal is bipolar with sequential 1s of opposite polarity. Consecutive nonzero signals of the same polarity represent a BPV.

DS1 signals use one of three transmission formats:

AMI. This is a bipolar signal with sequential 1s of opposite polarity (AMI). The Mux/Demux equipment ensures no more than eight consecutive 0s. To maintain a minimum pulse density of 12.5%, one bit out of eight must be reserved for pulse density maintenance. This limits a DS0 data channel to 56 kb/s for data transmission.

B8ZS. (*bipolar with eight-zero suppression*). This is a bipolar signal with a sequence of 1s of opposite polarity (AMI). The sequence 000VB0VB is inserted for eight consecutive 0s. Since this requires the signal to be buffered for 8 bits, this format introduces a nominal delay of about 5 μs. Since no in-channel bits are used, it supports 64 kb/s clear channel transmission per DS0.

ZBTSI. (*zero byte time slot interchange*). This format processes the data stream to remove excess zeros. It requires data channel in ESF format (so an overhead channel is available). It introduces approximately 1.5-ms delay. It is rarely used today.

DS3 signals use bipolar with three-zero substitution (B3ZS). Each block of three consecutive zero signals is replaced by 00V or B0V (V represents a bipolar violation and B represents normal bipolar signaling). The choice of substitution block is made so that the polarity of consecutive V signal elements alternates to avoid introducing a DC component into the signal (number of B pulses between consecutive V pulses is odd).

E1, E2, and E3 signals use high density bipolar of order three (HDB3). Each block of four consecutive zero signals is replaced by 000V or B00V. The choice of substitution block is made so that the polarity of consecutive V signal elements alternates to avoid introducing a DC component into the signal (number of B pulses between consecutive V pulses is odd).

E4, SONET, and SDH signals use coded mark inversion (CMI). This is an NRZ 100% duty cycle code with two signaling voltage levels of the same amplitude but opposite polarity. Binary 1 is represented by either voltage level being sustained for one full signaling time interval. Successive binary 1s use alternate voltage levels. For binary 1, there is a positive transition at the start of the binary unit time interval if in the preceding time interval the signal level was low. For binary 1, there is a negative transition at the start of the binary unit time interval if the preceding last binary 1 signal level was high. Binary 0 is represented by both voltage levels, each being sustained consecutively for half a signaling time interval. For binary 0, there is always a positive transition at the midpoint of the signaling time interval.

DS1 signals come in two formats as shown in Figure 6.19 and Figure 6.20 (American National Standards Institute (ANSI), 1995a).

DS3 signals have one general format as shown in Figure 6.21 (American National Standards Institute (ANSI), 1995a).

DS3s can be operated in one of four ways:

M13. This is the original format. It supports DS2 mapping. It does not support C-bit alarm, status, or loopback features.

C-Bit Parity. This mode supports end to end performance management (PM) and control by redefining DS2 stuffing bits (but it is typically implemented differently by different manufactures). Owing to

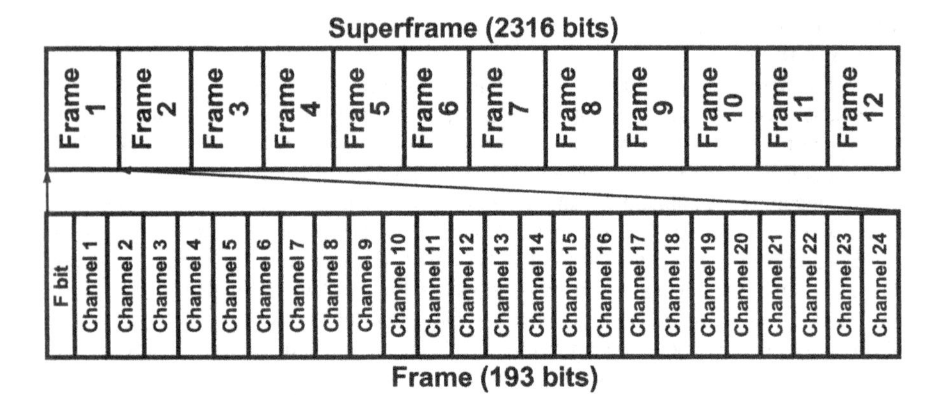

Frame

Frame 125 μs long
Frame = 24 DS0 channels (8 bits) + one F (Frame) bit

Superframe ("D4")

Composed of 12 frames
F bits used for framing (sequence: 100011011100)
Two signaling channels, A & B
Signaling bits robbed from 8th bit of each 6th DS0 byte
Signaling typically used for E & M signaling

Figure 6.19 DS1 superframe format.

proprietary implementations, it often has interoperability issues with different vendors' equipment. It is incompatible with M13 format.

Clear channel. This format usually retains only DS3 framing. It is used for encrypted data or video transmission. Lack of zero suppression and nonstandard pulses can cause compatibility issues.

Syntran. This is an obsolete synchronous format rarely used today.

SONET and SDH TDM formats (American National Standards Institute (ANSI), 1993; American National Standards Institute (ANSI), 1995b; International Telecommunication Union—Telecommunication Standardization Sector (ITU-T), 2003) are very popular for medium- and long-distance data transport (Fig. 6.22).

The North American SONET format is defined (American National Standards Institute (ANSI), 1995c; American National Standards Institute (ANSI), 1997) for four data rates as shown in Figure 6.23.

Within the lowest speed format, STS-1 or STM-0, various synchronous virtual tributaries (VT-X) are defined for encapsulating plesiochronous and packet signals. For SONET STS-1, four VTs are defined for DS1 signals (Fig. 6.24).

Locked Mode. This is an obsolete mode that locks all VTs together (no VT pointer processing) within the STS-1. The DS1 must be locked to the STS-1 device.

Floating Byte Synchronous Mode. This mode pointer processes (moves VT forward or backward) one byte (8 bits) at a time. DS0 grooming (drop and insert) requires slip buffers. The DS1 must be synchronized to the STS-1 network.

Floating Bit Synchronous Mode. This mode provides single bit pointer processing. DSO grooming can be performed without slip buffers. The DS1 must be synchronized to the STS-1 network.

Floating Asynchronous Mode. This mode provides multiple bit pointer processing. The VT is synchronous with the DS1 but asynchronous with the STS-1. This mode supports DS0 grooming

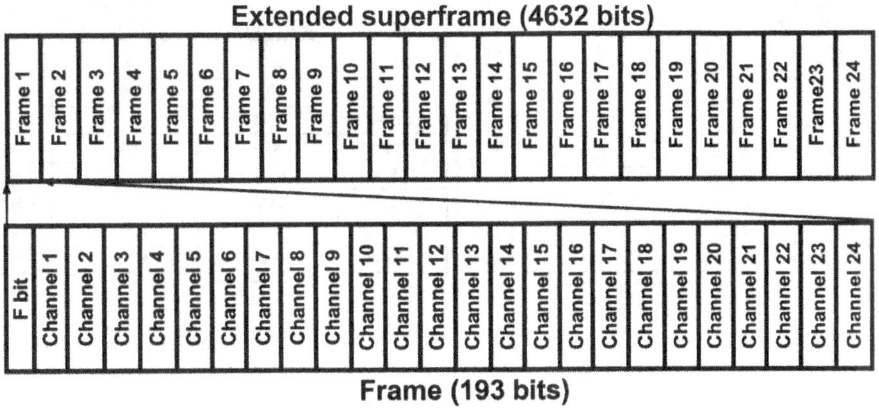

Extended superframe (4632 bits)

Frame 1 | Frame 2 | Frame 3 | Frame 4 | Frame 5 | Frame 6 | Frame 7 | Frame 8 | Frame 9 | Frame 10 | Frame 11 | Frame 12 | Frame 13 | Frame 14 | Frame 15 | Frame 16 | Frame 17 | Frame 18 | Frame 19 | Frame 20 | Frame 21 | Frame 22 | Frame23 | Frame 24

F bit | Channel 1 | Channel 2 | Channel 3 | Channel 4 | Channel 5 | Channel 6 | Channel 7 | Channel 8 | Channel 9 | Channel 10 | Channel 11 | Channel 12 | Channel 13 | Channel 14 | Channel 15 | Channel 16 | Channel 17 | Channel 18 | Channel 19 | Channel 20 | Channel 21 | Channel 22 | Channel 23 | Channel 24

Frame (193 bits)

Frame

Frame 125 μs long
Frame = 24 DS0 channels (8 bits) + one F bit

Extended Superframe

Composed of 24 frames
Four signaling modes
F bits used for 8 kb/s overhead channel
- 2kb/s framing channel (sequence: 001011)
- 2kb/s CRC-6 word for PM
- 4kb/s facility data link (FDL)

Supports end to end PM using FDL and CRC-6

Figure 6.20 DS1 extended superframe format.

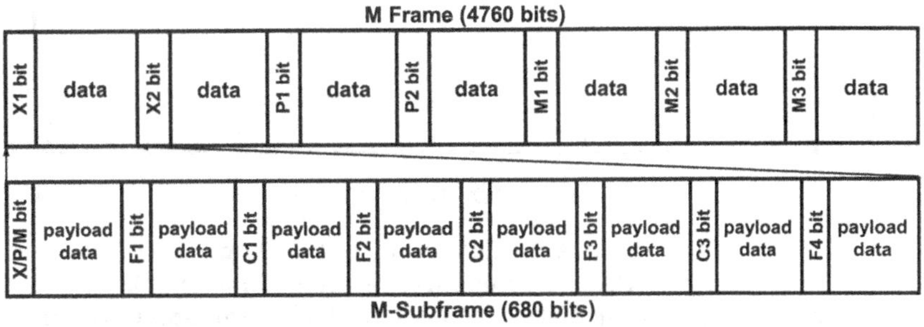

M Frame (4760 bits)

X1 bit | data | X2 bit | data | P1 bit | data | P2 bit | data | M1 bit | data | M2 bit | data | M3 bit | data

X/P/M bit | payload data | F1 bit | payload data | C1 bit | payload data | F2 bit | payload data | C2 bit | payload data | F3 bit | payload data | C3 bit | payload data | F4 bit | payload data

M-Subframe (680 bits)

M-Subframe

Composed of payload data from 4 DS1s
F bits used for M-subframe alignment (sequence:1001)
C bits reserved for DS2 pulse stuffing for standard M13
C bits used for alarm, status and loop backs for C-bit parity

M Frame

Composed of 7 M-subframes
X bits used low speed message channel
P bits used for parity check of previous M-frame
M bits used for M-frame alignment (sequence: 010)

Figure 6.21 DS3 framing format.

Line rate (Mb/s)	SONET signal	PDH signal T1s (1544 kb/s)	VF channels
1.728	VT1.5	1	24
2.304	VT2	-	30
3.456	VT3	2	48
6.912	VT6	4	96
51.84	STS-1	28	672
155.52	STS-3	84	2,016
622.08	STS-12	336	8,064
2488.32	STS-48	1344	32,256
9953.28	STS-192	5376	129,024

Ref: ANSI T1.102/105

Line rate Mb/s	SDH signal	PDH signal E1s (2048 kb/s)	VF channels
2.048	VC-12	1	30
34.368	VC-3	16	480
51.84	STM-0	21	630
139.264	VC-4	64	1,920
155.52	STM-1	63	1,890
622.08	STM-4	252	7,560
2488.32	STM-16	1,088	30,240
9953.28	STM-64	4,032	120,960

Ref: ITU-T G.707

Figure 6.22 North American Synchronous Optical Network (SONET) and European Synchronous Digital Hierarchy (SDH).

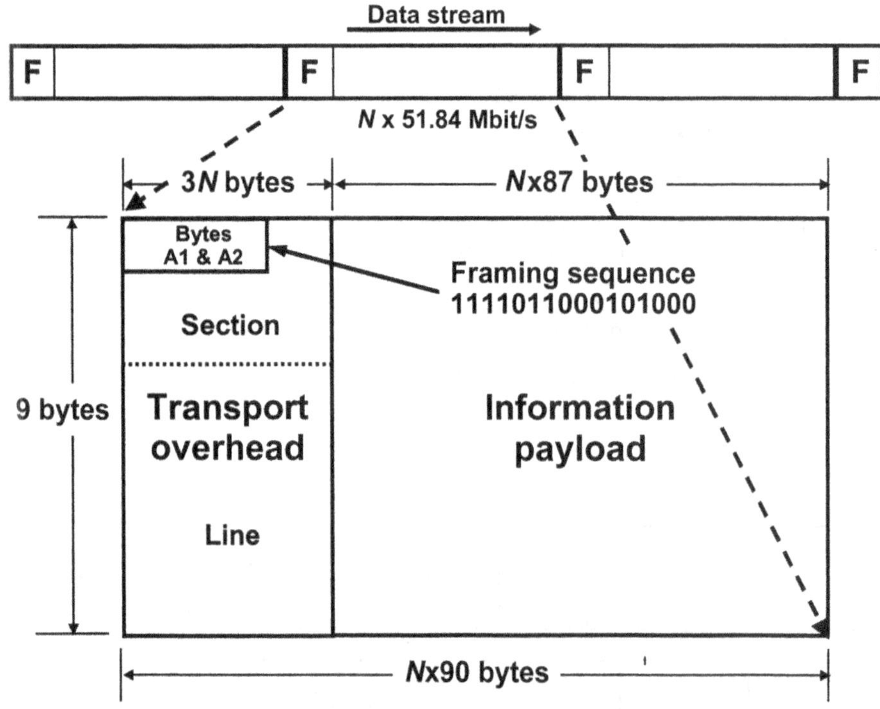

Figure 6.23 North American SONET framing format.

without slip buffers or network synchronization of DS1. Since this mode completely decouples the operation of DS1s from the SONET network, it is the preferred mode in use today.

As with plesiochronous networks employing drop and insert functionality, all DS0s within the DS1 must be mutually synchronized (American National Standards Institute (ANSI), 1999; International Telecommunication Union—Telecommunication Standardization Sector (ITU-T), 2003; International Telecommunication Union—Telecommunication Standardization Sector (ITU-T), 2000a).

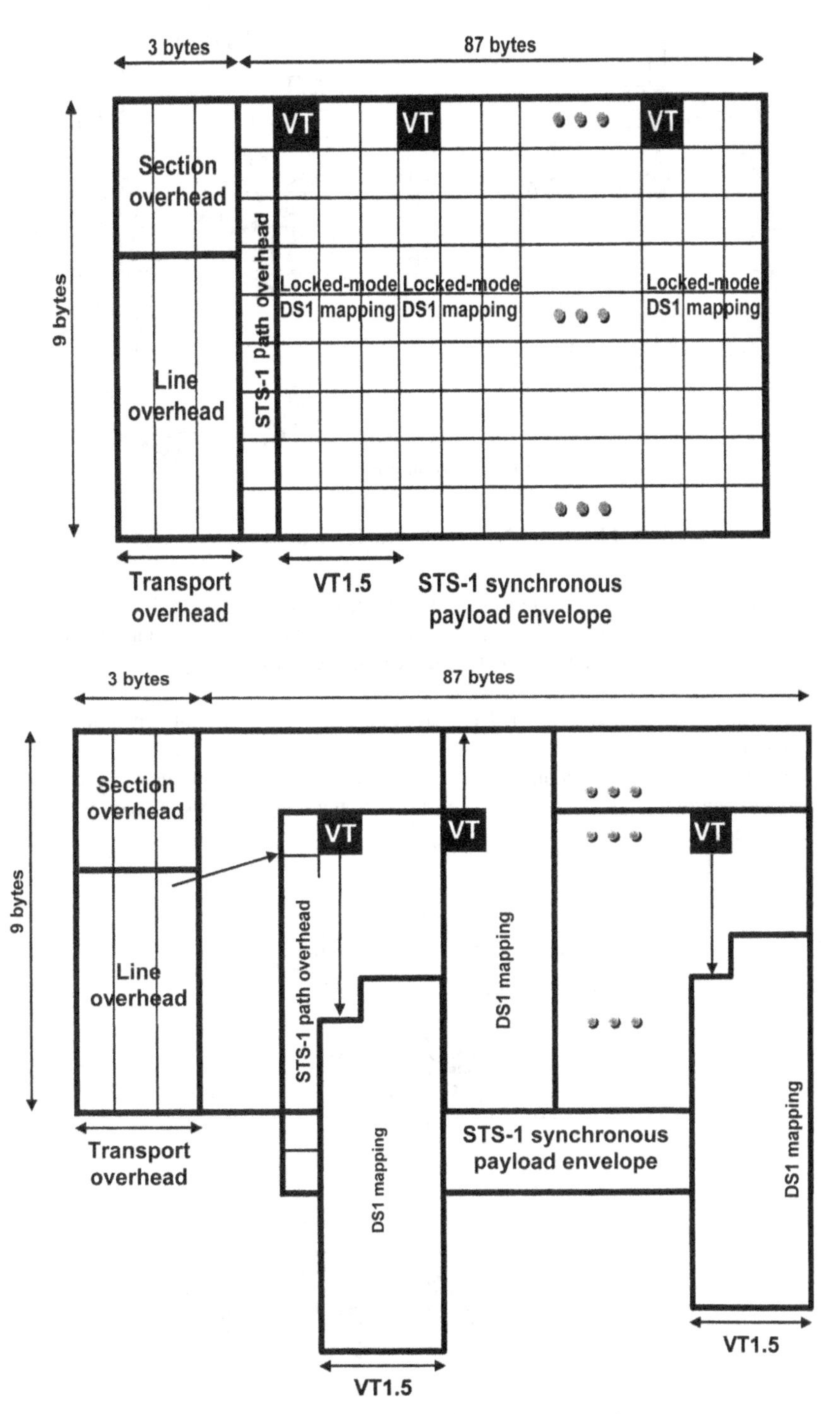

Figure 6.24 STS-1 locked and floating VT 1.5 framing formats.

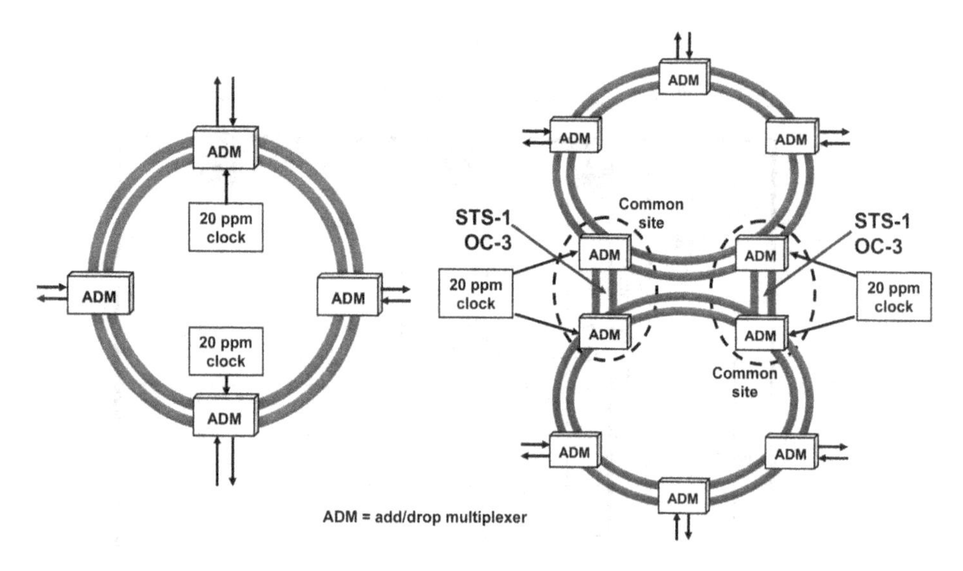

Figure 6.25 Simple synchronous networks.

By their very nature, the network elements (NEs) (typically add/drop multiplexers) in SONET or SDH networks must be mutually synchronized (American National Standards Institute (ANSI), 1996). If the network is small and one set of sync clocks can see all NEs, a relatively low cost clock is adequate (Fig. 6.25).

If no single pair of clocks can see all NEs (and the interconnects are all synchronous), high quality clocks are required (Fig. 6.26).

Interconnects also affect the required sync clock quality. If the interconnects between synchronous networks is plesiochronous, then no mutual synchronization among networks is required and low quality clocks are adequate (Fig. 6.27).

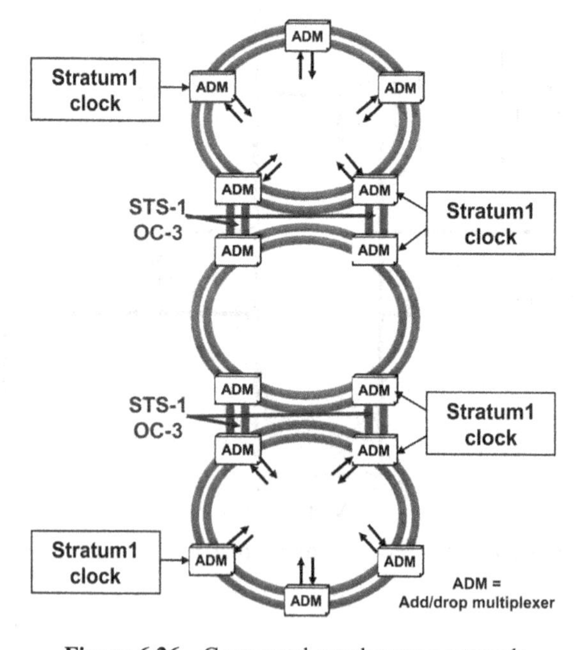

Figure 6.26 Compound synchronous network.

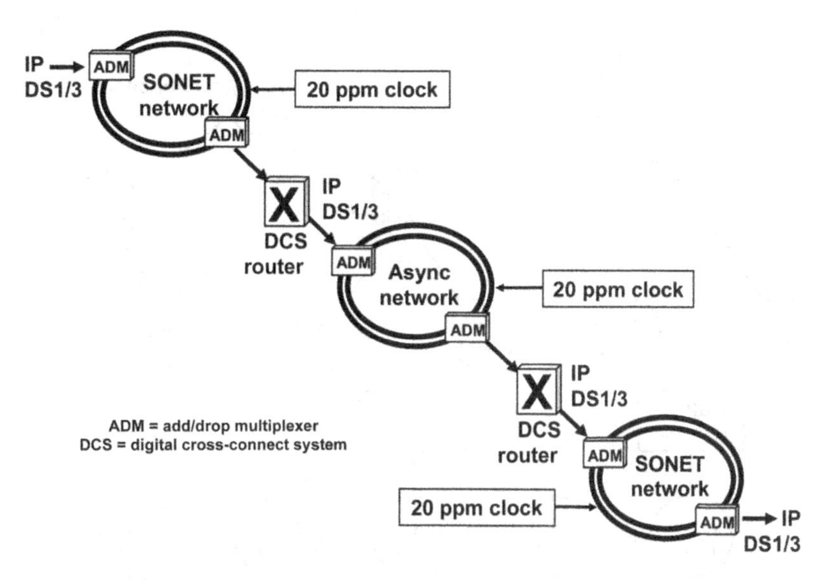

Figure 6.27 Plesiochronously connected synchronous networks.

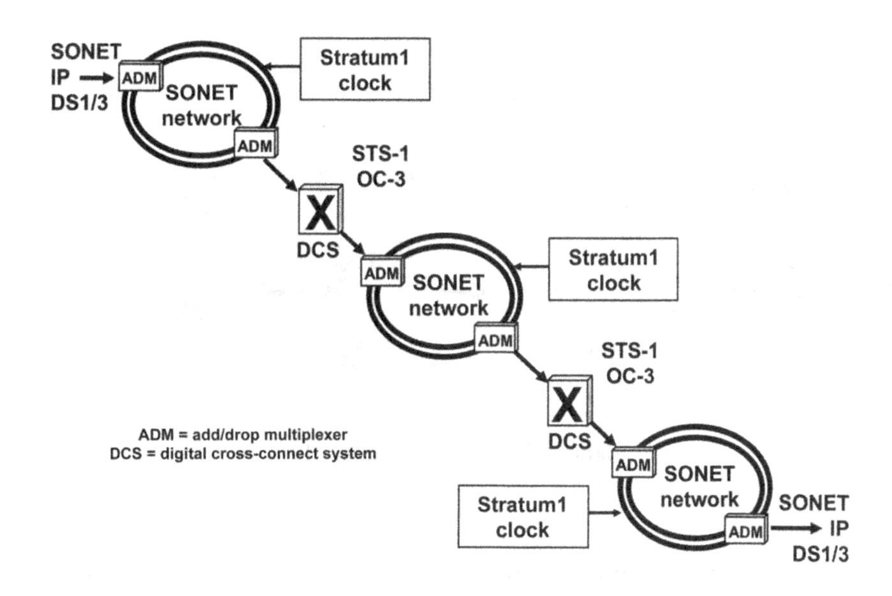

Figure 6.28 Synchronously connected synchronous networks.

However, if the networks are interconnected synchronously, all networks require a high quality clock (since they all cannot use the same one) (Fig. 6.28). If external synchronization clocks are required, one of two architectures is usually used (Bregni, 2002; Okimi and Fukinuki, 1981) (Fig. 6.29). It is sometimes useful to synchronize a chain of clocks (typically through cascaded NEs or BITS shelves) (Fig. 6.30). The number of clocks in the series (cascaded) should be limited to avoid excessive wander and network breathing (International Telecommunication Union—Telecommunication Standardization Sector (ITU-T), 2000d; International Telecommunication Union—Telecommunication Standardization Sector (ITU-T), 2000e; International Telecommunication Union—Telecommunication Standardization Sector (ITU-T), 2002b). G.703 (International Telecommunication Union—Telecommunication Standardization

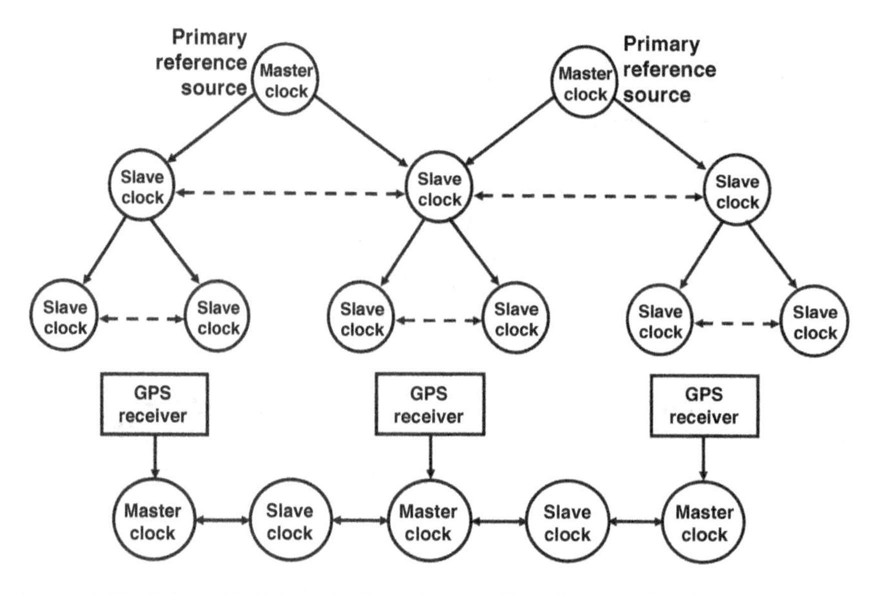

Figure 6.29 Hierarchical (root-leaf) and independent (flat) synchronization networks.

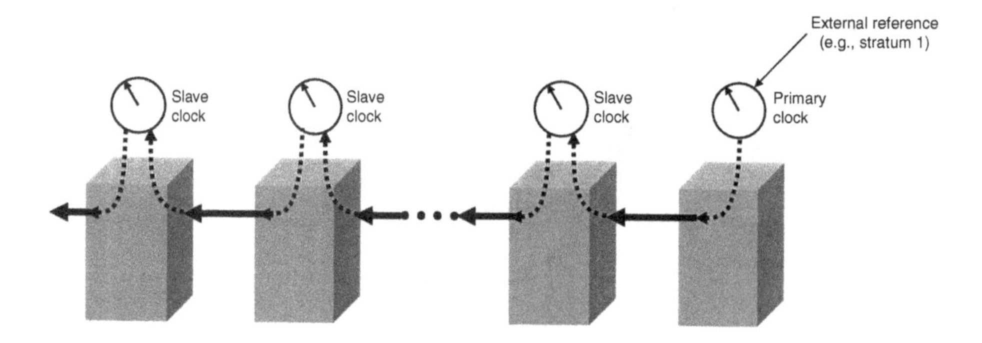

Figure 6.30 Synchronization chain

Sector (ITU-T), 2001) suggests that the maximum number of cascaded clocks is 60 although many industry sources suggest informally that this number should be on the order of 20 or 30.

For simple junction stations, cross-wiring synchronization signal between NEs is usually adequate. For larger sites, a building-integrated synchronization supply (BITS) shelf with appropriate quality clock is usually required to mutually synchronize the various NEs. For channel banks and small cross-connects, the synchronization source is typically a 64-kb/s composite clock. For higher speed devices, a DS1 or E1 synchronization signal is usually used.

For large networks, redundant BITS shelf (DS1 or E1) synchronization sources may be used to increase sync reliability. DS1 or E1 sources of synchronization may have Sync Messaging (source synchronization quality signals) (American National Standards Institute (ANSI), 1996; American National Standards Institute (ANSI), 1994; International Telecommunication Union—Telecommunication Standardization Sector (ITU-T), 2003) embedded within the DS1 or E1 signals. For small networks, sync messaging is implemented in the individual NEs. For large networks, sync messaging is implemented in the BITS shelf.

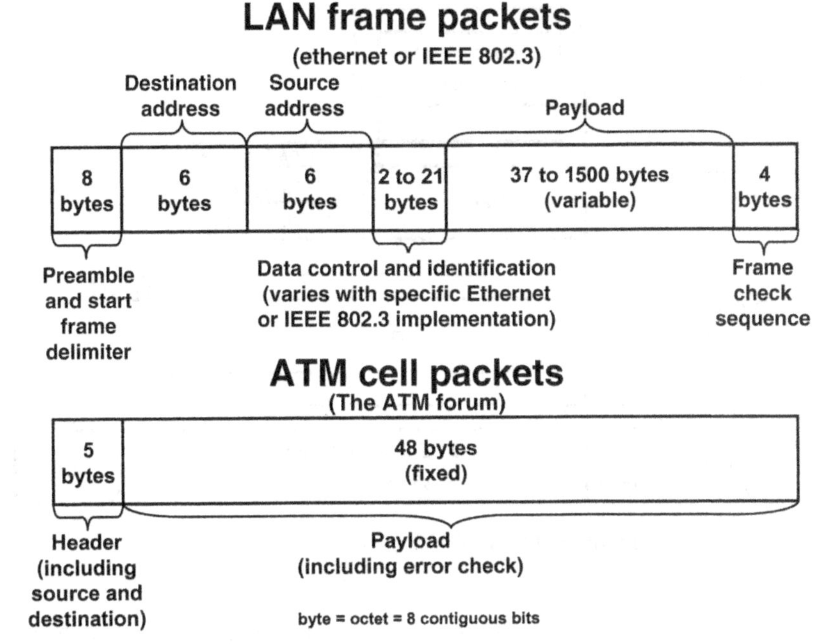

Figure 6.31 Popular packet formats

Most SONET or SDH microwave radios are SONET or SDH compatible, not compliant. That means they transport the SONET or SDH signals but probably do not support the synchronization and alarm and performance monitoring standards. They appear as active fiber (i.e., do not interface with SONET or SDH network management). Since they are loop-timed from the incoming synchronous signal, they do not require external synchronization.

Today, most user interfaces are based on a PSN. The most popular modern forms are LAN and ATM (Fig. 6.31).

Ethernet LAN packet interfaces (Institute of Electrical and Electronics Engineers, 1999–2008) are becoming the de facto interface for many new user data circuits (Fig. 6.32).

Line rate (Mb/s)	Signal designation	Media
10	10BASE2/5	Coax
	10BASE - T	Copper pairs
	10BASE - FB/FL/FP	Optical fibers
100	100BASE - X	Generic
	100BASE - T2/T4/TX	Copper pairs
	100BASE - FX	Optical fibers
1000	1000BASE - X	Generic
	1000BASE - CX/T	Copper pairs
	1000BASE - LX/SX	Optical fibers
10000	10GBASE - CX/T	Copper pairs
	10GBASE - ER/EW/LR/ LX4/LW/SR/SW/SX	Optical fibers

Ref: IEEE 802.3( )

Figure 6.32 Local area network (LAN) "Ethernet" physical interfaces.

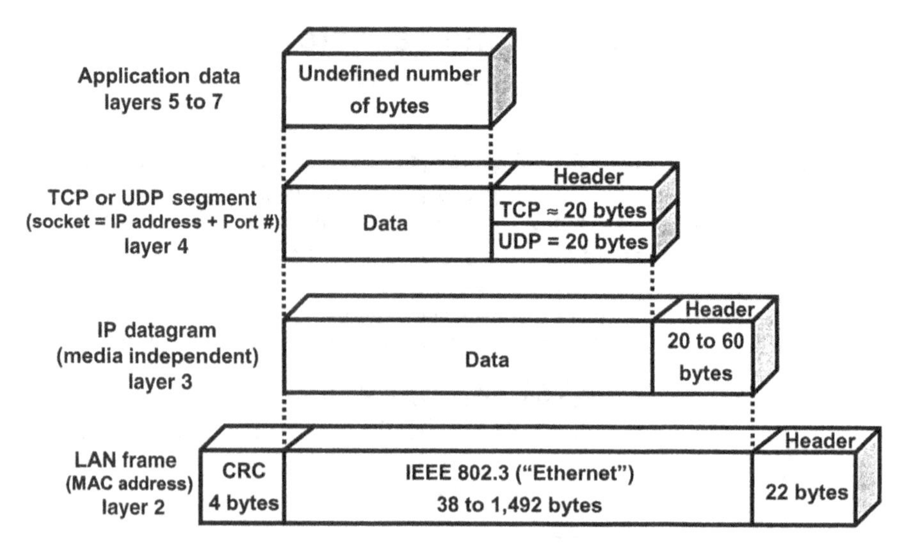

Figure 6.33 Typical message encapsulation.

The above formats are the direct physical interface to the user. The user message must be encapsulated so it can be routed to the far end destination. Encapsulation and routing occurs by layer (Fig. 6.33).

Routing of packets among nodes is usually described on the basis of an IP four-layer (Internet Engineering Task Force (IETF), 1989) or OSI seven-layer model (International Telecommunication Union—Telecommunication Standardization Sector (ITU-T), 1994) (Fig. 6.34).

The OSI model supports the following layers:

7. Application Layer: NNTP, SIP, SSI, DNS, FTP, Gopher, HTTP, NFS, NTP, SMPP, SMTP, SNMP, Telnet, DHCP, Netconf, RTP, SPDY
6. Presentation Layer: MIME, XDR, TLS, SSL
5. Session Layer: NetBIOS, SAP, L2TP, PPTP, Named Pipes
4. Transport Layer: TCP, UDP, SCTP, DCCP, SPX
3. Network Layer: IP (IPv4, IPv6), ICMP, IPsec, IGMP, IPX, AppleTalk

TCP/UDP/IP protocol stack	OSI seven layer stack	Stack services
Processes (applications and services)	**7 Application Layer**	**Application services** (file transfers, transaction services)
	6 Presentation Layer	**Communication services** (format conversions, encryption)
	5 Session Layer	**Connection managment**
Host to host (UDP or TCP)	**4 Transport Layer**	**Transparent data transfer** (flow control and error recovery)
Internet (IP)	**3 Network Layer**	**Routing, addressing segmenting**
Network access Medium Access Control (MAC) (IEEE 802.X)	**2 Data Link Layer**	**Transmission and flow control**
	1 Physical Layer	**Physical connection**

Figure 6.34 Message encapsulation layers.

2. Data Link Layer: ATM, SDLC, HDLC, ARP, CSLIP, SLIP, GFP, PLIP, IEEE 802.3, Frame Relay, ITU-T G.hn DLL, PPP, X.25, Network Switch
1. Physical Layer: EIA/TIA-232, EIA/TIA-449, ITU-T V-Series, I.430, I.431, POTS, PDH, SONET/SDH, PON, OTN, DSL, IEEE 802.3, IEEE 802.11, IEEE 802.15, IEEE 802.16, IEEE 1394, ITU-T G.hn PHY, USB, Bluetooth, Hubs

MPLS operates between traditional definitions of Layer 2 (Data Link Layer) and Layer 3 (Network Layer) in the OSI model layer and is often referred to as a *Layer 2.5 protocol.*

The TCP/IP model (RFC 1122) supports four layers:

4. Application Layer: BGP, DHCP, DNS, FTP, HTTP, IMAP, IRC, LDAP, MGCP, NNTP, NTP, POP, RIP, RPC, RTP, SIP, SMTP, SNMP, SSH, Telnet, TLS/SSL, XMPP
3. Transport Layer: TCP, UDP, DCCP, SCTP, RSVP, ECN
2. Internet Layer: IP (IPv4 • IPv6), ICMP, ICMPv6, IGMP, IPsec
1. Link Layer: ARP/InARP, NDP, OSPF, Tunnels (L2TP), PPP, Media Access Control (Ethernet • DSL • ISDN • FDDI)

Routing between nodes is via OSI Layer 1, Layer 2, or Layer 3 (Fig. 6.35).

The four- and seven-layer models are historical and neither exactly fits modern packet networks. In practice, the layers are described rather loosely.

The latest Ethernet-based transport products (routers and radios) will have native Ethernet interfaces and emulated (pseudowire) interfaces for TDM (DS1, DS3, E1, E3, SONET, SDH, and Frame Relay) and ATM circuits. Many organizations including Internet Engineering Task Force (IETF RFCs 3985 through 5287), International Telecommunication Union-Telecommunication Standardization Sector (ITU-T Y series), and Metropolitan Ethernet Forum (MEF 3 and 8) have recommendations for pseudowire circuit emulation.

Many next generation radios are native Ethernet transport products. They carry IP signals as direct inputs and outputs. They carry TDM signals (e.g., DS1 and DS3) as emulated (pseudowire) circuits. Since the TDM signals must be encapsulated, transmitted as packets, and then reassembled at the receive end, increased delay (relative to a conventional TSM radio) will occur. The point to point TSM signal will have been transported by a series of separate packets. These packets are routed and received individually. TDM signals have maximum allowable wander and jitter requirements. This requires that the TDM

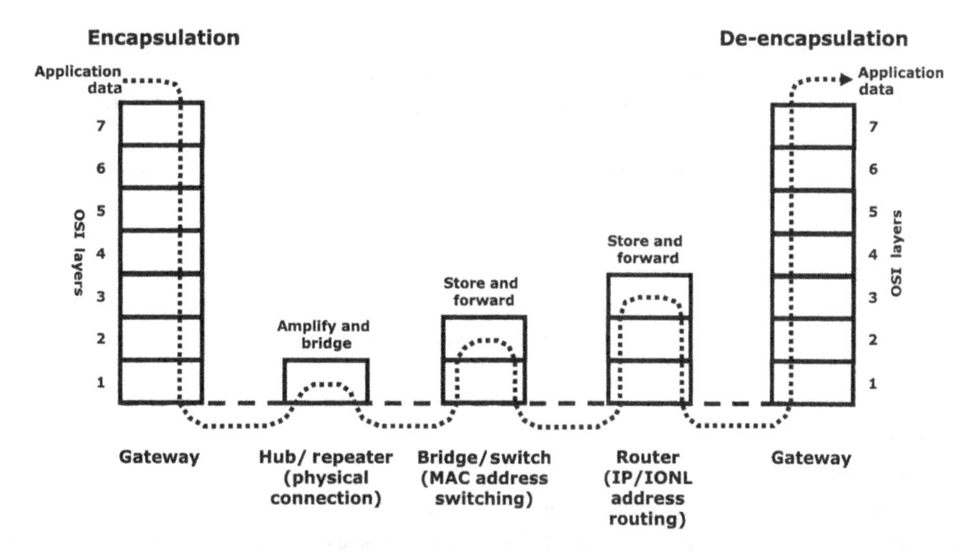

Figure 6.35 Typical message routing.

signal be buffered and then clocked out at the same rate as it was received at the packet source. Since the packet network is not synchronous, recovering the TDM signal clock requires additional consideration. See Chapter 3.16 for further discussion on this topic.

Standards for quality and availability of IP circuits are still in the early stages of development (Song et al., 2007). ITU-R Report F2058 (International Telecommunication Union—Radiocommunication Sector (ITU-R), 2006). The most significant factors mentioned are bandwidth, delay, and lost packets. ITU-T Recommendation Y.1540 (International Telecommunication Union—Telecommunication Standardization Sector (ITU-T), 2002a) defines QoS/CoS parameters. Currently, parameters of interest include successful packet transfer, errored packets, lost packets, spurious packets, average packet delay, and packet delay variation. See Section 3.16 for further discussion on this topic.

Error performance of the radio network can impact the PSN. During the time that multipath or rain attenuation is occurring, the received signal may have repetitive errors. For typical routers, a single error to the IP header will cause the entire packet to be discarded. A single radio error results in a signal gap resulting in a frame loss for the TDM circuit. Significant repetitive errors from a fading radio will cause some routers to disable the transmission port from that radio. The router views that connection as a defective "flapping" port. It is important that the radio receiver recompute the IP packet CRC error checking block to avoid these error-related issues being propagated.

6.7 OPERATIONS AND MAINTENANCE

Telecommunications operations companies provide services. These services must be initiated, administered, and maintained. The primary goal of network management is to support telecommunications services by maintaining efficient, reliable operation both when the network is under stress because of overload or failure and when it is changed by the introduction of new equipment and services. At the same time, network management must increase the performance of the network in terms of the quality and quantity of service provided to its end users.

Network management is a recursive, three-step process, applied to several functions necessary to network operation:

- Data analysis and interpretation
- Situation assessment
- Planning and response generation.

The ISO telecommunications management network model describes five functions of network management: fault, configuration, accounting (or administration), performance, and security (FCAPS) (International Telecommunication Union—Telecommunication Standardization Sector (ITU-T), 2000b):

Fault Management. A set of functions that enables the detection, isolation, and correction of abnormal operation of the network or its elements:
- Alarm surveillance
- Failure localization
- Testing.

Configuration Management. A set of functions to exercise control over, identify, collect data from, and provide data to NEs:
- Status and Control
- Installation
- Provisioning
- Network overall ongoing operations and management.

Accounting (Billing) Management. A set of functions that enables the use of the network service to be measured and the cost for such use to be determined and assigned to an appropriate subscriber.

PM. A set of functions to evaluate and report on the behavior of telecommunications equipment and the effectiveness of the network or its elements:

- Status and control
- Performance monitoring.

Security Management. A set of functions that protects telecommunications networks and systems from denial of service and the unauthorized disclosure of information, modification of information, or access to resources.

Over the years, various specialized electronic systems (typically computer-based) have been developed to support these activities. Collectively, these software and hardware functions, as well as their inter- and intracommunication circuits, have been termed the *telecommunications management network* (International Telecommunication Union—Telecommunication Standardization Sector (ITU-T), 2000c) (Fig. 6.36).

We now focus on network managers most closely involved with the actual NEs (e.g., radios, add/drop multiplexers, cross-connects, and routers) and the element managers and network management equipment that directly interact with them.

Element managers are the computer-based products which directly configure and provision the NEs. They may be simple craft interfaces or sophisticated systems providing end-to-end or system-wide connections, path restoration, or root cause analysis. Typically, they are very tightly coupled to the NE and are vendor-unique. They provide the specialized, evolving, interactive, unique functionality of a particular NE.

Element managers and other network managers provide one or more of the basic five TMN functions (International Telecommunication Union—Telecommunication Standardization Sector (ITU-T), 2000b): performance, fault, configuration, accounting, and security management. Several NMs may operate in parallel or the structure of NMs may be tiered. Of particular interest to network operators are fault management and PM.

6.7.1 Fault Management

The basic function of a *fault management* system is to allow one or more users to perform commands at and receive alarms or status events from a remote location device or unit. Typical fault management functions include the following:

- Accepting and acting on fault detection notifications
- Tracing and identifying fault locations
- Correcting faults
- Carrying out diagnostic tests
- Maintaining and examining fault history reports.

ITU-T suggests (International Telecommunication Union—Telecommunication Standardization Sector (ITU-T), 1992) that the purpose of fault management is to minimize both the occurrence and impact of failures and to ensure that, in the case of failure, the right notification sends the right personnel to the right place with the right equipment and the right information at the right time to perform the right action.

A computer running-specialized management software and connected to a data communication network acts as the local interface for the user. The remote location monitoring unit may be a discrete device or may be included in the function of the remote transmission equipment. The remote equipment is often called a *network element* (*NE*). The communication channel between the computer(s) and the monitoring units is called the data communications channel (DCC), data communications network (DCN), embedded communications channel (ECC), embedded operations channel (EOC), or simply the telemetry channel. This channel may be a physically separate channel (over an external network) or a virtual channel within the NE payload.

Communication over the telemetry channel is via a predefined language called a *protocol.* The protocol defines the dictionary and syntax of words appropriate for command and responses. If the words are short

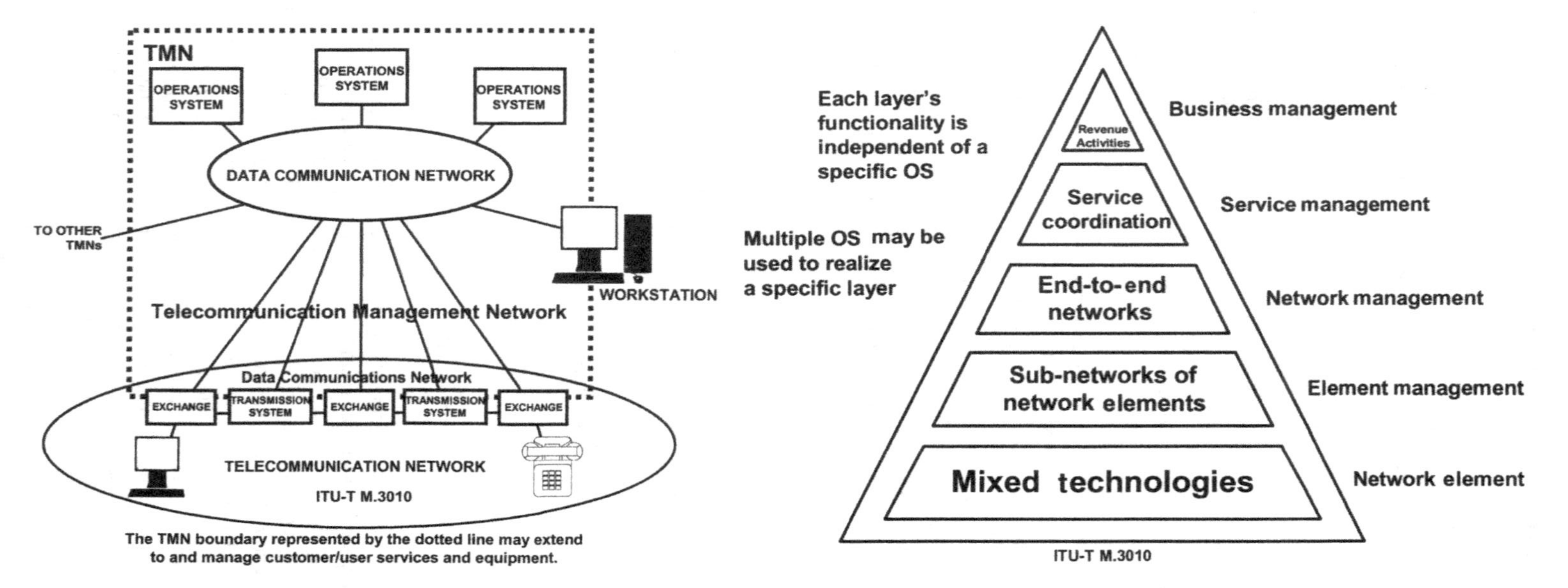

Figure 6.36 Telecommunications management network and its functional management layers.

or they are composed of ASCII characters, communication is generally asynchronous. If the words are long, synchronous communication is required. The source and termination of a data channel is termed *data terminal equipment* (DTE). The master and remote units are DTE devices. The equipment forming the transmission channel is called data circuit-terminating equipment (DCE). Radios and modems are examples of DCE. If the transmission channel is synchronous, the DCE supplies the transmit and receive data clocks.

Proper operation of the fault management system assumes error-free data transmission between the master and the remote units. Loss of communication or corrupted data can cause errors in the telemetry channel. Most protocols have error-checking methods that preclude bad data words from being reported as valid or causing a remote control. Errors can also cause the data connection between the origination point and the termination point to be lost. This condition is the usual cause of "no report" status. Multiple NEs with the same protocol address can create confusing or unpredictable behavior. In some protocols, this can generate a "no report" condition. For OSI-based systems (TL1 in North America and CMISE/CMIP in Europe), loss of data due to buffer overflow or long delays in message transfer due to alarm storms is common. Similarly, IP-based systems (SNMP) have similar problems although the predominant issue is loss of data due to data collisions. For both OSI- and IP-based systems, the network manager must have specialized algorithms to mitigate these issues.

Two types of fault management architectures are common. The first is based on a *peer–peer* relationship between the user computer and the remote units. In this architecture, any remote unit can (in theory) communicate with any fault management master at any time. The computer used to support this architecture is termed a *manager*. Examples of this type of network are SNMP, CMISE, and TL1 (Fig. 6.37).

Although simple in concept, multiple remote units attempting to contact multiple masters at the same time significantly stress the speed and reliability of the telemetry network. The advantages of this architecture are that interaction is possible between all network devices (peer-to-peer communication) and it is easily scaled to large networks. The limitations of this architecture include potential congestion, loss of communication, uncontrolled transmission delays, and unknown reliability. The use of relatively large (verbose) standardized alarm messages makes this system relatively slow. SNMP's use of UDP protocol on a data channel shared with other communications makes alarm messaging inherently unreliable. Since messages are never expected unless an alarm occurs, the manager must determine the loss of communication to a site or the site's functional state. Loss of alarm messages and failure to learn of site and equipment failures are common issues. The manager must mitigate these.

The second architecture is based on a *master–slave* relationship between the user computer and the remote units. The computer used to support this architecture is termed a *master*. Examples of this are vendor proprietary protocols such as MCS-11 and FarScan (Fig. 6.38).

In this architecture, the master polls one remote unit at a time. The advantage of this approach is that telemetry channel activity is minimized and data collisions are eliminated (unless multiple NEs share the same name/address). The master always knows whether the remote unit or site is present and operating. Use of compacted, predefined packets makes this approach relatively fast. Usually, this architecture uses

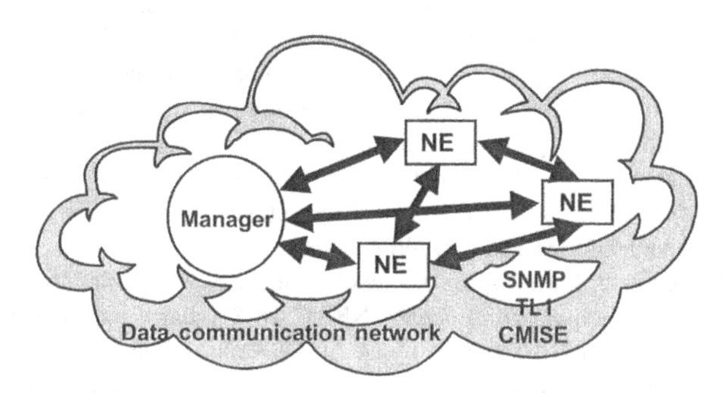

Figure 6.37 Peer–peer network management.

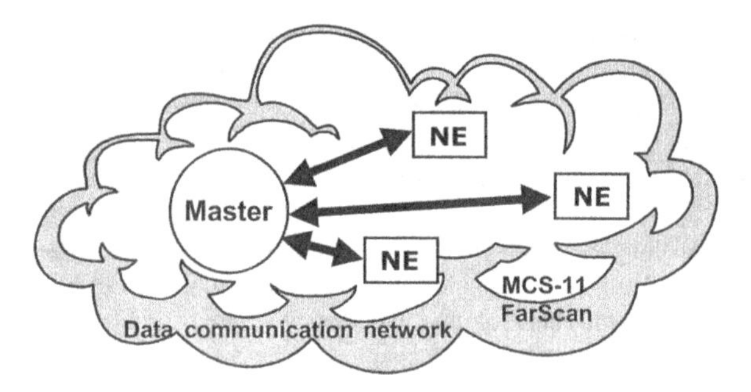

Figure 6.38 Master–slave architecture.

a dedicated communications channel with error-detecting protocols. This maximizes alarm reliability as well as fast detection of site or communication channel failure. This architecture is optimum for moderate-sized networks. However, for very large networks, the master may take a long time to determine the status of all NEs. An additional limitation is that communication between remotes is not possible. The only allowed communication is between remotes and the master. Use of multiple active masters is difficult.

Facilities and equipment are the two basic entities monitored by fault managers. *Facilities* are the signals carried or supported by the transmission equipment. These are of specific interest to the *customer.* Performance monitoring and performance thresholds are most commonly associated with facilities. *Equipment* is the actual hardware that is used to support the transportation of the customer facilities. Equipment is of particular interest to the *network operator.* Alarms and status (nonfailure conditions or events) are most commonly associated with equipment.

6.7.2 Alarms and Status

Traditionally, alarm or event messages have been classified as binary alarms or status conditions. An alarm is a binary state indicative of equipment failure. A status is a binary state that represents a condition of equipment not associated with failure. Typically, alarms or status conditions are indicated by contact closures. An "off" condition is an open circuit or battery voltage and an "on" condition is closure to ground. If the alarm sensing circuitry receives the binary alarms as contact closures, the alarm (status) points are termed *parallel.* If the alarm sensing circuitry accepts the alarm (or status) data as a preprocessed serial binary data stream [such as telemetry byte oriented serial (TBOS)], the alarms (status) are termed *serial.*

Alarms are unipolar if they are only reported when an event occurs. Bipolar alarms report the transition from "no event" to "event" status as well as the transition for "event" to "no event" status. If alarms are not released until they have been reported over the fault management telemetry channel, they have been "stretched" or "latched." Alarm integration is a process by which an alarm is not declared until it has been continuously present for a predefined period of time. The time period is typically 2.5 s in terminal equipment and a few tens of milliseconds in transport equipment. Virtual alarms (derived alarms) are user-defined alarms based on logical combinations of other alarms.

The Bell System and Telcordia (previously Bellcore) define three basic alarms. Notification of an upstream facility outage is an alarm indication signal (AIS) or blue signal. It is used to suppress alarms downstream from the failure. A local unprotected facility outage is a red signal. A downstream facility outage is a remote defect indication (RDI) or yellow signal.

Telcordia has further defined alarm types and levels (Telcordia Technologies, 2000a). Alarm types are service affecting (SA) and nonservice affecting (NSA). An SA alarm indicates an equipment failure which causes loss of the transported (baseband) signal. An NSA alarm indicates that an equipment failure has occurred but that functionality was automatically restored by backup equipment. Critical alarms are SA alarms that indicate failures that could affect many users or considerable bandwidth. Telcordia defines critical alarms as failures that require immediate corrective action independent of the time of the day (early Telcordia documents suggested that a critical alarm was a failure of more than 5 DS1s). Major alarms are SA alarms that affect fewer users or less bandwidth. Telcordia defines major alarms as failures that require immediate attention (early Telcordia documents suggested that a major alarm was a failure of 1–5 DS1s). Minor alarms are NSA alarms or SA alarms that indicate a failure that affects few customers or little bandwidth. Telcordia defines minor alarms as NSA failures (early Telcordia documents suggested that a minor alarm was an NSA failure or an SA failure that affected <1 DS1 of bandwidth). Nonalarmed is a nonalarm condition, status, or event. Typically, it is not reported or stored.

6.7.3 Performance Management

Customers expect telecommunications service providers to provide high quality (highly accurate), dependable (high availability) data transmission service. PM parameters are the result of the operator's evaluation of the effectiveness of the network or its NEs to provide this service. Typical PM functions include the following:

- Gathering statistical information
- Determining system performance under natural and artificial conditions
- Altering system states or modes of operation on the basis of system performance
- Maintaining and examining logs of system states or performance histories.

6.7.3.1 Payload PM Parameters

Bit Error Ratio (*Rate*) (*BER*). This is the ratio of errored received digits divided by the total number of digits transmitted over a defined period of time. This parameter is used in North America but no longer used in ITU-T and ITU-R Recommendations.

Residual Bit Error Ratio (*RBER*). This is the bit error ratio in the absence of fading or external interference. It is the result of inherent system errors, environment, aging effects, cross talk, and local interference. This parameter is used in North America but no longer used in ITU-T and ITU-R Recommendations.

BBER. This is an ITU-T and ITU-R term for the ratio of errored blocks that are not contained within an SES to the total number of transmitted blocks.

Dribble. This is a colloquialism for occasional errors (typically the BER is better than 10^{-9}).

Hit. This is a colloquialism for short duration loss of signal (LOS) (typically the loss is <10 s).

Errored Second (ES). An ES is a second during which at least one of the following has occurred: coding violations, slips, out-of-frame indications, or LOF alignment indications. An important reason for counting ES is that the quality statements in data service tariffs are generally given in terms of percentage of Error Free Seconds (EFSs). In addition, by combining the count of ES and the effective BER, a measure of the error distribution can be obtained. Distinguishing between bursty and randomly distributed errors can be important in diagnosing facility problems. An ES may commence with the detection of a coding violation and end 1 s later (synchronous ES) or the 1 s interval may be timed independently of the errors (asynchronous ES). ESs are to be measured only when service is available.

Burst Errored-Second (*BES*). This is a Telcordia digital signal performance parameter that indicates the second in which at least 100 errors occurred.

Consecutive Severely Errored-Seconds (*CSESs*). This is a Bell System digital signal performance parameter that describes a sequence of SES for 3–10 consecutive seconds. During this time, the digital signal is severely impaired. After 10 s of SES, the signal is deemed unavailable.

Severely Errored Second (SES). In North America, SES is a second that contains more that N coding violations. The value of N will vary with frame size and bit rate and should be chosen to correspond to an appropriate bit error ratio. Recently, the appropriate bit error ratio has been under review. Currently, the most common choices are 10^{-3} for asynchronous systems or 10^{-6} for SONET and radio systems. The assumption is generally made that errors are randomly distributed (not true for systems employing forward error correction or Viterbi decoders). An SES may commence with the detection of a coding violation and end 1 s later (synchronous SES) or the 1-s interval may be timed independently of the errors (asynchronous SES). This count may be used to determine problems for particular types of services. It may be used as a measure of facility outage duration. In conjunction with ES and effective BER, the SES count yields additional information on error distributions. Like the case of ESs, SESs should be measured during available time only. For ITU-T and ITU-R Recommendations, SES is a second with a minimum of 30% errored blocks or a severely disturbed period (e.g., loss of pointer or LOF).

Degraded Minute (*DM*). This is an obsolete ITU-T digital signal performance parameter that indicates a minute with an average BER of 10^{-6} or worse.

Available. For ITU-T and Telcordia cable-based equipment, the transmission system has operated for the last 10 consecutive seconds with no SESs. After SES-free transmission has occurred for those 10 seconds, they are counted as available time. North American radio systems use a near-instantaneous BER threshold to define availability.

Unavailable. For ITU-T and Telcordia cable-based equipment, a transmission system is considered unavailable if the transmission system has experienced SESs, LOF, or LOS for the last 10 consecutive seconds. After those 10 s have occurred, those seconds are counted as part of the unavailable time. North American radio systems use a near instantaneous BER threshold to define unavailability.

Unavailable Seconds (*UASs*). Service is said to become unavailable if the transmission system has experienced SESs, LOF, or LOS for the last 10 consecutive seconds. Once the service becomes unavailable, it remains unavailable until 10 consecutive non-SESs occur, after which it becomes available again. The parameter UAS measures the duration for which the service was unavailable, in seconds. Unavailable time is a measure used in data service tariff specifications. More detailed information may be needed to adequately describe the performance of a data channel. For example, if a data channel has a bit error ratio of 10^{-4} and there is only a facility alarm when a UAS event occurs, the degraded performance may not be detected, although this level of service may not be acceptable to a data customer. In effect, UAS can be used to determine availability but not QoS.

Currently, a couple of issues remain unresolved. The first is lack of universality of performance measurement. All NEs do not measure the same performance parameters the same way. Stallings (1993) noted that different LAN bridges measured packet counts, a basic LAN performance parameter, quite differently. That is not an isolated issue. Another issue is event clock synchronization. IP networks have time services available for most networks. Having managers and NEs capable of utilizing these services to synchronize their local clocks is critical to synchronized event and performance data. SONET networks have no defined time services. Synchronization over the relatively slow speed embedded communications channels induces delay dependent on the NE's distance for the time-setting computer. Perhaps a worse issue is lack of NE clock accuracy. Some clocks are tied to the SONET signal and are quite accurate. Others have such poor accuracy that they drift several minutes a year. All of these issues significantly complicate performance measurement comparisons among various NEs.

There are a couple of ambiguities in all current PM definitions. The first is whether the parameters are measured into or out of the transmission equipment. For small equipment, such as radios or add/drop multiplexers, it is usually only measured on the signal entering the equipment. For large equipment (such as cross-connects) which may encompass several racks or bays and in which the signal may be highly processed, the parameters are often both into and out of the equipment. Another ambiguity is the concept of "into" and "out of" the equipment. The terms *transmit and receive* are used frequently. However, these terms have widely varying meanings to different people. Telcordia defines a transmit signal as the baseband signal *into* either the signal originating or the terminating equipment (such as a SONET add/drop multiplexer). A receive signal is defined as the baseband signal *out of* the originating or terminating equipment. For transmission equipment connected to the drop side (contrasted with line side) signals, this means the transmit signal is leaving the signal carrying transmission equipment and the receive signal is entering it. Many equipment manufacturers do not conform to this protocol. In addition, drop side interconnects have obvious ambiguity. This leads to considerable confusion when different operators use transmit and receive to describe a signal's direction. Agreement between operators is necessary to minimize misunderstandings.

Performance monitoring parameters are very useful both for verifying quality and availability of service and for equipment maintenance. In addition, thresholds of performance measurements are useful for defining when particular maintenance actions (automatic or manual) need to occur. These performance thresholds are usually alarms derived from the PM values.

Telcordia sets equipment protection switching thresholds (Telcordia Technologies, 1998) on the basis of BER. For all transmission systems, two automatic switching thresholds are suggested. The lower threshold occurs in the range of 10^{-3} to 10^{-4}. The higher threshold is user selectable in the range 10^{-3} to 10^{-7} for digital radio systems and 10^{-6} to 10^{-9} for fiber optics systems. Network operators often set the higher threshold to correspond to the BER point at which 100% of ESs is experienced at the customer signal data rate. For equipment without error correction (i.e., a data stream with random equally probable errors), this BER is just the reciprocal of the binary signal transmission rate (e.g., 6.5×10^{-7} for DS1, 2.2×10^{-8} for DS3, and 6.4×10^{-9} for OC3). If error correction is used, errors come in bursts and the 100% ESs occurs at a lower BER. Tests have shown that this lower bit rate can be two or three orders of magnitude lower than the theoretical. Actual equipment measurements are needed to accurately estimate this. Experiments have confirmed that BER is essentially constant for all downstream signals (assuming it is not changed by external interference or error correction) at all levels of multiplexing (even when the originating signal had error bursts). Percentage of ESs obviously does change with data rate. The 100% ES point must be measured at the data rate of interest.

Telcordia defines (Telcordia Technologies, 2000a) several background ("residual") error criteria.

Residual BER. The BER at all interface levels between DS1 and DS3 shall be less than 2×10^{-10} (excluding all BESs). During a BES, neither the number of bit errors nor the number of bits is counted. This requirement applies to every channel in a normal operating environment.

Burst Errored Seconds (BESs). The frequency of BESs shall average no more than four per day at all interface levels between DS1 and DS1 (excluding burst errors caused by hard equipment failures or protection switching). This requirement applies to every channel in a normal operating environment.

DS1 ESs. For systems interfacing at the DS1 level, the long-term percentage of ESs (measured at the DS1 rate) shall not exceed 0.04%. This is equivalent to 99.96 EFSs. It is equivalent to no more that 10 ESs during a 7-h one-way test. This requirement applies to every DS1 channel in a normal operating environment. It is also an acceptance criterion.

DS3 ESs. For systems interfacing at the DS3 level, the long-term percentage of ESs (measured at the DS3 rate) shall not exceed 0.4%. This is equivalent to 99.6 EFSs. It is equivalent to no more that 29 ESs during a 2-h one-way test. This requirement applies to every DS3 channel in a normal operating environment. It is also an acceptance criterion.

Telcordia defines (Telcordia Technologies, 2000b) specific digital network signal alarm points:

Minor Alarm.

- BER exceeded 10^{-6} measured over a few minutes to 1 h
- Out of frame count exceeded 17 measured over 24 h
- Slip count exceeded 4 measured over 24 h.

Major Alarm.

- BER exceeded 10^{-3} measured over a few seconds
- Out of frame count exceeded 511 measured over 24 h
- Slip count exceeded 255 measured over 24 h.

The criticality of the alarm is generally a function of the service offering and the customer. For many DS1 customers, 10^{-6} is roughly the 100% ESs point, and represents very poor performance. The 10^{-3} point is generally considered the out-of-service point.

ES is a common criterion for DS1, DS3, and OC-3 data rates. Error counts are usually used at higher data rates (because the ES criterion imposes an unrealistic BER threshold).

Log Viewer

2007-06-02 16:00:00 From/To 2007-08-02 16:00:00

Results | Search Rules

Local Time	Event	Severity	Description	Location		SA	User	Acked	Net	NE	HW Name
2007-06-05 23:16:27	On	MJ	B RX Frame Loss	080 : MDR-8000-3 to 504 ch1	: B Side Alarms		mfortys	2007-06-05 23:17:30	Polling Engine 1	RDS C6A	P 43
2007-06-05 23:27:08	On	MN	A RX Frame Loss	080 : MDR-8000-3 to 504 ch1	: A Side Alarms				Polling Engine 1	RDS C6A	P 35
2007-06-05 23:31:30	Off	MN	A RX Frame Loss	080 : MDR-8000-3 to 504 ch1	: A Side Alarms			2007-06-05 23:31:30	Polling Engine 1	RDS C6A	P 35
2007-06-05 23:37:50	On	MN	A RX Frame Loss	080 : MDR-8000-3 to 504 ch1	: A Side Alarms				Polling Engine 1	RDS C6A	P 35
2007-06-05 23:52:56	Off	MN	A RX Frame Loss	080 : MDR-8000-3 to 504 ch1	: A Side Alarms			2007-06-05 23:52:56	Polling Engine 1	RDS C6A	P 35
2007-06-05 23:59:16	On	MN	A RX Frame Loss	080 : MDR-8000-3 to 504 ch1	: A Side Alarms				Polling Engine 1	RDS C6A	P 35
2007-06-06 00:03:38	Off	MN	A RX Frame Loss	080 : MDR-8000-3 to 504 ch1	: A Side Alarms			2007-06-06 00:03:38	Polling Engine 1	RDS C6A	P 35
2007-06-06 00:20:44	On	MN	B RX Channel Fail	080 : MDR-8000-3 to 504 ch1	: B Side Alarms				Polling Engine 1	RDS C6A	P 42
2007-06-06 06:51:00	On	MN	A RX Frame Loss	080 : MDR-8000-3 to 504 ch1	: A Side Alarms				Polling Engine 1	RDS C6A	P 35
2007-06-06 07:06:03	Off	MN	B RX Channel Fail	080 : MDR-8000-3 to 504 ch1	: B Side Alarms			2007-06-06 07:06:03	Polling Engine 1	RDS C6A	P 42
2007-06-06 07:16:48	Off	MN	A RX Frame Loss	080 : MDR-8000-3 to 504 ch1	: A Side Alarms			2007-06-06 07:16:48	Polling Engine 1	RDS C6A	P 35
2007-06-06 07:23:10	On	MN	A RX Frame Loss	080 : MDR-8000-3 to 504 ch1	: A Side Alarms				Polling Engine 1	RDS C6A	P 35
2007-06-06 07:27:32	Off	MN	A RX Frame Loss	080 : MDR-8000-3 to 504 ch1	: A Side Alarms			2007-06-06 07:27:32	Polling Engine 1	RDS C6A	P 35
2007-06-06 08:11:23	On	MN	A RX Frame Loss	080 : MDR-8000-3 to 504 ch1	: A Side Alarms		tech01	2007-06-06 12:52:47	Polling Engine 1	RDS C6A	P 35
2007-06-06 08:31:53	Off	MN	A RX Frame Loss	080 : MDR-8000-3 to 504 ch1	: A Side Alarms			2007-06-06 08:31:53	Polling Engine 1	RDS C6A	P 35
2007-06-06 08:38:15	On	MN	B RX Channel Fail	080 : MDR-8000-3 to 504 ch1	: B Side Alarms		tech01	2007-06-06 12:52:47	Polling Engine 1	RDS C6A	P 42
2007-06-06 14:48:18	Off	MN	B RX Channel Fail	080 : MDR-8000-3 to 504 ch1	: B Side Alarms			2007-06-06 14:48:18	Polling Engine 1	RDS C6A	P 42
2007-06-06 14:53:42	Off	MJ	B RX Frame Loss	080 : MDR-8000-3 to 504 ch1	: B Side Alarms			2007-06-06 14:53:42	Polling Engine 1	RDS C6A	P 43
2007-07-16 13:59:54	On	MJ	B RX Frame Loss	080 : MDR-8000-3 to 504 ch1	: B Side Alarms		tech01	2007-07-16 14:00:44	Polling Engine 1	RDS C6A	P 43
2007-07-16 13:59:54	On	MN	B RX Channel Fail	080 : MDR-8000-3 to 504 ch1	: B Side Alarms		tech01	2007-07-16 14:00:44	Polling Engine 1	RDS C6A	P 42
2007-07-16 13:59:54	On	MN	A RX Frame Loss	080 : MDR-8000-3 to 504 ch1	: A Side Alarms		tech01	2007-07-16 14:00:44	Polling Engine 1	RDS C6A	P 35
2007-07-16 13:59:54	On	MN	A RX Channel Fail	080 : MDR-8000-3 to 504 ch1	: A Side Alarms		tech01	2007-07-16 14:00:44	Polling Engine 1	RDS C6A	P 34
2007-07-16 14:00:53	Off	MN	A RX Channel Fail	080 : MDR-8000-3 to 504 ch1	: A Side Alarms			2007-07-16 14:00:53	Polling Engine 1	RDS C6A	P 34
2007-07-16 14:00:53	Off	MN	A RX Frame Loss	080 : MDR-8000-3 to 504 ch1	: A Side Alarms			2007-07-16 14:00:53	Polling Engine 1	RDS C6A	P 35
2007-07-16 14:00:53	Off	MN	B RX Channel Fail	080 : MDR-8000-3 to 504 ch1	: A Side Alarms			2007-07-16 14:00:53	Polling Engine 1	RDS C6A	P 42
2007-07-16 14:00:53	Off	MJ	B RX Frame Loss	080 : MDR-8000-3 to 504 ch1	: B Side Alarms			2007-07-16 14:00:53	Polling Engine 1	RDS C6A	P 43
2007-07-16 14:25:40	On	MJ	B RX Frame Loss	080 : MDR-8000-3 to 504 ch1	: B Side Alarms		tech01	2007-07-16 14:40:57	Polling Engine 1	RDS C6A	P 43
2007-07-16 14:25:40	On	MN	B RX Channel Fail	080 : MDR-8000-3 to 504 ch1	: B Side Alarms		tech01	2007-07-16 14:40:57	Polling Engine 1	RDS C6A	P 42
2007-07-16 14:25:40	On	MN	A RX Frame Loss	080 : MDR-8000-3 to 504 ch1	: A Side Alarms		tech01	2007-07-16 14:40:57	Polling Engine 1	RDS C6A	P 35
2007-07-16 14:25:40	On	MN	A RX Channel Fail	080 : MDR-8000-3 to 504 ch1	: A Side Alarms		tech01	2007-07-16 14:40:57	Polling Engine 1	RDS C6A	P 34
2007-07-16 14:27:37	Off	MN	A RX Channel Fail	080 : MDR-8000-3 to 504 ch1	: A Side Alarms			2007-07-16 14:27:37	Polling Engine 1	RDS C6A	P 34
2007-07-16 14:27:37	Off	MN	A RX Frame Loss	080 : MDR-8000-3 to 504 ch1	: A Side Alarms			2007-07-16 14:27:37	Polling Engine 1	RDS C6A	P 35
2007-07-16 14:27:37	Off	MN	B RX Channel Fail	080 : MDR-8000-3 to 504 ch1	: B Side Alarms			2007-07-16 14:27:37	Polling Engine 1	RDS C6A	P 42
2007-07-16 14:27:37	Off	MJ	B RX Frame Loss	080 : MDR-8000-3 to 504 ch1	: B Side Alarms			2007-07-16 14:27:37	Polling Engine 1	RDS C6A	P 43
2007-07-27 17:37:52	On	MJ	B RX Frame Loss	041 : MDR-7000 to 507	: B Side Alarms		mfortys	2007-07-28 00:19:44	Polling Engine 3	RDS C6A	P 43
2007-07-27 17:39:50	Off	MJ	B RX Frame Loss	041 : MDR-7000 to 507	: B Side Alarms			2007-07-27 17:39:50	Polling Engine 3	RDS C6A	P 43

36 records, 1 selected

Figure 6.39 Example of an alarm report. *Source*: Reprinted with permission of Fial Incorporated.

6.8 MAINTAINING THE NETWORK

Performance monitoring is a powerful tool for service introduction and verification. Owing to nonuniform implementations of performance determination, time synchronization, and the impact of maintenance actions, its use for long-term network performance monitoring is not always useful.

Virtually all NEs provide alarms. Most fault managers provide alarm reports with filtering by time, NE type, alarm level, and other important attributes. Figure 6.39 shows a typical alarm log filtered by a time and alarm name (channel fail and frame loss).

These reports are quite helpful in finding "dead" equipment and doing "after the fact" fault analysis. These reports are especially helpful in finding standing failures or degradations of off-line subelements of hot standby equipment that may not be identified by simply using equipment summary alarms wired to other local equipment for fault reporting. However, these reports are of little help in finding transient events. If the fault manager does a good job of retrieving the alarm "on," alarm "off," and NE "no report" conditions, significant ancillary data can be obtained by processing the alarm logs. The following reports can be created from an alarm log file of the network fault manager. While it is helpful to have a fault manager that has these reports built in, an industrious software programmer can create these reports from exported comma delimited alarm reports ("csv" files) from the fault manager.

For large, dynamic digital systems, it is not unusual to experience several tens of thousands of alarms a day but to have only a few hundred alarms active ("standing") at any one time. Special reports to pinpoint the transient but repetitive alarms are very helpful in finding those "nuisance" alarms. Figure 6.40 is an example of a simple report that collapses all network alarms to a single line with summary statistics.

This report takes all the alarms of a prefiltered alarm report and summarizes the alarm transitions ("toggles" total number and average) and percentage of off time (pseudo "availability"). The number of alarm toggles indicates the activity (instability) of the alarm and the percentage of off time indicates the level of concern. For this report, it is helpful to be able to sort on any of the fields including toggles and off time. By sorting on toggles (from most to least), the operator can focus on unstable devices. By sorting on–off time (from least to most), the operator can focus on network impact. Common problems identified include "flapping" open door alarms, over cycling dehydrators, cycling input signals, and various transient equipment degradations.

It is important to keep track of the fault manager outages. If the fault manager is "out of service" and not receiving alarms, NEs will appear alarm free when they are not. The "cumulative missing days"

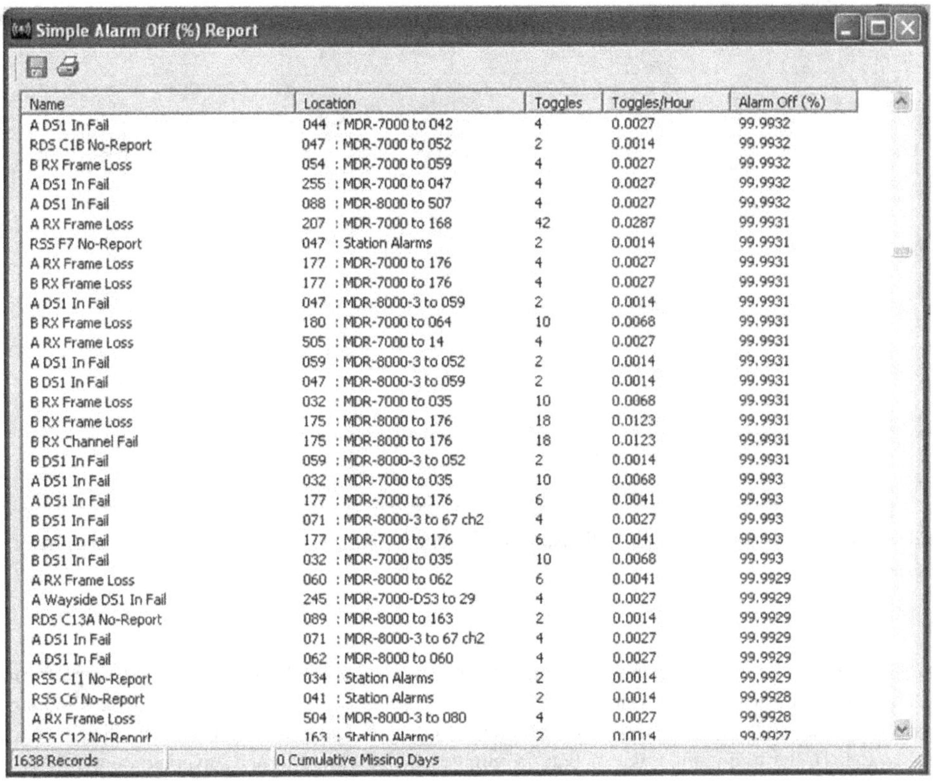

Simple Alarm Off (%) Report

Name	Location	Toggles	Toggles/Hour	Alarm Off (%)
A DS1 In Fail	044 : MDR-7000 to 042	4	0.0027	99.9932
RDS C1B No-Report	047 : MDR-7000 to 052	2	0.0014	99.9932
B RX Frame Loss	054 : MDR-7000 to 059	4	0.0027	99.9932
A DS1 In Fail	255 : MDR-7000 to 047	4	0.0027	99.9932
A DS1 In Fail	088 : MDR-8000 to 507	4	0.0027	99.9932
A RX Frame Loss	207 : MDR-7000 to 168	42	0.0287	99.9931
RSS F7 No-Report	047 : Station Alarms	2	0.0014	99.9931
A RX Frame Loss	177 : MDR-7000 to 176	4	0.0027	99.9931
B RX Frame Loss	177 : MDR-7000 to 176	4	0.0027	99.9931
A DS1 In Fail	047 : MDR-8000-3 to 059	2	0.0014	99.9931
B RX Frame Loss	180 : MDR-7000 to 064	10	0.0068	99.9931
A RX Frame Loss	505 : MDR-7000 to 14	4	0.0027	99.9931
A DS1 In Fail	059 : MDR-8000-3 to 052	2	0.0014	99.9931
B DS1 In Fail	047 : MDR-8000-3 to 059	2	0.0014	99.9931
B RX Frame Loss	032 : MDR-7000 to 035	10	0.0068	99.9931
B RX Frame Loss	175 : MDR-8000 to 176	18	0.0123	99.9931
B RX Channel Fail	175 : MDR-8000 to 176	18	0.0123	99.9931
B DS1 In Fail	059 : MDR-8000-3 to 052	2	0.0014	99.9931
A DS1 In Fail	032 : MDR-7000 to 035	10	0.0068	99.993
A DS1 In Fail	177 : MDR-7000 to 176	6	0.0041	99.993
B DS1 In Fail	071 : MDR-8000-3 to 67 ch2	4	0.0027	99.993
B DS1 In Fail	177 : MDR-7000 to 176	6	0.0041	99.993
B DS1 In Fail	032 : MDR-7000 to 035	10	0.0068	99.993
A RX Frame Loss	060 : MDR-8000 to 062	6	0.0041	99.9929
A Wayside DS1 In Fail	245 : MDR-7000-DS3 to 29	4	0.0027	99.9929
RDS C13A No-Report	089 : MDR-8000 to 163	2	0.0014	99.9929
A DS1 In Fail	071 : MDR-8000-3 to 67 ch2	4	0.0027	99.9929
A DS1 In Fail	062 : MDR-8000 to 060	4	0.0027	99.9929
RSS C11 No-Report	034 : Station Alarms	2	0.0014	99.9929
RSS C6 No-Report	041 : Station Alarms	2	0.0014	99.9928
A RX Frame Loss	504 : MDR-8000-3 to 080	4	0.0027	99.9928
RSS C12 No-Report	163 : Station Alarms	2	0.0014	99.9927

1638 Records | 0 Cumulative Missing Days

Figure 6.40 Example of a simple, collapsed alarm report. *Source*: Reprinted with permission of Fial Incorporated.

item at the bottom of the above report provides that information. It keeps track of the "system up time" of the fault manager and reports cumulative "out of service time" during the reporting period.

A related issue is "no reports" (condition where the NE is unreachable owing to a failure of that NE or failure of other portions of the network that causes communication to that NE to be lost). It is important to keep track of when an NE has been unreachable (in "no report" status). Although that is not an alarm, it is a significant event. The "no report" NE is in an indeterminate status. In addition to individual NE alarms, it is important to have a simple collapsed report for all NEs displaying cumulative "no report" statistics. If the NE was in "no report," lack of alarm activity does not necessarily indicate a functional status.

The simple, collapsed alarm report is helpful for finding equipment instability. If the equipment is redundant and the alarms can differentiate between the two types of equipment (e.g., A side and B side of a redundant NE), dual collapsed (simultaneous A side and B side) alarms can be useful for finding facility failures (loss of transmission capability) (Fig. 6.41).

While equipment and facility problems are important, many radio problems are propagation or maintenance related. Six derived alarm reports (path alarm duration summary, path alarm duration detail, path alarm occurrence, path "no report" duration summary, path "no report" duration detail, and path "no report" occurrence) are especially helpful in highlighting these issues. Figure 6.42 shows the alarm performance of nine paths that were taken from an actual large microwave radio network. In this figure, the reports are sorted so that both directions ("go" and "return") of a path are next to each other (A, B, and A&B in one direction followed by A, B, and A&B in the other direction). If the statistics or duration of events are significantly different by direction, the event is probably not related to propagation.

The alarms duration report (Fig. 6.42) shows a summary of a user-defined alarm event. In these cases, the user-defined Boolean combination of receiver (RX), channel failure (Fail), or receiver frame

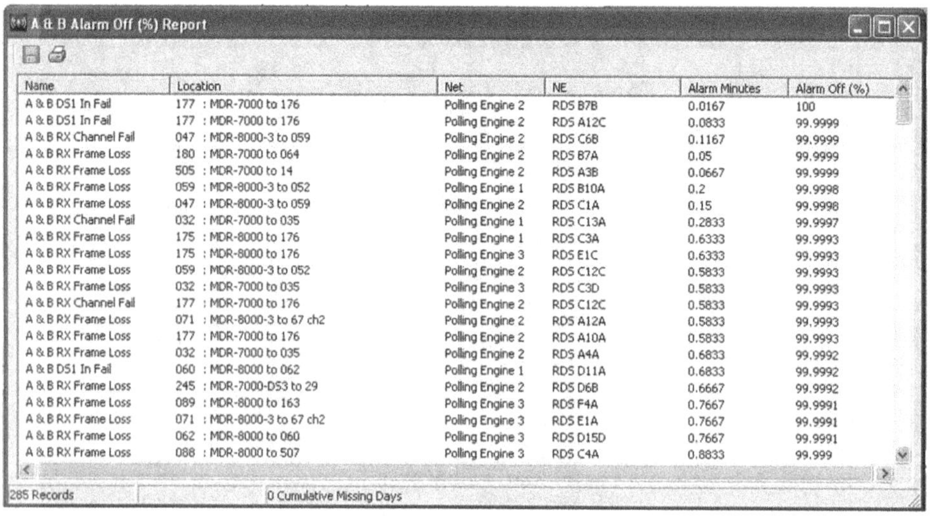

A & B Alarm Off (%) Report

Name	Location	Net	NE	Alarm Minutes	Alarm Off (%)
A & B DS1 In Fail	177 : MDR-7000 to 176	Polling Engine 2	RDS B7B	0.0167	100
A & B DS1 In Fail	177 : MDR-7000 to 176	Polling Engine 2	RDS A12C	0.0833	99.9999
A & B RX Channel Fail	047 : MDR-8000-3 to 059	Polling Engine 2	RDS C6B	0.1167	99.9999
A & B RX Frame Loss	180 : MDR-7000 to 064	Polling Engine 2	RDS B7A	0.05	99.9999
A & B RX Frame Loss	505 : MDR-7000 to 14	Polling Engine 2	RDS A3B	0.0667	99.9999
A & B RX Frame Loss	059 : MDR-8000-3 to 052	Polling Engine 1	RDS B10A	0.2	99.9998
A & B RX Frame Loss	047 : MDR-8000-3 to 059	Polling Engine 2	RDS C1A	0.15	99.9998
A & B RX Channel Fail	032 : MDR-7000 to 035	Polling Engine 1	RDS C13A	0.2833	99.9997
A & B RX Frame Loss	175 : MDR-8000 to 176	Polling Engine 1	RDS C3A	0.6333	99.9993
A & B RX Frame Loss	175 : MDR-8000 to 176	Polling Engine 3	RDS E1C	0.6333	99.9993
A & B RX Frame Loss	059 : MDR-8000-3 to 052	Polling Engine 2	RDS C12C	0.5833	99.9993
A & B RX Frame Loss	032 : MDR-7000 to 035	Polling Engine 3	RDS C3D	0.5833	99.9993
A & B RX Channel Fail	177 : MDR-7000 to 176	Polling Engine 2	RDS C12C	0.5833	99.9993
A & B RX Frame Loss	071 : MDR-8000-3 to 67 ch2	Polling Engine 2	RDS A12A	0.5833	99.9993
A & B RX Frame Loss	177 : MDR-7000 to 176	Polling Engine 2	RDS A10A	0.5833	99.9993
A & B RX Frame Loss	032 : MDR-7000 to 035	Polling Engine 2	RDS A4A	0.6833	99.9992
A & B DS1 In Fail	060 : MDR-8000 to 062	Polling Engine 1	RDS D11A	0.6833	99.9992
A & B RX Frame Loss	245 : MDR-7000-DS3 to 29	Polling Engine 2	RDS D6B	0.6667	99.9992
A & B RX Frame Loss	089 : MDR-8000 to 163	Polling Engine 3	RDS F4A	0.7667	99.9991
A & B RX Frame Loss	071 : MDR-8000-3 to 67 ch2	Polling Engine 3	RDS E1A	0.7667	99.9991
A & B RX Frame Loss	062 : MDR-8000 to 060	Polling Engine 3	RDS D15D	0.7667	99.9991
A & B RX Frame Loss	088 : MDR-8000 to 507	Polling Engine 3	RDS C4A	0.8833	99.999

285 Records | 0 Cumulative Missing Days

Figure 6.41 Example of a dual collapsed alarm report. *Source*: Reprinted with permission of Fial Incorporated.

loss is interpreted as a radio channel outage. The report monitors individual radio system performance and entire receiver performance (individual A or B side outages for nonstandby radios or simultaneous A&B outages for hot standby radios). The number of outages (Outages), the maximum single outage in minutes (Max), the minimum single outage in minutes (Min), the arithmetic average outage (Avg), the median outage (Med), and total outage time in minutes (Total) are all listed. A large difference between Max and Min and between Avg and Med is indicative of unusual behavior. In these reports, hardware configurations are indicated by NA (nonstandby, A side equipped), NB (nonstandby, B side equipped), hot standby nondiversity (HSB), and hot standby space diversity (HSB-SD). For the space diversity configurations, diversity improvement (Div) can be calculated (ratio of the lower outage time of the A and B side equipments to the total A&B outage time). If the diversity improvement is not in the double digits, the outages are not being protected by the diversity configuration (and probably are not multipath related).

By simple observation, the B receive equipment on the "go" direction of paths 4 and 6 have clearly had abnormal problems. In general, diversity action is good in both directions of paths 4, 7, and 8. There is something wrong with diversity action on one direction of paths 1, 5, and 9. Diversity is poor in both directions of paths 2, 5, and 6. Outages are not balanced on paths 4 and 5. The B receiver has died on the "go" direction of path 6. Overall, radio outages by direction are roughly similar. If a path were experiencing persistent interference, there would usually be a significant difference in outages by direction.

A related report (Fig. 6.43) sorts alarm events by event duration and provides a total of events of a labeled duration (M = minutes, s = seconds). Multipath outages are typically tens of seconds "wide" (actually they are of much shorter duration but appear this wide because of alarm retrieval latency). Rain outages are typically a few minutes "wide." Long-duration events are usually associated with nonpropagation events (e.g., interference, equipment failure, or maintenance action). The events longer than 5 or 10 min are unquestionably unusual events requiring additional investigation.

In Figure 6.43, outage event durations are measured and the number of event durations are summed by time "bin" width. The left side of the report shows the number of events with 2 s (0M2s), 3 s (0M3s),..., to 30 min or greater (30+M). Typical multipath fading occurs in the few-seconds duration time bins. However, there are also several unusual events (23M, 24M, and 30+M). These are not normal events. Note that both A and B sides of a radio in one direction (but not the other direction) have the same duration of failure. This suggests human intervention and requires further investigation.

Alarm Type = (RX Channel Fail) OR (RX Frame Loss)

Report Start Date = 1/2/20XX
Report Start Time = 8:48:00
Report End Date = 1/8/20XX
Report End Time = 8:48:00

Average Availability(%) = 99.9910
Average Availability(fraction) = 0.99991
Cumulative Missing Days = 0.0

Site	Direction	Rcv ID	Path	HW	Outages	Max	Min	Avg	Med	Total	Div
112	MDR-8000-3 to 505 Ch1	B2A	204	HSB-SD	A = 26	0.7	0.1	0.2	0.2	4.6	
112	MDR-8000-3 to 505 Ch1	B2A	204	HSB-SD	B = 32	3.2	0.1	0.4	0.2	12.7	
112	MDR-8000-3 to 505 Ch1	B2A	204	HSB-SD	A & B = 0	0.0	0.0	0.0	0.0	0.0	-
505 Ch1	MDR-8000-3 to 112	B3A	204	HSB-SD	A = 33	0.7	0.1	0.2	0.2	6.4	
505 Ch1	MDR-8000-3 to 112	B3A	204	HSB-SD	B = 52	0.9	0.1	0.3	0.2	14.7	
505 Ch1	MDR-8000-3 to 112	B3A	204	HSB-SD	A & B = 5	0.3	0.1	0.2	0.2	0.8	8
112	MDR-8000-3 to 505 Ch2	B2B	305	HSB-SD	A = 16	0.9	0.1	0.5	0.5	7.3	
112	MDR-8000-3 to 505 Ch2	B2B	305	HSB-SD	B = 24	3.1	0.1	0.9	0.5	21.0	
112	MDR-8000-3 to 505 Ch2	B2B	305	HSB-SD	A & B = 2	0.5	0.5	0.5	0.5	0.9	8
505 Ch2	MDR-8000-3 to 112	B3B	305	HSB-SD	A = 14	13.4	0.1	1.5	0.5	20.3	
505 Ch2	MDR-8000-3 to 112	B3B	305	HSB-SD	B = 22	13.8	0.1	1.0	0.4	22.1	
505 Ch2	MDR-8000-3 to 112	B3B	305	HSB-SD	A & B = 2	13.4	0.5	6.9	0.5	13.8	2
512	MDR-8000-3 to 005	D1A	558	HSB	A = 98	59.5	0.1	1.5	0.5	150.4	
512	MDR-8000-3 to 005	D1A	558	HSB	B = 199	59.5	0.1	1.1	0.5	216.1	
512	MDR-8000-3 to 005	D1A	558	HSB	A & B = 95	59.5	0.1	1.3	0.5	127.6	
005	MDR-8000-3 to 512	D2A	558	HSB	A = 118	24.4	0.0	0.6	0.4	67.1	
005	MDR-8000-3 to 512	D2A	558	HSB	B = 192	24.4	0.0	0.7	0.4	129.1	
005	MDR-8000-3 to 512	D2A	558	HSB	A & B = 108	24.4	0.0	0.6	0.3	61.8	
512	MDR-8000-3 to 147	D1B	559	HSB-SD	A = 13	1.1	0.1	0.4	0.4	5.5	
512	MDR-8000-3 to 147	D1B	559	HSB-SD	B = 23	4.0	0.1	0.7	0.5	17.0	
512	MDR-8000-3 to 147	D1B	559	HSB-SD	A & B = 0	0.0	0.0	0.0	0.0	0.0	-
147	MDR-8000-3 to 512	D2B	559	HSB-SD	A = 11	0.9	0.1	0.4	0.4	4.0	
147	MDR-8000-3 to 512	D2B	559	HSB-SD	B = 68	397.8	0.1	12.2	0.9	829.9	
147	MDR-8000-3 to 512	D2B	559	HSB-SD	A & B = 1	0.1	0.1	0.1	0.1	0.1	57
080 Ch2	MDR-8000-3 to 504	G2D	456	HSB-SD	A = 36	157.2	0.1	5.1	0.5	184.8	
080 Ch2	MDR-8000-3 to 504	G2D	456	HSB-SD	B = 45	163.8	0.1	5.6	0.4	250.9	
080 Ch2	MDR-8000-3 to 504	G2D	456	HSB-SD	A & B = 35	157.2	0.1	5.2	0.5	183.4	1
504	MDR-8000-3 to 080 Ch2	G1E	345	HSB-SD	A = 119	2.8	0.1	0.5	0.4	57.7	
504	MDR-8000-3 to 080 Ch2	G1E	345	HSB-SD	B = 151	6.9	0.1	0.8	0.4	113.3	
504	MDR-8000-3 to 080 Ch2	G1E	345	HSB-SD	A & B = 12	0.9	0.1	0.3	0.2	4.0	14
080	MDR-8000-3 to 071 Ch3	H13A	102	HSB-SD	A = 20	4.2	0.1	0.6	0.4	11.8	
080	MDR-8000-3 to 071 ch3	H13A	102	HSB-SD	B = 30	8640.0	8640.0	8640.0	8640.0	8640.0	
080	MDR-8000-3 to 071 Ch3	H13A	102	HSB-SD	A & B = 20	4.2	0.1	0.6	0.4	11.8	1
071 Ch3	MDR-8000-3 to 080	H14A	102	HSB-SD	A = 21	192.3	0.1	11.3	0.4	237.9	
071 Ch3	MDR-8000-3 to 080	H14A	102	HSB-SD	B = 36	0.9	0.1	0.3	0.3	12.2	
071 Ch3	MDR-8000-3 to 080	H14A	102	HSB-SD	A & B = 20	0.5	0.1	0.3	0.2	5.3	2
062	MDR-8000 to 063	H7C	583	HSB-SD	A = 25	1.2	0.1	0.3	0.2	7.9	
062	MDR-8000 to 063	H7C	583	HSB-SD	B = 25	0.9	0.1	0.3	0.2	7.1	
062	MDR-8000 to 063	H7C	583	HSB-SD	A & B = 0	0.0	0.0	0.0	0.0	0.0	-
063	MDR-8000 to 062	H8C	583	HSB-SD	A = 10	0.5	0.1	0.2	0.1	1.6	
063	MDR-8000 to 062	H8C	583	HSB-SD	B = 13	0.5	0.1	0.2	0.2	2.3	
063	MDR-8000 to 062	H8C	583	HSB-SD	A & B = 1	0.1	0.1	0.1	0.1	0.1	28
504	MDR-8000 to 076	G1A	776	HSB-SD	A = 72	1.8	0.1	0.4	0.3	29.3	
504	MDR-8000 to 076	G1A	776	HSB-SD	B = 61	17.2	0.1	0.8	0.4	48.5	
504	MDR-8000 to 076	G1A	776	HSB-SD	A & B = 2	0.5	0.5	0.5	0.5	0.9	32
076	MDR-8000 to 504	G2A	776	HSB-SD	A = 74	0.9	0.1	0.3	0.2	24.5	
076	MDR-8000 to 504	G2A	776	HSB-SD	B = 77	1.4	0.1	0.4	0.3	30.6	
076	MDR-8000 to 504	G2A	776	HSB-SD	A & B = 3	0.4	0.1	0.2	0.2	0.7	36
504	MDR-8000-3 to 080 Ch1	G1C	865	HSB-SD	A = 67	3.1	0.1	0.6	0.3	38.7	
504	MDR-8000-3 to 080 Ch1	G1C	865	HSB-SD	B = 53	5.1	0.1	0.6	0.4	31.2	
504	MDR-8000-3 to 080 Ch1	G1C	865	HSB-SD	A & B = 2	0.2	0.2	0.2	0.2	0.4	89
080 Ch1	MDR-8000-3 to 504	G2C	865	HSB-SD	A = 160	17.1	0.1	0.6	0.2	88.7	
080 Ch1	MDR-8000-3 to 504	G2C	865	HSB-SD	B = 130	17.1	0.1	0.5	0.2	61.1	
080 Ch1	MDR-8000-3 to 504	G2C	865	HSB-SD	A & B = 59	17.1	0.1	0.5	0.1	29.8	2

Figure 6.42 Alarm duration summary report. *Source*: Reprinted with permission of Fial Incorporated.

The alarms occurrence report (Fig. 6.44) shows a summary of a user-defined alarm event sorted by time of the outage event (as before, logically derived). The number in each hour "bin" is the percentage of time the outages occurred within that one hour "window." Multipath events usually occur at night. Multipath and rain outages typically occur approximately equally in both directions of the path.

Paths 1, 6, 7, 8, and 9 appear fairly typical for multipath fading. Path 2 has "go" fading occurring in the very early morning and "return" fading in the late afternoon. Most of the fading of path 3 occurs in the middle of the day. Path 4 has fading occurring at all times of the day. Path 5 has "go" fading before midnight and "return" fading after midnight. Path 6 has receiver B of the "go" direction dead all the

Site	Direction	0M2s	0M3s	0M4s	0M5s	0M6s	0M7s	0M8s	0M9s	0M10s	0M11s	0M12s	0M13s	0M14s	<-->	23M	24M	<-->	30+M
112	MDR-8000-3 to 505 Ch1			5	6			1	5	3			1						
112	MDR-8000-3 to 505 Ch1			2	3	1		1	8				1	3					
112	MDR-8000-3 to 505 Ch1																		
505 Ch1	MDR-8000-3 to 112			3	5	1			10	1		1	2	2					
505 Ch1	MDR-8000-3 to 112			2	6			1	12	1	1		4	1					
505 Ch1	MDR-8000-3 to 112				2				1		1								
112	MDR-8000-3 to 505 Ch2				3														
112	MDR-8000-3 to 505 Ch2				1					1									
112	MDR-8000-3 to 505 Ch2																		
505 Ch2	MDR-8000-3 to 112					2													
505 Ch2	MDR-8000-3 to 112					1	2					1	2	2					
505 Ch2	MDR-8000-3 to 112																		
512	MDR-8000-3 to 005				1	11	7	3	1			1	3	1		1			1
512	MDR-8000-3 to 005				1	17	16	5	2	1	2	3	7	1		1			1
512	MDR-8000-3 to 005				1	11	7	3	1			1	3	1		1			1
005	MDR-8000-3 to 512	1		2	2	21	12	4	2		2		4	1			1		
005	MDR-8000-3 to 512			2	4	23	11	5	1	1	1	8	7	5			1		
005	MDR-8000-3 to 512	1		2	2	21	12	4	2		2		4	1			1		
512	MDR-8000-3 to 147					2					1	1							
512	MDR-8000-3 to 147					2				1			2						
512	MDR-8000-3 to 147																		
147	MDR-8000-3 to 512			1	1	1		1											
147	MDR-8000-3 to 512			1		2	2												6
147	MDR-8000-3 to 512			1															
080 Ch2	MDR-8000-3 to 504					5	3	1											1
080 Ch2	MDR-8000-3 to 504					8	2	1				1		1					2
080 Ch2	MDR-8000-3 to 504					5	3	1											1
504	MDR-8000-3 to 080 Ch2				2	11	13		3		2	8	3	5					
504	MDR-8000-3 to 080 Ch2				1	19	6	4	3	1	1	8	10	3					
504	MDR-8000-3 to 080 Ch2					2			2					2					
080	MDR-8000-3 to 071 Ch3					3	2						1	1					
080	MDR-8000-3 to 071 ch3																		1
080	MDR-8000-3 to 071 Ch3					3	2						1	1					
071 Ch3	MDR-8000-3 to 080				1	4	2	2						1					2
071 Ch3	MDR-8000-3 to 080				1	4	2	3	1		1		4	1					
071 Ch3	MDR-8000-3 to 080				1	4	2	2					1	1					
062	MDR-8000 to 063			1	4				5	1			5						
062	MDR-8000 to 063			3	2			2	6				3	1					
062	MDR-8000 to 063																		
063	MDR-8000 to 062			2	3			1	2										
063	MDR-8000 to 062			1	4			1	2	1			2						
063	MDR-8000 to 062				1														
504	MDR-8000 to 076					8	10	3	3	4		2	1	1					
504	MDR-8000 to 076				1	6	5	2	1	1	1	1	1						
504	MDR-8000 to 076																		
076	MDR-8000 to 504					8	11	3	1	1	3	4	4	3					
076	MDR-8000 to 504					9	9	3				5	5	2					
076	MDR-8000 to 504					1						1							
504	MDR-8000-3 to 080 Ch1			3	16		1	1	5	4				3					
504	MDR-8000-3 to 080 Ch1			5	3				7		1	1	1	2					
504	MDR-8000-3 to 080 Ch1								1			1							
080 Ch1	MDR-8000-3 to 504			14	24	5		1	13	10	4	3	7	4					
080 Ch1	MDR-8000-3 to 504			17	30	5		1	4	5	2	3	2	5					
080 Ch1	MDR-8000-3 to 504			9	23	6			1	5	2	3							

Figure 6.43 Alarm duration detail report. *Source*: Reprinted with permission of Fial Incorporated.

time. Path 3 has no diversity so is probably a high fade margin radio on a relatively short path. The day time outages (which are fairly intense based on the first report (alarm duration summary) and very long based on the second report (alarm duration)) should be investigated. Human intervention or some other unusual circumstances seem likely.

Sometimes all the path "availabilities" are averaged for the entire network for long-term trending (Fig. 6.45). These summary reports are quite valuable but can be misleading for times when alarms cannot be retrieved for a radio. If communication to a radio is lost because of path outage, the far end radio may appear operational from an alarm perspective. However, if the fault manager supports (background) polling, it will record that radio (and that simplex path) as being in "no report" status. If all of the above reports are replicated by replacing the outage alarms (Boolean combination of receiver (RX), channel failure (Fail), or receiver frame loss) with "no report" condition, another equally important view of the network is provided. Although these reports are more difficult to interpret ("no report" represents an unknown status rather than an outage), they provide valuable insight into the operational effectiveness of the network.

Fault managers often "stretch" alarm and "no report" durations owing to delays in alarm polling, report buffering, or delayed retrieval after initial alarm was lost. Stretch factors of 10 to 100 have been observed on large active networks. Time stamping of events is usually at the network manager (even if the NE has an internal clock, the network managers do not always use that alarm raise or alarm clear time). Some NEs have no alarm numbering or alarm history files to facilitate lost alarm retrieval. For polled slave–master, it often takes several seconds (or minutes) to retrieve an alarm clear event. For peer–peer alarm systems, alarm clear messages are often lost and must be retrieved after the fact. For these reasons, the above "availability" calculations are only relative numbers and should not be taken

Site	Direction	Midnight	1AM	2AM	3AM	4AM	5AM	6AM	7AM	8AM	9AM	10AM	11AM	Noon	1PM	2PM	3PM	4PM	5PM	6PM	7PM	8PM	9PM	10PM	11PM
112	MDR-8000-3 to 505 Ch1	0	0	13	0	0	8	45	15	2	0	11	0	0	0	0	0	0	0	0	0	4	0	0	4
112	MDR-8000-3 to 505 Ch1	14	1	2	0	1	6	12	6	0	9	9	0	0	0	0	0	0	0	0	0	0	0	5	36
112	MDR-8000-3 to 505 Ch1	0	0	0	0	0	0	0	0	0	0	0	0	0	0	0	0	0	0	0	0	0	0	0	0
505 Ch1	MDR-8000-3 to 112	0	1	9	0	0	8	55	15	0	0	8	0	0	0	0	0	0	0	0	0	0	3	0	1
505 Ch1	MDR-8000-3 to 112	4	0	12	1	16	1	19	14	7	9	14	0	0	0	0	0	0	1	0	0	0	0	2	0
505 Ch1	MDR-8000-3 to 112	0	0	19	0	0	0	81	0	0	0	0	0	0	0	0	0	0	0	0	0	0	0	0	0
112	MDR-8000-3 to 505 Ch2	19	25	0	0	0	13	6	19	6	0	0	0	0	0	0	0	0	0	0	12	0	0	0	0
112	MDR-8000-3 to 505 Ch2	28	17	4	0	2	7	32	4	1	0	0	0	0	0	0	0	0	0	0	0	0	0	0	4
112	MDR-8000-3 to 505 Ch2	50	0	0	0	0	50	0	0	0	0	0	0	0	0	0	0	0	0	0	0	0	0	0	0
505 Ch2	MDR-8000-3 to 112	9	2	0	0	2	3	13	0	4	0	0	0	0	0	0	0	0	0	49	17	0	0	0	0
505 Ch2	MDR-8000-3 to 112	6	8	0	4	3	3	1	3	10	0	0	0	0	0	0	0	0	0	45	17	0	0	0	0
505 Ch2	MDR-8000-3 to 112	3	0	0	0	0	0	0	0	0	0	0	0	0	0	0	0	0	0	72	24	0	0	0	0
512	MDR-8000-3 to 005	0	3	2	1	0	3	0	3	4	5	3	0	24	28	18	2	4	0	0	0	0	0	0	0
512	MDR-8000-3 to 005	1	3	2	2	2	4	2	5	8	19	4	0	17	12	12	1	3	0	0	0	0	0	1	0
512	MDR-8000-3 to 005	0	3	2	2	0	3	0	3	5	6	3	0	28	19	20	1	4	0	0	0	0	0	0	0
005	MDR-8000-3 to 512	2	5	1	0	2	6	4	6	10	13	3	0	1	2	37	5	2	0	0	1	0	1	0	1
005	MDR-8000-3 to 512	5	4	2	3	3	7	6	7	14	15	4	0	0	1	20	3	2	0	0	0	0	1	1	1
005	MDR-8000-3 to 512	2	4	1	0	1	5	3	6	11	12	3	0	1	2	41	4	2	0	0	0	0	1	0	1
512	MDR-8000-3 to 147	25	8	8	0	0	4	0	39	17	0	0	0	0	0	0	0	0	0	0	0	0	0	0	0
512	MDR-8000-3 to 147	5	0	3	1	0	0	1	6	0	3	2	0	0	44	5	0	18	3	0	0	0	5	3	1
512	MDR-8000-3 to 147	0	0	0	0	0	0	0	0	0	0	0	0	0	0	0	0	0	0	0	0	0	0	0	0
147	MDR-8000-3 to 512	11	34	14	0	0	0	0	37	0	3	0	0	0	0	0	0	0	0	0	0	0	2	0	0
147	MDR-8000-3 to 512	0	0	0	0	2	12	9	5	7	7	7	10	17	11	10	2	0	0	0	0	0	0	0	0
147	MDR-8000-3 to 512	0	0	0	0	0	0	0	0	0	0	0	0	0	0	0	0	0	0	0	0	0	100	0	0
080 Ch2	MDR-8000-3 to 504	0	0	0	0	1	0	0	0	0	0	0	0	1	2	1	0	2	0	0	0	2	25	32	33
080 Ch2	MDR-8000-3 to 504	0	0	3	19	1	0	0	0	0	0	0	0	1	2	1	0	2	0	0	0	1	21	24	25
080 Ch2	MDR-8000-3 to 504	0	0	0	0	1	0	0	0	0	0	0	0	1	2	0	0	2	0	0	0	2	25	33	33
504	MDR-8000-3 to 080 Ch2	13	7	2	7	12	13	9	13	7	5	6	0	0	0	0	0	0	0	0	0	1	0	1	5
504	MDR-8000-3 to 080 Ch2	16	13	7	1	18	7	10	10	5	3	3	0	0	0	0	0	0	0	0	0	0	1	2	4
504	MDR-8000-3 to 080 Ch2	86	0	2	0	0	0	0	11	0	0	0	0	0	0	0	0	0	0	0	0	0	0	0	0
080	MDR-8000-3 to 071 Ch3	0	4	0	1	2	0	38	6	1	37	4	0	0	0	0	0	0	0	0	0	0	0	7	0
080	MDR-8000-3 to 071 ch3	4	4	4	4	4	4	4	4	4	4	4	4	4	4	4	4	4	4	4	4	4	4	4	4
080	MDR-8000-3 to 071 Ch3	0	4	0	1	2	0	38	6	1	37	4	0	0	0	0	0	0	0	0	0	0	0	7	0
071 Ch3	MDR-8000-3 to 080	0	2	25	38	27	4	0	0	1	3	0	0	0	0	0	0	0	0	0	0	0	0	0	0
071 Ch3	MDR-8000-3 to 080	2	22	1	13	0	3	8	4	15	13	18	0	0	0	0	1	0	0	0	0	0	0	0	1
071 Ch3	MDR-8000-3 to 080	4	34	2	9	0	4	2	0	17	20	8	0	0	0	0	0	0	0	0	0	0	0	0	0
062	MDR-8000 to 063	0	0	0	0	0	3	6	21	44	25	0	0	0	0	0	0	0	0	0	0	0	0	2	0
062	MDR-8000 to 063	0	0	0	1	11	12	21	25	20	8	2	0	0	0	0	0	0	0	0	0	0	0	0	0
062	MDR-8000 to 063	0	0	0	0	0	0	0	0	0	0	0	0	0	0	0	0	0	0	0	0	0	0	0	0
063	MDR-8000 to 062	0	0	0	0	0	0	16	46	13	19	0	0	0	0	0	0	5	0	0	0	0	0	0	0
063	MDR-8000 to 062	0	0	0	0	10	0	7	33	34	0	6	0	0	0	0	0	4	0	0	0	0	0	7	0
063	MDR-8000 to 062	0	0	0	0	0	0	0	0	0	0	0	0	0	0	0	0	100	0	0	0	0	0	0	0
504	MDR-8000 to 076	0	1	7	6	18	19	8	4	17	16	1	0	0	0	0	0	0	0	0	0	0	0	0	1
504	MDR-8000 to 076	0	1	1	2	3	8	8	5	7	10	0	0	2	0	30	17	1	1	1	0	0	0	1	1
504	MDR-8000 to 076	0	0	0	0	0	0	0	0	0	100	0	0	0	0	0	0	0	0	0	0	0	0	0	0
076	MDR-8000 to 504	0	0	7	8	5	26	9	17	7	10	1	0	0	0	0	0	0	0	0	0	0	1	4	3
076	MDR-8000 to 504	1	3	7	2	12	18	16	8	18	12	2	0	0	0	0	0	0	0	0	0	0	0	0	0
076	MDR-8000 to 504	0	0	15	0	0	55	0	30	0	0	0	0	0	0	0	0	0	0	0	0	0	0	0	0
504	MDR-8000-3 to 080 Ch1	4	12	9	1	2	19	18	28	5	1	0	0	0	0	0	0	0	0	0	0	0	1	1	0
504	MDR-8000-3 to 080 Ch1	2	6	5	0	9	48	26	4	1	0	0	0	0	0	0	0	0	0	0	0	0	0	0	0
504	MDR-8000-3 to 080 Ch1	0	57	0	0	0	0	43	0	0	0	0	0	0	0	0	0	0	0	0	0	0	0	0	0
080 Ch1	MDR-8000-3 to 504	4	6	7	2	5	9	21	10	7	1	3	0	0	16	9	0	0	0	1	0	0	0	0	0
080 Ch1	MDR-8000-3 to 504	7	5	2	2	4	12	7	8	7	4	4	0	0	23	12	0	0	0	2	0	0	0	0	0
080 Ch1	MDR-8000-3 to 504	4	0	0	2	0	0	2	1	4	2	7	0	0	47	25	0	0	0	4	1	0	0	0	0

Figure 6.44 Alarm occurrence report. *Source*: Reprinted with permission of Fial Incorporated.

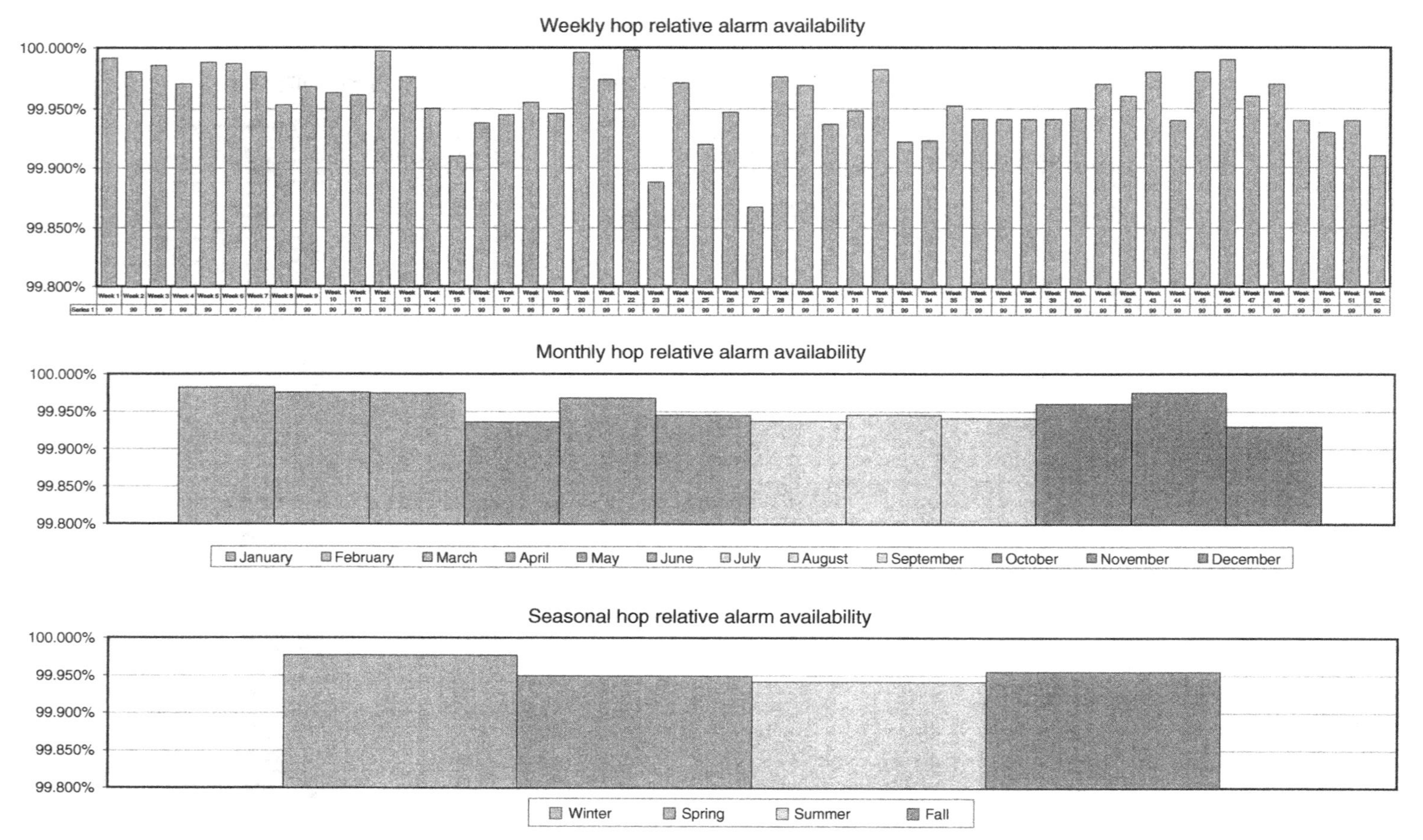

Figure 6.45 Relative path performance report. *Source*: Reprinted with permission of Fial Incorporated.

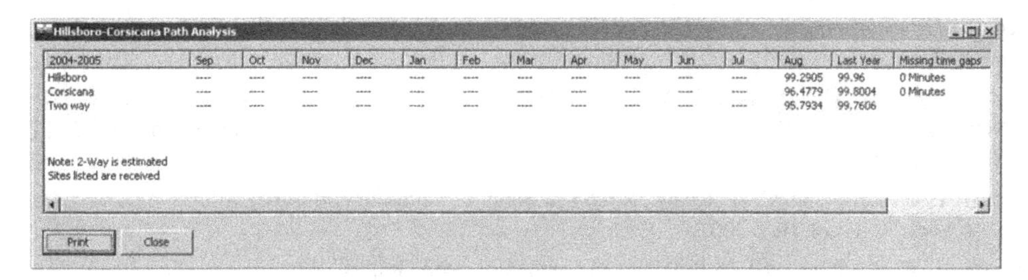

2004-2005	Sep	Oct	Nov	Dec	Jan	Feb	Mar	Apr	May	Jun	Jul	Aug	Last Year	Missing time gaps
Hillsboro	----	----	----	----	----	----	----	----	----	----	----	99.2905	99.96	0 Minutes
Corsicana	----	----	----	----	----	----	----	----	----	----	----	96.4779	99.8004	0 Minutes
Two way	----	----	----	----	----	----	----	----	----	----	----	95.7934	99.7606	

Figure 6.46 Path availability report. *Source*: Reprinted with permission of Fial Incorporated.

literally. However, they are very powerful for comparing paths, validating basic operation, and finding unusual events that require further investigation, which would be difficult to determine by traditional methods.

Some modern radios provide performance data such as SESs, not just on a receiver basis but on a path (facility) basis. If the network management system can retrieve that data, the availability of the path can be calculated. Figure 6.46 is an example of outage seconds (facility SESs) captured on both receivers at both ends of a path.

The one-way path availability is calculated for each receiver. The two one-way path availabilities are multiplied together to determine the two-way availability. This result is displayed by month and year. This greatly simplifies identification of poorly performing paths.

REFERENCES

American National Standards Institute (ANSI), Standards Committee T1—Telecommunications, ANSI T1.102, *Digital Hierarchy—Electrical Interfaces*, 1993.

American National Standards Institute (ANSI), Standards Committee T1—Telecommunications, T1 Technical Report No. 33, *A technical report on synchronization network management using synchronization status messages*, 1994.

American National Standards Institute (ANSI), Standards Committee T1—Telecommunications, ANSI T1.107, *Telecommunications—Digital hierarchy - Format specifications*, 1995a.

American National Standards Institute (ANSI), Standards Committee T1—Telecommunications, ANSI T1.105, *Telecommunications—Synchronous Optical Network (SONET) - Basic Description including Multiplex Structure, Rates, and Formats*, 1995b.

American National Standards Institute (ANSI), Standards Committee T1—Telecommunications, ANSI T1.105.03a, *Telecommunications -Synchronous Optical Network (SONET) - Jitter at Network Interfaces - DS1 Supplement* 1), 1995c.

American National Standards Institute (ANSI), Standards Committee T1—Telecommunications, ANSI T1.105.09, *Telecommunications—Synchronous Optical Network (SONET) - Network element timing and synchronization*, 1996.

American National Standards Institute (ANSI), Standards Committee T1—Telecommunications, ANSI T1.105.03b, *Telecommunications -Synchronous Optical Network (SONET) - Jitter at Network Interfaces - DS3 Wander Supplement*, 1997.

American National Standards Institute (ANSI), Standards Committee T1—Telecommunications, ANSI T1.101, *Synchronization Interface Standard*, 1999.

Aoun, B., Boutaba, R., Iraqil, Y. and Kenward, G., "Gateway Placement Optimization in Wireless Mesh Networks with QoS Constraints," *IEEE Journal on Selected Areas in Communications*, Vol. 24, pp. 2127–2136, November 2006.

Bragg, A. W., "Which Network Design Tool is Right for You?," *IT Professional*, Vol. 2, pp. 23–31, September/October 2000.

Bregni, S., *Synchronization of Digital Telecommunications Networks*. New York: John Wiley & Sons, Inc., 2002.

Chattopadhyay, N. G., Morgan, T. W. and Raghuram, A., "An Innovative Technique for Backbone Network Design," *IEEE Transactions on Systems, Man and Cybernetics*, Vol. 19, pp. 1122–1132, October 1989.

Cormen, T. H., Leiserson, C. E., Rivest, R. L. and Stein, C., Introduction to Algorithms, Second Edition. New York: McGraw-Hill, 2001.

Dijkstra, E. W., "A Note on Two Problems in Connexion with Graphs," *Numerische Mathematic*, Vol. 1, pp. 269–271, December 1959.

Doshi, B. and Harshavardhana, P., "Broadband Network Infrastructure of the Future: Roles of Network Design Tools in Technology Deployment Strategies," *IEEE Communications Magazine*, Vol. 36, pp. 60–71, May 1998.

Dutta, A., Kubat, P. and Liu, H., "Topological Design of Collector Rings in Metropolitan Networks," *Proceedings of the Fourth International Workshop on Design of Reliable Communication Networks*, pp. 282–287, October 2003.

Dutta, A. and Mitra, S., "Integrating Heuristic Knowledge and Optimization Models for Communication Network Design," *IEEE Transactions on Knowledge and Data Engineering*, Vol. 5, pp. 999–1017, December 1993.

Gersht, A. and Weihmayer, R., "Joint Optimization of Data Network Design and Facility Selection," *IEEE Journal on Selected Areas in Communications*, Vol. 8, pp. 1667–1681, December 1990.

Gupta, P. and Kumar, P. R., "The Capacity of Wireless Networks," *IEEE Transactions on Information Theory*, Vol. 46, pp. 388–404, March 2000.

Gupta, P. and Kumar, P. R., "Internets in the Sky: Capacity of 3-D Wireless Networks," *Proceedings of the 39th IEEE Conference on Decision and Control*, pp. 2290–2295, December 2000.

Institute of Electrical and Electronics Engineers, IEEE 802.3 Working Group, "*IEEE Std 802.3(), Standards for Ethernet based LANs*," 1999–2008.

International Telecommunication Union—Radiocommunication Sector (ITU-R), "*Report ITU-R F.2058, Design Techniques Applicable to Broadband Fixed Wireless Access Systems Conveying Internet Protocol Packets or Asynchronous Transfer Mode Cells*," 2006.

International Telecommunication Union—Telecommunication Standardization Sector (ITU-T), "*Recommendation M20, Maintenance Philosophy for Telecommunication Networks*," 1992.

International Telecommunication Union—Telecommunication Standardization Sector (ITU-T), "*Recommendation G.702, Digital Hierarchy Bit Rates*," 1993.

International Telecommunication Union—Telecommunication Standardization Sector (ITU-T), "*Recommendation X.200, Information Technology - Open Systems Interconnection - Basic Reference Model: The Basic Model*," 1994.

International Telecommunication Union—Telecommunication Standardization Sector (ITU-T), "*Recommendation G.704, Synchronous frame structures used at 1544, 6312, 2048, 8448 and 44 736 kbit/s hierarchical levels*," 1998.

International Telecommunication Union—Telecommunication Standardization Sector (ITU-T), "*Recommendation G.803, Architecture of transport networks based on the synchronous digital hierarchy (SDH)*," 2000a.

International Telecommunication Union—Telecommunication Standardization Sector (ITU-T), "*Recommendation M3400, TMN management functions*," 2000b.

International Telecommunication Union—Telecommunication Standardization Sector (ITU-T), "*Recommendation M3010, Principles for a telecommunications management network*," 2000c.

International Telecommunication Union—Telecommunication Standardization Sector (ITU-T), "*Recommendation G.823, The control of jitter and wander within digital networks which are based on the 2048 kbit/s hierarchy*," 2000d.

International Telecommunication Union—Telecommunication Standardization Sector (ITU-T), "*Recommendation G.824, The control of jitter and wander within digital networks which are based on the 1544 kbit/s hierarchy*," 2000e.

International Telecommunication Union—Telecommunication Standardization Sector (ITU-T), "*Recommendation G.703, Physical/electrical characteristics of hierarchical digital interfaces)*," 2001.

International Telecommunication Union—Telecommunication Standardization Sector (ITU-T), "*Recommendation Y.1540, Internet Protocol Data Communication Service—IP Packet Transfer and Availability Performance Parameters*," 2002a.

International Telecommunication Union—Telecommunication Standardization Sector (ITU-T), "*Recommendation Y.1541, Network Performance Objectives for IP-based Services*," 2002b.

International Telecommunication Union—Telecommunication Standardization Sector (ITU-T), "*Recommendation G.707, Network node interface for the synchronous digital hierarchy (SDH)*," 2003.

Internet Engineering Task Force (IETF), "*Request for Comments (RFC) 1122, Requirements for Internet Hosts—Communication Layers*," 1989.

Ju, H.-J. and Rubin, I. A., "Backbone Topology Synthesis for Multiradio Mesh Networks," *IEEE Transactions on Selected Areas in Communications*, Vol. 24, pp. 2116–2126, November 2006.

Jun, J. and Sichitiu, M. L., "The Nominal Capacity of Wireless Mesh Networks," *IEEE Wireless Communications*, Vol. 10, pp. 8–14, October 2003.

Kasch, W. T., Ward, J. R. and Andrusenko, J., "Wireless Network Modeling and Simulation Tools for Designers and Developers," *IEEE Communications Magazine*, Vol. 47, pp. 120–134, March 2009.

Knuth, D. E., *The Art of Computer Programming*, Volume 1, *Third Edition*. Boston: Addison-Wesley, 1997.

Ko, K.-T., Tang, K.-S., Chan, C.-Y., Man, K.-F. and Kwong, S., "Using Genetic Algorithms to Design Mesh Networks," Computer, Vol. 30, pp. 56–61, August 1997.

Kubat, P., Smith, J. M. and Yum, C., "Design of Cellular Networks with Diversity and Capacity Constraints," *IEEE Transactions on Reliability*, Vol. 49, pp. 165–175, June 2000.

Luss, H., Rosenwein, M. B. and Wong, R. T., "Topological Network Design for SONET Ring Architecture," *IEEE Transactions on Systems, Man, and Cybernetics—Part A: Systems and Humans*, Vol. 28, pp. 780–790, November 1998.

McGregor, P. V. and Shen, D., "Network Design: An Algorithm for the Access Facility Location Problem," *IEEE Transactions on Communications*, Vol. 25, pp. 61–73, January 1977.

Okimi, K. and Fukinuki, H., "Master–slave Synchronization Techniques," *IEEE Communications Magazine*, Vol. 19, pp. 12–21, May 1981.

Ruston, L. and Sen, P., "Rule-Based Network Design: Application to Packet Radio Networks," *IEEE Network Magazine*, Vol. 3, pp. 31–39, July 1989.

Song, J., Chang, M. Y., Lee, S. S. and Joung, J., "Overview of ITU-T NGN QoS Control," *IEEE Communications Magazine*, Vol. 45, pp. 116–132, September 2007.

Soni, S., Narasimhan, S. and LeBlanc, L. J., "Telecommunication Access Network Design with Reliability Constraints," *IEEE Transactions on Reliability*, Vol. 53, pp. 532–541, December 2004.

Stallings, W., "Digital Signaling Techniques," *IEEE Communications Magazine*, Vol. 22, pp. 21–25, December 1984.

Stallings, W., *SNMP, SNMPv2 and CMIP, The Practical Guide to Network Management Standards*. New York: Addison-Wesley, pp. 113–122, 1993.

Telcordia Technologies, GR-499-CORE, Issue 2, *Transport Systems Generic Requirements (TSGR): Common Requirements*, 1998.

Telcordia Technologies, GR-253-CORE, Issue 3, *Synchronous Optical Network (SONET) Transport Systems: Common Generic Criteria*, 2000a.

Telcordia Technologies, *Special Report SR-2275, Telcordia Notes on the Networks*, 2000b.

7

HYPOTHETICAL REFERENCE CIRCUITS

Reference circuits were overviewed in Chapter 4. This chapter goes into more detail as to how the path objectives of those standards were derived. It is assumed that the reader has already read Chapter 4 and fully understands the concepts of quality and availability as used in North American and ITU recommendations.

The purpose of hypothetical reference circuits is to establish nominal circuit configurations that meet defined end to end objectives. Even though actual circuits may be significantly different than the reference circuits, it is assumed that actual circuits, if designed according to the reference circuit objectives, will meet the defined end to end objectives. The purpose of this chapter is to reduce the plethora of possible reference circuits to simple objectives. Path designers can make decisions as to applicability and then use the appropriate objectives in their designs. It is assumed that the following objectives will be used as a guideline. However, it is expected that the system designer will tailor the objectives to the particular architecture of their system.

In the process of reviewing these objectives, it may be noticed that the North American availability objectives are highly developed but quality objectives are less developed. ITU quality objectives are highly developed and availability objectives are less developed.

7.1 NORTH AMERICAN (NA) AVAILABILITY OBJECTIVES

Historically, the Bell System defined the availability objectives for all Bell System Long Lines radio networks. Although these standards are no longer actively managed, they are generally regarded as the "gold standard" for North American path design. Modern path designers sometimes deviate from these standards but they still offer a significant reference for high quality path design.

In North America, Telcordia currently sets the telecommunications standards for all public-switched telephone networks. Radio systems expected to be placed into those networks are expected to conform to the Telcordia objectives.

7.1.1 NA Bell System Hypothetical Reference Circuit-Availability Objectives

In North America, the two most generally accepted hypothetical reference circuits for fixed point to point microwave radios are those of the Bell System (former AT&T before divestiture in 1984) Bell Labs and Telcordia (formerly Bellcore). Availability objectives were based on those reference circuits.

Digital Microwave Communication: Engineering Point-to-Point Microwave Systems, First Edition. George Kizer.

Bell System Bell Labs originally developed performance objects for two hypothetical reference circuits (American Telephone and Telegraph Bell System Staff, 1984; Jansen and Prime, 1971; Vigants and Pursley, 1979). One was a long-haul 4000-mile (6400-km) linear system composed of 150 equal-distance radio paths ("hops"). The average path length was 26.7 miles (42.9 km). Multiline equipment and path protection was used for every three consecutive hops ("span"). High speed (DS3 interconnect) cross-connects were assumed at each third hop and low speed (DS1 interconnect) cross-connects were assumed every nine hops (except for one span) for a total of 51 digital terminals. The other was a short-haul 250-mile (400-km) linear system composed of 10 equidistant radio paths. The average path length was 25 miles (40.2 km). Low speed digital interconnects (DS1) were assumed (Fig. 7.1).

All digital circuits had a one-way background ("residual") error objective of 0.5 ESs per day per mile one way, measured at the DS1 level.

Regardless of the hypothetical reference circuit, the end to end transmission circuit-availability objective was 99.98%. This was equivalent to an end to end two-way outage objective of 0.02%. Outage time was allocated among the following outage mechanisms:

- 50% for media (propagation outages because of multipath, rain and upfades);
- 25% for obstruction fades (ObsFade);
- 25% for equipment and maintenance induced failures.

Outage time was the time when the transmission circuit was performing worse than a hypothetical customer requirement. Historically, this was a BER of 10^{-3} in either direction of transmission (a two-way objective). Later this BER objective was raised to 10^{-6} for most fixed point to point microwave radio applications.

A year is 31,557,600 s long (365.25 × 24 × 60 × 60). The 0.02% two-way outage allocation is therefore 6312 s.

The long-haul (4000-mile hypothetical reference circuit) objectives are the following:

$$\text{Propagation media/outage objective} = (0.50 \times 6312\ \text{s})/4000\ \text{miles}$$
$$= 0.7890\ \text{s/mile two way}$$

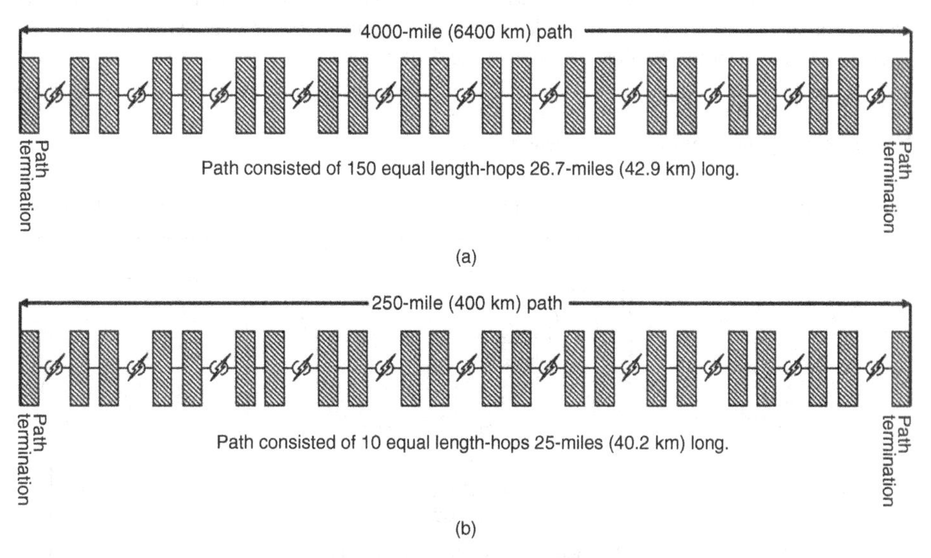

Figure 7.1 Bell System hypothetical reference circuits. Overall objective is 99.98% average annual two-way availability end to end. Path objectives include hop diversity but not multiplex/demultiplex or path-protective switching equipment. (a) Long haul and (b) short haul. A typical radio hop is 26.7 miles long (long haul) or 25 miles long (short haul). *Source*: Bell System Technical Journal, pp. 2085–2116, September 1971 and pp. 1779–1796, October 1979.

$$\text{ObsFade and equipment outage objective} = (0.25 \times 6312\ \text{s})/4000\ \text{miles}$$
$$= 0.3945\ \text{s/mile two way}$$

Typical path propagation outage objective = 0.789 × 26.7 = 21 s/year two way.

This is approximately equivalent to a two-path availability objective of 99.9999%.

For high frequency (≫10 GHz) microwave radio paths dominated by rain fading, the one-way and two-way path objectives are the same. For low frequency (≪10 GHz) microwave radio paths dominated by multipath fading, the one-way outage objective is about half of the two-way outage objective. For the typical low frequency path, this is a 99.99995% one-way availability objective.

The short-haul (250-mile hypothetical reference circuit) objectives are the following:

$$\text{Propagation media/outage objective} = (0.50 \times 6312\ \text{s})/250\ \text{miles}$$
$$= 12.6\ \text{s/mile two way}$$

$$\text{ObsFade and equipment outage objectives} = (0.25 \times 6312\ \text{s})/250\ \text{miles}$$
$$= 6.31\ \text{s/mile two way}$$

Typical path propagation outage objective = 12.6 × 25 = 316 s year two way.

This is approximately equivalent to a two-path availability objective of 99.999%.

For high frequency (≫10 GHz) microwave radio paths dominated by rain fading, the one-way and two-way path objectives are the same. For low frequency (≪10 GHz) microwave radio paths, the one-way outage objective is about half of the two-way outage objective. For the typical low frequency path, this is a 99.9995% one-way availability objective.

7.1.2 NA Telcordia Hypothetical Reference Circuit-Availability Objectives

Most Telcordia performance objectives (Telcordia (Bellcore) Staff, 2009) are based on a 250-mile hypothetical reference circuit (Fig. 7.2).

Telcordia has defined the hypothetical reference circuit (Telcordia (Bellcore) Staff, 2009) as a short-haul (250 miles) configuration of equipment for intraLATA interoffice transport systems. This circuit defines the performance for a source DS1 to sink DS1 data connection. For SONET systems, the performance allocation corresponds to an average downtime of 105 min/year/DS1 channel over 250 miles.

Outage time (unavailability) is allocated among the following outage mechanisms:

- 75% for transmission media (radio path or fiber);
- 25% for equipment (terminals and regenerators).

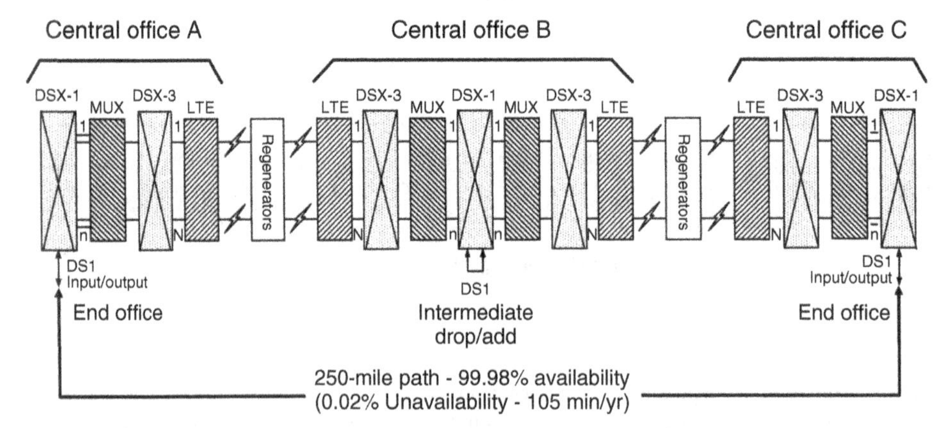

Figure 7.2 Telcordia short-haul hypothetical reference circuit, average 250-mile path. The conventional assumption is that each radio hop is 25 miles long. Path objectives include but not multiplex/demultiplex or path-protective switching equipment. *Source*: Telcordia GR-499-CORE.

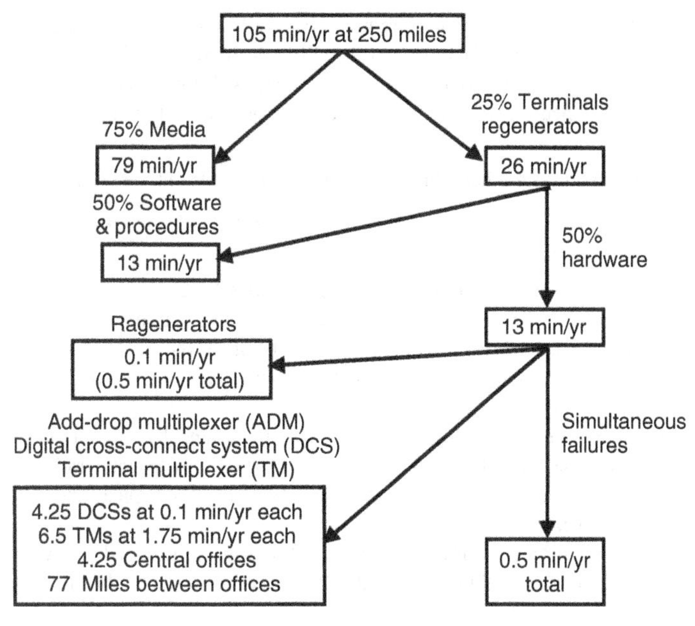

Figure 7.3 Telcordia outage allocations.

The hardware outage time is further subdivided (Fig. 7.3). The hypothetical reference circuit is a linear system by definition. Failures as a result of terminal or regenerator equipment are assumed to be because of hardware or software. No distinction is made between software failure and procedural (operational) errors. The hardware and software failure (unavailability) allocation is divided equally. Hardware outage is further subdivided into simultaneous failures (e.g., failures of the power system), regenerators, cross-connects, and terminal equipment. These hypothetical system objectives are applied to a wide range of transmission equipments (even if they are not SONET based) such as the comparable ITU-SDH Recommendations. Although the circuit objectives are labeled as SONET in application, no distinction is made between synchronous and plesiochronous (asynchronous) terrestrial point to point technology.

From the hypothetical reference circuit diagram, the above objectives apply to hot standby equipment. The above allocations are based on a number of assumptions. The unavailability of the operations systems (network management equipment that reports equipment failures) is 28 min/year (equally divided between hardware and software/procedural failures). Working spares and trained maintenance personnel are available. For plug-in replaceable interoffice transmission equipment, the MTTR is assumed to be 2 h for office equipment or 4 h for remote (unattended) sites. Loop fiber equipment is assumed to have a 4-h MTTR if in the central office and 6 h if remote. If the equipment is not plug-in replaceable (hardwired), the MTTR is assumed to be 48 h. The actual MTTR value dramatically affects system availability.

Cross-connects are electronic digital cross-connect systems. The terminal multiplexer (SONET add/drop multiplexer configured for terminal operation) unavailability is assumed to be calculated on the basis of a two-way DS1 to high speed line failure. The regenerator unavailability is assumed to be calculated on the basis of a high speed interface failure. Keep in mind that the above equipment objective is for hardware outages, in systems approximately 250 miles from end to end. The Telcordia assumption is that central offices are about 80 miles apart and about 6.5 terminals or add/drop multiplexers are in series in the system.

On the basis of various assumptions, Telcordia makes the following (two-way hot standby linear) hardware failure allocations for a short-haul system SONET equipment. Cross-connects should achieve 0.1-min unavailability per year. Assuming a 2-h MTTR, this is an MTBF of 1.1×10^7 h. Terminal multiplexers should achieve 1.75-min unavailability per year. Assuming a 4-h MTTR, this is an MTBF of 1.2×10^6 h. Regenerators should achieve 0.1-min unavailability per year. Assuming a 4-h MTTR, this is an MTBF of 2.1×10^7 h. High capacity radios are usually considered regenerators. Low capacity radios, as they include a multiplexing function, are usually considered terminals.

The media is assigned 75% of the outage allocation, that is, 79-min/year two way. The short-haul (250-mile hypothetical reference circuit) objectives are therefore the following:

$$\begin{aligned}\text{Propagation media/outage objective} &= (79 \times 60 \text{ s})/250 \text{ miles} \\ &= 19.0 \text{ s/mile two way}\end{aligned}$$

The reference circuit would have 10 radio hops each 25 miles long.

Typical path propagation outage objective = 19.0 × 25 = 475-s/year two way

This is approximately equivalent to a two-path availability objective of 99.999%.

Notice that this outage objective is the same as the combined AT&T short-haul outage objectives for propagation and obstruction fading.

For high frequency (≫10 GHz) microwave radio paths dominated by rain fading, the one-way and two-way path objectives are the same. For low frequency (≪10 GHz) microwave radio paths, the one-way outage objective is about half of the two-way outage objective. For the typical path, this is a 99.9995% one-way availability objective.

With the introduction of metropolitan area networks, Telcordia introduced the Metro 50-mile hypothetical reference circuit. The Metro (50-mile hypothetical reference circuit) objectives are the following:

$$\begin{aligned}\text{Propagation media/outage objective} &= (79 \times 60 \text{ s})/50 \text{ miles} \\ &= 94.8 \text{ s/mile two way}\end{aligned}$$

The reference circuit would have two radio hops each 25 miles long.

Typical path propagation outage objective = 94.8 × 25 = 2370 s/year two way.

This is approximately equivalent to a two-way availability objective of 99.99%.

For high frequency (>10 GHz) microwave radio paths dominated by rain fading, the one-way and two-way path objectives are the same. For low frequency (≪10 GHz) microwave radio paths, the one-way outage objective is about half of the two-way outage objective. For the typical path, this is a 99.995% one-way availability objective.

High frequency radios are often used in metro systems. For a 10-mile (16.1-km) path, the two-way availability is 99.997%. For a 16.7-mile (26.8-km) path, the two-way availability is 99.995% (Fig. 7.4).

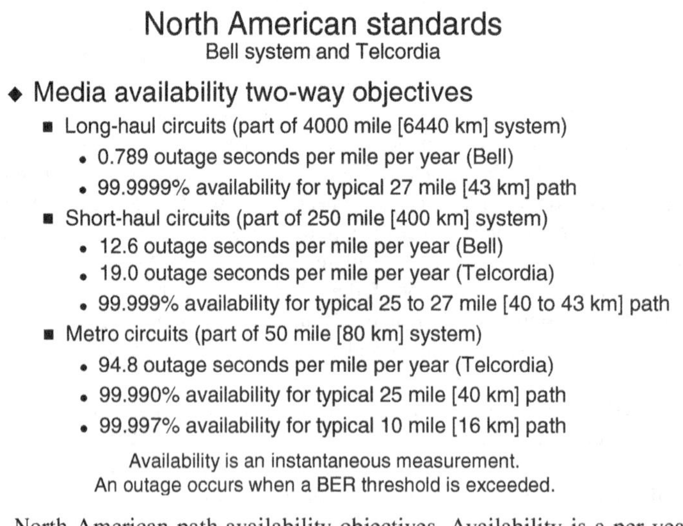

North American standards
Bell system and Telcordia

- Media availability two-way objectives
  - Long-haul circuits (part of 4000 mile [6440 km] system)
    - 0.789 outage seconds per mile per year (Bell)
    - 99.9999% availability for typical 27 mile [43 km] path
  - Short-haul circuits (part of 250 mile [400 km] system)
    - 12.6 outage seconds per mile per year (Bell)
    - 19.0 outage seconds per mile per year (Telcordia)
    - 99.999% availability for typical 25 to 27 mile [40 to 43 km] path
  - Metro circuits (part of 50 mile [80 km] system)
    - 94.8 outage seconds per mile per year (Telcordia)
    - 99.990% availability for typical 25 mile [40 km] path
    - 99.997% availability for typical 10 mile [16 km] path

Availability is an instantaneous measurement.
An outage occurs when a BER threshold is exceeded.

Figure 7.4 North American path availability objectives. Availability is a per year objective.

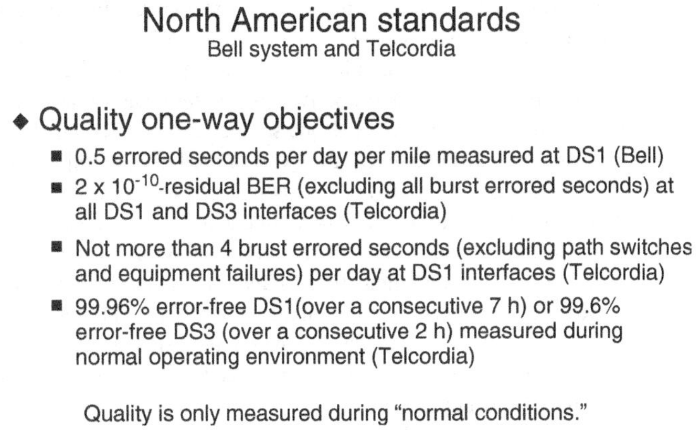

North American standards
Bell system and Telcordia

- Quality one-way objectives
  - 0.5 errored seconds per day per mile measured at DS1 (Bell)
  - 2×10^{-10}-residual BER (excluding all burst errored seconds) at all DS1 and DS3 interfaces (Telcordia)
  - Not more than 4 brust errored seconds (excluding path switches and equipment failures) per day at DS1 interfaces (Telcordia)
  - 99.96% error-free DS1(over a consecutive 7 h) or 99.6% error-free DS3 (over a consecutive 2 h) measured during normal operating environment (Telcordia)

Quality is only measured during "normal conditions."

Figure 7.5 North American path quality objectives. Typically measured over a short period of time.

7.2 NORTH AMERICAN QUALITY OBJECTIVES

Telcordia defines (Telcordia (Bellcore) Staff, 2009) several quality criteria (Fig. 7.5).

7.2.1 Residual BER

The BER at all interface levels between DS1 and DS3 shall be less than 2×10^{-10} (excluding all burst errored seconds). During a burst errored second, neither the number of bit errors nor the number of bits is counted. This requirement applies to every channel in a normal operating environment.

7.2.2 Burst Errored Seconds

The frequency of burst errored seconds (seconds with at least 100 errors) shall average no more than 4 per day at all interface levels between DS1 and DS1 (excluding burst errors caused by hard equipment failures or protection switching). This requirement applies to every channel in a normal operating environment.

7.2.3 DS1 Errored Seconds

For systems interfacing at the DS1 level, the long-term percentage of ES (measured at the DS1 rate) shall not exceed 0.04%. This is equivalent to 99.96 error-free seconds. It is equivalent to no more than 10 ES during a 7-h one-way test. This requirement applies to every DS1 channel in a normal operating environment. It is also an acceptance criterion.

7.2.4 DS3 Errored Seconds

For systems interfacing at the DS3 level, the long-term percentage of ES (measured at the DS3 rate) shall not exceed 0.4%. This is equivalent to 99.6 error-free seconds. It is equivalent to no more than 29 ES during a 2-h one-way test. This requirement applies to every DS3 channel in a normal operating environment. It is also an acceptance criterion.

7.3 INTERNATIONAL OBJECTIVES

All member nations of the United Nations are bound by treaty obligation to conform to the standards of the ITU for all international telecommunication equipment and services. For radio systems, these standards are outlined in the ITU radio regulations ("Red Book") (ITU-R, 2008). The ITU recommendations support these regulations. While the radio regulations are mandatory and the recommendations are *voluntary*, the distinction is seldom recognized. This UN agency is the de facto international telecommunications authority. However, its regulations and recommendations legally apply only to circuits or transmission that intentionally or unintentionally cross national borders. All telecommunication services totally contained within a nation are viewed as "internal matters" and not subject to the ITU. In the United States, for example, the FCC is allowed [except within 35 miles (56.3 km) of the US borders] to set independent fixed point to point microwave radio standards (Part 101) but is compelled to establish fixed satellite standards (Part 25) in conformance with ITU regulations and recommendations. In Europe, telecommunications agencies (PTT agencies) of most countries have their own internal standards, which are often much stricter than those of the ITU. The European Telecommunications Standards Institute (ETSI) sets local equipment standards for the EU countries. These standards may extend well beyond ITU recommendations, similar to Telcordia in the United States. The ITU standards can be regarded as the minimal requirements for telecommunications systems. However, many telecommunications agencies extend or modify those standards to achieve their own internal goals.

The ITU-T (previously CCITT) is responsible for setting telecommunications standards for all cross-border international telecommunications systems. The ITU-R (previously CCIR) is responsible for developing recommendations in support of the ITU-T recommendations. Digital cable systems were deployed nationally and internationally beginning in the 1960s. These were followed by digital, fixed, point to point microwave radio relay systems in the 1970s. In the 1980s, ITU began development of digital telecommunications system recommendations (ITU-T, 1988; ITU-T, 1984).

The first overall quality standard was ITU-T G.821 (ITU-T, 2002a), adopted in 1980. It introduced the following quality parameters:

BER. Ratio of errors to total transmitted digits averaged over a suitable time interval.

ES. 1 s with any data error.

DM. 1 min with average BER worse than 10^{-6}.

SESs. 1 s with average BER worse than 10^{-3}.

The above parameters could be either synchronous (time period beginning with the first error) or asynchronous (measurement based on an arbitrary clock interval). Either definition was allowable (leading to some confusion when comparisons were made along a communications path). Parameters were only to be measured along the communications path (off-line equipment quality was not defined). Quality parameters were only to be made during intervals of availability ($<$10 consecutive SESs). After the adoption of G.826, BER and DM were eliminated as performance standards and no longer are included in the standard.

The hypothetical reference connection (HRX) for G.821 was delineated as high, medium, and low grade sections (Fig. 7.6).

The high grade portion of the G.821 HRX was used as the hypothetical reference digital path (HDRP) for most radio standards (Fig. 7.7).

Several digital radio standards (F.557, F.634, F.695, F.696, F.697, and F.1400) were developed in support of G.821 and F.556-1 (ITU-R, 1986; ITU-R, 1997a; ITU-R, 1990; ITU-R, 1997b; ITU-R, 1997c; ITU-R, 1999). These standards have been superseded and are no longer used to design new systems. However, they have been retained in the active recommendations to support legacy systems' operation and expansion. This is sometimes confusing to designers unfamiliar with the ITU recommendations.

In the early 1990s, ITU-T developed a new series of standards, G.826, G.827, and G.828 (ITU-T, 2002b; ITU-T, 2003; ITU-T, 2000), to support the recently established SDH. While depending on SDH, these standards were intended to be applied to plesiochronous ("asynchronous") as well as synchronous networks.

ITU-T G.828 introduced a new standard for error-performance parameters and a new hypothetical reference path. It provided the framework for the new quality (G.826) and availability (G.827) standards. G.828 introduced the following quality parameters:

Errored block (*EB*). A data block in which one or more bits are in error or has at least one defect such as a loss of pointer.

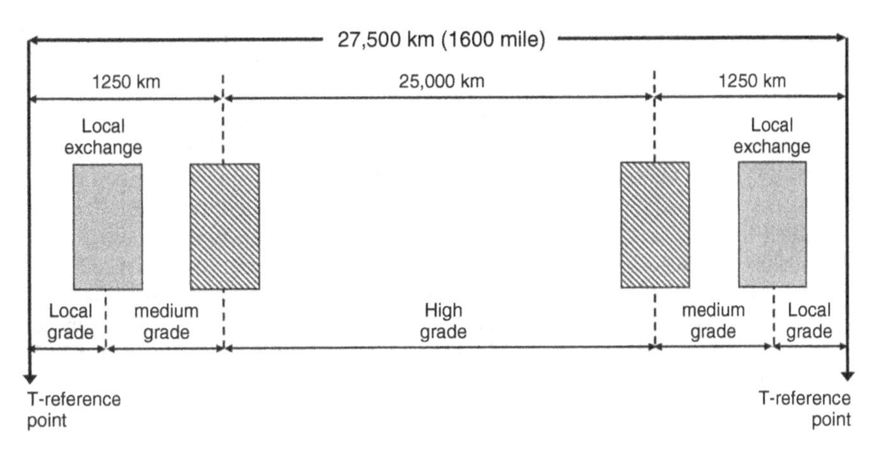

Figure 7.6 Legacy ITU-T hypothetical reference connection. No performance allocations are set from the T-reference point to (subscriber) terminal equipment. The local and medium-grade apportionments are block allowances (constant regardless of actual length). The high grade allowance is allocated on the basis of length using a hypothetical reference digital link which shall be based on a number of 280-km sections. *Source*: ITU-T G.821.

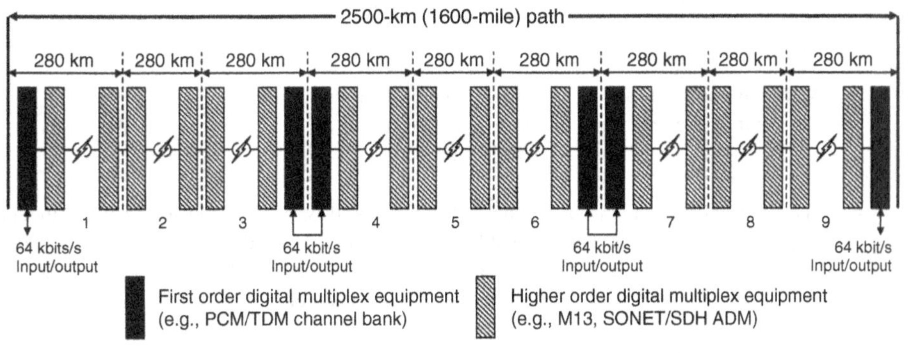

Figure 7.7 Legacy ITU-R hypothetical reference digital path for high-grade performance. Reference path does not include multiplex/demultiplex or (path) protective switching equipment. *Sources*: ITU-T G.821 and ITU-R F.1556-1.

ES. A second with one or more EBs.

SESs. 1 s with a minimum of 30% EBs or a severely disturbed period.

Background Block Error (*BBE*). An EB that is not contained within an SES.

Error-performance objectives (EPOs) were established using ratios of the above quantities to the total number of seconds (or total number of blocks) during the transmission period of interest (1 month).

ITU-T G.826 created new, significantly more stringent quality (error performance) standards for modern digital networks. ITU-R developed several quality standards in support of G.826. Most of them have been withdrawn (F.1092, F.1189, F.1397, and F.1491). Currently, the only ITU-R quality (error performance) recommendation is F.1668.1 (ITU-R, 2007a). This recommendation is very clear. In its first paragraph it states, "This Recommendation [F.1668.1] provides updated information on error-performance objectives for real digital fixed wireless links used in 27 500 km hypothetical reference paths and connections. It is the only Recommendation defining error-performance objectives for all real digital fixed wireless links. Performance events and objectives for connections using equipment designed prior to approval of ITU-T Recommendation G.826 in December 2002 are given in ITU-T Recommendation

G.821 and Recommendations ITU-R F.634, ITU R F.696 and ITU R F.697 (ITU-R, 1997d; ITU-R, 1997b; ITU-R, 1997c). Recommendations ITU R F.1397 and ITU R F.1491 are superseded by this Recommendation."

ITU-T G.827 introduced new, significantly more stringent availability standards for modern digital networks. ITU-R developed two availability standards in support of G.827, which have been subsequently withdrawn (F.1492 and F.1493). Currently, the only ITU-R availability recommendation is F.1703 (ITU-R, 2005). This recommendation is very clear. In its first paragraph, it states, "This Recommendation provides updated information on availability objectives for real digital fixed wireless links used in 27 500 km hypothetical reference paths taking into account ITU T Recommendation G.827 (approved in 2003). It is the only Recommendation defining availability objectives for all real digital fixed wireless links. Recommendations ITU R F.1492 and ITU R F.1493 are superseded by this Recommendation. The applicability of Recommendations ITU R F.557, ITU R F.695, ITU R F.696, and ITU R F.697 is limited to systems designed prior to the approval of this Recommendation." (Fig. 7.8 and Fig. 7.9)

Recommendations ITU-R F.592-4 (ITU-R, 2007b) and ITU-R F.594-4 (ITU-R, 1997e) are supporting recommendations.

7.3.1 International Telecommunication Union Availability Objectives

7.3.1.1 Legacy Availability Objectives The legacy availability standards (F.557, F.695, F.696, F.697, and F.1400) are still used for expansion of existing systems. We will start with the outage objectives and derive a nominal radio path objective. Historically, typical radio paths were often assumed to be 40 km (25 miles) or 46.7 km (29 miles). We will use 46.7 km as the typical distance.

High Grade Circuit (F.557 and F.695)

$$\text{Unavailability for path and equipment} = U \text{ [path (media) and equipment]}$$
$$U \text{ (path and equipment)} = 0.003 \times (L/2500);$$
$$L = \text{path length (km)}.$$

Let us make the typical assumption that half the unavailability is allocated to the path (media) and half to the equipment (If the equipment is considered perfect and the media gets the entire unavailability allocation, the results only change by a factor of 2.). This has the added advantage that it gives us a result that is applicable to both the path and the equipment separately.

U (path) $= 0.0015 \times (L/2500)$.

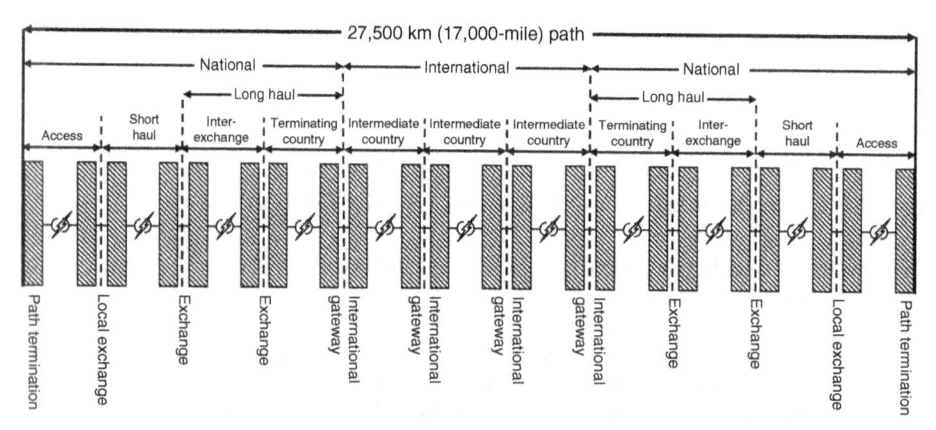

Figure 7.8 Section definitions for quality and availability objectives. Modern ITU-R hypothetical reference path. The technology or media of each section is not explicitly defined. The actual number of spans/hops of equipment is not explicitly defined. Path objectives do not include multiplex/demultiplex or protective switching equipment. *Sources*: ITU-T G.801, G.826, G.827 and G.828 and ITU-R F.1668 and F.1703.

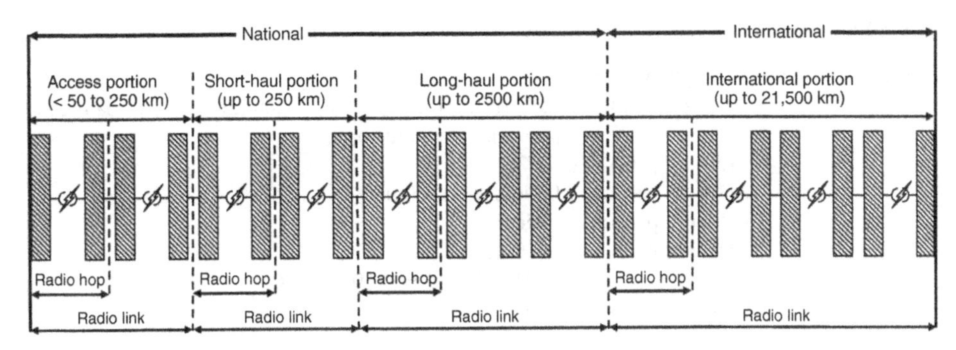

Figure 7.9 Radio path definitions used in the ITU HRP for quality and availability objectives. *Sources*: ITU-R F.1688 and F.1703.

Multiplying by seconds in a year yields the following:

U (s/km/year) = 31,557,600 × 0.0015/2500 = 18.93;
U (s/mile/year) = 30.47.

For a typical path of 46.7 km,

Unavailability = 0.00002802;
Availability = 99.997%.

Medium Grade Circuit (F.696)

Class 1: 280-km section

Unavailability for path and equipment = 0.033%;
U (path) = 0.0165% (using the common assumption of equal path and equipment availability);
U (46.7-km path) = 0.0165 × (46.7/280) = 0.002752%;
A (46.7-km path) = 99.997%;
U (s/km/year) = 0.000165 × 31,557,600/280 = 18.60;
U (s/mile/year) = 29.93.

Class 2: 280-km section

Unavailability for path and equipment = 0.05%;
U (path) = 0.025% (using the common assumption of equal path and equipment availability);
U (46.7-km path) = 0.025 × (46.7/280) = 0.004170%;
A (46.7-km path) = 99.996%;
U (s/km/year) = 0.00025 × 31,557,600/280 = 28.18;
U (s/mile/year) = 45.34.

Class 3: 50-km section

Unavailability for path and equipment = 0.05%;
U (path) = 0.025% (using the common assumption of equal path and equipment availability);
U (50-km path) = 0.025%;
U (10-km path) = 0.025 × (10/50) = 0.005%;
A (50-km path) = 99.975%;
A (10-km path) = 99.995%;
U (s/km/year) = 0.00025 × 31,557,600/50 = 157.8;
U (s/mile/year) = 253.9.

Class 4: 50-km section

Unavailability for path and equipment = 0.1%;
U (path) = 0.05% (using the common assumption of equal path and equipment availability);
U (50-km path) = 0.05%;
U (10-km path) = 0.05 × (10/50) = 0.01%;
A (50-km path) = 99.950%;
A (10-km path) = 99.990%;
U (s/km/year) = 0.0005 × 31,557,600/50 = 315.6;
U (s/mile/year) = 507.9.

Local-Grade Circuit (F.697) Per Annex 1, there is no availability objective for this circuit.
These results are summarized in Figure 7.10.

7.3.1.2 Current Availability Objectives Recommendation F.1703 uses two parameters to quantify availability: availability ratio (AR), the ratio of available time to total operating time, and outage intensity (OI), the number of outages in the total operating time. OI is the average duration of each outage during the total operating time. The operating time is universally taken as 1 year (e.g., Section 3, Annex 1 by International Telecommunication Union–Radiocommunication Sector (ITU-R, 2005)). The parameter Mo, the mean time between outage, is inversely related to OI and also considered.

The availability parameters are stated as one way. Since most of the outages that affect availability are inherently the same one way and two way, the assumption will be made that these one-way numbers may be used as two-way numbers. For those wishing more conservative results, the traditional way to convert one-way parameters to two way is to consider the two directional outages as statistically independent and employ Bayes' theorem to calculate the result:

$$A(\text{availability, \%, two way}) = 100\left[\left(\frac{A\ (\text{availability, \%, one way})}{100}\right)\right]^2 \tag{7.1}$$

$$\text{Mo (two way)} = 2 \times \text{Mo (one way)} \tag{7.2}$$

International legacy standards
ITU-R (ITU-T G.821)

- availability two-way objectives
  (path or equipment)
  - High grade circuit, F.557-4 & F.695
    - 99.997% for 46.7-km hop in 2500-km HDRP
  - Medium grade circuit, F.696-2
    - Class 1280-km radio section
      - 99.997% for 46.7-km hop
    - Class 2280-km radio section
      - 99.996% for 46.7-km hop
    - Class 350-km radio sub section
      - 99.98% for 50-km hop
      - 99.995% for 10-km hop
    - Class 450-km radio sub section
      - 99.95% for 50-km hop
      - 99.99% for 10-km hop
  - Local grade link, F.697-2
    - Commonly used for high frequency (>17-GHz) hops
    - No availability objectives

It is assumed that equipment and media availability are equal
Availability measured using 10 s on/off window.
Total measurement period under study but probably greater than 1 year.
Typical path degradations are rain and long term interference.

Figure 7.10 Legacy international path availability objectives.

$$\text{OI (two way)} = \frac{\text{OI (one way)}}{2} \tag{7.3}$$

For actual systems, it is assumed that the path designer will calculate availabilities for that system architecture. However, it is desirable to calculate availabilities for "typical" paths so that we get a feel for typical results and can compare them with the "typical" path performance of other standards. Therefore, an assumption must be made as to what a typical path is. For older G.821 systems, the conventional path was often taken as 46.7 km (29 miles). For the current G.826 standards, it is tempting to use 50 km (31 miles) as the typical path as it divides evenly into the radio section limits. However, this path length is somewhat longer than a typical 6- or 7-GHz path. For the following calculations, a typical path (when applicable) will be assumed to be 41.7 km (26 miles) as it is more typical of 6- or 7-GHz path lengths and also divides (approximately) evenly into the radio section limits.

Availability and OI as calculated by F.1703 include the effects of both the equipment and the path (transmission media). If we make the conventional assumption that the contribution of each is equal, the product of the equipment and path availabilities (expressed as fractions) is equal to the hop availability (expressed as a fraction). This leads to the following relationships:

$$A \text{ (path availability)} = A \text{ (radio hop availability)}^{1/2} \tag{7.4}$$

$$\text{OI (path outage intensity)} = \frac{\text{OI (radio hop outage intensity)}}{2} \tag{7.5}$$

$$\text{Mo (path mean time between outage)} = \text{Mo (radio hop Mo)} \times 2 \tag{7.6}$$

Obviously, the assumptions affect the results. However, regardless of the assumptions, the following give order-of-magnitude results that are helpful in gaining a perspective on the recommendations.

First let us address Availability Ratio (AR):

$$\text{AR} = 1 - \left[B_n \left(\frac{L_{\text{link}}}{2500}\right) + C_n\right] \tag{7.7}$$

AR = availability ratio (fraction) for the entire radio link;
L_{link} = radio link length (km);
B_n = parameter based on radio link type and length;
C_n = parameter based on radio link type and length.

Access Section $50\,\text{km} < L_{\text{link}} \leq 250\,\text{km}$.

AR (entire radio link) = 1 − 0.0005 = 0.9995 (block allocation not based on length).

If the entire section is one hop,

$$A \text{ (hop availability)} = 99.9500\% \tag{7.8}$$

$$A \text{ (path availability)} = 100 \times (0.999500)^{(1/2)} = 99.9750\% \tag{7.9}$$

If the section has six identical cascaded hops,

$$A \text{ (hop availability)} = 100\left(\frac{99.9500}{100}\right)^{(1/6)} = 99.9917\% \tag{7.10}$$

$$A \text{ (path availability)} = 100 \times (0.999917)^{(1/2)} = 99.9958\% \tag{7.11}$$

Short-Haul Section $50\,\text{km} < L_{\text{link}} \leq 250\,\text{km}$.

AR (entire radio link) = 1 − 0.0004 = 0.9996 (block allocation not based on length).

If the entire section is one hop,

$$A\ (\text{hop availability}) = 99.9600\% \tag{7.12}$$

$$A\ (\text{path availability}) = 99.9800\% \tag{7.13}$$

If the section has six identical cascaded hops,

$$A\ (\text{hop availability}) = 100\left(\frac{99.9600}{100}\right)^{(1/6)} = 99.9933\% \tag{7.14}$$

$$A\ (\text{path availability}) = 99.9966\% \tag{7.15}$$

National or International Section $50\,\text{km} < L_{\text{link}} \leq 250\,\text{km}$.

$$\text{AR (entire radio link)} = 1 - \left(\left[0.0019\left(\frac{L_{\text{link}}}{2500}\right)\right] + 0.00011\right) \tag{7.16}$$

If the entire section is one hop of 41.7 km (26 miles),

$$\text{AR} = 1 - \left\{\left[0.0019\left(\frac{50}{2500}\right)\right] + 0.00011\right\} = 1 - 0.000148 = 0.999852$$

$$A\ (\text{hop availability}) = 99.9852\% \tag{7.17}$$

$$A\ (\text{path availability}) = 99.9926\% \tag{7.18}$$

If the section is 250 km long and composed of six cascaded hops of 41.7 km (26 miles) length each,

$$\text{AR} = 1 - \left\{\left[0.0019\left(\frac{250}{2500}\right)\right] + 0.00011\right\} = 1 - 0.0003 = 0.9997 \tag{7.19}$$

$$A\ (\text{hop availability}) = 100 \times (0.9997)^{(1/6)} = 99.9950\% \tag{7.20}$$

$$A\ (\text{path availability}) = 99.9975\% \tag{7.21}$$

$250\,\text{km} < L_{\text{link}} \leq 2500\,\text{km}$.

$$\text{AR (entire radio link)} = 1 - \left\{\left[0.003\left(\frac{L_{\text{link}}}{2500}\right)\right] + 0\right\} \tag{7.22}$$

If the section is 250 km long and composed of six cascaded hops of 41.7 km (26 miles) length each:

$$\text{AR} = 1 - \left(\frac{0.003 \times 250}{2500}\right) = 1 - 0.0003 = 0.9997 \tag{7.23}$$

$$A(\text{hop availability}) = 100 \times (0.9997)^{(1/6)} = 99.9950\%$$

$$A(\text{path availability}) = 99.9975\% \tag{7.24}$$

If the section is 2500 km long and composed of 60 cascaded hops of 41.7 km (26 miles) length each,

$$\text{AR} = 1 - \left(\frac{0.003 \times 2500}{2500}\right) = 1 - 0.003 = 0.997 \tag{7.25}$$

$$A\ (\text{hop availability}) = 100 \times (0.997)^{(1/60)} = 99.9950\% \tag{7.26}$$

$$A\ (\text{path availability}) = 99.9975\% \tag{7.27}$$

International Section AR (entire radio link) $= 1 - [(0.003 \times (L_{\text{link}}/2500)) + 0]$.
$2500\,\text{km} < L_{\text{link}} \leq 21{,}500\,\text{km}$.

If the section is 2500 km long and composed of 60 cascaded hops of 41.7 km (26 miles) length each,

$$\text{AR} = 1 - \left(\frac{0.003 \times 2500}{2500}\right) = 1 - 0.003 = 0.997 \tag{7.28}$$

$$A \text{ (hop availability)} = 100 \times (0.997)^{(1/60)} = 99.9950\% \tag{7.29}$$

$$A \text{ (path availability)} = 99.9975\% \tag{7.30}$$

If the section is 7500 km long and composed of 180 cascaded hops of 41.7 km (26 miles) length each,

$$\text{AR} = 1 - \left(\frac{0.003 \times 7500}{2500}\right) = 1 - 0.009 = 0.991 \tag{7.31}$$

$$A \text{ (hop availability)} = 100 \times (0.991)^{(1/180)} = 99.9950\% \tag{7.32}$$

$$A \text{ (path availability)} = 99.9975\% \tag{7.33}$$

If the section is 21,500 km long and composed of 516 cascaded hops of 41.7 (41.667) km (26 miles) length each,

$$\text{AR} = 1 - \left(\frac{0.003 \times 21{,}500}{2500}\right) = 1 - 0.0258 = 0.9741 \tag{7.34}$$

$$A \text{ (hop availability)} = 100 \times (0.9741)^{(1/516)} = 99.9950\% \tag{7.35}$$

$$A \text{ (path availability)} = 99.9975\% \tag{7.36}$$

Next, let us address Outage Intensity (OI) and Mean Time Between Outage (Mo).

$$\text{OI} = \left[D_{\text{n}}\left(\frac{L_{\text{link}}}{2500}\right) + E_{\text{n}}\right] \tag{7.37}$$

OI = outage intensity (number of outages per year) for the entire radio link.

In F.1703, 1 year is 525,960 min (365.25 days) (ITU-R, 2005).

$$\text{Mo} = \frac{\text{(minutes in 1 year)}}{\text{OI}} = \frac{525{,}960}{\text{OI}} \tag{7.38}$$

Mo = mean time between outages (min) for the entire radio link;
L_{link} = radio link length (km);
D_{n} = parameter based on radio link type and length;
E_{n} = parameter based on radio link type and length.

Access Section $50\,\text{km} < L_{\text{link}} \leq 250\,\text{km}$.

$$\text{OI (entire radio link)} = 100 \text{ (block allocation not based on length)} \tag{7.39}$$

If the entire section is one hop,

$$\text{OI (hop outage intensity)} = 100 \text{ outages/year} \tag{7.40}$$

$$\text{OI (path outage intensity)} = \frac{100}{2} = 50 \text{ outages/year} \tag{7.41}$$

$$\text{Mo(hop mean time between outage)} = \frac{525{,}960}{100}$$

$$= 5260 \text{ min (3.7 days)} \quad (7.42)$$

$$\text{Mo (path mean time between outage)} = 10{,}520 \text{ min(7.3 days)} \quad (7.43)$$

If the section has six identical cascaded hops,

$$\text{OI (hop outage intensity)} = \frac{100}{6} = 17 \text{ outages/year} \quad (7.44)$$

$$\text{OI (path outage intensity)} = 8 \text{ outages/year} \quad (7.45)$$

$$\text{Mo(hop mean time between outage)} = \frac{525{,}960}{(100/6)}$$

$$= 31{,}558 \text{ min (21.9 days)} \quad (7.46)$$

$$\text{Mo (path mean time between outage)} = 63{,}116 \text{ min (43.8 days)} \quad (7.47)$$

Short-Haul Section 50 km < $L_{\text{link}} \leq 250$ km.
OI (entire radio link) = 120 (block allocation not based on length).
If the entire section is one hop,

$$\text{OI (hop outage intensity)} = 120 \text{ outages/year} \quad (7.48)$$

$$\text{OI (path outage intensity)} = 60 \text{ outages/year} \quad (7.49)$$

$$\text{Mo(hop mean time between outage)} = \frac{525{,}960}{120}$$

$$= 4383 \text{ min (3.0 days)} \quad (7.50)$$

$$\text{Mo (path mean time between outage)} = 8766 \text{ min (6.1 days)} \quad (7.51)$$

If the section has six identical cascaded hops,

$$\text{OI (hop outage intensity)} = \frac{120}{6} = 20 \text{ outages/year} \quad (7.52)$$

$$\text{OI (path outage intensity)} = 10 \text{ outages/year} \quad (7.53)$$

$$\text{Mo(hop mean time between outage)} = \frac{525{,}960}{120/6}$$

$$= 26{,}298 \text{ min (18.3 days)} \quad (7.54)$$

$$\text{Mo (path mean time between outage)} = 52{,}596 \text{ min (36.5 days)} \quad (7.55)$$

National (Long-Haul or Terminating Country) or International Section
50 km < $L_{\text{link}} \leq 250$ km.

$$\text{OI (entire radio link)} = 150\left(\frac{L_{\text{link}}}{2500}\right) + 50 \quad (7.56)$$

If the entire section is one hop of 41.7 km (26 miles),

$$\text{OI (hop outage intensity)} = 150\left(\frac{50}{2500}\right) + 50 = 50 \text{ outages/year} \quad (7.57)$$

$$\text{OI (path outage intensity)} = 25 \text{ outages/year} \quad (7.58)$$

$$\text{Mo (hop mean time between outage)} = \frac{525{,}960}{50}$$

$$= 10{,}519 \ \text{min (7.3 days)} \tag{7.59}$$

$$\text{Mo (path mean time between outage)} = 21{,}038 \ \text{min (14.6 days)} \tag{7.60}$$

If the section is 250 km long and composed of six cascaded hops of 41.7 km (26 miles) length each,

$$\text{OI (hop outage intensity)} = \left[\frac{150\,(250/2500) + 50}{6}\right]$$

$$= 11 \ \text{outages/year} \tag{7.61}$$

$$\text{OI (path outage intensity)} = 5 \ \text{outages/year} \tag{7.62}$$

$$\text{Mo (hop mean time between outage)} = \frac{525{,}960}{10.8333}$$

$$= 48{,}550 \ \text{min (34 days)} \tag{7.63}$$

$$\text{Mo (path mean time between outage)} = 97{,}100 \ \text{min (67.4 days)} \tag{7.64}$$

$250\,\text{km} < L_{\text{link}} \leq 2500\,\text{km}$.

$$\text{OI (entire radio link)} = 100\left(\frac{L_{\text{link}}}{2500}\right) + 55 \tag{7.65}$$

If the section is 250 km long and composed of six cascaded hops of 41.7 km (26 miles) length each,

$$\text{OI (hop outage intensity)} = \left[\frac{100\,(250/2500) + 55}{6}\right]$$

$$= 11 \ \text{outages/year} \tag{7.66}$$

$$\text{OI (path outage intensity)} = 5 \ \text{outages/year} \tag{7.67}$$

$$\text{Mo (hop mean time between outage)} = \frac{525{,}960}{10.8333}$$

$$= 48{,}550 \ \text{min (33.7 days)} \tag{7.68}$$

$$\text{Mo (path mean time between outage)} = 97{,}100 \ \text{min (67.4 days)} \tag{7.69}$$

If the section is 2500 km long and composed of 60 cascaded hops of 41.7 km (26 miles) length each,

$$\text{OI (hop outage intensity)} = \left[\frac{100\,(2500/2500) + 55}{60}\right]$$

$$= 3 \ \text{outages/year} \tag{7.70}$$

$$\text{OI (path outage intensity)} = 1 \ \text{outage/year} \tag{7.71}$$

$$\text{Mo (hop mean time between outage)} = \frac{525{,}960}{2.58333}$$

$$= 203{,}597 \ \text{min (141 days)} \tag{7.72}$$

$$\text{Mo (path mean time between outage)} = 407{,}194 \ \text{min (283 days)} \tag{7.73}$$

International Section

$$\text{OI (entire radio link)} = 100\left(\frac{L_{\text{link}}}{2500}\right) + 55 \tag{7.74}$$

$2500\,\text{km} < L_{\text{link}} \leq 21{,}500\,\text{km}$.

If the section is 2500 km long and composed of 60 cascaded hops of 41.7 km (26 miles) length each,

$$\begin{aligned}\text{OI (hop outage intensity)} &= \left[\frac{100\,(2500/2500) + 55}{60}\right] \\ &= 3 \text{ outages/year} \end{aligned} \tag{7.75}$$

OI (path outage intensity) = 1 outage/year.

$$\begin{aligned}\text{Mo (hop mean time between outage)} &= \frac{525{,}960}{2.58333} \\ &= 203{,}597 \text{ min (141 days)} \end{aligned} \tag{7.76}$$

$$\text{Mo (path mean time between outage)} = 407{,}194 \text{ min (283 days)} \tag{7.77}$$

If the section is 7500 km long and composed of 180 cascaded hops of 41.7 km (26 miles) length each,

$$\begin{aligned}\text{OI (hop outage intensity)} &= \left[\frac{100\,(7500/2500) + 55}{180}\right] \\ &= 2 \text{ outages/year} \end{aligned} \tag{7.78}$$

$$\text{OI (path outage intensity)} = 1 \text{ outage/year} \tag{7.79}$$

$$\begin{aligned}\text{Mo (hop mean time between outage)} &= \frac{525{,}960}{1.97222} \\ &= 266{,}684 \text{ min (185 days)} \end{aligned} \tag{7.80}$$

$$\text{Mo (path mean time between outage)} = 533{,}368 \text{ min (370 days)} \tag{7.81}$$

If the section is 21,500 km long and composed of 516 cascaded hops of 41.7 (41.667) km (26 miles) length each,

$$\begin{aligned}\text{OI (hop outage intensity)} &= \left[\frac{100\,(21{,}500/2500) + 55}{516}\right] \\ &= 2 \text{ outages/year} \end{aligned} \tag{7.82}$$

$$\text{OI (path outage intensity)} = 1 \text{ outage/year} \tag{7.83}$$

$$\begin{aligned}\text{Mo (hop mean time between outage)} &= \frac{525{,}960}{1.77326} \\ &= 296{,}606 \text{ min (206 days)} \end{aligned} \tag{7.84}$$

$$\text{Mo (path mean time between outage)} = 593{,}212 \text{ min (412 days)} \tag{7.85}$$

These results are summarized in Figure 7.11.

7.4 INTERNATIONAL TELECOMMUNICATION UNION QUALITY OBJECTIVES

7.4.1 Legacy Quality Objectives

The legacy quality (error performance) recommendations (F.634, F.696, and F.697) are still used for expansion of existing systems (The legacy F.1400 Recommendation was developed strictly for wireless access to the public-switched telephone network and is not applicable for general use.). The recommendations are based on the traditional digital transmission models (ITU-T, 1984) and plesiochronous ("asynchronous") digital hierarchy (ITU-T, 1988). The digital levels mentioned in the recommendations

International standards
G.827 and F.1703

- Availability two-way objectives (path or equipment)
  - International digital links (50–21,500 km)
    - Availability = 99.9926% to 99.9975%
    - Outage intensity = 25 to 1 outage per year
  - National digital links
    - Terminating country (50–2500 km)
      - Availability = 99.9926% to 99.9975%
      - Outage intensity = 25 to 1 outage per year
    - Long haul (50–2500 km)
      - Availability = 99.9926% to 99.9975%
      - Outage intensity = 25 to 1 outage per year
    - Short haul (50–250 km)
      - Availability = 99.9800% to 99.9966%
      - Outage intensity = 60 to 10 outage per year
    - Access (50–250 km, typically < 50 km)
      - Availability = 99.9750% to 99.9958%
      - Outage intensity = 50 to 8 outage per year

Availability objectives are per year.
Availability measured using 10 s on/off window.
It is assumed that equipment and media availability are equal.
Typical path degradations are rain and long term interference.

Figure 7.11 International path availability objectives.

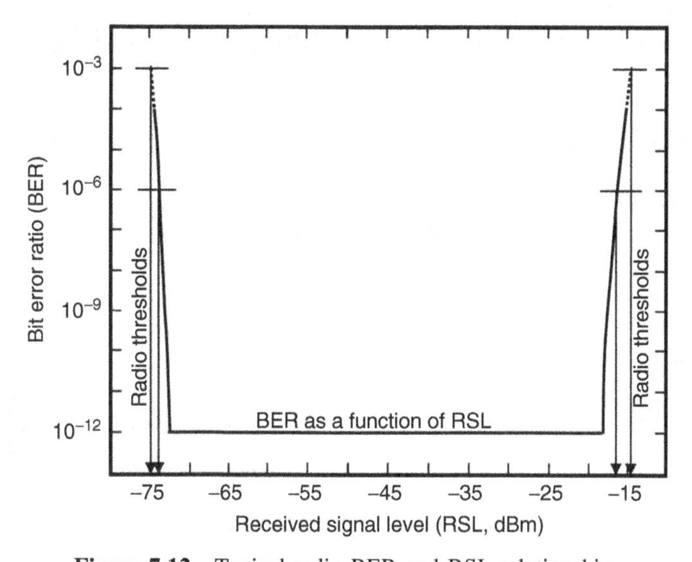

Figure 7.12 Typical radio BER and RSL relationship.

are Level 1 (DS1 or E1), Level 2 (DS2 or E2), or Level 3 (DS3 or E3). We will start with the outage objectives and derive a nominal radio path objective.

The following objectives include both ES and SES objectives. Under normal conditions, digital radios run essentially error free. Only when the radios have very low or very large RSLs do they create errors. The transition from error-free service to severely errored performance is quite rapid (Fig. 7.12).

The ES performance requirement is basically a network operations requirement. If operational personnel are careful, the radio network runs error free until it is degraded because of propagation or interference events. Therefore, from a path-engineering perspective, only the SES objectives are related to path engineering.

7.4.1.1 High Grade Circuit (F.634-4)

An SES represents a radio path outage.
Severely Errored Seconds Ratio (SESR) = ratio of SESs to the total number of seconds in a month.

$$\text{SESR} = 0.00054\left(\frac{L}{2500}\right) \tag{7.86}$$

L = path length (km).

Assume the typical 46.7-km radio path.

$$\begin{aligned}\text{Availability, one way (46.7-km path)} &= 100 - \left[100 \times 0.00054\left(\frac{46.7}{2500}\right)\right]\\ &= 99.9990\%\end{aligned} \tag{7.87}$$

$$\text{Availability, two way (46.7-km path)} = 100 \times (0.999990)^2 = 99.9980\% \tag{7.88}$$

7.4.1.2 Medium Grade Circuit (F.696-2)

Class 1: 280-km section.

$$\text{SESR} = 0.00006\left(\frac{L}{280}\right) \tag{7.89}$$

L = path length (km).

Assume the typical 46.7-km radio path.

$$\begin{aligned}\text{Availability, one way (46.7-km path)} &= 100 - \left[100 \times 0.00006\left(\frac{46.7}{280}\right)\right]\\ &= 99.9990\%\end{aligned} \tag{7.90}$$

$$\text{Availability, two way (46.7-km path)} = 100 \times (0.999990)^2 = 99.9980\% \tag{7.91}$$

Class 2: 280-km section.

$$\text{SESR} = 0.000075\left(\frac{L}{280}\right) \tag{7.92}$$

L = path length (km).

Assume the typical 46.7-km radio path.

$$\begin{aligned}\text{Availability, one way (46.7-km path)} &= 100 - \left[100 \times 0.000075\left(\frac{46.7}{280}\right)\right]\\ &= 99.9987\%\end{aligned} \tag{7.93}$$

$$\text{Availability, two way (46.7-km path)} = 100 \times (0.999987)^2 = 99.9975\% \tag{7.94}$$

Class 3: 50-km section.

$$\text{SESR} = 0.00002\left(\frac{L}{50}\right)$$

$$\text{Availability, one way (50-km path)} = 100 - \left[100 \times 0.00002\left(\frac{50}{50}\right)\right] = 99.9980\%$$

$$\text{Availability, two way (50-km path)} = 100 \times (0.999980)^2 = 99.9960\%$$

$$\text{Availability, one way (10-km path)} = 100 - \left[100 \times 0.00002\left(\frac{10}{50}\right)\right] = 99.9996\%$$

$$\text{Availability, two way (10-km path)} = 100 \times (0.999996)^2 = 99.9992\% \tag{7.95}$$

L = path length (km).

Class 4: 50-km section.

$$\text{SESR} = 0.00005\left(\frac{L}{50}\right)$$

$$\text{Availability, one way (50-km path)} = 100 - \left[100 \times 0.00005\left(\frac{50}{50}\right)\right] = 99.9950\%$$

$$\text{Availability, two way (50-km path)} = 100 \times (0.999950)^2 = 99.9900\%$$

$$\text{Availability, one way (10-km path)} = 100 - \left[100 \times 0.00005\left(\frac{10}{50}\right)\right] = 99.9990\%$$

$$\text{Availability, two way (10-km path)} = 100 \times (0.999990)^2 = 99.9980\% \qquad (7.96)$$

L = path length (km).

7.4.1.3 *Local-Grade Circuit (F.697-2)*

$$\text{SESR} = 0.00015 \text{ (regardless of path length)}$$

$$\text{Availability, one way} = 100 - (100 \times 0.00015) = 99.9850\%$$

$$\text{Availability, two way} = 100 \times (0.999850)^2 = 99.9700\% \qquad (7.97)$$

These results are summarized in Figure 7.13.

7.4.1.4 Fixed Wireless Access to Public-Switched Telephone Network (F.1400) This recommendation addresses fixed wireless access (FWA) between a user network and the public-switched telephone network (PSTN). This recommendation, specifically, is not applicable to new services such as Internet, packet data, and multimedia applications over FWA networks.

International legacy standards
ITU-R (ITU-T G.821)

- Path one-way quality objectives
  - High grade circuit, F.634-4
    - Part of 2500-km HDRP
    - 99.9990% free of SES for 46.7-km hop
  - Medium grade circuit, F.696-2
    - Class 1280-km radio subsection
      - 99.9990% free of SES for 46.7-km hop
    - Class 2280-km radio subsection
      - 99.9987% free of SES for 46.7-km hop
    - Class 350-km radio subsection
      - 99.9980% free of SES for 50-km hop
      - 99.9996% free of SES for 10-km hop
    - Class 450-km radio subsection
      - 99.9950% free of SES for 50-km hop
      - 99.9990% free of SES for 10-km hop
  - Local grade link, F.697-2
    - Commonly used for high frequency (>17 GHz) links
    - 99.9850% free of SES for hop

Quality objectives defined as error performance during worst month.
Performance is only measured when circuit is available (using 10 s on/off window).
Typical path degradation is multipath fading short term interference.

Figure 7.13 Legacy international path quality objectives.

It defines three types of services:

Type 1. Analog transmission (3.1-kHz audio signals).

Type 2. Low speed digital transmission [data rates below the primary rate (DS1 or E1)].

Type 3. High speed digital transmission [data rates at or above the primary rate (DS1 or E1)].

Quality (error-performance) objectives for Type 2 systems are to be based on G.821 (superseded by G.826) and F.697 (superseded by F.1668). Quality objectives for Type 3 systems are to be based on F.1189 (now withdrawn).

Availability for Type 2 systems are block allocations of 99.99% for medium quality applications and 99.999% for high quality applications. Availability objectives for Type 3 systems are not defined.

This recommendation, while still in force, has limited scope and is based on obsolete recommendations. It should not be used for new systems.

7.4.2 Current Quality Objectives

Radio quality (error-performance) objectives are described in F.1668-1 (ITU-R, 2005). Error Performance Objectives (EPOs) are defined for ESRs, SESRs, and BBERs. These parameters were defined above in the discussion of G.828. As noted above, when radios are operating under normal conditions, they run essentially error free [Owing to testing time limitations, radios with low speed (1.5 or 2 Mb/s) tributaries are generally tested to confirm that the bit error ratio is $<10^{-10}$ and for radios with high speed (>45 Mb/s) tributaries that the bit error ratio is $<10^{-12}$.]. The radios do not contribute errors to the ESR or BBER tests. When the radio RSL reaches an extremely low or high value, the radio errors increase dramatically and the path becomes unavailable (and errors contributing to ESR and BBER are not counted). Digital system errors during normal operating conditions are limited by network operational considerations (test and maintenance activities by operational staff) and not predictable from radio or path characteristics. Therefore, we limit our discussion to the parameter that is directly related to radio path performance, SESR.

Eliminating BBER and ESR criterions from practical consideration for radio systems is not new. Shafi and Smith (1993) pointed out that for typical 50-km (31-mile) links, a BER of 10^{-12} for a high speed (155 Mb/s) radio and 10^{-10} for a low speed (1.5/2 Mb/s) radio would meet the ESR objectives. They also observed that if the ESR objective was met, the BBER objective would also be met.

SESR is the fraction of the available time (typically a month) during which SESs occur on a link. The SESR must not exceed the objective specified for that link. F.1668-1 gives limits for international (intermediate countries) and national (terminating countries, interexchange, short-haul and access) radio networks. Limits are subdivided into those for synchronous (SDH as defined in G.828) and plesiochronous ("asynchronous") (PDH as defined in G.826). The SESR limits for SDH and PDH are exactly the same for a given network.

For national access networks,

$$\mathrm{EPO} = 0.002\,C \tag{7.98}$$

$C =$ a user-defined parameter such that $0.075 \leq C \leq 0.085$.

For national short-haul networks,

$$\mathrm{EPO} = 0.002\,B \tag{7.99}$$

$B =$ a user-defined parameter such that $0.075 \leq B \leq 0.085$.

For national interexchange (long-haul) networks,

$$\mathrm{EPO} = 0.002\,A \tag{7.100}$$

$$A = (A_1 + 0.002)\frac{L_{\mathrm{link}}}{100} \quad \text{for } 50\text{ km} \leq L_{\mathrm{link}} \leq 100\text{ km} \tag{7.101}$$

$$A = A_1 + \left(0.00002\,L_{\mathrm{link}}\right) \quad \text{for } 100\text{ km} < L_{\mathrm{link}} \tag{7.102}$$

A_1 = a user-defined parameter such that $0.01 \leq A_1 \leq 0.02$.

For international and national (long-haul) terminating country networks,

$$\text{EPO} = B_\text{n}\left(\frac{L_\text{link}}{2500}\right) + C_\text{n} \qquad (7.103)$$

EPO = SESR for the entire radio link;
L_link = radio link length (km);
B_n = equation based on radio link type, link length, and a factor B_R;
B_R = a user defined parameter such that $0 \leq B_\text{R} \leq 1$;
C_n = parameter (fraction) based on radio link type and length.

As mentioned in the notes (Be sure to read all notes and annexes in recommendations!), the minimum L_link for EPO calculations is 50 km. The period of evaluation is 1 month. The EPOs are one-way (each direction) objectives.

In the following sections, the parameter NPO (non-EPO) will be the quantity 100 [1 − (EPO for the hop)]. It will have the same look and feel as an availability number but will represent the percentage of time that the hop one-way performance meets or exceeds the hop EPO.

7.4.2.1 Access Section 50 km < L_link ≤ 250 km.

$$\text{EPO (for the entire radio link)} = 0.002\ C \qquad (7.104)$$

C = a user-defined parameter such that $0.075 \leq C \leq 0.085$.

If the entire section is one hop,

$$\text{EPO} = 0.002 \times 0\ 075 = 0.00015 \quad \text{for } C = 0.075 \qquad (7.105)$$

$$\text{EPO} = 0.002 \times 0\ 085 = 0.00017 \quad \text{for } C = 0.085$$

$$\text{NPO (hop\% of no SES)} = 100(1 - \text{EPO}) = 99.985\% \quad \text{for } C = 0.075 \qquad (7.106)$$

$$\text{NPO (hop\% of no SES)} = 100(1 - \text{EPO}) = 99.983\% \quad \text{for } C = 0.085$$

If the section has six identical cascaded hops,

$$\text{EPO (entire link)} = 0.002 \times 0\ 075 = 0.00015 \quad \text{for } C = 0.075 \qquad (7.107)$$

$$\text{EPO (entire link)} = 0.002 \times 0\ 085 = 0.00017 \quad \text{for } C = 0.085$$

$$\text{NPO (hop\% of no SES)} = 100(1 - \text{EPO})^{(1/6)} = 99.9975\% \quad \text{for } C = 0.075 \qquad (7.108)$$

$$\text{NPO (hop\% of no SES)} = 100(1 - \text{EPO})^{(1/6)} = 99.9972\% \quad \text{for } C = 0.085$$

7.4.2.2 Short-Haul Section 50 km < L_link ≤ 250 km.

$$\text{EPO} = 0.002\ B$$

B = a user-defined parameter such that $0.075 \leq C \leq 0.085$.

If the entire section is one hop,

$$\text{EPO} = 0.002 \times 0\ 075 = 0.00015 \quad \text{for } C = 0.075 \qquad (7.109)$$

$$\text{EPO} = 0.002 \times 0\ 085 = 0.00017 \quad \text{for } C = 0.085$$

$$\text{NPO (hop\% of no SES)} = 100(1 - \text{EPO}) = 99.985\% \quad \text{for } C = 0.075$$

$$\text{NPO (hop\% of no SES)} = 100(1 - \text{EPO}) = 99.983\% \quad \text{for } C = 0.085 \tag{7.110}$$

If the section has six identical cascaded hops,

$$\text{EPO (entire link)} = 0.002 \times 0\ 075 = 0.00015 \quad \text{for } C = 0.075 \tag{7.111}$$

$$\text{EPO (entire link)} = 0.002 \times 0\ 085 = 0.00017 \quad \text{for } C = 0.085$$

$$\text{NPO (hop\% of no SES)} = 100(1 - \text{EPO})^{(1/6)} = 99.9975\% \quad \text{for } C = 0.075 \tag{7.112}$$

$$\text{NPO (hop\% of no SES)} = 100(1 - \text{EPO})^{(1/6)} = 99.9972\% \quad \text{for } C = 0.085$$

7.4.2.3 *National Interexchange Section*

$50\,\text{km} < L_{\text{link}} \leq 100\,\text{km}$.

$$\text{EPO} = 0.002\ A \tag{7.113}$$

$$A = (A_1 + 0.002)\frac{L_{\text{link}}}{100} \tag{7.114}$$

A_1 = a user-defined parameter such that $0.01 \leq A_1 \leq 0.02$.

If the entire section is one hop of 41.7 km (26 miles),

$$\text{EPO} = 0.002(0.01 + 0.002)\frac{50}{100} = 0.000012 \quad \text{for } A_1 = 0.01 \tag{7.115}$$

$$\text{EPO} = 0.002(0.02 + 0.002)\frac{50}{100} = 0.000022 \quad \text{for } A_1 = 0.02$$

$$\text{NPO (hop\% of no SES)} = 100(1 - \text{EPO}) = 99.9988\% \quad \text{for } A_1 = 0.01 \tag{7.116}$$

$$\text{NPO (hop\% of no SES)} = 100(1 - \text{EPO}) = 99.9978\% \quad \text{for } A_1 = 0.02$$

If the section is 83.4 km long and composed of two cascaded hops of 41.7 km (26 miles) length each,

$$\text{EPO} = 0.002(0.01 + 0.002)\frac{83.4}{100} = 0.000020 \quad \text{for } A_1 = 0.01 \tag{7.117}$$

$$\text{EPO} = 0.002(0.02 + 0.002)\frac{83.4}{100} = 0.000037 \quad \text{for } A_1 = 0.02$$

$$\text{NPO (hop\% of no SES)} = 100(1 - \text{EPO})^{(1/2)} = 99.9990\% \quad \text{for } A_1 = 0.01 \tag{7.118}$$

$$\text{NPO (hop\% of no SES)} = 100(1 - \text{EPO})^{(1/2)} = 99.9982\% \quad \text{for } A_1 = 0.02$$

$100\,\text{km} < L_{\text{link}} \leq 2500\,\text{km}$.

$$\text{EPO} = 0.002\ A$$

$$A = A_1 + (0.00002\ L_{\text{link}})$$

A_1 = a user-defined parameter such that $0.01 \leq A_1 \leq 0.02$.

If the section is 125.1 km long and composed of three cascaded hops of 41.7 km (26 miles) length each,

$$\text{EPO} = 0.002[0.01 + (0.00002 \times 125.1)] = 0.000025 \quad \text{for } A_1 = 0.01 \tag{7.119}$$

$$\text{EPO} = 0.002[0.02 + (0.00002 \times 125.1)] = 0.000045 \quad \text{for } A_1 = 0.02$$

$$\text{NPO (hop\% of no SES)} = 100(1 - \text{EPO})^{(1/3)} = 99.9992\% \quad \text{for } A_1 = 0.01 \tag{7.120}$$

$$\text{NPO (hop\% of no SES)} = 100(1 - \text{EPO})^{(1/3)} = 99.9985\% \quad \text{for } A_1 = 0.02$$

If the section is 2500 km long and composed of 60 cascaded hops of 41.7 km (26 miles) length each,

$$\text{EPO} = 0.002\ [0.01 + (0.00002 \times 2500)] = 0.00012 \quad \text{for } A_1 = 0.01 \tag{7.121}$$

$$\text{EPO} = 0.002[0.02 + (0.00002 \times 2500)] = 0.00014 \quad \text{for } A_1 = 0.02$$

$$\text{NPO (hop\% of no SES)} = 100(1 - \text{EPO})^{(1/3)} = 99.9998\% \quad \text{for } A_1 = 0.01 \tag{7.122}$$

$$\text{NPO (hop\% of no SES)} = 100(1 - \text{EPO})^{(1/2)} = 99.9998\% \quad \text{for } A_1 = 0.02$$

7.4.2.4 *National Terminating Country Section*

$50\,\text{km} < L_{\text{link}} \leq 1000\,\text{km}$.

$$\text{EPO} = 0.0001\ (1 + B_R)\left(\frac{L_{\text{link}}}{2500}\right), 0 \leq B_R \leq 1$$

If the entire section is one hop of 41.7 km (26 miles),

$$\text{EPO} = 0.0001\ (1 + 0)\left(\frac{50}{2500}\right) = 0.000002 \quad \text{for } B_R = 0 \tag{7.123}$$

$$\text{EPO} = 0.0001(1 + 1)\left(\frac{50}{2500}\right) = 0.000004 \quad \text{for } B_R = 1$$

$$\text{NPO (hop\% of no SES)} = 100 \times (1 - \text{EPO}) = 99.9998\% \quad \text{for } B_R = 0 \tag{7.124}$$

$$\text{NPO (hop\% of no SES)} = 100 \times (1 - \text{EPO}) = 99.9996\% \quad \text{for } B_R = 1$$

If the section is 500 km long and composed of 12 cascaded hops of 41.7 km (26 miles) length each,

$$\text{EPO} = 0.0001(1 + 0)\left(\frac{500}{2500}\right) = 0.00002 \quad \text{for } B_R = 0 \tag{7.125}$$

$$\text{EPO} = 0.0001(1 + 1)\left(\frac{500}{2500}\right) = 0.00004 \quad \text{for } B_R = 1$$

$$\text{NPO (hop\% of no SES)} = 100(1 - \text{EPO})^{(1/12)} = 99.9998\% \quad \text{for } B_R = 0 \tag{7.126}$$

$$\text{NPO (hop\% of no SES)} = 100(1 - \text{EPO})^{(1/12)} = 99.9997\% \quad \text{for } B_R = 1$$

$500\,\text{km} < L_{\text{link}} \leq 2500\,\text{km}$.

$$\text{EPO} = \left[0.0001\left(\frac{L_{\text{link}}}{2500}\right)\right] + (0.00002\ B_R), 0 \leq B_R \leq 1$$

If the section is 500 km long and composed of 12 cascaded hops of 41.7 km (26 miles) length each,

$$\text{EPO} = \left[0.0001\left(\frac{500}{2500}\right)\right] = 0.00002 \quad \text{for } B_R = 0 \tag{7.127}$$

$$\text{EPO} = \left[0.0001\left(\frac{500}{2500}\right)\right] + 0.00002 = 0.00004 \quad \text{for } B_R = 1$$

$$\text{NPO (hop\% of no SES)} = 100(1 - \text{EPO})^{(1/12)} = 99.9998\% \quad \text{for } B_R = 0 \tag{7.128}$$

$$\text{NPO (hop\% of no SES)} = 100(1 - \text{EPO})^{(1/12)} = 99.9997\% \quad \text{for } B_R = 1$$

If the section is 2500 km long and composed of 60 cascaded hops of 41.7 km (26 miles) length each,

$$\text{EPO} = \left[0.0001\left(\frac{2500}{2500}\right)\right] = 0.0001 \quad \text{for } B_R = 0 \tag{7.129}$$

$$\text{EPO} = \left[0.0001\left(\frac{2500}{2500}\right)\right] + 0.00002 = 0.00012 \quad \text{for } B_R = 1$$

$$\text{NPO (hop\% of no SES)} = 100(1 - \text{EPO})^{(1/60)} = 99.9998\% \quad \text{for } B_R = 0 \tag{7.130}$$

$$\text{NPO (hop\% of no SES)} = 100(1 - \text{EPO})^{(1/60)} = 99.9998\% \quad \text{for } B_R = 1$$

7.4.2.5 International Section 50 km < $L_{\text{link}} \leq$ 1000 km.

$$\text{EPO} = 0.0001\ (1 + B_R)\left(\frac{L_{\text{link}}}{2500}\right),\ 0 \leq B_R \leq 1 \tag{7.131}$$

If the entire section is one hop of 41.7 km (26 miles),

$$\text{EPO} = 0.0001(1 + 0)\left(\frac{50}{2500}\right) = 0.000002 \quad \text{for } B_R = 0 \tag{7.132}$$

$$\text{EPO} = 0.0001(1 + 1)\left(\frac{50}{2500}\right) = 0.000004 \quad \text{for } B_R = 1$$

$$\text{NPO (hop\% of no SES)} = 100(1 - \text{EPO}) = 99.9998\% \quad \text{for } B_R = 0 \tag{7.133}$$

$$\text{NPO (hop\% of no SES)} = 100(1 - \text{EPO}) = 99.9996\% \quad \text{for } B_R = 1$$

If the section is 1000 km long and composed of 24 cascaded hops of 41.7 km (26 miles) length each,

$$\text{EPO} = 0.0001(1 + 0)\left(\frac{1000}{2500}\right) = 0.00004 \quad \text{for } B_R = 0 \tag{7.134}$$

$$\text{EPO} = 0.0001(1 + 1)\left(\frac{1000}{2500}\right) = 0.00008 \quad \text{for } B_R = 1$$

$$\text{NPO (hop\% of no SES)} = 100(1 - \text{EPO})^{(1/24)} = 99.9998\% \quad \text{for } B_R = 0 \tag{7.135}$$

$$\text{NPO (hop\% of no SES)} = 100(1 - \text{EPO})^{(1/24)} = 99.9997\% \quad \text{for } B_R = 1$$

1000 km < $L_{\text{link}} \leq$ 21,500 km.

$$\text{EPO} = \left[0.0001\left(\frac{L_{\text{link}}}{2500}\right)\right] + (0.00004\ B_R),\ 0 \leq B_R \leq 1 \tag{7.136}$$

International standards
G.826, G.828 and F.1668

- Path quality one-way objectives
  - International digital links
    - 50–21,500 km
    - 99.9996% to 99.9998% free of SESs
  - National digital links
    - Terminating country (50–2500 km)
      - 99.9996% to 99.9998% free of SESs
    - Inter-exchange (50–2500 km)
      - 99.9978% to 99.9998% free of SESs
    - Short haul (50–250 km)
      - 99.9830% to 99.9975% free of SESs
    - Access (50–250 km, typically <50 km)
      - 99.9830% to 99.9975% free of SESs

Quality objectives defined as error performance during worst month.
Performance is only measured when circuit is available (using 10 s on/off window).
Typical path degradation is multipath fading and short term interference.

Figure 7.14 International path quality objectives.

If the section is 1000 km long and composed of 24 cascaded hops of 41.7 km (26 miles) length each,

$$\text{EPO} = 0.0001\left(\frac{1000}{2500}\right) = 0.00004 \qquad \text{for } B_R = 0 \tag{7.137}$$

$$\text{EPO} = \left[0.0001\left(\frac{1000}{2500}\right)\right] + 0.00004 = 0.00008 \qquad \text{for } B_R = 1$$

$$\text{NPO (hop\% of no SES)} = 100(1 - \text{EPO})^{(1/24)} = 99.9998\% \qquad \text{for } B_R = 0 \tag{7.138}$$

$$\text{NPO (hop\% of no SES)} = 100(1 - \text{EPO})^{(1/24)} = 99.9997\% \qquad \text{for } B_R = 1$$

If the section is 21,500 km long and composed of 516 cascaded hops of 41.7 km (26 miles) length each,

$$\text{EPO} = 0.0001\left(\frac{21{,}500}{2500}\right) = 0.00086 \qquad \text{for } B_R = 0 \tag{7.139}$$

$$\text{EPO} = \left[0.0001\left(\frac{21{,}500}{2500}\right)\right] + 0.00004 = 0.00090 \qquad \text{for } B_R = 1$$

$$\text{NPO (hop\% of no SES)} = 100(1 - \text{EPO})^{(1/516)} = 99.9998\% \qquad \text{for } B_R = 0 \tag{7.140}$$

$$\text{NPO (hop\% of no SES)} = 100(1 - \text{EPO})^{(1/516)} = 99.9998\% \qquad \text{for } B_R = 1$$

These results are summarized in Figure 7.14.

7.5 ERROR-PERFORMANCE RELATIONSHIP AMONG BER, BBER, AND SESs

The international basis of circuit availability is the SES. Until the early 1990s, an SES was based on a BER of 10^{-3}. With the adoption of G.828, the BER definition was dropped and an Errored Block (EB) definition was introduced.

A related dilemma is circuit error performance. Historically, error performance was measured using the bit error ratio. With the adoption of G.828, the BER definition was dropped and the concept of Background Block Error Ratio (BBER) was introduced. While this has obvious advantages for in-service monitoring of SDH (and SONET) networks, it is expected to be applied to plesiochronous ("asynchronous") networks as well. At present, it is not clear how these standards are to be applied to variable length IP blocks.

Most plesiochronous, some synchronous, and most radio systems measure error performance using BER. The relationship between BBER and BER and the point at which an SES occurs is of considerable interest to network operators.

$$\text{BER} = \frac{\text{(errored bits per second)}}{\text{(bits transmitted per second)}}$$
$$\text{BBER} = \left[\frac{\text{(errored blocks per second)}}{\text{(blocks transmitted per second)}}\right] \tag{7.141}$$

To relate BER to BBER, we need to know the number of blocks transmitted per second and at least first-order statistics about the block errors. In general, the statistics of block errors will not be known. However, if the radio channel is operating normally, errors will be infrequent. We make the assumption that demodulator errors will occur once per EB. Today, all radios have scramblers, forward error correction, and often TCM convolutional encoding. All of these processes extend a single decoded error into a burst of errors. The number of errors per burst is highly dependent on the actual equipment. Error bursts on the order of 10–20 per error event (International Telecommunication Union–Radiocommunication Sector (ITU-R), 2003) have been documented. Shafi and Smith (1993) suggest that error bursts could be as large as 30 per error event. We name the number of burst errors per demodulated error event as the *length of error bursts* (LEB). On the basis of these definitions and assumptions we have the following:

$$\text{BER} = \frac{\text{(errored bits per second)}}{\text{(bits transmitted per second)}}$$
$$= \frac{[\text{(blocks/second)} \times \text{BBER} \times \text{LEB}]}{[\text{(blocks/second)} \times \text{(bits/block)}]} \tag{7.142}$$
$$= \frac{[\text{BBER} \times \text{LEB}]}{[\text{(blocks/second)} \times \text{(bits/block)}]} \tag{7.143}$$

ITU solves the blocks per second and bits per block issue by assuming all systems are synchronous. For plesiochronous data, it is assumed that traffic is mapped into an appropriate virtual container (VC-N). The Table 7.1 is based on G. 828 (ITU-T, 2000), Tables 1, B.1, and B.4, and F.1605 (ITU-R, 2003), Table 1 (the value 19,940 listed twice in Table 1 of F.1605 should be 19,440) (Table 7.1).

The BER that relates to an SES is also of interest to operators. An SES occurs when 30% (or more) of the transmitted blocks contain an error. On the basis of that definition and an assumption of one error event per block, the following chart was developed (Table 7.2).

TABLE 7.1 Definition of Standardized Data Blocks

Transported Digital Signal	Path Signal Type	Path Bit Rate, Mb/s	Blocks per Second	Bits per Block
DS1	VC-11	1.664	2000	832
E1	VC-12	2.240	2000	1120
DS3	VC-3	48.960	8000	6120
E3	VC-3	48.960	8000	6120
STS-1	STM-0	51.840	8000	6480
STM-0	STM-0	51.840	8000	6480
3 x DS3	VC-4	150.336	8000	18,792
3 x E3	VC-4	150.336	8000	18,792
STS-3	STM-1	155.520	8000	19,440
STM-1	STM-1	155.520	8000	19,440

TABLE 7.2 Relationship Between BER and SESs

Transported Digital Signal	Path Signal Type	Errored Blocks/Second	BER (LEB = 1)	BER (LEB = 30)
DS1	VC-11	600	3.6×10^{-4}	1.1×10^{-2}
E1	VC-12	600	2.7×10^{-4}	8.0×10^{-3}
DS3	VC-3	2400	4.9×10^{-5}	1.5×10^{-3}
E3	VC-3	2400	4.9×10^{-5}	1.5×10^{-3}
STS-1	STM-0	2400	4.6×10^{-5}	1.4×10^{-3}
STM-0	STM-0	2400	4.6×10^{-5}	1.4×10^{-3}
3 × DS3	VC-4	2400	1.6×10^{-5}	4.8×10^{-4}
3 × E3	VC-4	2400	1.6×10^{-5}	4.8×10^{-4}
STS-3	STM-1	2400	1.5×10^{-5}	4.6×10^{-4}
STM-1	STM-1	2400	1.5×10^{-5}	4.6×10^{-4}

It appears that, in practice, the point at which the BBER SES threshold is reached is somewhere between 10^{-5} and 10^{-2}. From a practical perspective, the current 10^{-6} and 10^{-3} radio thresholds currently used for SES are not that different from the new thresholds. The difference between 10^{-6} and 10^{-3} is only 1 dB change in received signal level for a modern radio. In fact, owing to use of forward error correction and convolutional coding, modern radios transition abruptly from near error free operation to out of service for very little change in received signal level. Both multipath and rain fading are so fast that from a practical perspective, all the above definitions of SES are nearly the same when viewed from an operational outage time perspective.

REFERENCES

American Telephone and Telegraph Bell System Staff, "Radio Engineering, Microwave Radio, Digital Systems," *AT&T Western Electric Practices, Section 940-300-130, Issue 2*, pp. 28–42, March 1984.

International Telecommunication Union–Radiocommunication Sector (ITU-R), "Recommendation F.556-1, Hypothetical Reference Digital Path for Radio-Relay Systems which may form part of an Integrated Services Digital Network with a Capacity above the Second Hierarchical Level," pp. 1–2, 1986.

International Telecommunication Union–Radiocommunication Sector (ITU-R), "Recommendation F.695, Availability Objectives for Real Digital Radio-relay Links forming part of a High-Grade Circuit within an Integrated Services Digital Network," pp. 1–2, 1990.

International Telecommunication Union–Radiocommunication Sector (ITU-R), "Recommendation F.557-4, Availability Objective for Radio-Relay Systems over a Hypothetical Reference Circuit and a hypothetical Reference Digital Path," pp. 1–4, 1997a.

International Telecommunication Union–Radiocommunication Sector (ITU-R), "Recommendation F.696-2, Error Performance and Availability Objectives for Hypothetical Reference Digital Sections forming part or all of the Medium-Grade Portion of an Integrated Services Digital Network Connection at a bit rate below the Primary rate Utilizing Digital Radio-Relay Systems," pp. 1–4, 1997b.

International Telecommunication Union–Radiocommunication Sector (ITU-R), "Recommendation F.697-2, Error Performance and Availability Objectives for the Local-Grade portion at each end of an Integrated Services Digital Network Connection at a bit rate below the Primary Rate utilizing Digital Radio-Relay Systems," pp. 1–4, 1997c.

International Telecommunication Union–Radiocommunication Sector (ITU-R), "Recommendation F.634-4, Error Performance Objectives for Real Digital Radio-Relay Links forming part of the high-grade portion of International Digital Connections at a bit rate below the primary rate within an Integrated Services Digital Network," pp. 1–7, 1997d.

International Telecommunication Union–Radiocommunication Sector (ITU-R), "Recommendation F.594-4, Error Performance Objectives of the Hypothetical Reference Digital Path for Radio-Relay Systems providing Connections at a bit rate below the Primary Rate and forming Part or all of the High Grade Portion of an Integrated Services Digital Network," pp. 1–3, 1997e.

International Telecommunication Union–Radiocommunication Sector (ITU-R), "Recommendation F.1400, Performance and Availability Requirements and Objectives for Fixed Wireless Access to Public Switched Telephone Network," pp. 1–5, 1999.

International Telecommunication Union–Radiocommunication Sector (ITU-R), "Recommendation F.1605, Error performance and availability estimation for synchronous digital hierarchy terrestrial fixed wireless systems," pp. 1–12, 2003.

International Telecommunication Union–Radiocommunication Sector (ITU-R), "Recommendation F.1703, Availability objectives for real digital fixed wireless links used in 27 500 hypothetical reference paths and connections," pp. 1–12, 2005.

International Telecommunication Union–Radiocommunication Sector (ITU-R), "Recommendation F.1668-1, Error performance objectives for real digital fixed wireless links used in 27 500 km hypothetical reference paths and connections," pp. 1–14, 2007a.

International Telecommunication Union–Radiocommunication Sector (ITU-R), "Recommendation F.592-4, Vocabulary of Terms for the Fixed Service," pp. 1–15, 2007b.

International Telecommunication Union–Radiocommunication Sector (ITU-R), *Radio Regulations*, as adopted by World Radiocommunication Conference, Geneva, 2007 (WRC-07), 2008.

International Telecommunication Union–Telecommunication Standardization Sector (ITU-T), "Recommendation G.801, Digital Transmission Models," pp. 1–6, 1984.

International Telecommunication Union–Telecommunication Standardization Sector (ITU-T), "Recommendation G.702, Digital Hierarchy Bit Rates," pp. 1–4, 1988.

International Telecommunication Union–Telecommunication Standardization Sector (ITU-T), "Recommendation G.828, Error Performance Parameters and Objectives for International, Constant Bit Rate Synchronous Digital Paths," pp. 1–17, 2000.

International Telecommunication Union–Telecommunication Standardization Sector (ITU-T), "Recommendation G.821, Error Performance of an International Digital Connection Operating at a Bit Rate below the Primary Rate and Forming a Part of an Integrated Services Digital Network," pp. 1–7, 2002a.

International Telecommunication Union–Telecommunication Standardization Sector (ITU-T), "Recommendation G.826, End-to-End Error Performance Parameters and Objectives for International, Constant Bit-rate Digital Paths and Connections," pp. 1–24, 2002b.

International Telecommunication Union–Telecommunication Standardization Sector (ITU-T), "Recommendation G.827, Availability Performance Parameters and Objectives for End-to-End International Constant Bit-rate Digital Paths," pp. 1–18, 2003.

Jansen, R. M. and Prime, R. C., "TH-3 Microwave Radio System Considerations," *Bell System Technical Journal*, Vol. 50, pp. 2085–2116, September 1971.

Shafi, M. and Smith, P., "The Impact of G.826," *IEEE Communications Magazine*, Vol. 31, pp. 56–62, September 1993.

Telcordia (Bellcore) Staff, *Telcordia Generic Requirements GR-499-CORE, Transport Systems Generic Requirements (TSGR): Common Requirements, Issue 4*, pp. 2–1 to 4–3, November 2009.

Vigants, A. and Pursley, M. V., "Transmission Unavailability of Frequency-Diversity Protected Microwave FM Radio Systems Caused by Multipath Fading," *Bell System Technical Journal*, Vol. 58, pp. 1779–1796, October 1979.

8

MICROWAVE ANTENNA THEORY

At microwave frequencies, the antennas with the largest physical area have the most gain. These antennas are called *aperture antennas* because their defining feature is a large physical area or aperture. For the aperture antennas (passive reflectors, circular parabolic reflector, and square antennas) commonly used at microwave frequencies, antenna pattern, beamwidth, and illumination efficiency are primarily functions of the shape of the aperture (physical antenna radiation surface), the radiation pattern of the power distributed across the antenna aperture (antenna illumination), and the operating frequency (Hollis, 1970). If two antennas have the same aperture shape, aperture illumination, and the same normalized dimension D/λ (or W/λ), they are electrically identical. As Silver (1949) observed, for linear polarization, in the plane of measurement of the E field (i.e., plane of polarization), the radiation pattern is determined by the aperture dimension in that plane only. Antenna illumination efficiency (η) is a function of antenna shape and illumination density. It is a critical function in determining far field gain and near field power density.

Most commercial microwave antennas use power illumination that is spread with equal (constant) phase across the aperture. Side lobes are reduced by tapering the (equal phase) illumination to reduce illuminated power near the edge of the antenna. This produces the greatest boresight gain (sometimes called *directivity*) for a given illumination efficiency. Increased side lobe reduction can be achieved by varying the phase as well as the amplitude of the aperture illumination (currently there is much theoretical work using antennas composed of many small antenna "patches" that are used to vary spot amplitude and phase). This technique is termed *superdirectivity* (in the past this was referred to using the misleading term, *supergaining*). Theoretically, any side lobe pattern can be produced by an aperture antenna but the boresight gain and operating bandwidth will be reduced when compared to equal phase illumination. All antennas considered in this chapter use equal phase illumination across the aperture.

Illumination efficiency is misunderstood by many engineers (especially when they estimate near field power density). Illumination efficiency does not represent a power loss. It represents loss of antenna boresight gain relative to an antenna with energy equally distributed across the aperture of the antenna. As the antenna illumination efficiency is lowered, more transmit energy is located near the center of the antenna (and boresight power is decreased). The relationship between illumination efficiency and antenna gain and beamwidth is investigated for common circular and square aperture antennas.

The characteristics of an antenna are different at different distances from it. Very near the antenna (a wavelength or less), the dominant energy is nonradiating "reactive" fields. This is called the *reactive near field region*. Further away but still near the antenna is the radiating near field (or Fresnel) region. In

Digital Microwave Communication: Engineering Point-to-Point Microwave Systems, First Edition. George Kizer.

this region, the energy is radiating (it produces the conventional antenna radiation pattern when viewed from the far field). Most of the energy is directly in front of the antenna. The energy in front of the antenna is almost independent of the distance from the antenna and is maximum along the antenna boresight (direction perpendicular to the center of the antenna). Much farther away from the antenna is the far field (or Fraunhofer) region. In this region, energy is distributed in a pattern that is dependent on the observation angle from the boresight. The pattern relative to the boresight is independent of the distance from the antenna. In this region, at a given angle relative to the boresight, energy is a function of the distance (from the antenna) squared. At distances far from the antenna, antenna gain and side lobe (pattern) characteristics are of primary interest. At distances very near the antenna, the power density is of primary interest as this is related to public safety.

The nominal transition point from radiating near field to far field (Friis, 1946) is traditionally $2\ D^2/\lambda$ where D is the widest dimension of the antenna perpendicular to the line of propagation. This equation came from Fresnel's research into optics. As we shall see later, this transition estimation is only a very rough guideline. It ignores shape and surface illumination. The far field transition distance is based on a maximum deviation from parallel of $\pi/16$ radians when viewed from a point directly in front of the antenna. For electrically small antennas with the widest dimension smaller than the wavelength, the radiating near field region may not exist. This equation is a conservative estimator (much less than 1 dB error) for antenna gain measurements. For close coupling loss between two antennas, this estimate is somewhat optimistic (errors can be as large as a decibel). As we shall see later, this estimator is not particularly accurate for near field calculations.

As measurements are moved closer to the antenna from the radiating near field transition point, the measured (peak) power near the center of the antenna is relatively constant with distance from the antenna. This is radiating electromagnetic energy with the electric and magnetic fields orthogonal to each other as well as to the direction of propagation. Obviously, these fields cannot exist at the aperture antenna surface (an electric field cannot exist on a perfectly conductive surface in a direction parallel to that surface). Very close to the antenna aperture (disk or square surface), an additional nonradiating ("reactive") field must exist to satisfy Maxwell's equations. Most sources suggest that this field exists only as far as λ (Hansen, 1964; Mikki and Antar, 2011), $\lambda/2\pi$ (Balanis, 2008) or $0.62\sqrt{D^3/\lambda}$ (Balanis, 2008) in front of the antenna. Hansen (1964) noted that the very near field (including reactive terms) was a series of functions of $(1/r)^n$ where r is the aperture maximum dimension divided by wavelength and $n \geq 1$. The far field solution was the term for $n = 1$. He noted that for a small circular antenna, $D/\lambda = 1$, the reactive field extended to D, the aperture diameter. For a larger antenna, $D/\lambda = 10$, the reactive field only extended to $0.1D$ (Hansen, 1964, Fig. 17). Numerical calculations (Laybros et al., 2005) for worst case (measurements perpendicular to the center of fully illuminated) circular and square antenna shapes ("apertures") have replicated Hansen's results. If we consider reactive near field power to be 10 dB less than the radiating near field power, Figure 15 and Figure 19 of (Laybros et al., 2005) show that for circular antennas the reactive near field is significant only at less than D in front of the antenna. For square antennas, it is significant only at much less than λ in front of the antenna. Many circular antennas have a radome or shroud that completely encloses the reactive field. For square antennas, the reactive field is essentially only on the surface of the antenna. Therefore, for practical electromagnetic exposure estimates to be considered later, only the radiating near field needs to be considered.

Over the last couple of decades, interest in limiting human exposure to radio frequency energy has increased substantially. Fixed point-to-point microwave and satellite radio transmission systems continue to be installed throughout the world. Typically, these systems, operating at frequencies greater than 1 GHz, employ transmit antennas (including reflectors) which may be near people. The near field power density of these antennas determines the radio frequency exposure level of those people. Safety organizations and government agencies attempt to estimate radio energy power near these antennas and compare those power levels with established limits (Cleveland et al., 1997; IEEE International Committee on Electromagnetic Safety (SCC39), 2005). Owing to the complexity of theoretical computations, most practicing engineers use simple, inaccurate methods to estimate near field power densities. This chapter provides simple yet more accurate near field power density estimation methods.

Fixed point-to-point microwave systems require frequency planning to minimize interference between different radio systems. The **far field radiation pattern** (relative to the maximum power or "boresight") is a critical function used in the frequency planning process. The majority of the radiated power of the antenna is concentrated into a single major lobe whose angular width is termed the *antenna beamwidth*.

While some authors refer to antenna beamwidth as the angular width from boresight to a predefined power level (sometimes termed *half angle beamwidth*), this chapter uses beamwidth as the total angular width between two consecutive occurrences of the defined power criterion. The far field radiation pattern is directly related to the illumination over the surface of an aperture antenna. The far field radiation pattern is the Fourier transform of the distribution of the electric field across the aperture plane of that transmit antenna (Hansen, 1964).

8.1 COMMON PARAMETERS

The following definitions or relationships are used:

$$D/\lambda = \text{Normalized diameter of circular antenna}$$
$$= 1.0167\ D(\text{ft})\ f(\text{GHz}) = 3.3356\ D(\text{m})\ f(\text{GHz})$$

$$W/\lambda = \text{Normalized width of square antenna}$$
$$= 1.0167\ W(\text{ft})\ f(\text{GHz}) = 3.3356\ W(\text{m})\ f(\text{GHz})$$

$$\Delta = \frac{d}{\left(\frac{2D^2}{\lambda}\right)} = \text{Normalized distance parameter for circular antenna}$$
$$= \frac{0.49179\ d(\text{ft})}{[D(\text{ft})^2\ f(\text{GHz})]}$$
$$= \frac{0.14990\ d(\text{m})}{[D(\text{m})^2 f(\text{GHz})]}$$

$$\Delta = \frac{d}{\left(\frac{2W^2}{\lambda}\right)} = \text{Normalized distance parameter for square antenna}$$

$$\Delta_{dB} = 10\ \log(\Delta)$$

$$\Delta = 1\ (\Delta_{dB} = 0) = \text{Normalized distance at nominal far field crossover point}$$

$$\Delta G(\text{dB}) = \text{Close antenna gain variation due to reflections}$$

d = distance from the center of the antenna to the point of interest;
D = diameter of the circular antenna (aperture);
W = width of the rectangular antenna (aperture with $\zeta = 0$);
ζ = angular rotation of a rectangular antenna (aperture);
$\zeta = 0$ when one side of the rectangular aperture is parallel to the plane of the Earth;
λ = radio free space wavelength.

$$\lambda(\text{ft}) = \frac{0.98357}{f}(\text{GHz})$$

$$\lambda(\text{m}) = \frac{0.29980}{f}(\text{GHz})$$

f = radio signal frequency;
$G(\text{dB})$ = antenna far field maximum ("boresight") gain relative to an isotropic radiator;
$g(\text{power ratio}) = 10^{G/10}$ = antenna isotrophic power gain;
$L_{NF}(\text{dB})$ = near field coupling loss (a negative decibel number);
= composite loss experienced by two close antennas relative to far field gain (overall gain = far field gain of one antenna + far field gain of second antenna + near field coupling loss);
p = power delivered to the antenna;
P_{dB} = antenna far field gain (a function of transmission angle) normalized to antenna maximum gain (at boresight).

S (power density) = power/area

$$P_{\mathrm{NNF}}(\mathrm{dB}) = 10\ \log\left[\frac{S\,(\Delta)}{S(\Delta = 1)}\right] = \text{Normalized near field power}$$

$$P_{\mathrm{NNFL}}(\mathrm{dB}) = 10\ \log\left[\frac{S\,(\Delta \ll 0.1)}{S(\Delta = 1)}\right]$$

= Normalized near field power limit(measured very near the antenna)

= Limit value to far left on normalized near field power graphs

η (power ratio) = antenna illumination efficiency (between 0.1 and 1.0);
θ = half beamwidthangle of power reception relative to antenna boresight;
θ_{ndB} = angle measured between boresight and (either) n dB value of P_{dB} (referenced to boresight);
ϕ = conventional (two-sided) antenna beamwidth;
= angle between two similar relative power levels;
ϕ_{ndB} = angle measured between the two n dB values of P_{dB} (referenced to boresight).

$$\phi_{\mathrm{ndB}} = 2\theta_{\mathrm{ndB}}$$

$$u = \left(\frac{D}{\lambda}\right)\sin\theta \cong \left(\frac{D}{\lambda}\right)\theta\ \text{(radians)} = \left(\frac{p}{180}\right)\left(\frac{D}{\lambda}\right)\theta\ \text{(degrees)}$$

The above approximation for u is accurate within 1% for $\theta < 14^\circ$ and within 10% for $\theta < 45^\circ$.

$L_{\mathrm{NF}}(\mathrm{dB})$ = near field coupling loss;
I_0 = modified Bessel function of the first kind and zero order;
I_1 = modified Bessel function of the first kind and first order;
J_0 = Bessel function of the first kind and zero order;
J_1 = Bessel function of the first kind and first order;
sqrt(x) = square root of x;
log(x) = common (base 10) logarithm of x;
sin(x) = sine function of x;
cos(x) = cosine function of x;
C = Fresnel cosine integral;
S = Fresnel sine integral (only used with passive square reflectors, not to be confused with the power density function).

8.2 PASSIVE REFLECTORS

Fixed point-to-point radio systems sometimes employ passive reflectors to change the directions of a radio wave without active devices. This is usually an approximately circular or rectangular flat surface used as a radio wave "mirror" to redirect radio wave energy. These reflectors are elliptical or rectangular flat panels that appear as circles or squares when viewed from the radio path (their geometric projection on the path is a circle or a square.). Since these passive antennas are (at least approximately) in the far field of the transmit and receive antennas, energy is distributed across the antenna uniformly.

The theoretical results for passive reflectors are developed assuming the passives to be oriented directly orthogonal to the path of transmission (uniform phase illumination). If the passive is rotated in a plane which includes the line of maximum power transmission (the radio path), a linear phase error is introduced onto the aperture illumination. This causes the main beam to rotate and broaden relative to the uniform phase case. From the point of view of an observer, the result is exactly the same as a passive reflector oriented directly orthogonal to the path of transmission but which has the area of the projection of the original passive into that orthogonal plane of reference.

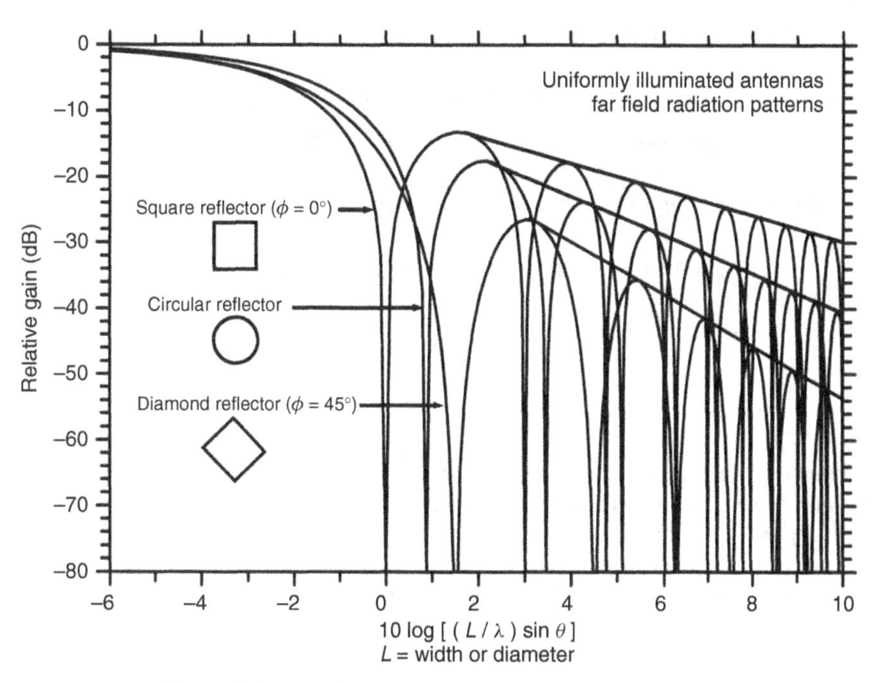

Figure 8.1 Passive reflector far field radiation patterns.

8.2.1 Passive Reflector Far Field Radiation Pattern

For simple antennas such are (rectangular or circular) passive reflectors (which have uniform illumination), the antenna patterns (Fig. 8.1) are well known (Silver, 1949).

For a *circular reflector*, the radiation pattern is given by

$$P_{\text{dB}} = 10 \log \left[\frac{2J_1(\pi u)}{(\pi u)} \right]^2 \tag{8.1}$$

$$u = \left(\frac{D}{\lambda} \right) \sin\theta \cong \left(\frac{D}{\lambda} \right) \theta \text{ (radians)}$$

θ = angle of power measurement relative to boresight;
D = diameter of the circular reflector.

For $u \cong (\pi/180)(D/\lambda)\theta$ (degrees) the following relationships apply:

$$\theta_{1\text{dB}} = 34.7 \left(\frac{\lambda}{D} \right) \text{degrees}$$

$$\theta_{3\text{dB}} = 59.0 \left(\frac{\lambda}{D} \right) \text{degrees}$$

$$\theta_{10\text{dB}} = 99.6 \left(\frac{\lambda}{D} \right) \text{degrees}$$

$$\theta_{20\text{dB}} = 215 \left(\frac{\lambda}{D} \right) \text{degrees (This is the angle to the widest 20 dB measurement.)}$$

The first side lobe is −17.6 dB below the peak boresight power.

The envelope of the circular reflector radiation pattern is

$$P_{\mathrm{dB}} = 20\ \log\left[\{\sin(2.680u_W)\}/(2.680u_W)\right], \quad u_W < 0.775 \tag{8.2}$$

$$P_{\mathrm{dB}} = 4.06 - 30\ \log(\pi u_W), \quad u_W >= 0.775 \tag{8.3}$$

For a *rectangular reflector*, the radiation pattern is given by

$$P_{\mathrm{dB}} = 10\ \log\left\{\left[\{\sin(\pi u_{\mathrm{H}})\}/(\pi u_{\mathrm{H}})\right]^2\left[\{\sin(\pi u_{\mathrm{W}})\}/(\pi u_{\mathrm{W}})\right]^2\right\} \tag{8.4}$$

$$u_{\mathrm{H}} = \left(\frac{H}{\lambda}\right)\sin\theta\ \cos\phi$$

$$u_{\mathrm{W}} = \left(\frac{W}{\lambda}\right)\sin\theta\ \cos\phi$$

H = height of the rectangular reflector (with $\phi = 0$);
W = width of the rectangular reflector (with $\phi = 0$);
θ = angle of power measurement relative to boresight.

For a *rectangular reflector* parallel to the Earth ($\phi = 0$):

$$P_{\mathrm{dB}} = 10\ \log\left[\{\sin(\pi u_{\mathrm{W}})\}/(\pi u_{\mathrm{W}})\right]^2 \tag{8.5}$$

W = width of the rectangular reflector (measured in the pattern plane of interest).

For $u_{\mathrm{W}} \cong \left(\frac{\pi}{180}\right)\left(\frac{W}{\lambda}\right)\theta$ (degrees) the following relationships apply:

$$\theta_{1\mathrm{dB}} = 30.0\left(\frac{\lambda}{D}\right)\text{degrees}$$

$$\theta_{3\mathrm{dB}} = 50.7\left(\frac{\lambda}{D}\right)\text{degrees}$$

$$\theta_{10\mathrm{dB}} = 84.6\left(\frac{\lambda}{D}\right)\text{degrees}$$

$$\theta_{20\mathrm{dB}} = 307\left(\frac{\lambda}{D}\right)\text{degrees (This is the angle to the widest 20 dB measurement.)}$$

The first side lobe is −13.3 dB below the peak boresight power.
The second side lobe is −17.8 dB below the peak boresight power.

The envelope of the square reflector radiation pattern is

$$P_{\mathrm{dB}} = 20\ \log\left[\{\sin(\pi u_{\mathrm{W}})\}/(\pi u_{\mathrm{W}})\right], \quad u_{\mathrm{W}} < 0.500 \tag{8.6}$$

$$P_{\mathrm{dB}} = -20\ \log(\pi u_{\mathrm{W}}), \quad u_{\mathrm{W}} >= 0.500 \tag{8.7}$$

For a *diamond reflector* ($W = H$ and $\phi = 45°$):

$$P_{\mathrm{dB}} = 10\ \log\left[\left\{\left[\sin\frac{\pi u_{\mathrm{W}}}{\mathrm{sqrt}\,(2)}\right]\Big/\left[\frac{\pi u_{\mathrm{W}}}{\mathrm{sqrt}\,(2)}\right]\right\}^4\right] \tag{8.8}$$

W = width of any side of the diamond

For $u_W \cong \left(\frac{\pi}{180}\right)\left(\frac{W}{\lambda}\right)\theta$ (degrees) the following relationships apply:

$$\theta_{1\text{dB}} = 30.1 \left(\frac{\lambda}{D}\right) \text{degrees}$$

$$\theta_{3\text{dB}} = 51.7 \left(\frac{\lambda}{D}\right) \text{degrees}$$

$$\theta_{10\text{dB}} = 90.3 \left(\frac{\lambda}{D}\right) \text{degrees}$$

$$\theta_{20\text{dB}} = 120 \left(\frac{\lambda}{D}\right) \text{degrees}$$

The first side lobe is −26.5 dB below the peak boresight power.

The envelope of the diamond reflector radiation pattern is

$$P_{\text{dB}} = 40 \log\left[\{\sin(2.221\, u_W)\}/(2.221\, u_W)\right], \quad u_W < 0.70 \tag{8.9}$$

$$P_{\text{dB}} = 6.02 - 40 \log(\pi u_W), \quad u_W >= 0.700 \tag{8.10}$$

For the preceding reflector discussions, the shape of the above reflector is the shape projected onto the path of radio transmission. For example, a square or circular reflector is physically a rectangle or ellipse, respectively. All dimensions are for the projected shape, not the physical shape. If the passive reflector is rotated in the plane of pattern measurement, projected width = [cosine (Θ) x physical width] where Θ is the angle between a line perpendicular to the face of the reflector and the line of radio wave propagation. The angle Θ is half the angle formed at the reflector by the incoming and outgoing radio paths.

8.2.2 Passive Reflector Near Field Power Density

For circular and square passive reflectors, the maximum (center of the reflector) near field power density is well known (Bickmore and Hansen, 1959) (Fig. 8.2).

8.2.2.1 For the Circular Reflector:

$$P_{\text{NNF}} = 10 \log \left[\frac{S(\Delta)}{S(\Delta = 1)}\right] = 10 \log\left\{13.14 \left[1 - \cos\left(\frac{\pi}{8\Delta}\right)\right]\right\} \tag{8.11}$$

8.2.2.2 For the Square Reflector:

$$\begin{aligned} P_{\text{NNF}} &= 10 \log \left(\frac{S(\Delta)}{S(\Delta = 1)}\right) \\ &= 20 \log \left\{4.05 \left[C^2 \left\{\frac{1}{[2\,\text{sqrt}(\Delta)]}\right\} + S^2 \left\{\frac{1}{[2\,\text{sqrt}(\Delta)]}\right\}\right]\right\} \end{aligned} \tag{8.12}$$

The circular reflector has an infinite number of peaks of 14.2 dB with $\Delta = 0.125$ ($\Delta_{\text{dB}} = -9.031$) being the farthest from the reflector. The square reflector has a peak of 11.2 dB at $\Delta = 0.1704$ ($\Delta_{\text{dB}} = -7.685$). The square reflector peaks become smaller as measurements are made closer to the reflector. The limiting value is 6.13 dB.

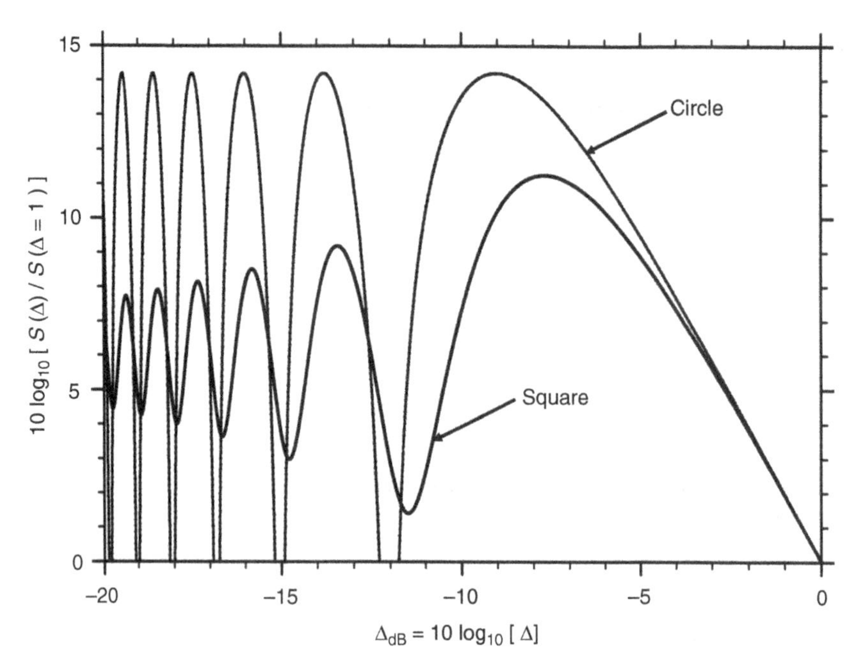

Figure 8.2 Passive reflector near field power density.

The general case for circular and square antennas is numerically derived subsequently. The reflectors are the limiting cases of $\eta = 1.0$ (100% illumination efficiency). As will be noted later, the abovementioned formulas are actually limiting cases where $D/\lambda \gg 10$ or $W/\lambda \gg 10$. For moderate to large antennas, they accurately predict the peak power density but do not always predict its exact location. More accurate numerical methods follow.

8.3 CIRCULAR (PARABOLIC) ANTENNAS

Commercial microwave radio transmit and receive antennas are usually large, circular-shaped parabolic reflectors using waveguide-driven feedhorns to illuminate the reflecting surface. The radio antennas are characterized by a circular aperture with energy distributed over the aperture in such a way that off-axis ("side lobe") energy is minimized while energy directly in front of the antenna ("boresight") remains large. The analytical analysis difficulty is related to the nonuniform distribution of energy across the antenna aperture.

Most commercial microwave transmit antennas place most of the energy into the center of the antenna. The energy power is tapered toward a finite ("pedestal") value at the edge of the antenna. The purpose of this tapered power illumination is to reduce the spurious side lobe responses (reduce spurious radiation and reception). However, the undesired effect is the reduction of the gain of the antenna (relative to a uniformly illuminated antenna). This gain reduction is called a *power efficiency factor* η.

8.3.1 Circular (Parabolic) Antenna Far Field Radiation Pattern

Aperture antenna analysis requires precise knowledge of the illumination of the aperture. Previously, attempts (Sciambi, 1965; Silver, 1949) to define it required multiple variables with no knowledge on how to vary them to achieve a typical antenna illumination. Hansen (1976a) solved this problem by using a pedestal parabolic antenna illumination defined by the following function:

$$P_{\text{I}}(\text{dB}) = 20 \log \left\{ \frac{I_0\left[\pi\, \mathbf{H}\, \text{sqrt}\left(1 - \psi^2\right)\right]}{I_0(\pi \mathbf{H})} \right\} \tag{8.13}$$

ψ = normalized radial distance from the center of the antenna (between 0 and 1).

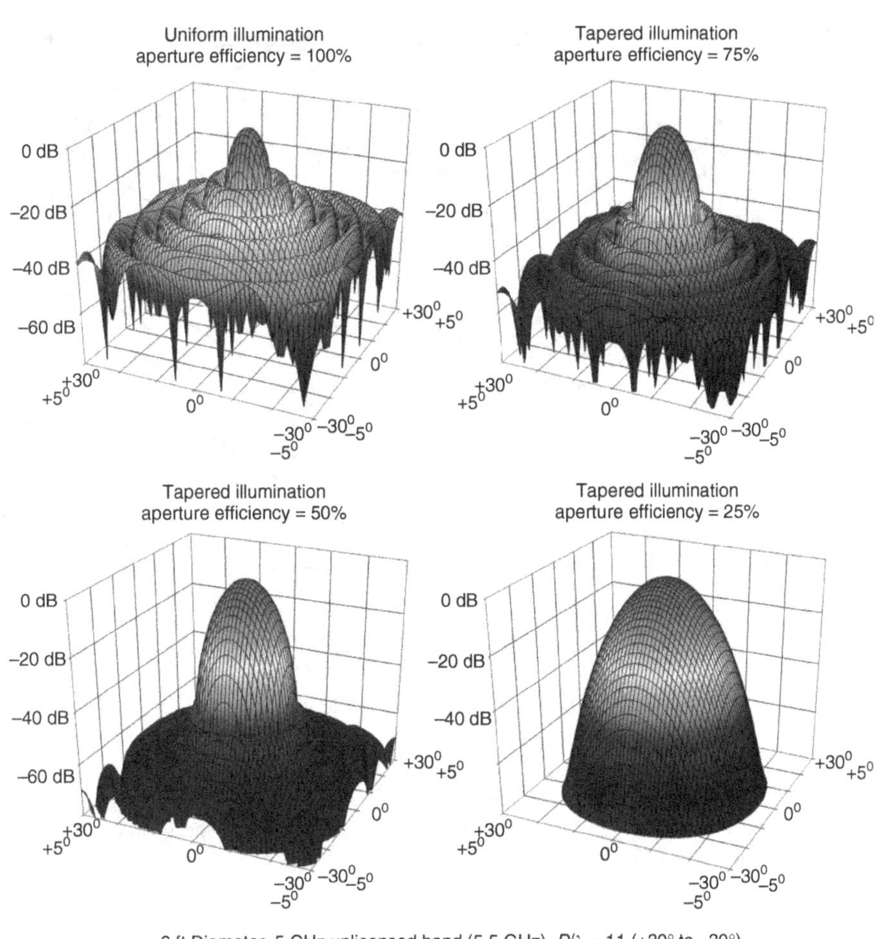

Figure 8.3 Circular antenna far field radiation patterns.

This is the distinct advantage of making the illumination a function of only one parameter **H** (which can be related to antenna illumination efficiency).

Based on this circular antenna illumination, Hansen derived the far field antenna pattern:

$$P_{\mathrm{dB}} = 20\ \log\left[\frac{\{\mathbf{H}\ \mathbf{I}_1\left(\pi\,\mathrm{sqrt}\left[\mathbf{H}^2 - u^2\right]\right)\}}{\{\mathrm{sqrt}[\mathbf{H}^2 - u^2]\ \mathbf{I}_1[\pi\mathbf{H}]\}}\right], \quad \text{for } u \le \mathbf{H}$$

$$P_{\mathrm{dB}} = 20\ \log\left[\frac{\{\mathbf{H}\ \mathbf{J}_1\left(\pi\,\mathrm{sqrt}\left[u^2 - \mathbf{H}^2\right]\right)\}}{\{\mathrm{sqrt}[u^2 - \mathbf{H}^2]\ \mathbf{I}_1[\pi\mathbf{H}]\}}\right], \quad \text{for } u > \mathbf{H}$$

$$u = \left(\frac{D}{\lambda}\right)\ \sin(\theta) \tag{8.14}$$

Examples of circular far field antenna patterns are presented in Figure 8.3.

For similarly shaped antennas with the same illumination, the far field antenna pattern is a function of $u = \left(\frac{D}{\lambda}\right)\ \sin(\theta)$. The function sin (θ) is a nearly linear function of θ for small values of θ. For $\theta < 0.7854$ rad (45°), $\theta = \sin(\theta)$ with accuracy of better than 10%. Therefore, for values of θ up to 45°, the antenna side lobe value is directly related to the $[D/\lambda]\theta$ as shown in Figure 8.3.

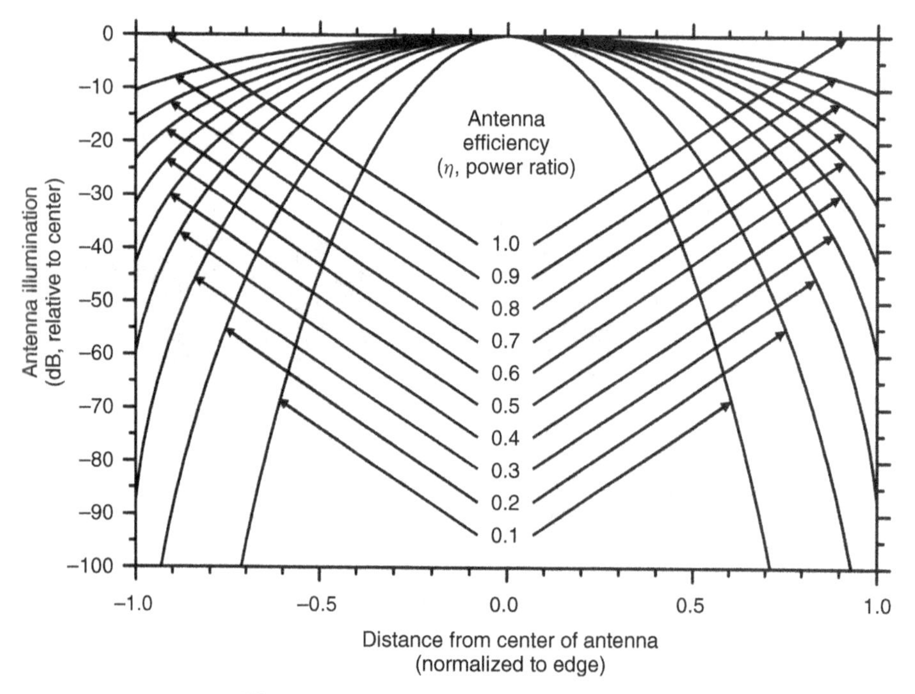

Figure 8.4 Circular antenna illumination.

For some frequency planning studies, FCC Category A and B antenna envelopes are used to limit off-boresight antenna patterns. These limits begin at −20 or −25 dB below the boresight value. All that is needed to complete them is a near boresight envelope. A simple envelope formula that is accurate (within 1 dB over the range of 0 to −25 dB and $0.35 \leq \eta \leq 0.65$) is

$$P_{\mathrm{dB}} = C_{\mathrm{fac}} \log\left[\frac{\sin\,(u)}{(u)}\right] \tag{8.15}$$

$$C_{\mathrm{fac}} = -24.6752 + 221.881\eta - 103.738\eta^2 + 96.6907\eta^3$$

H is a function related to antenna power efficiency, η, defined by the following equation:

$$\eta = \frac{[4I_1{}^2(\pi\mathbf{H})]}{\{\pi^2\mathbf{H}^2[I_0{}^2(\pi\mathbf{H}) - I_1{}^2(\pi\mathbf{H})]\}} \tag{8.16}$$

The following curve-fitted equation, accurate for $0.1 \leq \eta \leq 1.0$, makes **H** a function of η (Fig. 8.4):

$$\begin{aligned}\mathbf{H} = {} & (10789.06678518257 - 21713.73557586721\eta \\ & + 11772.42148035185\eta^2 - 847.7526882273281\eta^3) \\ & /(1 + 8454.692672442079\eta - 14801.06001940764\eta^2 \\ & + 5699.562595386604\eta^3 + 652.2024360393683\eta^4)\end{aligned} \tag{8.17}$$

Hansen's antenna illumination function is similar to actual illumination of commercial antennas. This is demonstrated by the close match between the calculated antenna pattern and the actual commercial antenna patterns near the boresight (Fig. 8.5).

The antenna pattern off-boresight is a function of antenna feedhorn/support structure scattering, phase pattern changes of the feedhorn, and reflector illumination variation. Therefore, deviation of the actual performance from the theoretical for angles away from the boresight is to be expected. It is interesting

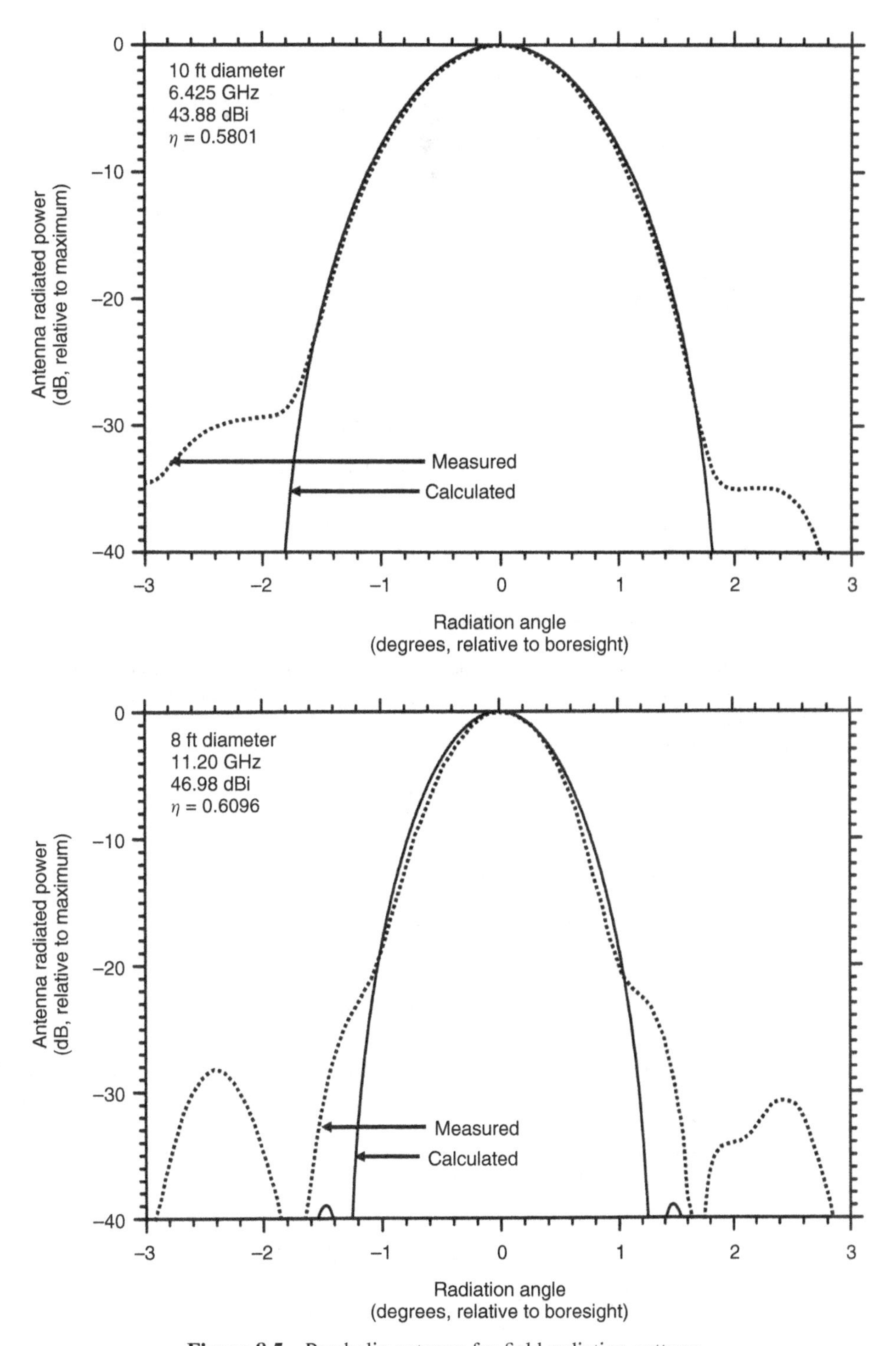

Figure 8.5 Parabolic antenna far field radiation patterns.

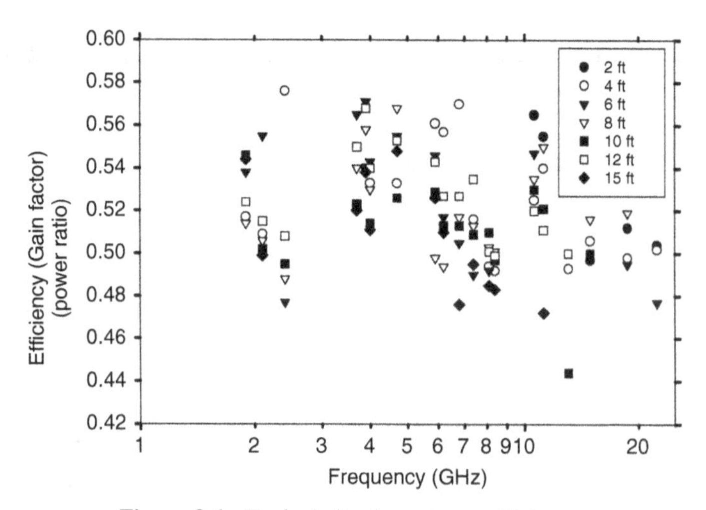

Figure 8.6 Typical circular antenna efficiency.

to note that the side lobe level and null angles of practical antennas are often quite different than those calculated theoretically.

8.3.2 Circular (Parabolic) Antenna Efficiency

While antenna illumination efficiency is seldom provided for commercial microwave antennas, all manufactures list antenna diameter and isotropic gain for a given frequency. Antenna illumination efficiency, η, for commercial antennas may be estimated from this data.

The power gain of an aperture antenna (Kizer, 1990) is given by:

$$g = \frac{4\pi A\eta}{\lambda^2} \tag{8.18}$$

This yields the following equations for antenna efficiency:

$$\begin{aligned} 10\log(\eta) &= -10.1 + G(\text{dB}) - 20\log[f(\text{GHz})] - 20\log[D(\text{ft})] \\ &= -20.4 + G(\text{dB}) - 20\log[f(\text{GHz})] - 20\log[D(\text{m})] \end{aligned} \tag{8.19}$$

$$\eta = \text{antenna relative power efficiency (0 to 1)} \tag{8.20}$$

This formula works because the parabolic antenna itself is nearly lossless and loss of gain is essentially all due to illumination. Typical values of η for various typical commercial antennas were derived using these formulas (Fig. 8.6):

While circular antenna efficiency is easily inferred from size and gain, it can also be inferred from the common 3-dB beamwidth specification, $\phi_{3\text{dB}}$, using the following formulas:

$$\begin{aligned} \eta &= \frac{(N0 + N1\beta + N2\beta^2 + N3\beta^3 + N4\beta^4)}{(1 + D1\beta + D2\beta^2 + D3\beta^3 + D4\beta^4)}, \quad 0.1 \leq \eta \leq 1.0 \\ \beta &= \left[\left(\frac{D}{\lambda}\right)\sin\left(\frac{\phi_{3\text{dB}}}{2}\right)\right], \quad 1.68 \geq \beta \geq 0.515 \end{aligned} \tag{8.21}$$

$N0 = 2.30135726199837$;
$N1 = -9.718502599996146$;
$N2 = 10.71969385496424$;
$N3 = 0.2532903862637571$;

$N4 = 0.0;$
$D1 = -4.929022134884171;$
$D2 = 18.21074703351663;$
$D3 = -44.98768585675341;$
$D4 = 43.03842941363276.$

This formula may be needed if for some reason the antenna is electrically lossy.

8.3.3 Circular (Parabolic) Antenna Beamwidth

Circular antenna efficiency, η, can be related to antenna beamwidth ϕ_n using the following formulas:

$$\left(\frac{D}{\lambda}\right)\sin\theta_n = \frac{(N0 + N1\eta + N2\eta^2 + N3\eta^3 + N4\eta^4)}{(1 + D1\eta + D2\eta^2 + D3\eta^3 + D4\eta^4)}, \quad 0.1 \le \eta \le 1.0 \tag{8.22}$$

$$\phi_{n\text{dB}} = 2\ \arcsin\left\{\frac{\left[\left(\frac{D}{\lambda}\right)\ \sin\theta_n\right]}{\left(\frac{D}{\lambda}\right)}\right\}$$

For $\theta_{3\text{dB}}$ the following values are used:

$N0 = 7.972180065637231;$
$N1 = 137.7187514674955;$
$N2 = -65.72554929123112;$
$N3 = -169.5462549241622;$
$N4 = 92.52937419842578;$
$D1 = 90.30209868556027;$
$D2 = 299.7227877858257;$
$D3 = -615.0321426632636;$
$D4 = 229.7331500526253.$

$$1.68 \ge \left(\frac{D}{\lambda}\right)\sin\theta_{3\text{dB}} \ge 0.515$$

For $\theta_{10\text{dB}}$ the following values are used:

$N0 = 10.82186120921261;$
$N1 = 88.12091114204749;$
$N2 = -134.7746923842901;$
$N3 = -12.95576841796835;$
$N4 = 49.0194213930968;$
$D1 = 48.19536883990477;$
$D2 = 35.09170209420454;$
$D3 = -191.3604015533953;$
$D4 = 107.3398772721432.$

$$3.05 \ge \left(\frac{D}{\lambda}\right)\sin\theta_{10\text{dB}} \ge 0.869$$

For $\theta_{20\text{dB}}$ the following values are used:

$N0 = 14.57321339163523;$
$N1 = 90.54344327554624;$
$N2 = -182.6759854406716;$
$N3 = 57.45550532477225;$
$N4 = 20.32003841569955;$
$D1 = 42.19500334507131;$
$D2 = -1.350679649101605;$

$D3 = -108.9283729566957$;
$D4 = 67.28268578480228$.

$$4.28 \geq \left(\frac{D}{\lambda}\right) \sin\theta_{20\text{dB}} \geq 1.0$$

The above formula calculates the beamwidth of the primary antenna beam (it ignores side lobes). For circular antennas of very high efficiency ($1.0 \geq \eta > 0.978$), the first side lobe can exceed −20 dB relative power (for $\eta = 1.0$, the first side lobe power peaks at −17.6 dB). For $1.0 \geq \eta > 0.978$, $\theta_{20\text{dB}}$ calculated by the above formula should be expanded by a multiplicative factor of 1.73 to 1.49, respectively, if the first side lobe power is considered.

$\phi_{3\text{dB}}$ is a commonly used criterion. For $(D/\lambda) \geq 1$ and $\eta = 0.5$, the formula may be simplified to:

$$\theta_{3\text{dB}} = \frac{88.0}{\left(\frac{D}{\lambda}\right)} \text{degrees} \tag{8.23}$$

The actual beamwidth values vary approximately ±10% from the above values over the η range of 0.4–0.6.

The 3-dB angle formula may be compared to other commonly used historical values:

$$\theta_{3\text{dB}} = \frac{61}{\left(\frac{D}{\lambda}\right)} \text{degrees (TIA Subcommittee TR-14.7, 2005), converted to normalized format;}$$

$$\theta_{3\text{dB}} = \frac{70}{\left(\frac{D}{\lambda}\right)} \text{ degrees (TIA Subcommittee TR-14.7, 1996);}$$

$$\theta_{3\text{dB}} = \frac{71}{\left(\frac{D}{\lambda}\right)} \text{degrees (White, 1970), converted to normalized format;}$$

$$\theta_{3\text{dB}} = \frac{71}{\left(\frac{D}{\lambda}\right)} \text{degrees (Reintjes and Coate, 1952);}$$

$$\theta_{3\text{dB}} = \frac{84}{\left(\frac{D}{\lambda}\right)} \text{ degrees (Silver, 1949), Table 6.2, 0.56 gain (efficiency) factor.}$$

This difference may be explained by the observation that earlier antennas studied used significantly higher efficiency than modern antennas. An emphasis on frequency reuse has caused most modern antennas to be designed with lower efficiency to reduce off-boresight radiation. An exception to this is the recent introduction of very small antennas into unlicensed bands where high efficiency antennas are used (at the expense of frequency reuse).

Another way to look at parabolic antenna beamwidth is the beamwidth expansion as a function of illumination efficiency. As the efficiency is reduced, the antenna beamwidth expands.

$$\theta_{1\text{dB}} = 34.66\, E_{\text{X1}} \left(\frac{\lambda}{D}\right) \text{degrees}$$

$$\theta_{3\text{dB}} = 58.90\, E_{\text{X3}} \left(\frac{\lambda}{D}\right) \text{degrees}$$

$$\theta_{20\text{dB}} = 124.7\, E_{\text{X20}} \left(\frac{\lambda}{D}\right) \text{degrees}$$

$E_{\text{X}(\,)}$ = antenna expansion factor;
θ = angle of power measurement relative to boresight;
D = diameter of the circular reflector.

Note that the θ_{20dB} formula uses a factor of 124.7 rather than the previous 215.2. The previous formula considered side lobe values. This factor only considers main lobe values. For efficiencies greater than 0.978, the parabolic antenna first side lobe peak value will exceed −20 dB.

The E_{X1}, E_{X3}, and E_{X20} represent expansion factors of the full illumination beamwidth. They clearly show the beamwidth penalty of lower antenna efficiency. These factors may be calculated (to at least four significant figures) using the formulas below (Fig. 8.7):

$C11 = 11.47467471422744;$
$C12 = 96.4836471787781;$
$C13 = -149.130810770249;$
$C14 = -7.726291805073669;$
$C15 = 49.11001513993868;$
$C16 = 49.56175002248294;$
$C17 = 34.99249700379614;$
$C18 = -207.6119399582662;$
$C19 = 135.3743821944512;$
$C110 = -13.10545450684395.$

$$E_{X1} = \frac{(C11 + C12\eta + C13\eta^2 + C14\eta^3 + C15\eta^4)}{(1 + C16\eta + C17\eta^2 + C18\eta^3 + C19\eta^4 + C110\eta^5)} \tag{8.24}$$

$C31 = 11.65806521303201;$
$C32 = 96.51565982372452;$
$C33 = -152.1423956242208;$
$C34 = -4.217102204230871;$
$C35 = 48.37581221497451;$
$C36 = 49.10391941747248;$
$C37 = 32.82051399322144;$
$C38 = -201.8020843245464;$

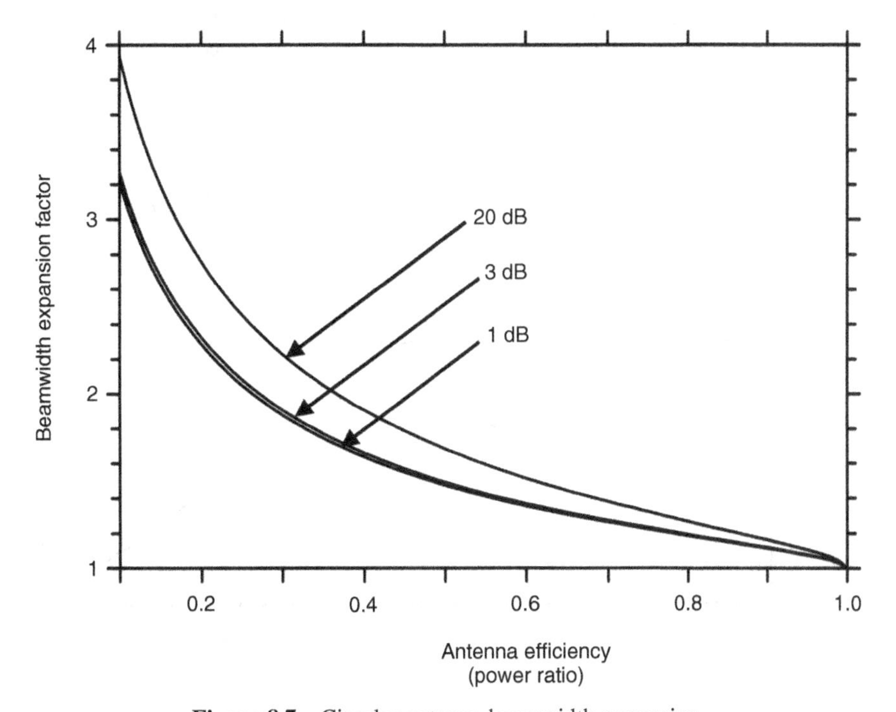

Figure 8.7 Circular antenna beamwidth expansion.

$C39 = 130.4562878319081;$
$C310 = -11.38859718015969.$

$$E_{X3} = \frac{(C31 + C32\eta + C33\eta^2 + C34\eta^3 + C35\eta^4)}{(1 + C36\eta + C37\eta^2 + C38\eta^3 + C39\eta^4 + C310\eta^5)} \tag{8.25}$$

$C201 = 12.68923557512405;$
$C202 = 61.4965668903022;$
$C203 = -163.3946591887904;$
$C204 = 100.3483444646388;$
$C205 = -11.09614160640891;$
$C206 = 37.0644387489422;$
$C207 = -24.55611931753965;$
$C208 = -69.50012243371823;$
$C209 = 67.85296330438491;$
$C2010 = -11.81781396481483.$

$$E_{X20} = \frac{(C201 + C202\eta + C203\eta^2 + C204\eta^3 + C205\eta^4)}{(1 + C206\eta + C207\eta^2 + C208\eta^3 + C209\eta^4 + C2010\eta^5)} \tag{8.26}$$

8.3.4 Circular (Parabolic) Antenna Near Field Power Density

The near field power density of an antenna is of considerable interest for evaluating potential health hazards near an antenna. Calculating near field energy is relatively difficult. Over the years, several attempts have been made to approximate the near field power density of a circular parabolic antenna. Since it is well known that the maximum near field power is in a straight line in front of the center of the antenna, previous estimations calculated that power (near field on axis power density).

One estimate (Saad et al., 1971) of near field on axis power density used Bickmore and Hansen's results (Bickmore and Hansen, 1959) for a circular antenna with power linearly tapered (electric field parabolically tapered) with maximum at the antenna center and zero at the edge (Fig. 8.8).

$$\begin{aligned} P_{\text{NNF}} &= 10\ \log\left[\frac{S(\Delta)}{S(\Delta = 1)}\right] \\ &= 10\ \log\left[26.1\left(1 - \left[(2\delta)\ \sin\left(\frac{1}{\delta}\right)\right] + \left\{(2\delta^2)\left[1 - \cos\left(\frac{1}{\delta}\right)\right]\right\}\right)\right] \end{aligned} \tag{8.27}$$

$$\delta = \frac{8\Delta}{\pi}.$$

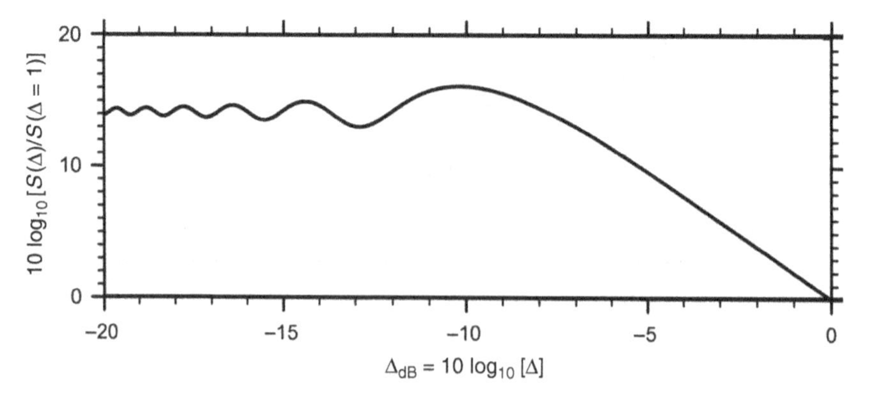

Figure 8.8 Tapered antenna near field power density.

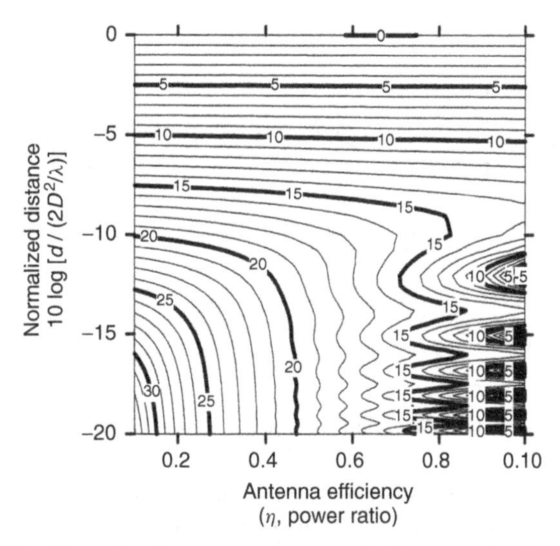

Figure 8.9 Hansen's circular antenna near field power density estimate.

This approach predicts a power peak (16.17 dB for $\Delta = 0.09612$ with limiting value of 14.17 dB). This peak is not observed by actual measurements (Medhurst, 1959). With this broad taper, illumination power is spread out more than is typical for a commercial antenna (as evidenced by poorer side lobe performance). This result is not indicative of commercial antennas.

Hansen (1976b) derived the near field on axis power density for circular microwave transmit antennas on the basis of his pedestal illumination function **H**. His result (after correcting typographical errors) is the following:

$$S = \frac{1}{\Delta^2}\left|\int_0^1 I_0\left(\pi \mathbf{H}\sqrt{1-p^2}\right)\left[e^{j\frac{\pi}{8\Delta}\left(1-p^2\right)}\right] p\mathrm{d}p\right|^2 \tag{8.28}$$

Applying Euler's identity, this equation becomes

$$S = \frac{1}{\Delta^2}\left\{\left[\int_0^1 I_0\left(\pi \mathbf{H}\sqrt{1-p^2}\right)\cos\left[\frac{\pi}{8\Delta}\left(1-p^2\right)\right] p\mathrm{d}p\right]^2 + \left[\int_0^1 I_0\left(\pi \mathbf{H}\sqrt{1-p^2}\right)\sin\left[\frac{\pi}{8\Delta}\left(1-p^2\right)\right] p\mathrm{d}p\right]^2\right\} \tag{8.29}$$

Numerical integration provides the results for $P_{\mathrm{NNF}} = 10 \log [S(\Delta)/S(\Delta = 1)]$ as shown in Figure 8.9.

Near field power density is a function of antenna size (D/λ) and illumination efficiency. Hansen's formula represents the limiting case for a very large antenna. If only the peak value is of interest (but its location Δ is unimportant), Hansen's formula is adequate for $D/\lambda \gg 10$. If null values are of interest, Hansen's approximation requires $D/\lambda \gg 100$.

8.3.5 General Near Field Power Density Calculations

As one goes closer and closer to an antenna, he or she enters the near field of the antenna. In this region, the measured power density is no longer the far field value. When the observer is very close to the antenna, the measured power is due to evanescent electromagnetic fields orthogonal to the face of the antenna. This is the reactive region of the antenna. The measured energy in this region is complicated

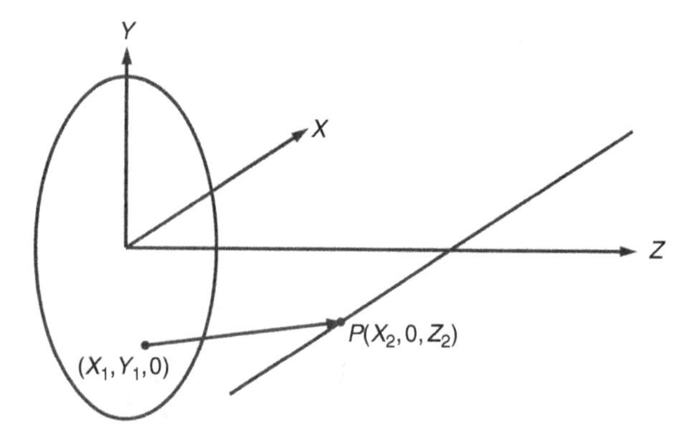

Figure 8.10 Antenna power density calculation geometry.

to estimate analytically. Fortunately, we seldom need to calculate this energy because it only exists very close to the face of the aperture antenna. As the observer retreats from the face of the antenna, he or she enters a near field region where the energy radiates from the antenna but is still not predicted by the far field assumptions. This is the Fresnel region of the antenna. Fortunately, this energy can be predicted by optical analogy. The Huygens–Fresnel principle states that the amplitude of the electromagnetic wave at any given point equals the superposition of the amplitudes of all secondary wavelets emitted from the entire wave front before that point. For our purposes, the wave front of interest is the aperture of an antenna.

Assume an antenna aperture described by points x_1 and y_1 in an X–Y plane (for circular parabolic antennas, this is a circular planar surface directly in front of the antennas where the wave front from the parabolic reflective surface is totally in phase across the entire virtual aperture). Assume an orthogonal Z-axis centered on the center of the (virtual) antenna aperture. This defines a horizontal X–Z plane intersecting the antenna at $z = 0$. A point (x, y, z) on the antenna is $(x_1, y_1, 0)$. A point in space in front of the antenna (but variable in the X–Z plane) is $(x_2, 0, z_2)$ (Fig. 8.10).

Silver (1949) describes the near field energy of an aperture antenna U_P at point $P(x_2, 0, z_2)$ in very general terms (page 170, Eq. 5).

$$U_p = \frac{1}{4\pi}\int_A F(x, y)\frac{e^{-jkr}}{r}\left[\left(jk + \frac{1}{r}\right)(i_z \bullet r_1) + jk(i_z \bullet s)\right]\mathrm{d}x\mathrm{d}y \tag{8.30}$$

This is based on Huygen's concept of estimating a field at a point of interest by summing the contributions of all energy radiating from the aperture antenna. The factor i_z is a unit vector normal to the aperture. The factor r_1 is a unit vector in line with a ray from a point on the aperture $(x_1, y_1, 0)$ to a point of interest $P(x_2, 0, z_2)$. To achieve maximum antenna boresight gain, the phase of the antenna aperture illumination must have constant phase of unity. We will assume that our antenna has this attribute (which all commercial antennas do). Assuming the antenna (field) illumination function $F(x, y)$ has constant unity phase $(i_2 \bullet s = 1)$, replacing k with $2\pi/\lambda$ and applying Euler's identity, Silver's results become

$$U_p = \left(\frac{\pi}{\lambda^2}\right)\int_A F(x, y)\left(\frac{1}{\delta}\right)(B + jC)\mathrm{d}x\mathrm{d}y \tag{8.31}$$

where the integral is over the entire aperture (A) of the antenna.

$F(x, y)$ = the field (square root of power density) illumination (**H**) function;

$$B = (1 + \cos\phi)\sin\delta + \left[\frac{(\cos\phi)}{\delta}\right]\cos\delta;$$

$$C = (1 + \cos\phi)\cos\delta - \left[\frac{(\cos\phi)}{\delta}\right]\sin\delta;$$

$\delta = \left(\frac{2\pi r}{\lambda}\right)$;

r = distance from a point on the antenna $(x_1, y_1, 0)$ to a point P $(x_2, 0, z_2)$ in free space;
= sqrt[$(x_2-x_1)^2 + y_1{}^2 + z_2{}^2$] with sqrt[ ] as the square root function;

ϕ = the angle formed by Z-axis and a ray from a point on the antenna to a point P in free space.

$\cos(\phi) = i_z \bullet r_1 = \frac{z_2}{r}$

$|U_p|^2$ = near field power density at free space point $(x_2, 0, z_2)$.

Since the results will be normalized to the far field transition point, $F(x, y)$ can be taken as simply the Hansen illumination function **H** redefined as a function of radial distance. Both, the real and imaginary components are integrated separately and power summed (sum of squares) to arrive at the composite power. The results of the integration are normalized to the power at the nominal far field transition point.

This integral is challenging to numerically integrate owing to the oscillating sine and cosine functions. It requires considerable area granularity, especially for calculations away from antenna boresight. Integrating a rectangular aperture is straightforward. The circular aperture requires a differential sector area (formed from the difference of two sector areas). The area of a sector slice defined by two radii, R/x and $(R-1)/x$, and angle α (radians) is $\alpha(2R-1)/(2x^2)$. The x, y, and z distances are normalized by dividing the distances by $(2D^2/\lambda)$ where D is the diameter of the circle or the width of the square and λ is the wavelength. For convenience, the x and y coordinates are normalized to (d/D) so that $d/D = \pm 0.5$ represents the edges of the antenna. The (d/D) values are then divided by $(2D/\lambda)$. See Appendix 8.A.1 for a numerical approximation of this integration.

Silver's formula calculates the radiating power in the Fresnel and Fraunhofer regions. It does not predict the nonradiating reactive fields very near the antenna. Hansen (1964) noted that this energy is restricted to the area no greater than one wavelength (λ) from the face of the antenna (where radial distance is measured to the closest part of the aperture). This result was reconfirmed by Laybros et al. (2005). Silver's formula is also based on discarding high order terms of a series expansion (Fresnel field approximation) involving $1/\delta$. Comparing Silver's formula with Hansen's results, which include higher order terms (Hansen, 1964), shows that Silver's formula has negligible error (<10%) for radial distances as close as the reactive field transition point. The Silver equation should be quite accurate for all practical distances from the antenna.

The most popular antenna (Comsearch, 1999, Unpublished report) for the lower 6-GHz fixed point-to-point microwave frequency band (FCC, 2004) has a 10-ft diameter. For the 11-GHz band, the most popular antenna (Comsearch, 1999, Unpublished report) diameter is 6 ft (followed very closely by 10 ft). These popular frequencies and diameters yield the following parameters:

$$\text{For lower 6 GHz,} \quad \frac{D}{\lambda} = \left[\frac{10}{\left(\frac{0.98357}{6.175}\right)}\right] = 62.8 \tag{8.32}$$

$$\text{For 11 GHz,} \quad \frac{D}{\lambda} = \left[\frac{6}{\left(\frac{0.98357}{11.2}\right)}\right] = 68.3 \tag{8.33}$$

For many of the following illustrations, $D/\lambda = 66$, the approximate average of the above, is used.

The following near field power densities were calculated and normalized to power density at the far field transition point $\{d/[2(D/\lambda)^2] = 1\}$. For Figure 8.11, Figure 8.12, Figure 8.13, Figure 8.14, Figure 8.15, and Figure 8.16, the antenna edge is [x/diameter] = ±0.5. The center of the antenna is [x/diameter] = 0.

For high efficiency antennas, the near field power density varies considerably near the antenna (Fig. 8.17). The variation smoothes out considerably as the efficiency is reduced. In all cases, the highest power density (for unity phase antennas) is on a line at the center of the antenna and orthogonal to the face of the antenna.

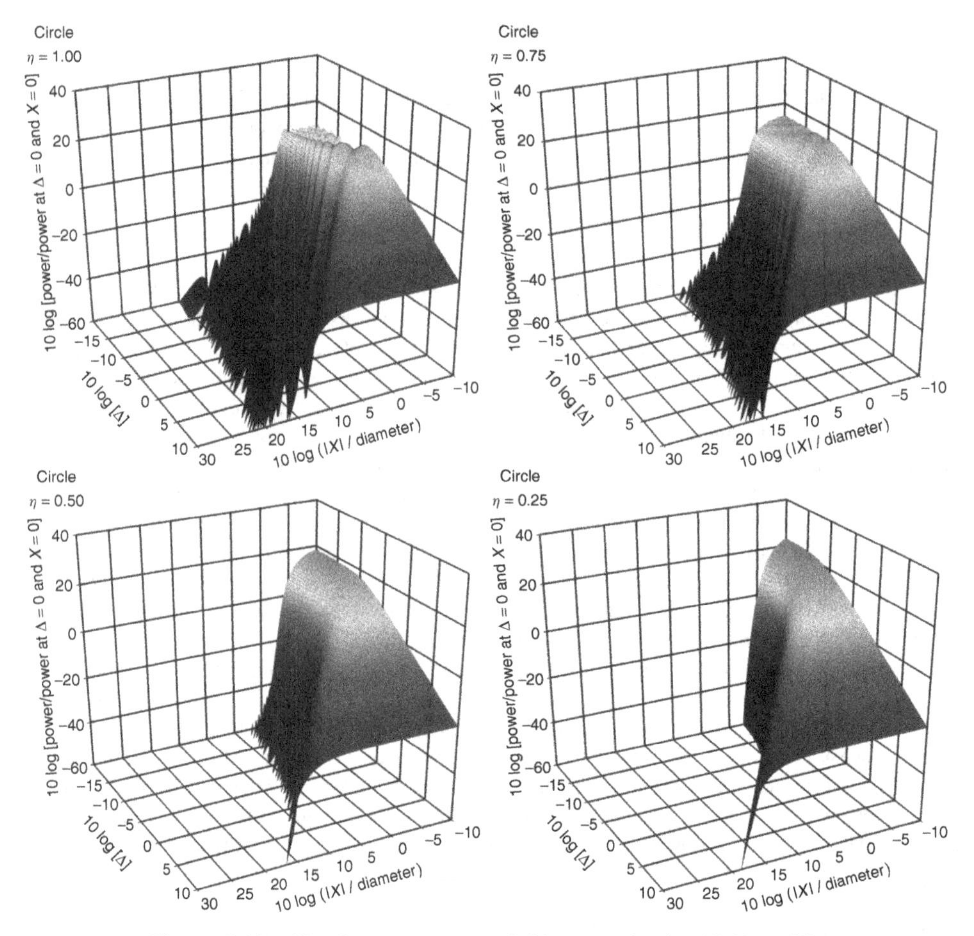

Figure 8.11 Circular antenna near field power density ($D/\lambda = 66$).

For high efficiency antennas, near field power density is a function of antenna size. For fully illuminated ($\eta = 1$) circular antennas with diameter $D = 2n\lambda$ (n an integer), the near field intensity as a function of distance in front of the center of the antenna has $n - 1$ maximums and n minimums (Laybros et al., 2005). Using numerical procedures, the near field boresight power density is calculated for illumination efficiencies, η, between 0.1 and 1 and for values of (D/λ) of 1, 2.5, 5, 10, 100, and 250.

When these calculations are made for efficiencies η above about 0.5, the near field power densities oscillate with periods that are a function of (D/λ). The peaks are the same values for the different values of (D/λ) but occur for different values of normalized distance Δ. For values of (D/λ) approaching 1, the near field power densities are smaller than larger values of (D/λ). This effect will be ignored in the following estimates. Worst case, these calculations will be a little conservative for those antennas. However, these antennas have very little near field region anyway. Figure 8.18 graphs the worst case (peak) near field power density as a function of efficiency and distance.

The limiting near and far field power relationships to Δ_{dB} are well defined. As the power density transitions from one to the other, the relationship is more complicated. The main use of near field power density calculations is to estimate human exposure. Using a simple, relatively conservative method is appropriate. For simple near field estimations, somewhere between the deep near field and the far field, one can calculate $[2\Delta_{dB}]$ and P_{NNFL}(dB). The smaller of the two values represents a simple straight line approximation to the near field power density.

$$10 \log_{10} \left[\frac{S(\Delta)}{S(\Delta = 1)} \right] = -2\Delta_{dB} \tag{8.34}$$

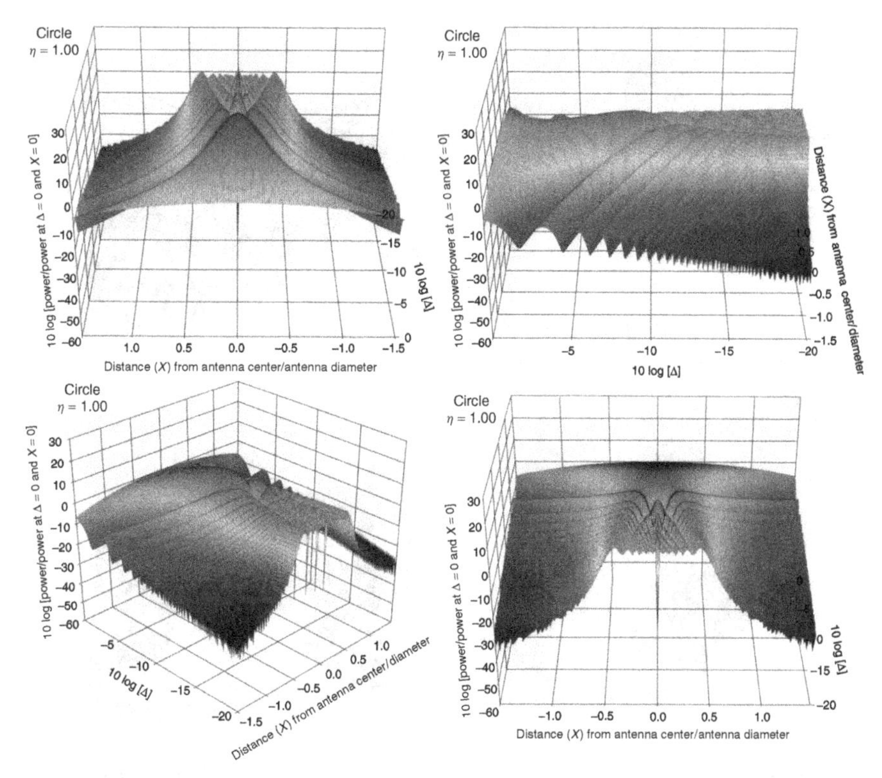

Figure 8.12 Circular antenna near field power density ($\eta = 1.0$, $D/\lambda = 66$).

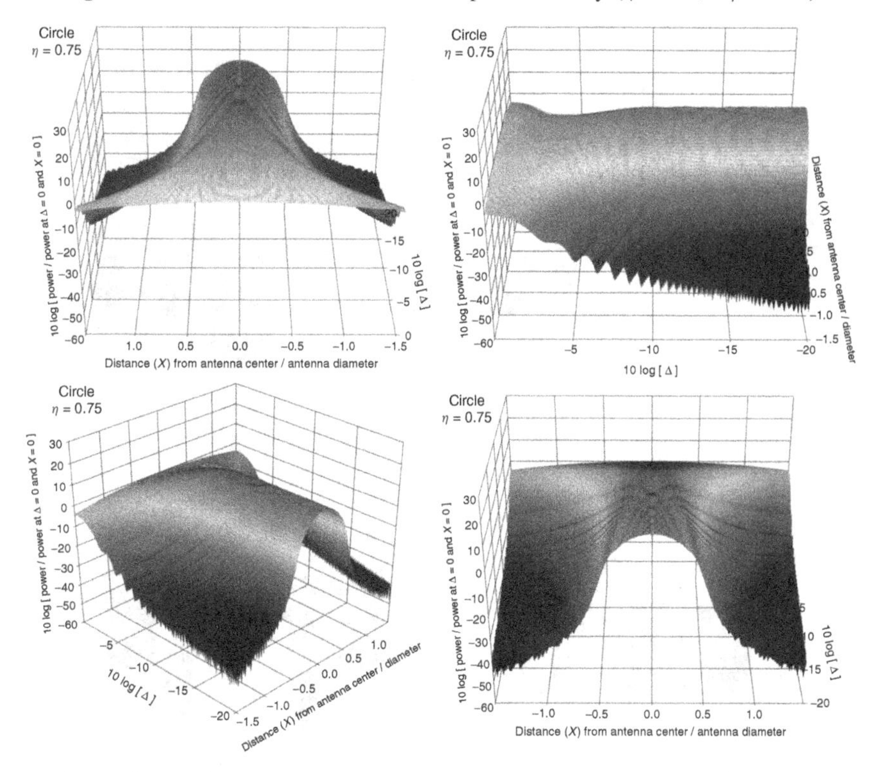

Figure 8.13 Circular antenna near field power density ($\eta = 0.75$, $D/\lambda = 66$).

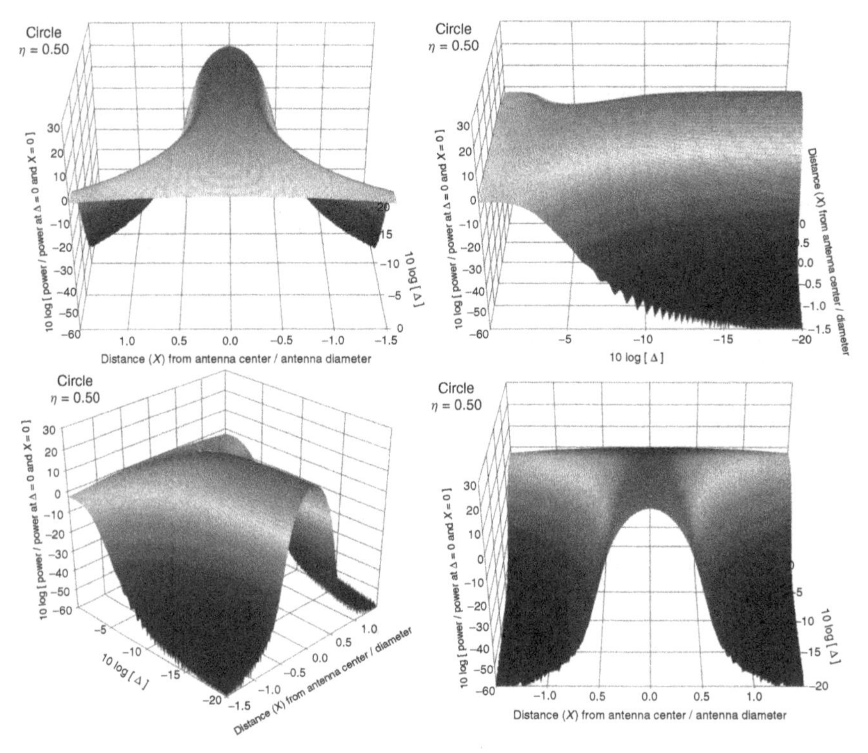

Figure 8.14 Circular antenna near field power density ($\eta = 0.5$, $D/\lambda = 66$).

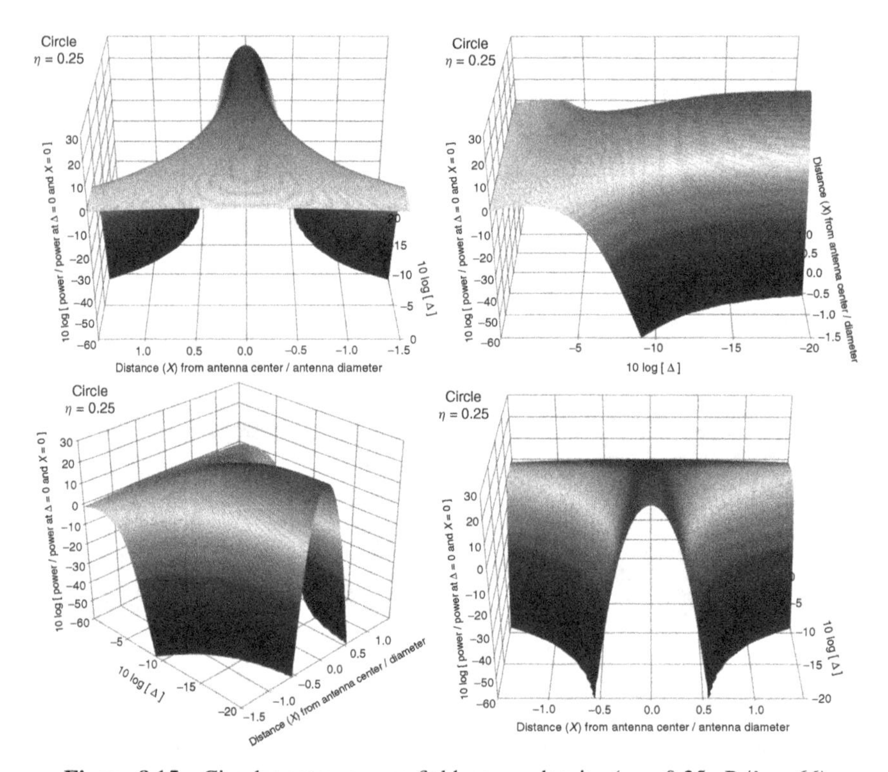

Figure 8.15 Circular antenna near field power density ($\eta = 0.25$, $D/\lambda = 66$).

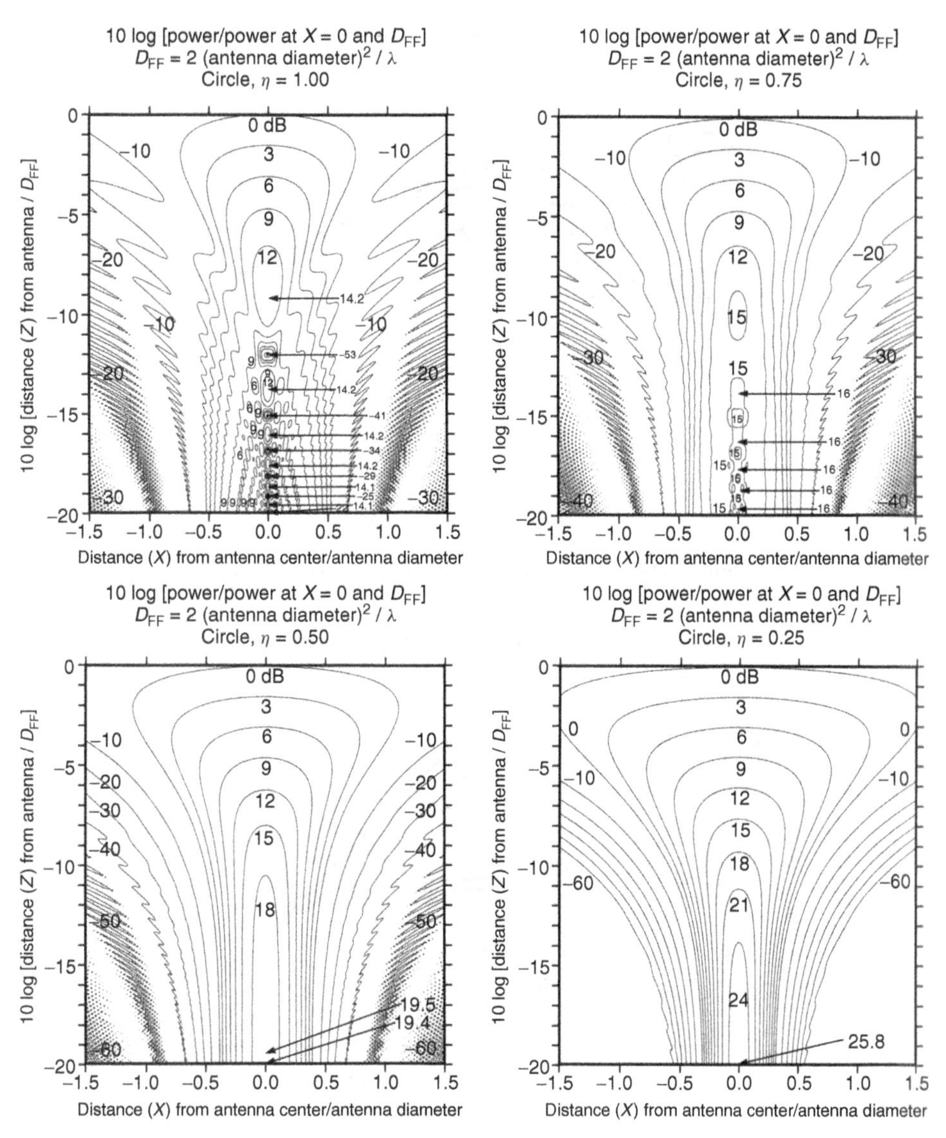

Figure 8.16 Circular antenna near field power density contours ($D/\lambda = 66$).

or

$$10 \log_{10} \left[\frac{S(\Delta)}{S(\Delta = 1)} \right] = A + B\eta + \frac{C}{\eta} + D\eta^2 + \frac{E}{\eta^2} + F\eta^3 \tag{8.35}$$

$A = 40.430453$;
$B = -61.480406$;
$C = -0.46691971$;
$D = 55.376708$;
$E = 0.04791274$;
$F = -19.805638$.

Both formulas are evaluated. Eqn. 8.34 describes a diagonal line for the far field power. Eqn. 8.35 defines a horizontal line for the near field power. The lower power is the power estimate normalized to the far field crossover power.

This approach is very accurate in the near field and far field regions and conservative in between. It also has the merit of being quite easy to use.

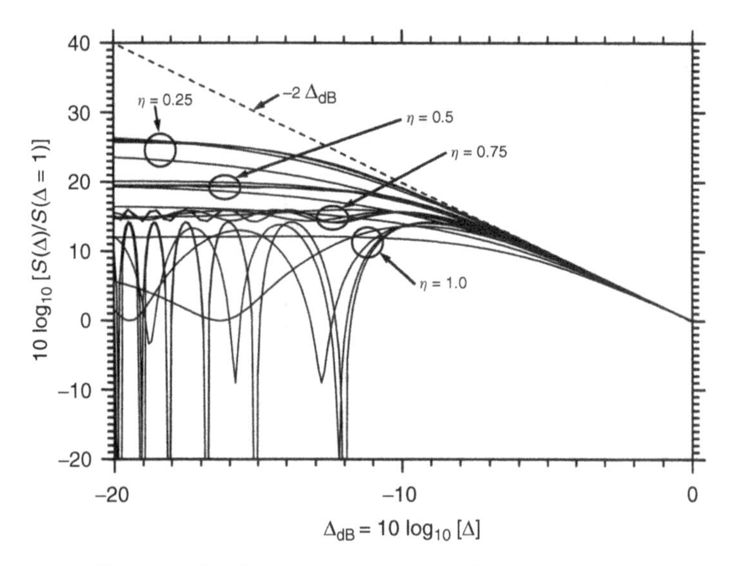

Figure 8.17 Circular antenna near field power density.

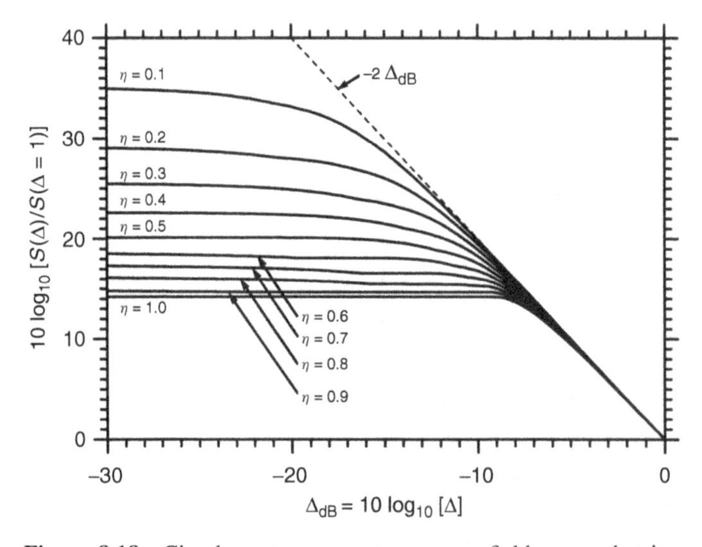

Figure 8.18 Circular antenna worst case near field power density.

8.3.6 Circular Antenna Near Field Power Density Transitions

It is well known (Collin and Zucker, 1969; Editorial Board, 1968; Silver, 1949) that near a circular antenna most of the emitted power is contained in a cylinder centered on the center of the antenna. In this (Fresnel or near field) region, power is essentially constant with distance from the antenna (with the exception of $\eta \approx 1$). Far away from the antenna (Fraunhofer or far field region), power varies as distance squared from the antenna. The area between these two regions has been designated the transition region. In this region, power versus distance from the antenna transitions between the near field and far field limiting conditions. Various criteria have been suggested for the limits between the near field and transition as well as between the transition and far field regions.

Sources (Cleveland et al., 1997; Editorial Board, 1968) have suggested the transition point between the Fresnel zone and the transition region as $D^2/(4\lambda)$ or $D^2/(2\lambda)$ with the latter formula being traditional. The $D^2/(2\lambda)$ criterion represents phase error ("astigmatism") at the aperture edge of $\lambda/4$ radians. Historically (Friis, 1946), the formula $2D^2/\lambda$ has been used as the transition point for the far field region. The

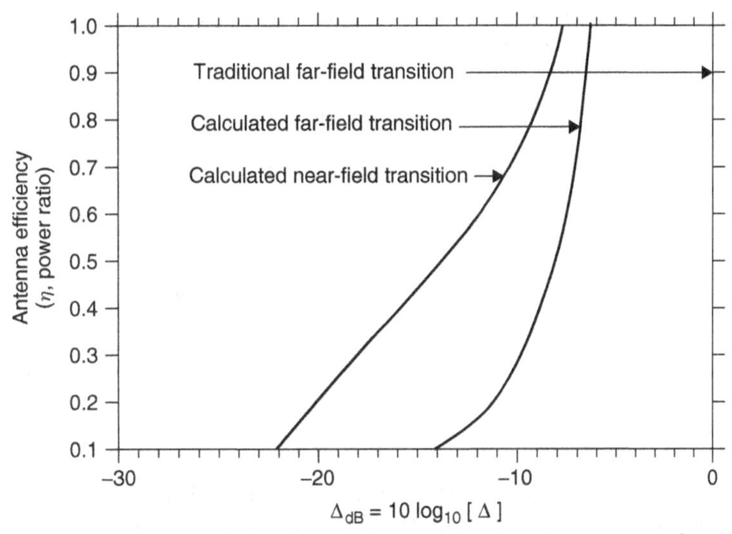

Figure 8.19 Circular antenna near field transition regions.

$2D^2/\lambda$ criterion represents phase error at the aperture edge of $\lambda/16$ radians. Since the traditional far field transition distance formula $2D^2/\lambda$ fails to account for the effective reduction in antenna size as illumination efficiency η is reduced, some engineers have suggested $2\eta D^2/\lambda$ as an "updated" far field transition distance formula.

None of these limits have any performance basis. They are artifacts from the era of Huygens (1678) and Fresnel (1816). They ignore the effects of aperture shape and illumination. While $2D^2/\lambda$ is fairly conservative for single antenna power density measurements, it is less conservative when considering other near field situations such as power transfer between two antennas and antenna pattern measurements.

Figure 8.19 plots the radiating near and far field limits as a function of illumination efficiency, η. The far field limit is the point at which the far field equation power is 1 dB greater than the calculated power. The near field is defined as the point at which the final peak value is achieved.

If the far field transition is defined as the point at which the circular antenna boresight gain is 1 dB less than the far field gain for that distance, the following formula is much more accurate in predicting the transition point:

$$\Delta_{\text{dB}}(\text{1 dB far field transition distance}) = -10.46 + 8.730\eta - 4.116\eta^2 - \frac{0.4638}{\eta} \tag{8.36}$$

8.3.7 Circular Antenna Far Field Reference Power

Near field power density results are normalized to antenna power S at $\Delta = 1$ (the historical "far field transition point"). This is the power density value at the far field transition distance, $d = (2D^2/\lambda)$. It is necessary to determine that power. Microwave radio antenna gain is normally specified relative to an isotropic radiator (a hypothetical antenna that radiates uniformly in all directions).

$$S(\text{power density}) = \frac{pg}{\text{surface area of sphere with radius } d} = \frac{pg}{4\pi d^2} \tag{8.37}$$

At the far field transition point ($\Delta = 1$), $d = (2\,D^2)/\lambda$.
Therefore

$$S(\Delta = 1) = \frac{pg\lambda^2}{16\pi D^4} \tag{8.38}$$

Since most power density safety limits are specified in milliwatts (mW) per centimeter (cm), p is transmitter power in milliwatts, D and λ are in centimeters, and f is the radio frequency.

$$\lambda(\text{cm}) = \text{Free space wavelength} = \frac{29.98}{f(\text{GHz})} \tag{8.39}$$

$$D(\text{cm}) = \text{Aperture diameter} = 30.48 D(\text{ft}) \tag{8.40}$$

Antenna isotropic gain, g (Kizer, 1990), is given by

$$g = \frac{4\pi A\eta}{\lambda^2} = \frac{\pi^2 D^2 \eta}{\lambda^2} \tag{8.41}$$

Substituting this into Equation 8.37 yields the far field (to near field transition) reference power:

$$S(\Delta = 1) = \frac{\pi p \eta}{16\, D^2} \tag{8.42}$$

The illumination efficiency, η, can be determined as described earlier as circular (parabolic) antenna efficiency. The final result is to use this value to calculate the near field power.

$$\left[\frac{S(\Delta)}{S(\Delta = 1)}\right] = 10^{\{10 \log [S(\Delta)/S(\Delta=1)]\}/10} \tag{8.43}$$

$$S(\Delta) = \left[\frac{S(\Delta)}{S(\Delta = 1)}\right] S(\Delta = 1) \tag{8.44}$$

Worst case near field power is calculated at the worst case location (boresight at the reactive near field to Fresnel field transition point of $d = \lambda$). Therefore $S(\Delta)$ is calculated at the following Δ.

$$\Delta = \frac{1}{\left[2\left(\frac{D}{\lambda}\right)^2\right]} \tag{8.45}$$

8.4 SQUARE FLAT PANEL ANTENNAS

Small square panel antennas (parallel to the Earth or rotated 45° to appear as a diamond) are beginning to appear in several unlicensed microwave bands. These antennas use phased array techniques to achieve very thin depth while achieving significant gain. These antennas are composed of a series of "patch" dipoles spread across a square dielectric panel. Owing to frequency-dependent dielectric losses, these antennas are usually limited to operation no higher than 6 GHz.

These radio antennas are characterized by a square aperture with energy distributed over the aperture in such a way that off-axis ("side lobe") energy is minimized while energy directly in front of the antenna ("boresight") remains large. The analytical difficulty is related to the distribution of energy across the antenna aperture.

As with circular antenna, we will use Hansen's (Hansen, 1976a) pedestal parabolic antenna illumination defined by the following function:

$$P(\text{dB}) = 20 \log \left[\frac{I_0\left[\pi \mathbf{H} \text{sqrt}\left(1 - \psi^2\right)\right]}{I_0(\pi \mathbf{H})}\right] \tag{8.46}$$

ψ = normalized radial distance from the center of the antenna (between 0 and 1).

While Hansen used this illumination function in the context of parabolic antennas, it must be modified to use in square antenna calculations. If we assume ψ is zero at the center of the square and 1 at the corner, we may numerically calculate antenna power efficiency η for any given value of **H**.

$$\text{Antenna gain} = \text{Gm} = \text{Antenna gain relative to 100\% illumination}$$

$$= \frac{4\pi|\int_A F(x, y)dxdy|^2}{\lambda^2 \int_A |F(x, y)|^2 dxdy} \tag{8.47}$$

$$\text{Reference gain} = \text{Go} = \text{Antenna gain with } F(x, y) = 1 \text{ (100\% illumination)}$$

$$= \pi^2 \left(\frac{D}{\lambda}\right)^2 \text{ for circle of diameter } D$$

$$= 4\pi \left(\frac{W}{\lambda}\right)^2 \text{ for square of width } W \tag{8.48}$$

$F(\psi)$ = antenna illumination function = $I_0[\pi \mathbf{H} \text{sqrt}\,(1 - \psi^2)]/I_0(\pi \mathbf{H})$;
η (power efficiency) = Gm/Go.

The integrals are over the illuminated antenna aperture. The values of η were calculated for various **H** values. **H** is related to square antenna power efficiency η by the following curve-fitted equation, accurate for $0.1 \leq \eta \leq 1.0$:

$$\begin{aligned}\mathbf{H} = {} & (-1979.796930437251 + 5081.776118785949\eta \\ & - 4381.289862237859\eta^2 + 1279.304377692569\eta^3) \\ & /(1 - 1010.234447176994\eta + 2522.983250131826\eta^2 \\ & - 2209.833454513456\eta^3 + 694.1627969027892\eta^4)\end{aligned} \tag{8.49}$$

Notice that this **H** is applied to a square antenna. It is not the same **H** used for circular antennas.

Currently, there is no closed form solution for the far field antenna pattern. It must be calculated using the Silver method described in Section 8.3.5 for circular antennas.

$$U_p = \left(\frac{\pi}{\lambda^2}\right) \int_{A_1} F_1(x, y) \left(\frac{1}{\delta}\right) (B + jC) dxdy \tag{8.50}$$

where the integral is over the entire aperture of the transmit antenna

$$|U_p|^2 = \text{Near-field power density at free space point } (x_2, y_2, z_2) \tag{8.51}$$

If $z_2 \gg 2W^2/\lambda$ then

$$P_{dB} = 20\ \log|U_p|(x_2, y_2, z_2) - 20\ \log|U_p|(0, 0, z_2) \tag{8.52}$$

P_{dB} = normalized far field antenna pattern.

See Appendix 8.A.2 for a numerical approximation of this integration.

Hansen's antenna illumination function is similar to the actual illumination of commercial antennas. This is demonstrated by the close match between the calculated antenna pattern and actual square commercial antenna envelopes near the boresight (Fig. 8.20).

The actual antenna pattern off-boresight is a function of the granularity of the phased arrays being used to approximate a smooth illumination function. For these antennas, theoretical performance will typically not be achieved for angles well off boresight.

Currently, there is no known closed form solution for the normalized far field antenna pattern, P_{dB}. The double integral must be numerically integrated to determine the far field antenna pattern. Figure 8.21 illustrates the results.

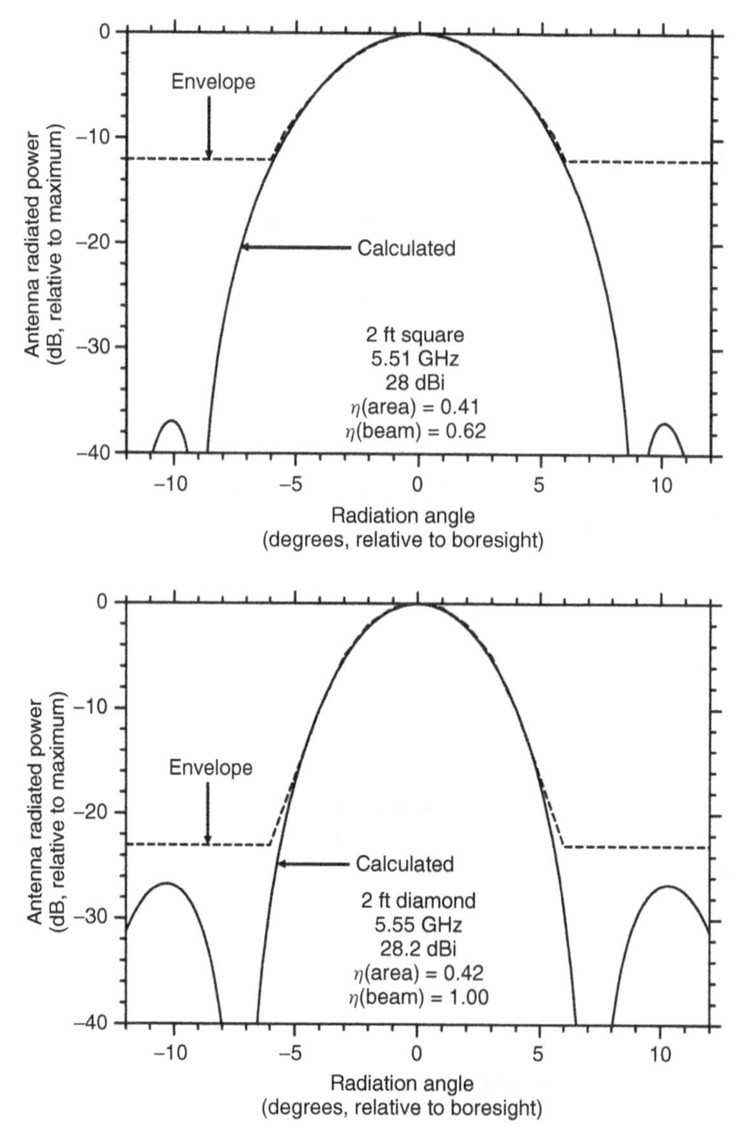

Figure 8.20 Square antenna far field radiation pattern.

For similarly shaped antennas with the same illumination, the far field antenna pattern is a function of $u = (D/\lambda)\sin(\theta)$. $\sin(\theta)$ is a nearly linear function of θ for small values of θ. For $\theta < 0.7854$ rad (45°), $\theta = \sin(\theta)$ with an accuracy of better than 10%. Therefore, for values of θ up to 45°, the antenna side lobe value is directly related to the $(D/\lambda)\theta$ as shown in Figure 8.21.

8.4.1 Square Antenna Beamwidth

Far field antenna theory provides the following relationships:

$$
\begin{aligned}
G(\text{dB}) &= 11.1 + 20\ \log\ f(\text{GHz}) + 10\ \log\ \text{area}(\text{ft}^2) + 10\ \log \eta \\
&= +21.5 + 20\ \log\ f(\text{GHz}) + 10\ \log\ \text{area}(\text{m}^2) + 10\ \log \eta
\end{aligned}
\tag{8.53}
$$

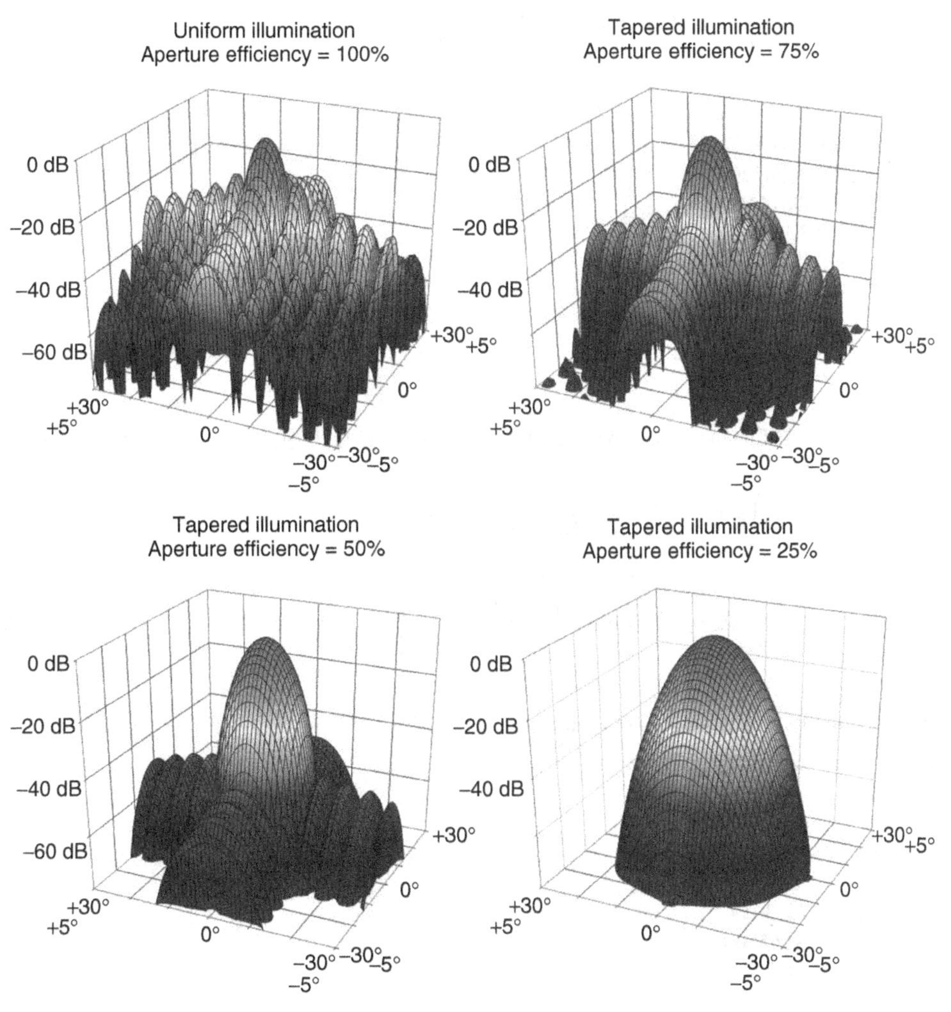

Figure 8.21 Square antenna far field radiation patterns.

From this we may infer η from size, gain, and frequency:

$$\begin{aligned}\eta &= 10^{\alpha/10}\\ \alpha &= G(\text{dB})-11.1-20\ \log\ f(\text{GHz})-20\ \log\ W(\text{ft})\\ &= G(\text{dB})-21.5-20\ \log\ f(\text{GHz})-20\ \log\ W(\text{m})\end{aligned} \tag{8.54}$$

While square antenna efficiency could, in theory, be inferred from size, gain, and frequency, most square antennas are small and relatively lossy. The efficiency of interest, η, is the radiation efficiency that is directly related to the radiation pattern characteristics. That efficiency is more accurately estimated directly from the 3-dB beamwidth specification, $\phi_{3\text{dB}}$, using the following formulas:

$$\eta = \frac{(N0 + N1\beta + N2\beta^2 + N3\beta^3 + N4\beta^4)}{(1 + D1\beta + D2\beta^2 + D3\beta^3 + D4\beta^4)}, \quad 0.1 \le \eta \le 1.0$$

$$\beta = \left[\left(\frac{W}{\lambda} \right) \sin \left(\frac{\phi_{3\text{dB}}}{2} \right) \right], \quad 1.49 \geq \beta \geq 0.447 \tag{8.55}$$

$N0 = -0.4468979109577574;$
$N1 = 2.705347403057084;$
$N2 = -5.689139811168476;$
$N3 = 5.017375871680245;$
$N4 = -1.037085334383484;$
$D1 = -7.914244751535077;$
$D2 = 24.1096714821637;$
$D3 = -33.58979930453501;$
$D4 = 18.85685129777957.$

Square antenna efficiency η can be related to antenna beamwidth, ϕ_n, using the following formulas:

$$\left(\frac{W}{\lambda} \right) \sin \theta_n = \frac{(N0 + N1\eta + N2\eta^2 + N3\eta^3 + N4\eta^4)}{(1 + D1\eta + D2\eta^2 + D3\eta^3 + D4\eta^4)}, \quad 0.1 \leq \eta \leq 1.0$$

$$\phi_{n\text{dB}} = 2 \arcsin \left\{ \frac{\left[\left(\frac{W}{\lambda} \right) \sin \theta_n \right]}{\left(\frac{W}{\lambda} \right)} \right\} \tag{8.56}$$

For $\theta_{3\text{dB}}$ the following values are used:

$N0 = 3.683863445307736;$
$N1 = 5.948533059131917;$
$N2 = -12.00998709023712;$
$N3 = 2.602565857888975;$
$N4 = 0.0;$
$D1 = 20.0539872236591;$
$D2 = -20.764915655005;$
$D3 = 0.214464192754996;$
$D4 = 0.0;$

$$1.49 \geq (W/\lambda), \quad \sin \theta_{3\text{dB}} \geq 0.447.$$

For $\theta_{10\text{dB}}$ the following values are used:

$N0 = 872.9332198184043;$
$N1 = 45161.34589980058;$
$N2 = 48795.11666508175;$
$N3 = -104521.2404994755;$
$N4 = 12668.83297462799;$
$D1 = 11923.81527760833;$
$D2 = 102924.4164651547;$
$D3 = -85685.52877384194;$
$D4 = -25261.64181058896;$

$$2.71 \geq (W/\lambda), \quad \sin \theta_{10\text{dB}} \geq 0.763.$$

For $\theta_{20\text{dB}}$ the following values are used:

$N0 = 7.229987491461088;$

$N1 = 78.30645073634908;$
$N2 = 1419.187443833135;$
$N3 = -1283.159662762055;$
$N4 = -187.1848795537545;$
$D1 = 9.486125571531771;$
$D2 = 483.034865668168;$
$D3 = 674.8234257427976;$
$D4 = -1133.084099849063;$

$$3.81 \geq (W/\lambda), \quad \sin\theta_{20dB} \geq 0.976.$$

Beamwidth is nominally related to square antenna rotation. Ignoring rotation angle in the formulas introduces the following error:

For θ_{3dB}, error is 0.0%, 0.2%, 0.5%, and 0.9% for $\eta = 0.1$, 0.5, 0.75, and 1.0 respectively. For θ_{10dB}, error is 0.0%, 0.7%, 1.7%, and 3.9% for $\eta = 0.1$, 0.5, 0.75, and 1.0 respectively. For θ_{20dB}, error is 0.0%, 1.2%, 3.7%, and 7.5% for $\eta = 0.1$, 0.5, 0.75, and 1.0 respectively.

θ_{3dB} is a commonly used criterion. For $(W/\lambda) \geq 1$ and $\eta = 1.0$, the formula may be simplified to the following (with no more than 10% error):

$$\theta_{3dB} = \frac{51.2}{\left(\frac{W}{\lambda}\right)} \text{degrees} \tag{8.57}$$

The actual beamwidth increases 52% if η is reduced to 0.5.

8.4.2 Square Near Field Power Density

Bickmore and Hansen (1959) calculated the near field power density for a uniformly illuminated ($\eta = 1$) square antenna. This result gave no guidance for nonuniform illumination ($\eta = 1$).

An approach (Editorial Board, 1968) commonly used is to take the transmitter power p at the input to the antenna, multiply it by 4, and divide by the antenna area, A. This approach overestimates the near field power for the uniformly illuminated antenna ($\eta = 1$) by 3.5 dB. It gives no guidance for illumination factors less than unity ($\eta < 1$). For the 50% illumination efficiency ($\eta = 0.5$) case, this approach underestimates the power density by 2 dB.

Another approach (Cleveland, Sylvar and Ulcek, 1997, Eq. 13) estimates the same near field power density and the previous approach. However, it underestimates the power density by 5 dB for the 50% illumination efficiency ($\eta = 0.5$) case.

To date, accurate near field power density calculation methods are not available. This limitation will be overcome by numerically calculating the power density using the same approach as that used for circular antennas (but with the **H** factor normalized for square antennas).

Near field power densities were calculated and normalized to power density at the far field transition point ($d/[2(W/\lambda)^2] = 1$). For Figure 8.22, Figure 8.23, Figure 8.24, Figure 8.25, Figure 8.26, Figure 8.27, Figure 8.28, Figure 8.29, Figure 8.30, Figure 8.31, Figure 8.32, and Figure 8.33, the antenna edge is (x/width) $= \pm 0.5$. The center of the antenna is (x/width) $= 0$. Patterns are calculated for both the square (0° rotation relative to the Earth) and diamond (45° rotation) orientations.

Rotating the square antenna 45° in the vertical plane reduces the energy off the boresight. For high efficiency antennas, the near field power density varies considerably near the antenna. The variation smoothes out considerably as the efficiency is reduced. In all cases, the highest power density (for unity phase antennas) is in a line at the center of the antenna and orthogonal to the face of the antenna.

Using the numerical procedures outlined earlier with $z_2 \leq 2W^2/\lambda$, the near field boresight power density was calculated for illumination efficiencies η between 0.1 and 1 and for values of (D/λ) between 1 and 250 (Fig. 8.34).

For square antennas, antennas boresight gain as well as boresight near field power density are independent of antenna rotation. When these calculations are made for efficiencies, η, above about 0.5, the near

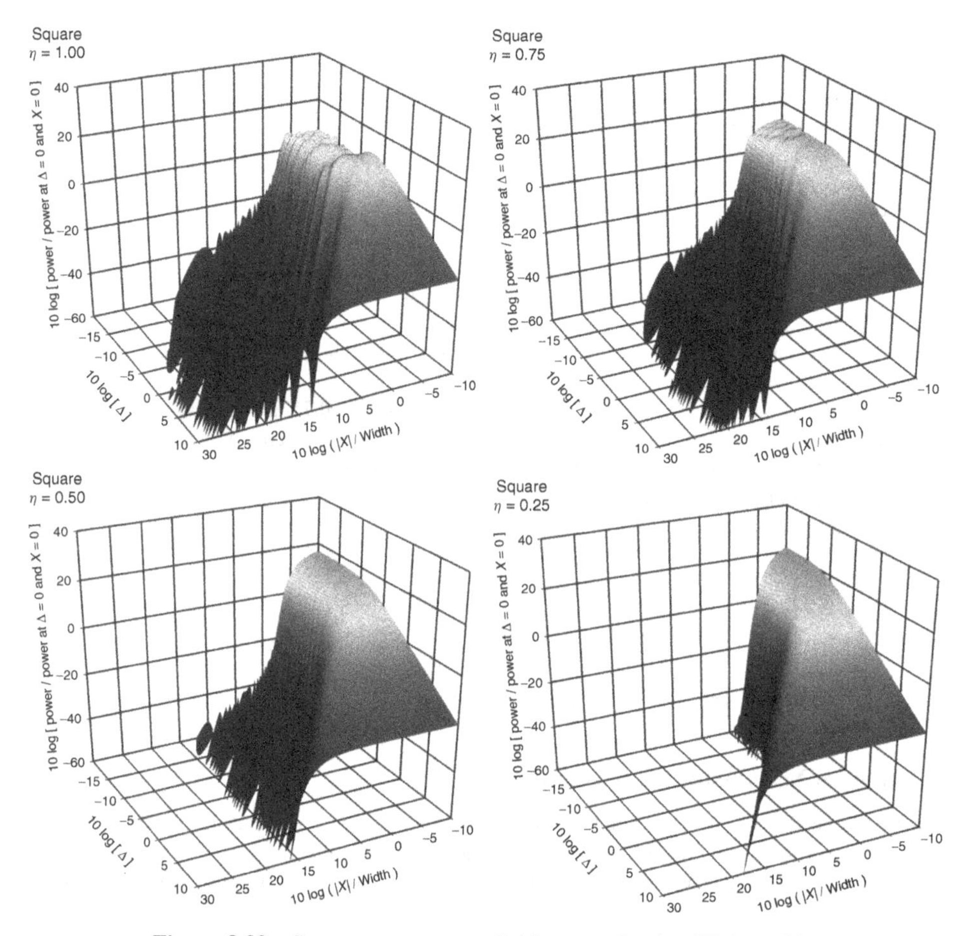

Figure 8.22 Square antenna near field power density ($W/\lambda = 66$).

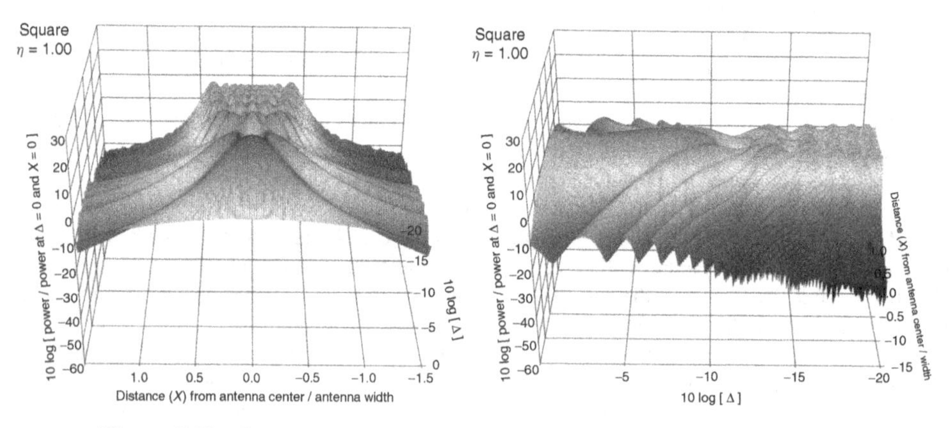

Figure 8.23 Square antenna near field power density ($\eta = 1.0$, $W/\lambda = 66$).

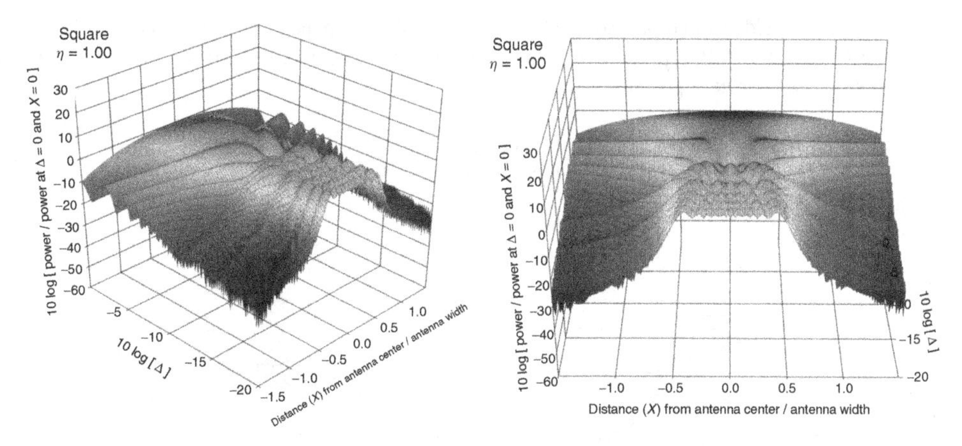

Figure 8.23 (*Continued*)

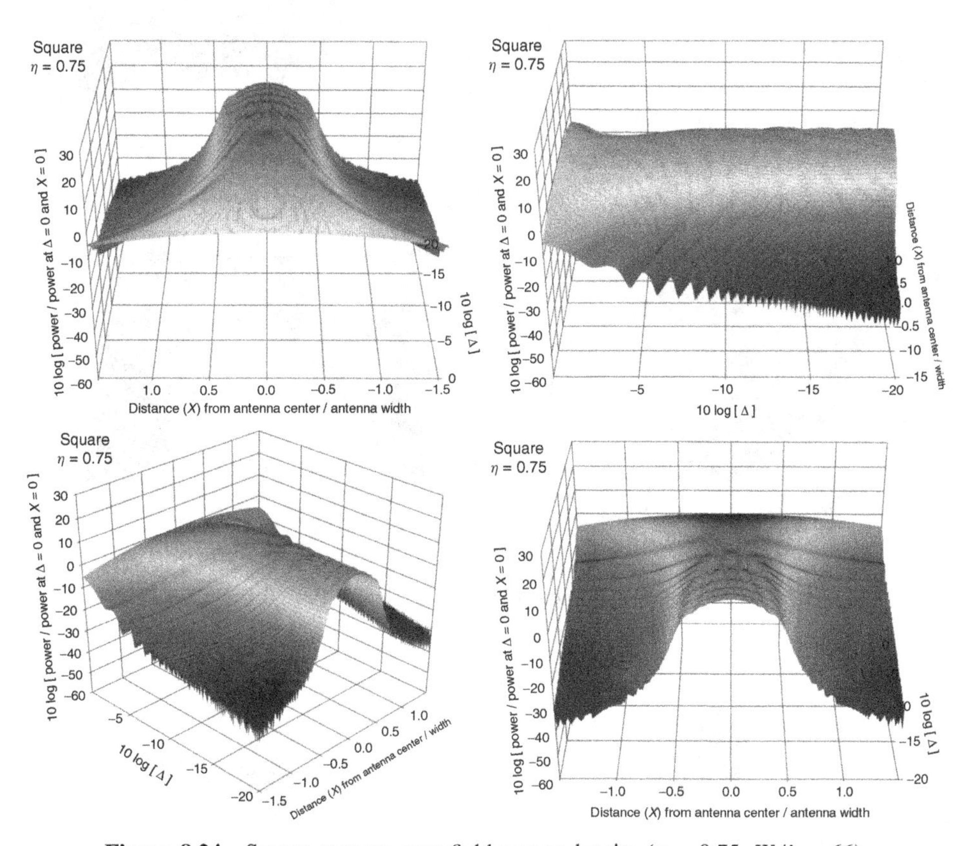

Figure 8.24 Square antenna near field power density ($\eta = 0.75$, $W/\lambda = 66$).

field power densities oscillate with periods as a function of (D/λ). The peaks are the same for the different values of (D/λ) but occur for different values of normalized distance, Δ. For values of (D/λ) approaching 1, the near field power densities are smaller than larger values of (D/λ). This effect will be ignored in the following estimates. In the worst case, these calculations will be a little conservative for those antennas. Figure 8.35 graphs the worst case (peak) near field power density as a function of efficiency and distance.

For simple near field estimations somewhere between the deep near field and the far field, one can calculate $(-2\Delta_{\text{dB}})$ and $P_{\text{NNFL}}(\text{dB})$. The smaller of the two values represents a simple straight line approximation to the near field power density.

$$10\ \log_{10}\left[\frac{S\,(\Delta)}{S(\Delta = 1)}\right] = -2\ \Delta_{\text{dB}} \tag{8.58}$$

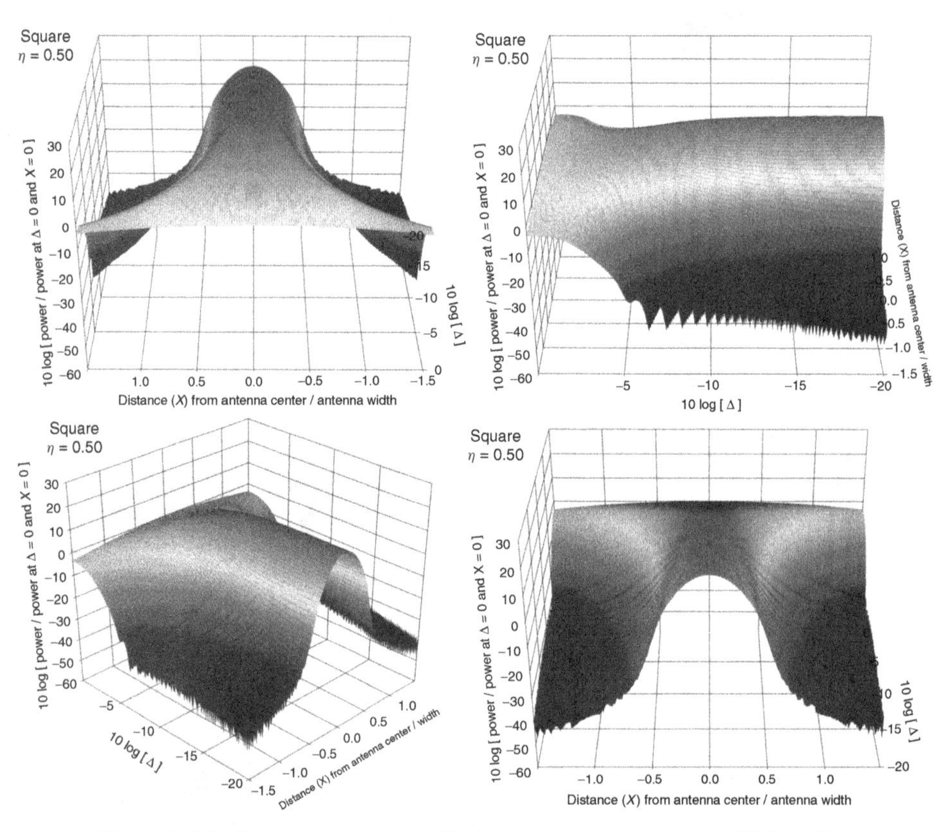

Figure 8.25 Square antenna near field power density ($\eta = 0.5$, $W/\lambda = 66$).

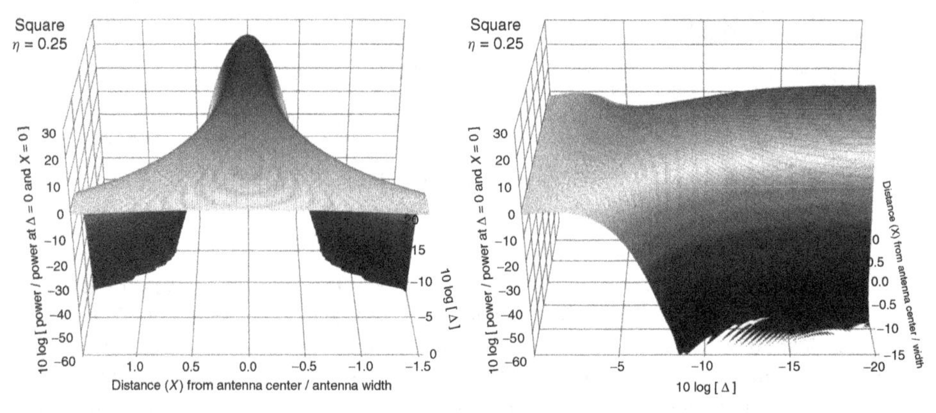

Figure 8.26 Square antenna near field power density ($\eta = 0.25$, $W/\lambda = 66$).

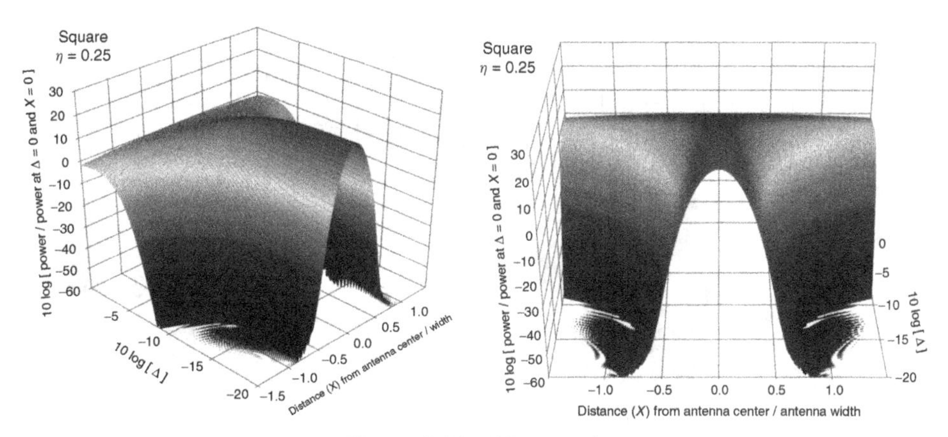

Figure 8.26 *(Continued)*

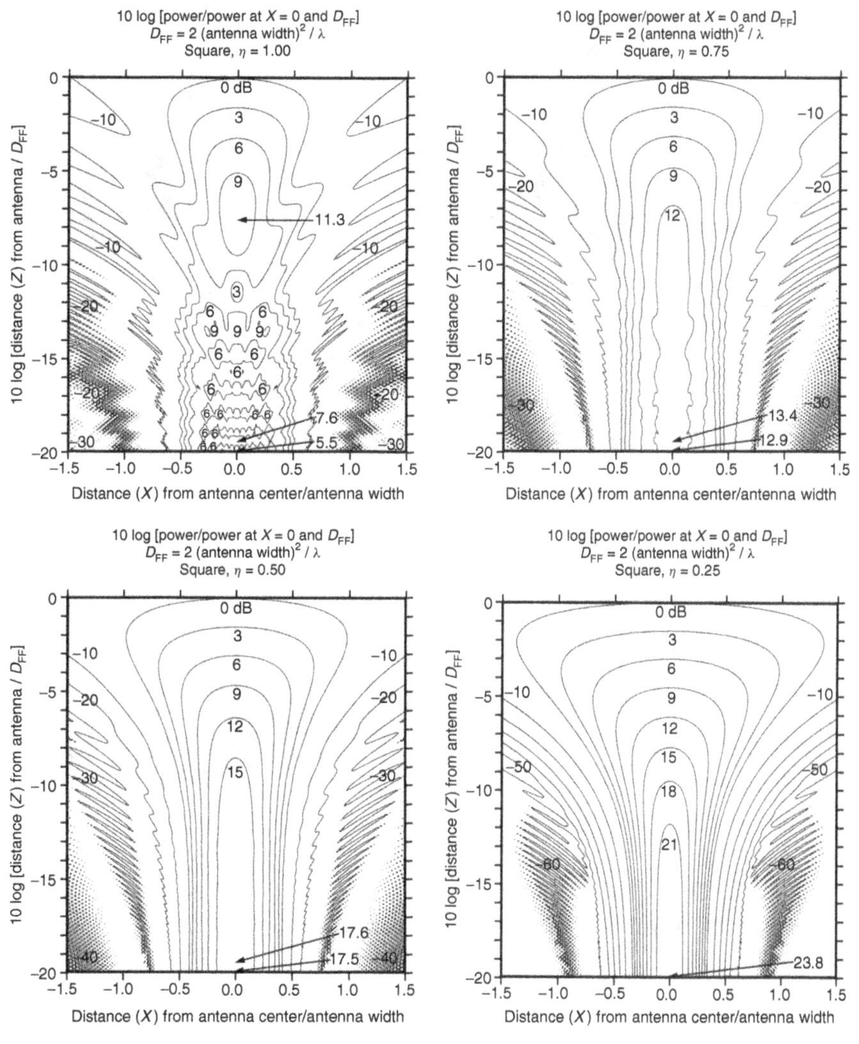

Figure 8.27 Square antenna near field power density ($W/\lambda = 66$).

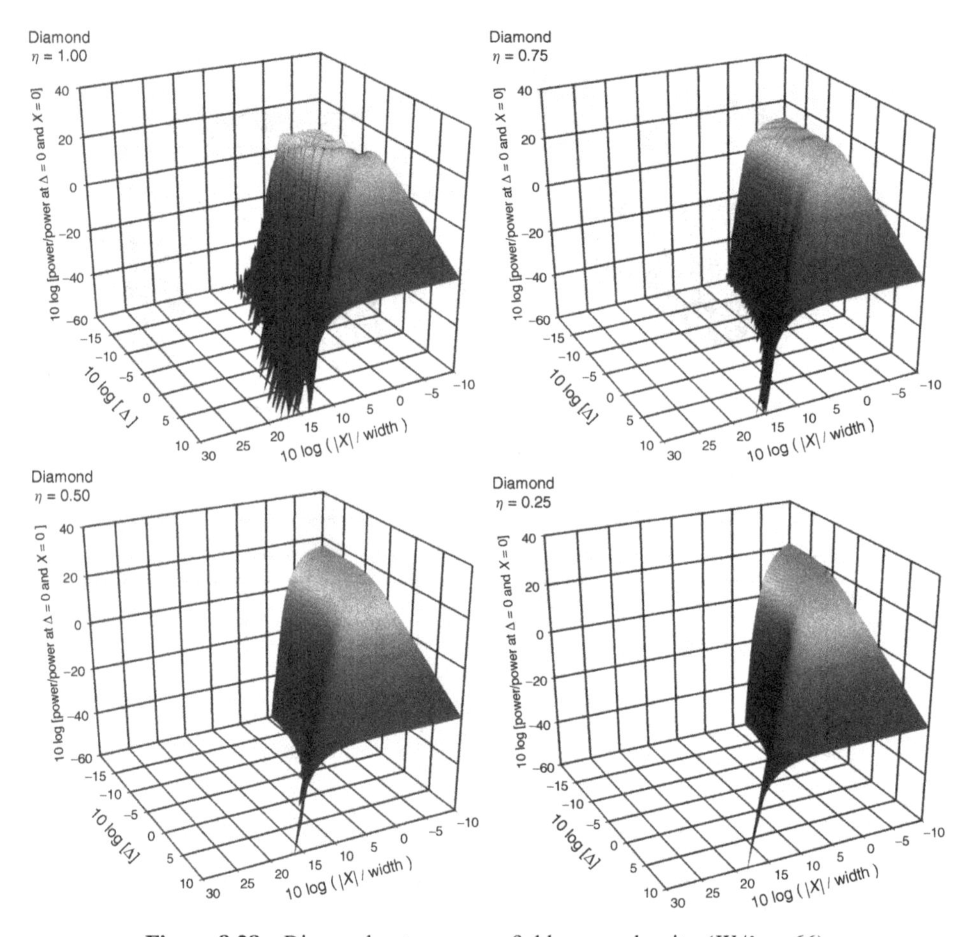

Figure 8.28 Diamond antenna near field power density ($W/\lambda = 66$).

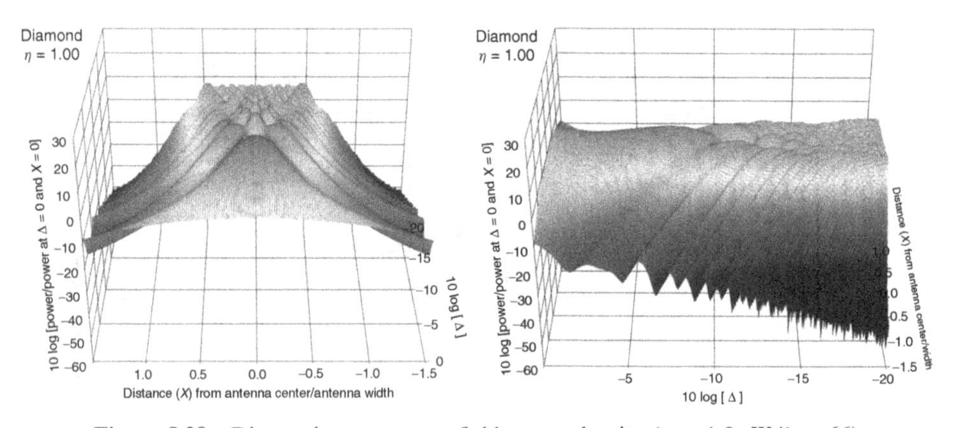

Figure 8.29 Diamond antenna near field power density ($\eta = 1.0$, $W/\lambda = 66$).

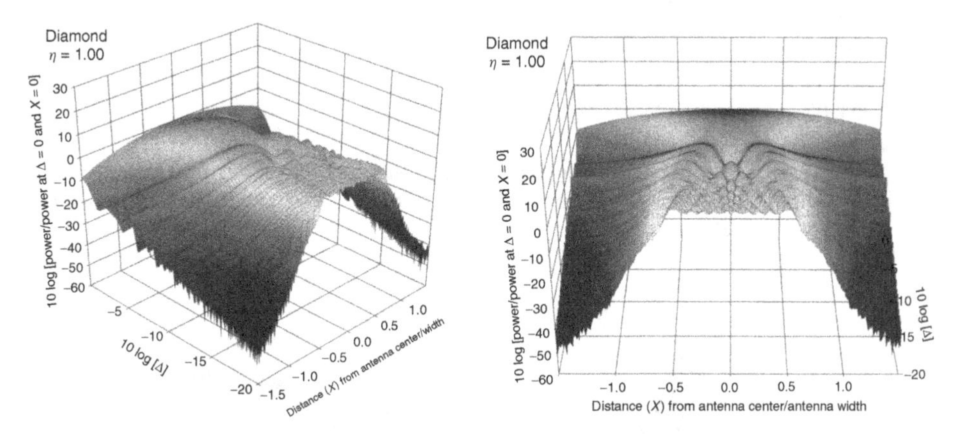

Figure 8.29 (*Continued*)

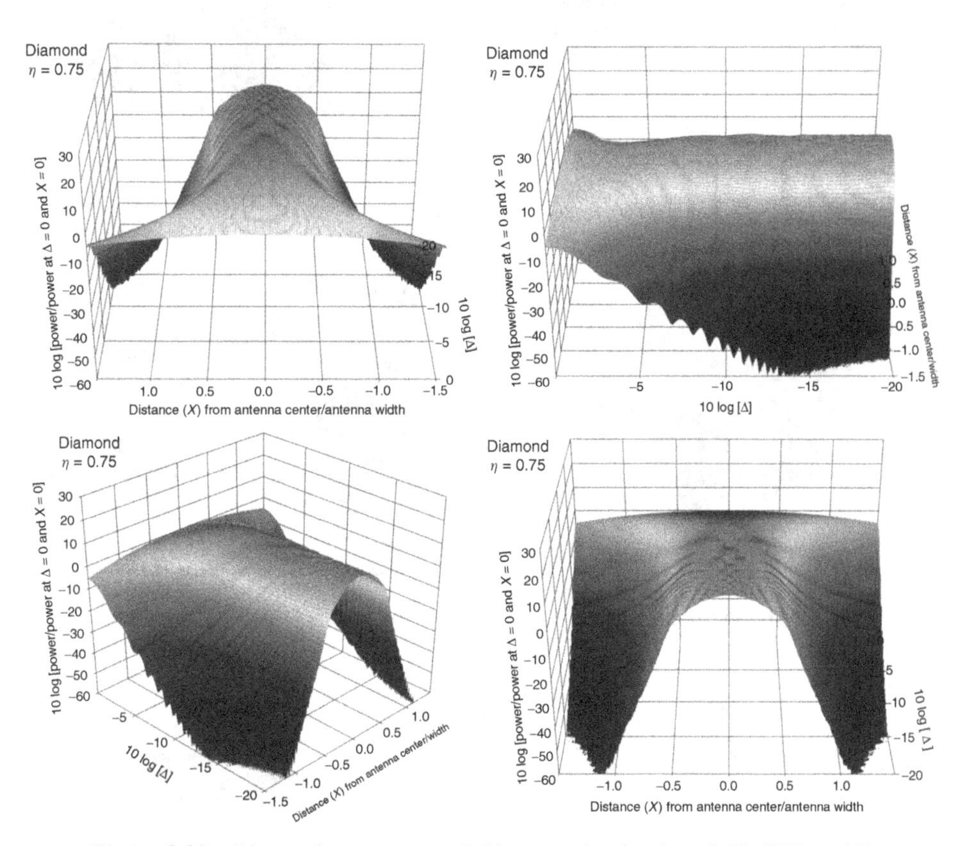

Figure 8.30 Diamond antenna near field power density ($\eta = 0.75$, $W/\lambda = 66$).

or

$$10 \log_{10}\left[\frac{S(\Delta)}{S(\Delta = 1)}\right] = A + B\eta + \frac{C}{\eta} + D\eta^2 + \frac{E}{\eta^2} + F\eta^3 \tag{8.59}$$

$A = 34.223061$;
$B = -58.288613$;

$C = 0.51017224;$
$D = 64.124471;$
$E = -0.013593334;$
$F = -29.354905.$

Both formulas are evaluated. Eqn. 8.58 describes a diagonal line for the far field power. Eqn. 8.59 defines a horizontal line for the near field power. The lower power is the power estimate normalized to the far field crossover power.

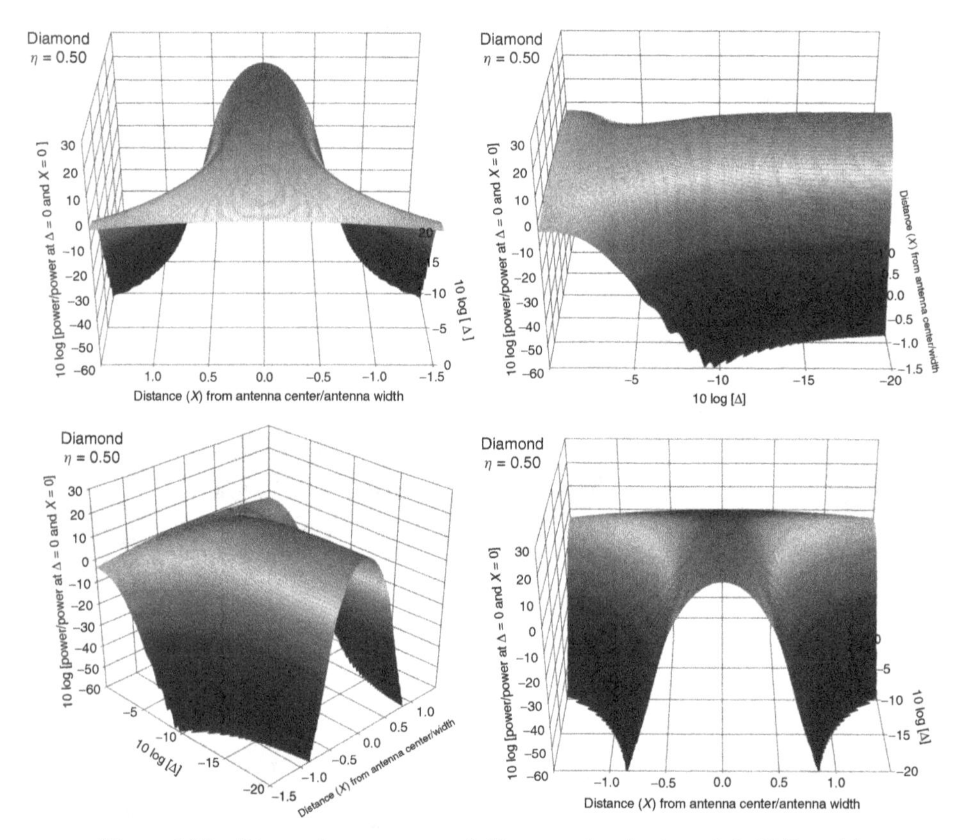

Figure 8.31 Diamond antenna near field power density ($\eta = 0.5$, $W/\lambda = 66$).

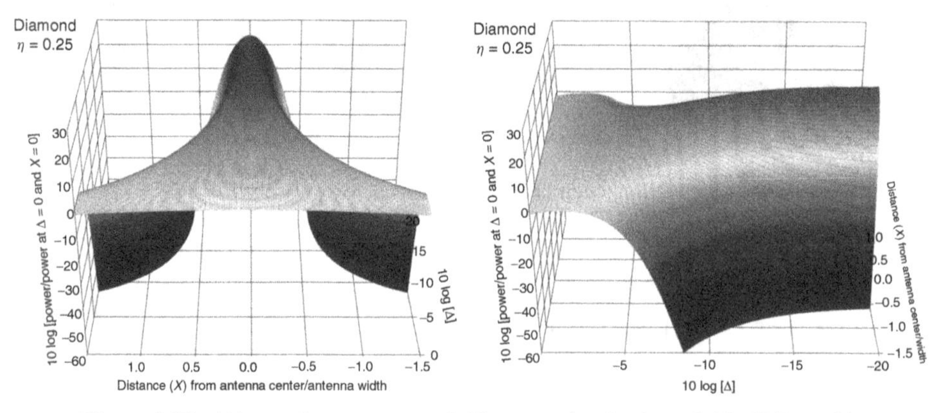

Figure 8.32 Diamond antenna near field power density ($\eta = 0.25$, $W/\lambda = 66$).

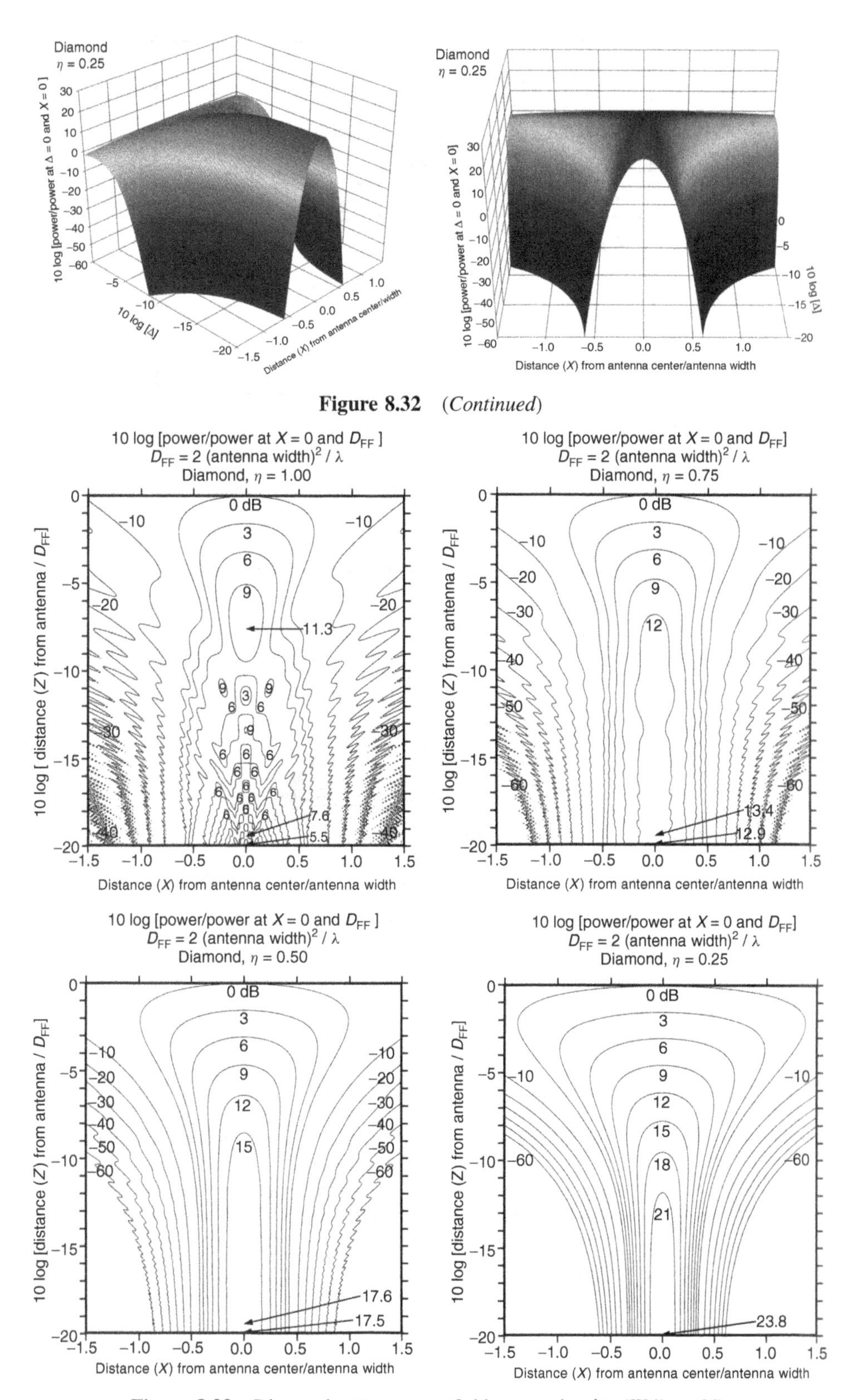

Figure 8.32 (*Continued*)

Figure 8.33 Diamond antenna near field power density ($W/\lambda = 66$).

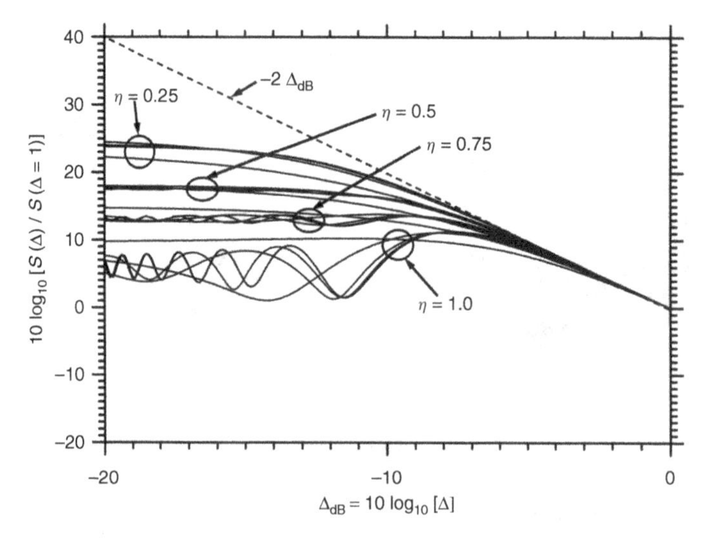

Figure 8.34 Square antenna near field power density.

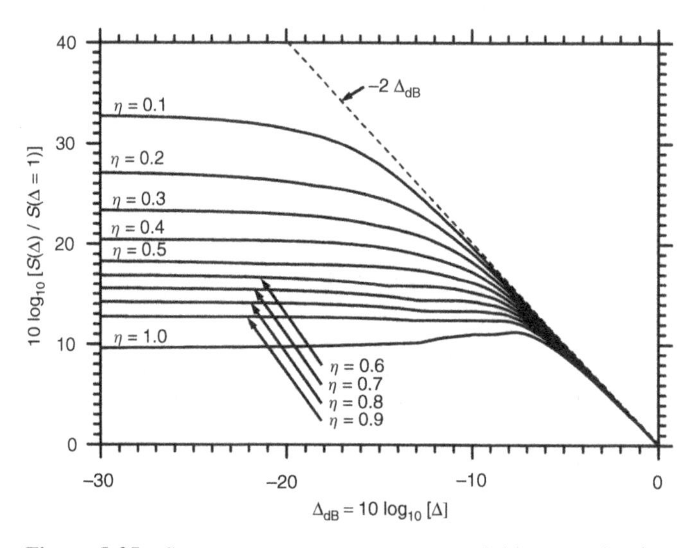

Figure 8.35 Square antenna worst case near field power density.

8.4.3 Square Antenna Far Field Reference Power

Near field power density results are normalized to antenna power S at $\Delta = 1$ (essentially the historical "far field transition point"). This is the power density value at the far field transition distance, $d = (2W^2/\lambda)$. It is necessary to determine that power. Microwave radio antenna gain is normally specified relative to an isotropic radiator (a hypothetical antenna that radiates uniformly in all directions).

$$S \text{ (power density)} = \frac{pg}{\text{Surface area of sphere with radius } d} = \frac{pg}{4\pi d^2} \tag{8.60}$$

At the far field transition point ($\Delta = 1$), $d = (2W^2)/\lambda$.

Therefore,

$$S(\Delta = 1) = \frac{pg\lambda^2}{16\pi W^4} \tag{8.61}$$

Since most power density safety limits are specified in milliwatts (mW) per centimeter (cm), p is in milliwatts, D and λ are in centimeters, and f is the radio frequency.

$$\lambda(\text{cm}) = \frac{29.98}{f}(\text{GHz}) \tag{8.62}$$

$$W(\text{cm}) = 30.48W(\text{ft}) \tag{8.63}$$

Antenna isotropic gain, g, (Kizer, 1990) is given by

$$g = \frac{4\pi A\eta}{\lambda^2} = \frac{4\pi W^2\eta}{\lambda^2} \tag{8.64}$$

Substituting this into Equation 8.60 yields the far field (to near field transition) reference power:

$$S(\Delta = 1) = \frac{p\eta}{4W^2} \tag{8.65}$$

The illumination efficiency, η, can be estimated on the basis of results in the paragraph on square antenna beamwidth. The final result is to use this value to calculate the nearfield power.

$$\left[\frac{S(\Delta)}{S(\Delta = 1)}\right] = 10^{\{10 \log [S(\Delta)/S(\Delta=1)]\}/10} \tag{8.66}$$

$$S(\Delta) = \left[\frac{S(\Delta)}{S(\Delta = 1)}\right] S(\Delta = 1) \tag{8.67}$$

Worst case near field power is calculated at the worst case location (boresight at the reactive near field to Fresnel field transition point of $d = \lambda$). Therefore $S(\Delta)$ is calculated at the following Δ.

$$\Delta = \frac{1}{\left[2\left(\frac{W}{\lambda}\right)^2\right]} \tag{8.68}$$

Example calculations are in 8.6.2.

8.4.4 Square Near Field Power Density Transitions

Figure 8.36 shows curves representing a 1 dB deviation for the far field or near field limits based on the previous numerical results for square antennas (strictly speaking, the traditional far field transition point uses (sqrt (2) W instead of W but we will ignore this nuance). The far field limit is the conventional inverse of the distance squared solution. The near field is defined as the constant (peak) value very close to the antenna. For the square aperture, the very near field reaches a peak and then becomes slightly less ("fades") as the measurement is made closer to the antenna. The peak value is the value used for the near field transition calculations.

As with circular antennas, the historical transition distance formulas are of limited value as they ignore the effect of aperture shape and illumination efficiency, η. If the far field transition is defined as the point at which the square antenna boresight gain is 1 dB less than the far field gain for that distance, the following formula may be used to estimate that limit:

$$\Delta_{\text{dB}} \text{ (far field transition distance)} = -8.544 + 6.188\eta - 1.954\eta^2 - \frac{0.5349}{\eta} \tag{8.69}$$

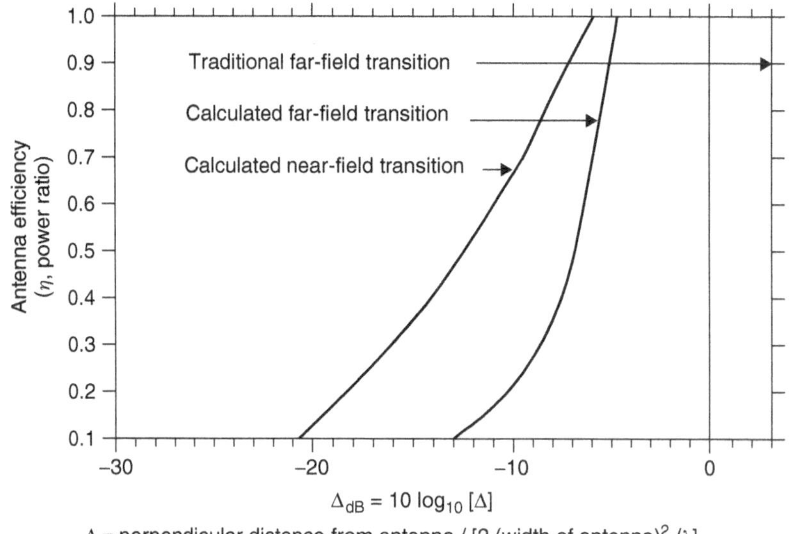

Figure 8.36 Square antenna near field transition regions.

8.5 REGULATORY NEAR FIELD POWER DENSITY LIMITS

The FCC (Cleveland et al., 1997) requires an uncontrolled exposure (general population) limit of 1 mW/cm^2 and controlled exposure (occupational) limit of 5 mW/cm^2 for frequencies greater than 1.5 GHz. A recent study (Zhadobov et al., 2009) reaffirmed that high frequency rf energy no greater than the FCC 1-mW/cm^2 limit poses no health hazard. The IEEE (IEEE International Committee on Electromagnetic Safety (SCC39), 2005) limit in an uncontrolled exposure (Table 9, general public) environment is 1 mW/cm^2 for frequencies between 2 and 100 GHz. The IEEE (IEEE International Committee on Electromagnetic Safety (SCC39), 2005) controlled exposure (Table 8, environments controlled per IEEE Std. C95.7-2005) limit is 10 mW/cm^2 for frequencies between 3 and 100 GHz The OSHA limit (United States Federal Government, 2004) is 10 mW/cm^2. For those wishing to understand the history of these limits, Steneck's book (Steneck, 1984) is highly recommended. Guy (1984) wrote an excellent historical overview of the biological effect of microwave energy. The IEEE standard (IEEE International Committee on Electromagnetic Safety (SCC39), 2005) provides an extensive summary of the research used to establish the exposure limits. Sanchez-Hernandez (2009) provides a modern overview of the subject.

The following examples plot the energy around an antenna were the worst case energy is the regulatory limit of 1 mw/cm. The circular antenna examples assume that the antenna (with $D/\lambda = 66$) has been supplied with exactly the amount of power to create 1 mW (1000 μW/cm^2) worst case near field power density (Fig. 8.37 and Fig. 8.38). Likewise, the square antenna examples assume that the antenna (with $W/\lambda = 66$) has been supplied with exactly the amount of power to create 1 mW (1000 μW/cm^2) worst case near field power density (Fig. 8.39, Fig. 8.40, and Fig. 8.41).

8.6 PRACTICAL NEAR FIELD POWER CALCULATIONS

Currently, several different methods are being used to calculate worst case antenna near field power density. A common system engineering approach (Editorial Board, 1968) for circular and square antennas estimates the power density directly in front of the antenna as $4P/A$ (P is the power at antenna input and A is antenna area). This approach calculates the correct near field power density for the fully illuminated

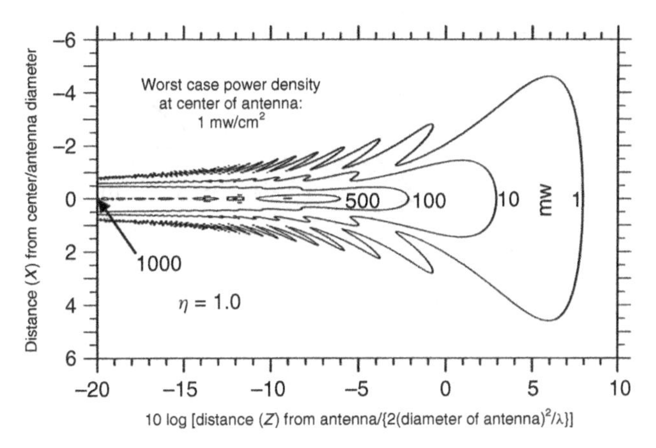

Figure 8.37 Circular antenna power density contours.

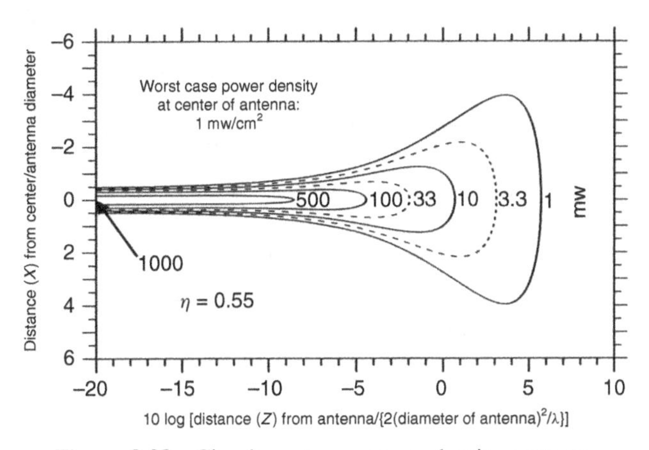

Figure 8.38 Circular antenna power density contours.

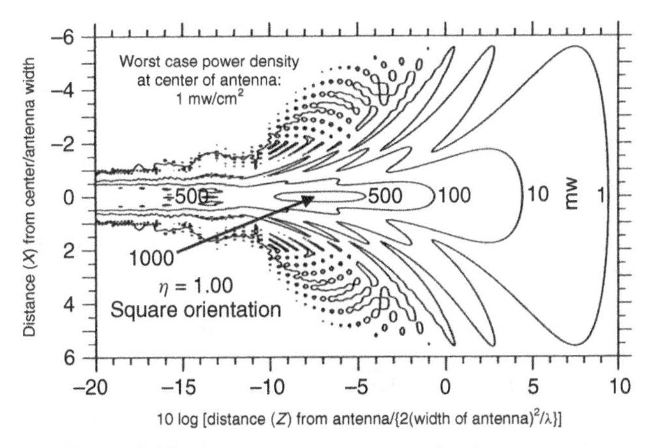

Figure 8.39 Square antenna power density contours.

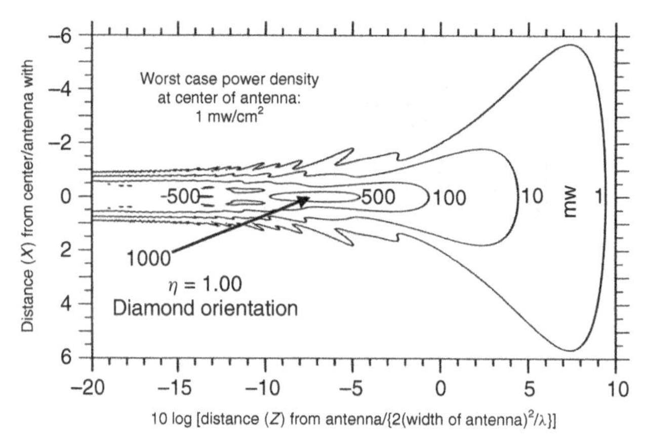

Figure 8.40 Square antenna power density contours.

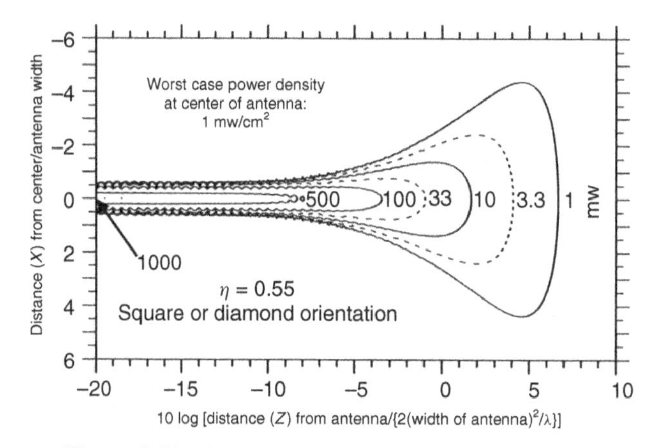

Figure 8.41 Square antenna power density contours.

($\eta = 1$) circular antenna but overestimates the power density for a fully illuminated square antenna by 3.5 dB. It gives no guidance for lower efficiency ($\eta < 1$) antennas. A variation of this approach (Cleveland et al., 1997, Eq. 13) attempts to compensate for illumination efficiency by multiplying the $4P/A$ result by η, the antenna illumination efficiency factor. This misinterprets the η factor as a power loss rather than as a power density concentration factor. Although it gives the same result as the first approach for uniform illumination, it underestimates the power density for the $\eta = 0.5$ case (typical for commercial parabolic antennas) by 6 dB for a circular antenna and 5 dB for a square antenna.

Bickmore and Hansen (1959) calculated antenna near field power density for a fully illuminated square antenna. Their result gives no guidance for illumination efficiencies η other than unity. They (Bickmore and Hansen, 1959) also calculated power density for a circular antenna with uniform or linearly tapered antenna illumination power. These antenna illuminations are not typical of commercial antennas and either *overestimate* or *underestimate* the actual near field power value of an antenna. The Bickmore and Hansen results are sometimes used for circular antennas (Saad et al., 1971).

Currently, the most accurate method for circular antennas is Hansen's work (Hansen, 1976b). Unfortunately, the published integral formula has typographical errors and the numerical results provided are not typical of commercial antennas. When the typographical errors are corrected, Hansen's approach is accurate for the limiting case of large circular antennas. However, it is very difficult to calculate, only represents the limiting large antenna case, and places the peak and null values in the wrong location for small- and medium-sized antennas.

The antenna near field power density simplified calculations in the previous paragraphs overcome these limitations. They provide simple, easy to use, accurate formulas. Since they are straight line worst case estimates, they are conservative in the transition region between the near and far fields. They also smooth over the power nulls so the engineer does not need to worry about them.

8.6.1 A Parabolic Antenna Near Field Power Example Calculation

Let us consider a 30-dBm 6.2-GHz microwave transmitter feeding a 10-ft circular parabolic antenna with an isotropic gain of 43 dBi.

$p(\text{mW}) = 10^{30/10} = 1000$ mW;
$g(\text{pr}) = 10^{43/10} = 19{,}953$;
$\lambda(\text{cm}) = 29.98/f(\text{GHz}) = 29.98/6.2 = 4.835$ cm.

$$D(\text{cm}) = 30.48\ D(\text{ft}) = 30.48\text{x}10 = 304.8\ \text{cm}$$

$$10\ \log(\eta) = -10.1 + G(\text{dB}) - 20\ \log[f(\text{GHz})] - 20\ \log[D(\text{ft})]$$

$$= -10.1 + 43.0 - 20\ \log(6.2) - 20\ \log(10) = -2.948$$

$$\eta = 10^{[\ 10\ \log\ (h)]\ /10} = 0.5072$$

The power density at the far field transition ($\Delta = 1$) point $d = (2\ D^2)/\lambda$ is given by

$$S(\Delta = 1) = \frac{pg}{4\pi d^2} = \frac{[p(\text{mW})\ g\lambda(\text{cm})^2]}{[16\pi D(\text{cm})^4]}$$

$$= \frac{[1000\text{x}19{,}953(4.835)^2]}{[16\text{x}\pi(304.8)^4]} = 0.001075\ \text{mW/cm}^2 \tag{8.70}$$

From the near field limit equation, the normalized near field power density near antenna limit is given by

$$P_{\text{NNFL}}(\text{dB}) = 10\ \log\left[\frac{S(D \ll 0.1)}{S(D = 1)}\right]$$

$$= A + B\eta + \frac{C}{\eta} + D\eta^2 + \frac{E}{\eta^2} + F\eta^3$$

$$= 40.43 - (61.48\ \text{x}\ 0.5072) - \left(\frac{0.4669}{0.5072}\right)$$

$$+ (55.38\ \text{x}\ 0.5072^2) + \left(\frac{0.04791}{0.5072^2}\right) - (19.81\text{x}\ 0.5072^3)$$

$$= 20.19 \tag{8.71}$$

$A = 40.430453$;
$B = -61.480406$;
$C = -0.46691971$;
$D = 55.376708$;
$E = 0.04791274$;
$F = -19.805638$.

$$\frac{S(\Delta \ll 0.1)}{S(\Delta = 1)} = 10^{\text{PNNFL}/10} = 104.5(\text{power ratio}) \tag{8.72}$$

The (worst case) near field power density is the given by the following product:

$$S(\Delta \ll 0.1) = \left[\frac{S(\Delta \ll 0.1)}{S(\Delta = 1)}\right]\ \text{x}\ [S(\Delta = 1)]$$

$$= 0.001075\text{x}104.5 = 0.1123\ \text{mW/cm}^2 \tag{8.73}$$

This is significantly lesser than the FCC (Cleveland et al., 1997), IEEE (IEEE International Committee on Electromagnetic Safety (SCC39), 2005), and OSHA (United States Federal Government, 2004) limits for uncontrolled areas of 1, 4, and 10 mW/cm^2, respectively.

As will become obvious as one gains experience with these calculations, the highest near field power densities are for small antennas, not large ones. This is counter-intuitive but clear when it is realized that the same power concentrated in a smaller area must produce higher near field power densities.

8.6.2 Safety Limits

It is desirable to estimate the power into the antenna that does not exceed the worst case (FCC) statutory power density limit.

8.6.2.1 Circular Antenna

$$\begin{aligned} S(\Delta \ll 0.1) &= (10^{\text{PNNFL(dB)}/10})\text{x}[S(\Delta = 1)] = 1 \\ &= (10^{\text{PNNFL(dB)}/10})\frac{(\pi \eta p)}{(16D^2)} \end{aligned} \tag{8.74}$$

$$p(\text{mW}) = \frac{16(30.48^2 D^2)}{[\pi \eta (10^{\text{PNNFL(dB)}/10})]}, \quad D(\text{ft}) \tag{8.75}$$

$$\begin{aligned} P(\text{dBm}) &= 36.75 + 20\ \log\,[D(\text{ft})] - 10\ \log(\eta) - \text{PNNFL(dB)} \\ &= 47.07 + 20\ \log[D(\text{m})] - 10\ \log(\eta) - \text{PNNFL(dB)} \end{aligned} \tag{8.76}$$

For the typical $\eta = 0.55$ (55% illumination efficiency) and D in ft,

$$\begin{aligned} P_{\text{NNFL}}(\text{dB}) &= 40.43 - (61.48\text{x}0.55) - \left(\frac{0.4669}{0.55}\right) \\ &\quad + (55.38\text{x}0.55^2) + \left(\frac{0.04791}{0.55^2}\right) - (19.81\text{x}0.55^3) \\ &= 19.38 \end{aligned} \tag{8.77}$$

$$\begin{aligned} P(\text{dBm}) &= 36.75 + 20\ \log\,[D(\text{ft})] + 2.60 - 19.38 \\ &= 19.97 + 20\ \log\,[D(\text{ft})] \end{aligned} \tag{8.78}$$

8.6.2.2 Square Antenna

$$\begin{aligned} S(\Delta \ll 0.1) &= (10^{\text{PNNFL(dB)}/10})\text{x}[S(\Delta = 1)] = 1 \\ &= (10^{\text{PNNFL(dB)}/10})\frac{(\eta p)}{(4W^2)} \end{aligned} \tag{8.79}$$

$$p(\text{mW}) = \frac{4(30.48^2 W^2)}{[\eta (10^{\text{PNNFL(dB)}/10})]}, \quad W \text{ in ft} \tag{8.80}$$

$$\begin{aligned} P(\text{dBm}) &= 35.70 + 20\ \log[W(\text{ft})] - 10\ \log(\eta) - \text{PNNFL(dB)} \\ &= 46.02 + 20\ \log[W(\text{m})] - 10\ \log(\eta) - \text{PNNFL(dB)} \end{aligned} \tag{8.81}$$

For the typical $\eta = 1.0$ (100% illumination efficiency) and W in ft

$$P_{\mathrm{NNFL}}(\mathrm{dB}) = 34.22 - (58.29\mathrm{x}1.00) + \left(\frac{0.51}{1.00}\right) + (64.12\mathrm{x}1.00^2) - \left(\frac{0.01}{1.00^2}\right) - (29.35\mathrm{x}1.00^3) = 11.20 \quad (8.82)$$

$$P(\mathrm{dBm}) = 35.70 + 20\ \log[W(\mathrm{ft})] + 0.0 - 11.20 = 24.50 + 20\ \log[W(\mathrm{ft})] \quad (8.83)$$

The above may be summarized in Table 8.1:

Table 8.1 illustrates the nonintuitive result that the smaller the antenna, the greater the worst case near field power density. Also, the worst case power density is not a function of frequency or antenna gain—just size and illumination efficiency (how densely the energy is spread across the aperture of the antenna). Table 8.1 applies to virtually all commercial antennas that use fixed point-to-point microwave radio applications.

The above results assume that a near field exists. For very small antennas, the near field may not exist. The far field may begin near the transition from the reactive near field (less than one wavelength from a square antenna). This situation exists for very small square antennas currently used in the unlicensed bands in the United States. Again, calculations are made for these small antennas in Table 8.2.

Very small antennas (with (D or W)/λ < 2) may have no radiating near field and the maximum radiating energy may occur in the far field. We will automatically consider this if we always perform the near field power density calculations at $d = \lambda$.

TABLE 8.1 Maximum Antenna Input Power (dBm) Not Exceeding the FCC Power Density Limit of 1 mW/cm²

Diameter or Width, ft	Parabolic Antenna (55% Efficiency)	Square Antenna (100% Efficiency)
0.25	7.9	12.4
0.5	13.9	18.5
1	19.9	24.5
2	26.0	30.5
3	29.5	34.0
4	32.0	36.5
6	35.5	40.0
8	38.0	42.5
10	39.9	44.5
12	41.5	46.1
15	43.5	48.0

TABLE 8.2 Maximum Antenna Input Power (dBm) Not Exceeding the FCC Power Density Limit of 1 mW/cm² for Very Small Panel Antennas

Width, ft	2.4 GHz	5.2 GHz	5.8 GHz
0.25	26.2 dBm	12.8 dBm	12.4 dBm
0.5	20.2 dBm	18.5 dBm	18.5 dBm
1	24.5 dBm	24.5 dBm	24.5 dBm

8.7 NEAR FIELD ANTENNA COUPLING LOSS

Antennas and passive reflectors are used to transmit radio frequency signals between transmitting and receiving locations. Usually, these locations are so far apart that for a constant transmitted power the received power is reduced as a function of distance (between transmit and receive antennas) squared (Friis, 1946) (the "far field" case). As the antennas are moved closer together, the received energy is no longer entirely perpendicular to the receiving antenna's aperture. The phase errors associated with this situation cause a loss in received power when compared to the far field case. Owing to the complexity of this situation, numerical techniques will be used to estimate this near field antenna coupling loss.

8.7.1 Antenna to Antenna Near Field Coupling Loss

When antennas are placed near each other, their boresight gain may be reduced. This is due to the astigmatism introduced by the nearness of the two apertures. The parallel ray assumption of the far field gain derivation becomes progressively in error as the antennas come closer together. The amount of this gain loss is difficult to estimate. This problem has been attempted by several researchers (Hu, 1958; Pace, 1969; Robieux, 1959; Takeshita and Sakurai, 1964). Sometimes, only complex integral results were offered. Typically, antenna illumination functions of second order exponential or various parabolic functions were assumed. These functions did not represent practical antenna illumination functions. Usually, results were obtained from numerical integration over a restricted range. Pace (1969) provided a truncated series approximation that provided satisfactory results for the uniform illumination with one antenna much larger than the other. For nonuniform illumination and similar sized antennas, his results diverged too rapidly to be of practical use. Kay (1960) used Bessel functions with the edge pedestal. He offered a range of results. The relation between the results and actual antennas was not obvious.

The close-coupled antenna loss problem will be approached in a general way. The close-coupled loss will be the difference in calculated composite gain of both antennas relative to the composite far field gain of the two antennas. Hansen's one-parameter illumination function will be used. Assume a left transmitting antenna aperture described by points x_1 and y_1 in an X–Y plane. Assume an orthogonal Z-axis centered on the center of the antenna aperture. This defines a horizontal X–Z plane intersecting the antenna at $z = 0$. A point (x,y,z) on this antenna is $(x_1, y_1, 0)$. Assume a right receiving antenna aperture centered on the Z-axis and parallel to the left antenna aperture. A point on that antenna aperture is (x_2, y_2, z_2) (Fig. 8.42).

Using the Huygens–Fresnel principle and the aperture-field method, Kay (1960) and Silver (1949) showed that the power transmitted between two antennas in free space is the integral of the energy field created by the transmitting antenna at each point of the receiving antenna times the illumination function of the receiving antenna at that point.

Silver (1949) described the transmitting antenna induced near field U_p created at point P (x_2, y_2, z_2) in very general terms. Assuming the transmit and receive antenna (field) illumination functions F_1 and

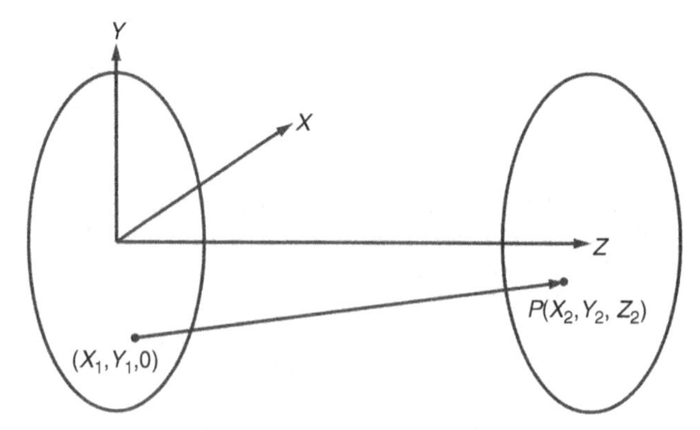

Figure 8.42 Antenna coupling loss geometry.

F_2 have constant unity phase and applying Euler's identity, Silver's results may be used to determine the total energy received (U_r):

$$U_r = \text{constant} \int_{A_2} F_2(x_2 y_2) \int_{A_1} F_1(x_1, y_1) \left(\frac{1}{\delta}\right)(B + jC)\mathrm{d}x_1 \mathrm{d}y_1 \mathrm{d}x_2 \mathrm{d}y_2 \tag{8.84}$$

The integrals are over the entire aperture of both antennas.

A_1 and A_2 represent the areas of the respective antenna apertures;
$F_1(x_1, y_1)$ = the field illumination function of the transmit antenna;
$F_2(x_2, y_2)$ = the field illumination function of the receive antenna;
$B = (1 + \cos\phi)\ \sin\delta + [(\cos\phi)/\delta]\cos\delta$;
$C = (1 + \cos\phi)\cos\delta - [(\cos\phi)/\delta]\sin\delta$;
$\delta = (2\pi r/\lambda)$.
r = distance from a point on the antenna to a point in free space;
= $\mathrm{sqrt}[(x_2 - x_1)^2 + (y_2 - y_1)^2 + {z_2}^2]$ with sqrt [ ] as the square root function;
ϕ = angle formed by Z-axis and a ray from a point on the antenna to a point in free space;
$\cos(\phi) = z_2/r$;
U_r = total energy received by right antenna;
$|U_r|^2$ = total received power.

By comparing the calculated received power to the expected far field result, the composite antenna to antenna coupling loss may be derived. To test the validity of the methodology, calculations ($\eta = 0.5$) were compared with measured values (Yang, R. F. H. 1964, Private communication).

From Figure 8.43, we can see that the calculated and measured values are in reasonable agreement. Using this approach, the coupling loss between two circular parabolic and two square antennas was calculated (Fig. 8.44).

Before moving forward, a word of caution is in order. Silver (1949) observed that as antennas are located close together, they interact because of reflections from the antenna themselves. He observed that two similar parabolic antennas positioned at the traditional far field transition point could experience 0.25-dB composite gain variation due to reflection interaction. Closer antennas would experience more. In general, Silver's results for two similar parabolic antennas equated (Kizer, 1990) to a potential gain variation ΔG of the following:

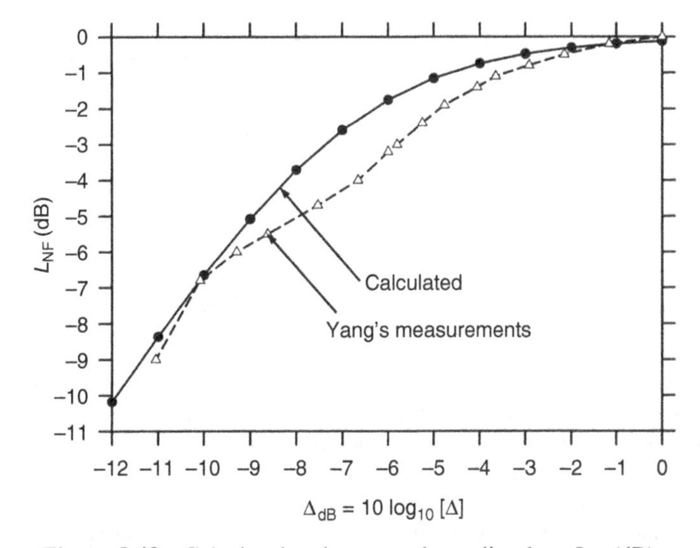

Figure 8.43 Calculated and measured coupling loss L_{NF}(dB).

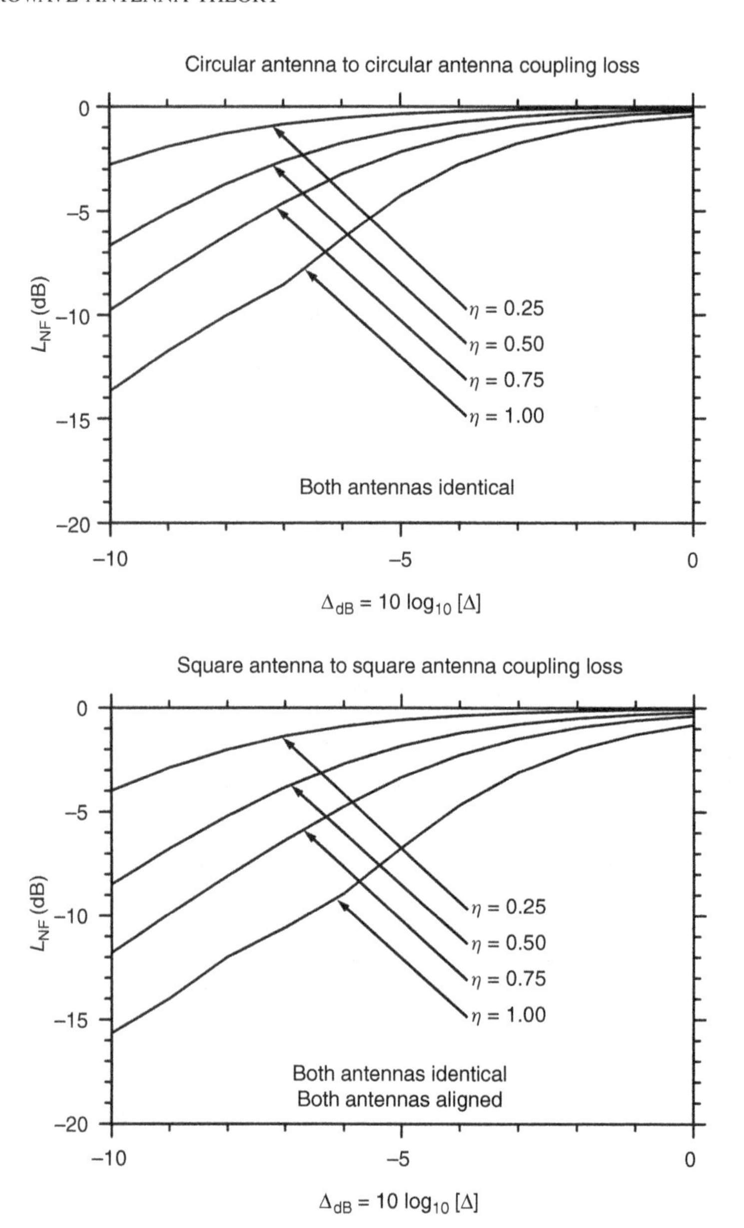

Figure 8.44 Coupling loss L_{NF}(dB) between identical antennas.

$$\Delta G(\text{dB}) = \text{Close antenna gain variation} = 20 \log\left[1 + \left(\frac{0.03}{\Delta^2}\right)\right] \tag{8.85}$$

This interaction is so large that it can significantly affect closely coupled antenna results. This variation in the near field suggests that if closely coupled antennas are required, the use of significantly different-sized antennas may be necessary to avoid significant interaction.

Multiple reflections from closely coupled antennas can produce significant variation in gain but can also produce multiple reflected signals similar to those produced by poor return loss in waveguide (or coax) transmission lines. This suggests that these echo signals can cause unexpected residual bit error degradation in digital receivers. For this reason, very closely coupled antennas are not recommended.

Since reflectors are tilted relative to the active antenna (and therefore would not reflect much energy back toward the active antenna), this interaction should be much reduced relative to face to face antennas.

For the following sections, the parameter $\Delta(\text{dB}) = 10 \log [d/(2D^2/\lambda)] = 10 \log[d/(2W^2/\lambda)]$ is calculated using the D or W dimension for the *larger* antenna.

Figure 8.45 and Figure 8.46 display the relationship between coupling loss and the difference in size of antennas.

The following definitions or relationships are used to estimate near field coupling loss between antennas:

$$\Delta\text{dB} = 10 \log(\Delta), \quad -10 \le D = \Delta_{\text{dB}} \le 0,$$

$D = \Delta_{\text{dB}}$ = normalized distance between antennas;
$\Delta = d/(2D^2/\lambda)$ = normalized distance parameter for the larger circular antenna;
$\Delta = d/(2D^2/\lambda)$ = normalized distance parameter for the larger square antenna;
R = ratio of smaller antenna width/larger antenna width;

$$0 \le R \le 1;$$

$N = \eta$ = illumination efficiency;

$$0.25 \le \eta \le 1.0;$$

L_{NF} = near field antenna to antenna coupling loss (dB);
= value to be added to far field free space loss.

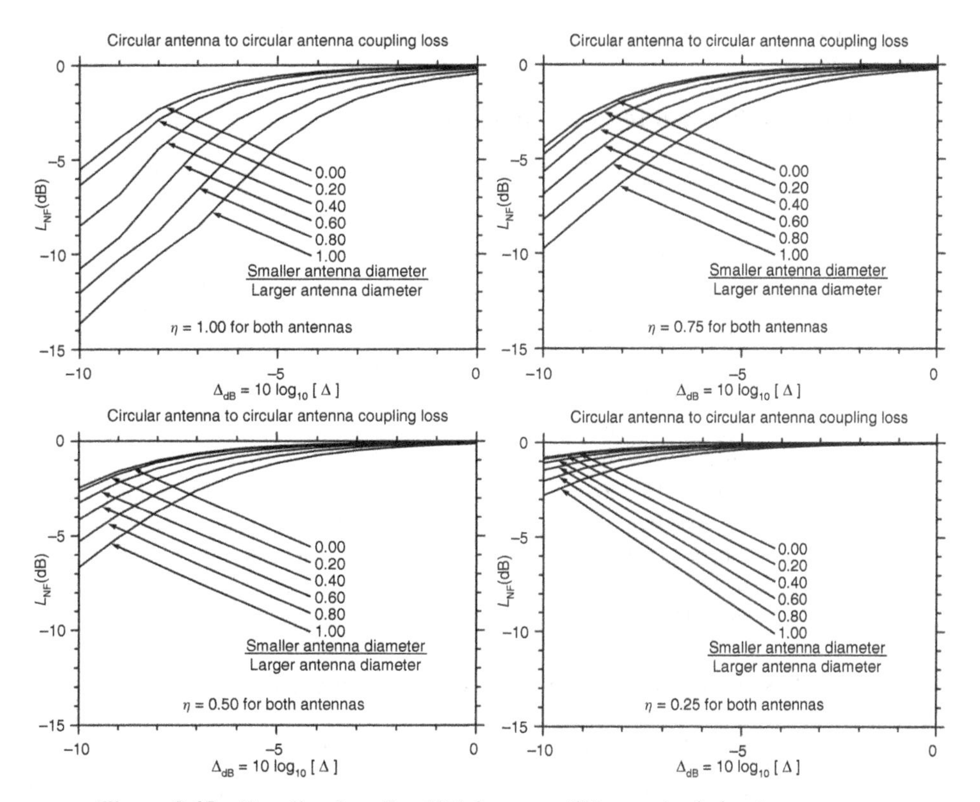

Figure 8.45 Coupling loss L_{NF}(dB) between different-sized circular antennas.

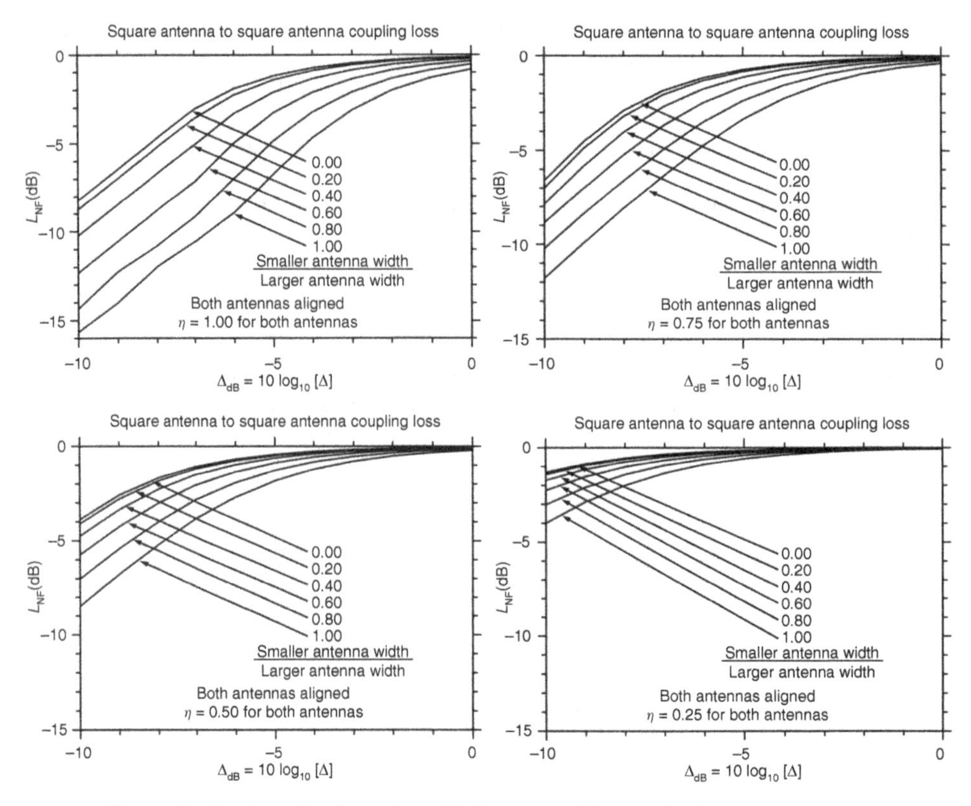

Figure 8.46 Coupling loss L_{NF}(dB) between different-sized square antennas.

8.7.2 Coupling Loss between Identical Antennas

$$L_{\mathrm{NF}} = C1 + C2 \times D + C3 \times N + C4 \times D^2 + C5 \times N^2 + C6 \times D \times N$$
$$+ C7 \times D^3 + C8 \times N^3 + C9 \times D \times N^2 + C10 \times D^2 \times N \tag{8.86}$$

Circular (Parabolic) Antennas	Square (Aligned) Antennas
$C1 = 0.5688330523739922$	$C1 = 0.246159837605786$
$C2 = -0.2843725447552475$	$C2 = -0.8316031494312162$
$C3 = -5.913339006420615$	$C3 = -6.285507130745876$
$C4 = -0.005544709076135127$	$C4 = -0.07920675867559671$
$C5 = 12.46763586356988$	$C5 = 15.99365443163881$
$C6 = 0.2328735832473644$	$C6 = 1.544481166675684$
$C7 = 0.00194088480963481$	$C7 = -0.0002143259518259517$
$C8 = -7.561944604459402$	$C8 = -10.80646533812489$
$C9 = 0.299596446278667$	$C9 = -0.1245258794169019$
$C10 = -0.0905373781148429$	$C10 = -0.02108477297350535$

8.7.3 Coupling Loss between Different-Sized Circular Antennas

$$L_{\mathrm{NF}} = C1 + C2 \times D + C3 \times R + C4 \times D^2 + C5 \times R^2 + C6 \times D \times R$$
$$+ C7 \times D^3 + C8 \times R^3 + C9 \times D \times R^2 + C10 \times D^2 \times R \tag{8.87}$$

$N = \eta = 1.00$	$N = \eta = 0.75$
$C1 = -0.7499780289155289$	$C1 = -0.2714292776667776$
$C2 = -0.5133628527953528$	$C2 = 0.02152531477781479$
$C3 = 3.911939824296074$	$C3 = 3.675637157518407$
$C4 = -0.07520881757131757$	$C4 = 0.0321249222999223$
$C5 = -6.863575712481961$	$C5 = -10.12064281204906$
$C6 = 0.6943731306193806$	$C6 = -0.1223080099067599$
$C7 = 0.002120276482776482$	$C7 = 0.006847157472157472$
$C8 = 3.416898148148148$	$C8 = 6.74849537037037$
$C9 = 0.08539001623376626$	$C9 = 0.334229301948052$
$C10 = -0.0209452047952048$	$C10 = -0.0458997668997669$

$N = \eta = 0.50$	$N = \eta = 0.25$
$C1 = 0.1013892010767011$	$C1 = 0.1093747294372294$
$C2 = 0.2001590773115773$	$C2 = 0.142861466033966$
$C3 = -0.6023601345413846$	$C3 = -0.5683255501443001$
$C4 = 0.05432371378621378$	$C4 = 0.03395236985236985$
$C5 = 0.4445470328282829$	$C5 = 0.6277401244588743$
$C6 = -0.4858658146020646$	$C6 = -0.3156246699134199$
$C7 = 0.005887286324786324$	$C7 = 0.002853418803418803$
$C8 = -0.01127946127946136$	$C8 = -0.192550505050505$
$C9 = 0.3789476461038961$	$C9 = 0.1949728084415584$
$C10 = -0.05209527139527139$	$C10 = -0.02924848484848485$

8.7.4 Coupling Loss between Different-Sized Square Antennas

$$L_{\mathrm{NF}} = C1 + C2 \times D + C3 \times R + C4 \times D^2 + C5 \times R^2 + C6 \times D \times R + C7 \times D^3 + C8 \times R^3 + C9 \times D \times R^2 + C10 \times D^2 \times R \quad (8.88)$$

$N = \eta = 1.00$	$N = \eta = 0.75$
$C1 = -0.9729305264180264$	$C1 = -0.3371434274059274$
$C2 = -0.8812381138306138$	$C2 = -0.1846516780441781$
$C3 = 4.742492976699226$	$C3 = 0.887764439033189$
$C4 = -0.2101820207570208$	$C4 = -0.01862155622155622$
$C5 = -8.833426902958152$	$C5 = -1.338570752164502$
$C6 = 1.227099754828505$	$C6 = 0.1831897564935065$
$C7 = -0.005192715617715618$	$C7 = 0.006001243201243201$
$C8 = 4.532586279461279$	$C8 = 0.6045138888888888$
$C9 = 0.1829265422077922$	$C9 = 0.3801099837662338$
$C10 = 0.0529536297036297$	$C10 = -0.004622077922077921$

$N = \eta = 0.50$	$N = \eta = 0.25$
$C1 = -0.0139240342990343$	$C1 = 0.1224828477078477$
$C2 = 0.1031087360787361$	$C2 = 0.169524861989862$
$C3 = -0.2748664266289267$	$C3 = -0.7053111198986198$
$C4 = 0.03496463952713952$	$C4 = 0.04124757742257742$
$C5 = 0.08828057359307374$	$C5 = 0.8015615981240979$
$C6 = -0.343841555944056$	$C6 = -0.3901483508158508$
$C7 = 0.006170593758093758$	$C7 = 0.003830542605542605$
$C8 = 0.08970959595959589$	$C8 = -0.2554187710437709$
$C9 = 0.4123693181818182$	$C9 = 0.2612280844155844$
$C10 = -0.0415508991008991$	$C10 = -0.03715820845820846$

For values of η between the values mentioned in the above tables, calculate the values for $\eta = 1.00$, 0.75, 0.50, and 0.25 and use two-dimensional cubic interpolation (see Appendix A) to find the intermediate value.

8.7.5 Parabolic Antenna to Passive Reflector Near Field Coupling Loss

Often, terrain obstructs a direct path between a transmitter and a receiver. In that case, passive reflectors ("beam benders") are sometimes used (Norton, 1962) to provide a direct transmission path. Passive reflector systems are most efficient when the reflector is near either the transmit or receive antenna. However, as the reflector is moved close to the transmit or receive antenna, near field coupling loss becomes important. Near field coupling loss for conventional reflector systems can be calculated as with the dual parabolic antennas. However, one antenna (the passive reflector) has uniform illumination ($F_2(x_2, y_2) = 1$) and the aperture is in the shape of the reflector (typically square) as it is projected onto the transmission path (i.e., if the reflector is at a 45° angle with the transmission path, a projected square would be a physical rectangle).

This problem has been solved in the past (Drexler, 1954; Jakes, 1953; Kizer, 1990; Medhurst, 1959) for "periscope" reflectors (no longer legal in the United States commercial bands (FCC, 2004)). Previous results were limited in range and also assumed antenna illumination functions inconsistent with current lower side lobe ("high performance") antennas. Calculations using the appropriate illumination assumptions are compared to previous results in Figure 8.47.

The methodology is consistent with previous results. General results have been calculated (Fig. 8.48 and Fig. 8.49) for the most common modern antennas and reflector combinations. Although both circular and square reflectors have been used commercially in the past, current practice is to use rectangular ("square") reflectors. The following results will be for square reflectors. However, if the actual reflectors are rectangular, using the reflector dimension in the plane of the signal polarization to derive an equivalent square reflector achieves essentially the same result. Be sure to remember that the reflector size in the following results is the geometric projection of the reflector on the path, not the physical size (usually the reflector height is not changed but the projected width is). See Section 8.2.1 for projected width.

8.7.5.1 Square Antenna and Square Reflector

The following definitions or relationships are used to estimate near field coupling loss between an antenna and a reflector:

$$\Delta_{dB} = 10 \log(\Delta), \quad -8 \leq \Delta_{dB} \leq 0$$

$D = \Delta_{dB}$ = the normalized distance between the antenna and the reflector;
$\Delta = d/(2W^2/\lambda)$ = the normalized distance parameter for the reflector width.

$$0 \leq R \leq 1$$

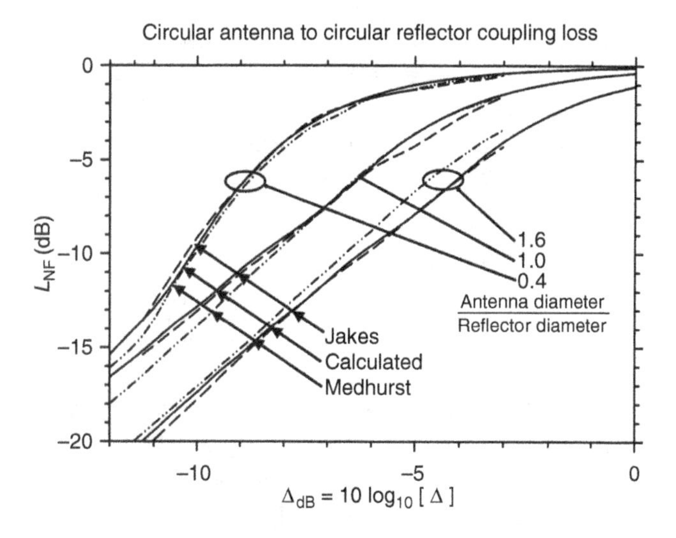

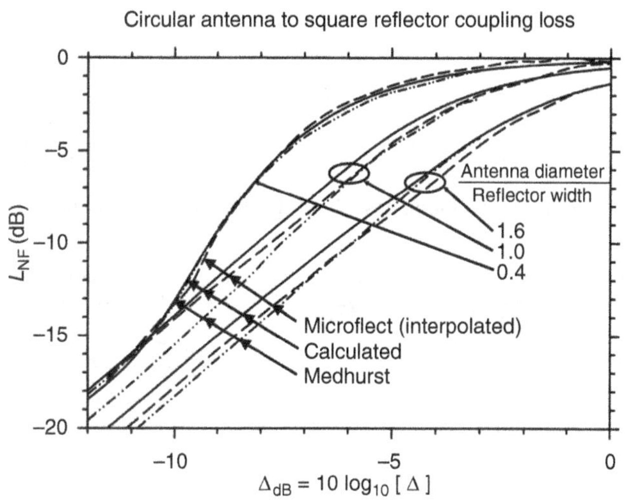

Figure 8.47 Comparison with previously published results.

R = the ratio of antenna width/reflector width.

$$0.25 \leq \eta \leq 1.0$$

$N = \eta$ = illumination efficiency;

L_{NF} = near field antenna to reflector coupling loss (dB);

= value to be added to far field free space loss.

8.7.6 Coupling Loss for Circular Antenna and Square Reflector

$$\begin{aligned} L_{NF} = {} & C1 + C2 \times D + C3 \times R + C4 \times D^2 + C5 \times R^2 + C6 \times D \times R \\ & + C7 \times D^3 + C8 \times R^3 + C9 \times D \times R^2 + C10 \times D^2 \times R \end{aligned} \tag{8.89}$$

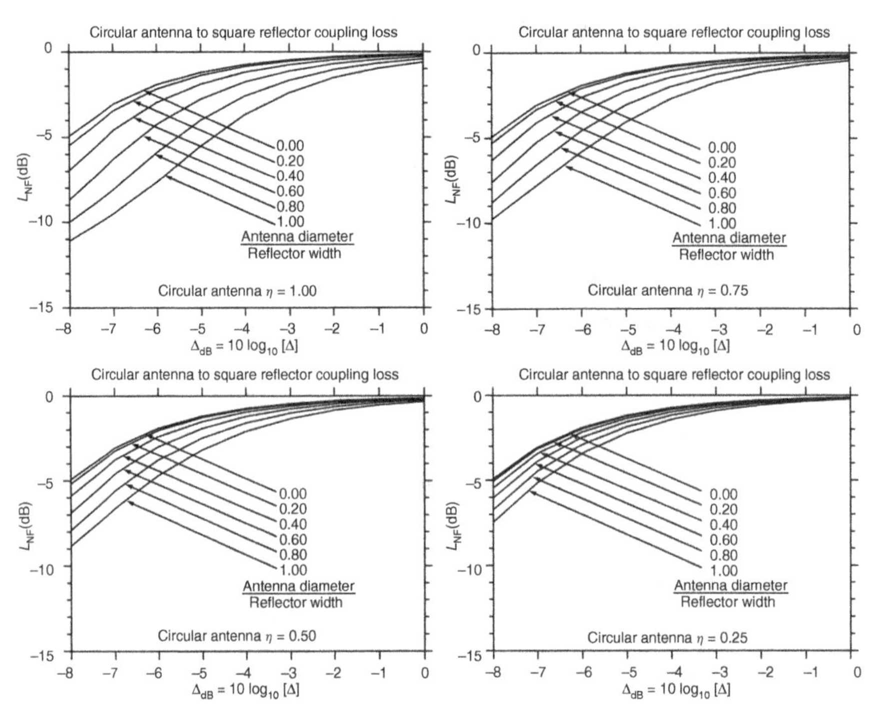

Figure 8.48 Antenna reflector coupling loss L_{NF}(dB) for circular (parabolic) antenna and square reflector.

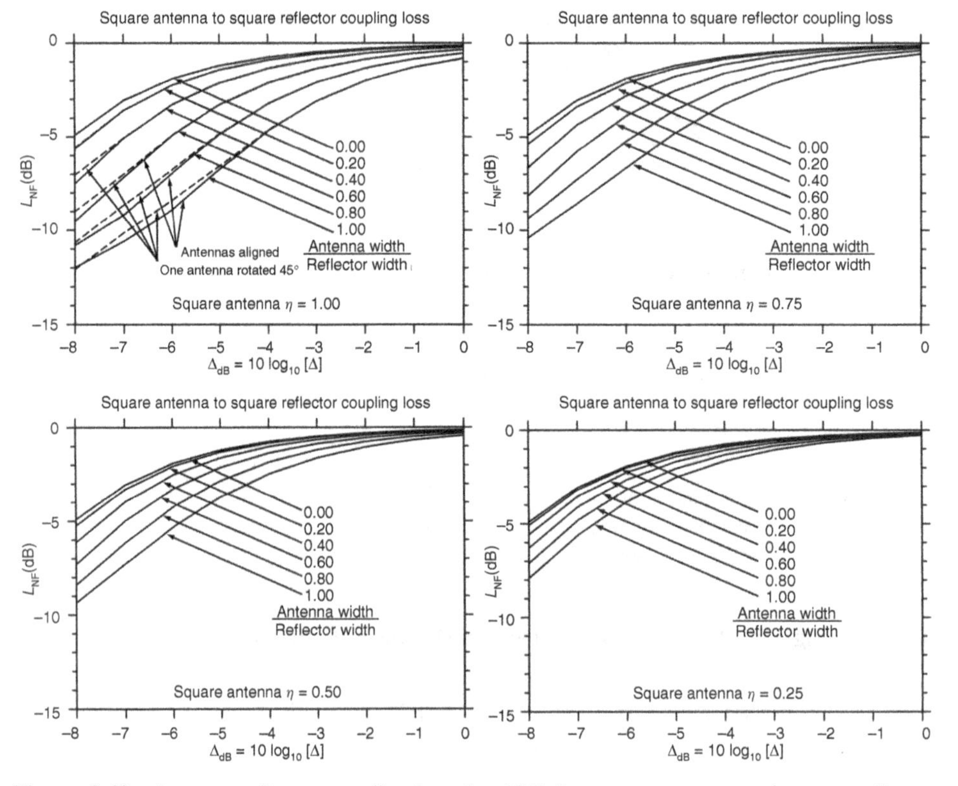

Figure 8.49 Antenna reflector coupling loss L_{NF}(dB) for square antenna and square reflector.

$N = \eta = 1.00$	$N = \eta = 0.75$
$C1 = -0.4566947811447811$	$C1 = -0.2431097426647427$
$C2 = -0.2020482792780412$	$C2 = 0.06178604780801208$
$C3 = 2.191165239698573$	$C3 = 1.058045280984448$
$C4 = -0.02878061052703909$	$C4 = 0.03989009482580911$
$C5 = -4.37696097883598$	$C5 = -2.486101521164021$
$C6 = 0.3237634547000618$	$C6 = -0.03516842648423002$
$C7 = 0.007908487654320989$	$C7 = 0.01286877104377104$
$C8 = 2.162461419753087$	$C8 = 1.312422839506173$
$C9 = 0.3109449404761905$	$C9 = 0.3230811011904762$
$C10 = -0.03379225417439703$	$C10 = -0.04815576685219542$

$N = \eta = 0.50$	$N = \eta = 0.25$
$C1 = -0.131068519320186$	$C1 = -0.05018120811287484$
$C2 = 0.1926641022755904$	$C2 = 0.2815940879028379$
$C3 = 0.4079589168136392$	$C3 = -0.09241433875794949$
$C4 = 0.07328302583659727$	$C4 = 0.09614733044733044$
$C5 = -1.326173390652558$	$C5 = -0.2993532848324522$
$C6 = -0.2101255144557823$	$C6 = -0.2734471239177489$
$C7 = 0.01526261223344557$	$C7 = 0.01694427609427609$
$C8 = 0.7796810699588479$	$C8 = 0.2686085390946506$
$C9 = 0.3159635416666667$	$C9 = 0.2477760416666666$
$C10 = -0.05155578231292517$	$C10 = -0.04260757575757575$

8.7.7 Coupling Loss for Square Antenna and Square Reflector (Both Aligned)

$$L_{\mathrm{NF}} = C1 + C2 \times D + C3 \times R + C4 \times D^2 + C5 \times R^2 + C6 \times D \times R + C7 \times D^3 + C8 \times R^3 + C9 \times D \times R^2 + C10 \times D^2 \times R \quad (8.90)$$

$N = \eta = 1.00$	$N = \eta = 0.75$
$C1 = -0.6466395109828443$	$C1 = -0.3507770017636684$
$C2 = -0.4242813704505371$	$C2 = -0.06543854400009162$
$C3 = 3.395700560098338$	$C3 = 1.723228713590936$
$C4 = -0.08242216209716208$	$C4 = 0.008616482855768571$
$C5 = -6.342335537918872$	$C5 = -3.605354938271605$
$C6 = 0.8051073917748917$	$C6 = 0.206446064471243$
$C7 = 0.004274172278338946$	$C7 = 0.01070726711560045$
$C8 = 2.923791152263375$	$C8 = 1.771116255144033$
$C9 = 0.1849970238095238$	$C9 = 0.2753497023809524$
$C10 = -0.006447186147186151$	$C10 = -0.03658038033395176$

$N = \eta = 0.50$	$N = \eta = 0.25$
$C1 = -0.1980865496232163$	$C1 = -0.07070867564534233$
$C2 = 0.1141195494056208$	$C2 = 0.2586083355092878$
$C3 = 0.8217887665009888$	$C3 = 0.03945184971407202$
$C4 = 0.05379672232529374$	$C4 = 0.09023423177351747$
$C5 = -2.054483906525573$	$C5 = -0.5848569223985893$
$C6 = -0.07965355287569573$	$C6 = -0.2614715762213976$
$C7 = 0.01389718013468013$	$C7 = 0.01650486812570146$
$C8 = 1.094843106995885$	$C8 = 0.4112782921810701$
$C9 = 0.3029285714285714$	$C9 = 0.2718556547619048$
$C10 = -0.04733208101422387$	$C10 = -0.04616696042053185$

For values of η between the values above, calculate the values for η = 1.00, 0.75, 0.50, and 0.25, and use two-dimensional cubic interpolation (see Appendix A) to find the intermediate value.

8.7.8 Back-to-Back Square Passive Reflector Near Field Coupling Loss

For severe path obstruction cases (e.g., going over a mountain), two close-coupled rectangular reflectors are sometimes used (Cappuccini and Gasparini, 1958) to achieve a radio path. Yang (1957) assumed two rectangular reflectors, which appear as squares when projected onto a plane perpendicular to the radio path. He solved this problem in integral form. A general closed form solution is easily achieved. The following factor, Y(dB), accounts for the combined gain of two flat square reflectors relative to the gain of a single reflector the size of the smaller reflector:

These formulas are valid for $R_W \geq 1$ and for all Δ.

d = separation between the two reflectors;
a^2 = projected area of the smaller reflector;
b^2 = projected area of the larger reflector;
Projected area = Physical area x [cosine (1/2 path included path angle at reflector)];
a = width of the smaller reflector assuming reflector projection is square;
b = width of the larger reflector assuming reflector projection is square;
λ = free space wavelength of radio wave;

a, b, d, and λ are in the same linear units;

$$\Delta = d/(2a^2/\lambda);$$
$$\Delta\text{dB} = 10 \log(\Delta);$$
$$R_W = \text{sqrt}(b^2/a^2) = b/a;$$
$$k = (1/2)\,\text{sqrt}(1/\Delta);$$
$$p = k(R_W - 1);$$
$$q = \; = k(R_W + 1).$$

Y(dB) = Combined gain of both reflectors relative to gain of single smaller reflector

$$= 20 \log[(U^2 + V^2)\,(2\Delta)] = \alpha_3 \text{ in Yang's derivation (Yang, 1957).} \tag{8.91}$$

$$U = \int_p^q C(t)\; dt, \quad C = \text{Fresnel cosine integral} \tag{8.92}$$

$$V = \int_p^q S(t)\; dt, \quad S = \text{Fresnel sine integral} \tag{8.93}$$

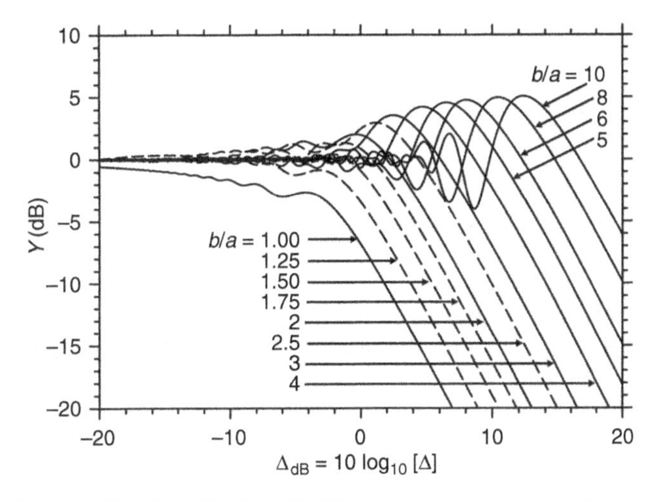

Figure 8.50 Coupling loss Y(dB) between two square reflectors.

Integrating formally,

$$U = qC(q) - pC(p) + \left(\frac{1}{\pi}\right)\left[\sin\left(\frac{\pi p^2}{2}\right) - \sin\left(\frac{\pi q^2}{2}\right)\right] \tag{8.94}$$

$$V = qS(q) - pS(p) + \left(\frac{1}{\pi}\right)\left[\cos\left(\frac{\pi q^2}{2}\right) - \cos\left(\frac{\pi p^2}{2}\right)\right] \tag{8.95}$$

Note that the sign of p was changed (for convenience) relative to Yang's derivation. Since $C(-p) = -C(p)$ and $S(-p) = -S(p)$, the resultant integral values are unchanged. The values for C and S are calculated using standard polynomial approximations for the Fresnel integrals (Abramowitz and Stegun, 1968, Sections 7.3.9, 7.3.10, 7.3.32, and 7.3.33).

$$C(z) = 0.5 + f(z)\ \sin\left(\frac{\pi z^2}{2}\right) - g(z)\ \cos\left(\frac{\pi z^2}{2}\right)$$

$$S(z) = 0.5 - f(z)\ \cos\left(\frac{\pi z^2}{2}\right) - g(z)\ \sin\left(\frac{\pi z^2}{2}\right)$$

$$f(z) = \frac{(1 + 0.926z)}{(2 + 1.792z + 3.104z^2)}$$

$$g(z) = \frac{1}{(2 + 4.142z + 3.492z^2 + 6.670z^3)}$$

For $\Delta \gg 1$ (reflectors greatly separated), $Y(\text{dB}) \cong -6 + 40 \log R_W - 20 \log \Delta$.
For $\Delta \ll 1$ (reflectors so close they appear as one), $Y(\text{dB}) \cong 0$.
The above results are plotted in Fig. 8.50.

8.A APPENDIX

8.A.1 Circular Antenna Numerical Power Calculations

```
REM Circular (Parabolic) Antenna Power Calculation
REM CALCULATE THE POWER AT A POINT X2, Y2=0,  Z2 = d / (2 D^2 / LAMBDA)
```

```
REM QuickBasic Program
REM # indicates a double precision variable

PI# = 3.141592653589793#
E# = 2.718281828459045#

CLS
OPEN "C:\OUTPUT.CSV" FOR OUTPUT AS #1

'DOVERLAMBDA# = normalized antenna diameter
'EFFEC# = antenna illumination efficiency
INPUT "Input D/Lambda (1 to 200) and Efficiency (.25 to 1.00):", DOVERLAMBDA#, EFFEC#

GOSUB 8000      ' CALCULATE H

X# = PI# * H#
GOSUB 9000
ILLUMREF# = I0#
PNORM# = 1!

'///////////////////////////////////////////////////////////////////////
Use this section for near field calculations
SPEEDUP = 1
Z2# = 1!
X2# = 0!
Y2# = 0!
GOSUB 3000
PNORM# = NPOWER#

FOR K = 0 TO 120

DSTEP# = 10! - (K / 4)
DELTA# = 10! ^ (DSTEP# / 10!)

FOR J = 0 TO 100 STEP 1

AFACTOR# = J / 100!

X2# = AFACTOR# / (2! * DOVERLAMBDA#)
Z2DB# = 10! * LOG(DELTA#) / LOG(10!)
Z2# = DELTA#

SPEEDUP = 1
IF DSTEP# >= -8! AND ASTEP# <= 15! THEN SPEEDUP = 4
IF DSTEP# >= 0! AND ASTEP# <= 15! THEN SPEEDUP = 8
IF ASTEP# > 15! AND (ASTEP# - DSTEP#) <= 17 THEN SPEEDUP = 8

GOSUB 3000

PRINT
PRINT USING " ###.## ,  ###.#### ,  ####.## "; AFACTOR#; Z2DB#; PDBNORM#
      'Width, Depth, Power
PRINT #1, USING " ###.## ,  ###.#### ,  ####.## "; AFACTOR#; Z2DB#; PDBNORM#

NEXT J
```

```
NEXT K
'///////////////////////////////////////////////////////////////////////////
Use this section for far field calculations
SPEEDUP = 5
Z2# = 100!
X2# = 0!
Y2# = 0!
GOSUB 3000
PNORM# = NPOWER#

FOR KK = 0 TO 60 STEP 1
FOR JJ = 0 TO 60 STEP 1

THEANGLE1# = -30! + CDBL(JJ / 1) 'ANGLE IN FAR FIELD PATTERN
THEANGLERADS1# = (PI# / 180!) * THEANGLE1#
X2# = Z2# * TAN(THEANGLERADS1#)

THEANGLE2# = -30! + CDBL(KK / 1) 'ANGLE IN FAR FIELD PATTERN
THEANGLERADS2# = (PI# / 180!) * THEANGLE2#
Y2# = Z2# * TAN(THEANGLERADS2#)

GOSUB 3000

PRINT
PRINT USING " ###.#### ,   ###.#### ,   #####.#### ";
      THEANGLE1#; THEANGLE2#; PDBNORM#
PRINT #1, USING " ###.######## ,   ###.######## ,   #####.######## ";
      THEANGLE1#; THEANGLE2#; PDBNORM#
NEXT JJ
NEXT KK
'///////////////////////////////////////////////////////////////////////////

999 CLOSE #1
PRINT
PRINT USING "D/Lambda =####, Effeciency =##.##"; DOVERLAMBDA#; EFFEC#
PRINT
INPUT "Press Enter key:"; Enter
END


3000 REM INTEGRATE CIRCLE (UNITY DIAMETER) * ANTENNA ILLUMINATION
        AT POINT Z2#

SUMREAL# = 0!
SUMIMAG# = 0!
RADIALSTEPS = 400 / SPEEDUP
ANGULARSTEPS = 720 / SPEEDUP
FOR R = 1 TO RADIALSTEPS
FOR A = 1 TO ANGULARSTEPS
SANGLE# = (PI# / 180!) * (360! / CDBL(ANGULARSTEPS))     'SECTOR ANGLE IN RADIANS
RADIUS# = CDBL(R / (RADIALSTEPS * 2))                     'NORMALIZED RADIUS
SECTORAREA# = (SANGLE# / 2!) * CDBL(((2 * R) - 1) / (RADIALSTEPS * RADIALSTEPS * 4))
RADIUSLOWER# = CDBL((R - 1) / (RADIALSTEPS * 2))
RHO# = (RADIUS# + RADIUSLOWER#) / 2!
X# = PI# * H# * SQR(1! - (RHO# * RHO# * 4!))
```

```
GOSUB 9000
ILLUM# = I0# / ILLUMREF#

ANGLE# = (PI# / 180!) * (360! * CDBL(A / ANGULARSTEPS)) 'ANGLE IN RADIANS
        AT POINT ON ANTENNA
X1# = RHO# * COS(ANGLE#)
Y1# = RHO# * SIN(ANGLE#)
X1# = X1# / (2! * DOVERLAMBDA#)
Y1# = Y1# / (2! * DOVERLAMBDA#)
RAY# = (SQR(((X2# - X1#) ^ 2) + ((Y2# - Y1#) ^ 2) + (Z2# ^ 2)))
COSPHI# = Z2# / RAY#
REAL# = ((1 + COSPHI#) * SIN(2! * PI# * RAY# * (2! * (DOVERLAMBDA# ^ 2))))
        + ((COSPHI# / (2! * PI# * RAY# * (2! * (DOVERLAMBDA# ^ 2))))
        * COS(2 * PI# * RAY# * (2! * (DOVERLAMBDA# ^ 2))))
IMAG# = ((1 + COSPHI#) * COS(2! * PI# * RAY# * (2! * (DOVERLAMBDA# ^ 2))))
        - ((COSPHI# / (2! * PI# * RAY# * (2! * (DOVERLAMBDA# ^ 2))))
        * SIN(2 * PI# * RAY# * (2! * (DOVERLAMBDA# ^ 2))))
SUMREAL# = SUMREAL# + (SECTORAREA# * ILLUM#
        * (1! / (RAY# * (2! * (DOVERLAMBDA# ^ 2)))) * REAL#)
SUMIMAG# = SUMIMAG# + (SECTORAREA# * ILLUM#
        * (1! / (RAY# * (2! * (DOVERLAMBDA# ^ 2)))) * IMAG#)

IF INKEY$ <> "" THEN GOTO 999
NEXT A
PRINT USING "####.####   ####.####   ####.##  #####   ####";AFACTOR#; Z2DB#;
PDBNORM#; R; RADIALSTEPS
NEXT R
NPOWER# = ((SUMIMAG# ^ 2) + (SUMREAL# ^ 2)) / PNORM#
PDBNORM# = 10! * LOG(NPOWER#) / LOG(10!)

3001 RETURN

8000 REM EFFEC# TO H# CONVERSION
REM INPUT: X#
REM OUTPUT: H#
REM USES: HA# - HH#
HA# = 10789.06678518257#
HB# = 8454.692672442079#
HC# = -21713.73557586721#
HD# = -14801.06001940764#
HE# = 11772.42148035185#
HF# = 5699.562595386604#
HG# = -847.7526882273281#
HH# = 652.2024360393683#

H# = (HA# + HC# * EFFEC# + HE# * (EFFEC# ^ 2) + HG# * (EFFEC# ^ 3)) /
        (1 + HB# * EFFEC# + HD# * (EFFEC# ^ 2) + HF# * (EFFEC# ^ 3)
        + HH# * (EFFEC# ^ 4))

RETURN

9000 REM BESSEL FUNCTION SUBROUTINE
REM INPUT: X#
REM OUTPUT: J0#, J1#, I0#, I1#, C#, S#
REM USES: A0# - H0#, A1# - H1#, E#, F#, G#, PI#, T#
```

```
REM USES: JFA0# - JFF0#, JTA0# - JTF0#, JF0#, JT0#
REM USES: JFA1# - JFF1#, JTA1# - JTF1#, JF1#, JT1#

X# = ABS(X#)
PI# = 3.141592654#
E# = 2.718281828#

REM BESSEL FUNCTIONS
REM Abramowitz and Stegun, SECTION 9.4 PAGES 369-370

REM MODIFIED BESSEL FUNCTION I0
REM SECTION 9.8 PAGE 378
9300 IF X# <= 3.75 GOTO 9400
    IF X# > 3.75 GOTO 9500
9400 REM 0 <= X# <= 3.75
T# = X# / 3.75
A0# = 3.5156229#
B0# = 3.0899424#
C0# = 1.2067492#
D0# = .2659732#
E0# = .0360768#
F0# = .0045813#
I0# = 1! + A0# * T# ^ 2 + B0# * T# ^ 4 + C0# * T# ^ 6 + D0# * T# ^ 8 + E0# * T# ^ 10
        + F0# * T# ^ 12
GOTO 9600
9500 REM 3.75 < X#
T# = 3.75 / X#
A0# = .01328592#
B0# = .00225319#
C0# = .00157565#
D0# = .00916281#
E0# = .02057706#
F0# = .02635537#
G0# = .01647633#
H0# = .00392377#
I0# = .39894228# + A0# * T# ^ 1 + B0# * T# ^ 2 - C0# * T# ^ 3 + D0# * T# ^ 4
        - E0# * T# ^ 5 + F0# * T# ^ 6 - G0# * T# ^ 7 + H0# * T# ^ 8
I0# = I0# * E# ^ X# / SQR(X#)
GOTO 9600

9600 RETURN
```

8.A.2 Square Antenna Numerical Power Calculations

```
REM Square Antenna Near Field Power Calculation
REM CALCULATE THE POWER AT A POINT X2, Y2=0,  Z2 = d / (2 D^2 / LAMBDA)
REM QuickBasic Program

PI# = 3.141592653589793#
E# = 2.718281828459045#

CLS
OPEN "C:\OUTPUT.CSV" FOR OUTPUT AS #1
```

```
ROTATION# = 0!    ' 0.0 OR 45.0         'ANTENNA ROTATION IN DEGREES
ROTATIONRADS# = (PI# / 180!) * ROTATION#

'DOVERLAMBDA# = NORMALIZED ANTENNA DIAMETER
'EFFEC# = ANTENNA POWER EFFECIENCY (FRACTION)
INPUT "Input D/Lambda (1 to 200) and Effeciency (.25 to 1.00):",DOVERLAMBDA#,EFFEC#

GOSUB 8000      ' CALCULATE H
X# = PI# * H#
GOSUB 9000
ILLUMREF# = I0#
PNORM# = 1!

'//////////////////////////////////////////////////////////////////////
Use this section for near field calculations
SPEEDUP = 1
Z2# = 1!
X2# = 0!
Y2# = 0!
GOSUB 3000
PNORM# = NPOWER#

FOR K = 0 TO 120

DSTEP# = 10! - (K / 4)
DELTA# = 10! ^ (DSTEP# / 10!)

SPEEDUP = 1
IF DELTA# > .15 THEN SPEEDUP = 4
IF DELTA# > 1! THEN SPEEDUP = 8

FOR J = 0 TO 100 STEP 1

AFACTOR# = J / 100!

X2# = AFACTOR# / (2! * DOVERLAMBDA#)
Z2DB# = 10! * LOG(DELTA#) / LOG(10!)
Z2# = DELTA#
GOSUB 3000

PRINT
PRINT USING " ###.#### ,   ###.#### ,   ####.## "; AFACTOR#; Z2DB#; PDBNORM#
          'Width, Depth, Power
PRINT #1, USING " ###.#### ,   ###.#### ,   ####.## "; AFACTOR#; Z2DB#; PDBNORM#

NEXT J
NEXT K
'//////////////////////////////////////////////////////////////////////
Use this section for far field calculations
SPEEDUP = 10
Z2# = 100!
X2# = 0!
Y2# = 0!
GOSUB 3000
```

```
PNORM# = NPOWER#

FOR KK = 0 TO 60 STEP 1
FOR JJ = 0 TO 60 STEP 1

THEANGLE1# = −30! + CDBL(JJ / 1) 'ANGLE IN FAR FIELD PATTERN
THEANGLERADS1# = (PI# / 180!) * THEANGLE1#
X2# = Z2# * TAN(THEANGLERADS1#)

THEANGLE2# = −30! + CDBL(KK / 1) 'ANGLE IN FAR FIELD PATTERN
THEANGLERADS2# = (PI# / 180!) * THEANGLE2#
Y2# = Z2# * TAN(THEANGLERADS2#)

GOSUB 3000

PRINT
PRINT USING " ###.#### ,   ###.#### ,   #####.#### "; THEANGLE1#;
        THEANGLE2#; PDBNORM#
PRINT #1, USING " ###.######## ,   ###.######## ,   #####.######## "; THEANGLE1#;
        THEANGLE2#; PDBNORM#

NEXT JJ
NEXT KK
'////////////////////////////////////////////////////////////////////////////

999 CLOSE #1
PRINT
PRINT USING "D/Lambda =####, Efficiency =##.##";DOVERLAMBDA#,EFFEC#
PRINT
INPUT "Press Enter key:";Enter
END

3000 REM INTEGRATE SQUARE * ANTENNA ILLUMINATION AT POINT Z2#

SUMREAL# = 0!
SUMIMAG# = 0!
STEPS = 800 / SPEEDUP
RSTEPS# = CDBL(STEPS)
FOR X1S = 1 TO STEPS
FOR Y1S = 1 TO STEPS
AREA# = 1! / (RSTEPS# * RSTEPS#)
X1# = (CDBL(X1S) - (.5 + (RSTEPS# / 2!))) / RSTEPS#
Y1# = (CDBL(Y1S) - (.5 + (RSTEPS# / 2!))) / RSTEPS#

RHO# = (SQR(X1# ^ 2 + Y1# ^ 2)) / SQR(.5)
X# = PI# * H# * SQR(1! - (RHO# * RHO#))
GOSUB 9000
ILLUM# = I0# / ILLUMREF#

X1OLD# = X1#
Y1OLD# = Y1#
X1# = (X1OLD# * COS(ROTATION#)) - (Y1OLD# * SIN(ROTATION#))
Y1# = (X1OLD# * SIN(ROTATION#)) + (Y1OLD# * COS(ROTATION#))

X1# = X1# / (2! * DOVERLAMBDA#)
```

```
Y1# = Y1# / (2! * DOVERLAMBDA#)

RAY# = (SQR(((X2# - X1#) ^ 2) + ((Y2# - Y1#) ^ 2) + (Z2# ^ 2)))
COSPHI# = Z2# / RAY#
REAL# = ((1 + COSPHI#) * SIN(2! * PI# * RAY# * (2! * (DOVERLAMBDA# ^ 2))))
          + ((COSPHI# / (2! * PI# * RAY# * (2! * (DOVERLAMBDA# ^ 2))))
          * COS(2 * PI# * RAY# * (2! * (DOVERLAMBDA# ^ 2))))
IMAG# = ((1 + COSPHI#) * COS(2! * PI# * RAY# * (2! * (DOVERLAMBDA# ^ 2))))
          - ((COSPHI# / (2! * PI# * RAY# * (2! * (DOVERLAMBDA# ^ 2))))
          * SIN(2 * PI# * RAY# * (2! * (DOVERLAMBDA# ^ 2))))
SUMREAL# = SUMREAL# + (AREA# * ILLUM# * (1! / (RAY#
          * (2! * (DOVERLAMBDA# ^ 2)))) * REAL#)
SUMIMAG# = SUMIMAG# + (AREA# * ILLUM# * (1! / (RAY#
          * (2! * (DOVERLAMBDA# ^ 2)))) * IMAG#)
IF INKEY$ <> "" THEN GOTO 999
NEXT Y1S
PRINT USING "####.####   ####.####   ####.##   ####.###   ####.### ";AFACTOR#; Z2DB#;
PDBNORM#; (X1S / STEPS); (Y1S / STEPS)
NEXT X1S
NPOWER# = ((SUMIMAG# ^ 2) + (SUMREAL# ^ 2)) / PNORM#
PDBNORM# = 10! * LOG(NPOWER#) / LOG(10!)

3001 RETURN

8000 REM EFFEC# TO H# CONVERSION
REM SQUARE ANTENNA
REM INPUT: X#
REM OUTPUT: H#
REM USES: HA# - HH#
HA# = −1979.796930437251#
HB# = −1010.234447176994#
HC# = 5081.776118785949#
HD# = 2522.983250131826#
HE# = −4381.289862237859#
HF# = −2209.833454513456#
HG# = 1279.304377692569#
HH# = 694.1627969027892#

H# = (HA# + HC# * EFFEC# + HE# * (EFFEC# ^ 2) + HG# * (EFFEC# ^ 3)) /
      (1 + HB# * EFFEC# + HD# * (EFFEC# ^ 2) + HF# * (EFFEC# ^ 3)
      + HH# * (EFFEC# ^ 4))

RETURN
9000 REM BESSEL FUNCTION SUBROUTINE
REM INPUT: X#
REM OUTPUT: J0#, J1#, I0#, I1#, C#, S#
REM USES: A0# - H0#, A1# - H1#, E#, F#, G#, PI#, T#
REM USES: JFA0# - JFF0#, JTA0# - JTF0#, JF0# , JT0#
REM USES: JFA1# - JFF1#, JTA1# - JTF1#, JF1# , JT1#

X# = ABS(X#)
PI# = 3.141592654#
E# = 2.718281828#

REM MODIFIED BESSEL FUNCTION I0
```

```
REM Abramowitz and Stegun, SECTION 9.8 PAGE 378
9300 IF X# <= 3.75 GOTO 9400
     IF X# > 3.75 GOTO 9500
9400 REM 0 <= X# <= 3.75
T# = X# / 3.75
A0# = 3.5156229#
B0# = 3.0899424#
C0# = 1.2067492#
D0# = .2659732#
E0# = .0360768#
F0# = .0045813#
I0# = 1! + A0# * T# ^ 2 + B0# * T# ^ 4 + C0# * T# ^ 6 + D0# * T# ^ 8 + E0# * T# ^ 10
      + F0# * T# ^ 12
GOTO 9600
9500 REM 3.75 < X#
T# = 3.75 / X#
A0# = .01328592#
B0# = .00225319#
C0# = .00157565#
D0# = .00916281#
E0# = .02057706#
F0# = .02635537#
G0# = .01647633#
H0# = .00392377#
I0# = .39894228# + A0# * T# ^ 1 + B0# * T# ^ 2 - C0# * T# ^ 3 + D0# * T# ^ 4
     - E0# * T# ^ 5 + F0# * T# ^ 6 - G0# * T# ^ 7 + H0# * T# ^ 8
I0# = I0# * E# ^ X# / SQR(X#)
GOTO 9600

9600 RETURN
```

8.A.3 Bessel Functions

While Bessel functions appear commonly in antenna pattern calculations, they are seldom available as subroutines for calculators or computers. The following approximations (Abramowitz and Stegun, 1968) are useful in calculating these functions.

$I_0(X)$ = I0 = modified Bessel function of the first kind and zero order
$I_1(X)$ = I1 = modified Bessel function of the first kind and first order
$J_0(X)$ = J0 = Bessel function of the first kind and zero order
$J_1(X)$ = J1 = Bessel function of the first kind and first order

X >= 0

SQR(X) = (positive) square root of X
COS(X) = cosine of X (radians)

```
IF 0 <= X <= 3.75
T = X / 3.75
A0 = 3.5156229
B0 = 3.0899424
C0 = 1.2067492
D0 = 0.2659732
E0 = 0.0360768
F0 = 0.0045813
```

```
I0 = 1.0 + A0 * T ^ 2 + B0 * T ^ 4 + C0 * T ^ 6 + D0 * T ^ 8
        + E0 * T ^ 10 + F0 * T ^ 12

IF X > 3.75
T = 3.75 / X
E =  2.7182818
A0 = 0.01328592
B0 = 0.00225319
C0 = -0.00157565
D0 = 0.00916281
E0 = -0.02057706
F0 = 0.02635537
G0 = -0.01647633
H0 = 0.00392377
Gzero = 0.39894228 + A0 * T ^ 1 + B0 * T ^ 2 + C0 * T ^ 3 + D0 * T ^ 4
           + E0 * T ^ 5 + F0 * T ^ 6 + G0 * T ^ 7 + H0 * T ^ 8
I0 = Gzero * (E ^ X) / SQR(X)
```

```
IF 0<= X <= 3.75
T = X / 3.75
A0 = 0.87890594
B0 = 0.51498869
C0 = 0.15084934
D0 = 0.02658733
E0 = 0.00301532
F0 = 0.00032411
Gone = 0.5 + A0 * T ^ 2 + B0 * T ^ 4 + C0 * T ^ 6 + D0 * T ^ 8
          + E0 * T ^ 10 + F0 * T ^ 12
I1 =Gone *  X

IF X > 3.75
T = 3.75 / X
E =  2.7182818
A0 = -0.03988024
B0 = -0.00362018
C0 = 0.00163801
D0 = -0.01031555
E0 = 0.02282967
F0 = -0.02895312
G0 = 0.01787654
H0 = -0.00420059
Gtwo = 0.39894228 + A0 * T ^ 1 + B0 * T ^ 2 + C0 * T ^ 3 + D0 * T ^ 4
          + E0 * T ^ 5 + F0 * T ^ 6 + G0 * T ^ 7 + H0 * T ^ 8
I1 = Gtwo * (E ^ X) / SQR(X)
```

```
IF 0 <= X <= 3.0
T = X / 3.0
A0 = -2.2499997
B0 = 1.2656208
C0 = -0.3163866
D0 = 0.0444479
E0 = -0.0039444
F0 = 0.00021
J0 = 1.0 + A0 * T ^ 2 + B0 * T ^ 4 + C0 * T ^ 6 + D0 * T ^ 8
        + E0 * T ^ 10 + F0 * T ^ 12
```

```
IF X > 3.0
T = 3.0 / X
A0 = −0.00000077
B0 = −0.0055274
C0 = −0.00009512
D0 = 0.00137237
E0 = −0.00072805
F0 = 0.00014476
Fzero = 0.79788456 + A0 * T ^ 1 + B0 * T ^ 2 + C0 * T ^ 3
          + D0 * T ^ 4 + E0 * T ^ 5 + F0 * T ^ 6
A0 = −0.04166397
B0 = −0.00003954
C0 = 0.00262573
D0 = −0.00054125
E0 = −0.00029333
F0 = 0.00013558
Pzero = −0.78539816 + A0 * T ^ 1 + B0 * T ^ 2 + C0 * T ^ 3
          + D0 * T ^ 4 + E0 * T ^ 5 + F0 * T ^ 6
P0 = X + Pzero
J0 = Fzero * COS(P0) / SQR(X)
```

```
IF 0 <=  X <= 3.0
T = X / 3.0
A0 = −0.56249985
B0 = 0.21093573
C0 = −0.03954289
D0 = 0.00443319
E0 = −0.00031761
F0 = 0.00001109
Hone = 0.5 + A0 * T ^ 2 + B0 * T ^ 4 + C0 * T ^ 6 + D0 * T ^ 8
          + E0 * T ^ 10 + F0 * T ^ 12
J1 = Hone * X

IF X > 3.0
T = 3.0 / X
A0 = 0.00000156
B0 = 0.01659667
C0 = 0.00017105
D0 = −0.00249511
E0 = 0.00113653
F0 = −0.00020033
Fone = 0.79788456 + A0 * T ^ 1 + B0 * T ^ 2 + C0 * T ^ 3
          + D0 * T ^ 4 + E0 * T ^ 5 + F0 * T ^ 6
A0 = 0.12499612
B0 = 0.0000565
C0 = −0.00637879
D0 = 0.00074348
E0 = 0.00079824
F0 = −0.00029166
Pone = −2.35619449 + A0 * T ^ 1 + B0 * T ^ 2 + C0 * T ^ 3
          + D0 * T ^ 4 + E0 * T ^ 5 + F0 * T ^ 6
P1 = X + Pone
J1 = Fone * COS(P1) / SQR(X)
```

REFERENCES

Abramowitz, M. and Stegun, I., *Handbook of Mathematical Functions (NBS AMS 55, seventh printing with corrections)*. Washington, DC: U. S. Government Printing Office, pp. 301–302, 369–370 and 378, 1968.

Balanis, C. A., *Modern Antenna Handbook*. Hoboken: John Wiley & Sons, Inc., pp. 3–56, 2008.

Bickmore, R. W. and Hansen, R. C., "Antenna Power Densities in the Fresnel Region," *Proceedings of the IRE*, Vol. 47, pp. 2119–2120, December 1959.

Cappuccini, F. and Gasparini, F., "Passive Repeaters Using Double Flat Reflectors," *Proceedings of the IRE*, pp. 784–785, April 1958.

Cleveland, Jr., R. F., Sylvar, D. M. and Ulcek, J. L., *Evaluating Compliance with FCC Guidelines for Human Exposure to Radiofrequency Electromagnetic Fields, OET Bulletin 65*. Washington, DC: Federal Communications Commission, Office of Engineering and Technology, pp. 26–30, 1997.

Collin, R. E. and Zucker, F. J., *Antenna Theory, Part 2*, New York: McGraw-Hill, p. 50, 1969.

Drexler, J., "An Experimental Study of a Microwave Periscope," *Proceedings of the IRE*, p. 10–22, June 1954.

Editorial Board, *Reference Data for Radio Engineers*, *Fifth Edition*, Indianapolis: H. W. Sams & Co., pp. 25–47, 25–48, 1968.

FCC, "Part 101 – Fixed Microwave Services, Subpart C – Technical Standards," *Code of Federal Regulations, Title 47, Chapter I, Federal Communications Commission, Part 101, paragraph 101*, Washington, DC, 2004.

Friis, H. T., "A Note on a Simple Transmission Formula," *Proceedings of the IRE*, Vol. 34, pp. 254–256, May 1946.

Guy, A. W., "History of Biological Effects and Medical Applications of Microwave Energy," *IEEE Transactions on Microwave Theory and Techniques*, Vol. 32, pp. 1182–1200, September 1984.

Hansen, R. C., *Microwave Scanning Antennas*, Volume I, *Apertures*. New York: Academic Press, pp. 10–11, 24–33, 1964.

Hansen, R. C., "A One-Parameter Circular Aperture Distribution with Narrow BeamWidth and Low Sidelobes," *IEEE Transactions on Antennas and Propagation*, Vol. 24, pp. 477–480, July 1976a.

Hansen, R. C., "*Circular-Aperture Axial Power Density*," *Microwave Journal*, Vol. 19, pp. 50–52, February 1976b.

Hollis, J. S., Lyon, T. J. and Clayton, Jr., L., *Microwave Antenna Measurements*. Atlanta: Scientific-Atlanta, 1970.

Hu, M., "Near-Zone Power Transmission Formulas," *IRE National Convention Record, Part 8*, pp. 128–135, 1958.

IEEE International Committee on Electromagnetic Safety (SCC39), *IEEE Standard for Safety Levels with Respect to Human Exposure to Radio Frequency Electromagnetic Fields, 3 kHz to 300 GHz, IEEE C95.1-2005*, New York: The Institute of Electrical and Electronics Engineers, pp. 23–27, 2005.

Jakes, Jr., W. C., "A Theoretical Study of an Antenna-Reflector Problem," *Proceedings of the IRE*, pp. 272–274, February 1953.

Kay, A. F., "Near-Field Gain of Aperture Antennas," *IRE Transactions on Antennas and Propagation*, Vol. 8, pp. 586–593, November 1960.

Kizer, G. M., *Microwave Communication*. Ames: Iowa State University Press, pp. 335–339, 1990.

Laybros, S., Combes, P. F. and Mametsa, H. J., "The "Very-Near-Field" Region of Equiphase Radiating Apertures," *IEEE Antennas and Propagation Magazine*, Vol. 47, pp. 50–66, August 2005.

Medhurst, R. G., "Passive Microwave Mirrors," *Electronic and Radio Engineers*, pp. 443–449, July 1959.

Mikki, S. M. and Antar, Y. M. M., "A Theory of Antenna Electromagnetic Near Field – Part I and Part II," *IEEE Transactions on Antennas and Propagation*, Vol. 59, pp. 4706–4724, December 2011.

Norton, M. L., "Microwave System Engineering Using Large Passive Reflectors," *IRE Transactions on Communications Systems*, Vol. 10, pp. 304–311, September 1962.

Pace, J. R., "Asymptotic Formulas for Coupling Between Two Antennas in the Fresnel Region," *IEEE Transactions on Antennas and Propagation*, Vol. 17, pp. 285–291, May 1969.

Reintjes, J. F. and Coate, G. T., *Principles of Radar, Third Edition*, New York: McGraw-Hill, p. 950, 1952.

Robieux, J., "Near-Zone Power Transmission Formulas," *Proceedings of the IRE*, p. 1161, June 1959.

Saad, T. S., Hansen, R. C. and Wheeler, G. J., *Microwave Engineers' Handbook*, Volume 2. Dedham: Artech House, p. 34, 1971.

Sanchez-Hernandex, D. A., *High Frequency Electromagnetic Dosimetry*. Norwood: Artech House, 2009.

Sciambi, A. F., "The Effect of the Aperture Illumination on the Circular Aperture Antenna Pattern Characteristics," *The Microwave Journal*, Vol. 8, pp. 79–84, August 1965.

Silver, S., *Microwave Antenna Theory and Design*. New York: McGraw-Hill, pp. 169–180, 195, 587–592, 1949.

Steneck, N. H., *The Microwave Debate*. Cambridge: The MIT Press, 1984.

Takeshita, S. and Sakurai, N., "Path Loss in the Fresnel Region," *Electronics and Communications in Japan*, Vol. 47, pp. 29–39, August 1964.

TIA Subcommittee TR-14.7, *Structural Standards for Steel Antenna Towers and Antenna Supporting Structures, ANSI/TIA/EIA-222-F*, Arlington: Telecommunications Industry Association, p. 74, 1996.

TIA Subcommittee TR-14.7, *Structural Standard for Antenna Supporting Structures and Antennas, ANSI/TIA/EIA-222-G*, Arlington: Telecommunications Industry Association, p. 204, 2005 (Addendum 1, 2007).

United States Federal Government, "OSHA Radiation Protection Guide," *Code of Federal Regulations, 29 CFR Part 1910, Subpart G, Standard 1910.97, subparagraph (a) (2) (i)*, Washington, DC: U. S. Government Printing Office, 2004.

White, R., *Engineering Considerations for Microwave Communications Systems*, San Carlos: Lenkurt Electric, p. A5, 1970.

Yang, R. F. H., "Passive Repeater Using Double Flat Reflectors," *IRE National Convention Record, Part 1*, pp. 36–41, 1957.

Zhadobov, M., Nicolaz, C. N., Sauleau, R., Desmots, F., Thouroude, D., Michel, D. and Le Drean, Y., "Evaluation of the Potential Biological Effects of the 60-GHz Millimeter Waves Upon Human Cells," *IEEE Transactions on Antennas and Propagation*, pp. 2949–2955, October 2009.

9

MULTIPATH FADING

The basics of microwave radio transmission were developed in the 1930s and 1940s. In the late 1940s and early 1950s, many companies including Western Union and AT&T began deploying microwave radio systems. As the systems were deployed, it became apparent that the received signal level for moderate-to-long microwave paths was not constant (Fig. 9.1). For long-distance (20–30 miles) microwave transmission, the received signal level continuously varied up or down by a small amount. While this "breathing" increased with path length, it was not a significant operational factor (although it complicated determining average received signal level (RSL) level).

However, it soon became clear (Bullington, 1950, 1957; Durkee, 1948; England et al., 1938; Friis, 1948) overall performance of fixed point to point microwave systems was limited by short-duration, large, received signal power variations called *fading* (Fig. 9.2).

As a microwave signal travels over a relatively long path, the received signal may have propagated over several slightly different paths between the transmitter and the receiver. If the path differences exceed a significant portion of the radio wave wavelength, the combination of received signals, when added together at the receive antenna, can produce a composite signal larger or smaller than the signal that would be present if only one transmission path existed. The different paths are caused by slight time- and space-dependent variation in the atmospheric refractive index. This multipath fading is also called *scintillation*.

The Bell System research organization, Bell Laboratories, was tasked with investigating the signal-fading phenomenon. Bell Laboratories discovered (Barnett, 1974a; Bullington, 1950, 1957; Burrows et al., 1935; Clark, 1968; Crawford and Jakes, 1952; England et al., 1933; Friis, 1948; Jakes, 1974; Schelleng et al., 1933; Vigants, 1971; Young, 1952) that short-term signal power fading was both diurnal and seasonal (Fig. 9.3).

For linearly polarized microwave signals, both a copolarized and a cross-polarized signal can propagate along the radio path. The cross-polarized signal is discriminated by the receive antenna. This discrimination is termed as cross-polarization discrimination (XPD). As a linearly polarized signal propagates along a path, part of that signal is transferred to the opposite polarization. The exact mechanism that creates this cross-polarized signal is not well understood. It is believed to be caused by reflections from the Earth, or from elevated ducting layer multipath (either directly or via cross-polarization pattern differences for signals of different angles of arrival) (Mottl, 1977; Sakagami Morita, 1979). Both the copolarized and cross-polarized signals fade with the same statistical characteristics (Barnett, 1974b;

Digital Microwave Communication: Engineering Point-to-Point Microwave Systems, First Edition. George Kizer.

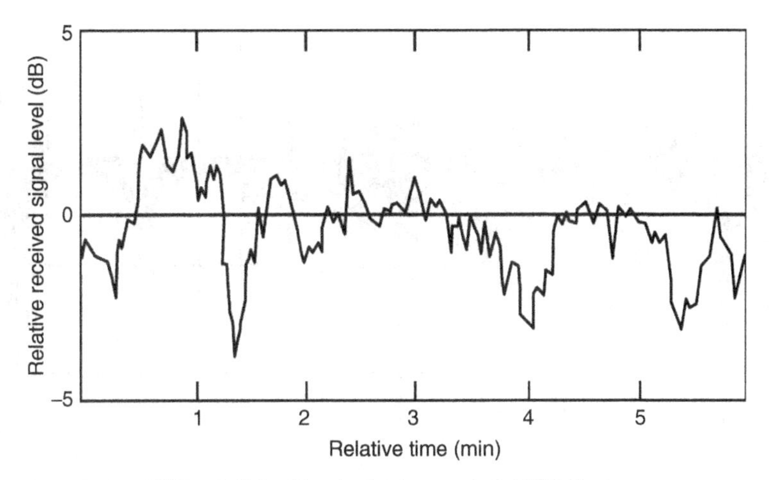

Figure 9.1 Nominal microwave path fading.

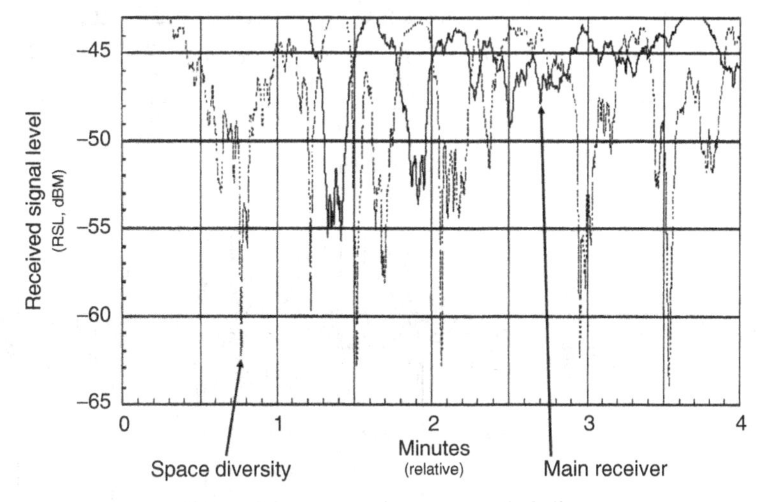

Figure 9.2 Deep microwave path fading.

Lin, 1977). However, the fading on orthogonal polarizations is completely uncorrelated (Barnett, 1974b). Since the cross-polarized signal is only suppressed a few tens of decibels by the receive antenna, sometimes the cross-polarized signal is as strong as the main signal (0-dB XPD) (Fig. 9.4).

If the radio path is only operating on one polarization, the appearance of an extra signal on the opposite polarization is not a factor. However, if both polarizations are being used for communication, receiver cross-polarization interference cancellation (XPIC) is required to provide satisfactory operation.

Most atmospheric multipath fading occurs at night in the summer or early fall. (An exception is reflective multipath caused by surface reflections. This multipath is greatest during the day when solar-induced thermal updrafts vary the reflection path length.) Conventional atmospheric multipath fading is associated with stable air masses such as warm, quiet air that facilitates atmospheric layering. High pressure cells and clear skies are associated with deep fading. Unstable air masses usually break stratification and reduce fading. The passing of weather fronts, wind, rain, and snow usually reduce fading if they occur along most of the radio path. Schiavone (1983) observed that the best predictor of atmospheric multipath was monthly wind speed (the higher the wind, the lower the fading). He also observed that fading increased for monthly water vapor pressures greater than 10 mbar. Outside low latitude coastal

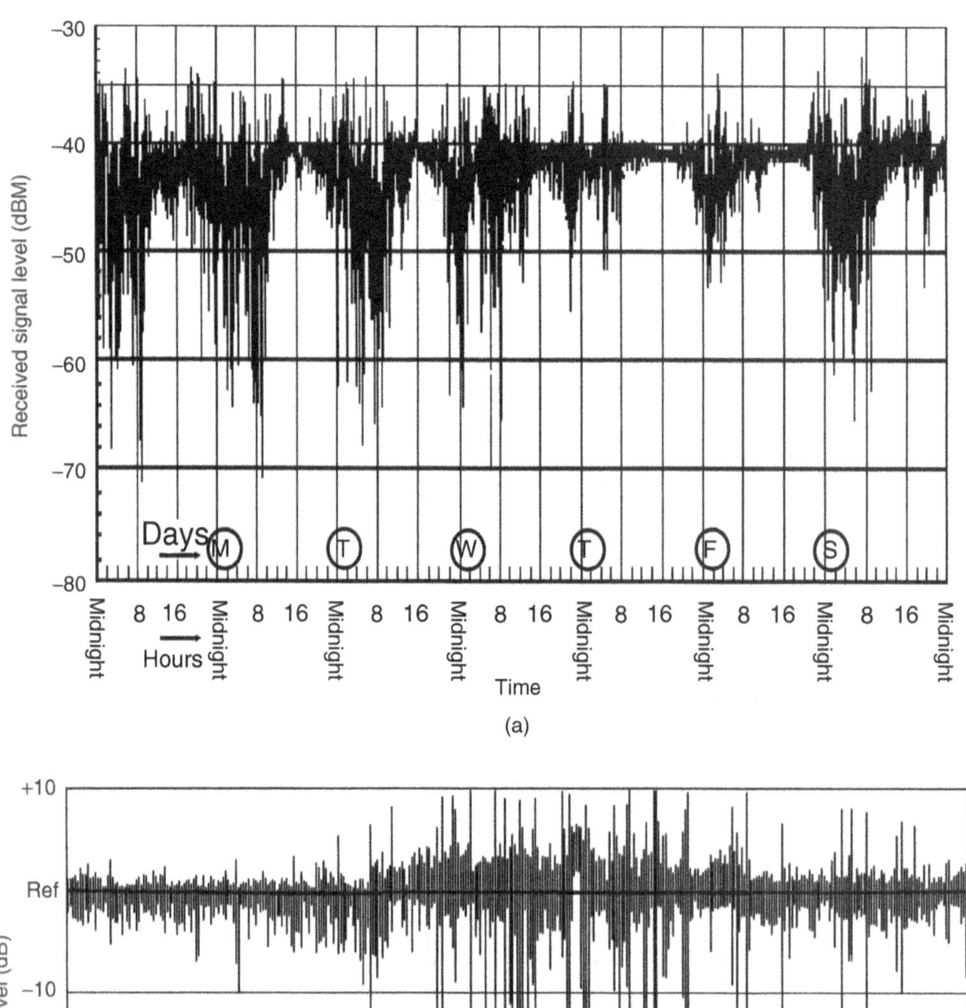

Figure 9.3 (a) Diurnal and (b) seasonal fading.

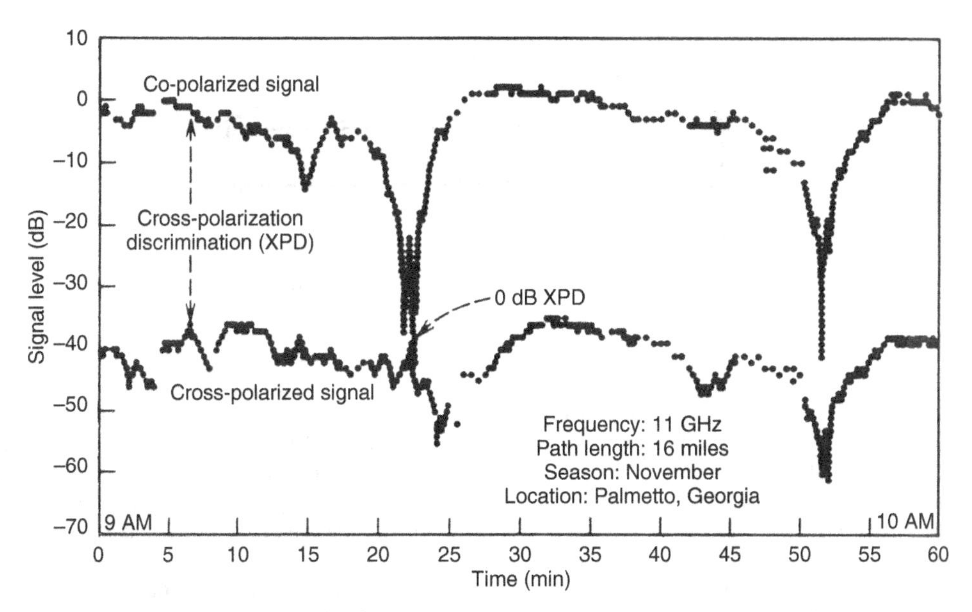

Figure 9.4 Co- and cross-polarization fading. *Source*: Lin (1977). Reprinted with permission of Alcatel-Lucent USA, Inc.

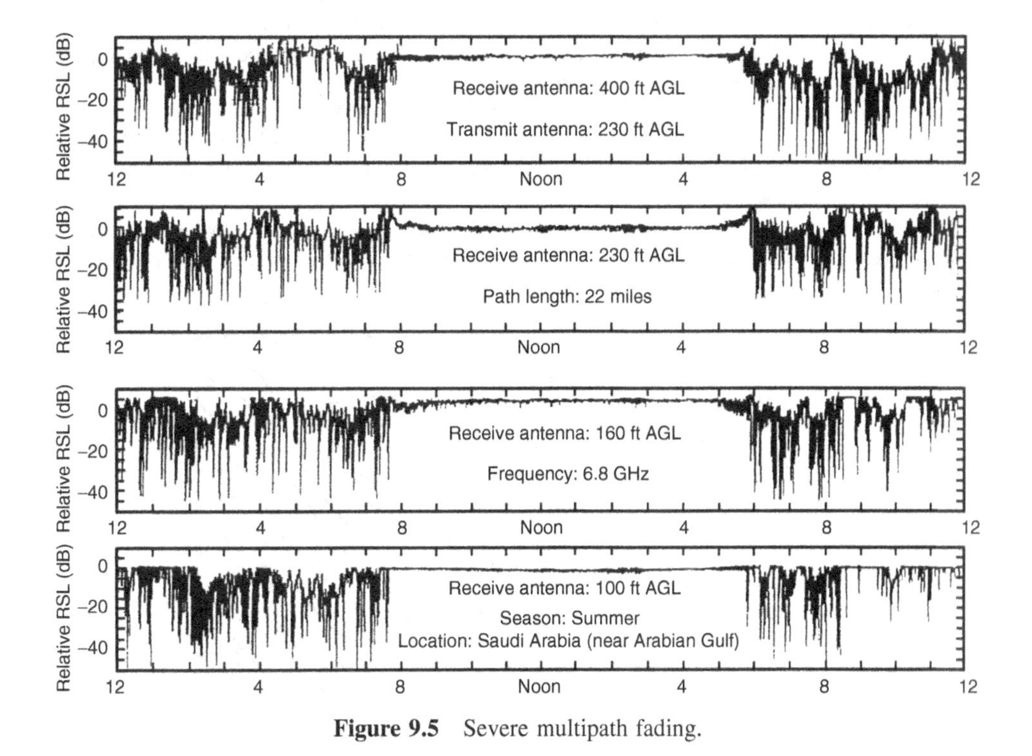

Figure 9.5 Severe multipath fading.

regions, fading occurred most often during humid seasons with low precipitation. Year to year fading variability is greatest in early autumn. Seasonal fading depends strongly on geographical location and therefore on local weather patterns. One of the worst propagation areas is the Saudi Arabian peninsula.

Figure 9.5 is an example of severe multipath fading (coupled with ducting). The flat topping of the bottom two plots was due to receiver limiting of upfading received signals. During daylight hours, the

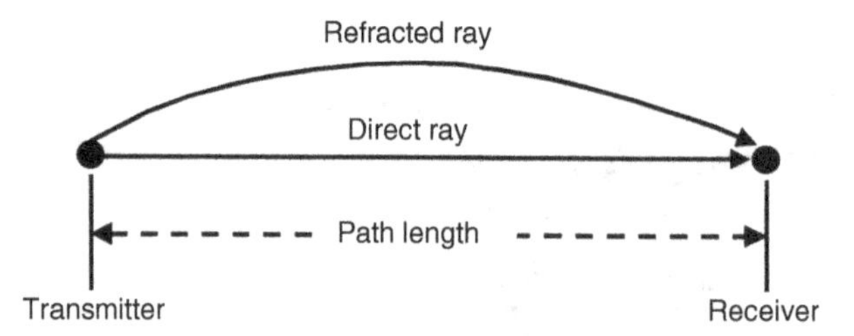

Figure 9.6 Basic multipath fading model.

thermal drafts from intense surface heating stirred up the atmosphere. At night, quiet relatively cool air developed layers that fostered multipath fading.

Investigators (Crawford and Jakes, 1952; De Lange, 1952; Kaylor, 1953; Rummler, 1979; Stephansen, 1981) measured the characteristics of several typical paths and discovered that the received signal variation was caused by the transmitted signal being split into several different signals traveling slightly different paths to the receiver. The received signals were replicas of the transmitted signal but delayed in time (Fig. 9.6).

This splitting and delay was hypothesized (England et al., 1938) to be due to refractive or reflective "layers" in the atmosphere (created by small vertical variations in the refractive index of air due to slight changes in air temperature and humidity). Owing to this delay, when the multiple signals were combined at the receive antenna, phase variations caused the delayed signals to add or subtract from the "normal" signal producing a wide variation in received signal level (RSL). This variation was both in time and vertical location of the receiving antenna. Figure 9.7 shows moderate fading produced during a quiet day where solar heating caused layers to drift up through the path.

Since this signal variation was produced by the combination of multiple signals traversing slightly different paths between the transmitter and the receiver, the phenomenon was named *multipath fading* (Monsen, 1980; Rummler et al., 1986; Siller, 1984). Figure 9.8 illustrates normal and simplified multipath transmission in the atmosphere.

The effects of deep multipath fading were moderated by designing microwave paths such that during normal propagation conditions, the received signal power was significantly greater than necessary to

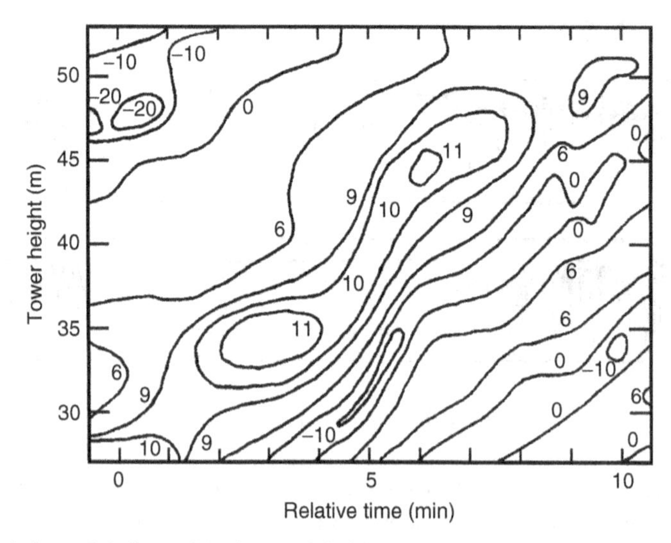

Figure 9.7 Variation of fading with time and height. *Source*: Adapted from Giloi, H., "A Study of Field Strength Height Profiles Caused by Multipath Fading," *IEEE Transactions on Antennas and Propagation*, p. 1384, December 1985. Reprinted with permission of IEEE.

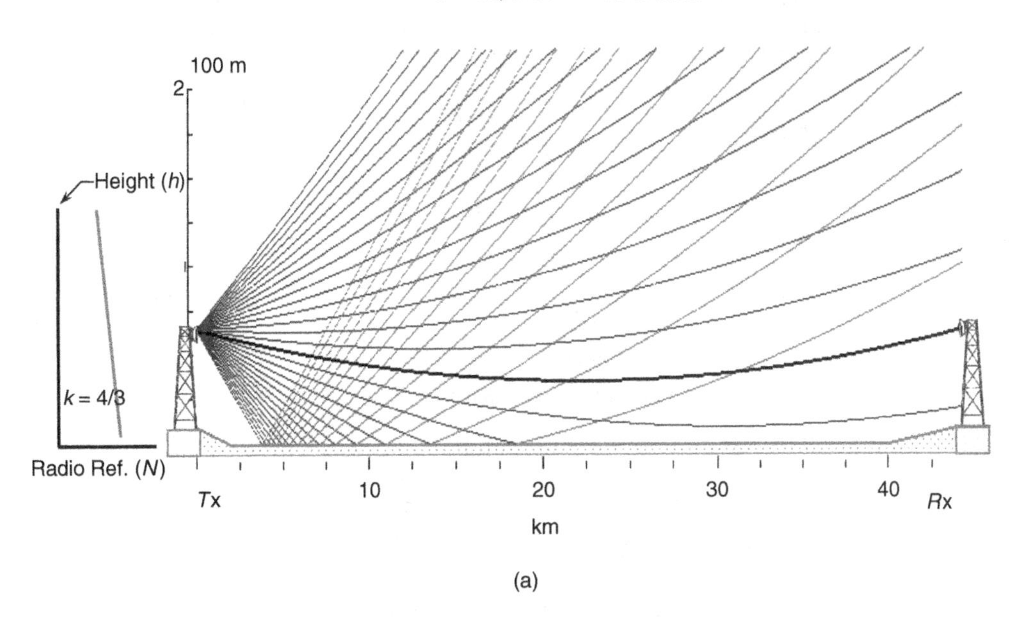

(a)

Within superrefractive layer $k = -1.1$; $dn/dh = -300$ N-units/km

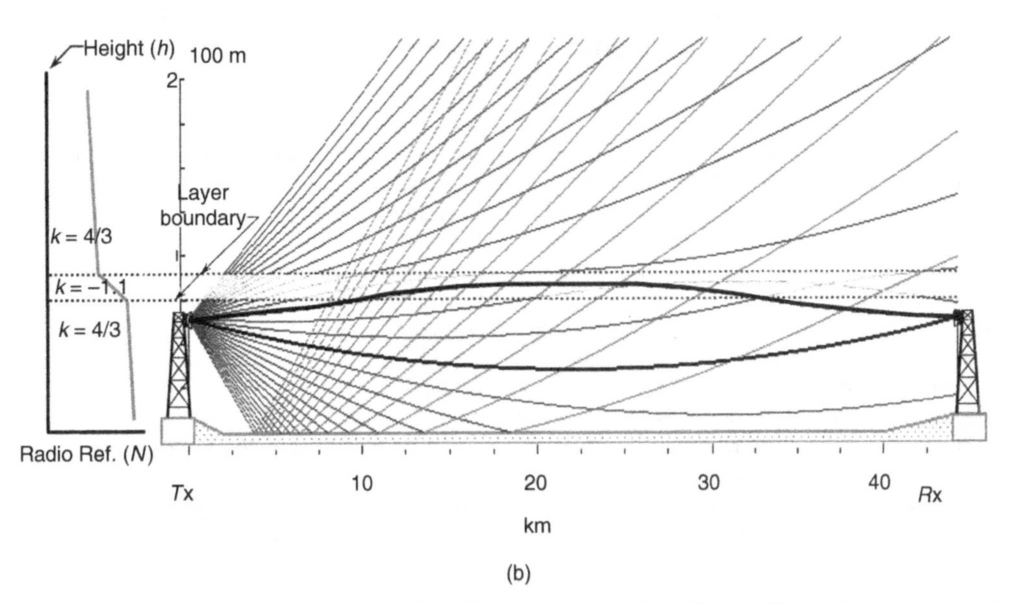

(b)

Figure 9.8 (a) Normal and (b) multipath radio wave propagation. *Source*: Illustrations courtesy of Eddie Allen. Used with permission.

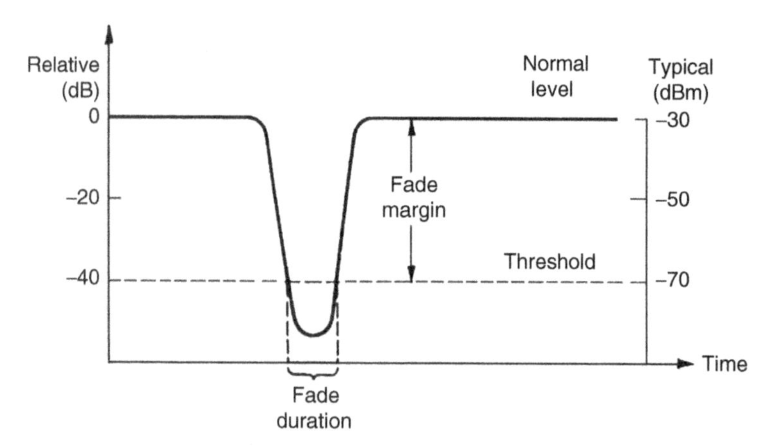

Figure 9.9 Thermal fade margin.

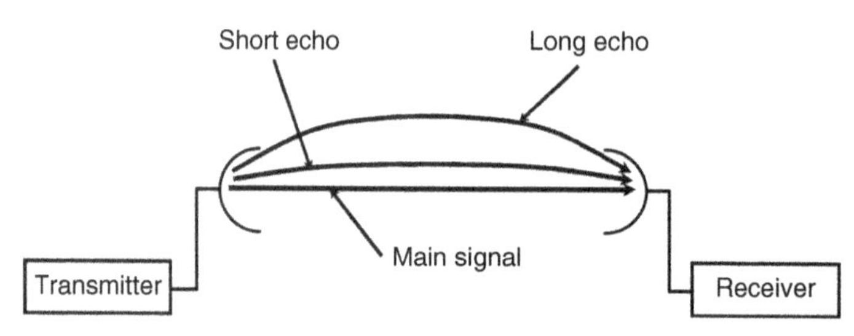

Figure 9.10 Ruthroff multipath model.

achieve acceptable communications. The difference between the normal RSL and the unacceptable RSL was termed as *fade margin*. To differentiate it from DFM, discussed later, this fade margin is often termed as *flat fade margin* (*FFM*) or *thermal fade margin* (as it is limited by the thermal noise generated in the front end of the microwave receiver) (Fig. 9.9).

Early researchers (Crawford and Jakes, 1952; De Lange, 1952; Kaylor, 1953), measuring paths in New Jersey and Iowa from 22 to 31 miles long, observed several different delayed signals. They observed signals delayed from significantly less than 1 ns to delays of 1, 3, 5, 9, and 12 ns. (The results are very similar to those reported from Denmark (Stephansen, 1981)). These delays represented path length extensions of approximately 1, 3, 5, 9, and 12 ft (longer than the nominal 22- to 31-mile path lengths). For radio signals with wavelengths of 2 inches. (6 GHz) to 2/3 inch. (18 GHz), these randomly varying path lengths would cause randomly varying RSLs when the different path (and relative phase) signals were combined at the receive antenna.

In the early 1970s, Ruthroff (1971) of Bell Laboratories derived a detailed theory of microwave radio path signal delay based on atmospheric stratified layers with differing refractive indexes (Fig. 9.10).

He showed that fading was due to only two delayed signals, one with delay less than 1 ns (short echo) and another typically with delay 5–10 ns (long echo). Reports of multiple echoes of varying delay were attributed to a long delayed signal echo with time-varying delay due to short-term variation in the height of the refractive atmospheric layer. Other researchers (Martin, 1985; Pickering and DeRosa, 1979) have confirmed the theoretical validity of the three-ray (main plus two different delayed signals) model as the most probable propagation case. Webster and Merritt demonstrated (1990) that 82% of the time path propagation is characterized as one, two, or three rays (Fig. 9.11).

Webster's time-lapse measurements (Webster, 1991) clearly show the existence of two or three rays during deep multipath fading. The relative amplitude of the direct and delayed signals varied considerably but when fading occurred, they were all nearly the same amplitude. As will be noted later, the short echo

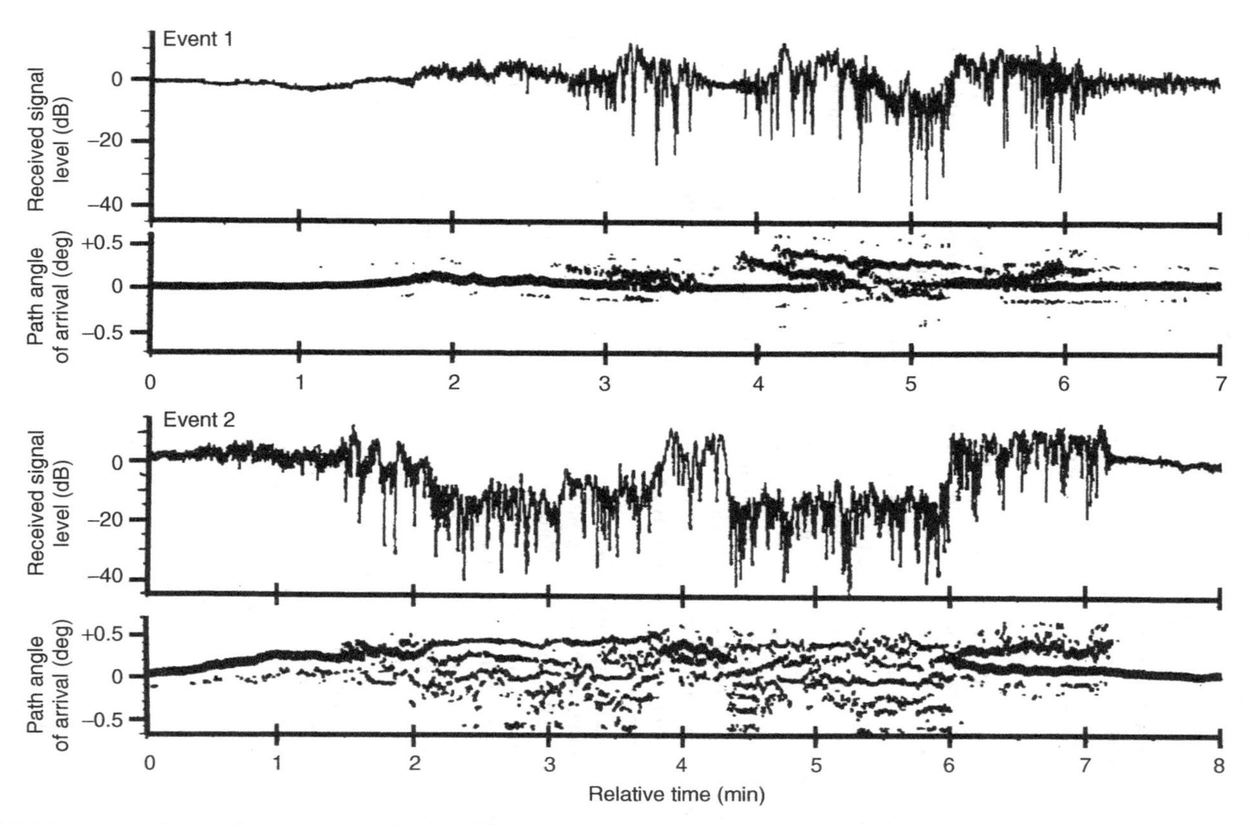

Figure 9.11 Multiple propagation paths during multipath fading. *Source*: Adapted from Webster, A. R., "Multipath Angle-of-Arrival Measurements on Microwave Line-of-Sight Links," *IEEE Transactions on Communications*, pp. 798–803, June 1991. Reprinted with permission of IEEE.

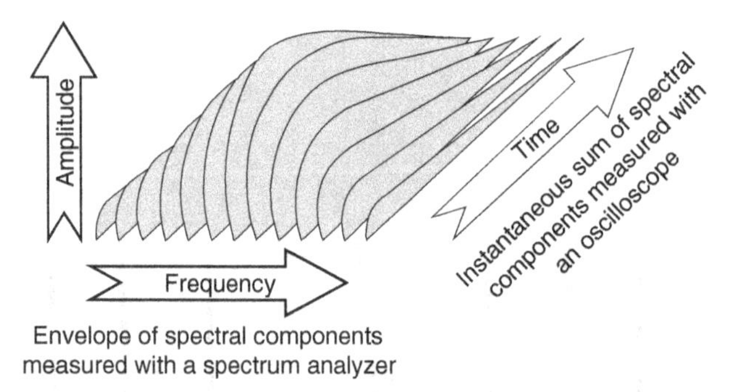

Figure 9.12 Relationship of frequency and time domain measurements.

was responsible for broadband received signal loss. The longer echo was responsible for significant signal amplitude distortion (but insignificant power loss). Ruthroff (1971) and Sasaki and Akiyama (1979) demonstrated theoretically that for horizontal paths, the longer delay time increases as the third power of the distance between the transmitting and receiving antennas. Sasaki and Akiyama's (1979) results suggest that the delay may decrease as the path is slanted. However, Lane (1991) noted that delay might actually increase due to off-path surface reflections if the slanted path has excessive clearance.

When considering the delayed echo, if the direct signal is the dominant signal, the fading produced is termed *minimum phase*. If the delayed signal is dominant, the fading produced is termed as *nonminimum phase*. Consider the transfer function of a two-path network, which could represent two propagation paths through the atmosphere. If the first path (having the shorter delay) is stronger than the second (longer delay) path, the transfer function is minimum phase. If the second path is stronger than the first, the transfer function is nonminimum phase. This is not a theoretical problem or a mathematical abstraction. It is a real effect that can be verified with a network analyzer. If a pair of signal strengths with a fixed relative delay produces a transmission minimum at a given frequency, interchanging the delay associated with the two paths will clearly reverse the sign of the delay distortion in the region near the minimum.

When fading is light, the predominant fading is minimum phase (ITU-R, 2003). When deep fading is occurring, both minimum phase and nonminimum phase fades are approximately equally likely (Giger and Barnett, 1981; ITU-R, 2003). Often the path alternates between minimum phase and nonminimum phase fades several times during a fading period. During this time, the main and long delayed signals are within a few tenths of a decibel of the same power level. A slight change in the power of either signal can cause the change between a minimum phase and nonminimum phase fade.

A signal or path can be viewed in the time domain or the frequency domain (Fig. 9.12).

If the radio signal is viewed in the frequency domain, the received signal observed on a spectral analyzer is the spectrum of the transmitter to which is added the frequency domain transfer function (Fig. 9.13) of the path. The frequency domain distortion (amplitude vs frequency power transfer function) caused by a single echo is easily determined (Kizer, 1990):

$$\begin{aligned} A(\text{dB}) &= \text{channel amplitude versus frequency response} \\ &= \frac{\text{received power at } f}{\text{transmitted power at } f} \\ &= 10\log[1 + r^2 - 2r\cos(\phi + 2\pi f \tau 10^{-3})] \end{aligned} \tag{9.1}$$

f = measurement frequency (MHz) relative to channel center frequency;
ϕ = time-varying phase shift;
τ = $1000/\Delta f$ (MHz);
= echo delay (ns);
Δf = frequency difference between consecutive power nulls (or peaks);
r^2 = (echo power)/(main signal power);
E = echo relative amplitude (dB) 10 log (r^2);
= $10\log(r^2)$;

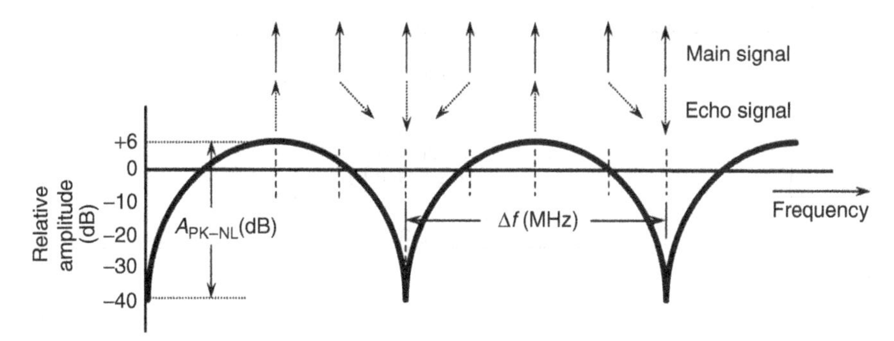

Figure 9.13 General frequency domain distortion (path frequency domain transfer function).

$$= 20\log\left\{\frac{\left[\left(10^{A_{PK-NL}/20}\right)-1\right]}{\left[\left(10^{A_{PK-NL}/20}\right)+1\right]}\right\};$$

$= $ echo relative amplitude (dB) 10 log (r^2);

$A_{PK-NL} = $ maximum power (peak, dB)−minimum power (null, dB).

9.1 FLAT AND DISPERSIVE FADING

The short echo (relative delay typically <1 ns) usually varies in and out of phase with the direct signal fairly quickly. Its effect is to produce a short-duration, broad, frequency-independent signal power depression near the operating frequency (or more rarely, a slight signal enhancement). This produces the well-known *flat fade*. The overall received signal power loss due to the longer delayed signal echo of a *dispersive fade* is usually relatively insignificant. Both of these effects are illustrated in Fig. 9.14.

For typical dispersive fade delays, the frequency distortion is quite narrow and overall received signal power depression is slight (<10 dB) (Fig. 9.14). This explains Kaylor's comment (Kaylor, 1953) that "... deep selective fading is ordinarily accompanied by a [broadband] 6 to 10 dB signal depression ...." This was demonstrated experimentally (Liniger, 1984) on a 29-mile (47 km) upper 6-GHz path in Switzerland (Fig. 9.15).

The 0-dB relative amplitude response represented channel response without fading. During dispersive fading, the overall channel amplitude versus frequency response was depressed approximately 10 dB while a dispersive notch traveled through the channel at about 10 MHz/s.

Figure 9.16 is an example of a 30-mile lower 6-GHz path in southern California. The radio is a SONET OC3 radio operating in a nominal 30-MHz channel. The time sequence pictures (beginning at upper left and finishing at lower right) are of the received signal spectrum during a dispersive fade event. The spectrum analyzer vertical display is 5 dB per division. The horizontal display is 5 MHz per division. The dispersive fade notch moved across the spectrum of the digital radio from left (lower frequency) to right (higher frequency). The total elapsed time for the sequence was 45 s.

A significant exception is the abnormally long echo caused by off-path reflections (Fig. 9.17). This echo produces no significant in-band power loss but causes considerable FM echo distortion noise and digital radio intersymbol interference.

Multipath fading is usually caused by a short- or long-delayed echo signal (depending on the current atmospheric conditions). The short delay is usually less than 1 ns and causes flat fading. The long delay is typically 6–18 ns with 9 or 10 ns (Rummler, 1979, 1980) (not the often-quoted 6.3 ns) being typical. The long echo causes dispersive fading.

In the mid-1970s, digital radios began to be used on fixed point to point microwave paths. By the 1980s, these were the dominant long-distance microwave radios. While flat fades affected these radios, it was noticed that serious radio receiver outages were occurring at times when flat fading was not occurring (Giger and Barnett, 1981). This issue was addressed by various administrations worldwide. The

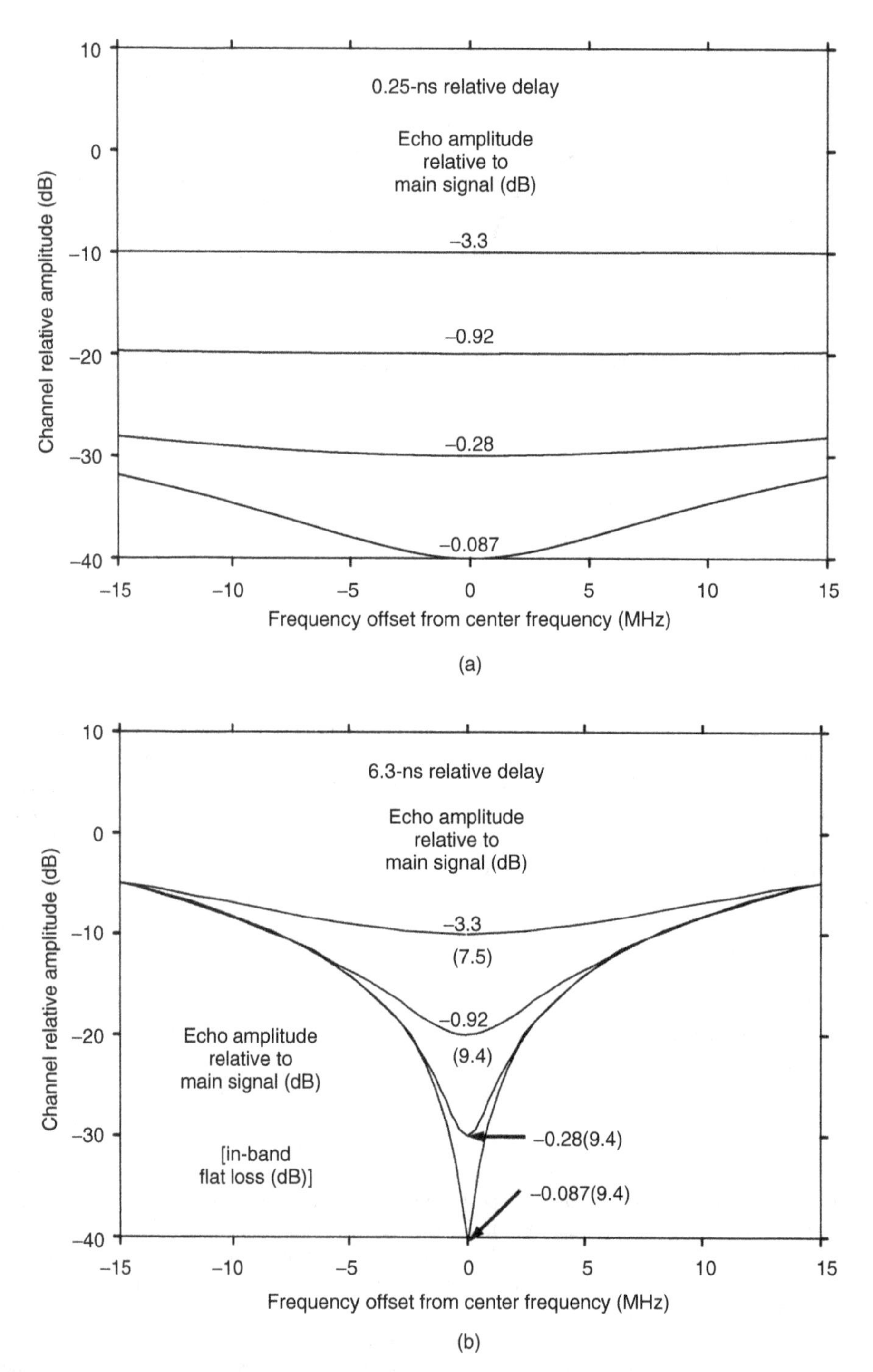

Figure 9.14 Frequency domain distortion: (a) 0.25- and (b) 6.3-ns echos. Typically, the spectrum shifts left or right over time.

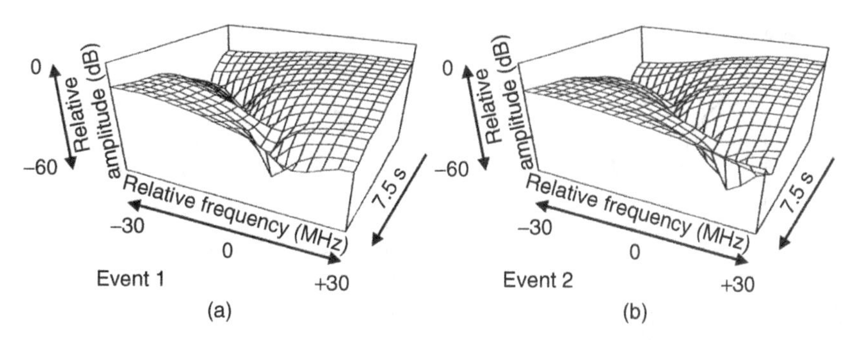

Figure 9.15 (a) Channel amplitude versus (b) frequency response during dispersive fading. *Source*: Adapted from Liniger, M., "Sweep Measurements of Multipath Effects on Cross-Polarized RF-Channels Including Space Diversity," *IEEE Global Telecommunications Conference (Globecom) Record, Volume III*, pp. 45.7.1–45.7.5, November 1984. Reprinted with permission of IEEE.

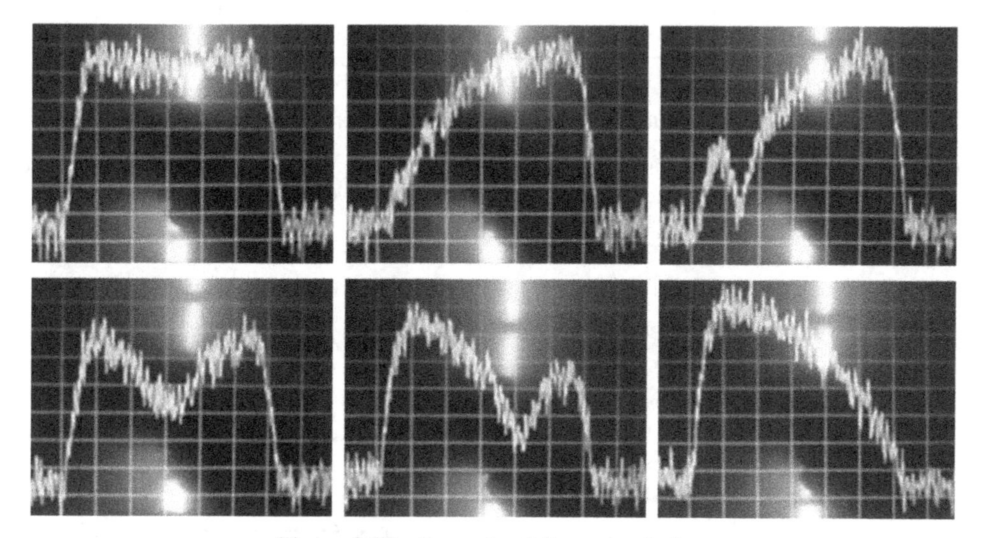

Figure 9.16 Example of dispersive fading.

mechanism for this type of fading was the long path echo that could be ignored in older FM transmission radios. The long echo caused negligible received signal power loss. However, it was significantly delayed in time from the direct signal and therefore caused considerable distortion in the time domain signal of the digital radio (Fig. 9.18).

The long echo delay, typically in the 6- to 12-ns range, is still short relative to a digital radio's signaling interval. It has much more effect on high speed digital signals (that are relatively closely spaced in time).

The effect of the long echo combining with the direct signal is to distort the signal and cause the transmitted digital pulse to broaden or disperse in time. The outages caused by this distortion were termed *dispersive fading* to distinguish them from the more common flat fading.

For the (unusual) case of very long echoes [echo delay as long or longer than the radio signal time spacing (baud interval)] caused by off-axis terrain reflections, the echo can be considered as an interfering signal. In this case, the approximate power level and relative delay can be calculated and the interference treated with conventional interference (C/I and T/I) calculations (TIA, 1994). For the more typical case of relatively short echoes, the effect of the echo is directly related to in-band frequency response distortion (Fig. 9.19).

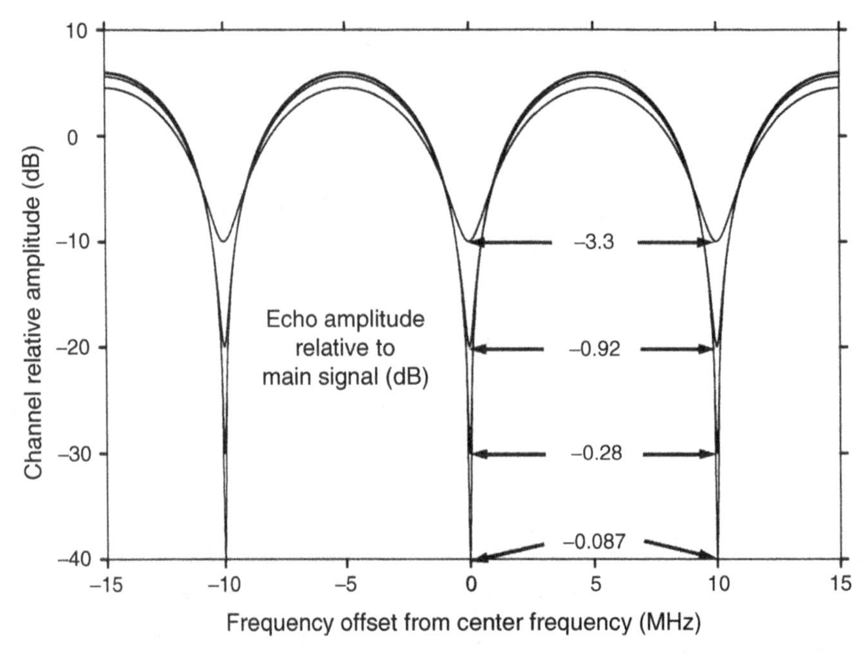

Figure 9.17 Frequency domain distortion (effect of unusually long echo of 100 ns). Could be caused by off-path reflection. Typically, the spectrum shifts left or right over time.

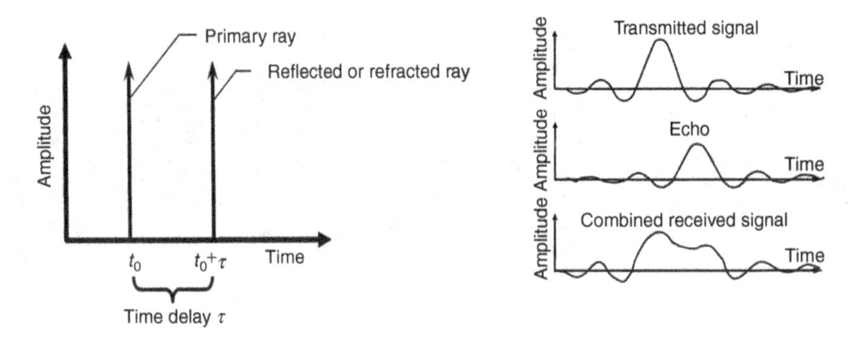

Figure 9.18 Time domain echo distortion. The widened received signal represents signal dispersion.

Many approaches have been taken to characterize dispersive multipath fading (Rummler et al., 1986). Rummler of Bell Laboratories characterized dispersive fading (Lundgren and Rummler, 1979; Rummler, 1979, 1980, 1981, 1982, 1988) with a model that has become the industry standard. As he could not measure the effect of signal distortion in the time domain, he measured the frequency band distortion (narrow signal spectrum nulls or "notches") and related that distortion to an inferred long delay echo. The parameter he focused on was the difference between the depth of the notch relative to the highest power value within the radio channel. This value, in decibels, he termed *in-band power distortion* (IBPD).

He determined that the typical in-band frequency domain distortion statistical characteristics (notably, the critical IBPD time below level characteristic of $e^{-B/\tau}$ as noted in Rummler's research) could be adequately characterized by assuming a two-path model (Rummler, 1981, 1982). A direct connection between the transmitter and the receiver would be used to simulate one atmospheric path. Another connection would provide a connection between transmitter and receiver with a nominal echo delay (in milliseconds) of about one divided by six times the radio channel bandwidth (in megahertz). For typical 30-MHz microwave radio channels, a long echo delay of 6.31 (usually rounded off to 6.3) ns

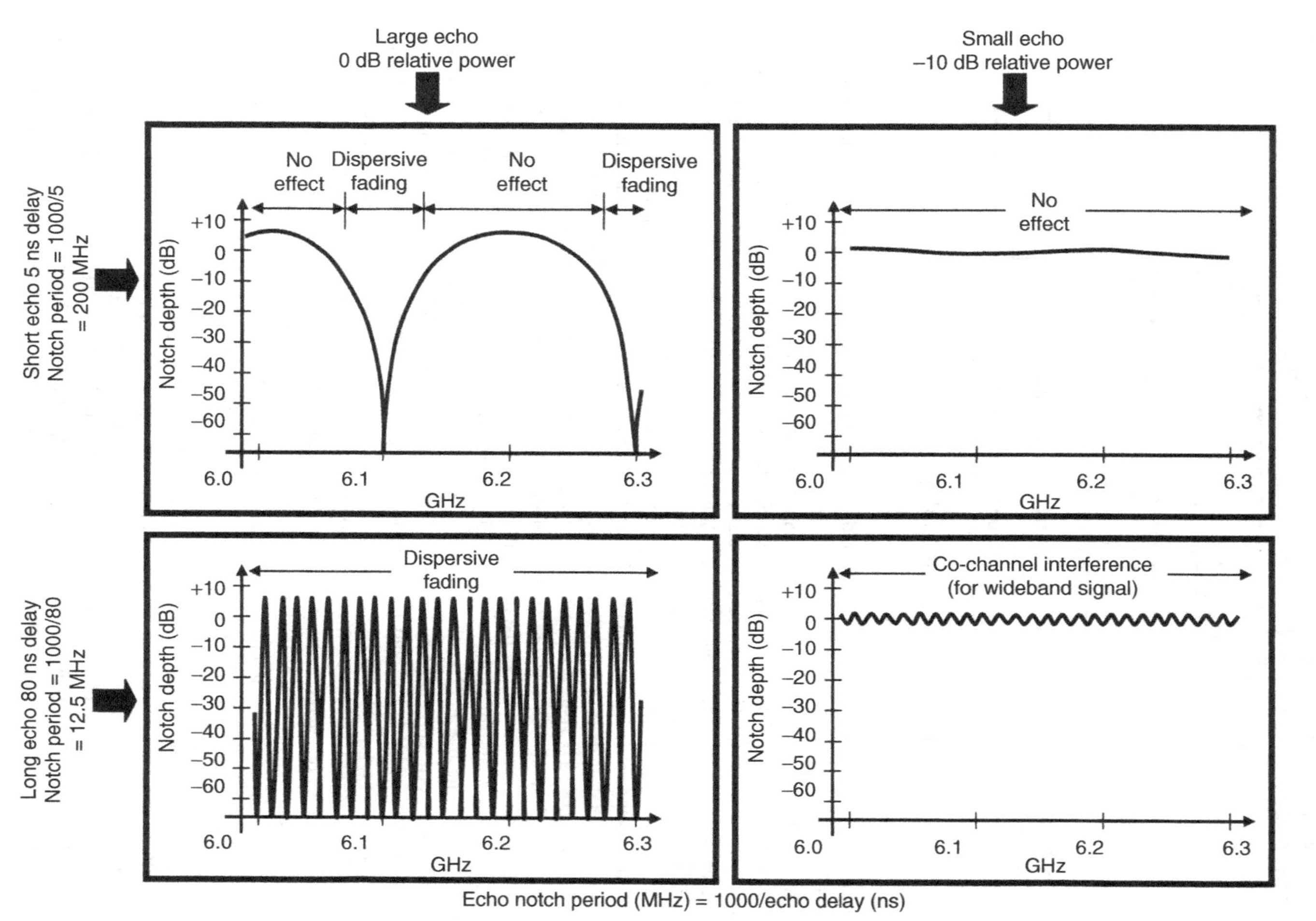

Figure 9.19 Dispersive fade frequency domain distortion.

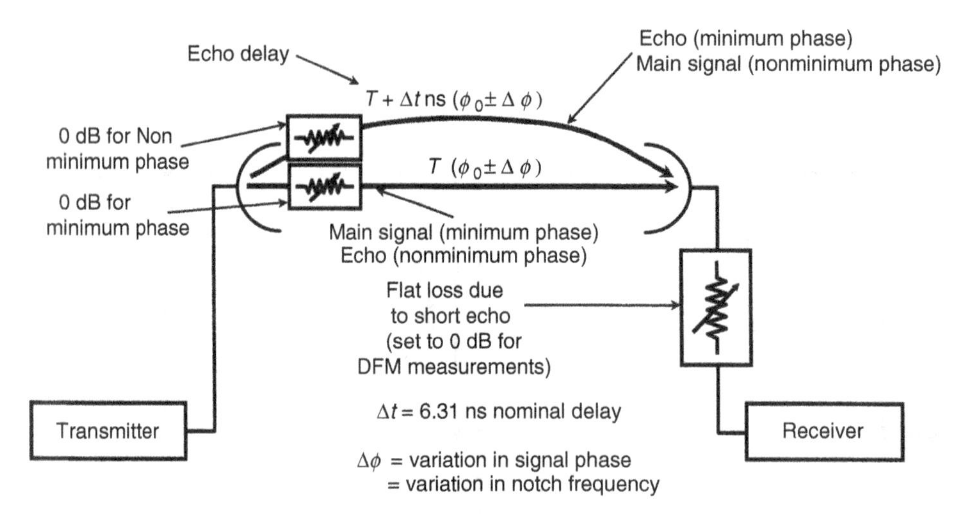

Figure 9.20 Rummler dispersive fading model.

was chosen. Short echo flat fading was considered to be adequately characterized by a simple variable attenuator (Fig. 9.20).

Rummler never claimed 6.3 ns was the typical long echo delay for his radio path. Actually, he measured echo delays from 5.6 to 18.4 ns (Rummler, 1979). Stephansen (1981) reported echo delays up to 15 ns for paths shorter than 80 km. The average echo delay Rummler measured on his 26.4-mile test hop was 9.1 (Rummler, 1979) or 10 (Rummler, 1980) ns. What he did state (Rummler, 1981) was that if τ was the delay in nanoseconds and B was the radio channel bandwidth in megahertz, $\tau \leq 1000/(6B)$ was necessary to adequately characterize the majority of the shapes of the observed in-band distortion (Other researchers have suggested slightly other formulas, e.g., Xiong (2006) suggests 10 rather than 6.). Too long a reference delay underestimates short delays and too short a reference delay underestimates long delays. Rummler chose a 6.31-ns delay to statistically characterize the in-band distortion measured on his 30-MHz test hop. Rummler observed (1979) that using a 6.31-ns delay in his model achieved not more than 1-dB amplitude error for 98% of the observations. The 6.31-ns delay was a good estimator except for (relatively rare) long delays (of course, this simplified three-ray model does not estimate upfades). Many of his results, including the applicability of the τ (ns) $\geq$ {1000/[6B (MHz)]} criterion for a wideband radio channel, have been reproduced by others (Sylvain and Lavergnat, 1985) with a wide range of radio channel bandwidths (and delay τ) and path characteristics. Many engineers mistakenly assume 6.3 ns characterizes the path's typical delay despite Rummler's clear statement that this delay represented a nonphysical model (Rummler, 1979). Rummler (1979, paragraph 3.2) points out some of the purposeful artificial manipulations of reference delay necessary to obtain good statistics using such a simple model.

Keep in mind that the purpose of this long echo delay τ is to estimate the statistics of the majority of the IBPD events. It is not meant to characterize the mean path delay (Mean path delay was discussed at length in the work by Rummler (1979) (Section V).). A wide range of delays can be used to characterize the IBPD. While 6.3 ns is the most popular delay, results similar to Rummler's have been obtained using reference delays from 2.6 to 3.5 ns (Caskey and Moreland, 1988; Sandberg, 1980; Sylvain and Lavergnat, 1985). Greenfield (1984) reproduced Rummler's statistics on a long path that must have been characterized by very long delay. Caskey and Moreland (1988) got the same statistics using reference long echo delays of both 4.2 and 6.3 ns on a 40-MHz wide radio path. As has been shown many times, IBPD amplitude versus time statistics are (relatively) independent of reference delay as well as path length and physical characteristics. Median long echo path delay (which determines the appropriate path signature curve) is not.

Other researchers have expanded Rummler's two-ray model to a three-ray model where one of the rays is the delay reference, one ray is very short delay (on the order of 0.5 ns) and one ray is much longer (on the order of 6–16 ns). The short delay ray was used to simulate flat fading. In Rummler's model, this just represented as an attenuation of the main signal (upfades were ignored). The long delay ray would lag or lead the main reference ray delay depending on whether minimum or nonminimum phase dispersive fading was being simulated. This model provides insight into both flat fading and dispersive fading and, unlike Rummler's model, it does exhibit upfades. It has been studied extensively (Goldman, 1991; Shafi, 1987). However, it has not produced new practical results.

Engineering techniques using flat fade margin (FFM) were well established and engrained by the mid-1970s (Vigants, 1975). Although several new methodologies were suggested for characterizing dispersive fading, they typically were only useful in an academic environment. The various new methodologies, while interesting, offered no clear engineering guidelines. Most engineers were comfortable with fade margin and wanted to continue to use that concept. Dupuis et al. (1979) and Rummler (1982b) introduced the concept of dispersive fade margin (DFM). This approach has become the North American industry standard method of characterizing digital radio sensitivity to long-path echo distortion.

The basic idea of path performance analysis (from the Vigants North American perspective) is that path outage time can be estimated using a fade margin (FM, dB) as follows:

$$\text{Seconds of fading} = \frac{2.5 \times 10^{-6}\, CRfD^3T_0\, 10^{-\mathrm{FM}/10}}{I} \tag{9.2}$$

The seconds of fading are calculated first for flat fading using the FFM and then for dispersive fading using the DFM. The overall path performance is estimated by the sum of both sets of calculated seconds of fading. In these calculations, C is climate factor, R is path roughness factor, f is operating frequency in gigahertz, D is path distance, T_0 is the seasonal fading time, and I is diversity improvement factor. See Section 16.5 for more details on this formula.

Given the similarity in fading mechanisms that produce flat and dispersive fading, one would expect the fading time formulas to be similar. Vigants (1984) showed experimentally that $10^{-\mathrm{DFM}/10}$ varies as a function of distance cubed. Sasaki and Akiyama (1979) showed experimentally and theoretically that both short and long multipath delays vary as a function of distance cubed. To date, no experiments have demonstrated the effect of frequency or climate and roughness factors.

The effect of DFM is to set an upper limit on path performance that cannot be improved by the typical approaches such as increasing transmit power or antenna gain. The examples in Figure 9.21 were calculated at 6 GHz using space diversity improvement described by Lin et al. (1988), Kizer (1995), and TIA (1994).

Rummler and Lundgren demonstrated that flat fading and dispersive fading were uncorrelated (1979, Fig. 4). Therefore, the effects of each could be determined and added to get a composite fading estimate for the path. Flat fading was well characterized. What Rummler and Lundgren did was to determine the formulas necessary to calculate the probability of dispersive fading. It was a short step to using that information to create a DFM (discussed later).

DFM is a hypothetical fade margin synthesized by using the Rummler multipath model test setup (with zero flat fade attenuation) to measure a receiver's sensitivity to dispersive fading. The test setup measures receiver BER as a function of path echo delay phase (frequency of in-band null or "notch" relative to channel center frequency) and amplitude of the echo relative to the undelayed signal. The resulting receiver signature curve, often called as *W curve* (or an M curve if the decibel values are negative), is measured for a defined BER (typically 10^{-3} or 10^{-6}). The curve is then used to calculate a DFM.

Figure 9.22 plots the frequency versus notch depth caused by a 6.3-ns echo that causes a 10^{-3} BER in a typical wideband receiver. The derivation of DFM is based on procedures described by Rummler and Lundgren (1979) and Rummler (1982b). The basic procedure is to derive a curve of notch depth for a given BER over the band pass of the receiver. Although a receiver BER of 10^{-3} was used, Rummler and Lundgren pointed out (Lundgren and Rummler, 1979) that their method was applicable in the BER range of 10^{-3} to 10^{-6}. They provided a formula for estimating the time duration of a given notch depth. The estimated time of receiver outage is the integral of the notch curve values multiplied by the Rummler–Lundgren relationship. The outage time can then be directly related to average yearly or monthly outage by dividing by the appropriate time. Taking this, the percentage of outage time can then be converted to a hypothetical "fade margin" using conventional FFM formulas.

In 1987, Martin (1987) outlined a concept for integrating the receiver signature curve to provide an indicator of probability of error. In 1989, Bellcore (now Telcordia) published a formula (Bellcore (Telcordia) Staff, 1989) that did all the necessary steps to relate the signature to DFM.

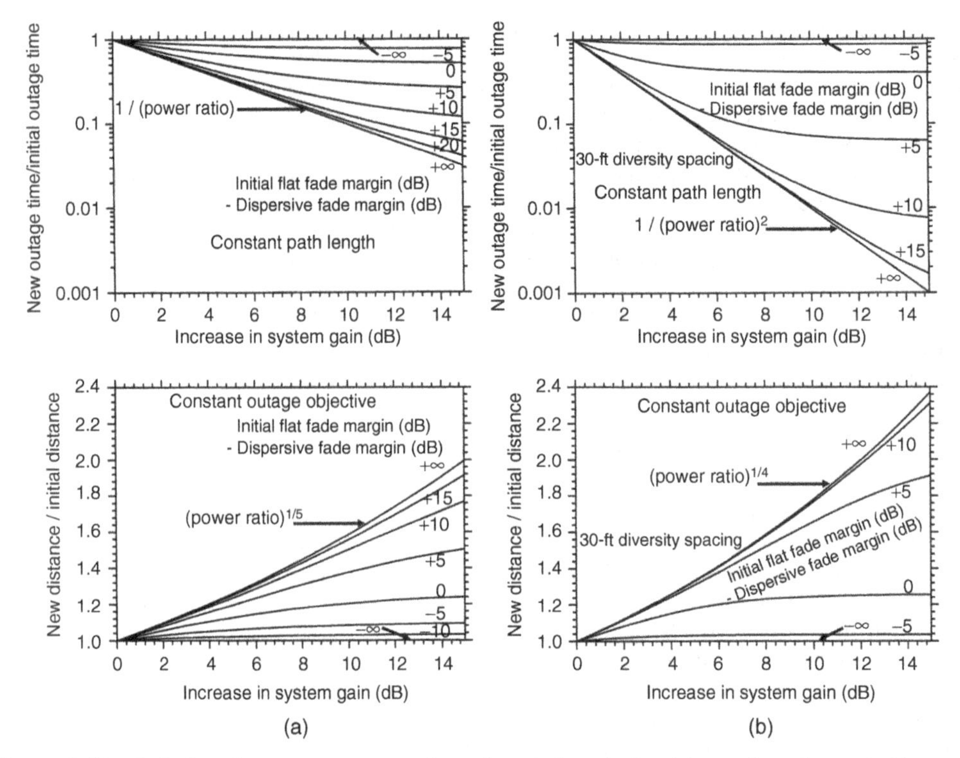

Figure 9.21 Path length versus system gain and path length for (a) nondiversity and (b) space diversity paths (multipath-dominated 6-GHz paths).

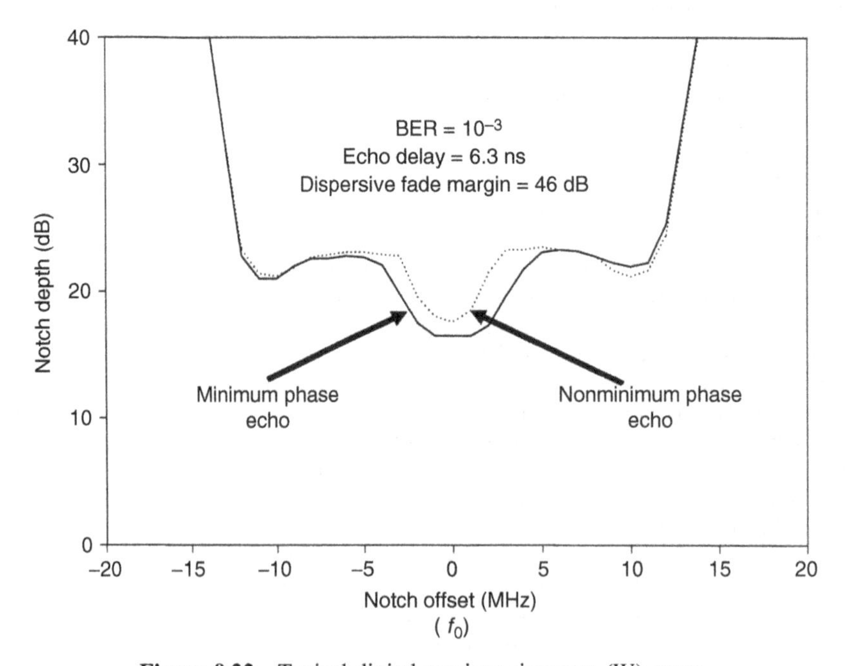

Figure 9.22 Typical digital receiver signature (W) curve.

$$\mathrm{DFM(dB)} = 17.6 - 10\log\left(\frac{\mathrm{Sw}}{158.4}\right) = 39.6 - 10\log(\mathrm{Sw}) \tag{9.3}$$

$$\mathrm{Sw} = \int_{-39.6}^{+39.6} [\mathrm{e}^{-Bn/3.8} + \mathrm{e}^{-Bm/3.8}]\mathrm{df}(MHz) \tag{9.4}$$

Bn = decibel value of B at frequency f (MHz) for a nonminimum phase W curve;
Bm = decibel value of B at frequency f (MHz) for a minimum phase W curve.

A more general form of this formula is given later.

Test sets that estimate DFM use equally spaced frequency offset values of Bn and Bm with simple rectangular integration to approximate Sw

$$\mathrm{Sw} \approx \sum_{J=1}^{K} [\mathrm{e}^{-\mathrm{Bn}(J)/3.8} + \mathrm{e}^{-\mathrm{Bm}(J)/3.8}]\Delta f \tag{9.5}$$

Δf = (highest frequency − lowest frequency)/(K − 1).

Assume a typical 30-MHz channel lower 6-GHz digital radio. Further assume $K = 16$ equally (frequency offset) spaced measurements of Bn(J) and Bm(J) across the radio channel. Those normalized measurements would be taken at frequencies −15, −13, −11 MHz, ..., +11, +13, and +15 MHz (relative to channel center frequency).

$$\Delta f = \frac{+15-(-15)}{16-1} = 2\text{ MHz for the above example} \tag{9.6}$$

Typically measured Bn and Bm are between 10 and 40 dB. If, for this example, Bn(J) = Bm(J) = 24.9, then DFM ≈ 50 dB. At present, DFM estimates exceeding 60 dB are not very repeatable.

The Telcordia formula is the standard formula in North America for determining DFM. The advantage of this approach is the DFM can be used in the path design calculations exactly similar to those for FFM. It is powerful in that it allows different similar bandwidth radios to be compared on their ability to discriminate against dispersive fading. One disadvantage is it only accurately characterizes dispersive fading for relatively short paths. (Typical current DFM calculations use 6.3-ns delay to characterize the path regardless of length or characteristics while Rummler showed that his typical 26.4-mile path was characterized by a median delay of 9.1 ns (Rummler, 1981).) It fails to accommodate the characteristics of actual paths. Modifying W curves to accommodate different path delays is well understood. Estimating anticipated path delay or dispersive fading is not. An attempt was made to relate actual path dispersive fading to a reference fading (Ranade and Greenfield, 1983) by modifying the dispersive fading outage time but that approach was never finalized.

On the basis of current research, the best estimate seems to be to calculate DFM for the standard 6.3-ns W curve and then modify it for an estimated 9.1-ns average long echo for a 26.4-mile path (Rummler, 1979). That is accomplished (as noted in Appendix 1) by subtracting 3.6 dB = 22.8 log (9.1/6.3) from the DFM calculated for the 6.3-ns W curve to account for W-curve variation with path delay.

$$\text{Seconds of dispersive fading} = \frac{2.5\times10^{-6}\,CRf\,D^3T_0\,10^{-\mathrm{DFM_m}/10}}{I} \tag{9.7}$$

$$\mathrm{DFM_m}(9.1\text{ ns}) = \text{DFM }(6.31\text{ ns})-\text{W curve effect of delay}-\text{probability effect of delay}$$
$$= \text{DFM } -3.6 \tag{9.8}$$

Ruthroff (1971) theorized and Vigants (1984) demonstrated that path delay is a function of distance cubed (D^3). ITU (ITU-R, 2003) suggests a $D^{1.3}$ to $D^{1.5}$ path variation. However, Vigants' results (Vigants, 1984) suggest the ITU results are the result of unique aspects of propagation or analysis. On the basis of current research the D^3 factor in the seconds of dispersive fading formula provides the best estimate for fading variation due to path length.

Although diversity improvement I is well understood (with the exception of angle diversity) for flat fading, there is no industry agreement (nationally or internationally) as to how to characterize diversity improvement for dispersive fade calculations. While the Bell Laboratories results (Lin et al., 1988) for dispersive fading space diversity improvement have been accepted by TIA (1994), they have not achieved widespread use.

An issue not currently addressed by industry methodologies is receiver hysteresis. All published W curves are based on "static" measurements (BER after the receiver has recovered from any anomalous performance). It is well known (Lundgren and Rummler, 1979) that dispersive events occur quickly and can cause momentary loss of synchronization for large BERs. "Dynamic" measurements more accurately represent this actual performance. Nevertheless, these "dynamic" measurements, described in Bellcore specifications (Bellcore (Telcordia) Staff, 1989) and typical DFM measurement equipment manuals, are generally not available. The measured differences between static and dynamic DFMs range from less than 1 dB to several decibels depending on the particular receiver (primarily a function of the receiver time domain equalizer convergence).

Another significant issue is the extreme variability from year to year of total time of IBPD fading. Variations in year-to-year fading time for a given radio path of two to one (Rummler, 1981) or three to one (Ranade and Greenfield, 1983) are not uncommon.

DFM is a powerful equipment comparison parameter. It is also a useful, although somewhat inexact, path design tool.

9.A APPENDIX

9.A.1 Fading Statistics

Lin (1971, 1972) (based in part on Rice's work (Rice, 1944, 1945, 1948, 1958)) determined the following general relationships for path flat multipath fading under very general conditions:

$$P(v \leq L) \propto L^{2M} \quad = \text{probability that the received signal is less than or equal to } L \text{ for an overland path} \tag{9.A.1}$$

For $M = 1$, an order of magnitude outage probability change (e.g., 10 s of outage to 1 s of outage) represents a 10-dB change in fade depth (30- to 40-dB fade margin).

For $M = 2$, an order of magnitude outage probability change (e.g., 10 s of outage to 1 s of outage) represents a 5-dB change in fade depth (30- to 35-dB fade margin).

$$T(L) \propto L \quad = \text{average duration of a single received signal level fade to level } L \tag{9.A.2}$$

$L = \sqrt{p_{\text{FM}}} << 1$;
$\quad =$ inverse of fade margin expressed as a voltage ratio;
$p_{\text{FM}} = 10^{-\text{FFM}/10}$;
$\quad =$ inverse of FFM expressed as a power ratio;
FFM $=$ flat fade margin (dB) expressed as a positive number greater than 20;
$M =$ order of diversity;
$\quad = 1/2$ for highly reflective over water path;
$\quad = 1$ for typical nondiversity path;
$\quad = 2$ for space or frequency diversity path;
$\quad = 4$ for quad diversity path.

Barnett (1974a) and Vigants (1969, 1971) determined that for a typical 28.5-mile path the average fade duration at 4, 6, or 11 GHz is given by the following equation:

$$\begin{aligned} T &= 410\ L \text{ for nondiversity receivers} \\ &= 205\ L \text{ for space or frequency diversity receivers} \\ &= \text{average duration of multipath fade in seconds} \end{aligned} \tag{9.A.3}$$

For a 40-dB fade margin, the average nondiversity fade lasts about 4 s. The diversity fade lasts half that time. If the fade margin is less than 40 dB or the path is shorter, the outage will be longer; if the fade margin is greater or the path longer, it will be less.

Ranade (1987) determined that for a typical 28.5-mile 6-GHz path the probability of a dispersive fade is given by the following equation:

$$P_i(t > \tau) = 100[e^{-0.85\sqrt{\tau}}] \tag{9.A.4}$$

$P_i(t > \tau)$ = probability (%) that an individual dispersive fading event lasts longer than τ seconds. The median (expected or "average") event is 2/3 s long.

The above results assume the fade is at least 20 dB below the nominal RSL. Olsen and Segal (1992) derived an estimate of shallow fading from 0 to 20 dB below median level.

9.A.2 DFM Equation Derivation

Flat fading is estimated using the well-known formula (Vigants, 1975):

$$\text{Seconds of flat fading} = \frac{2.5 \times 10^{-6} CRf D^3 T_0 10^{-\text{FFM}/10}}{I} \tag{9.A.5}$$

C = climate factor (typical value of 1);
R = path roughness factor (typical value of 1);
f = operating frequency in gigahertz;
D = path length in miles;
T_0 = number of seconds of active fading in the time period of interest;
FFM = flat fade margin in decibels;
I = diversity improvement factor.

An equivalent formula for dispersive fading would be:

$$\text{Seconds of dispersive fading} = \frac{2.5 \times 10^{-6} CRf D^3 T_0 10^{-\text{DFM}/10}}{I} \tag{9.A.6}$$

DFM = dispersive fade margin in decibels.

As before, see Section 16.5 for more details on this formula. Bill Rummler's research indicated that T_0 is essentially the same as for flat fading. We will defer the diversity improvement issue. What is needed, then, is a formula for dispersive fading using the following relationships:

$$\text{Po Pod} = 2.5 \times 10^{-6} CRf D^3 10^{-\text{DFM}/10} \tag{9.A.7}$$

Po = fraction of time dispersive fading occurs during active fading;
Pod = probability of dispersive fading outage (radio BER threshold exceeded) during dispersive fading.

Rummler and Lundgren showed (Lundgren and Rummler, 1979) that the probability of dispersive fading outage during dispersive fading was given by:

$$\text{Pod} = \int_{-\pi}^{\pi} \int_{B}^{\infty} p(\phi) p(x) \mathrm{d}x \mathrm{d}\phi \tag{9.A.8}$$

Pod = probability of radio BER threshold being exceeded;
$p(\phi)$ = probability of a frequency notch of phase ϕ occurring;

$p(x)$ = probability of a notch of depth x at phase ϕ causing a radio BER threshold to be exceeded (dispersive fading causing an outage).

2π represents the angular phase shift ϕ of the delayed echo between two consecutive distortion nulls. As noted previously, the frequency difference between consecutive nulls is [1000/(delay in nanoseconds)] = 1000/6.31 = 158.5 MHz. Therefore, a phase shift of 2π represents a shift in null location of 158.5 MHz.

Rummler and Lundgren's empirical data showed that virtually all dispersive fading occurred over the ϕ range of $+\pi/2$ (+90°) to $-\pi/2$ (−90°) and the probability of ϕ was constant over that range. (Sylvain and Lavergnat (1985) reported that the probability was constant over an even larger range of ϕ.) This represents a probable frequency range of +39.6 to −39.6 MHz for null frequencies. Rummler and Lundgren's 6-GHz empirical data also showed that

$$\int_{Bc}^{\infty} p(x)\mathrm{d}x = \mathrm{e}^{-B/3.8}, \tag{9.A.9}$$

where B is the decibel value of a W curve (value of radio threshold for a given BER threshold) for a notch at frequency f. (Many researchers (Caskey and Moreland, 1988; Greenfield, 1984; Mayrargue, 1984; McKay and Shafi, 1988; Sandberg, 1980; Sylvain and Lavergnat, 1985) have reproduced the Rummler/Lundgren $\mathrm{e}^{-B/3.8}$ relationship for a wide range of radio bandwidths and path lengths and physical characteristics).

Rummler assumed that the B values for nonminimum phase and minimum phase dispersion were the same. Although nonminimum phase and minimum phase fading are approximately equally likely (Gardina and Vigants, 1984), actual radio receiver testing shows B values for the same notch can be different depending on the type of fading. Therefore, the above formula needs to be modified to account for this.

$$\mathrm{e}^{-B/3.8} = 1/2\,[\mathrm{e}^{-\mathrm{Bn}/3.8} + \mathrm{e}^{-\mathrm{Bm}/3.8}] \tag{9.A.10}$$

Bn = decibel value of B at frequency f for a nonminimum phase echo;
Bm = decibel value of B at frequency f for a minimum phase echo.

Taking the above into account, we may rewrite the probability of dispersive fading formula as the following:

$$\mathrm{Pod} = \int_{-\pi/2}^{+\pi/2} p(\phi)\{0.5[\mathrm{e}^{-\mathrm{Bn}/3.8} + \mathrm{e}^{-\mathrm{Bm}/3.8}]\}\mathrm{d}\phi \tag{9.A.11}$$

Converting the phase shift to frequency shift and recognizing that $p(\phi)$ is constant over the integral range and the probability integral must be zero for B equal to zero allows us to rewrite the formula as:

$$\mathrm{Pod} = 0.5(1/79.25) \int_{-39.6}^{+39.6} [\mathrm{e}^{-\mathrm{Bn}/3.8} + \mathrm{e}^{-\mathrm{Bm}/3.8}]\,\mathrm{d}f(\mathrm{MHz}) \tag{9.A.12}$$

$$\mathrm{Pod} = (1/158.5)\mathrm{Sw} \tag{9.A.13}$$

In 1981, Rummler published (Rummler, 1980) two measured values for the dispersive fading period in an active fading month. Averaging those values and dividing by the number of seconds in a (fading) month (number of seconds in 365.25 days/12) yields Po, the fraction of time dispersive fading is occurring:

$$\mathrm{Po} = \frac{\left[\frac{1}{2}\,(8{,}060 + 15{,}250)\right]}{2{,}629{,}800} = 0.004432 \tag{9.A.14}$$

Going back to our original relationships,

$$2.5 \times 10^{-6} CRf D^3 10^{-\mathrm{DFM}/10} = \mathrm{Po\ Pod} = (0.004432)\left(\frac{1}{158.5}\right)\mathrm{Sw} \tag{9.A.15}$$

Applying the values of $C = R = 1$, $f = 6.034$ (the center of the frequency band used for measurements) and $D = 26.4$, the values appropriate for Rummler's research path, results in the following:

$$10^{-\mathrm{DFM}/10} = \left(\frac{1}{62.6}\right)\left(\frac{\mathrm{Sw}}{158.5}\right) \tag{9.A.16}$$

Taking -10 times the common logarithm of both sides gives the result as:

$$\mathrm{DFM\ (dB)} = 17.97 - 10\log\left(\frac{\mathrm{Sw}}{158.5}\right) = 39.6 - 10\log(\mathrm{Sw}) \tag{9.A.17}$$

$$\mathrm{Sw} = \int_{-39.6}^{+39.6} [e^{-\mathrm{Bn}/3.8} + e^{-\mathrm{Bm}/3.8}]\,df\,(\mathrm{MHz}) \tag{9.A.18}$$

Bn = decibel value of B at frequency f for a nonminimum phase W curve;
Bm = decibel value of B at frequency f for a minimum phase W curve.

Bellcore's DFM formula originally (Ranade, 1985) was essentially the same as the above (with 17.97 replaced by 17.96). Later, Bellcore revised the DFM formula (Bellcore (Telcordia) Staff, 1989) to replace 17.96 with 17.6 and the integration limits were changed from 39.6 to 39.5.

The above formulas can be generalized to any long delay by generalizing Sw:

$$\mathrm{Sw} = \int_{F_{\mathrm{min}}}^{F_{\mathrm{max}}} [e^{-\mathrm{Bn}/3.8} + e^{-\mathrm{Bm}/3.8}]df\,(\mathrm{MHz}) \tag{9.A.19}$$

F_{max} (MHz) = +250/delay (ns);
F_{min} (MHz) = +250/delay (ns).

F_{max} and F_{min} account for the frequency difference between consecutive nulls for various delays. In addition, the probability density function $p(\phi)$ must be changed from unity to $[39.6/F_{\mathrm{max}}]$ to account for probability variation in path delay. Making these charges results in the following *general DFM formula*:

$$\mathrm{DFM_m(dB)} = 39.6 - 10\log\left(\frac{39.6}{F_{\mathrm{max}}}\right) - 10\log(\mathrm{Sw}) \tag{9.A.20}$$

$$\mathrm{Sw} = \int_{F_{\mathrm{min}}}^{F_{\mathrm{max}}} [e^{-\mathrm{Bn}/3.8} + e^{-\mathrm{Bm}/3.8}]\,df\,(\mathrm{MHz}) \tag{9.A.21}$$

F_{max} (MHz) = +250/delay (ns);
F_{min} (MHz) = +250/delay (ns).

The $\mathrm{DFM_m}$ is used to note that the DFM has been modified to accommodate actual path delay for the reference path.

Note that this DFM *is calculated for the reference path* distance (26.4 miles), reference path operating frequency (6.034 GHz), and unity path roughness and climate factor. Changes in operating path characteristics must be accommodated by the various C, R, f, and D parameters in the seconds of dispersive fading formula. The implied assumption is that the radio path is relatively flat without pronounced reflective surfaces. Paths with excessive clearance or other unusual characteristics (such as paths over water) are not modeled well by this approach. There is no industry-accepted way to improve the fade outage estimation model.

As long as $[F_{\max} - F_{\min}] > $ [W-curve effective bandwidth Δf_{eff}], the above formulas apply. Since the integral is limited by receiver bandwidth Δf_{eff}, the only effect on the integral of varying path delay is the variation in the W curve (addressed in the main body of this paper). If $[F_{\max} - F_{\min}] \leq$ [W-curve effective bandwidth Δf_{eff}], the delayed signal is at least one radio symbol delayed from the original radio symbol. Since transmitters have scramblers (to smooth the transmit spectrum), the delayed signal will be uncorrelated with the main received signal. It will appear to be an interfering signal similar to the desired signal. For this long delayed signal situation, C/I and T/I techniques (TIA, 1994) would be appropriate.

9.A.3 Characteristics of Receiver Signature Curves and DFM

Several W-curve general characteristics are known. The BER W curve can be characterized by an average notch depth B_{avg} over an effective bandwidth Δf_{eff} (Fig. 9.A.1).

This rectangular approximation of the W curve allows the Bellcore (now Telcordia) DFM equation to be simplified:

$$\text{DFM(dB)} = 39.6 - 10\log(\text{Sw}) \tag{9.A.22}$$

$$\text{Sw} = 2\Delta f_{\text{eff}}\, e^{-B_{\text{avg}}/3.8} \tag{9.A.23}$$

Δf_{eff} (MHz) = effective receiver bandwidth;
B_{avg} (dB) = average height of W curve over Δf_{eff} (averaging both minimum and nonminimum W curves equally).

Straightforward manipulation yields the following simplification:

$$\text{DFM(dB)} = 36.6 - 10\log(\Delta f_{\text{eff}}) + 1.14B_{\text{avg}} \tag{9.A.24}$$

If the minimum and nonminimum W curves are significantly different, the above formula may be used for each and the results averaged.

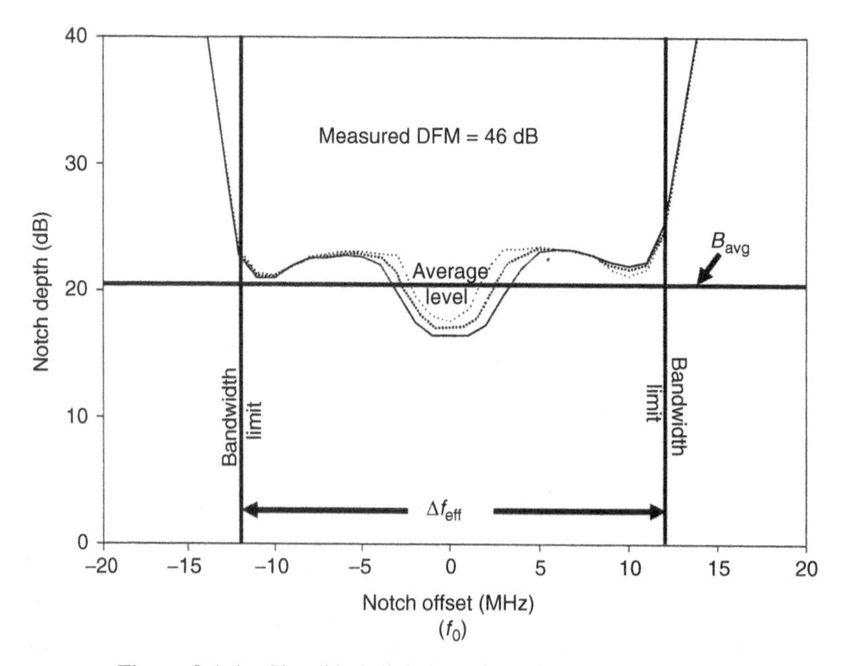

Figure 9.A.1 Simplified digital receiver signature (W) curve.

In the above figure, the estimated $\text{DFM(dB)} \cong 36.6 - 10\log(24) + 1.14(20.5) = 46.2\,\text{dB}$.

For ITU-R P.530 dispersive fading calculations, the effective receiver bandwidth and average W-curve height are required. If the Bellcore DFM and effective receiver bandwidth are known, the above formulas may be used to infer the average W-curve height B_{avg}:

$$B_{avg} = -32.1 + 0.88\ \text{DFM(dB)} + 8.8\log(\Delta f_{eff}) \tag{9.A.25}$$

The receiver effective bandwidth Δf_{eff} for W curves is essentially the same as the receiver noise bandwidth. This is about the same numerically as the radio baud rate. For typical radios, this is usually half way between the transmit signal 3-dB bandwidth and the 99% bandwidth. This value may be inferred from a normalized receiver CW T/I curve or transmit spectrum plot.

Several general characteristics of the W curve are well known (Campbell and Coutts, 1982; Emshwiller, 1978; ITU-R, 2003).

Emshwiller (1978) showed that the voltage amplitude of a data point on the W curve is directly proportional to the echo delay for PSK receivers. Giger (1991) suggests the result is much more general. Unpublished tests have confirmed the property for QAM and TCM receivers. Therefore, for W curves plotted in positive decibel, the curve moves up or down directly as −20 log [echo delay (ns)/reference delay (6.3 ns)]. It is assumed that echo delay is significantly less than the radio symbol time spacing (inverse of the baud rate). Echo delays approaching or exceeding the radio symbol time spacing would be characterized by co-channel C/I measurements.

For a particular receiver configuration, sensitivity to dispersive fades will not change with data rate. However, the width of the receiver band pass and the W curve are directly proportional to the radio (symbol) signaling rate (reciprocal of the baud time period).

Campbell and Coutts (1982), Emshwiller (1978), Giger (1991), Campbell and Coutts (1982), and ITU-R (2003) show that the shape and width of the W curve are independent of echo delay.

Unpublished laboratory measurements demonstrate that the W-curve depth is inversely related to the BER threshold by a factor of half the difference in receiver thresholds (RTs).

With the above W-curve information, dispersive fading performance could be estimated for various radio modulation formats. Normalized W curves could be measured. Specific radio curves would be obtained by scaling the normalized curves by radio signaling (baud) rate and shifting the W curves to account for coding and forward error correction gains (for the BER of interest). These synthesized curves could then be used to estimate dispersive fade threshold.

The previously mentioned DFM equation approximation coupled with general knowledge of W curves allows some generalization to be made:

DFM decreases as the radio signaling rate increases.

$$\text{DFM(dB)} + 22.8\ \log\ (\text{signaling rate}) = \text{a constant} \tag{9.A.26}$$

$\text{DFM(dB)} = \text{DFM}_{ref}\text{(dB)} - 22.8\log(\text{signaling rate/reference signaling rate})$;

DFM_{ref} = dispersive fade margin for receiver with reference signaling rate.

DFM decreases as effective receiver bandwidth increases.

$$\text{DFM(dB)} + 10\log(\Delta f) = \text{a constant} \tag{9.A.27}$$

$\text{DFM(dB)} = \text{DFM}_{ref}\text{(dB)} - 10\log(\Delta f/\Delta f_{ref})$;

DFM = dispersive fade margin for receiver with effective bandwidth of Δf;

DFM_{ref} = dispersive fade margin for receiver with effective bandwidth of Δf_{ref}.

DFM decreases as the echo delay increases.

$$\text{DFM(dB)} + 22.8\log[\text{echo delay (ns)}] = \text{a constant} \tag{9.A.28}$$

$\text{DFM(dB)} = \text{DFM}_{ref}\text{(dB)} - 22.8\log[\text{echo delay(ns)/reference delay(6.3 ns)}]$;

DFM_{ref} = dispersive fade margin for receiver with reference delay.

REFERENCES

Barnett, W. T., "Multipath Propagation at 4, 6 and 11 GHz," *Bell System Technical Journal*, pp. 321–361, June 1974a.

Barnett, W. T., "Deterioration of Cross-Polarization Discrimination During Rain and Multipath Fading at 4 GHz," *IEEE International Conference on Communications*, pp. 12D-1 to 12D-3, June 1974b.

Bellcore (Telcordia) Staff, Bellcore (Telcordia) Technical Reference TR-TSY-000752, Microwave Digital Radio Systems Criteria, pp. 7–13, October 1989.

Bullington, K., "Radio Propagation Variations at VHF and UHF," Proceedings of the IRE, pp. 27–32, January 1950.

Bullington, K., "Radio Propagation Fundamentals," *Bell System Technical Journal*, pp. 593–626, May 1957.

Burrows, C. R., Hunt, L. E. and Decino, A., "Ultra-Short-Wave Propagation: Mobile Urban Transmission characteristics," *Bell System Technical Journal*, pp. 253–272, April 1935.

Campbell, J. C. and Coutts, R. P., "Outage Prediction of Digital Radio Systems," *Electronics Letters*, pp. 1071–1072, December 9 1982.

Caskey, M. D. and Moreland, K. W., "Multipath Fading Effects in an 40 MHz Channel," *IEEE International Conference on Communications*, Vol. 3, pp. 1633–1640, June 1988.

Clark, R. H., "A Statistical Theory of Mobile Radio Reception," *Bell System Technical Journal*, pp. 957–1000, July-August 1968.

Crawford, A. B. and Jakes, W. C., Jr., "Selective Fading of Microwaves," *Bell System Technical Journal*, pp. 68–90, January 1952.

De Lange, O. E., "Propagation Studies at Microwave Frequencies by Means of Very Short Pulses," *Bell System Technical Journal*, pp. 91–103, January 1952.

Dupuis, P., Joindot, M., Leclert, A. and Rooryck, M., "Fade Margin of High Capacity Digital Radio System," *IEEE International Conference on Communication*, pp. 48.6.1–48.6.5, June 1979.

Durkee, A. L., "Results of Microwave Propagation Tests on a 40-Mile Overland Path," *Proceedings of the IRE*, pp. 197–205, February 1948.

Emshwiller, M., "Characterization of the Performance of PSK Digital Radio Transmission in the Presence of Multipath Fading," *IEEE International Conference on Communication*, pp. 47.3.1–47.3.6, June 1978.

England, C. R., Crawford, A. B. and Mumford, W. W., "Some Results of a Study of Ultra-Short-Wave Transmission Phenomena," *Bell System Technical Journal*, pp. 197–227, April 1933.

England, C. R., Crawford, A. B. and Mumford, W. W., "Ultra-Short-Wave Transmission and Atmospheric Irregularities," *Bell System Technical Journal*, pp. 489–519, October 1938.

Friis, H. T., "Microwave Repeater Research," *Bell System Technical Journal*, pp. 183–246, April 1948.

Gardina, M. F. and Vigants, A., "Measured Multipath Dispersion of Amplitude and Delay at 6 GHz in a 30 MHz Bandwidth," *IEEE International Conference on Communications*, Vol. 3, pp. 1433–1436, May 1984.

Giger, A. J., *Low-Angle Microwave Propagation: Physics and Modeling*. Norwood: Artech House, pp. 160–174, 1991.

Giger, A. J. and Barnett, W. T., "Effects of Multipath Propagation on Digital Radio," *IEEE Transactions on Communications*, pp. 1345–1352, September 1981.

Goldman, H., "Mathematical Analysis of the Three-Ray Dispersive Fading Channel Model," *IEE Proceedings*, pp. 87–94, April 1991.

Greenfield, P. E., "Digital Radio Performance on a Long, Highly Dispersive Fading Path," *IEEE International Conference on Communications*, pp. 1451–1454, May 1984.

International Telecommunication Union—Radiocommunication Sector (ITU-R), "*Effects of Multipath Propagation on the Design and Operation of Line-of-Sight Digital Radio-Relay Systems, ITU-R Recommendation 1093–1*," *ITU-R Recommendations.* Geneva: International Telecommunication Union, September 2003.

Jakes, W. C., *Microwave Mobile Communications*. New York: John Wiley & Sons, Inc., pp 11–131, 1974.

Kaylor, R. L., "A Statistical Study of Selective Fading of Super-High Frequency Radio Signals," *Bell System Technical Journal*, pp. 1187–1202, September 1953.

Kizer, G. M., *Microwave Communication*. Ames: Iowa State University Press, pp 225–226, 1990.

Kizer, G. M., "Microwave Radio Communication," *Handbook of Microwave Technology,* Volume 2, *Applications*. Ishii, T. K., Editor. San Diego: Academic Press, pp 449–504, 1995.

Lane, R. U., "Special Transmission Design Techniques for Difficult 45 Mbit/s Digital Microwave Links," ENTELEC Presentation, March 1991.

Lin, S. H., "Statistical Behavior of a Fading Signal," *Bell System Technical Journal*, pp. 3211–3270, December 1971.

Lin, S. H., "Statistical Behavior of Deep Fades of Diversity Signals," *IEEE Transactions on Communications*, pp. 1100–1107, December 1972.

Lin, S. H., "Impact of Microwave Depolarization During Multipath Fading on Digital Radio Performance," *Bell System Technical Journal*, pp. 645–674, May 1977.

Lin, S. H., Lee, T. C. and Gardina, M. F., "Diversity Protections for Digital Radio—Summary of Ten-Year Experiments and Studies," *IEEE Communications Magazine*, pp. 51–64, February 1988.

Liniger, M., "Sweep Measurements of Multipath Effects on Cross-Polarized RF-Channels Including Space Diversity," *IEEE Global Telecommunications Conference (Globecom) Record*, Vol. 3, pp. 45.7.1–45.7.5, November 1984.

Lundgren, C. W. and Rummler, W. D., "Digital Radio Outage Due to Selective Fading—Observation vs Prediction From Laboratory Simulation," *Bell System Technical Journal*, pp. 1073–1100, May-June 1979.

Martin, A. L., "Phase Distortions of Multipath Transfer Functions," *IEEE International Conference on Communications*, Vol. 3, pp. 46.2.1–46.2.5, June 1985.

Martin, A. L., "Dispersion Signatures, a Statistically Based, Dynamic, Digital Microwave Radio System Measurement Technique," *IEEE Journal on Selected Areas in Communications*, pp. 427–436, April 1987.

Mayrargue, S., "Parametric Characterization of Channel Transfer Functions during Multipath Fading Events," *IEEE Global Telecommunications Conference (Globecom) Record*, Vol. III, pp. 1471–1474, November 1984.

McKay, R. G. and Shafi, M., "Multipath Propagation Measurements on an Overwater Path in New Zealand," *IEEE Transactions on Communications*, pp. 781–788, July 1988.

Monsen, P., "Fading Channel Communications," *IEEE Communications Magazine*, pp. 16–25, January 1980.

Mottl, T. O., "Dual-Polarized Channel Outages During Multipath Fading," *Bell System Technical Journal*, pp. 675–701, May-June 1977.

Olsen, R. L. and Segal, B., "New Techniques for Predicting the Multipath Fading Distribution on VHF/UHF/SHF Terrestrial Line-of-Sight Links in Canada," *Canadian Journal of Electrical and Computer Engineering*, pp. 11–23, January 1992.

Pickering, L. W. and DeRosa, J. K., "Refractive Multipath Model for Line-of-Sight Microwave Relay Links," *IEEE Transactions on Communications*, pp. 1174–1182, August 1979.

Ranade, A., "Statistics of the Time Dynamics of Dispersive Multipath Fading and its Effects on Digital Microwave Radios," *IEEE International Conference on Communications*, Vol. 3, pp. 47.7.1–47.7.4, June 1985.

Ranade, A., "Frequency and Duration of Dispersive Fades at 6 GHz," *IEEE International Conference on Communications*, Vol. 1, pp. 10.7.1–10.7.5, June 1987.

Ranade, A. and Greenfield, P. E., "An Improved Method of Digital Radio Characterization from Field Measurements," *IEEE International Conference on Communications*, pp. C2.6.1–C2.6.5, June 1983.

Rice, S. O., "Mathematical Analysis of Random Noise, Parts I and II," *Bell System Technical Journal*, pp. 282–332, July 1944.

Rice, S. O., "Mathematical Analysis of Random Noise, Part III," *Bell System Technical Journal*, pp. 46–156, January 1945.

Rice, S. O., "Statistical Properties of a Sine Wave Plus Random Noise," *Bell System Technical Journal*, pp. 109–157, January 1948.

Rice, S. O., "Distribution of the Duration of Fades in Radio Transmission: Gaussian Noise Model," *Bell System Technical Journal*, pp. 581–635, May 1958.

Rummler, W. D., "A New Selective Fading Model: Application to Propagation Data," *Bell System Technical Journal*, pp. 1037–1071, May-June 1979.

Rummler, W. D., "Time- and Frequency-Domain Representation of Multipath Fading on Line-of-Sight Microwave Paths," *Bell System Technical Journal*, pp. 763–795, May-June 1980.

Rummler, W. D., "More on the Multipath Fading Channel Model," *IEEE Transactions on Communications*, pp. 346–352, March 1981.

Rummler, W. D., "A Simplified Method for the Laboratory Determination of Multipath Outage of Digital Radios in the Presence of Thermal Noise," *IEEE Transactions on Communications*, pp. 487–494, March 1982.

Rummler, W. D., "A Comparison of Calculated and Observed Performance of Digital Radio in the Presence of Interference," *IEEE Transactions on Communications*, pp. 1693–1700, July 1982b.

Rummler, W. D., "Characterizing the Effects of Multipath Dispersion on Digital Radios," *IEEE Global Telecommunications Conference (Globecom)*, pp. 52.5.1–52.5.7, June 1988.

Rummler, W. D., Coutts, R. P. and Liniger, M. "Multipath Fading Channel Models for Microwave Digital Radio," *IEEE Communications Magazine*, pp. 30–42, November 1986.

Ruthroff, C. L., "Multiple-Path Fading on Line-of-Sight Microwave Radio Systems as a Function of Path Length and Frequency," *Bell System Technical Journal*, pp. 2375–2398, September 1971.

Sakagami, S. and Morita, K., "Cross Polarization Discrimination Characteristics During Multipath Fading at 4 GHz," *Antennas and Propagation Society International Symposium*, pp. 821–824, June 1979.

Sandberg, J., "Extraction of Multipath Parameters from Swept Measurements on a Line-of-Sight Path," *IEEE Transactions on Antennas and Propagation*, pp. 743–750, November 1980.

Sasaki, O. and Akiyama, T., "Multipath Delay Characteristics on Line-of-Sight Microwave Radio Systems," *IEEE Transactions on Communications*, pp. 1876–1886, December 1979.

Schelleng, J. C., Burrows, C. R. and Ferrell, E. B., "Ultra-Short Wave Propagation," *Bell System Technical Journal*, pp. 125–161, April 1933.

Schiavone, J. A., "Microwave Radio Meteorology: Seasonal Fading Distributions," *Radio Science*, pp. 369–380, May-June 1983.

Shafi, M., "Statistical Analysis/Simulation of a Three Ray Model for Multipath Fading with Applications to Outage Prediction," *IEEE Journal on Selected Areas in Communications*, pp. 389–401, April 1987.

Siller, C. A., Jr., "Multipath Propagation," *IEEE Communications Magazine*, pp. 6–15, February 1984.

Stephansen, E. T., "Clear-air Propagation on Line-of-Sight Radio Paths: A Review," *Radio Science*. pp. 609–629, September-October 1981.

Sylvain, M. and Lavergnat, J., "Modeling the Transfer Function of a 55 MHz Wide Radio Channel During Multipath Propagation," *IEEE International Conference on Communications Record*, Vol. 3, pp. 1541–1546, June 1985.

Telecommunications Industry Association (TIA), TIA/EIA Telecommunications Systems Bulletin 10-F (TSB10-F), Interference Criteria for Microwave Systems. Washington, DC: Telecommunications Industry Association, June 1994.

Vigants, A., "The Number of Fades and Their Durations on Microwave Line-of-Sight Links With and Without Space Diversity," *IEEE International Conference on Communications (ICC) Proceedings*, pp. 3.7–3.11, June 1969.

Vigants, A., "Number and Duration of Fades at 6 and 4 GHz," *Bell System Technical Journal*, pp. 815–841, March 1971.

Vigants, A., "Space-Diversity Engineering," *Bell System Technical Journal*, pp. 103–142, January 1975.

Vigants, A., "Temporal Variability of Distance Dependence of Amplitude Dispersion and Fading," *IEEE International Conference on Communications*, Vol. 3, pp. 1447–1450, May 1984.

Webster, A. R., "Multipath Angle-of-Arrival Measurements on Microwave Line-of-Sight Links," *IEEE Transactions on Communications*, pp. 798–803, June 1991.

Webster, A. R. and Merritt, T. S., "Multipath Angle-of-Arrival on a Terrestrial Microwave Link," *IEEE Transactions on Communications*, pp. 25–30, January 1990.

Xiong, F., Digital Modulation Techniques, Second Edition. Boston: Artech House, pp 547–550, 2006.

Young, W. R. Y., Jr., "Comparison of Mobile Radio Transmission at 150, 450, 900 and 3700 Mc," *Bell System Technical Journal*, pp. 1068–1085, November 1952.

10

MICROWAVE RADIO DIVERSITY

Atmospheric multipath (Chapter 9) is a fundamental limitation to the performance of a fixed point to point microwave system. The one-way (transmitter-to-receiver transmission path) atmospheric multipath fading outage time may be estimated by the following (Vigants, 1975; Rummler et al., 1986; Committee 14–11, 1994):

$$T(\mathrm{s}) = T_{\mathrm{o}}\, P \tag{10.1}$$

$T(\mathrm{s})$ = number of fading seconds in a time period of interest (a year or the "worst" month);
T_{o} = seconds in a time period of interest (year or month);
P = probability of heavy fading during the time period of interest.

Two forms of multipath commonly occur: flat fading and dispersive fading. Dispersive fading is almost always accompanied by some flat fading. However, deep fade levels are uncorrelated with dispersive effects (Rummler, 1979, 1988). This means that the deepest flat fade events do not occur at the same time as the most dispersive fading events. Thus the composite results of the two types of fading may be treated additively.

$$T_{\mathrm{FF}}(\mathrm{s}) = T_{\mathrm{o}}\, P_{\mathrm{FF}} \tag{10.2}$$

$T_{\mathrm{FF}}(\mathrm{s})$ = seconds of dispersive fading.

$$\begin{aligned} T_{\mathrm{DF}}(\mathrm{s}) &= T_{\mathrm{o}}\, P_{\mathrm{DF}} \\ P_{\mathrm{FF}} &= \frac{(R \times 10^{-\mathrm{FFM}/10})}{I_{\mathrm{FF}}} \\ P_{\mathrm{DF}} &= \frac{(R \times 10^{-\mathrm{DFM}/10})}{I_{\mathrm{DF}}} \end{aligned} \tag{10.3}$$

$T_{\mathrm{DF}}(\mathrm{s})$ = seconds of dispersive fading;
R = fade occurrence factor (function of terrain roughness, operating frequency, path length, and climate);

Digital Microwave Communication: Engineering Point-to-Point Microwave Systems, First Edition. George Kizer.

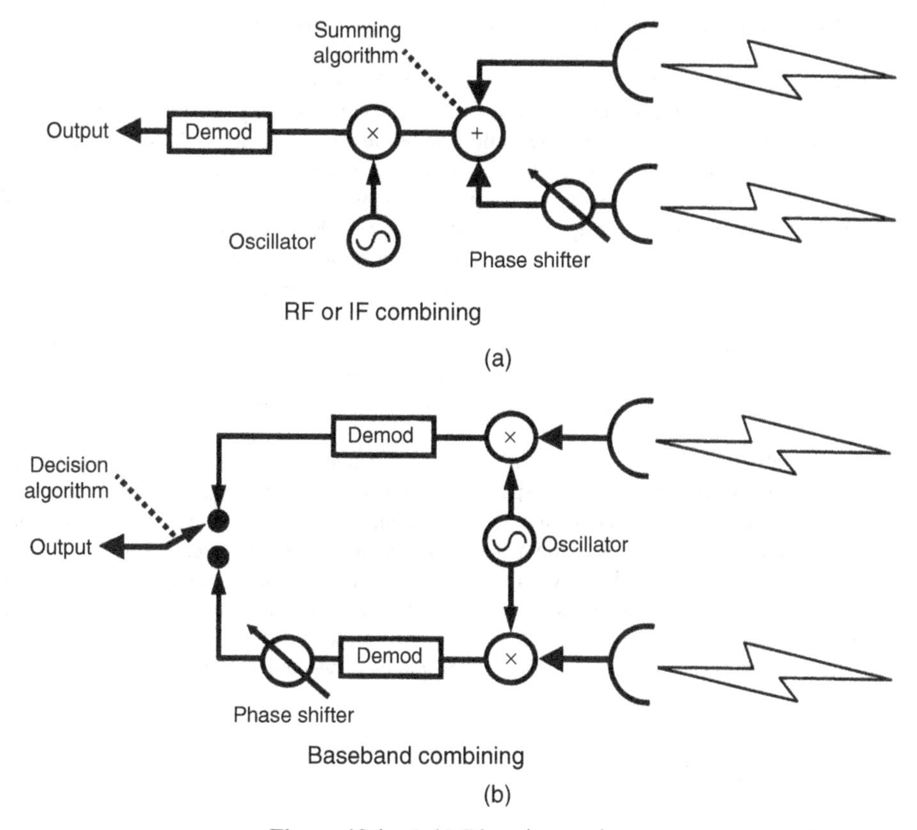

Figure 10.1 (a,b) Diversity receivers.

FFM = path flat fade margin (dB);
DFM = path dispersive fade margin (dB);
I_{FF} = flat fading improvement factor;
I_{DF} = dispersive fading improvement factor.

Sometimes, the performance of a simple path (transmitter to a single receiver) is adequate and diversity is not used (and the diversity improvement factors I_{FF} and I_{DF} above are 1). However, for many (long or difficult) paths, a diversity configuration is required to achieve acceptable performance. The basic idea of diversity is to use two different antennas or frequencies and combine their signals in such a way as to improve overall radio system performance (Fig. 10.1).

For older analog FM microwave systems, RF or IF analog combining (maximal ratio squared or switching) was popular in diversity systems. This form of combining used the overall RSL as the primary estimate of signal quality. With digital systems, however, it was realized that when dispersive fading was occurring, the channel could be severely distorted but the RSL could be essentially unchanged (Anderson et al., 1979; Barnett, 1979; Giger and Barnett, 1981). An early measure of dispersive fading was in band power distortion (IBPD), a measure of dispersive fading in-band frequency notch depth. However, this was difficult to measure and not all notches were equally troublesome.

Owing to its inability to simply and cost-effectively address dispersive fading, RF or IF analog combining was abandoned in favor of baseband switching. This gave the digital receiver time domain equalizers the opportunity to improve the received signal before a quality decision was made. This was viewed as a significant improvement because now decisions would be made on the actual quality of the receiver output, regardless of the path distortion. However, this introduced a new decision: when should the decoded data output be switched from one receiver to the other? Unlike the RF or IF combining (or switching), which was done on the basis of overall power, digital baseband quality was based on BER. The desire was to achieve essentially error-free operation [generally viewed as BER $< 10^{-9}$ for low speed (DS1 or E1) systems or BER $< 10^{-12}$ for high speed (>100 Mb/s) systems] to the greatest extent possible. However, if a switch was not made until an error was sensed on the primary receiver (yet the secondary receiver was running error free), this was generally viewed as not acceptable. Considerable effort has been expended in attempting to determine when to switch from one receiver to the other in an attempt to achieve error-free operation. An obvious choice is receiver BER. However, this measurement requires an integration time interval. When multipath fading is significant, RSL and IBPD are changing at rates greater than 100 dB/s. By the time a significant BER has been sensed, the receiver output is significantly degraded. In an attempt to anticipate receiver degradation, four parameters are usually considered: RSL (a fast indicator of flat fading), time domain equalizer stress (a fast indicator of dispersive fading), eye closure (a fast indicator of general performance), and BER before forward error correction (a slow indicator of general performance). How to optimize these measurements remains an elusive challenge. The degree to which this problem has been solved determines the degree to which practical diversity improvement approaches the theoretical.

Diversity is usually achieved using two vertically spaced antennas (space diversity), multiple transmitter frequencies (frequency diversity), both space and frequency diversity (quad diversity), or reception using two different antenna patterns (angle diversity). Frequency diversity is common for US government and international radio systems. The US commercial radio systems are prohibited (Mosley) (with some exceptions) from using frequency diversity unless the path employs at least three transmitter–receiver pairs with different payloads for every protection channel. Exceptions include a waiver due to practicality (Mosley), use in collapsed ring architecture (Mosley, 2000), use of lower priority (preemptable) traffic on the protection channels (Knerr, 1998), and use of IP radios connected to a router.

10.1 SPACE DIVERSITY

Frequency diversity was the first diversity used by fixed point to point microwave systems. It was quite popular and had the desirable feature of allowing operation on one channel while doing maintenance on the other. However, pressure by the FCC forced the use of space diversity (two receive antennas vertically spaced receiving the same frequency signal, Fig. 10.2) in the United States in the 1960s. In the United States, space diversity engineering was pioneered by Collins Radio (now Alcatel-Lucent) (Albertson, 1964), Lenkurt Electric (now Aviat Networks) (White, 1968), and Bell Labs (now Alcatel-Lucent) (Vigants, 1967). Space diversity is more expensive than frequency diversity (owing to the additional antenna and coaxial cable or waveguide). However, frequency diversity requires wide frequency spacing to be effective. Space diversity has become the most common form of commercial path diversity in the United States.

Figure 10.3 shows RSL recordings of a properly aligned (see below) 24-mile-path over Lake Pontchartrain in Louisiana. Despite the significant short-term reductions in RSL ("fading"), the radio output was error free during the entire commissioning test period.

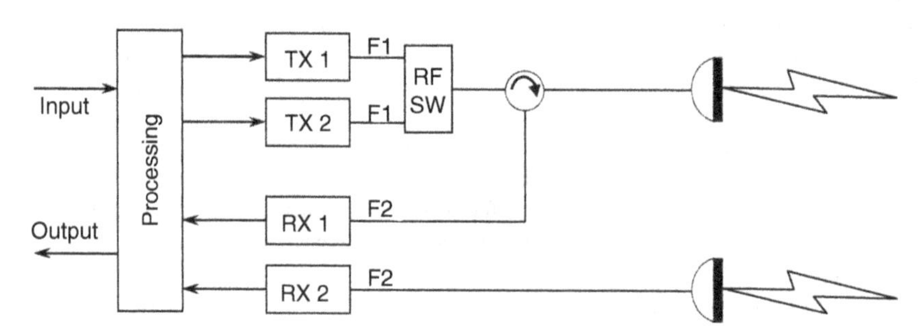

Figure 10.2 Space diversity radio configuration.

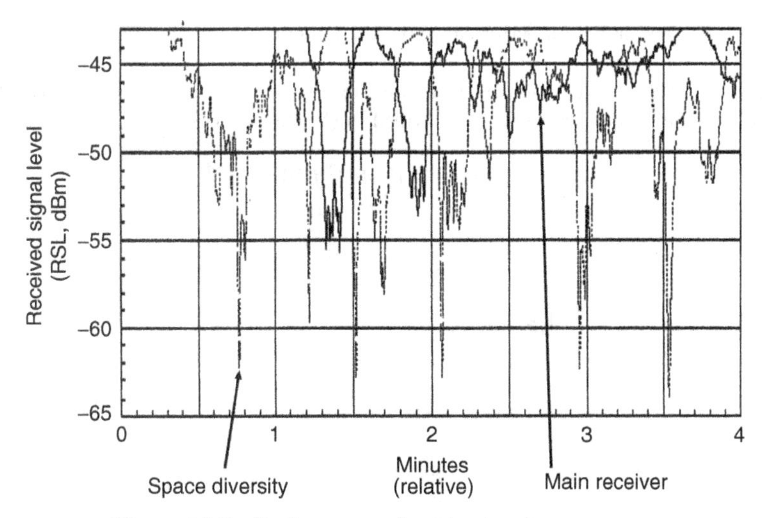

Figure 10.3 Performance of a 24-mi path over water.

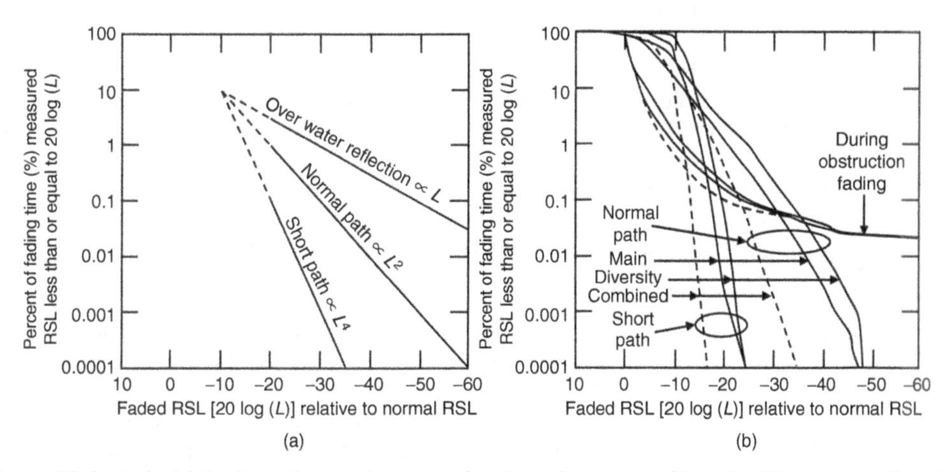

Figure 10.4 (a,b) Lin's theoretical and measured path performance with and without space diversity.

The objective of space diversity is to improve this path performance when needed. Barnett (1972) published experimentally derived characteristics of multipath fading. Lin derived a theoretical basis for the characteristics of multipath fading (Lin, 1971, 1972) with and without diversity.

Lin's nondiversity models (Lin, 1971) are depicted in Fig. 10.4a. This chart should not be taken literally. The curves will move up or down depending on path characteristics. As Lin noted, the overwater curve can also apply to highly reflective overland paths. Vigants (1968) suggests Lin's results only apply for fades deeper than 20 dB below normal RSL.

Figure 10.4b shows actual nondiversity and diversity path-fading statistics. The combined curve is theoretical performance based on instantaneous hysteresis-free switching between the better of the two receiver RSLs. Notice that during obstruction fading, diversity has no significant effect. The obstruction fading curve is similar to the curve of a rain fading event.

Lin (1972) showed that diversity causes an overwater path (amount of fading $\propto L$, the RMS voltage level of the faded signal relative to its unfaded level) to behave like a normal path (fading $\propto L^2$, the power of the fading signal relative to its unfaded power). He also showed that diversity causes a normal path (fading $\propto L^2$) to behave even better (fading $\propto L^4$). This result is shown in the measured data above.

Space diversity improves short path performance but the empirical data above suggests the improvement still of order L^4.

Flat fading improvement was extensively investigated by Vigants (1967, 1968, 1969, 1970, 1975). The following flat fading improvement was described in (Vigants, 1975):

$$
\begin{aligned}
I_{\text{FF-SD}} &= \frac{7 \times 10^{-5}\,\eta\,[s(\text{ft})]^2\, f\, 10^{\text{P}/10}\, 10^{\text{FFM}/10}}{D\ (\text{miles})} \\
&= \frac{1.2 \times 10^{-3}\,\eta\,[s(\text{m})]^2 f\, 10^{\text{P}/10}\, 10^{\text{FFM}/10}}{D\ (\text{km})} \qquad (10.4) \\
& 1 \le I_{\text{FF-SD}} \le 200 \\
& 0 \le s\,(\text{ft}) \le 50 \\
& 0 \le s\,(\text{m}) \le 15
\end{aligned}
$$

$I_{\text{FF-SD}}$ = flat fading improvement factor for space diversity;
FFM = flat fade margin (for the primary antenna);
D = path distance;
s = vertical separation of receiving antenna centers;
f = radio operating frequency (GHz);
η = switching hysteresis efficiency;
= $2/[(10^{\text{H}/10}) + 1/(10^{\text{H}/10})]$;
H = switching hysteresis (dB);
= decibel value by which the RSL of the second receiver must be greater than the RSL of the first receiver before the output signal will be switched from the first receiver to the second receiver;
$P1$ = isotropic boresight gain (dB) of primary receive antenna;
$P2$ = isotropic boresight gain (dB) of space diversity receive antenna;
P = difference between space diversity antenna gain and the main antenna gain;
= $P2 - P1$.

Vigants limits both the amount of diversity improvement and the maximum vertical distance between vertical antennas. Some engineers have questioned this. "Isn't more better," they asked. This issue was addressed experimentally by Lee and Lin (1988) (Fig. 10.5).

They demonstrated empirically that there is a limit to diversity improvement as the antennas are separated vertically (although this limit will probably vary with path length and characteristics). Rummler demonstrated theoretically (Rummler, 1983) that 6-GHz space diversity improvement (when averaged over all possible paths) does not increase for antenna separations greater than 46 ft (Keep in mind that

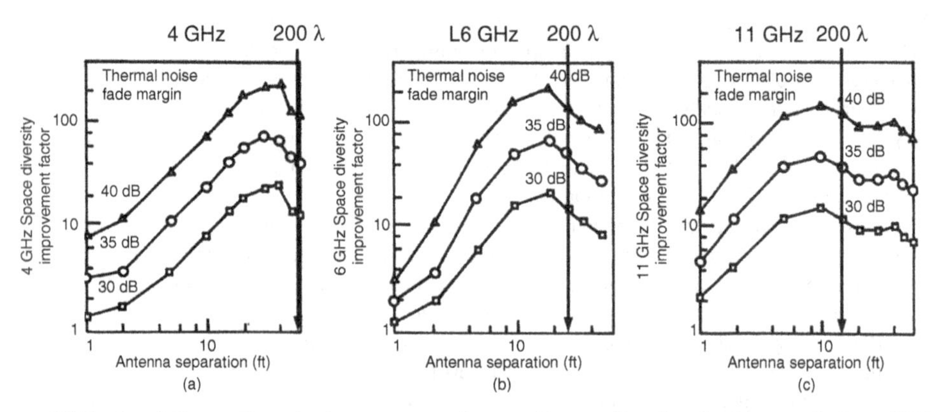

Figure 10.5 (a–c) Space diversity improvement factors. *Source*: Reprinted with permission of IEEE.

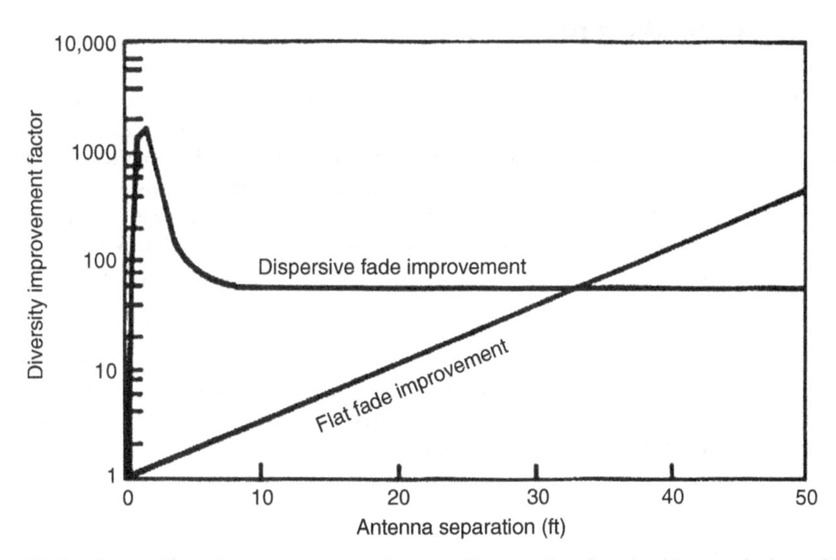

Figure 10.6 Space diversity improvement factors. *Source*: Reprinted with permission of IEEE.

Rummler was discussing nonreflective paths. Reflective paths require more detailed engineering to achieve appropriate diversity vertical spacing.).

Early in the investigations of digital radio outages, it was recognized that using the FFM space diversity improvement factors underestimated dispersive fading diversity improvement (Greenstein and Shafi, 1972; Wang, 1979; Smith and Cormack, 1984). Dispersive fading space diversity improvement was investigated extensively (Rummler, 1982; Lee and Lin, 1986; Lin et al., 1988), both experimentally and by simulation. Although graphical results were published (Lin et al., 1988), formulas were not (Fig. 10.6).

For a typical 6-GHz 26-mile path, simulations by Lin, Lee and Gardina (1988) showed dispersive space diversity improvement was a maximum for vertical separation of only 1 ft. Improvement was less but constant for separations exceeding 8 ft. Obviously, this is quite different from flat fading.

Private discussions between Rockwell International and Bell Labs staff members (Evans, M. and Knight, W., private communication with Bell Labs staff, 1989) resulted in the following conservative estimates (ignoring the short vertical spacing effect because short spacing is inconsistent with flat fading diversity improvement engineering) for dispersive fading improvement:

$$
\begin{aligned}
I_{\mathrm{DF-SD}} &= \frac{0.09\, f\, 10^{\mathrm{DFM}/10}}{D\ (\text{miles})} \\
&= \frac{0.14\, f\, 10^{\mathrm{DFM}/10}}{D\ (\text{km})} \\
1 \leq\ & I_{\mathrm{DF-SD}} \leq 200
\end{aligned}
\tag{10.5}
$$

$I_{\mathrm{DF-SD}}$ = dispersive fading improvement factor for space diversity;
DFM = dispersive fade margin (dB) (Chapter 9).

The above results are published in (Kizer, 1995; Knight, 2006). DFMs may be calculated with (dynamic) or without hysteresis (static) (Bellcore (Telcordia) Staff, 1989). It is suggested the use of DFMs with hysteresis be used to more nearly characterize actual receiver action. The method of switching between receivers varies with manufacturer. At present, the effect of hysteresis in switching between receivers during dispersive fading is not known.

10.2 DUAL-FREQUENCY DIVERSITY

Figure 10.7 diagrams a typical frequency diversity radio. The circles with curved arrows represent circulators that transfer the signal entering one port to the next port. The curved arrows indicate the direction of signal transfer. Each transmitter (TX) and receiver (RX) is assumed to have an RF filter that reflects all signals with frequency different from the frequency of operation for that transmitter or receiver.

Figure 10.8 shows receiver signals for a frequency diversity path. The path was 100 miles from a site on the top of a mountain 1 mile above ground level to another site with an antenna 200 ft above ground level. White (1968) reported excellent diversity improvement for crossband upper 6- and 12-GHz (an obsolete band) operation.

Dual (two)-frequency diversity (typically designated as $1 + 1$, $1:1$, or 1×1) receiver configurations are popular for international and US government projects. Flat fading frequency diversity improvement was initially investigated experimentally by Barnett (1970) and later theoretically by Vigants and Pursley (1979). The Vigants and Pursley work, which extends Barnett's results, are examined in detail below. Using those results for the dual-frequency case results in the following:

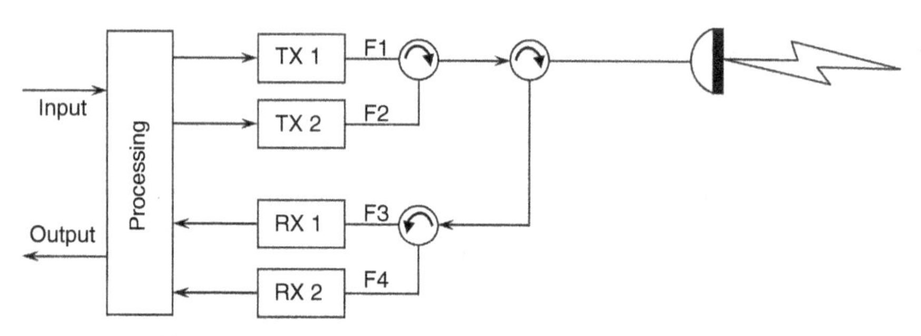

Figure 10.7 Dual-frequency diversity radio configuration.

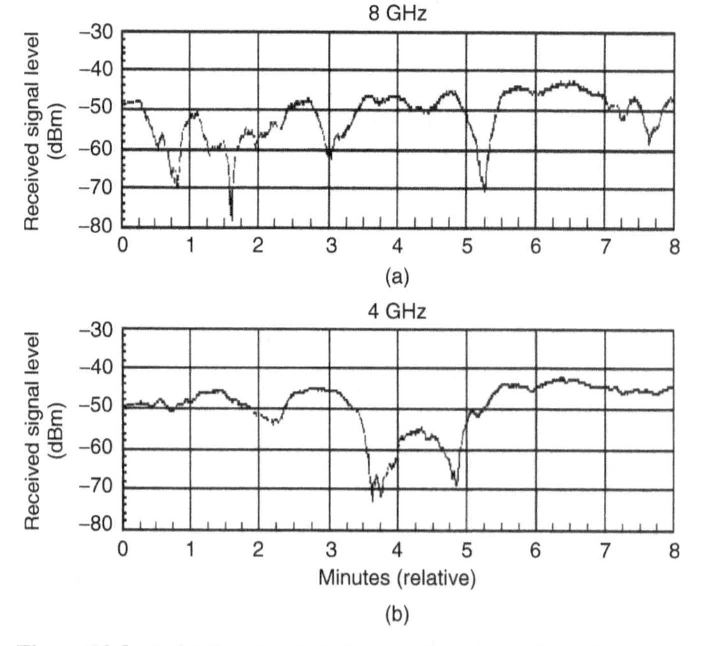

Figure 10.8 (a,b) Crossband frequency diversity path performance.

$$
\begin{aligned}
I_{FF-FD} &= \frac{100\, f_{avg}\, \eta\, 10^{FFM/10}}{[D\,(\text{miles})\, G]} \\
&= \frac{161\, f_{avg}\, \eta\, 10^{FFM/10}}{[D\,(\text{km})\, G]} \\
1 \le I_{FF-FD} &\le 200
\end{aligned}
\tag{10.6}
$$

I_{FF-FD} = flat fading improvement factor for frequency diversity;
D = path distance;
f_{avg} = average operating frequency;
= $(f_1 + f_2)/2$;
f_2 = higher channel frequency (GHz);
f_1 = lower channel frequency (GHz);
FFM = effective flat fade margin;
= RSLPrdB (dBm) − receiver threshold (dBm);
$P1$ = receive power of f_1 channel (mW);
$P2$ = receive power of f_2 channel (mW);
Pr = reference power (mW);
= $(P1 + P2)/2$;
PrdB = 10 log (Pr) (dBm);
RSLPrdB = received signal level (dBm) for PrdB transmit power (dBm);
η = switching hysteresis efficiency;
= $2/[(10^{H/10}) + (1/10^{H/10})]$;
H = switching hysteresis (dB);
= decibel value by which the RSL of the second receiver must be greater than the RSL of the first receiver before the output signal will be switched from the first receiver to the second receiver.

The hysteresis factor η was developed (Vigants, 1975) after the original frequency diversity improvement factors were developed (Vigants and Pursley, 1979). However, the derivation (Vigants, 1975) suggests it would be equally valid for frequency diversity.

Solving the multiline equations below for the simple dual-channel frequency diversity (1 + 1, 1 : 1) yields the following:

$$
\begin{aligned}
G &= \left\{ \frac{2\left[f_{avg}\right]3}{[(f_2 - f_1) P_a P_b]} \right\} \\
&= \left\{ \frac{(f_1 + f_2)^3}{[4\,(f_2 - f_1)\, P_a P_b]} \right\} \\
P_a &= \frac{P1}{\text{Pr}} \\
P_b &= \frac{P2}{\text{Pr}}
\end{aligned}
\tag{10.7}
$$

As the frequency difference $(f_2 - f_1)$ increases, diversity improvement increases. For dual frequencies with similar receive power levels $(P_a = P_b)$, diversity improvement factor is maximized ($1/G$ minimized) when $f_2 = 2\, f_1$. Diversity improvement is within 85% of maximum for $f_2 = 1.5\, f_1$. Increasing f_2 beyond $f_2 = 2\, f_1$ slowly degrades diversity improvement. For systems where large frequency separation is not possible, at least 5% frequency separation is recommended (White, 1968).

Early in the investigations of digital radio outages, it was recognized that using the FFM frequency diversity improvement factors underestimated dispersive fading diversity improvement (Cellerino et al.,

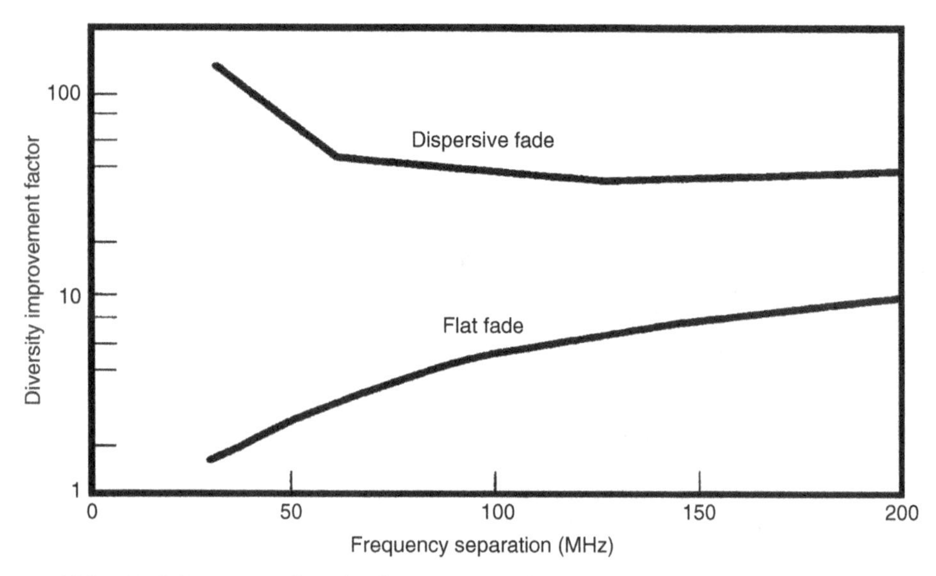

Figure 10.9 Dual-frequency diversity improvement factors. *Source*: Reprinted with permission of IEEE.

1985; Dirner and Lin, 1985). Dispersive fading frequency diversity improvement was investigated extensively (Kostal, 1985; Lee and Lin, 1985; Lin et al., 1988) both experimentally and by simulation. Although graphical results were published (Lee and Lin, 1985; Lin et al., 1988), formulas were not (Fig. 10.9).

Private discussions between Rockwell International and Bell Labs staff members (Evans, M. and Knight, W., private communication with Bell Labs staff, 1989) resulted in the following conservative estimates (ignoring the increasing improvement for separation <50 MHz) for dispersive fading improvement:

$$
\begin{aligned}
I_{\mathrm{DF-FD}} &= \frac{468 \times 10^{\mathrm{DFM}/15}}{[(f_{\mathrm{avg}})^2 D(\mathrm{miles})]} \\
&= \frac{753 \times 10^{\mathrm{DFM}/15}}{[(f_{\mathrm{avg}})^2 D(\mathrm{km})]} \\
1 \leq\ & I_{\mathrm{DF-FD}} \leq 200
\end{aligned}
\tag{10.8}
$$

$I_{\mathrm{DF-FD}}$ = dispersive fading improvement factor for frequency diversity;
DFM = dispersive fade margin (dB) (Chapter 9).

The above results have been published in (Kizer, 1995; Knight, 2006). As DFM is not a function of receive signal level, using different receive power levels only applies to flat fading improvement factor. DFMs may be calculated with (dynamic) or without hysteresis (static) (Bellcore (Telcordia) Staff, 1989). It is suggested the use of DFMs with hysteresis be used to more nearly characterize actual receiver action. The method of switching between receivers varies with manufacturer. At this time, the effect of hysteresis in switching between receivers during dispersive fading is not known.

It should be noted that frequency diversity is generally less effective than space diversity (typically by a factor of 3–10).

10.3 QUAD (SPACE AND FREQUENCY) DIVERSITY

Combining dual-channel space and frequency diversity produces a powerful diversity improvement receiver configuration (Fig. 10.10).

Figure 10.11 illustrates the receive signal levels for a quad-diversity path. This example was a 100 mile path in Germany from the top of the Zugspitze (1 mile above ground level) to a site with antennas 200 ft above ground level. It ran error free for a year of commissioning testing.

This configuration has long been known (Florman, 1968) to be a significant improvement over space or frequency diversity alone. It was used in the Bell System (Vigants and Pursley, 1979) for paths needing improved performance. Diversity improvement is estimated (Kostal, 1985) by multiplying the frequency and space diversity improvement factors calculated above.

$$I_{\mathrm{FF-QD}} = I_{\mathrm{FF-SD}} \times I_{\mathrm{FF-FD}} \tag{10.9}$$

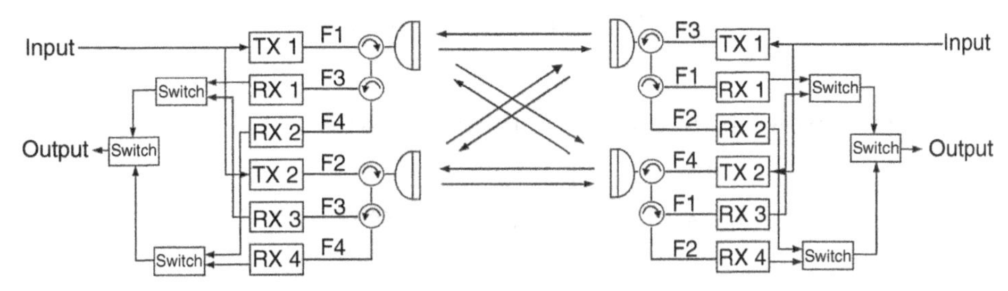

Figure 10.10 Quad diversity.

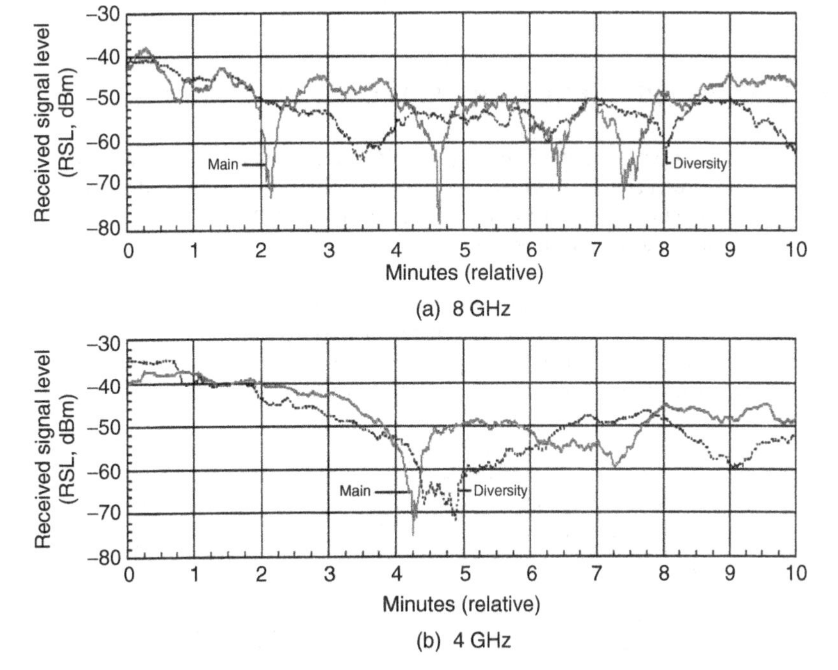

Figure 10.11 (a,b) The 100-mile quad diversity path.

I_{FF-QD} = quad diversity flat fading diversity improvement factor.

$$I_{DF-QD} = I_{DF-SD} \times I_{DF-FD} \tag{10.10}$$

I_{DF-QD} = quad diversity dispersive fading diversity improvement factor.

This configuration is the one most commonly used to fix difficult paths such as long overwater hops.

10.4 HYBRID DIVERSITY

This configuration is a variation of quad diversity (Fig. 10.12).

Diversity improvement is estimated (Knight, 2006) by selecting the larger of the frequency and space diversity improvement factors calculated above.

$$I_{FF-HD} = I_{FF-SD} \text{ or } I_{FF-FD}, \text{ whichever is larger} \tag{10.11}$$

I_{FF-HD} = hybrid diversity flat fading diversity improvement factor.

$$I_{DF-HD} = I_{DF-SD} \text{ or } I_{DF-FD}, \text{ whichever is larger} \tag{10.12}$$

I_{DF-HD} = hybrid diversity dispersive fading diversity improvement factor.

This configuration is typically used when diversity is needed but space diversity cannot be installed at a site. White (1968) reports excellent performance with this configuration.

10.5 MULTILINE FREQUENCY DIVERSITY

Figure 10.13 diagrams a typical multiple radio (multiline) frequency diversity configuration. The circles with curved arrows represent circulators that transfer the signal entering a port to the next port. The curved arrows indicate the direction of signal transfer. Each transmitter (TX) and receiver (RX) is assumed to have an RF filter (represented by rectangles) that reflects all signals with frequency different from the frequency of operation for that transmitter or receiver.

For paths requiring large transmission capacity, use of multiple transmitter–receiver pairs on the same transmission path is common. These transmitter–receiver pairs (working or regular channels) are usually nonstandby radios. Equipment and path protection is achieved by one or two additional nonstandby transmitter–receiver pairs (protection channels). This type of radio transmission system is called *multiline*,

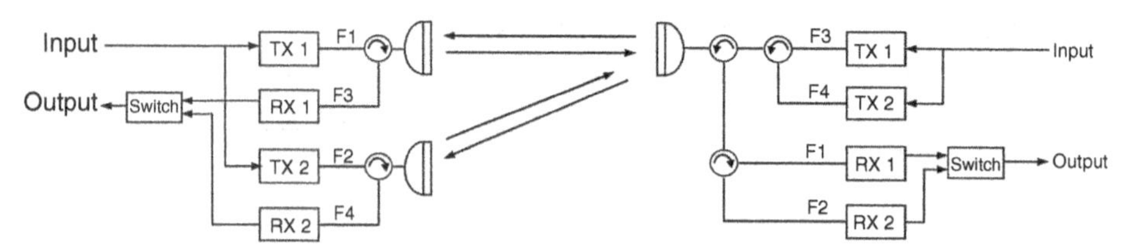

Figure 10.12 Hybrid diversity.

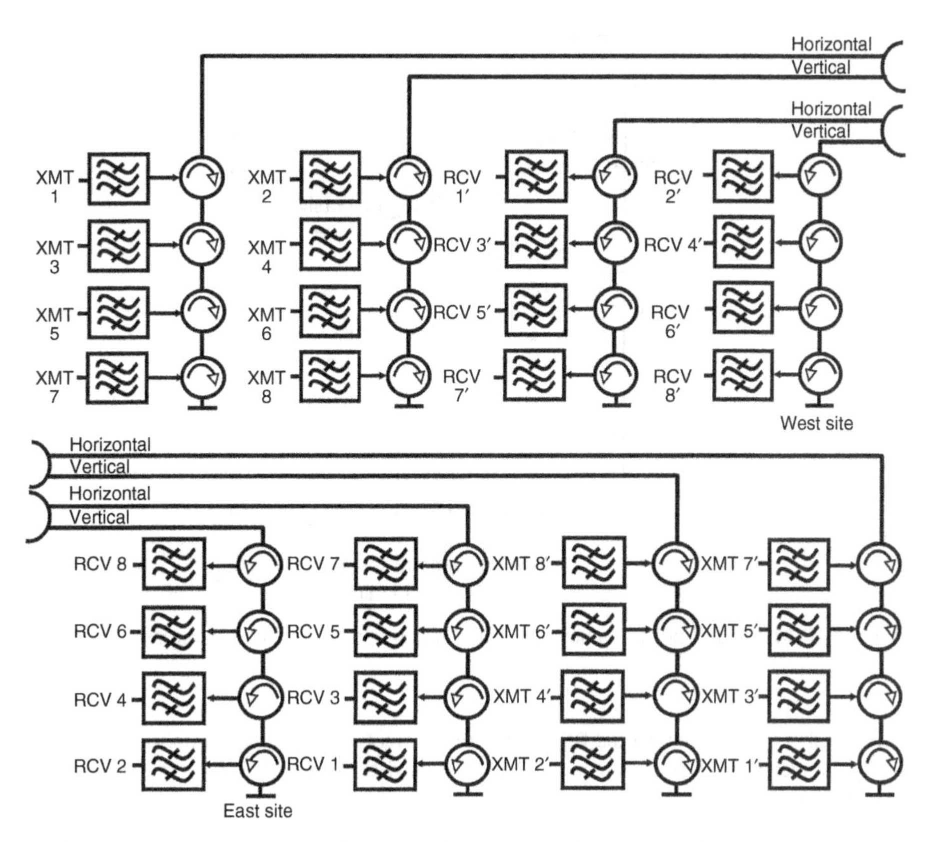

Figure 10.13 Multiline frequency diversity radio configuration—transmitters and receivers. Transmitters and receivers are reverse stacked to equalize losses due to circulators and filter reflections.

$U + N$ ("U plus N"), $U{:}N$ ("U for N"), or $U \times N$ ("U by N") where U is the number of protection channels and N is the number of working channels.

For systems with more than one working channel ("multiline"), Vigants and Pursley (1979) gave the following results for flat fading:

$$
\begin{aligned}
I_{\mathrm{FF-FD}} &= \frac{100 \times f_{\mathrm{avg}} \times \eta \times 10^{\mathrm{FFM}/10}}{D\,(\mathrm{miles})\,G} \\
&= \frac{161 \times f_{\mathrm{avg}} \times \eta \times 10^{\mathrm{FFM}/10}}{D\,(\mathrm{km})\,G} \qquad (10.13) \\
1 &\le I_{\mathrm{FF-FD}} \le 200
\end{aligned}
$$

$I_{\mathrm{FF-FD}}$ = flat fading improvement factor for frequency diversity;
FFM = effective flat fade margin (assuming all RSLs the same);
f_{avg} = average channel frequency $[(f_1 + \cdots + f_{N+U})/(N+U)]$ (GHz);
D = path distance;
G = diversity improvement factor (discussed below);
η = switching hysteresis efficiency;
$= 2/[(10^{\mathrm{H}/10}) + 1/(10^{\mathrm{H}/10})]$;
H = switching hysteresis (dB);

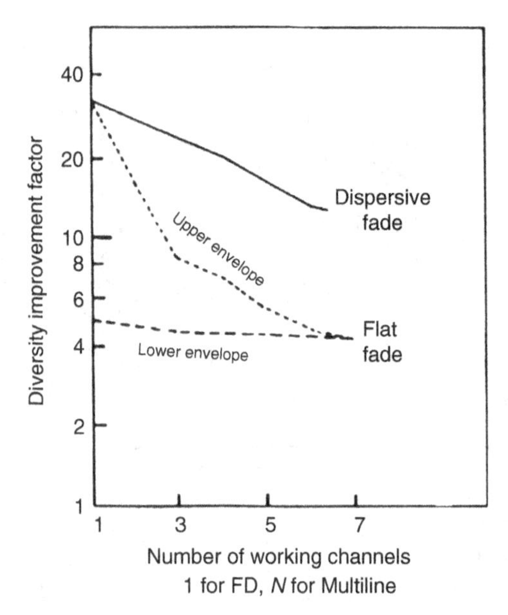

Figure 10.14 Multiline frequency diversity. *Source*: Reprinted with permission of IEEE.

= decibel value by which the RSL of the second receiver much be greater than the RSL of the first receiver before the output signal will be switched from the first receiver to the second receiver.

The hysteresis factor η was developed (Vigants, 1975) after the original frequency diversity improvement factors were developed (Vigants and Pursley, 1979). However, the derivation (Vigants, 1975) suggests it would be equally valid for frequency diversity.

Early in the investigations of digital radio outages, it was recognized that using the FFM diversity improvement factors underestimated dispersive fading diversity improvement (Greenstein and Shafi, 1972; Cellerino et al., 1985; Dirner and Lin, 1985). Dispersive fading frequency diversity improvement was investigated extensively (Kostal, 1985; Lee and Lin, 1985; Lin et al., 1988) both experimentally and by simulation. Although graphical results were published (Lee and Lin, 1985; Lin et al., 1988), formulas were not (Fig. 10.14).

Private discussions between Rockwell International and Bell Labs staff members (Evans, M. and Knight, W., private communication with Bell Labs staff, 1989) resulted in the following conservative estimates for dispersive fading improvement:

$$I_{\text{DF-FD}} = \frac{9324 \times f_{\text{avg}} \times 10^{\text{DFM}/15}}{D(\text{miles})\, G} = \frac{15{,}000 \times f_{\text{avg}} \times 10^{\text{DFM}/15}}{D(\text{km})\, G} \quad (10.14)$$

$$1 \leq I_{\text{DF-FD}} \leq 200$$

$I_{\text{DF-FD}}$ = dispersive fading improvement factor for frequency diversity;
DFM = dispersive fade margin (dB) (Chapter 9).

The above results are published in (Kizer, 1995; Knight, 2006). DFMs may be calculated with (dynamic) or without hysteresis (static) (Bellcore (Telcordia) Staff, 1989). It is suggested the use of

DFMs with hysteresis be used to more nearly characterize actual receiver action. The method of switching between receivers varies with manufacturer. The effect of hysteresis in switching between receivers during dispersive fading is not known at present.

The G factor is quite difficult to calculate for the general multiline case. For the single-protection channel ($U = 1$) with all radio receivers receiving the same receive signal power, Vigants offered an approximation (Vigants, 1975).

$$G = \frac{2\ (f_{\text{avg}})^3}{(\Delta f_{\text{ch}}\ \Delta f_{\text{eq}})} \tag{10.15}$$

N = the number of working radio channels;
U = the number of protection channels (typically 1 or 2);
Δf_{ch} = channel spacing (frequency difference between consecutive channels) (GHz);
$\Delta f_{\text{eq}} \approx N/\{N + [(N-1)/2] + [(N-2)/3] + \cdots + [1/N\}$.

Unfortunately, Vigants' approximation underestimates the value of Δf_{eq} for all $N > 1$. Nevertheless, this formula has widespread acceptance within the microwave radio community. Note that Δf_{eq} in Vigants' article (Vigants, 1975) is equivalent to $\Delta f_{\text{ch}} \Delta f_{\text{eq}}$ shown in the formula for G (Eq. 10.15).

The more general case (including an improvement to Vigant's 1:N approximation) was described later by Vigants and Pursley (1979). The basic assumption of the Vigants and Pursley algorithm is that radio channel outage is directly related to operating frequency. Both flat and dispersive fading are believed to conform to this relationship (Barnett, 1972; Rummler, 1988).

The $M = N + U$ consecutive radio channels have center frequencies $f_1, f_2, f_3, \ldots, f_x, \ldots, f_M$. The channel indices Cf_x are $1, 2, 3, \ldots, M \cdot P$ designates a unique pair of channels a and b drawn from the above pool of frequencies such that $f_b > f_a$. Merging equations (Smith and Cormack, 1984; Dirner and Lin, 1985) from the Vigants and Pursley (1979) paper yields the following:

$$G = \frac{1}{N} \sum_{i=1}^{N} \sum_{k=1}^{J(U+i)} \left\{ \frac{\left[(-1)^{(i-1)} (U+i) C_{(U-1)}^{(U+i-2)}\right]}{\left[\sum_{P=1}^{I(U+i)} \frac{(P_a P_b \delta_p)}{f_p^2}\right]} \right\} \tag{10.16}$$

$$C_n^m = \text{binomial coefficient} = \frac{m!}{(m-n)!n!} = \frac{m\,(m-1)\cdots(m-n-1)}{n!}, m \geq n$$

$$I(U+i) = C_2^{U+i}$$

$$J(U+i) = C_{(U+i)}^{M}$$

$$P_a = \frac{\text{receive power of channel a}}{\text{reference power}}$$

$$P_b = \frac{\text{receive power of channel b}}{\text{reference power}}$$

$$\delta_P = \frac{(f_b - f_a)}{f_P}$$

$$f_P = \frac{(f_b + f_a)}{2}$$

$$\Delta f_{\text{ch}} = f_{a+1} - f_a = \text{channel bandwidth}$$

$$f_b > f_a$$

Channel center frequencies f_a and f_b are chosen in such a way that all possible pairs P of frequencies are chosen without repetition.

Notice no mention is made of the frequency of the protection channel(s). We are dealing with average performance of protected channels, not individual channels. Unprotected lower frequency channels will always have lower path-related outage than higher frequency unprotected channels. As Vigants and Pursley observed (Vigants and Pursley, 1979), "... the [frequency] location of the protection channel affects the spread of the individual service failure times. The facility [protected transmission channel] service failure time and, therefore, the service failure time of an average [protected transmission] channel are not affected by the [frequency] location of the protection channel."

With the exception of the US 4-GHz frequency plan, throughout the world, all frequency plans are defined in groups of two contiguous sets of frequencies (low sub-band and high sub-band). One sub-band is assigned to transmitter frequencies and the other is assigned to receiver frequencies. Each sub-band is composed of several consecutive channels of the same bandwidth. If we assume channels are chosen from such a frequency plan, the average frequency of any channel pair f_P is nearly the same as the average frequency of all the channels. As we are choosing from groups of equally spaced channels, the frequency difference between channel pairs is an integral number of channel bandwidths. This allows considerable computational simplification using the following definitions:

$$\delta_P = \Delta f_{ch} \frac{(Cf_b - Cf_a)}{f_P}$$

f_P = average channel frequency (sum of all channel frequencies/M);
Cf_a = the channel index of f_a and Cf_b is the channel index of f_b.

$$G = \frac{1}{N}\sum_{i=1}^{N}\sum_{k=1}^{J(U+i)}\left(\frac{\text{Numerator}}{\text{Denominator}_1}\right) \tag{10.17}$$

$$\text{Numerator} = (-1)^{(i-1)}(U+i)\,C_{(U-1)}^{(U+i-2)}$$

$$\text{Denominator}_1 = \frac{\Delta f_{ch}}{f_P^3}\left\{\sum_{d_1=1}^{U+i-1}\sum_{d_2=d_1+1}^{U+i}[Cf_b(i,d_2) - Cf_a(i,d_1)]\{P_b[Cf_b(i,d_2)]P_a[Cf_a(i,d_1)]\}\right\}$$

We may now redefine the G factor calculation as follows:

$$G = \frac{2f_P^3}{\Delta f_{ch}\,\Delta f_{eq}} \tag{10.18}$$

$$\Delta f_{eq} = \frac{1}{\left[\frac{1}{2}\frac{1}{N}\sum_{i=1}^{N}\sum_{k=1}^{J(U+i)}\left(\frac{\text{Numerator}}{\text{Denominator}_2}\right)\right]}$$

$$\text{Numerator} = (-1)^{(i-1)}(U+i)\,C_{(U-1)}^{(U+i-2)}$$

$$\text{Denominator}_2 = \sum_{d_1=1}^{U+i-1}\sum_{d_2=d_1+1}^{U+i}\{[Cf_b(i,d_2) - Cf_a(i,d_1)][P_b(Cf_b(i,d_2))P_a(Cf_a(i,d_1))]\}$$

The indices (effectively normalized channel frequencies) $Cf(i,d_0) = A\%(i, d_0)$ may be calculated using the following BASIC algorithm (developed by Brad Wick of Rockwell International):

```
1000 R = U& + I& REM [U% = U and I& = I in Vigants and Pursley article]
REM [All variables are integers]
REM [All variables with % suffix are short (2 bytes) integers]
REM [All variables with & suffix are long (4 bytes) integers]
OPEN AIDATA$ FOR OUTPUT AS #1 REM [open file AIDATA$ to record A%]
FOR NI = 1 TO R
A%(NI) = NI
IF NI = R THEN A%(NI) = R - 1
B%(NI) = M& - R + NI
NEXT NI
NN& = 0
1100 A%(R) = A%(R) + 1
FOR NL = 1 TO R
PRINT #1, A%(NL) REM [record the value A% = Cf(i,d0), a channel index]
NEXT NL
NN& = NN& + 1
IF A%(R) >= B%(R) THEN
FOR NP = 1 TO R
ND = (R + 1) - NP
IF A%(ND) <B%(ND) THEN GOTO 1200 REM [reset A%]
NEXT NP
GOTO 1300 REM [terminate the subroutine]
1200 REM [reset A%]
A%(ND) = A%(ND) + 1
FOR NQ = ND TO (R - 1)
A%(NQ + 1) = A%(NQ) + 1
NEXT NQ
A%(R) = A%(R - 1)
END IF
GOTO 1100
1300 CLOSE #1 REM [close file AIDATA$ of A% values for a given i value]
REM [file AIDATA$ remains available for reading]
RETURN REM [end of subroutine]
```

These algorithms were used to calculate Δf_{eq} for U from 1 to 4 and N from 1 to 20 assuming all channels have the same RSL. Curve fitting the results gave the following:

$$G = \frac{2(f_{avg})^3}{(\Delta f_{ch}\,\Delta f_{eq})} \tag{10.19}$$

f_{avg} = average channel frequency $[(f_1 + \cdots + f_{N+U})/N + U]$ (GHz);
N = the number of working radio channels;
U = the number of protection channels (typically 1 or 2);
Δf_{ch} = channel spacing (frequency difference between consecutive channels) (GHz).

$$\Delta f_{eq} = 2\,\frac{(C1 + C2\;N + C3\;N^2 + \;C4\;N^3)}{(1 + C5\;N + \;C6\;N^2 + \;C7\;N^3)} \tag{10.20}$$

For 1:N protection (U = 1)	For 2:N protection (U = 2)
$C1 = 0.5975701514570466$	$C1 = 1.833935974042738$
$C2 = 0.7428445800439416$	$C2 = 1.37635098350984$
$C3 = 0.3082229552151934$	$C3 = 0.3351140480008432$
$C4 = 0.04384794647757599$	$C4 = 0.03839451275666113$
$C5 = 1.583046763318658$	$C5 = 1.311603244602883$
$C6 = 0.6973150200993524$	$C6 = 0.3335083601043749$
$C7 = 0.1046098514355536$	$C7 = 0.04273458297033073$

For 3:N protection (U = 3)	For 4:N protection (U = 4)
$C1 = 3.738277399105584$	$C1 = 6.320916126203917$
$C2 = 2.032342560033778$	$C2 = 2.636255876978117$
$C3 = 0.3648839616080699$	$C3 = 0.3680260657151042$
$C4 = 0.03197283771075967$	$C4 = 0.03080961423005372$
$C5 = 1.216758924321869$	$C5 = 1.161986088953397$
$C6 = 0.2272356326313417$	C6 = 0.1606827250777983
$C7 = 0.02299512674588805$	$C7 = 0.01632182640267676$

Although the above formulas were based on N up to 20, the shape of the curves (Fig. 10.15) suggests that these formulas would be quite accurate for considerably larger values of N.

For frequency plans with consecutive channel frequencies, the basic assumption of using an average channel frequency rather than the specific channel pair average frequency introduces no observable error for fixed microwave radio-frequency bands (excluding the unusual US 4-GHz band with offset and interleaved channels). If double precision arithmetic is used, the above calculations using average channel

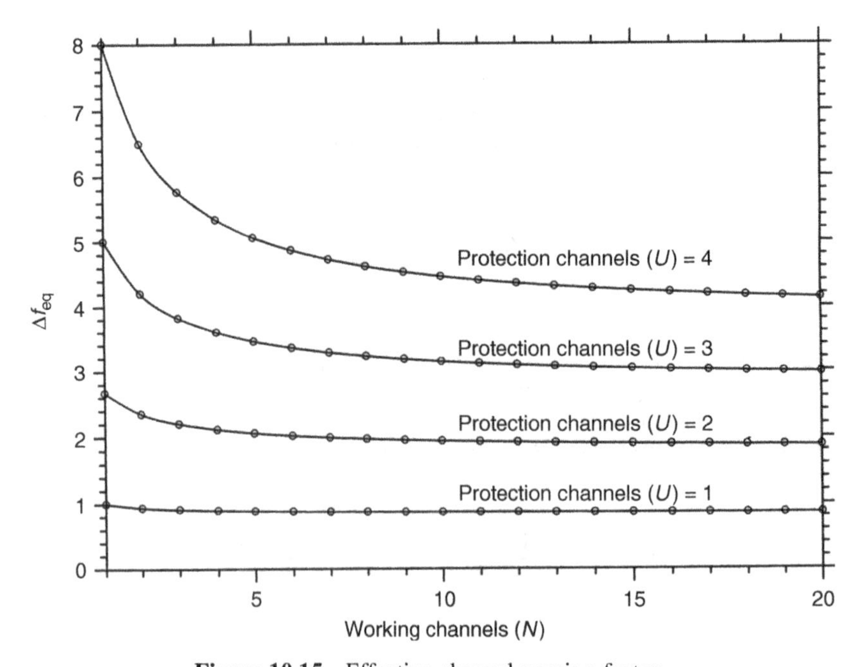

Figure 10.15 Effective channel spacing factor.

pair frequency gets exactly the same results as Vigants and Pursley (1979, Table 3, 6 GHz 1 × 7 and 2 × 6 values, $f_{avg} = 6.049$ and $\Delta f_{ch} = 0.02965$).

The effect of different receive powers for different channels is complicated to compute. However, under normal circumstances (assuming normal engineering practice of reverse-stacking transmitters and receivers on opposite ends of the path to equalize path loss), all channels on a path have the same receive signal power (if they are in the same frequency band).

10.6 CROSSBAND MULTILINE

With the congestion of frequency bands, use of more than one frequency band might be required to achieve the desired overall transmission capacity. Use of this form of crossband frequency diversity complicates the calculations. Assume all radio channels have the same bandwidth Δf_{ch}. Group the radio channel center frequencies into two groups: high frequency and low frequency.

$$f_{hi} = \text{average channel center frequency of high frequency group}$$

$$f_{lo} = \text{average channel center frequency of low frequency group}$$

$$R_{hilo} = \frac{f_{hi}}{f_{lo}}$$

$$P_{lo} = \text{receive power of low frequency channels/reference power}$$

$$P_{hi} = \text{receive power of high frequency channels/reference power}$$

$$G = \frac{2\,f_{lo}^3}{\Delta f_{ch}\,\Delta f_{eq}} \tag{10.21}$$

$$\Delta f_{eq} = \frac{1}{\left[\dfrac{1}{2N}\displaystyle\sum_{i=1}^{N}\sum_{k=1}^{J(U+i)}\left(\frac{\text{Numerator}}{\text{Denominator}_3}\right)\right]}$$

$$\text{Numerator} = (-1)^{(i-1)}(U+i)\,C_{(U-1)}^{(U+i-2)}$$

$$\text{Demoninator}_3 = \sum_{d_1=1}^{U+i-1}\sum_{d_2=d_1+1}^{U+i}\left[F_f P_b(Cf_b(i,d_2))P_a(Cf_a(i,d_1))\right]$$

Notice that the reference average frequency is F_{lo}.

If f_a and f_b are both in the low frequency group

$$F_f = \left[\frac{(Cf_b - Cf_a)}{1}\right]P_{lo}{}^2$$

If f_a is in the low frequency group and f_b is in the high frequency group

$$F_f = \frac{(R_{hilo}-1)\left(\frac{F_{lo}}{\Delta f_{ch}}\right)}{\left[\frac{(R_{hilo}+1)}{2}\right]^3}P_{lo}P_{hi}$$

If f_a and f_b are both in the high frequency group

$$F_f = \frac{(Cf_b - Cf_a)}{(R_{hilo})^3}P_{hi}{}^2$$

In the United States, the most likely candidates for crossband frequency diversity are the lower 6- and the 11-GHz frequency bands. For short paths or paths in favorable rain areas (e.g., the western United States), crossband diversity is practical. For these bands,

Δf_{ch} = channel bandwidth = 0.030 GHz;
F_{hi} = average frequency of high frequency band = 11.2 GHz;
F_{lo} = average frequency of low frequency band = 6.175;
R_{hilo} = ratio of average channel frequencies = 1.814.

We will assume the same parabolic dish transmit and receive antennas for both bands. For parabolic dish antennas, antenna gain (dB) increases as a function of frequency 20 log (R_{hilo}). Path loss (dB) increases as a function of 20 log (R_{hilo}). As there is only one path loss but two antennas per path, we might expect a net receive signal gain of 20 log (1.814) = 5.2 dB for the 11-GHz channels relative to the lower 6 GHz. However, transmission line (waveguide) losses are much higher and typical transmit powers are a couple of decibel lower for 11 GHz radios. Receiver combined noise figures and input filter loss are also typically 1–2 dB worse at 11 GHz. For purposes of this example, we will assume essentially the same effective receive signal power for both lower 6 and 11 GHz.

The lower 6-GHz band has 8 duplex 30-MHz channels and the 11-GHz band has 13. We will consider U:N systems with U from 1 to 4 and N up to 16. Conventional 1:N multiline systems can only have a maximum of eight channels can be in the lower 6-GHz band (Fig. 10.16). Cross-polarization cancellation (XPC) multiline systems can have up to 16 channels in the lower 6-GHz band (but they usually operate as two independent 1 : 7 systems).

The results are graphed in Figure 10.16. It is obvious that the optimum number of channels to be operated in the high band is U, the number of protect channels. Placing the protect channels in the upper band significantly improves the frequency diversity improvement factor.

The question still remains regarding the effect of operating the higher frequency channels with lower receive signal power than the lower frequency channels. We will consider a few systems that meet the FCC-mandated one protection channel per three working channels. We will again consider the lower 6 GHz (8 channels of 30 MHz) and 11 GHz (13 channels of 30 MHz) as our potential bands (Table 10.1).

The higher band relative power (Table 10.1) is the relative receiver power each of the higher band channels needs so that the multiline channel performance is the same as if all channels were in the lower 6-GHz band (which is not even possible for the larger multiline systems). For most of the systems, the higher band can have significantly lower receive power and still have better frequency diversity performance relative to an all lower band solutions. Even the higher capacity systems need only modest power increase for the high band channels.

Use of crossband frequency diversity appears practical. Placing at least the protection channels of a crossband diversity system in the higher band significantly increases frequency diversity improvement relative to a single frequency band case. Of course, the assumption is made that the path is short enough that rain attenuation availability is satisfactory.

10.7 ANGLE DIVERSITY

The purpose of angle diversity antennas is to mitigate the destructive effects of multipath propagation without using a vertically spaced diversity antenna on the microwave tower (Fig. 10.17). Multipath

TABLE 10.1 Higher Band Relative Power

Multiline System	Channels in Higher Band	Higher Band Relative Power, dB
1 : 4	1	−15.4
1 : 8	1	−14.2
2 : 8	2	−12.1
3 : 12	7	+2.3
4 : 12	8	+1.9

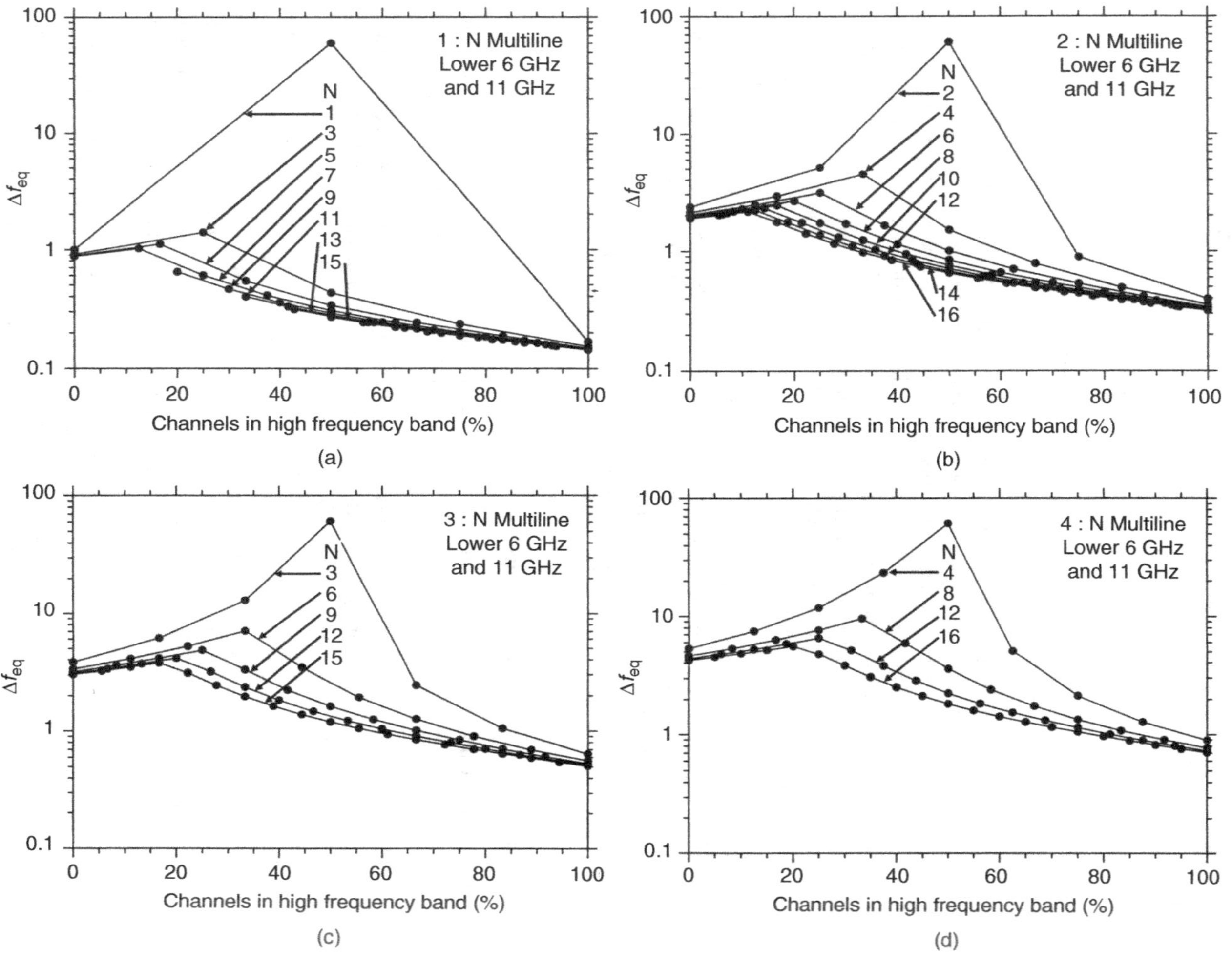

Figure 10.16 (a–d) Dual-band operation.

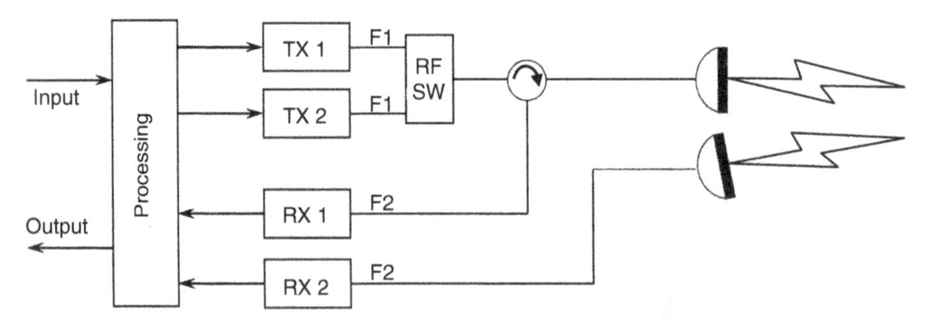

Figure 10.17 Angle diversity.

propagation occurs when the transmitted signal arrives at the receiver having traveled two or more paths. These multiple receive signals, all replicas of the original transmitted signal, have slightly different relative phase because they travel slightly different paths from transmitter to receiver. When they combine at the receive antenna, their differences in relative phase cause the composite received signal to increase (upfade) or decrease (downfade) depending on the relative amplitude and phase of the incoming signals. Inasmuch as the different transmission paths are changing slightly as the atmosphere changes, the fading is dynamic. Secondary signals are caused by refraction due to layered variations in the atmospheric refractive index or by reflections form the earth's surface. This causes the secondary signal to arrive at a slightly different angle ("angle of arrival") (Webster and Scott, 1987; Webster and Merritt, 1990; Webster, 1983, 1991) than the normal signal. Most nondiversity microwave paths over reflective terrain are intentionally near the ground to minimize the effect of ground reflections. Secondary signals, which cause multipath fading, normally have a greater angle of arrival than the primary signal. Changing the relative amplitude of the main and secondary signals through antenna beam shaping can result in a significantly different composite signal power. Angle diversity antennas attempt to take advantage of that observation.

10.7.1 Angle Diversity Configurations

Angle diversity can be achieved using two different antenna configurations (Allen, 1989a): Two separate antennas mounted side by side or a single antenna with dual-beam feedhorn. These can be operated in several ways.

For the configuration with *two separate antennas*, the most common alignment is for one antenna to be bore-sighted on the primary (normal) transmitted signal and the second antenna to have an elevation angle slightly greater than the primary antenna (Fig. 10.18).

The second antenna is aligned such that the primary transmitted signal arrives at the first low side null of the main beam. The second antenna theoretically will only see signals that have a higher angle of arrival than the primary signal. This second antenna will have no received signal during normal propagation conditions but should have a significant signal when the secondary signals are causing cancellation of the primary signal at the first (normally aligned) antenna. While this is optimum alignment, because it forces the second antenna receiver to be in constant alarm (due to low RSL), this alignment is not popular with maintenance staff.

If the transmission path has excessive terrain clearance (e.g., mountaintop-to-mountaintop path), the receive antenna may experience significant secondary signals from below the normal receive path angle. In this case, the second antenna is either pointed above or below the main antenna, depending on whether the secondary signal is expected to be (high angle) normal atmospheric multipath or (low angle) ground reflections (or low reflective atmospheric layer refractions). Sometimes, this decision must be made based on experience.

During normal propagation conditions, with the second antenna aligned on the first low side null of the upper beam as specified above, the secondary antenna may not be receiving a significant signal (causing the diversity radio to be constantly in alarm). For this reason, some operators change the alignment of the second antenna back toward the main signal path (moving slightly out of the null) to keep the diversity

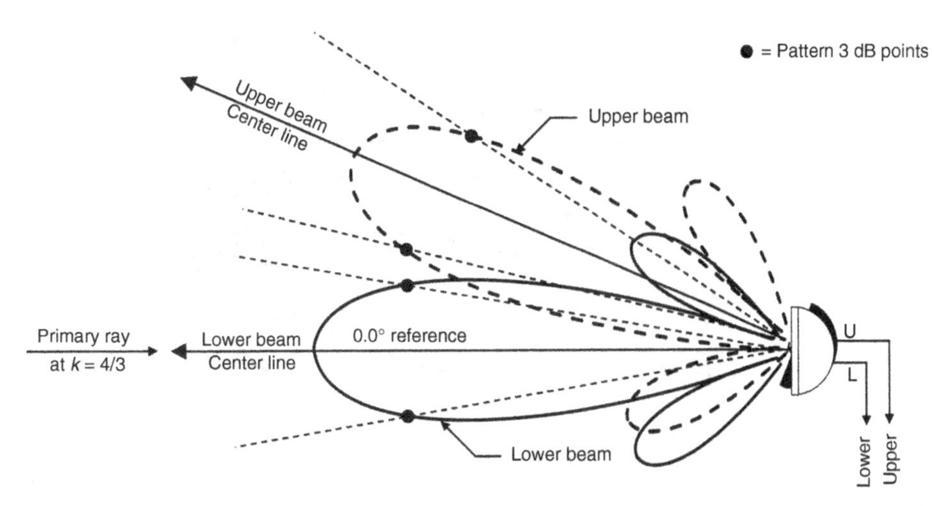

Figure 10.18 Angle diversity using two antennas. Drawing courtesy of Eddie Allen. Used with permission.

receiver from being in constant alarm. While this does not provide optimum multipath protection, it has the operational advantage of reducing receiver alarms.

Another angle diversity configuration is a *single antenna with dual-antenna feedhorns*. As before, the diversity (upper pattern) antenna feedhorn is aligned such that the primary signal arrives at the first low side null of the upper beam (Fig. 10.19).

Usually, this causes the main antenna received signal to be slightly less than normal (typically about 1 dB). This procedure will also slightly degrade cross-polarization discrimination on the main feed.

Many microwave paths experience signal reflections from the ground as well as atmospheric refractions from above the path. Although the preceding case is the most typical angle diversity installation, it is not optimum for this situation. A common attempt to cover secondary signals from above and below the normal propagation path is to align the antenna so that both ports have the same signal power during normal propagation (Fig. 10.20).

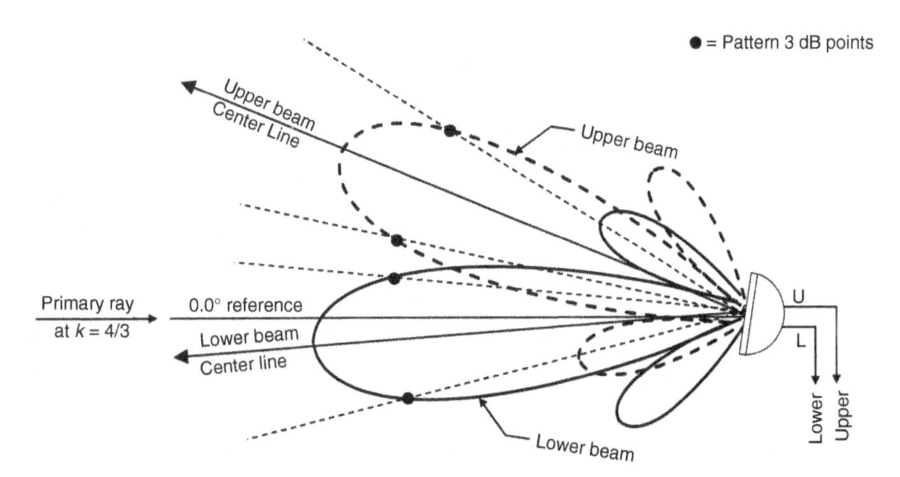

Figure 10.19 Angle diversity using one antenna with dual pattern. Drawing courtesy of Eddie Allen. Used with permission.

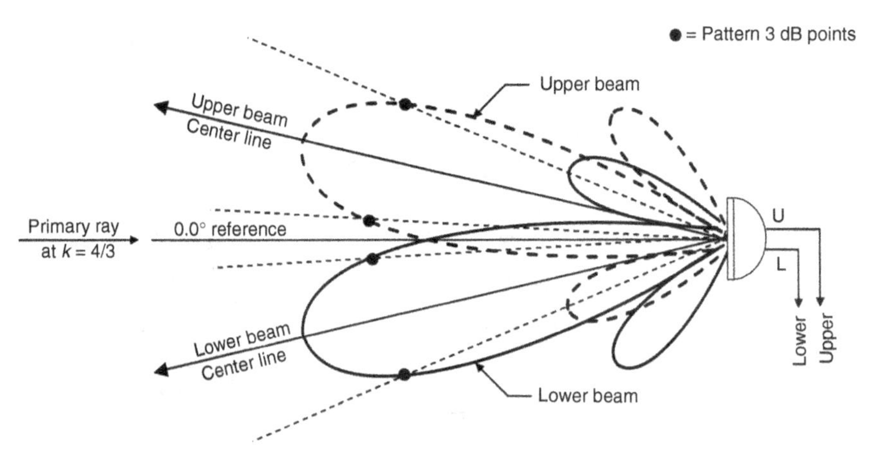

Figure 10.20 Angle diversity using one antenna with dual pattern. Drawing courtesy of Eddie Allen. Used with permission.

While this might appear desirable, actual path testing (Allen, 1989a, 1989b) has shown that this design performs quite poorly. This alignment is not effective for any type of fading. If multipath signals are expected both above and below the main path, the use of an RF hybrid (below) produces much better results.

The RF hybrid approach takes the two feedhorn patterns and produces a synthetic pattern formed by the sum or difference in the two signals. The antenna is aligned so that the sum port is centered on the main path and the difference port signal is at a null on the primary signal (Fig. 10.21).

While effective, this approach causes the diversity receiver to be in alarm under normal conditions. This is sometimes a maintenance concern. With this configuration, the sum port of the RF hybrid has 2–3 dB less gain than a nonangle diversity antenna of the same size. This type of antenna is fine for a receiver but is somewhat limited for use with transmitters in frequency-congested areas. Using the hybrid, the sum signal antenna pattern fails FCC Category A antenna pattern standards. If a Category A transmit antenna is required to coordinate the station, the hybrid cannot be used for transmitting.

The hybrid angle diversity approach is by far the most difficult to install and maintain of all the angle diversity configurations. However, it is the only dual-angle diversity configuration that can minimize fading due to both atmospheric multipath and ground reflections.

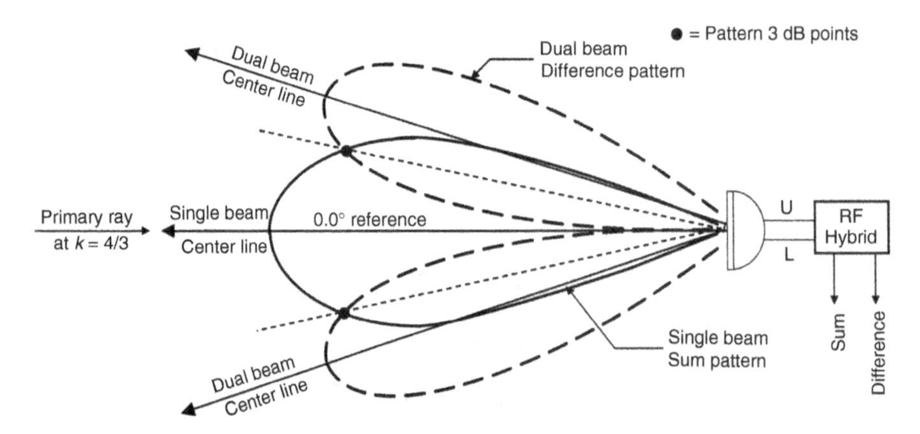

Figure 10.21 Angle diversity using one antenna and a hybrid. Drawing courtesy of Eddie Allen. Used with permission.

10.7.2 Angle Diversity Performance

From the early 1980s to the early 1990s, many investigators (Gardina and Lin, 1985; Hubbard, 1985; Malaga and Parl, 1985; Balaban et al., 1987; Dekan et al., 1987; Lin et al., 1987; Allen, 1988a, 1988b, 1992a; Noerpel, 1988; Alley et al., 1987, 1990; Valentin et al., 1987, 1990) tested angle diversity on a large number of microwave radio paths. People implemented angle diversity in different ways. There was no agreement on engineering or installing these systems. Likewise, there is no agreement as to what angle diversity improvement to expect. Lin (1988) suggested "If properly designed, angle diversity performance can be far superior to space diversity" but failed to mention what "properly designed" meant. Allen (1992b pointed out that "... in some cases [angle diversity] performs only slightly better than an unprotected antenna ...." Some researchers (Alley et al., 1987; Balaban et al., 1987; Lin et al., 1987) claimed angle diversity was generally better than space diversity. Others suggested (Mohamed et al., 1989; Valentin et al., 1989; Allen, 1991; Glauner, 1993) just the opposite. Some (Allen, 1988b; Lin, 1988; Satoh and Sasaki, 1989) said sometimes it was better and sometimes it was worse. Some (Satoh and Sasaki, 1989; Alley et al., 1990) suggested angle diversity improvement was a function of path conditions but they could not quantify what those were. Some (Allen, 1988a; Lin, 1988; Mohamed et al., 1989) said angle diversity worked better on paths dominated by flat (thermal) multipath fading. Others (Lin et al., 1988; Valentin et al., 1990; Vergeres et al., 1990; Di Zenobio et al., 1992) said it worked better on paths dominated by dispersive multipath fading. Giger (1991) suggests angle diversity can be effective for paths dominated by either long dispersive paths or strong near end ground reflections (i.e., for situations where the secondary signal has a significantly different angle of arrival relative to the normal signal). Angle diversity can only be optimized for one of these conditions.

Alley et al. (1987) of AT&T Bell Labs stated "The variation of SFF [Single Frequency Fading] improvement factors and SES [Severely Errored Seconds] improvement factors with time for a given [angle diversity] antenna configuration illustrates the difficulty in making precise statements concerning the performance of [angle] diversity antenna systems. ... The upper and lower 81% confidence limits for the SES improvement factor were shown to vary from ±150% to ±250% of the mean SES improvement factor for various [angle] diversity configurations after 5 months of data collection on a hop which suffers extremely heavy multipath fading ..." Nearly everyone agreed that angle diversity could provide significant path diversity protection but there was little agreement as to how much improvement to expect and under what configuration or path conditions. Although initially a strong supporter of angle diversity, AT&T Long Lines (of the Bell System) eventually avoided the use of angle diversity except when no other alternative was practical.

Giger (1991), of AT&T Bell Labs, developed an angle diversity improvement estimate model. His model predicted "Angle diversity is, therefore, equivalent to space diversity where the two antennas are separated by a distance equal to half the antenna diameter D. All the results obtained for space diversity now also apply to angle diversity." It is not clear under what path conditions this model applies. However, this model suggests that angle diversity will generally perform much worse than the typical space diversity implementation.

Rummler, also of AT&T Bell Labs, and Shafi (Rummler and Dhafi, 1989) observed, "As yet there are no algebraic formulas which permit the estimation of improvement by using angle diversity. However, recent studies by Lin have shown that improvements in performance over conventional space diversity are possible. It is clear ... that the treatment of angle diversity lacks completeness, and further work in required. In particular, the dependence of the observed improvement factors on the nature of the path has yet to be determined." Danielsson and Johansson (1993) also stated "... new models for [angle diversity] improvement calculation is [sic] needed."

Glauner (1993) and Danielsson and Johansson (1993) compared angle diversity to other diversity methods. Unfortunately, their derivations required knowledge of the angle of arrival of the receive signals—and in general that information is not known. Glauner (1993) stated that average hop angle diversity improvement was in the range 5–20. This is similar to typical frequency diversity improvement. Lin et al. (1988) state "... antenna angle diversity are very effective against dispersion caused outages but are less effective against thermal noise-caused [flat fading] outages." These results have lead some designers (Knight, 2006) to treat angle diversity as a typical space diversity configuration and use dispersive diversity improvement as calculated for space diversity but divide flat fading diversity improvement by a factor of 10.

Despite all the research, currently angle diversity has to be considered somewhat experimental. It can provide effective path protection but how much and under what circumstances is not clear. It appears it should be used only if conventional space diversity is not practical.

10.A APPENDIX

10.A.1 Optimizing Space Diversity Vertical Spacing

Space diversity systems have been engineered for the last 40 years. For years, the engineering rule of thumb (White, 1968) was to place the diversity antenna about 200 wavelengths below (or above) the primary antenna. This was the origin of the popular 30-ft spacing for 6-GHz systems and 20 ft for 11-GHz systems (Fig. 10.A.1).

Lee and Lin's (1988) results (Fig. 10.6) support this approach. Thousands of systems have been successfully deployed using this simple methodology.

Nevertheless, space diversity antenna engineering remains an art. White's classic paper (White, 1968) describes the path design issues in some detail and is still well worth reading. Paths come in many types. Many paths with rough terrain need no special engineering. The 200 wavelength rule works just fine. If the path has a reflection but there is a couple of decibel of antenna discrimination against the reflected signal (a common condition for highly inclined paths), special analysis is not necessary. Deep fades can only occur when the reflected signal power is closer than a decibel to the main signal power (Chapter 9). In some cases, the reflection path can be blocked by moving the towers or antenna heights in such a way as to shield the antenna from reflections using terrain or trees. On short paths, one or both antennas may be tilted up (Hartman and Smith, 1977) to put the reflection into an antenna null. Of course, this degrades cross-polarization discrimination and makes the path more susceptible to multipath fading.

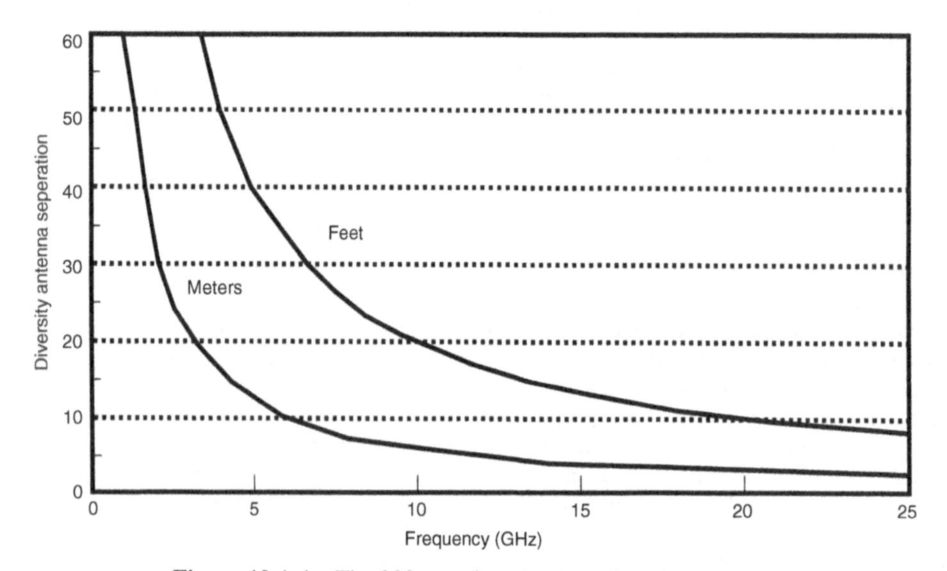

Figure 10.A.1 The 200-wavelength space diversity spacing.

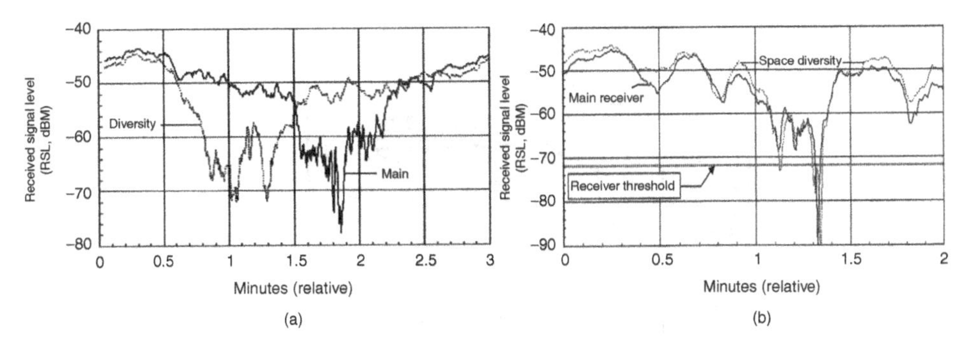

Figure 10.A.2 (a) Good and (b) bad 8-GHz main and space diversity path performance.

The usual way of designing paths is to build them relatively near the ground, so ground reflections are not a problem (although the path may be subject to obstruction fading, Chapter 12). In the normal case, all the designer has to deal with is atmospheric reflections arriving from above the normal path. For path with one or both antennas very high (e.g., on a mountain or bluff), normal path design methods do not apply. As the path has excessive clearance, normal methods of optimizing the path do not work (reflection points due terrain or atmospheric layers are dozens of Fresnel zones below the main path and the even and odd zone radii nearly overlap). The natural inclination is to put the antenna near the edge of the terrain. In fact, just the opposite is needed. For long, high paths, move the antenna as far back as practical to allow terrain to shield reflection from terrain (and atmospheric layer) reflections. If the reflections are blocked, typical 200 wavelength multipath protection spacing is adequate. If the reflected signals cannot be blocked (e.g., the antennas are on the sides of mountains), significant fading can be expected and quad diversity is recommended.

For paths over water or flat terrain, engineering for reflections is mandatory. Figure 10.A.2 illustrates performance of two long paths over flat land: One space diversity path is improperly designed; the other is designed and performing properly.

As you will note in Chapter 13, all flat surfaces reflect radio waves nearly parallel to the Earth (due to different dielectric coefficients and Snell's Law). That is true whether the surface is a pure conductor (Lake Tahoe) or a pure dielectric (salt flats) or anything in between. Even rough terrain with rice paddies, pine trees, or wheat fields giving off large amounts of water vapor can appear as reflective surfaces when there is no wind. If the terrain is flat and the reflection path is not blocked, assume the path is reflective.

The following procedure is suggested for diversity antenna spacing on reflective paths. In general, the method is to place the antennas in the best locations for low order Fresnel zone reflections. The intent is to place one antenna so it has an enhanced signal due to the reflection and the other antenna has a diminished signal [White calls this *complementary diversity* (White, 1968).]. This optimizes the antennas over a defined range of K factors. Then, if necessary, the spacing is changed slightly to account for large negative K factors.

Figure 10.A.3a: Determine the propagation area and the appropriate antenna clearances section 12.4.1. Draw the path profile using K factors appropriate for the propagation area. $K =$ Infinity plots the measured terrain heights. Use equation 5.36 to distort terrain height for other K factors. For this example, we are assuming an overwater path in a poor propagation area (such as Lake Pontchartrain in Louisiana).

Figure 10.A.3b: Position the two main antennas so they meet the clearance criteria for the selected propagation area (in this case, the main to main antennas have grazing clearance for $K = 1/2$ and first Fresnel zone clearance for $K = 4/3$). In this example, the lowest antenna positions are constrained by the center of the path ($K = 1/2$) and a tree on the far right side of the path ($K = 4/3$).

Figure 10.A.4: Position the two space diversity antennas so they have the appropriate clearance (0.6 Fresnel zone for $K = 4/3$). In this example, the space diversity (lower) antenna heights are constrained by trees on either side of the path. We have now established the starting point for spacing optimization.

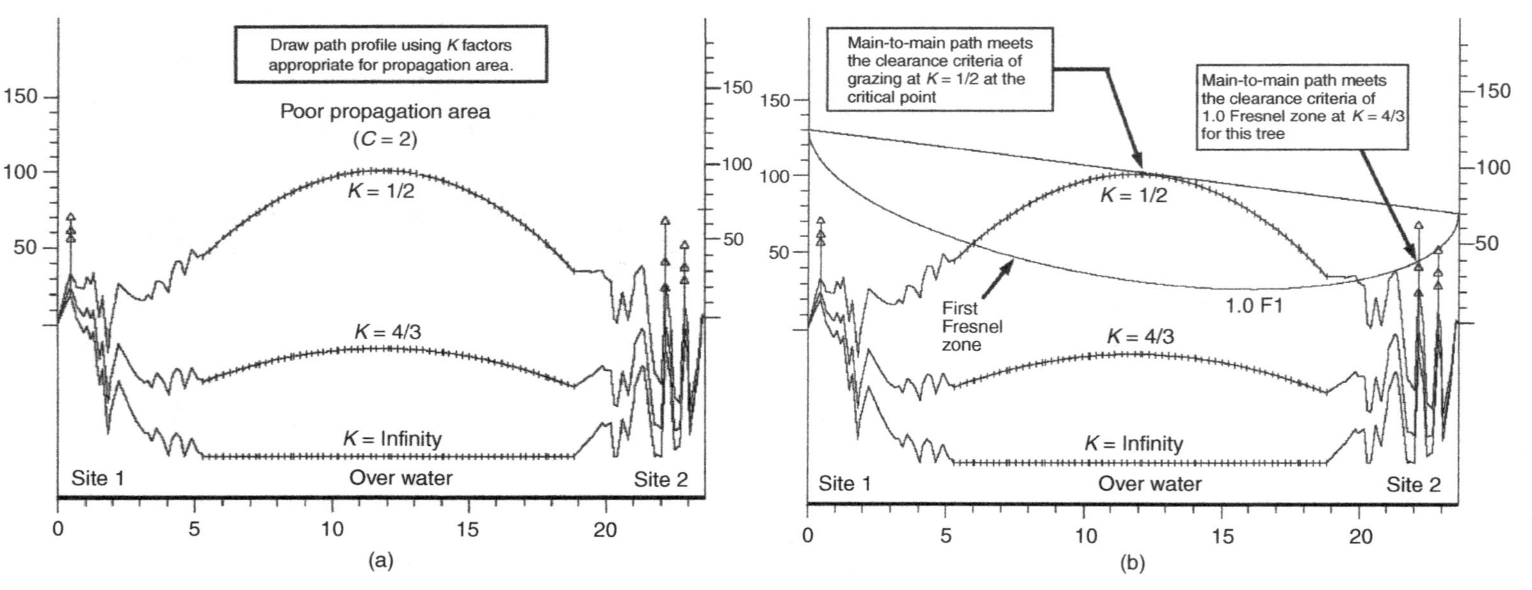

Figure 10.A.3 (a,b) Space diversity antenna placement.

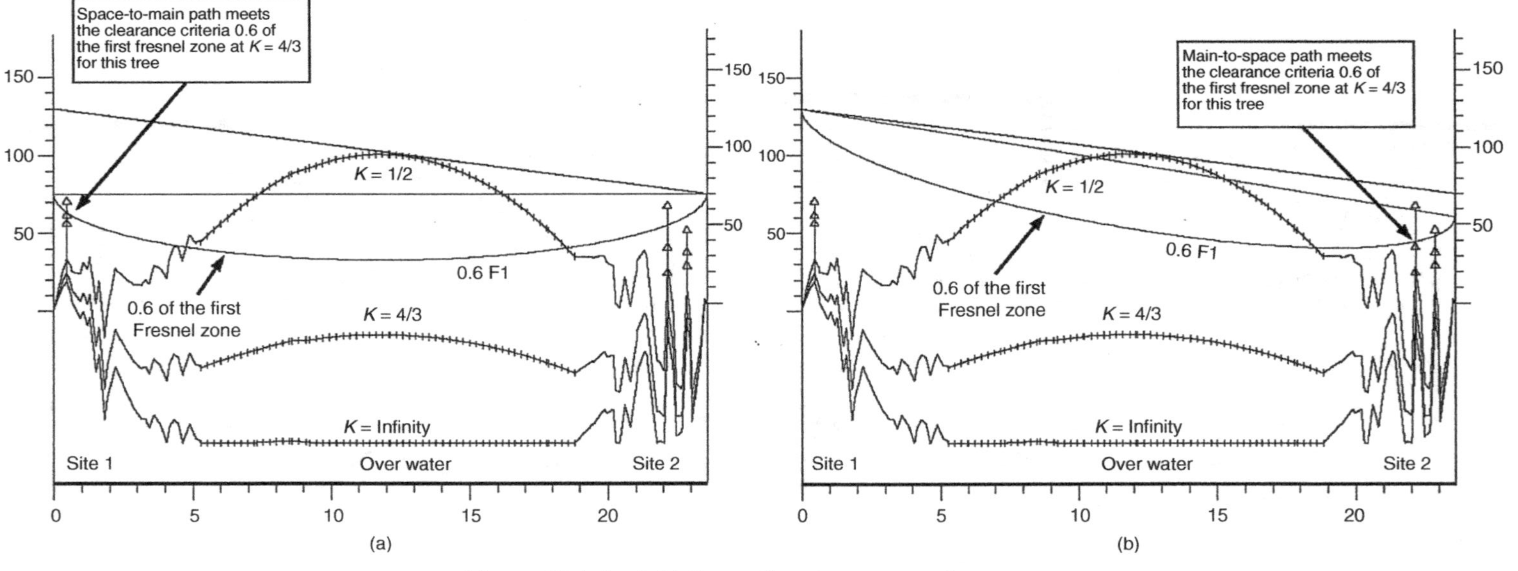

Figure 10.A.4 (a,b) Space diversity antenna placement.

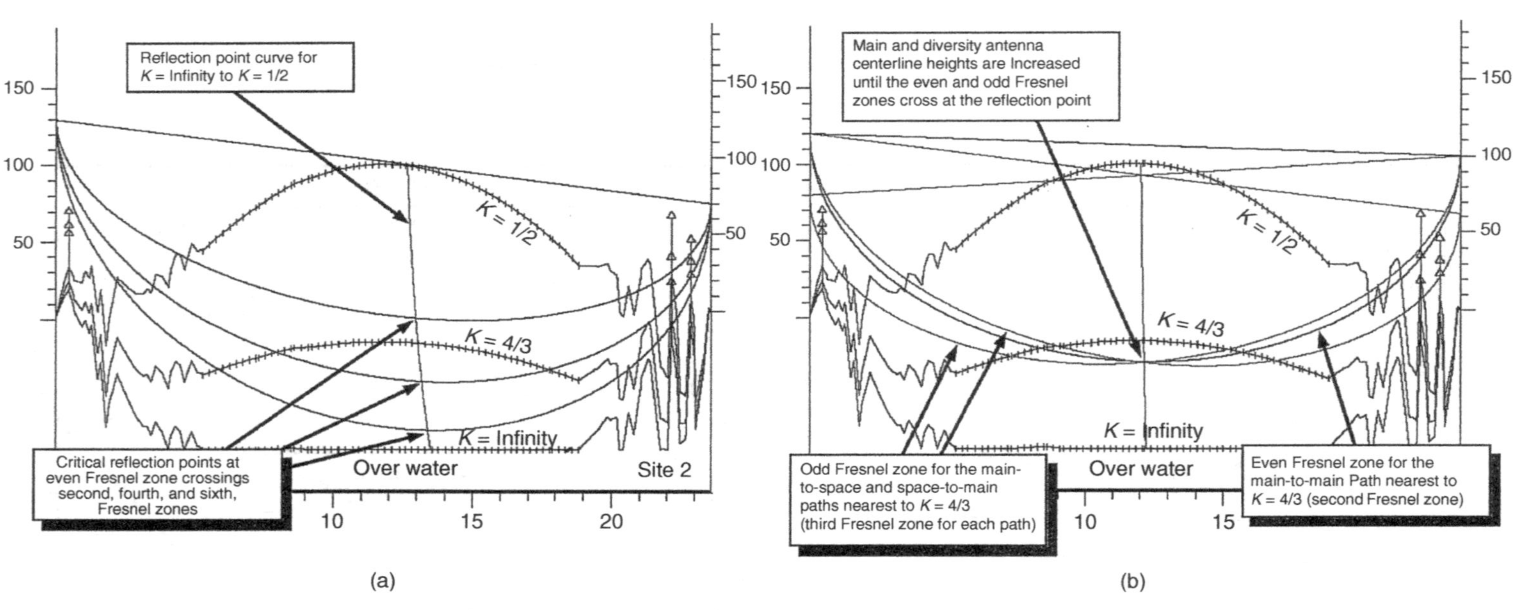

Figure 10.A.5 (a,b) Space diversity antenna placement.

Figure 10.A.5a: Determine the reflection points on the modified earth for the main to main antenna path for the various K factors from $K \ll 1$ to K = infinity. The formula for the reflection point is in section 13.4 as well as Appendix A. The reflection point formula is based on the height difference between the antenna and the reflection point, not absolute height. This issue is usually resolved by computer program by curve fitting the terrain near the center of the path and finding an iterated solution. If the transmit and receive antennas are at different elevations, the reflection point will vary with atmospheric refractivity (D factor). Over the ocean, it will vary with the tides.

Draw a curve through the reflection points. This will be a straight line if the path is parallel to the Earth or a curve if the path is inclined. Next locate the low order (typically second, fourth, and sixth) even Fresnel radii. These will be critical points of interest. These will probably occur for K factors different than those for which the path profile was drawn.

If there are several reflection points, pick the one that has an even-order Fresnel radius near the $K = 4/3$ earth surface as this will be the controlling case.

Figure 10.A.5b: Plot the odd-order Fresnel zone (typically first or third) for the main-to-diversity path that crosses the closest to the critical point (even-order Fresnel zone) for the main to main paths. Adjust the diversity antenna as necessary to cause the even-order main to main Fresnel zone reflection point to coincide with the main-to-diversity odd-order Fresnel zone reflection point. It may be necessary to move both main and diversity antennas in some situations.

Figure 10.A.6a: Check the overlap of the low order even main to main Fresnel zones (second, fourth, and sixth) with the low order odd main-to-diversity Fresnel zones. Adjust for the best compromise. When this is finished, the main and diversity antennas will have been placed at locations were the reflection path to one antenna will enhance the received signal while the reflective path to the other will cause signal loss. However, as the K factor changes (due to weather-induced atmospheric refractivity changes), the signal enhancement and loss will change but be complementary. Be careful to maintain the original clearance criteria. The objective is to maintain the clearance criteria and achieve complementary diversity.

Figure 10.A.6b and Figure 10.A.7 illustrate this process using a different set of criteria and terrain dominated by trees.

10.A.2 Additional Optimization

At this point, the antennas have been optimized for the typical range of K values of $K \ll 1$ to K = infinity. This is adequate for most paths. However, White (1968) observed that in very poor propagation areas (where ducting is a possibility), such as Florida, K factors as great as -0.5 have been observed. (A K factor of infinity represents a refractive index gradient of -157 N units/km, a "flat-earth" condition. A negative K factor indicates a refractive index gradient more negative than -157 N/km.) In this case the Earth appears concave. The spacing between Fresnel zones becomes very small. The even and odd radii are so close that the above methodology is ineffective (a similar problem to that of mountaintop-to-mountaintop paths with excessive clearance).

When complementary diversity has been achieved, one antenna has a better than normal signal and the other has a worse than normal. Over a wide range of K values, a null condition on one antenna will be associated with a near maximum condition on the other. As K increases past infinity, the intervals between successive Fresnel zones diminish. Eventually, as K becomes large and negative, the point may be reached at which the vertical spacing between the two antennas causes the reflection point to be at two adjacent even Fresnel zones. When this occurs, both antennas will experience a null signal.

To avoid this case, White suggests optimizing the path as indicated above. Then determine the reflection point for the worst case K factor (e.g.,-0.5). Move the diversity antenna slightly up so that both antennas are no longer receiving even Fresnel zone reflections for that K value. While this slightly deoptimizes the path, it essentially guarantees that both antennas will not experience a null simultaneously for negative K values ("less" than infinity).

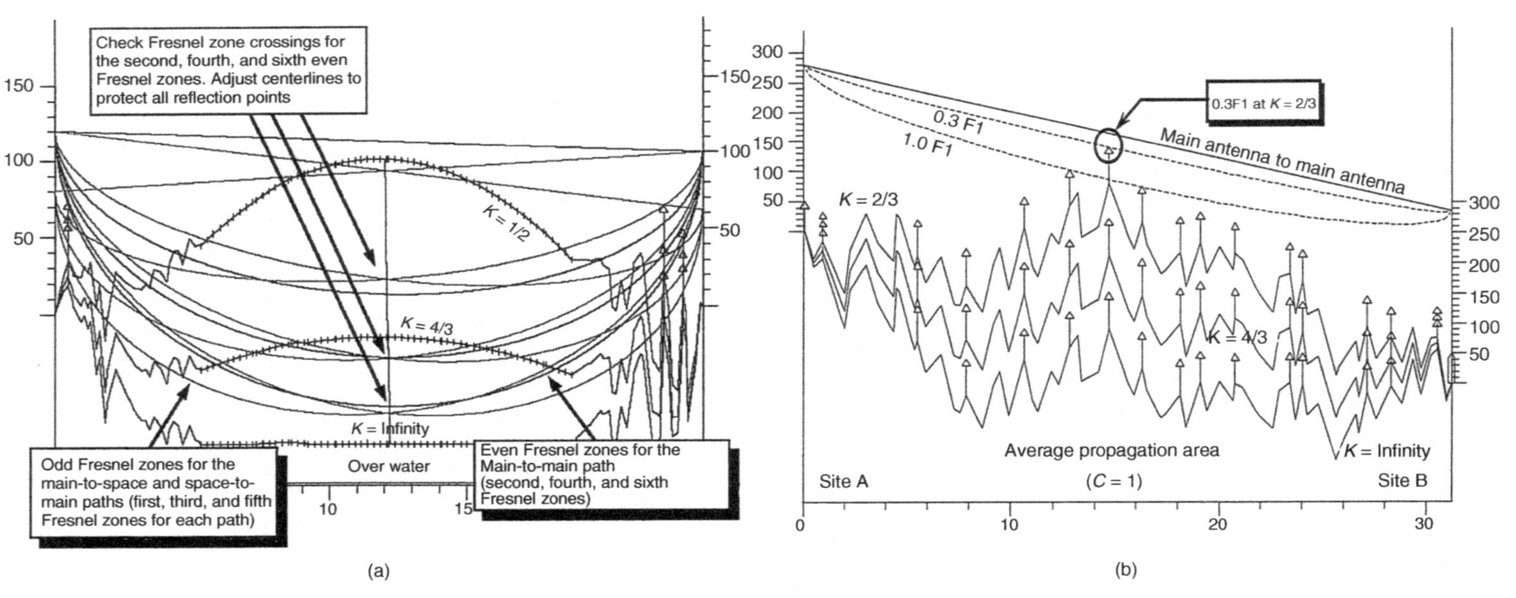

Figure 10.A.6 (a,b) Space diversity antenna placement.

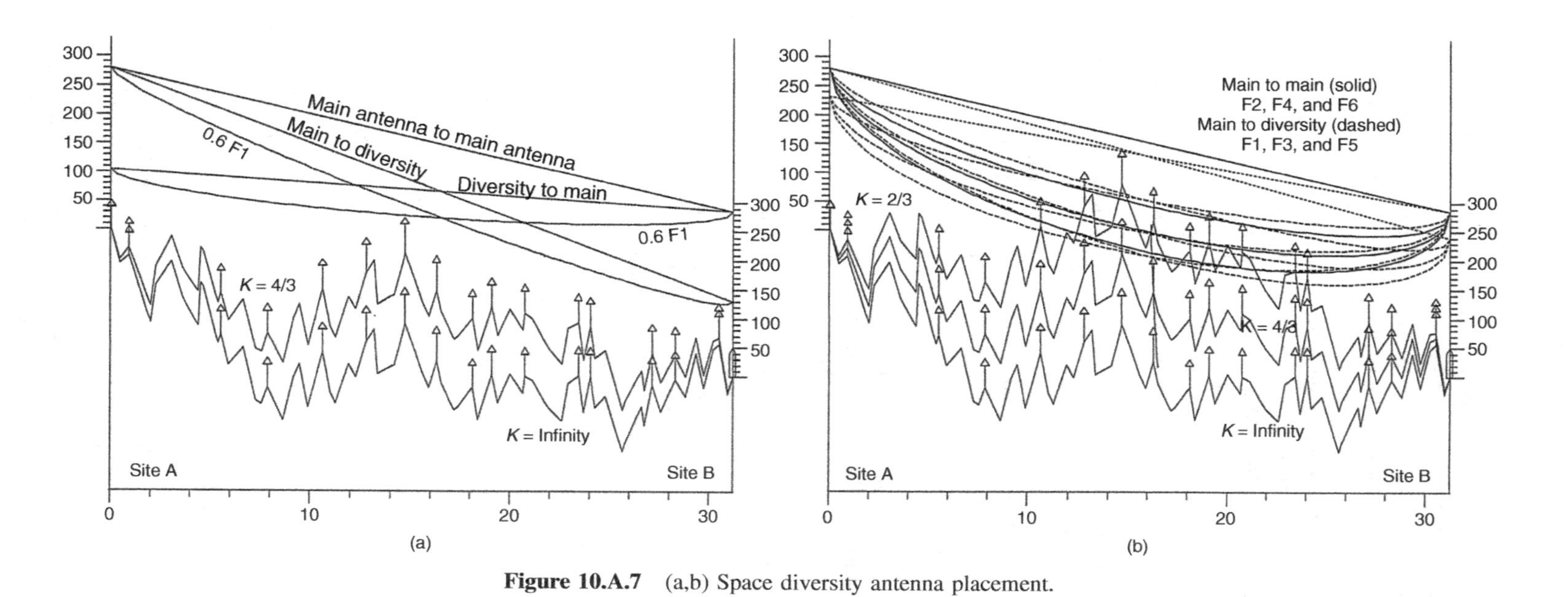

Figure 10.A.7 (a,b) Space diversity antenna placement.

REFERENCES

Albertson, J. N., "Space Diversity on the Microwave Systems of the Southern Pacific," *Wire and Radio Communications*, pp. 83–87, November 1964.

Allen, E. W., "Angle Diversity Test Results at 6 GHz Using a Single-Aperture Dual-Beam Antenna," *Rockwell International Transmission Systems Engineering Symposium*, pp. B6.51–B6.73, October 1988a.

Allen, E. W., "Angle Diversity Test Using a Single Aperture Dual Beam Antenna," *IEEE International Conference on Communications (ICC) Proceedings*, Vol. 3, pp. 50.1.1–50.1.7, June 1988b.

Allen, E. W., "Angle Diversity at 6 GHz: Methods of Alignment and Test Results," *IEEE International Conference on Communications (ICC) Proceedings*, Vol. 2, pp. 24.1.1–24.1.6, June 1989a.

Allen, E. W., "The Effects of Atmospheric Conditions on Angle Diversity Performance at 6 GHz," *IEEE International Conference on Communications (ICC) Proceedings*, Vol. 2, pp. 24.5.1–24.5.6, June 1989b.

Allen, E. W., "More on Angle Diversity vs Space Diversity," *Bellcore National Radio and Wireless Engineers Conference Record*, pp. EWA1–EWA14, 1991.

Allen, E. W., "Path Protection Using Angle Diversity Antenna Systems," Energy Telecommunications and Electrical Association *(ENTELEC) Conference Proceedings*, pp. 189–205, March 1992a.

Allen, E. W., "Angle Diversity vs Space Diversity at 6 GHz: Results of a 3.5 Year Test," *International Telecommunications Symposium (Taipei, Taiwan) Proceedings*, pp. ITS-92AD/EWA 1–ITS-92AD/EWA 13, February 1992b.

Alley, G. D., Peng, W. C., Robinson, W. A. and Lin, E. H., "The Effect on Error Performance of Angel Diversity in a High Capacity Digital Microwave Radio System," *IEEE Global Telecommunications Conference (Globecom) Conference Record*, Vol. 2, pp. 31.4.1–31.4.6, November 1987.

Alley, G. D., Bainchi, C. H. and Robinson, W. A., "Angle Diversity and Space Diversity Experiments on the Salton/Brawley Hop," *IEEE Global Telecommunications Conference (Globecom) Conference Record*, Vol. 2, pp. 504.5.1–504.5.11, December 1990.

Anderson, C. W., Barber, S. G. and Patel, R. N., "The Effect of Selective Fading on Digital Radio," *IEEE Transactions on communications*, Vol. 27, pp. 1870–1876, December 1979.

Balaban, P., Sweedyk, E. A. and Axeling, G. S., "Angle Diversity with Two Antennas: Model and Experimental Results," *IEEE International Conference on Communications (ICC) Proceedings*, Vol. 2, pp. 23.7.1–23.7.7, June 1987.

Barnett, W. T., "Microwave Line-of-Sight Propagation With and Without Frequency Diversity," *Bell System Technical Journal*, Vol. 49, pp. 1827–1871, October 1970.

Barnett, W. T., "Multipath Propagation at 4, 6 and 11 GHz," *Bell System Technical Journal*, Vol. 51, pp. 321–361, February 1972.

Barnett, W. T., "Multipath Fading Effects on Digital Radio," *IEEE Transactions on communications*, Vol. 27, pp. 1842–1848, December 1979.

Bellcore (Telcordia) Staff, *Bellcore (Telcordia) Technical Reference TR-TSY-000752, Microwave Digital Radio Systems Criteria*, pp. 7–13, October 1989.

Cellerino, G., D'Avino, P. and Moreno, L., "Frequency Diversity Protection in Digital Radio with Hitless Switching," *IEEE International Conference on Communications (ICC) Conference Record*, Vol. 3, pp. 47.2.1–47.2.5, June 1985.

Committee 14–11, "Interference Criteria for Microwave Systems," *TIA/EIA Telecommunications Systems Bulletin 10-F*, pp. B-1–B-8, June 1994.

Danielsson, B. and Johansson, U., "Measured Improvements Using Angle and Space Diversity on a Terrestrial Microwave Radio Link," *IEE Radio Relay Systems Conference Publication No. 386*, pp. 215–220, October 1993.

Dekan, P. M., Berg, J. H. and Evans, M., "Aperture Diversity Using Similar Antennas," *IEEE International Conference on Communications (ICC) Proceedings*, Vol. 2, pp. 23.6.1–23.6.4, June 1987.

Di Zenobio, D., Santella, G., Candeo, S. and Mandich, D., "Angle Diversity and Space Diversity: Experimental Comparison," *IEEE Global Telecommunications Conference (Globecom) Conference Record*, Vol. 3, pp. 1851–1857, December 1992.

Dirner, P. L. and Lin, S. H., "Measured Frequency Diversity Improvement for Digital Radio," *IEEE Transactions on Communications*, Vol. 33, pp. 106–109, January 1985.

Florman, E. F., "Comparison of Space-Polarization and Space-Frequency Diversities," *IEEE Transactions on Communication Technology*, Vol. 16, pp. 283–288, April 1968.

Gardina, M. F. and Lin, S. H., "Measured Performance of Horizontal Space Diversity on a Microwave Radio Path," *IEEE Global Telecommunications Conference (Globecom) Conference Record*, Vol. 3, pp. 1104–1107, December 1985.

Giger, A. J., *Low-Angle Microwave Propagation: Physics and Modeling*. Boston: Artech House, pp. 214–218 and 231, 1991.

Giger, A. J. and Barnett, W. T., "Effects of Multipath Propagation on Digital Radio," *IEEE Transactions on Communications*, Vol. 29, pp. 1345–1352, September 1981.

Glauner, M., "Equivalence Relations Between Angle Diversity and Other Diversity Methods. Analytical Model and Practical Results," *IEE Radio Relay Systems Conference Publication No. 386*, pp. 261–266, October 1993.

Greenstein, L. J. and Shafi, M., "Outage Calculation Methods for Microwave Digital Radio," *IEEE Communications Magazine*, Vol. 25, pp. 30–39, February 1972.

Hartman, W. J. and Smith, D., "Tilting Antennas to Reduce Line-of-Sight Microwave Link Fading," *IEEE Transactions on Antennas and Propagation*, Vol. 25, pp. 642–645, September 1977.

Hubbard, R. W., "Angle Diversity Reception for LOS Digital Microwave Radio," *IEEE Military Communications (Milcom) Conference Proceedings*, pp. 19.6.1–19.6.7, October 1985.

Kizer, G. M., "Microwave Radio Communication," *Handbook of Microwave Technology*, *Volume* 2, Ishii, T. K., Editor. San Diego: Academic Press, pp. 449–504, 1995.

Knerr, A. R., FCC Private Letter to M. Blomstrom, Reference: PS&PWD-LTAB-647, Gettysburg, PA, September 3 1998.

Knight, W., *Microwave Path Design Reference Manual*. Plano (north Dallas): Alcatel-Lucent, A.1–C.12, June 2006.

Kostal, H., "Diversity Protection in Microwave Radio," *IEEE International Conference on Communications (ICC) Conference Record*, Vol. 2, pp. 31.7.1–31.7.5, June 1985.

Lee, T. C. and Lin, S. H., "More on Frequency Diversity for Digital Radio," *IEEE Global Telecommunications Conference (Globecom) Conference Record*, pp. 36.7.1–36.7.4, 1985.

Lee, T. C. and Lin, S. H., "A Model of Space Diversity Improvement for Digital Radio," *International Union of Radio Science, Open Symposium, Wave Propagation, University of New Hampshire*, pp. 7.3.1–7.3.4, July 1986.

Lee, T. C. and Lin, S. H., "A Model of Space Diversity for Microwave Radio Against Thermal Noise Caused Outage During Multipath Fading," *IEEE Global Telecommunications Conference (Globecom) Record*, Vol. 3, pp. 44.2.1–44.2.7, November 1988.

Lin, S. H., "Statistical Behavior of a Fading Signal," *Bell System Technical Journal*, Vol. 50, pp. 3211–3270, December 1971.

Lin, S. H., "Statistical Behavior of Deep Fades of Diversity Signals," *IEEE Transactions on Communications*, Vol. 20, pp. 1100–1107, December 1972.

Lin, S. H., "Measured Relative Performance of Antenna Pattern Diversity, Antenna Angle Diversity and Vertical Space Diversity in Mississippi," *IEEE Global Telecommunications Conference (Globecom) Conference Record*, Vol. 3, pp. 44.1.1–44.1.7, November 1988.

Lin, E. H., Giger, A. J. and Alley, G. D., "Angle Diversity on Line-of-Sight Microwave Paths Using Dual-Beam Dish Antennas," *IEEE International Conference on Communications Proceedings*, Vol. 2, pp. 23.5.1–23.5.11, June 1987.

Lin, S. H., Lee, T. C. and Gardina, M. F., "Diversity Protections for Digital Radio—Summary of Ten-Year Experiments and Studies," *IEEE Communications Magazine*, Vol. 26, pp. 51–64, February 1988.

Malaga, A. and Parl, S. A., "Experimental Comparison of Angle and Space Diversity for Line-of-Sight Microwave Links," *IEEE Military Communications (Milcom) Conference Proceedings*, pp. 19.5.1–19.5.8, October 1985.

Mohamed, S. A., Richman, G. D. and Huish, P. W., "Further Results of Angle and Space Diversity Measurements on a Line-of-Sight Radio Link," *IEEE Global Telecommunications Conference (Globecom) Conference Record*, Vol. 1, pp. 2.4.1–2.4.6, November 1989.

Mosley, R. A., Director, FCC Order, "Request for Ruling that Part 101 Frequency Diversity Restrictions Are Not Applicable to Collapsed Ring Architecture for Microwave Systems," Reference: DA 99–1502, Washington, DC, January 21 2000.

Mosley, R. A., Director, *Code of Federal Regulations (CFR), Title 47 - Telecommunication, Chapter I, Federal Communications Commission, Part 101, paragraph 103(c)*. Washington: Office of the Federal Register, 2013 (updated regularly).

Noerpel, A. R., "Pattern Diversity Using a Single Horn Reflector Antenna," *IEEE Global Telecommunications Conference (Globecom) Conference Record*, Vol. 3, pp. 44.5.1–44.5.6, November 1988.

Rummler, W. D., "A New Selective Fading Model: Application to Propagation Data," *Bell System Technical Journal*, Vol. 58, pp. 1037–1071, May–June 1979.

Rummler, W. D., "A Statistical Model of Multipath Fading on a Space Diversity Radio Channel," *IEEE International Conference on Communications Proceedings (ICC)*, Vol. 1, pp. 3B.4.1–3B.4.6, June 1982.

Rummler, W. D., "A Rationalized Model for Space and Frequency Diversity Line-of-Sight Radio Channels," *IEEE International Conference on Communications Proceedings*, Vol. 3, pp. E2.7.1–E2.7.5, June 1983.

Rummler, W. D., "Characterizing the Effects of Multipath Dispersion on Digital Radios," *IEEE Globecom Conference Record*, pp. 52.5.1–52.5.7, 1988.

Rummler, W. D. and Dhafi, M., "Route Design Methods," *Terrestrial Digital Microwave Communications*. Ivanek, F., Editor. Norwood: Artech House, pp. 326–329, 1989.

Rummler, W. D., Coutts, R. P. and Liniger, M., "Multipath Fading Channel Models for Microwave Digital Radio," *IEEE Communications Magazine*, Vol. 24, pp. 30–42, November 1986.

Satoh, A. and Sasaki, O., "Tilted-Beam, Beam-Width and Space-Diversity Improvements on Various Paths," *IEEE Global Telecommunications Conference (Globecom) Conference Record*, Vol. 1, pp. 2.3.1–2.3.5, November 1989.

Smith, D. R. and Cormack, J. J., "Improvement in Digital Radio due to Space Diversity and Adaptive Equalization," *IEEE Global Telecommunications Conference (Globecom) Conference Record*, Vol. 3, pp. 45.6.1–45.6.6, November 1984.

Valentin, R., Metzger, K., Giloi, H. G. and Dombek, K. P., "Effects of Angle Diversity on the Performance of Line-of-Sight Digital Radio Systems," *IEEE Global Telecommunications Conference (Globecom) Conference Record*, Vol. 2, pp. 31.5.1–31.5.4, November 1987.

Valentin, R., Giloi, H. G. and Metzger, K., "Space Versus Angle Diversity—Results of System Analysis Using Propagation Data," *IEEE International Conference on Communications (ICC) Proceedings*, Vol. 2, pp. 24.2.1–24.2.5, June 1989.

Valentin, R., Giloi, H. G. and Metzger, K., "More on Angle Diversity for Digital Radio Links," *IEEE Global Telecommunications Conference (Globecom) Conference Record*, Vol. 2, pp. 504.3.1–504.3.5, December 1990.

Vergeres, D., Jordi, P. and Loembe, A., "Simultaneous Error Performance of Antenna Pattern Diversity and Vertical Space Diversity on a 64 QAM Radio Link," *IEEE Global Telecommunications Conference (Globecom) Conference Record*, Vol. 2, pp. 504.4.1–504.4.5, December 1990.

Vigants, A., "Variations of Space-Diversity Performance on Line-of-Sight Links," *Proceedings of the IEEE*, Vol. 55, pp. 595–596, April 1967.

Vigants, A., "Space-Diversity Performance as a Function of Antenna Separation," *IEEE Transactions on Communication Technology*, Vol. 16, pp. 831–836, December 1968.

Vigants, A., "The Number of Fades and Their Durations on Microwave Line-of-Sight Links With and Without Space Diversity," *IEEE International Conference on Communications Conference Record (ICC)*, pp. 3-7-3-11, June 1969.

Vigants, A., "The Number of Fades in Space-Diversity Reception," *Bell System Technical Journal*, Vol. 49, pp. 1513-1530, September 1970.

Vigants, A., "Space-Diversity Engineering," *Bell System Technical Journal*, Vol. 54, pp. 103-142, January 1975.

Vigants, A. and Pursley, M. V., "Transmission Unavailability of Frequency-Diversity Protected Microwave FM Radio Systems Caused by Multipath Fading," *Bell System Technical Journal*, Vol. 58, pp. 1779-1796, October 1979.

Wang, Y. Y., "Simulation and Measured Performance of a Space Diversity Combiner for 6 GHz Digital Radio," *IEEE Transactions on Communications*, Vol. 27, pp. 1896-1907, December 1979.

Webster, A. R., "Angles-of-Arrival and Delay Times on Terrestrial Line-of-Sight Microwave Links," *IEEE Transactions on Antennas and Propagation*, Vol. 31, pp. 12-17, January 1983.

Webster, A. R., "Multipath Angle-of-Arrival Measurements on Microwave Line-of-Sight Links," *IEEE Transactions on Antennas and Propagation*, Vol. 39, pp. 798-803, June 1991.

Webster, A. R. and Merritt, T. S., "Multipath Angles-of-Arrival on a Terrestrial Microwave Links," *IEEE Transactions on Communications*, Vol. 38, pp. 25-30, January 1990.

Webster, A. R. and Scott, A. M., "Angle-of-Arrival and Tropospheric Multipath Microwave Propagation," *IEEE Transactions on Antennas and Propagation*, Vol. 35, pp. 94-99, January 1987.

White, R. F., "Space Diversity on Line-of-Sight Microwave Systems," *IEEE Transactions on Communications Technology*, Vol. 16, pp. 119-133, February 1968.

11

RAIN FADING

Rain rate data is used for an amazingly wide range of applications. It is used to design space launch vehicles, storm sewers, bridges and a host of other unrelated applications. This chapter applies rain rate data to high frequency microwave radio path design.

Airborne atmospheric water can be in the gaseous, liquid, or solid state. Liquid or solid water is rain, snow, or ice. These cloud or precipitation particles are known as *hydrometeors*. Gaseous and solid water have little effect on microwave radio communications below about 40 GHz (International Telecommunication Union—Radiocommunication Sector (ITU-R), 1999). However, liquid water (rain) can have a significant effect on microwave frequency radio waves.

Rain is composed of many conductive water droplets. As with all electrical conductors in an RF environment, rain does not absorb much radio energy; it primarily reflects it. The result of this reflection is to reduce the received signal power at the desired receiver. However, this redirected energy can significantly increase RF interference to receive antennas pointed in other directions. Rain can cause loss of signal polarization discrimination in addition to the basic rain loss.

For linear polarization and a given rain path attenuation, cross-polarization discrimination (XPD) increases with frequency (Fig. 11.1) (Chu, 1974). Differential phase shift is the dominant factor in the rain-induced cross-polarization at frequencies below about 10 GHz. For higher frequencies, differential attenuation becomes increasingly important. The result of this relationship is that rain-induced cross-polarization is relatively independent of rain rate (Chu, 1974). A surprising result is that rain can have a significant effect for low microwave frequencies. Although rain path attenuation at 4 and 6 GHz is usually not significant, XPD degradation can be so. Barnett (1974) observed a 4-GHz path that only experienced 4 dB of rain-induced path loss suffered a 21-dB reduction in XPD (from a normal value of 27.5 dB to a value of 6.5 dB due to rain).

Unlike multipath fading, for terrestrial paths, rain fading is essentially independent of vertical placement of transmit or receive antenna. Rain fading is dependent on horizontal location. As deep rain fading in North America is localized near high intensity rain cells, geographic diversity has been proposed (Bergmann, 1977; Crane, 1989; Kanellopoulos and Koukoulas, 1987; Lin et al., 1980; Osborne, 1971; Paulson et al., 2006) to mitigate its effect. Lin et al. (1980) noted that terminals (and presumably paths) separated by 20 km could achieve diversity improvement of 10 in some cases. International Telecommunication Union—Radiocommunication Sector (ITU-R) (2009) suggests that parallel paths with separation of 4–8 km have significantly less diversity improvement. To date, inadequate practical statistics are available

Digital Microwave Communication: Engineering Point-to-Point Microwave Systems, First Edition. George Kizer.

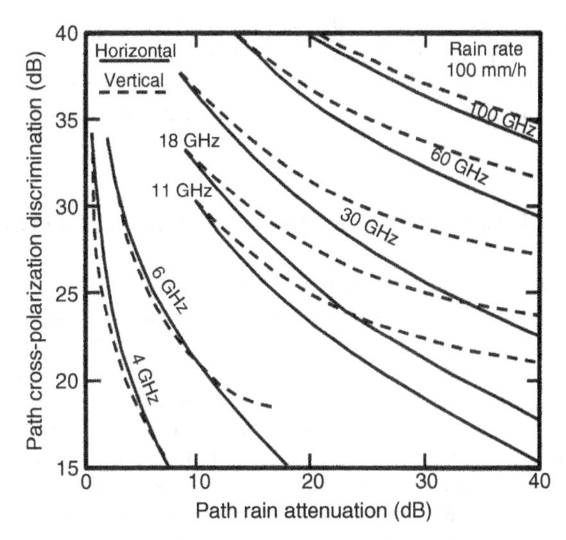

Figure 11.1 Rain-induced cross-polarization discrimination. Source: Adapted from Chu, T. S., "Rain-Induced Cross-Polarization at Centimeter and Millimeter Wavelengths," *Bell System Technical Journal*, pp. 1557–1579, October 1974.

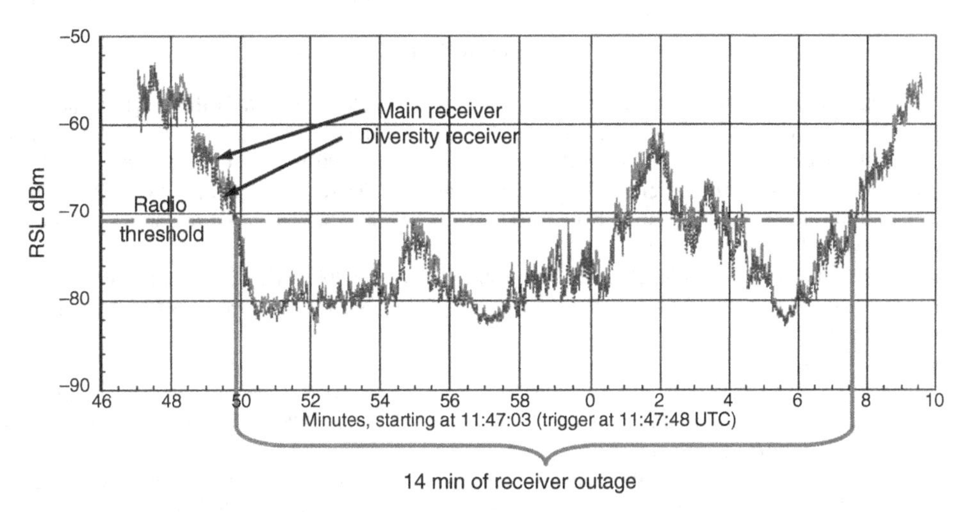

Figure 11.2 Rain radio-fade event.

to provide generalized engineering methods. However, the technique has been shown to offer availability improvement in some situations.

From the microwave radio path perspective, the effect of rain is a short-term reduction of received signal level termed a *rain fade*. Figure 11.2 depicts an intense rain-fading event. This fading occurred on an 8-GHz path in central Italy. While signal depression may last for several minutes, the duration over which the receiver is out of service is typically a few minutes (but occasionally much longer outages occur). Lin (1973) observed that the median rain-fade duration in North America is about 2.5 min for a moderate receiver fade margin. Crane (1996) suggests the median fade duration is 2 min. The rain-induced received signal fade is a function of the aggregate effect of rain intensity throughout the entire radio path between the transmitter and receiver. The overall radio fading is estimated by predicting the rain intensity at one location (point rain intensity), estimating its RF attenuation, and modifying that attenuation to account for the rain intensity statistics of the entire path. The radio path length is lengthened or shortened as necessary so that the radio (flat or thermal) fade margin is not exceeded more than a desired period of time (typically specified as an availability percentage). Each rain event is a few minutes long. The function of the path designer is to specify the path length so that the sum of the path outages, averaged over a statistically significant period (at least 10 to 20 years), does not exceed a predefined objective.

11.1 POINT (SINGLE-LOCATION) RAIN LOSS (FADE) ESTIMATION

Reflective rain losses are estimated following the general procedures developed by Goldstein et al. (1951). Size and shape determines terminal velocity. Size distribution was estimated by Laws and Parsons (1944) and terminal velocity by Gunn and Kinzer (1949). These results are in excellent agreement with Medhurst's results (Medhurst, 1965) for rain rates of 5 mm/h or greater. Raindrop photographs show a flat or concave shape (Jones, 1959). Most researchers assume the shape to be elliptical as described by the Pruppacher and Pitter model (Pruppacher and Pitter, 1971). Crane observed (Crane, 1975) that other raindrop shapes, although possible, were of little significance. Marshall and Palmer (1948) discovered the basic exponential relationship of (single-location) rain rate to rain attenuation. Olsen et al. (1978) extended those results to the basic KR^{α} equation for radio signal loss as a function of frequency, polarization, rain rate, and temperature. Vertically polarized signals experience slightly less attenuation than do horizontally polarized signals. Rain loss is also slightly temperature dependent. Attenuation is greater for low frequencies at low temperatures and just the opposite for higher temperatures. Attenuation variation over the temperature range 5–20 °C is only 4% (Lin, 1973). Most rain attenuation models assume a rain temperature of 20 °C. This leads to the following formula for single-location rain attenuation:

$$\gamma_R = KR^{\alpha}\,(\mathrm{dB/km}) \tag{11.1}$$

The parameters K and α are a function of frequency and polarization. R is the single-location rain rate in millimeters per hour not exceeded with probability p. Crane's 1980 model (Crane, 1980) ignores the effect of polarization. He is silent on the parameters in his 1996 model (Crane, 1996). As all path designers accept the polarization effect on the parameters, the ITU-R values (International Telecommunication Union—Radiocommunication Sector (ITU-R), 2005) for K and α, listed in Table 11.1, are used universally, regardless of rain model.

ITU-R Recommendation P.838 changed to the above values in 2005. These are significantly different from previously published values. This can lead to differences in results from different path design software programs.

Values of γ_R are plotted in Figure 11.3.

Values for other frequencies may be calculated using the ITU-R Recommendation (International Telecommunication Union—Radiocommunication Sector (ITU-R), 2005) or by interpolating using the formulas discussed in the following.

Accurate K and α variables are the basis of all radio rain path loss calculations. It is highly desirable to achieve at least four significant figures for these variables. Interpolation of Table 11.1 is challenging because the shape of the curves are changing rapidly in the first few tens of gigahertz. The optimum linear interpolation formula (Appendix A) changes on the basis of frequency of interest. Any simple interpolation choice will fail to achieve high accuracy over all frequencies of interest. For rain path loss calculations figures, it is recommended that the formulas and values from ITU-R Recommendation P.838-3 (International Telecommunication Union—Radiocommunication Sector (ITU-R), 2005) be used to calculate frequency-specific values:

$$\log_{10}(K) = L_K + M_K \log_{10}(F) + \sum_{n=1}^{4} A_K(n) \exp\left\{-\left[\frac{B_K(n) + \log_{10}(F)}{C_K(n)}\right]^2\right\} \tag{11.2}$$

$K = 10^{\log_{10}(K)}$;

$$\alpha = L_\alpha + M_\alpha \log_{10}(F) + \sum_{n=1}^{5} A_\alpha(n) \exp\left\{-\left[\frac{B_\alpha(n) + \log_{10}(F)}{C_\alpha(n)}\right]^2\right\};$$

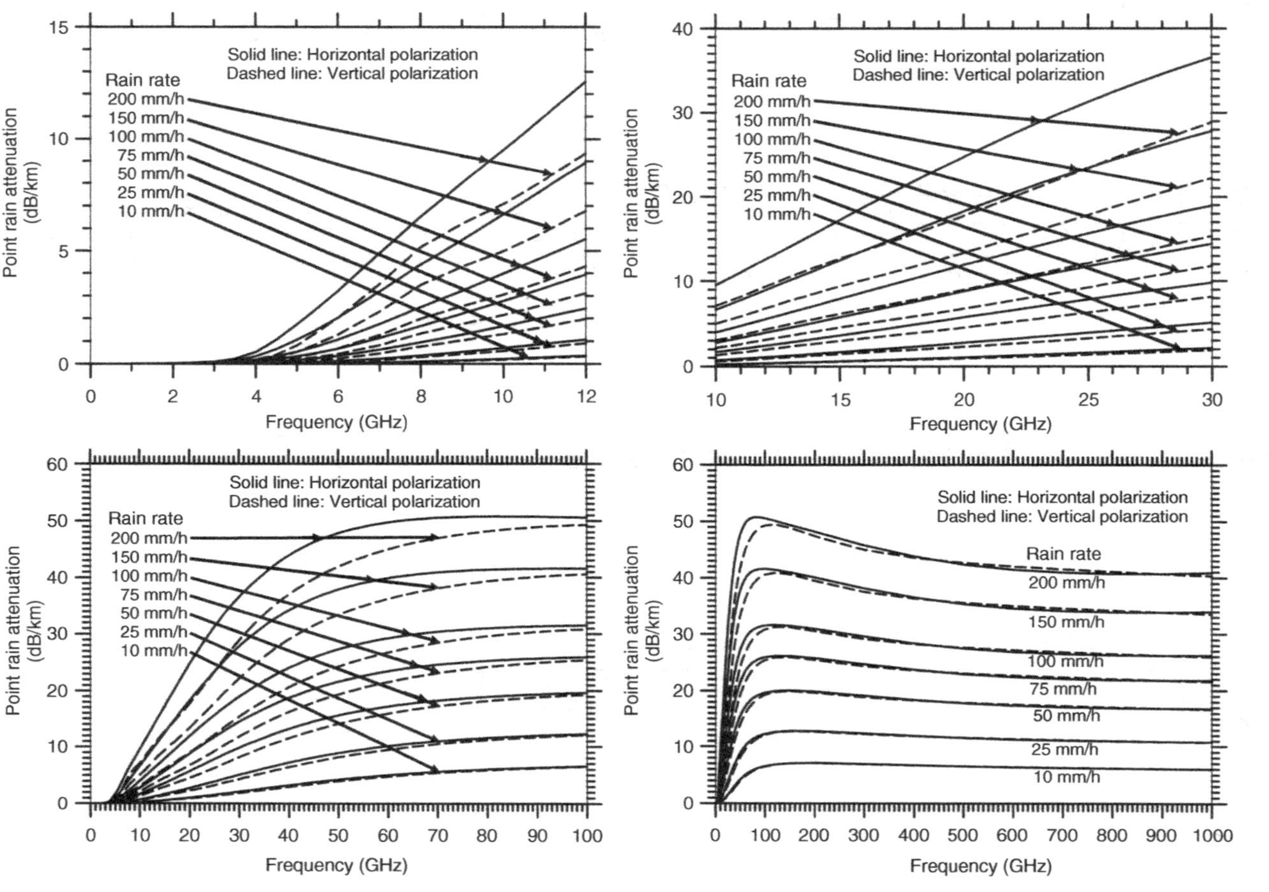

Figure 11.3 Rain attenuation per ITU-RP.838-3.

TABLE 11.1 Rain Attenuation Factors (ITU-R Rec. P.838-3)

Frequency	K_h	α_h	K_v	α_v
1	2.590E-05	0.9691	3.080E-05	0.8592
1.5	4.430E-05	1.0185	5.740E-05	0.8957
2	8.470E-05	1.0664	9.980E-05	0.9490
2.5	1.321E-04	1.1209	1.464E-04	1.0085
3	1.390E-04	1.2322	1.942E-04	1.0688
3.5	1.155E-04	1.4189	2.346E-04	1.1387
4	1.071E-04	1.6009	2.461E-04	1.2476
4.5	1.340E-04	1.6948	2.347E-04	1.3987
5	2.162E-04	1.6969	2.428E-04	1.5317
5.5	3.909E-04	1.6499	3.115E-04	1.5882
6	7.056E-04	1.5900	4.878E-04	1.5728
7	1.915E-03	1.4810	1.425E-03	1.4745
8	4.115E-03	1.3905	3.450E-03	1.3797
9	7.535E-03	1.3155	6.691E-03	1.2895
10	1.217E-02	1.2571	1.129E-02	1.2156
11	1.772E-02	1.2140	1.731E-02	1.1617
12	2.386E-02	1.1825	2.455E-02	1.1216
13	3.041E-02	1.1586	3.266E-02	1.0901
14	3.738E-02	1.1396	4.126E-02	1.0646
15	4.481E-02	1.1233	5.008E-02	1.0440
16	5.282E-02	1.1086	5.899E-02	1.0273
17	6.146E-02	1.0949	6.797E-02	1.0137
18	7.078E-02	1.0818	7.708E-02	1.0025
19	8.084E-02	1.0691	8.642E-02	0.9930
20	9.164E-02	1.0568	9.611E-02	0.9847
21	0.1032	1.0447	0.1063	0.9771
22	0.1155	1.0329	0.1170	0.9700
23	0.1286	1.0214	0.1284	0.9630
24	0.1425	1.0101	0.1404	0.9561
25	0.1571	0.9991	0.1533	0.9491
26	0.1724	0.9884	0.1669	0.9421
27	0.1884	0.9780	0.1813	0.9349
28	0.2051	0.9679	0.1964	0.9277
29	0.2224	0.9580	0.2124	0.9203
30	0.2403	0.9485	0.2291	0.9129
31	0.2588	0.9392	0.2465	0.9055
32	0.2778	0.9302	0.2646	0.8981
33	0.2972	0.9214	0.2833	0.8907
34	0.3171	0.9129	0.3026	0.8834
35	0.3374	0.9047	0.3224	0.8761
36	0.3580	0.8967	0.3427	0.8690
37	0.3789	0.8890	0.3633	0.8621
38	0.4001	0.8816	0.3844	0.8552
39	0.4215	0.8743	0.4058	0.8486
40	0.4431	0.8673	0.4274	0.8421
41	0.4647	0.8605	0.4492	0.8357
42	0.4865	0.8539	0.4712	0.8296
43	0.5084	0.8476	0.4932	0.8236
44	0.5302	0.8414	0.5153	0.8179
45	0.5521	0.8355	0.5375	0.8123

TABLE 11.1 (*Continued*)

Frequency	K_h	α_h	K_v	α_v
46	0.5738	0.8297	0.5596	0.8069
47	0.5956	0.8241	0.5817	0.8017
48	0.6172	0.8187	0.6037	0.7967
49	0.6386	0.8134	0.6255	0.7918
50	0.6600	0.8084	0.6472	0.7871
51	0.6811	0.8034	0.6687	0.7826
52	0.7020	0.7987	0.6901	0.7783
53	0.7228	0.7941	0.7112	0.7741
54	0.7433	0.7896	0.7321	0.7700
55	0.7635	0.7853	0.7527	0.7661
56	0.7835	0.7811	0.7730	0.7623
57	0.8032	0.7771	0.7931	0.7587
58	0.8226	0.7731	0.8129	0.7552
59	0.8418	0.7693	0.8324	0.7518
60	0.8606	0.7656	0.8515	0.7486
61	0.8791	0.7621	0.8704	0.7454
62	0.8974	0.7586	0.8889	0.7424
63	0.9153	0.7552	0.9071	0.7395
64	0.9328	0.7520	0.9250	0.7366
65	0.9501	0.7488	0.9425	0.7339
66	0.9670	0.7458	0.9598	0.7313
67	0.9836	0.7428	0.9767	0.7287
68	0.9999	0.7400	0.9932	0.7262
69	1.0159	0.7372	1.0094	0.7238
70	1.0315	0.7345	1.0253	0.7215
71	1.0468	0.7318	1.0409	0.7193
72	1.0618	0.7293	1.0561	0.7171
73	1.0764	0.7268	1.0711	0.7150
74	1.0908	0.7244	1.0857	0.7130
75	1.1048	0.7221	1.1000	0.7110
76	1.1185	0.7199	1.1139	0.7091
77	1.1320	0.7177	1.1276	0.7073
78	1.1451	0.7156	1.1410	0.7055
79	1.1579	0.7135	1.1541	0.7038
80	1.1704	0.7115	1.1668	0.7021
81	1.1827	0.7096	1.1793	0.7004
82	1.1946	0.7077	1.1915	0.6988
83	1.2063	0.7058	1.2034	0.6973
84	1.2177	0.7040	1.2151	0.6958
85	1.2289	0.7023	1.2265	0.6943
86	1.2398	0.7006	1.2376	0.6929
87	1.2504	0.6990	1.2484	0.6915
88	1.2607	0.6974	1.2590	0.6902
89	1.2708	0.6959	1.2694	0.6889
90	1.2807	0.6944	1.2795	0.6876
91	1.2903	0.6929	1.2893	0.6864
92	1.2997	0.6915	1.2989	0.6852
93	1.3089	0.6901	1.3083	0.6840

TABLE 11.1 (*Continued*)

Frequency	K_h	α_h	K_v	α_v
94	1.3179	0.6888	1.3175	0.6828
95	1.3266	0.6875	1.3265	0.6817
96	1.3351	0.6862	1.3352	0.6806
97	1.3434	0.6850	1.3437	0.6796
98	1.3515	0.6838	1.3520	0.6785
99	1.3594	0.6826	1.3601	0.6775
100	1.3671	0.6815	1.3680	0.6765
120	1.4866	0.6640	1.4911	0.6609
150	1.5823	0.6494	1.5896	0.6466
200	1.6378	0.6382	1.6443	0.6343
300	1.6286	0.6296	1.6286	0.6262
400	1.5860	0.6262	1.5820	0.6256
500	1.5418	0.6253	1.5366	0.6272
600	1.5013	0.6262	1.4967	0.6293
700	1.4654	0.6284	1.4622	0.6315
800	1.4335	0.6315	1.4321	0.6334
900	1.4050	0.6353	1.4056	0.6351
1000	1.3795	0.6396	1.3822	0.6365

TABLE 11.2 *K* Factors (ITU-R Rec. P.838-3)

N	Horizontal $A_K(N)$	Horizontal $B_K(N)$	Horizontal $C_K(N)$	Vertical $A_K(N)$	Vertical $B_K(N)$	Vertical $C_K(N)$
1	−5.33980	0.10008	1.13098	−3.80595	−0.56934	0.81061
2	−0.35351	−1.26970	0.45400	−3.44965	0.22911	0.51059
3	−0.23789	−0.86036	0.15354	−0.39902	−0.73042	0.11899
4	−0.94158	−0.64552	0.16817	0.50167	−1.07319	0.27195

Horizontal	Horizontal	Vertical	Vertical
L_K	M_K	L_K	M_K
0.71147	−0.18961	0.63297	−0.16398

F = frequency (GHz);
$\exp(X) = e^X$

The constants necessary to calculate the above are listed in Table 11.2 and Table 11.3.

Use of the above formulas eliminates K and α variation due to interpolation method, a problem common among various path loss calculation programs.

11.2 PATH RAIN-FADE ESTIMATION

High frequency radio signal rain loss is a function of raindrop size and distribution. These are a complex function of rain rate, wind, and type of rain. Rain in temperate climates is usually spatially characterized as *stratiform* and *convective*. Stratiform precipitation covers a large horizontal region and is relatively mild rain. Convective rain is very intense rain typically associated with thunderheads. Tropical regions have monsoon and cyclone rain in addition to the temperate climate rain types. Monsoon rain is usually

TABLE 11.3 α Factors

N	Horizontal $A_\alpha(N)$	Horizontal $B_\alpha(N)$	Horizontal $C_\alpha(N)$	Vertical $A_\alpha(N)$	Vertical $B_\alpha(N)$	Vertical $C_\alpha(N)$
1	−0.14318	−1.82442	−0.55187	−0.07771	−2.338400	−0.762840
2	0.29591	−0.77564	0.19822	0.56727	−0.955450	0.540390
3	0.32177	−0.63773	0.13164	−0.20238	−1.145200	0.268090
4	−5.37610	0.96230	1.47828	−48.29910	−0.791669	0.116226
5	16.17210	3.29980	3.43990	48.58330	−0.791459	0.116479

Horizontal	Horizontal	Vertical	Vertical
L_α	M_α	L_α	M_α
−1.95537	0.67849	0.83433	−0.053739

intense convective rain followed by stratiform precipitation. Cyclone rain is a series of circular bands of intense rain centered on a calm rain-free center zone. Cyclone rain may extend several hundred kilometers. Most rain-fading models only include estimates of stratiform and convective rain (Capsoni et al., 2009). Crane (1996) refers to stratiform rain as *debris rain* and convective rain as *volume cell* rain. The ITU-R model is unique in that it assumes a predefined rain rate curve (Capsoni et al., 2009).

Many microwave radio path rain attenuation methods have been used in the past. The earliest in wide use was by Rice and Holmberg (1973) (with enhancements by Dutton and Dougherty (1979)). Lin's method (Lin, 1977; Lin et al., 1980; Lin and Pursley, 1985) was used extensively in the Bell System. At present, the Crane model (Crane, 2003) is most popular in North America and ITU-R (International Telecommunication Union—Radiocommunication Sector (ITU-R), 2009) is most popular internationally. The Crane and ITU-R models are different in both methodology and philosophy. The ITU-R model is based on parameter curve fits to rain databases. Two rain data shapes (one for latitudes beyond $\pm 30^\circ$ and one for latitudes between $\pm 30^\circ$) are assumed. Crane also represents curve fits to rain data. However, a significance difference is that Crane fits the data to a model based on physical processes (stratiform and convective rain).

Most rain databases are accumulated over 20 years. A rain rate of 0.01% would represent 1052 1-min integrated samples or 210 5-min integrated samples. For rain rates of 0.001%, the number of samples is 105 and 21, respectively. This is the practical limit of statistical significance. Yet rain rates lower than 0.0001% are required for North American high frequency path design. Direct measurement cannot yield statistically significant data for such low rain probabilities. However, using a model (based upon extreme value theory) with parameters curve-fitted using statistically significant data, rain rates for very low probabilities (which cannot be measured) may be estimated. The ITU-R method is limited to rain unavailability of 0.01 % or greater (which represents the practical limit of statistically significant rain rate measurements). As the Crane model is based on curve fits to rain models, the Crane model does not have this limitation.

Radio path engineering attempts to design radio paths so that the path (thermal or flat) fade margin M is not exceeded more than a specified percentage of the time (i.e., to achieve an annual average path availability of p). This is done by solving the following equation:

$$M = \gamma_{\mathrm{R}} d_{\mathrm{eff}} \tag{11.3}$$

M = path flat fade margin (dB);
γ_{R} = single-location ("point") rain attenuation (dB/km);
d_{eff} = effective path length (km);
= $d\,F_{\mathrm{pp}}$;
F_{pp} = point rain rate to path average rain rate conversion factor;
d = actual (physical) path length (km).

In general, radio path design involves defining a path availability and then determining the necessary fade margin M to achieve it.

The following approach is used for the *ITU-R method* (International Telecommunication Union—Radiocommunication Sector (ITU-R), 2009; International Telecommunication Union—Radiocommunication Sector (ITU-R), 2007a; International Telecommunication Union—Radiocommunication Sector (ITU-R), 2005) of rain path attenuation estimation.

In general, the ITU-R methods (837-5 and 1144-4) (International Telecommunication Union—Radiocommunication Sector (ITU-R), 2007a; International Telecommunication Union—Radiocommunication Sector (ITU-R), 2007b) use worldwide-measured parameters to infer single-location average point rain rates for probability $p = 0.01\%$. Rain rates for other values of p are then estimated using the following equations (International Telecommunication Union—Radiocommunication Sector (ITU-R), 2009). Data to calculate the 0.01% rain rate may be downloaded from http://www.itu.int/ITU-R/index.asp?category=study-groups&rlink=rsg3-soft ware-ionospheric&lang=en. Under the tab "Software and data sets concerning Tropospheric Propagation," download the software for "Rainfall rate model (Rec. P.837)." The data is in the zipped folder.

Point rain attenuation is calculated according to IAW ITU-R 838-3 (International Telecommunication Union—Radiocommunication Sector (ITU-R), 2005). The point rain attenuation parameters changed in 2005 and the point rain rate calculation method changed in 2007. Both of these are significantly different from previous recommendations. This can cause significant differences in path design software programs. For rain databases with integration time longer than 1 min, International Telecommunication Union—Radiocommunication Sector (ITU-R) (2007a) proposes a conversion factor from longer integration time measurements to 1-min integration time measurements. It should be remembered that this factor only applies to point rain rates for probability $p = 0.01\%$.

The ITU-R rain calculation starts by solving the rain attenuation equation for $p = 0.01\%$.

$$A_{0.01} = \text{rain attenuation for } p = 0.01\% = \gamma_{\mathrm{R}} d_{\mathrm{eff}} \tag{11.4}$$

$\gamma_{\mathrm{R}} = K(R_{0.01})^{\alpha}$ (dB/km);

K and α are defined by International Telecommunication Union—Radiocommunication Sector (ITU-R) (2005) (as previously mentioned);

$R_{0.01}$ = rain rate not exceeded for $p > 0.01\%$;

d = actual path distance (km);

$d_{\mathrm{eff}} = d\{1/[1 + (d/d_0)]\}$ (dimensionless multiplicative factor).

The preceding formula for d_{eff} is exactly the same as described in the ITU-R Recommendation (International Telecommunication Union—Radiocommunication Sector (ITU-R), 2009) except d has been added to facilitate calculating M.

$d_0 = 35/e^{F_1}$, reference distance (km);

$F_1 = 0.015 R_{0.01}$ if $R_{0.01} \leq 100$;

$F_1 = 1.5$ if $R_{0.01} > 100$;

$R_{0.01}$ = single-location rain rate exceeded for 0.01% of the time;

= rain rate calculated per International Telecommunication Union—Radiocommunication Sector (ITU-R) (2007a) if long-term measurements are not available.

Bill Knight (Knight, 1999) observed that the above formula has evolved over the years. In 1986, the formula for d_0 was $d_0 = 0.045$. In 1990, the formula became $d_0 = 35/e^{F_1}$.but without any limits on F_1. In 1995, the limit based on $R_{0.01}$ was added. In the early 2000s, two rain curve shapes were defined (one for latitude beyond ±30° and one for latitude ±30° or less).

e = Euler's number $\cong 2.71828182846$;

A_{p} = relative attenuation that is exceeded no more than probability p (%);

= actual path fade margin of interest = M.

For latitudes equal to or greater than 30° (North or South),

$$\frac{A_p}{A_{0.01}} = \frac{0.12}{p^{F_2}}$$
$$F_2 = 0.546 + 0.043 \log_{10}(p) \tag{11.5}$$

For latitudes less than 30° (North or South),

$$\frac{A_p}{A_{0.01}} = \frac{0.07}{p^{F_3}}$$
$$F_3 = 0.855 + 0.139 \log_{10}(p) \tag{11.6}$$

If the desired path unavailability p is known and the related path fade margin A_p is desired, the task is defined. However, if the desired fade margin A_p is known and the task is to define the related path unavailability p, a little more work is required. Dividing both sides of the $A_p/A_{0.01}$ equation by 0.12 or 0.07, then taking the logarithm of both sides, rearranging terms, and solving the quadratic equation yields the following (where sqrt(x) is the square root of x).

For latitudes equal to or greater than 30° (North or South),

$$p = 10^{F_4}$$
$$F_4 = -6.349 + \text{sqrt}\left[40.31 - 23.26 \log_{10}\left(\frac{A_p}{0.12 A_{0.01}}\right)\right] \tag{11.7}$$

For latitudes less than 30° (North or South),

$$p = 10^{F_5}$$
$$F_5 = -3.076 + \text{sqrt}\left[9.459 - 7.194 \log_{10}\left(\frac{A_p}{0.07 A_{0.01}}\right)\right] \tag{11.8}$$

The above ITU-R formulas for $A_p/A_{0.01}$ are only to be calculated for $0.001\% \leq p \leq 1\%$ (availabilities of 99.999%–99%) (International Telecommunication Union—Radiocommunication Sector (ITU-R), 2009).

For latitudes equal to or greater than 30° (North or South), the allowable range is $0.1200 \leq (A_p/A_{0.01}) \leq 2.138$. The formulas fail for $(A_p/A_{0.01}) > 6.489$. Real path designs often require $p < 0.001\%$. For these cases, most designers use these equations for $(A_p/A_{0.01}) \leq 6.489$. $A_p/A_{0.01}$ is limited to a maximum value of 6.489 ($p = 0.0000005\%$). This is a more than adequate range for real designs.

For latitudes less than 30° (North or South), the allowable range is $0.06994 \leq (A_p/A_{0.01}) \leq 1.443$. The formulas fail for $(A_p/A_{0.01}) > 1.445$. Real path designs often require $p < 0.001\%$. For these cases, the formulas may not be used because they fail. To deal with this, it is suggested that for $(A_p/A_{0.01}) > 1.443$, the path designer use the above formulas for latitudes equal to or greater than 30° (North or South) but with $A_p/A_{0.01}$ replaced by 1.482 $(A_p/A_{0.01})$. $A_p/A_{0.01}$ is limited to a maximum value of 4.378 ($p = 0.0000005\%$).

The following algorithm may be used to calculate path outage probability p based on a path fade margin M:

Step One: Determine the point rain attenuation for probability $p = 0.01\%$.

M = path fade margin (dB)
Latitude = latitude (decimal) near center of path
Determine K and Alpha for operating frequency using ITU-R formulas
Determine local rain rate R01 (mm/hr) not exceeded more than 0.01% of time
KRAlpha = K * (R01 ^ Alpha)
REMARK A^B = A^B

Step Two: Find the effective path distance.

```
PathLenKM = path length (kilometers)
RD0 = R01
IF R01 > 100 THEN RD0 = 100
D0 = 35 / (EXP(0.015 * RD0))
REMARK EXP(x) = e^x
Pt2Path = 1 / (1 + (PathLenKM / D0))
PathEff = Pt2Path * PathLenKM
```

Step Three: Given path fade margin M, determine the expected probability p (%) that M is not exceeded.

```
A01 = KRAlpha * PathEff
Ratio = M / A01

IF ABS(Latitude) >= 30 THEN REMARK For latitudes equal to or greater than 30
     (North or South):
  IF Ratio > 6.48901 THEN Ratio = 6.48901
  PART = 23.26 * (LOG(Ratio / 0.12) / LOG(10))
  REMARK LOG(x) = natural log of x = log_e(x)
  REMARK LOG(x)/LOG(10) = common log of x = log_10(x)
  FP = -6.34901 + SQR(40.31 - PART)
  REMARK SQR(x) = square root of x
  P = 10 ^ FP
END IF

IF ABS(Latitude) < 30 THEN  REMARK For latitudes less than 30 (North or South):
  IF Ratio > 4.37801 THEN Ratio = 4.37801
  IF Ratio <= 1.443 THEN
     PART = 7.194 * (LOG(Ratio / 0.07) / LOG(10))
     FP = -3.076 + SQR(9.459 - PART)
     P = 10 ^ FP
END IF
  IF Ratio > 1.443 THEN
     PART = 23.26 * (LOG((1.482 * Ratio) / 0.12) / LOG(10))
     FP = -6.349 + SQR(40.31 - PART)
     p = 10 ^ FP
  END IF
END IF
```

Step Four: Determine the outage time associated with p.

```
OutagePerCent = p
AvailabilityPerCent = 100−OutagePerCent
MinutesInYear = 365.25 * 24 * 60
OutageMinutes = MinutesInYear * (OutagePerCent / 100)
OutageSeconds = 60 * OutageMininutes
```

The ITU-R rain attenuation formulas are simple to use but do not conform to the standard path attenuation formula. If we wish to compare the ITU-R method with the Crane method, we must relate the above formulas to the standard formula $\gamma_R = KR^\alpha$ (dB/km).

$$\gamma_R = K(R_{0.01})^\alpha \left(\frac{A_p}{A_{0.01}}\right) = K\left[\left(\frac{A_p}{A_{0.01}}\right)^{1/\alpha} R_{0.01}\right]^\alpha \text{ (dB/km)}$$

$$R = \left(\frac{A_p}{A_{0.01}}\right)^{1/\alpha} R_{0.01} = \text{ITU-R rain rate(mm/h)} \quad (11.9)$$

We wish to use a standard fade margin formula $M = \gamma_R d_{eff}$ with terms as previously defined. As the ITU-R method for calculating fade margin M is not written in the standard $\gamma_R d_{eff}$ format, the assignment of pieces to that formula is artificial. As the path unavailability decreases, rain rate increases but we would expect path attenuation to stay constant or decrease. Therefore, the factor $A_p/A_{0.01}$ was assigned to the rain rate factor γ_R rather than the effective path length factor d_{eff} as $(A_p/A_{0.01})$ matches the action of γ_R (they both increase with decreasing path unavailability). This leaves us with the curious result that ITU-R point rain rate is frequency dependent (through the α factor's frequency dependency). For example, a point rain rate of 50 mm/h 0.01% of the time becomes a 0.001% point rain rate of 87 (86) mm/h at 8 GHz, 96 (94) mm/h at 11 GHz, 107 (101) mm/h at 18 GHz, 110 (105) mm/h at 23 GHz, and 122 (119) mm/h at 38 GHz for a vertically (horizontally) polarized signal.

To derive an approximation of the ITU-R rain rates (simplified ITU-R rain rates) to compare with the Crane city data, the approximation $\alpha \cong 1$ are used. While this approximation introduces significant error for many frequencies, for the popular US 18- and 23-GHz vertical polarization frequencies, this approximation is surprisingly good. It introduces no more than 22 % error for high rain rates (200 mm/h) and significantly less for lower rain rates. As noted above, for moderate rain rates over the frequency range 11–23 GHz, the approximation is accurate within 10%. This allows the following redefinition to facilitate comparisons.

$$R \cong \left(\frac{A_p}{A_{0.01}}\right) R_{0.01} = \text{approximate ("simplified") ITU-R single-location rain rate} \quad (11.10)$$

This approximate relationship for R is used below for comparison with Crane rain rate measurements. The resulting rain curves are termed the *simplified ITU-R rain curves*.

The following approach is used for the *Crane method* (Crane, 1980,1996) of rain path attenuation estimation:

Crane uses actual rain rate measurements to define average rain rates R for a geographic area called a *zone* (Crane, 1980,1996) or actual city rain rates (Crane, 2003). The Crane method of rain path attenuation has been defined several times by Crane. The primary versions are the following:

Crane (1980). This paper introduced the concepts of point rain attenuation and point-to-path conversion factor. End to end path rain attenuation = Point rain attenuation × Point-to-path conversion factor

Point rain attenuation = αR^{β} (dB/km) where α and β are factors based on frequency and polarization. Crane's α and β factors ignored polarization and therefore are not used by the industry. The industry universally uses the ITU-R factors (P.838 with α and β being relabeled as K and α respectively.). R is rain rate in millimeters per hour at a single location (point rain rate). Crane used various alphabetically labeled land areas (zones), which had predefined rain rates. (This approach was very similar to the ITU-R P.837-1 methodology but the zones and rain rates were different.)

Point-to-path conversion factor (also termed the *path reduction factor*) was a complicated function of rain rate, path length, and RF.

Crane (1996). In this model, the point rain attenuation factor was relabeled as KR^{α} to conform to the ITU-R definition. Crane did not address how to determine K and α but the industry approach is to use ITU-R factors (initially P.838 and now P.838-3). As before, the definition of R was unchanged. However, Crane updated the zones used to determine R. He introduced two new zones: B1 and B2. He also updated the zone rain rates (typically, they increased relative to the 1980 rates).

Although Crane changed the appearance of the formulas of the point-to-path conversion factor, a little mathematical manipulation will show that the 1996 formulas are exactly the same as those in his 1980 model. One typo needs correction. A minus sign must be placed in the exponent of formula (3.3) so it represents a lognormal distribution.

Crane (2003). In this model, only point rain attenuation was addressed. In Appendix 5.2, Crane determined K and α using an obscure 1981 source. The industry ignores this and uses ITU-R factors (P.838 initially and now P.838-3). The big change was to introduce location-specific rain rates for R. (This is similar to the ITU-R P.837-5, which calculates R on a specific longitude and latitude basis.) He listed rates for 109 cities. (The data for two cities, Greeley, CO, and Boise, ID, are flawed and are not used) Rain data by Segal (Canada) and Kizer [NOAA (National Oceanic and Atmospheric Administration) data for other US cities] extended the city data list to 279 cities in North America (see Appendix, 11.A).

The Crane method of rain loss calculation is an iterative approach that uses two factors: point rain attenuation and point-to-path conversion factor. The Crane point-to-path conversion factor has never changed. Other than changing labels to conform to ITU-R usage and relying on ITU-R values, the only difference in the point rain attenuation factor among the three models is the method of calculating R. In the 1980 model, one table of zone R values is used. In the 1996 model, an updated set of zone R values is used. In the 2003 model, the R values are based on specific cities.

Basically, the three Crane models define what R (rain rate) values one uses for the calculation. Industry practice is to use the ITU-R 838-X model to determine the values of K and α. The basic methodology of using those values has never changed (although the K and α. values have). The values of the older ITU-R 838 are generally smaller than the current ITU-R 838 values. The new factors result in larger rain attenuation for a given rain rate. This, in turn, results in shorter path distances for a given calculated outage probability.

The rain path attenuation equations are defined in the work by Crane (1980), Eqs. 3–7 for Crane 1980 and Equation 4.7 and Equation 4.8 and the following unnumbered equations of the work by Crane (1996). While the two sets of equations appear different, a little manipulation will show that they are exactly the same. They are equivalent to the following equations. In general, the following notation follows Crane 96 but the roles of d and D are reversed.

$$M = \gamma_{\mathrm{R}} d_{\mathrm{eff}} \tag{11.11}$$

M = radio path fade margin not exceeded with probability p;
$\gamma_{\mathrm{R}} = KR(p)^{\alpha}$ (dB/km), $\alpha R(p)^{\beta}$ in Crane 80, and $KR(p)^{\alpha}$ in Crane 96;
$R(p)$ = rain rate (mm/h) not exceeded with probability p;

d_{eff} is relatively complex to calculate.

Crane considers path attenuation to be influenced by two factors—one due to intense rain cells and one due to more diffused debris rain. These factors are built into Crane's two-component path attenuation model.

G_1 = effective path length due to component 1

$$= \frac{e^{U\alpha d} - 1}{U\alpha} \tag{11.12}$$

$G_{1\mathrm{L}}$ = outer limit of effective path length due to component 1

$$= \frac{e^{U\alpha D} - 1}{U\alpha} \tag{11.13}$$

G_2 = effective path length due to component 2

$$= \frac{e^{w\alpha}(e^{c\alpha d} - e^{c\alpha D})}{c\alpha} \tag{11.14}$$

$G_{2\mathrm{L}}$ = outer limit of effective path length due to component 2

$$= \frac{e^{w\alpha}[e^{c\alpha(22.5)} - e^{c\alpha D}]}{c\alpha} \tag{11.15}$$

d = path distance (km) = 1.609 × path distance (miles), D in Crane 80 and 96;
$b = 2.3/R^{0.17}$;
$w = \ln(b) = 0.83 - 0.17\ln(R)$, B in Crane 96;
$c = 0.026 - 0.03\ \ln(\mathrm{R})$;

$D = 3.8 - 0.6\ \ln(R) =$ rain cell diameter, $R \leq 550$ mm/h, d in Crane 80 and $\delta(R)$ in Crane 96;
$U = [\ln(be^{cD})]/D = [\ln(b)/D] + c = (w/D) + c$;
$e^{w\alpha} = b^{\alpha}$;
$\ln(x) = \log_e(x)$.

Using the above Crane model, effective path distance is defined as follows:

For $0 < d \leq D$,

$$d_{\text{eff}} = G_1.$$

For $D < d \leq 22.5$,

$$d_{\text{eff}} = G_{1L} + G_2.$$

For $22.5 < d$,

$$d_{\text{eff}} = G_{1L} + G_{2L}$$

For $d > 22.5$, probability of occurrence p is replaced by a modified probability of occurrence p_m where $p_m = (d/22.5) \times p$. The rain rate value R is not changed, only its probability of occurrence. This has the effect of increasing rain outage time as the path length increases beyond 22.3 km.

In practice, the fade margin of the path is determined from system parameters. The above formulas are used to iterate to an appropriate rain rate. Then the probability of rain rate occurrence is calculated (usually using a lookup table) to determine the unavailability of the path due to rain.

The following algorithm may be used to calculate path outage probability p based on a path fade margin M.

```
Step One: Determine rain rate R (mm/h) associated with fade margin M (dB)
          Determine K and α for operating frequency using ITU-R formulas

          PathLenKM = path length in kilometers

             L = PathLenKM
             IF PathLenKM > 22.5 THEN L = 22.5
             R = 0.001
             Rstep = 20
          10 B = 2.3 / (R ^ (0.17))
             REMARK A^B = A^B
             C = 0.026 - (0.03 * LOG(R))
             D = 3.8 - (0.6 * LOG(R))
             U = (LOG(B) / D) + C
             REMARK LOG(x) = natural log of x = log_e(x)
             KRAlpha = K * (R ^ Alpha)
             IF L <= D THEN PathEff = ((EXP(U * Alpha * L)) - 1!) / (U * Alpha)
             IF L > D THEN
                NM1 = ((EXP(U * Alpha * D)) - 1!) / (U * Alpha)
                NM2 = (B ^ Alpha) * ((EXP(C * Alpha * L)) - (EXP(C * Alpha * D))) /
                        (C * Alpha)
                REMARK EXP(x) = e^x
                PathEff = (NM1 + NM2)
             END IF
             TrialMargin = K * (R ^ Alpha) * PathEff
             IF TrialMargin < M THEN GOTO 20
             R = R - Rstep
             Rstep = Rstep / 2!
          20 IF Rstep < .01 THEN GOTO 30
```

```
        R = R + Rstep
        GOTO 10
    30 REMARK R has been determined
```

Step Two: Determine the outage time associated with R

Determine outage probability p (%) associated with R (mm/h) by interpolating a lookup table of rain rates for the given area.

```
IF PathLenKM > 22.5 THEN p = p * PathLenKM / 22.5

OutagePerCent = p
AvailabilityPerCent = 100 – OutagePerCent
MinutesInYear = 365.25 * 24 * 60
OutageMinutes = MinutesInYear * (OutagePerCent / 100)
OutageSeconds = 60 * OutageMininutes
```

11.3 POINT-TO-PATH LENGTH CONVERSION FACTOR

Two factors differentiate the ITU-R and Crane rain models: point rain rate and point-to-path conversion factor F_{pp}. First consider the point-to-path conversion factor F_{pp}.

Figure 11.4 shows the point-to-path attenuation factors F_{pp} based on the results of the above paragraphs. The Crane factor increases with path distance for low point rain rates. Crane observed that if the rain rate at one location is low, it is likely to be higher at another location on the path. However, if the rain rate at a location is high, it is likely that the most intense rain is at that location. The Crane factor also has frequency dependency. This effect has been observed by others (Hodge, 1977; Kheirallah et al., 1980).

Obviously, there is a significant difference between the point-to-path conversion factors using the Crane or ITU-R methods. A direct comparison can be a little misleading because the ITU-R conversion factor is calculated for the rain rate at unavailability 0.01% while the Crane conversion is calculated at the unavailability rate of interest. The ITU-R conversion factor is limited to $R \leq 100$ mm/h. In general, the Crane conversion factor is greater than the ITU-R factor. The Crane conversion factor is significantly greater than the ITU-R factor for very low rain rates (<25 mm/h) and for rain rates near 100 mm/h. For other rates, they are closer.

Deciding which point-to-path rain attenuation conversion factor is more realistic is not easy. Lin (Lin, 1975) reviewed the statistics of 96 rain gauges located in a grid with 1.3-km spacing. The incidence of 100 mm/h rain was higher by a factor of 5 for the upper 25% of rain rates as opposed to the lowest 25%. In another study of four rain gauges spaced in a square with 1-km sides, Lin noted that for rain rates greater than 80 mm/h, rain rates varied by a factor of 3. He observed "... on a short-term basis, the relationship between the path rain attenuation distribution and the [point] rain rate distribution measured by a single rain gauge is *not unique*." Different paths in the same area will experience different average rain attenuation when averaged over the same time period. Therefore, it can be inferred short-term measurements could be found to support either method—and actual short term performance could deviate significantly from either. Only long-term average performance can be expected to yield stable statistics. Lin and the other Bell Labs researchers terminated their studies without reaching a definitive conclusion regarding point-to-path rain statistics. To date, there is no study in North America to support one method over the other.

11.4 SINGLE-LOCATION RAIN RATE R

Having determined the basic radio rain loss equations, the next step is to determine the rain rate at or near a particular location. The current ITU-R and Crane methods use measurements at specific locations to estimate average rain rate distributions as a function of time.

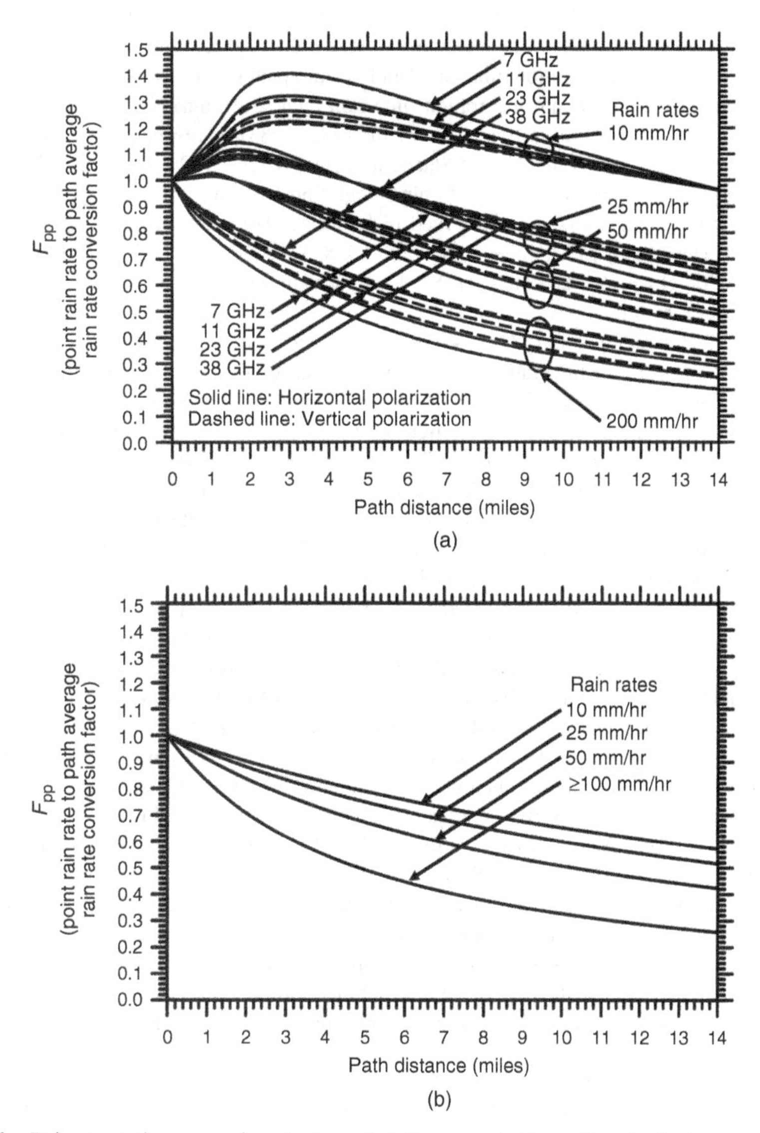

Figure 11.4 Point-to-path conversion factors. (a) Crane point-to-path rain factor and (b) ITU-R point-to-path rain factor.

Rain rates are measured at meteorological stations by capturing rain in a small container. The rain container (gauge) is measured and emptied periodically. The gauge repetition rate (integration time) may be a few or several minutes. The measured rain rate is the accumulated depth of rain divided by the integration time. Shorter integration time produces relatively large "instantaneous" rates. Longer integration time smoothens out the peaks and produces lower rates. Rain rates vary from measurement to measurement. Averaging over several years is required to obtain stable statistics.

It should be noted that short-term rainfall rate determines radio path outage time, not total annual amount of rain that falls in an area. The northwest coast of the United States is a primary example of a very wet region where there are virtually no rain-related path outages. Large-scale climatological factors that seem to bear some relation to high rain rates are number of thunderstorms, late summer humidity, and total July precipitation. Rough terrain and mountains contribute to the formation of thunderstorms. Intense storms develop on the plains but show highly variable rain rates. In mountainous regions, precipitation tends to increase with altitude. When moist winds are lifted by a mountain chain, the windward slope tends to have very heavy rainfall, while areas on the other side may be quite dry.

Crane and Lin use 5-min integration time rain data for radio path fading estimation. Rice and Holmberg and ITU-R specify the use of 1-min integration time rain data. Unfortunately, very few meteorological facilities use such a short integration time (Tattelman, 1982, 1983). Five-minute integration is the most common short-term integration in the United States. It is clear (Crane, 1996; Jones and Sims, 1978) there is no significant difference between 1- and 5-min integration for low rain rates (typically <75 mm/h). For higher rain rates, researchers have observed differences. Various researchers (Crane, 1996; Emiliani et al., 2009; Hershfield, 1972; Jones and Sims, 1978; Lin, 1976,1977; Rice and Holmberg, 1973; Tattelman, 1982; Tattelman and Grantham, 1985) have measured results that differ by location. There is little commonality of result.

International Telecommunication Union—Radiocommunication Sector (ITU-R) (2007a) proposes a conversion factor from 5-min integration time measurements to 1-min integration time measurements. Emiliani et al. (2009, 2010) compared the ITU-R conversion method to other methods. The ITU-R method had the most error but all better methods required location-specific data, which is generally unavailable. This is similar to the results of Lin (1977), which also required site-specific information. As Emiliani et al. noted (Emiliani et al., 2010), the ITU-R conversion factor is only valid for the 0.01% rain rate, the only actual rain rate of interest for the ITU-R rain attenuation calculation method (International Telecommunication Union—Radiocommunication Sector (ITU-R), 2009; Jung et al., 2007). A general 5- to 1-min conversion method is not currently known.

Crane (1996) disagrees with the use of the ITU-R conversion factor to convert longer integration time rain rates to 1-min integration rain rates. "The data obtained from MIT [Massachusetts Institute of Technology] do not support such a recommendation [using the ITU-R conversion factor]. If observations have an integration time of 5 min or less, they should be used to represent instantaneous rain-rate distributions." In this chapter, Crane's hypothesis is used to consider 1- and 5-min rain rate distributions essentially the same. While this introduces some error, variation with location is a much larger factor and seems to dominate actual rain rate estimations. Given that a general conversion factor is unknown, and 1 min integration data is rare in North America, this decision seems inevitable.

The lowest value of p (path unavailability) for the ITU-R model is 0.001% (99.999% path availability, often referred to as *five* 9*'s* availability). This range of availability is adequate for many high frequency radio path designs whose path performance is dominated by rain. However, some designers want to design very high reliability paths. For the western United States, rain rates are so low, multipath can be a significant factor on 11-GHz paths. For those paths, rain rates for unavailability smaller than 0.001% are needed. Extending the existing data to smaller unavailability numbers was not addressed in the early Crane papers (Crane, 1975, 1980, 1989) and is expressly disallowed in the ITU-R model (lack of uniformity in extrapolating the Crane data to low unavailabilities being one reason different path design packages yielding different results for these low rates). Extending the Crane model to any arbitrary unavailability (outage time) is now easily accomplished using the procedures outlined below. Both the ITU-R and Crane models produce average rain rate distributions. Some path designers would prefer "worst-case" average values. Path designers would like to understand how much variation there is within each of the rain zones. They would also like to understand how different the ITU-R and Crane models are. Fortunately, those issues may now be addressed.

Crane divided the world into several rain zones. The latest version was published in 1996 (Crane, 1996); the name Crane 96 is usually used to refer to this version of the Crane zone model. Those zones for North America are recreated in Figure 11.5 and Figure 11.6.

The maps in Figure 11.7 and Figure 11.8 show four regions, C1, C2, F1, and F2, while Crane only uses two regions, C (for C1 and C2) and F (for F1 and F2). The reason for this update is explained later.

The dots on the maps show the meteorological stations used to gather Canadian (Segal (1979)) and US (NOAA) rain data. Crane's published 1996 zone diagram (Crane, 1996) of the regions was too granular to accurately place several cities (including Buffalo, Halifax, Houston, Juneau, Rochester, Saint John, Seattle, Tacoma, Vancouver, and Wichita). Where necessary, the city rain rate distributions were compared to the rain zone averages and the cities were matched to the zone more nearly matching their rain statistics. For reasons that are addressed below, in this chapter, the Crane C and F zones were

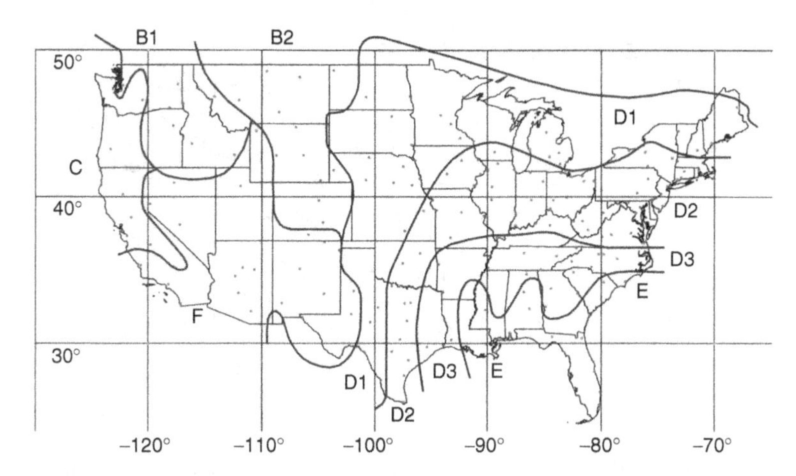

Figure 11.5 Original Crane 96 rain zones for the lower 48 states in the United States.

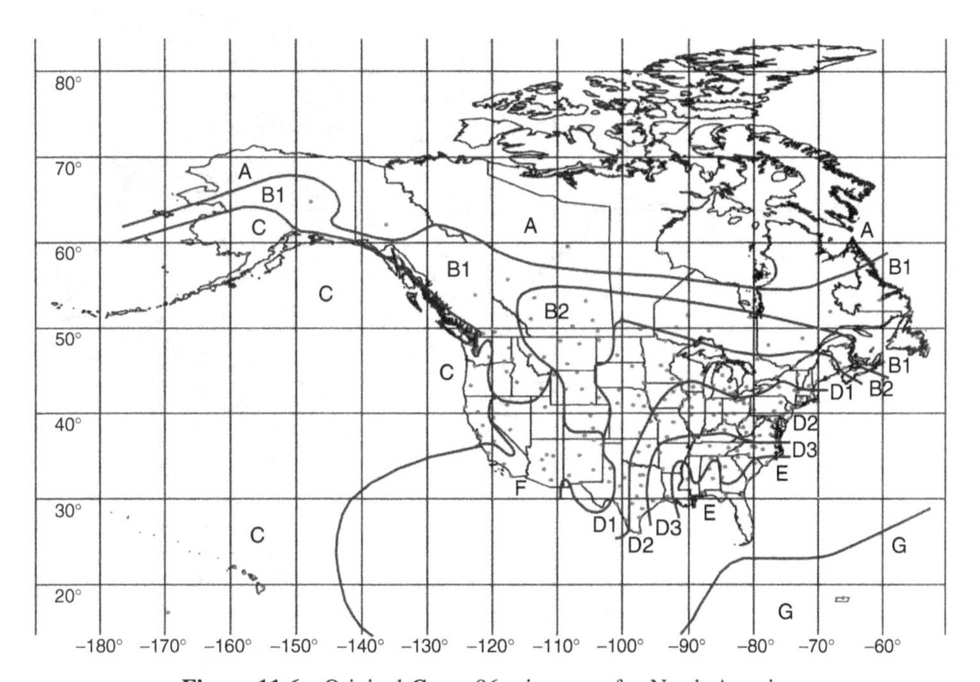

Figure 11.6 Original Crane 96 rain zones for North America.

subdivided into smaller zones (C1, C2, F1, and F2). The zones for various cities are listed in the Appendix at the end of this chapter.

Crane's rain rate model ((Crane, 1996, Eqs. 3.1–3.4), or (Crane, 2003, Eq. 5.3)) uses a two component distribution:

$$P(r \geq R) = \text{probability } r \text{ exceeds } R(\%)$$

$$= P_{\text{V}}(r \geq R) + P_{\text{W}}(r \geq R) \tag{11.16}$$

$$P_{\text{V}}(r \geq R) = \text{volume cell component (intense rain cell contribution)}$$

$$= P c e^{(-R/R_{\text{C}})} \tag{11.17}$$

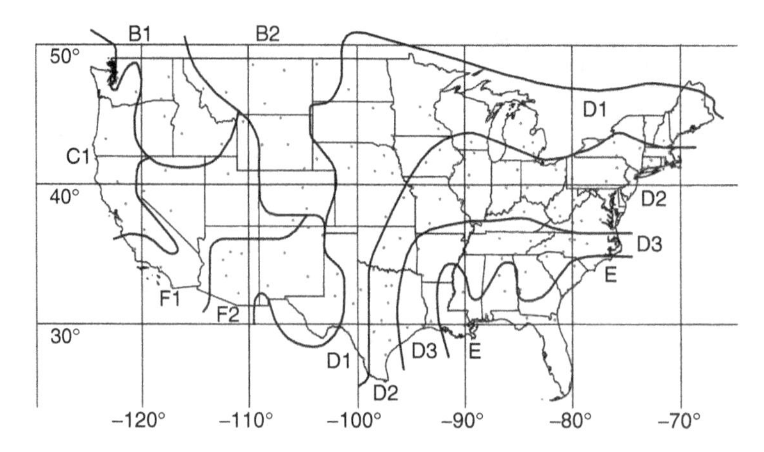

Figure 11.7 Updated Crane 96 rain zones for the lower 48 states in the United States.

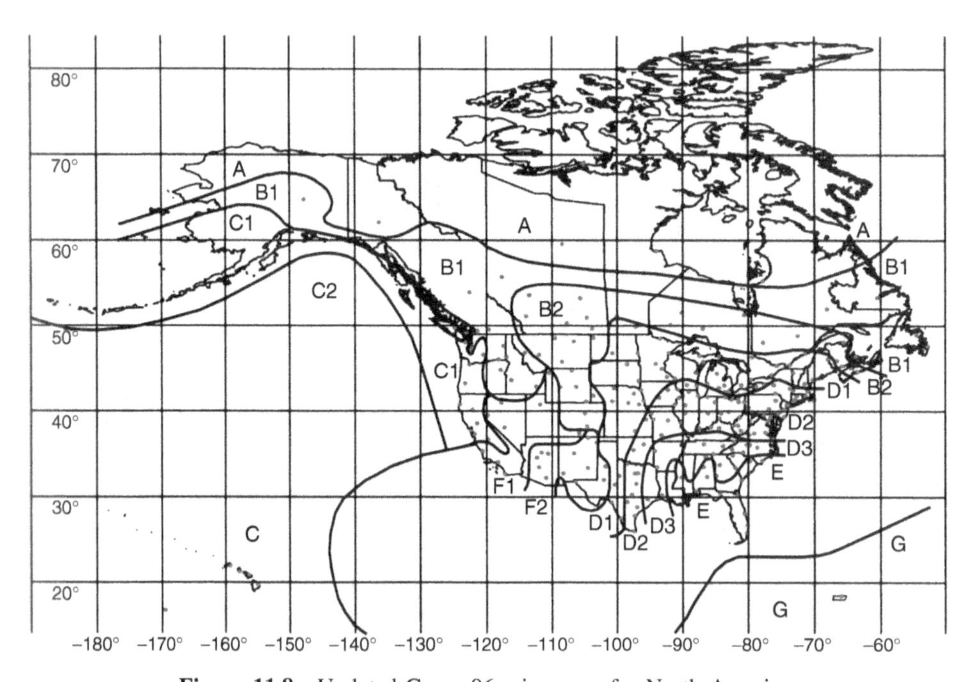

Figure 11.8 Updated Crane 96 rain zones for North America.

$$P_W(r \geq R) = \text{debris component(low density widely distributed rain contribution)}$$

$$= P_D Q\left[\frac{\ln(R) - \ln(R_D)}{SD}\right] \tag{11.18}$$

r = observed rain rate (mm/h);
R = specified rain rate (mm/h);
e = Euler's constant ≈2.718281828459045;
$\ln(x)$ = natural logarithm of x;
P_C = probability of a cell (different for month or year average);

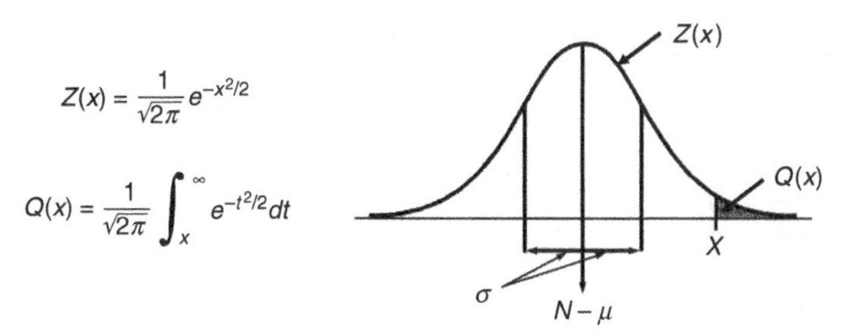

Figure 11.9 Normal distribution parameters.

P_D = probability of debris (different for month or year average);
R_C = average cell rain rate (constant for month or year average);
R_D = median rain rate in debris (constant for month or year average);
SD = standard deviation of natural logarithm of rain rate (constant for month or year average);
$Q(x)$ = Gaussian cumulative distribution function (Fig. 11.9).

Using a standard numerical approximation for Q (Abramowitz and Stegun, 1968), Eqs. 26.2.1 and 26.2.17) yields the following:

$$Q(x) \approx Z[(B1 \times T) + (B2 \times T^2) + (B3 \times T^3) + (B4 \times T^4) + (B5 \times T^5)]$$

$$T = \frac{1}{1 + (P \times X)}$$

$$Z = \frac{e^{[-(X \times X)/2]}}{(2\pi)^{(1/2)}}$$

$$X = \frac{N - \mu}{\sigma}$$

$$N = \text{variable value} = \ln(R)$$

$$\mu = \text{mean} = \ln(R_D)$$

$$\sigma = \text{standard deviation} = \text{SD}$$

$$P = 0.2316419$$

$$B1 = 0.319381530$$

$$B2 = -0.356563782$$

$$B3 = 1.781477937$$

$$B4 = -1.821255978$$

$$B5 = 1.330274429$$

Crane (2003, Appendix 5.1) lists annual and monthly values of the various parameters for 109 cities in the United States. Two cities, Greely, CO, and Boise, ID, have flawed annual parameters and are not used.

Table 11.4 shows the annual parameters (Crane, 1996, Table 3.2) for the model for the Crane zones.

The new zones C1, C2, F1, and F2 are explained in the following. Using the formulas and parameters in Table 11.4, the average Crane rain zone rain rates have been calculated (Table 11.5).

Interpolation between values can be accurately determined using the following formulas, any convenient base logarithm LOG(x) and the data in Table 11.6.

TABLE 11.4 Crane 1996 Zone Annual Rain Rate Parameters

Zone	P_C	R_C	P_D	R_D	S_D
A	0.000088	3.450	2.270	0.205	1.490
B	0.023000	14.300	10.800	0.178	1.440
B1	0.009000	11.100	12.000	0.092	1.610
B2	0.023000	17.500	12.000	0.181	1.480
C[a]	0.033000	19.800	12.000	0.293	1.310
C1[b]	0.001380	30.379	1.282	1.562	0.926
C2[b]	0.103568	24.300	31.470	0.300	0.678
D1	0.026000	23.200	8.190	0.463	1.340
D2	0.031000	14.300	9.270	0.475	1.480
D3	0.048000	17.000	4.000	1.970	1.210
E	0.220000	23.200	5.250	2.020	1.250
F[a]	0.004800	8.260	6.950	0.099	1.810
F1[b]	0.001300	31.379	1.282	1.562	0.926
F2[b]	0.024568	25.700	31.470	0.300	0.678
G	0.028000	50.500	9.820	1.820	1.200
H	0.048000	35.400	7.000	2.470	1.490

[a] Not recommended.
[b] New zone.

$$
\begin{aligned}
U1 &= \text{outage time 1} \\
U2 &= \text{outage time 2} \\
U &= \text{outage time of interest} \\
X1 &= \text{LOG}(U1) \\
X2 &= \text{LOG}(U2) \\
X &= \text{LOG}(U) \\
Y1 &= \text{rain rate at } U1 \\
Y2 &= \text{rain rate at } U2 \\
M &= \frac{Y2 - Y1}{X2 - X1} \\
Y &= [(X - X2)M] + Y2 = \text{rain rate at } U
\end{aligned}
\tag{11.19}
$$

It is possible to solve the fade margin equation $M = \gamma_R d_{eff}$ (mentioned at the beginning of this chapter and discussed in more detail below) to determine the expected path length for different availabilities and frequencies. Results were derived for the path availabilities of 99.99% (99.99, "four 9's"), 99.999% (99.999, "five 9's"), and 99.9999% (99.9999, "six 9's"). Three Crane rain zones were considered: C1 (low rain rates, Fig. 11.10), D2 (moderate rain rates, Fig. 11.11), and E (high rain rates, Fig. 11.12). The Crane 96 and ITU-R methods (averaged for all the cities in the Crane zone) were used for the three zones.

The graph (a) uses the Crane F_{pp} factor; the graph (b) uses the ITU-R F_{pp} factor. The underlying calculations used the following frequencies: "11 GHz" = 11.5 GHz, "18 GHz" = 19.5 GHz, "23 GHz" = 22.5 GHz, and "38 GHz" = 39.65 GHz. These frequencies are the center of the upper sub-band of each FCC frequency plan. These curves are compared in the graphs in Figure 11.13, Figure 11.14, and Figure 11.15.

For low rain rates (e.g., rain zone C1), for all availabilities less than 99.9999% the ITU-R method is more optimistic than the Crane method (often by a very large amount). It should be remembered that for 11 GHz in low rain rate zones, rain fading is not a significant factor; atmospheric multipath is the primary path length limitation.

For moderate to high rain rate regions, with the exception of 11 GHz, for all availabilities between 99.99% and 99.999%, the ITU-R method is more optimistic than Crane. For all availabilities between 99.9999% and 99.999%, just the opposite is true. Again, 11 GHz is a minor exception.

TABLE 11.5 Crane 1996 Average Annual Rain Rates by Zone

Zone	Outage Time, %	A	B	B1	B2	C	C1	C2	D1	D2	D3	E	F	F1	F2	G	H
Outage Time, min/yr		Rain Rate, mm/h															
0.053	0.000010	154.1	171.7	205.7	217.7	178.3	151.3	224.7	258.8	527.6	493.6	652.1	474.1	154.2	200.6	550.9	2641.2
0.074	0.000014	138.3	156.4	184.4	197.6	168.6	141.1	216.5	237.9	476.6	452.9	597.3	418.1	143.7	192.0	511.4	2381.5
0.105	0.000020	123.0	141.9	163.9	178.6	159.0	130.4	207.8	218.2	427.3	412.8	543.4	365.2	132.7	182.8	473.3	2130.7
0.158	0.000030	107.4	127.5	143.2	159.8	148.7	118.3	198.0	198.5	376.6	370.9	487.2	312.4	120.2	172.4	434.3	1873.5
0.210	0.000040	97.4	118.6	129.9	148.1	141.7	109.8	191.0	186.0	343.9	343.3	450.3	279.1	111.4	165.0	409.0	1707.7
0.263	0.000050	90.2	112.2	120.4	139.8	136.4	103.2	185.6	177.0	320.2	323.2	423.3	255.5	104.6	159.3	390.5	1588.0
0.316	0.000060	84.7	107.3	113.1	133.5	132.2	97.9	181.1	170.0	301.9	307.4	402.2	237.5	99.1	154.6	376.1	1495.5
0.368	0.000070	80.2	103.4	107.2	128.5	128.7	93.5	177.4	164.3	287.2	294.6	385.1	223.2	94.5	150.6	364.3	1421.0
0.421	0.000080	76.5	100.2	102.4	124.3	125.7	89.6	174.1	159.6	274.9	283.8	370.8	211.5	90.6	147.2	354.4	1359.1
0.473	0.000090	73.4	97.4	98.4	120.8	123.1	86.3	171.3	155.5	264.4	274.6	358.6	201.5	87.1	144.2	345.9	1306.4
0.526	0.00010	70.7	95.0	94.9	117.7	120.7	83.3	168.7	152.0	255.4	266.5	347.9	193.0	84.1	141.5	338.4	1260.7
0.736	0.00014	62.6	87.7	84.5	108.4	113.4	74.1	160.5	141.1	228.2	242.1	315.6	167.9	74.5	132.8	315.3	1123.9
1.05	0.00020	54.8	80.5	74.8	99.4	105.7	64.8	151.9	130.3	202.1	218.3	284.3	144.4	65.0	123.6	292.1	992.9
1.58	0.00030	47.0	72.9	65.2	89.8	97.2	55.2	142.0	118.6	175.6	193.6	252.2	121.2	55.2	113.2	267.0	859.9
2.10	0.00040	42.1	67.8	59.1	83.4	91.2	49.1	135.0	110.6	158.7	177.5	231.6	106.9	49.0	105.8	249.9	775.0
2.63	0.00050	38.6	63.9	54.7	78.7	86.6	44.9	129.6	104.6	146.5	165.9	216.9	96.7	44.7	100.1	237.0	714.1
3.16	0.00060	35.9	60.9	51.4	74.9	82.9	41.7	125.2	99.8	137.2	156.8	205.7	89.1	41.5	95.4	226.7	667.4
3.68	0.00070	33.7	58.3	48.7	71.7	79.8	39.2	121.4	95.9	129.8	149.5	196.7	83.1	39.1	91.4	218.1	630.0
4.21	0.00080	31.9	56.2	46.5	69.0	77.1	37.2	118.2	92.4	123.6	143.4	189.4	78.1	37.1	88.0	210.7	599.0
4.73	0.00090	30.4	54.3	44.6	66.7	74.7	35.6	115.3	89.5	118.4	138.2	183.1	74.0	35.4	85.0	204.3	572.7
5.26	0.0010	29.1	52.6	42.9	64.7	72.6	34.2	112.8	86.8	113.9	133.7	177.7	70.5	34.0	82.3	198.7	550.0
7.36	0.0014	25.3	47.4	37.9	58.2	65.9	30.2	104.6	78.6	100.6	120.2	161.7	60.1	30.1	73.6	181.0	482.5
10.5	0.0020	21.7	42.1	33.1	51.7	58.8	26.5	95.9	70.1	88.1	107.2	146.5	50.7	26.5	64.5	162.8	418.7
15.8	0.0030	18.1	36.4	28.1	44.5	51.0	23.0	86.1	60.9	75.6	93.9	130.9	41.6	23.0	54.0	143.1	354.9
21.0	0.0040	15.8	32.4	24.8	39.7	45.6	20.8	79.1	54.6	67.7	85.4	120.8	36.0	20.8	46.6	129.7	314.7

(Continued)

TABLE 11.5 ***(Continued)***

Zone	Outage Time, %	A	B	B1	B2	C	C1	C2	D1	D2	D3	E	F	F1	F2	G	H
Outage Time, min/yr		Rain Rate, mm/h															
26.3	0.0050	14.3	29.5	22.5	36.1	41.5	19.3	73.6	49.8	62.1	79.1	113.4	32.1	19.2	40.9	119.7	286.3
31.6	0.0060	13.1	27.2	20.7	33.3	38.2	18.1	69.2	46.1	57.8	74.3	107.6	29.2	18.1	36.2	111.8	264.7
36.8	0.0070	12.1	25.4	19.3	30.9	35.5	17.1	65.5	43.1	54.4	70.4	102.8	27.0	17.1	32.3	105.4	247.5
42.1	0.0080	11.4	23.8	18.1	29.0	33.2	16.3	62.2	40.5	51.5	67.2	98.8	25.1	16.3	28.8	100.0	233.4
47.3	0.0090	10.7	22.5	17.1	27.4	31.2	15.7	59.4	38.3	49.1	64.4	95.3	23.6	15.7	25.8	95.4	221.6
52.6	0.010	10.2	21.3	16.2	25.9	29.5	15.1	56.8	36.4	47.0	62.0	92.2	22.3	15.1	23.1	91.3	211.5
73.6	0.014	8.5	17.9	13.7	21.7	24.4	13.4	48.6	30.8	40.7	54.6	82.8	18.5	13.3	14.5	79.2	181.7
105	0.020	7.0	14.7	11.3	17.8	19.8	11.7	40.0	25.5	34.8	47.5	73.3	15.1	11.7	5.6	67.7	154.1
158	0.030	5.6	11.7	9.1	14.0	15.4	10.0	30.1	20.5	28.7	40.1	63.1	11.8	10.0	3.1	56.3	127.0
210	0.040	4.7	9.9	7.7	11.9	13.0	8.9	23.1	17.4	24.9	35.4	56.2	9.9	8.9	2.7	49.2	110.1
263	0.050	4.1	8.7	6.8	10.4	11.4	8.1	17.7	15.4	22.2	31.9	51.1	8.5	8.1	2.5	44.3	98.3
316	0.060	3.7	7.8	6.1	9.3	10.2	7.4	13.3	13.9	20.2	29.3	47.0	7.6	7.4	2.3	40.6	89.3
368	0.070	3.3	7.1	5.6	8.5	9.3	6.9	9.5	12.7	18.6	27.2	43.7	6.8	6.9	2.2	37.7	82.2
421	0.080	3.0	6.6	5.2	7.8	8.6	6.5	6.3	11.8	17.2	25.4	40.9	6.2	6.5	2.2	35.3	76.4
473	0.090	2.8	6.1	4.8	7.3	8.1	6.2	4.0	11.0	16.1	23.9	38.5	5.7	6.2	2.1	33.3	71.5
526	0.10	2.6	5.8	4.5	6.9	7.6	5.8	3.1	10.4	15.2	22.6	36.4	5.3	5.8	2.0	31.6	67.3
736	0.14	2.0	4.7	3.7	5.6	6.3	4.9	2.3	8.5	12.4	18.8	30.2	4.1	4.9	1.8	26.7	55.2
1052	0.20	1.5	3.8	2.9	4.5	5.1	4.0	1.9	6.9	10.0	15.3	24.3	3.1	4.0	1.7	22.2	44.0
1578	0.30	1.1	2.9	2.2	3.4	4.0	3.1	1.6	5.3	7.6	11.9	18.5	2.2	3.1	1.5	17.8	33.4
2104	0.40	0.8	2.4	1.8	2.8	3.4	2.5	1.5	4.4	6.2	9.7	15.0	1.7	2.5	1.4	15.2	27.1
2630	0.50	0.6	2.1	1.5	2.4	2.9	2.0	1.4	3.8	5.3	8.3	12.7	1.4	2.0	1.3	13.3	22.8
3156	0.60	0.5	1.8	1.3	2.1	2.6	1.7	1.3	3.3	4.6	7.2	11.0	1.2	1.7	1.2	11.9	19.7
3682	0.70	0.4	1.6	1.2	1.9	2.4	1.4	1.2	3.0	4.1	6.3	9.6	1.0	1.4	1.2	10.8	17.3
4208	0.80	0.4	1.5	1.0	1.7	2.1	1.2	1.2	2.7	3.7	5.7	8.6	0.9	1.2	1.1	9.9	15.4
4734	0.90	0.3	1.3	0.9	1.6	2.0	1.0	1.1	2.4	3.3	5.1	7.7	0.8	1.0	1.1	9.1	13.8
5260	1.0	0.3	1.2	0.9	1.4	1.8	0.8	1.1	2.2	3.0	4.6	7.0	0.7	0.8	1.1	8.5	12.5

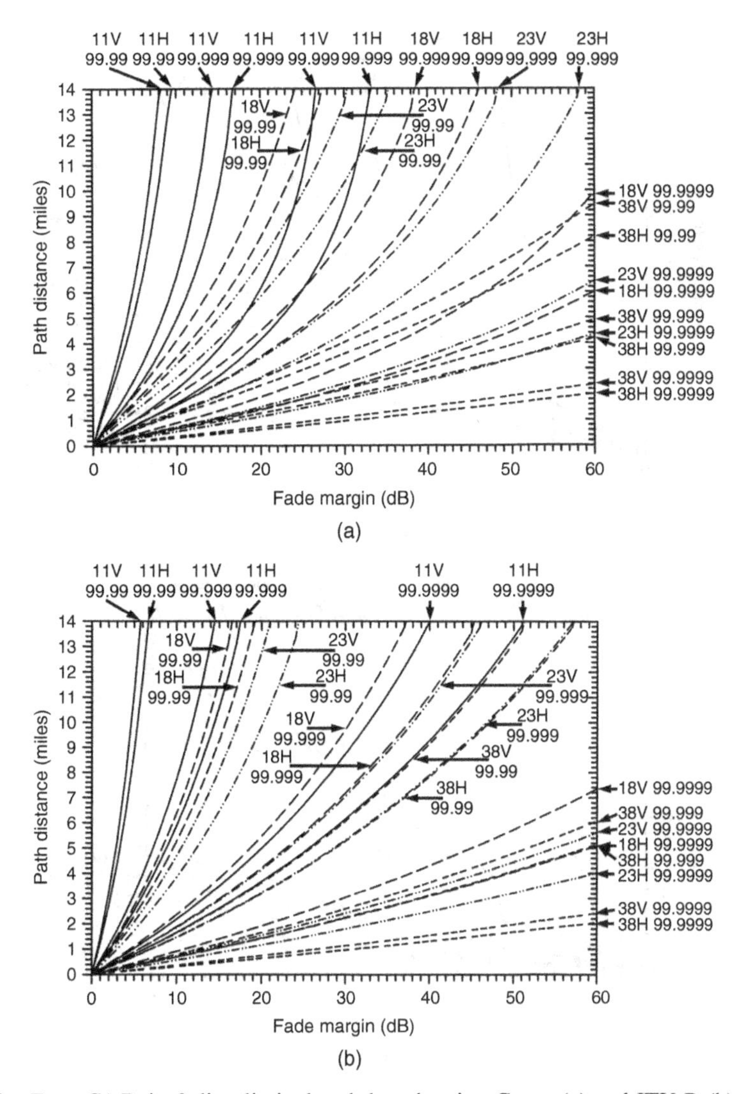

Figure 11.10 Zone C1 Rain-fading-limited path length using Crane (a) and ITU-R (b) point-to-path factors.

For high rain rates, the ITU-R method is more optimistic for 99.99% availability. For 99.999% both methods are about the same. For 99.9999% availability, the ITU-R method is slightly more pessimistic than the Crane method.

One should remember that the ITU-R recommendation is only applicable to rain rates related to availabilities no greater than 99.999%. The point-to-path factor is not applicable to higher rain rates and is probably the reason the ITU-R model is more conservative than the Crane model for the very high availabilities.

11.5 CITY RAIN RATE DATA FOR NORTH AMERICA

Tattelman and Grantham (1983a,1983b) attempted to gather 1-min integration rain rates worldwide. However, their results were limited to rain rates between 0.01% and 1%. Even then they had to resort to conversion methodologies due to lack of short-term integration data. One-minute integration rain rate data

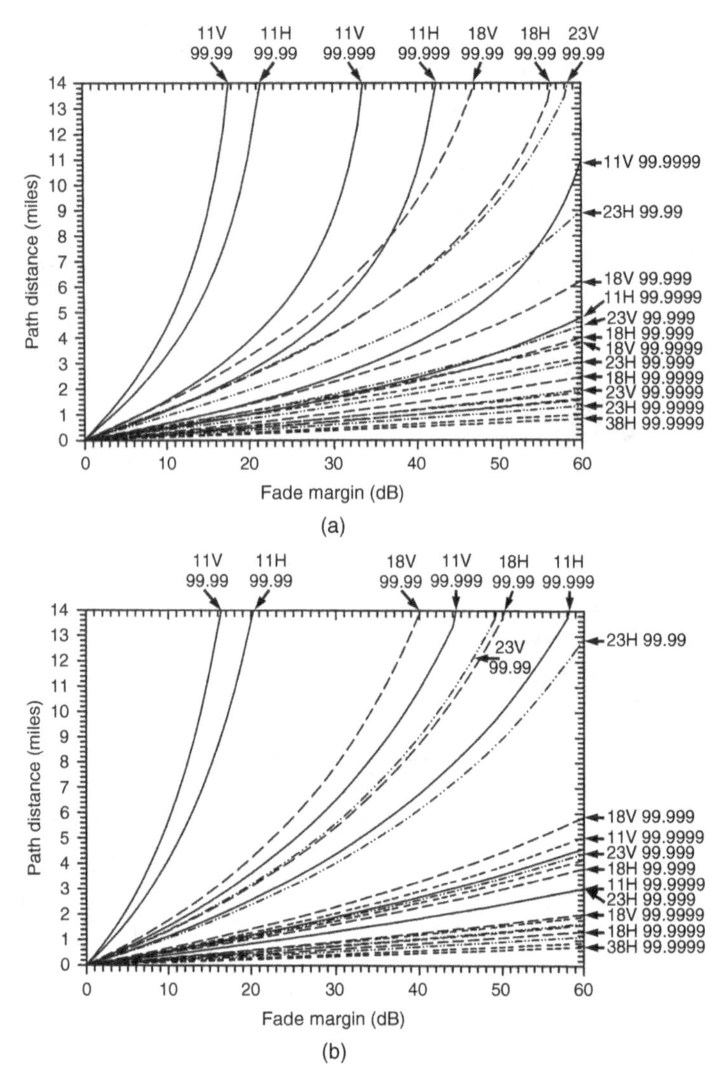

Figure 11.11 Zone D2 Rain-fading-limited path length using Crane (a) and ITU-R (b) point-to-path factors.

for cities in Canada were published by Segal (1979) in 1979. Five-minute integration rain data for the United States were accumulated by Lin (Kizer, 2008; Lin, 1977) (Lin, S. H., "11 GHz path engineering curves for 226 U. S. locations—rain attenuation outages calculated from 20 and 50 Year rain data," unpublished internal Bell Laboratories document, 1977) at Bell Labs in 1977 and distributed informally by Bell Labs about that time to several engineering organizations but never officially published by the Bell System. That data was limited to outage times of 0.0002% to 0.02%. At some locations 5-min data was not available and conversion methodologies were used to create the 5-min rain curves from much longer rain-averaged data. In addition, some site curves have significant data anomalies.

NOAA annual precipitation for the 48 United States shows an increase over the last 30 years. This suggests that data more modern than the Lin data might exhibit higher rain rates (Fig. 11.16).

In fact, modern data (taken in the last 30 years since the Lin data was finished) does show significantly higher rain rates at some cities (Miami and Miami Beach being worst-case examples) than the historical Lin Bell Labs data.

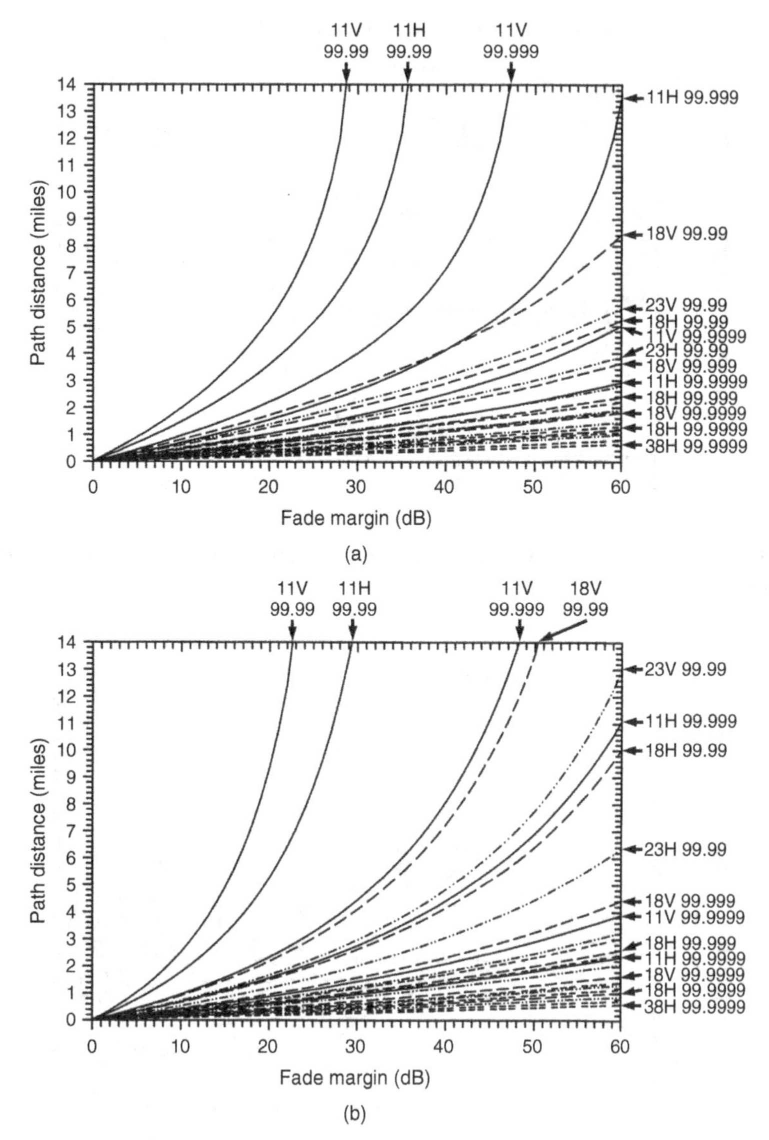

Figure 11.12 Zone E rain-fading-limited path length using Crane (a) and ITU-R (b) point-to-path factors.

Fortunately, significant modern rain data have been recorded for the United States. Data for several United States cities were published by Crane (2003). Crane's data was in the form of parameters (Appendix, 11.A) that permitted calculation (as described earlier) of the cell and debris components of the composite rain rate (See Fig. 11.17 for examples).

Unpublished public domain NOAA data was used to create average rain curves for several additional locations. The Segal and NOAA data was curve-fitted to conform to the Crane model parameters. Those consolidated parameters are listed in Appendix 11.A. Use of these new data (termed *city data* in this chapter) overcomes the limitations of previous data and provides ability to model average rain rate distributions for any desired path outage time at specific sites.

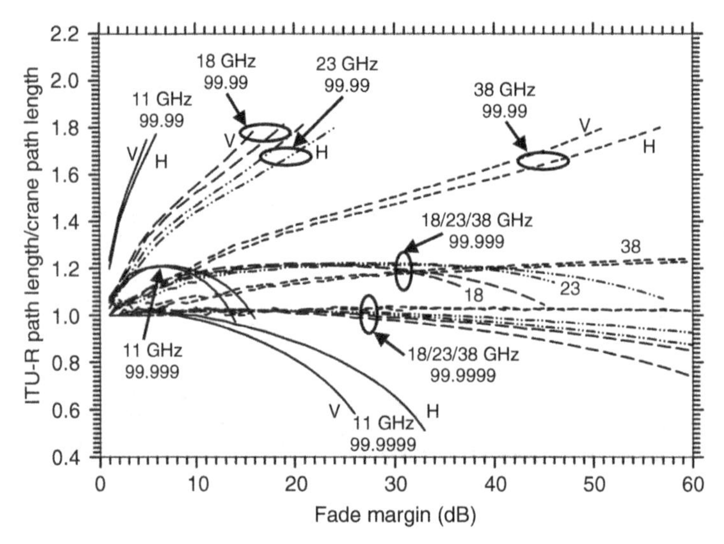

Figure 11.13 Comparison of zone C1 rain-fading-limited path length using Crane and ITU-R point-to-path factors.

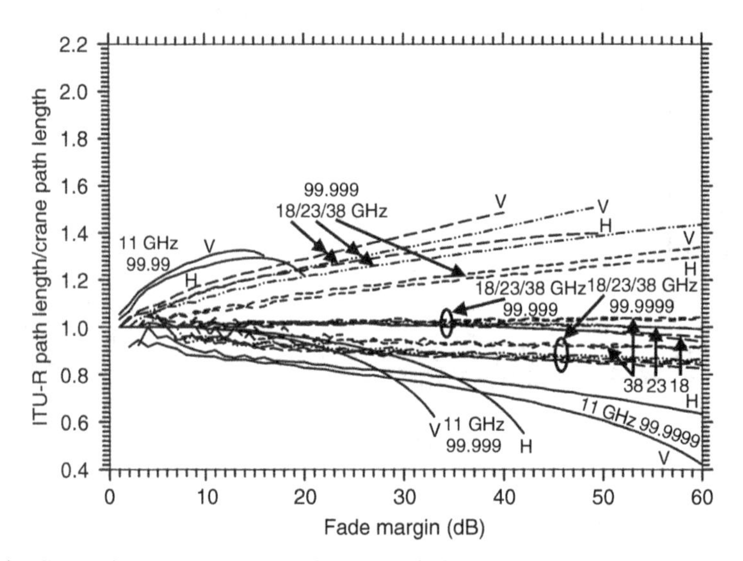

Figure 11.14 Comparison of zone D2 rain-fading-limited path length using Crane and ITU-R point-to-path factors.

The following are examples of Crane-modeled North American city rain curves (city data) and calculated simplified ITU-R rain rates for the same cities. Notice the difference in curve shapes, the city value differences and the limits of the ITU-R curves. The graphs in Figure 11.18a, Figure 11.19a, Figure 11.20a, Figure 11.21a, Figure 11.22a, Figure 11.23a, Figure 11.24a, Figure 11.25a, Figure 11.26a, and Figure 11.27a plot city data at the extremes of the rain zone as well as the rain zone average (of all cities in that zone) and the Crane rain zone curve defined by the parameters in Table 11.2. The graphs in Figure 11.18b, Figure 11.19b, Figure 11.20b, Figure 11.21b, Figure 11.22b, Figure 11.23b, Figure 11.24b, Figure 11.25b, Figure 11.26b, and Figure 11.27b plot the ITU-R rain rates for the same cities shown on the city data graphs. They also show the city average rate and Crane rain zone curves.

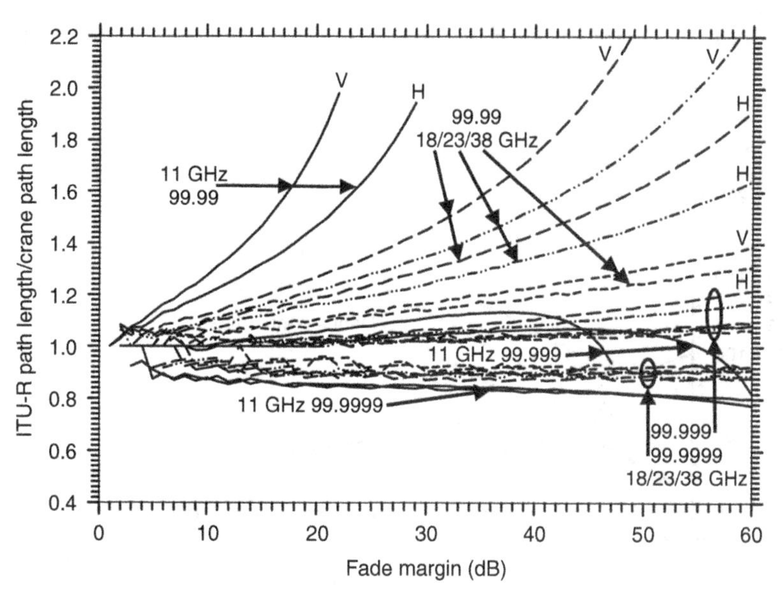

Figure 11.15 Comparison of zone E rain-fading-limited path length using Crane and ITU-R point-to-path factors.

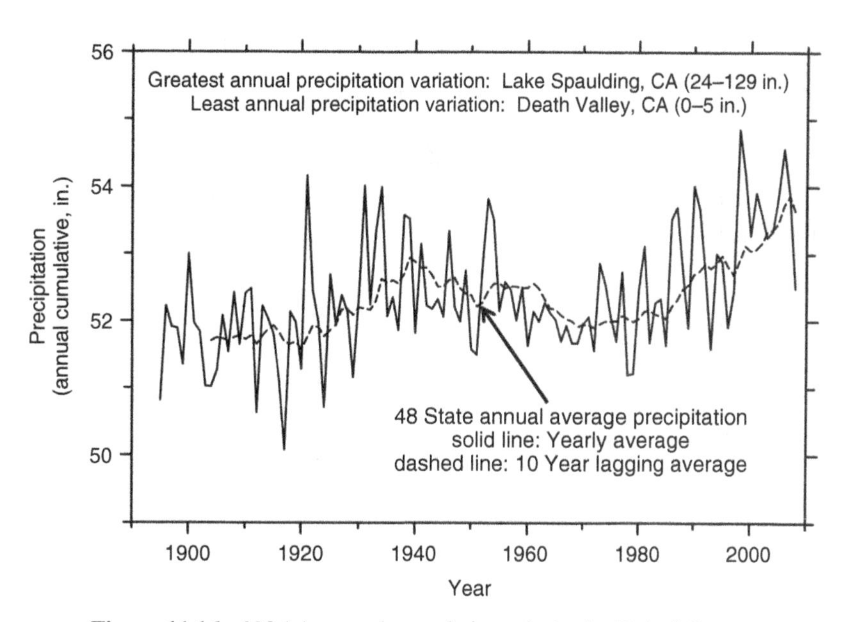

Figure 11.16 NOAA annual cumulative rain in the United States.

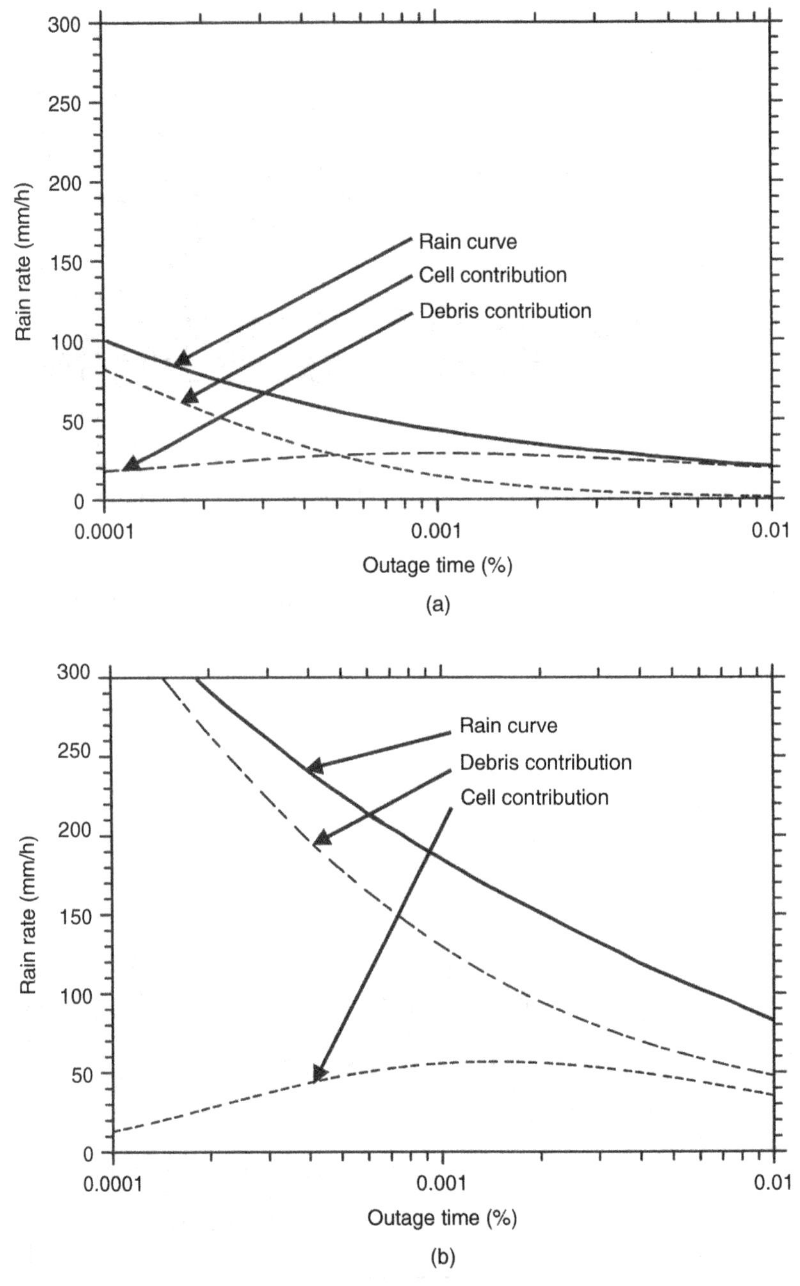

Figure 11.17 Crane rain rate curves for San Francisco (a) and Miami (b).

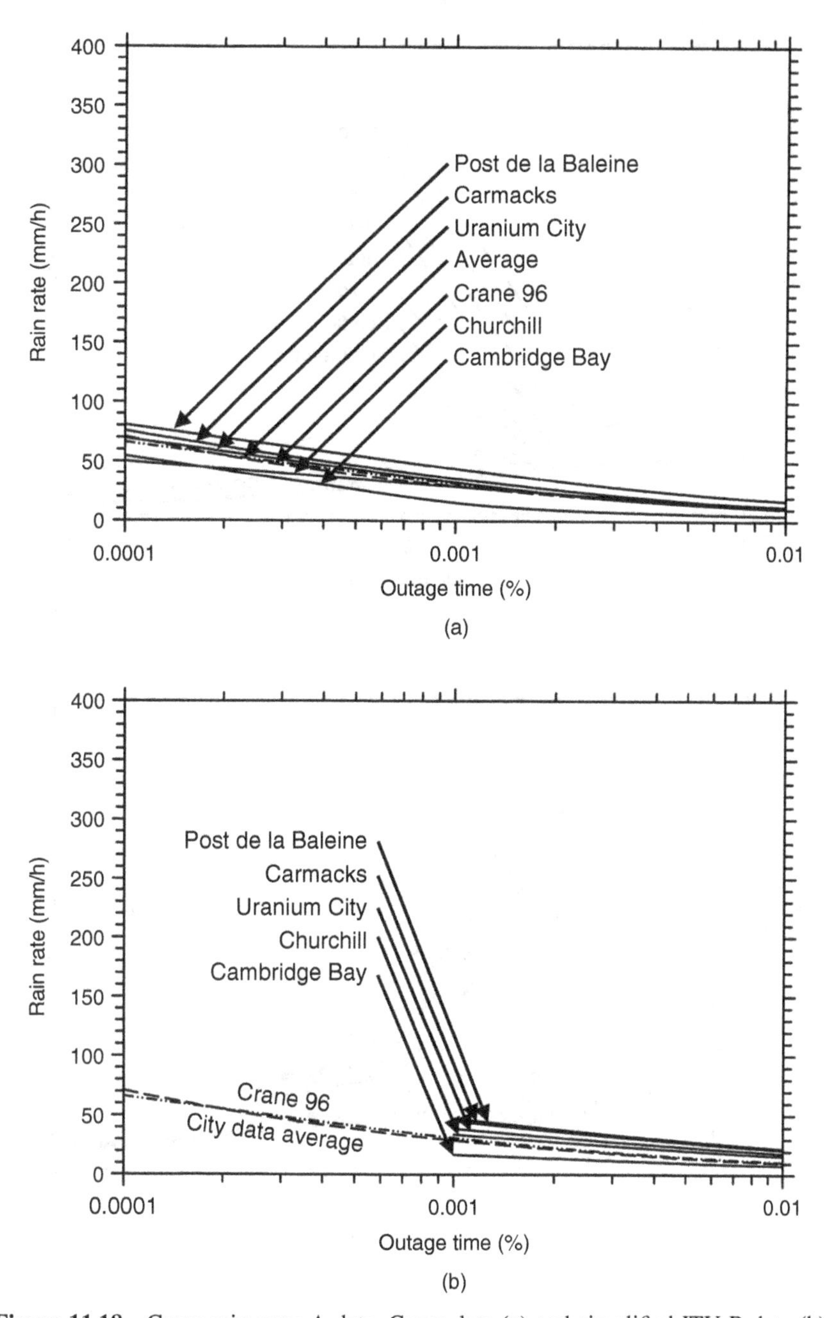

Figure 11.18 Crane rain zone A data, Crane data (a) and simplified ITU-R data (b).

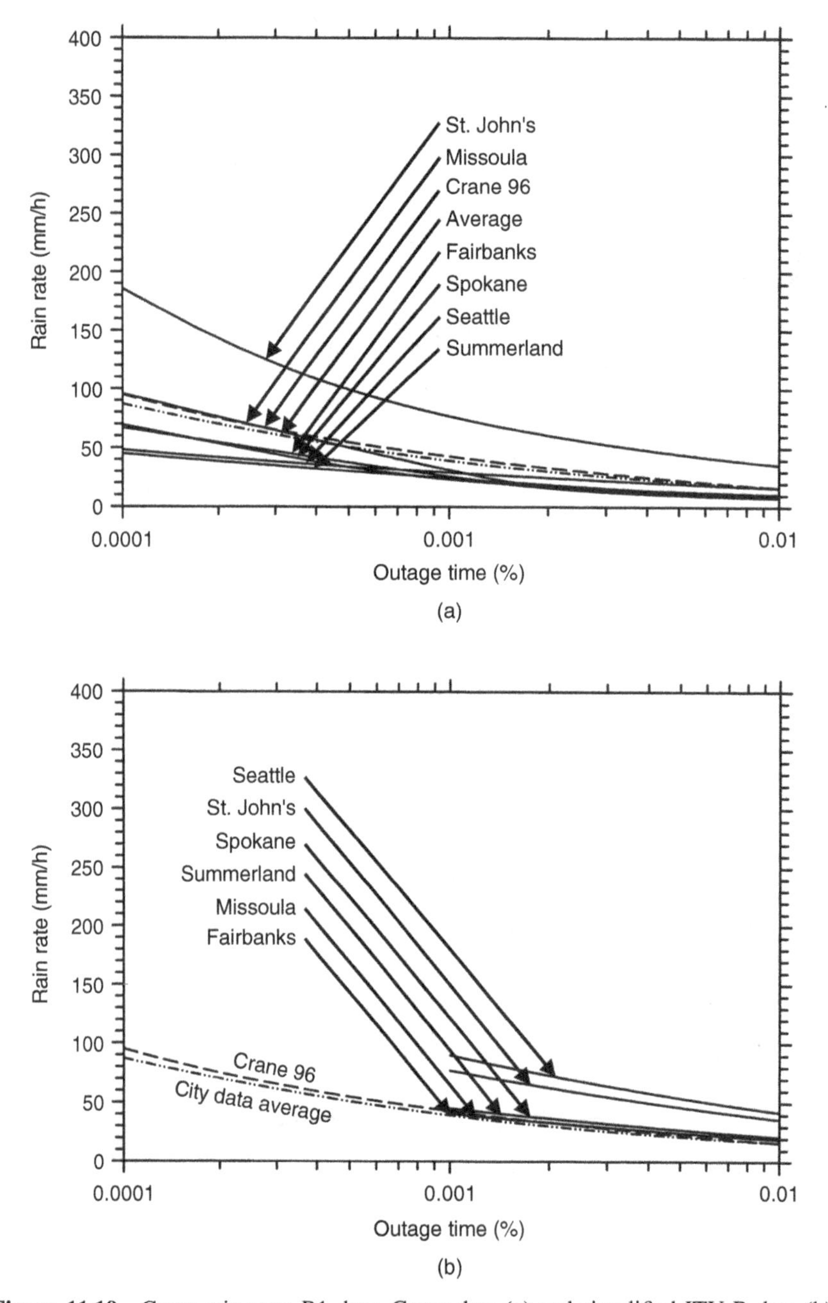

Figure 11.19 Crane rain zone B1 data, Crane data (a) and simplified ITU-R data (b).

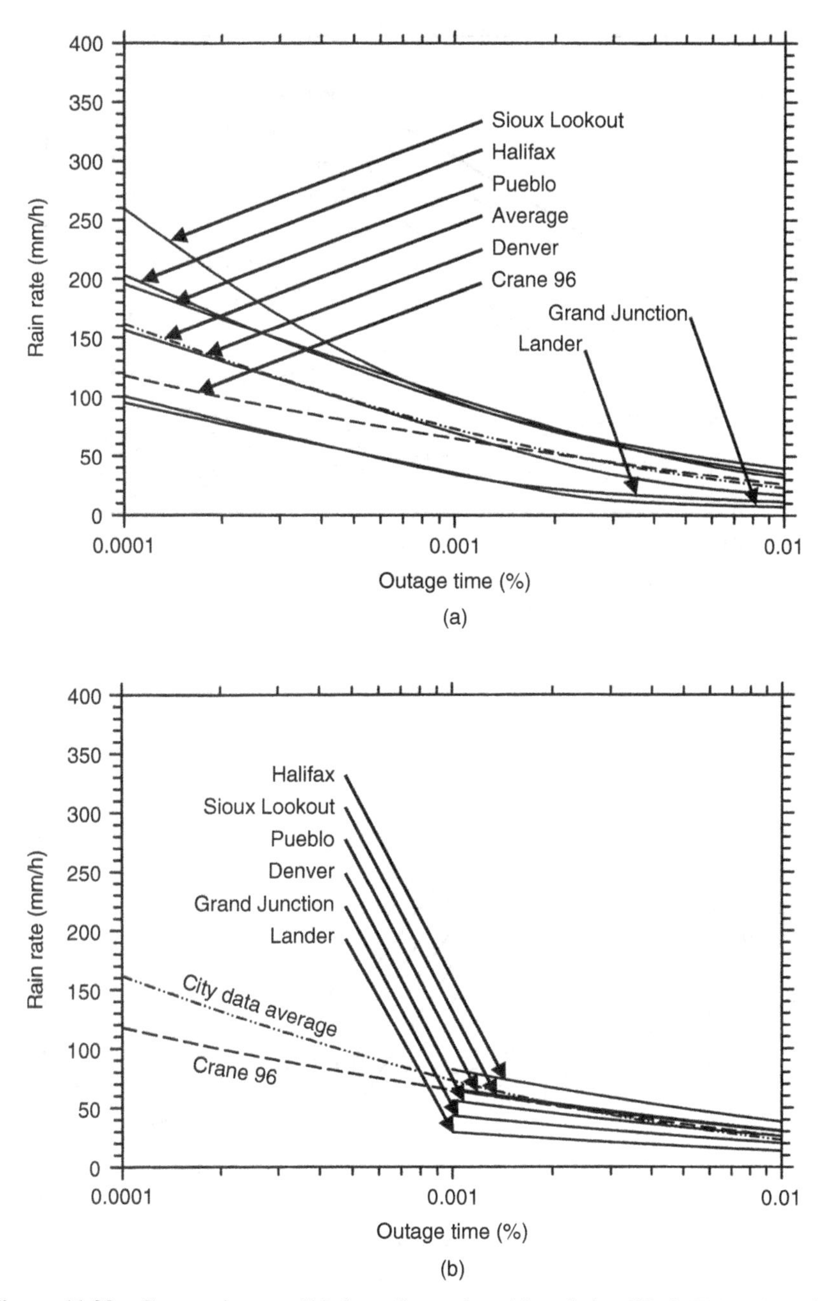

Figure 11.20 Crane rain zone B2 data, Crane data (a) and simplified ITU-R data (b).

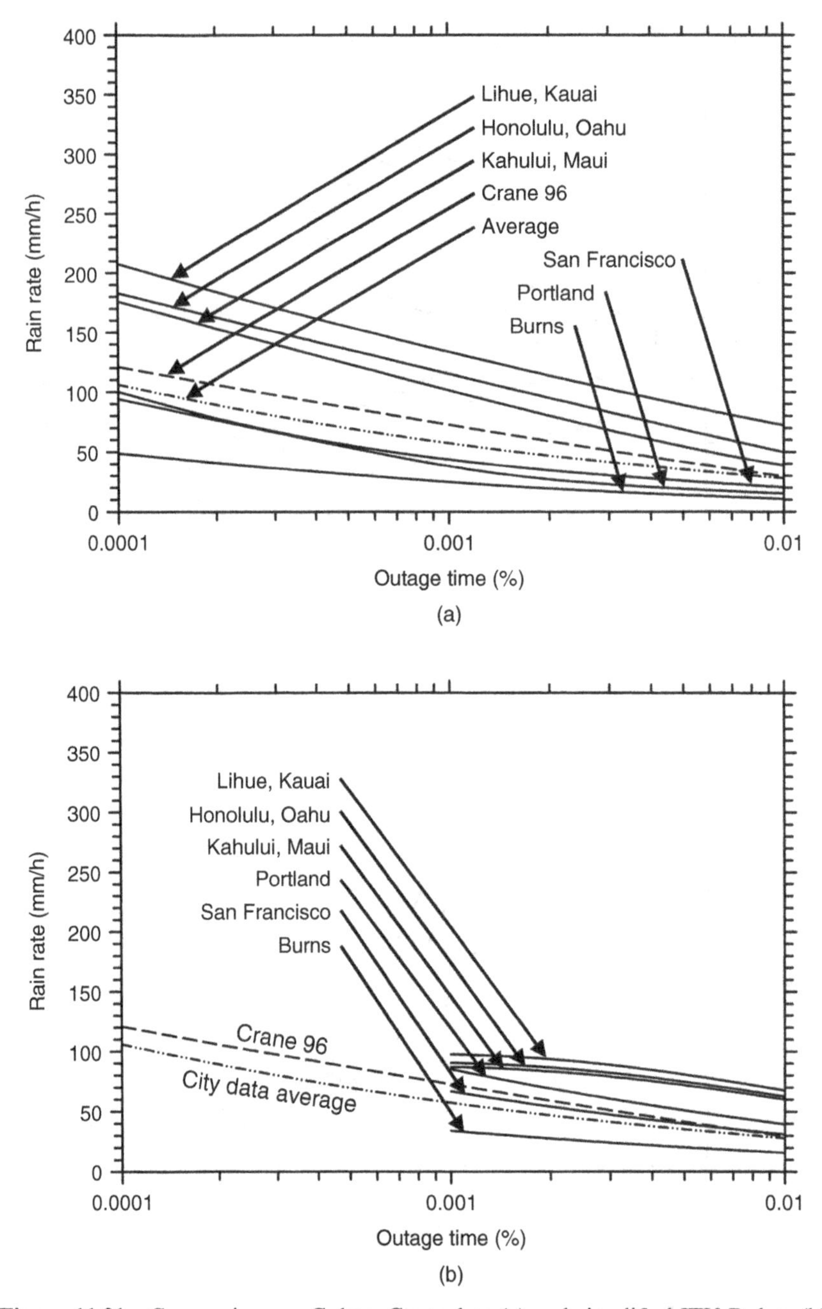

Figure 11.21 Crane rain zone C data, Crane data (a) and simplified ITU-R data (b).

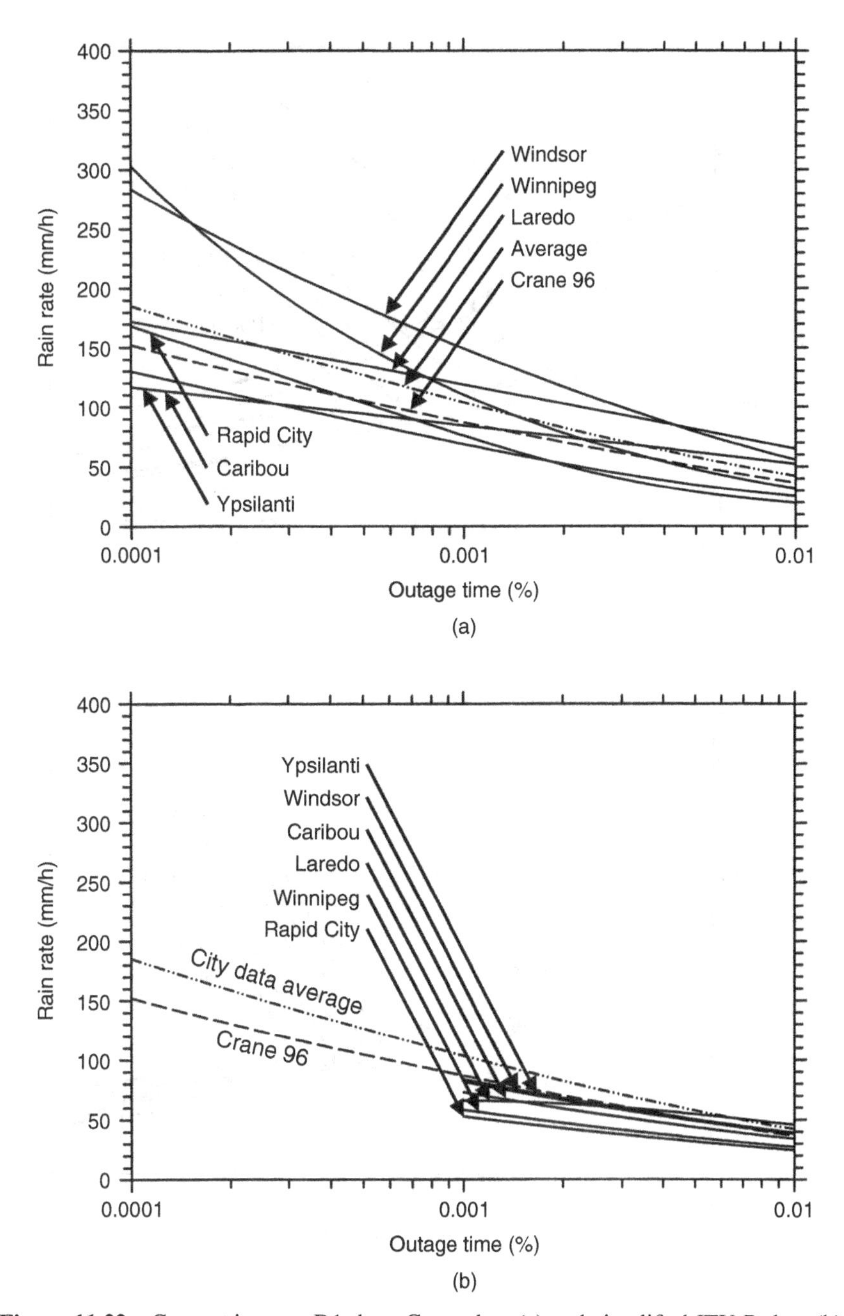

Figure 11.22 Crane rain zone D1 data, Crane data (a) and simplified ITU-R data (b).

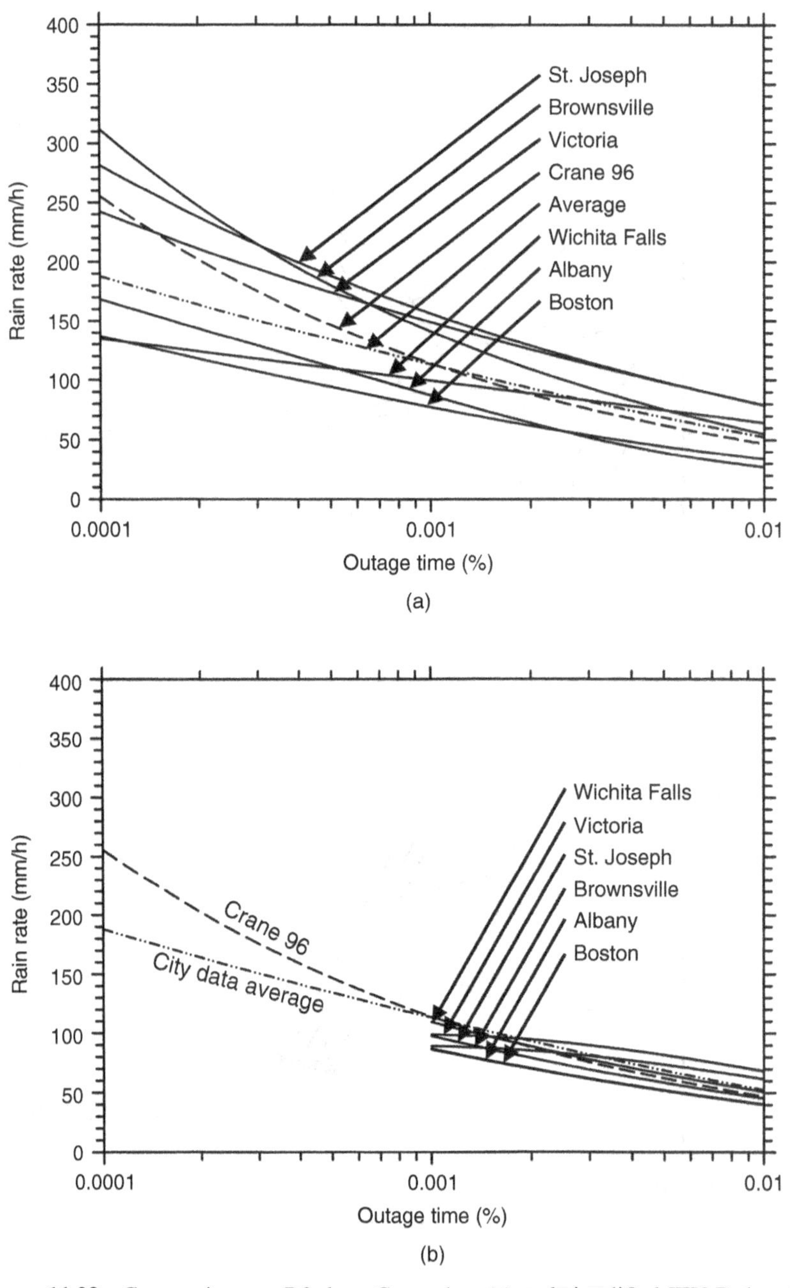

Figure 11.23 Crane rain zone D2 data, Crane data (a) and simplified ITU-R data (b).

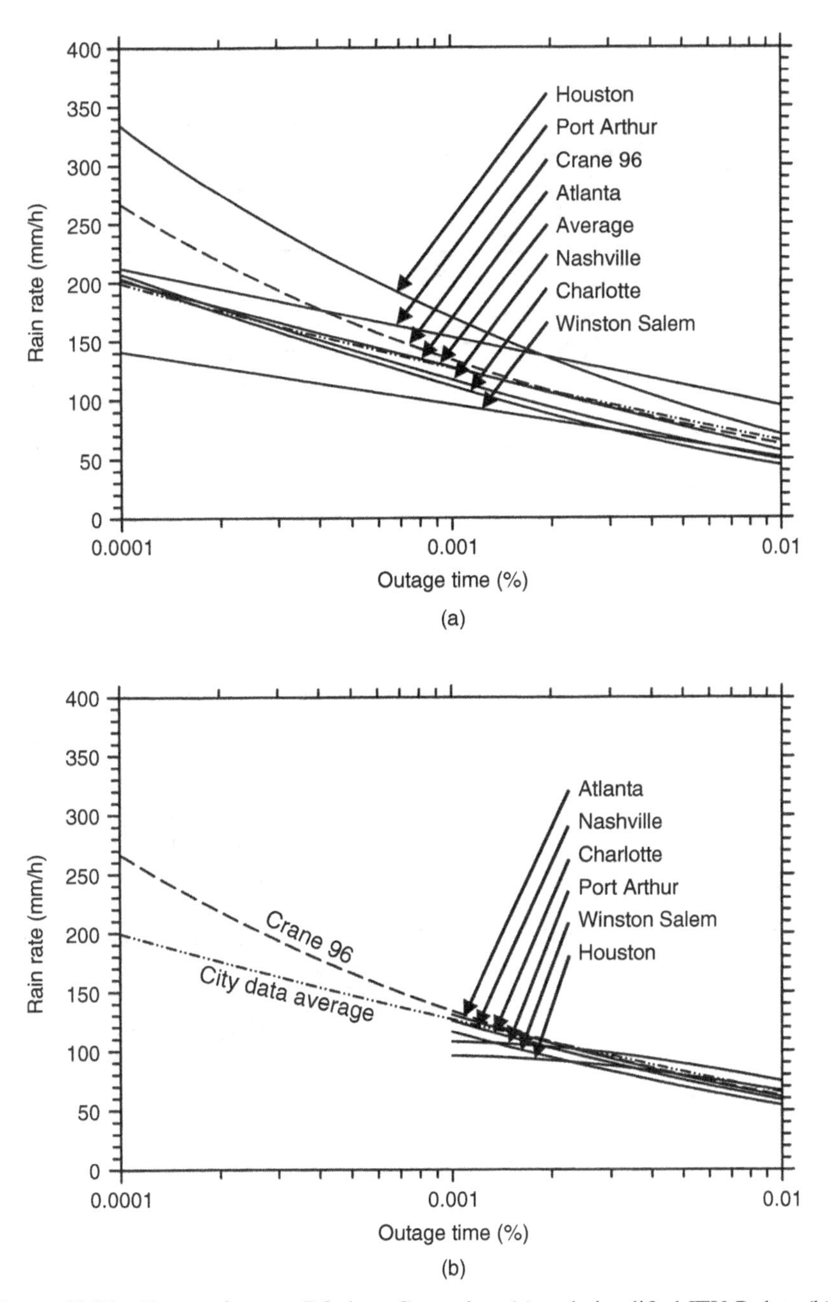

Figure 11.24 Crane rain zone D3 data, Crane data (a) and simplified ITU-R data (b).

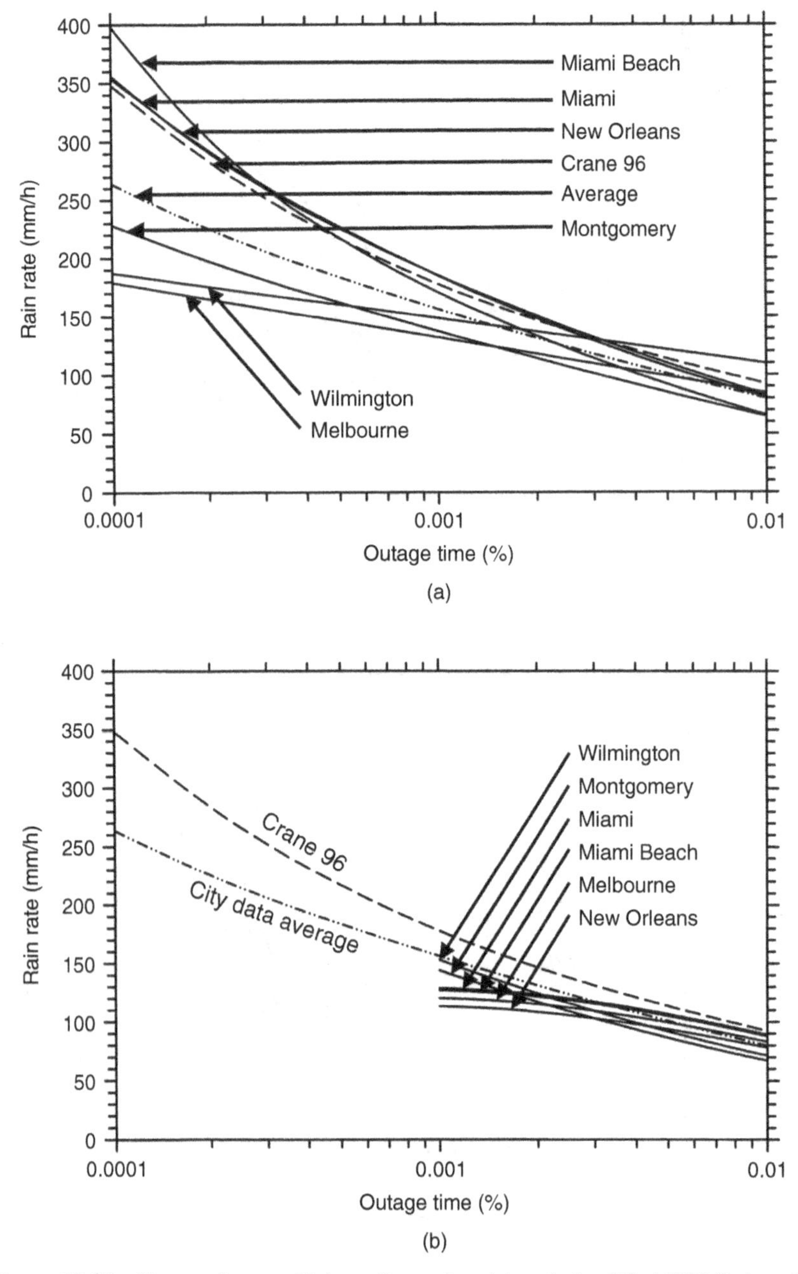

Figure 11.25 Crane rain zone E data, Crane data (a) and simplified ITU-R data (b).

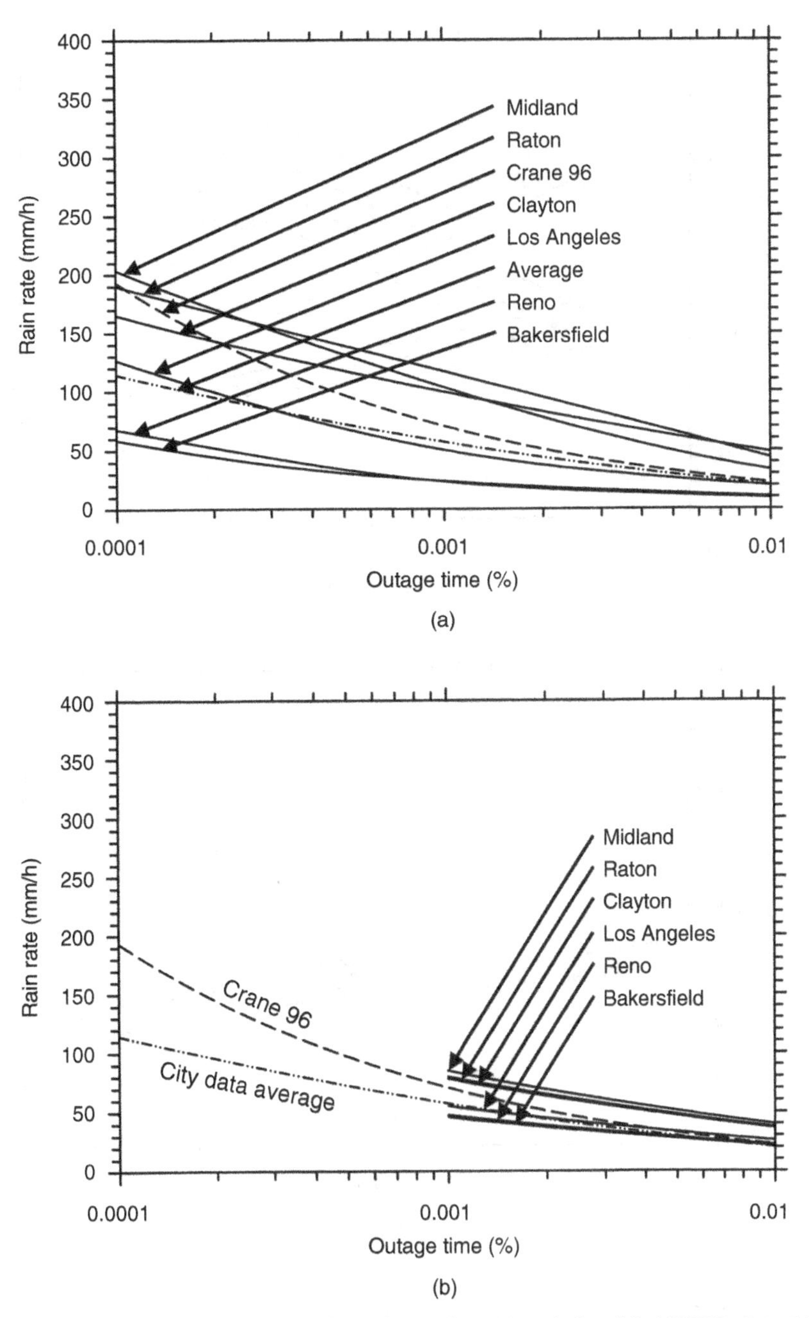

Figure 11.26 Crane rain zone F data, Crane data (a) and simplified ITU-R data (b).

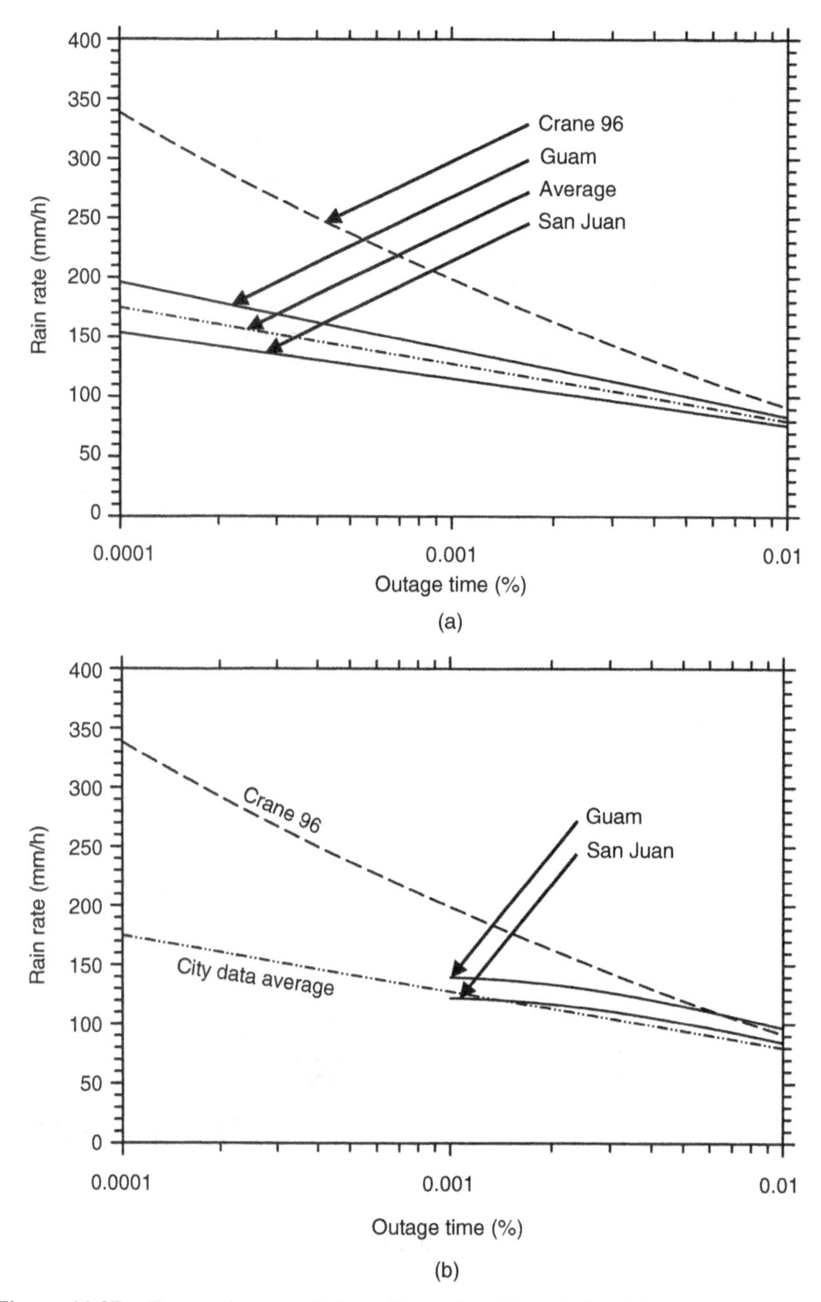

Figure 11.27 Crane rain zone G data, Crane data (a) and simplified ITU-R data (b).

Table 11.6 and Table 11.7 are the Crane model city average and city maximum (highest city value within the zone) rain rates, respectively, for all cities within a given zone.

The average and maximum simplified ITU-R rain rates for all cities with a given zone are given in Table 11.8 and Table 11.9, respectively.

TABLE 11.6 Crane 2003 City Data Average Rain Rates

Average Outage Time, min/yr	Outage Time, %	A	B1	B2	C	C1	C2	D1	D2	D3	E	F	F1	F2	G
		Rain Rate, mm/h													
0.053	0.000010	105.1	159.5	285.0	175.0	153.4	232.4	291.8	282.1	295.5	456.2	186.3	147.6	224.9	222.4
0.074	0.000014	99.0	146.8	263.9	163.5	141.2	222.9	273.6	266.4	278.9	420.2	174.7	137.6	211.7	215.4
0.105	0.000020	92.7	134.3	242.9	151.9	129.0	212.9	255.5	250.6	262.4	385.2	162.8	127.3	198.4	208.1
0.158	0.000030	85.8	121.2	220.5	139.4	115.9	201.8	236.2	233.6	244.9	349.3	149.9	115.9	184.0	199.7
0.210	0.000040	80.9	112.4	205.5	130.9	107.2	194.1	223.2	222.0	233.2	326.1	141.0	107.9	174.1	193.8
0.263	0.000050	77.2	105.9	194.3	124.5	100.7	188.2	213.5	213.4	224.5	309.3	134.3	101.8	166.7	189.2
0.316	0.000060	74.3	100.7	185.4	119.5	95.5	183.3	205.8	206.5	217.6	296.4	128.9	97.0	160.8	185.4
0.368	0.000070	71.8	96.5	178.0	115.3	91.3	179.3	199.4	200.7	211.9	286.0	124.4	92.9	155.9	182.2
0.421	0.000080	69.6	92.8	171.7	111.7	87.7	175.8	194.0	195.8	207.1	277.3	120.5	89.4	151.7	179.5
0.473	0.000090	67.7	89.7	166.3	108.7	84.6	172.7	189.3	191.6	202.9	269.9	117.2	86.3	148.0	177.0
0.526	0.00010	66.0	86.9	161.5	105.9	81.9	170.0	185.1	187.8	199.2	263.5	114.2	83.6	144.8	174.8
0.736	0.00014	60.7	78.5	146.7	97.5	73.6	161.3	172.1	176.0	187.7	244.2	104.8	75.1	134.6	167.9
1.05	0.00020	55.1	70.0	131.6	89.0	65.4	152.1	158.7	163.9	175.9	225.4	95.3	66.5	124.0	160.6
1.58	0.00030	48.9	61.0	115.4	80.0	56.8	141.8	144.1	150.5	163.0	205.9	84.8	57.2	112.4	152.2
2.10	0.00040	44.5	55.2	104.4	74.0	51.2	134.5	134.0	141.3	154.0	193.1	77.7	51.0	104.3	146.3
2.63	0.00050	41.1	50.9	96.2	69.5	47.3	128.9	126.3	134.3	147.2	183.6	72.3	46.5	98.1	141.7
3.16	0.00060	38.4	47.7	89.7	66.1	44.2	124.4	120.2	128.6	141.7	176.1	68.1	43.1	93.2	137.9
3.68	0.00070	36.2	45.0	84.4	63.3	41.8	120.5	115.1	123.9	137.1	169.9	64.7	40.4	89.0	134.7
4.21	0.00080	34.3	42.8	80.0	60.9	39.8	117.2	110.7	119.9	133.2	164.7	61.8	38.1	85.4	132.0
4.73	0.00090	32.6	40.9	76.2	58.9	38.1	114.3	106.9	116.3	129.7	160.2	59.3	36.2	82.3	129.6
5.26	0.0010	31.2	39.3	72.9	57.2	36.7	111.7	103.5	113.2	126.6	156.2	57.1	34.6	79.5	127.4
7.36	0.0014	26.9	34.5	62.8	51.9	32.6	103.5	93.0	103.3	117.0	143.7	50.4	30.0	70.7	120.4
10.5	0.0020	22.9	30.1	53.2	46.8	28.8	94.8	82.3	93.1	106.9	131.1	43.8	25.9	61.7	113.1
15.8	0.0030	19.0	25.8	43.7	41.4	25.0	85.2	70.8	82.0	95.8	117.3	37.0	22.1	51.9	104.7
21.0	0.0040	16.7	23.1	37.8	37.9	22.7	78.4	63.2	74.4	88.1	107.9	32.5	19.8	45.2	98.8
26.3	0.0050	15.0	21.2	33.7	35.3	21.0	73.3	57.6	68.7	82.3	100.7	29.2	18.2	40.2	94.2

(Continued)

TABLE 11.6 ***(Continued)***

Average Outage Time, min/yr	Outage Time, %	A	B1	B2	C	C1	C2	D1	D2	D3	E	F	F1	F2	G
		Rain Rate, mm/h													
31.6	0.0060	13.9	19.8	30.6	33.3	19.8	69.2	53.2	64.2	77.6	95.0	26.6	17.0	36.3	90.4
36.8	0.0070	12.9	18.7	28.1	31.6	18.8	65.7	49.7	60.6	73.8	90.3	24.6	16.0	33.2	87.2
42.1	0.0080	12.2	17.7	26.1	30.2	17.9	62.8	46.8	57.5	70.4	86.3	22.9	15.2	30.5	84.5
47.3	0.0090	11.6	16.9	24.4	29.0	17.2	60.3	44.3	54.8	67.6	82.8	21.4	14.5	28.3	82.0
52.6	0.010	11.1	16.2	22.9	27.9	16.6	58.1	42.1	52.5	65.0	79.8	20.2	14.0	26.4	79.9
73.6	0.014	9.6	14.2	18.8	24.7	14.7	51.2	35.8	45.5	57.2	70.3	16.5	12.3	20.8	72.9
105	0.020	8.3	12.3	15.2	21.5	12.9	44.3	29.8	38.7	49.2	60.8	13.5	10.6	16.3	65.6
158	0.030	7.0	10.3	11.9	18.2	11.1	37.1	24.0	31.9	40.7	50.9	10.6	8.9	12.3	57.2
210	0.040	6.2	9.1	10.0	16.0	9.9	32.3	20.4	27.5	35.1	44.5	8.9	7.9	9.9	51.3
263	0.050	5.6	8.2	8.7	14.4	9.0	28.7	17.8	24.4	30.9	39.8	7.9	7.1	8.7	46.7
316	0.060	5.2	7.6	7.7	13.2	8.4	26.0	15.9	22.1	27.6	36.3	7.1	6.5	7.8	42.9
368	0.070	4.8	7.0	7.0	12.2	7.9	23.7	14.5	20.2	25.0	33.4	6.6	6.0	7.1	39.7
421	0.080	4.5	6.5	6.4	11.3	7.4	21.7	13.4	18.7	22.7	31.1	6.1	5.6	6.5	37.0
473	0.090	4.3	6.2	5.9	10.6	7.1	20.0	12.4	17.5	20.8	29.1	5.7	5.3	6.0	34.6
526	0.10	4.1	5.8	5.5	10.0	6.7	18.6	11.6	16.4	19.2	27.4	5.3	5.0	5.6	32.4
736	0.14	3.5	4.9	4.3	8.0	5.8	14.0	9.4	13.4	14.4	22.2	4.3	4.1	4.5	25.5
1052	0.20	2.9	4.0	3.2	6.4	4.9	10.5	7.5	11.0	10.5	17.3	3.4	3.3	3.4	18.1
1578	0.30	2.3	3.1	2.2	5.1	4.0	7.8	5.7	8.6	7.2	12.7	2.4	2.5	2.4	9.9
2104	0.40	2.0	2.6	1.7	4.2	3.4	6.2	4.7	7.3	5.5	9.9	1.9	2.0	1.8	7.3
2630	0.50	1.7	2.3	1.4	3.6	3.0	5.0	4.0	6.3	4.8	7.9	1.5	1.6	1.3	5.4
3156	0.60	1.6	2.0	1.2	3.1	2.7	4.1	3.5	5.6	4.3	6.6	1.2	1.4	1.1	3.8
3682	0.70	1.4	1.8	1.0	2.8	2.5	3.6	3.1	5.0	3.9	5.8	1.0	1.2	0.9	2.5
4208	0.80	1.3	1.6	0.8	2.6	2.3	3.3	2.7	4.6	3.5	5.2	0.8	1.0	0.7	1.4
4734	0.90	1.2	1.4	0.7	2.4	2.1	3.0	2.5	4.2	3.3	4.8	0.7	0.9	0.6	0.9
5260	1.0	1.1	1.3	0.6	2.2	1.9	2.8	2.2	3.9	3.0	4.4	0.6	0.7	0.4	0.7

TABLE 11.7 Crane 2003 City Data Maximum Rain Rates

Average Outage Time, min/yr	Outage Time, %	A	B1	B2	C	C1	C2	D1	D2	D3	E	F	F1	F2	G
		Rain Rate, mm/h													
0.053	0.000010	132.9	389.5	628.1	300.3	229.8	300.3	724.4	640.0	625.8	960.3	387.4	219.4	387.4	252.2
0.074	0.000014	124.8	352.6	564.6	288.1	204.7	288.1	659.4	592.0	583.3	871.7	359.8	208.6	359.8	245.8
0.105	0.000020	114.0	304.4	482.6	271.4	172.6	271.4	574.3	528.1	526.4	756.6	322.7	193.1	322.7	236.5
0.158	0.000030	103.0	271.7	403.5	253.9	145.5	253.9	490.8	463.7	469.1	644.6	285.2	175.9	285.2	226.2
0.210	0.000040	95.9	250.6	355.0	242.2	134.2	242.2	438.6	422.7	432.5	575.3	261.2	164.0	261.2	218.9
0.263	0.000050	92.2	234.5	323.4	233.4	125.6	233.4	401.7	393.2	406.3	526.5	244.0	154.8	244.0	213.4
0.316	0.000060	89.2	221.6	306.5	226.5	118.8	226.5	373.6	370.5	386.1	489.7	230.7	147.3	230.7	208.8
0.368	0.000070	86.6	210.8	292.4	220.8	113.1	220.8	351.3	352.2	369.8	460.5	221.6	141.1	221.6	205.0
0.421	0.000080	84.4	201.6	280.3	215.9	108.2	215.9	332.9	337.0	356.3	436.6	214.9	135.7	214.9	201.7
0.473	0.000090	82.5	193.6	269.6	211.6	104.0	211.6	317.4	324.1	344.8	416.6	209.0	131.0	209.0	198.8
0.526	0.00010	80.8	186.5	260.2	207.9	100.9	207.9	304.0	312.9	334.9	399.5	203.8	126.8	203.8	196.2
0.736	0.00014	75.2	164.3	230.1	195.9	91.6	195.9	263.8	278.9	304.3	348.7	187.3	113.3	187.3	187.7
1.05	0.00020	69.5	143.0	200.3	183.9	83.8	183.9	237.6	247.3	275.5	303.8	171.0	99.8	171.0	179.0
1.58	0.00030	63.0	121.6	168.6	170.6	76.9	170.6	213.3	215.3	245.5	262.3	155.0	85.2	155.0	169.1
2.10	0.00040	58.4	108.4	148.0	161.5	72.2	161.5	197.0	199.6	225.8	242.0	145.9	75.5	145.9	162.1
2.63	0.00050	54.8	99.3	133.4	154.5	68.8	154.5	184.7	188.3	211.3	227.2	138.8	68.5	138.8	156.6
3.16	0.00060	51.9	92.7	122.4	148.9	66.0	148.9	175.0	179.5	200.0	215.5	133.0	63.1	133.0	152.2
3.68	0.00070	49.5	87.5	113.8	144.2	63.7	144.2	167.0	172.3	190.7	206.0	128.1	58.9	128.1	148.4
4.21	0.00080	47.4	83.3	108.0	140.1	61.8	140.1	160.1	166.2	182.9	198.0	123.9	55.5	123.9	145.1
4.73	0.00090	45.6	79.8	103.3	136.6	60.1	136.6	154.2	161.0	176.1	191.1	120.2	52.7	120.2	142.3
5.26	0.0010	43.9	76.9	99.1	133.5	58.6	133.5	149.0	156.5	170.3	185.3	116.8	50.3	116.8	139.7
7.36	0.0014	38.8	68.4	86.2	123.7	54.1	123.7	132.7	142.6	152.2	167.7	106.1	46.3	106.1	131.5
10.5	0.0020	33.7	60.6	73.2	113.5	49.6	113.5	116.4	128.9	135.6	150.5	94.8	42.6	94.8	122.8
15.8	0.0030	28.3	53.0	61.6	102.4	44.7	102.4	98.9	114.6	125.3	132.2	82.0	38.8	82.0	112.8
21.0	0.0040	24.9	48.2	55.3	94.7	41.5	94.7	87.3	105.2	117.9	125.1	72.8	36.2	72.8	105.8
26.3	0.0050	22.6	44.7	51.0	89.0	39.2	89.0	80.9	98.3	112.3	121.3	65.8	34.3	65.8	100.4

(*Continued*)

TABLE 11.7 (*Continued*)

Average Outage Time, min/yr	Outage Time, %	A	B1	B2	C	C1	C2	D1	D2	D3	E	F	F1	F2	G
		Rain Rate, mm/h													
31.6	0.0060	20.9	42.1	47.8	84.4	37.3	84.4	76.7	93.0	107.6	118.2	60.0	32.8	60.0	95.9
36.8	0.0070	19.6	40.0	45.2	80.7	35.8	80.7	73.1	88.6	103.7	115.6	55.3	31.5	55.3	92.1
42.1	0.0080	18.6	38.2	43.1	77.5	34.5	77.5	70.0	85.1	100.3	113.4	52.7	30.5	52.7	88.9
47.3	0.0090	17.7	36.7	41.3	74.7	33.4	74.7	67.2	82.0	97.3	111.4	50.4	29.6	50.4	86.0
52.6	0.010	17.0	35.4	39.8	72.3	32.4	72.3	64.8	79.3	94.6	109.6	48.5	28.8	48.5	83.4
73.6	0.014	15.0	31.4	35.3	65.2	29.5	65.2	57.0	71.1	86.0	104.0	42.5	26.3	42.5	75.2
105	0.020	13.2	27.6	31.1	59.2	26.5	59.2	48.6	63.0	76.9	98.0	36.6	23.8	36.6	66.5
158	0.030	11.4	23.6	26.8	52.4	23.5	52.4	39.2	54.6	66.5	91.2	30.5	21.0	30.5	57.8
210	0.040	10.4	21.0	24.0	47.6	21.5	47.6	32.9	49.2	59.7	86.3	26.7	19.2	26.7	52.9
263	0.050	9.6	19.1	22.0	43.8	20.0	43.8	29.7	45.2	55.2	82.6	24.0	17.8	24.0	49.2
316	0.060	9.0	17.7	20.4	40.8	18.9	40.8	27.1	42.2	51.6	79.5	21.9	16.7	21.9	46.1
368	0.070	8.5	16.5	19.2	38.2	17.9	38.2	25.1	39.7	48.7	76.9	20.2	15.7	20.2	43.5
421	0.080	8.1	15.5	18.1	35.9	17.1	35.9	23.7	37.7	46.3	74.7	18.8	15.0	18.8	41.3
473	0.090	7.7	14.7	17.2	33.9	16.5	33.9	22.6	35.9	44.3	72.7	17.6	14.3	17.6	39.3
526	0.10	7.4	13.9	16.5	32.2	15.9	32.2	21.5	34.4	42.6	70.9	16.6	13.7	16.6	37.5
736	0.14	6.5	11.7	14.1	27.6	14.1	27.6	18.5	29.9	37.5	65.2	13.7	11.9	13.7	31.9
1052	0.20	5.6	9.5	11.8	23.7	12.3	23.7	15.9	25.6	32.8	59.2	10.9	10.1	10.9	25.9
1578	0.30	4.7	7.2	9.5	19.8	10.5	19.8	13.4	21.5	28.3	52.4	8.2	8.2	8.2	19.0
2104	0.40	4.1	5.7	7.9	17.2	9.3	17.2	11.8	19.4	25.5	47.6	7.0	7.0	6.5	14.2
2630	0.50	3.6	4.9	6.8	15.4	8.5	15.4	10.7	17.8	23.5	43.8	6.1	6.1	5.3	10.4
3156	0.60	3.3	4.3	5.9	14.0	7.8	14.0	9.8	16.5	22.0	40.8	5.5	5.5	4.3	7.4
3682	0.70	3.0	3.8	5.2	12.8	7.3	12.8	9.1	15.5	20.8	38.2	4.9	4.9	3.6	4.8
4208	0.80	2.7	3.4	4.6	11.9	6.8	11.9	8.5	14.7	19.9	35.9	4.5	4.5	3.0	2.6
4734	0.90	2.5	3.1	4.1	11.1	6.4	11.1	8.0	14.0	19.0	33.9	4.1	4.1	2.4	1.5
5260	1.0	2.3	2.9	3.6	10.3	6.1	10.3	7.6	13.4	18.3	32.2	3.8	3.8	1.9	1.2

TABLE 11.8 Simplified ITU-R Average Rain Rates

Average Rain Time, min/yr	Unavailability, %	A	B1	B2	C	C1	C2	D1	D2	D3	E	F	F1	F2	G
		Rain Rate, mm/h													
5.26	0.0010	36.6	61.5	55.8	80.3	74.7	95.3	76.6	97.7	127.2	140.8	61.2	46.0	76.5	130.7
7.36	0.0014	33.2	55.7	50.5	74.8	67.7	93.9	69.6	88.8	115.9	131.3	55.5	41.7	69.2	128.9
10.5	0.0020	29.7	49.9	45.3	69.0	60.7	91.2	62.6	80.0	104.5	121.0	49.7	37.3	62.1	125.2
15.8	0.0030	26.1	43.8	39.7	62.3	53.2	86.5	55.1	70.5	92.3	109.3	43.6	32.8	54.5	118.7
21.0	0.0040	23.7	39.8	36.1	57.6	48.4	82.4	50.1	64.2	84.2	101.1	39.6	29.8	49.5	113.1
26.3	0.0050	21.9	36.8	33.4	54.0	44.8	78.7	46.5	59.6	78.2	94.9	36.7	27.6	45.8	108.1
31.6	0.0060	20.6	34.6	31.4	51.2	42.0	75.6	43.7	56.0	73.5	89.8	34.4	25.9	43.0	103.7
36.8	0.0070	19.5	32.7	29.7	48.8	39.8	72.8	41.4	53.0	69.6	85.6	32.6	24.5	40.7	99.9
42.1	0.0080	18.6	31.2	28.3	46.7	37.9	70.2	39.4	50.6	66.4	82.0	31.0	23.3	38.8	96.4
47.3	0.0090	17.8	29.8	27.1	44.9	36.3	68.0	37.8	48.5	63.7	78.8	29.7	22.3	37.1	93.3
52.6	0.010	17.1	28.7	26.0	43.3	34.9	65.9	36.3	46.6	61.3	76.1	28.6	21.5	35.7	90.4
73.6	0.014	15.0	25.2	22.9	38.5	30.7	59.2	32.0	41.1	54.0	67.5	25.1	18.9	31.4	81.2
105	0.020	13.1	21.9	19.9	33.6	26.7	52.0	27.8	35.7	47.0	59.0	21.8	16.4	27.3	71.4
158	0.030	11.1	18.6	16.9	28.5	22.6	44.1	23.6	30.3	39.8	50.0	18.5	13.9	23.1	60.6
210	0.040	9.8	16.5	15.0	25.1	20.0	38.8	20.9	26.8	35.3	44.1	16.4	12.3	20.5	53.2
263	0.050	8.9	15.0	13.6	22.7	18.2	34.8	19.0	24.4	32.0	39.9	14.9	11.2	18.6	47.8
316	0.060	8.2	13.8	12.6	20.9	16.8	31.8	17.5	22.5	29.5	36.6	13.8	10.4	17.2	43.6
368	0.070	7.7	12.9	11.7	19.4	15.7	29.3	16.3	21.0	27.5	34.1	12.9	9.7	16.0	40.2
421	0.080	7.3	12.2	11.0	18.2	14.8	27.3	15.4	19.7	25.9	31.9	12.1	9.1	15.1	37.4
473	0.090	6.9	11.5	10.5	17.2	14.0	25.5	14.6	18.7	24.5	30.1	11.5	8.6	14.3	35.1
526	0.10	6.6	11.0	10.0	16.3	13.3	24.0	13.9	17.8	23.3	28.5	10.9	8.2	13.7	33.0
736	0.14	5.6	9.4	8.5	13.7	11.4	19.6	11.8	15.2	19.9	24.0	9.4	7.0	11.7	27.0
1052	0.20	4.7	7.9	7.2	11.3	9.6	15.7	10.0	12.7	16.7	19.8	7.9	5.9	9.8	21.5
1578	0.30	3.9	6.5	5.9	9.0	7.9	11.9	8.1	10.4	13.6	15.7	6.5	4.9	8.1	16.3
2104	0.40	3.3	5.6	5.1	7.6	6.8	9.6	7.0	9.0	11.7	13.3	5.6	4.2	7.0	13.2
2630	0.50	3.0	5.0	4.5	6.6	6.1	8.1	6.2	8.0	10.4	11.6	5.0	3.7	6.2	11.2
3156	0.60	2.7	4.5	4.1	5.9	5.5	7.0	5.7	7.2	9.4	10.4	4.5	3.4	5.7	9.7
3682	0.70	2.5	4.2	3.8	5.4	5.1	6.2	5.2	6.6	8.6	9.5	4.2	3.1	5.2	8.6
4208	0.80	2.3	3.9	3.5	5.0	4.7	5.6	4.9	6.2	8.0	8.7	3.9	2.9	4.8	7.7
4734	0.90	2.2	3.7	3.3	4.6	4.4	5.1	4.5	5.8	7.5	8.1	3.6	2.7	4.5	7.0
5260	1.0	2.1	3.5	3.1	4.3	4.2	4.6	4.3	5.5	7.1	7.6	3.4	2.6	4.3	6.4

TABLE 11.9 Simplified ITU-R Maximum Rain Rates

Average Outage Time, min/yr	Outage Time,%	A	B1	B2	C	C1	C2	D1	D2	D3	E	F	F1	F2	G
		Rain Rate, mm/h													
5.26	0.0010	47.4	92.9	82.1	105.8	105.8	104.2	98.9	119.8	159.6	161.1	94.0	72.8	94.0	139.4
7.36	0.0014	42.9	84.1	74.4	102.8	95.8	102.8	89.6	108.5	144.6	145.9	85.1	66.0	85.1	137.5
10.5	0.0020	38.5	75.4	66.7	99.8	85.9	99.8	80.3	97.2	129.6	130.8	76.3	59.1	76.3	133.5
15.8	0.0030	33.8	66.2	58.5	94.7	75.4	94.7	70.5	89.7	113.7	116.9	66.9	51.9	66.9	126.6
21.0	0.0040	30.7	60.1	53.1	90.2	68.4	90.2	64.0	85.4	103.3	111.3	60.8	47.1	60.8	120.6
26.3	0.0050	28.4	55.7	49.2	86.2	63.4	86.2	59.3	81.7	95.7	106.4	56.3	43.7	56.3	115.3
31.6	0.0060	26.7	52.2	46.2	82.7	59.5	82.7	55.6	78.4	89.7	102.1	52.8	41.0	52.8	110.6
36.8	0.0070	25.2	49.4	43.7	79.6	56.3	79.6	52.6	75.4	84.9	98.3	50.0	38.8	50.0	106.5
42.1	0.0080	24.0	47.1	41.6	76.9	53.6	76.9	50.1	72.8	80.9	94.9	47.6	36.9	47.6	102.8
47.3	0.0090	23.0	45.1	39.8	74.4	51.3	74.4	48.0	70.5	77.5	91.8	45.6	35.4	45.6	99.5
52.6	0.010	22.1	43.3	38.3	72.1	49.4	72.1	46.2	68.3	74.5	89.0	43.8	34.0	43.8	96.4
73.6	0.014	19.5	38.1	33.7	64.8	43.4	64.8	41.0	61.4	66.9	80.0	38.6	29.9	38.6	86.6
105	0.020	16.9	33.2	29.3	56.9	37.7	56.9	36.1	54.0	58.8	70.3	33.5	26.0	33.5	76.1
158	0.030	14.3	28.1	24.8	48.3	32.0	48.3	30.6	45.7	49.9	59.6	28.4	22.0	28.4	64.6
210	0.040	12.7	24.9	22.0	42.4	28.4	42.4	26.9	40.2	43.8	52.4	25.2	19.5	25.2	56.7
263	0.050	11.5	22.6	20.0	38.1	25.8	38.1	24.2	36.1	39.4	47.1	22.9	17.7	22.9	51.0
316	0.060	10.7	20.9	18.5	34.8	23.8	34.8	22.2	32.9	35.9	42.9	21.1	16.4	21.1	46.5
368	0.070	10.0	19.5	17.2	32.1	22.2	32.1	20.8	30.4	33.5	39.6	19.7	15.3	19.7	42.9
421	0.080	9.4	18.4	16.2	29.8	20.9	29.8	19.6	28.3	31.6	36.8	18.6	14.4	18.6	39.9
473	0.090	8.9	17.4	15.4	27.9	19.8	27.9	18.5	26.5	29.9	34.5	17.6	13.7	17.6	37.4
526	0.10	8.5	16.6	14.7	26.3	18.9	26.3	17.7	24.9	28.5	32.5	16.8	13.0	16.8	35.2
736	0.14	7.2	14.2	12.5	21.5	16.2	21.5	15.1	20.4	24.4	26.6	14.3	11.1	14.3	28.8
1052	0.20	6.1	12.0	10.6	17.1	13.6	17.1	12.7	16.2	20.5	21.1	12.1	9.4	12.1	22.9
1578	0.30	5.0	9.8	8.7	13.0	11.1	13.0	10.4	12.6	16.8	17.0	9.9	7.7	9.9	17.4
2104	0.40	4.3	8.5	7.5	10.5	9.6	10.5	9.0	10.9	14.5	14.7	8.6	6.6	8.6	14.1
2630	0.50	3.8	7.5	6.7	8.9	8.6	8.9	8.0	9.7	13.0	13.1	7.6	5.9	7.6	11.9
3156	0.60	3.5	6.9	6.1	7.8	7.8	7.7	7.3	8.8	11.8	11.9	6.9	5.4	6.9	10.3
3682	0.70	3.2	6.3	5.6	7.2	7.2	6.8	6.7	8.1	10.9	11.0	6.4	5.0	6.4	9.1
4208	0.80	3.0	5.9	5.2	6.7	6.7	6.1	6.3	7.6	10.1	10.2	5.9	4.6	5.9	8.2
4734	0.90	2.8	5.5	4.9	6.3	6.3	5.5	5.9	7.1	9.5	9.6	5.6	4.3	5.6	7.4
5260	1.0	2.7	5.2	4.6	5.9	5.9	5.1	5.5	6.7	9.0	9.0	5.3	4.1	5.3	6.8

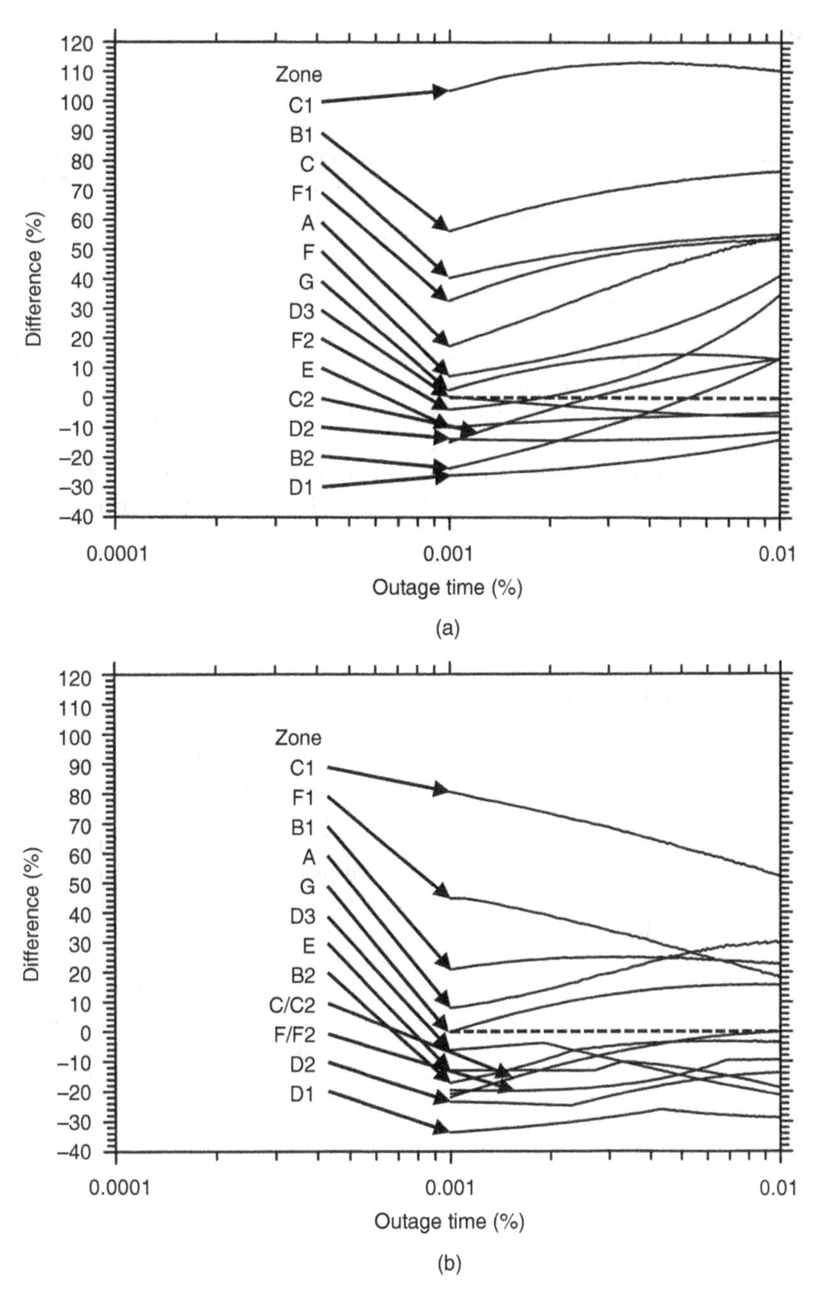

Figure 11.28 Differences in the Crane data and simplified ITU-R average (a) and maximum (b) rain rates.

We may compare the difference of the zone averages or maximum values in the charts in Figure 11.28. The average (a) curve graphs 100 (ITU-R average—city data average)/(city data average). The maximum (b) curve graphs 100 (ITU-R maximum—city data maximum)/(city data maximum). Clearly, the ITU-R data tends to overestimate rain rates on the western coast of the United States.

11.6 NEW RAIN ZONES

A typical approach of high frequency microwave path design is to use the Crane 96 rain zone rain rates for any location within the zone. However, this introduces error that varies from city to city (Fig. 11.29).

For example, in zone D2 (Fig. 11.29a) using the zone rain rate to calculate point outage rate of 0.001% will be about right for Kansas City but will overestimate outage time in Boston by a factor of 4 and underestimate it in St. Joseph by a factor of 3. In zone C, (Fig. 11.29b) using the zone rain rate for 0.001% outage overestimates outage for Burn by a factor of 20 but underestimates outage for a site in Kauai by a factor of 20. The excessive estimation error in zones C and F will be addressed below. The error in using this simplified "one-size-fits-all" zone approach rather than using a specific location rain rate can be calculated for the cities in Appendix 11.A (See Table 11.10).

All rain zones could be subdivided to improve zone-wide error performance. However, two Crane 96 rain zones, C and F, are of particular interest primarily because of the large number of high frequency paths designed on the west coast of the United States. These zones have significantly large variation between site-specific rain rates and zone rain rates. Zone C covers the Pacific Islands as well as the northwestern coast of the United States. It overestimates rain on the US coast and underestimates the rain in the Pacific Islands. Zone F covers the lower western coast of the United States as well as west Texas. It underestimates Texas rain and overestimates west coast rain. Using the F and C zone rates invariably causes western US coast paths to be overdesigned. By subdividing the F and C zones as depicted earlier, the error in using the zone rain rate is significantly reduced and more economical designs result (Fig. 11.30, Fig. 11.31, Fig. 11.32, and Fig. 11.33).

11.7 WORST-MONTH RAIN RATES

For ITU-R style quality (error performance) calculations, worst-case month performance is calculated. Rain outages usually last longer than 10 s, so rain performance contributes to yearly availability calculations, not the monthly quality calculations. However, some engineers desire average worst-case monthly rain rates. Until now, that data was not available. Crane (2003) provided average annual and monthly rain rate parameters for 109 US cities. Using the data and the Crane two-component rain rate model, monthly and annual rate rates were calculated for the cities. The worst monthly rain rate was located for each rain rate value (different months were worst case for different rain rates). The ratio of average worst month to annual rain rates were calculated by city and then merged by Crane zone (Fig. 11.34).

The noisy results to the right of the figure are due to low rain rates and rain rate calculations only carried to 0.1 mm/h. For unavailabilities greater than approximately 0.05%, the city-to-city ratio variability increases considerably. Calculated values are given in Table 11.11.

The number of cities in each zone, whose rain data has been presented, is rather small. A more statistically significant result is obtaining by merging all the rain data. That results in the "all zones" data. The following equation is a curve fit to the "all zones" data:

$$
\begin{aligned}
Y &= \frac{A + BX + CX^2 + DX^3 + EX^4 + FX^5}{1 + GX + HX^2 + JX^3 + KX^4 + LX^5} \\
A &= 1.18867960530659 \\
B &= 1748.639400422324 \\
C &= 19665.34928770431 \\
D &= 74175.75060507632 \\
E &= -20874.54217398086 \\
F &= -24559.47954713423 \\
G &= 1066.250147796061 \\
H &= 22242.85620371334
\end{aligned}
\tag{11.20}
$$

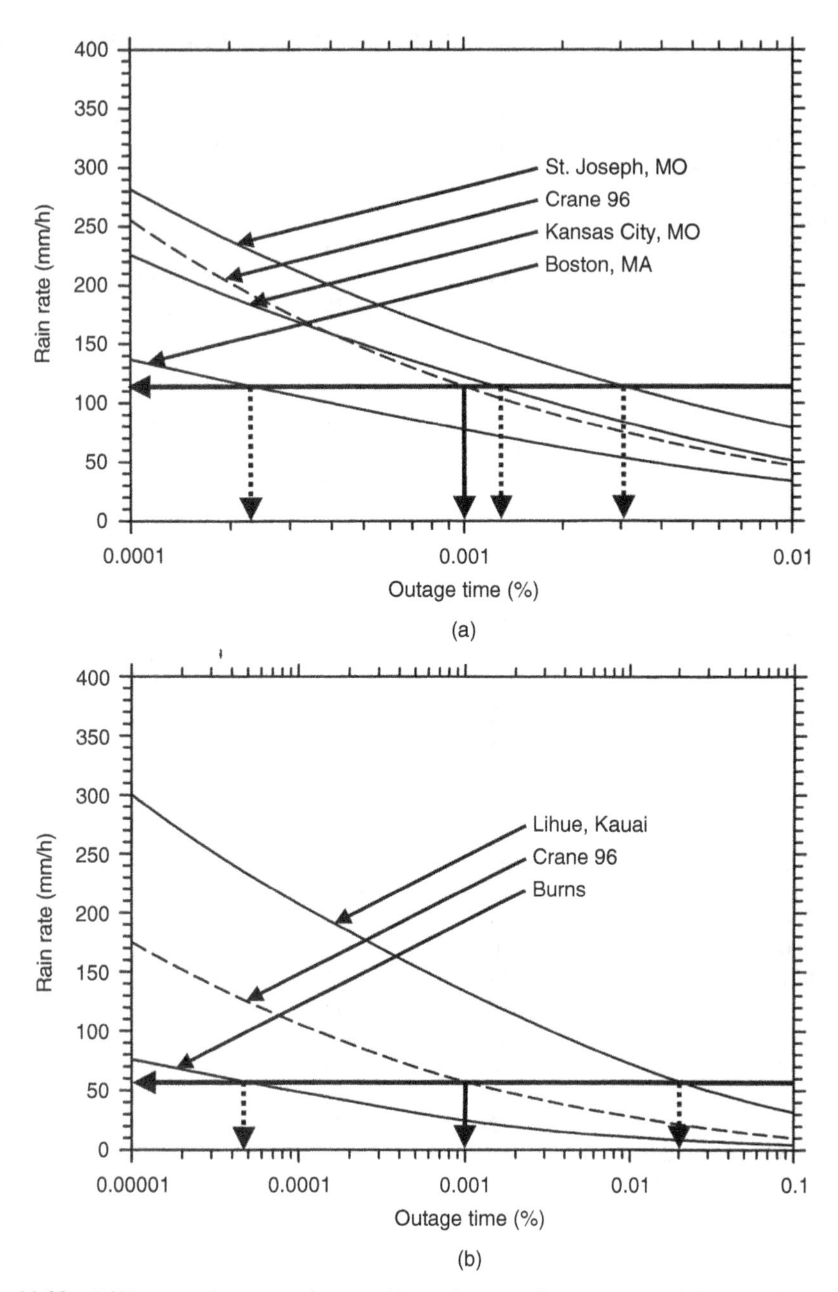

Figure 11.29 Differences between city-specific and zone rain rates, zone D2 (a) and zone C (b).

$$J = 9932.63602429962$$

$$K = 23961.37871718089$$

$$L = -37527.65786180049$$

Y = worst month rain rate/yearly average rain rate;
X = unavailability (0.0001–1%).

TABLE 11.10 Upper and Lower Bounds of Differences Between Zone Rain Rates and Crane Rain Rates

Zone	Outage Time, %	Minimum Difference, %	Maximum Difference, %	Lower Average Difference, %	Upper Average Difference, %	Average Difference, %
A	0.0001	−29	14	−18	10	−7
	0.00032	−26	35	−20	21	4
	0.001	−48	51	−24	28	7
	0.0032	−59	56	−29	28	5
	0.01	−55	67	−55	24	8
B1	0.0001	−52	97	−26	26	−8
	0.00032	−6	86	−25	25	−6
	0.001	−46	79	−32	20	−8
	0.0032	−55	90	−31	24	−8
	0.01	−52	119	−23	39	0
B2	0.0001	−19	121	−9	48	37
	0.00032	−27	85	−14	37	28
	0.001	−47	53	−19	27	13
	0.0032	−73	38	−25	23	−2
	0.01	−72	54	−28	25	−11
C	0.0001	−60	72	−32	41	−12
	0.00032	−63	76	−42	46	−18
	0.001	−73	84	−49	54	−21
	0.0032	−80	102	−51	68	−18
	0.01	−84	145	−47	84	−5
C1	0.0001	−41	21	−16	13	−2
	0.00032	−33	41	−13	25	3
	0.001	−43	71	−22	25	7
	0.0032	−55	96	−21	39	9
	0.01	−70	115	−26	46	10
C2	0.0001	−12	23	−10	12	1
	0.00032	−11	20	−7	13	0
	0.001	−10	18	−7	10	−1
	0.0032	−21	19	−11	9	−1
	0.01	−32	27	−16	21	2
D1	0.0001	−23	100	−11	29	22
	0.00032	−16	79	−10	27	21
	0.001	−21	72	−11	25	19
	0.0032	−37	62	−13	25	16
	0.01	−46	78	−16	31	16
D2	0.0001	−47	23	−28	11	−26
	0.00032	−38	23	−17	10	−14
	0.001	−32	37	−11	11	−1
	0.0032	−32	52	−10	18	9
	0.01	−42	69	−13	28	12
D3	0.0001	−47	26	−27	25	−25
	0.00032	−38	27	−17	27	−15
	0.001	−28	27	−11	10	−5
	0.0032	−21	34	−9	14	2
	0.01	−28	53	−11	21	5
E	0.0001	−48	15	−31	6	−24
	0.00032	−37	4	−23	4	−18
	0.001	−26	4	−17	3	−12
	0.0032	−23	1	−13	0	−10
	0.01	−30	19	−17	12	−14

TABLE 11.10 (*Continued*)

Zone	Outage Time, %	Minimum Difference, %	Maximum Difference, %	Lower Average Difference, %	Upper Average Difference, %	Average Difference, %
F	0.0001	−69	6	−44	4	−41
	0.00032	−69	30	−40	20	−29
	0.001	−67	66	−39	31	−19
	0.0032	−65	99	−37	46	−11
	0.01	−57	117	−37	61	−9
F!	0.0001	−30	51	−17	21	−1
	0.00032	−32	55	−15	29	4
	0.001	−32	48	−16	26	2
	0.0032	−37	71	−20	38	−4
	0.01	−37	91	−26	39	−7
F2	0.0001	−28	44	−14	32	2
	0.00032	−30	37	−16	27	−1
	0.001	−33	42	−20	26	−3
	0.0032	−40	52	−24	32	−4
	0.01	−51	110	−31	59	14
G	0.0001	−55	−42	−48	0	−48
	0.00032	−49	−36	−43	0	−43
	0.001	−42	−30	−36	0	−36
	0.0032	−32	−21	−26	0	−26
	0.01	−16	−9	−13	0	−13

In most cases, only one month is the worst month. It usually occurs in the summer but often it is in the winter on the west coast. Generally, regions with the lowest average rain rates have the highest monthly to annual rain rate ratios. Also, the monthly to annual ratio tends to be lower for very high rain rates but is relatively high for low rain rates.

Generalizations regarding the occurrence of the worst month are difficult for a given city, as the site-to-site variation is great. The following observations may be made for merged data.

All US Rain Zones

Any month may be the worst month.

February can be the worst month only for $1 \geq U\ (\%) \geq 0.2$ (low rain rates).

The worst month is never October for $0.02 \geq U\ (\%) \geq 0.004$ (moderate rain rates).

US Rain Zone B1

The worst months are November or December and May, June, July, or August.

US Rain Zone B2

Worst months are June and July for $0.025 \geq U\ (\%) \geq 0.0001$.

Worst months are May and June for $0.13 \geq U\ (\%) \geq 0.025$.

Worst months are March and May for $1.0 \geq U\ (\%) \geq 0.13$.

Worst months are often March and October for $1.0 \geq U\ (\%) \geq 0.03$.

US Rain Zone C

The worst months are January, May, November, and December.

The worst months are March, August and October for $0.0024 \geq U\ (\%) \geq 0.0001$.

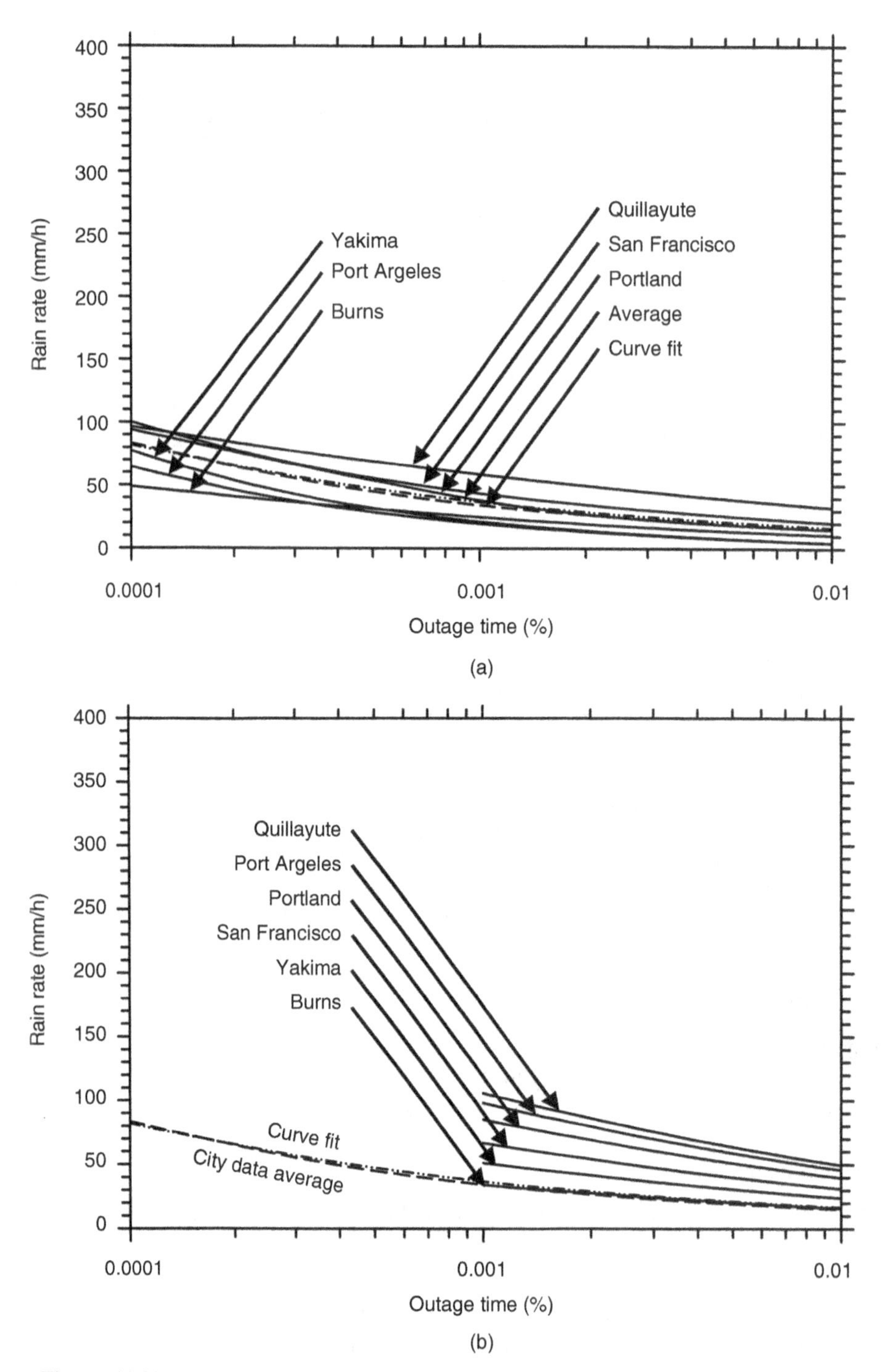

Figure 11.30 New rain zone C1 data, Crane data (a) and simplified ITU-R data (b).

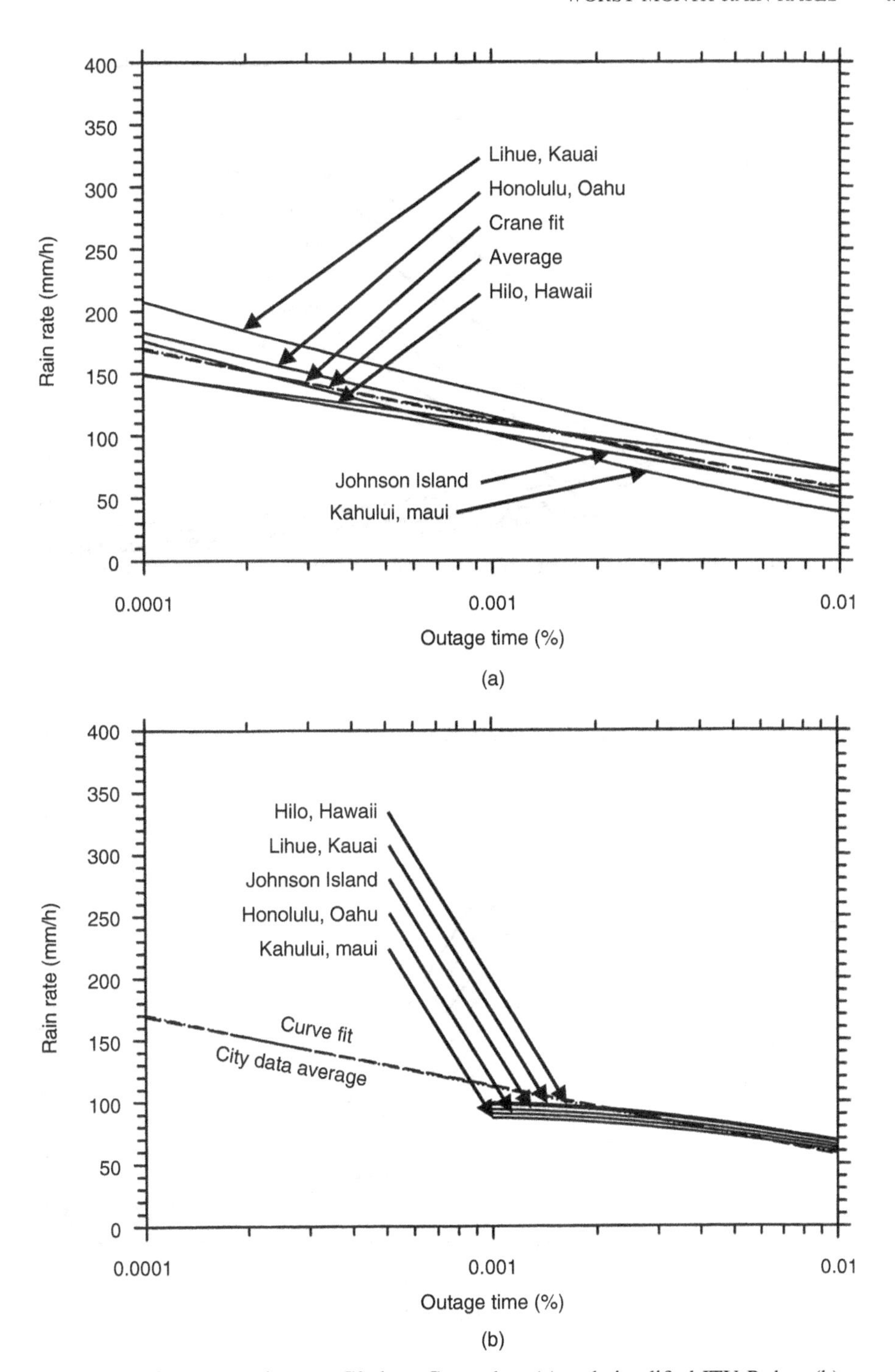

Figure 11.31 New rain zone C2 data, Crane data (a) and simplified ITU-R data (b).

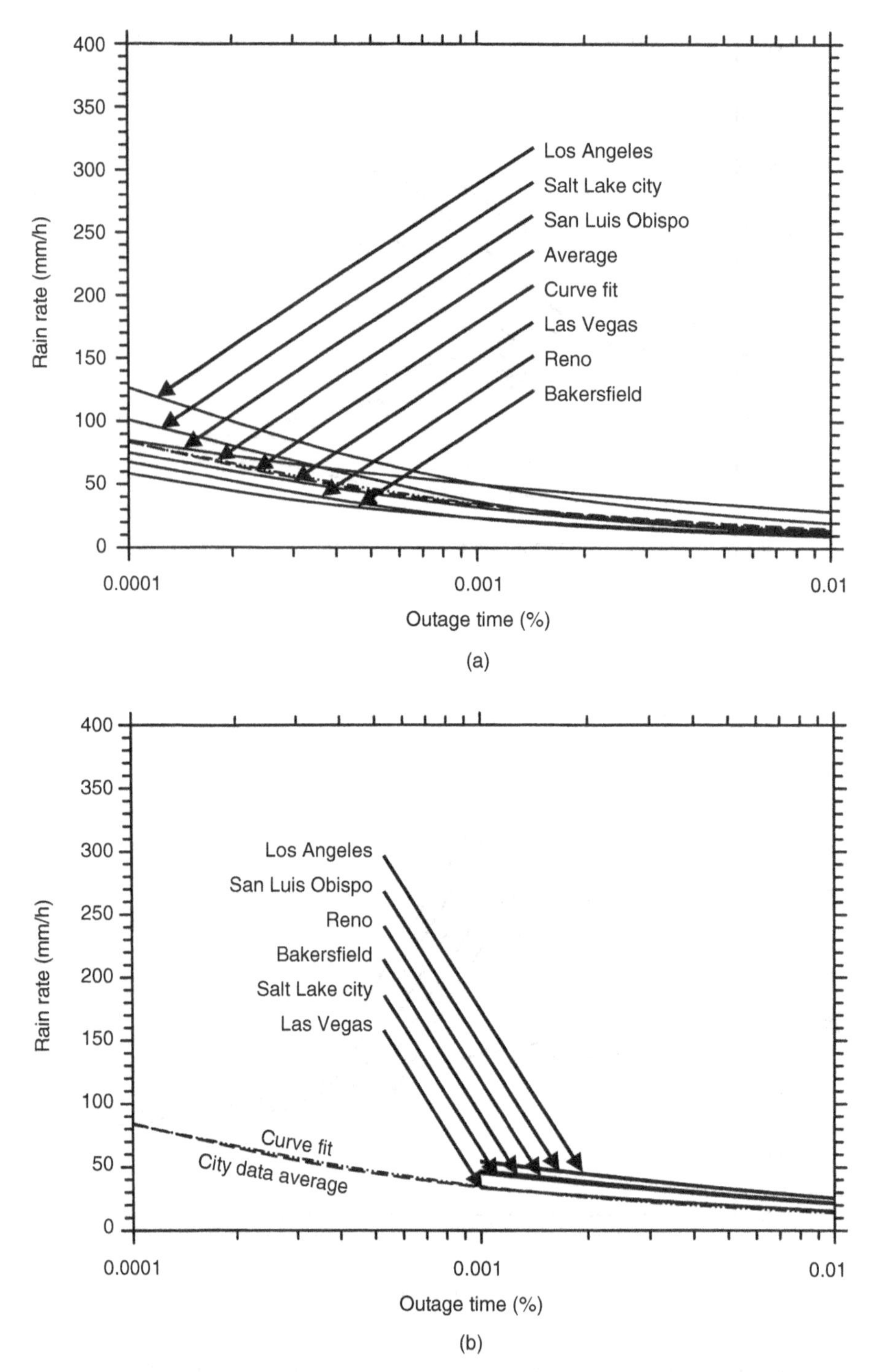

Figure 11.32 New rain zone F1data, Crane data (a) and simplified ITU-R data (b).

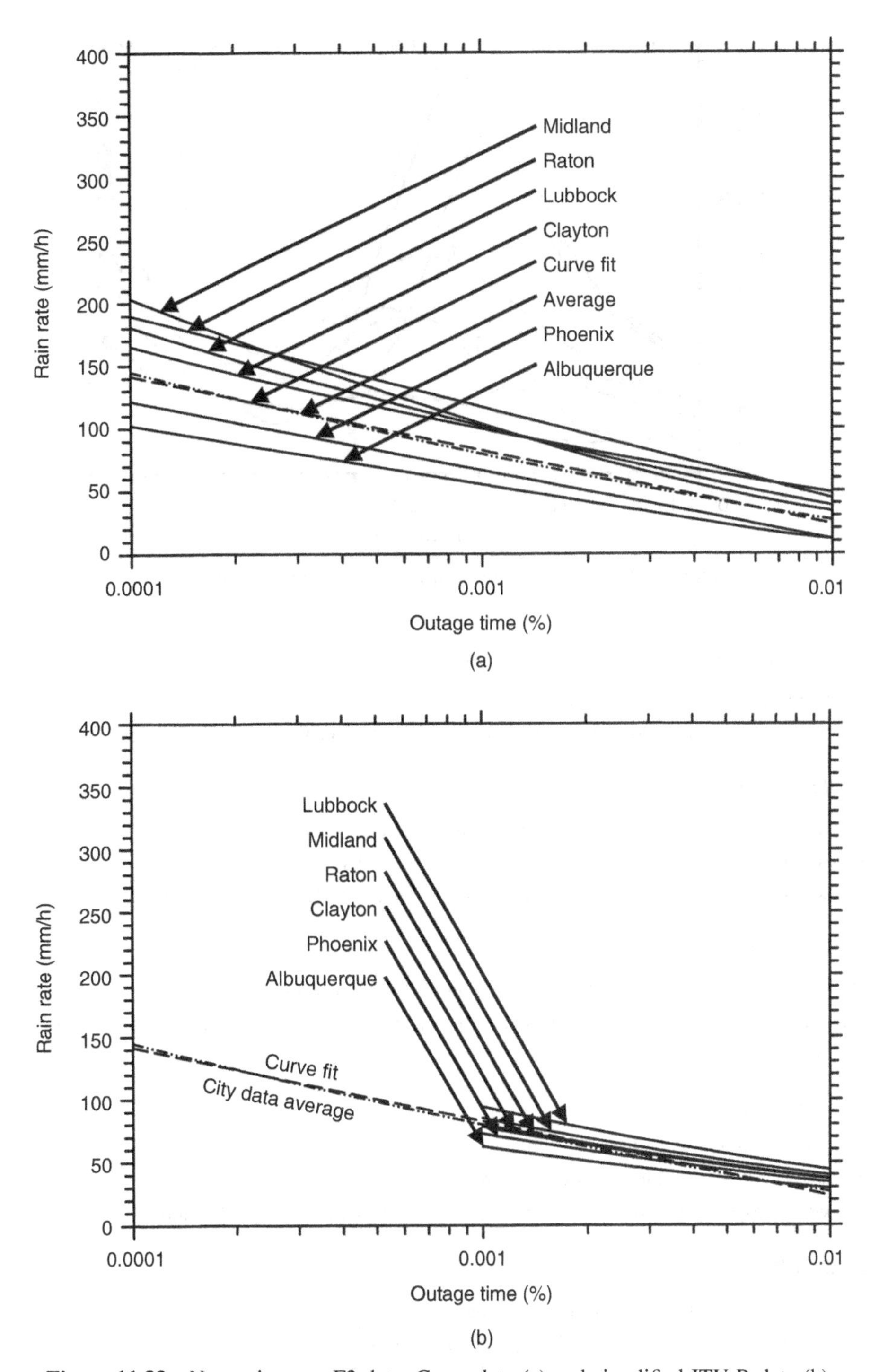

Figure 11.33 New rain zone F2 data, Crane data (a) and simplified ITU-R data (b).

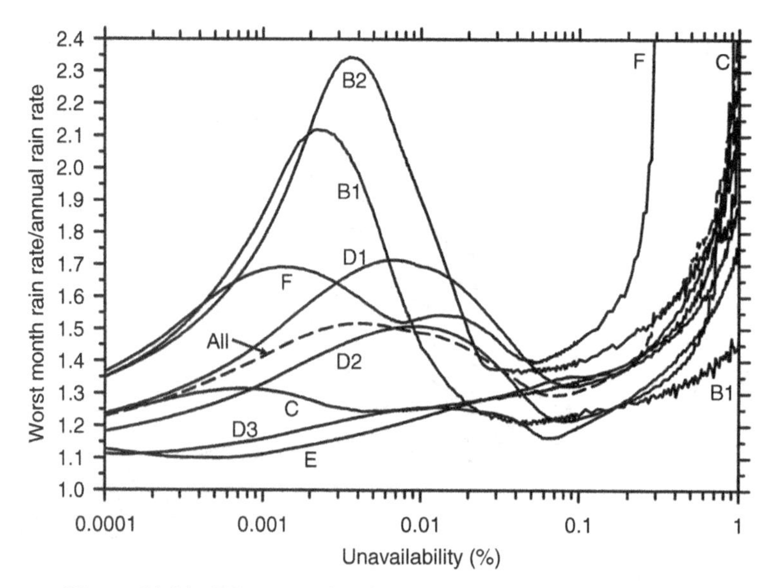

Figure 11.34 Worst-month rain rate divided by annual rain rate.

US Rain Zone D1

The worst months are June and July.

The worst months are April, May, September, and November for $1.0 \geq U\ (\%) \geq 0.07$.

US Rain Zone D2

The worst months are May, June, August, and September.

The worst month is also July for $0.088 \geq U\ (\%) \geq 0.0001$.

The worst months are also March, April, and November for $1.0 \geq U\ (\%) \geq 0.06$.

US Rain Zone D3

The worst months are March, April, or May and July, August, or September.

The worst months are also January, February, March, September, and December for $1.0 \geq U\ (\%) \geq 0.2$.

US Rain Zone E

A worst month is always July.

A worst month is also June for $1.0 \geq U\ (\%) \geq 0.0002$.

The worst months are also March or April and August, September, or December.

US Rain Zone F

The worst months are March, May, and August.

A worst month is June except for $0.7 \geq U\ (\%) \geq 0.044$.

A worst month is July except for $0.035 \geq U\ (\%) \geq 0.022$.

A worst month is January except for $0.00023 \geq U\ (\%) \geq 0.0001$.

For the above U (%) is the probability of the rain rate being exceeded [$P(r \geq R)$]. What constitutes a "worst month" varies with outage probability U (%). A rain zone may have several "worst months" if all those months have the same rain statistics for a given U (%).

TABLE 11.11 Worst Month Rain Rate Divided by Annual Rain Rate

Outage Minutes	Unavailability, %	All Zones	B1	B2	C	D1	D2	D3	E	F
0.53	0.00010	1.229	1.350	1.349	1.234	1.236	1.182	1.113	1.128	1.365
0.74	0.00014	1.247	1.384	1.381	1.251	1.257	1.196	1.111	1.118	1.404
1.05	0.00020	1.269	1.432	1.422	1.269	1.281	1.213	1.115	1.109	1.452
1.58	0.00030	1.299	1.501	1.479	1.289	1.314	1.235	1.122	1.102	1.519
2.10	0.00040	1.324	1.564	1.530	1.301	1.341	1.252	1.130	1.101	1.568
2.63	0.00050	1.344	1.622	1.576	1.308	1.365	1.268	1.135	1.101	1.605
3.16	0.00060	1.361	1.675	1.619	1.312	1.387	1.282	1.141	1.101	1.631
3.68	0.00070	1.377	1.725	1.661	1.314	1.407	1.295	1.146	1.103	1.652
4.21	0.00080	1.390	1.772	1.702	1.313	1.427	1.307	1.151	1.106	1.665
4.73	0.00090	1.402	1.815	1.741	1.312	1.445	1.318	1.155	1.109	1.676
5.26	0.00100	1.413	1.858	1.779	1.311	1.462	1.329	1.159	1.112	1.683
7.36	0.00140	1.450	2.005	1.925	1.300	1.520	1.365	1.174	1.126	1.691
10.52	0.00200	1.484	2.108	2.125	1.280	1.583	1.404	1.192	1.140	1.679
15.78	0.00300	1.511	2.085	2.320	1.256	1.647	1.446	1.212	1.158	1.638
21.04	0.00400	1.518	1.984	2.337	1.246	1.684	1.471	1.226	1.173	1.600
26.30	0.00500	1.516	1.861	2.280	1.247	1.705	1.488	1.235	1.185	1.566
31.56	0.00600	1.509	1.744	2.196	1.246	1.713	1.498	1.242	1.195	1.539
36.82	0.00700	1.502	1.641	2.105	1.247	1.715	1.504	1.248	1.203	1.524
42.08	0.00800	1.497	1.566	2.032	1.249	1.710	1.507	1.251	1.211	1.521
47.34	0.00900	1.492	1.504	1.964	1.251	1.704	1.508	1.254	1.219	1.526
52.60	0.01000	1.488	1.444	1.903	1.252	1.694	1.508	1.255	1.225	1.533
73.63	0.01400	1.470	1.346	1.710	1.263	1.668	1.495	1.256	1.247	1.542
105.19	0.02000	1.437	1.267	1.491	1.275	1.610	1.460	1.249	1.268	1.527
157.79	0.03000	1.384	1.231	1.374	1.288	1.514	1.387	1.237	1.290	1.468
210.38	0.04000	1.342	1.212	1.373	1.296	1.439	1.320	1.217	1.306	1.418
262.98	0.05000	1.314	1.207	1.375	1.305	1.386	1.270	1.186	1.318	1.403
315.58	0.06000	1.298	1.210	1.379	1.311	1.352	1.236	1.164	1.334	1.405
368.17	0.07000	1.295	1.221	1.380	1.318	1.334	1.220	1.164	1.343	1.418
420.77	0.08000	1.296	1.226	1.389	1.327	1.323	1.214	1.174	1.349	1.431
473.36	0.09000	1.302	1.237	1.389	1.334	1.321	1.217	1.187	1.352	1.449
525.96	0.10000	1.308	1.232	1.406	1.339	1.323	1.223	1.196	1.352	1.464
	Cities in zone data	107	4	5	11	19	32	10	9	17

11.8 POINT RAIN RATE VARIABILITY

A potentially significant factor is rain loss associated with wet radomes and passive reflectors. Losses of less than 1 dB to over 14 dB per antenna (depending on frequency, antenna covering material, and rain rate) have been documented (Anderson, 1975; Blevis, 1965; Burgueno et al., 1987; Crane, 2002; Effenberger and Strickland, 1986; Lin, 1973, 1975; Rummler, 1987). Rummler (1987) proposes a single-radome losses estimate for all rain rates greater than 10 mm/h: wet radome loss (dB) = {4 × sqrt [*f*/11]}, where *f* is the radio operating frequency in gigahertz. He assumes only one radome is wet at a time. Lin (1975) estimated wet-radome loss at 11 GHz as 1.5 dB but assumed both radomes to be wet. Andrew Corporation (Skarpiak, 1988) made 11-GHz measurements of wet 8-ft radomes composed of polymer-coated fabric (Teglar™) and fiberglass (Hypalon™). The fabric radome experienced about 1 dB of loss and the fiber glass radome experienced 5 dB of loss when either was lightly sprinkled with water. The radomes experienced 2 and 6 dB of loss, respectively, when completely covered with a sheet of water. (Wet-radome loss, if used, should only be applied to rain-fade margin. Multipath fading does not typically occur during rain, so adding wet-radome loss as an additional overall system loss is inappropriate.) Recent measurements suggest modern radomes clear rain quickly and have little residual effect. Some authorities suggest wet radomes are no longer a significant factor.

Microwave path design is based on long-term point rain rates that are based on long-term measured rain rates averaged over at least 20 years. These long-term averages represent our best estimate of the expected performance on a path for any given year. However, given the highly variable nature of rain, the long-term averages are rarely (if ever) observed over any given year. Rain intensity for a given location is simply highly variable.

Point rain rate tables or charts used for microwave path loss calculations list rates not exceeded more than a percentage of time when measured with the stated gauge-integration time. Long-term rain rate data gathered from a single rain gauge requires a very long time base (several years) to yield stable statistics. If the time base is not sufficiently long, the short-term results will underestimate or overestimate the long-term large-sample average in an unpredictable way. Data taken over a period of less than 10 years is generally unreliable for moderate to high rain rates. Incidence of high rain rates at a single point is so low in the western United States that much longer time bases (a few decades) are required to obtain stable statistics.

Regardless of the rain data used, exactly the same data would probably not be obtained if measurements were made over the next couple of decades. The data represents the best current estimate of the point rain rate that would not be exceeded over any 1 year *on the average*. The actual rain rate measured over any one specific year will be different from the average value—sometimes greater and sometimes smaller. Osborne (1977) observed that the worst-case 1-year rain rates can exceed long-term averages by a multiplicative factor of 2 to 20. Lin (1977) observed this variation in rain rate data taken at Newark, NJ (Fig. 11.35).

Worst-case month or day rates can exceed long-term averages by extremely large factors. The following example illustrates this situation. Rain rates were measured in Philadelphia during a particularly rainy year (September 2008 to August 2009) (Fig. 11.36).

Actual measured rain rates are contrasted with long-term monthly and yearly average rain rates measured by Crane (2003) at the same location (Fig. 11.37).

The measured rain statistics are dominated by a rare but very intense (3150 mm/h) rain event lasting 10 min. This "once in a hundred years" event significantly distorts short-term statistics. There is no practical method to limit the worst-case outage time of radio paths with path loss dominated by rain

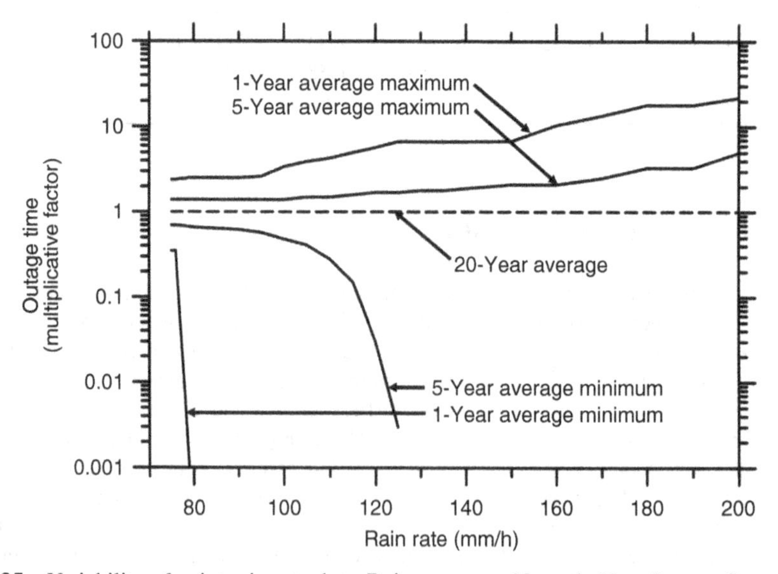

Figure 11.35 Variability of point rain rate data. Rain averages, Newark, New Jersey. *Source*: Adapted from Figure 7, Lin, S. H., "Nationwide Long-Term Rain Statistics and Empirical Calculation of 11-GHz Microwave Rain Attenuation," *Bell System Technical Journal*, pp. 1581–1604, November 1977.

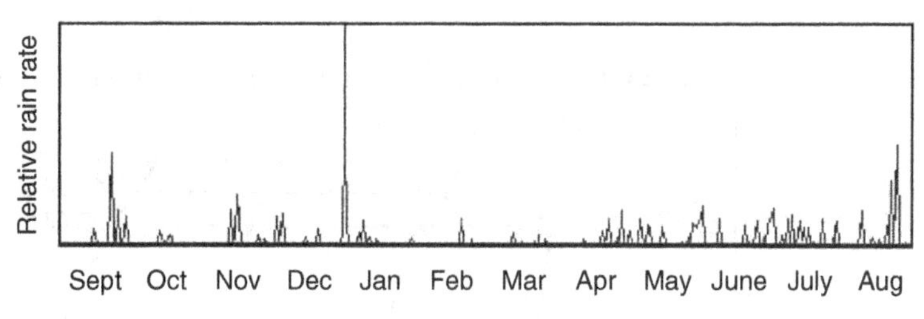

Figure 11.36 Short-term rain rates measured in Philadelphia.

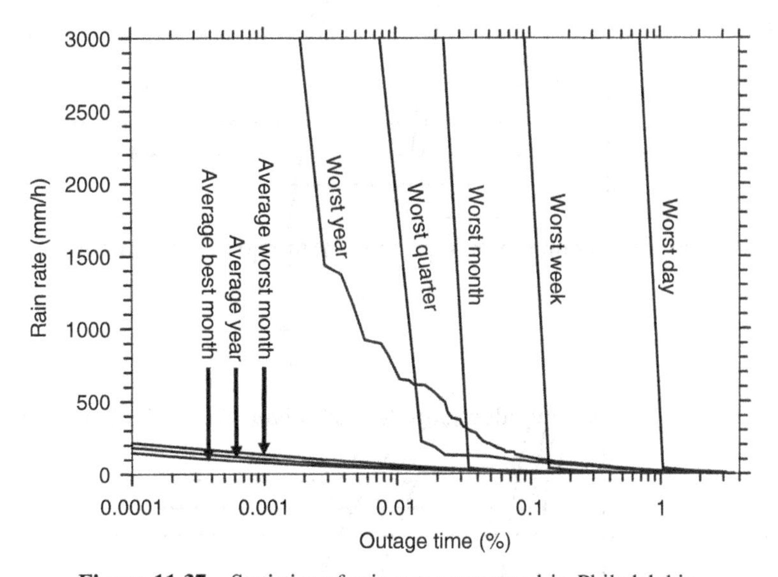

Figure 11.37 Statistics of rain rates measured in Philadelphia.

attenuation. Engineering paths based on worst-case statistics would lead to uneconomical and impractical radio systems. All rain-loss-dominated high frequency microwave radio paths are susceptible to rare, intense rain rate events. Under those circumstances, the path will experience an outage regardless of the path design. If the path is designed using long-term rain data, the expectation is that the composite path outage time *averaged over the measurement time* will approach the expected result as the measurement time approaches 20 years. For any given year, the outage time can significantly exceed the path design value.

11.9 EXAMPLES OF RAIN-LOSS-DOMINATED PATH DESIGNS

Before leaving the topic, let us take a look at the results of the above rain rate estimates applied to standard, commercial, high frequency microwave radios. We will assume the Crane 96 rain model and the Vigants–Barnett multipath model (only significant for 11-GHz paths or very long higher frequency paths, Fig. 11.38 and Fig. 11.39).

It is obvious that for a given path-availability objective, the path can be much longer in the Northwest United States than in the Southeast. The Midwest is in the middle both graphically and in performance. It is interesting that the performance of a high capacity (256 QAM) radio in Chicago is similar to that of a moderate capacity (32 QAM) radio in Miami.

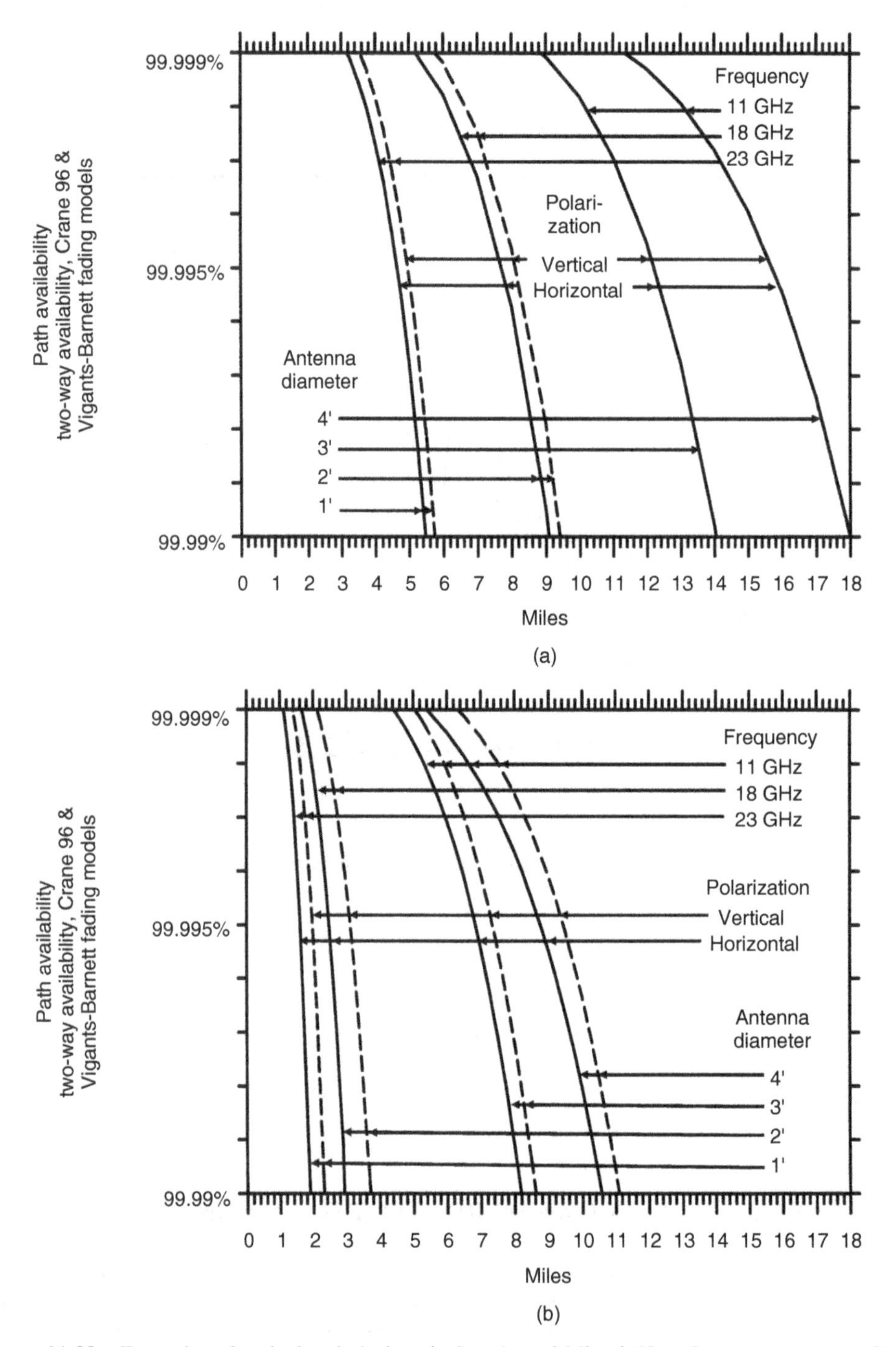

Figure 11.38 Examples of typical path designs in Seattle and Miami. How far you can go: (a) Crane rain zone B1 (Seattle) 32 QAM 100 Mb/s and (b) Crane rain zone E (Miami) 32 QAM 100 Mb/s.

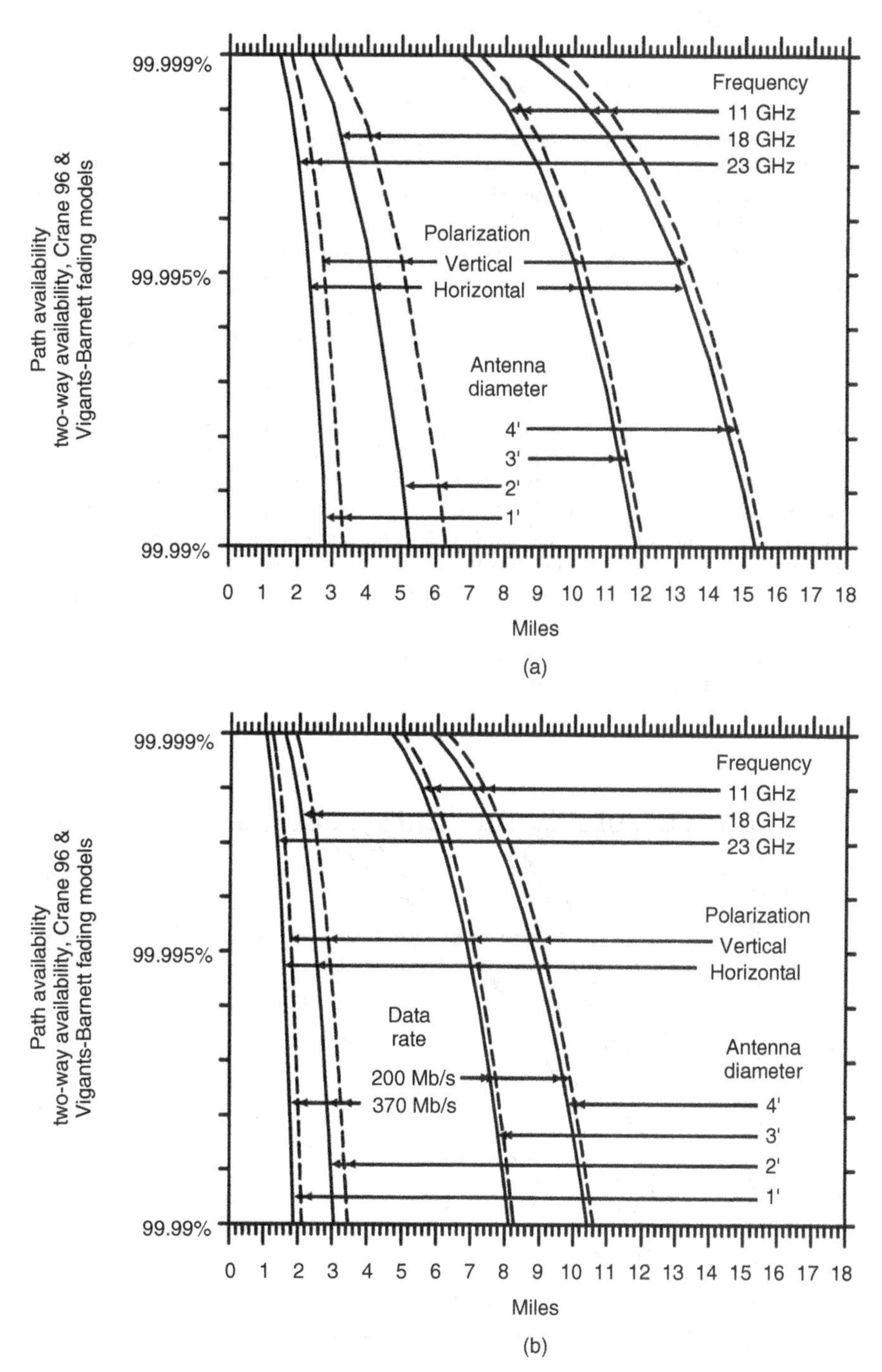

Figure 11.39 Examples of typical path designs in Chicago. How far you can go Crane rain zone D2 (Chicago): (a) 32 QAM 100Mb\s and (b) 256 QAM.

11.10 CONCLUSIONS

The path designer must make a decision regarding what model should be used for point rain rate and point-to-path conversion factor.

The ITU-R city rain rates for the various zones, when averaged, are relatively similar to Crane zone averages for all zones except A, B1, and C1. However, the ITU-R method is significantly different than Crane for specific locations such as Houston, Miami, New Orleans, San Francisco, and Seattle. Overall, the ITU-R method gives fairly accurate results but not as good as actual measurements (as would be expected). It appears to be a good resource if no other data is available.

It should be remembered that the simplified ITU-R rain curves ignore the frequency dependency. Actual calculations will be slightly different for frequencies other than 18- and 23-GHz vertical polarization. The ITU-R point rain rate versus unavailability curve change shape dramatically at latitude 30°. While there is some shape change in the city data rain rate curves, in general, the Crane site-specific data does not support this dramatic change in shape. The ITU-R recommendation (International Telecommunication Union—Radiocommunication Sector (ITU-R), 2009) recommends using actual long-term data when available from local sources and only using the ITU-R formulas when the local data is not available. Since Crane location-specific data is available, use of the ITU-R formulas for 0.01% point rain rate estimates is unnecessary when using the ITU-R path design method in the United States or Canada.

The path designer should remember that the ITU-R point rain rate calculation method is only valid over the outage range 0.1% to 0.001% (availabilities of 99.9–99.999%). Attempts to extend the range to smaller outages (greater availability) should be resisted. Comparing the shapes of the curves in the previous graphs make the reason clear.

Having addressed the source of rain rate data, we return to the original issue. The path designer is faced with the following equation (as defined above) to solve for path availability based on available fade margin:

$$M = \gamma_{\mathrm{R}} d_{\mathrm{eff}} \tag{11.21}$$

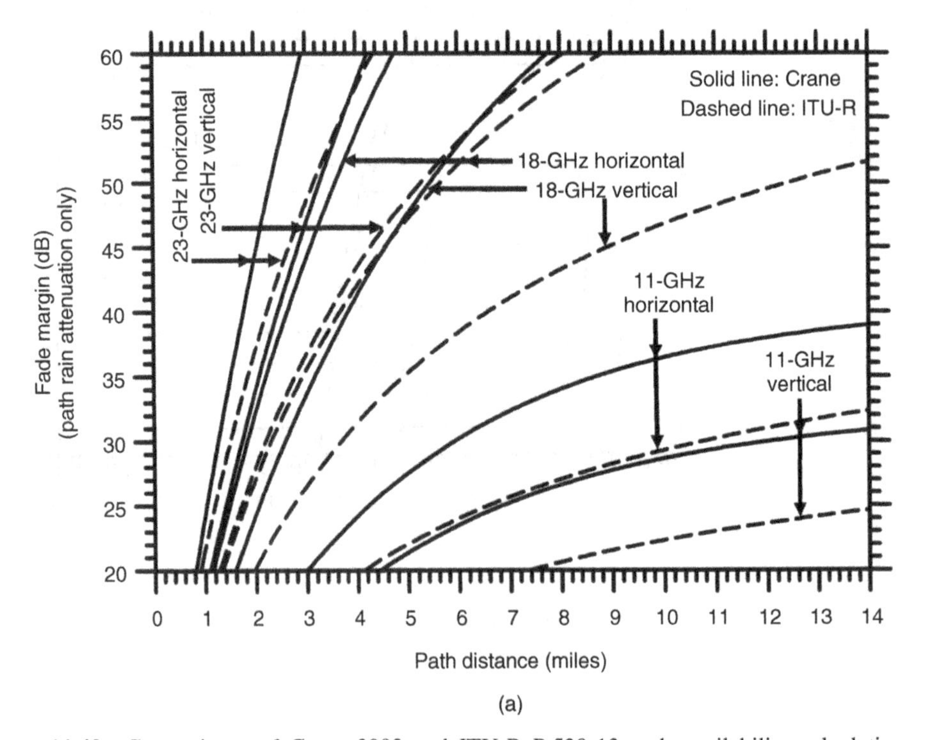

Figure 11.40 Comparisons of Crane 2003 and ITU-R P.530-13 path availability calculations. (a) Crane rain zone D2 (Chicago) 99.999% path availability and (b) Crane rain zone E (Miami) 99.999% path availability.

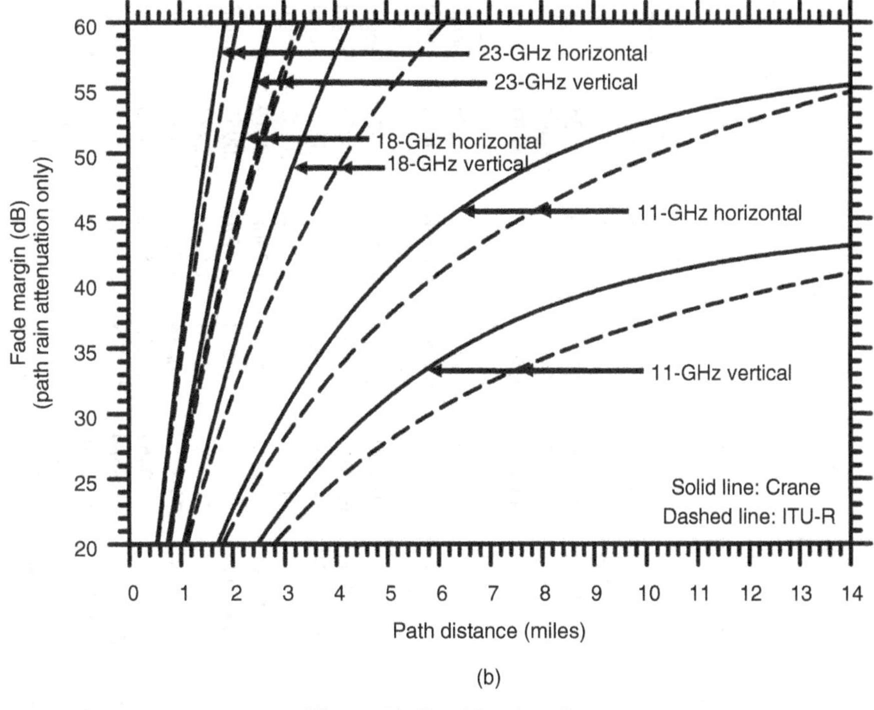

(b)

Figure 11.40 (*Continued*)

Comparing the Crane and ITU-R rain fade estimation methods is difficult. Figures 11.13 to 11.15 and 11.28 compare the point to path conversion factor and point rain rate differences separately. Actual examples offer additional perspective on the differences.

For high rain rate zones, high frequencies, and high path availabilities, the difference is noticeable. However, for lower rain rate zones, lower frequencies, and lower path availabilities, the difference becomes more pronounced. It should be remembered that for longer paths and lower path availabilities, multipath also is an influence. Multipath moderates the significant difference in methods at 11 GHz. It is clear the Crane method is much more conservative than the ITU-R method for many practical situations. Studies that could prove which method is more accurate to date have been inconclusive.

11.A APPENDIX

11.A.1 North American City Rain Data Index

TABLE 11.A.1 North American City Rain Data Index

Zone	City/ Location	State/ Province	Country	Latitude, degrees	Longitude, degrees	P_C	R_C	P_D	R_D	S_D	Source
B2	Calgary	Alberta	Canada	51.10	−114.02	0.006087	40.0663	0.7467	1.1516	1.1002	Segal
B2	Edmonton	Alberta	Canada	53.57	−113.52	0.004355	44.4817	0.2049	2.7328	1.2530	Segal
B1	Watino	Alberta	Canada	55.72	−117.62	0.005172	21.2994	0.1154	1.9073	1.1535	Segal
C1	Comox	British Columbia	Canada	49.72	−124.90	0.001312	34.0793	4.8487	0.7093	0.9995	Segal
B1	Hope	British Columbia	Canada	49.38	−121.43	0.000490	38.1462	4.5643	1.2324	0.8607	Segal
B1	Mission	British Columbia	Canada	49.15	−122.27	0.012234	16.8470	1.7869	1.9592	0.8281	Segal
B1	Prince George	British Columbia	Canada	53.88	−122.67	0.056328	6.0368	0.1533	1.1198	1.5296	Segal
B1	Summerland	British Columbia	Canada	49.57	−119.65	0.011386	9.2159	0.7295	0.8388	0.9693	Segal
B1	Vancouver	British Columbia	Canada	49.18	−123.17	0.007962	15.5055	2.6879	1.5878	0.7465	Segal
A	Churchill	Manitoba	Canada	58.75	−94.07	0.019353	9.4569	2.3259	1.3164	0.7722	Segal
B2	Dauphin	Manitoba	Canada	51.10	−100.05	0.003005	68.4388	1.4031	1.6831	1.0889	Segal
D1	Winnipeg	Manitoba	Canada	49.90	−97.23	0.042771	4.3482	0.6182	1.1257	1.5559	Segal
B2	Fredericton	New Brunswick	Canada	45.92	−66.62	0.007021	31.3335	0.4335	2.7166	1.1437	Segal
B2	Saint John	New Brunswick	Canada	45.32	−65.88	0.007894	40.0102	1.5742	3.2808	0.9019	Segal
B1	Gander	Newfoundland	Canada	48.95	−54.57	0.001536	46.7400	4.5628	0.7796	1.0957	Segal
B1	Goose Bay	Newfoundland	Canada	53.32	−60.42	0.010207	17.0258	0.2070	1.5680	1.2284	Segal
B1	St. John's	Newfoundland	Canada	47.62	−52.75	0.001116	73.3137	1.0845	4.2605	0.8881	Segal
B1	Stephenville	Newfoundland	Canada	48.53	−58.55	0.018877	16.0462	1.7222	2.5216	0.8060	Segal
A	Cambridge Bay	Northwest Territories	Canada	69.10	−105.12	0.002111	17.8684	4.6244	0.2389	1.0167	Segal
B2	Halifax	Nova Scotia	Canada	44.63	−63.50	0.004374	52.7376	1.8347	3.9795	0.8772	Segal
B2	Kentville	Nova Scotia	Canada	45.07	−64.48	0.037992	22.5736	1.2506	2.6334	0.9119	Segal
B1	Sydney	Nova Scotia	Canada	46.17	−60.05	0.020230	17.7430	1.7538	3.0250	0.8032	Segal

B2	Central Patricia	Ontario	Canada	51.50	−90.15	0.029789	14.0173	0.2834	0.7474	1.6943	Segal
B2	Geraldton	Ontario	Canada	49.68	−86.95	0.022601	11.6170	0.2931	0.8107	1.5679	Segal
D1	Kingston	Ontario	Canada	44.23	−76.48	0.003796	43.1368	3.8717	1.2423	1.1806	Segal
D1	London	Ontario	Canada	43.03	−81.15	0.001270	81.6708	1.1633	2.0196	1.2460	Segal
B2	Moosonee	Ontario	Canada	51.27	−80.65	0.001834	60.6367	1.5526	2.7184	0.8981	Segal
D1	North Bay	Ontario	Canada	46.37	−79.42	0.019447	34.0000	3.1586	1.3459	1.1105	Segal
D1	Ottawa	Ontario	Canada	45.38	−75.72	0.010498	29.6379	1.3232	1.5692	1.2816	Segal
D1	Sault Ste. Marie	Ontario	Canada	46.48	−84.50	0.011721	26.9663	1.6180	1.3345	1.2532	Segal
B2	Sioux Lookout	Ontario	Canada	50.12	−91.90	0.001388	93.5399	0.8104	3.0329	1.0709	Segal
D1	Toronto	Ontario	Canada	43.68	−79.63	0.005877	42.6782	4.1412	1.0275	1.2549	Segal
D1	Windsor	Ontario	Canada	42.27	−82.97	0.011972	46.3000	1.0567	2.4059	1.2522	Segal
B1	Summerside	Prince Edward Island	Canada	46.43	−63.83	0.012484	16.6759	2.1114	2.3381	0.7856	Segal
B2	Caplan	Quebec	Canada	48.10	−65.65	0.043405	10.9421	0.2253	1.4929	1.4672	Segal
B1	Gagnon	Quebec	Canada	51.95	−68.13	0.012535	18.1854	0.7591	2.3450	0.9260	Segal
D1	Montreal	Quebec	Canada	45.47	−73.75	0.017809	36.1263	3.1996	1.3784	1.1145	Segal
B2	Normandin	Quebec	Canada	48.85	−72.53	0.012093	4.0385	0.3456	2.5198	1.1446	Segal
A	Post de la Baleine	Quebec	Canada	55.28	−77.77	0.014411	16.1882	2.4104	1.9329	0.7584	Segal
D1	Quebec	Quebec	Canada	46.80	−71.38	0.015969	30.7690	2.7200	1.3976	1.1732	Segal
B2	Val d’Or	Quebec	Canada	48.05	−77.78	0.010434	34.8249	0.6871	2.7484	1.0551	Segal
B2	Prince Albert	Saskatchewan	Canada	53.22	−105.68	0.000505	35.5518	0.2499	2.3373	1.2540	Segal
B2	Regina	Saskatchewan	Canada	50.93	−104.67	0.000607	41.1939	0.3292	3.1280	1.1879	Segal
B2	Swift Current	Saskatchewan	Canada	50.27	−107.73	0.001909	36.5455	0.2345	2.2020	1.2751	Segal
A	Uranium City	Saskatchewan	Canada	59.57	−108.48	0.006779	16.4236	1.5593	1.1444	0.8406	Segal
B2	Weyburn	Saskatchewan	Canada	49.67	−103.85	0.011035	34.5071	0.2704	1.3409	1.2707	Segal
A	Carmacks	Yukon	Canada	62.10	−136.30	0.004072	16.4680	0.2739	1.1575	1.1863	Segal

(*Continued*)

TABLE 11.A.1 ***(Continued)***

Zone	City/ Location	State/ Province	Country	Latitude, degrees	Longitude, degrees	P_C	R_C	P_D	R_D	S_D	Source
G	Guam		Pacific Ocean	13.31	144.49	0.304239	24.4303	7.5490	0.0489	0.8789	Kizer
C2	Johnston Island		Pacific Ocean	16.44	−169.31	0.137406	20.6951	1.6599	0.0325	0.9005	Kizer
C2	Wake Island		Pacific Ocean	19.17	166.39	0.219446	20.1992	5.9667	0.0492	0.8521	Kizer
E	Birmingham	Alabama	United States	33.52	−86.82	0.670921	20.7511	100.8637	0.0159	0.9847	Kizer
D3	Huntsville	Alabama	United States	34.65	−86.77	0.024000	36.3820	2.8780	2.4010	1.1140	Crane
E	Mobile	Alabama	United States	30.68	−88.25	0.021000	42.1120	2.6770	2.7140	1.2080	Crane
E	Montgomery	Alabama	United States	32.30	−86.40	0.052520	30.9560	2.0110	2.0290	1.1840	Crane
B1	Fairbanks	Alaska	United States	64.82	−147.83	0.004000	18.1310	2.8120	0.7690	0.8280	Crane
B1	Juneau	Alaska	United States	58.21	−134.35	0.004521	11.6578	2.4178	1.7176	0.7550	Kizer
F2	Flagstaff	Arizona	United States	35.13	−111.67	0.014000	25.6950	2.2920	1.3670	0.9960	Crane
F2	Phoenix	Arizona	United States	33.27	−111.59	0.015943	23.9358	0.4572	0.1378	1.2536	Kizer
F2	Prescott	Arizona	United States	34.57	−112.43	0.017920	22.3603	1.2668	2.3345	0.9743	Kizer
F2	Tucson	Arizona	United States	32.13	−111.00	0.014370	27.4780	0.7980	1.2230	1.1050	Crane
F2	Winslow	Arizona	United States	35.02	−110.43	0.016339	24.2191	0.2419	0.0282	1.4297	Kizer
F1	Yuma	Arizona	United States	32.67	−114.67	0.001800	37.9070	0.2850	1.2110	1.1170	Crane
D3	Fort Smith	Arkansas	United States	35.20	−94.22	0.074676	27.6924	34.6163	2.5514	0.9098	Kizer
D3	Little Rock	Arkansas	United States	34.73	−92.23	0.064100	28.0140	2.0020	1.4980	1.2240	Crane
D3	Texarkana	Arkansas	United States	33.27	−94.00	0.210428	24.6742	31.1085	1.6854	0.8715	Kizer
F1	Bakersfield	California	United States	35.42	−119.00	0.001000	24.1460	0.8130	1.3290	0.9060	Crane
C1	Eureka	California	United States	40.80	−124.17	0.003000	21.7740	4.3400	1.6400	0.9250	Crane
C1	Fresno	California	United States	36.78	−119.67	0.001000	30.3790	1.2820	1.5620	0.9260	Crane
F1	Los Angeles	California	United States	33.93	−118.33	0.002000	40.9750	0.9820	2.0330	0.9610	Crane
C1	Red Bluff	California	United States	40.11	−122.12	0.002235	29.7513	3.6267	1.5419	0.9591	Kizer
C1	Sacramento	California	United States	38.35	−121.30	0.085218	4.4258	0.3261	1.2512	1.2110	Kizer

F1	San Diego	California	United States	32.82	−117.17	0.005000	24.1540	0.9630	1.5540	0.9730	Crane
C1	San Francisco	California	United States	37.62	−122.33	0.001000	39.9570	2.0540	1.7510	0.9420	Crane
C1	San Jose	California	United States	37.33	−121.88	0.157839	4.6853	2.5241	1.0679	1.0488	Kizer
F1	San Luis Obispo	California	United States	35.33	−120.67	0.377203	6.1135	4.0528	1.6615	0.9691	Kizer
F1	Alamosa	Colorado	United States	37.45	−105.83	0.003000	26.0330	1.1570	0.9920	0.9690	Crane
B2	Denver	Colorado	United States	39.77	−104.83	0.006000	38.0900	2.0250	1.1310	0.9880	Crane
B2	Grand Junction	Colorado	United States	39.12	−108.50	0.004000	25.6760	1.3200	1.2050	0.6970	Crane
B2	Pueblo	Colorado	United States	38.14	−104.38	0.008418	42.9956	0.7330	2.6775	1.0270	Kizer
D2	Bridgeport	Connecticut	United States	41.12	−73.12	0.031651	30.0938	3.3573	1.4140	1.0933	Kizer
D2	Hartford	Connecticut	United States	41.93	−72.68	0.015000	32.6140	3.8270	1.7390	1.0000	Crane
D2	New Haven	Connecticut	United States	41.34	−72.96	0.066009	21.7820	9.2882	1.4066	0.9869	Kizer
D2	Wilmington	Delaware	United States	39.77	−75.54	0.062517	16.6092	23.6044	2.7072	0.9185	Kizer
D2	Washington	District of Columbia	United States	38.95	−77.45	0.027780	32.2260	2.7200	1.4970	1.1280	Crane
E	Jacksonville	Florida	United States	30.48	−81.70	0.027000	40.5400	2.2090	2.2220	1.2440	Crane
E	Melbourne	Florida	United States	28.04	−80.37	6.765880	16.8243	36.3133	0.2019	0.6785	Kizer
E	Miami	Florida	United States	25.80	−80.30	0.036000	38.7350	1.9100	2.4910	1.2760	Crane
E	Miami Beach	Florida	United States	25.47	−80.08	0.024584	37.4200	34.2400	0.0716	1.9020	Kizer
E	Tallahassee	Florida	United States	30.38	−84.37	0.032000	42.4020	2.2300	2.9190	1.1950	Crane
E	Tampa	Florida	United States	27.97	−82.53	0.072950	30.4160	1.0820	1.3280	1.3490	Crane
E	West Palm Beach	Florida	United States	26.41	−80.06	0.568568	23.8402	31.4702	0.2995	0.6777	Kizer
D3	Atlanta	Georgia	United States	33.65	−84.43	0.051020	31.8380	2.3000	1.7790	1.0650	Crane
D3	Augusta	Georgia	United States	33.28	−81.58	0.392381	20.9192	67.3086	0.0301	0.9140	Kizer
D3	Columbus	Georgia	United States	32.31	−84.57	0.367155	24.7719	74.4775	0.0586	0.8599	Kizer
E	Savannah	Georgia	United States	32.13	−81.20	0.068840	28.8850	1.9410	1.2690	1.3630	Crane

(Continued)

TABLE 11.A.1 (*Continued*)

Zone	City/ Location	State/ Province	Country	Latitude, degrees	Longitude, degrees	P_C	R_C	P_D	R_D	S_D	Source
C2	Hilo/Hawaii	Hawaii	United States	19.43	−155.03	0.676588	16.8243	36.3133	0.2019	0.6785	Kizer
C2	Honolulu/Oahu	Hawaii	United States	21.19	−157.52	0.047826	29.4518	3.1264	1.9518	0.9654	Kizer
C2	Kahului/Maui	Hawaii	United States	20.54	−156.26	0.019498	32.3083	1.8287	1.9720	1.0460	Kizer
C2	Lihue/Kauai	Hawaii	United States	21.59	−159.20	0.052152	29.3437	5.2401	4.6216	0.8949	Kizer
B1	Boise	Idaho	United States	43.34	−116.13	0.115296	3.7231	0.4245	1.4257	1.1120	Kizer
B1	Lewiston	Idaho	United States	46.37	−117.02	0.052287	4.1817	0.2095	1.0344	1.3220	Kizer
B1	Pocatello	Idaho	United States	42.55	−112.34	0.097505	5.8819	0.4142	1.8336	1.1429	Kizer
D3	Cairo	Illinois	United States	37.21	−89.18	0.144785	26.1233	0.3647	0.0173	0.9922	Kizer
D2	Chicago	Illinois	United States	42.00	−87.88	0.020000	35.1250	3.0060	1.3380	1.1710	Crane
D2	Moline	Illinois	United States	41.28	−90.31	0.086781	29.6207	4.9305	1.9408	1.0068	Kizer
D2	Peoria	Illinois	United States	40.40	−89.41	0.148852	23.4514	13.3984	0.0497	0.8509	Kizer
D2	Rockford	Illinois	United States	42.12	−89.06	0.029255	33.3454	2.0558	3.7931	1.0162	Kizer
D2	Springfield	Illinois	United States	39.85	−89.68	0.061480	22.8500	2.3110	0.9080	1.2440	Crane
D2	Evansville	Indiana	United States	38.03	−87.32	0.129439	18.6766	3.6387	2.7085	1.0890	Kizer
D2	Fort Wayne	Indiana	United States	41.00	−85.20	0.029000	27.4670	3.0520	1.3030	1.1310	Crane
D2	Indianapolis	Indiana	United States	39.73	−86.27	0.015000	41.0550	3.1690	1.6140	1.0980	Crane
D2	South Bend	Indiana	United States	41.42	−86.20	0.024725	38.1273	4.9752	2.3628	1.0362	Kizer
D2	Burlington	Iowa	United States	40.49	−91.10	0.008229	38.8423	7.4321	2.9735	1.0138	Kizer
D2	Davenport	Iowa	United States	41.58	−90.64	0.147884	21.2581	2.8060	1.2504	0.9274	Kizer
D2	Des Moines	Iowa	United States	41.53	−93.65	0.014000	40.7020	2.6190	1.4520	1.1470	Crane
D2	Dubuque	Iowa	United States	42.24	−90.42	0.062441	29.7820	9.6180	2.3455	0.9851	Kizer
D1	Sioux City	Iowa	United States	42.23	−96.23	0.062326	28.0267	0.8233	0.0066	1.0663	Kizer
D2	Waterloo	Iowa	United States	42.33	−92.24	0.016079	46.7144	3.7153	2.5694	1.0762	Kizer
D1	Goodland	Kansas	United States	39.37	−101.67	0.008000	46.4740	1.5690	1.1720	1.2130	Crane

D2	Topeka	Kansas	United States	39.04	−95.38	0.081510	25.1486	6.2310	1.4252	0.9668	Kizer
D2	Wichita	Kansas	United States	37.65	−97.43	0.011000	38.8500	1.8700	1.7860	1.1920	Crane
D2	Lexington	Kentucky	United States	38.02	−84.36	0.222253	22.7233	12.1034	0.0263	0.9297	Kizer
D2	Louisville	Kentucky	United States	38.18	−85.73	0.018000	37.5610	3.1440	1.8000	1.0620	Crane
D3	Lake Charles	Louisiana	United States	30.07	−93.14	0.328451	25.3438	100.8637	2.2818	0.8703	Kizer
E	New Orleans	Louisiana	United States	29.98	−90.25	0.012000	46.0420	2.5670	2.9680	1.2060	Crane
D3	Shreveport	Louisiana	United States	32.47	−93.82	0.070690	28.4100	1.6710	1.3080	1.3220	Crane
D1	Caribou	Maine	United States	46.87	−68.02	0.011000	27.1540	5.1850	1.1560	1.0050	Crane
D1	Portland	Maine	United States	43.65	−70.30	0.007000	35.1160	4.3600	1.6530	1.0130	Crane
D2	Baltimore	Maryland	United States	39.28	−76.61	0.066890	29.7385	6.2229	1.6972	0.9594	Kizer
D2	Frederick	Maryland	United States	39.60	−77.67	0.021691	23.8343	5.3109	2.6014	1.0497	Kizer
D2	Boston	Massachusetts	United States	42.37	−71.03	0.012000	26.1540	4.0550	1.6720	1.0260	Crane
D2	Nantucket	Massachusetts	United States	41.15	−70.04	0.032849	24.3704	3.2687	1.3779	1.0804	Kizer
D2	Pittsfield	Massachusetts	United States	42.45	−73.25	0.043419	26.1465	3.7025	0.8695	1.0657	Kizer
D1	Detroit	Michigan	United States	42.23	−83.33	0.015000	35.2270	3.2950	1.2740	1.0810	Crane
D1	Escanaba	Michigan	United States	45.81	−86.91	0.025786	30.4902	5.7815	2.4153	1.0096	Kizer
D1	Flint	Michigan	United States	42.58	−83.45	0.060029	29.5122	2.5126	0.0045	1.0919	Kizer
D2	Grand Rapids	Michigan	United States	42.88	−85.52	0.013000	36.8170	3.8190	1.3080	1.1320	Crane
D1	Houghton Lake	Michigan	United States	44.22	−84.41	0.028338	25.4871	12.5769	1.9073	0.9733	Kizer
D1	Lansing	Michigan	United States	42.47	−84.35	0.050183	27.7247	5.7620	1.8733	1.0044	Kizer
D1	Marquette	Michigan	United States	46.33	−87.23	0.035060	27.3292	3.1044	0.6834	1.1124	Kizer
D1	Muskegon	Michigan	United States	43.10	−86.14	0.099535	19.5634	6.5078	0.0132	0.9266	Kizer
D1	Sault Ste. Marie	Michigan	United States	46.47	−84.35	0.015000	27.2530	5.6410	0.7760	1.2720	Crane
D1	Ypsilanti	Michigan	United States	42.25	−83.62	0.418046	13.9707	7.8378	0.0049	1.0857	Kizer

(Continued)

TABLE 11.A.1 (*Continued*)

Zone	City/ Location	State/ Province	Country	Latitude, degrees	Longitude, degrees	P_C	R_C	P_D	R_D	S_D	Source
D1	Duluth	Minnesota	United States	46.83	−92.18	0.011000	39.1530	3.4670	1.1650	1.1020	Crane
D1	International Falls	Minnesota	United States	48.57	−93.38	0.017000	28.9160	3.0740	0.9130	1.1770	Crane
D1	Minneapolis	Minnesota	United States	44.88	−93.22	0.011000	38.2700	2.8530	1.1630	1.2020	Crane
D1	Rochester	Minnesota	United States	43.54	−92.30	0.018446	36.5642	5.4178	2.2431	1.0436	Kizer
D1	St. Cloud	Minnesota	United States	45.33	−94.03	0.008146	44.9906	2.9819	2.3209	1.1349	Kizer
E	Jackson	Mississippi	United States	32.32	−90.08	0.074400	30.6000	1.7670	1.5690	1.2620	Crane
D2	Columbia	Missouri	United States	38.58	−92.22	0.114076	25.3279	10.9647	2.3966	0.9151	Kizer
D2	Kansas City	Missouri	United States	39.32	−94.72	0.013000	34.8180	2.6880	1.7160	1.2160	Crane
D2	Springfield	Missouri	United States	37.14	−93.23	0.141471	22.3692	22.2614	2.7314	0.8342	Kizer
D2	St. Joseph	Missouri	United States	39.46	−94.55	0.023245	15.2134	4.7528	4.0929	1.0331	Kizer
D2	St. Louis	Missouri	United States	38.75	−90.37	0.020000	32.4040	2.5590	1.6540	1.1340	Crane
B2	Billings	Montana	United States	45.48	−108.33	0.013383	25.4820	1.3421	0.4244	0.9450	Kizer
B2	Glasgow	Montana	United States	48.13	−106.37	0.023728	22.8424	0.2623	0.5156	1.2075	Kizer
B2	Glasgow	Montana	United States	48.22	−106.67	0.004000	39.4000	1.7370	0.8780	1.0970	Crane
B2	Havre	Montana	United States	48.33	−109.46	0.011012	24.6284	0.6284	0.8471	1.1195	Kizer
B2	Helena	Montana	United States	46.61	−111.96	0.007048	19.9737	0.1846	0.8803	1.4161	Kizer
B1	Missoula	Montana	United States	46.93	−114.17	0.003000	28.0530	2.7010	0.9860	0.7880	Crane
D1	Grand Island	Nebraska	United States	40.97	−98.32	0.014000	36.6700	1.8950	1.2680	1.2540	Crane
D1	Lincoln	Nebraska	United States	40.51	−96.45	0.155108	21.4321	3.7676	0.0155	0.9759	Kizer
D1	Norfolk	Nebraska	United States	41.59	−97.26	0.050082	29.2416	5.9444	2.0415	1.0125	Kizer
D1	North Platte	Nebraska	United States	41.13	−100.67	0.007000	43.6150	1.8210	1.2600	1.2190	Crane
D1	Omaha	Nebraska	United States	41.22	−96.01	0.041009	30.8208	3.8946	2.1681	1.0984	Kizer
B2	Scottsbluff	Nebraska	United States	41.87	−103.67	0.004000	45.1460	1.9690	1.0450	1.1620	Crane

D1	Valentine	Nebraska	United States	42.52	−100.33	0.049476	25.0537	6.9579	1.7879	0.9800	Kizer
F1	Elko	Nevada	United States	40.83	−115.83	0.001000	20.5970	1.5100	1.0350	1.0500	Crane
F1	Ely	Nevada	United States	39.28	−114.83	0.001000	32.5610	1.7180	1.0430	0.9370	Crane
F1	Las Vegas	Nevada	United States	36.08	−115.17	0.002000	23.0770	0.5260	1.2010	1.0560	Crane
F1	Reno	Nevada	United States	39.50	−119.83	0.002000	22.4980	1.2050	1.1620	0.8570	Crane
F1	Tonopah	Nevada	United States	38.04	−117.05	0.001012	30.6524	0.4622	1.0774	1.1341	Kizer
F1	Winnemucca	Nevada	United States	40.90	−117.81	0.053880	3.1392	0.2061	0.9366	1.2938	Kizer
D1	Concord	New Hampshire	United States	43.20	−71.50	0.017000	24.0890	3.8310	1.4560	0.9980	Crane
D2	Atlantic City	New Jersey	United States	39.38	−74.42	0.078239	24.6017	2.9120	0.0110	0.9887	Kizer
D2	Newark	New Jersey	United States	40.43	−74.11	0.057240	14.1767	5.2727	2.5365	1.0284	Kizer
D2	Trenton	New Jersey	United States	40.22	−74.73	0.030871	30.5850	3.5600	1.9733	1.0798	Kizer
F2	Albuquerque	New Mexico	United States	35.03	−106.67	0.014740	20.4780	0.6900	1.1940	0.8040	Crane
F2	Clayton	New Mexico	United States	36.27	−103.09	0.033004	23.3434	1.2829	3.8078	0.9759	Kizer
F2	Jornada	New Mexico	United States	32.62	−106.73	0.010000	24.4690	0.4170	2.4010	0.9130	Crane
F2	Raton	New Mexico	United States	36.90	−104.47	0.039721	31.7264	0.2681	0.0052	1.1051	Kizer
F2	Roswell	New Mexico	United States	33.30	−104.50	0.017000	22.1000	1.0120	1.8130	1.2570	Crane
F2	Silver City	New Mexico	United States	32.75	−108.25	0.020783	23.4736	0.3720	0.4059	1.2429	Kizer
D2	Albany	New York	United States	42.75	−73.80	0.011000	35.6420	3.7760	1.5020	0.9630	Crane
D2	Binghamton	New York	United States	42.05	−76.06	0.027068	23.6612	2.5118	1.9442	1.1386	Kizer
D1	Buffalo	New York	United States	42.93	−78.73	0.046370	19.9170	4.7130	0.9940	1.0080	Crane
D2	New York	New York	United States	40.65	−73.78	0.019000	33.7200	2.9860	1.8250	1.0180	Crane
D1	Rochester	New York	United States	43.12	−77.68	0.033082	22.5150	3.9414	2.7889	0.9748	Kizer
D2	Syracuse	New York	United States	43.12	−76.12	0.013000	33.8230	5.0070	1.1230	1.0860	Crane

(Continued)

TABLE 11.A.1 (*Continued*)

Zone	City/ Location	State/ Province	Country	Latitude, degrees	Longitude, degrees	P_C	R_C	P_D	R_D	S_D	Source
D3	Asheville	North Carolina	United States	35.26	−82.29	0.037094	25.3316	2.3974	1.9362	1.1658	Kizer
D3	Charlotte	North Carolina	United States	35.22	−80.93	0.011000	40.1670	2.9100	2.0390	1.0820	Crane
D3	Greensboro	North Carolina	United States	36.06	−79.57	0.172668	21.7292	2.4694	0.0084	1.0627	Kizer
D3	Raleigh	North Carolina	United States	35.87	−78.78	0.016000	36.9330	2.7700	1.9180	1.0970	Crane
E	Wilmington	North Carolina	United States	34.16	−77.54	0.613502	20.5353	101.4532	0.0338	0.9199	Kizer
D3	Winston Salem	North Carolina	United States	36.10	−80.25	0.137239	19.5058	3.4048	0.0097	1.0346	Kizer
D1	Bismarck	North Dakota	United States	46.77	−100.83	0.018210	26.1970	1.7430	0.8880	1.0610	Crane
D1	Fargo	North Dakota	United States	46.90	−96.80	0.008000	43.6290	2.3370	0.8350	1.3900	Crane
B2	Williston	North Dakota	United States	48.12	−103.39	0.024404	28.6605	0.1619	0.3669	1.2325	Kizer
D2	Akron	Ohio	United States	40.55	−81.26	0.041598	23.9099	3.4154	2.0669	1.1222	Kizer
D2	Cincinnati	Ohio	United States	39.06	−84.26	0.138231	23.4580	0.2799	0.0130	1.0432	Kizer
D2	Cleveland	Ohio	United States	41.42	−81.87	0.030890	25.5830	3.7410	1.1770	1.0720	Crane
D2	Columbus	Ohio	United States	40.00	−82.88	0.021630	35.1400	3.2640	1.3570	1.0820	Crane
D2	Dayton	Ohio	United States	39.54	−84.13	0.063164	25.0586	7.1867	2.3049	0.9581	Kizer
D2	Sandusky	Ohio	United States	41.39	−82.65	0.063610	22.8906	1.7720	0.5866	1.2371	Kizer
D2	Toledo	Ohio	United States	41.35	−83.48	0.070535	24.0601	11.0137	1.9889	0.9217	Kizer
D2	Youngstown	Ohio	United States	41.15	−80.40	0.092621	21.6041	21.9363	2.5064	0.9084	Kizer
D2	Oklahoma City	Oklahoma	United States	35.40	−97.60	0.022000	36.8300	1.6230	1.8530	1.2130	Crane
D2	Tulsa	Oklahoma	United States	36.12	−95.53	0.085745	26.1616	12.3446	2.3986	1.0009	Kizer
B1	Baker	Oregon	United States	44.84	−117.81	0.034333	4.8604	0.2328	1.2051	1.2712	Kizer
C1	Burns	Oregon	United States	43.58	−119.00	0.005000	11.8330	2.1220	0.9010	0.9280	Crane
C1	Eugene	Oregon	United States	44.12	−123.17	0.005000	22.6320	5.1960	1.7470	0.8940	Crane

C1	Medford	Oregon	United States	42.38	−122.83	0.002000	26.3090	2.8950	1.2700	0.8860	Crane
C1	Portland	Oregon	United States	45.60	−122.67	0.004000	25.4250	5.4860	1.3590	0.8090	Crane
C1	Roseburg	Oregon	United States	43.12	−123.21	0.037219	4.7938	0.1912	1.0646	1.3610	Kizer
D2	Harrisburg	Pennsylvania	United States	40.13	−76.51	0.050140	19.5264	3.3012	1.8629	1.0982	Kizer
D2	Philadelphia	Pennsylvania	United States	39.88	−75.25	0.015000	34.0770	3.2610	1.7080	1.0930	Crane
D2	Pittsburgh	Pennsylvania	United States	40.50	−80.22	0.019498	32.3083	1.8287	1.9720	1.0460	Crane
D2	Wilkes Barre	Pennsylvania	United States	41.20	−75.44	0.048577	24.5318	4.7569	1.7192	1.0455	Kizer
G	San Juan	Puerto Rico	United States	18.26	−66.00	0.930309	16.8243	36.3133	0.2019	0.6785	Kizer
D2	Providence	Rhode Island	United States	41.72	−71.43	0.046664	23.5200	3.0158	1.0557	1.0928	Kizer
E	Charleston	South Carolina	United States	32.78	−79.93	0.239285	27.5924	101.1732	0.0028	1.1430	Kizer
D3	Columbia	South Carolina	United States	33.95	−81.12	0.042860	32.3290	2.2700	1.7610	1.2110	Crane
E	Florence	South Carolina	United States	34.11	−79.43	0.100404	30.9873	20.9810	2.2175	0.9237	Kizer
D3	Greenville-Spartanburg	South Carolina	United States	34.88	−82.22	0.253773	20.3494	28.0836	0.0146	0.9658	Kizer
D1	Aberdeen	South Dakota	United States	45.44	−98.41	0.046552	30.3674	3.8989	1.2848	1.0889	Kizer
D1	Rapid City	South Dakota	United States	44.05	−103.00	0.006000	40.8730	2.1380	1.0370	1.0680	Crane
D1	Sioux falls	South Dakota	United States	43.57	−96.73	0.034770	23.7640	2.0170	0.9000	1.2480	Crane
D3	Bristol	Tennessee	United States	36.28	−82.24	0.099804	24.9314	5.6079	0.0084	1.0130	Kizer
D3	Chattanooga	Tennessee	United States	35.02	−85.12	0.192426	21.5225	16.2158	0.0217	0.9221	Kizer
D3	Knoxville	Tennessee	United States	35.49	−83.59	0.276434	19.3723	101.3824	0.0242	0.9021	Kizer
D3	Memphis	Tennessee	United States	35.05	−90.00	0.047850	30.3520	2.2330	1.9800	1.1650	Crane
D3	Nashville	Tennessee	United States	36.12	−86.68	0.014000	37.2410	3.1200	2.0450	1.1060	Crane
D1	Abilene	Texas	United States	32.42	−99.68	0.019000	36.3090	1.1150	1.6640	1.2150	Crane
D1	Amarillo	Texas	United States	35.23	−101.67	0.009000	43.4110	1.3100	1.4980	1.2460	Crane

(Continued)

TABLE 11.A.1 (*Continued*)

Zone	City/ Location	State/ Province	Country	Latitude, degrees	Longitude, degrees	P_C	R_C	P_D	R_D	S_D	Source
D2	Austin	Texas	United States	30.28	−97.70	0.025000	38.3820	1.5380	1.5660	1.2810	Crane
D2	Brownsville	Texas	United States	25.90	−97.43	0.010000	28.9740	1.5200	1.8070	1.3470	Crane
D2	Corpus Christi	Texas	United States	27.77	−97.51	0.334617	21.2124	27.4150	0.0735	0.7753	Kizer
D2	Dallas-Fort Worth	Texas	United States	32.90	−97.03	0.015000	34.1890	1.6840	2.4070	1.1690	Crane
D1	Del Rio	Texas	United States	29.37	−101.00	0.009000	43.6260	0.9400	1.3490	1.3440	Crane
F2	El Paso	Texas	United States	31.80	−106.33	0.010000	24.4690	0.4170	2.4010	0.9130	Crane
D3	Houston	Texas	United States	29.97	−95.35	0.008000	48.1800	2.4400	2.3710	1.2500	Crane
D1	Laredo	Texas	United States	27.34	−99.30	0.161371	23.2980	0.0100	2.1650	0.8278	Kizer
F2	Lubbock	Texas	United States	33.65	−101.83	0.021000	29.7270	1.0270	1.3630	1.2560	Crane
F2	Midland	Texas	United States	31.95	−102.17	0.008000	40.7020	0.9510	1.4880	1.2610	Crane
D3	Port Arthur	Texas	United States	29.57	−94.01	0.406342	25.5252	101.4259	0.0502	0.8653	Kizer
D1	San Angelo	Texas	United States	31.21	−100.30	0.089851	21.9579	6.8986	1.3853	0.9588	Kizer
D2	San Antonio	Texas	United States	29.53	−98.47	0.077960	28.0067	4.6310	1.9997	1.0932	Kizer
D2	Victoria	Texas	United States	28.52	−96.56	0.035851	32.1962	6.8454	3.8890	0.9772	Kizer
D2	Waco	Texas	United States	31.37	−97.14	0.120863	27.4259	13.2000	3.5529	0.8452	Kizer
D2	Wichita Falls	Texas	United States	33.59	−98.30	0.678455	15.2618	0.5685	0.1936	0.6871	Kizer
F1	Salt Lake City	Utah	United States	40.78	−112.00	0.003000	29.6540	2.2790	1.2810	0.8670	Crane
D1	Burlington	Vermont	United States	44.47	−73.15	0.021150	27.7030	4.0690	1.1480	1.0110	Crane
D2	Lynchburg	Virginia	United States	37.20	−79.12	0.175727	15.3406	8.6397	1.1811	1.1779	Kizer
D2	Norfolk	Virginia	United States	36.90	−76.20	0.025180	33.4200	2.9250	1.7100	1.1520	Crane
D2	Richmond	Virginia	United States	37.52	−77.33	0.029860	33.7510	2.7280	1.5990	1.1510	Crane
D2	Roanoke	Virginia	United States	37.32	−79.97	0.012000	41.5070	3.1570	1.8340	1.0080	Crane

D2	Wallops Island	Virginia	United States	37.93	−75.48	0.004000	48.0510	3.0380	1.7090	1.1920	Crane
C1	Blaine	Washington	United States	49.00	−122.75	0.005000	16.4690	3.9470	2.0600	0.7670	Crane
C1	Port Angeles	Washington	United States	48.06	−123.32	0.009511	3.5598	0.0999	0.4671	1.5948	Kizer
C1	Quillayute	Washington	United States	47.95	−124.50	0.005000	16.4690	10.3240	1.8350	0.9200	Crane
B1	Seattle	Washington	United States	47.45	−122.33	0.003000	10.7450	5.3600	1.3900	0.8410	Crane
B1	Spokane	Washington	United States	47.63	−117.50	0.002000	23.1020	3.1280	1.1050	0.8120	Crane
B1	Tacoma	Washington	United States	47.25	−122.43	0.000030	142.1950	6.8700	0.8648	1.0150	Kizer
B1	Walla Walla	Washington	United States	46.10	−118.33	0.001000	34.8530	2.8440	1.1600	0.8510	Crane
C1	Yakima	Washington	United States	46.34	−120.33	0.009235	3.9003	0.0929	0.3925	1.7221	Kizer
D2	Beckley	West Virginia	United States	37.48	−81.07	0.071197	22.8728	2.6072	0.0066	1.0113	Kizer
D2	Charleston	West Virginia	United States	38.23	−81.35	0.091094	28.1972	2.0865	0.0128	0.9941	Kizer
D2	Elkins	West Virginia	United States	38.53	−79.51	0.133100	20.5668	2.6795	0.0191	0.9549	Kizer
D2	Huntington	West Virginia	United States	38.23	−82.33	0.148852	23.4514	13.3984	0.0497	0.8509	Kizer
D2	Parkersburg	West Virginia	United States	39.21	−81.26	0.060347	26.9060	6.6306	2.4323	0.9592	Kizer
D1	Green Bay	Wisconsin	United States	44.29	−88.08	0.015508	33.4129	2.9546	1.9170	1.1273	Kizer
D1	La Crosse	Wisconsin	United States	43.52	−91.15	0.026722	28.4508	3.1320	1.9947	1.1429	Kizer
D2	Madison	Wisconsin	United States	43.13	−89.33	0.031620	31.7040	2.2920	1.2300	1.0770	Crane
D2	Milwaukee	Wisconsin	United States	43.07	−88.03	0.034144	26.7960	10.8231	2.7907	0.9155	Kizer
B2	Casper	Wyoming	United States	42.54	−106.28	0.024617	19.6086	0.4181	1.1438	1.1149	Kizer
B2	Cheyenne	Wyoming	United States	41.15	−104.83	0.008000	31.2690	2.1080	0.9050	1.1200	Crane
B2	Lander	Wyoming	United States	42.82	−108.67	0.003000	29.4180	2.0840	1.1970	0.8420	Crane
B2	Sheridan	Wyoming	United States	44.77	−107.00	0.003000	33.1510	2.6510	1.0170	0.9030	Crane
F1	Yellowstone Park	Wyoming	United States	44.50	−110.50	0.157839	4.6853	2.5241	1.0679	1.0488	Kizer

Much of the tabular data from this chapter is available at the Wiley web site mentioned in the Preface.

REFERENCES

Abramowitz, M., and Stegun I., *Handbook of Mathematical Functions*, (NBS AMS 55, seventh printing with corrections), 931–933. Washington, DC: U.S. Government Printing Office, 1968.

Anderson, I., "Measurements of 20-GHz Transmission Through a Radome in Rain," *IEEE Transactions on Antennas and Propagation*, Vol. 23, pp. 619–622, September 1975.

Barnett, W. T., "Deterioration of Cross-Polarization Discrimination During Rain and Multipath Fading at 4 GHz," *IEEE International Conference on Communications*, Vol. 4, pp. 12D-1–12D-3, June 1974.

Bergmann, H. J., "Satellite Site Diversity: Results of a Radiometer Experiment at 13 and 18 GHz," *IEEE Transactions on Antennas and Propagation*, Vol. 25, pp. 483–489, July 1977.

Blevis, B. C., "Losses Due to Rain on Radomes and Antenna Reflecting Surfaces," *IEEE Transactions on Antennas and Propagation*, Vol. 13, pp. 175–176, January 1965.

Burgueno, A., Austin, J., Vilar, E. and Puigcerver, M., "Analysis of Moderate and Intense Rainfall Rates Continuously Recorded Over half a Century and Influence on Microwave Communications Planning and Rain-Rate Data Acquisition," *IEEE Transactions on Communications*, Vol. 35, pp. 382–395, April 1987.

Capsoni, C., Luini, L., Paraboni, A., Riva, C. and Martellucci, A., "A New Prediction Model of Rain Attenuation that Separately Accounts for Stratiform and Convective Rain," *IEEE Transactions on Antennas and Propagation*, Vol. 57, pp. 196–204, January 2009.

Chu, T. S., "Rain-Induced Cross-Polarization at Centimeter and Millimeter Wavelengths," *Bell System Technical Journal*, Vol. 53, pp. 1557–1579, October 1974.

Crane, R. K., "Analysis of the Effects of Water on the ACTS Propagation Terminal Antenna," *IEEE Transactions on Antennas and Propagation*, Vol. 50, pp. 954–965, July 2002.

Crane, R. K., "Attenuation Due to Rain—A Mini-Review," *IEEE Transactions on Antennas and Propagation*, Vol. 23, pp. 750–752, September 1975.

Crane, R. K., "Prediction of Attenuation by Rain," *IEEE Transactions on Communications*, Vol. 28, pp. 1717–1733, September 1980.

Crane, R. K., "A Two-Component Rain Model for the Prediction of Site Diversity Performance," *Radio Science*, Vol. 24, pp. 641–665, September-October 1989.

Crane, R.K. 1996. *Electromagnetic Wave Propagation Through Rain*, 107–184. New York: John Wiley & Sons, Inc.

Crane, R. K., *Propagation Handbook for Wireless Communication System Design*. Boca Raton: CRC Press, pp 225–301, 2003.

Dutton, E. J. and Dougherty, H. T., "Year-to-Year Variability of Rainfall for Microwave Applications in the U.S.A," *IEEE Transactions on Communications*, Vol. 27, pp. 829–832, May 1979.

Effenberger, J. A. and Strickland, R. R., "The Effects of Rain on a Radome's Performance," *Microwave Journal*, pp. 261–272, May 1986.

Emiliani, L. D., Luini, L. and Capsoni, C., "Analysis and Parameterization of Methodologies for the Conversion of Rain-Rate Cumulative Distributions from Various Integration Times to One Minute," *IEEE Antennas and Propagation Magazine*, Vol. 51, pp. 70–84, June 2009.

Emiliani, L. D., Luini, L. and Capsoni, C., "On the Optimum Estimation of 1-Minute Integrated Rainfall Statistics from Data with Longer Integration Time," *European Conference on Antennas and Propagation 2010 (EuCAP 2010)*, April 2010.

Goldstein, H., Kerr, D. E. and Bent, A. E., "Meteorological Echoes," *Propagation of Short Radio Waves*. Kerr, D. E. Editor. New York: McGraw-Hill, pp. 588–640, 1951.

Gunn, R. and Kinzer, G. D., "The Terminal Velocity of Fall for Water Droplets in Stagnant Air," *Journal of Meteorology*, Vol. 6, pp. 243–248, August 1949.

Hershfield, D. M., "Estimating the Extreme-Value 1-Minute Rainfall," *Journal of Applied Meteorology*, Vol. 11, pp. 936–940, September 1972.

Hodge, D. B., "Frequency Scaling of Rain Attenuation," *IEEE Transactions on Antennas and Propagation*, Vol. 25, pp. 446–447, May 1977.

International Telecommunication Union—Radiocommunication Sector (ITU-R), "ITU-R Recommendation P.840-3, Attenuation Due to Clouds and Fog," *ITU-R Recommendations*. Geneva: International Telecommunication Union, 1999.

International Telecommunication Union—Radiocommunication Sector (ITU-R), "ITU-R Recommendation P.838-3, Specific Attenuation Model for Rain for Use in Prediction Methods," *ITU-R Recommendations*. Geneva: International Telecommunication Union, 2005.

International Telecommunication Union—Radiocommunication Sector (ITU-R), "ITU-R Recommendation P.837-5, Characteristics of Precipitation for Propagation Modeling," *ITU-R Recommendations*. Geneva: International Telecommunication Union, 2007a.

International Telecommunication Union—Radiocommunication Sector (ITU-R), "ITU-R Recommendation P.1144-4, Guide to the Application of the Propagation Methods of Radio communications Study Group 3," *ITU-R Recommendations*. Geneva: International Telecommunication Union, 2007b.

International Telecommunication Union—Radiocommunication Sector (ITU-R), "ITU-R Recommendation P.530-13, Propagation Data And Prediction Methods Required for the Design of Terrestrial Line-of-sight Systems," *ITU-R Recommendations*. Geneva: International Telecommunication Union, 2009.

Jones, D. M. A., "The Shape of Raindrops," *Journal of Applied Meteorology*, Vol. 16, pp. 504–510, June 1959.

Jones, D. M. A. and Sims, A. L., "Climatology of Instantaneous Rainfall Rates," *Journal of Applied Meteorology*, Vol. 17, pp. 1135–1140, August 1978.

Jung, M., Han, I., Choi, M., Lee, J. and Pack, J., "Study on the Empirical Prediction of 1-min Rain Rate Distribution from Various Integration Time Data," *Proceedings of the Korea-Japan Microwave Conference (KJMW)*, pp. 89–92, November 2007.

Kanellopoulos, J. D. and Koukoulas, S. G., "Analysis of the Rain Outage Performance of Route Diversity Systems," *Radio Science*, Vol. 22, pp. 549–565, July-August 1987.

Kheirallah, H. N., Knight, J. P., Olsen, R. L., McCormick, K. S. and Segal, B. "Frequency Dependence of Effective Path Length in Prediction of Rain Attenuation Statistics," *Electronics Letters*, Vol. 16, pp. 448–450, June 1980.

Kizer, G., "Improvements in Fixed Point-to-Point Microwave Radio Path Design," *IEEE Wireless Transmission Symposium (WTS) 2008*, pp. 354–359, April 2008.

Knight, W., "Rain Outage Models," *Fixed Wireless Transmission Engineering Seminar*, Plano: Alcatel Network Systems, Slide. 9, June 1999.

Laws, J. O. and Parsons, D. A., "The Relation of Raindrop-Size to Intensity," *American Geophysical Union Transactions of 1943, part II (April, 1943)*, Vol. 24, pp. 452–460, January 1944.

Lin, S. H., "Statistical Behavior of Rain Attenuation," *Bell System Technical Journal*, Vol. 52, pp. 557–581, April 1973.

Lin, S. H., "A Method for Calculating Rain Attenuation Distributions on Microwave Paths," *Bell System Technical Journal*, Vol. 54, pp. 1051–1086, July-August 1975.

Lin, S. H., "Dependence of Rain-Rate Distribution on Rain-Gauge Integration Time," *Bell System Technical Journal*, Vol. 55, pp. 135–141, January 1976.

Lin, S. H., "Nationwide Long-Term Rain Statistics and Empirical Calculation of 11-GHz Microwave Rain Attenuation," *Bell System Technical Journal*, Vol. 56, pp. 1581–1604, November 1977.

Lin, S. H. and Pursley, M. V., "Rain Attenuation Model for Short Microwave Radio Paths," *IEEE Global Telecommunications Conference (Globecom) Conference Record*, pp. 9.4.1–9.4.4, December 1985.

Lin, S. H., Bergmann, J. J. and Pursley, M. V., "Rain Attenuation Distributions on Earth-Satellite Paths—Summary of 10 Year Experiments and Studies," *Bell System Technical Journal*, Vol. 59, pp. 183–228, February 1980.

Marshall, J. S. and Palmer, W. McK., "The Distribution of Raindrops with Size," *Journal of Meteorology (Atmospheric Science)*, Vol. 5, pp. 165–166, August 1948.

Medhurst, R. G., "Rainfall Attenuation of Centimeter Waves: Comparison of Theory and Measurement," *IEEE Transactions on Antennas and Propagation*, Vol. 13, pp. 550–564, July 1965.

Olsen, R. L., Rogers, D. V. and Hodge, D. B., "The aR^b Relation in the Calculation of Rain Attenuation," *IEEE Transactions on Antennas and Propagation*, Vol. 26, pp. 318–329, March 1978.

Osborne, T. L., "Rain Outage Performance of Tandem and Path Diversity 18-GHz Short Hop Radio Systems," *Bell System Technical Journal*, Vol. 50, pp. 59–79, January 1971.

Osborne, T. L., "Applications of Rain Attenuation Data to 11-GHz Radio Path Engineering," *Bell System Technical Journal*, Vol. 56, pp. 1605–1627, November 1977.

Paulson, K. S., Watson, R. J. and Usman, I. S., "Diversity Improvement Estimation From Rain Radar Databases using Maximum Likelihood Estimation," *IEEE Transactions on Antennas and Propagation*, Vol. 54, pp. 168–174, January 2006.

Pruppacher, H. R. and Pitter, R. L., "A Semi-empirical determination of the shape of cloud and raindrops," *Journal of Atmospheric Sciences*, Vol. 28, pp. 86–94, 1971.

Rice, P. L. and Holmberg, N. R., "Cumulative Time Statistics of Surface-Point Rainfall Rates," *IEEE Transactions on Communications*, Vol. 21, pp. 1131–1136, October 1973.

Rummler, W. D., "Advances in Microwave Radio Route Engineering for Rain," *IEEE Conference on Communications (ICC) Proceedings*, Vol. 1, pp. 10.8.1–10.8.5, June 1987.

Segal, B., *High-Intensity Rainfall Statistics for Canada, CRC Report No. 1329-E*. Ottowa: Communications Research Centre, Department of Communications, Canada, 1979.

Skarpiak, C., "Teglar/Hypalon Gain and Return Loss Comparison," *private letter to M. Evans*. Orland Park, Illinois: Andrew Corporation, 20 July 1988.

Tattelman, P. and Grantham, D. D., *A Survey of Techniques for Estimating Short-duration Precipitation Rate Statistics, AFGL-TR-0357*. Hanscom AFB: Air Force Geophysics laboratory, Air Force Systems Command, 1982.

Tattelman, P. and Grantham, D. D., *Northern Hemisphere Atlas of 1-Minute Rainfall Rate, AFGL TR-83-0267*. Hanscom AFB: Air Force Geophysics laboratory, Air Force Systems Command, 1983a.

Tattelman, P. and Grantham, D. D., *Southern Hemisphere Atlas of 1-Minute Rainfall Rate, AFGL TR-83-0285*. Hanscom AFB: Air Force Geophysics laboratory, Air Force Systems Command, 1983b.

Tattelman, P. and Grantham, D. D., "A Review of Models for Estimating 1 min Rainfall Rates for Microwave Attenuation Calculations," *IEEE Transactions on Communications*, Vol. 33, pp. 361–372, April 1985.

12

DUCTING AND OBSTRUCTION FADING

12.1 INTRODUCTION

As an unconstrained electromagnetic wave (such as light or a radio signal) travels between transmit and receive locations, it loses power. The first type of loss, of course, is free space loss. Free space loss is the loss of power that occurs as the wave front expands with distance without encountering any obstructions, reflections, or atmospheric effects. The electromagnetic wave power is the (cross) product of (mutually orthogonal) electric and magnetic fields. Each field expands linearly with distance, so the resultant power decreases as a function of distance squared. For propagation in the atmosphere near the Earth, the wave may be trapped in or reflected (by a duct) or blocked (obstructed) during rare but severe refractive index variations from normal conditions. This results in path transmission losses that can be significantly less or greater than free space loss calculations would predict.

The atmosphere above the earth's surface up to about 1.5 miles is reasonably static (except for slow changes) and the conventional standard (exponential atmospheric refractive index) model (Bean and Dutton, 1966) is quite accurate. Below a height of about 1.5 miles, below this height the atmosphere is highly influenced by the earth's surface. This transition point is termed the *atmospheric boundary layer* and its study is known as *boundary layer meteorology*. The area closest to the Earth is termed the *surface layer*. The surface layer is generally defined as the first 100 m above the terrain. This layer has the greatest influence on terrestrial microwave radio paths. The surface layer is relatively unstable and highly influenced by temperature and humidity microvariations of the earth's surface. The linear atmospheric refractive index model for a "normal" atmosphere does not always apply.

All electromagnetic waves (including light, heat, and radio waves) travel in straight lines through any isotropic, homogeneous medium with a velocity $v = c/n$, where c is the velocity of light in a vacuum and n is the index of refraction of the medium. The velocity of the wave and its direction of travel changes with a change in the index of refraction. While the index of refraction of air is ordinarily considered to be unity, its actual value is slightly greater and varies with the state of the atmosphere.

$$n = \text{index of refraction} \cong 1.000319 \tag{12.1}$$

$$\begin{aligned} N &= \text{refractivity} = (n-1)10^6 \\ &= \text{dry component} + \text{wet component} \end{aligned} \tag{12.2}$$

Digital Microwave Communication: Engineering Point-to-Point Microwave Systems, First Edition. George Kizer.

dry component $= [77.6p]/[273 + T]$;
wet component $= [3.73 \times 10^5 e_S H_R]/[273 + T]^2$;
$p =$ atmospheric pressure (mbar);
$=$ 1.33 (pressure in millimeters of mercury);
$=$ 33.9 (pressure in inches of mercury);
$T =$ temperature (°C);
$H_R =$ relative humidity (%)/100;
$e_S =$ saturation vapor pressure (mbar) for temperature T;
$= 6.11 \times 10^{[(7.50T)/(273 + T)]}$.

The above equation is a simplification derived by Bean and Dutton (1966). It applies to microwave radio frequencies less than 40 GHz. Above 40 GHz, atmospheric oxygen and water vapor resonances add a reactive component to the refractive index. For light waves, only the dry component applies. A standard atmosphere (Bean and Dutton, 1966; International Telecommunication Union—Radiocommunication Sector (ITU-R), 2005) can be defined as a hypothetical atmosphere in which the properties are chosen to fit average conditions. The US standard atmosphere is one that averages summer and winter conditions over the United States. For the US standard atmosphere, p is 1013 mbar, T is +15 °C, and H_R is 60%. This yields a refractivity of air at sea level of 319 for radio waves and 273 for light. The height variation of refractivity N through the normal atmosphere can be approximated by the following (per convention, using metric units) (Bean and Dutton, 1966):

$$\begin{aligned} N &= N_S e^{-C_h} \\ &= N_S e^{-h/h_0} \end{aligned} \tag{12.3}$$

$N =$ refractivity at height h;
$N_S =$ surface value of refractivity;
$h =$ height above mean sea level;
$h_0 =$ reference height (mean sea level);
$C_h = \ln[N_S/(N_S + \Delta N)]$;
$\ln(x) = \log_e(x)$;
$\Delta N =$ mean change of N in the first kilometer of height;
$N' = dN/dh$;
$=$ rate of change of reactivity;
$= [N_S/h_0]e^{-h/h_0}$.

While the rate of change of the atmospheric refractivity is roughly exponential for terrestrial microwave paths nearly tangential to the Earth, this rate of change may be approximated as linear (normally −63 N units/mile or −39 N units/km at sea level). This approximation is quite accurate for radio paths within the first mile above sea level.

Under normal atmospheric conditions, radio and light waves curve toward the Earth. If the Earth were replaced by a sphere of radius aK (where a is the physical radius of the Earth and K is a function of refractivity), a radio or light wave launched parallel to the Earth would remain parallel to this modified "Earth." For a radio wave nearly parallel to the Earth, the following approximates K:

$$\begin{aligned} K &\approx \frac{1}{[1 + a(dn/dh)]} \\ &= \frac{253}{253 + [dN/dh(\text{N units/mile})]} \\ &= \frac{157}{157 + [dN/dh(\text{N units/km})]} \end{aligned}$$

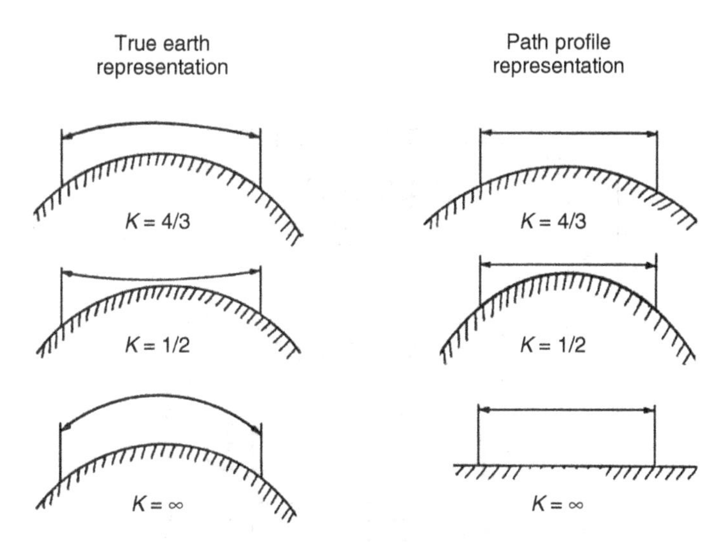

Figure 12.1 Path profiles as a function of atmospheric refractivity.

$$= \text{typically } \frac{6}{5}(\text{average}) \text{ to } \frac{7}{5}(\text{midday}) \text{ for light}$$

$$= \text{typically } \frac{4}{3}(\text{average}) \text{ for radio waves } <40\,\text{GHz} \quad (12.4)$$

a = physical earth radius;
K = effective earth radius factor.

Microwave paths are typically designed and analyzed using a plot of the terrain between the transmitter and the receiver with a line representing the maximum power of the radio wave front as it moves from transmitter to receiver. The radio wave path will bend up or down depending on the K factor (atmospheric refractivity). The historical convention is to always plot the radio wave path as a straight line and to move the Earth up or down (adding terrain height using equation 5.36) as a function of the K factor to preserve the vertical distance between the radio wave and the Earth at any location on the path (Fig. 12.1).

For a normal K factor of 4/3 ($dN/dh = -40$ N units/km), the radio wave bends down slightly (or the Earth bulges slightly) as it moves from transmitter to receiver. As the K factor increases (or becomes negative), the radio wave bends down more (or the Earth flattens or becomes concave). As the K factor decreases, the radio wave bends up (or the Earth bulges more) (Bean and Dutton, 1966; International Telecommunication Union—Radiocommunication Sector (ITU-R), 2007a; International Telecommunication Union—Radiocommunication Sector (ITU-R), 2005).

A normal atmosphere contains well-mixed, homogeneous air that decreases in temperature as height is increased. Mountainous areas ("high, dry, and windy") such as the Alps, Andes, Atlas, Himalayas, and the Rockies are good propagation areas. Difficult propagation areas usually have high humidity and/or temperature and poorly mixed low altitude atmosphere (often due to little air turbulence with abnormal temperature and/or humidity conditions). Tropical coastal areas and most desert regions fall into this category. In these areas, unusual conditions can cause an inversion of the normal, slightly decreasing, refractive index with height. Figure 12.2 (Samson, 1975) illustrates the extremes of radio wave atmospheric refractivity.

12.1.1 Power Fading

Variations in the atmospheric refractive index can cause the radio path of propagation to vary. Variation of atmospheric refraction can cause change in the apparent angle of arrival of the LOS ratio ray, particularly in the vertical plane. This can cause an effective reduction in gain in the receive antenna (Kizer, 1990) called *power fading*.

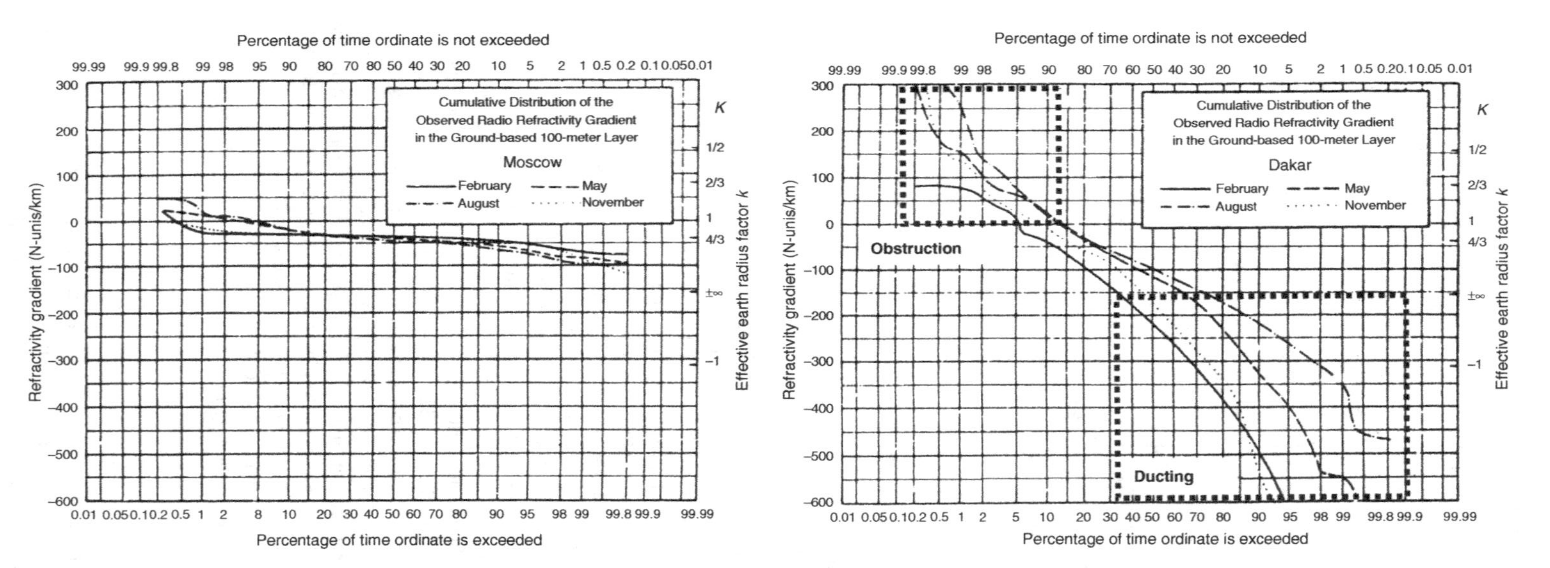

Figure 12.2 Refractive index in Moscow and Dakar, Africa. *Source*: From Samson, C. A., *Refractivity Gradients in the Northern Hemisphere*, Office of Telecommunications Report 75–59. Washington, DC: U.S. Department of Commerce, 1975.

Dougherty (Dougherty, 1968; Kizer, 1990) has shown that the variation in angle of arrival can be calculated from the following:

$$\theta(K_2) - \theta(K_1) = -0.00722\ D\ \text{(miles)} \left[\frac{1}{K_2} - \frac{1}{K_1}\right]$$

$$= -0.00449\ D\ \text{(km)} \left[\frac{1}{K_2} - \frac{1}{K_1}\right] \qquad (12.5)$$

$\theta(K)$ = angle of arrival for a given K value;
D = distance between the transmitter and the receiver.

Angle of arrival variation of as much as 0.75° has been observed (Kizer, 1990). Angle of arrival variation seems to increase with path distance. For this reason, it is recommended that high gain antennas (with narrow beamwidths) be avoided on long paths. If high gain antennas must be used, use space diversity with a smaller antenna (with wider beamwidth). This will reduce conventional space diversity improvement (slightly) but reduce power fading (significantly).

12.2 SUPERREFRACTION (DUCTING)

If a layer of air with unusually rapid decrease of water vapor with height and/or increase in temperature with height exists near the Earth, electromagnetic refractivity decreases very quickly with height. This condition, called *superrefraction* or *ducting*, causes the curvature of radio rays to exceed the curvature of the earth's surface. This effect can trap or reflect radio waves. Superrefraction (ducting) is usually associated with refractive indexes less than −157 N units/km ($K < 0$). Figure 12.3 (Bean et al., 1966) illustrates locations where ducting can be expected.

Superrefraction is formed when cool moist air appears below warm dry air (a temperature or humidity inversion). This can occur in many ways. It can occur when low moist air moves from a lake onto a desert (advection) or evaporation occurs from vegetation or irrigated areas (especially near warm mountain slopes). Solar heating can cause high moisture layers near bodies of water. Subsidence is a term to describe a slow settling of high altitude dry air from a high pressure system. The air mass is warmed by compression and entraps a cooler, moist air mass supported by surface moisture. Subsidence can cause surface or elevated ducts. Nocturnal radiation cooling can create a temperature inversion of heavy cold air below lighter warm air. Sometimes, a low fog layer is indicative of an inversion. However, a fog layer has little effect in ducting without a temperature inversion. In general, strong or gusty winds will breakup superrefactive layers by stirring the atmosphere. However, low level sea breezes can increase the layering.

Optically, superrefraction is called a *superior mirage*. Under these conditions, an image below the optical horizon appears inverted above the actual object. An image above or at the optical horizon appears erect above the actual object.

Ducts may be divided into two basic types: ground or elevated. Figure 12.4 illustrates normal propagation as well as propagation through an elevated duct.

Notice that if both transmit and receive antennas are in the duct, several different paths exist between the transmitter and receiver. Owing to different propagation delays, these signals arriving at the receiver cause significant variation in received signal level. Propagation conditions can be expected to vary because of variations in refractivity over time. Figure 12.5 is an example of this.

The underside of a ground-based duct is in contact with the earth's surface while the underside of an elevated duct is above the earth's surface and overlies another layer of air. Prolonged fading or signal enhancement may result from propagation through the duct. Signals may be trapped within the duct and propagated far beyond the normal radio horizon. This can cause the received signal level to be much greater than normal ("upfade"). It can also cause the received signal to go away (Fig. 12.6).

Ducting has been observed in many areas of the world (Anderson and Gossard, 1955; Day and Trolese, 1950; Dougherty and Dutton, 1981; Dougherty and Hart, 1976; Dougherty, 1968, 1979; Dutton, 1982; Fruchtenicht, 1974; Hubbard, 1979; Ikegami, 1959; Katzin et al., 1947; Liebe and Giommestad, 1978; Mahmoud et al., 1987; Schiavone, 1982). Ducts are usually stationary but sometimes

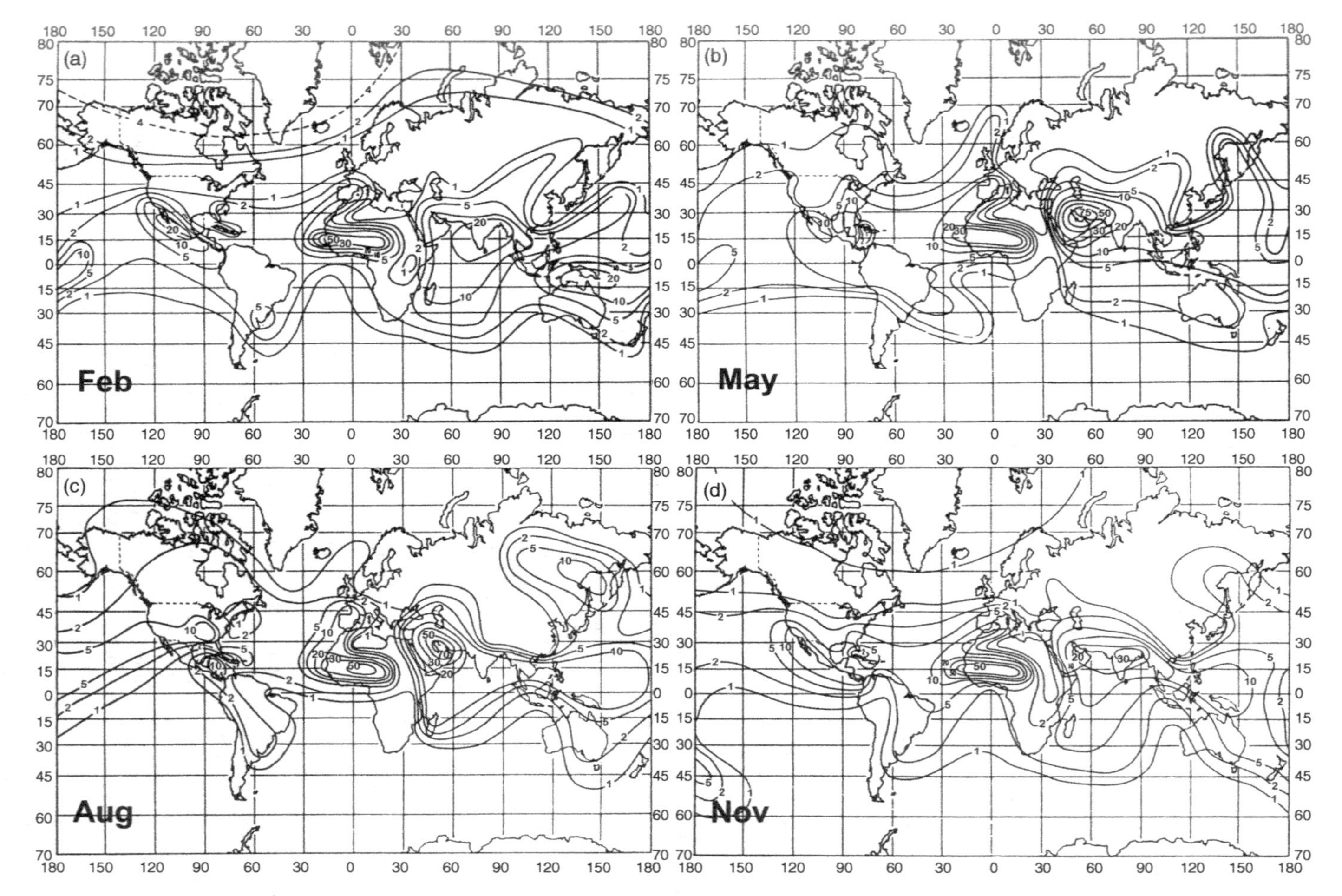

Figure 12.3 Percentage of time $N' \leq -157$ N/km, (a) February, (b) May, (c) August, and (d) November. *Source*: From Bean, B. R., Cahoon, B. A., Samson, C. A. and Thayer, G. D., *A World Atlas of Atmospheric Radio Refractivity*, ESSA Monograph 1. Washington, DC: U.S. Department of Commerce, 1966.

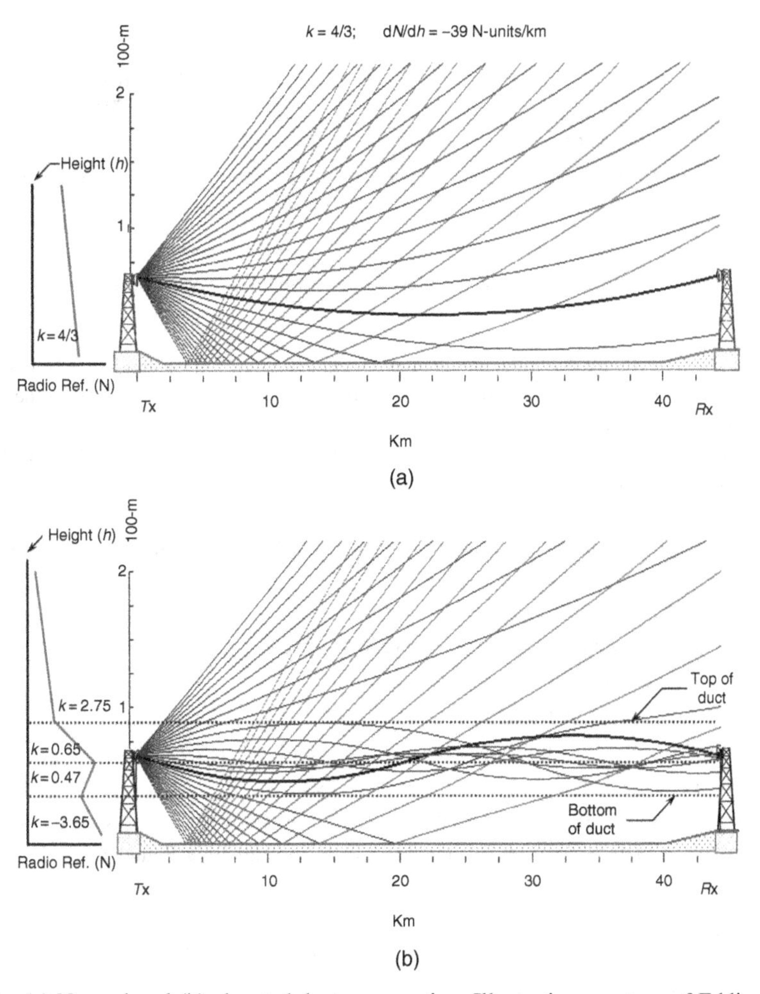

Figure 12.4 (a) Normal and (b) elevated duct propagation. Illustration courtesy of Eddie Allen. Used with permission.

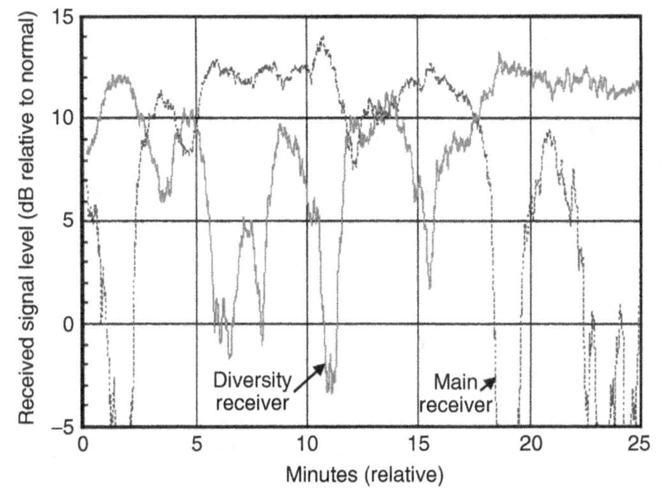

Figure 12.5 Propagation in a duct.

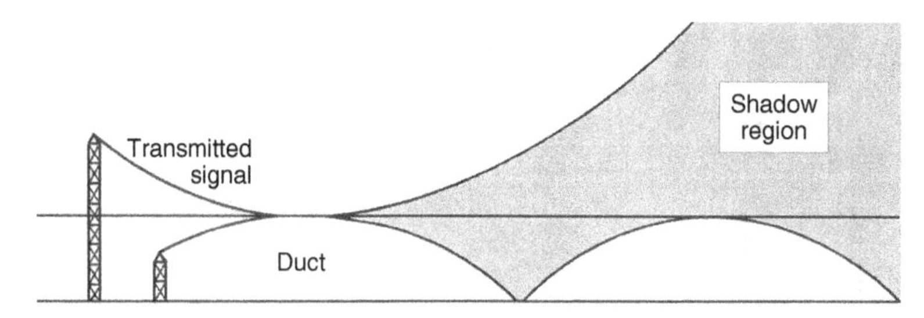

Figure 12.6 Example of duct propagation variation.

they move up through the lower atmosphere. It is especially common on land–sea transitions or along coastlines in tropical areas. It is common on the seas. It is also common at night and in the early morning hours in desert areas (Day and Trolese, 1950).

Ducting has been observed in the following areas (Bean et al., 1960, 1966; Dougherty and Dutton, 1981; Dougherty and Hart, 1976; Samson, 1975, 1976):

- Continental west coast areas at latitudes 20°–35° (N or S) in the summer and 10°–25° (N or S) in winter;
- Tropical ocean areas subjected to trade winds;
- Siberia, the Canadian interior, and the Antarctic in winter;
- Eastern Mediterranean and Black Sea during summer;
- Southeast Arabia during summer;
- Northern Arabian Sea (including the Gulf of Aden and the Persian Gulf);
- In Africa between Dakar, Senegal, and Fort Lamy, Chad;
- India, Bay of Bengal, Southeast Asia, Indonesia, and the northern tip of Australia;
- Southwest coast of North America (Southern California and Baja Peninsula), the California coast and a portion of the North Pacific;
- Gulf of Mexico and Caribbean area;
- Northwest coast of Africa and western Mediterranean;
- Antarctica in winter.

One of the worst propagation areas is the Saudi Arabian peninsula. Figure 12.7 is an example of severe ducting. The flat topping of the bottom two plots was due to receiver limiting of upfading received signals. The upper two paths began the day with nominal fading but ended the day with significant fading.

A particularly severe form of fading produced by surface ducts has been termed *blackout fading* (Laine, 1975). Low clearance paths traversing areas supporting superrefractive ground-based layers can experience complete loss of signal for periods exceeding a day. The fades are sudden, catastrophic, and not frequency dependent (within the same frequency band). An invisible, rising atmospheric layer (sometimes associated with visible steam fog formed over warm water or moist ground) may intercept and trap the signal. The failure is sometimes preceded by reflection fades and obstruction fading.

Blackout fading has been observed in the following areas:

- Roswell, New Mexico
- Imperial Valley and southern coast, California
- Cape Hatteras, North Carolina
- Pacific Northwest, United States
- In Texas near Houston, San Antonio, Lubbock, and the Oklahoma Panhandle
- Syrian and Arabian coasts
- Australian coasts and North West Cape.

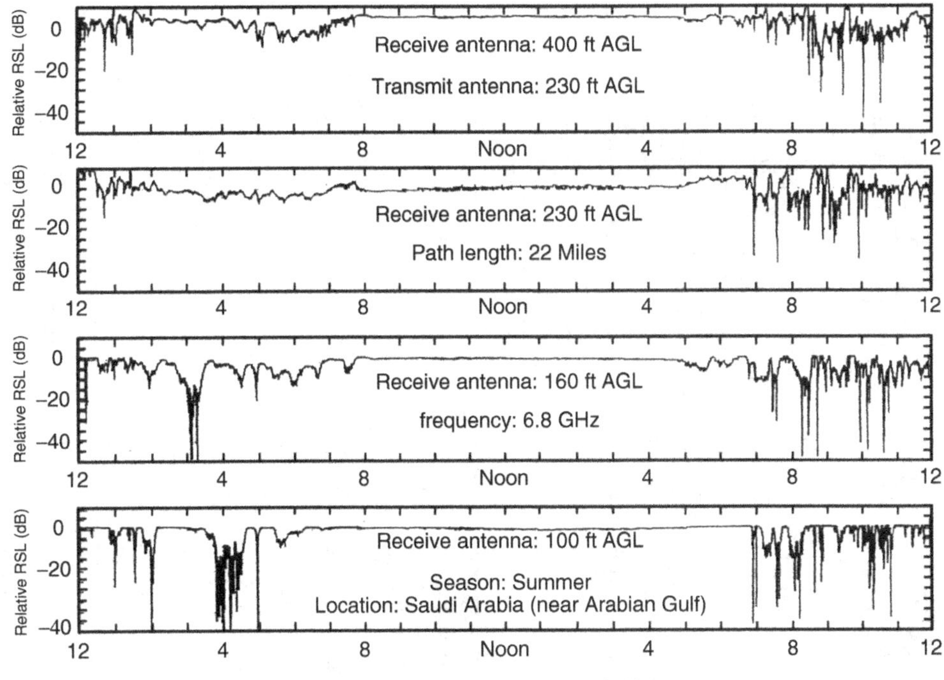

Figure 12.7 Example of duct fading.

At present, there are few path engineering guidelines for propagation in ducting environments. Dougherty (Dougherty and Dutton, 1981; Dougherty and Hart, 1976; Dougherty, 1979) and International Telecommunication Union—Radiocommunication Sector (ITU-R) (2003) offer methods for predicting probability of ducting but provide no path engineering guidelines for mitigating it.

The best defense against ducting fades is essentially the same as for obstruction fading: tall towers on short paths. If that is not possible, other approaches can help. A high–low path with large launch angles (at least 1°) is one approach. Orienting an antenna slightly up or down can help. Another approach is to use space diversity with main to main and diversity-to-diversity radio paths as parallel to the terrain as possible. Using smaller antennas (to increase radio vertical beamwidth) may help. Sometimes, very high main and very low diversity antennas are effective for unusual situations such as ducting over large bodies of water. At present, path engineering in a ducting environment is based on experience more than science.

12.3 SUBREFRACTION (EARTH BULGE OR OBSTRUCTION)

If refractivity increases with height near the Earth (owing to unusually high humidity or temperature gradients), a positive refractive gradient condition called *subrefraction* (also termed *obstruction* or *bulge*) exists. In this case, there is upward bending of the electromagnetic wave rather than the normal, slightly downward bending. This bending may become so extreme that the radio wave cannot reach the receive antenna (Fig. 12.10b). This phenomenon is known as *obstruction fading*. Using short paths and/or tall towers usually moderates this condition. Subrefraction (obstruction or bulge) is usually associated with refractive indexes greater than 0 N units/km ($0 < K < 1$). Figure 12.8 (Bean et al., 1966) illustrates locations where obstruction fading can be expected:

Subrefraction happens when cool moist air appears above warm dry air. It can occur when a high level moist weather front moves from a lake onto a desert. Solar heating of the Earth can cause subrefraction. In most cases, surface wind must be low or nonexistent for this condition to occur (high or gusty wind stirs up the atmosphere, breaking up abnormal layers). However, high level wind from the sea can increase subrefraction. Advection, the movement of one air type over another, is common in coastal areas. In

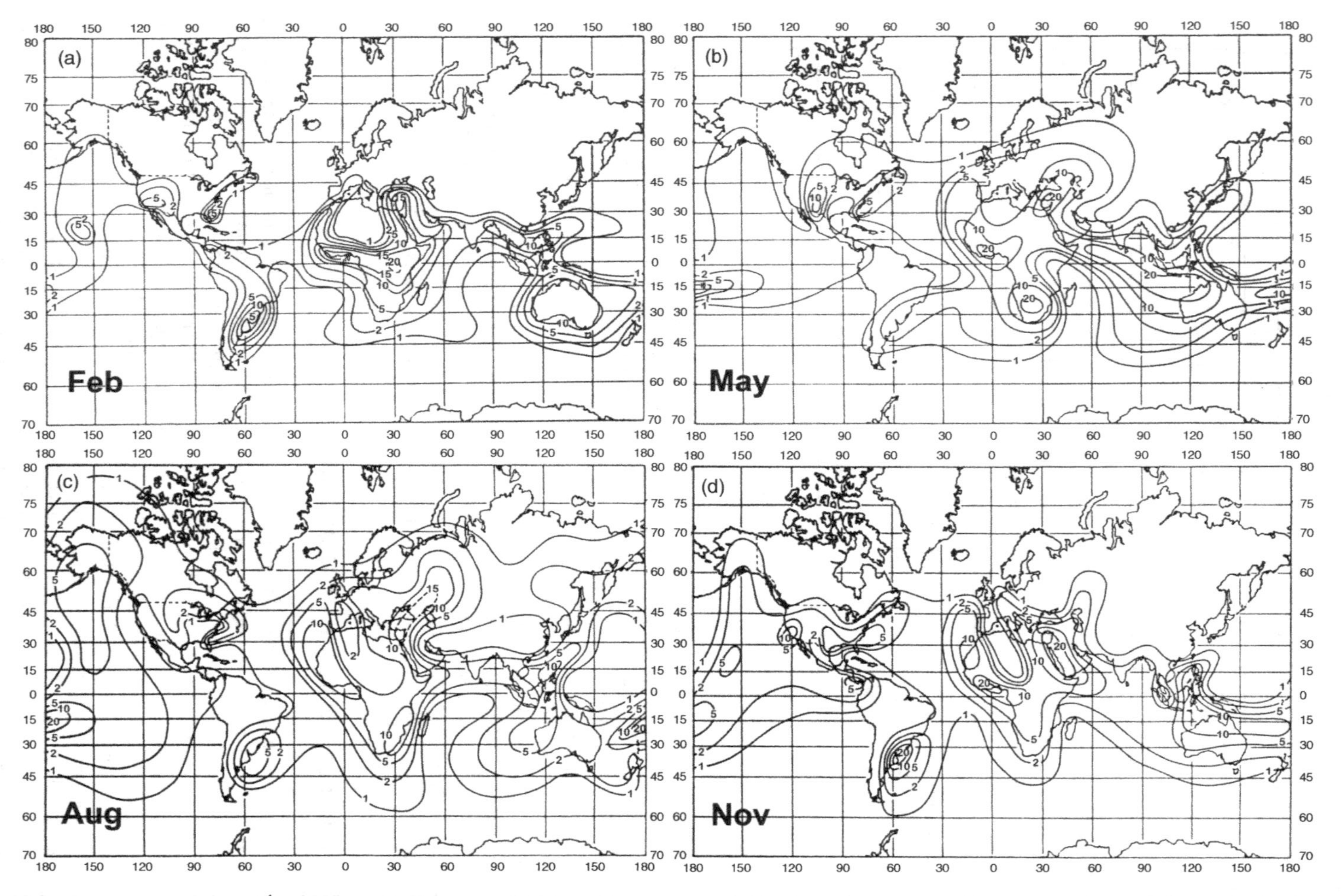

Figure 12.8 Percentage of time $N' \geq 0$ N/km, (a) February, (b) May, (c) August, and (d) November. *Source*: From Bean, B. R., Cahoon, B. A., Samson, C. A. and Thayer, G. D., *A World Atlas of Atmospheric Radio Refractivity*, ESSA Monograph 1. Washington, DC: U.S. Department of Commerce, 1966.

these areas, humid air originating from an ocean or lake moves inland over existing dry air. Nocturnal radiation cooling can produce ground fog. Air refractivity with fog is smaller than air refractivity with the same water in vapor form. Relatively flat areas, areas within 200 miles of a large water mass, and fine weather conditions with little or no wind have all been correlated to subrefraction.

Optically, subrefraction is called an *inferior mirage*. Under these conditions, an image above the optical horizon appears inverted below the actual object. "Pools of water" on hot highways is a common example of this.

Obstruction fading has been observed in the following areas (Samson, 1975, 1976; Bean et al., 1960, 1966; Lee, 1985, 1986; Vigants, 1972; Achariyapaopan, 1986):

- Large desert steppe regions such as the Sahara, the Australian interior, the southwestern United States desert (such as Las Vegas), and the Asian land area southeast of the Caspian Sea
- High plateau areas
- Southeast coast of the United States (Texas to North Carolina)
- Great Lakes area in the United States
- Hawaiian Islands
- South Africa
- Southeast coast of South America
- Southern California
- Northern Indian Ocean
- Isthmus of Panama
- Eastern Mediterranean
- Red Sea
- Indonesian and Southwest Pacific area
- Ivory Coast and Ghana lowlands of Africa.

While most North American transmission engineers are conditioned to expect obstruction fading near large bodies of water (e.g., south of the Great Lakes and near the coast of the Gulf Coast), many are unaware that the worst obstruction fading occurs in desert areas (the worst measured obstruction fading conditions in North America occur near Las Vegas). Figure 12.9 is an example of severe desert obstruction fading. Notice that the top path does not have obstruction fading. Although the top path is high enough to avoid obstruction fading, it is being subjected to severe ducting.

12.4 MINIMIZING OBSTRUCTION FADING

Notice that the top path does not have obstruction fading. Although the top path is high enough to avoid obstruction fading, it is being subjected to severe ducting. The performance of terrestrial fixed point to point microwave radio paths is greatly influenced by three propagation phenomena: rain, multipath, ducting, and obstruction. The remainder of this chapter will focus on obstruction. Obstruction fading is caused by a large positive gradient in the atmospheric refractivity. The effect is to cause the radio wave path ("ray") to bend so much that no radio path is possible between the transmitter and the receiver (Fig. 12.10). The effect of obstruction fading is to cause both the main and diversity received signal levels to be depressed for several minutes. These fades can last over an hour (until the positive gradient air layer has been stirred up or pushed aside) (Fig. 12.11).

12.4.1 Path Clearance (Antenna Vertical Placement) Criteria

High frequency (>13 GHz) antennas have such a narrow radiation pattern and the paths are so short that as long as the transmitter-to-receiver path has adequate clearance (to avoid power blockage), excessive vertical path clearance is not an issue. The assumption is that the paths are too short to be affected by refractive index changes and the antenna beamwidths are too small to illuminate significant terrain. These

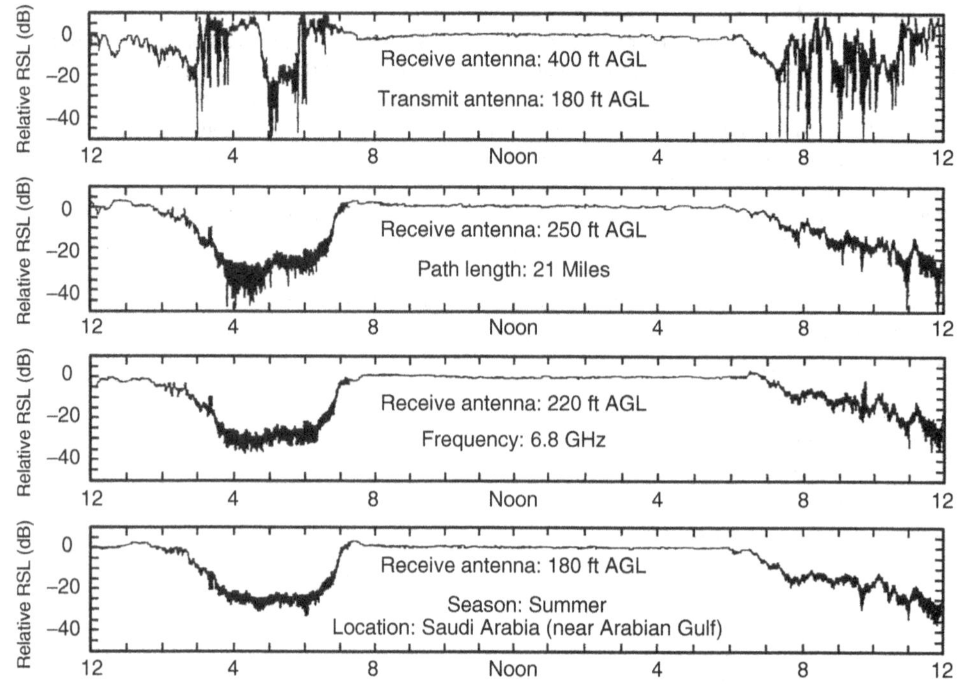

Figure 12.9 Example of obstruction fading.

paths are usually designed for a 1.0 F_1 clearance at $K = 4/3$ where F_1 is the first Fresnel zone height. Often the vertical path clearance is much greater than this. These paths require an inspection to verify that path or building reflections are not significant. If they pass that check, the path is considered adequate.

For longer low frequency (<13 GHz) paths, the terrain between the transmitter and receiver are illuminated with radio energy. Reflections from the terrain can affect the received signal. In addition, changes in the refractive index along the path can affect the received signal. The microwave path designer's objective is to place the transmission antennas at such a height that satisfactory path performance is achieved (obstruction fading is not excessive).

Reflections from the terrain require the path to be relatively low. Avoiding obstruction fading requires the antennas to be raised. In the past, this compromise was attempted by using the following "rules of thumb." There are several currently in use.

12.4.1.1 ITU-R Path Clearance Criteria

For Frequencies Above 2 GHz and Below 13 GHz The guidelines for main transmit antenna to main receive antenna path clearance (Table 12.1) (International Telecommunication Union—Radiocommunication Sector (ITU-R), 2007bb):

1. Determine the antenna heights to achieve 1.0 F_1 clearance at $K = 4/3$. F_1 is the first Fresnel zone radius defined in the following.
2. Determine a K value using Figure 2 in the work in recommendation P. 530-12 by International Telecommunication Union—Radiocommunication Sector (ITU-R), (2007bb). The K value Figure 2 may be found using the following formula:

$$K = 0.80379426 + (0.0029438097D) - \left(\frac{6.1701657}{D}\right) - (0.000013619815D^2) + \left(\frac{10.375301}{D^2}\right) + (0.000000022844768D^3) \quad (12.6)$$

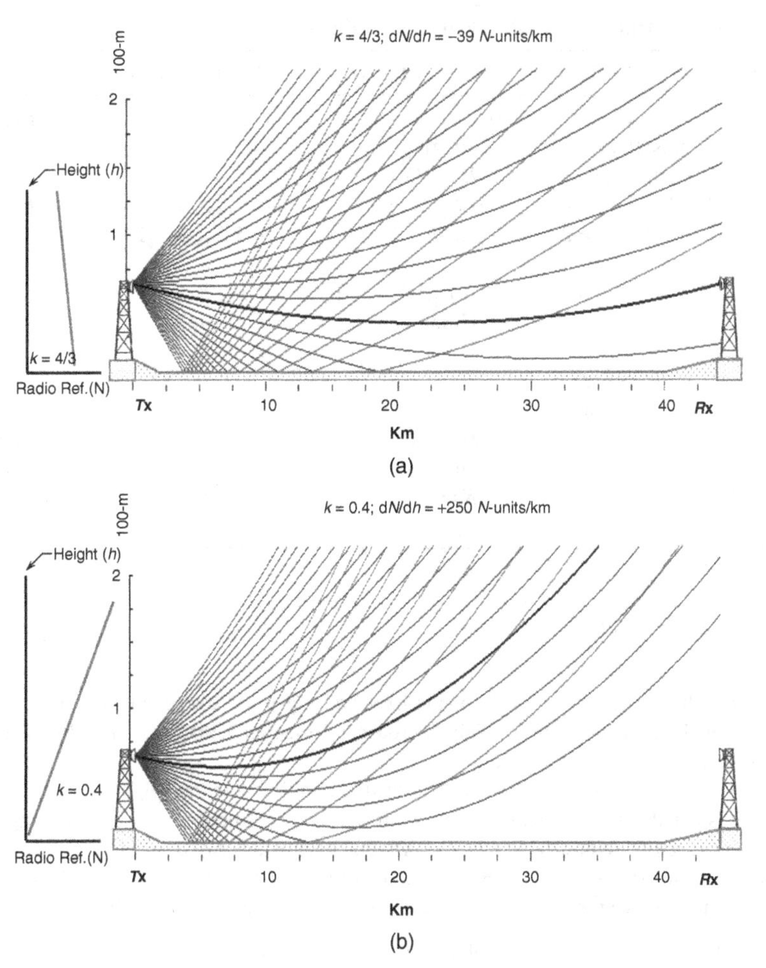

Figure 12.10 (a) Normal propagation and (b) obstruction fading. Illustrations courtesy of Eddie Allen. Used with permission.

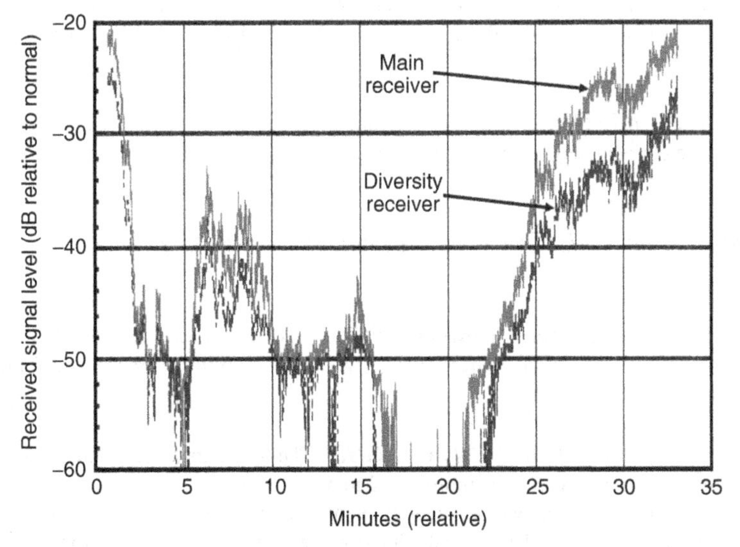

Figure 12.11 Obstruction fading.

TABLE 12.1 North American Path Clearance Guidelines

Climate Areas	GTE Lenkurt	Bell Labs	Alcatel-Lucent	Aviat Networks
Good	0.6 F_1, $K=1$ plus 10 ft (light route)	0.6 F_1, $K=1$ Grazing, $K=2/3$	0.6 F_1, $K=1$	0.6 F_1, $K=1$
Average	1.0 F_1, $K=4/3$ 0.3 F_1, $K=2/3$ (heavy route)	1.0 F_1, $K=4/3$ 0.3 F_1, $K=2/3$	0.6 F_1, $K=4/3$ 0.3 F_1, $K=2/3$	0.6 F_1, $K=1$
Poor	[1.0 F_1, $K=4/3$] Grazing, $K=1/2$ (very difficult heavy route)	[1.0 F_1, $K=4/3$] Grazing, $K=1/2$	1.0 F_1, $K=4/3$ Grazing, $K=1/2$	1.0 F_1, $K=4/3$ 0.3 F_1, $K=2/3$
Very Poor		[1.0 F_1, $K=4/3$] Grazing, $K=5/12$	1.0 F_1, $K=4/3$ Grazing, $K=4/10$	Grazing, $K=1/2$ or grazing, $K=1$ above 150 ft duct

D = path distance (km), $20 \leq D \leq 200$.

The above K value is described as the value expected not to be exceeded more than 0.01% of the time in a continental temperate climate. This approach is essentially that described by Wheeler (1977). A more general result is described by Figure 12.12.

For temperate climates (average propagation areas), determine antenna heights for 0.0 F_1 (grazing) if there is a single isolated path obstruction or 0.3 F_1 if the path obstruction is extended along a portion of the path. Path clearance is determined using the K value calculated above.

For tropical climates (poor propagation areas), determine antenna heights for 0.6 F_1 for path lengths greater than 30 km. Path clearance is determined using the K value calculated above.

Use the higher of the antenna clearances obtained from steps 1 and 2.

The steps for main transmit antenna to diversity receive antenna path clearance:

Use $K=4/3$.

Determine antenna heights for 0.3–0.0 F_1 if there are one or two isolated obstacles in the path.

Determine antenna heights for 0.6–0.3 F_1 if the path obstruction is extended along a portion of the path.

For Frequencies Above 13 GHz Add 1.0 F_1 clearance to the above values. This is due to the degradation of obstacle height accuracy at these frequencies. In general, diversity antennas are not used in this frequency range.

For Frequencies Below 2 GHz Use the above main transmit antenna to diversity receive antenna criteria.

12.4.1.2 North American Path Clearance Criteria Several criteria are currently in use in North America (Knight, 2004; Laine, 2004; Latter, 1980; Vigants, 1975; White, 1970). The following are the most popular:

For Frequencies Above 2 GHz and Below 13 GHz The guidelines for main transmit antenna to main receive antenna path clearance are summarized in Table 12.1.

Where two criteria apply, use the worse (greater antenna heights) case. The criteria within the brackets [ ] of Table 12.1 do not appear in the original source guidelines but were added for completeness. The above clearances are for the main antenna to main antenna path. For the main antenna to diversity antenna path, see below.

As Schiavone observed (Schiavone, 1981), significant obstruction conditions require homogeneous vertical gradients over a horizontal distance of the typical path length of 25 miles (40 km). Mojoli (1979) noted that abnormal refractivity conditions are relatively localized. Samson (1975) observed "... the more extreme positive refractivity gradients near the surface are greatly influenced by local conditions of terrain, moisture sources, etc., and are not likely to extend over wide areas." White (1970) pointed out that "... the unusual conditions causing these extreme values [of K factor] are unlikely to occur over more than a small part of the path at any given instant." For longer paths, path-averaged refractivity conditions tend to move toward the expected K value of 4/3. With the obvious exception of paths near and parallel to the ocean, relaxing the K factor obstruction value for longer paths is appropriate (In the United States, for paths near and parallel to the ocean or land to island/oil platform paths, think twice before using 6-GHz paths longer than 20 miles if high reliability is important!). Alcatel-Lucent applies a K factor correction for paths greater than 25 miles. These correction factors (based on work done by Teletra, an Italian company acquired by Alcatel-Lucent) are obtained using the following formulas.

For $K = 1$,

$$K = 1.3525442 - (0.0025870465\,D) - \left(\frac{13.402012}{D}\right) + (0.000018703584\,D^2) + \left(\frac{148.40133}{D^2}\right) - (0.000000045263631\,D^3) \tag{12.7}$$

For $K = \frac{2}{3}$,

$$K = 1.0561294 - (0.0026434873\,D) - \left(\frac{14.930824}{D}\right) + (0.000025692173\,D^2) + \left(\frac{161.86573}{D^2}\right) - (0.00000007235408\,D^3) \tag{12.8}$$

For $K = 0.5$,

$$K = 0.82999889 - (0.0015503296\,D) - \left(\frac{13.665673}{D}\right) + (0.0000172027\,D^2) + \left(\frac{152.99299}{D^2}\right) - (0.000000046838057\,D^3) \tag{12.9}$$

For $K = 0.4$,

$$K = 0.64549929 - (0.00116752\,D) - \left(\frac{11.930858}{D}\right) + (0.000017371414\,D^2) + \left(\frac{156.48897}{D^2}\right) - (0.000000053622664\,D^3) \tag{12.10}$$

D = path distance (miles) where $25 \leq D \leq 150$.

For those wishing to apply the above criteria in an ITU-R manner, consideration should be given to the work of Mojoli (1979). He showed both theoretically and experimentally the moderation of K factor as path length is increased. Figure 12.12 depicts his results for K factor not exceeding more than 0.01% of the time.

The results have been normalized to a path length of 25 miles. The relaxation factor is multiplied by the K factor of the propagation area of interest. The curve labeled as ITU-R is the Figure 2 curve of ITU-R Recommendation P.530-12 normalized to 25 miles. The ITU-R K value at 25 miles is 0.75.

The steps for main transmit antenna to diversity receive antenna path clearance:

For the main to space diversity antenna path, path clearance is significantly relaxed. Typically 0.6 F_1 at $K = 4/3$ is considered adequate. For nonreflective paths, main receive and space diversity receive antenna spacing of 200 wavelengths ("200 λ") provides adequate receive signal de-correlation. This is the basis of the typical 30-ft separation for 6 GHz and 20-ft separation for 11 GHz. For reflective paths, spacing of the receive antennas should be optimized to minimize simultaneous fading on both antennas

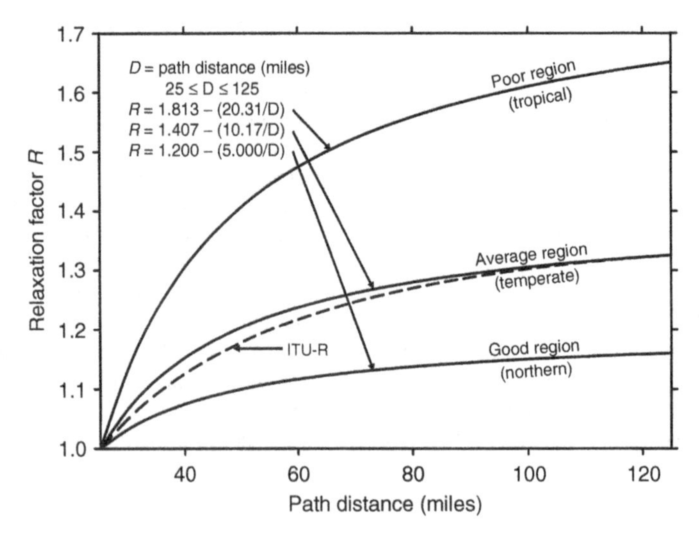

Figure 12.12 *K* factor relaxation for long paths

(Chapter 10). In addition, a minimum clearance of 10 ft for stationary objects and 30 ft for trees below the lower antenna aperture to a distance of 500 ft is recommended.

For Frequencies Above 13 GHz There is no definitive standard for this frequency range. The most popular criterion is to engineer path clearance of 1.0 F_1 for $K = 4/3$ for all areas. In general, diversity antennas are not used in this frequency range.

For Frequencies Below 2 GHz Engineer path clearance of 0.6 F_1 for $K = 4/3$ for all areas.

There are obviously different views on what the path clearance criteria should be. There is no objective evidence to guide the engineer as to which set of rules to use but the choice has significant economic

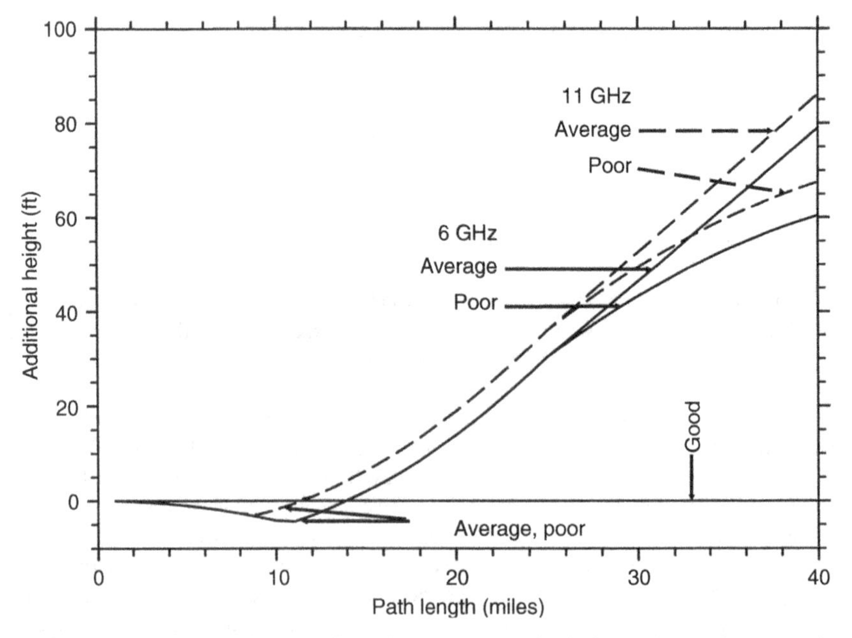

Figure 12.13 Additional tower height using Alcatel-Lucent criteria instead of Aviat Networks criteria.

impact. In North America, the most popular choices for low frequency paths are the Alcatel-Lucent and the Aviat Networks (previously Harris Stratex) criterions. Assuming a flat earth, the following chart shows the additional tower height required if the Alcatel-Lucent (ALU) path criterions are used rather than the Aviat (A) criterions (for main transmitter antenna to main receiver antenna) (Fig. 12.13).

There is no objective evidence indicating which criteria are more accurate. It is not clear what constitutes good, average, poor, or very poor regions for a path design. We will address these issues after we develop an obstruction fading model next.

12.5 OBSTRUCTION FADING MODEL

To date, all path designs have been based on Fresnel zone based criteria (as noted in the previous section). While this has considerable historical precedence, it is not obvious. As paths increase in length, the Fresnel zone near the center of the path (typically the obstruction controlling point) becomes smaller. Decreasing terrain clearance as path length increases seems counterintuitive. That issue was addressed by Vigants (1981) and Schiavone (1981). Unlike the traditional Fresnel zone criteria approach, they developed a model of obstruction fading from a physical and metrological perspective.

The effects of weather-related phenomena such as rain and multipath fading have been estimated based on direct measurement. Unfortunately, obstruction fading is highly variable in both geographic location and time of occurrence. A database of obstruction fading events does not exist and would be difficult to create. However, Schiavone observed that obstruction fading can be related to meteorological variations for which databases already exist.

Schiavone developed a model on the basis of estimates of the mean μ and standard deviation σ of the daily and nightly rate of change of refractivity gradient N' based on season and location in the continental United States (CONUS). The atmospheric refractivity N' mean (μ) had been previously researched by Bean et al. (1966) (Fig. 12.14).

Schiavone used Bean et al.'s empirical data for mean μ. The refractivity standard deviation σ was more difficult to estimate. To determine this, he developed models based on the major mechanisms that produce dynamic changes to the lower atmosphere. While several mechanisms are known to produce inverted (positive refractivity gradient) water vapor vertical profiles (ground fog, nocturnal temperature inversion, dry air convection, upland rainfall evaporation), these processes are so localized they are not anticipated to extend over a wide enough area to cause microwave path obstruction. Nocturnal advection of moist air at water body–land interfaces is a possible mechanism but is unlikely to extend more than a few tens of miles inland. Schiavone proposed large-scale advection of moist air as the mechanism for most microwave path obstruction fading. Most obstruction fading (wide-based positive refractivity gradients) occurs during fine weather condition with calm winds. During the day, solar radiation causes convective mixing, which produces a well-mixed lower atmosphere. At night, radiative cooling of the earth's surface produces a relative thin, density-stable boundary layer. For calm conditions near a large body of water, at night, winds typically blow from the water onto the land. This causes moist air to be advected over the top of surface-based dry air. However, measurements show that elevated moist layers occur at sites too far from large bodies of water to have them account for the inverted moisture. To account for this situation, Schiavone developed two models: a coastal region submodel and an interior region submodel.

The coastal submodel was based on six parameters: frequency of temperature inversion, air homogeneity, water body proximity, wind direction, air moisture capacity, and surface moisture. Methods of Hosler (1961), Bean et al. (1966), and the USIGS (1970) were applied to the methodologies of Ikegami et al. (1968) and Akiyama and Sasaki (1979). This model was expected to extend from the coast (of a large body of water) into the interior for the distance air can be expected to penetrate during 12 h at a height of 100 m. The coastal submodel was applied to areas near the Gulf Coast.

The interior submodel was based on five parameters: frequency of temperature inversion, air homogeneity, a combined water body proximity and wind direction factor, air moisture capacity, and surface moisture. The interior submodel was applied to the Pacific coast and the rest of the United States not covered by the coastal submodel.

Schiavone (1981) developed a method of estimating the positive refractivity parameters based on published data (Samson, 1975) and his coastal and interior submodels for the CONUS. He observed

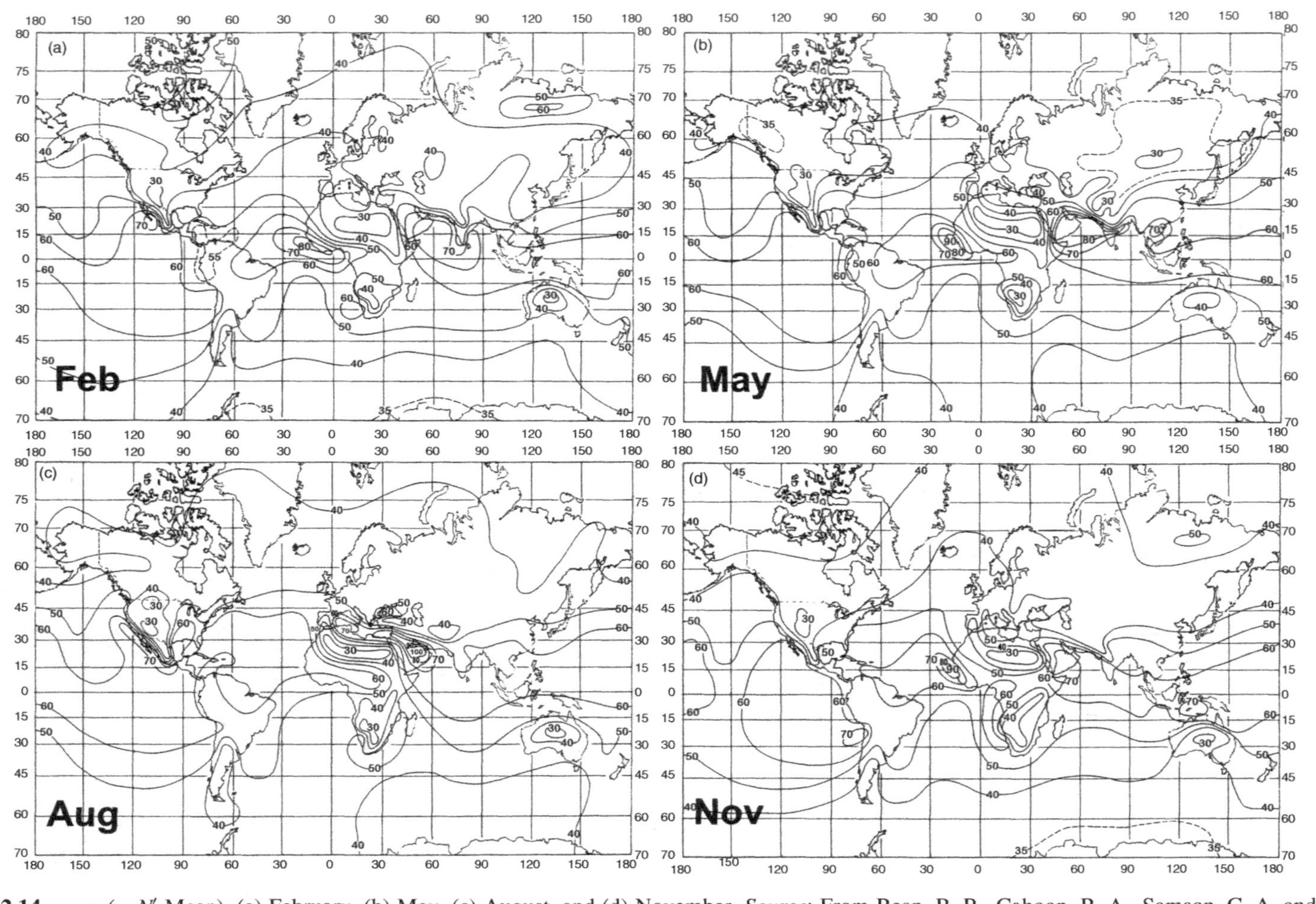

Figure 12.14 $-\mu$ ($-N'$ Mean), (a) February, (b) May, (c) August, and (d) November. *Source*: From Bean, B. R., Cahoon, B. A., Samson, C. A. and Thayer, G. D., *A World Atlas of Atmospheric Radio Refractivity*, ESSA Monograph 1. Washington, DC: U.S. Department of Commerce, 1966.

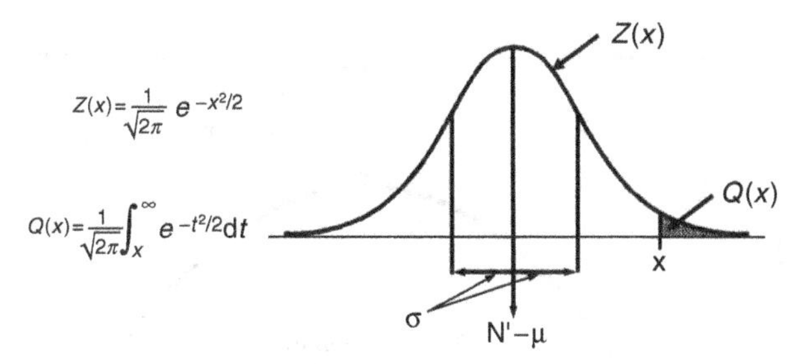

Figure 12.15 Basic normal distribution parameters.

that positive refractivity gradients are caused by a combination of two different normally distributed mechanisms: mixed atmosphere and stratified atmosphere. Typically, during the day, the mixed atmospheric refractivity is controlled by a distribution of air with seasonal mean μ and a median standard deviation σ_m of 15 N units/km. Intense gradients due to stratified air typically occur during the night and are characterized by the seasonal mean μ and a seasonal standard deviation σ_s that varies by geographic location. The daytime gradient distribution occurs about 80% of the time and the nightly more extreme gradient distribution occurs about 20% of the time.

Using the normal distribution (Fig. 12.15) assumption, the seasonal probability $P(N',\text{Season})$ of refractivity exceeding $N'(N' \geq x)$ is given by the following expressions:

$$P(N', \text{Season}) = 0.8\ Q(x, \mu, \sigma_m) + 0.2\ Q(x, \mu, \sigma_s), \quad \sigma_m = 15 \tag{12.11}$$

$$K = \frac{157}{157 + N'} \tag{12.12}$$

$$N'(\text{N units/km}) = \mu + X\sigma \tag{12.13}$$

μ = mean (average value) of N';
σ_m = standard deviation of N' in mixed atmosphere = 15;
σ_s = standard deviation of N' in stratified atmosphere;
$x = (N' - \mu)/\sigma$;
Probability of N' = $Z(x,\mu,\sigma)$ = Gaussian probability density function;
Total area of $Z(x)$ = 1.0;
Probability of $N' \geq x$ = $Q(x,\mu,\sigma)$ = Gaussian cumulative distribution function.

The total time the refractivity gradient N' exceeds a value x is given by the following expressions:

$$P_{\text{average}} = \left[\frac{P\left(N', \text{February}\right) + P(N', \text{May}) + P(N', \text{August}) + P(N', \text{November})}{4}\right]$$

$$[\text{Annual time } N' \geq x] = P_{\text{average}} \times (\text{total number seconds in a year})$$

While daytime refractivity statistical parameters are essentially geographically invariant (and generally insignificant), nightly parameters vary greatly by location.

12.6 OBSTRUCTION FADING ESTIMATION

Vigants (1981) used Schiavone's refractivity model to estimate obstruction fading. He observed that when a path is obstructed (path clearance is less than grazing), the obstruction fade level could be estimated using the following relationships (Fig. 12.16):

$$M = \text{Obstruction fade level (dB)} \geq 10$$

$$= +10 + 20\left(\frac{E}{F_1}\right) \tag{12.14}$$

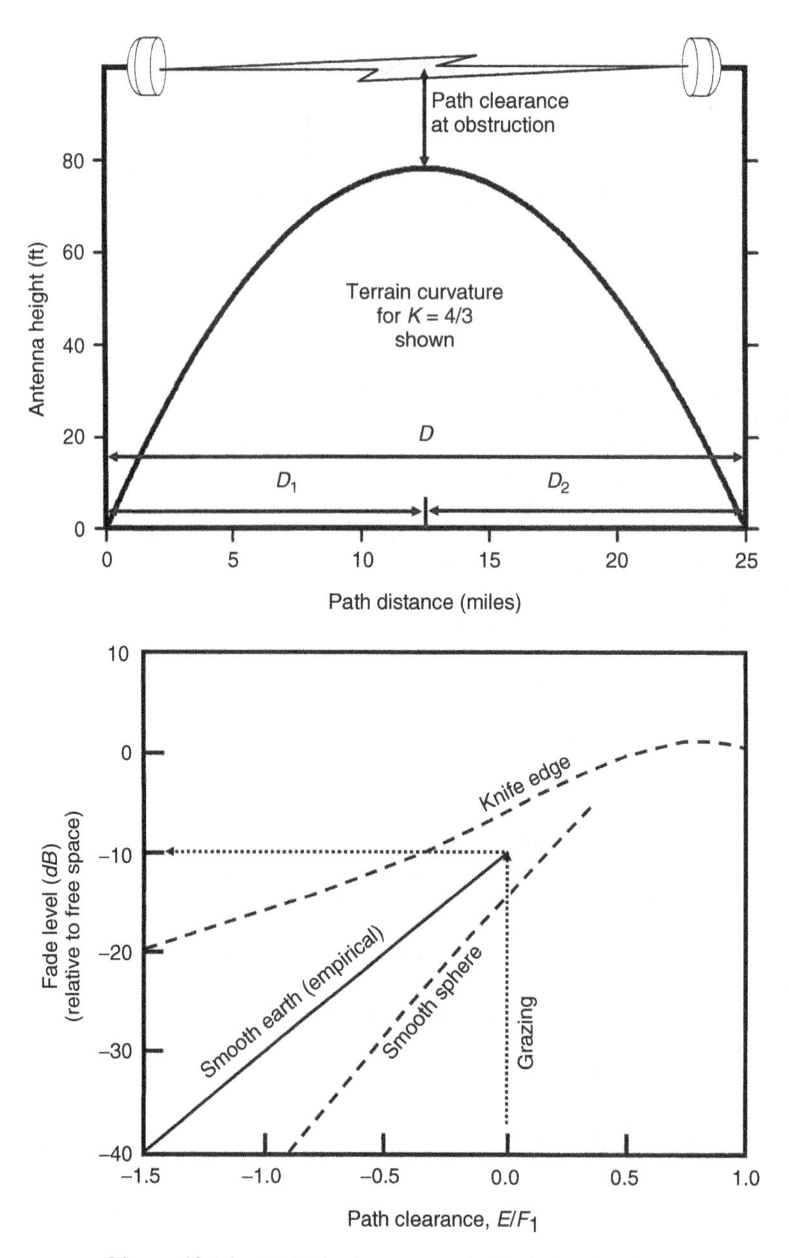

Figure 12.16 (a) Path clearance and (b) obstruction loss.

E = obstruction height that causes M dB fade greater than 0;
F_1 = first Fresnel zone height at potential obstruction;
F_1 (ft) = 72.1 sqrt{[$n \times d_1$(miles) $\times d_2$(miles)]/[F(GHz) $\times D$(miles)]};
F_1 (m) = 17.3 sqrt{[$n \times d_1$(km) $\times d_2$(km)]/[F(GHz) $\times D$(km)]};
D_1 = distance from one end of the path to the reflection;
D_2 = distance from other end of the path to the reflection;
D = total path distance = $D_1 + D_2$;
F = frequency of radio wave.

Combining the results of Schiavone and Vigants yields the following algorithm for estimating obstruction fading outage time:

```
/Input path data/
Input M /fade margin in dB ≥ 10/
Input TAH /transmit antenna height in feet above mean sea level/
Input OH /obstruction height in feet above mean sea level/
Input RAH /receive antenna height in feet above mean sea level/
Input D /path distance in miles/
Input D1 /distance from transmit antenna to obstruction in miles/
D2 = D - D1 /distance from receive antenna to obstruction in miles/
Input F /radio transmitting center frequency in GHz/

/Calculate basic parameters/
F1 = 72.1 * SQR((D1 * D2) / (F * D))
      /first Fresnel zone radius in feet, SQR = square root function/
E = (F1 / 20.) * (M - 10.) /obstruction height that blocks the path/
PH = TAH + ((RAH - TAH) * (D1 / D)) /height of path at obstruction/
H = PH + E - OH /earth bulge required to produce M dB fade level/
K = (D1 * D2) / (1.5 * H) /K factor needed to produce earth bulge/
N = (157. / K)−157 /refractivity gradient N' required to produce earth bulge/

/Input refractive gradient statistics/
Input N6 /February mean atmospheric refractivity gradient/
Input N7 /May mean atmospheric refractivity gradient/
Input N8 /August mean atmospheric refractivity gradient/
Input N9 /November mean atmospheric refractivity gradient/
Input N10 /February atmospheric refractivity gradient standard deviation/
Input N11 /May atmospheric refractivity gradient standard deviation/
Input N12 /August atmospheric refractivity gradient standard deviation/
Input N13 /November atmospheric refractivity gradient standard deviation/

/Calculate refractive gradient probability/
/February/
      MEAN = N6
      STDDEV = 15.
      GOSUB 1000 /go to subroutine 1000 and return/
PROBMIXED = Q
      MEAN = N6
      STDDEV = N10
      GOSUB 1000
PROBSTRATIFIED = Q
PFEB = (.8 * PROBMIXED) + (.2 * PROBSTRATIFIED)

/May/
      MEAN = N7
      STDDEV = 15.
      GOSUB 1000
PROBMIXED = Q
      MEAN = N7
      STDDEV = N11
      GOSUB 1000
PROBSTRATIFIED = Q
PMAY = (.8 * PROBMIXED) + (.2 * PROBSTRATIFIED)

/August/
      MEAN = N8
      STDDEV = 15.
```

```
    GOSUB 1000
PROBMIXED = Q
    MEAN = N8
    STDDEV = N12
    GOSUB 1000
PROBSTRATIFIED = Q
PAUG = (.8 * PROBMIXED) + (.2 * PROBSTRATIFIED)

/November/
    MEAN = N9
    STDDEV = 15.
    GOSUB 1000
PROBMIXED = Q
    MEAN = N9
    STDDEV = N13
    GOSUB 1000
PROBSTRATIFIED = Q
PNOV = (.8 * PROBMIXED) + (.2 * PROBSTRATIFIED)

PROB = (PFEB + PMAY + PAUG + PNOV) / 4.
SECS = 365.25 * 24. * 60. * 60.  /seconds in year/
OUTAGE = SECS * PROB /obstruction fading time in seconds/
END

1000 /Subroutine to calculate normal curve Q value/
PI = 3.1415926536 /constant pi/
EC = 2.7182818285 /constant e/
X = (N - MEAN) / STDDEV
ZX = (EC ^ (−(X * X) / 2.)) / SQR(2. * PI)
R = .2316419
T = 1. / (1. + (R * X))
B1 = .319381530
B2 = −.356563782
B3 = 1.781477937
B4 = −1.821255978
B5 = 1.330274429
Q = ZX * ((B1 * T) + (B2 * T ^ 2) + (B3 * T ^ 3) + (B4 * T ^ 4) + (B5 * T ^ 5))
    /unavailability probability/
RETURN
```

In the above algorithm, comments are placed between slants (//). The above subroutine for $Q(x)$ is based on Equations 26.2.1 and 26.2.17 of the work by Abramowitz and Stegun (1968) and the relationship $Q(x) = 1 - P(x)$. Abreu's tight upper bound (Abreu, 2012) on the Q-function also is an excellent approximation for Q.

Comparison of actual obstruction fading outages with the predictions of the above algorithm suggests that the above algorithm is somewhat optimistic. Of course, some of this optimism can be attributed to the fact that atmospheric parameters are measured only two fixed times a day (which seldom coincide with poor propagation events). However, the algorithm is most sensitive to refractivity gradient standard deviation. Typically, the refractivity gradient is greatest for one season in the year. The algorithm averages the refractivity gradients for all four seasons before estimating outage. If all season gradients were similar or outage time were a linear function of refractivity gradient, calculating an average result would be appropriate. However, usually, refractivity gradient is much greater for one season than the others and outage is a complicated exponential function of refractivity gradient. Therefore, averaging refractivity gradient before calculating would underestimate seasonal outage time. A more accurate approach might be to calculate outage time for each season and sum the results.

12.7 BELL LABS SEASONAL PARAMETER CHARTS

To facilitate practical application, Schiavone (of Bell Labs) produced charts (Kizer, 2008) (Schiavone, J. A., "Obstruction fading (OBSFAD) refractivity gradient curves for the United States," unpublished internal Bell Laboratories document, 1981) illustrating the seasonal variation of the mean and standard deviation of the nightly refractivity parameters. He started with Bean et al.'s results (Bean et al., 1966) for mean $N'(\mu)$. For the standard deviation of $N'(\sigma)$, he resorted to models. He started with measured data accumulated and processed by Samson 1975 (Fig. 12.17).

Samson provided refractivity gradient probability distributions for 22 cities in North America. Schiavone only focused on the continental United States (17 cities). He eliminated data from Denver, El Paso, New York, and Long Beach because those sites had significantly higher occurrence of obstruction (large, positive refractivity gradients) than his model predicted. He believed refractivity at those sites was caused by unusual local effects.

Schiavone compared his models with Samson's $N' \geq 100$ N units/km data. The match was excellent for the 17 cities. He used his refractivity models to fill in the areas not covered by Samson's data. He also took advantage of Vigant's experience with the actual performance of Bell System microwave paths to enhance the model.

The result of Schiavone's refractivity estimation efforts was a series of eight charts (Kizer, 2008), showing geographically for the four seasons his estimates for the standard deviation σ_s for the nighttime stratified atmosphere and the daily mean value μ for the atmospheric refractivity gradient N'. These charts and the above obstruction calculation model have been used in the Bell System to engineer point to point microwave paths subject to obstruction fading.

In the 1980s, Bell Labs gave Schiavone's refractivity charts to the transmission engineering department of the Collins Microwave Group of Rockwell International. That group (and its Alcatel and Alcatel-Lucent successors) has been using the Schiavone charts and the Schiavone–Vigants algorithm for decades. The eight refractivity charts have been included in their microwave transmission engineer manual (Knight, 2004) and distributed to consultants and customers since the 1980s. The experience of engineers using the charts is generally that obstruction fading as predicted by the Schiavone charts and the Schiavone–Vigants algorithm is insignificant in much of the interior of the United States. The model is quite valuable in the Gulf Coast and East Coast areas where obstruction fading is prevalent. The model clearly shows the pronounced effect of path length in that area of the country. However, when compared to actual experience, the model underestimates obstruction fading time in the Gulf Coast and southern Texas

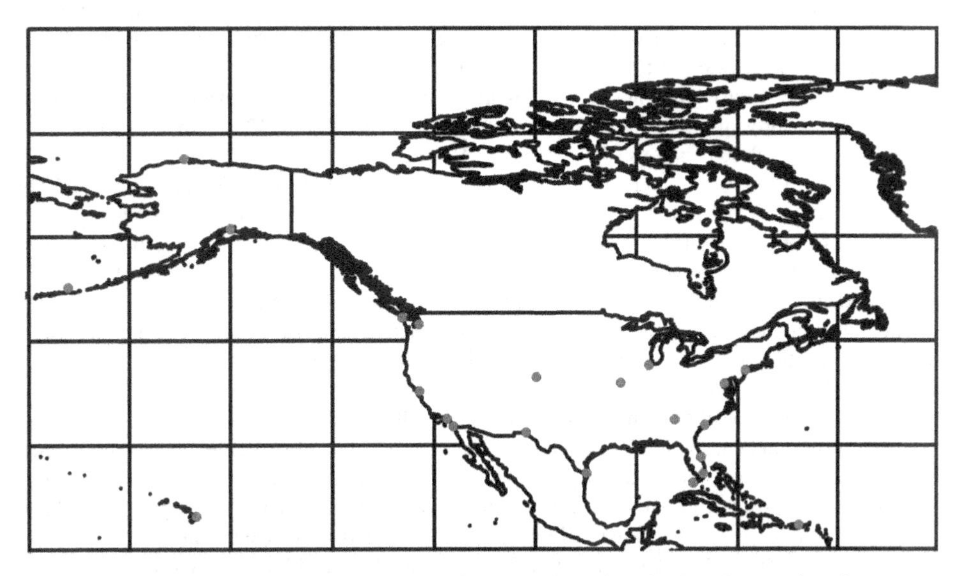

Figure 12.17 Samson's 1975 refractivity gradient data for North America sites.

areas. It also fails to predict obstruction fading near the Great Lakes, in the desert near Las Vegas, and in Southern California. As long as these limitations are understood, the Schiavone–Vigants algorithm remains a useful path-engineering tool.

12.8 REFRACTIVITY DATA LIMITATIONS

The underestimation of obstruction fading time could be caused by several factors in the refractivity gradient measurement. First, the time of measurement of atmospheric refractivity data is not optimum for observing obstruction fading. All refractivity gradient data is taken by a radiosonde only twice a day. Before the International Geophysical Year (IGY) of 1957–1958, radiosonde data was usually taken at 03:00 and 15:00 UTC (Coordinated Universal Time, also known as *GMT*, *Greenwich Mean Time* or *Z*, *Zulu time*). After the IGY, the convention was to take the data at 00:00 (midnight) and 12:00 (noon) UTC. Local standard time in the United States is 5, 6, 7, or 8 h earlier depending on the time zone. None of the times are ideal for observing nighttime positive refractive indexes. As Samson (1975) noted, "Variations in refractivity gradients tend to be closely related to the local or sun-referenced time, because of the influence of heating and cooling of the earth's surface on the stratification of air layers near the ground. For example, extreme gradients of refractivity often occur near sunrise, when nocturnal temperature inversions tend to be most pronounced, and near sunset, when there is a rapid shift from gain to loss of heat in the air layer near the ground. RAOBs [radiosonde observations], however, are taken at standard times which do not always coincide with the local time when extreme gradients are most likely to occur." He also noted (Samson, 1976), "Consequently, observations made during these [standardized measurement] periods may not be representative of the conditions existing even one hour earlier or later." Obstruction fading is generally a relatively rare event and conventional radiosonde data can be expected to underestimate it.

It should also be appreciated that radiosonde measurements are made from a rapidly rising balloon. The temperature and humidity are reported separately. The radiosonde toggles between the two measurements. They are not made simultaneously. The balloon passes through the first 100 m of atmosphere in about 30 s. Samson (1976) observed, "... the time interval over which a single observation might be expected to be representative will vary with the local climatic conditions as well as the time of observation ... it is difficult to determine the statistical significance of these data ..." (Samson, 1976). Therefore, the low atmosphere data is inaccurate owing both to cycling between measurements and the fact that the data of interest is sampled very briefly at a nonoptimum time. Any latency in the measurements will also skew the data. Bean et al. (1966, Table E-1) illustrated that high refractivity gradients occur in the first 100 m. Even slightly higher elevations have significantly lower gradients. This emphasizes that obstruction fading is not expected for microwave paths in mountainous terrain.

Atmospheric refractivity gradients (N') are highly location dependent. For example, the following site pairs have significantly different measured refractivity gradients (Table 12.B.1) even though they are geographically quite close: Flint and Mt. Clemens (Michigan), Quillayute (Tatoosh Island) and Seattle (Washington), and Long Beach and San Diego (California). The significant differences are primarily the difference in local geography (terrain, wind, and sources of humidity). The highest refractivity gradients in North America were measured—not in the Gulf Coast—but in Las Vegas. Obviously, this was not a humidity factor but unusual heating and cooling (inverting the normal temperature vs height profile) with low wind conditions (supporting layering). Given the limited number of North American and worldwide atmospheric measurements, the current data can only represent a geographic area in an approximate way. Considerable variation at nearby (but unmeasured) sites is to be expected.

The variability of measured data was highlighted by the Samson (Samson, 1975 and 1976) and Segal and Barrington (1977) results. Samson provided two results for the same location. One data set was labeled Quillayute and the other Tatoosh Island but they had the same geographic coordinates. One (Quillayute) showed moderated refractivity gradients (similar to Port Hardy above it). The other (Tatoosh Island) showed very low refractivity gradients (similar to Seattle below it). Given the geographic location, the Quillayute data set was reported in the Table 12.B.1. Samson as well as Segal and Barrington reported results for Caribou, Maine. The results are quite different. The Segal and Barrington results are reproduced in the Table 12.B.1. Samson as well as Segal and Barrington reported results for Fairbanks, Alaska. The results are very different. Segal and Barrington report results similar to those for all the other cities in Alaska. Samson reports very high refractive gradients for Fairbanks—much different than for any other city in Alaska. The Segal and Barrington results were reported in the Table 12.B.1. Fairbanks sits on a plain near mountains. The conditions are such that obstruction fading could occur under the right conditions. However, under most conditions, it appears Fairbanks should behave like the other cities in Alaska. Nevertheless, the significant difference in different measured data sets for similar locations highlights the considerable variability of (rare) obstruction fading events., this phenomenon is difficult to measure.

The data reported in Appendix 12.A and Appendix 12.B and the various models and charts represent the best-available published knowledge. However, the geographic and measurement time limitations of the current data limit our knowledge of much of the world. This data provides useful guidance for estimating path performance but it is still clearly very approximate. If high accuracy is required, long-term refractivity index measurements must be made at appropriate daily times over several entire seasons. Only then can one be confident of the predicted results.

12.9 REVIEWING THE BELL LABS SEASONAL PARAMETER CHARTS

The Schiavone (Bell Labs) refractive index charts (Kizer, 2008) (Schiavone, J. A., "Obstruction fading (OBSFAD) refractivity gradient curves for the United States," unpublished internal Bell Laboratories document, 1981) were created in the early 1980s. It is of interest to evaluate them on the basis of currently available data and experience. First let us consider the four refractive index mean (μ) charts. These charts are labeled winter, spring, summer, and fall. They represent the months February, May, August, and November, respectively. We compare them to the data obtained by Bean et al. (1966, Figures B-3, B-6, B-9 and B-12 respectively reproduced above as Figure 12.12). The winter and spring charts are exact matches to the Bean data (with the exception that the upper 40 contour in the winter chart is mislabeled 30). The summer chart is similar to that of Bean but the two separate 30 contours are merged and the upper left portion of the 40 contour is shifted to the left. The fall chart is somewhat curious. It matches the Bean Figure B-11 for October (rather than Figure B-12 for November). The reasons for these discrepancies are unclear.

When the Schiavone charts were prepared, additional data (Samson, 1976; Segal and Barrington, 1977) was available (for locations shown in Figure 12.18). For unknown reasons, the additional data was not used. This data—plus 25 years of additional experience—can be used to evaluate the accuracy of the four refractive index standard deviation charts.

In general, the charts accurately estimate refractivity in much of the United States. However, there are some notable exceptions. Obstruction fading in the Great Lakes area is well known to be nearly as extreme as in the Gulf Coast but the charts do not reflect this phenomenon. The charts do show a standard deviation as high as 70 N units/km in a small area south of the lakes but neither mirrors the intensity nor the geographic area of known obstruction fading (Lee, 1985, 1986). They also miss much of the area in Southern California and the Texas panhandle, areas known to exhibit obstruction fading. Lastly, they totally ignore the Las Vegas desert area, the location of the greatest subrefraction measurements in North America. Schiavone recognized that his interior model was less accurate than his coastal model. The results of the charts, while somewhat optimistic, do a fair job of estimating obstruction fading in the Gulf Coast and Florida geographic areas. We now compare the coastal model with the chart results in that area where the charts seem most effective.

Schiavone (1981, Fig. 8) noted the relationship between the model-calculated refractivity and the Samson measurements (Samson, 1975). For the cities Brownsville, Charleston, Cocoa Beach, and Miami the ratio of (calculated refractivity/measured refractivity) was approximately 1.5, 1.1, 1, and 1, respectively, for refractivity measurements of 100 N/km (Schiavone's value for comparison). Obstruction fading is dominated by the largest refractive index standard deviation and, in the United States, that almost always occurs during one season for a given location. Therefore, we focus on the worst-case value. For the cities in the above order, the worst-case (smallest probability) 100 N/km values were found at 99.5% in August, 99.2% in November, 99.2% in November, and 99.95% (interpolated) in February, respectively. These yield model refractivity standard deviations of $1.5 \times (100/2.576) = 58$ (in August), $1.1 \times (100/2.409)$

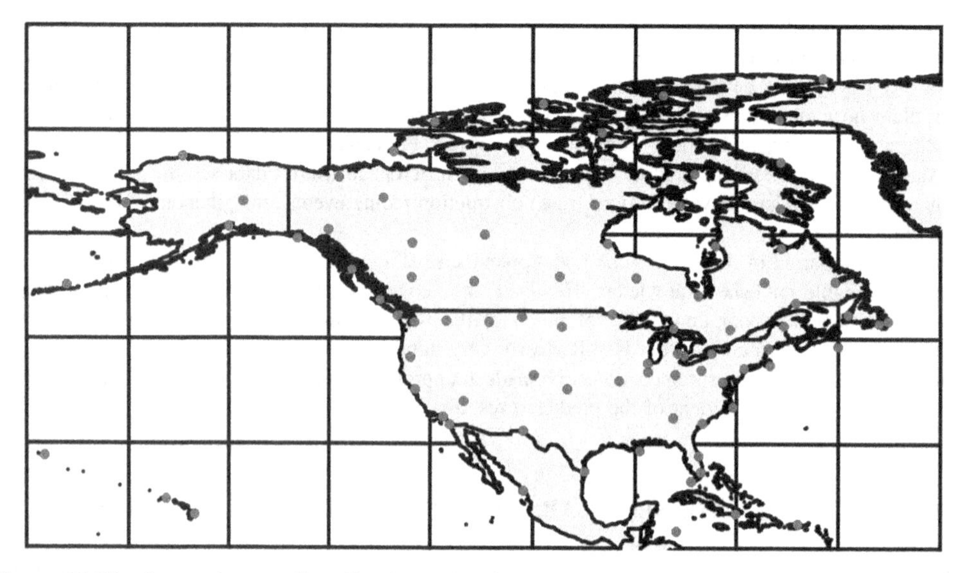

Figure 12.18 Samson's as well as Segal and Barrington's refractivity data for North America sites.

= 46 (in November), (100/2.409) = 42 (in November), and (100/3.291) = 30 (in February). The charts provide refractivity standard deviations of 100, 105, 120, and 65 for the cities and all of these (worst-case) values are in November. Schiavone did not describe how he differentiated among seasons. However, an inspection of the Samson as well as the Segal and Barrington North American data discloses significant refractivity standard deviations in all seasons. Observing the total standard deviations for each season shows that the average worst-case season is August (summer) for both North American and international sites. Individual site inspection yields similar results. Schiavone's charts show the greatest (most severe) standard deviation values in November (fall). The difference is puzzling. While the chart values seem useful in the Gulf Coast area, they do not appear to be based on the Schiavone coastal model (as the model was claimed to match the Samson data but the charts are significantly different than the Samson data). Presumably the charts are based on the empirical data of Vigants.

Overall, the Schiavone interior and coastal seasonal refractivity parameter models appear to be of limited use. The Schiavone refractivity standard deviation charts are also limited in their accuracy. However, they appear useful in the important Gulf Coast and Florida areas.

The Schiavone–Vigants obstruction fading model is based on general meteorological models and has no knowledge of local geographic and meteorological anomalies. The model assumes a Gaussian distribution for the refractivity parameters and actual data shows that assumption to be a very rough approximation. However, when used with reasonable mean and standard deviation parameters, it provides reasonable results.

12.10 OBSTRUCTION FADING PARAMETER ESTIMATION

Conventional path clearance criteria work well for good and average propagation areas. For poor propagation areas, the Schiavone–Vigants obstruction fading model should be considered to augment the path design. Under no circumstances should the Schiavone–Vigants algorithm be the only tool to determine the actual clearance criteria or antenna centerline heights. In some (particularly good propagation) areas, the estimated clearance will be too optimistic. The conventional clearance criteria should be used and the model used to determine if additional clearance is needed in areas where poor propagation is considered likely. In the United States, poor propagation can be expected for paths on or near the southeast coast, Southern California, selected inland areas or the Great Lakes region. The model illustrates why 6 GHz paths longer than 20 miles are not recommended on the Gulf Coast.

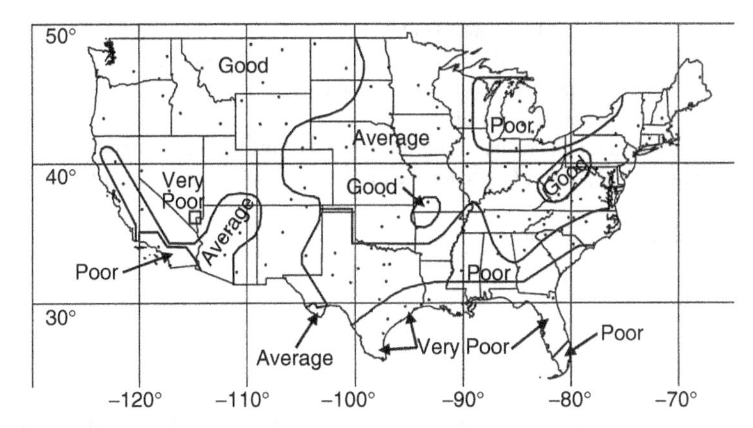

Figure 12.19 Obstruction fading propagation areas (based on data provided by Bill Knight).

From experience, Bill Knight of Alcatel-Lucent proposed the following unpublished criteria (Knight, W., private communications with George Kizer in 1994) for obstruction propagation areas:

Good : $\sigma_s \leq 45$, for all σ_s

Average : $C \leq 1$ and $45 < \sigma_s$, for at least one σ_s and $\sigma_s < 98$ for all σ_s

Poor : $C > 1$ and $45 < \sigma_s$, for at least one σ_s and $\sigma_s < 98$ for all σ_s

Very Poor : $\sigma_s \geq 98$, for at least one σ_s

C = climate factor;
σ_s = standard deviation of N' in stratified atmosphere.

On the basis of these criteria, Knight developed estimates of the propagation areas of the United States (Fig. 12.19). It should be realized that all refractive data is taken at relatively low elevation sites. This data definitely does not apply to sites on mountains (or those high California hills). Obstruction fading virtually never occurs at high elevation (e.g., Sierra Nevada mountain chain) but moderate fading may be experienced at ground levels (e.g., Sacramento, near the Sierra Nevada Mountains).

As the available refractivity gradient data is fairly sparse, it is important to pick a location similar to the area of interest. For example, the Denver site data is located in an unusual geographic area—just east of the Rocky Mountains. Prevailing winds going east over the mountains tends to trap air on the east side of the Rockies and produce the unusual propagation at Denver. Obstruction fading and ducting are common in that area. Denver data should be used for any similar location (such as Colorado Springs) but would not be appropriate for other areas.

12.11 EVALUATING PATH CLEARANCE CRITERIA

With the above Bell Labs obstruction fading estimation algorithm, we may now evaluate the performance of the common path clearance criteria. The three most commonly used path clearance criteria were evaluated (using the Bean et al. data) for three propagation areas: Good (Seattle, Washington), Average (Columbia, Missouri), and Very Poor (Cocoa Beach, Florida).

To simplify the comparison, the terrain was considered flat and both antenna elevations were the same height. Space diversity antennas were not considered. The following definitions apply:

F = 6.175 = radio operating frequency (GHz);
Path fade margin = 40 dB;
F_1 = first Fresnel zone radius at the middle of the path (ft);

D = total path distance (miles)

$D_1 = D_2 = D/2$ = distance from either antenna to the center of the path;

K = atmospheric refraction K factor;

Antenna height (ft) = clearance + terrain;

Clearance = path clearance in feet imposed by Fresnel zone criterion;

Midpath terrain height = $(D_1 \times D_2)/(1.5 \times K)$ = earth bulge (ft)at center of path for defined K factor.

Alcatel-Lucent Clearance Criteria:

Good Propagation area:

Clearance = $0.6 \times F_1$

$K = 1$

$$\text{If } D > 25 \text{ then } K = 1.3525442 - (0.0025870465 \times D) - \left(\frac{13.402012}{D}\right) + (0.000018703584 \times D^2) + \left(\frac{148.40133}{D^2}\right) - (0.000000045263631 \times D^3) \quad (12.15)$$

Average Propagation area:

Clearance = $0.3 \times F_1$

$K = 2/3$

$$\text{If } D > 25 \text{ then } K = 1.0561294 - (0.0026434873 \times D) - \left(\frac{14.930824}{D}\right) + (0.000025692173 \times D^2) + \left(\frac{161.86573}{D^2}\right) - (0.00000007235408 \times D^3) \quad (12.16)$$

Very Poor Propagation area:

Clearance = 0.0(grazing)

$K = 0.4$

$$\text{If } D > 25 \text{ then } K = 0.64549929 - (0.00116752 \times D) - \left(\frac{11.930858}{D}\right) + (0.000017371414 \times D^2) + \left(\frac{156.48897}{D^2}\right) - (0.000000053622664 \times D^3) \quad (12.17)$$

Aviat Networks (formerly Harris Stratex) Clearance Criteria:

Good or average propagation area

Clearance = $0.6 \times F_1$

$K = 1$

Very poor propagation area

Clearance = 0.0 (grazing)

$K = 0.5$

ITU-R Clearance Criteria:

Clearance = $0.3 \times F1$

If Poor Propagation and $D \geq 18.6$ then Clearance $= 0.6 \times F_1$

PD = $D \times 1.609344$

If PD < 20 then PD $= 20$

If PD > 200 then PD = 200

D = path distance in miles

PD = path distance in kilometers

$$K = 0.80379426 + (0.0029438097 \times \text{PD}) - \left(\frac{6.1701657}{\text{PD}}\right) - (0.000013619815 \times \text{PD}^2)$$
$$+ \left(\frac{10.375301}{PD^2}\right) + (0.000000022844768 \times \text{PD}^3) \quad (12.18)$$

Using the above criteria and the Schiavone–Vigants obstruction fading time algorithm described previously, the results in Figure 12.20 were obtained.

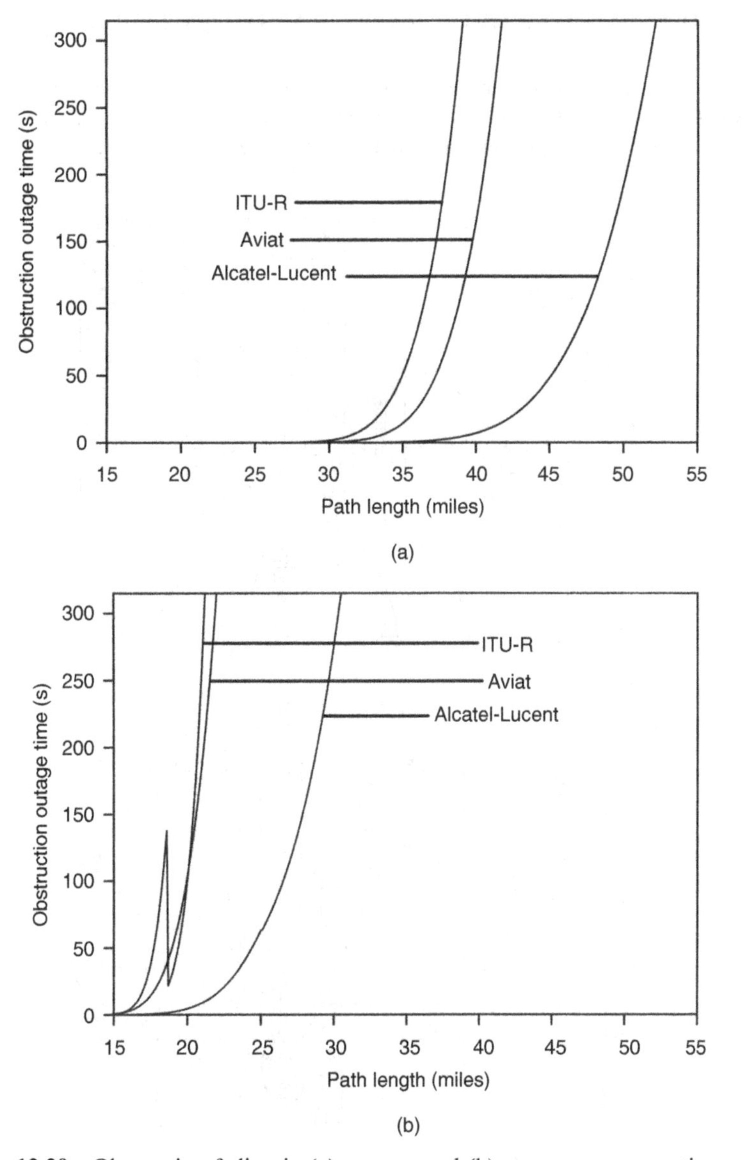

Figure 12.20 Obstruction fading in (a) average and (b) very poor propagation areas.

As reference, if the only propagation outage is due to obstruction, 315 s of outage represents a two-way path availability of 99.999%. For 32-s outage, the two-way availability is 99.9999%. There was no significant obstruction fading estimated for the good propagation area. For average areas, obstruction fading is not a significant issue for typical path lengths. The results also suggest that the ITU-R poor propagation area path clearance factor of 0.6 F_1 should be imposed for all distance paths in poor propagation areas. Path performance clearly rapidly degrades as path clearance is reduced. These results show the improved performance of additional path clearance. It is also noticeable that the Schiavone–Vigants obstruction algorithm is a little optimistic (probably because of optimistic refractivity data). In very poor propagation areas, it is well known that obstruction fading limits 6-GHz paths to a maximum distance of about 20 miles. These calculations suggest a path using the Alcatel-Lucent clearance criteria could exceed 25 miles. That is not good engineering practice. However, the relative effects of path clearance are clearly shown by the model.

12.A APPENDIX: NORTH AMERICAN REFRACTIVITY INDEX CHARTS

Refractive index values are highly dependent on geographic location. However, general guidelines for path design would be helpful. Digitized Bean et al. (1966) obstruction fading data was used to create generalized charts of refractivity gradient mean μ for North America (Fig. 12.A.1).

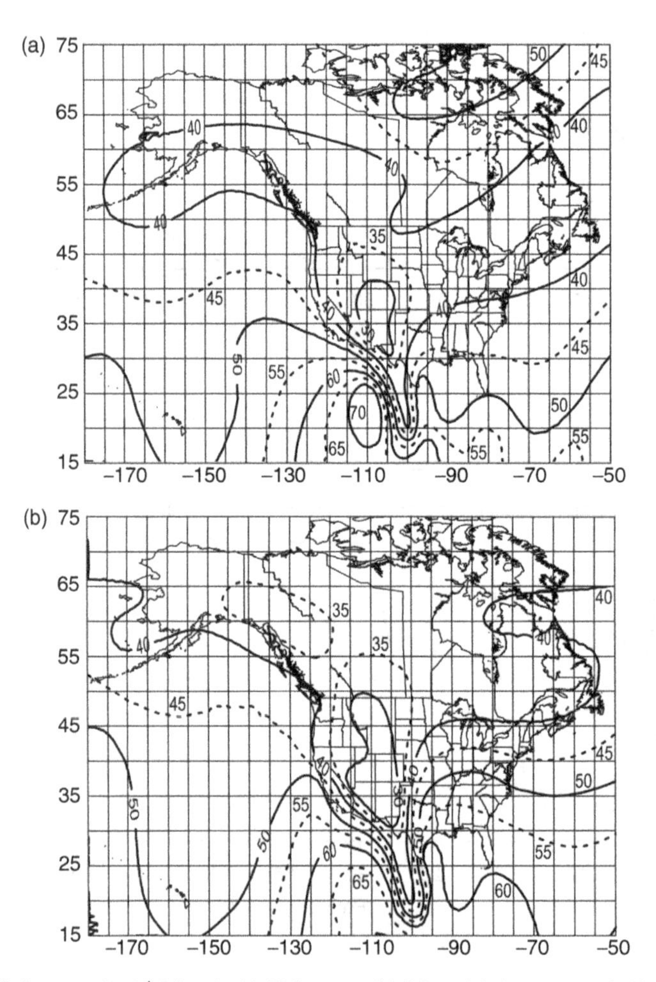

Figure 12.A.1 $-\mu$ ($-N'$ Mean), (a) February, (b) May, (c) August, and (d) November.

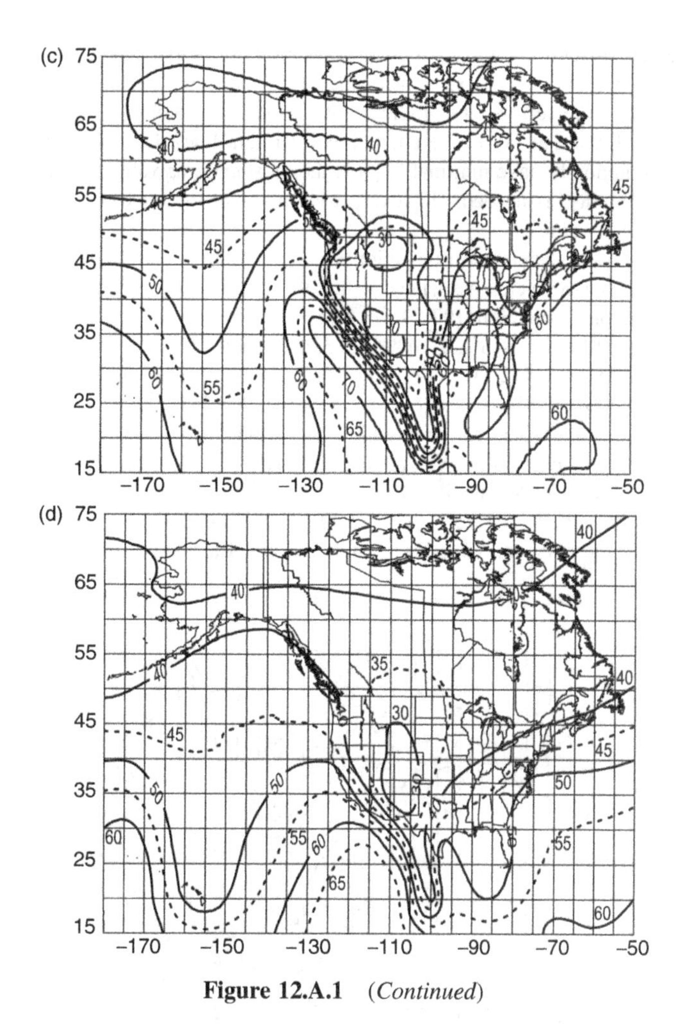

Figure 12.A.1 *(Continued)*

The obstruction fading data of Samson, as also that of Segal and Barrington were used to create generalized charts of refractivity gradient standard deviation σ_s (Fig. 12.A.2). In the United States, refractivity values not available from measurement were determined based on experience. In Canada, additional data was created using triangular interpolation of the Segal and Barrington data. Ocean values were based on Samson's Ship V data.

Detailed databases used to create the figures are available from the Wiley web site mentioned in the preface.

12.B APPENDIX: WORLDWIDE OBSTRUCTION FADING DATA

The data of Samson (1975, 1976) and Segal and Barrington (1977) were analyzed to determine seasonal mean and standard deviation of positive atmospheric refractivity gradients. The geographic locations of the data sites are shown in Figure 12.B.1.

The Samson as well as the Segal and Barrington sources were used to determine the appropriate atmospheric refractivity factors. Typically, the sources provide four graphs (February, May, August, and November) of refractivity gradient measured at various locations around the world. The vertical (ordinate or Y) axis is refractivity gradient in N units/km. Horizontal (abscissa or X) axis is percentage of time the ordinate (X) value is not exceeded (see Fig. 12.2 as an example). Two data values are needed: mean μ and standard deviation σ_s. The value of mean μ is easily read from the graph. It is the refractivity gradient for 50% of the time.

The standard deviation requires a couple of processing steps. First, the most positive refractivity gradient R for a large probability P is read from the chart. This step requires some judgment. From the charts, it is clear that the refractivity gradient data is only approximately normally distributed (if the process were normal, the data would be charted as a straight line). However, the normal process is assumed. For this assumption to have minimal impact, the most extreme valid data point R (value for the largest cumulative probability) is needed. However, it is important to notice the shape of the curve and avoid picking "data noise" values.

After R and μ are chosen, the standard deviation σ_s is simply $(R - \mu)/N_{SD}$ where N_{SD} is the number of standard deviations (from the mean) associated with a value sampled with probability P on a two-sided normally distributed sample population. (97.50% = 1.960 SD, 98.00% = 2.054 SD, 98.20% = 2.097 SD, 98.50% = 2.170 SD, 98.80% = 2.257 SD, 99.00% = 2.326 SD, 99.20% = 2.409 SD, 99.30% = 2.457 SD, 99.50% = 2.576 SD, 99.55% = 2.612 SD, 99.60% = 2.652 SD, 99.65% = 2.697 SD, 99.70% = 2.748 SD, 99.75% = 2.807 SD, 99.80% = 2.878 SD, 99.81% = 2.894 SD, 99.90% = 3.090 SD, 99.95% = 3.291 SD). Let us use Fig. 12.2b (Dakar, Africa) for an example. For August, we read that $\mu = -100$ N units/km. For $P = 99\%$, $R = 250$ N units/km. We determine that $N_{SD} = 2.326$ for $P = 99\%$. Therefore, $\sigma_s = (250 - [-100])/2.326 \approx 150$ N units/km. Clearly, long microwave paths in Dakar are impractical.

Worldwide refractivity parameters are listed in Table 12.B.1. Additional site data (Kizer) was interpolated on the basis of the data and experience of Segal and Barrington, as well as that of Samson.

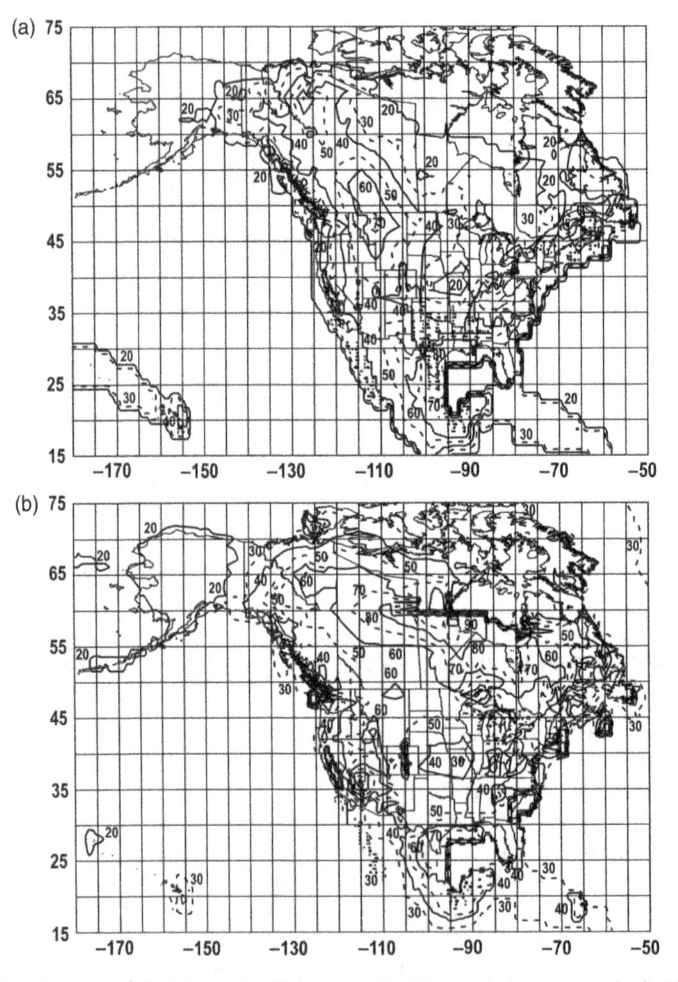

Figure 12.A.2 σ_s (N' Std Dev), (a) February, (b) May, (c) August, and (d) November.

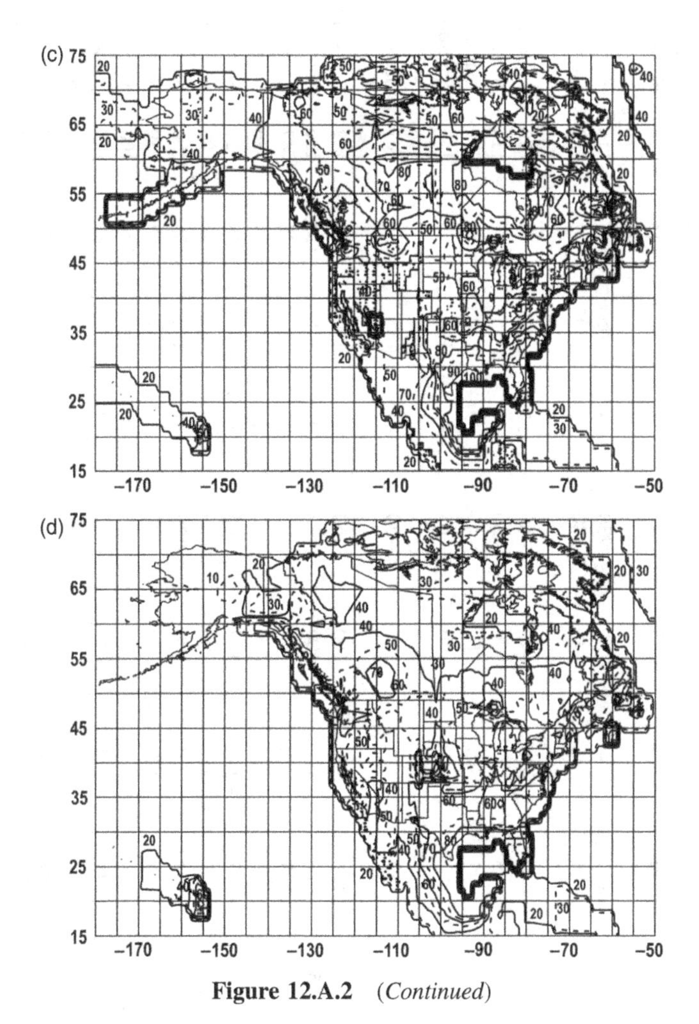

Figure 12.A.2 (*Continued*)

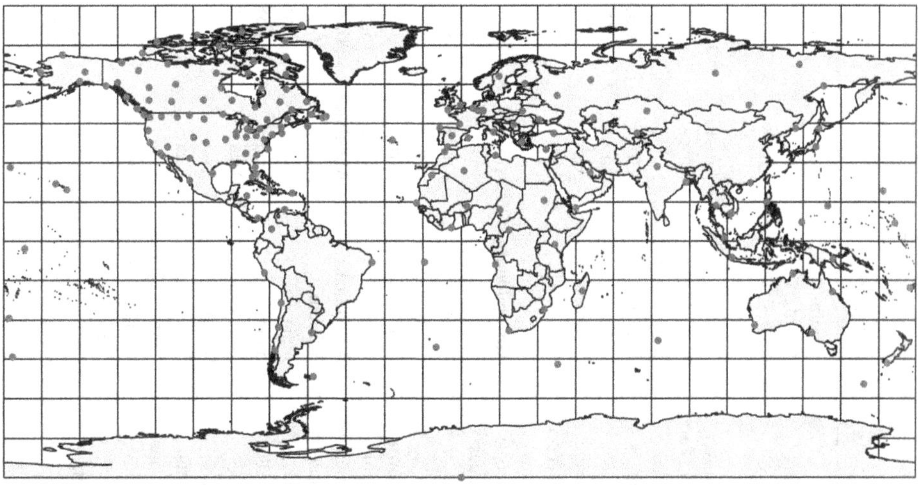

Figure 12.B.1 Refractivity gradient data sites.

TABLE 12.B.1 Obstruction Fading Data

City	State/ Provence/Country	Country/ Region	Latitude	Longitude	Feb-Mean	May-Mean	Aug-Mean	Nov-Mean	Feb-SD	May-SD	Aug-SD	Nov-SD	Source
Aoulef	Algeria	Africa	26.97	1.08	−35	−40	−45	−40	19	25	27	25	Samson
Luanda	Angola	Africa	−8.82	13.22	−40	−50	−50	−40	83	54	33	43	Samson
Bangui	Central African Republic	Africa	4.38	18.57	−50	−75	−75	−75	111	56	43	58	Samson
Fort Lamy	Chad	Africa	12.13	15.03	−150	−100	−150	−230	64	105	107	116	Samson
Abidjan (Port Bouet)	Ivory Coast (Côte d'Ivoire)	Africa	5.25	−3.93	−55	−40	−40	−30	76	49	52	89	Samson
Nairobi	Kenya	Africa	−1.30	36.75	−40	−50	−40	−50	56	23	64	58	Samson
Tripoli	Libya	Africa	32.90	13.28	−50	−75	−80	−50	40	55	92	115	Samson
Fort Trinquet	Mauritania	Africa	25.23	−11.62	−45	−45	−60	−45	29	43	159	71	Samson
Kenitra (Port Lyautey)	Morocco	Africa	34.30	−6.60	−50	−50	−50	−65	43	104	43	130	Samson
Lourenco Marques	Mozambique	Africa	−25.92	32.57	−50	−25	−30	−50	43	35	31	61	Samson
Niamey	Niger	Africa	13.48	2.17	−115	−90	−160	−160	85	100	73	112	Samson
Dakar	Senegal	Africa	14.73	−17.50	−220	−120	−100	−175	104	146	150	126	Samson
Cape Town	South Africa	Africa	−33.97	18.60	−35	−50	−50	−20	27	35	25	47	Samson
Khartoum	Sudan	Africa	15.60	32.55	−25	−45	−60	−30	45	43	54	56	Samson
Amundsen Scott	Antarctica	Antarctica	−90.00	0.00	−65	−150	−165	−65	17	43	49	21	Samson
Muharraq	Bahrain	Asia	26.27	50.62	−65	−210	−290	−80	29	87	130	109	Samson
Calcutta	India	Asia	22.65	88.45	−105	−70	−70	−125	134	120	36	104	Samson
New Delhi	India	Asia	28.58	77.20	−85	−120	−70	−105	28	37	17	39	Samson
Djakarta	Indonesia	Asia	−6.18	106.83	−60	−60	−60	−60	100	100	100	100	Samson
Atyrau	Kazakhstan	Asia	47.12	51.92	−45	−35	−35	−45	9	82	67	125	Samson
Karaganda	Kazakhstan	Asia	49.80	73.13	−40	−40	−40	−40	16	19	23	16	Samson
Verkhoyansk	North Yakut	Asia	67.55	133.38	−90	−35	−45	−75	21	4	21	16	Samson
Singapore	Republic of Singapore	Asia	1.35	103.90	−70	−80	−70	−70	31	155	134	27	Samson
Bangkok	Thailand	Asia	13.73	100.50	−65	−100	−50	−95	41	89	21	43	Samson
Lviv (L'vov)	Ukraine	Asia	49.82	23.95	−30	−40	−45	−45	4	54	27	21	Samson
Odessa	Ukraine	Asia	46.48	30.63	−35	−45	−50	−45	12	31	90	10	Samson

Tashkent	Uzbekistan	Asia	41.33	69.30	−40	−55	−55	−55	8	58	80	17	Samson
Saigon	Viet Nam	Asia	10.82	106.67	−60	−70	−80	−80	112	72	50	90	Samson
Tan An	Viet Nam	Asia	10.53	106.38	−75	−75	−75	−75	153	153	153	153	Samson
Aden	Yemen	Asia	12.83	45.02	−55	−75	−65	−40	138	123	107	155	Samson
Lajes	Azores	Atlantic	38.75	−27.08	−50	−55	−55	−50	16	34	74	42	Samson
Stanley	Falkland Islands	Atlantic	−51.70	−57.87	−50	−45	−50	−50	17	10	12	27	Samson
Thule	Greenland	Atlantic	76.52	−68.73	−80	−30	−30	−60	19	14	66	21	Samson
Ascension Island		Atlantic	−7.97	−14.40	−50	−50	−50	−50	49	33	26	23	Samson
Gough Island		Atlantic	−40.32	−9.90	−30	−35	−30	−30	13	32	13	9	Samson
Adelaide	Australia	Australia	−34.93	138.58	−30	−50	−50	−30	41	83	43	25	Samson
Darwin	Australia	Australia	−12.43	130.87	−55	−30	−30	−30	34	34	89	146	Samson
Perth	Australia	Australia	−31.95	115.82	−25	−30	−40	−30	61	120	80	106	Samson
Townsville	Australia	Australia	−19.27	146.75	−55	−40	−40	−40	43	35	21	43	Samson
Calgary	Alberta	Canada	51.10	−114.02	−38	−32	−40	−35	68	53	60	65	Kizer
Edmonton	Alberta	Canada	53.32	−113.58	−35	−32	−41	−34	71	52	55	71	Segal and Barrington
Watino	Alberta	Canada	55.72	−117.62	−39	−35	−42	−38	51	56	53	50	Kizer
Comox	British Columbia	Canada	49.72	−124.90	−40	−40	−52	−41	29	67	73	64	Kizer
Fort Nelson	British Columbia	Canada	58.83	−122.58	−39	−33	−39	−37	51	53	55	39	Segal and Barrington
Hope	British Columbia	Canada	49.38	−121.43	−39	−38	−48	−39	34	33	56	53	Kizer
Mission	British Columbia	Canada	49.15	−122.27	−40	−39	−50	−40	27	31	50	48	Kizer
Port Hardy	British Columbia	Canada	50.68	−127.37	−39	−39	−46	−40	48	55	75	73	Segal and Barrington
Prince George	British Columbia	Canada	53.88	−122.67	−35	−33	−39	−35	41	44	27	41	Segal and Barrington
Summerland	British Columbia	Canada	49.57	−119.65	−39	−36	−44	−38	42	39	53	48	Kizer
Vancouver	British Columbia	Canada	49.18	−123.17	−40	−40	−51	−40	24	43	42	46	Kizer
Churchill	Manitoba	Canada	58.75	−94.07	−45	−38	−41	−38	26	103	88	30	Segal and Barrington
Dauphin	Manitoba	Canada	51.10	−100.05	−40	−35	−41	−35	31	54	55	44	Kizer

(*continued*)

TABLE 12.B.1 (*Continued*)

City	State/ Provence/Country	Country/ Region	Latitude, degrees	Longitude, degrees	Feb-Mean	May-Mean	Aug-Mean	Nov-Mean	Feb-SD	May-SD	Aug-SD	Nov-SD	Source
The Pas	Manitoba	Canada	53.97	−101.10	−39	−35	−40	−36	18	50	73	29	Segal and Barrington
Winnipeg	Manitoba	Canada	49.90	−97.23	−40	−36	−42	−35	37	60	59	49	Kizer
Frederiction	New Brunswick	Canada	45.92	−66.62	−39	−40	−49	−40	40	23	53	32	Kizer
Saint John	New Brunswick	Canada	45.32	−65.88	−39	−40	−51	−41	48	21	62	35	Kizer
Argentia	Newfoundland	Canada	47.23	−54.02	−34	−37	−45	−39	40	44	33	47	Segal and Barrington
Gander	Newfoundland	Canada	48.95	−54.57	−39	−40	−48	−40	40	29	33	35	Kizer
Goose Bay	Newfoundland	Canada	53.32	−60.42	−38	−36	−40	−35	24	69	107	59	Segal and Barrington
St. John's	Newfoundland	Canada	47.62	−52.75	−38	−42	−52	−41	12	43	43	39	Segal and Barrington
Stephenville	Newfoundland	Canada	48.53	−58.55	−34	−35	−45	−34	21	44	102	82	Segal and Barrington
Cambridge Bay	Northwest Territories	Canada	69.10	−105.12	−49	−38	−39	−43	16	34	59	36	Kizer
Alert	Northwest Territory	Canada	82.50	−62.33	−49	−37	−36	−44	15	31	37	13	Segal and Barrington
Baker Lake	Northwest Territory	Canada	64.30	−96.00	−48	−37	−38	−41	16	51	38	23	Segal and Barrington
Clyde	Northwest Territory	Canada	70.45	−68.55	−40	−36	−36	−36	16	54	69	41	Segal and Barrington
Coppermine	Northwest Territory	Canada	67.83	−115.12	−46	−36	−37	−39	17	53	63	38	Segal and Barrington
Coral Harbour	Northwest Territory	Canada	64.20	−83.37	−45	−37	−38	−40	14	40	40	23	Segal and Barrington
Eureka	Northwest Territory	Canada	80.00	−85.93	−52	−37	−36	−49	16	26	42	16	Segal and Barrington
Fort Smith	Northwest Territory	Canada	60.02	−111.97	−41	−34	−38	−37	28	80	94	33	Segal and Barrington

Frobisher Bay	Northwest Territory	Canada	63.75	−68.55	−44	−37	−35	−37	19	60	47	20	Segal and Barrington
Hall Beach	Northwest Territory	Canada	68.78	−81.25	−46	−39	−39	−42	14	45	59	25	Segal and Barrington
Inuvik	Northwest Territory	Canada	68.30	−133.48	−47	−35	−35	−41	17	38	64	34	Segal and Barrington
Isachsen	Northwest Territory	Canada	78.78	−103.53	−53	−38	−38	−47	17	41	58	25	Segal and Barrington
Mould Bay	Northwest Territory	Canada	76.23	−119.33	−51	−37	−36	−43	17	38	38	33	Segal and Barrington
Norman Wells	Northwest Territory	Canada	65.30	−126.85	−45	−35	−38	−42	71	71	57	49	Segal and Barrington
Resolute	Northwest Territory	Canada	74.68	−94.92	−49	−38	−37	−44	15	45	42	44	Segal and Barrington
Sachs Harbour	Northwest Territory	Canada	71.98	−125.28	−49	−39	−38	−42	15	73	69	46	Segal and Barrington
Halifax	Nova Scotia	Canada	44.63	−63.50	−39	−41	−54	−42	50	23	73	35	Kizer
Kentville	Nova Scotia	Canada	45.07	−64.48	−39	−41	−53	−41	50	22	72	35	Kizer
Sable Island	Nova Scotia	Canada	43.93	−60.03	−37	−39	−56	−39	63	82	96	90	Segal and Barrington
Sydney	Nova Scotia	Canada	46.17	−60.05	−39	−41	−52	−41	48	33	96	35	Kizer
Central Patricia	Ontario	Canada	51.50	−90.15	−41	−38	−45	−37	27	64	66	41	Kizer
Geraldton	Ontario	Canada	49.68	−86.95	−40	−38	−47	−37	19	45	58	38	Kizer
Kingston	Ontario	Canada	44.23	−76.48	−39	−41	−46	−40	25	50	65	58	Kizer
London	Ontario	Canada	43.03	−81.15	−39	−43	−48	−39	43	50	77	48	Kizer
Moosonee	Ontario	Canada	51.27	−80.65	−40	−37	−43	−37	21	48(72)	66(93)	23(46)	Segal and Barrington
North Bay	Ontario	Canada	46.37	−79.42	−39	−40	−47	−38	25	52	58	57	Kizer
Ottawa	Ontario	Canada	45.38	−75.72	−39	−40	−46	−39	26	52	61	56	Kizer

(continued)

TABLE 12.B.1 *(Continued)*

City	State/ Provence/Country	Country/ Region	Latitude, degrees	Longitude, degrees	Feb-Mean	May-Mean	Aug-Mean	Nov-Mean	Feb-SD	May-SD	Aug-SD	Nov-SD	Source
Sault Ste. Marie	Ontario	Canada	46.48	−84.50	−39	−40	−49	−38	27	48	60	57	Kizer
Sioux Lookout	Ontario	Canada	50.12	−91.90	−40	−38	−46	−37	32	62	70	45	Kizer
Toronto	Ontario	Canada	43.68	−79.63	−39	−42	−47	−39	45	51	92	55	Kizer
Trout Lake	Ontario	Canada	53.83	−89.87	−41	−36	−41	−36	31	73	55(88)	38	Segal and Barrington
Windsor	Ontario	Canada	42.27	−82.97	−39	−44	−50	−40	31	48	77	48	Kizer
Summerside	Prince Edward Island	Canada	46.43	−63.83	−39	−40	−50	−40	50	22	92	35	Kizer
Caplan	Quebec	Canada	48.10	−65.65	−38	−39	−47	−39	13	30	64	39	Kizer
Fort Chimo	Quebec	Canada	58.10	−68.42	−41	−37	−39	−36	16	45	67	20	Segal and Barrington
Gagnon	Quebec	Canada	51.95	−68.13	−38	−39	−44	−38	18	50	58	47	Kizer
Inoucdjouac	Quebec	Canada	58.45	−78.12	−43	−38	−41	−36	22	45(68)	85	57	Segal and Barrington
Maniwaki	Quebec	Canada	46.37	−75.98	−37	−38	−47	−36	29	54	63	53	Segal and Barrington
Montreal	Quebec	Canada	45.47	−73.75	−39	−40	−47	−39	42	47	53	53	Kizer
Nitchequon	Quebec	Canada	53.20	−70.90	−39	−35	−40	−35	15	50	49	47	Segal and Barrington
Normandin	Quebec	Canada	48.85	−72.53	−38	−39	−46	−38	24	53	57	40	Kizer
Post de La Baleine	Quebec	Canada	55.28	−77.77	−40	−39	−44	−38	25	72	84	38	Kizer
Quebec	Quebec	Canada	46.80	−71.38	−38	−39	−47	−39	37	34	41	40	Kizer
Sept Iles	Quebec	Canada	50.22	−66.27	−37	−37	−43	−36	20	67	53	50	Segal and Barrington
Val d'Or	Quebec	Canada	48.05	−77.78	−39	−39	−46	−38	25	58	52	53	Kizer
Prince Albert	Saskatchewan	Canada	53.22	−105.68	−39	−34	−42	−35	39	51	63	46	Kizer
Regina	Saskatchewan	Canada	50.93	−104.67	−40	−33	−39	−34	43	55	52	43	Kizer

Swift Current	Saskatchewan	Canada	50.27	−107.73	−38	−31	−34	−33	55	58	53	52	Kizer
Uranium City	Saskatchewan	Canada	59.57	−108.48	−40	−36	−41	−39	29	82	84	35	Kizer
Weyburn	Saskatchewan	Canada	49.67	−103.85	−40	−33	−37	−34	45	56	41	50	Kizer
Carmacks	Yukon	Canada	62.10	−136.30	−40	−34	−40	−40	40	49	46	28	Kizer
Whitehorse	Yukon	Canada	60.72	−135.07	−35	−32	−29	−33	29	55	45	16	Segal and Barrington
Guantanamo	Cuba	Caribbean	19.90	−75.15	−50	−50	−50	−50	38	35	35	21	Samson
Curacao	Netherlands Antilles	Caribbean	12.18	−68.98	−60	−60	−60	−60	10	26	21	26	Samson
San Juan	Puerto Rico	Caribbean	18.43	−66.00	−65	−75	−75	−75	25	60	34	39	Samson
Swan Island		Caribbean	17.40	−83.93	−55	−70	−70	−70	19	23	40	28	Samson
Balboa	Panama	Central America	8.93	−79.57	−60	−60	−60	−60	31	40	31	47	Samson
Hong Kong	Hong Kong Special Administrative Region	China	22.30	114.17	−45	−65	−70	−50	12	17	45	17	Samson
Bruxelles	Belgium	Europe	50.90	4.48	−50	−50	−50	−50	66	66	66	66	Samson
Gibraltar	British Colony	Europe	36.15	−5.35	−50	−60	−65	−60	122	120	124	122	Samson
Camborne	England	Europe	50.22	−5.32	−50	−50	−50	−50	16	31	23	19	Samson
Cardington	England	Europe	52.10	−0.42	−45	−45	−45	−45	104	104	104	104	Samson
Helsinki	Finland	Europe	60.32	24.97	−40	−45	−45	−40	12	7	12	14	Samson
Bordeaux	France	Europe	44.85	−0.70	−55	−70	−75	−65	31	43	56	19	Samson
Bitburg	Germany	Europe	49.95	6.52	−35	−35	−35	−35	9	19	33	23	Samson
Gross Rohrheim	Germany	Europe	49.72	8.47	−55	−55	−55	−55	74	74	74	74	Samson
Wiesbaden	Germany	Europe	50.05	8.33	−35	−35	−35	−35	29	21	54	43	Samson
Athens	Greece	Europe	37.97	23.72	−40	−50	−50	−40	25	33	145	29	Samson
Rome	Italy	Europe	41.80	12.60	−40	−55	−55	−40	45	33	91	83	Samson
La Coruna	Spain	Europe	43.37	−8.42	−40	−50	−50	−50	126	132	134	113	Samson
Madrid	Spain	Europe	40.40	−3.68	−35	−35	−35	−35	23	50	29	52	Samson
Ostersund	Sweden	Europe	63.18	14.62	−40	−40	−50	−40	6	12	35	10	Samson
Beograd	Yugoslavia	Europe	44.78	20.53	−40	−50	−55	−40	47	98	134	66	Samson
Tananarive	Madagascar (Malagasy Republic)	Indian Ocean	−18.90	47.53	−80	−75	−75	−75	71	21	37	21	Samson
Marion Island		Indian Ocean	−46.88	37.87	−40	−40	−40	−40	8	12	39	14	Samson

(continued)

TABLE 12.B.1 *(Continued)*

City	State/ Provence/Country	Country/ Region	Latitude, degrees	Longitude, degrees	Feb-Mean	May-Mean	Aug-Mean	Nov-Mean	Feb-SD	May-SD	Aug-SD	Nov-SD	Source
Nouvelle Amsterdam Island		Indian Ocean	−37.83	77.57	−65	−50	−50	−50	33	27	41	17	Samson
Sapporo	Hokkaido	Japan	43.05	141.33	−35	−35	−45	−35	5	23	30	17	Samson
Tokyo (Tateno)	Honshu	Japan	36.05	140.13	−50	−50	−50	−50	21	56	35	19	Samson
Nicosia	Cyprus	Mediterranean	35.15	33.28	−25	−40	−75	−20	26	61	116	119	Samson
Palma	Majorca	Mediterranean	39.60	2.70	−40	−30	−40	−40	173	168	173	162	Samson
Samsun	Turkey	Mideast	41.28	36.33	−40	−50	−50	−45	17	19	34	17	Samson
Mazatlan	Mexico	North America	23.18	−106.43	−80	−70	−60	−70	47	26	42	36	Samson
Ship C		North Atlantic Ocean	52.75	−35.50	−40	−40	−45	−40	10	28	36	40	Samson
Ship K		North Atlantic Ocean	45.00	−16.00	−40	−50	−50	−50	17	52	69	63	Samson
Ship V		North Pacific Ocean	34.00	164.00	−60	−60	−70	−60	14	23	12	12	Samson
Ship M		North Sea	66.00	2.00	−40	−40	−45	−40	7	12	54	9	Samson
Nandi	Fiji Islands	Pacific	−17.75	177.45	−65	−60	−30	−60	62	101	49	140	Samson
Raoul Island	Kermadec Islands	Pacific	−29.25	−177.92	−50	−40	−40	−50	16	23	50	29	Samson
Guam	Mariana Islands	Pacific	13.55	144.83	−65	−60	−50	−55	52	49	54	33	Samson
Majuro Island	Marshall Islands	Pacific	7.08	171.38	−55	−55	−65	−55	4	6	10	4	Samson
Lae	New Guinea	Pacific	−6.75	146.98	−65	−65	−65	−85	21	13	24	52	Samson
Invercargill	New Zealand	Pacific	−46.42	168.32	−50	−50	−50	−50	12	19	43	10	Samson
Canton Island	Phoenix Islands	Pacific	−2.77	−171.72	−70	−70	−70	−70	24	36	24	24	Samson
Chatham Island		Pacific	−43.97	−176.55	−50	−50	−50	−60	10	10	10	14	Samson
Koror Island		Pacific	7.33	134.48	−25	−25	−35	−30	4	23	6	2	Samson
Macquarie Island		Pacific	−54.50	158.95	−40	−40	−40	−40	6	9	13	6	Samson
Midway Island (Atoll)		Pacific	28.22	−177.37	−50	−50	−60	−50	35	12	23	12	Samson
Wake Island (Atoll)		Pacific	19.28	166.65	−60	−75	−85	−85	56	80	133	26	Samson
Clark Field	Luzon	Philippines	15.13	120.58	−45	−55	−65	−65	35	86	89	68	Samson
Okhotsk	Khabarovsk Krai	Russia	59.37	143.20	−50	−45	−50	−45	8	21	35	12	Samson
Syktyvkar	Komi Krai	Russia	61.67	50.85	−35	−45	−45	−35	4	52	45	4	Samson

Tura	Krasnoyarsk Krai	Russia	64.27	100.23	−70	−40	−40	−55	16	8	10	10	Samson
Vladivostok	Primorsky Krai	Russia	43.12	131.90	−30	−45	−55	−40	19	17	31	10	Samson
Chita	Zabaykalsky Krai	Russia	52.08	113.48	−55	−30	−35	−35	8	4	30	4	Samson
Moscow		Russia	55.82	37.62	−35	−35	−45	−35	10	21	33	12	Samson
Buenos Aires	Argentina	South America	−34.83	−58.53	−45	−50	−45	−40	153	161	45	32	Samson
Ezeiza	Argentina	South America	−34.83	−58.53	−40	−50	−50	−25	168	62	32	41	Samson
Recife	Brazil	South America	−8.12	−34.92	−55	−55	−55	−55	24	17	19	43	Samson
Antofagasta	Chile	South America	−23.47	−70.43	−60	−50	−50	−50	14	26	14	21	Samson
Puerto Montt	Chile	South America	−41.48	−72.85	−50	−50	−50	−50	26	127	14	5	Samson
Quintero	Chile	South America	−32.78	−71.53	−45	−50	−50	−50	41	54	43	26	Samson
Bogota	Columbia	South America	4.70	−74.15	−60	−65	−60	−70	41	39	34	34	Samson
Lima	Peru	South America	−12.10	−77.03	−65	−50	−50	−50	25	14	10	9	Samson
Birmingham	Alabama	United States	33.52	−86.82	−43	−55	−62	−46	40	45	65	67	Kizer
Huntsville	Alabama	United States	34.73	−86.58	−42	−54	−62	−45	44	45	57	67	Kizer
Mobile	Alabama	United States	30.73	−88.05	−45	−57	−62	−47	55	57	98	79	Kizer
Montgomery	Alabama	United States	32.38	−86.32	−44	−56	−62	−47	40	45	79	69	Kizer
Anchorage	Alaska	United States	61.17	−149.98	−45	−45	−45	−45	12	19	36	19	Samson
Annette Island	Alaska	United States	55.00	−131.58	−37	−37	−42	−36	45	57	46	44	Segal and Barrington
Barrow	Alaska	United States	71.30	−156.78	−45	−40	−45	−45	14	12	57	12	Samson
Fairbanks	Alaska	United States	64.82	−147.87	−43	−36	−39	−39	13	11	34	1	Segal and Barrington
Juneau	Alaska	United States	58.36	−134.58	−40	−35	−40	−40	26	35	46	39	Kizer
Nome	Alaska	United States	64.50	−165.43	−55	−50	−50	−50	10	12	30	7	Samson
Shemya	Alaska	United States	52.72	−174.10	−40	−40	−40	−40	13	11	73	19	Samson
Yakutat	Alaska	United States	59.50	−139.68	−37	−38	−40	−37	39	42	49	45	Segal and Barrington
Flagstaff	Arizona	United States	35.17	−111.67	−30	−27	−27	−31	45	68	48	55	Kizer

(continued)

TABLE 12.B.1 (*Continued*)

City	State/ Provence/Country	Country/ Region	Latitude, degrees	Longitude, degrees	Feb-Mean	May-Mean	Aug-Mean	Nov-Mean	Feb-SD	May-SD	Aug-SD	Nov-SD	Source
Phoenix	Arizona	United States	33.45	−112.07	−34	−32	−31	−38	45	39	46	53	Kizer
Prescott	Arizona	United States	34.58	−112.35	−32	−29	−28	−35	44	64	47	51	Kizer
Tucson	Arizona	United States	32.22	−110.97	−36	−35	−35	−40	48	44	44	55	Kizer
Winslow	Arizona	United States	35.03	−110.67	−29	−26	−27	−29	44	65	50	55	Kizer
Yuma	Arizona	United States	32.67	−114.67	−42	−42	−50	−48	40	34	57	37	Kizer
Fort Smith	Arkansas	United States	35.39	−94.32	−40	−52	−56	−42	39	45	46	63	Kizer
Little Rock	Arkansas	United States	34.73	−92.23	−41	−54	−61	−43	37	45	69	69	Kizer
Texarkana	Arkansas	United States	33.49	−93.93	−42	−54	−61	−44	40	45	74	69	Kizer
Bakersfield	California	United States	35.42	−119.00	−41	−43	−53	−49	34	50	52	51	Kizer
Eureka	California	United States	40.80	−124.18	−42	−40	−47	−47	25	30	40	45	Kizer
Fresno	California	United States	36.73	−119.78	−40	−42	−48	−48	50	72	64	59	Kizer
Los Angeles (Long Beach)	California	United States	33.82	−118.15	−50	−40	−40	−50	50	30	23	61	Samson
Red Bluff	California	United States	40.17	−122.25	−40	−38	−40	−46	45	65	60	57	Kizer
Sacramento	California	United States	38.58	−121.50	−40	−40	−45	−48	41	61	58	55	Kizer
San Diego	California	United States	32.73	−117.17	−60	−45	−50	−60	23	9	10	42	Samson
San Francisco (Oakland)	California	United States	37.73	−122.20	−50	−50	−50	−50	19	34	16	55	Samson
San Jose	California	United States	37.33	−121.88	−42	−45	−55	−51	23	35	39	49	Kizer
San Luis Obispo	California	United States	35.33	−120.67	−43	−47	−62	−53	29	32	43	47	Kizer
Alamosa	Colorado	United States	37.42	−105.87	−28	−27	−32	−27	39	42	41	51	Kizer
Denver	Colorado	United States	39.77	−104.88	−35	−35	−50	−50	30	66	73	30	Samson
Grand Junction	Colorado	United States	39.12	−108.50	−29	−27	−32	−28	48	59	43	53	Kizer
Pueblo	Colorado	United States	38.27	−104.62	−29	−29	−34	−28	26	38	51	27	Kizer
Bridgeport	Connecticut	United States	41.22	−73.29	−39	−44	−49	−45	54	61	57	58	Kizer
Hartford	Connecticut	United States	41.77	−72.68	−39	−43	−49	−44	52	57	53	55	Kizer
New Haven	Connecticut	United States	41.34	−72.96	−39	−44	−50	−45	53	62	57	57	Kizer
Washington	District of Columbia	United States	38.85	−77.03	−40	−40	−40	−40	18(54)	42	49	49	Samson
Wilmington	Delaware	United States	39.75	−75.55	−40	−46	−48	−46	54	44	52	54	Kizer

Cocoa Beach	Florida	United States	28.23	−80.60	−50	−65	−65	−65	59	57	57(114)	112	Samson
Jacksonville	Florida	United States	30.33	−81.67	−45	−56	−60	−49	59	60	117	87	Kizer
Key West	Florida	United States	24.55	−81.75	−50	−50	−50	−50	14	14	14	19	Samson
Melbourne	Florida	United States	28.08	−80.60	−47	−58	−60	−50	59	57	113	112	Kizer
Miami	Florida	United States	25.82	−80.28	−50	−75	−70	−55	36	42	38(76)	21(63)	Samson
Miami Beach	Florida	United States	25.78	−80.21	−49	−59	−60	−50	36	42	74	64	Kizer
Tallahassee	Florida	United States	30.45	−84.28	−45	−57	−60	−48	56	57	100	76	Kizer
Tampa	Florida	United States	27.97	−82.53	−47	−58	−60	−49	59	57	118	112	Kizer
West Palm Beach	Florida	United States	26.68	−80.11	−48	−59	−60	−50	52	53	88	97	Kizer
Atlanta	Georgia	United States	33.65	−84.43	−55	−65	−85	−65	38	26	30(60)	69	Samson
Augusta	Georgia	United States	33.42	−82.05	−43	−54	−60	−48	51	51	94	67	Kizer
Columbus	Georgia	United States	32.50	−84.88	−43	−56	−61	−47	40	41	80	69	Kizer
Savannah	Georgia	United States	31.96	−81.03	−44	−55	−60	−49	65	83	106	67	Kizer
Hilo (Hawaii)	Hawaii	United States	19.73	−155.07	−50	−65	−75	−60	43	40	59	80	Samson
Honolulu/Oahu	Hawaii	United States	21.32	−157.92	−49	−49	−58	−48	30	27	24	41	Kizer
Kahului/Maui	Hawaii	United States	20.90	−156.43	−49	−49	−58	−48	38	33	36	61	Kizer
Lihue (Kauai)	Hawaii	United States	21.98	−159.35	−50	−60	−60	−50	30	24	24	30	Samson
Boise	Idaho	United States	43.62	−116.20	−35	−31	−32	−34	50	46	40	52	Kizer
Lewiston	Idaho	United States	46.38	−116.96	−37	−32	−34	−35	46	41	48	47	Kizer
Pocatello	Idaho	United States	42.87	−112.45	−32	−30	−33	−31	58	59	40	51	Kizer
Cairo	Illinois	United States	37.21	−89.18	−40	−51	−61	−42	36	45	56	66	Kizer
Chicago	Illinois	United States	41.87	−87.68	−39	−45	−55	−39	24	46	63	60	Kizer
Joliet	Illinois	United States	41.50	−88.17	−45	−70	−70	−70	12(24)	29(44)	60	47(60)	Samson
Moline	Illinois	United States	41.80	−89.98	−39	−45	−54	−38	23	46	67	57	Kizer
Peoria	Illinois	United States	40.70	−89.60	−39	−46	−56	−39	22	43	70	51	Kizer
Rantoul	Illinois	United States	40.30	−88.15	−40	−50	−50	−40	26	45	54	30	Samson
Rockford	Illinois	United States	42.28	−89.10	−39	−44	−54	−38	25	50	69	60	Kizer
Springfield	Illinois	United States	39.80	−89.65	−39	−48	−57	−40	21	39	68	40	Kizer
Evansville	Indiana	United States	38.05	−87.54	−40	−50	−60	−42	44	46	62	57	Kizer

(*continued*)

TABLE 12.B.1 (*Continued*)

City	State/ Provence/Country	Country/ Region	Latitude, degrees	Longitude, degrees	Feb-Mean	May-Mean	Aug-Mean	Nov-Mean	Feb-SD	May-SD	Aug-SD	Nov-SD	Source
Fort Wayne	Indiana	United States	41.07	−85.13	−39	−46	−55	−40	25	50	70	60	Kizer
Indianapolis	Indiana	United States	39.77	−86.17	−39	−48	−58	−41	22	38	67	46	Kizer
South Bend	Indiana	United States	41.68	−86.25	−39	−45	−55	−39	25	50	70	60	Kizer
Burlington	Iowa	United States	40.80	−91.22	−39	−46	−54	−38	18	38	64	51	Kizer
Davenport	Iowa	United States	41.58	−90.64	−39	−45	−54	−38	21	43	66	55	Kizer
Des Moines	Iowa	United States	41.58	−93.62	−38	−44	−50	−37	24	42	61	55	Kizer
Dubuque	Iowa	United States	42.55	−90.85	−39	−44	−53	−38	23	45	62	54	Kizer
Sioux City	Iowa	United States	42.50	−96.40	−37	−41	−46	−35	30	44	58	50	Kizer
Waterloo	Iowa	United States	42.50	−92.33	−39	−43	−52	−37	24	43	61	52	Kizer
Dodge City	Kansas	United States	37.77	−99.97	−50	−50	−40	−50	21	33	40	14	Samson
Goodland	Kansas	United States	39.37	−101.67	−32	−35	−38	−31	42	45	51	61	Kizer
Topeka	Kansas	United States	39.05	−95.67	−38	−46	−47	−37	17	32	60	45	Kizer
Wichita	Kansas	United States	37.68	−97.33	−37	−46	−43	−37	21	35	60	39	Kizer
Lexington	Kentucky	United States	38.05	−84.50	−40	−50	−60	−43	45	41	57	59	Kizer
Louisville	Kentucky	United States	38.19	−85.74	−40	−50	−60	−42	44	42	61	55	Kizer
Burrwood	Louisiana	United States	28.97	−89.37	−60	−65	−65	−60	31(62)	47(71)	9(99)	17(85)	Samson
Lake Charles	Louisiana	United States	30.23	−93.22	−46	−58	−65	−48	62	57	98	83	Kizer
New Orleans	Louisiana	United States	29.95	−90.07	−46	−58	−64	−47	62	58	98	85	Kizer
Shreveport	Louisiana	United States	32.47	−93.82	−43	−56	−62	−45	40	45	80	69	Kizer
Caribou	Maine	United States	46.87	−68.02	−36	−36	−45	−36	8	16	31	12	Segal and Barrington
Portland	Maine	United States	43.65	−70.32	−34	−37	−47	−36	50	54	68	65	Segal and Barrington
Baltimore	Maryland	United States	39.28	−76.62	−40	−47	−48	−46	54	42	51	50	Kizer
Frederick	Maryland	United States	39.60	−77.67	−40	−47	−48	−45	54	44	50	49	Kizer

Boston	Massachusetts	United States	42.35	−71.05	−39	−43	−51	−43	51	36	59	54	Kizer
Nantucket	Massachusetts	United States	41.25	−70.07	−45	−50	−55	−45	43	83	59	54	Samson
Pittsfield	Massachusetts	United States	42.45	−73.25	−39	−43	−48	−43	51	55	50	54	Kizer
Detroit	Michigan	United States	42.38	−83.03	−39	−44	−50	−39	32	47	79	49	Kizer
Escanaba	Michigan	United States	45.81	−86.91	−39	−40	−50	−38	25	50	63	58	Kizer
Flint	Michigan	United States	42.97	−83.73	−35	−37	−44	−36	38	57	65	45(60)	Segal and Barrington
Grand Rapids	Michigan	United States	42.88	−85.52	−39	−44	−53	−39	28	50	70	60	Kizer
Houghton Lake	Michigan	United States	44.42	−84.83	−39	−42	−51	−38	28	50	70	58	Kizer
Lansing	Michigan	United States	42.73	−84.55	−39	−44	−52	−39	34	52	68	60	Kizer
Marquette	Michigan	United States	46.54	−87.48	−39	−40	−50	−37	22	43	56	46	Kizer
Mt. Clemens	Michigan	United States	42.60	−82.82	−40	−45	−45	−40	17(34)	27(41)	95	29(48)	Samson
Muskegon	Michigan	United States	43.23	−86.27	−39	−43	−53	−38	25	50	70	60	Kizer
Sault Ste. Marie	Michigan	United States	46.47	−84.37	−36	−35	−45	−36	29	47	61	58	Segal and Barrington
Ypsilanti	Michigan	United States	42.25	−83.62	−39	−44	−51	−39	33	51	70	56	Kizer
Duluth	Minnesota	United States	46.81	−92.19	−39	−39	−48	−36	32	52	59	52	Kizer
International Falls	Minnesota	United States	48.57	−93.38	−36	−35	−41	−35	41	62	88	50	Segal and Barrington
Minneapolis	Minnesota	United States	44.98	−93.27	−39	−40	−49	−36	31	51	57	55	Kizer
Rochester	Minnesota	United States	44.02	−92.47	−39	−41	−51	−37	27	47	60	51	Kizer
St. Cloud	Minnesota	United States	45.58	−94.17	−39	−39	−47	−36	33	52	54	54	Kizer
Jackson	Mississippi	United States	32.30	−90.18	−44	−56	−63	−46	40	45	82	69	Kizer
Columbia	Missouri	United States	38.97	−92.37	−40	−55	−65	−65	10	28	49	49	Samson
Kansas City	Missouri	United States	39.08	−94.49	−38	−47	−50	−38	14	31	61	50	Kizer
Springfield	Missouri	United States	37.22	−93.30	−39	−50	−55	−40	21	35	43	50	Kizer
St. Joseph	Missouri	United States	39.75	−94.83	−38	−46	−49	−37	18	34	60	52	Kizer
St. Louis	Missouri	United States	38.70	−90.33	−39	−49	−58	−40	21	38	61	48	Kizer
Billings	Montana	United States	45.78	−108.50	−35	−29	−28	−30	60	60	49	48	Kizer
Glasgow	Montana	United States	48.22	−106.62	−36	−32	−35	−34	60	64	59	41	Segal and Barrington

(continued)

TABLE 12.B.1 (*Continued*)

City	State/ Provence/Country	Country/ Region	Latitude, degrees	Longitude, degrees	Feb-Mean	May-Mean	Aug-Mean	Nov-Mean	Feb-SD	May-SD	Aug-SD	Nov-SD	Source
Great Falls	Montana	United States	47.48	−111.37	−32	−31	−30	−32	74	58	81	55	Segal and Barrington
Havre	Montana	United States	48.55	−109.58	−37	−30	−29	−32	63	59	63	51	Kizer
Helena	Montana	United States	46.60	−112.03	−35	−29	−28	−32	70	57	67	53	Kizer
Missoula	Montana	United States	46.83	−114.17	−36	−29	−29	−33	57	47	61	47	Kizer
Grand Island	Nebraska	United States	40.97	−98.33	−36	−41	−43	−34	27	40	54	40	Kizer
Lincoln	Nebraska	United States	40.82	−96.70	−37	−43	−46	−36	25	38	60	45	Kizer
Norfolk	Nebraska	United States	42.08	−97.47	−37	−41	−45	−35	30	42	54	48	Kizer
North Platte	Nebraska	United States	41.17	−100.80	−34	−36	−41	−32	41	47	49	47	Kizer
Omaha	Nebraska	United States	41.25	−96.00	−37	−43	−47	−36	25	40	60	48	Kizer
Scottsbluff	Nebraska	United States	41.83	−103.67	−32	−31	−37	−30	47	49	49	51	Kizer
Valentine	Nebraska	United States	42.70	−100.53	−35	−36	−42	−32	40	47	48	30	Kizer
Elko	Nevada	United States	40.83	−115.83	−34	−31	−31	−34	48	51	40	53	Kizer
Ely	Nevada	United States	39.28	−114.83	−33	−30	−31	−33	51	50	40	48	Kizer
Las Vegas	Nevada	United States	36.08	−115.15	−30	−25	−25	−50	64	106	126	33	Samson
Reno	Nevada	United States	39.52	−119.82	−38	−36	−35	−41	30	36	40	55	Kizer
Tonopah	Nevada	United States	38.08	−117.25	−36	−32	−32	−38	44	47	40	55	Kizer
Winnemucca	Nevada	United States	40.98	−117.75	−36	−33	−31	−36	40	47	40	45	Kizer
Concord	New Hampshire	United States	43.20	−71.53	−39	−42	−49	−42	50	48	54	56	Kizer
Atlantic City	New Jersey	United States	39.37	−74.58	−40	−46	−50	−47	54	37	61	56	Kizer
Newark	New Jersey	United States	40.73	−74.17	−40	−45	−49	−45	51	54	53	59	Kizer
Trenton	New Jersey	United States	40.22	−74.73	−40	−45	−48	−46	54	46	53	57	Kizer
Albuquerque	New Mexico	United States	35.08	−106.65	−28	−25	−30	−27	42	43	40	62	Kizer
Clayton	New Mexico	United States	36.35	−103.17	−31	−31	−34	−30	37	45	71	69	Kizer
Jornada	New Mexico	United States	32.62	−106.73	−28	−26	−30	−29	56	46	38	49	Kizer
Raton	New Mexico	United States	36.90	−104.47	−29	−28	−33	−28	26	31	46	37	Kizer
Roswell	New Mexico	United States	33.40	−104.52	−30	−27	−31	−29	48	45	53	69	Kizer
Silver City	New Mexico	United States	32.75	−108.25	−29	−27	−28	−31	53	39	40	52	Kizer

Albany	New York	United States	42.65	−73.75	−39	−43	−48	−42	51	53	50	53	Kizer
Binghamton	New York	United States	42.10	−75.92	−39	−43	−47	−42	50	52	51	52	Kizer
Buffalo	New York	United States	42.93	−78.73	−35	−38	−46	−37	53	61	105	48	Segal and Barrington
New York	New York	United States	40.77	−73.87	−35	−45	−50	−35	31(47)	57	52	61	Samson
Rochester	New York	United States	43.17	−77.60	−39	−42	−46	−40	33	50	79	50	Kizer
Syracuse	New York	United States	43.05	−76.15	−39	−42	−46	−41	25	50	60	49	Kizer
Asheville	North Carolina	United States	35.60	−82.55	−41	−53	−59	−47	53	42	57	63	Kizer
Cape Hatteras	North Carolina	United States	35.27	−75.55	−50	−65	−65	−65	57	56	21(63)	35(70)	Samson
Charlotte	North Carolina	United States	35.22	−80.85	−41	−53	−58	−48	47	45	75	68	Kizer
Greensboro	North Carolina	United States	36.07	−79.78	−41	−52	−56	−48	46	45	72	69	Kizer
Raleigh	North Carolina	United States	35.78	−78.63	−41	−51	−57	−49	45	45	80	70	Kizer
Wilmington	North Carolina	United States	34.23	−77.92	−42	−53	−60	−51	65	75	97	67	Kizer
Winston Salem	North Carolina	United States	36.10	−80.25	−41	−52	−56	−48	49	44	69	66	Kizer
Bismarck	North Dakota	United States	46.77	−100.75	−37	−34	−40	−36	48	54	56	41	Segal and Barrington
Fargo	North Dakota	United States	46.84	−96.88	−39	−37	−43	−34	41	54	49	47	Kizer
Williston	North Dakota	United States	48.60	−103.67	−40	−33	−35	−33	49	57	42	47	Kizer
Akron	Ohio	United States	41.08	−81.52	−39	−46	−49	−41	26	44	70	48	Kizer
Cincinnati	Ohio	United States	39.18	−84.47	−40	−49	−57	−42	29	37	57	58	Kizer
Cleveland	Ohio	United States	41.44	−81.64	−39	−45	−49	−41	28	47	72	48	Kizer
Columbus	Ohio	United States	39.97	−83.00	−40	−48	−53	−42	34	41	60	56	Kizer
Dayton	Ohio	United States	39.87	−84.12	−40	−45	−80	−45	10	27	49	59	Samson
Sandusky	Ohio	United States	41.39	−82.65	−39	−45	−50	−40	27	50	70	53	Kizer
Toledo	Ohio	United States	41.65	−83.55	−39	−45	−51	−40	29	50	70	56	Kizer
Youngstown	Ohio	United States	41.10	−80.65	−39	−46	−48	−41	20	32	69	39	Kizer
Oklahoma City	Oklahoma	United States	35.47	−97.52	−38	−48	−43	−39	37	44	61	60	Kizer

(*continued*)

TABLE 12.B.1 (*Continued*)

City	State/ Provence/Country	Country/ Region	Latitude, degrees	Longitude, degrees	Feb-Mean	May-Mean	Aug-Mean	Nov-Mean	Feb-SD	May-SD	Aug-SD	Nov-SD	Source
Tulsa	Oklahoma	United States	36.15	−96.00	−39	−50	−48	−40	30	44	60	51	Kizer
Baker	Oregon	United States	44.87	−117.83	−37	−33	−33	−36	45	50	40	48	Kizer
Burns	Oregon	United States	43.58	−119.00	−37	−34	−32	−37	38	44	40	53	Kizer
Eugene	Oregon	United States	44.12	−123.17	−40	−38	−37	−42	18	32	40	41	Kizer
Medford	Oregon	United States	42.38	−122.87	−45	−50	−50	−45	14	49	56	49	Samson
Portland	Oregon	United States	45.52	−122.68	−40	−38	−38	−41	17	24	40	36	Kizer
Roseburg	Oregon	United States	43.25	−123.33	−40	−38	−38	−43	19	36	40	44	Kizer
Harrisburg	Pennsylvania	United States	40.27	−76.88	−40	−46	−47	−45	54	47	50	50	Kizer
Philadelphia	Pennsylvania	United States	39.96	−75.22	−40	−46	−48	−46	54	44	51	55	Kizer
Pittsburgh	Pennsylvania	United States	40.50	−80.22	−35	−55	−60	−45	6	21	17	19	Samson
Wilkes Barre	Pennsylvania	United States	41.25	−75.88	−39	−44	−47	−44	54	52	50	54	Kizer
San Juan	Puerto Rico	United States	18.44	−66.00	−51	−59	−60	−59	27	60	34	39	Kizer
Providence	Rhode Island	United States	41.83	−71.42	−39	−43	−51	−44	52	43	58	54	Kizer
Charleston	South Carolina	United States	32.90	−80.03	−50	−60	−50	−50	47(71)	56(84)	47(106)	66	Samson
Columbia	South Carolina	United States	33.95	−81.12	−42	−54	−59	−49	46	47	85	69	Kizer
Florence	South Carolina	United States	34.18	−79.80	−42	−53	−59	−50	66	77	99	67	Kizer
Greenville-Spartanburg	South Carolina	United States	34.85	−82.40	−42	−53	−60	−47	45	41	69	70	Kizer
Aberdeen	South Dakota	United States	45.50	−98.50	−39	−37	−42	−34	40	51	49	49	Kizer
Rapid City	South Dakota	United States	44.05	−103.00	−35	−32	−37	−31	48	50	43	50	Kizer
Sioux Falls	South Dakota	United States	43.55	−96.73	−38	−40	−45	−35	33	47	54	52	Kizer
Bristol	Tennessee	United States	36.50	−82.17	−41	−52	−58	−46	58	41	49	56	Kizer
Chattanooga	Tennessee	United States	35.05	−85.32	−42	−53	−62	−45	46	38	56	68	Kizer
Knoxville	Tennessee	United States	35.97	−83.92	−41	−53	−61	−45	58	45	54	59	Kizer
Memphis	Tennessee	United States	35.17	−89.93	−41	−54	−62	−44	40	45	76	69	Kizer
Nashville	Tennessee	United States	36.17	−86.78	−41	−53	−62	−44	58	42	52	65	Kizer
Abilene	Texas	United States	32.47	−99.72	−38	−45	−39	−40	40	45	74	69	Kizer

Amarillo	Texas	United States	35.23	−101.67	−33	−36	−34	−33	40	45	74	69	Kizer
Austin	Texas	United States	30.27	−97.75	−44	−53	−53	−46	62	54	94	72	Kizer
Brownsville	Texas	United States	25.92	−97.47	−60	−60	−60	−60	40(80)	40(60)	63(106)	49(98)	Samson
Corpus Christi	Texas	United States	27.80	−97.40	−48	−58	−59	−50	80	71	99	85	Kizer
Dallas	Texas	United States	32.78	−96.80	−42	−52	−52	−43	40	45	76	69	Kizer
Del Rio	Texas	United States	29.38	−100.92	−38	−42	−37	−41	42	46	85	69	Kizer
El Paso	Texas	United States	31.80	−106.40	−50	−35	−60	−60	61	70	19(38)	37	Samson
Fort Worth	Texas	United States	32.75	−97.33	−41	−52	−49	−43	40	45	76	69	Kizer
Houston	Texas	United States	29.75	−95.37	−46	−58	−63	−48	62	61	98	84	Kizer
Laredo	Texas	United States	27.49	−99.46	−43	−47	−43	−45	80	71	99	79	Kizer
Lubbock	Texas	United States	33.58	−101.85	−33	−36	−34	−34	40	45	74	69	Kizer
Midland	Texas	United States	32.00	−102.08	−33	−36	−34	−36	40	45	72	69	Kizer
Port Arthur	Texas	United States	29.90	−93.93	−46	−58	−65	−48	62	58	98	85	Kizer
San Angelo	Texas	United States	31.47	−100.43	−37	−43	−38	−40	40	45	74	69	Kizer
San Antonio	Texas	United States	29.43	−98.48	−43	−51	−51	−46	69	62	95	74	Kizer
Victoria	Texas	United States	28.83	−97.00	−47	−57	−60	−49	80	71	98	81	Kizer
Waco	Texas	United States	31.55	−97.13	−43	−53	−53	−45	40	45	83	69	Kizer
Wichita Falls	Texas	United States	33.92	−98.48	−38	−48	−42	−40	40	45	74	69	Kizer
Salt Lake City	Utah	United States	40.75	−111.88	−31	−30	−33	−31	54	57	41	52	Kizer
Burlington	Vermont	United States	44.48	−73.22	−39	−41	−47	−40	50	50	51	59	Kizer
Lynchburg	Virginia	United States	37.42	−79.15	−40	−50	−52	−47	54	39	57	58	Kizer
Norfolk	Virginia	United States	36.85	−76.28	−41	−50	−55	−50	46	45	70	70	Kizer
Richmond	Virginia	United States	37.53	−77.43	−40	−49	−51	−48	52	43	64	63	Kizer
Roanoke	Virginia	United States	37.27	−79.95	−40	−50	−53	−47	54	39	51	56	Kizer
Wallops Island	Virginia	United States	37.93	−75.48	−40	−48	−52	−49	45	45	68	70	Kizer
Blaine	Washington	United States	49.00	−122.75	−40	−40	−51	−40	24	37	43	45	Kizer
Port Angeles	Washington	United States	48.06	−123.32	−40	−40	−52	−41	27	73	67	63	Kizer
Quillayute	Washington	United States	48.38	−124.73	−40	−40	−40	−40	55	83	85	73	Samson
Seattle	Washington	United States	47.45	−122.30	−45	−45	−45	−55	14	16	16	23	Samson

(continued)

TABLE 12.B.1 (*Continued*)

City	State/ Provence/Country	Country/ Region	Latitude, degrees	Longitude, degrees	Feb-Mean	May-Mean	Aug-Mean	Nov-Mean	Feb-SD	May-SD	Aug-SD	Nov-SD	Source
Spokane	Washington	United States	47.62	−117.65	−36	−30	−31	−36	47	49	52	53	Segal and Barrington
Tacoma	Washington	United States	47.25	−122.43	−40	−39	−46	−41	17	23	27	33	Kizer
Walla Walla	Washington	United States	46.08	−118.17	−37	−33	−35	−36	43	40	44	48	Kizer
Yakima	Washington	United States	46.60	−120.52	−39	−36	−38	−39	25	25	33	31	Kizer
Beckley	West Virginia	United States	37.75	−81.17	−40	−50	−54	−45	58	36	41	50	Kizer
Charleston	West Virginia	United States	38.35	−81.63	−40	−50	−53	−44	58	43	40	51	Kizer
Elkins	West Virginia	United States	38.92	−79.83	−40	−48	−49	−45	56	32	41	48	Kizer
Huntington	West Virginia	United States	38.33	−82.38	−40	−50	−55	−44	54	48	41	55	Kizer
Parkersburg	West Virginia	United States	39.25	−81.41	−40	−48	−51	−43	58	41	43	50	Kizer
Green Bay	Wisconsin	United States	44.52	−88.02	−39	−41	−52	−38	25	50	70	58	Kizer
La Crosse	Wisconsin	United States	43.81	−91.24	−39	−42	−51	−37	25	46	60	53	Kizer
Madison	Wisconsin	United States	43.07	−89.38	−39	−43	−53	−38	25	49	65	59	Kizer
Milwaukee	Wisconsin	United States	43.03	−87.90	−39	−43	−53	−38	25	50	70	60	Kizer
Casper	Wyoming	United States	42.85	−106.32	−32	−29	−33	−29	46	53	42	51	Kizer
Cheyenne	Wyoming	United States	41.13	−104.82	−31	−30	−35	−29	25	36	49	28	Kizer
Lander	Wyoming	United States	42.83	−108.75	−32	−29	−32	−29	56	56	40	54	Kizer
Sheridan	Wyoming	United States	44.77	−107.00	−34	−29	−30	−30	52	59	40	45	Kizer
Yellowstone Park	Wyoming	United States	44.50	−110.50	−34	−29	−30	−30	65	58	40	54	Kizer

The additional sites were chosen to match the rain cities in Chapter 11. In some of the data below, standard deviation values were listed as xx(zz). The xx values are those derived from the source refractivity charts. In these cases, the derived estimates are believed to be too small. The (zz) values are estimates believed to be more appropriate for the geographic area based on comparison with data of other nearby locations.

The data contained in Table 12.B.1 is available at the Wiley web site mentioned in the Preface.

REFERENCES

Abramowitz, M., and I. Stegun 1968. *Handbook of Mathematical Functions (NBS AMS 55, seventh printing with corrections)*, 931–933. Washington, DC: U.S. Government Printing Office.

Abreu, G., "Very Simple Tight Bounds on the Q-Function," *IEEE Transactions on Communications*, Vol. 60, pp. 2415–2420, September 2012.

Achariyapaopan, T., "A Model of Geographic Variation of Multipath Fading Probability," *Bellcore National Radio Engineers Conference Record*, pp. TA1–TA16, 1986.

Akiyama, T. and Sasaki, O. "Statistical Distribution and Maximum Critical Value for Refractivity Gradient Variation," *Review of the Radio Communication Systems Laboratories (NTT)*, Vol. 27, pp. 841–848, 1979.

Anderson, L. J. and Gossard, E. E., "Prediction of Oceanic Duct Propagation from Climatological Data," *IRE Transactions on Antennas and Propagation*, Vol. 3, pp. 163–167, October 1955.

Bean, B.R., and E.J. Dutton 1966. *Radio Meteorology*, 4–9, 49–172. Washington, DC: U.S. Government Printing Office.

Bean, B.R., J.D. Horn, and A.M. Ozanich Jr. 1960. *Climatic Charts and Data of the Radio Refractive Index for the United States and the World, National Bureau of Standards Monograph 22*. Washington, DC: U.S. Government Printing Office.

Bean, B.R., B.A. Cahoon, C.A. Samson, and G.D. Thayer 1966. *A World Atlas of Atmospheric Radio Refractivity, Environmental Science Services Administration Monograph 1*. Washington, DC: U.S. Government Printing Office.

Day, J. P. and Trolese, L. G., "Propagation of Short Radio Waves Over Desert Terrain," *Proceedings of the IRE*, Vol. 38, pp. 165–175, February 1950.

Dougherty, H. 1968. *A Survey of Microwave Fading Mechanisms: Remedies and Applications, Environmental Science Services Administration Technical Report ERL 69-WPL4*, 4–32 March. Washington, DC: U.S. Department of Commerce.

Dougherty, H. T., "Recent Progress in Duct Propagation Predictions", *IEEE Transactions on Antennas and Propagation*, Vol. 27, pp. 542–548, July 1979.

Dougherty, H., and E.J. Dutton 1981. *The Role of Elevated Ducting for Radio Service and Interference Fields, NTIA Report 81–69*. Washington, DC: U.S. Department of Commerce March.

Dougherty, H.T., and B.A. Hart 1976. *Anomalous Propagation and Interference Fields, Office of Telecommunications Report 76–107*, 20–31 December. Bolder: Institute of Telecommunications Sciences, U.S. Department of Commerce.

Dutton, E. J., "A Note on the Distribution of Atmospherically Ducted Signal Power Near the Earth's Surface," *IEEE Transactions on Communications*, Vol. 30, pp. 301–303, January 1982.

Fruchtenicht, H. W., "Notes on Duct Influences on Line-of-Sight Propagation," *IEEE Transactions on Antennas and Propagation*, Vol. 22, pp. 295–302, March 1974.

Hosler, C. R., "Low-level Inversion Frequency in the Contiguous United States," *Monthly Weather Review*, Vol. 89, pp. 319–339, September 1961.

Hubbard, R.W. 1979. *Investigation of Digital Microwave Communications in a Strong Meteorological Ducting Environment, NTIA Report 79–24*. Washington, DC: U.S. Department of Commerce August.

Ikegami, F., "Influence of an Atmospheric Duct on Microwave Fading," *IEEE Transactions on Antennas and Propagation*, Vol. 7, pp. 252–257, July 1959.

Ikegami, F., Akiyama, T., Aoyagi, S. and Yoshida, H., "Variation of Radio Refraction in the Lower Atmosphere," *IEEE Transactions on Antennas and Propagation*, Vol. 16, pp. 194–200, March 1968.

International Telecommunication Union—Radiocommunication Sector (ITU-R), "ITU-R Recommendation P.453-9, The Radio Refractive Index: Its Formula and Refractivity Data," *ITU-R Recommendations*. Geneva: International Telecommunication Union, 2003.

International Telecommunication Union—Radiocommunication Sector (ITU-R), "ITU-R Recommendation P.835-4, Reference Standard Atmospheres," *ITU-R Recommendations*. Geneva: International Telecommunication Union, 2005.

International Telecommunication Union—Radiocommunication Sector (ITU-R), "ITU-R Recommendation P.834-6, Effects of Tropospheric Refraction on Radiowave Propagation," *ITU-R Recommendations*. Geneva: International Telecommunication Union, 2007a.

International Telecommunication Union—Radiocommunication Sector (ITU-R), "ITU-R Recommendation P.530-12, Propagation Data And Prediction Methods Required for the Design of Terrestrial Line-of-sight Systems," *ITU-R Recommendations*. Geneva: International Telecommunication Union, 2007b.

Katzin, M., Bauchman, R. W. and Binnian, W., "3- and 9- Centimeter Propagation in Low Ocean Ducts," *Proceedings of the IRE*, Vol. 35, pp. 891–905, September 1947.

Kizer, G.M. 1990. *Microwave Communication*, 423–424. Ames: Iowa State University Press.

Kizer, G., "Improvements in Fixed Point-to-Point Microwave Radio Path Design," *IEEE Wireless Transmission Symposium (WTS) 2008*, pp. 354–359, April 2008.

Knight, W., *Microwave Path Design Reference Manual*, page 4.2 (path clearance criteria) and Figures 2.4.5—2.4.8 and 2.5.5—2.5.8 (refractivity gradient charts), Alcatel North America, June 2004.

Laine, R. U., "Blackout Fading in Ling-of-Sight Microwave Links," *PIEA-PESA-PEPA Conference Proceedings*, April 1975.

Laine, R., "Digital Microwave Link Engineering (slide 148)," *Digital Microwave Systems Applications Seminar*, Vol. I, Harris Corp., February 2004.

Latter, R. F., "Guidelines for the Selection of Antenna Centerlines Based on Obstruction Fading Performance Objectives," *AT&T Long Lines Radio Systems Memorandum—Microwave Propagation-Obstruction Fading Predictions*, Section I, March 25 1980.

Lee, J. L., "Refractivity Gradient and Microwave Fading Observations in Northern Indiana", *IEEE Global Telecommunications Conference (Globecom) Conference Record*, Vol. 3, pp. 36.8.1–36.8.5, December 1985.

Lee, J. L., "Observed Atmospheric Structure Causing Degraded Microwave Propagation in the Great Lakes Area", *IEEE Global Telecommunications Conference (Globecom) Conference Record*, Vol. 3, pp. 1548–1552, December 1986.

Liebe, H. J. and Giommestad, G. G., "Calculation of Clear Air EHF Refractivity," *Radio Science*, Vol. 13, pp. 245–251, March-April 1978.

Mahmoud, S. F., Boghdady, H. N. and El-Sayed, O. L., "Analysis of Multipath Fading in the Presence of an Elevated Atmospheric Duct," *Proceedings of the IEE*, Vol. 134, pp. 71–76, February 1987.

Mojoli, L. F., "A New Approach to the Visibility Problems in Line of Sight Hops," *IEEE Global Telecommunications Conference (Globecom) Conference Record*, Vol. 1, pp. 16.4.1–16.4.6, December 1979.

Samson, C.A. 1975. *Refractivity Gradients in the Northern Hemisphere, Office of Telecommunications Report 75–59*. Washington, DC: U.S. Government Printing Office.

Samson, C.A. 1976. *Refractivity and Rainfall Data for Radio Systems Engineering, Office of Telecommunications Report 76–105*. Washington, DC: U.S. Government Printing Office.

Schiavone, J. A., "Prediction of Positive Refractivity Gradients for Line-of-Sight Microwave Radio Paths," *Bell System Technical Journal*, Vol. 60, pp. 803–822, July-August 1981.

Schiavone, J. A., "Microwave Radio Meteorology: Fading by [Duct] Beam Focusing," *IEEE International Conference on Communications (ICC) Conference Record*, Vol. 3, pp. 7B.1.1–7B.1.5, June 1982.

Segal, B., and R.E. Barrington 1977. *The Radio Climatology of Canada Tropospheric Refractivity Atlas for Canada, Communications Research Centre Report No. 1315-E*. Ottawa: Department of Communications December.

U.S. Department of the Interior Geological Survey 1970. *The National Atlas of the United States of America*. Washington, DC: U.S. Department of the Interior Geological Survey, plate 62.

Vigants, A., "Observations of 4 GHz Obstruction Fading," *IEEE International Conference on Communications (ICC) Conference Record*, pp. 28.1–28.2, June 1972.

Vigants, A., "Space-Diversity Engineering," *Bell System Technical Journal*, Vol. 54, pp. 103–142, January 1975.

Vigants, A., "Microwave Radio Obstruction Fading," *Bell System Technical Journal*, Vol. 60, pp. 785–801, July-August 1981.

Wheeler, H. A., "Microwave Relay Fading Statistics as a Function of a Terrain Clearance Factor," *IEEE Transactions on Antennas and Propagation*, Vol. 25, pp. 269–273, March 1977.

White, R. 1970. *Engineering Considerations for Microwave Communications Systems*, 43, 51. San Carlos: GTE Lenkurt.

13

REFLECTIONS AND OBSTRUCTIONS

A radio signal between a transmitter and a receiver usually travels through free space. However, radio transmitters employ practical antennas that emit significant energy in directions other than directly between the transmitter and the receiver. A radio signal traveling near the Earth can be subjected to a number of propagation obstacles. The most common is the earth's surface. The combined effect at the receive antenna of the direct signal and the signal reflected from the Earth can be important.

If the Earth is a smooth flat surface, the reflection from the Earth will appear at the receive antenna along with the direct signal. The interaction of those two signals will cause an increase or decrease in the received signal level that is dependent on the vertical placement of the receive (or transmit) antenna. That can be described as a series of secondary lobes superimposed on the antenna's normal propagation pattern (Fig. 13.1a) or as a variation in received signal level called *obstruction gain* (Fig. 13.1b). (F_1 is the first Fresnel zone defined below.) Figure 13.1b assumes perfect reflection from the obstruction. Obstruction gain is discussed at the end of this chapter.

The effect of the Earth on the radio signal reflected from the Earth is a combination of at least three factors: scattering, divergence, and reflection. The radio engineer must determine, on the basis of transmit and receive antenna patterns and surface conditions, whether not reflected energy is a significant factor. Many sources suggest evaluating only the reflecting earth surface within the first Fresnel zone (see the following). While this is generally accurate, there are cases where it is not true (Beckmann and Spizzichino, 1987). The areas close to the transmit and receive antennas are especially important but other reflective surfaces can be significant. Actual path examination by an experienced person is a practical approach.

13.1 THEORETICAL ROUGH EARTH REFLECTION COEFFICIENT

If the earth's surface is perfectly flat, the angle of the reflected signal is the same as the angle of the incident signal. Snell's law of optical reflection (the angle of incidence is equal to the angle of reflection) applies. The path is symmetric; the path traveled by the incident and reflected signal can be reversed (an incident signal following the original reflected signal's path will be reflected along the original incident signal's path). As the surface becomes rough, the situation begins to change (Rice, 1951). When the incident signal strikes the rough surface, some energy is directed at angles different than predicted by

Digital Microwave Communication: Engineering Point-to-Point Microwave Systems, First Edition. George Kizer.

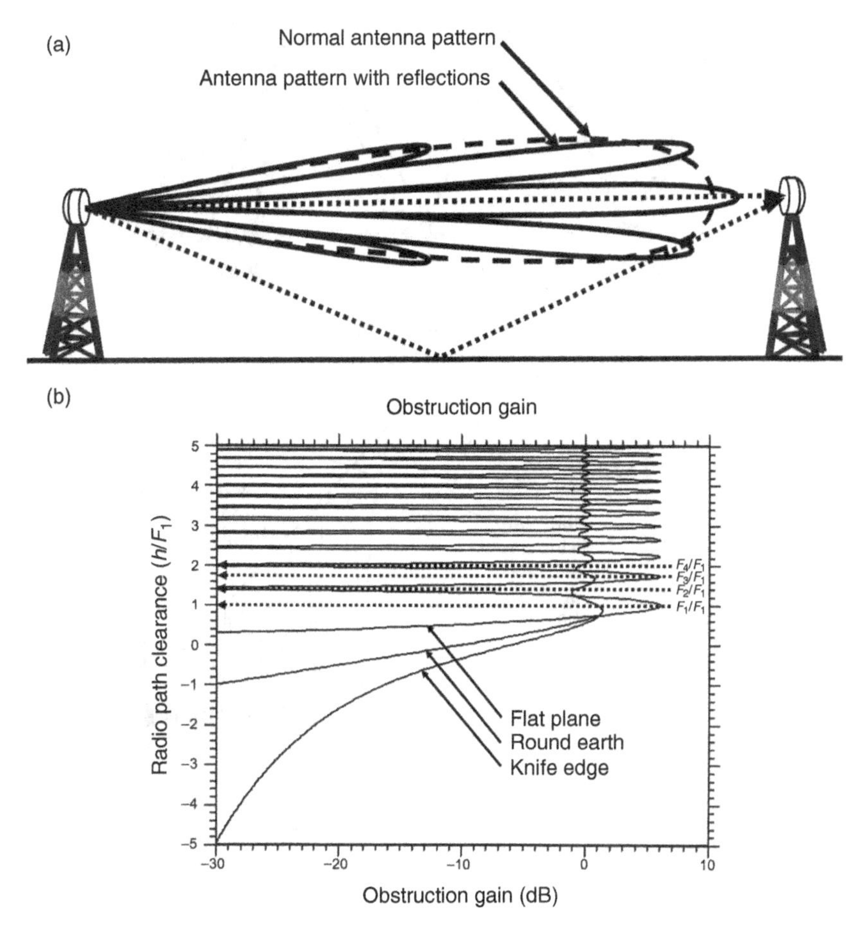

Figure 13.1 (a,b) Reflection lobes and obstruction gain.

Snell's law. As the roughness increases, the potential reflection point diffuses the incident signal and little directed energy is reflected. The potential reflection point is termed a *glistening area*. Under these conditions, path reflection symmetry is totally lost. All signals are diffused regardless of their direction of arrival. If the earth's surface is smooth, the reflected signal is coherent (specular). Scattered signal loss is negligible and the reflected signal power is only a function of divergence and reflection. The criterion for transition from rough to smooth reflection is not well defined. Various sources suggested coherent reflection occurs when surface height irregularities h are shorter than $\lambda/(8 \sin \phi)$ (Ishimaru, 1978; Long, 2001), or $\lambda/(16 \sin \phi)$ (Reed and Russell, 1953) where λ is the wavelength and ϕ is the grazing angle. The Rayleigh criterion, $\lambda/(8 \sin \phi)$, limits reflected path phase differences to no more than $\pi/2$. Landee et al. (1957) suggest that to maintain reflection magnitude within 1 dB of theoretical, parabolic and flat reflectors must maintain RMS surface error less than $\lambda/14$ and $\lambda/28$, respectively. As will be noted below, these commonly used criteria are misleading on microwave paths. Coherent reflection (at reduced levels) occurs for all degrees of terrain roughness (Fig. 13.3).

The energy reflected in the direction of a (coherent) specular reflected signal (a signal that obeys Snell's law of reflection) (Fig. 13.2) is determined by the reflection coefficient (ratio of coherent reflected power to incident power) for the surface. This coefficient is not a function of polarization.

Estimating scattered energy from rough surfaces is complicated (Bullington, 1954; Liao and Sarabandi, 2010; Norton, 1952; Rice, 1951). Accurate estimates of this scattered energy as a function of surface roughness are not known in general. However, limiting cases (Beckmann and Spizzichino, 1987; Boithias, 1987) for the reflection coefficient have been derived. If the surface is perfectly smooth, the

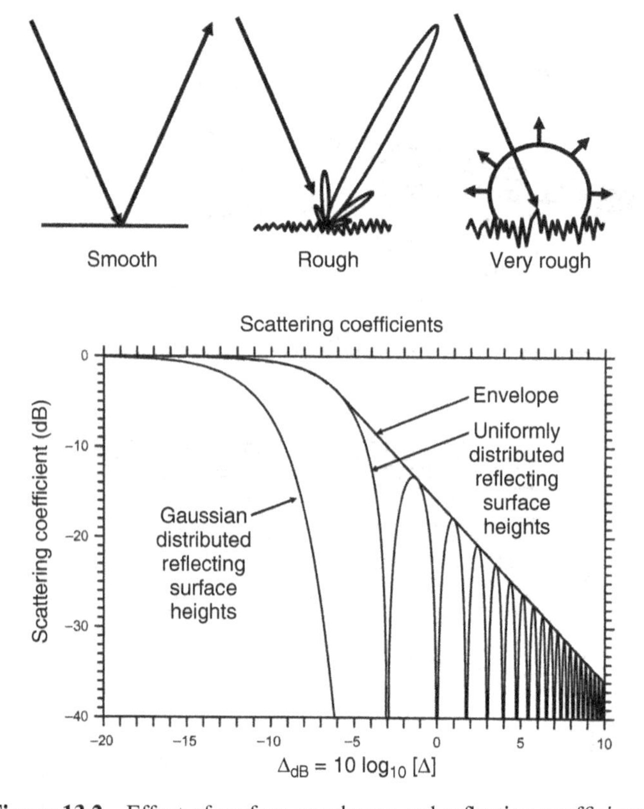

Figure 13.2 Effect of surface roughness and reflection coefficients.

reflection coefficient is 1 (0 dB). If the surface heights near the reflection can be characterized as a group of small areas with a Gaussian height distribution, the Gaussian model applies. If the surface heights near the reflection can be characterized as a group of small areas with a uniformly distributed heights, the uniform model applies.

$$\Delta = \Delta h \ \sin \ \frac{\phi}{\lambda} \cong 0.01745 \ \frac{\Delta h \ \phi\,(^\circ)}{\lambda} \quad \text{for small grazing angle } \phi$$

$$\text{since } \sin \ \phi \cong \ \phi(\text{rad}) \ \text{within } 10\% \text{ for } \phi \leq 0.785 \ (45^\circ) \ \text{ or } 1\% \text{ for } \phi \leq 0.262 \ (15^\circ) \quad (13.1)$$

Δ = effective surface height;
Δh = surface actual height;
ϕ = grazing angle;
λ = radio wave free space wavelength.

13.1.1 Gaussian Model

Beckmann and Spizzichino (1987) propose the following:

$$C_S \ (\text{dB}) = 10 \ \log \ \left(e^{-16\pi^2\Delta^2}\right) = -685.810 \ \Delta^2 \quad (13.2)$$

C_S (dB) = reflection (scattering) coefficient (dB);
Δh = standard deviation of the normal distribution of reflecting surface heights.

Boithias (1987) proposes the following:

$$C_S\,(\text{dB}) = 10\log\left[I_0\left(16\pi^2\Delta^2\right)e^{16\pi^2\Delta^2}\right]$$
$$\approx 5\log\left\{\frac{\left[1+\left(16\pi^2\Delta^2\right)\right]}{\left[1+2.35\left(16\pi^2\Delta^2\right)+2\pi\left(16\pi^2\Delta^2\right)^2\right]}\right\} \tag{13.3}$$

C_S (dB) = reflection coefficient (dB);
I_0 = modified Bessel function of the first kind of order zero.

13.1.2 Uniform Model

Beckmann and Spizzichino (1987) propose the following:

$$C_S\,(\text{dB}) = 10\log\frac{\left[\sin^2(2\pi\Delta)\right]}{(2\pi\Delta)^2} \tag{13.4}$$

C_S (dB) = reflection coefficient;
Δh = maximum difference of uniformly distributed reflecting surface heights.

$$\text{Reflection coefficient envelope (dB)} = 20\log\frac{[\sin(2\pi\Delta)]}{(2\pi\Delta)},\ \Delta_{\text{dB}} \le -5.8$$
$$\text{Reflection coefficient envelope (dB)} = -16.0 - 2\,\Delta_{\text{dB}},\ \Delta_{\text{dB}} > -5.8 \tag{13.5}$$
$$\Delta_{\text{dB}} = 10\log(\Delta)$$

Note that Δh is based on difference in heights and not on absolute height.

In the above sources, the reflection coefficient is called a *scattering coefficient.*

Bullington (1954) noted that variations from complete randomness tend to fill in nulls predicted by theoretical calculations. Rice (1951) observed that mildly rough surface reflections were Rayleigh distributed. Liao and Sarabandi (2010) observed that these approximations were derived for roughness on the order of the radio wavelengths or smaller. They also do not account for the shadowing that occurs for very small grazing angles.

13.2 SCATTERING FROM EARTH TERRAIN

For more general cases, surface roughness can be characterized by the Rayleigh roughness parameter.

$$P = 4\pi\left(\frac{\sigma}{\lambda}\right)\sin\theta \tag{13.6}$$

P = Rayleigh roughness parameter;
= effective terrain roughness;
σ = RMS surface height (measured from crest to trough);
λ = radio wave free space wavelength;
θ = grazing angle of incident signal relative to the mean surface plane.

By convention, if $P \le \pi/2$, the surface is deemed smooth. If $P \ge 2\pi$, the surface is considered rough. The relative reflected signal level is defined by the terrain reflection coefficient.

R = reflection coefficient
= (voltage) amplitude of the reflected signal relative to the amplitude of the incident signal
$20 \log_{10}(R)$ = reflected signal power/incident signal power (dB).

With the exception of large bodies of water, actual earth terrain seldom conforms to any regular shape. The reflection coefficient R is related to the Rayleigh roughness parameter but the relationship is difficult to define. This relationship is usually assumed to conform approximately to one of the following models:

Exponential

$$R = \frac{1}{\left[1 + \left(\frac{P^2}{2}\right)\right]} \tag{13.7}$$

Pseudoexponential

$$R = \frac{1}{\left[1 + \left(\frac{2P^2}{3}\right)\right]^{(3/4)}} \tag{13.8}$$

Normal

$$R = \exp\left(\frac{-P^2}{2}\right) \tag{13.9}$$

$$\exp(x) = e^x$$

Longley–Rice Empirical

$$R = \exp\left(\frac{-P}{2}\right) \tag{13.10}$$

Weiner (1982) summarized the results of various researchers who measured scattering from actual terrain, terrain models, and the ocean. Weiner's data is summarized in Figure 13.3.

The normal model is commonly used in theoretical research papers. The Longley–Rice empirical model is often used in practical, irregular terrain estimations. However, on the basis of the Weiner data, the pseudoexponential model appears to most closely match experimental results. Regardless of the model used, the difference between the model prediction and actual results can be several decibels.

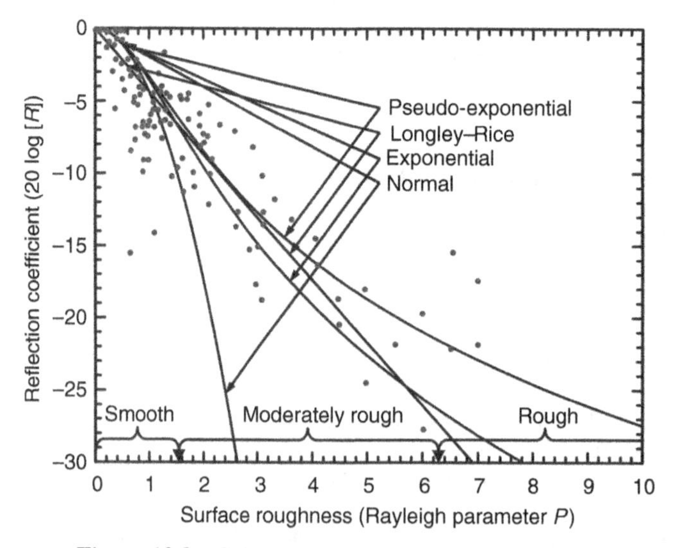

Figure 13.3 Coherent scattered signal power levels.

13.3 PRACTICAL EARTH REFLECTION COEFFICIENT

Flat terrain is highly reflective. As we shall see later in this chapter, all smooth terrain, whether dry dielectric or wet conductor, is reflective. Sometimes, potential reflections can be avoided by tilting one or both antennas (Hartman and Smith, 1977) (recognizing that this degrades CPD) or by placing antennas so that terrain, buildings, or trees block potential reflections. However, most practical situations do not permit a theoretical evaluation of the earth's reflection coefficient. Bullington (1954) observed "... that the antenna heights at which the maxima and minima occur [on a path in central Nebraska] varied with the time of day so no fixed combination could be optimum all of the time. Moreover, it was found that moving the tower only about 100 ft at one terminal reduced the reflection coefficient on this path from 0.72 (−3 dB) to about 0.55 (−5 dB), which indicates that the magnitude of the reflection coefficient cannot be predicted accurately from the gross features of the path profile."

Bullington (1954) continued, "A rigorous determination of the reflection from a rough surface is a difficult problem. In fact, it is virtually impossible except for certain idealized profiles. In general, it is not possible to represent an actual profile by a reasonable number of parameters. Considerable judgment is required in formulating the problem and the end result may be no more than a range of possible answers." This situation was addressed by Laine (1978). He related reflection coefficient to typical earth surface types. Table 14.1 is adapted from his Table 1.

In Table 14.1, the reflection coefficient (E_R/E_S) is the ratio of the reflected signal RMS voltage divided by the normal signal RMS voltage. The signal-to-reflection power is normal signal power (dBm) minus reflected signal power (dBm). The maximum upfade column shows the maximum increase in normal received signal level that can be caused by the level of reflection shown on that line. The maximum downfade column shows the minimum decrease in normal received signal level that can be caused by the level of reflection shown on that line. Note that if the reflected signal is at least 1 dB less than the desired signal, the worst-case downfade is −20 dB. If the reflected signal is 3 dB less than the desired signal, the worst-case downfade is less than −10 dB. If the reflected signal is −20 dB, the downfade will be an insignificant 1 dB.

The objective of the path designer is to recognize the cases where reflections are likely and attempt to reduce (using terrain or antenna discrimination) the reflection power to less than 20 dB of the desired signal at the receive antenna. (If the reflection is stronger, the anticipated free space loss will not be achieved.)

In addition to the surfaces listed in Table 13.1, there are invisible reflective surfaces that sometimes occur. Plants (such as wheat and pine trees) and agricultural area (such as rice paddies and irrigated fields) put out significant moisture that can form an invisible reflective surface when the air is quiet.

For mountaintop-to-mountaintop paths, invisible layers are almost always present over quiet valleys. These layers, if not anticipated, can cause significant reflective fading. As the layers are many Fresnel zones below the main signal path, they are difficult to combat with space diversity. An effective method of reducing their effect is to move the mountaintop antennas far back from the edge of the site, so potential illumination of the layers is blocked.

13.4 REFLECTION LOCATION

If the terrain is relatively smooth, potential reflections must be considered. The location of a reflection on a smooth earth was derived by Boithias (1987) (Fig. 13.4).

$$h_1 \geq h_2$$

h_1 = physical height of the antenna at one end of the path *above the reflection point physical height*;
h_2 = physical height of the antenna at the other end of the path *above the reflection point physical height*.

$$D = d_1 + d_2$$

TABLE 13.1 Lane's Practical Earth Reflection Coefficients

Type of Terrain	Reflection Coefficient (E_R/E_S)	Signal-to-Reflection Power, dB	Maximum Upfade, dB	Maximum Downfade, dB
Smooth Water or	−1.00	0.0	6.0	−∞
Salt Flat or	−0.99	0.1	6.0	−40
Desert	−0.98	0.2	5.9	−34
	−0.97	0.3	5.9	−30.5
	−0.95	0.4	5.8	−26.0
	−0.93	0.6	5.7	−23.1
	−0.91	0.8	5.6	−20.9
Rough Water or	−0.89	1.0	5.5	−19.2
Cotton or	−0.84	1.5	5.3	−15.9
Low Grass	−0.79	2.0	5.1	−13.6
	−0.75	2.5	4.9	−12.0
	−0.71	3.0	4.7	−10.8
Sagebrush or	−0.63	4.0	4.2	−8.6
High Grass	−0.50	6.0	3.5	−6.0
Partially Wooded	−0.40	8.0	2.9	−4.4
	−0.32	10.0	2.4	−3.3
	−0.24	12.0	1.9	−2.4
	−0.20	14.0	1.6	−1.9
	−0.16	16.0	1.3	−1.5
	−0.13	18.0	1.1	−1.2
Heavily Wooded	−0.10	20.0	0.8	−0.9
Forest	−0.05	26.0	0.4	−0.4
	−0.01	40.0	0.1	−0.1

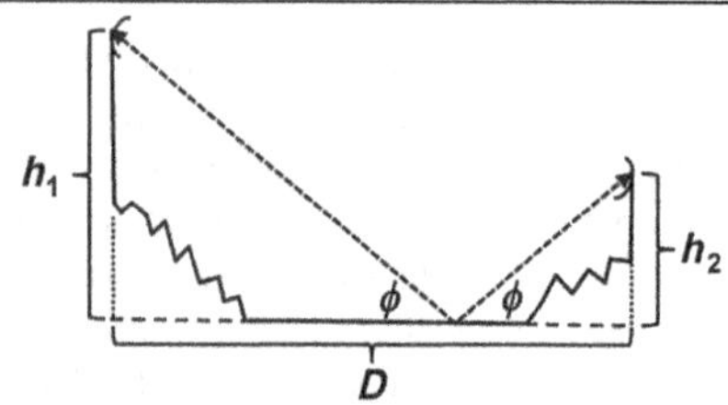

Figure 13.4 Flat earth path profile reflection geometry (flat earth, $K = \infty$).

D = distance from one end of the path to the other end;
= total path distance;
d_1 = distance from one end of the path to the reflection point;
d_2 = distance from the other end of the path to the reflection point;
B = normalized distance from the center of the path to the reflection point (−1 to 1).

$$C = \frac{(h_1 - h_2)}{(h_1 + h_2)} \geq 0$$

$$M = \frac{D^2}{[4Ka(h_1 + h_2)]}$$

$$d_1 = \left(\frac{D}{2}\right)(1 + B)$$

$$d_2 = D - d_1$$

K = equivalent earth radius factor;
a = physical earth radius $\cong$ 6367 km $\cong$ 3957 (statute) miles.

All the above equations are in the same units of distance. B (reflection location factor) is the solution to the following equation:

$$BM^3 - BM - B + C = 0 \tag{13.11}$$

For ducting conditions ($K \leq 0$), three solutions exist (two near the ends of the path and one near the center). For normal conditions ($K > 0$), only the following solution exists:

$$B = 2\sqrt{\frac{M+1}{3M}}\cos\left\{\left(\frac{\pi}{3}\right) + \left[\frac{1}{3}\text{Arc}\cos\left(\frac{3C}{2}\sqrt{\frac{3M}{(M+1)^3}}\right)\right]\right\} \tag{13.12}$$

$$\phi\,(\text{rad}) = \left[\frac{(h_1 + h_2)}{D}\right]\left[1 - M\left(1 + B^2\right)\right]$$

Boithias's flat earth results apply to overwater paths and paths in dessert terrain where the earth is flat. However, in most other cases, the terrain near the reflection point is inclined relative to the surface of a smooth earth. Boithias's flat earth results can be extended to this case by conceptually rotating the terrain, so the reflecting plane becomes parallel to the Earth. This allows B and ϕ to be calculated using Boithias's formulas. However, for the geometry to be correct, the antenna heights must be adjusted so that they are measured relative to the extension of the flat surface that has been conceptually rotated to be parallel to the earth's surface (Fig. 13.5).

For small angles and short distances (relative to the earth's radius), the flat, inclined terrain case could be rotated (in the vertical plane in line with the path) so as to be congruent to the flat earth case if h_1 and h_2 are redefined as follows:

h_1 = physical height of the antenna at one end of the path *above the extended flat terrain physical height*;
h_2 = physical height of the antenna at the other end of the path *above the extended flat terrain physical height.*

With these h_1 and h_2 redefinitions, the flat terrain reflection results apply to the Boithias flat, inclined terrain case.

As the path profile will have flat terrains of varying sizes, the obvious question is, "What is the minimum size of flat terrain that will produce a significant reflection?" Assume a reflected radio path where we are only concerned with the energy contained within a perpendicular radius (h/F_1) of the main reflected path. F_1 is the first Fresnel zone radius (See the end of this chapter for an explanation of Fresnel zones.) and h is a number yet to be determined. Norton (1952) observed that "... only the curvature of the reflecting surface in the plane of propagation will appreciably influence the intensity of the ground-reflected wave, and we need not even investigate the surface curvature normal to the propagation path."

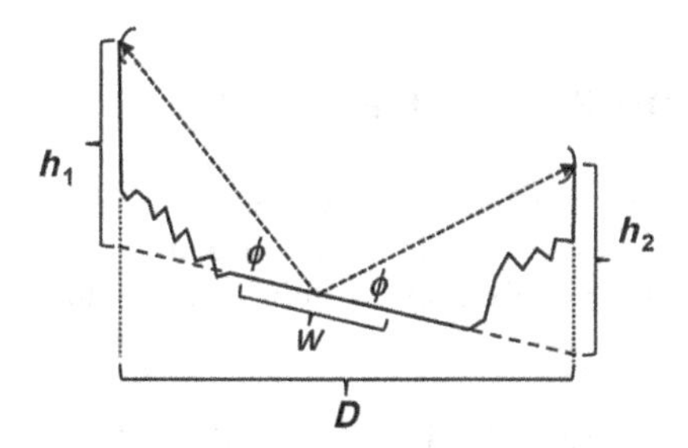

Figure 13.5 Smooth inclined earth path profile reflection geometry (flat earth, $K = \infty$).

In other words, we are only concerned with the energy that would be reflected by the terrain within an ellipse with relatively small off-axis (normal to direction of propagation) total width of 2 (h/F_1) and potentially large on-path (in line with the direction of propagation) total width of $W = 2\left[\left(h/F_1\right)/\sin\phi\right]$, where ϕ is the radio path (launch) angle relative to the flat earth. The reflected path energy contained within the (h/F_1) radius was calculated by Hristov (2000). He showed that if $h \approx 1/2$ the reflected energy is the same as for h infinitely large (full earth reflecting surface). For $h > 1/2$, the power varies up and down but eventually converges to the $h \approx 1/2$ value. Like Norton, Hristov also observed that for an inclined surface, the Fresnel zone radius is stretched along the radio path. From this, we may infer that the minimum flat terrain on path width that has potential for significant effect would be on the order of at least $W = 2[(1/2)(F_1 \sin\phi)] = F_1/\sin\phi$. This essentially indicates that the significant energy in the radio wave is contained within one Fresnel zone of the ray between the transmitter and the receiver. (This region is often called the *Fresnel volume* in other technologies such as seismic imaging and sonar detection.)

Hristov's calculations showed that energy from the area greater than approximately the first Fresnel zone's projection onto the Earth is not significant for uniform terrain. He derived the projected dimensions of the Fresnel zone onto a flat earth ((Hristov, 2000), Eqs. 3.13 and 3.14). They simplify to the following:

$$L_n = \frac{d\sqrt{1 + \frac{(4h_1 h_2)}{(n\lambda d)}}}{1 + \frac{(h_1 + h_2)^2}{(n\lambda d)}} \tag{13.13}$$

L_n = projected length of Fresnel zone projected onto the smooth earth in the direction of the radio path = 2 (semimajor axis) = $2a_n$ as per the book by Hristov (2000).

$$W_n = \frac{\sqrt{1 + \frac{(4h_1 h_2)}{(n\lambda d)}}}{\sqrt{1 + \frac{(h_1 + h_2)^2}{(n\lambda d)}}} \tag{13.14}$$

W_n = projected width of Fresnel zone projected onto the smooth earth in the direction of the radio path = 2(semiminor axis) = $2b_n$ as per the book by Hristov (2000);
n = Fresnel zone integer designation (1 for first Fresnel zone);
d = path length between transmitter and receiver;
$\approx d_o + (n\lambda/2) \approx d_o + (1/4)$ as per the book by Hristov (2000);
h_1 = transmit antenna height above reflection point height;
h_2 = receive antenna height above reflection point height;
λ = radio wavelength;
λ(ft) = 0.98357/F (GHz);
λ(m) = 0.29980/F (GHz);
F = radio operating frequency.

D, h_1, h_2, and λ are all in the same distance units.

13.5 SMOOTH EARTH DIVERGENCE FACTOR

The divergence factor accounts for the power divergence of the signal reflected by a round surface (assuming perfect reflection). This coefficient is not a function of polarization and is only significant for small grazing angles. The factor is given by the following formulas (Hall, 1979; Kizer, 1990) for reflection from a smooth earth:

$$C_D = \text{sqrt}\left\{1 + \frac{(2d_1 d_2)}{(K\,r\,d\,\sin(\phi))}\right\} \tag{13.15}$$

C_D = reflection coefficient (voltage ratio);
= Abs (reflected signal magnitude/incident signal magnitude);
ϕ = grazing angle at reflection = angle of incidence = angle of reflection;
d_1 = distance from one end of the path to the location of interest;
d_2 = distance from the other end of the path to the location of interest;
$d = d_1 + d_2$;
R_e = earth's equatorial radius = 6378 km = 3963 miles;
R_p = earth's polar radius = 6357 km = 3950 miles;
E^2 = Eccentricity$^2 = (R_e{}^2 - R_p{}^2)/R_e{}^2$;
ψ = latitude at reflection point;
r = Earth's radius at reflection point = $R_p/\text{sqrt}[1 - (E^2 \cos \psi)]$;
K = equivalent earth radius factor (typically 4/3);
$C_D(\text{dB}) = 20 \log [C_D \text{ (voltage ratio)}]$;
$\phi(°) = (\frac{180}{\pi})\phi(\text{rad})$.

The above formula, based on optical analogy, was derived by Kerr (1951). For small ϕ, the above divergence factor formula will significantly overestimate reflection loss. Long (2001) suggests that the formula is accurate as long as the difference between the main and reflected signal exceeds $\lambda/4$. This is equivalent to requiring the path to have a minimum first Fresnel zone clearance (h/F_1) of [1/sqrt(2)] where h is the path height above the reflection point, F_1 is the first Fresnel zone height (for the definition of F_1, see below), and sqrt is square root]. Conventional fixed microwave line-of-sight radio path design limits the minimum path clearance (above the Earth) to approximately this value.

Although this formula is commonly available in path calculation programs, in many practical cases, it overestimates reflection path loss (primarily because of path blockage and roughness). Actual microwave paths are rarely smooth reflective surfaces. The use of this formula is only recommended for line-of-sight microwave paths over water or flat terrain.

13.6 REFLECTIONS FROM OBJECTS NEAR A PATH

One of the tasks of the path designer is to determine potential reflections or obstructions along the path between the transmit and receive antennas. Potentially reflecting surfaces on buildings and water towers are fairly common. As buildings are usually composed of flat surfaces, treating them as passive reflectors is relatively easy. However, structures with more complex shapes are much more difficult to analyze.

In theory, if the structure is in the far field of both antennas and is smaller than the first Fresnel zone, it will be uniformly illuminated (in amplitude and phase) by the transmit antenna (If the structure is larger across than the first Fresnel zone, the illumination will have a significant phase change across the surface of the structure and the reflection loss will be greater than calculated by the following method.). "All" that needs to be done is to integrate the E field across the structure accounting for the reflected ray phase change at each point on the surface of the structure, divide the result by the structure area, and take 20 times the common logarithm of the result. This is generally easier said than done.

However, for simple structures this is practical. If a surface is flat, the two-way (reception and retransmission) gain is the following:

$$\text{Two-way flat reflector gain (dB)} = 22.2 + 40 \log F\,(\text{GHz}) + 20 \log A(\text{ft}^2) + 20 \log \cos\left(\frac{C}{2}\right) \quad (13.16)$$

$$42.9 + 40 \log F\,(\text{GHz}) + 20 \log A(\text{m}^2) + 20 \log \cos\left(\frac{C}{2}\right)$$

F = frequency of radio wave;
A = reflector area;

C = angle between incoming and outgoing radio signal paths.

If the surface is a simple cylinder or sphere, we can, in principle, calculate the divergence factor to account for the curvature of the object's surface. The object's reflective gain would be the following:

$$\text{Two-way flat reflector gain (dB)} = \text{Two-way flat reflector gain (dB)} + \text{divergence factor (dB)} \qquad (13.17)$$

Assume a circular cylinder with normalized radius of R_n = cylinder radius/λ (wavelength). The height of the cylinder is unimportant from a divergence perspective; only its radius of curvature is significant. The divergence factor can be simplified to the following integral:

$$\text{Cylinder divergence factor} = 20\log\left\{\int_0^{+1} \cos\left[2\pi\left(2R_n\sqrt{(1-x^2)}\right)\mathrm{d}x\right]\right\} \qquad (13.18)$$

Assume a circular sphere with normalized radius of R_n = sphere's radius/λ (wavelength). The divergence factor can be simplified to the following integral:

$$\text{Sphere divergence factor} = 20\log\left\{\int_0^{+1} r\cos\left[2\pi\left(2R_n\sqrt{(1-r^2)}\right)\mathrm{d}r\right]\right\} \qquad (13.19)$$

Both of these integrals are difficult to integrate formally. Likewise, numerical integration is challenging because of the rapid oscillation of the cosine term for large values of R_n. The cylinder divergence factor integral was numerically integrated. The results were a smooth curve that, when curve fitted, resulted in the following equation:

$$\text{Cylinder divergence factor (dB)} = -12 - 10\ \log\ (R_n) \qquad (13.20)$$

$$R_n = \frac{\text{cylinder's radius}}{\lambda}\ \text{(wavelength)}$$

$$= 1.017\ \text{cylinder's radius (ft)}\ f\ \text{(GHz)}$$

$$= 3.336\ \text{cylinder's radius (m)}\ f\ \text{(GHz)}$$

$$f = \text{operating frequency}$$

The spherical divergence factor integral was numerically integrated. The results varied widely depending on R_n. If R_n was an integer or an integer plus 1/2, the divergence factor was $-\infty$. If R_n was an integer plus 1/4, a peak value was calculated. If R_n was an integer plus 3/4, a near-peak value was calculated. If we limit the results to just the peak values when R_n was an integer plus 1/4, a smooth curve was generated. Curve-fitting this curve resulted in the following equation:

$$\text{Sphere divergence factor (dB)} = -32 - 40\ \log\ (R_n) \qquad (13.21)$$

$$R_n = \frac{\text{sphere's radius}}{\lambda}\ \text{(wavelength)}$$

$$= 1.017\ \text{cylinder's radius (ft)}\ f\ \text{(GHz)}$$

$$= 3.336\ \text{cylinder's radius (m)}\ f\ \text{(GHz)}$$

$$f = \text{operating fequency}$$

These results are graphed in Figure 13.6.

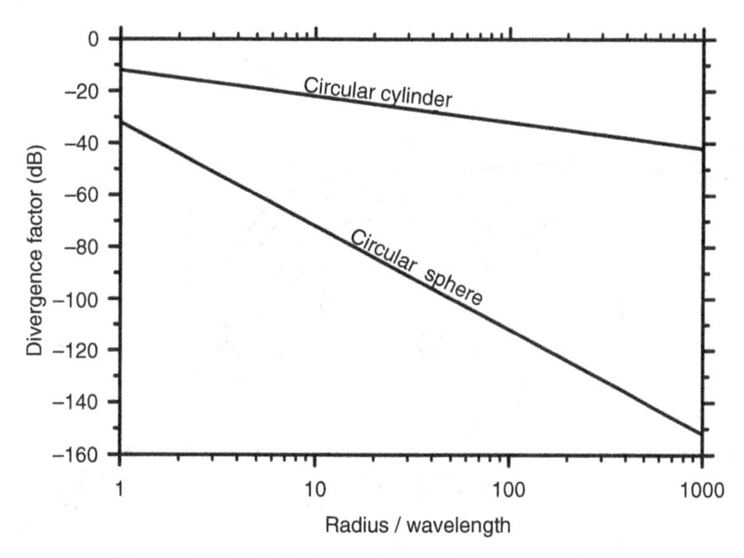

Figure 13.6 Cylinder and sphere divergence factors.

13.7 FRESNEL ZONES

Path clearance is often described in terms of Fresnel zones, a concept introduced by Augustin Fresnel (1788–1827) in his study of optics. Fresnel zones are families of ellipsoidal boundaries described by points at which a radio wave, if reflected at that point toward the receiver, would travel n times half a wavelength further than by a direct route between the transmitter and the receiver (extra radio path length is $n(\lambda/2)$ where the factor n must be a positive integer). Fresnel zone values may be calculated from the following formulas (Fig. 13.7, Fig. 13.8, and Fig. 13.9):

$$F_n \text{ (ft)} = 72.1 \text{ sqrt} \left\{ \frac{[n \times d_1 \text{ (miles)} \times d_2 \text{ (miles)}]}{[F \text{ (GHz)} \times D \text{ (miles)}]} \right\}$$

$$\mathrm{F}_n \text{ (m)} = 17.3 \text{ sqrt} \left\{ \frac{[n \times d_1 \text{ (km)} \times d_2 \text{ (km)}]}{[F \text{ (GHz)} \times D \text{ (km)}]} \right\} \quad (13.22)$$

$$F_n = F_1 \text{ sqrt } (n)$$

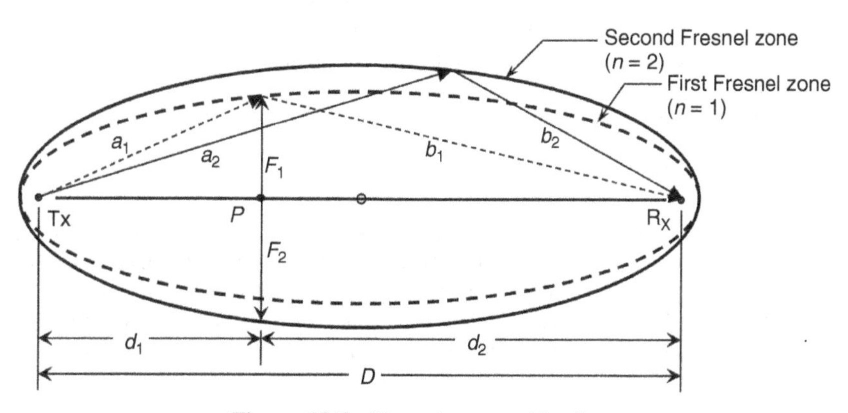

Figure 13.7 Fresnel zones, side view.

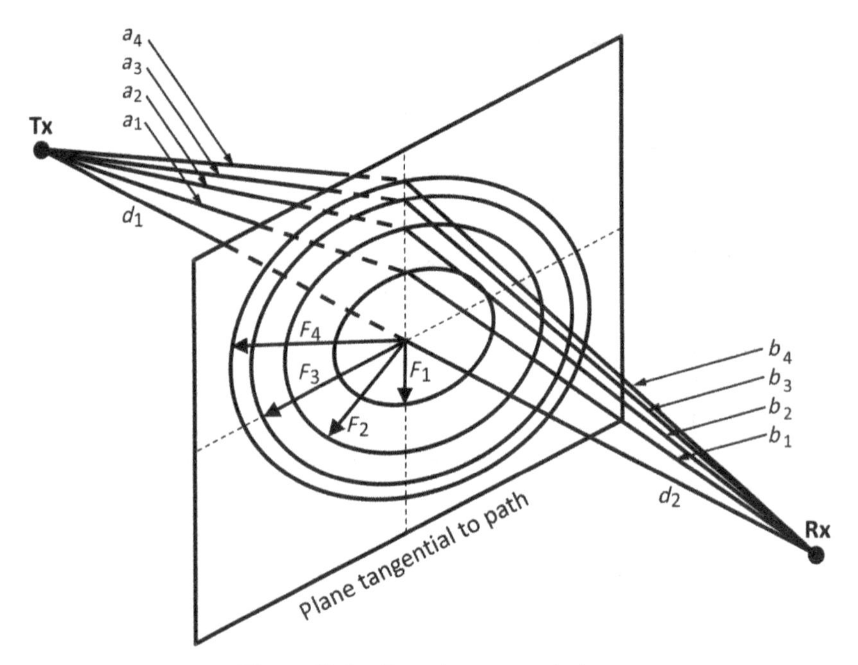

Figure 13.8 Fresnel zones, end view.

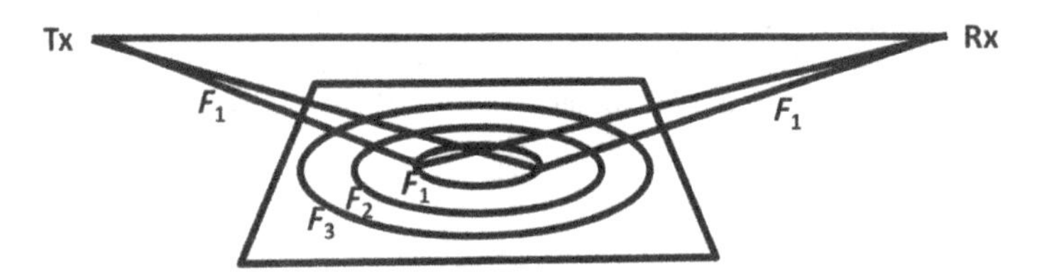

Figure 13.9 Fresnel zones, flat surface reflections.

n = Fresnel zone number (an integer);
d_1 = distance from one end of the path to the reflection;
d_2 = distance from the other end of the path to the reflection;
a_n = distance from one end of the path to the reflection;
b_n = distance from the other end of the path to the reflection;
$a_n + b_n = D + \left[n\left(\frac{\lambda}{2}\right)\right]$;
λ = radio wavelength;
D = total path distance = $d_1 + d_2$;
F = frequency of radio wave.

In fixed microwave radio path design, path clearance h is usually described in the normalized parameter h/F_1. The radio path clearance h is measured on a line perpendicular to the main radio signal path ("beam"). For typical, nearly horizontal radio paths, the convention is to measure the path clearance perpendicular to the Earth. In optics, Fresnel zone clearance is defined as $(h/F_1)^2$. Fixed microwave radio engineering follows the convention of Bullington (1957) and defines Fresnel zone clearance as h/F_1.

13.8 ANTENNA LAUNCH ANGLE (TRANSMIT OR RECEIVE ANTENNA TAKEOFF ANGLE)

$$\theta = \arctan \phi_1 - \arcsin \phi_2$$
$$\phi_1 = \frac{d}{D}$$
$$= \frac{1.894 \times 10^{-4}\, d\ (\mathrm{ft})}{D\ (\mathrm{miles})}$$
$$= \frac{1.000 \times 10^{-3}\, d\ (\mathrm{m})}{D\ (\mathrm{km})}$$
$$\phi_2 = \frac{\left[\mathrm{sqrt}\ (d^2 + D^2)\right]}{(2\ Ka)}$$
$$= \frac{\left[\mathrm{sqrt}\ \left(D^2\ (\mathrm{miles}) + 3.587 \times 10^{-8}\, d^2\ (\mathrm{ft})\right)\right]}{7913\ K}$$
$$= \frac{\left[\mathrm{sqrt}\ \left(D^2\ (\mathrm{km}) + 1.000 \times 10^{-6}\, d^2\ (\mathrm{m})\right)\right]}{12{,}735\ K}$$
$$d = h_F - h_N \tag{13.23}$$

θ = antenna launch angle;
h_F = height of the far-end antenna above mean sea level;
h_N = height of the near end antenna above mean sea level;
D = distance between near and far-end antennas;
K = equivalent earth radius factor;
a = earth radius $\cong$ 3957 (statute) miles $\cong$ 6367 km.

For d much smaller than D {nearly horizontal paths such that [d (ft)/D (miles) < 900] or [d (m)/D (km) < 170]} the following has less than 1% error:

$$\theta\ (\mathrm{rad}) = \frac{\left[d - \left(\frac{D^2}{(2\ Ka)}\right)\right]}{D}$$
$$\theta\ (^\circ) = \frac{\left(\frac{d(\mathrm{ft})}{92.15}\right) - \left[\frac{D^2\ (\mathrm{miles})}{(138.1\ K)}\right]}{D\ (\mathrm{miles})} \tag{13.24}$$
$$\theta\ (^\circ) = \frac{\left(\frac{d(\mathrm{m})}{17.45}\right) - \left[\frac{D^2\ (\mathrm{km})}{(222.3\ K)}\right]}{D(\mathrm{km})}$$

Angles are positive if above the horizon and negative if below.

Note that this formula can be used to determine the angle of arrival of a reflected signal after the reflection point has been determined. It can also be used to estimate beam angle movement (to estimate power fading due to antenna pattern) over an expected K-factor range.

13.9 GRAZING ANGLE

At the reflection point, Snell's Law dictates that the angle of incidence is equal to the angle of reflection. The reflection point grazing angle is given by the following formulas (Bell Laboratories Staff, 1964):

$$\phi(^\circ) = \frac{\left[h_F\,(\text{ft}) - \left(\frac{0.6667\, D_F^2\,(\text{miles})}{K}\right)\right]}{[92.15\, D_F\,(\text{miles})]}$$

$$\phi(^\circ) = \frac{\left[h_F\,(\text{m}) - \left(\frac{0.07843\, D_F^2\,(\text{km})}{K}\right)\right]}{[17.45\, D_F\,(\text{km})]} \tag{13.25}$$

Angles are positive if above the horizon and negative if below.

h_F = height of the antenna at one end of the path *above the reflection point height*;
D_F = distance between the reflection point and far-end antenna;
K = equivalent earth radius factor (nominally 4/3 for normal atmospheric conditions).

13.10 ADDITIONAL PATH DISTANCE

On the basis of geometry of a nearly horizontal radio path, the additional path distance of a reflected radio wave is given by the following equations:

$$d = \left(\frac{\lambda}{2}\right)\left(\frac{h}{F_1}\right)^2 \tag{13.26}$$

d = additional reflected path distance;
h = perpendicular distance from main radio signal path ("beam") to obstruction reflection point;
F_1 = first Fresnel zone distance;
λ = free space radio wavelength.

13.11 ESTIMATING THE EFFECT OF A SIGNAL REFLECTED FROM THE EARTH

For a signal reflected from the Earth, estimating the reflected signal relative power and potential impact can be done in three steps:

1. *Determine Loss at Transmit Antenna.* Determine the takeoff angle of the main signal and the path toward the reflection point. The reflection point can be calculated using the formulas in Section 13.4 above. The takeoff angle can be calculated using the formula above. The transmit antenna loss L_t (dB) < 0 is determined using the antenna pattern and determining the pattern loss at the angle equal to the difference between the direct and reflected path takeoff angles (as the antenna pattern is the same in the vertical plane as in the horizontal plane).
2. *Determine Reflection Loss.* The loss at the reflection point L_{rp} (dB) < 0 is estimated using the above procedures.
3. *Determine Loss at Receive Antenna.* Determine the receive antenna loss L_r (dB) < 0 in the same way as at the transmit end.

As the main and reflected paths are nearly the same length and the reflection point is larger than a Fresnel zone, accounting for free space loss and reflection gain is unnecessary.

$$P_{rel}\,(\text{dB}) = L_t + L_{rp} + L_r < 0 \tag{13.27}$$

P_{rel} (dB) = power of received reflected signal relative to power of main received signal.

$$V_{rel}\ (\text{dB}) = 10^{\frac{P_{rel}}{20}} < 1 \tag{13.28}$$

V_{rel} (dB) = voltage of received reflected signal relative to voltage of main received signal.

$$P_{max}\ (\text{dB}) = 20\ \log\ (1\ +\ V_{rel}) \tag{13.29}$$

P_{max} (dB) = maximum received signal power level relative to direct received signal power.

$$P_{min}\ (\text{dB}) = 20\ \log\ (1\ -\ V_{rel}) \tag{13.30}$$

P_{min} (dB) = minimum received signal power level relative to direct received signal power.

13.12 FLAT EARTH OBSTRUCTION PATH LOSS

The composite power of the combined direct and reflected radio wave reflected by the (nearly) flat earth (reflection) obstruction (ignoring antenna discrimination) for a nearly horizontal path (ignoring reflection phase shift other than π radians) may be estimated by the following formulas:

$$X = \frac{h}{F_1} \geq 0$$

h = perpendicular distance from main radio signal path ("beam") to the obstruction reflection point;
F_1 = first Fresnel zone distance.

$$P\ (\text{dB}) = 20\ \log\left\{\left|1 + C_{comp}\ \cos\left[\pi\left(1 + X^2\right)\right]\right|\right\} \tag{13.31}$$

P = (received signal power with obstruction/received signal power without obstruction);
C_{comp} = composite reflected signal voltage.

13.13 SMOOTH EARTH OBSTRUCTION LOSS

If the radio wave is relatively close to a smooth earth (a very large sphere) obstruction, the results of Vigants (1981) augmented by those of Bullington (1957) and Rice (1954), to allow obstruction loss to be estimated by the following formulas:

$$X = \frac{h}{F_1} \leq 0.75$$

h = perpendicular distance from main beam to obstruction;
F_1 = first Fresnel zone distance;
P = (received signal power with obstruction/received signal power without obstruction).

$$P(\text{dB}) = -10 + 20X, \qquad X \leq 0 \tag{13.32}$$

$$P(\text{dB}) = -10 + 20X - 6.665X^2, \qquad 0 \leq X \leq 0.75 \tag{13.33}$$

Smooth earth obstruction is not a function of polarization or earth conductivity. At present, the International Telecommunication Union—Radiocommunication Sector (ITU-R) (2009a) suggests that if obstruction irregularities do not exceed the following value for Δh, the obstruction surface can be considered smooth:

$$\Delta h = 0.04\left[R\lambda^2\right]^{1/3} \text{ (m)} \tag{13.34}$$

R = obstacle curvature radius (m);
λ = radio signal wavelength (m).

Figure 13.1b graphs the obstruction gain for flat earth, knife edge (see below) and rounded earth. Note that free space propagation loss is achieved for h/F_1 values of 0.550, 0.634, and 0.707 for flat earth, knife edge, and round earth obstruction, respectively. This is the basis of 0.6 first Fresnel zone clearance criterion for normal propagation conditions ($K = 4/3$).

13.14 KNIFE-EDGE OBSTRUCTION PATH GAIN

If the radio wave faces a knife-edge obstruction, the standard Fresnel integral characterization (Bacon, 2003) coupled with standard Fresnel integral approximations (Abramowitz and Stegun, 1968) allows knife-edge obstruction loss to be estimated by the following formulas (Fig. 13.10).

Before using the following formulas, be sure to adjust vertical elevations for radio path refractivity (add $[d_1(\text{miles})\, d_2(\text{miles})]/[1.500\, K]$ or $[d_1(\text{km})\, d_2(\text{km})]/[12.74\, K]$ to elevations as described in Eqn. 5.36).

$$X = \frac{h}{F_1}$$

$$G_{KE} < 0 \text{ for } X < 0.55$$

$$A = 2^{\frac{1}{2}}\,|X|$$

$$F = \frac{(1 + 0.9260\, A)}{(2 + 1.792\, A + 3.104\, A^2)}$$

$$G = \frac{1}{(2 + 4.142\, A + 3.492\, A^2 + 6.670\, A^3)}$$

$$S = \sin\left(\frac{\pi\, A^2}{2}\right)$$

$$C = \cos\left(\frac{\pi\, A^2}{2}\right)$$

$$\text{CFI} = 0.5 + (F\, S) - (G\, C)$$

$$\text{SFI} = 0.5 - (F\, C) - (G\, S)$$

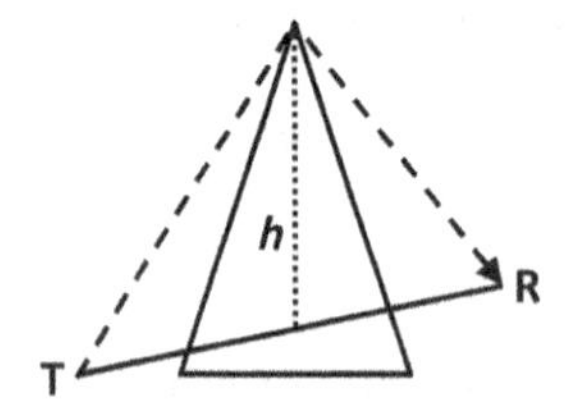

Figure 13.10 Knife-edge diffraction.

$$G_{KE}\ (\mathrm{dB}) = 10\ \mathrm{Log}_{10}\ [0.25\ +\ 0.5(\mathrm{SFI}+\mathrm{CFI}) + 0.5\ (\mathrm{SFI}^2 +\ \mathrm{CFI}^2)],\quad X > 0$$

$$G_{KE}\ (\mathrm{dB}) = 10\ \mathrm{Log}_{10}\ [0.25\ -0.5(\mathrm{SFI}+\mathrm{CFI}) + 0.5\ (\mathrm{SFI}^2 +\ \mathrm{CFI}^2)],\quad X \leq 0 \qquad (13.35)$$

h = perpendicular distance from the main beam to the obstruction;
> 0 if the path is below the knife edge (path is obstructed);
< 0 if the path is above the knife edge (path clears the obstruction);
F_1 = first Fresnel zone distance calculated at the obstruction;
G_{KE} = knife-edge diffraction gain (dB) = -Knife-edge diffraction loss (dB);
= received signal power with knife-edge obstruction (dBm) - received signal power without knife-edge obstruction (dBm);
= 10 $\log_{10}$ (received signal power with knife-edge obstruction/received signal power without knife-edge obstruction).

A curved edge can be considered a knife edge if the following condition applies:

$$|\phi| \leq \frac{\lambda}{4r}$$

ϕ = angle (radians) formed by a horizontal plane passing through the obstruction's edge and the ray that hits the obstruction's edge;
λ = wavelength of the radio wave;
r = radius of curvature of the obstruction's edge.

RF screens are a typical example of a knife-edge obstruction. Knife-edge obstruction is not a function of polarization. As Preikschat (1964) noted, knife-edge obstruction loss is not a function of the knife-edge conductivity. As long as the RF signal is blocked, the obstruction can be conductive or an RF absorber. Often, several consecutive knife edges are needed to block secondary reflections to active theoretical obstruction loss. Adding strips and openings to the top of the knife edge can enhance blockage loss and moderate loss variation.

13.15 ROUNDED-EDGE OBSTRUCTION PATH GAIN

If the radio wave faces a rounded-edge obstruction, the obstruction attenuation is greater than that of a knife edge (Fig. 13.11). As with knife-edge diffraction, be sure to add atmospheric refractivity correction (Eqn. 5.36) to the profile heights before beginning this analysis. This approach is appropriate for hills and ridges but is not accurate for large bulges on the order of the Earth itself.

$$\begin{aligned} G_{RE} &= G_{KE} + G_{RO} \\ &\leq 0 \\ &= +\ 6 +\ 20\ \log_{10}(mn) - 7.2\ m^{1/2} +\ (2 - 17\ n)\ m - 3.6\ m^{3/2} +\ 0.8\ m^2 \text{ if } mn > 4 \\ &= -\ 7.2\ m^{1/2} +\ (2 - 12.5\ n)m - 3.6\ m^{3/2} +\ 0.8\ m^2 \text{ if } mn\ \leq\ 4 \end{aligned}$$

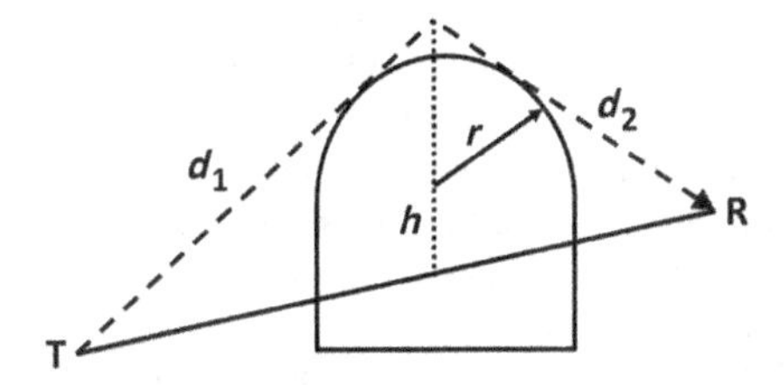

Figure 13.11 Rounded-edge diffraction.

$$m = r\frac{\left[\frac{(d_1+d_2)}{(d_1 d_2)}\right]}{\left(\frac{\pi r}{\lambda}\right)^{1/3}} \tag{13.36}$$

$$n = h\frac{\left(\frac{\pi r}{\lambda}\right)^{2/3}}{r}$$

G_{RE} = 10 $\log_{10}$ (received signal power with rounded edge obstruction/received signal power without rounded edge obstruction);
G_{KE} = knife-edge gain (dB) as calculated above;
G_{RO} = additional rounded obstruction gain (dB)—additional rounded obstruction loss (dB) per the work by International Telecommunication Union—Radiocommunication Sector (ITU-R) (2009a);
h = perpendicular distance from the main beam to the obstruction;
$h \geq 0$ (path is obstructed or grazing);
d_1 = distance from the transmit antenna to path ray intersection above obstruction;
d_2 = distance from the receive antenna to path ray intersection above obstruction;
r = obstruction radius of curvature;
λ = radio free space wavelength;
= 0.98357/F (GHz) (ft);
= 0.29980/F (GHz) (m);
F = radio operating frequency (GHz).

The primary challenge in using this method is to estimate the radius of curvature r. Rice et al. (1965) recommend a smooth crest radius between the points (horizons) determined by the rays from the path terminals. Assis (1971), Chrysdale (1958), and International Telecommunication Union—Radiocommunication Sector (ITU-R) (2009a) prefer using the radius of curvature to a parabola fitted to the obstacle profile in the vicinity of the top. The ratio ($\pi\ r\ /\ \lambda$) is critical to the obstruction gain (loss) value.

13.16 COMPLEX TERRAIN OBSTRUCTION LOSSES

Generally, fixed point to point microwave paths are designed so they are unobstructed for most propagation conditions (International Telecommunication Union—Radiocommunication Sector (ITU-R), 2009b). Occasionally, a short obstructed path may be designed if the obstruction loss is tolerable. Short obstructed paths do have the advantage of being more stable than a similar length unobstructed path (Carlson and Waterman, 1966, 1967; Dickson et al., 1953). Long-distance obstructed paths are of considerable interest to those performing frequency interference studies. In North American fixed point to point networks, these studies typically utilize NSMA procedures (Working Group 2, 2000) or the ITS Longley–Rice model (Hufford et al., 1982; Longley, 1978; Longley and Rice, 1968; Rice et al., 1965) (Hufford, G. A., "Memorandum to Users of the ITS Irregular Terrain Model," Unpublished (Hufford_1985_Memo.pdf at http://flattop.its.bldrdoc.gov/itm.html), January 30 1985; Hufford, G. A., "The ITS Irregular Terrain Model, Version 1.2.2, The Algorithm," Unpublished (itm_alg.pdf at http://flattop.its.bldrdoc.gov/itm.html), Undated). Internationally, ITU-R Recommendations P.526 (point to point links) (International Telecommunication Union—Radiocommunication Sector (ITU-R), 2009a) and P.1546 (point-to-area links) (International Telecommunication Union—Radiocommunication Sector (ITU-R), 2009c) are popular.

Complex obstructed paths are generally modeled as one or more rounded cylinder (Assis, 1971; Chrysdale, 1958; Rice, 1954) or knife edge (Anderson and Trolese, 1958, 1959; Bullington, 1947; Deygout, 1966, 1991; Durgin, 2008; Epstein and Peterson, 1953; Giovaneli, 1984; Millington et al., 1962; Mokhtari, 1999; Vogler, 1981) obstructions. If a knife edge is on top of a rounded obstruction, the knife-edge loss dominates the results (Wait and Spies, 1968). In some situations, cascaded knife edges may also include reflections (e.g., this often occurs when buildings are modeled as dual knife edges) (Komijani et al., 2010; Meeks, 1982; Zhao and Vainikainen, 2001).

Assis observed (2011) that the problem of diffraction by terrain irregularities is well known to be quite complex. The general solution requires the numerical solution to a three dimensional integral equation (Giovaneli, 1984; Tzaras and Saunders, 2000). As this is generally impractical, simplified methods (included those listed in the previous paragraph) are available.

The most popular simplified methods are those of Bullington (1947), Deygout (1966, 1991), Epstein-Peterson (1953), Longley–Rice (Hufford et al., 1982; Longley, 1978; Longley and Rice, 1968; Rice et al., 1965) (Hufford, G. A., "Memorandum to Users of the ITS Irregular Terrain Model," Unpublished (Hufford_1985_Memo.pdf at http://flattop.its.bldrdoc.gov/itm.html), January 30 1985; Hufford, G. A., "The ITS Irregular Terrain Model, Version 1.2.2, The Algorithm," Unpublished (itm_alg.pdf at http://flattop.its.bldrdoc.gov/itm.html), Undated), and ITU-R P.526 (International Telecommunication Union—Radiocommunication Sector (ITU-R), 2009a). For all these methods, the heights of each terrain profile are first corrected for earth curvature and atmospheric refractivity (Eqn. 5.36). Then the significant obstacles are determined using basic, stretched-string analysis (Liniger et al., 2006).

The Bullington method initially determines the position and height of an equivalent knife-edge obstacle from the intersection between the two straight lines defining the horizons from each terminal. The loss due to the equivalent knife-edge obstacle is taken as the additional path loss due to diffraction (Liniger et al., 2006).

The Deygout method involves the calculation of the loss due to each obstacle in the absence of all others. The obstacle providing the highest loss subdivides the path into two, to which the procedure is recursively applied. The individual losses are again added to determine additional path loss due to diffraction (Liniger et al., 2006).

In the Epstein–Peterson method, the loss for each obstacle is calculated by assuming imaginary terminals placed at the top of the immediately adjacent obstacles to its left and right (or at the top of the real terminals, for the first and the last obstacles). The individual contributions are then added to determine the total path loss due to diffraction (Liniger et al., 2006).

The Longley–Rice model computes transhorizon loss on the basis of climate type, frequency, polarization, ground conductivity and dielectric constant, atmospheric refractivity at ground level, transmitter and receiver antenna ground clearances, and the terrain profile between the transmit and receive antennas. A roughness factor is calculated from the terrain profile. A straight line is fitted to the terrain profile and the effective heights of transmit and receive antennas above this line are determined. The model locates the horizon viewed from transmit and receive antennas and calculates elevation angles and ranges to those horizons. The model calculates two-ray (multipath) interference, double knife-edge diffraction loss, spherical earth diffraction loss, and tropospheric scatter loss (Meeks, 1983).

The ITU-R P.526 method incorporates aspects of both the Deygout and Epstein–Peterson methods (Liniger et al., 2006).

Regarding simplified methods, Assis noted, "... the accuracy [of these simplified methods] is closely related to specific topographical features." He continues, "Numerical results derived from the solution currently adopted by [ITU-R] Recommendation 526–10 to this case [paths between smooth earth and that of isolated obstacles] do not fit well when compared to experimental data." As simplified methods are necessary for most practical applications, the obvious question is, "how good are the simplified methods in estimating path loss on real paths?" The Deygout method is well respected because its results are similar to the theoretical Millington method (Giovaneli, 1984). However, it is very sensitive to the number and closeness of obstructions used (Tzaras and Saunders, 2000).

Surface irregularities or finite obstruction width can affect loss. If the obstruction is rough, path loss increases relative to the smooth obstruction case (Assis and Cerqueira, 2007; Shkarofsky et al., 1958). If the obstruction has finite width (perpendicular to the path), path loss is usually less than the infinite width case (Bachynski and Kingsmill, 1962; Chang et al., 2000; Davis, 2002; Durgin, 2009). The effect of hills is shadow loss that follows a normal probability distribution (Bullington, 1950). Trees and vegetation have dimensions that are of the same order of magnitude as the radio wavelength and can act as parasitic radiators whose loss is Rayleigh distributed (Bullington, 1950). Foreground reflections can affect the result but it is difficult to predict owing to the granular nature of most terrain profile data.

We are dealing with radio wavelengths of the order of a few centimeters but our terrain profiles provide terrain elevations with samples 10–100 m apart. It is not obvious how accurately these different diffraction models predict real-world results. Several attempts have been made to address this concern (Costa et al., 2011; IEEE Vehicular Society Committee, 1988; Kholod et al., 2007; Liniger et al., 2006; Longley, 1976, 1978; Longley and Reasoner, 1970; Silva et al., 2006; Weiner, 1986).

TABLE 13.2 Average Model Error

Obstacles	0	1	2	3	4	5
Bullington, dB	6.0	6.6	−8.0	−3.4	−1.5	4.5
ITU-R P.526, dB	8.4	3.0	12.0	11.5	13.5	15.0
Epstein–Peterson, dB	6.0	6.6	4.5	−2.4	−7.5	−16.3
Longley–Rice, dB	8.0	1.3	6.1	7.6	9.2	10.1
Deygout, dB	6.0	6.6	4.6	3.5	6.0	4.0

TABLE 13.3 Average Standard Deviation of Model Error

Obstacles	0	1	2	3	4	5
Bullington, dB	10.5	11.9	17.5	16.1	14.3	13.7
ITU-R P.526, dB	10.5	13.6	16.6	16.1	13.7	13.3
Epstein–Peterson, dB	10.5	11.9	12.6	13.1	13.6	16.1
Longley–Rice, dB	10.8	13.4	14.3	15.4	12.2	12.4
Deygout, dB	10.5	11.9	12.8	13.5	13.2	16.7

Costa et al. (2011) compared model results with actual path loss measurements for over 9000 paths taken from the ITU-R Study Group 3 and Institute of Telecommunications Sciences databases. The model estimates were based on terrain profiles created from the 200- to 300-m resolution data (Tables 13.2 and 13.3). If the error value was positive, the model estimated obstruction loss less than the measured value by the amount indicated (the model was optimistic). If the error value was negative, the model overestimated the obstruction loss by the amount indicated (the model was pessimistic).

Liniger et al. (2006) compared model results with about 500 measurements of 19 TV transmitters operating at 12 different sites. Paths were chosen to avoid effects such as vegetation or nearby obstacles not considered in the models. The model estimates were based on terrain profiles created from the 100-m resolution USGS Shuttle Radar Topography Mission (STRM) database. Typical transmitter height was 10 m (Tables 13.4 and 13.5).

TABLE 13.4 Average Model Error

Obstacles	0	1	2	3	4	5
Bullington, dB	4	5	−1	5	7	8
ITU-R P.526-9, dB	3	2	−16	−18	−17	−22
Epstein–Peterson, dB	4	4	−10	−16	−24	−36
Longley–Rice, dB	−1	−13	−17	−9	−13	−17
Deygout, dB	2	2	−15	−23	−29	−45

TABLE 13.5 Average Standard Deviation of Model Error

Obstacles	0	1	2	3	4	5
Bullington, dB	12.5	12.3	13.5	13.5	12.0	11.7
ITU-R P.526-9, dB	13.1	13.4	14.0	13.3	14.0	13.2
Epstein–Peterson, dB	12.5	12.3	14.0	13.5	13.1	13.8
Longley–Rice, dB	14.3	14.3	18.7	17.1	14.7	13.7
Deygout, dB	14.5	14.6	16.2	17.2	16.1	16.8

It is clear that the models are only approximately accurate. The expected variation between the model-estimated loss and the actual observed loss is several decibels. This variability has been observed (Bullington, 1950; IEEE Vehicular Society Committee, 1988; Longley, 1976; Longley and Reasoner, 1970) and has been attributed (IEEE Vehicular Society Committee, 1988) to three categories: variation in time, location characteristics (variation in height, obstruction, or nearby reflections), and situation (unknown or unknowable) factors.

Longley (1976) studied mean signal variability for many different types of nonurban obstructed paths (including paths with single and multiple obstructions of various shapes and roughness). She noted that transmission loss variation increased rapidly with distance for small distances but distance quickly became irrelevant for moderate-to-long distances. The most significant factors of mean signal variation were either frequency or terrain roughness. On the basis of averages of 800 paths up to 120 km long, operating at frequencies up to 9 GHz, the following relationships were observed:

$$\sigma_L = 6.6 + 9.5\, F^{(1/2)} - 1.5\, F \tag{13.37}$$

For the above, path roughness was ignored. A more accurate estimate included both path roughness and frequency.

$$\sigma_L = 6.0 + 0.55 \left(\frac{\Delta h}{\lambda}\right)^{\left(\frac{1}{2}\right)} - 0.004 \left(\frac{\Delta h}{\lambda}\right) \text{ for } \left(\frac{\Delta h}{\lambda}\right) \leq 4700 \tag{13.38}$$

$$\sigma_L = 24.9 \text{ for } \left(\frac{\Delta h}{\lambda}\right) > 4700 \tag{13.39}$$

σ_L = standard deviation of the observed path loss mean relative to the estimated mean using the Longley–Rice model (dB);
F = radio operating frequency (GHz);
λ = radio free space wavelength (m);
Δh = interdecile range of terrain elevations (m);
= RMS terrain elevation calculated from regularly spaced terrain elevation samples relative to a straight line fitted to the terrain elevation samples (the lowest and highest 10% values are eliminated before creating the RMS value) (Table 13.6).

Longley found no observable change in standard deviation with increasing distance. Except for rugged mountainous areas, 90% of all measurements fell within 2 dB of the predicted standard deviation values using the above ($\Delta h/\lambda$) formulas. Longley's estimates of path loss variability are similar to the values observed by Costa et al. Significant average (mean) receive signal level variation from model prediction appears inevitable for obstructed paths.

TABLE 13.6 Estimates of Δh Values

Type of Terrain	Δh, m
Water or very smooth plains	0–5
Smooth plains	5–20
Slightly rolling hills	20–40
Rolling hills	40–80
Typical value	90
Hills	80–150
Typical mountains	150–300
Rugged mountains	300–700
Very rugged mountains	>700

13.A APPENDIX

13.A.1 Smooth Earth Reflection Coefficient

The reflection coefficient is the relative magnitude of signal reflected from the Earth relative to the signal before reflection, ignoring divergence factor. This coefficient is highly dependent on polarization and grazing angle. However, it is essentially unity power ratio (0 dB) with phase reversal for small grazing angles. The earth's reflection coefficient is calculated using the following formulas (Hall, 1979):

$$R = \text{reflection coefficient}$$

$$= \text{reflected signal (voltage) magnitude/incident signal magnitude} \quad (13.A.1)$$

$$R_{\mathrm{H}} = \text{reflection coefficient for horizontal polarization}$$

$$= \frac{\{\sin(\phi) - \mathrm{sqrt}[\eta^2 - \cos^2(\phi)]\}}{\{\sin(\phi) + \mathrm{sqrt}[\eta^2 - \cos^2(\phi)]\}} \quad (13.A.2)$$

$$R_{\mathrm{V}} = \text{reflection coefficient for vertical polarization}$$

$$= \frac{\{\eta^2 \sin(\phi) - \mathrm{sqrt}[\eta^2 - \cos^2(\phi)]\}}{\{\eta^2 \sin(\phi) + \mathrm{sqrt}[\eta^2 - \cos^2(\phi)]\}} \quad (13.A.3)$$

η^2 = complex refractive index of reflecting surface;
$= \varepsilon_{\mathrm{r}} - j[\sigma/(\omega\varepsilon_{\mathrm{o}})] = \varepsilon_{\mathrm{r}} - j60\lambda\sigma$ (a function of frequency);
ε_{r} = permittivity (dielectric constant relative to air);
σ = conductivity (S/m);
F = frequency (GHz);
ϕ (rad) = grazing angle = angle of incidence = angel of reflection;
λ = wavelength (m) = 0.2998/F (GHz) = 299.8/F (MHz).

13.A.2 Procedure for Calculating R_{H} AND R_{V}

The deceptively simple formulas for reflective coefficient are surprisingly difficult to calculate. The square root function and ambiguity of the arctangent function make further simplification of the equations impractical. However, a computer procedure is relatively easy to outline:

$$R = \text{reflection coefficient} = \text{reflected signal magnitude/incident signal magnitude}$$

$$R_{\mathrm{H}} = \text{reflection coefficient for horizontal polarization} \quad (13.A.4)$$

$$= \frac{\{\sin(\phi) - \mathrm{sqrt}[\eta^2 - \cos^2(\phi)]\}}{\{\sin(\phi) + \mathrm{sqrt}[\eta^2 - \cos^2(\phi)]\}}$$

$$R_{\mathrm{V}} = \text{reflection coefficient for vertical polarization}$$

$$= \frac{\{\eta^2 \sin(\phi) - \mathrm{sqrt}[\eta^2 - \cos^2(\phi)]\}}{\{\eta^2 \sin(\phi) + \mathrm{sqrt}[\eta^2 - \cos^2(\phi)]\}} \quad (13.A.5)$$

η^2 = (normalized admittance)^2 = complex refractive index of reflecting surface;
$= \varepsilon_{\mathrm{r}} - j\sigma/(\omega\varepsilon_0) = \varepsilon_{\mathrm{r}} - j60\lambda\sigma$ (a function of frequency);
ε_{r} = permittivity (dielectric constant relative to air);
σ = conductivity (S/m);
F = frequency (GHz);
ϕ (rad) = grazing angle = angle of incidence = angel of reflection;
λ = wavelength (m) = 0.2998/F (GHz).

$R = \varepsilon_r - \cos^2(\phi)$
$I = - 60 * \lambda * \sigma$
Go to Subroutine 1 and return
Go to Subroutine 3 and return
M = Ms
$\theta = \theta s$
Go to Subroutine 2 and return
CR = R
CI = I
For Vertical Polarization:
    $DR = \varepsilon_r$
    $DI = - 60\ \lambda\ \sigma$
For Horizontal Polarization:
    DR = 1
    DI = 0
$R = [DR * \sin(\phi)] - CR$
$I = [DI * \sin(\phi)] - CI$
Go to Subroutine 1 and return
Mn = M
$\theta n = \theta$
$R = (DR * \sin(\phi)) + CR$
$I = (DI * \sin(\phi)) + CI$
Go to Subroutine 1 and return
Md = M
$\theta d = \theta$
Go to Subroutine 4 and return
If $\theta > \pi$ then $\theta \Rightarrow \theta - 2\pi$
If $\theta < - \pi$ then $\theta \Rightarrow \theta + 2\pi$
$R_{H \text{ or } V} = M\ e^{j\theta}$
$R_{H \text{ or } V}$ Magnitude (dB) = 20 Log (M)
$R_{H \text{ or } V}$ Phase (degrees) = $(180 / \pi)\ \theta$(radians)

Subroutine 1:
$R + j\ I \Rightarrow M\ e^{j\theta}$
$M = \text{Sqrt}(\ R^2 + I^2\ )$
$\theta = \text{Arctan}(\ I / R\ ),\ - \pi / 2 \leq \theta \leq + \pi / 2$
If $R < 0$ and $I > 0$ then $\theta \Rightarrow \theta + \pi$
If $R < 0$ and $I < 0$ then $\theta \Rightarrow \theta - \pi$

Subroutine 2:
$M\ e^{j\theta} \Rightarrow R + j\ I$
$R = M \cos(\theta)$
$I = M \sin(\theta)$

Subroutine 3
$\text{Sqrt}\ [\ M\ e^{j\theta}\ ] = Ms\ e^{j\theta s}$
$Ms = \text{Sqrt}(M)$
$\theta s = \theta / 2$

Subroutine 4
$[\ Mn\ e^{j\theta n}\ ] / [\ Md\ e^{j\theta d}\ ] = M\ e^{j\theta}$
$M = Mn / Md$
$\theta = \theta n - \theta d$

The reflection coefficient equations require knowledge of the earth's permittivity and conductivity. While there are many sources for this data for low frequencies (e.g., related to power distribution and radio broadcasts), there are few sources for this data at microwave frequencies. An extensive source is ITU-R Recommendation 527–3 (International Telecommunication Union—Radiocommunication Sector (ITU-R), 1992). This source provides earth conductivity data in graphical format. This format, while useful, does not lend itself to automated calculation. To facilitate these calculations, Recommendation 527-3 data was curve fitted over the range of 100 kHz to 100 GHz. A plot of the curve fits and the numerical equations follow.

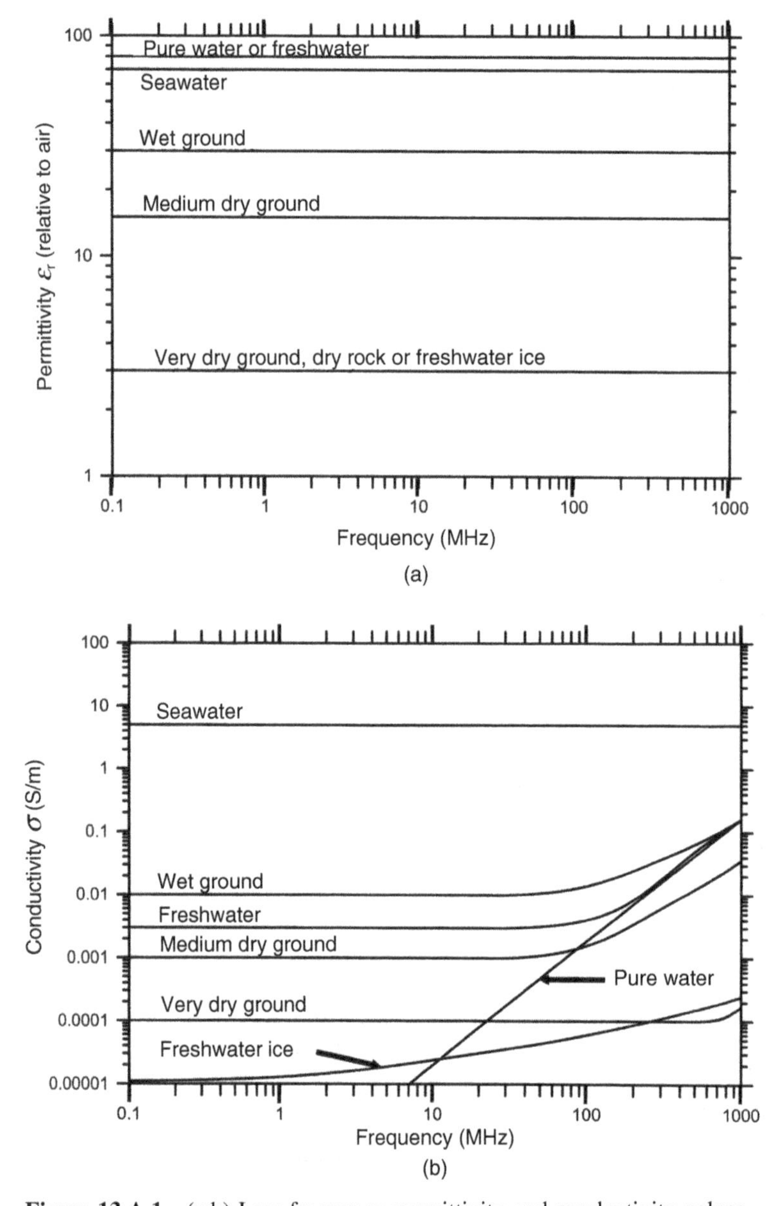

Figure 13.A.1 (a,b) Low frequency permittivity and conductivity values.

13.A.3 Earth Parameters for Frequencies Between 100 kHz and 1 GHz

General formulas for calculating ε_r and σ (Fig. 13.A.1)

$$R_{\text{exp}} = \frac{(\text{Cr1} + \text{Cr2}\ F + \text{Cr3}\ F^2 + \text{Cr4}\ F^3 + \text{Cr5}\ F^4)}{(1 + \text{Cr6}\ F + \text{Cr7}\ F^2 + \text{Cr8}\ F^3 + \text{Cr9}\ F^4)} \tag{13.A.6}$$

$$\sigma = 10^{R_{\text{exp}}}$$

F = radio wave frequency (MHz).

For Fresh Water	For Pure Water
For $0.1 \leq F \leq 1000$ $\varepsilon_r = 80$ For $0.1 \leq F \leq 30$ $\sigma = 0.003$ For $30 \leq F \leq 1000$ (use the above formula and following coefficients) $Cr1 = -2.614441823488561$ $Cr2 = -0.0004668855275106429$ $Cr3 = 1.308404511022228 \times 10^{-6}$ $Cr4 = 0.0$ $Cr5 = 0.0$ $Cr6 = 0.001326412261081864$ $Cr7 = 1.804647661107618 \times 10^{-6}$ $Cr8 = -1.789022600542104 \times 10^{-9}$ $Cr9 = 0.0$	For $0.1 \leq F \leq 1000$ $\varepsilon_r = 80$ For $0.1 \leq F \leq 1000$ (use the above formula and following coefficients) $F\log = \log_{10}(F)$: $Cr1 = -6.69304383571563$ $Cr2 = 2.003990023349996 \times \frac{F\log}{F}$ $Cr3 = -1.006287589534218 \times 10^{-2} \times \frac{F\log^2}{F^2}$ $Cr4 = -3.021202859233963 \times 10^{-3} \times \frac{F\log^3}{F^3}$ $Cr5 = 1.188748132191522 \times 10^{-3} \times \frac{F\log^4}{F^4}$ $Cr6 = 0.0$ $Cr7 = 0.0$ $Cr8 = 0.0$ $Cr9 = 0.0$
For Wet Ground	**For Medium Dry Ground**
For $0.1 \leq F \leq 1000$ $\varepsilon_r = 30$ For $0.1 \leq F \leq 30$ $\sigma = 0.01$ For $30 \leq F \leq 1000$ (use the above formula and following coefficients) $Cr1 = -2.017850105148183$ $Cr2 = 0.0$ $Cr3 = -3.589310575656583 \times 10^{-5}$ $Cr4 = 0.0$ $Cr5 = -3.199977514317552 \times 10^{-12}$ $Cr6 = 0.0$ $Cr7 = 2.787058734413573 \times 10^{-5}$ $Cr8 = 0.0$ $Cr9 = 2.278280532564988 \times 10^{-11}$	For $0.1 \leq F \leq 1000$ $\varepsilon_r = 15$ For $0.1 \leq F \leq 30$ $\sigma = 0.001$ For $30 \leq F \leq 1000$ (use the above formula and following coefficients) $Cr1 = -3.023718934870313$ $Cr2 = 0.0$ $Cr3 = -4.896992146047716 \times 10^{-5}$ $Cr4 = 0.0$ $Cr5 = -8.439863059581116 \times 10^{-12}$ $Cr6 = 0.0$ $Cr7 = 2.512593711118078 \times 10^{-5}$ $Cr8 = 0.0$ $Cr9 = 1.572552265268755 \times 10^{-11}$
For Very Dry Ground	**For Pure or Fresh Water Ice (−10 °C)**
For $0.1 \leq F \leq 1000$ $\varepsilon_r = 3$ For $0.1 \leq F \leq 500$ $\sigma = 0.0001$ For $500 \leq F \leq 1000$ (use above formula and following coefficients) $Cr1 = -3.995131826587921$ $Cr2 = 0.0$ $Cr3 = 5.621093748833045 \times 10^{-6}$	For $0.1 \leq F \leq 1000$ $\varepsilon_r = 3$ For $0.1 \leq F \leq 1000$ (use above formula and following coefficients) $Cr1 = -4.967089460343868$ $Cr2 = -0.7638338206326642$ $Cr3 = -0.003857304863292114$ $Cr4 = -1.378230458708166 \times 10^{-7}$ $Cr5 = 0.0$

Cr4 = 0.0	Cr6 = 0.1708111349047694
Cr5 = $-3.740241438751809 \times 10^{-12}$	Cr7 = 0.001023209735555101
Cr6 = 0.0	Cr8 = $1.242615497159902 \times 10^{-7}$
Cr7 = $-1.421760479745282 \times 10^{-6}$	Cr9 = 0.0
Cr8 = 0.0	
Cr9 = $9.815379602688738 \times 10^{-13}$	

For Dry Rock	For Seawater
For $0.1 \le F \le 1000$	For $0.1 \le F \le 1000$
$\varepsilon_r = 3$	$\varepsilon_r = 70$
For $0.1 \le F \le 1000$	For $0.1 \le F \le 1000$
$\sigma =$ (σ for very dry ground) $\times 10^{-6}$	$\sigma = 5.0$

13.A.4 Earth Parameters for Frequencies Between 1 GHz and 100 GHz

General formulas for calculating ε_r and σ (Fig. 13.A.2)

$$\varepsilon_r = \frac{(\text{Ce1} + \text{Ce2}\, F + \text{Ce3}\, F^2)}{(1 + \text{Ce4}\, F + \text{Ce5}\, F^2)} \tag{13.A.7}$$

$$\sigma = \frac{(\text{Cr1} + \text{Cr2}\, F + \text{Cr3}\, F^2)}{(1 + \text{Cr4}\, F + \text{Cr5}\, F^2)} \tag{13.A.8}$$

F = radio wave frequency (GHz).

For Pure or Freshwater	For Seawater
For $1 \le F \le 7$	For $1 \le F \le 3$
$\varepsilon_r = 80$	$\varepsilon_r = 70$
For $7 \le F \le 100$ (use the above formula and following coefficients)	For $3 \le F \le 100$ (use the above formula and following coefficients)
Ce1 = 80.52092160517354	Ce1 = 76.55562349662371
Ce2 = $-4.971024269121863 \times 10^{-1}$	Ce2 = $-7.83255339930585 \times 10^{-1}$
Ce3 = $1.978751495586252 \times 10^{-2}$	Ce3 = $2.169668074841692 \times 10^{-2}$
Ce4 = $-2.473058671438465 \times 10^{-2}$	Ce4 = $1.402674394563023 \times 10^{-2}$
Ce5 = $3.069772306422429 \times 10^{-3}$	Ce5 = $2.498108458270239 \times 10^{-3}$
For $1 \le F \le 100$ (use the above formula and following coefficients)	For $1 \le F \le 100$ (use the above formula and following coefficients)
Cr1 = $-1.040961919734746 \times 10^{-1}$	Cr1 = 4.171579918614342
Cr2 = $1.203114992817814 \times 10^{-1}$	Cr2 = $7.701236770774619 \times 10^{-1}$
Cr3 = $1.587606230105653 \times 10^{-1}$	Cr3 = $5.437239720515787 \times 10^{-2}$
Cr4 = $1.960874690996541 \times 10^{-4}$	Cr4 = $-6.889537462455094 \times 10^{-3}$
Cr5 = $2.175207124202258 \times 10^{-3}$	Cr5 = $8.624043908144374 \times 10^{-4}$

For Wet Ground	For Medium Dry Ground
For $80 \le F \le 100$	For $1 \le F \le 5$
$\varepsilon_r = 4$	$\varepsilon_r = 15$
	For $80 \le F \le 100$
	$\varepsilon_r = 4$

For $1 \leq F \leq 80$ (use the above formula and following coefficients)	For $5 \leq F \leq 80$ (use the above formula and following coefficients)
$Ce1 = 31.97815090233879$	$Ce1 = 11.04307913651391$
$Ce2 = 1.220006423195523$	$Ce2 = 8.042739723856723 \times 10^{-2}$
$Ce3 = 8.601249801091193 \times 10^{-2}$	$Ce3 = 4.938818643114097 \times 10^{-2}$
$Ce4 = 8.350803061140581 \times 10^{-2}$	$Ce4 = -1.017535215238241 \times 10^{-1}$
$Ce5 = 2.563127582423338 \times 10^{-2}$	$Ce5 = 1.416381797171442 \times 10^{-2}$
For $1 \leq F \leq 100$ (use the above formula and following coefficients)	For $1 \leq F \leq 100$ (use the above formula and following coefficients)
$Cr1 = -5.250013032638385 \times 10^{-2}$	$Cr1 = -7.213138125174718 \times 10^{-3}$
$Cr2 = 1.907805647742724 \times 10^{-1}$	$Cr2 = 2.067147943106849 \times 10^{-2}$
$Cr3 = 2.81814167590337 \times 10^{-2}$	$Cr3 = 2.398607708989673 \times 10^{-2}$
$Cr4 = 4.218810534444952 \times 10^{-2}$	$Cr4 = 3.953994735601837 \times 10^{-2}$
$Cr5 = 1.150155569466024 \times 10^{-3}$	$Cr5 = 8.474075518630894 \times 10^{-4}$
For Very Dry Ground	**For Pure or Freshwater Ice (−10 °C)**
For $1 \leq F \leq 100$	For $1 \leq F \leq 100$
$\varepsilon_r = 3$	$\varepsilon_r = 3$
For $1 \leq F \leq 100$ (use the above formula and following coefficients)	For $1 \leq F \leq 100$ (use the above formula and following coefficients)
$Cr1 = 7.063715531814967 \times 10^{-4}$	$Cr1 = 2.672701152839106 \times 10^{-5}$
$Cr2 = -1.121337658125633 \times 10^{-3}$	$Cr2 = 2.295305938496779 \times 10^{-4}$
$Cr3 = 5.814679174877295 \times 10^{-4}$	$Cr3 = 1.110066276134025 \times 10^{-5}$
$Cr4 = -4.494535598458532 \times 10^{-3}$	$Cr4 = 8.938918373780938 \times 10^{-2}$
$Cr5 = 5.329704350479167 \times 10^{-4}$	$Cr5 = -3.69777442471754 \times 10^{-5}$
For Dry Rock	
For $1 \leq F \leq 100$:	
$\varepsilon_r = 3$	
For $1 \leq F \leq 100$:	
$\sigma = (\sigma \text{ for very dry ground}) \times 10^{-6}$	

13.A.5 Comments on Conductivity and Permittivity

The various water properties were measured at $20^{\circ}C$. Ice properties were measured at $-10^{\circ}C$. Ice from pure water or freshwater have the same conductivity values. Dry rock values were scaled from low frequency values using very dry ground as a model. Conductivity values of ice (function of temperature) and dry rock (function of composition and dampness) vary widely. However, reflection coefficients as calculated below for ice and dry rock are quite insensitive to large changes in actual conductivity values of this type of Earth. Rock and Ground have similar properties. A little moisture changes their properties drastically and similarly. For most applications, very dry ground and dry rock can be used interchangeably. Below 100 kHz, the permittivity of ice increases. For very low frequencies, pure water and ice conductivity values continue to decrease. Most other conductivity and permittivity values remain constant at low frequencies. The above formulas may be used down to 10 kHz with little error (with the exception of ice permittivity).

13.A.6 Reflection Coefficients

Using the above conductivity values and calculation procedures, the amplitude and phase response of the reflection coefficients were calculated for various types of Earth (Fig. 13.A.3, Fig. 13.A.4, Fig. 13.A.5,

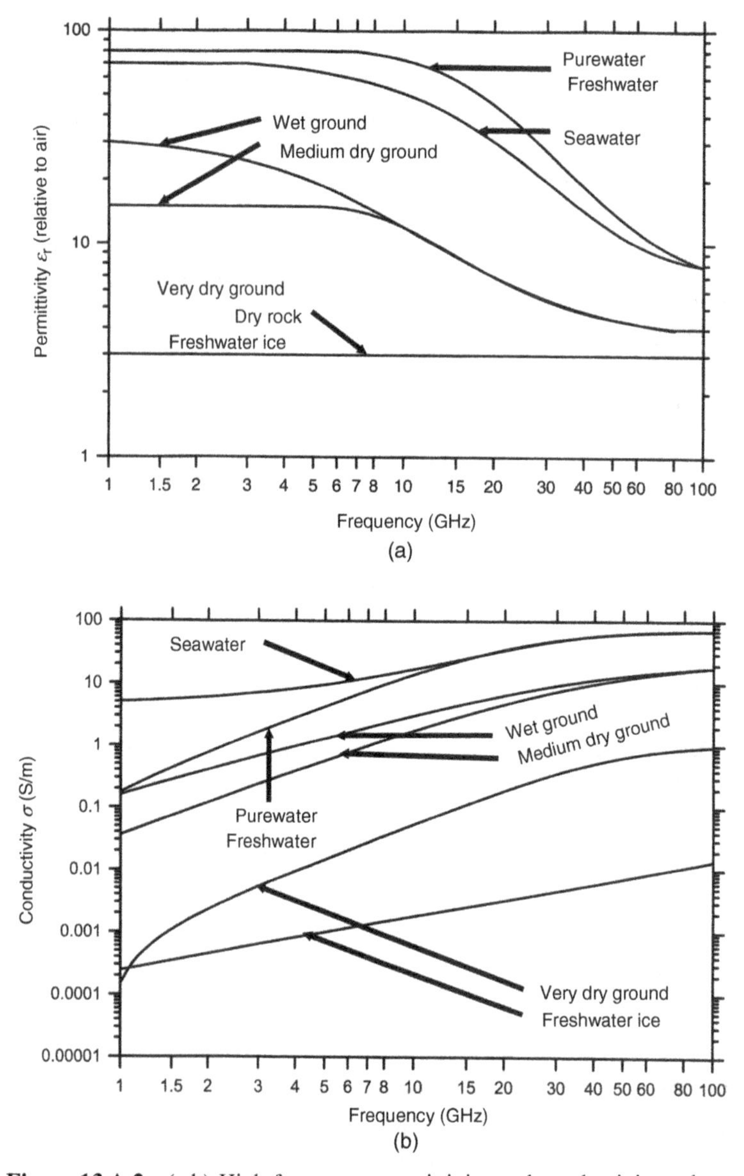

Figure 13.A.2 (a,b) High frequency permittivity and conductivity values.

Fig. 13.A.6, Fig. 13.A.7, Fig. 13.A.8, Fig. 13.A.9, Fig. 13.A.10, Fig. 13.A.11, Fig. 13.A.12, Fig. 13.A.13, Fig. 13.A.14, Fig. 13.A.15, Fig. 13.A.16, Fig. 13.A.17, and Fig. 13.A.18).

The angle relative to the Earth is the grazing angle of the reflection. For horizontal polarization, the reflected signal is essentially always reversed in phase 180° but relatively unchanged in amplitude. For vertical polarization, the reflected signal experiences a relatively small amplitude (null) at or near a grazing angle whose value depends on the conductivity values of the Earth and frequency of the radio signal. If the reflective surface is nonconductive, the angle is called *Brewster's angle*. If the reflective surface is conductive, the angle is called a *pseudo-Brewster angle*.

The vertical polarization null (minimum) value and angle of the (pseudo-)Brewster angle as a function of type of Earth and frequency are shown in Figure 13.A.19 and Figure 13.A.20, respectively.

For vertical polarization, the phase shift at the (pseudo-)Brewster angle is 90°. These results show that for most types of earth (pseudo-)Brewster angles occur at larger grazing angles as frequency increases.

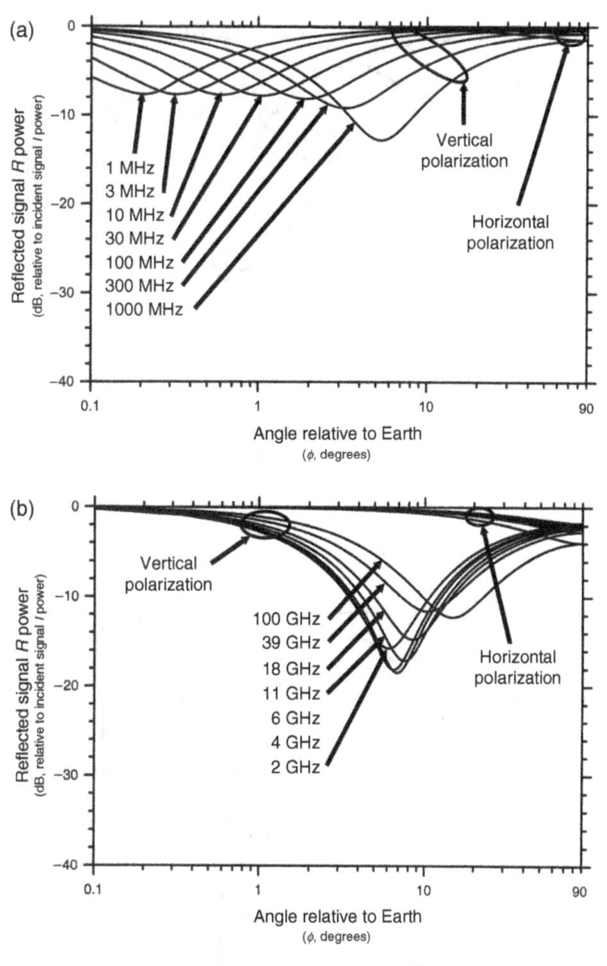

Figure 13.A.3 (a,b) Low and high frequency reflection power for seawater.

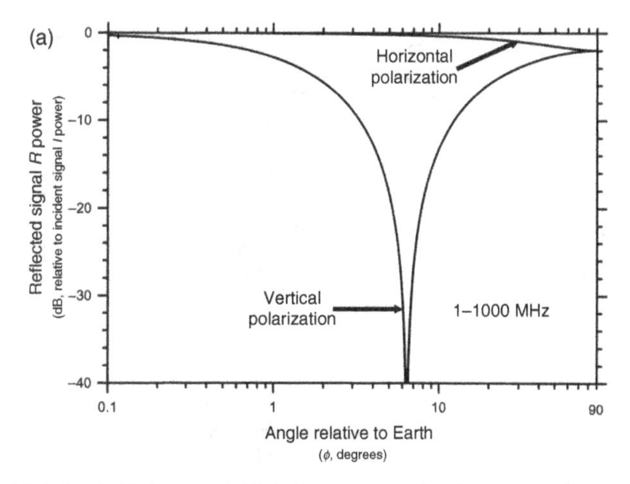

Figure 13.A.4 (a,b) Low and high frequency reflection power for pure water.

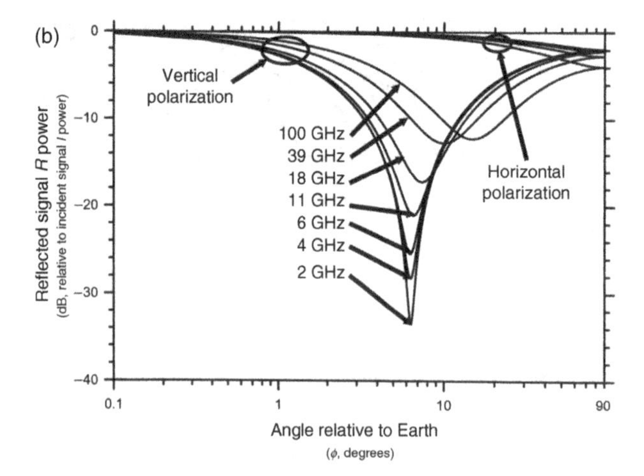

Figure 13.A.4 (*Continued*)

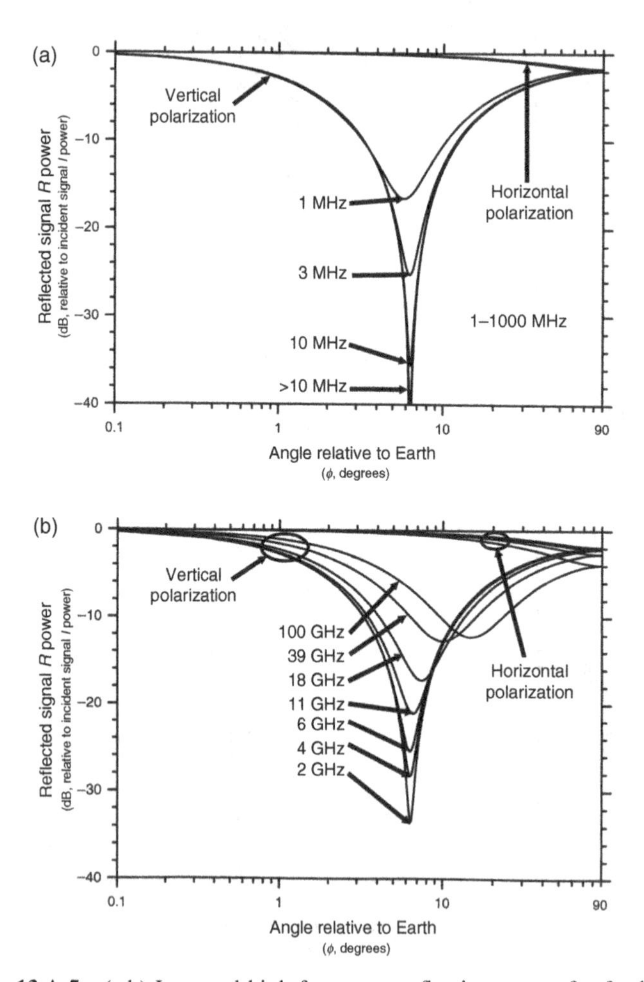

Figure 13.A.5 (a,b) Low and high frequency reflection power for freshwater.

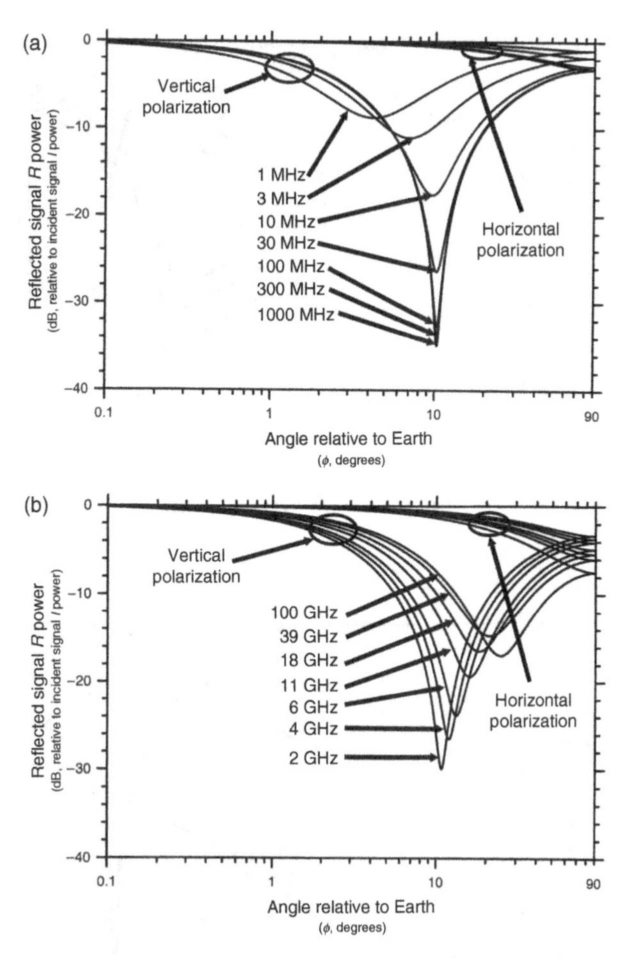

Figure 13.A.6 (a,b) Low and high frequency reflection power for wet ground.

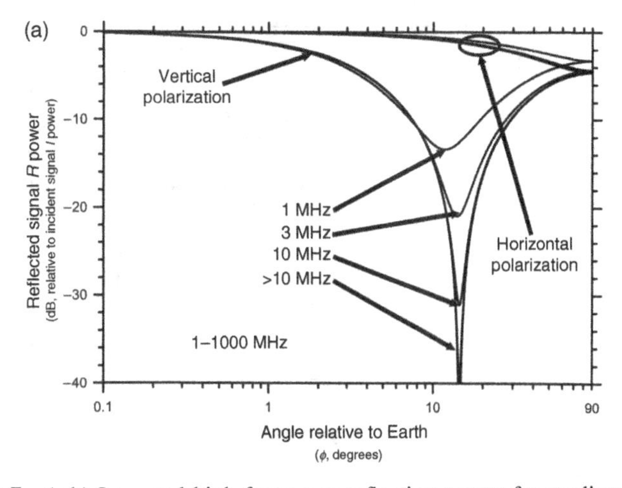

Figure 13.A.7 (a,b) Low and high frequency reflection power for medium dry ground.

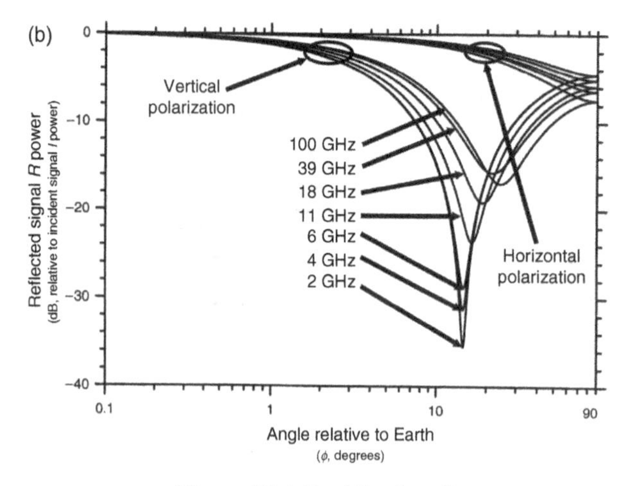

Figure 13.A.7 (*Continued*)

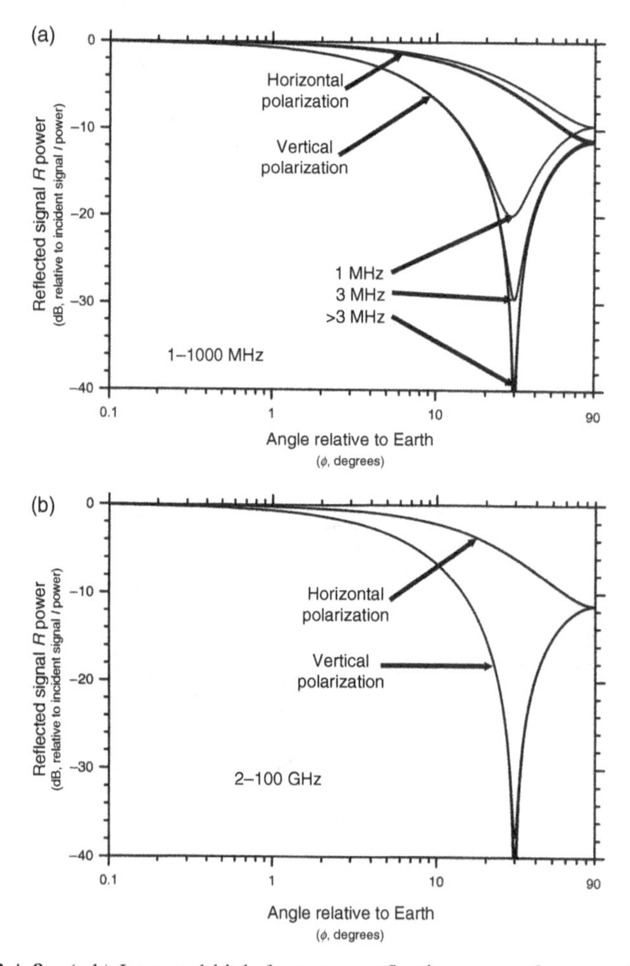

Figure 13.A.8 (a,b) Low and high frequency reflection power for very dry ground.

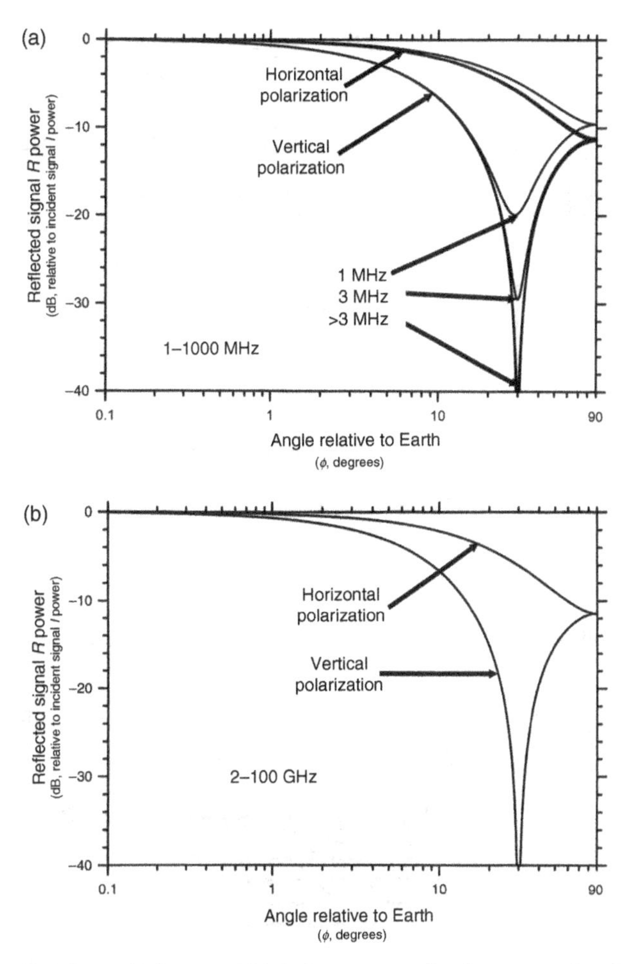

Figure 13.A.9 (a,b) Low and high frequency reflection power for dry rock.

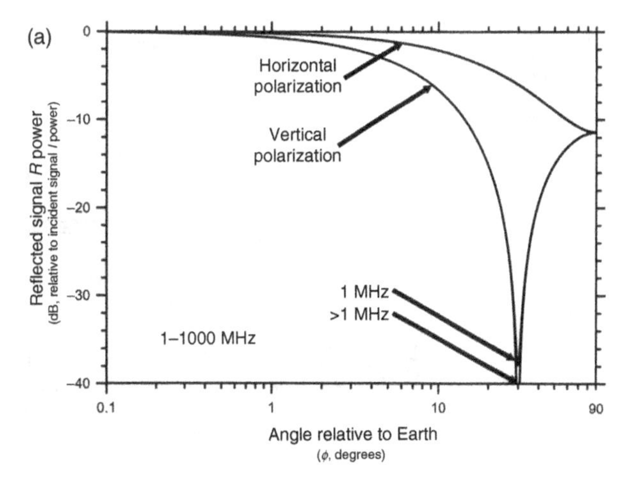

Figure 13.A.10 (a,b) Low and high frequency reflection power for fresh water ice.

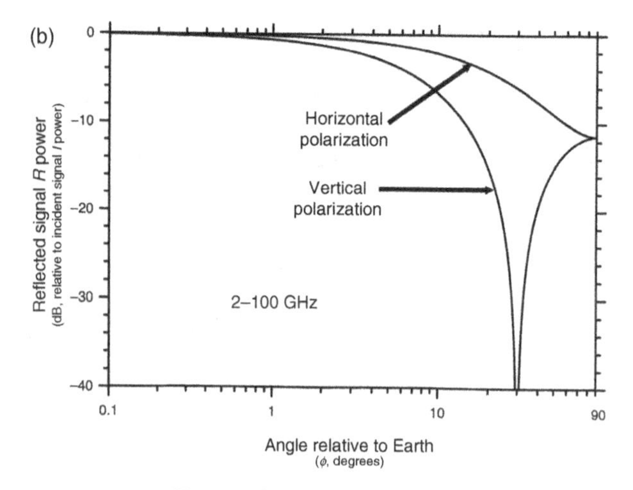

Figure 13.A.10 (*Continued*)

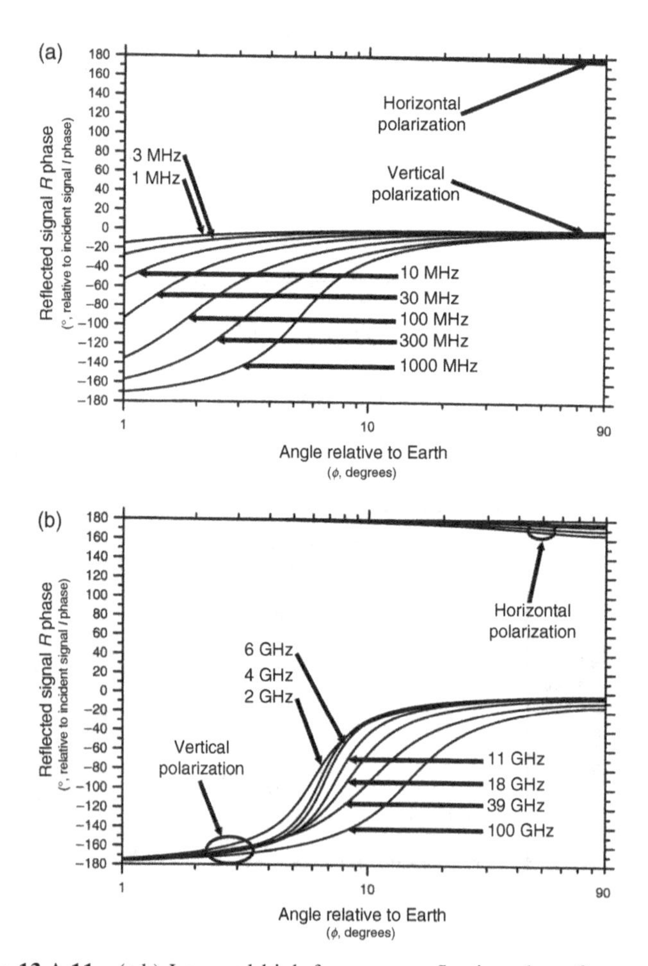

Figure 13.A.11 (a,b) Low and high frequency reflection phase for sea water.

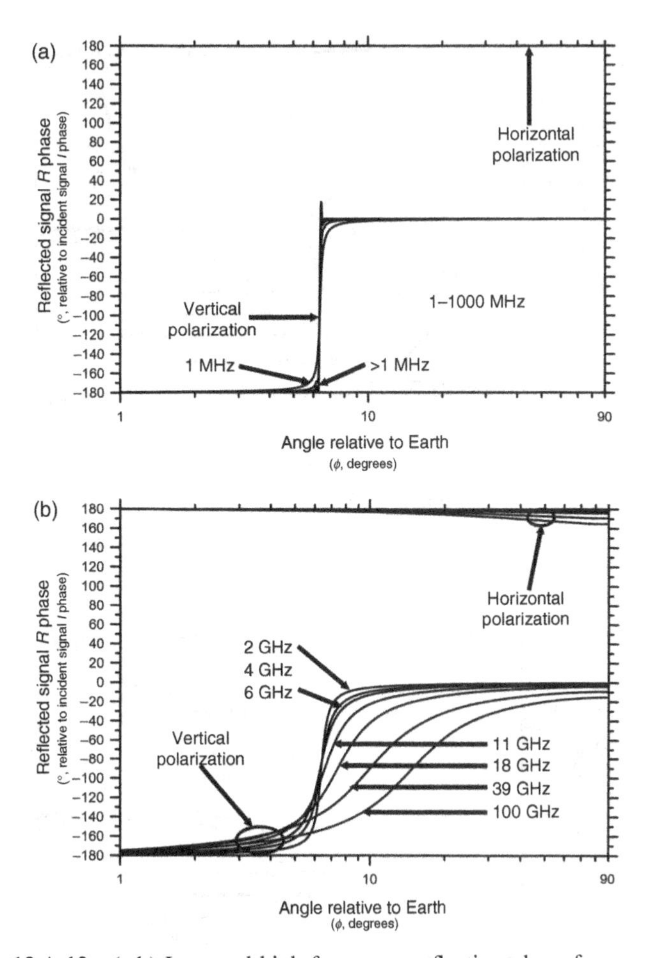

Figure 13.A.12 (a,b) Low and high frequency reflection phase for pure water.

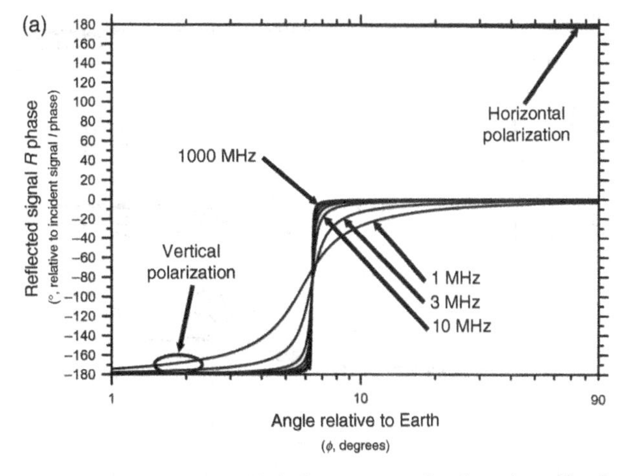

Figure 13.A.13 (a,b) Low and high frequency reflection phase for fresh water.

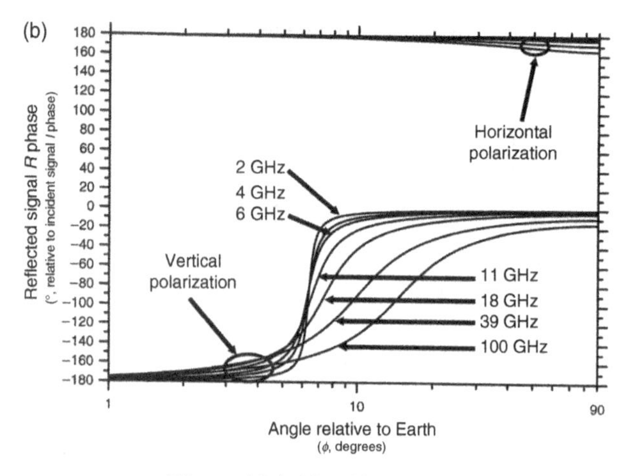

Figure 13.A.13 (*Continued*)

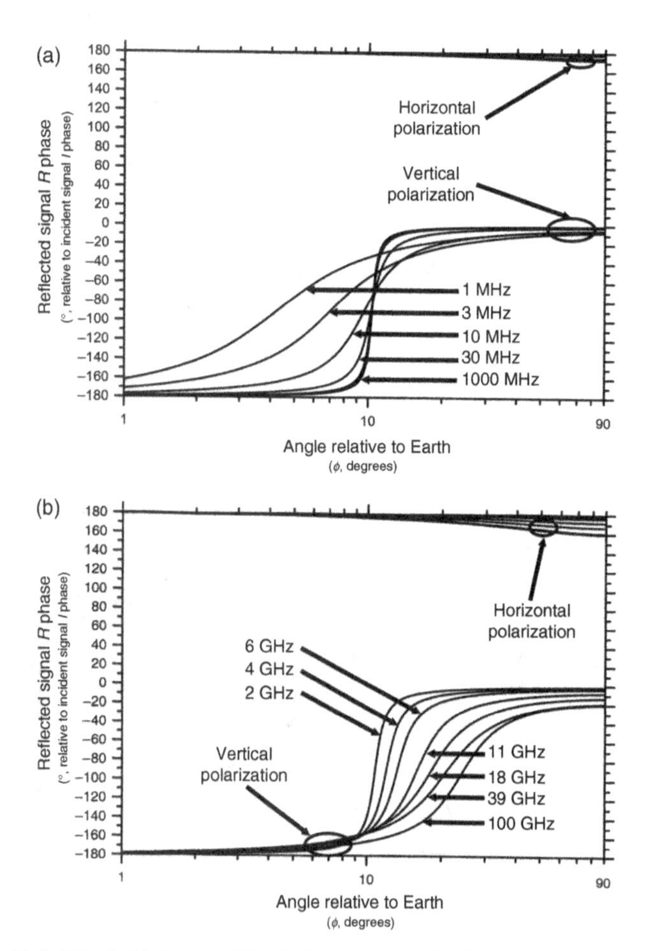

Figure 13.A.14 (a,b) Low and high frequency reflection phase for wet ground.

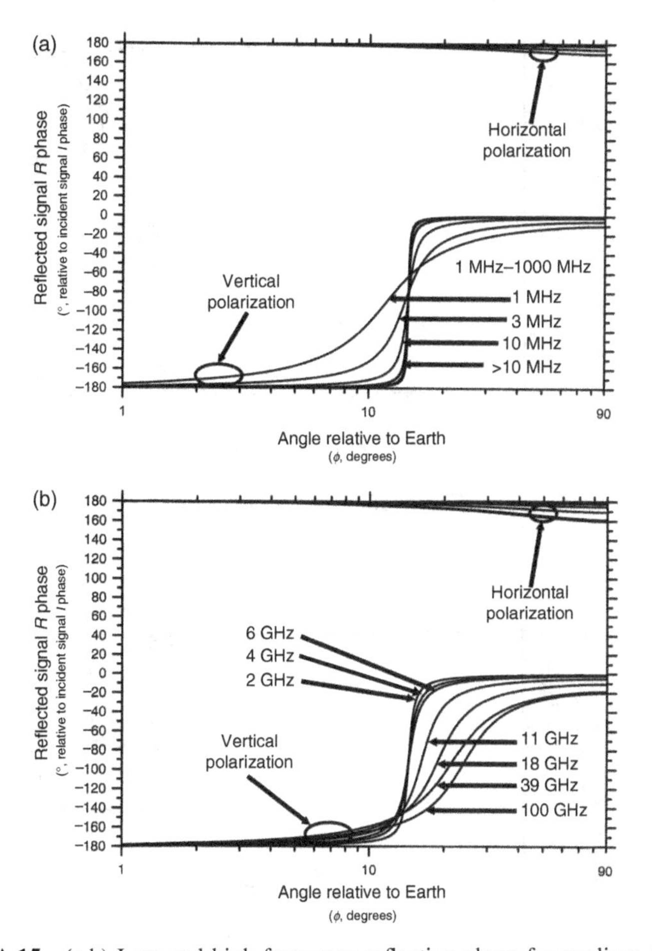

Figure 13.A.15 (a,b) Low and high frequency reflection phase for medium dry ground.

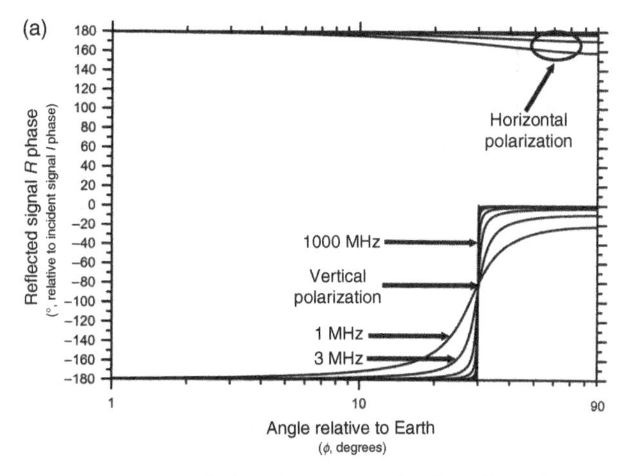

Figure 13.A.16 (a,b) Low and high frequency reflection phase for very dry ground.

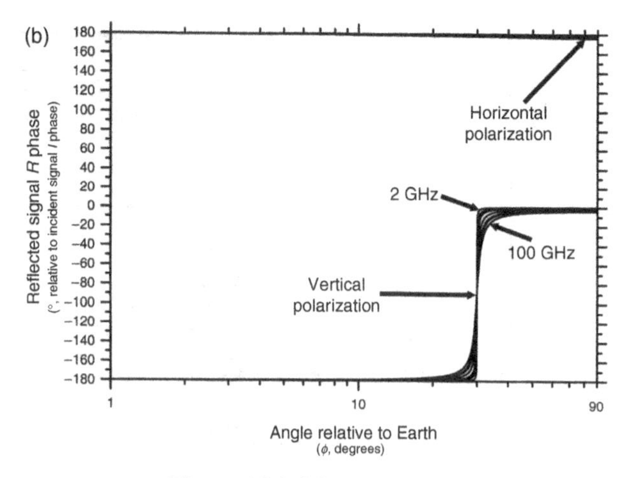

Figure 13.A.16 (*Continued*)

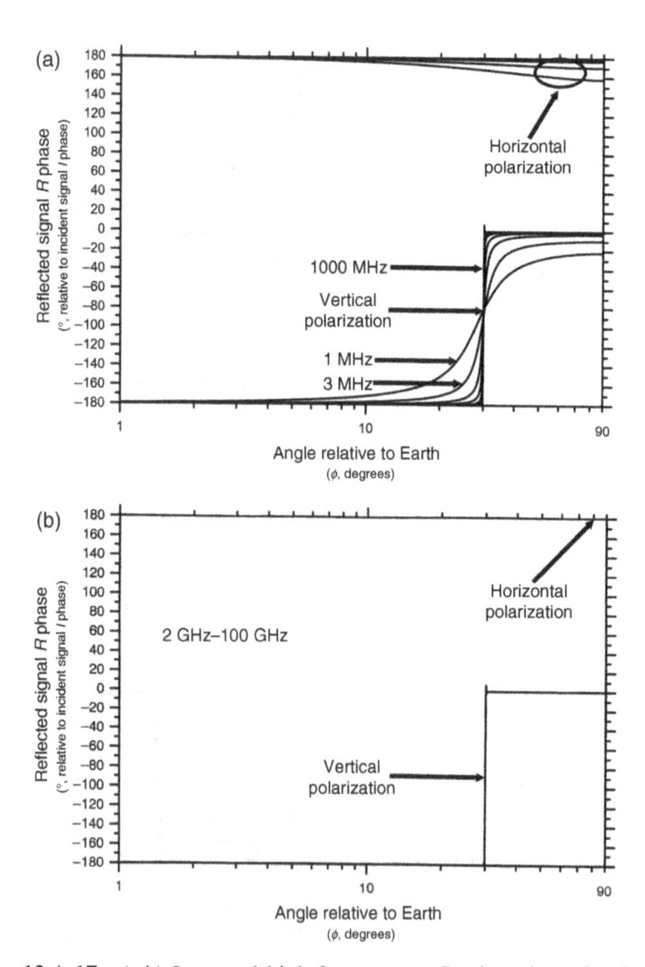

Figure 13.A.17 (a,b) Low and high frequency reflection phase for dry rock.

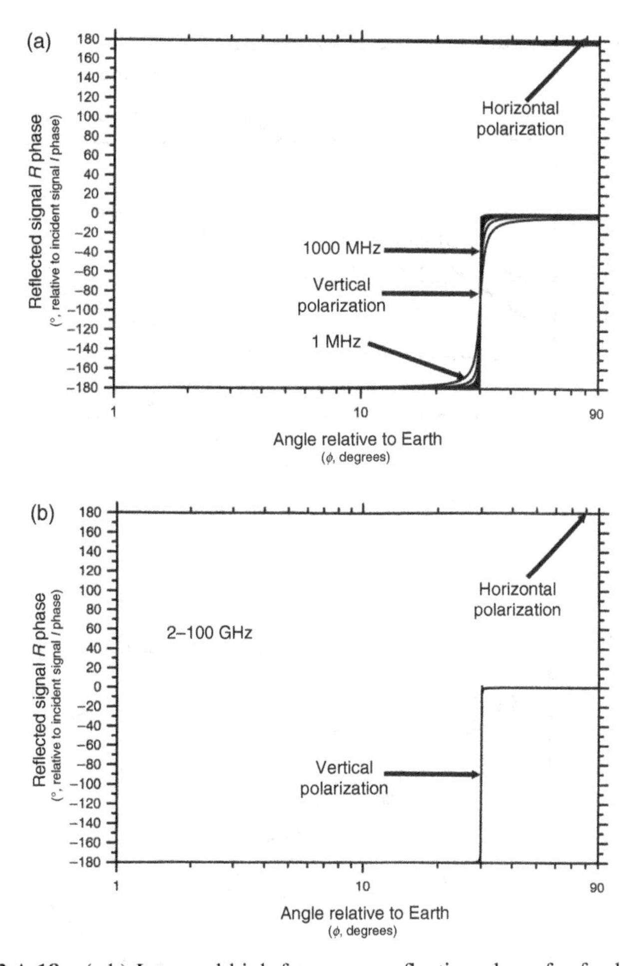

Figure 13.A.18 (a,b) Low and high frequency reflection phase for fresh water ice.

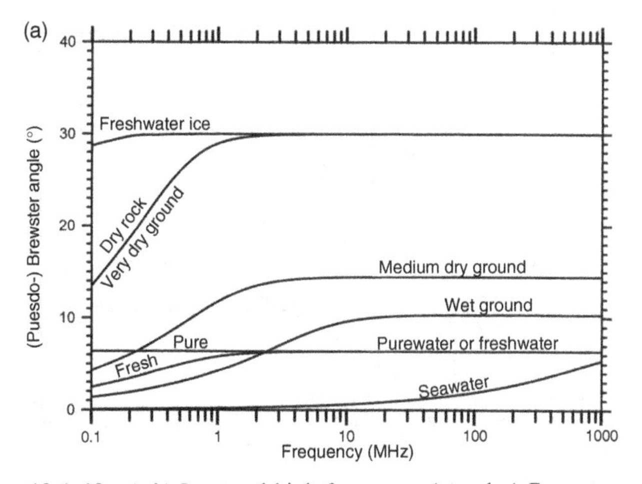

Figure 13.A.19 (a,b) Low and high frequency (puesdo-) Brewster angles.

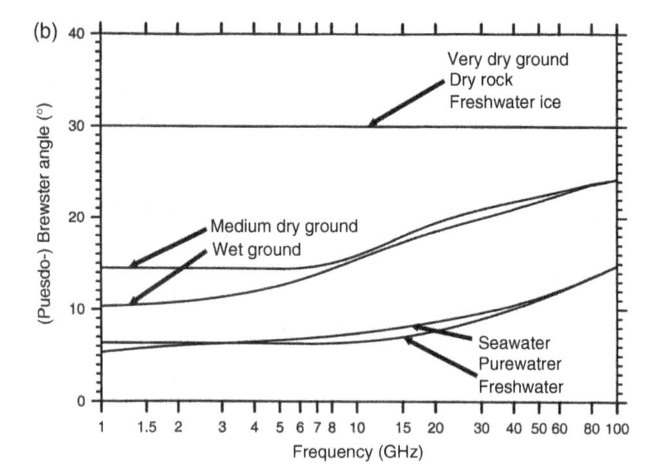

Figure 13.A.19 (*Continued*)

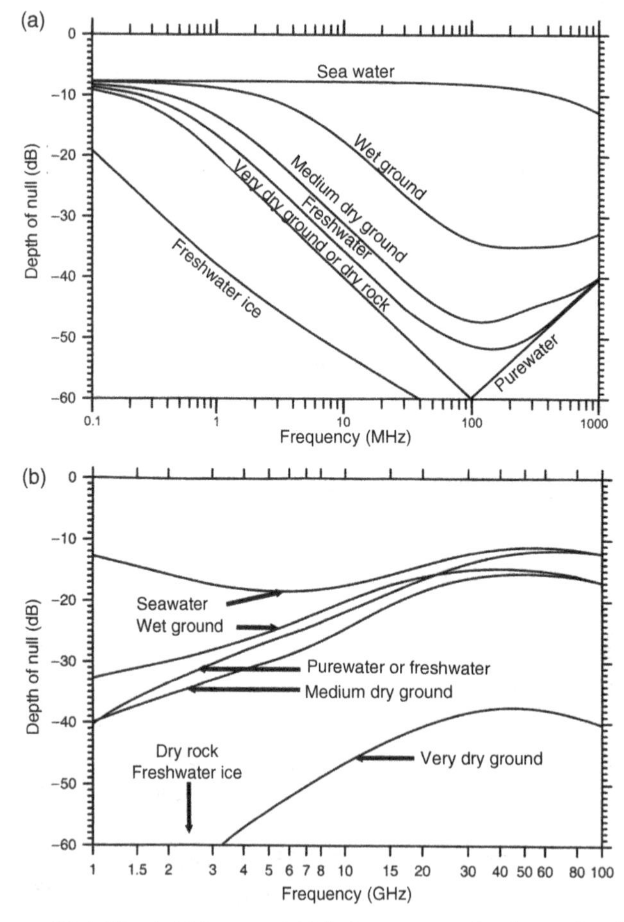

Figure 13.A.20 (a,b) Low and high frequency reflection null values.

REFERENCES

Abramowitz, M. and Stegun, I., *Handbook of Mathematical Functions (NBS AMS 55, seventh printing with corrections).* Washington, DC: U.S. Government Printing Office, pp. 300–302, 1968.

Anderson, L. J. and Trolese, L. G., "Simplified Method for Computing Knife Edge Diffraction in the Shadow Region," *IRE Transactions on Antennas and Propagation*, Vol. 6, pp. 281–286, July 1958.

Anderson, L. J. and Trolese, L. G., "Modification of Simplified Method for Computing Knife Edge Diffraction in the Shadow Region," *IRE Transactions on Antennas and Propagation*, Vol. 7, p. 198, April 1959.

Assis, M. S., "A Simplified Solution to the Problem of Multiple Diffraction over Rounded Obstacles," *IEEE Transactions on Antennas and Propagation*, Vol. 19, pp. 292–295, March 1971.

Assis, M. S., "Radio Wave Diffraction by Terrain Irregularities," *URSI General Assembly and Scientific Symposium*, p. 1, 2011.

Assis, M. S. and Cerqueira, J. L., "Diffraction by Terrain Irregularities: A Review and New Results," *Proceedings of the Second European Conference on Antennas and Propagation*, pp. 1–6, November 2007.

Bachynski, M. P. and Kingsmill, M. G., "Effect of Obstacle Profile on Knife-Edge Diffraction," *IRE Transactions on Antennas and Propagation*, Vol. 10, pp. 201–205, March 1962.

Bacon, D., "Introduction to Diffraction," *Propagation of Radiowaves, Second Edition.* Barclay, L., Editor. London: The Institution of Electrical Engineers, pp. 129–144, 2003.

Beckmann, P. and Spizzichino, A., *The Scattering of Electromagnetic Waves from Rough Surfaces*. Norwood: Artech House, pp. 10–16, 244–248, 1987.

Bell Laboratories Staff, *Radio Engineering, Microwave Radio, Propagation Path Testing*, New York: American Telephone and Telegraph Company, Section 940-310-104, Issue 1, 14–33, September 1964.

Boithias, L., *Radio Wave Propagation*. New York: McGraw-Hill, pp. 56–62, 1987.

Bullington, K., "Radio Propagation at Frequencies Above 30 Megacycles," *Proceedings of the IRE*, Vol. 35, pp. 1122–1136, October 1947.

Bullington, K., "Radio Propagation Variations at VHF and UHF," *Proceedings of the IRE*, Vol. 38, pp. 27–32, January 1950.

Bullington, K., "Reflection Coefficients of Irregular Terrain," *Proceedings of the IRE*, Vol. 42, pp. 1258–1262, August 1954.

Bullington, K., "Radio Propagation Fundamentals," *Bell System Technical Journal*, Vol. 36, pp. 593–626, May 1957.

Carlson, A. and Waterman, A., "Microwave Propagation over Mountain-Diffraction Paths," *IEEE Transactions on Antennas and Propagation*, Vol. 14, pp. 489–496, July 1966.

Carlson, A. and Waterman, A., "Further Comment on Microwave Propagation over Mountain-Diffraction Paths," *IEEE Transactions on Antennas and Propagation*, Vol. 15, p. 716, September 1967.

Chang, S. K., Kim, S. W., Kim, B. K., Hong, U. S., Baek, C. H. and Choi, B. J., "A Propagation Loss Prediction Model for Various Types of Finite Width Obstacles using Numerical Approach," *Asia-Pacific Microwave Conference*, pp. 1584–1588, December 2000.

Chrysdale, J. H., "Comparison of Some Experimental Terrain Diffraction Losses with Predictions based on Rice's Theory for Diffraction by a Parabolic Cylinder," *IRE Transactions on Antennas and Propagation*, Vol. 6, pp. 293–295, July 1958.

Costa, E., Silva, M. A. N. and Liniger, M., "Prediction of Diffraction Effects Due to Irregular Terrain on Radio Wave Propagation in the VHF and UHF Bands," *URSI General Assembly and Scientific Symposium*, pp. 1–4, 2011.

Davis, B. A., "Diffraction by a Randomly Rough Knife Edge," *IEEE Transactions on Antennas and Propagation*, Vol. 50, pp. 1769–1778, December 2002.

Deygout, J., "Multiple Knife-Edge Diffraction of Microwaves," *IEEE Transactions on Antennas and Propagation*, Vol. 14, pp. 480–489, July 1966.

Deygout, J., "Correction Factor for Multiple Knife-Edge Diffraction," *IEEE Transactions on Antennas and Propagation*, Vol. 39, pp. 1256–1258, August 1991.

Dickson, F. H., Egli, J. J., Herbstreit, J. W. and Wickizer, G. S., "Large Reductions of VHF Transmission Loss and Fading by the Presence of a Mountain Obstacle in Beyond-line-of-sight Paths," *Proceedings of the IRE*, Vol. 41, pp. 967–969, August 1953.

Durgin, G. D., "Practical Geometrical Behavior of Knife-Edge Diffraction," *Antennas and Propagation Society International Symposium*, pp. 1–4, July 2008.

Durgin, G. D., "The Practical Behavior of Various Edge-Diffraction Formulas," *IEEE Antennas and Propagation Magazine*, Vol. 51, pp. 24–35, June 2009.

Epstein, J. and Peterson, D. W., "An Experimental Study of Wave Propagation at 850 MHz," *Proceedings of the IRE*, Vol. 41, pp. 595–611, May 1953.

Giovaneli, C. L., "An Analysis of Simplified Solutions for Multiple Knife-Edge Diffraction," *IEEE Transactions on Antennas and Propagation*, Vol. 32, pp. 297–301, March 1984.

Hall, M. P. M., *Effects of the Troposphere on Radio Communication*. New York: P. Peregrinus (on behalf of the IEE), pp. 86, 88–91, 95–97, 1979.

Hartman, W. J. and Smith, D., "Tilting Antennas to Reduce Line-of-Sight Microwave Link Fading," *IEEE Transactions on Antennas and Propagation*, Vol. 25, pp. 642–645, September 1977.

Hristov, H.D. 2000. *Fresnel Zones in Wireless Links, Zone Plate Lenses and Antennas*, Boston: Artech House, pp. 48–50.

Hufford, G. A., Longley, A. G. and Kissick, W. A., "A Guide to the Use of the ITS Irregular Terrain Model in the Area Prediction Mode," *NTIA Report 82–100 (NTIS Pub PB82217977)*. Washington, DC: U.S. Government Printing Office, April 1982.

IEEE Vehicular Society Committee on Radio Propagation, "Coverage Prediction for Mobile Radio Systems Operating in the 800/900 MHz Frequency Range," *IEEE Transactions on Vehicular Technology*, pp. 3–72, February 1988.

International Telecommunication Union—Radiocommunication Sector (ITU-R), "ITU-R Recommendation P.527-3, Electrical Characteristics of the Surface of the Earth," *ITU-R Recommendations*. Geneva: International Telecommunication Union, 1992.

International Telecommunication Union—Radiocommunication Sector (ITU-R), "Recommendation P.526-11, Propagation by Diffraction," pp. 1–37, 2009a.

International Telecommunication Union—Radiocommunication Sector (ITU-R), "Recommendation P.530-13, Propagation Data and Prediction Methods Required for the Design of Terrestrial Line-of-sight Systems," pp. 3–4, 2009b.

International Telecommunication Union—Radiocommunication Sector (ITU-R), "Recommendation P.1546-4, Method for Point to Area Prediction for Terrestrial Service in the Frequency Range 30 MHz to 3000 MHz," pp. 3–4, 2009c.

Ishimaru, A., *Wave Propagation and Scattering in Random Media, Volume* 2. New York: Academic Press, pp. 463–492, 1978.

Kerr, D.E., *Propagation of Short Radio Waves*. New York: McGraw-Hill, pp. 404–406, 1951.

Kholod, A., Rohner, M. and Liniger, M., "Comparison of Different Diffraction models Using ITRU-R Study Group 3 Database," *Proceedings of the Second European Conference on Antennas and Propagation*, pp. 1–3, 2007.

Kizer, G., *Microwave Communication*. Ames: Iowa State University Press, pp. 360–362, 612–616, 1990.

Komijani, J., Mirkamamali, A. and Nateghi, J., "Combining Multiple Knife-Edge Diffraction and Ground Reflections for Terrain Path Loss Calculation," *Proceedings of the Fourth European Conference on Antennas and Propagation*, pp. 1–3, April 2010.

Laine, R. U., "Antenna Decoupling as a Major Cause of Nocturnal Fading in Microwave Links," *PIEA Conference Proceedings*, April 1978.

Landee, R.W., D.C. Davis, and A.P. Albrecht 1957. *Electronic Designers' Handbook*. New York: McGraw-Hill, pp. 17–2–17–3.

Liao, D. and Sarabandi, K., "On the Effective Low-Grazing Reflection Coefficient of Random Terrain Roughness for Modeling Near-Earth Radiowave Propagation," *IEEE Transactions on Antennas and Propagation*, Vol. 58, pp. 1315–1324, April 2010.

Liniger, M., Marghitola, M., Rohner, M., Silva, M. A. N. and Costa, E., "Wave Propagation Models: Comparison of Prediction Results with Measurements," *International Conference on Communication Technology*, pp. 1–5, 2006.

Long, M. W., *Radar Reflectivity of Land and Sea*. Boston: Artech, pp. 116–117, 295–297, 2001.

Longley, A. G., "Local Variability of Transmission Loss - Land Mobile and Broadcast Systems," *OT Report 76–87 (NTIS Pub PB254472)*. Washington, DC: U.S. Government Printing Office, May 1976.

Longley, A. G., "Radio Propagation in Urban Areas," *OT Report 78–144 (NTIS Pub PB281932)*. Washington, DC: U.S. Government Printing Office, April 1978.

Longley, A. G. and Reasoner, R. K., "Comparison of Propagation Measurements with Predicted Values in the 20 to 10,000 MHz Range," *ESSA Tech Report ERL 148 - ITS 97 (NTIS Pub AD703579)*. Washington, DC: U.S. Government Printing Office, January 1970.

Longley, A. G. and Rice, P. L., "Prediction of Tropospheric Radio Transmission over Irregular Terrain, A Computer Method-1968," *ESSA Tech Report ERL 79 - ITS 67 (NTIS Pub AD676874)*. Washington, DC: U.S. Government Printing Office, July 1968.

Meeks, M. L., "A Propagation Experiment Combining Reflection and Diffraction," *IEEE Transactions on Antennas and Propagation*, Vol. 30, pp. 318–321, March 1982.

Meeks, M. L., "VHF Propagation over Hilly, Forested Terrain," *IEEE Transactions on Antennas and Propagation*, Vol. 31, pp. 483–489, May 1983.

Millington, G., Hewitt, R. and Immirzi, F. S., "Double Knife-edge Diffraction in Field Strength Predictions," *Proceedings of the IEE, Part C, Monograph 507E*, Vol. 109, pp. 419–429, March 1962.

Mokhtari, H., "A Comprehensive Double Knife-Edge Diffraction Computation Method Based on the Complete Fresnel Theory and a Recursive Series Expansion Method," *IEEE Transactions on Vehicular Technology*, Vol. 48, pp. 589–592, March 1999.

Norton, K. A., "Transmission Loss of Space Waves Propagated Over Irregular Terrain," *Transactions of the IRE Professional Group on Antennas and Propagation*, Vol. 3, pp. 152–166, August 1952.

Preikschat, F. K., "Screening Fences for Ground Reflection Reduction," *The Microwave Journal*, Vol. 7, pp. 46–50, August 1964.

Reed, H. R. and Russell, C. M., *Ultra High Frequency Propagation*. New York: John Wiley & Sons, Inc., pp. 23–182, 229–238, 1953.

Rice, S. O., "Reflection of Electromagnetic Waves from Slightly Rough Surfaces," *The Theory of Electromagnetic Waves*. New York: Interscience Publishers, pp. 351–378, 1951.

Rice, S. O., "Diffraction of Plane Radio Waves by a Parabolic Cylinder," *Bell System Technical Journal*, Vol. 33, pp. 417–504, March 1954.

Rice, P. L., Longley, A. G., Norton, K. A. and Barsis, A. P., *National Bureau of Standards [NBS] Technical Note 101, Volumes I and II (NTIS Pubs AD687820 and AD687821)*. Washington, DC: U.S. Government Printing Office, May 1965 (revised May 1966 and January 1967).

Shkarofsky, I. P., Neugebauer, H. E. J. and Bachynski, M. P., "Effect of Mountains with Smooth Crests on Wave Propagation," *IRE Transactions on Antennas and Propagation*, Vol. 6, pp. 341–348, October 1958.

Silva, M. A. N., Costa, E. and Liniger, M., "Digital Elevation Data and Their Use for Improved Broadcast Coverage and Frequency Utilization," *First European Conference on Antennas and Propagation*, pp. 1–7, 2006.

Tzaras, C. and Saunders, S. R., "Comparison of Multiple-Diffraction Models for Digital Broadcasting Coverage Prediction," *IEEE Transactions on Broadcasting*, Vol. 46, pp. 221–226, September 2000.

Vigants, A., "Microwave Radio Obstruction Fading," *Bell System Technical Journal*, Vol. 60, pp. 785–822, July–August 1981.

Vogler, L. E., "The Attenuation of Electromagnetic Waves by Multiple Knife-Edge Diffraction," *NTIA Report 81–86 (NTIS publication PB 82–139239)*, Institute for Telecommunication Sciences: Boulder, CO, 1981.

Wait, J. R. and Spies, K. P., "Radio Propagation Over a Cylindrical Hill Including the Effect of a Surmounted Obstacle," *IEEE Transactions on Antennas and Propagation*, Vol. 16, pp. 700–705, November 1968.

Weiner, M. M., "Comparison of the Longley-Rice Semi-Empirical Model with Theoretical Models for Coherent Scatter," *US Air Force Systems Command Electronic Systems Division Report ESD-TR-82-133 (NTIS publication ADA 114644)*, Mitre Corporation, Bedford, Massachusetts, 1982.

Weiner, M. M., "Use of the Longley-Rice and Johnson-Gierhart Tropospheric Radio Propagation Programs:0.02-20 GHz," *IEEE Journal on Selected Areas in Communications*, Vol. 4, pp. 297–307, March 1986.

Working Group 2, *OHLOSS [Over the Horizon Loss] Path Loss Computation with OHLOSS Tutorial, Recommendation WG 2.99.052*. Washington, DC: National Spectrum Managers Association, pp. 1–39, October 2000.

Zhao, X. and Vainikainen, P., "Multipath Propagation Study Combining Terrain Diffraction and Reflection," *IEEE Transactions on Antennas and Propagation*, Vol. 49, pp. 1204–1209, August 2001.

14

DIGITAL RECEIVER INTERFERENCE

Fixed point to point microwave radio systems use transmitters and receivers deployed miles apart to transport high speed digital signals. The reliability of the transmission is directly related to the path fade margin (the difference between the normal received signal power and the lowest received signal power that still supports receiver operation). In the absence of external interference, the lowest operational received power (receiver threshold) is determined by the receiver's front end (Gaussian) noise. External interference can cause the receiver threshold to occur at a larger (stronger) received power, thereby reducing the effective path fade margin (and path reliability). Three techniques are commonly used to manage this interference.

14.1 COMPOSITE INTERFERENCE ($\Delta T/T$) CRITERION

The composite interference criterion is a simple interference criterion used when few system details are known. It avoids the need to determine the type of desired and interfering signals. Knowledge of the spectral powers and receiver filtering is usually adequate. In this methodology, the operator first calculates the anticipated receiver front end Gaussian noise power (on the basis of noise figure and receiver bandwidth). Interference is limited to the (receiver-filtered) interference power that does not increase the composite receiver front end noise power beyond a defined limit (this is equivalent to reducing a radio receiver fade margin by the defined limit). Common limit values are 1 dB (interference power level 6 dB smaller than the Gaussian front end noise), 1/2 dB (relative interference power of 9 dB), and 1/4 dB (relative interference power of 12.25 dB). This methodology is commonly used in satellite and fixed microwave interference studies. It avoids any assumptions regarding modulation techniques but tacitly assumes that Gaussian noise plus interference is essentially Gaussian. The following results for T/I indicate that this assumption introduces an error of 2 dB for QAM victim receivers when (similar-modulation) interference power is 6 dB less than the Gaussian noise interference.

Digital Microwave Communication: Engineering Point-to-Point Microwave Systems, First Edition. George Kizer.

14.2 CARRIER-TO-INTERFERENCE RATIO (*C/I*) CRITERION

For licensed radio systems, frequency coordination among all licensed radio-system users is used to limit the impact of radio interference on receivers. Allowable external interference because of intersystem (foreign) radio transmitters is often described as a function of carrier-to-interference ratio (*C/I*) (Working Group 5, 1992).

$$(C/I)(\text{dB}) = \text{T/I (dB)} + \text{FM} - 7$$

$$(C/I)(\text{dB}) = 10 \log \left(\frac{\text{normal received signal (carrier) power}}{\text{receiver front-end (KTB) Gaussian noise}} \right) \tag{14.1}$$

The interference objective is given by

$$\begin{aligned}
I_{\text{coord}} &= \text{coordinated interference objective (dBm)} \\
&= \text{RSL}_{\text{min}} - (C/I)(\text{dB}) \\
&= \text{RSL}_{\text{norm}} - \text{FM(dB)} - (C/I)(\text{dB}) \\
\text{RSL}_{\text{min}} &= \text{receiver threshold (dBm, without interference)} \\
&= \text{lowest receive signal level that achieves a minimum defined} \\
&\quad \text{receiver error rate } \left(\text{typically } 10^{-6} \text{ or } 10^{-3} \text{ BER}\right) \\
\text{RSL}_{\text{norm}} &= \text{normal received signal level (dBm)} && (14.2) \\
\text{FM} &= \text{path fade margin (dB)} = \text{RSL}_{\text{norm}} - \text{RSL}_{\text{min}} && (14.3)
\end{aligned}$$

Note that for satellite systems, FM $\approx$ 0dB. For fixed point to point (relay) systems, 30 dB $<$ FM $<$ 50dB.

Historically *C/I* has been the parameter used for fixed point to point terrestrial microwave systems. It was used in North America for analog systems and is still popular internationally for satellite systems. Currently, ITU-R Recommendations S.741 and SF.766 are the primary *C/I* references. *C/I* requires knowledge of the typical radio RSL and the interference allowed into a victim receiver. Usually, the allowable interference power level into a victim receiver is for co-channel, adjacent channel, and next to adjacent channel interference. The interference objectives take into account the particular demodulation method of the receiver as well as the receiver filtering. The interference objectives are defined so that the victim receiver's effective fade margin is not adversely degraded. Estimation methodologies are well developed for (international) satellite systems but less mature for fixed point to point microwave systems (whose frequency coordination is usually defined and performed by local administrations and not the IRU-R).

14.3 MEASURING *C/I*

Currently there is no industry standard for measuring *C/I* for fixed point to point microwave systems. One might expect it to be measured similarly to *T/I* as follows, but at present, it is not. The following describes the current industry methodology.

C/I is measured as follows:

1. A desired signal digital transmitter is connected to a victim digital receiver through an attenuator. The attenuator is adjusted until the receiver is operating at a nominal RSL (assume −40 dBm).
2. Foreign interference is introduced to the victim receiver at an increasing power level until the receiver exhibits a defined threshold (10^{-6} BER in North America). Assume this occurs at a −70 dBm foreign interference signal into the victim receiver.

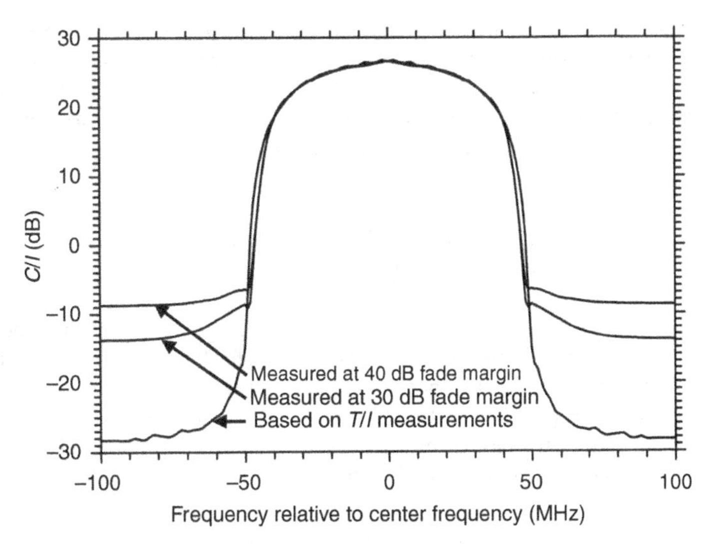

Figure 14.1 Typical *C/I*-FM curves.

3. The *C/I*-FM is the desired signal power minus the undesired signal power referenced to the input of the receiver. For this example, the receiver *C/I* would be $[-40\,\text{dBm} - (-70\,\text{dBm})] = +30\,\text{dB}$.

C/I and *T/I* are similar. However, *C/I* ignores the effect of receiver front end noise. For QAM receivers, if the desired- and interfering-transmitter signals are similar, *C/I* (dB)-FM $\approx$ *T/I* (dB) −7 dB.

The *C/I* measurement can be degraded depending on the assumed "normal" or reference RSL. For an interfering signal with frequency close to the desired signal frequency (e.g., co-channel interference), the *C/I* measurement is not a function of choice of "normal" RSL. However, for adjacent channel and next to adjacent channel interference where receiver filtering is important, the choice of reference RSL is important. If the composite (desired plus interfering signal) power at the receiver input comes within approximately 10 dB of the receiver's 10^{-6} BER overload point, the *C/I* measurements become degraded as a result of receiver preamplifier compression.

While all microwave receivers have broad front end filters, the critical filtering that defines adjacent channel *C/I* values occurs in the receiver after the receiver signal has passed through a front end amplifier. Measuring the effect of adjacent channel signals may require injecting a high power interfering signal that can overload the front end amplifier as well as the receiver analog to digital converter. This effect is illustrated in the following example of a like-modulation *C/I*-FM curve measured using a 256 QAM receiver operating in a 50 MHz radio channel (Fig. 14.1).

For QAM radios, a *C/I*-FM curve can be obtained from a *T/I* curve (see following text) by merely subtracting 7 dB from the *T/I* values—and as the *T/I* is a measure near radio threshold, the results are not sensitive to receiver front end overload.

14.4 ESTIMATING *C/I*

In some cases, specific interference objectives are not available and must be estimated. For analog FM receivers, receiver interference noise can be calculated numerically (Kizer, 1995; TIA/EIA, 1994). For digital receivers, interference objective estimation requires knowledge of the relationship between interference plus Gaussian noise power and receiver error rate as well as the filtering characteristics of the receiver. The filtering characteristics of the receiver are usually specified as the overall power suppression versus frequency for the receiver. The relationship between Gaussian noise power and receiver error rate (S/N or Eb/No vs bit error rate curve) is well known. The assumption is made that the composite Gaussian noise plus foreign interference is approximately Gaussian.

If the conventional assumption is made that the statistical characteristics of interference plus Gaussian noise are approximately Gaussian, the only other factor that is to be determined is the effect of receiver filtering on the interference presented to the receiver. If the normalized spectrum of the interfering signal $s(f)$ and the normalized composite receiver filtering $c(f)$ are known (s and c have nothing to do with S/N), the receiver filtering attenuation A_{Receiver} (dB) is given by

$$A_{\text{Receiver}} = 10\log\left[\int_{-\infty}^{+\infty} s\,(\tau - f)\,c(\tau)\mathrm{d}\tau\right] \leq 0 \tag{14.4}$$

A_{Receiver} is used to quantify the effect of receiver filtering in reducing the effective power of a wide, interfering signal. The term $c(f)$ is the normalized power transfer function (band-pass characteristic) of the victim receiver. If $C(f)$ is the receiver band-pass characteristic expressed in decibels and normalized to 0 dB for receiver center frequency ($f = 0$), $c(f) = 10^{C(f)/10}$. The term f is the frequency of interest relative to center frequency of the desired received signal. The term $s(f)$ is normalized spectral power density (power ratio) of the interfering signal. It is normalized so that its integral is unity.

$$\int_{-\infty}^{+\infty} s(\tau)d\tau = 1 \ \text{(unit area)} \tag{14.5}$$

The *C*/*I* objective may be estimated as

$$(C/I)(\text{dB}) \; - \; [\text{flat fade margin}] = (\text{S/N})(\text{dB}) + A_{\text{Receiver}}(\text{dB}) \tag{14.6}$$

(S/N)(dB) = desired receiver signal to Gaussian noise ratio based on a 10^{-6} or 10^{-3} bit error rate;
A_{Receiver} = receiver filtering attenuation (dB) of the interference signal appearing at the input of the victim receiver $\leq$0 dB.

14.5 THRESHOLD TO INTERFERENCE (*T*/*I*) CRITERION

In many situations (especially in bands where different radio services are used), receiver fade margins are either not known or are different for different radio users (or services) in the same frequency band. In these situations, it is desirable to determine interference objectives that limit receiver degradation to a defined level regardless of user fade margin. Interference is defined in such a way that it only decreases receiver performance by a specified amount. The parameter used in this case is threshold to interference (*T*/*I*). The threshold to interference (*T*/*I*) parameter is popular in North America (TIA/EIA, 1994) for frequency coordination of wideband digital fixed point to point microwave radios. At present, it is not popular internationally.

The interference objective is given by

$$\begin{aligned} I_{\text{coord}} &= \text{coordinated interference objective (dBm)} \\ &= \text{RSL}_{\text{min}} - (T/I)(\text{dB}) \\ &= \text{RSL}_{\text{norm}} - \text{FM(dB)} - (T/I)(\text{dB}) \end{aligned} \tag{14.7}$$

Typically, *T*/*I* is specified for similar signal interference that is co-channel, adjacent channel, or next to adjacent channel.

The use of *T*/*I* simplifies analysis of the effect of interference into a receiver. The victim receiver fade margin is reduced by the number of decibels by which the objective exceeded (as long as this is only a few decibels).

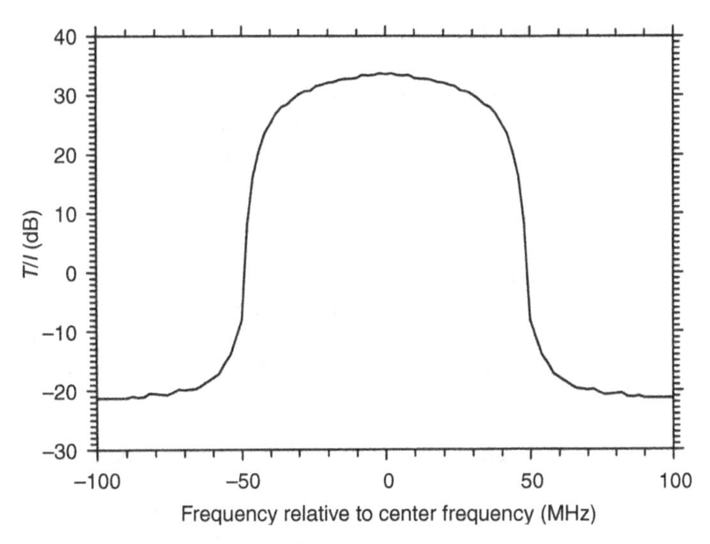

Figure 14.2 Typical *T/I* curve.

T/I is the decibel ratio of desired signal power (when the receiver is at threshold) and interfering power into the victim receiver (which degrades the receiver threshold by 1 dB). If the *T/I* is +30 dB and the receiver threshold signal power is −70 dBm, the maximum allowable interfering power is −100 dBm. If the −100 dBm interfering signal is present, the victim receiver threshold will be −69 dBm.

T/I is measured as follows:

1. A desired signal digital transmitter is connected to a victim digital receiver through an attenuator. The attenuator is increased until the receiver exhibits a defined threshold (10^{-6} BER in North America). Assume this occurs at a −70 dBm received signal level (RSL).
2. The attenuator is reduced 1 dB (the RSL is now −69 dBm). Foreign interference is introduced at an increasing power level until the BER is again the defined threshold(10^{-6}). Assume that the foreign signal power into the receiver is −100 dBm for this example.
3. The faded, desired RSL power (threshold without interference) to intersystem interference ratio (measured in decibels) which causes 10^{-6} BER in the victim receiver is called the *receiver's T/I value* for the specific interference tested. For this example, the receiver *T/I* would be [−70 dBm − (−100 dBm)] = +30 dB.

Figure 14.2 is an example of a like-modulation *T/I* curve for a 256-QAM receiver operating in a 50-MHz channel.

T/I is a function of receiver bandwidth, modulation format, and interfering signal spectral power and frequency. It is very similar to *C/I* except that it uses radio threshold rather than normal RSL.

14.6 WHY ESTIMATE *T/I*

In all cases, the most desirable way to determine *T/I* is via direct measurements. When frequency coordinators are estimating interference within a common user community, common transmitter and receiver measurements can be used to insure compliance with established interference limits. However, at present, many frequency bands are used with so many different capacity (and bandwidth) radios that actual interference measurements are often unavailable. In addition, regulatory agencies are often asked to share or subdivide frequency allocations among different radio services. Often, little is known about the characteristics of the new services except their proposed power and spectral density versus frequency. Practical methods are needed to deal with this real-world situation, where actual measurements are simply not possible or at least not practical.

14.7 *T/I* ESTIMATION—METHOD ONE

Sometimes, nothing is known about the victim receiver and the interference, other than the type of modulation and band-pass characteristics used by the receiver and the spectral power density characteristics (e.g., spectrum mask) of the interference. *T/I* may be estimated as

$$(T/I)(\text{dB}) = (\text{S/N})(\text{dB}) + A_{\text{Receiver}}(\text{dB}) + T(\text{dB}) \tag{14.8}$$

$(\text{S/N})(\text{dB})$ = average signal to Gaussian noise power (dB) required to achieve a 10^{-6} BER);
$T(\text{dB})$ = value depending on the statistical characteristics of the interfering signal;
$A_{\text{Receiver}}(\text{dB})$ = interfering signal attenuation due to receiver passband filtering ≤ 0 dB;
$\approx H_{\text{CW}}$ (dB) (described later) normalized to have value 0 dB for center frequency of the receiver

A_{Receiver} is used to quantify the effect of receiver filtering in reducing the effective power of a wide, interfering signal.

$$A_{\text{Receiver}} = 10\log\left[\int_{-\infty}^{+\infty} s(\tau - f)\, c(\tau)\text{d}\tau\right] \tag{14.9}$$

The term $c(f)$ is the normalized power transfer function (band-pass characteristic) of the victim receiver. If $C(f)$ is the receiver band-pass characteristic expressed in decibels and normalized to 0 dB for receiver center frequency ($f = 0$), $c(f) = 10^{C(f)/10}$.

The term $s(f)$ is normalized spectral power density of the interfering signal. It is normalized so that its integral is unity.

$$\int_{-\infty}^{+\infty} s(\tau)\text{d}\tau = 1 \quad \text{(unit area)} \tag{14.10}$$

Typical values of S/N and *T* are given in Table 14.B.1. The S/N versus 10^{-6} BER values assume receiver interference is simply the receiver front end Gaussian noise. These values are based on ITU-R Recommendations F.1101 (ITU-R, 1994), SF.766 (ITU-R, 1992), and Kizer (1995).

Coco (1988, p. 982, Eq. 3) has shown that BER performance of a QAM receiver subjected to an arbitrary, interfering signal is only a function of interfering power at the digital demodulator and the statistical characteristics (probability density function) of the interfering signal. We will use the *T* factor as a practical measure of the interfering signal's statistical characteristics. Interfering power will be determined by convolving the interfering signal with a measurement indicative of the power versus frequency response of the victim receiver. All research has been for co-channel interference. The assumption will be made that filtered interference statistical characteristics are the same as those of unfiltered interference. This is always true for Gaussian noise. Actual measurements imply that this is also true for modulated spectrums.

The continuous wave (CW) *T* values assume that the interfering signal is a sine wave. The similar modulation *T* values assume that the interference is a signal similar to the desired received signal but emitted from a foreign transmitter. The *T* values are based on results of Prabhu and Rosenbaum (phase-shift keying (PSK)) (Prabhu, 1969, 1982; Prabhu and Salz, 1981; Rosenbaum, 1969), Coco, Prabhu, Tobin, and Yao (QAM) (Coco, 1988; Prabhu, 1971, 1980, 1981, 1982; Yao and Tobin, 1976), and Kabal and Pasupathy (1975) (partial response (PR)).

Similar interference *T* values for quadrature partial response (QPR) were not located. Lucky et al. (1968) have shown that the basic Gaussian noise BER relationship holds over a wide range on interference cases for simple modulation formats. BER performance of linear quadrature systems such as QAM (including trellis coded modulation and multilevel modulation) and QPR may be calculated by investigating either quadrature and ignoring the other. This is the same as investigating the baseband equivalent (pulse amplitude modulation or PAM) of these systems (assuming perfect carrier detection and ignoring phase jitter). Proakis (1983, Eq. 4.2.137 and Eq. 6.2.61) demonstrated that at least for Gaussian noise

TABLE 14.1 Actual *T/I* Measurements

Receiver	S/N, dB No Interference	*T/I*, dB CW Interference	*T/I*, dB Similar Interference
16 QAM	20	25	27
64 QAM	26	30	32
256 QAM	32	36	38

interference, the performance of PR signaling (the baseband equivalent of QPR) was essentially the same as for the equivalent PAM signal (the baseband equivalent of QAM). Assuming similarity of this relationship for similar modulation interference, PAM results were used to estimate QPR-similar modulation interference.

Unfortunately, we cannot always predict the statistical characteristics of interfering signals. If they were similar to Gaussian noise, the *T* value would be 6 dB. If the interfering signal's "peak to average power ratio" is less than that of Gaussian noise (a CW sine wave being the most compact case), the *T* value would be smaller than 6 dB. For signal "peakier" than Gaussian noise, *T* values would be larger than 6 dB (high order QAM being an obvious example). What represents the statistical characteristics of "worst-case" interference into a receiver is difficult to predict. However, the working hypothesis used to create Table 14.B.1 was to assume that the worst-case *T* value was 1 dB greater than the similar modulation *T* value or 6 dB, whichever was greater. This assumption was based on informal discussions with various experienced frequency coordinators.

The sum of several interfering signals will not have the same statistical characteristics as a single interfering signal. As noted in ITU-R Recommendation SF.766 (ITU-R, 1992), Annex 21, paragraph 2.4.1, these cases tend to approach the Gaussian noise case (in accordance with the statistical central limit theorem). For Gaussian noise, the *T* value is 6 dB. Multiple interferences should tend to the value that is consistent with the preceding concept of worst-case *T* value.

Many modern digital radios use QAM with trellis or multilevel coding. Informal tests with these radios have shown that although these coding methods definitely improve the Gaussian and interference rejection, the difference between the signal-to-interference ratio for various types of interference (*T*-factors) remains the same as for plain QAM.

Table 14.B.1 was created from many difference sources. Overlap of results showed that different researchers arrived at different results for similar types of interferences. Differences of up to 1 dB were noticed. In addition, most current wideband digital radios use some form of QAM. Virtually all researchers assume coherent demodulation, while most practical implementations use differential demodulation (to achieve independence to constellation rotation). Usually, some form of error correction (e.g., Reed–Solomon coding) is added to recover the threshold degradation due to differential demodulation.

The preceding approach was tested using tests of digital microwave radio receivers. The following co-channel *T/I* measurements were made (Table 14.1).

The readings were compensated for receiver filtering loss with center frequency CW measurements being ignored (to ignore clock recovery issues). These results are in reasonable agreement with theoretical values as calculated by the preceding formula and Table 14.B.1.

14.8 *T/I* ESTIMATION—METHOD TWO

Sometimes, the victim receiver is well-characterized but little is known about the interfering signal except its spectral power density (e.g., its *spectrum mask*"). *T/I* may be estimated as follows:

A CW *T/I* curve (*T/I* vs interfering, unmodulated carrier frequency offset relative to desired signal center frequency) is typically used to determine the effect of FM signal interference into digital receivers. Since the CW signal is narrow in frequency, the CW *T/I* curve represents the signal rejection characteristics of a digital receiver. Of course, it also represents the sensitivity of the receiver demodulator to interference.

In theory, all one needs to determine a digital receiver's *T/I* for an arbitrary interfering signal is the receiver's CW *T/I*, a plot of power versus frequency for the interfering spectrum, and a co-channel *T/I* measurement using the interfering signal. To determine the *T/I* at the adjacent channel's frequency offset, one merely convolves the interfering spectrum with the digital receiver's CW *T/I* curve. Of course, the interfering spectrum would be offset by any frequency difference between the desired and interfering signals. The resultant measurement would then be offset by the difference between the co-channel *T/I* for the sine wave (CW signal) and the co-channel *T/I* for the arbitrary signal. The result is *T/I* for the arbitrary signal interference centered at the adjacent channel frequency.

If interference into a digital receiver is to be calculated, the following three items are required.

The first is a CW (sine wave interference) *T/I* curve C (dB). For purposes of the following calculations, the dB curve $C(f)$ is assumed to be converted to a relative power curve $c(f) = 10^{C(f)/10}$. The variable f is the (interference offset) frequency relative to the desired signal center frequency. Notice that this curve is not normalized. It typically is usually in the range of 21–32 dB at radio receiver center frequency ($f = 0$) for a QAM radio.

The second item is normalized spectral power density $s(f)$ of the interfering signal. It is normalized so that its integral is unity.

$$\int_{-\infty}^{+\infty} s(\tau)\mathrm{d}\tau = 1 \quad \text{(unit area)} \tag{14.11}$$

If the normalized spectrum is not available, the normalized spectrum will be assumed to be the (normalized) spectrum mask imposed by the appropriate regulatory agency.

The third item is a factor F (dB) that characterizes the differences in statistical characteristics between CW interference (plus Gaussian noise) and the interfering signal (plus Gaussian noise).

$$F(\mathrm{dB}) = T\text{-value(“worst-case” interference)} - T\text{-value(CW interference)} \tag{14.12}$$

The F value is used to convert the CW *T/I* to the *T/I* of a modulated spectrum. The F values from Table 14.B.1 vary from 1 to 4 dB. For very complex interference signals or for interference composed of many similar signals, the limiting case of Gaussian noise is approached and F is 2 dB.

Using the preceding three items, the interfering signal *T/I* may be calculated by convolving the normalized interfering spectrum $s(f)$ with the CW *T/I* curve, $c(f)$, and adding the statistical correction factor, F.

$$\frac{T}{I(f)} = F(\mathrm{dB}) + H_{\mathrm{CW}} \tag{14.13}$$

$$H_{\mathrm{CW}} = 10\log\left[\int_{-\infty}^{+\infty} s\,(\tau - f)\,c(\tau)\mathrm{d}\tau\right] \tag{14.14}$$

where f is the frequency of interest relative to center frequency of the desired received signal.

The preceding methodology was tested by laboratory measurements. QAM transmitters and receivers were used to measure adjacent channel *T/I*. Multiple frequency offset *T/I* data was measured. The broadband receiver *T/I* (solid lines) was measured on two different receivers (1 DS3 and 3 DS3 64 QAM) using two different bandwidth interfering transmitters (1 DS3 and 3 DS3 64 QAM). Simulated *T/I* curves (dotted lines) were calculated using the following method:

$$\text{Simulated modulated } (T/I)(\mathrm{dB}) = H_{\mathrm{CW}}(\mathrm{dB}) + 2\,\mathrm{dB} \tag{14.15}$$

The following graphs display the results (Fig. 14.3 and Fig. 14.4).

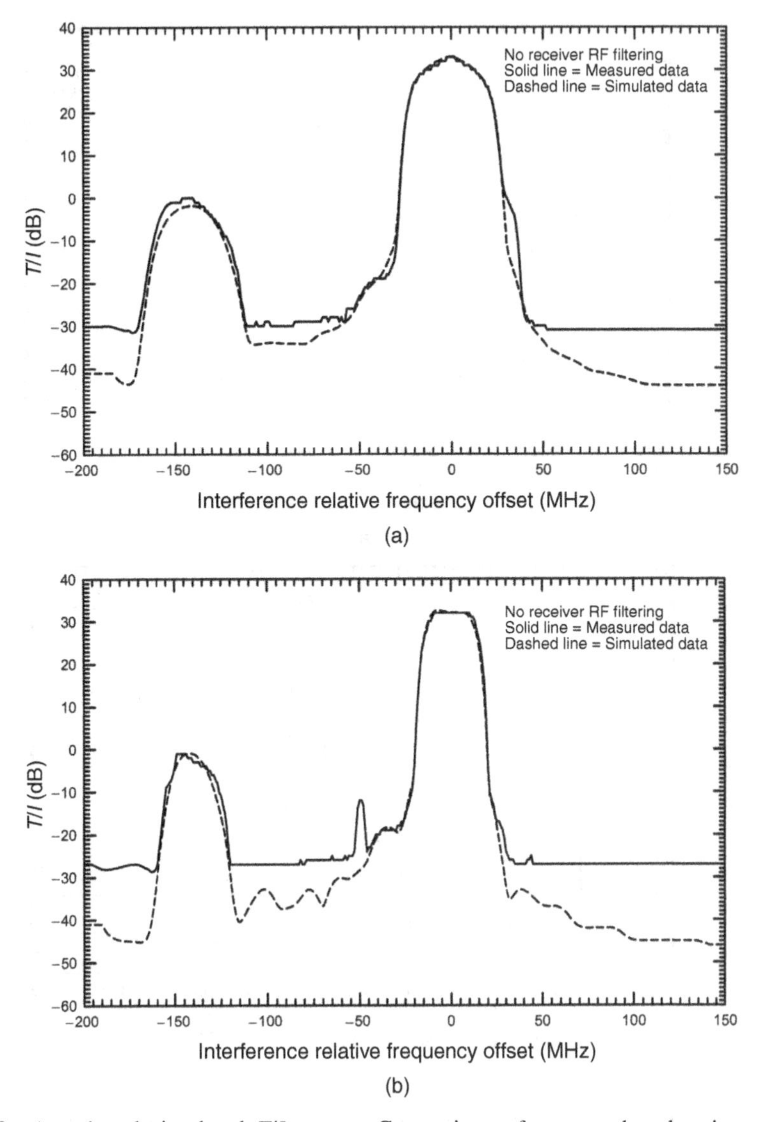

Figure 14.3 Actual and simulated *T/I* curves. Comparison of measured and estimated radio *T/I* curves: (a) 3 DS-3 64 QAM receiver, 3 DS-3 64 QAM interference; (b) 3 DS-3 64 QAM receiver, 1 DS-3 64 QAM interference.

The actual and simulated measurements were made on receivers without receiver RF filters so that values for large interfering signal frequency offsets would be measurable. In reality, (measurements made with receiver RF filters) *T/I*s performed on receivers with RF filters would cause the actual and simulated values to slope down dramatically for offsets far from the 0 MHz interference offset. To create realistic, simulated *T/I* curves, the interfering signal spectrum must first be filtered by the receiver RF filter bandpass characteristic. Then, the interfering signal spectral power density is convolved with the receiver CW *T/I* curve.

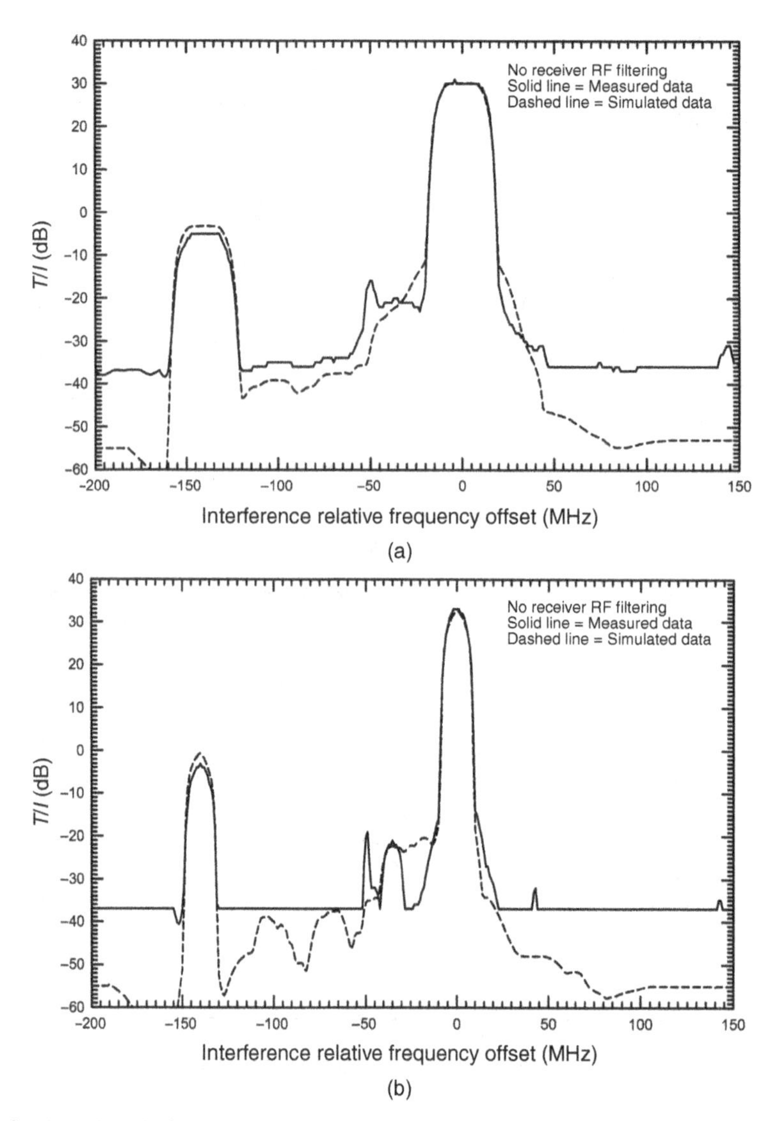

Figure 14.4 Actual and simulated *T/I* curves. Comparison of measured and estimated radio *T/I* curves: (a) 1 DS-3 64 QAM receiver, 3 DS-3 64 QAM interference; (b) 1 DS-3 64 QAM receiver, 1 DS-3 64 QAM interference.

The horizontal line of measured values represents measurement equipment limitations. The large peak to the left of the main peak is a heterodyne receiver-spurious response due to down-converting the RF signal to 70 MHz for further receiver processing. With the receiver RF filter present, this spurious response would not be observed. The simulated and actual measurements show good agreement within the limits of the measurement equipment.

The preceding methodology works well as long as receiver preamplifier compression (nonlinearity) is not a problem. Divergence between these estimates and actual measurements can be expected if measurements include receiver filters and the composite input to the receiver comes within approximately 10 dB of the receiver overload point.

14.9 CONCLUSION

The preceding methodology was developed by the TIA Fixed-Point-to-Point Microwave Committee TR 14–11, at the request of various PCS industry groups. The PCS groups needed a general interference coordination procedure for sharing the fixed point to point microwave bands. This methodology was initially used at 2 GHz and later expanded to all FCC-managed frequency bands to satisfy the requirements of TIA Bulletin 10-F (TIA/EIA, 1994). Despite the need for some assumptions, this methodology has worked well for calculating the interfering effect of a wide range of interfering signals into many of the receivers listed in Table 14.B.1 in real-world situations.

14.A APPENDIX

14.A.1 Basic 10^{-6} Threshold for Gaussian (Radio Front End) Noise Only

If receiver characteristics are known, a reasonable estimate of receiver threshold S/N may be made as follows:

$$\begin{aligned} \text{S/N for } 10^{-6} \text{ BER (dB)} &= 114 - 10\log(\text{bandwidth of receiver (MHz)}) \\ &\quad + \text{receiver } 10^{-6}\,\text{BER threshold (dBm)} \\ &\quad - \text{noise figure(dB)} + \text{coding gain(dB)} \end{aligned} \tag{14.A.1}$$

The bandwidth of the receiver is the baud rate of the radio. For current generation QAM-derived radios, this is typically 85% of the radio channel bandwidth.

The RSL threshold, without external interference, at which a given receiver achieves 10^{-6} BER, sets the fade margin of a radio path. This most basic threshold is required for initial path design. Sometimes, it is desirable to estimate the 10^{-6}-BER threshold for a new receiver (e.g., new data rate or frequency band). If the new radio has similar basic architecture to an existing radio with published parameters, the new radio's 10^{-6}-BER threshold may be estimated as follows:

$$\begin{aligned} 10^{-6}\,\text{Threshold(dBm)(new receiver)} &= 10^{-6} \text{ Threshold (dBm) (old receiver)} \\ &\quad + \text{(S/N) for } 10^{-6} \text{ threshold (dB) (new receiver)} \\ &\quad - \text{(S/N) for } 10^{-6}\text{threshold(dB)(old receiver)} \\ &\quad + \text{noise figure (new receiver)} \\ &\quad - \text{noise figure (old receiver)} \\ &\quad + 10 \log\left(\frac{\text{bandwidth of new receiver}}{\text{bandwidth of old receiver}}\right) \end{aligned} \tag{14.A.2}$$

Of course, the noise figures must be specified at the same physical wave guide location. If the receiver bandwidths are not available, substituting the respective radio's channel bandwidth will give essentially the same result.

If an existing radio is not available as a benchmark, a reasonable estimate may be made as follows:

$$\begin{aligned} 10^{-6} \text{ Threshold (dBm)} &= \text{S/N for } 10^{-6} \text{ threshold (dB)} \\ &\quad + \text{noise figure (dB)} \end{aligned}$$

$$+ 10 \log (\text{bandwidth of receiver (MHz)})$$

$$- \text{coding and FEC gain (dB)} - 114 \qquad (14.A.3)$$

Typical microwave radio receiver noise figures range from 2 to 5 dB. Coding and forward error correction (FEC) gain is typically 2–5 dB for a 10^{-6} BER.

14.A.2 Using a Spectrum Mask as a Default Spectrum Curve

In many practical situations, many different interfering signals are encountered. If band sharing proposals are being evaluated, interference guidelines will be established long before any interfering systems are defined. Usually, all that is known is that the interfering spectrum must conform to a spectrum mask. If that mask is normalized to integrated to unity (power) area, it can be used as a conservative power spectrum estimate.

The following is an example for fixed point to point microwave systems operating under the US FCC jurisdiction. The FCC rules 47 CFR 101.111 (FCC, 2004) define seven different spectrum masks (emission limitations). We choose the mask for digital fixed point to point operation below 15 GHz (47 CFR 101.111 (a) (2) (I) and (ii)). This defines the power A (dB) measured in a 4-kHz bandwidth as follows:

$$\begin{aligned} A(\text{dB}) &= -50.0 \text{ for } 50 < P \leq [68.75 - 12.5\log(B)] \\ &= +5 - 0.8P - 10\log(B) \text{ for } [68.75 - 12.5\log(B)] < P \leq [106.25 - 12.5\log(B)] \\ &= -80.0 \text{ for } [106.25 - 12.5\log(B)] < P \leq 250 \\ &= -43.0 \text{ for } P > 250 \end{aligned} \qquad (14.A.4)$$

A (dB) = power measured in a 4-kHz bandwidth relative to total transmitted power;
B = channel bandwidth (MHz);
P = absolute value of $[100(f - f_0)/B]$;
f_0 = channel center frequency (MHz);
f = frequency of interest (MHz);

Most digital radios operating below 15 MHz employ QAM (or a derivative). These modulators nearly fill the channel bandwidth with a uniform spectrum. If we assume that the spectrum within the radio channel is uniform (flat), we obtain the following relationship:

$$A(\text{dB}) = 10\log\left(\frac{4000}{B}\right) \quad \text{for } 0 < P < 50 \qquad (14.A.5)$$

When a measurement bandwidth other than 4 kHz is needed (e.g., 3.1 kHz), all the preceding values of A have the following factor added to them.

$$\text{Bandwidth correction factor} = 10\log\left(\frac{\text{measurement bandwidth (kHz)}}{4}\right) \qquad (14.A.6)$$

When a default interfering signal spectrum is required, the preceding A (dB) values are used as calculated. Of course, "log (x)" denotes the common (base 10 or Brigg's) logarithm of x in the preceding equation.

14.B APPENDIX: RECEIVER PARAMETERS

TABLE 14.B.1 Radio Receiver Interference Parameters

Receiver Type and Number of Symbols	S/N for 10^{-6} BER	*T* Value		
		CW or FDM-FM	Similar Modulation	Worst Case
CPSK				
2	13.6	5.4	5.4	6.4
4	13.8	5.1	5.1	6.1
8	19.3	5.0	5.0	6.0
16	25.1	4.9	4.9	6.0
DPSK				
2	15.9	5.8	5.8	6.8
4	16.1	6.1	6.1	7.1
8	22.1	6.4	6.4	7.4
16	28.1	5.8	5.8	6.8
QPR				
9	17.6	4.3	4.3	6.0
25	22.3	4.3	5.9	6.9
49	24.6	4.3	6.4	7.4
121	28.5	4.3	6.6	7.6
QAM (coherent demodulation)				
4	13.5	4.3	4.8	6.0
16	20.2	4.3	5.3	6.3
32	23.2	4.3	6.1	7.1
64	26.2	4.3	6.5	7.5
128	29.1	4.3	6.8	7.8
256	32.0	4.3	7.0	8.0
512	34.9	4.3	7.2	8.2
1024	37.7	4.3	7.3	8.3
TCM 2D				
32	20.9	4.3	6.1	7.1
128	27.2	4.3	6.8	7.8
256	30.3	4.3	7.0	8.0
512	33.3	4.3	7.2	8.2
TCM 4D				
32	19.9	4.3	6.1	7.1
128	26.2	4.3	6.8	7.8
256	29.3	4.3	7.0	8.0
512	32.3	4.3	7.2	8.2

Abbreviations: CPSK, Phase-shift keying, coherent demodulation; DPSK, phase-shift keying, differential demodulation; QPR, quadrature partial response; QAM, quadrature amplitude modulation; TCM 2D, Trellis coded modulation, two-dimensional; TCM 4D, Trellis coded modulation, four-dimensional.

REFERENCES

Coco, R. A., "Symbol Error Rate Curves for M-QAM Signals with Multiple Cochannel Interferers," *IEEE Transactions on Communications*, Vol. 36, pp. 980–983, August 1988.

FCC, "Part 101—Fixed Microwave Services, Subpart C—Technical Standards", *Code of Federal Regulations*, Title 47, Chapter I, *Federal Communications Commission*, Part 101, paragraph 111), Washington, D.C., 2004.

ITU-R, "Methods for determining the effects of interference on the performance and the availability of terrestrial radio-relay systems and systems in the fixed-satellite service", *Recommendation SF.766*, Geneva, 1992.

ITU-R, "Characteristics of digital fixed wireless systems below about 17 GHz", *Recommendation F.1101*, Geneva, 1994.

Kabal, P. and Pasupathy, S., "Partial-Response Signaling," *IEEE Transactions on Communications*, Vol. 23, pp. 921–934, September 1975.

Kizer, G. M., "Microwave Radio Communication," *Handbook of Microwave Technology*, Ishii, T. K., Editor. San Diego: Academic Press, pp. 449–504, 1995.

Lucky, R. W., Salz, J. and Weldon, Jr., E. J., *Principles of Data Communication*, New York: McGraw-Hill, pp. 125–127, 1968.

Prabhu, V. K., "Error Rate Considerations for Coherent Phase-Shift Keyed Systems with Co-Channel Interference," *Bell System Technical Journal*, Vol. 26, pp. 743–767, March 1969.

Prabhu, V. K., "Some Considerations of Error Bounds in Digital Systems," *Bell System Technical Journal*, Vol. 50, pp. 3127–3151, December 1971.

Prabhu, V. K., "The Detection Efficiency of 16-ary QAM," *Bell System Technical Journal*, Vol. 59, pp. 639–656, April 1980.

Prabhu, V. K., "Cochannel Interference Immunity of High Capacity QAM," *Electronics Letters*, Vol. 17, pp. 680–681, September 1981.

Prabhu, V. K., "Error Rate Bounds for Differential PSK," *IEEE Transactions on Communications*, Vol. 30, pp. 2547–2550, December 1982.

Prabhu, V. K., "Modified Chernoff Bounds for PAM Systems with Noise and Interference," *IEEE Transactions of Information Theory*, Vol. 28, pp. 95–99, January 1982.

Prabhu, V. K. and Salz, J., "On the Performance of Phase-Shift-Keying Systems", *Bell System Technical Journal*, Vol. 60, pp. 2307–2343, December 1981.

Proakis, J. G., *Digital Communications*, New York: McGraw-Hill, pp. 178–183, 346–349, 1983.

Rosenbaum, A. S., "PSK Error Performance with Gaussian Noise and Interference," *Bell System Technical Journal*, Vol. 48, pp. 413–442, February 1969.

TIA/EIA, "Interference Criteria for Microwave Systems," *TIA/EIA Telecommunications Systems Bulletin 10-F*, pp. B-1–B-8, June 1994.

Working Group 5, Report and Tutorial, Carrier to Interference Objectives, *Recommendation WG 5.92.008*, Washington: National Spectrum Managers Association, 1992.

Yao, K. and Tobin, R. M., "Moment Space Upper and Lower Error Bounds for Digital Systems with Intersymbol Interference," *IEEE Transactions on Information Theory*, Vol. 22, pp. 65–74, January 1976.

15

NETWORK RELIABILITY CALCULATIONS

Product quality is a key factor in making purchase decisions. Quality is often defined (Rausand and Hoyland, 2004) as "The totality of features and characteristics of a product or service that bear on the ability to satisfy stated or implied needs." Quality comparisons among similar products are subjective. However, everyone relates to initial cost and lifetime operating costs. The operating cost of an item of telecommunications equipment can be greater than the item's original purchase price. For this reason, customers are concerned with the number of times something must be repaired; it directly reflects in their operating costs. Repairs are related to the reliability of systems and the constituent equipment. Customer satisfaction is typically based on the total system being in an unfailed (or reliable) state at a satisfactory cost. Evaluating this cost involves reliability analysis.

Reliability is often defined (Rausand and Hoyland, 2004) as "The ability [stated as a probability] of an item to perform a required function, under given environmental and operational conditions and for a stated period of time" (Rausand and Hoyland, 2004), "the probability that an item will operate when needed," or "the average fraction of time that a system is expected to be in an operating condition" (Bellcore, 1986). Despite these definitions, the concept of reliability is subject to interpretation. Everyone agrees that failure to operate at all constitutes a failure. It is seldom clear if out of specification (degraded operation), module pull with no trouble found during repair, extraneous operation, or failure to change operating modes also constitute failure. Failure may be caused by the telecommunications equipment itself. However, cables and connectors, equipment backplanes, design deficiencies, manufacturing defects, maintenance personnel mistakes, and loss of system power are common causes of failure, which are difficult to quantify. It is also seldom clear if failure to monitor or manage the equipment via a remote network management function constitutes a failure.

In the following discussion, an inherent assumption is that the equipment under discussion are provided with uninterrupted support facilities, such as power, air conditioning and/or heating, and protected physical environment. Maintenance actions on or near the equipment are assumed not to affect the function of the equipment.

System failures are both transmission media and transmission equipment dependent. For radio systems, Bell Labs and Telcordia believe the media and equipment to be equally failure prone. For SONET optical systems, Telcordia believes the media (fiber cable) fails twice as often as the equipment. Other sources suggest that the media is significantly more failure prone than modern hardware and software.

Experience has shown that hardware-, software-, and operator-induced failures are all approximately equally likely. The current industry approach is to focus on catastrophic failure of a hardware component,

Digital Microwave Communication: Engineering Point-to-Point Microwave Systems, First Edition. George Kizer.

subsystem, or system as the most significant measure of the equipment's or system's reliability. This is obviously overly simplistic but reflects the current lack of standard measures of software or operator reliability.

15.1 HARDWARE RELIABILITY

Equipment availability is the percentage of total time the equipment performs its intended function. Telcordia (Telcordia Technologies, 2009) states "The availability of an entity is the probability that it can perform a required function at a given instant of time. It can be interpreted as the long-term fraction or percentage of time that it performs as intended." Equipment reliability is defined as the average number of equipment failures that occur per unit of time. Equipment availability can be improved up to a point by using more reliable components. Beyond this, the equipment is made more reliable by designing redundancy into the equipment. This greatly increases the time between (redundant) equipment failures, but, since the number of components is increased, the reliability (time between individual failures and related maintenance actions) of the equipment is reduced. Measures of both concepts are discussed.

For most electronic system reliability calculations, analysis is based on component failures without regard to the underlying physical failure mechanisms. This actuarial (as opposed to a physical or structural reliability) approach is based on six assumptions:

- A piece of equipment can be in only one of two states: failed or not failed.
- The failure of any component is independent of the failure of other components.
- The probability of equipment failure is quite small (allowing second-order probability factors to be ignored).
- The equipment is assumed to have aged beyond the infant mortality period (it is mature) but is not worn out.
- The equipment failure statistics are constant.
- The repaired equipment is "good as new."

It is not feasible to predict the lifetime or degradation rate for any individual electronic component. It is, however, possible to treat large populations of components probabilistically and thereby predict average statistical characteristics. Statistical measures of system reliability, based on such probabilistic analysis, form the basis for making equipment system reliability comparisons.

The following are the typical reliability definitions (the formula factors are defined later):

$$\begin{aligned} A &= \text{equipment availability} = \frac{\text{operational time}}{\text{total time}} \\ &= \frac{\text{operational time}}{(\text{operational time} + \text{outage time})} \\ &= \frac{\text{MTBF}}{(\text{MTBF} + \text{MTTR})} \\ &= \frac{1}{[1 + \lambda \times \text{MTTR})]} = \frac{1}{[1 + (\text{FITS} \times 10^{-9} \times \text{MTTR})]} \\ &= \frac{10^{+9}}{[10^{+9} + (\text{FITS} \times \text{MTTR})]} = \frac{1}{\left[1 + \left(\dfrac{\text{MTTR}}{\text{MTBF}}\right)\right]} \end{aligned} \tag{15.1}$$

$$\begin{aligned} U &= \frac{\text{MTTR}}{(\text{MTBF} + \text{MTTR})} \\ & 0 \le A \le 1 \text{ and } 0 \le U \le 1 \text{ and } A + U = 1 \end{aligned} \tag{15.2}$$

U = equipment unavailability = outage time/total time;
MTBF(hours) = mean time between (device two way) failure;
MTTR(hours) = mean time to restore (mean downtime).

$$\text{MTTR} = \text{RT} + \text{TT} + [(1 - \text{PS}) \times \text{TR}] + \text{MTR} \tag{15.3}$$

RT(hours) = mean time to detect, diagnose, and report an alarm to the appropriate repair person;
TT(hours) = mean travel time;
PS = probability of having a working spare module;
TR(hours) = mean time to obtain a spare module from an outside source if no spare is available locally;
MTR(hours) = mean time to replace (or repair) failed module and restore equipment.

These formulas assume that a reliable network operations center monitoring the network contacts the appropriate repair person promptly and without error. The repair person is assumed to be fully qualified, always available, and always have a working spare. That person performs maintenance without error. Each repair is assumed to make the system "as good as new." Sparing philosophy, maintenance of personnel training and staffing, site access, and effective fault management systems will significantly influence the actual MTTR achieved. These factors are critical to achieving the availability of which the network is capable. As can be inferred from Equation 15.1, MTBF (equipment quality) and MTTR (maintenance quality) are critical to high availability.

Some authors use the term *MTR* for the MTTR function. MTTR is then defined as the actual time to perform the repair (after being notified, obtaining the spare unit, and reaching the site). This is not common usage and the use of MTTR to encompass all restoral action time is preferred. In evaluating equipment specifications, it is critical that all specifications use the same MTTR value. If not, the comparisons can be quite misleading. To simplify comparing various vendors, Telcordia has suggested that MTTR be standardized to 2 h for central office equipment with separable modules, 4 h for remote site equipment with separable modules, and 48 h if the equipment must be replaced as a complete unit. ITU-T also uses the 4 h MTTR (International Telecommunication Union—Telecommunication Standardization Sector (ITU-T), 2003).

Rare equipment failures meeting the previous assumptions are modeled as a homogeneous Poisson process with exponential failure distribution. Subsystem (module) failure rate is assumed to be constant and defined by λ.

$$\lambda\,(\text{failures/hour}) = \frac{1}{\text{MTBF}} = \text{FITS} \times 10^{-9} \tag{15.4}$$

Device reliability $R(t)$ is the probability that the device will perform without failure over the time period 0 to t (h) when the device is operated within its intended environment. $R(t)$, the time integral of the failure probability density function $\lambda e^{-\lambda t}$, is also called the *survivor probability*.

$$R(t) = e^{-\lambda t} \tag{15.5}$$

e = Napier's constant (Euler's number) $\cong$ 2.7182818.

An interesting aspect of the Poisson distribution is, as it is exponential, most failures occur earlier in time than the MTBF. On average, half of the units will fail by 0.61 MTBF, the median failure time. By the MTBF time, it is expected that 63% (essentially 2/3) of the modules will have failed.

$$\begin{aligned}\text{FITS} &= \text{failures in time} = \text{failures per billion hours (per 114,077 years)}\\ &= \left[\frac{\text{average yearly outage time (min/year)}}{\text{525,960 (min/year)}}\right] \times \left(\frac{10^{+9}}{\text{MTTR}}\right)\\ &= \frac{[10^{+9} \times (1 - A)]}{(A \times \text{MTT}R)} = \frac{10^{+9}}{\text{MTBF}}\end{aligned} \tag{15.6}$$

$$\text{FITS(measured)} = \frac{\{[(\text{failures/year}) + 1] \times 10^{+9}\}}{[(\text{field population}) \times (\text{hours/year})]}$$

$$= \left[\frac{(\text{returns/year})}{\text{field population}}\right] \times \left(\frac{10^{+9}}{8766}\right) \tag{15.7}$$

$$\text{MTBF(hours)} = \text{mean time between (device two way) failure in hours}$$

$$= \text{integral over all time of the reliability function } R(t) = 1/\lambda$$

$$= \frac{(A \times \text{MTTR})}{(1 - A)} = \frac{10^{+9}}{\text{FITS}}$$

$$= \frac{8766}{\left[\frac{(\text{returns/year})}{\text{field population}}\right]}$$

$$= \left[\frac{525{,}960\ (\text{min/year})}{\text{average yearly outage time (min/year)}}\right] \times \text{MTTR(hours)} \tag{15.8}$$

MTBF represents the statistics of rare random failures of the entire population of similar devices. Mean time to failure (MTTF) or mean life (ML) are terms used to describe the average period until the device is worn out. They should not be confused with MTBF.

$$\text{MTBO(hours)} = \text{mean time between outage (subsystem or network element failure) in hours}$$

$$= \text{MTBF for a subsystem or network element}$$

$$= \frac{(A \times \text{MTTR})}{(1 - A)} = \frac{10^{+9}}{\text{FITS}} \text{(using system values rather than component values)}$$

$$= \left[\frac{525{,}960\ (\text{min/year})}{\text{average yearly system outage time (min/year)}}\right] \times \text{MTTR(hours)} \tag{15.9}$$

MTBO is not a term found in reliability literature. It is a colloquial term used to differentiate between the failure of a single module and the failure of a complete network element or communication system. Mathematically, MTBO is treated in the same manner as MTBF. If the failure of interest is of a complete (sub-)system, the term *MTBO* is preferred. If the failure of a (nonprotected) component or device is considered, the term *MTBF* is preferred. If failure of a complete subsystem is of interest, even if that failure does not cause an overall device failure (e.g., a module in a redundant device can fail, but the overall device function has not failed), the term *MTBF* is sometimes used to describe the subsystem.

The MTBO/MTBF concept works well to estimate equipment hardware availability. Overall hardware reliability can also be described using average annual maintenance actions or average annual downtime. Customers looking at life cycle cost of ownership ask for data on maintenance actions per year. Each action costs them for labor and additionally, if out of warranty, repair charges. Maintenance actions are really tied directly to the total failures in time (FITS) of the entire system, regardless of whether the cards are in series or parallel. In general, the more reliable the product is designed via redundancy, the more the total maintenance cost goes up. That is unfortunate but unavoidable.

$$\text{MA} = \text{(average annual) maintenance actions} = \text{(hours in a year)} \times \text{(failures per hour)}$$

$$= \frac{8766 \times (\text{grand total of FITS for all equipment subsystems under consideration})}{10^{+9}} \tag{15.10}$$

$$\text{DT} = \text{(average annual) outage (down) time (minutes per year)}$$

$$= \text{(hours in a year)} \times \text{(failures per hour)} \times \text{(outage time in minutes per failure)}$$

$$= 8766 \times (\text{FITS} \times 10^{-9}) \times (60 \times \text{MTTR})$$
$$= \frac{(525960 \times \text{MTTR} \times \text{FITS})}{10^{+9}} = \frac{(525960 \times \text{MTTR})}{\text{MTBF}} \quad (15.11)$$

Each outage lasts longer than the outage time or downtime. The equipment will only fail once every few years. When it does, the outage lasts the duration of the MTTR. The outage time is an average number, the outage time divided by the time between outages.

15.2 SYSTEM RELIABILITY

Most electronic systems are composed of electronic network elements (e.g., lightwave or radio terminals). These network elements have expectations for reliability (Telcordia Technologies, 1998, 2001, 2002, 2005, 2009). These in turn are composed of various electronic circuit packs or modules. The reliability of these packs or modules are calculated by industry-standardized methods (typically Telcordia's *Reliability Prediction Procedure for Electronic Equipment*, SR-332, the Department of Defense's *Reliability Prediction of Electronic Equipment*, MIL-HDBK-217, or ITU-T's "Parameters and Calculation Methodologies for Reliability and Availability of Fibre Optic Systems," G.911). When circuit packs or modules have multiple inputs and/or outputs, the module reliability may require the module to be subdivided into portions only affecting the signal flow of interest. Most manufactures keep track of actual module failures and develop module reliability estimates on the basis of this experience.

On the basis of predicted or measured circuit packs or modules, network element reliability is determined by estimating the availability. Many methods (Dovich, 1990; Foster et al., 1977; International Telecommunication Union—Telecommunication Standardization Sector (ITU-T), 1997; Telcordia Technologies, 2011; Tobias and Trindade, 1986; US Department of Defense, 1991; Ushakov and Harrison, 1994) are used: Markov modeling, Monte Carlo simulation, or graph theory (flow networks). A popular approach (suitable for static networks with rare and unrelated failures) is by analyzing the network element's reliability block diagram (Ascher and Feingold, 1984; Bellcore, 1986; Ireson, 1966; Klinger et al., 1990; Myers et al., 1964; Rosenheim, 1958; Sandler, 1963; US Bureau of Naval Weapons, 1964; Ushakov and Harrison, 1994). (This is very similar to fault tree analysis.) The reliability diagram is a block diagram showing the failure dependency of the various packs or modules. On the basis of the reliability diagram, analysis is made to determine the availability or unavailability of the entire network element. The following network element subsystems are often encountered.

15.2.1 Equipment in Series (System A, Fig. 15.1)

Consider n separate devices operated in cascade (all signal paths are in series)

$$A_{\text{total}} = A_1 \times A_2 \times A_3 \times \cdots \times A_n \quad (15.12)$$

A_{total} = the availability of all series devices considered as a group;
A_x = availability of the xth device.

$$\text{FITS}_{\text{total}} = \text{FITS}_1 + \text{FITS}_2 + \text{FITS}_3 + \cdots + \text{FITS}_n \quad (15.13)$$

$\text{FITS}_{\text{total}}$ = the FITS of all series devices considered as a group;
FITS_x = FITS of the xth device.

$$\text{MTBF}_{\text{total}} = \frac{1}{\left[\left(\frac{1}{\text{MTBF}_1}\right) + \left(\frac{1}{\text{MTBF}_2}\right) + \left(\frac{1}{\text{MTBF}_3}\right) + \cdots + \left(\frac{1}{\text{MTBF}_n}\right)\right]}$$
$$= \frac{\text{MTBF}_1 \times \text{MTBF}_2}{\text{MTBF}_1 + \text{MTBF}_2} \quad \text{for } n = 2 \quad (15.14)$$

$\text{MTBF}_{\text{total}}$ = the availability of all series devices considered as a single device;
MTBF_x = availability of the xth device.

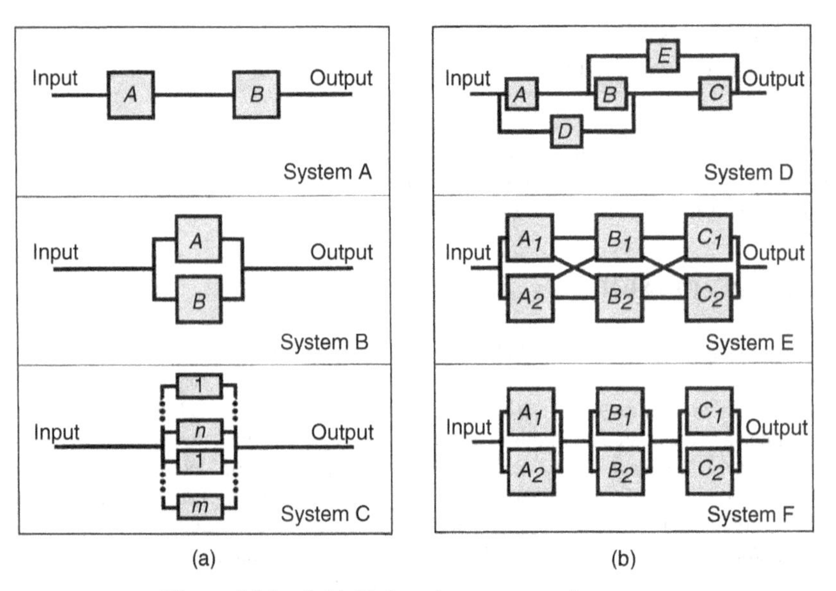

Figure 15.1 (a,b) Various interconnected systems.

15.2.2 Multiple Equipment in Parallel (Systems B and C, Fig. 15.1)

Consider n separate identical working devices (channels) operated with m separate identical standby (backup) devices (channels), where all n lines (signal paths through a working or standby device) must operate.

A_{total} = the availability of the total system consisting of all the working devices;
A_{w} = the availability of one of the working devices;
A_{p} = the availability of one of the standby (protection) devices;
U_{total} = the unavailability of the total system consisting of all the working devices;
U_{w} = the unavailability of one of the working devices;
U_{p} = the unavailability of one of the standby (protection) devices.

Consider $m = 1$ and $n = 1$ (typical hot standby or frequency diversity) case (System B, Fig. 15.1):

$$U_{\text{total}} = U_{\text{w}} \times U_{\text{p}} \tag{15.15}$$

$$\begin{aligned} A_{\text{total}} &= (1 - U_{\text{total}}) = (1 - [U_{\text{w}} \times U_{\text{p}}]) = (1 - [(1 - A_{\text{w}}) \times (1 - A_{\text{p}})]) \\ &= A_{\text{w}} + A_{\text{p}} - (A_{\text{w}} \times A_{\text{p}}) \end{aligned} \tag{15.16}$$

Consider $n > 1$, $m = 1$ (typical multiline) case (System C, Fig. 15.1):

$$\begin{aligned} A_{\text{total}} &= A_{\text{w}}{}^{n} + (n \times A_{\text{w}}{}^{(n-1)} \times A_{\text{p}}) - (n \times A_{\text{w}}{}^{n} \times A_{\text{p}}) \\ &= A_{\text{w}}{}^{n} + (n \times A_{\text{w}}{}^{(n-1)} \times A_{\text{p}})(1 - A_{\text{w}}) \end{aligned} \tag{15.17}$$

For the multiline and hot standby cases, usually the working and protect channels are similar. However, the A_{p} is usually the availability of the protection decision circuitry multiplied by the availability of the protection channel. Therefore, A_{w} and A_{p} are almost always different.

Consider the general case with all units identical ($m \geq 1, n \geq 1$) (System C, Fig. 15.1):

$$\begin{aligned} U_{\text{total}} &= \frac{[(n+m)! \times U^{(m+1)}]}{[n-1)! \times (m+1)!]} \\ &= U^2 \text{ for } n = m = 1 \text{ (hot standby or frequency diversity configuration)} \end{aligned} \tag{15.18}$$

U = the unavailability of one of the m or n identical units.

15.2.3 Nested Equipment (System D, Fig. 15.1)

$$A_{\text{total}} = (A_a \times A_e) + (A_c \times A_d) + (A_a \times A_b \times A_c) - [A_a \times A_b \times A_c \times A_d) + (A_a \times A_b \times A_c \times A_e) + (A_a \times A_c \times A_d \times A_e] + (A_a \times A_b \times A_c \times A_d \times A_e) \tag{15.19}$$

A_{total} = the availability of all devices as a group;
A_n = availability of the device n.

15.2.4 Meshed Duplex Configuration (Systems E and F, Fig. 15.1)

This configuration is common in high reliability computer and digital cross-connect systems (System E, Fig. 15.1). It is functionally equivalent (Klinger et al., 1990) to the reliability block diagram (System F, Fig. 15.1). It is evaluated by reducing the parallel components to equivalent series elements and then evaluating the reliability of the series units.

15.3 COMMUNICATION SYSTEMS

The evaluation of reliability of communication systems is similar to the evaluation of network elements. The system reliability block diagram of the system is drawn using the availability values of the network elements. Those values are determined from two-way MTBF and MTR values.

$$A = \text{equipment availability} = \frac{\text{MTBF}}{(\text{MTBF} + \text{MTTR})} \tag{15.20}$$

The system is then evaluated using the same techniques described earlier.

Linear systems are simply evaluated as series network elements (System A, Fig. 15.1).

For a single ring with single (nonredundant) inputs and outputs, availability is evaluated as two paths in parallel (System B, Fig. 15.1). One path is the normal (e.g., clockwise) branch and the other is the protection (e.g., counterclockwise) branch. Each of those branches is treated as series (linear) subsystems.

For more complex cases such as multiple, dual-node, interconnected rings, the rings and dual interconnecting paths are evaluated individually and the results are combined as series elements.

Mesh IP networks applied to radio networks usually have predefined paths between end sites to support virtual private networks. The objective of protection is to provide reliable communication in the event of failure of communication paths. Since path restoral must be fast and reliable (not influenced by other network failures), a mechanism that predefines restoral links is used. A typical routing protocol for this function is MPLS. Protection mechanisms are usually link protection or path protection. A typical assumption in analyzing these protection methods is that the effect of other network element failures and the network restoral of those paths will not affect restoral of the path of interest (i.e., the network is not oversubscribed).

Link protection precomputes an alternate detour for each link and recovers from a link failure by rerouting the traffic along the predetermined detour. The end to end path availability is computed by first treating each link and its alternate link as a parallel path network (System B, Fig. 15.1). The composite parallel path networks are then treated as individual elements in an end to end series element path (System A, Fig. 15.1).

Path protection assigns two paths, a primary and a backup, to each end to end connection. The traffic is switched onto the backup path in case of a primary path failure. The primary and backup paths need to be disjoint, so a single path element failure will not jeopardize both paths simultaneously. To evaluate this situation, first the availability of each path is determined, treating each path as a series of individual elements (System A, Fig. 15.1). Then the availability of the combined paths is determined using both paths as parallel elements (System B, Fig. 15.1).

Keep in mind that the availability of the path failure sensing element must be included in the backup path availability analysis. It should also be noted that path protection should not be used to improve end to end atmospheric propagation availability estimates. Path protection improves equipment and facility availability but is disruptive to the data path. Every time the physical path is changed, the end to end path delay is changed. This causes significant temporary disruption to data services (due to loss or repetition of data). This can induce significant temporary disruption to data service, which may be longer than the actual path change induced disruption.

15.4 APPLICATION TO RADIO CONFIGURATIONS

Radio system hardware reliability is usually specified as MTBO (system failure). The system planner wants to know what the equipment availability (or unavailability) is. Table 15.1 is a reference between the provided and desired quantities. If the MTBO is one or two way, the availability is one or two way, respectively. Notice that MTTR (a function of failure detection, reporting, spare availability and maintenance technician training, and reliability) has significant impact on achieved availability or required MTBF.

The previous formulas may be applied to determine the effect of different radio configurations on radio system hardware availability. Table 15.2 shows the increase in radio system availability when going from a nonstandby to a hot standby (or frequency diversity) configuration. Notice the significant increase in availability for even modestly reliable equipment. These results are shown graphically (Fig. 15.2).

Multiline radio systems, as they use an unprotected radio channel to protect multiple working radio channels, are inherently less reliable than hot standby or frequency diversity configures. Figure 15.3 shows the decrease in hardware reliability as working channels are increased in a one-for-N system. Cross-polarization multiline systems typically use two protection channels, one for vertical channels and the other for horizontal channels. These systems would be analyzed as one for N, where N is the number of vertical or horizontal working channels being protected.

15.5 SPARE UNIT REQUIREMENTS

Maintainability is often defined (Rausand and Hoyland, 2004) as "The ability of an item, understated conditions of use, to be restored to a state in which it can perform its required functions when maintenance is performed understated conditions and using prescribed procedures and resources." Usually, maintenance is performed by storing a number of repairable spare units (e.g., modules, cards, plug-ins, or blades) in reserve. These units are used to repair the system. The problem to be solved is to determine the number of spare units Nrequired to support a system of Q operational units with a probability of success P.

For complicated systems with interacting failure mechanisms, a Markov model analysis is necessary. For systems with rare failures occurring with a constant rate λ, the mean number of equipment failures is given by λt (with λ and t as previously defined). In this situation, a homogeneous Poisson process model (Rausand and Hoyland, 2004; Ushakov and Harrison, 1994) is generally applied. (The Erlang loss formula (Bellcore, 1986) achieves the same result for small failure rates.)

$$P_n = \frac{[(\lambda t)^n e^{-\lambda t}]}{n!} \quad \text{for } n = 0, 1, 2, 3, \ldots \text{ with } \lambda \text{ and } t \text{ as previously defined} \tag{15.21}$$

P_n = the probability that a failure occurs exactly n times in the time interval 0 to t;
S = the number of spares normally on hand;
P_S = the probability of a successful repair when a failure occurs;
= the probability that not more than n failures have occurred in the time interval 0 to t;
= the probability of having a spare available if S spares are normally available.

$$P_S = \sum_{n=0}^{S} P_n$$

$$R = \frac{\text{hours per year} \times Q}{\text{MTBF}} = \frac{8766 \times Q}{\text{MTBF}}$$

TABLE 15.1 Relationship Between Availability, Outage Time, and MTBF

Availability (Reliability), %	Unavailability (Outage), %	Average Outage Time per Year			MTBF (MTBO)					
		hours	minutes	seconds	MTTR = 2 hours	4 hours	8 hours	16 hours	24 hours	48 hours
0	100	8766	525,960	31,557,600	0	0	0	0	0	0
50	50	4383	262,980	15,778,800	2	4	8	16	24	48
80	20	1753	105,192	6,311,520	8	16	32	64	96	192
90	10	876.6	52,596	3,155,760	18	36	72	144	216	432
95	5	438.3	26,298	1,577,880	38	76	152	304	456	912
98	2	175.3	10,519	631,152	98	196	392	784	1,176	2,352
99	1	87.66	5,260	315,576	198	396	792	1,584	2,376	4,752
99.5	0.5	43.83	2,630	157,788	398	796	1,592	3,184	4,776	9,552
99.9	0.1	8.766	526.0	31,558	1,998	3,996	7,992	15,980	23,980	47,952
99.95	0.05	4.383	263.0	15,779	3,998	7,996	15,990	31,980	47,980	95,952
99.99	0.01	0.8766	52.60	3156	20,000	40,000	79,990	160,000	240,000	480,000
99.995	0.005	0.4383	26.30	1578	40,000	80,000	160,000	320,000	480,000	960,000
99.999	0.001	0.08766	5.260	315.6	200,000	400,000	800,000	1,600,000	2,400,000	4,800,000
99.9995	0.0005	0.04383	2.630	157.8	400,000	800,000	1,600,000	3,200,000	4,800,000	9,600,000
99.9999	0.0001	0.008766	0.5260	31.56	2,000,000	4,000,000	8,000,000	16,000,000	24,000,000	48,000,000
99.99995	0.00005	0.004383	0.2630	15.78	4,000,000	8,000,000	16,000,000	32,000,000	48,000,000	96,000,000
99.99999	0.00001	0.0008766	0.05260	3.156	20,000,000	40,000,000	80,000,000	160,000,000	240,000,000	480,000,000

Abbreviations: MTBF, mean time between failures (for component); MTBO, mean time between outages (for system); MTTR, mean time to restore (failed equipment); MTTR, mean time to detect, report, diagnose, acquire spares, travel, and repair failure. Year, 365.25 days.

TABLE 15.2 Decrease in Equipment Unavailability by Adding Protection

Working Channel Unprotected Availability	Unprotected Unavailability/Protected Unavailability: Protection Circuitry Perfect (Availability = 1)	Unprotected Unavailability/Protected Unavailability: Protection Circuitry Availability = Availability of Protection Channel
99	100	50
99.5	200	100
99.9	1,000	500
99.95	2,000	1,000
99.99	10,000	5,000
99.995	20,000	10,000
99.999	100,000	50,000
99.9995	200,000	100,000
99.9999	1,000,000	500,000
99.99995	2,000,000	1,000,000
99.99999	10,000,000	5,000,000

Working (normal) and protection channels are assumed to have the same availability.

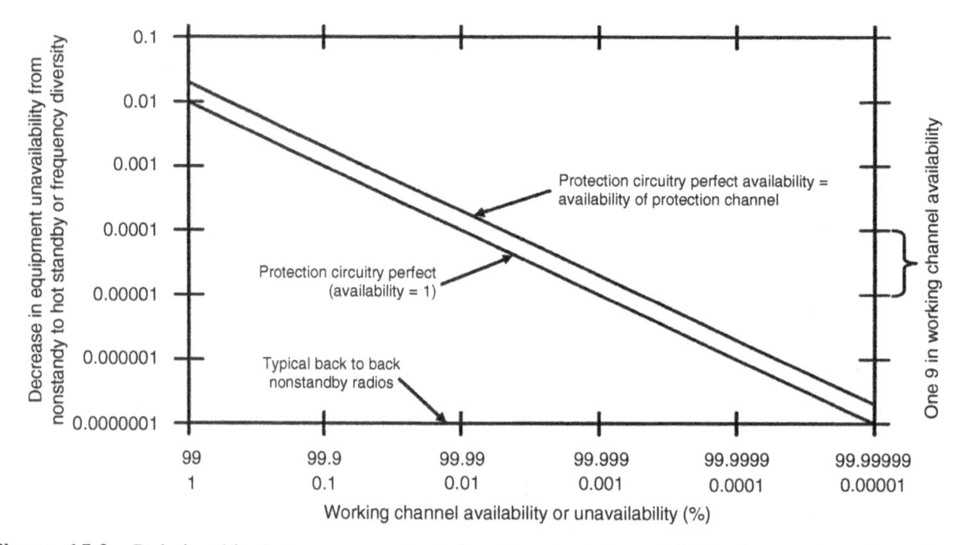

Figure 15.2 Relationship between unprotected and protected availability (protection channel and working channel both have same availability, factors listed are multiplicative factors applied to non-standby unavailability).

$$
\begin{aligned}
D &= \lambda t \\
&= \frac{R \times T}{\text{days per year}} \\
&= \frac{R \times T}{365.25}
\end{aligned} \tag{15.22}
$$

Q = quantity of units to be supported (number of units in population);
MTBF = mean time between failure of the unit being supported (hours);
T = time to restore (repair or replace) spare unit to spare reserves (days);

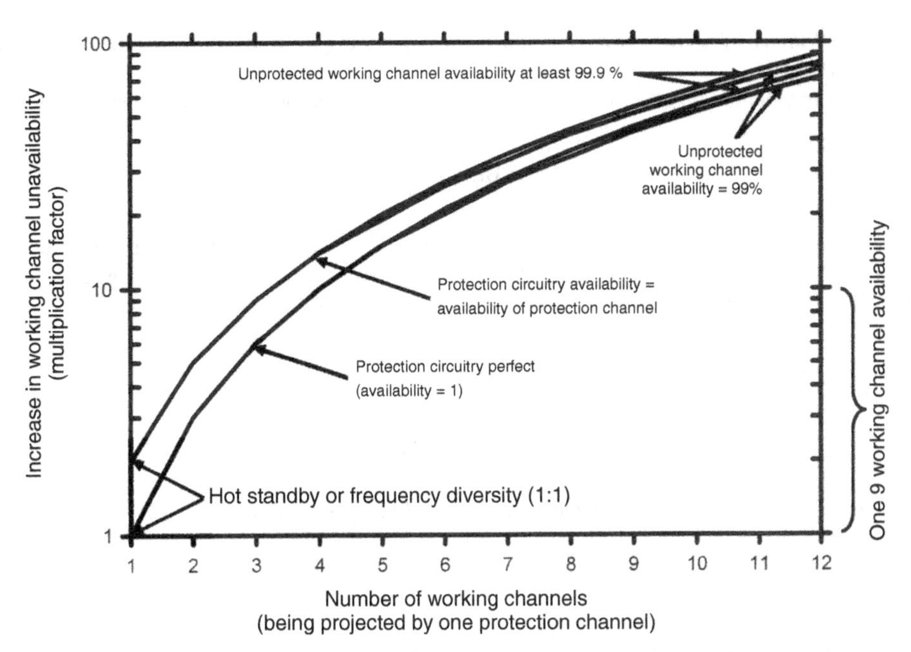

Figure 15.3 Multiline channel unavailability and working channels (reference unavailability is a 1 : 1 system with perfect protection circuitry; protection channel and all working channels have same availability).

R = average unit repairs (replacement of failures with spares) per year;
D = mean (average) demand (average failed units per year).

$$P_S(\%) = 100\left[\sum_{M=0}^{S} P_I\,(M)\right]$$

$$P_I(0) = e^{-D}$$

$$P_I(M) = \left(\frac{D}{M}\right)\,P_I(M-1) \tag{15.23}$$

Calculate P_S. Increment M in integers beginning at 1 until the desired probability of success P_S (%) is achieved or exceeded. Typical probabilities of success P range from 99% to 99.99%.

15.6 BER ESTIMATION

The fundamental quality of digital payloads, which may have a predefined number of states at any instant of time, is characterized by the probability of message error. BER estimation is a statistical measurement that attempts to measure current average error performance within a defined confidence range by measuring errors within a defined time interval. The classical approach is to use binomial distribution sampling theory applied to a predefined sample size [although there are some time advantages on using the slightly more complicated negative binomial sampling (Rice and Mazzeo, 2012)]. Direct binomial sampling theory yields the following results.

Assume N *digits* ($N > 10^4$) have been *transmitted* and E *digits* ($E \ll N$) were found to be *in error* (Jeruchim, 1984).

The median-unbiased maximum likelihood estimator (not exceeded 50% of the time) is as follows:

$$\text{BER} \cong \frac{E}{N} \tag{15.24}$$

If we wish to be more confident, then N must be further constrained. Estimate BER as per Section 15.24. We may be confident at the following levels that the BER falls within the range $[(2E)/N] \geq \text{BER} \geq [(0.5E)/N]$ if the following requirements for N are satisfied.

90% Confidence level biased-estimator (not exceeded 90% of the time):

$$N \geq \frac{6}{\text{BER}} \tag{15.25}$$

95% Confidence level biased-estimator (not exceeded 95% of the time):

$$N \geq \frac{8}{\text{BER}} \tag{15.26}$$

99% Confidence level biased-estimator (not exceeded 99% of the time):

$$N \geq \frac{15}{\text{BER}} \tag{15.27}$$

The above is the intuitive result. Less intuitive is the expected BER if N *digits* ($N > 10^4$) have been *transmitted* and *zero digits* were found to be *in error* (Crow, 1974).

Not finding an error in the sampled data is not to say, an error would not be found if the testing lasted long enough.

Median-unbiased maximum likelihood estimator (not exceeded 50% of the time) (Crow, 1974):

$$\text{BER} \cong \frac{1}{(3N)} \tag{15.28}$$

90% Confidence level biased-estimator (not exceeded 90% of the time):

$$\text{BER} \leq \frac{3.0}{N} \tag{15.29}$$

95% Confidence level biased-estimator (not exceeded 95% of the time):

$$\text{BER} \leq \frac{3.7}{N} \tag{15.30}$$

99% Confidence level biased-estimator (not exceeded 99% of the time):

$$\text{BER} \leq \frac{5.3}{N} \tag{15.31}$$

To be 95% confident that the BER is less than $1/N$, we must have received ($3.7\ N$) bits with no errors.

The above confidence levels are two sided. They also assume isolated errors from Gaussian noise sources. Confidence levels will be lower when error bursts occur.

15.6.1 Time to Transmit *N* Digits

Transmitted Digits (*N*)	10^{+9}	10^{+10}	10^{+11}	10^{+12}	10^{+13}	
DS0 (64 kb/s)	15,625	156,250	1,562,500	15,625,000	156,250,000	secs
	260.4	2,604.2	26,041.7	260,416.7	2,604,166.7	mins
	4.34	43.40	434.03	4,340.28	43,402.78	hours
DS1 (1.544 Mb/s)	648	6,477	64,767	647,668	6,476,684	secs
	10.8	107.9	1,079.4	10,794.5	107,944.7	mins
	0.18	1.80	17.99	179.9	1,799.08	hours
DS3 (44.736 Mb/s)	22	224	2,235	22,353	223,534	secs
	0.37	3.7	37.3	372.6	3,725.6	mins
	0.062	0.062	0.62	6.21	62.09	hours
OC-1/STS-1 (51.840 Mb/s)	19	193	1,929	19,290	192,901	secs
	0.32	3.2	32.2	321.5	3,215.0	mins
	0.054	0.054	0.54	5.36	53.58	hours
OC-3/STS-3 (155.520 Mb/s)	6.4	64	643	6,430	64,300	secs
	0.11	1.1	10.7	107.2	1,071.7	mins
	0.0018	0.018	0.18	1.79	17.86	hours
OC-12/STS-12 (622.080 Mb/s)	1.6	16	161	1,608	16,075	secs
	0.027	0.27	2.7	26.8	267.9	mins
	0.00045	0.0045	0.045	0.45	4.47	hours
OC-48/STS-48 (2488.320 Mb/s)	0.40	4.0	40	402	4,019	secs
	0.0067	0.067	0.67	6.7	67.0	mins
	0.00011	0.0011	0.011	0.11	1.12	hours
OC-192/STS-192 (9953.280 Mb/s)	0.10	1.0	10	100	1,005	secs
	0.0017	0.017	0.17	1.7	16.7	mins
	0.000028	0.00028	0.0028	0.028	0.28	hours

REFERENCES

Ascher, H. and Feingold, H., *Repairable Systems Reliability: Modeling, Inference, Misconceptions and Their Causes*. New York: Marcel Dekker, 1984.

Bellcore, *Reliability Manual, SR-TSY-000385, Issue 1*. Red Bank: Bellcore, June 1986.

Crow, E. L., *Confidence Limits for Digital Error Rates, OT Report 74–51*. Washington: US Department of Commerce/Office of Telecommunications, 1974.

Dovich, R. A., *Reliability Statistics*. Milwaukee: ASQ Quality Press, 1990.

Foster, J. W., Phillips, D. T. and Rogers, T. R., *Reliability, Availability and Maintainability*. College Station: Texas A&M University Press, 1977.

International Telecommunication Union—Telecommunication Standardization Sector (ITU-T), "Parameters and Calculation Methodologies for Reliability and Availability of Fibre Optic Systems, G.911," *ITU-T Recommendations*. Geneva: International Telecommunication Union, April 1997.

International Telecommunication Union—Telecommunication Standardization Sector (ITU-T), "Availability Performance Parameters and Objectives for End-to-End International Constant Bit-rate Digital Paths, G.827," *ITU-T Recommendations*. Geneva: International Telecommunication Union, September 2003.

Ireson, W. G., *Reliability Handbook*. New York: McGraw-Hill, 1966.

Jeruchim, M. C., "Techniques for Estimating the Bit Error Rate in the Simulation of Digital Communications Systems," *IEEE Transactions on Selected Areas in Communications*, Vol. 2, pp. 153–170, January 1984.

Klinger, D. J., Nakada, Y. and Menendez, M. A., *AT&T Reliability Manual*. New York: Van Nostrand Reinhold, 1990.

Myers, R. H., Wong, K. L. and Gordy, H. M., *Reliability Engineering for Electronic Systems*. New York: John Wiley & Sons, Inc., 1964.

Rausand, M. and Hoyland, A., *System Reliability Theory: Models, Statistical Methods, and Applications, Second Edition.* Hoboken: John Wiley & Sons, Inc., 2004.

Rice, M. and Mazzeo, B., "On the Superiority of the Negative Binomial Test Over the Binomial Test for Estimating the Bit Error Rate," *IEEE Transactions on Communications*, Vol. 60, pp. 2971–2981, October 2012.

Rosenheim, D. E., "Analysis of Reliability Improvement through Redundancy," *Proceedings of the New York Industry Conference on Reliability Theory*, pp. 119–142, June 1958.

Sandler, G. H., *System Reliability Engineering*. Englewood Cliffs: Prentice-Hall, 1963.

Telcordia Technologies, *LSSGR: Reliability, Section 12, GR-512-CORE, Issue 2.* Red Bank: Telcordia, January 1998.

Telcordia Technologies, *Generic Requirements for Assuring the Reliability of Components Used in Telecommunications Equipment, GR-357-CORE, Issue 1.* Red Bank: Telcordia, March 2001.

Telcordia Technologies, *Reliability and Quality Measurements for Telecommunications Systems (RQMS-Wireline), GR-929-CORE, Issue 8.* Red Bank: Telcordia, December 2002.

Telcordia Technologies, *Reliability and Quality Measurements for Telecommunications Systems (RQMS-Wireless), GR-1929-CORE, Issue 2.* Red Bank: Telcordia, February 2005.

Telcordia Technologies, *Transport Systems Generic requirements: Common Requirements, GR-499-CORE, Issue 4.* Red Bank: Telcordia, November 2009.

Telcordia Technologies, *Reliability Prediction Procedure for Electronic Equipment, SR-332, Issue 3.* Red Bank: Telcordia, February 2011.

Tobias, P. A. and Trindade, D., *Applied Reliability*. New York: Van Nostrand Reinhold, 1986.

US Bureau of Naval Weapons, *Handbook, Reliability Engineering, NAVWEPS 00-65-502*. Washington, DC: US Government Printing Office, June 1964.

US Department of Defense, *Reliability Prediction of Electronic Equipment, Mil-Hdbk-217F*. Washington, DC: US Government Printing Office, December 1991.

Ushakov, I. A. and Harrison, R. A., *Handbook of Reliability Engineering*. New York: John Wiley & Sons, Inc., 1994.

16

PATH PERFORMANCE CALCULATIONS

Microwave radio transmission engineers perform three main tasks:

1. They size the transmitter power amplifiers, determine the size of antennas, and pick the appropriate configuration of radio so that the radio system design meets appropriate performance objectives.
2. They engineer the placement of the microwave radio antennas to minimize obstruction and reflective fading.
3. They pick transmitter and receiver frequencies (perform frequency planning) in such a way that all existing and planned radio systems can operate with acceptable performance.

These tasks are covered in Chapters 2, 5, 9, 10, 11, and 12. Task 2 is a function of the physical path and relatively independent of the radio configuration (with the exception of diversity antennas if they are used). Minimizing obstruction fading is specifically covered in Chapter 12. Minimizing reflective fading is covered in the Appendix 10.A. For radio paths operating at frequencies greater than approximately 10 GHz in urban areas, additional terrain considerations can become important (International Telecommunication Union—Radiocommunication Sector (ITU-R), 2007a; International Telecommunication Union—Radiocommunication Sector (ITU-R), 2007b).

Task 1 is highly dependent on radio performance objectives. Performance objectives are estimated on the basis of either the North American or the ITU-R models and formulas. Chapters 4 and 7 detail the differences in these objectives. This chapter details the North American and ITU-R calculation methods. The implied assumption is that the paths are operating well below optical frequencies and therefore the Rayleigh or the Nakagami–Rice multipath (International Telecommunication Union—Radiocommunication Sector (ITU-R), 2007c) is the dominant multipath fading mechanism. As operating frequency increases, the dominant multipath fading mechanism is turbulent air (Hodges et al., 2006; International Telecommunication Union—Radiocommunication Sector (ITU-R), 2007d; International Telecommunication Union—Radiocommunication Sector (ITU-R), 2007e; International Telecommunication Union—Radiocommunication Sector (ITU-R), 2007f; Strohbehn, 1968; Strohbehn and Beran, 1969). This effect is not considered.

The performance of a telecommunications system is significantly influenced by many factors. The actions of operations personnel are quite important but difficult to model. The implied assumption is they are well trained and supported and their actions will not significantly detract from system performance. However, transmission path rerouting and other operator induced transient events are not unknown in real

Digital Microwave Communication: Engineering Point-to-Point Microwave Systems, First Edition. George Kizer.

networks. Another assumption is that external sources of noise are insignificant. But for the exception of when the sun passes directly in front of a receive antenna, this is generally true (International Telecommunication Union—Radiocommunication Sector (ITU-R), 2007g). Mother Nature also has an influence. Radio path performance is directly related to weather and unusual, unpredictable weather phenomenon can occur (consult http://www.weather.gov/, http://weather.unisys.com/ and http://www.wunderground.com/). We ignore rare exceptions and focus on relatively predictable system limitations. For microwave radio systems, performance objectives are defined by path, multipath, rain and obstruction fading, and external interference. Although other forms of degradation (operator actions, lightning, and influences of other equipment such as power systems) can occur in real systems, those events are considered rare and are usually not part of performance estimates.

16.1 PATH LOSS

In the absence of any external terrain or meteorological effects, the "free-space" transmission loss (Friis, 1946; International Telecommunication Union—Radiocommunication Sector (ITU-R), 1999a; Kizer, 1990; Kizer 1995) between a transmitter and a receiver is defined by the following:

$$\begin{aligned} L_{\mathrm{TR}} &= \text{transmission loss (dB)} \geq 0 \text{ dB} \\ &= P_{\mathrm{T}}(\mathrm{dB}) - P_{\mathrm{R}}(\mathrm{dB}) \\ &= 96.6 - G_{\mathrm{T}}(\mathrm{dB}) - G_{\mathrm{R}}(\mathrm{dB}) + 20 \log [f(\mathrm{GHz})] + 20 \log [d(\mathrm{miles})] \\ &= 92.4 - G_{\mathrm{T}} - G_{\mathrm{R}} + 20 \log [f(\mathrm{GHz})] + 20 \log [d(\mathrm{km})] \end{aligned} \tag{16.1}$$

P_{T} = RF power at the input to the transmitting antenna;
P_{R} = RF power at the output of the receiving antenna;
G_{T} = isotropic gain of the transmit antenna;
G_{R} = isotropic gain of the receive antenna;
f = operating frequency;
d = distance between the transmit and the receive antennas.

Free space loss (International Telecommunication Union—Radiocommunication Sector (ITU-R), 1994; Kizer, 1990) is transmission loss with the antenna gains removed.

$$\begin{aligned} L_{\mathrm{FSL}} &= \text{free space loss (dB)} \geq 0 \text{ dB} \\ &= L_{\mathrm{TR}} - G_{\mathrm{T}} - G_{\mathrm{R}} \\ &= 96.6 + 20 \log [f\ (\mathrm{GHz})] + 20 \log [d\ (\mathrm{miles})] \\ &= 92.4 + 20 \log [f\ (\mathrm{GHz})] + 20 \log [d\ (\mathrm{km})] \end{aligned} \tag{16.2}$$

Free space loss is the basic loss between the transmitter and receiver. It sets a fundamental limitation on the performance of the radio path. The primary task of the transmission engineer is to design a system in which free space loss is acceptable and the effects of external factors will not adversely change this loss by more than an acceptable amount. In reality, other propagation factors can affect path loss.

At the outset, it is important to mention the issues of antenna gain and operating frequencies. Most fixed point to point microwave radios operate in very wide (1/2–1 GHz) frequency allocations. Typical antennas have more than 1-dB variation across that frequency range. Likewise, free space loss can vary by over 1/2 dB over that frequency range. These variations will impact estimated path performance. Agreement among all interested parties as to the frequency at which performance estimates are to be made is important to minimize misunderstandings.

16.2 FADE MARGIN

Performance estimates are based on radio fade margin [decibel difference between the normal received signal level and the radio BER receiver threshold]. For North American radio systems, the radio threshold is usually a BER of 10^{-6} or occasionally 10^{-3}. (While this difference in threshold definition was significant for older radios, for modern radios using powerful error correction, there is little difference in threshold received signal level.) For ITU-R radio systems, the threshold is an SES. Today the ITU-T and ITU-R SES is defined as 30% or more errored blocks. Relating an SES threshold to a BER threshold requires knowledge of blocks per second, bits per block and the number of radio error bursts per error event (Section 7.5). Unlike the threshold in North America, ITU-R radio thresholds are highly variable (BERs of 10^{-5} to 10^{-2}) depending on the radio and baseband formats. Many manufactures simply ignore this variability. They only provide one threshold value (either 10^{-6} or 10^{-3}). Few provide the number of error bursts per error event. This complicates the user's estimation of path performance. This complication becomes less significant as radios use more error correction (which causes all thresholds to converge toward the same received signal level).

Fade margin is influenced by several engineering decisions, not just by equipment characteristics. Are "typical" or "guaranteed" radio transmit power and receiver threshold to be used? Is any "field margin" (a 1-, 2-, or 3-dB reduction of normal received signal level to account for unanticipated field installations issues) to be applied? Is any wet radome loss to be applied to one or both high frequency radio antennas? Are transmission lines to be assumed, estimated, or determined from site drawings? All these decisions will have significant influence on the fade margin, which drives the path performance calculations.

16.3 PATH PERFORMANCE

Path performance is specified as percentage of time that the radio path does not exceed a predefined performance objective. The performance is normally calculated for multipath fading and rain fading. As noted in Chapter 4, in North America, two-way multipath and two-way rain fading are accumulated together to obtain an overall two-way annual availability performance estimate. ITU-R has a one-way quality objective that is multipath performance measure over the "worst month." They have a two-way annual availability objective that includes the effects of rain-fading and equipment failures.

Multipath calculations are estimated one way. If a two-way estimate is needed, the outage time is doubled under the assumption that multipath fading at any instant each direction of transmission behaves independently. For rain fading, the fading is for both directions of transmission at the same time. Therefore, for rain fading, one-way outage time and two-way outage time are the same.

The one-way (transmitter-to-receiver transmission path) atmospheric multipath fading outage time may be estimated by the following:

$$T(s) = T_0 P \tag{16.3}$$

T = number of fading seconds in a time period of interest [a year (Vigants, 1975) or the "worst" month (International Telecommunication Union—Radiocommunication Sector (ITU-R), 1990; International Telecommunication Union—Radiocommunication Sector (ITU-R), 1992; International Telecommunication Union—Radiocommunication Sector (ITU-R), 2005a)];
T_0 = seconds in a time period of interest (year or month);
P = probability of heavy fading during the time period of interest.

While one could work with probabilities of outage, it is easier to deal with outage seconds, sum the appropriate different outages, and then convert this to an outage probability.

$$\text{Outage (\%)} = 100 \times \frac{T}{T_{\text{ref}}} \tag{16.4}$$

T_{ref} = reference time (1 year or 1 month);
T_{ref} = one year = 365.25 days/year × 24 × 60 × 60 s/day = 31,557,600 s/year;

T_{ref} = one month = 30.4375 days/month × 86,400 s/day = 2,629,800 s/month.

$$\text{Availability (\%)} = 100 - \text{Outage (\%)} \tag{16.5}$$

Flat, dispersive, rain fading, and obstruction fading can be calculated separately.
For the North American model,

quality is addressed as a bringing into service and maintenance issue and not directly calculated.
availability is two way and is based on all forms of fading (multipath, rain, obstruction, and short-term and long-term interferences).

For the ITU-R model,

quality is one way and is based on the one-way outage due to multipath and short-term interference.
availability is two way and is based on availability due to rain, obstruction fading, and long-term interference.

Which model should the path designer use? This is a question asked by many engineers. There are many opinions but few definitive answers. The North American (NA) methods were developed from paths in the United States. The ITU-R methods were developed from many sources but are generally viewed (at least by many in North America) as being based on European paths. Some North American engineers agree with Boithias (1987) that " ... results have been published suggesting that a 42 km link in the United States has the same statistical distribution as a 65 km link in Western European." Their position is North American methods are more stringent. Adherents for the ITU-R methods support their approach as being more general and modern. One study, namely, by Olsen et al. (2003), of this issue appears to support both claims. Ultimately, the choice of path propagation model becomes an engineering judgment issue.

16.4 ALLOWANCE FOR INTERFERENCE

In North America, the T_0/I coordination method (Committee TR14-11, 1994) is used for long-term interference estimation. Reducing path fade margin by 1 dB accounts for this interference. (The standard assumption is only one significant interference exposure.) If short-term interference is anticipated, path fade margin is reduced by 10 dB for 0.01% of the year (Working Group 9, 1985).

For ITU-R systems, an appropriate fade margin reduction (usually 1 or 0.5 dB) is used to account for long-term interference. Short-term interference is usually regulated by local authorities and fade margin reduction is tailored to those regulations.

16.5 NORTH AMERICAN (NA) PATH PERFORMANCE CALCULATIONS

The total path outage time for a digital radio receiver is the following:

$$T_{TOTAL} = \text{total two-way path outage(s)}$$

$$T_{TOTAL} = T_{FLAT} + T_{DISP} + T_{RAIN} + T_{UPFADE} + T_{OBST} \tag{16.6}$$

$$T_{FLAT} = 2 \times \text{(one-way flat multipath fading outage time)}$$

$$= 2\, T_{ref}\, P_{FF} \tag{16.7}$$

$$T_{DISP} = 2 \times \text{(one-way dispersive multipath fading outage time)}$$

$$= 2\, T_{ref}\, P_{DF} \tag{16.8}$$

$$T_{RAIN} = \text{rain fading outage}$$

$$= T_{\text{ref}}\ P_{\text{R}} \tag{16.9}$$

$$T_{\text{UPFADE}} = 2 \times \text{(one-way up fading outage-if calculated)}$$

$$= 2\ T_{\text{ref}}\ P_{\text{U}} \tag{16.10}$$

$$T_{\text{OBST}} = \text{obstruction outage time-if calculated}$$

$$= T_{\text{ref}}\ P_{\text{O}} \tag{16.11}$$

T_{ref} = reference time period (1 year);
= 31,557,600 s (1 year);
P_{FF} = probability of a one-way flat fade;
P_{DF} = probability of a one-way dispersive fade;
P_{R} = probability of a rain fade (one way = two way);
P_{U} = probability of a one-way upfade;
P_{O} = probability of an obstruction fade (one way = two way).

To calculate two-way outages, each outage, except rain and obstruction, is calculated and doubled on the assumption that fading each way (go/return) is uncorrelated. As rain and obstruction outages occur simultaneously in both directions of transmission, one-way and two-way outages are the same time. If the fade margins are different for the different directions (go/return) on the path, the rain and obstruction outages are calculated for the smaller fade margin.

The following calculations will be made on the basis of fade margin associated with flat and dispersive fading individually. Some designers merge the flat, dispersive fade margins into a composite fade margin (Rummler, 1982a) and perform the outage probability using that fade margin. This is technically correct for the nondiversity case. However, diversity improvement is different for flat and dispersive fading in both the North American and ITU-R methods. We treat flat and dispersive-fading effects separately.

If the path is operating in a US-licensed frequency band, frequency coordination has already taken place using the T/I methodology. If the T/I criteria have been met, the flat fade margin has not been degraded by more than 1 dB. For every decibel that this criterion has been exceeded, the flat fade margin has been reduced by that decibel amount. Therefore, interference is easily addressed by a modification of the flat fade margin. Interference has no effect on dispersive fading and therefore the dispersive fade margin is unchanged regardless of the level of interference. Of course, if the path is operating in an unlicensed band, the level of interference is unknown and uncontrolled.

16.5.1 Vigants–Barnett Multipath Fading (Barnett, 1972; Vigants, 1975)—NA

$$P_{\text{FF}} = \left[\frac{T_{\text{s}}}{T_{\text{ref}}}\right] \frac{R \times 10^{-\text{FFM}/10}}{I_{\text{FF}}} \tag{16.12}$$

$$P_{\text{DF}} = \left[\frac{T_{\text{s}}}{T_{\text{ref}}}\right] \frac{R_{\text{DF}}\ (R \times 10^{-\text{DFM}/10})}{I_{\text{DF}}} \tag{16.13}$$

FFM = path flat fade margin [positive number (dB)];
DFM = path-dispersive fade margin [positive number (dB)];
I_{FF} = flat-fading improvement factor;
I_{DF} = dispersive-fading improvement factor;
T_{S} = number of seconds in the heavy-fading season;
= $8 \times 10^6 \times (t/50)$;
T_{ref} = number of seconds in a year;
= $365.25 \times 24 \times 60 \times 60 = 31{,}557{,}600$;
t = average annual temperature (°F), $35\ °\text{F} \leq t \leq 75\ °\text{F}$;
R_{DF} = dispersive-fading dispersion ratio;

= 1 for typical paths;
= 0.5–1 for good paths;
= 1–3 for average paths (Ranade and Greenfield, 1983; Rummler, 1988);
= 5–7 for difficult paths;
= 9 for very difficult paths;
R = fade occurrence factor;
$= c(50/w)^{1.3}(f/4)D^3 \times 10^{-5}$;
c = climate and humidity factor;
c = 0.5 for dry climate;
c = 1 for average conditions;
c = 2 for coastal and over water paths.

After Vigants original results (Vigants, 1975), Vigants and others (Loso et al., 1992) suggested the following additional guidelines for c:

$c = 0.25$ for benign climates (mountainous, dry, or northerly conditions) such as the Rocky Mountains, Canada, and sections of Germany;
$c = 10$ for difficult United States (e.g., US Gulf Coast) and international areas;
$c = 100$ for worst-case conditions (areas of extreme heat and humidity such as the Red Sea, the Persian Gulf coastal plan, and equatorial climates).

These revised guidelines have yet to find acceptance within the North American engineering community.

The dispersive-fading outage time formula was based on measured fading on the Bell Labs reference path near Palmetto, Georgia. Other paths were observed to have different dispersive fading than that path (Greenfield, 1984; Ranade and Greenfield, 1983). Rummler (1988) suggested relating actual path-cumulative dispersive-fading time to the reference cumulative path-fading time by modifying the dispersive-fading outage time by the multiplicative dispersion factor R_{DF}. However, this approach was never researched further to determine general design guidelines. Some highly dispersive paths are paths with excessive path clearance over reflective terrain (including valleys with quiet air, trapped temperature, and/or humidity inversion layers) (Greenfield, 1984). Research (Ranade and Greenfield, 1983; Rummler, 1988) has noted that some paths experience a variation of three in year-to-year cumulative dispersive fading. In general, the relationship between path characteristics and dispersive fading is currently unknown. The above R_{DF} values (Gullett, 2001) are provisional.

w = terrain roughness factor (ft), $20 < w < 140$;
f = frequency (GHz);
D = path length (miles).

For metric calculations:

$$R = c\left(\frac{15.2}{w}\right)^{1.3} fD^3 \times 6 \times 10^{-7}$$

w = terrain roughness factor (m), $6 \le w \le 43$;
f = frequency (GHz);
D = path length (km).

The roughness factor w is calculated using n equally spaced elevations above a convenient reference plane (e.g., average mean sea level) taken from the design path (Fig. 16.1). For long paths, the elevation samples $\mathbf{x}_i$ are taken at 1 mile (Vigants, 1975) (North American method) or 1-km increments (Elevation samples are for bare earth; building or vegetation elevations are excluded from the samples.). If the path is short (e.g., less than 20 miles), some users divide the path into 20 equally spaced increments (Knight, 2006). Endpoints of the path are excluded from the calculation.

$$\overline{\mathbf{X}} = \left(\frac{1}{n}\right)\sum_{i=1}^{n} \mathbf{X}_i = \text{sample mean} \tag{16.14}$$

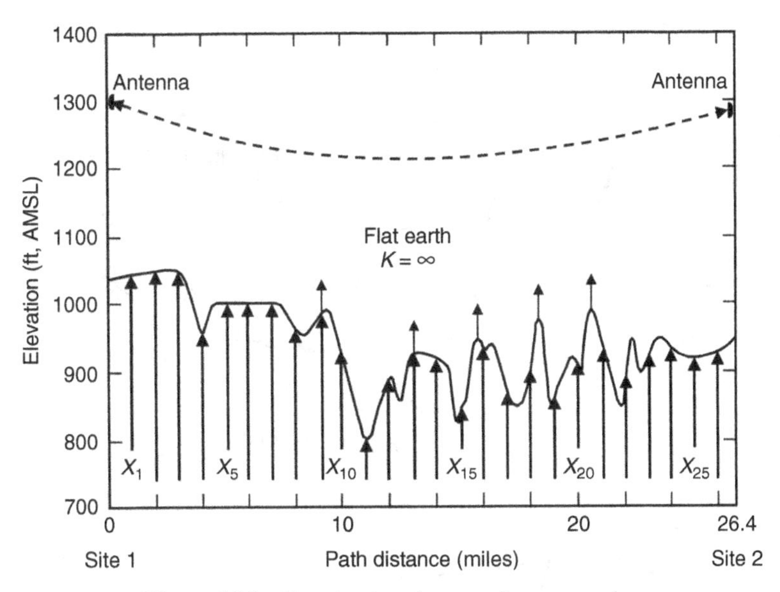

Figure 16.1 Terrain elevation roughness samples.

$$w = \sqrt{\left[\left(\frac{1}{n}\right)\sum_{i=1}^{n}(\mathbf{X}_i - \overline{\mathbf{X}})^2\right]} = \sqrt{\left[\left(\frac{1}{n}\right)\sum_{i=1}^{n}\mathbf{X}_i^2\right] - \overline{\mathbf{X}}^2} \quad (16.15)$$

Terrain data for the United States and Canada may be found at http://seamless.usgs.gov/index.php and http://www.geobase.ca/geobase/en/index.html, respectively.

The c-factor is determined heuristically. Several suggestions are in current use.

Figure 16.2a shows Vigants' suggestion (1975) for propagation areas in 1975. Figure 16.2b shows a chart updated by Bellcore (Achariyapaopan, 1986) (now Telecordia) in 1986. The Bellcore G-factor merges the c-factor and path roughness of the Vigants formula. Figure 16.2b includes terrain effects while Figure 16.2a does not. If the Gs 4 and 6 are both changed to 2 and 1/4 is changed to 1/2 in Figure 16.2b (to remove terrain effects) and they are relabeled as C, the Bellcore propagation areas would be good, average, and poor for Cs of 1/2, 1, and 2 respectively. The Bellcore map would then represent an updated version of the Vigants' map.

On the basis of the Bellcore update and his experience, Bill Knight of Alcatel-Lucent (ALU) created the data for Figure 16.3a. This chart modified the Great Lakes area and the Oklahoma/Texas border on the basis of experience. This was used by Alcatel-Lucent as an update to Vigants' original c-factor chart (Fig. 16.2a). Figure 16.2a, Figure 16.2b, and Figure 16.3a were originally developed for multipath propagation outage estimation. Multipath and obstruction fading are created by different processes. ALU design procedures reflect this difference by using different data to define abnormal propagation. (Figure 16.3b is based upon this data with modifications based upon conversations with Eddie Allen (Allen, E., private communication, 2010) and taking into account Samson's results discussed in Chapter 12.).

A distinction between multipath and obstruction areas is justified. However, using one criterion for multipath and another for obstruction fading (to set antenna heights—see Section 12.11) adds complication to the path design process. The largest differences between the climate factors (Figure 16.2a, Figure 16.2b, Figure 16.3a) and the abnormal propagation areas (Fig. 16.3b) are in the Average and Good areas. Simulation shows that obstruction fading is quite insensitive to antenna placement in Good and Average propagation areas. The climate factors and the abnormal propagation areas are quite similar in the Poor and Very Poor abnormal propagation areas. Many designers now use the climate C-factors to define abnormal propagation areas:

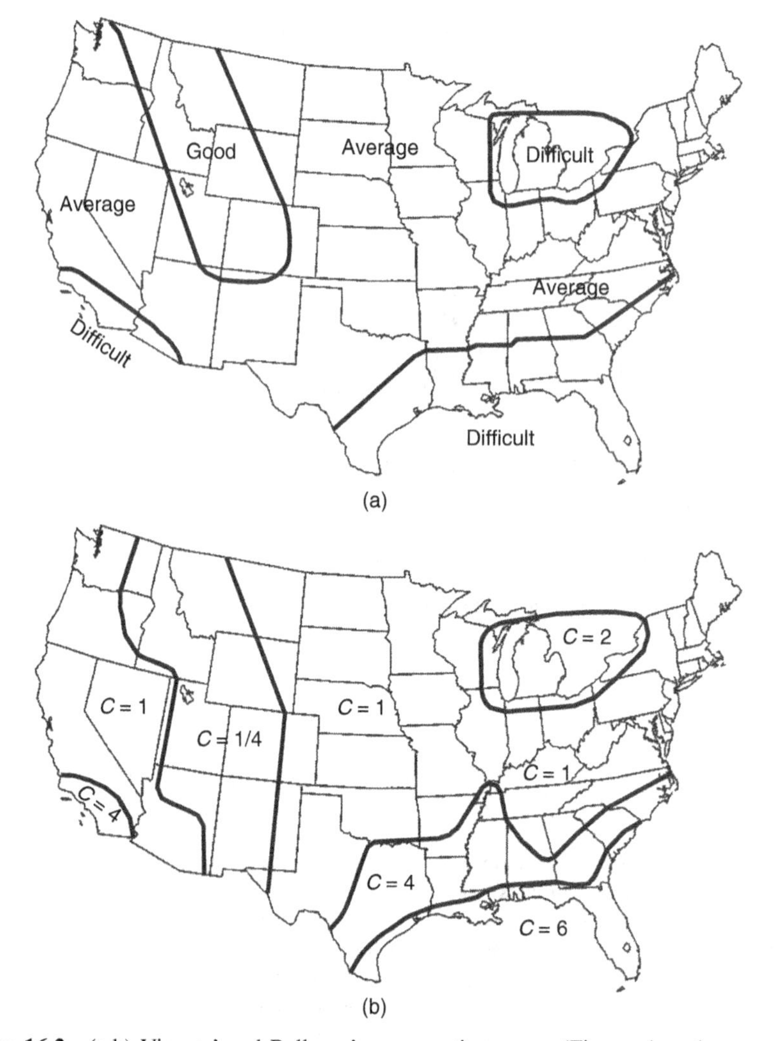

Figure 16.2 (a,b) Vigants' and Bellcore's propagation areas. (Figures based on sources).

$C = 0.5$ implies Propagation Area = Good

$C = 1.0$ implies Propagation Area = Average

$C = 2.0$ implies Propagation Area = Poor or Very Poor

Obstruction-fading outage time is calculated using the Schiavone algorithm (Chapter 12) for poor and very poor propagation. The conventional antenna clearance criteria should be used and the obstruction-fading model used to determine if additional clearance is needed in these areas.

Relatively recently, Vigants, Barnett, and others (Loso et al., 1992) have suggested that the c-factor for difficult United States (e.g., the US Gulf Coast) and international areas should be increased to 10. For worst-case conditions (areas of extreme heat and humidity such as the Red Sea or the Persian Gulf coastal plan or equatorial climates), they recommend a c-factor of 100. For benign climates (mountainous, dry or northerly conditions) such as the Rocky Mountains, Canada, and sections of Germany, they recommend c of 0.25. As yet, these suggestions have not been widely adopted in the design community.

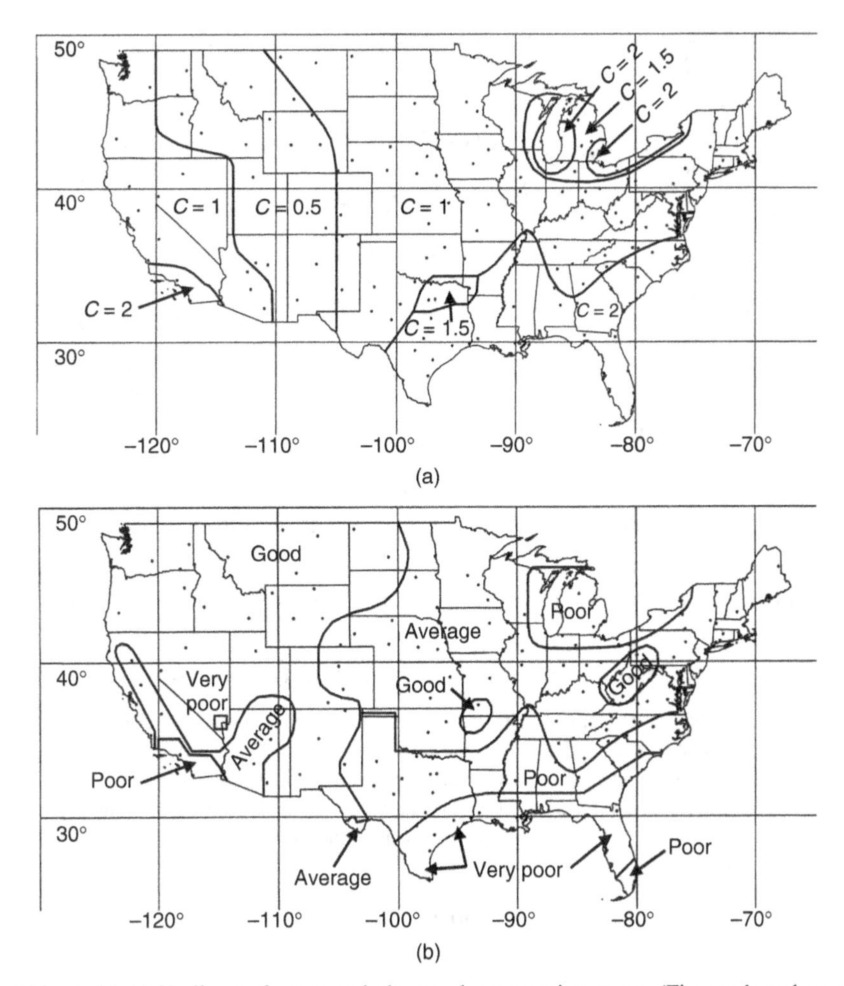

Figure 16.3 (a,b) ALU climate factors and abnormal propagation areas. (Figures based on sources).

Diversity improvement is addressed in Chapter 10. Flat and dispersive fade margins are described in Chapter 9. Some designers use a dispersion ratio (Rummler, 1988) to modify the dispersive fade margin. This is an attempt to account for the observation that some geophysically similar paths experience more dispersive fading than others. While this is an accepted issue, there is no industry agreement as to how to relate dispersion ratio to path characteristics. At present, this is addressed on an ad hoc basis.

Mean annual temperature at a location is a significant factor in the Vigants–Barnett multipath-fading model. The US NOAA maintains an extensive database of weather data, including temperature. NOAA's data (Figure 16.4) shows that the annual temperature of the United States varies significantly from year to year and from location to location.

Obviously, annual temperature must be averaged over 10–20 years to arrive at a statistically significant value. The following charts show a slight shift in temperature over the last 25 years (typically in the more northern states) (Fig. 16.5 and Fig. 16.6).

The ITU-R maintains a worldwide annual temperature data base. That data is located at http://www.itu.int/ITU-R/index.asp?category=study-groups&rlink=rsg3-software-ionospheric&lang=en. The data is in a zipped file under the tab "Annual mean surface temperature (Rec. P.1510)." At present, the data has an extraneous 360° (rightmost) column in the latitude and longitude files.

The ITU-R data (Fig. 16.7) is more granular than the NOAA data but otherwise the two sources are in general agreement.

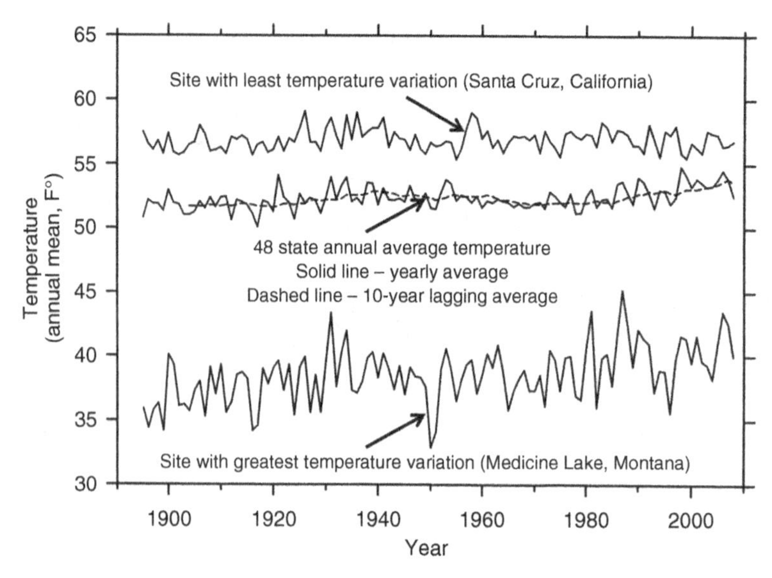

Figure 16.4 Annual temperature (°F) for the 48 states in the United States.

16.5.2 Cross-Polarization Discrimination Degradation Outages—NA

Some microwave systems are operated with transmit frequencies on both polarizations simultaneously. The operation of these systems is adversely impacted by cross-polarization degradation. Degradation of the received signal CPD can be caused by rotational misalignment (in any plane) of the transmit or receive antenna. Multipath and rain fading are also known to degrade CPD. At present, there is no North American industry agreement on how to estimate this degradation.

16.5.3 Space Diversity: Flat-Fading Improvement—NA

$$
\begin{aligned}
I_{\text{FF-SD}} &= \text{flat fading improvement factor for space diversity} \\
&= 7 \times 10^{-5}\, \eta\, [s\,(\text{ft})]^2\, \frac{f 10^{P/10}\, 10^{\text{FFM}/10}}{D\,(\text{miles})} \\
&= 1.2 \times 10^{-3}\, \eta\, [s(\text{m})]^2\, \frac{f\, 10^{P/10}\, 10^{\text{FFM}/10}}{D\,(\text{km})}
\end{aligned}
\tag{16.16}
$$

$1 \leq I_{\text{FF-SD}} \leq 200$
$0 \leq s(\text{ft}) \leq 50$
$0 \leq s(\text{m}) \leq 15$

FFM = flat fade margin (for the primary antenna);
D = path distance;
s = vertical separation of receiving antenna centers;
f = radio operating frequency (GHz);
η = switching hysteresis efficiency;
$= 2/[\{10^{H/10}\} + \{1/\{10^{H/10}\}\}]$;
H = switching hysteresis (dB);
= decibel value by which the RSL of the second receiver must be greater than the RSL of the first receiver before the output signal will be switched from the first receiver to the second receiver;

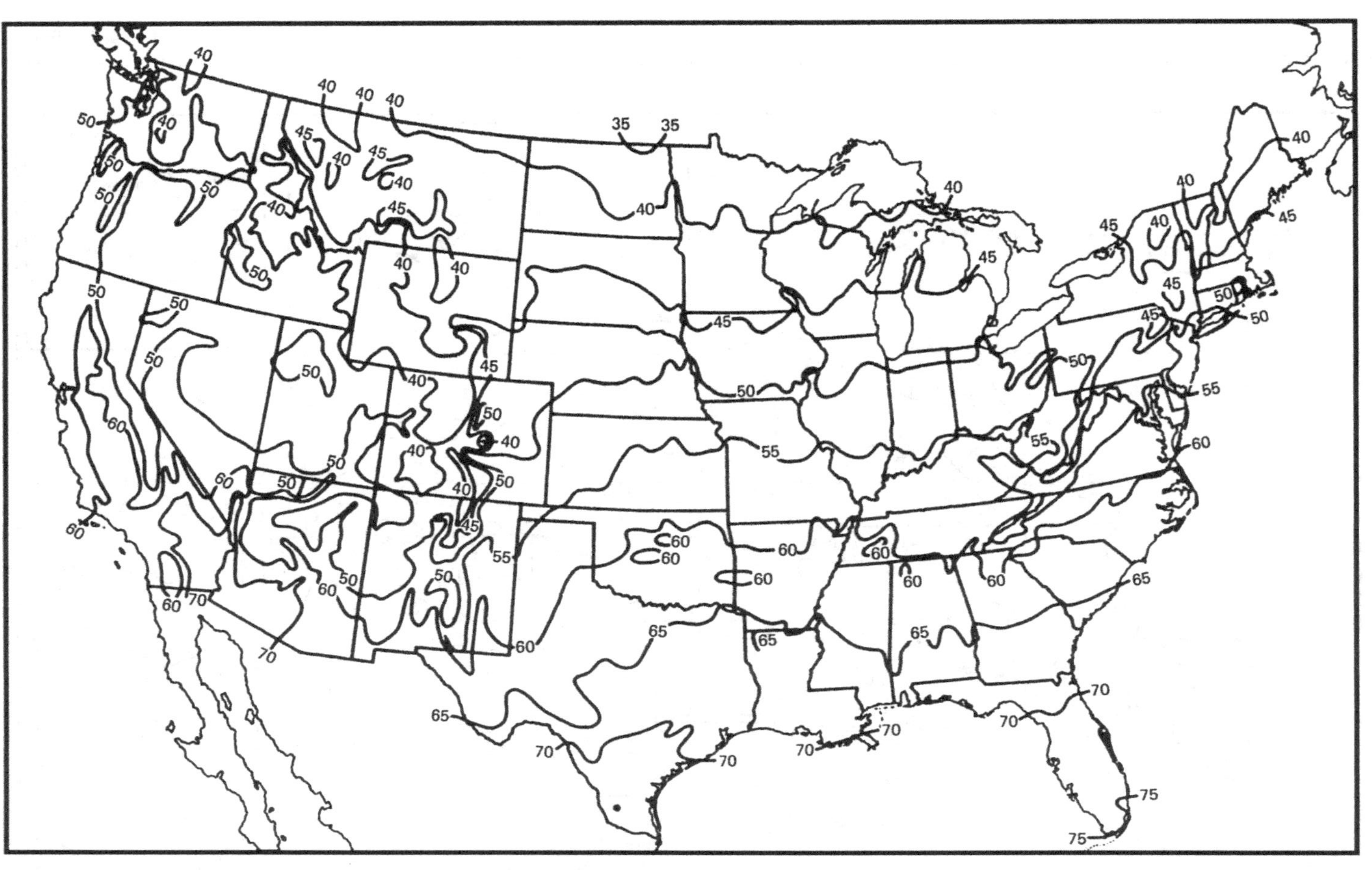

Figure 16.5 Annual temperature (°F). (*Source*: Vigants' 1975 data (Vigants, 1975). Reprinted with permission of Alcatel-Lucent USA, Inc).

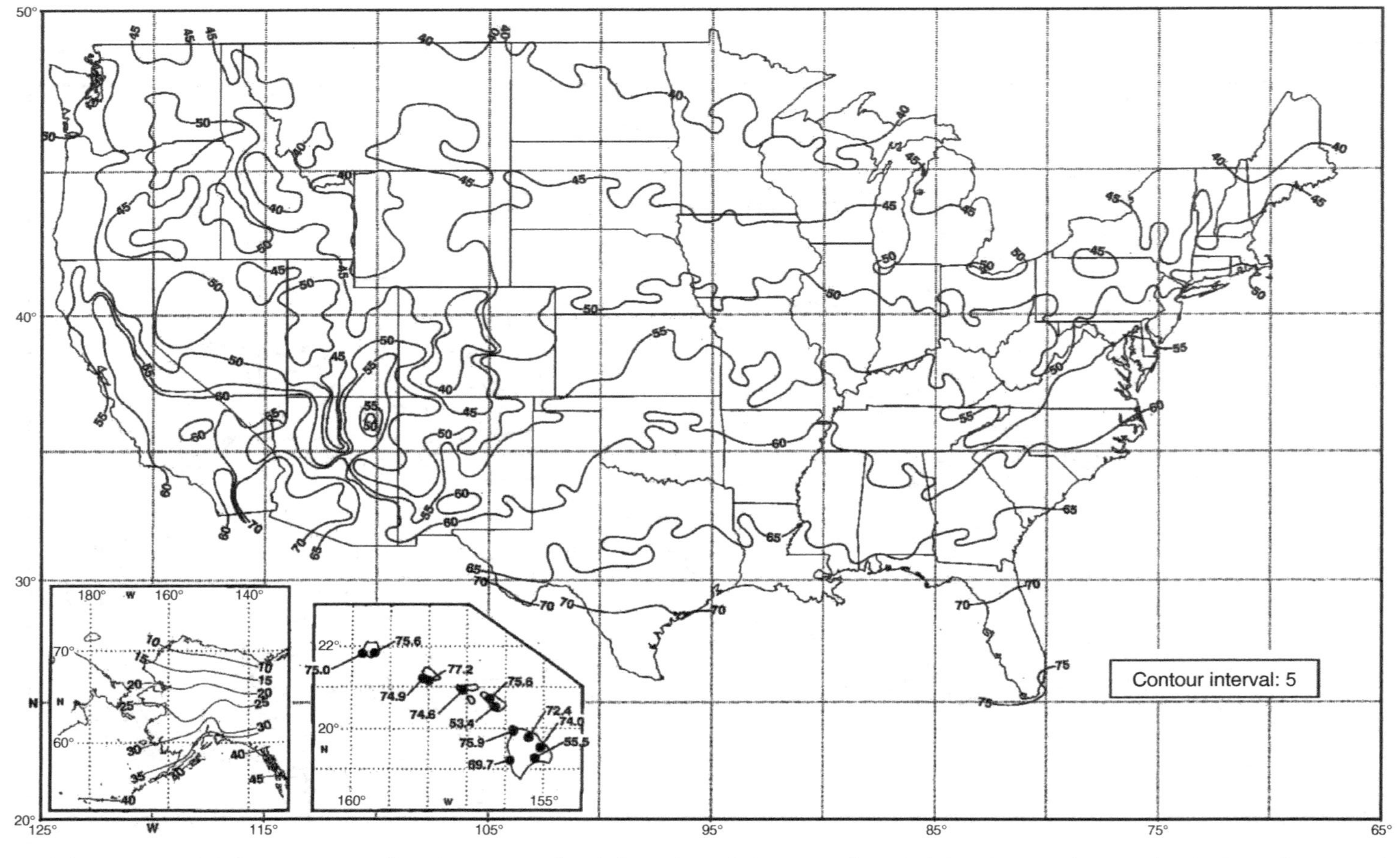

Figure 16.6 Current annual temperature (°F). (*Source*: National Oceanic and Atmospheric Administration, National Climatic Data Center, Asheville, North Carolina, 1961–1990 data).

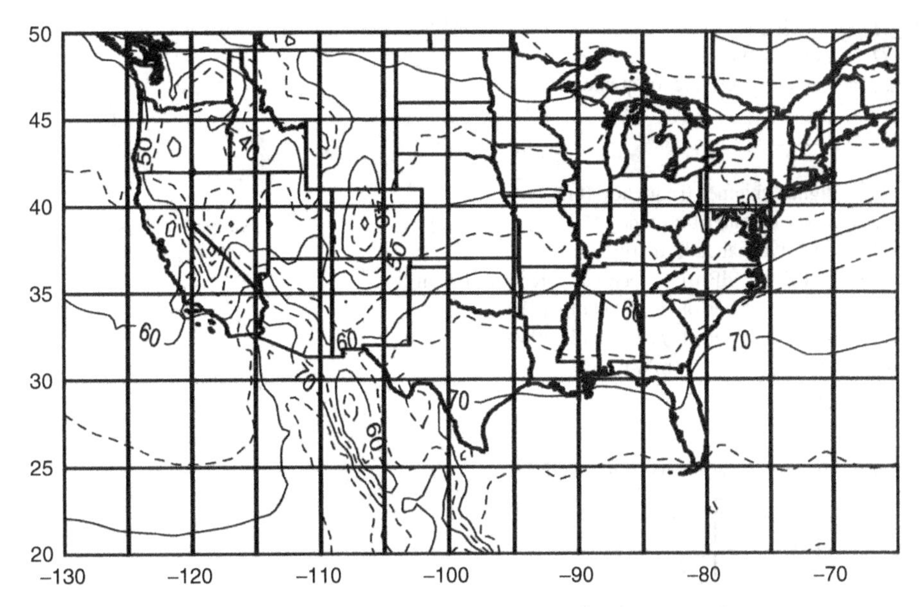

Figure 16.7 ITU-R mean annual temperature (°F) for the United States.

$P1$ = isotropic boresight gain (dB) of primary receive antenna;
$P2$ = isotropic boresight gain (dB) of space diversity receive antenna;
P = difference between space diversity antenna gain and the main antenna gain;
$= P2 - P1$.

16.5.4 Space Diversity: Dispersive-Fading Improvement—NA

Some engineers use the same diversity improvement factor for flat fading and dispersive fading. Early in the investigations of digital radio outages, it was recognized that using the flat fade margin diversity improvement factors underestimated dispersive-fading diversity improvement (Greenstein and Shafi, 1972). The following dispersive-fading improvement factors are based up Bell Labs investigations:

$$I_{\text{DF-SD}} = \text{dispersive fading improvement factor for space diversity}$$

$$= \frac{0.09 f\, 10^{\text{DFM}/10}}{D(\text{miles})}$$

$$= \frac{0.14 f\, 10^{\text{DFM}/10}}{D(\text{km})} \tag{16.17}$$

$1 \le I_{\text{DF-SD}} \le 200$

DFM = dispersive fade margin (dB) (Chapter 9).

16.5.5 Dual Frequency Diversity: Flat-Fading Improvement—NA

$$I_{\text{FF-FD}} = \text{flat fading improvement factor for frequency diversity}$$

$$= \frac{100 f_{\text{avg}} \eta 10^{\text{FFM}/10}}{D(\text{miles})\ G}$$

$$= \frac{161 f_{\text{avg}}\ \eta\ 10^{\text{FFM}/10}}{D(\text{km})\ G} \tag{16.18}$$

$1 \le I_{\text{FF-FD}} \le 200$

D = path distance;

f_{avg} = average operating frequency;

$$= \frac{(f_1 + f_2)}{2};$$

f_2 = higher channel frequency (GHz);

f_1 = lower channel frequency (GHz);

FFM = effective flat fade margin;

= RSLPrdB (dBm) − receiver threshold (dBm);

$P1$ = receive power of f_1 channel (mW);

$P2$ = receive power of f_2 channel (mW);

Pr = reference power (mW);

$$= \frac{P1 + P2}{2};$$

η = switching hysteresis efficiency;

$$= \frac{2}{\left[\left\{10^{(H/10)}\right\} + \left\{\frac{1}{\left[10^{(H/10)}\right]}\right\}\right]}$$

H = switching hysteresis (dB);

= decibel value by which the RSL of the second receiver must be greater than the RSL of the first receiver before the output signal will be switched from the first receiver to the second receiver;

$$G = \left\{\frac{2\left(f_{\text{avg}}\right)^3}{[(f_2 - f_1)\text{Pa Pb}]}\right\};$$

$$= \left\{\frac{(f_1 + f_2)^3}{[4(f_2 - f_1)\text{Pa Pb}]}\right\};$$

Pa = $P1$/Pr;

Pb = $P2$/Pr.

16.5.6 Dual Frequency Diversity: Dispersive-Fading Improvement—NA

Some engineers use the same diversity improvement factor for flat fading and dispersive fading. Early in the investigations of digital radio outages, it was recognized that using the flat fade margin diversity improvement factors underestimated dispersive-fading diversity improvement (Cellerino et al., 1985; Dirner and Lin, 1985). The following dispersive-fading improvement factors are based on Bell Labs investigations:

$$I_{\text{DF-FD}} = \text{dispersive fading improvement factor for frequency diversity}$$

$$= \frac{468 \times 10^{\text{DFM}/15}}{(f_{\text{avg}})^2 D(\text{miles})}$$

$$= \frac{753 \times 10^{\text{DFM}/15}}{(f_{\text{avg}})^2 D(\text{km})} \qquad (16.19)$$

$1 \le I_{\text{DF-FD}} \le 200$

DFM = dispersive fade margin (dB) (Chapter 9).

16.5.7 Quad (Space and Frequency) Diversity—NA

Quad diversity configurations are diagramed below in the ITU paragraphs on quad diversity. In North America, the standard quad diversity configuration has two transmitters and four receivers per radio terminal. Quad diversity improvement is estimated by multiplying the frequency and space diversity improvement factors calculated above.

$$I_{\text{FF-QD}} = \text{quad diversity flat fading diversity improvement factor}$$

$$= I_{\text{FF-SD}} \times I_{\text{FF-FD}} \tag{16.20}$$

$$I_{\text{DF-QD}} = \text{quad diversity dispersive fading diversity improvement factor}$$

$$= I_{\text{DF-SD}} \times I_{\text{DF-FD}} \tag{16.21}$$

16.5.8 Hybrid Diversity—NA

Diversity improvement is estimated by selecting the larger of the frequency and space diversity improvement factors calculated above.

$$I_{\text{FF-HD}} = \text{hybrid diversity flat fading diversity improvement factor}$$

$$= I_{\text{FF-SD}} \text{ or } I_{\text{FF-FD}}, \quad \text{whichever is larger} \tag{16.22}$$

$$I_{\text{DF-HD}} = \text{hybrid diversity dispersive fading diversity improvement factor}$$

$$= I_{\text{DF-SD}} \text{ or } I_{\text{DF-FD}}, \quad \text{whichever is larger} \tag{16.23}$$

This configuration is typically used when diversity is needed but space diversity cannot be installed at one of the sites.

16.5.9 Multiline Frequency Diversity—NA

For systems with more than one working channel and one or more protection channels ("multiline"), the following results apply:

$$I_{\text{FF-FD}} = \text{Flat fading improvement factor for frequency diversity}$$

$$= \frac{100 f_{\text{avg}} \eta 10^{\text{FFM}/10}}{D(\text{miles})G}$$

$$= \frac{161 f_{\text{avg}} \eta 10^{\text{FFM}/10}}{D(\text{km})G} \tag{16.24}$$

$1 \leq I_{\text{FF-FD}} \leq 200$

FFM = effective flat fade margin (assuming all RSLs to be the same);
f_{avg} = average channel frequency $[(f_1 + \cdots + f_{N+U})/(N+U)]$ (GHz);
D = path distance;
η = switching hysteresis efficiency;

$$= \frac{2}{\left[\left\{10^{(H/10)}\right\} + \left\{\frac{1}{\left[10^{(H/10)}\right]}\right\}\right]}$$

H = switching hysteresis (dB);
= decibel value by which the RSL of the second receiver must be greater than the RSL of the first receiver before the output signal will be switched from the first receiver to the second receiver.

$$I_{\text{DF-FD}} = \text{dispersive fading improvement factor for frequency diversity}$$

$$= \frac{9324\ f_{\text{avg}}\ 10^{\text{DFM}/15}}{D(\text{miles})\ G}$$

$$= \frac{15000\ f_{\text{avg}}\ 10^{\text{DFM}/15}}{D(\text{km})\ G} \tag{16.25}$$

$1 \leq I_{\text{DF-FD}} \leq 200$

DFM = dispersive fade margin (dB) (Chapter 9).

The G-factor is calculated as follows:

$$G = \frac{2(f_{\text{avg}})^3}{\Delta f_{\text{ch}} \Delta f_{\text{eq}}} \tag{16.26}$$

f_{avg} = average channel frequency $[(f_1 + \cdots + f_{N+U})/(N + U)]$ (GHz);
N = number of working radio channels;
U = number of protection channels (typically 1 or 2);
Δf_{ch} = channel spacing (frequency difference between consecutive channels) (GHz).

$$\Delta f_{\text{eq}} = \frac{2(C1 + C2N + C3N^2 + C4N^3)}{(1 + C5N + C6N^2 + C7N^3)} \tag{16.27}$$

For 1 : N protection ($U = 1$):	For 2 : N protection ($U = 2$):
$C1 = 0.5975701514570466$	$C1 = 1.833935974042738$
$C2 = 0.7428445800439416$	$C2 = 1.37635098350984$
$C3 = 0.3082229552151934$	$C3 = 0.3351140480008432$
$C4 = 0.04384794647757599$	$C4 = 0.03839451275666113$
$C5 = 1.583046763318658$	$C5 = 1.311603244602883$
$C6 = 0.6973150200993524$	$C6 = 0.3335083601043749$
$C7 = 0.1046098514355536$	$C7 = 0.04273458297033073$
For 3 : N protection ($U = 3$):	For 4 : N protection ($U = 4$):
$C1 = 3.738277399105584$	$C1 = 6.320916126203917$
$C2 = 2.032342560033778$	$C2 = 2.636255876978117$
$C3 = 0.3648839616080699$	$C3 = 0.3680260657151042$
$C4 = 0.03197283771075967$	$C4 = 0.03080961423005372$
$C5 = 1.216758924321869$	$C5 = 1.161986088953397$
$C6 = 0.2272356326313417$	$C6 = 0.1606827250777983$
$C7 = 0.02299512674588805$	$C7 = 0.01632182640267676$

16.5.10 Angle Diversity—NA

As noted in Chapter 10, at present, there is no industry agreement on how to estimate angle diversity improvement. There are theoretical methods but they require knowledge of the secondary path trajectory—and that varies in an unpredictable way from path to path (Rummler and Dhafi, 1989). Geiger (1991) suggests, "Angle diversity is, therefore, equivalent to space diversity where the two antennas are separated by a distance equal to half the antenna diameter D." This addresses flat fading but not dispersive fading (where the space diversity antenna spacing does not influence diversity improvement). In theory, angle diversity, when properly aligned (a difficult task), should be fairly effective against dispersive fading (as it is caused by a path with angle of arrival significantly different than the main path).

Some engineers simply use 10 for the flat-fading improvement and 20 (or an even larger number) for dispersive fading. A provisional approach would be to calculate flat- and dispersive-fading improvement for conventional space diversity (200 λ spacing) and divide the improvement factor by 10.

16.5.11 Upfading—NA

The mechanism that produces flat fading (signal cancellation or downfading) must also produce upfading (signal enhancement). Unpublished Bell Labs reports (Bell Laboratories, unpublished report on upfade experiments conducted at 4 and 6 GHz in Georgia, 1981) described this fading on the same path used to develop the flat-fading model. The upfades were lognormally distributed with a standard deviation of 3.5 dB. Upfades of 9.5 dB occured at the same rate as 30-dB downfades. Bill Knight and the author developed an unpublished method (based on the Bell Labs data and Abramowitz and Stegun (1968)) to estimate upfading.

$$\begin{aligned} U_{\mathrm{FM}} &= \text{upfade margin(dB)} \\ &= \mathrm{RSL}_{\mathrm{MAX}} - \mathrm{RSL} > 0 \end{aligned} \tag{16.28}$$

$\mathrm{RSL}_{\mathrm{MAX}}$ = receiver overload (dBm) specification (typically at 10^{-3} or 10^{-6} BER);
RSL = normal received signal level (dBm);
T_{FM} = thermal (flat fading) margin (dB) greater than 0;
$R_{10\log}$ = 10 log (upfade outage time/nondiversity downfade outage time).

$$\begin{aligned} R_{10\log} &= 10\log\left(\frac{\text{upfade outage time}}{\text{nondiversity downfade outage time}}\right) \\ &= 6.22 + T_{\mathrm{FM}} - (3.575 \times U_{\mathrm{FM}}) - (0.002718 \times U_{\mathrm{FM}}^3) \end{aligned} \tag{16.29}$$

$$\begin{aligned} R &= \frac{\text{upfade outage time}}{\text{nondiversity downfade outage time}} \\ &= 10^{(R_{10\log}/10)} \end{aligned} \tag{16.30}$$

$T_{\mathrm{UF}} = R \times T_{\mathrm{FF}}$
T_{UF} = upfade outage time (one way);
T_{FF} = flat-fading time (one way).

For systems with two receivers, the upfade margin is calculated for each receiver and the higher upfade margin is used. Upfading is much slower than downfading and it is not clear at this time whether diversity is effective for this type of fading. The current estimate is it is not.

16.5.12 Shallow Flat Fading—NA

All standard microwave-fading distributions assume a fade depth of at least 20 dB. Olsen and Segal (1992) developed an empirical formula for shallow multipath fading between 0 dB and approximately 25 dB in depth. Loso et al. (1992) adapted the method for use with any multipath-fading model.

$$\begin{aligned} P_{\mathrm{FF}} &= 1 - \frac{1}{e^E} \\ &= \text{probability of fade to at least } A \\ &= \frac{p_{\mathrm{w}}(\%)}{100} = 0.01 p_{\mathrm{w}} \end{aligned} \tag{16.31}$$

e = Euler's number $\approx$ 2.718281828459045;
$E = 1/10^{(qA/20)}$;

A = shallow fade depth (dB), $0\,\text{dB} \leq A \leq 25\,\text{dB}$;
$q = 2 + \{K[11]\,(q_t + R[11])\}$;
$K[11] = [1 + (0.3/10^{A/20})]/10^{0.016A}$;
$R[11] = 4.3[(1/10^{A/20}) + (A/800)]$;
$q_t = \{(r - 2)/K[30]\} - R[30]$;
$r = -0.8 \log \{-\ln (1 - P[30])\}$;
$K[30]$ = $K[11]$ evaluated for $A = 25$;
$R[30]$ = $R[11]$ evaluated for $A = 25$;
$P[30]$ = probability of a fade to at least 25 dB [P_{FF}] calculated using the conventional multipath-fading formula;
$\log(x)$ = common logarithm of x;
$\ln(x)$ = natural logarithm of x.

16.6 INTERNATIONAL TELECOMMUNICATION UNION—RADIOCOMMUNICATION SECTOR (ITU-R) PATH PERFORMANCE CALCULATIONS

Unlike the North American model where all path degradations are summed together, for the ITU-R model, different assessments are made.

Quality path degradations are those which individually last less than 10 consecutive seconds. These degradations are one-way effects of multipath and short-term interference. The period over which the objectives are estimated is the "worst month." "Worst month" is defined by ITU-R P.581-2 (International Telecommunication Union—Radiocommunication Sector (ITU-R), 1990). The note to this recommendation suggests that "worst month" should be interpreted as average "worst month" in an average year.

Availability degradations are those that individually last longer than 10 consecutive seconds. They are two-way degradations due to rain and/or obstruction fading in conjunction with interference. Acceptable levels of interference would not degrade availability without the concurrent occurrence of fading due to one of these mechanisms. The period over which the objectives are estimated is 1 year.

For quality, the following apply:

$$\begin{aligned} T_{\text{TOTAL-Q}} &= \text{total quality outage time} \\ &= T_{\text{FLAT}} + T_{\text{DISP}} + T_{\text{UPFADE}} \end{aligned} \tag{16.32}$$

T_{FLAT} = one-way flat-fading outage time;
= $T_{\text{ref}} P_{\text{FF}}$ (probability of flat fade outage);
T_{DISP} = one-way dispersive-fading outage time;
= $T_{\text{ref}} P_{\text{DF}}$ (probability of dispersive fade outage);
T_{UPFADE} = one-way upfading outage—if calculated;
= $T_{\text{ref}} P_{\text{U}}$ (probability of up fade outage);
T_{ref} = 2,629,800 s (1 month).

For availability, the following apply:

$$T_{\text{TOTAL-A}} = \text{total quality outage time}$$

$$= T_{\text{RAIN}} + T_{\text{OBST}} \tag{16.33}$$

$$T_{\text{RAIN}} = \text{rain fading outage}$$

$$= T_{\text{ref}}\, P_{\text{R}} \text{ (probability of rain fade outage)} \tag{16.34}$$

$$T_{\text{OBST}} = \text{obstruction outage time – if calculated}$$

$$= T_{\text{ref}}\, P_{\text{O}} \text{ (probability of obstruction fade outage)} \tag{16.35}$$

T_{ref} = 31,557,600 s (1 year).

16.6.1 Flat Fading—ITU-R

The ITU-R flat (nonselective)-fading model has evolved over the years (International Telecommunication Union—Radiocommunication Sector (ITU-R), 2006a; International Telecommunication Union—Radiocommunication Sector (ITU-R), 2006b; International Telecommunication Union—Radiocommunication Sector (ITU-R), 2009): As Lane (2008) observed, the early ITU-R fading model (CCIR Report 338) was essentially the Vigants–Barnett (United States) model (Barnett, 1972; Vigants, 1975):

$$P_{\mathrm{FF}} = KQfD^3 10^{-A} \tag{16.36}$$

The Morita (Japanese) model (Morita, 1970) was also supported.

$$P_{\mathrm{FF}} = KQf^{1.2}D^{3.5}10^{-A} \tag{16.37}$$

In 1995 ITU-R Recommendation P.530-5 introduced the Method 2 outage model:

$$P_{\mathrm{FF}} = \frac{K}{100} f^{0.93} d^{3.3} (1 + |\varepsilon_{\mathrm{p}}|)^{-1.1} \varphi^{-1.2} 10^{-A/10} \tag{16.38}$$

In 1999 ITU-R Recommendation P.530-7 introduced a new outage model:

$$P_{\mathrm{FF}} = \frac{K}{100} f^{0.89} d^{3.6} (1 + |\varepsilon_{\mathrm{p}}|)^{-1.4} 10^{-A/10} \tag{16.39}$$

This model was based on the work of Doble, Martin, Olsen, Segal, and Tjelta (Olsen and Segal, 1992; Olsen et al., 1986; Tjelta et al., 1990). In 2005 ITU-R Recommendation P.530-9 introduced the "Detailed Link Design" outage model based on the later work of Martin, Olsen, Segal, and Tjelta (Olsen and Tjelta, 1999; Olsen et al., 2003). This model was recently updated in P-530-13 (International Telecommunication Union—Radiocommunication Sector (ITU-R), 2009).

Standard Method:

$$\begin{aligned} P_{\mathrm{FF}} &= K d^{3.4} (1 + |\varepsilon_{\mathrm{p}}|)^{-1.03} f^{0.8} 10^{[-0.00076 h_{\mathrm{L}} - A/10]}/100 \\ &= 0.01 K d^{3.4} (1 + |\varepsilon_{\mathrm{p}}|)^{-1.03} f^{0.8} 10^{-0.00076 h_{\mathrm{L}}} 10^{-A/10} \end{aligned} \tag{16.40}$$

P_{FF} = probability of flat fade to at least A (dB) during the most active month ((International Telecommunication Union—Radiocommunication Sector (ITU-R), 2009), Eq. 7);
$= p_{\mathrm{w}}(\%)/100 = 0.01 p_{\mathrm{w}}$;
$K = (10 + s_{\mathrm{a}})^{-0.46} 10^{-4.4} 10^{-0.0027 \mathrm{dN}_1}$ ((International Telecommunication Union—Radiocommunication Sector (ITU-R), 2009), Eq. 4).

Quick-Planning Method ((International Telecommunication Union—Radiocommunication Sector (ITU-R), 2009), Eq. 8):

$$\begin{aligned} P_{\mathrm{FF}} &= K d^{3.1} (1 + |\varepsilon_{\mathrm{p}}|)^{-1.29} f^{0.8} 10^{-0.00089 h_{\mathrm{L}}} 10^{-A/10}/100, \\ &= \text{probability of flat fade to at least A during the most active month} \\ &= p_{\mathrm{w}}(\%)/100 = 0.01 p_{\mathrm{w}} \end{aligned} \tag{16.41}$$

$K = 10^{-4.6}\ 10^{-0.0027} \mathrm{d}N_1$, ((International Telecommunication Union—Radiocommunication Sector (ITU-R), 2009), Eq. 5);
p_{w} = probability (%) of flat fade to at least A;
$P_{\mathrm{FF}} = p_{\mathrm{w}}$ (%)/100 = probability (fraction) of flat fade to at least A;
d = path length (km) = 1.609 path length (miles);
$|\varepsilon_{\mathrm{p}}| = |h_{\mathrm{r}} - h_{\mathrm{e}}|/d$ = path tilt (radians)
h_{r} = emitting antenna height (m) = 0.3048 height (ft);

h_e = receiving antenna height (m) = 0.3048 height (ft);
f = operating frequency (GHz);
A = path fade margin (dB);
h_L = lower antenna height (m, above [mean] sea level).

dN_1 is the point refractivity gradient in the lowest 65 m of the atmosphere not exceeded for 1% of an average year. dN_1 is provided on a 1.5° grid in latitude and longitude in Recommendation ITU-R P.453. The correct value for the latitude and longitude at the path center should be obtained from the values for the four closest grid points by bilinear interpolation. The [Radio Refractivity (Rec. P.453)—Refractivity gradient in lowest 65 m] data can be downloaded from http://www.itu.int/ITU-R/index.asp?category=study-groups&rlink=rsg3-software-ionospheric&lang=en. Under the tab "Software concerning Tropospheric Propagation," download the software for "Radio Refractivity (Rec. P.453)—Refractivity gradient in lowest 65m." The data is in the zipped folder.

s_a is the area terrain roughness. It is defined as the standard deviation of terrain heights (m) within a 110 × 110 km area with a 30 arc second resolution (e.g., the Global GTOPO30 data). The area should be aligned with the longitude, such that the two equal halves of the area are on each side of the longitude that goes through the path center.

The GTOPO30 is not available from ITU-R. It is a global digital elevation model (DEM) resulting from a collaborative effort led by the staff at the US Geological Survey's EROS Data Center in Sioux Falls, South Dakota. Elevations in GTOPO30 are regularly spaced at 30 arc seconds (~1 km). It can be downloaded from http://edc.usgs.gov/products/elevation/gtopo30/gtopo30.html.

16.6.2 Dispersive Fading—ITU-R

P_S = probability of dispersive (frequency selective) fade to at least A during the most active month (A was defined in the previous section) (International Telecommunication Union—Radiocommunication Sector (ITU-R), 2006a; International Telecommunication Union—Radiocommunication Sector (ITU-R), 2006b; International Telecommunication Union—Radiocommunication Sector (ITU-R), 2009)

$$P_S = 2.15\eta\left(W_M \times 10^{-B_M/20}\frac{\tau_m^2}{|\tau_{r,M}|} + W_{NM} \times 10^{-B_{NM}/20}\frac{\tau_m^2}{|\tau_{r,NM}|}\right) \quad (16.42)$$

η = multipath activity factor = $1 - e^{-0.2(P_{FF})^{0.75}}$;
P_{FF} = probability of flat fading calculated in the previous section;
τ_m = mean dispersive time delay (ns) = $0.7(d/50)^{1.3}$;
$\tau_{r,M}$ = reference delay (ns) used to obtain the minimum phase signature curve;
$\tau_{r,NM}$ = reference delay (ns) used to obtain the nonminimum phase signature curve;
W_M = width (MHz) of minimum phase signature curve;
W_{NM} = width (MHz) of nonminimum phase signature curve;
B_M = average depth (dB) of minimum phase signature curve;
B_{NM} = average depth (dB) of nonminimum phase signature curve.

Typically, $W_M = W_{NM}$, $B_M \approx B_{NM}$ and $\tau_{r,M} = \tau_{r,NM}$. In this case, the outage probability formula simplifies to the following:

$$P_s = 4.3\eta\left(W_A \times 10^{-B_A/20}\frac{\tau_m^2}{|\tau_r|}\right) \quad (16.43)$$

τ_r = reference delay (ns) used to obtain the signature curve;
W_A = average width (MHz) of signature curve;
B_A = average depth (dB) of signature curve.

If we wish to use the Bellcore (Telcordia) dispersive fade margin, we may use the following relationship from Chapter 9:

$$B_{\mathrm{A}} = -32.1 + 0.88\mathrm{DFM} + 8.8\log(W_{\mathrm{A}})$$

$$\mathrm{DFM} = \text{Bellcore dispersive fade margin(dB)}$$

$$\tau_{\mathrm{r}} = 6.3 \text{ ns} \tag{16.44}$$

Let us consider a microwave receiver operating in a 30-MHz channel. A typical receiver would have a average signature curve width of about 80% of the operating channel ($W_{\mathrm{A}} = 24$ MHz). Using the above relationships yields the following results (Fig. 16.8).

In Figure 16.8, availability (%) $= 100 \times (1 - P_{\mathrm{FF}})$. It is clear that for typical Bellcore DFM values for wideband radios (45–55 dB), dispersive fading is a dominate outage mechanism. For North American calculations, dispersive fading is not a function of flat fading or of path length. Dispersive outages calculated for long, high availability paths using the ITU-R method are larger than those calculated using the North American methodology.

Let us investigate other relationships of the dispersive outage calculation.

$$e^X = 1 + x + \frac{x^2}{2} + \frac{x^3}{6} + \cdots \approx 1 + x, \quad \text{for } x \text{ very small} \tag{16.45}$$

P_{FF}, the probability of outage, is very small.

$$\eta = 1 - e^{-0.2(P_{\mathrm{FF}})^{0.75} \approx 1 - [1 - 0.2(P_{\mathrm{FF}})^{0.75}] = 0.2(P_{\mathrm{FF}})^{0.75}} \tag{16.46}$$

$$P_{\mathrm{s}} \approx 0.86 P_{\mathrm{FF}}^{0.75}\left(W_{\mathrm{A}} \times 10^{-B_{\mathrm{A}}/20}\frac{\tau_{\mathrm{m}}^2}{|\tau_{\mathrm{r}}|}\right) \tag{16.47}$$

The ITU-R dispersive outage is directly proportional to flat fading. In the North American method, they are independent.

$$P_{\mathrm{FF}} \propto p_{\mathrm{w}} \propto d^{3.4} f^{0.8} \text{ from the previous results} \tag{16.48}$$

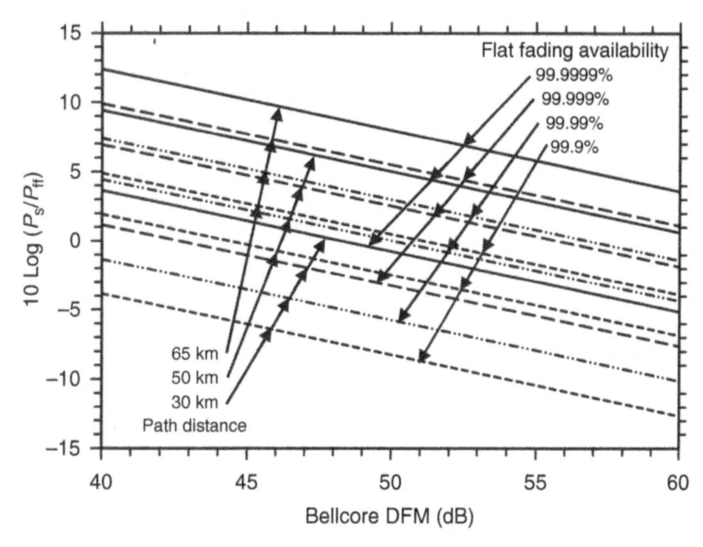

Figure 16.8 Calculated dispersive fading for 30-MHz radio channel.

$$P_S \propto (P_{FF})^{0.75} \propto (p_w)^{0.75} \propto d^{2.55} f^{0.6\ldots} \qquad (16.49)$$

The ITU-R dispersive outage is related to distance and frequency by exponential factors 2.55 and 0.6, respectively. The North American method relates outage to distance and frequency by exponential factors 3.0 and 1.0, respectively. Vigants (1984) demonstrated that dispersive fading was related to distance cubed. To date, the frequency dependence is experimentally inconclusive.

The mean delay τ_m used in the ITU-R method is generally less than 1 ns. While this value is consistent with delay associated with flat fading, it is significantly less than any delays associated with dispersive fading. Rummler (1981) suggests that dispersive delays must be at least {1000/[6 (receiver bandwidth in MHz)]} ns long. He measured dispersive delays of 5.6–18.4 ns on his 26-mile reference path (Chapter 9). There is no experimentally established relationship between flat-fading and dispersive-fading delays. The reason for using such short mean delay for calculating dispersive fading is not clear.

16.6.3 Cross-Polarization Discrimination Degradation Outages—ITU-R

Some microwave systems are operated with transmit frequencies on both polarizations simultaneously (International Telecommunication Union—Radiocommunication Sector (ITU-R), 2009). The operation of these systems is adversely impacted by cross-polarization degradation. Degradation of the receive signal cross-polarization discrimination (XPD) can be caused by rotational misalignment (in any plane) of the transmit or receive antenna. Multipath (clear air effects) and rain (precipitation) fading are also known to degrade CPD and occasionally cause propagation outages.

Cross-Polarization Discrimination Degradation for Multipath Fading

$$P_{XP} = \text{probability of outage due to clear air effects} = \frac{P_{FF}}{10^{M_{XPD}/10}} \qquad (16.50)$$

P_{FF} = probability of flat fading calculated in the previous two sections

$$M_{XPD} = \text{cross-polarization margin (dB) for a given bit error rate(BER)} = C - \left(\frac{C_O}{I}\right) + \text{XPIF} \qquad (16.51)$$

XPIF = cross-polarization canceller effect
= 20 dB for typical cross-polarization interference canceller (XPIC)
= 0 if cross-polarization interference canceller (XPIC) is not used

$\frac{C_O}{I}$ = carrier to interference ratio (dB) that produces a reference BER
= co-channel T/I (dB) for similar modulation if BER 10^{-6}

$C = \text{XPD}_O + Q$

$\text{XPD}_O = \text{XPD}_g + 5$ for $\text{XPD}_g \le 35$ dB
$= \text{XPD}_g$ for $\text{XPD}_g > 35$ dB

XPD_g = the lesser of the transmit or receive antenna's minimum boresite crosspolarization discrimination

$$Q = -10\log\left(\frac{k_{XP}\eta}{P_{FF}}\right)$$

η = multipath activity factor $= -e^{-0.2}(P_{FF})^{0.75}$

k_{XP} = 0.7 for one transmit antenna (or two antennas on the same vertical elevation)
$= 1 - 0.3\ \exp\left[-4\times 10^{-6}\left(\frac{S_t}{\lambda}\right)^2\right]$ for two (vertically spaced) transmit antennas

S_t = antenna vertical spacing (m);
λ = radio signal wavelength (m) = $0.29980/f$;
f = radio signal (center) frequency (GHz).

Cross-Polarization Discrimination Degradation for Rain Fading

$$P_{\mathrm{XPR}} = \text{probability of outage due to precipitation effects}$$
$$= 10^{(n-2)}$$
$$n = \frac{(-25.4 + \sqrt{161.23 - 4m})}{2}, \quad -3 \le n \le 0$$
$$m = 40 \text{ or } \left\{23.26 \log \left[\frac{A_{\mathrm{p}}}{0.12\ A_{0.01}}\right]\right\}, \quad \text{whichever is lesser}$$
$$= \text{path rain attenuation exceeded 0.01\% of the time (see Chapter 11)}$$

$$A_{\mathrm{p}} = 10^{(U - C_0/I + \mathrm{XPIF})/V}$$

$$U = 15 + 30 \log \mathrm{f}$$

$$U = 12.8\ f^{0.19} \text{ for } 8\ \mathrm{GHz} \le f \le 20\ \mathrm{GHz}$$

$$= 22.6 \text{ for } 20\ \mathrm{GHz} < \mathrm{f} \le 35\ \mathrm{GHz}$$

16.6.4 Upfading—ITU-R

All microwave radio receivers should have at least 10-dB upfade margin ("headroom") (International Telecommunication Union—Radiocommunication Sector (ITU-R), 2009).

$$P_{\mathrm{U}} = \text{probability of an upfade greater than 10 dB}$$
$$= \frac{p_{\mathrm{w}}}{100} \tag{16.52}$$
$$= \frac{\{100 - (10^{-0.4857}\ 10^{+0.05714A_{0.01}}\ 10^{-E/3.5})\}}{100}$$

p_{w} = probability of an upfade greater than 10 dB (%) ((International Telecommunication Union—Radiocommunication Sector (ITU-R), 2009), Eq. 19);
E = upfade power level (dB, relative to normal received signal level);
$A_{0.01}$ = fade margin (dB) that results in $P_{\mathrm{FF}} = 0.0001$ ($p_{\mathrm{w}} = 0.01\%$).

16.6.5 Shallow Flat Fading—ITU-R

All standard microwave fading distributions assume a fade depth of at least 20 dB (International Telecommunication Union—Radiocommunication Sector (ITU-R), 2009). Olsen and Segal (Olsen and Segal, 1992) developed an empirical formula for shallow multipath fading between 0 dB and approximately 25 dB in depth. A modified version of this method is detailed in ITU-R Recommendation P 530–12, para. 2.3.2. (International Telecommunication Union—Radiocommunication Sector (ITU-R), 2009)

$$p_{\mathrm{w}}[0] = 100 P_{\mathrm{FF}}[0] = \text{fading factor time axis intercept point}$$
$$= 100 P_{\mathrm{FF}}[0] \text{ calculated with } A = 0 \text{ using the flat fading formulas above for the Standard or Quick Planning Methods} \tag{16.53}$$

A = multipath fade depth (dB), $0\ \mathrm{dB} \le A$;
$A_{\mathrm{t}} = 25 + 1.2 \log (p_{\mathrm{w}}[0])$ = shallow fading transition point;

$P_t = (p_w[0]/100)/10^{A_t/10} = P_{FF}$ at shallow fading transition point.

For $A \geq A_t$, use the standard (deep) flat-fading formula for $p_w[11] = 100P_{FF}[11]$:

$$p_w[11] = p_w[0]\ 10^{-A/10} \tag{16.54}$$

For $A \leq A_t$, use the follow (shallow) flat-fading formulas (The following formulas are only valid if $p_w[0] < 2000$.):

$$\begin{aligned}
P_{FF} &= 1 - \left(\frac{1}{e^E}\right) \\
&= \text{probability of fade to at least } A \\
&= \frac{p_w(\%)}{100} = 0.01\ p_w \\
e &= \text{Euler's number} \approx 2.718281828459045 \\
E &= \frac{1}{10^{(qA/20)}} \\
A &= \text{shallow fade depth (dB)}, \quad 0\text{dB} \leq A \leq A_t \\
q &= 2 + \{K[11](q_t + R[11])\} = q_a \text{ in the ITU-R recommendation} \\
K[11] &= \frac{\left[1 + \left(\frac{0.3}{10^{A/20}}\right)\right]}{10^{0.016A}} \\
R[11] &= 4.3\left[\left(\frac{1}{10^{A/20}}\right) + \left(\frac{A}{800}\right)\right] \\
q_t &= \left[\frac{(r-2)}{K[A_t]}\right] - R[A_t] \\
r &= -\left(\frac{20}{A_t}\right)\log\{-\ln(1 - P_t)\} = q_a \text{ in the ITU-R recommendation}
\end{aligned} \tag{16.55}$$

$K[A_t] = K[11]$ evaluated for $A = A_t$;
$R[A_t] = R[11]$ evaluated for $A = A_t$;
$\log(x)$ = common logarithm on x;
$\ln(x)$ = natural logarithm of x.

16.6.6 Space Diversity Improvement—ITU-R

$$P_d = (P_{dns}^{0.75} + P_{ds}^{0.75})^{4/3} \tag{16.56}$$

P_d = total outage probability using space diversity (International Telecommunication Union—Radiocommunication Sector (ITU-R), 2009);
P_{dns} = probability of diversity nonselective (flat fading) outage;
P_{ds} = probability of diversity selective (dispersive fading) outage.

Flat-Fading Improvement

$$P_{dns} = \frac{P_{ns}}{I_{ns}}$$

$$P_{ns} = \frac{P_W}{100} = P_{FF} \text{ (calculated above in flat fading estimation)}$$

$$I_{ns} = [1 - e^{C}]\ 10^{-V/10}\ 10^{A/10}$$
$$C = -0.04\ S^{0.87}\ f^{-0.12}\ d^{0.48}\ (100\ P_{FF})^{-1.04}$$
$$V = |G_{main} - G_{diversity}| \tag{16.57}$$

I_{ns} = nonselective improvement factor;
S = vertical separation (center to center) of receiving antennas (m), $3\ m \leq S \leq 23\ m$;
f = operating frequency (GHz), $2\ \text{GHz} \leq f \leq 11\ \text{GHz}$;
d = path length (km), $25\ \text{km} \leq d \leq 240\ \text{km}$;
A = fade margin (dB);
G_{main} = main receive antenna gain (dBi);
G_{div} = diversity receive antenna gain (dBi).

$$k_{ns}^2 = 1 - \frac{I_{ns} \times P_{ns}}{\eta}$$

η = multipath activity factor $= 1 - e^{-0.2(P_{FF})^{0.75}}$.

Dispersive-Fading Improvement

$$P_{ds} = \frac{P_s^2}{\eta\ (1 - k_s^2)} \tag{16.58}$$

P_S = probability of nondiversity selective outage (see dispersive fading above).

$$k_s^2 = \begin{cases} 0.8238 & \text{for } r_w \leq 0.5 \\ 1 - 0.195\ (1 - r_w)^{0.109-0.13\log(1-r_w)} & \text{for } 0.5 < r_w \leq 0.9628 \\ 1 - 0.3957\ (1 - r_w)^{0.5136} & \text{for } r_w > 0.9628 \end{cases}$$

$$r_w = \begin{cases} 1 - 0.9746\ (1 - k_{ns}^2)^{2.170} & \text{for } k_{ns}^2 \leq 0.26 \\ 1 - 0.6921\ (1 - k_{ns}^2)^{1.034} & \text{for } k_{ns}^2 > 0.26 \end{cases}$$

$k_{ns}^2 = 1 - \frac{I_{ns} \times P_{ns}}{\eta}$ = nonselective correlation coefficient (to be used below).

16.6.7 Dual-Frequency Diversity Improvement—ITU-R

Calculate the same as for space diversity except use the following I_{ns} ((International Telecommunication Union—Radiocommunication Sector (ITU-R), 2009), Eq. 120).

$$I_{ns} = \left[\frac{80\ \Delta f}{(d\ f^2)}\right] 10^{A/10}$$
$$0 < \left(\frac{\Delta f}{f}\right) \leq 0.05 \tag{16.59}$$

I_{ns} = flat-fading improvement factor for frequency diversity;
Δf = frequency separation (GHz), $\Delta f \leq 0.5$ GHz;
f = operating frequency (GHz), $2\ \text{GHz} \leq f \leq 11\ \text{GHz}$;
d = path length (km), $30\ \text{km} \leq d \leq 70\ \text{km}$;
A = path (flat) fade margin (dB).

16.6.8 Quad (Space and Frequency) Diversity—ITU-R

ITU-R considers two different quad diversity configurations: (Fig. 16.9 and Fig. 16.10) (International Telecommunication Union—Radiocommunication Sector (ITU-R), 2009).

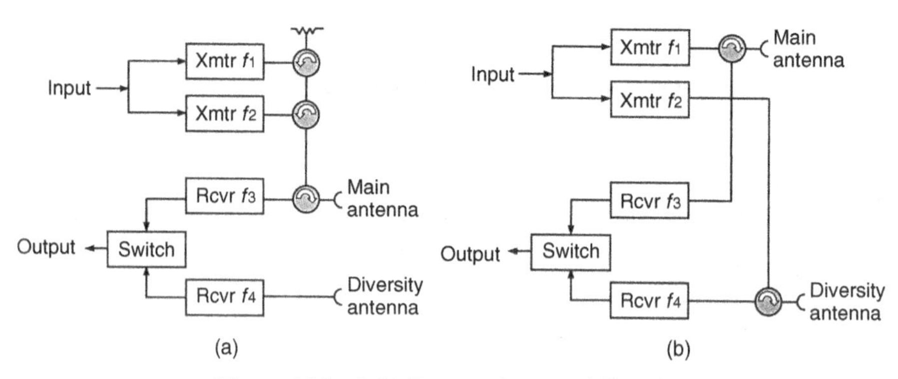

Figure 16.9 (a,b) Two receiver quad diversity.

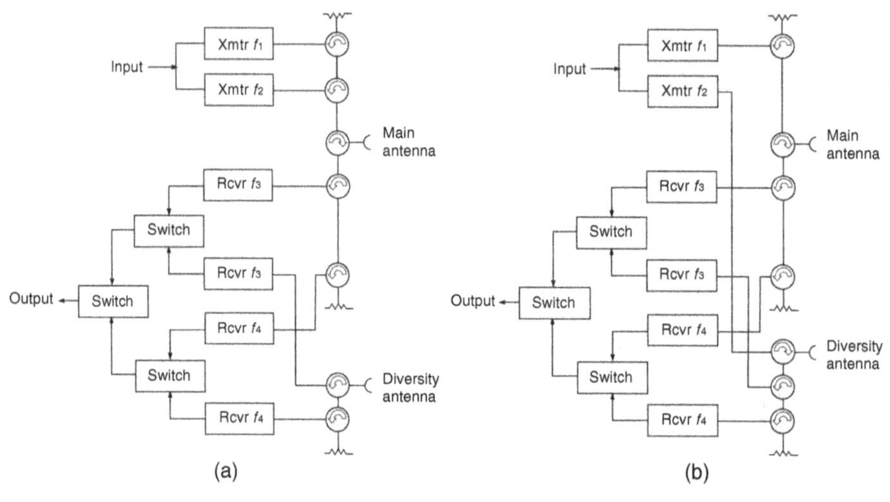

Figure 16.10 (a,b) Four receiver quad diversity.

16.6.8.1 Two Receivers

Calculate the same as for space diversity except substitute $k_{\text{ns}} = k_{\text{dns}}$.

$$k_{\text{dns}} = k_{\text{ns,s}} \times k_{\text{ns,f}} \tag{16.60}$$

k_{dns} = diversity nonselective correlation coefficient;
$k_{\text{ns,s}} = k_{\text{ns}}$ as calculated above for space diversity;
$k_{\text{ns,f}} = k_{\text{ns}}$ as calculated above for frequency diversity.

16.6.8.2 Four Receivers

$$P_{\text{d}} = (P_{\text{dns}}^{0.75} + P_{\text{ds}}^{0.75})^{4/3} \tag{16.61}$$

P_{d} = total outage probability using four receiver diversity.

Flat-Fading Improvement

$$m_{\text{ns}} = \eta^3 \ (1 - k_{\text{ns,s}}^2) \ (1 - k_{\text{ns,f}}^2)$$

$$P_{\text{dns}} = \frac{P_{\text{FF}}^4}{m_{\text{ns}}}$$

$$k_{\text{ns}}^2 = 1 - \sqrt{n} \ (1 - k_{\text{ns,s}}^2) \ (1 - k_{\text{ns,f}}^2) \tag{16.62}$$

m_{ns} = nonselective diversity parameter;
η = multipath activity factor (calculated above);
$k_{ns,s} = k_{ns}$ as calculated for space diversity above;
$k_{ns,f} = k_{ns}$ as calculated for frequency diversity above;
P_{dns} = probability of diversity nonselective outage.

Dispersive-Fading Improvement

$$P_{ds} = \left[\frac{P_s{}^2}{\eta\ (1 - k_{s,s}^2)}\right]^2 \tag{16.63}$$

P_{ds} = probability of diversity selective outage;
P_s = dispersive (selective) fading probability calculated above;
$k_{s,s} = k_s$ calculated for space diversity above.

16.6.9 Angle Diversity Improvement—ITU-R

$$P_d = \left(P_{dns}^{0.75} + P_{ds}^{0.75}\right)^{4/3} \tag{16.64}$$

P_d = total outage probability using angle diversity (International Telecommunication Union—Radiocommunication Sector (ITU-R), 2009).

Flat-Fading Improvement

$$P_{dns} = \frac{\eta Q_0}{10^{f/6.6}} \tag{16.65}$$

P_{dns} = probability of nonselective (flat fading) outage;
η = multipath activity factor (calculated above).

$$Q_0 = r\left(0.9399^{\mu_\theta} \times 10^{-24.58\ \mu_\theta^2}\right)\left[2.469^{1.879(\delta/\Omega)} \times 3.615^{\left[(\delta/\Omega)^{1.978}\ \ (\varepsilon/\delta)\right]} \times 4.601^{\left[(\delta/\Omega)^{2.152}\ \ (\varepsilon/\delta)^2\right]}\right]$$

$$r = q \text{ if } q \le 1$$

$$= 0.113\ \sin\left[150\left(\frac{\delta}{\Omega}\right) + 30\right] + 0.963 \text{ if } q > 1$$

$$q = 2505 \times 0.0437^{(\delta/\Omega)} \times 0.593^{(\varepsilon/\delta)}$$

$$\mu_\theta = 2.89 \times 10^{-5}\, G_m\, d\ (^\circ) = \text{estimated angle of arrival}$$

$$G_m = 157\,\frac{(1 - K)}{K}$$

ε = elevation angle (°) of upper (diversity) antenna (above horizon is negative angle);
δ = angular separation between the two antenna patterns (°);
Ω = half power beamwidth of the antenna patterns (°);
d = path distance (km);
G_m = average refractivity gradient (N units/km);
≈ −39 N units/km typically (equivalent to $K = 4/3$) ((Bean et al., 1966); (Bean et al., 1960); (Samson, 1976); (Samson, 1975); (Segal and Barrington, 1977)).

See Table 16.A.1 for average K values at specific locations.
Dispersive-Fading Improvement

$$P_{ds} = \frac{P_s^2}{\eta\ (1 - k_s^2)}$$

$$k_s^2 = 1 - (0.0763 \times 0.694^{\mu_\theta} \times 10^{23.3\mu_\theta^2})\ \delta\ (0.211 - 0.188\ \mu_\theta - 0.638\ \mu_\theta^2)^{\Omega} \tag{16.66}$$

P_{ds} = probability of selective (dispersive fading) outage;
P_S = probability of nonprotected selective outage (see above);
η = multipath activity factor (see above).

16.6.10 Other Diversity Improvements—ITU-R

Hybrid diversity and multiline frequency diversity improvements are not addressed.

16.7 RAIN FADING AND OBSTRUCTION FADING (NA AND ITU-R)

These topics are discussed is some detail in Chapters 11 and 12, respectively. Rain fading is addressed by both North American and ITU-R models. However, the ITU-R model ((International Telecommunication Union—Radiocommunication Sector (ITU-R), 2007h); (International Telecommunication Union—Radiocommunication Sector (ITU-R), 2005b)) usually results in much more optimistic results than does the North American Crane model. At present, obstruction fading has not been addressed by the ITU-R.

16.8 COMPARING THE NORTH AMERICAN AND THE ITU-R FLAT-FADING ESTIMATES

The North American (Vigants–Barnett) fading estimates and the ITU-R fading estimates models are quite different. A comparison of the results for a standard example will be made. The final version of the Bell Labs Vigants–Barnett fading model was created based on tests on a path near Atlanta, Georgia (Vigants, 1975) (Fig. 16.11). This path is often considered the "reference" North American path.

This 26.4-mile path was over scattered small clearings of oak–hickory–pine trees near the Chattahoochee River (Schiavone, 1982). A 260-ft tower was at the Atlanta end (latitude 33.84°, longitude −84.42°) and a 330-ft tower was at the Palmetto end (latitude 33.52°, longitude −84.67°) ((Rummler, 1982b); (Babler, 1972)). The main antenna at Palmetto was about 20 ft below the antenna at Atlanta. When space diversity was tested, a diversity antenna was placed 30 ft below the main antenna at the Palmetto end (Rummler, 1982b). For purposes of comparison, we assume an operating frequency of 6.15 GHz. The average roughness of the Atlanta–Palmetto path is 61.8 ft (average of the left-to-right calculated roughness of 63.1 ft and the right-to-left calculated roughness of 60.5 ft).

16.8.1 Vigants–Barnett Flat-Fading Estimation for Bell Labs Path

$$\begin{aligned} R &= c\left(\frac{50}{w}\right)^{1.3}\left(\frac{f}{4}\right) D^3 \times 10^{-5} \\ &= 1.5\left(\frac{50}{61.8}\right)^{1.3}\left(\frac{6.15}{4}\right) 26.4^3 \times 10^{-5} \end{aligned} \tag{16.67}$$

A c-factor of 1.5 was chosen as Atlanta straddles the 1 and 2 areas.

$$\begin{aligned} \frac{P_{FF}}{[T_s/T_{ref}]} &= \frac{0.3222}{10^{FFM/10}} \\ &= \text{one way outage probability during fading month} \end{aligned} \tag{16.68}$$

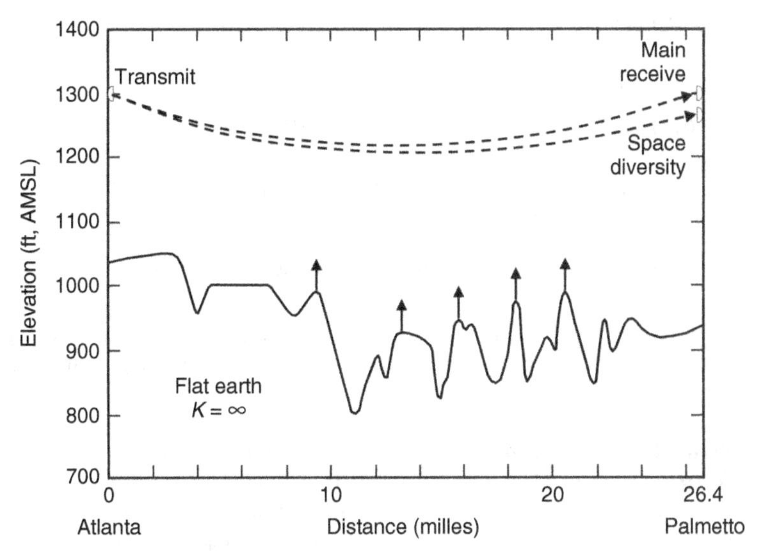

Figure 16.11 Path profile for Bell Labs Atlanta—Palmetto test path.

16.8.2 ITU-R Flat-Fading Estimation for Bell Labs Path

Standard Method:

$$|\varepsilon_p| = \frac{|h_r - h_e|}{d} = \frac{|20 \times 0.3048|}{(26.4/0.6214)} = 0.1435$$

$$s_a = 34.57 \text{ m} = 113.4 \text{ ft (calculated over the 110 km square)}$$

$$K = (10 + s_a)^{-0.46}\ 10^{-4.4}\ 10^{-0.0027dN_1}$$

$$= (34.57)^{-0.46}\ 10^{-4.4}\ 10^{-0.0027(-267)} = 0.00003651$$

$$P_{FF} = 0.01\ K\ d^{3.4}\ (1 + |\varepsilon_p|)^{-1.03}\ f^{0.8}\ 10^{-0.00076h_L} 10^{-A/10}$$

$$= 0.01\ K \left(\frac{26.4}{0.6214}\right)^{3.4} (1.1435)^{-1.03} \times\ 6.15^{0.8}\ 10^{-0.00076(1280/3.281)}\ 10^{-FFM/10} \tag{16.69}$$

$$= \frac{0.2360}{10^{FFM/10}} = \text{one way outage probability during fading month}$$

Owing to the complexity and time required to calculate s_a, many commercial software programs assume $w = s_a$. Making this assumption results in the following:

$$K = \left[10 + \left(\frac{61.8}{3.281}\right)\right]^{-0.46}\ 10^{-4.4}\ 10^{-0.0027\,(-267)} = 0.00004460$$

$$P_{FF} = 0.01\ K \left(\frac{26.4}{0.6214}\right)^{3.4} (1.1435)^{-1.03} \times\ 6.15^{0.8}\ 10^{-0.00076(1280/3.281)}\ 10^{-FFM/10}$$

$$= \frac{0.2884}{10^{FFM/10}} = \text{one way outage probability during fading month} \tag{16.70}$$

Quick-Planning Method:

$$K = 10^{-4.6}\ 10^{-0.0027(-267)} = 0.0001321$$

$$P_{\text{FF}} = 0.01\ K\ d^{3.1}\ (1 + |\varepsilon_{\text{p}}|)^{-1.29}\ f^{0.8}\ 10^{-0.00089 h_{\text{L}}} 10^{-A/10}$$

$$= 0.01\ K\ \left(\frac{26.4}{0.6214}\right)^{3.1}\ (1.1435)^{-1.29} \times\ 65^{0.8}\ 10^{-0.00089\,(1280/3.281)}\ 10^{-\text{FFM}/10}$$

$$= \frac{0.2383}{10^{\text{FFM}/10}} = \text{one way outage probability during fading month} \tag{16.71}$$

For the Bell Labs reference path, the ITU-R model produced results very similar to the Vigants–Barnett fading model. This is remarkable given the significantly different underlying methodologies of these models.

16.8.2.1 General Comparison of Vigants–Barnett and ITU-R Flat-Fading Estimation Methods in North America This comparison will be limited to flat fading as this is the most significant fading mechanism for most paths and this simplifies the comparison.

Typical frequency bands are divided into two sub-bands ("two-frequency plan"), one for transmission and one for reception. The following frequencies, the center frequency of the upper sub-band, will be used for fading estimation:

$$f = 6.3 \text{ for (lower) 6 GHz}$$

$$= 11.5 \text{ for 11 GHz}$$

For the Vigants–Barnett model, annual average temperature t varies from 35 °F (lowest fading) to 75 °F (greatest fading) with 50 °F being average.

For this model, annual climate and humidity factor c varies from 0.5 (lowest fading) to 2 (greatest fading) with 1 being average. The terrain roughness factor w varies from 140 (least fading) to 20 (greatest fading) with 50 being average.

We will consider the following hypothetical case:

Greatest Fading Values ("Worst Case", "hot and humid"):

$$c = 2$$

$$t = 75$$

$$w = 20$$

Typical Fading Values ("Typical"):

$$c = 1$$

$$t = 50$$

$$w = 50$$

Smallest Fading Values ("Best Case", "high and dry"):

$$c = 0.5$$

$$t = 35$$

$$w = 140$$

Diversity antenna separation in North America is typically about 200 wavelengths [30 ft (9 m) for 6 GHz and 20 ft (6 m) for 11 GHz]. We will use this spacing for all space diversity calculation. We will

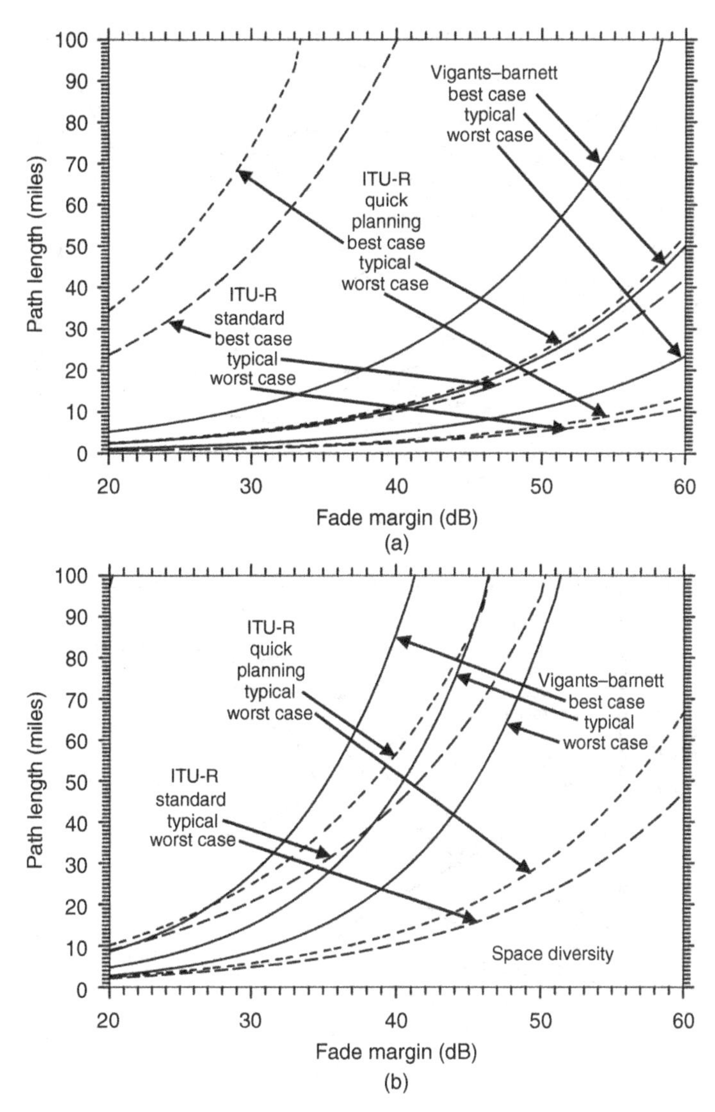

Figure 16.12 (a,b) Lower 6-GHz path lengths for 99.9999% availability (without and with space diversity).

assume the main and space diversity antennas are the same size ($P = 0$). We will also ignore hysteresis loss ($\eta = 1$), a common practice.

The ITU-R methodology uses site mean sea level. NOAA lists the elevation above sea level of 17,910 atmospheric measuring stations throughout the United States. The highest elevation in the 48 states is 3553 m (Coquille, Oregon); the lowest is −59 m (Death Valley). The mean elevation is 544 m. These elevations are assumed to be representative of North American site elevations. Antennas will be assumed to be nominally 300 ft (91 m) above the site elevation.

s_a is the ITU-R area terrain roughness in meters. We will assume the same values for w as for the Vigants–Barnett model (6–43, with 15 being typical).

dN_1 is the point refractivity gradient in the lowest 65 m of the atmosphere not exceeded for 1% of an average year. Over the United States, this value varies from −142 to −692 with median and mean value of −291.

Applying the above assumptions to the Vigants–Barnett and ITU-R models produces the following results (Fig. 16.12, Fig. 16.13, Fig. 16.14, Fig. 16.15, Fig. 16.16, Fig. 16.17).

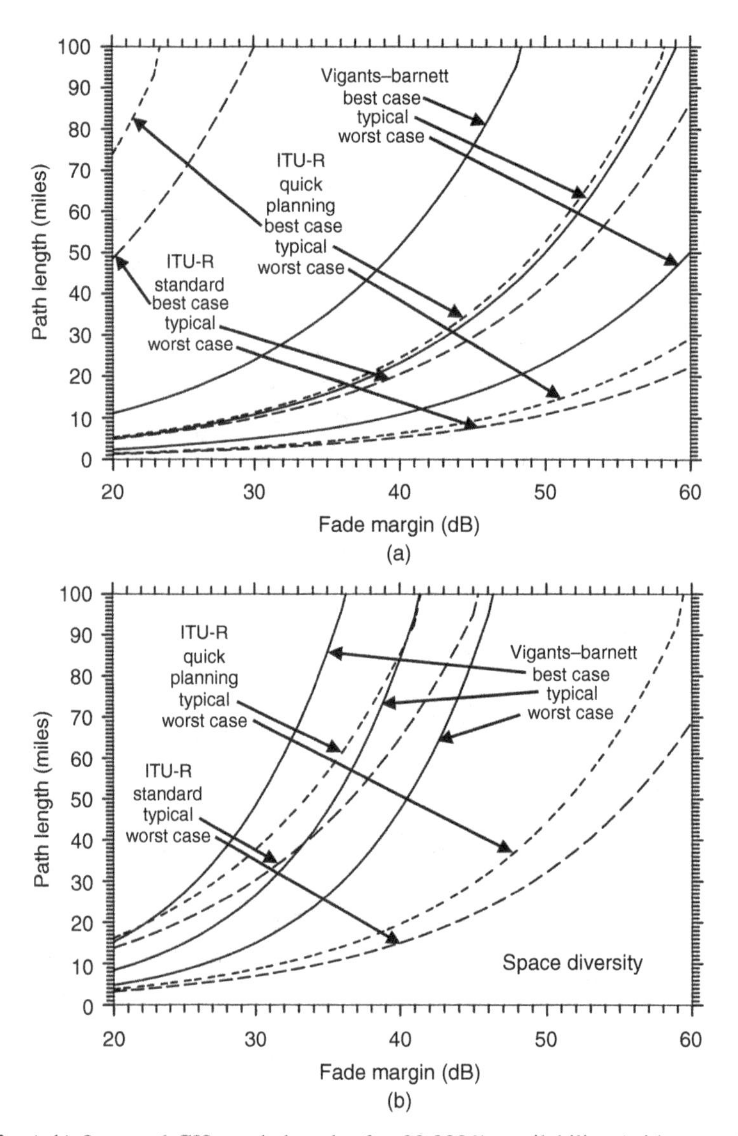

Figure 16.13 (a,b) Lower 6-GHz path lengths for 99.999% availability (without and with space diversity).

It is immediately obvious that lower 6-GHz paths perform better than 11-GHz paths for the same fade margin. However, this difference shrinks as path availability is reduced. Space diversity significantly improves path performance.

For the nondiversity case, both the ITU-R and Vigants–Barnett models achieve similar results. For the space diversity cases, the ITU-R model is more optimistic for low fade margins but similar to Vigants–Barnett for typical fade margins.

Of particular interest is the wide variation for the ITU-R model. Historically, each region used its own multipath calculation methodology: CCIR (now ITU-R) models were used in North West Europe, Morita for Japan, Vigants–Barnett for the United States, Pearson and Doble in the United Kingdom, and Dadenenko for the former Soviet Union (Olsen and Tjelta, 1999; Tjelta et al., 1990). When applied in their respective areas, the US and Japanese methods were considered superior to those of the CCIR (Olsen et al., 2003). In the late 1980s, the ITU-R (CCIR) began intensified efforts to develop "unified" fading models to accommodate worldwide conditions (Olsen and Tjelta, 1999; Tjelta et al., 1990). The updated

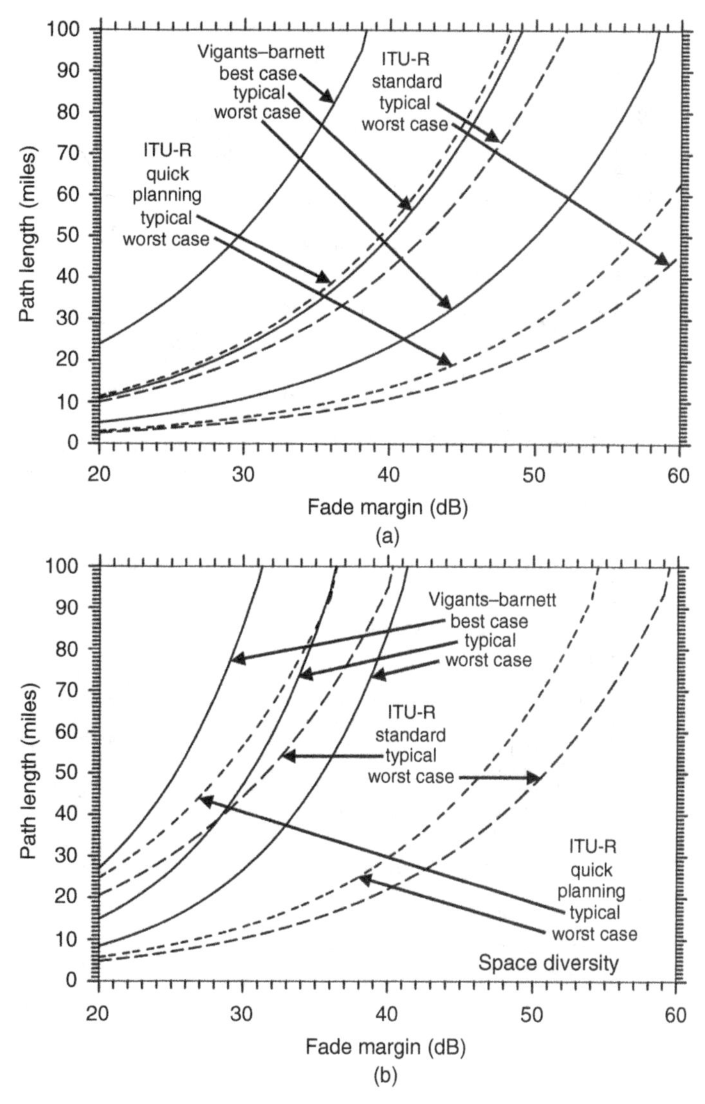

Figure 16.14 (a,b) Lower 6-GHz path lengths for 99.99% availability (without and with space diversity).

models were based on data of 238 links for 22 countries. However, no data was available for the United States (Olsen and Tjelta, 1999). When applied worldwide, the new models are superior to the historical regional methods (Olsen et al., 2003). The Vigants–Barnett model is slightly more conservative than the IRT-R model for all overland links (Olsen et al., 2003). This appears to be because the Vigants–Barnett model was developed for lower latitudes than most of the paths used to develop the ITU-R models (Olsen et al., 2003).

The standard deviation of error for the ITU-R methods was initially estimated to be 5.3 dB (Olsen et al., 2003). The latest outage model has estimated standard deviation of error of 5.2 dB (overland paths) to 7.3 dB (overwater paths). (International Telecommunication Union—Radiocommunication Sector (ITU-R), 2009). Standard- and quick-planning methods are claimed to obtain comparable results (standard deviation difference approximately 0.6 dB) except in high mountainous areas were the standard deviation difference may be several decibels. When compared to the ITU-R database of 239 paths in 22 countries, the Vigants–Barnett model has error with standard deviation of 6.8 dB (Olsen et al., 2003). When observed

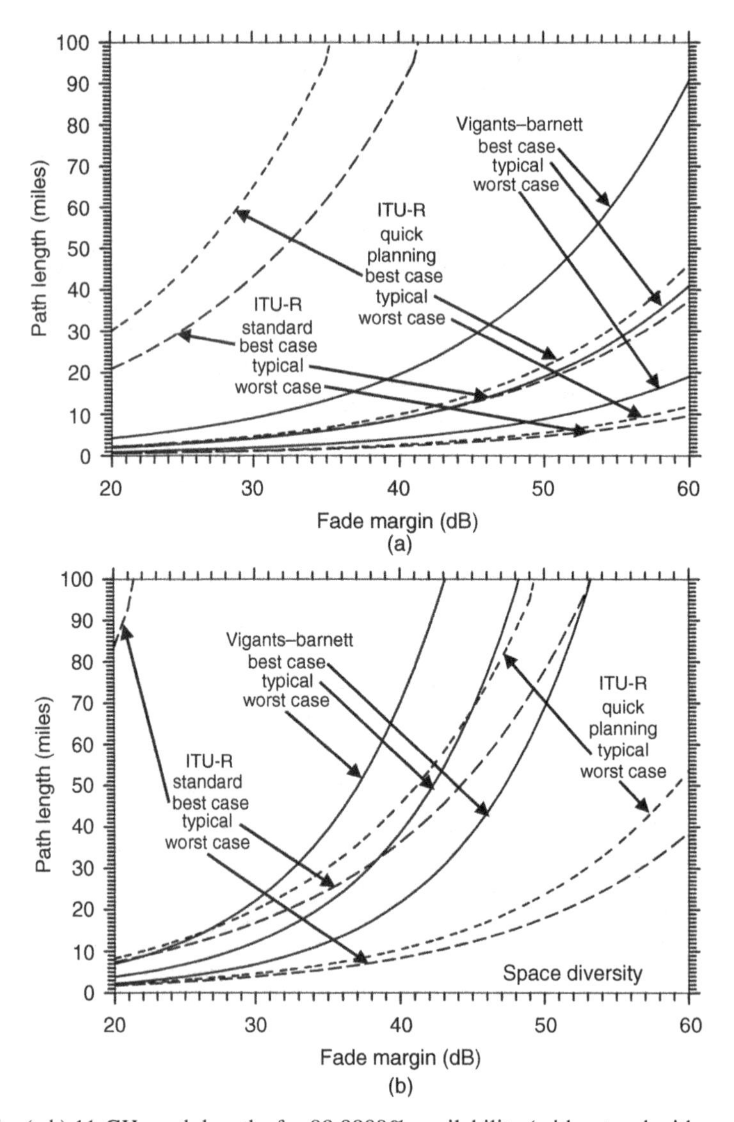

Figure 16.15 (a,b) 11-GHz path lengths for 99.9999% availability (without and with space diversity).

over 45 paths in the United States, the error standard deviation is estimated to be 10.2 dB (Achariyapaopan, 1986).

The following observations will apply to a typical 6-GHz nondiversity path with availability of 99.999% and a 40-dB fade margin. The ITU-R model has little sensitivity to path roughness s_a. In the standard model, when s_a changes from 15 (50 ft) to 43 m (140 ft) the path length increases 15%. For the Vigants–Barnett (V–B) model, the path length increases 56%. When s_a is reduced from 15 to 6 m (20 ft), the ITU-R path length decreases 11%; the V–B model path length decreases 33%. The ITU-R quick-planning model ignores path roughness. The current ITU-R standard model requires calculating path roughness over a 110 km square centered on the path. This is quite difficult and slow in practice. Using the conventional path roughness calculation (average roughness measured along the path) seems adequate.

Again the following observations will apply to a typical 6-GHz nondiversity path with availability of 99.999% and a 40-dB fade margin. The ITU-R standard model is sensitive to refractivity index gradient dN_1. Dividing the refractive index gradient by 2 (changing −291 to −146) increases the path length by 37%. Doubling the gradient (changing from −291 to −582) decreases the path length by 47%. The model is also sensitive to site elevation. Decreasing the antenna height by a factor of 2 (changing site plus antenna elevation of 544 + 91 m to 318 m) decreases the path length by 18%. Increasing antenna height

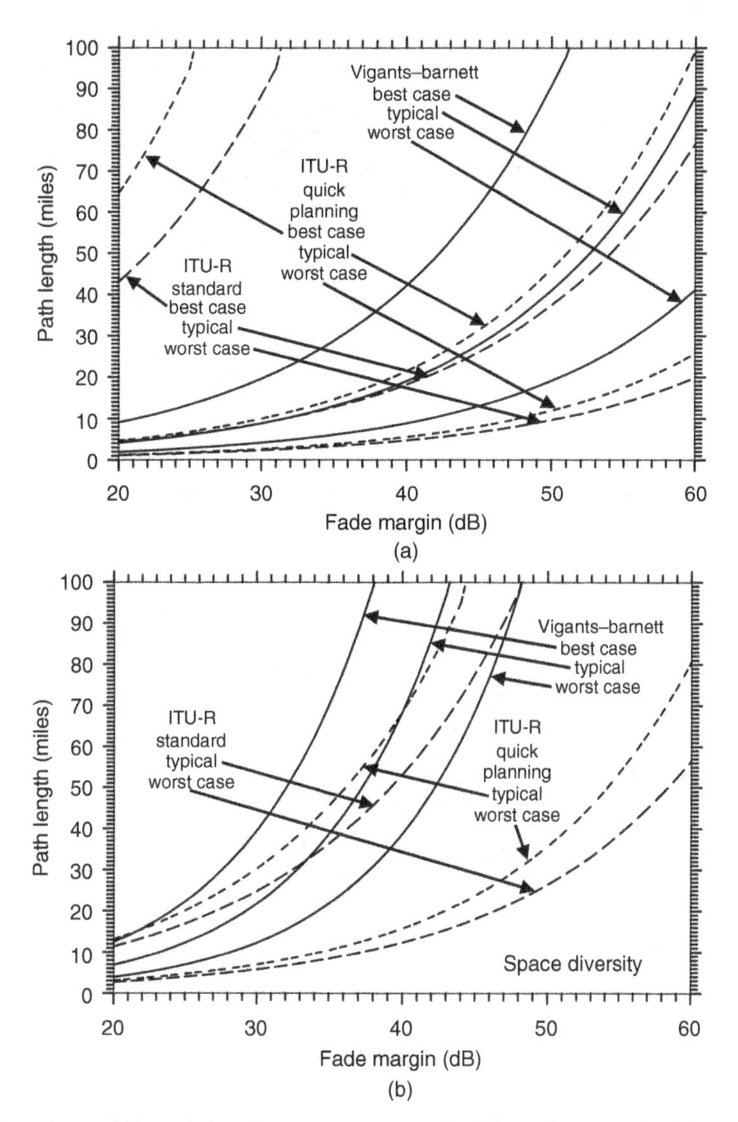

Figure 16.16 (a,b) 11-GHz path lengths for 99.999% availability (without and with space diversity).

by a factor of 2 (changing 635 m to 1270 m) increases the path length by 32%. As noted in the graphed results, this results in a much wider range of results than anticipated by the Vigants–Barnet model.

16.9 DIFFRACTION AND VEGETATION ATTENUATION

Diffraction attenuation (International Telecommunication Union—Radiocommunication Sector (ITU-R), 2005c) is of considerable importance in frequency coordination where estimates of interference from transmitters blocked by obstruction are important. The path loss for an obstructed path increases quickly as the obstruction increases (Section 12.6). This makes long paths, which may experience significant changes in atmospheric refractivity (e.g., K factor), subject to widely varying received signal levels. The objective of conventional path design is to avoid a diffracted path making the calculation unnecessary. Short paths (too short to be significantly influenced by refractivity changes) can be engineered and are quite stable. However, this is an exception to the general case.

Vegetation attenuation (International Telecommunication Union—Radiocommunication Sector (ITU-R), 2007i) calculations are important in estimating cellular phone coverage and interference from transmitters blocked by trees and other foliage. However, trees have long been known to be a significant

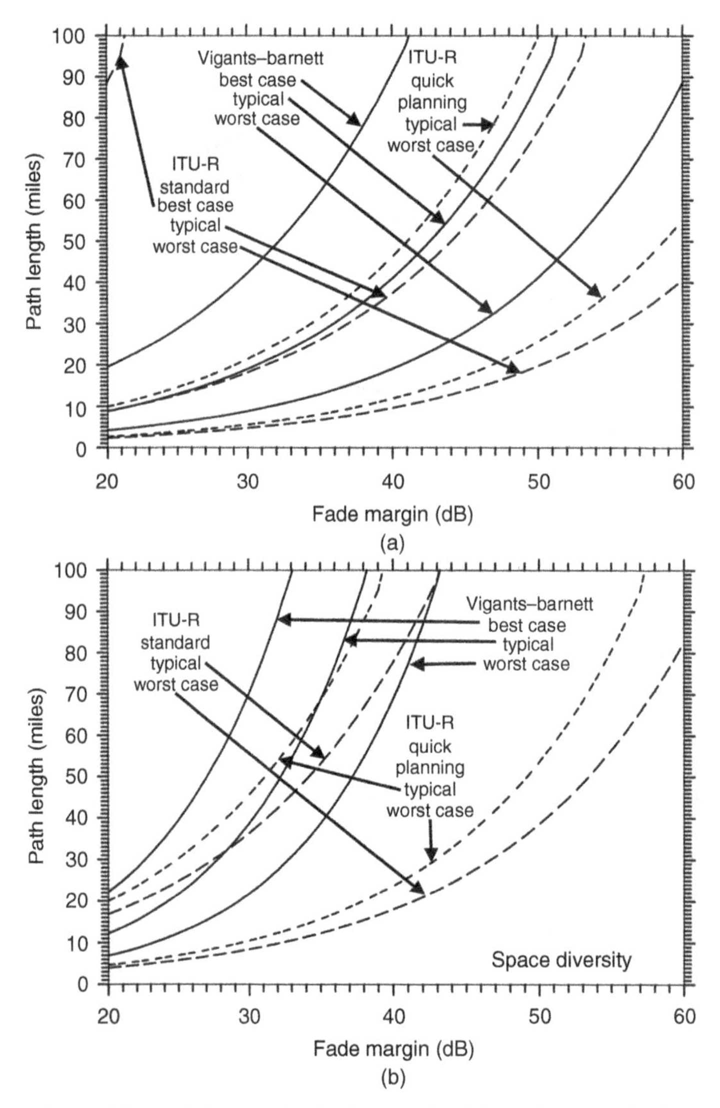

Figure 16.17 (a,b) 11-GHz path lengths for 99.99% availability (without and with space diversity).

obstruction (Friis, 1948) to fixed point to point microwave signals. The objective of path design is to provide clearance above trees. Therefore, for conventional path design, this calculation is unnecessary.

16.10 FOG ATTENUATION

For high microwave frequency signals, water droplets in fog and clouds can cause significant losses to the radio signal. For clouds or fog consisting entirely of small droplets, generally less than 0.01 cm, the Rayleigh approximation is valid for frequencies below 200 GHz and it is possible to express the attenuation in terms of the total water content per unit volume ((International Telecommunication Union—Radiocommunication Sector (ITU-R), 1999b); (International Telecommunication Union—Radiocommunication Sector (ITU-R), 2007j)).

$$\gamma_c = K_1 M \text{ (dB/km)}$$

$$
\begin{aligned}
M &\approx 0.009 + \left(\frac{136.1}{V^{1.4357}}\right) \\
V &= 0.3048 \times \text{visibility in ft} \\
K_1 &= \frac{0.819\,f}{\varepsilon''(1+\eta^2)}\ (\text{dB/km})/(\text{g/m}^3) \\
\eta &= \frac{2+\varepsilon'}{\varepsilon''} \\
\varepsilon'(f) &= \frac{\varepsilon_0-\varepsilon_1}{[1+(f/f_p)^2]} + \frac{\varepsilon_1-\varepsilon_2}{[1+(f/f_s)^2]} + \varepsilon_2 \\
\varepsilon''(f) &= \frac{f(\varepsilon_0-\varepsilon_1)}{f_p\,[1+(f/f_p)^2]} + \frac{f(\varepsilon_1-\varepsilon_2)}{f_s\,[1+(f/f_s)^2]} \\
\varepsilon_0 &= 77.6 + 103.3\,(\theta-1) \\
\varepsilon_1 &= 5.48 \\
\varepsilon_2 &= 3.51 \\
f_p &= 20.09 - 142\,(\theta-1) + 294\,(\theta-1)^2\ (\text{GHz}) \\
f_s &= 590 - 1500\,(\theta-1)\ (\text{GHz}) \\
\theta &= \frac{300}{T}
\end{aligned}
\qquad (16.72)
$$

γ_c = specific attenuation (dB/km) within the cloud;
K_1 = specific attenuation coefficient [(dB/km)/(g/m3)];
M = liquid water density in the cloud or fog (g/m3);
V = optical visibility (m), $5 \leq V \leq 500$ (International Telecommunication Union—Radiocommunication Sector (ITU-R), 1999b; Kizer, 1990);
f = operating frequency (GHz);
T = temperature (K).

Representative attenuation values are graphed in Fig. 16.18

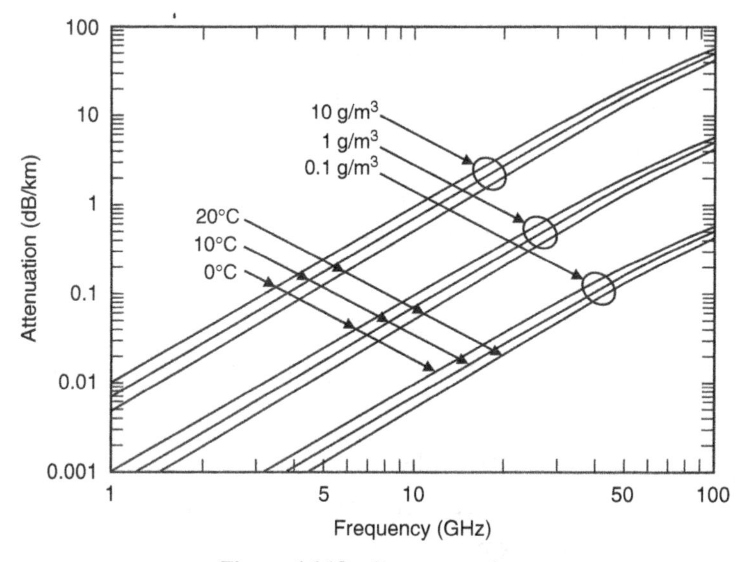

Figure 16.18 Fog attenuation.

16.11 AIR ATTENUATION

Atmospheric attenuation by absorption was first estimated by Liebe and Gimmestad (1978). Their work was simplified by Hause and Wortendyke (1979). This work was extended and improved by ITU-R Study Group 3. On the basis of this work, we may estimate atmospheric attenuation as follows (International Telecommunication Union—Radiocommunication Sector (ITU-R), 2007k; International Telecommunication Union—Radiocommunication Sector (ITU-R), 2007j):

Freq = frequency (GHz);
TempC = mean atmospheric temperature (°C);
Pres = mean atmospheric pressure [hectopascals (hPa) or millibars (mbar)];
Rho = water-vapor density (g/m^3).

$$\text{Rp} = \frac{\text{Pres}}{1013}$$

$$\text{Rt} = \frac{288}{(273 + \text{TempC})}$$

$$\text{EXP}\ (x) = e\hat{}x = 2.7182818\hat{}x$$

$$e\hat{}x = e^x$$

$$\text{LOG}(x) = \text{natural log of } x = \ln(x)$$

Atmospheric attenuation is primarily due to dry air (primarily oxygen and ozone) and water vapor resonances. For dry air, the attenuation AttnDryAir (dB/km) is given by the following equations:

$$\text{X11} = [\text{Rp}\hat{}(0.0717)] \times [\text{Rt}\hat{}(-1.8132)]$$

$$\text{X12} = \text{EXP}\{[(0.0156) \times (1 - \text{Rp})] + [(-1.6515) \times (1 - \text{Rt})]\}$$

$$\text{Xi1} = \text{X11} \times \text{X12}$$

$$\text{X21} = [\text{Rp}\hat{}(0.5146)] \times [\text{Rt}\hat{}(-4.6368)]$$

$$\text{X22} = \text{EXP}\{[(-0.1921) \times (1 - \text{Rp})] + [(-5.7416) \times (1 - \text{Rt})]\}$$

$$\text{Xi2} = \text{X21} \times \text{X22}$$

$$\text{X31} = [\text{Rp}\hat{}(0.3414)] \times [\text{Rt}\hat{}(-6.5851)]$$

$$\text{X32} = \text{EXP}\{[(0.213) \times (1 - \text{Rp})] + [(-8.5854) \times (1 - \text{Rt})]\}$$

$$\text{Xi3} = \text{X31} \times \text{X32}$$

$$\text{X41} = [\text{Rp}\hat{}(-0.0112)] \times [\text{Rt}\hat{}(0.0092)]$$

$$\text{X42} = \text{EXP}\{[(-0.1033) \times (1 - \text{Rp})] + [(-0.0009) \times (1 - \text{Rt})]\}$$

$$\text{Xi4} = \text{X41} \times \text{X42}$$

$$\text{X51} = [\text{Rp}\hat{}(0.2705)] \times [\text{Rt}\hat{}(-2.7192)]$$

$$\text{X52} = \text{EXP}\{[(-0.3016) \times (1 - \text{Rp})] + [(-4.1033) \times (1 - \text{Rt})]\}$$

$$\text{Xi5} = \text{X51} \times \text{X52}$$

$$\text{X61} = [\text{Rp}\hat{}(0.2445)] \times [\text{Rt}\hat{}(-5.9191)]$$

$$\text{X62} = \text{EXP}\{[(0.0422) \times (1 - \text{Rp})] + [(-8.0719) \times (1 - \text{Rt})]\}$$

$$\text{Xi6} = \text{X61} \times \text{X62}$$

$$X71 = [Rp\hat{}(-0.1833)] \times [Rt\hat{}(6.5589)]$$

$$X72 = EXP\{[(-0.2402) \times (1 - Rp)] + [(6.131) \times (1 - Rt)]\}$$

$$Xi7 = X71 \times X72$$

$$G541 = [Rp\hat{}(1.8286)] \times [Rt\hat{}(-1.9487)]$$

$$G542 = EXP\{[(0.4051) \times (1 - Rp)] + [(-2.8509) \times (1 - Rt)]\}$$

$$Gamma54 = 2.192 \times G541 \times G542$$

$$G581 = [Rp\hat{}(1.0045)] \times [Rt\hat{}(3.561)]$$

$$G582 = EXP\{[(0.1588) \times (1 - Rp)] + [(1.2834) \times (1 - Rt)]\}$$

$$Gamma58 = 12.59 \times G581 \times G582$$

$$G601 = [Rp\hat{}(0.9003)] \times [Rt\hat{}(4.1335)]$$

$$G602 = EXP\{[(0.0427) \times (1 - Rp)] + [(1.6088) \times (1 - Rt)]\}$$

$$Gamma60 = 15 \times G601 \times G602$$

$$G621 = [Rp\hat{}(0.9886)] \times [Rt\hat{}(3.4176)]$$

$$G622 = EXP\{[(0.1827) \times (1 - Rp)] + [(1.3429) \times (1 - Rt)]\}$$

$$Gamma62 = 14.28 \times G621 \times G622$$

$$G641 = [Rp\hat{}(1.432)] \times [Rt\hat{}(0.6258)]$$

$$G642 = EXP\{[(0.3177) \times (1 - Rp)] + [(-0.5914) \times (1 - Rt)]\}$$

$$Gamma64 = 6.819 \times G641 \times G642$$

$$G661 = [Rp\hat{}(2.0717)] \times [Rt\hat{}(-4.1404)]$$

$$G662 = EXP\{[(0.491) \times (1 - Rp)] + [(-4.8718) \times (1 - Rt)]\}$$

$$Gamma66 = 1.908 \times G661 \times G662$$

$$Delta1 = [Rp\hat{}(3.211)] \times [Rt\hat{}(-14.94)]$$

$$Delta2 = EXP\{[(1.583) \times (1 - Rp)] + [(-16.37) \times (1 - Rt)]\}$$

$$Delta = -0.00306 \times Delta1 \times Delta2$$

For Freq $\leq$ 54 GHz:

$$Ada1 = \frac{[7.2(Rt\hat{}2.8)]}{(Freq\hat{}2) + [0.34(Rp\hat{}2) \times (Rt\hat{}1.6)]}$$

$$Ada2 = \frac{(0.62 \times Xi3)}{[(54 - Freq)\hat{}(1.16 \times Xi1) + (0.83 \times Xi2)]}$$

$$AttnDryAir = \left[(Freq\hat{}2) \times \frac{(Rp\hat{}2)}{1000} \times (Ada1 + Ada2)\right] \quad (16.73)$$

For 54 GHz < Freq $\leq$ 60 GHz:

$$\text{Ada3} = \frac{(\text{LOG}(\text{Gamma54})) \times (\text{Freq} - 58) \times (\text{Freq} - 60)}{24}$$

$$\text{Ada4} = \frac{(\text{LOG}(\text{Gamma58})) \times (\text{Freq} - 54) \times (\text{Freq} - 60)}{8}$$

$$\text{Ada5} = \frac{(\text{LOG}(\text{Gamma60})) \times (\text{Freq} - 54) \times (\text{Freq} - 58)}{12}$$

$$\text{AttnDryAir} = \text{EXP}(\text{Ada3} - \text{Ada4} + \text{Ada5}) \tag{16.74}$$

For 60 GHz < Freq ≤ 62 GHz:

$$\text{AttnDryAir} = \text{Gamma60} + \left(\frac{(\text{Gamma62} - \text{Gamma60}) \times (\text{Freq} - 60)}{2}\right) \tag{16.75}$$

For 62 GHz < Freq ≤ 66 GHz:

$$\text{Ada6} = \frac{(\text{LOG}(\text{Gamma62})) \times (\text{Freq} - 64) \times (\text{Freq} - 66)}{8}$$

$$\text{Ada7} = \frac{(\text{LOG}(\text{Gamma64})) \times (\text{Freq} - 62) \times (\text{Freq} - 66)}{4}$$

$$\text{Ada8} = \frac{(\text{LOG}(\text{Gamma66})) \times (\text{Freq} - 62) \times (\text{Freq} - 64)}{8}$$

$$\text{AttnDryAir} = \text{EXP}(\text{Ada6} - \text{Ada7} + \text{Ada8}) \tag{16.76}$$

For 66 GHz < Freq ≤ 120 GHz:

$$\text{Ada9} = \left[\frac{3.02 \times (\text{Rt\^{}3.5})}{10{,}000}\right]$$

$$\text{Ada10A} = [0.283 \times (\text{Rt\^{}3.8})]$$

$$\text{Ada10B} = \{[(\text{Freq} - 118.75)\text{\^{}2}] + [2.91 \times (\text{Rp\^{}2}) \times (\text{Rt\^{}1.6})]\}$$

$$\text{Ada10} = \frac{\text{Ada10A}}{\text{Ada10B}}$$

$$\text{Ada11A} = 0.502 \times Xi6[1 - (0.0163 \times Xi7 \times (\text{Freq} - 66))]$$

$$\text{Ada11B} = [(\text{Freq} - 66)\text{\^{}}(1.4346 \times \text{Xi4})] + (1.15 \times Xi5)$$

$$\text{Ada11} = \frac{\text{Ada11A}}{\text{Ada11B}}$$

$$\text{AttnDryAir} = \left[\frac{(\text{Freq\^{}2})\,(\text{Rp\^{}2})}{1000}\right] \times (\text{Ada9} + \text{Ada10} + \text{Ada11}) \tag{16.77}$$

For 120 GHz < Freq ≤ 350 GHz:

$$\text{Ada12A} = \left(\frac{3.02}{10{,}000}\right)$$

$$\text{Ada12B} = 1 + \left[\frac{1.9\,(\text{Freq\^{}1.5})}{100{,}000}\right]$$

$$\text{Ada12} = \frac{\text{Ada12A}}{\text{Ada12B}}$$

$$\text{Ada13A} = 0.283(\text{Rt\^{}0.3})$$

$$\text{Ada13B} = [(\text{Freq} - 118.75)\text{\^{}2}] + [2.91 \times (\text{Rp\^{}2}) \times (\text{Rt\^{}1.6})]$$

$$\text{Ada13} = \frac{\text{Ada13A}}{\text{Ada13B}}$$

$$\text{Ada14} = \frac{(\text{Freq\^{}2})(\text{Rp\^{}2})(\text{Rt\^{}3.5})}{1000}$$

$$\text{AttnDryAir} = \text{Delta} + [\text{Ada14} \times (\text{Ada12} + \text{Ada13})] \quad (16.78)$$

For 350 GHz < Freq:

$$\text{AttnDryAir} = 0.030 \quad (16.79)$$

For water vapor, the attenuation AttnWetAir (dB/km) is given by the following:

$$\text{Eta1} = \{0.955 \times \text{Rp(Rt\^{}0.68)}\} + (0.006 \times \text{Rho})$$

$$\text{Eta2} = [0.735 \times \text{Rp(Rt\^{}0.5)}] + [0.0353(\text{Rt\^{}4}) \times \text{Rho}]$$

$$\text{Gfreq22} = 1 + \left[\frac{(\text{Freq} - 22)}{(\text{Freq} + 22)}\text{\^{}2}\right]$$

$$\text{Gfreq557} = 1 + \left[\frac{(\text{Freq} - 557)}{(\text{Freq} + 557)}\text{\^{}2}\right]$$

$$\text{Gfreq752} = 1 + \left[\frac{(\text{Freq} - 752)}{(\text{Freq} + 752)}\text{\^{}2}\right]$$

$$\text{Gfreq1780} = 1 + \left[\frac{(\text{Freq} - 1780)}{(\text{Freq} + 1780)}\text{\^{}2}\right]$$

$$\text{Awe1A} = 3.98 \times \text{Eta1[EXP(2.23(1 - Rt))]} \times \text{Gfreq22}$$

$$\text{Awe1B} = [(\text{Freq} - 22.235)\text{\^{}2}] + [9.42(\text{Eta1\^{}2})]$$

$$\text{Awe1} = \frac{\text{Awe1A}}{\text{Awe1B}}$$

$$\text{Awe2A} = 11.96 \times \text{Eta1[EXP(0.7(1 - Rt))]}$$

$$\text{Awe2B} = [(\text{Freq} - 183.31)\text{\^{}2}] + [11.14(\text{Eta1\^{}2})]$$

$$\text{Awe2} = \frac{\text{Awe2A}}{\text{Awe2B}}$$

$$\text{Awe3A} = 0.081 \times \text{Eta1[EXP(6.44(1 - Rt))]}$$

$$\text{Awe3B} = [(\text{Freq} - 321.226)\text{\^{}2}] + [6.29(\text{Eta1\^{}2})]$$

$$\text{Awe3} = \frac{\text{Awe3A}}{\text{Awe3B}}$$

$$\text{Awe4A} = 3.66 \times \text{Eta 1[EXP(1.6(1 - Rt))]}$$

$$\text{Awe4B} = [(\text{Freq} - 325.153)\text{\^{}2}] + [9.22(\text{Eta1\^{}2})]$$

$$\text{Awe4} = \frac{\text{Awe4A}}{\text{Awe4B}}$$

$$\text{Awe5A} = 25.37 \times \text{Eta1[EXP(1.09(1 - Rt))]}$$

$$\text{Awe5B} = [(\text{Freq} - 380)\text{\^{}2}]$$

$$\text{Awe5} = \frac{\text{Awe5A}}{\text{Awe5B}}$$

$$\text{Awe6A} = 17.4 \times \text{Eta1[EXP(1.46(1 - Rt))]}$$

$$\text{Awe6B} = [(\text{Freq} - 448)\text{\^{}2}]$$

$$\text{Awe6} = \frac{\text{Awe6A}}{\text{Awe6B}}$$

$$\text{Awe7A} = 844.6 \times \text{Eta1[EXP(0.17(1 - Rt))]} \times \text{Gfreq557}$$

$$\text{Awe7B} = [(\text{Freq} - 557)\hat{}2]$$

$$\text{Awe7} = \frac{\text{Awe7A}}{\text{Awe7B}}$$

$$\text{Awe8A} = 290 \times \text{Eta1}[\text{EXP}(0.41(1 - \text{Rt}))] \times \text{Gfreq752}$$

$$\text{Awe8B} = [(\text{Freq} - 752)\hat{}2]$$

$$\text{Awe8} = \frac{\text{Awe8A}}{\text{Awe8B}}$$

$$\text{Awe9A} = 83328 \times \text{Eta2}[\text{EXP}(0.99(1 - \text{Rt}))] \times \text{Gfreq1780}$$

$$\text{Awe9B} = [(\text{Freq} - 1780)\hat{}2]$$

$$\text{Awe9} = \frac{\text{Awe9A}}{\text{Awe9B}}$$

$$\text{Awe10A} = \text{Awe1} + \text{Awe2} + \text{Awe3} + \text{Awe4} + \text{Awe5}$$

$$\text{Awe10} = \text{Awe10A} + \text{Awe6} + \text{Awe7} + \text{Awe8} + \text{Awe9}$$

$$\text{AttnWetAir} = \frac{\text{Awe10}(\text{Freq}\hat{}2)(\text{Rt}\hat{}2.5)\text{Rho}}{10{,}000}$$

$$\text{AttnAir} = \text{total atmospheric attenuation (dB/km)}$$

$$= \text{AttnDryAir} + \text{AttnWetAir}$$

In the United States, typical water-vapor density is plotted in Figure 16.19 (International Telecommunication Union–Radiocommunication Sector (ITU-R) (2001b)").

If water-vapor density data is not available, it may be calculated from average temperature and relative humidity (Kizer, 1990):

$$\text{Rho} = \frac{217\ e_s\ H_R}{[100\ (\text{TempC} + 273)]}$$

$$e_s = 6.11\ 10^D$$

$$D = \left[\frac{7.50\ \text{TempC}}{(\text{TempC} + 273)}\right]$$

Rho = water-vapor density (g/m3);
H_R = relative humidity (%);
e_s = saturated vapor pressure (hectopascals [hPa] or millibars [mbar]) at air temperature TempC;
TempC = air temperature (degrees C).

Pres (mean pressure in hectopascals [hPa] or millibars [mbar]) ≈ 1013 at sea level. For other elevations in the midlatitudes, it varies linearly with altitude up to at least 2 miles (United States Air Force, 1960). In this region, seasonal variations are not significant. Variation with altitude may be estimated as follows:

$$\text{Press} = 1013 - 0.031505\ \text{Elev (ft)}$$

$$= 1013 - 0.10336\ \text{Elev (m)}$$

Elev = average path elevation above mean sea level (AMSL).

For the case of Pres = 1013 hPa (mbar), TempC = 15°C and Rho = 7.5 g/m^3, atmospheric attenuation (sum of dry air and water-vapor attenuation) is plotted in Figure 16.20 for frequencies up to 100 GHz. Below 350 GHz, besides background absorption, the most significant loss is due to a broad series of oxygen and ozone resonances from approximately 52 to 68 GHz and around 119 GHz as well as water resonances at 22, 183, and 323 GHz.

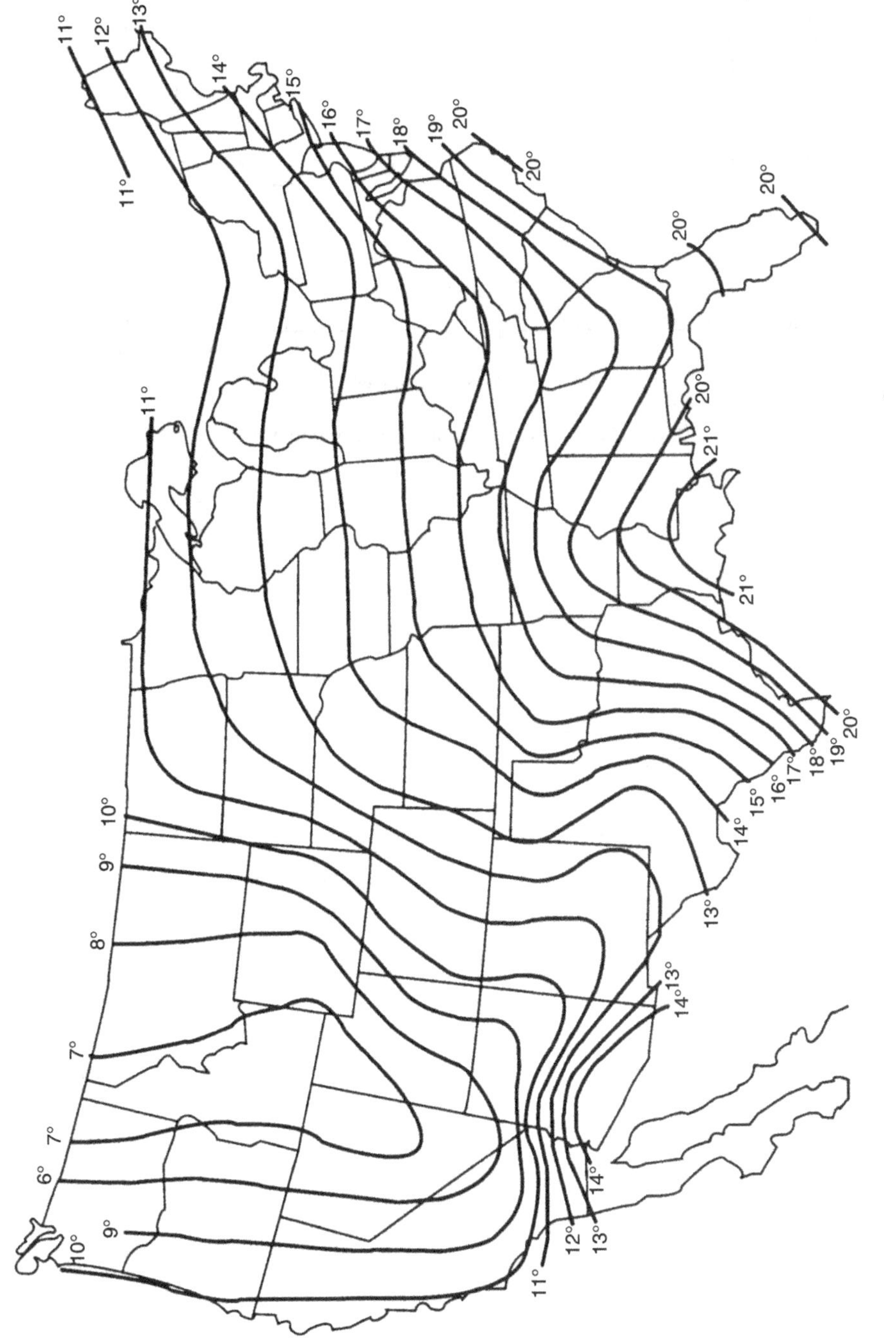

Figure 16.19 Average annual absolute humidity (g/m^3).

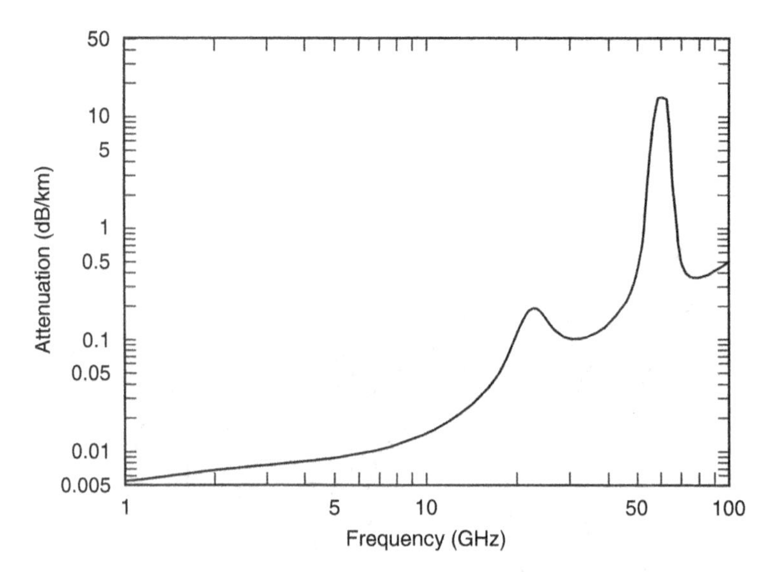

Figure 16.20 Typical atmospheric attenuation.

16.A APPENDIX

TABLE 16.A.1 Location *K* Factors

City	State/Province/ Country	Country/ Region	Latitude, degrees	Longitude, degrees	Feb-Mean	May-Mean	Aug-Mean	Nov-Mean	Average	Source
Aoulef	Algeria	Africa	26.97	1.08	1.29	1.34	1.40	1.34	1.34	Samson
Luanda	Angola	Africa	−8.82	13.22	1.34	1.47	1.47	1.34	1.40	Samson
Bangui	Central African Republic	Africa	4.38	18.57	1.47	1.91	1.91	1.91	1.78	Samson
Fort Lamy	Chad	Africa	12.13	15.03	22.43	2.75	22.43	−2.15	−314.00	Samson
Abidjan (Port Bouet)	Ivory Coast (Côte d'Ivoire)	Africa	5.25	−3.93	1.54	1.34	1.34	1.24	1.36	Samson
Nairobi	Kenya	Africa	−1.30	36.75	1.34	1.47	1.34	1.47	1.40	Samson
Tripoli	Libya	Africa	32.90	13.28	1.47	1.91	2.04	1.47	1.68	Samson
Fort Trinquet	Mauritania	Africa	25.23	−11.62	1.40	1.40	1.62	1.40	1.45	Samson
Kenitra (Port Lyautey)	Morocco	Africa	34.30	−6.60	1.47	1.47	1.47	1.71	1.52	Samson
Lourenco Marques	Mozambique	Africa	−25.92	32.57	1.47	1.19	1.24	1.47	1.33	Samson
Niamey	Niger	Africa	13.48	2.17	3.74	2.34	−52.33	−52.33	6.10	Samson
Dakar	Senegal	Africa	14.73	−17.50	−2.49	4.24	2.75	−8.72	48.31	Samson
Capetown	South Africa	Africa	−33.97	18.60	1.29	1.47	1.47	1.15	1.33	Samson
Khartoum	Sudan	Africa	15.60	32.55	1.19	1.40	1.62	1.24	1.34	Samson
Amundsen-Scott	Antarctica	Antarctica	−90.00	0.00	1.73	16.19	−14.67	1.74	3.49	Bean et al.
Muharraq	Bahrain	Asia	26.27	50.62	1.71	−2.96	−1.18	2.04	−36.94	Samson
Calcutta	India	Asia	22.65	88.45	3.02	1.80	1.80	4.91	2.43	Samson
New Delhi	India	Asia	28.58	77.20	2.18	4.24	1.80	3.02	2.53	Samson
Djakarta	Indonesia	Asia	−6.18	106.83	1.62	1.62	1.62	1.62	1.62	Samson
Atyrau	Kazakhstan	Asia	47.12	51.92	1.40	1.29	1.29	1.40	1.34	Samson
Karaganda	Kazakhstan	Asia	49.80	73.13	1.34	1.34	1.34	1.34	1.34	Samson
Verkhoyansk	North Yakut	Asia	67.55	133.38	2.34	1.29	1.40	1.91	1.64	Samson
Singapore	Republic of Singapore	Asia	1.35	103.90	1.80	2.04	1.80	1.80	1.86	Samson

(continued)

TABLE 16.A.1 (*Continued*)

City	State/Province/ Country	Country/ Region	Latitude, degrees	Longitude, degrees	Feb-Mean	May-Mean	Aug-Mean	Nov-Mean	Average	Source
Bangkok	Thailand	Asia	13.73	100.50	1.71	2.75	1.47	2.53	1.97	Samson
Lviv (L'vov)	Ukraine	Asia	49.82	23.95	1.24	1.34	1.40	1.40	1.34	Samson
Odessa	Ukraine	Asia	46.48	30.63	1.29	1.40	1.47	1.40	1.39	Samson
Tashkent	Uzbekistan	Asia	41.33	69.30	1.34	1.54	1.54	1.54	1.48	Samson
Saigon	Viet Nam	Asia	10.82	106.67	1.62	1.80	2.04	2.04	1.86	Samson
Tan An	Viet Nam	Asia	10.53	106.38	1.91	1.91	1.91	1.91	1.91	Samson
Aden	Yemen	Asia	12.83	45.02	1.54	1.91	1.71	1.34	1.60	Samson
Lajes	Azores	Atlantic	38.75	−27.08	1.47	1.54	1.54	1.47	1.50	Samson
Stanley	Falkland Islands	Atlantic	−51.70	−57.87	1.47	1.40	1.47	1.47	1.45	Samson
Thule	Greenland	Atlantic	76.52	−68.73	2.04	1.24	1.24	1.62	1.47	Samson
Ascension Island		Atlantic	−7.97	−14.40	1.47	1.47	1.47	1.47	1.47	Samson
Gough Island		Atlantic	−40.32	−9.90	1.24	1.29	1.24	1.24	1.25	Samson
Adelaide	Australia	Australia	−34.93	138.58	1.24	1.47	1.47	1.24	1.34	Samson
Darwin	Australia	Australia	−12.43	130.87	1.54	1.24	1.24	1.24	1.30	Samson
Perth	Australia	Australia	−31.95	115.82	1.19	1.24	1.34	1.24	1.25	Samson
Townsville	Australia	Australia	−19.27	146.75	1.54	1.34	1.34	1.34	1.39	Samson
Calgary	Alberta	Canada	51.10	−114.02	1.32	1.26	1.35	1.29	1.30	Kizer
Edmonton	Alberta	Canada	53.32	−113.58	1.29	1.26	1.35	1.28	1.29	Segal and Barrington
Watino	Alberta	Canada	55.72	−117.62	1.33	1.29	1.36	1.32	1.32	Kizer
Comox	British Columbia	Canada	49.72	−124.90	1.34	1.35	1.49	1.35	1.38	Kizer
Fort Nelson	British Columbia	Canada	58.83	−122.58	1.33	1.27	1.33	1.31	1.31	Segal and Barrington
Hope	British Columbia	Canada	49.38	−121.43	1.33	1.32	1.43	1.33	1.35	Kizer
Mission	British Columbia	Canada	49.15	−122.27	1.34	1.33	1.47	1.34	1.37	Kizer
Port Hardy	British Columbia	Canada	50.68	−127.37	1.33	1.33	1.41	1.34	1.35	Segal and Barrington
Prince George	British Columbia	Canada	53.88	−122.67	1.29	1.27	1.33	1.29	1.29	Segal and Barrington

Summerland	British Columbia	Canada	49.57	−119.65	1.33	1.30	1.39	1.32	1.33	Kizer
Vancouver	British Columbia	Canada	49.18	−123.17	1.34	1.34	1.49	1.35	1.38	Kizer
Churchill	Manitoba	Canada	58.75	−94.07	1.40	1.32	1.35	1.32	1.35	Segal and Barrington
Dauphin	Manitoba	Canada	51.10	−100.05	1.35	1.29	1.36	1.28	1.32	Kizer
The Pas	Manitoba	Canada	53.97	−101.10	1.33	1.29	1.34	1.30	1.31	Segal and Barrington
Winnipeg	Manitoba	Canada	49.90	−97.23	1.35	1.30	1.37	1.28	1.32	Kizer
Frederiction	New Brunswick	Canada	45.92	−66.62	1.33	1.34	1.46	1.34	1.36	Kizer
Saint John	New Brunswick	Canada	45.32	−65.88	1.33	1.35	1.48	1.35	1.37	Kizer
Argentia	Newfoundland	Canada	47.23	−54.02	1.28	1.31	1.40	1.33	1.33	Segal and Barrington
Gander	Newfoundland	Canada	48.95	−54.57	1.33	1.35	1.45	1.34	1.36	Kizer
Goose Bay	Newfoundland	Canada	53.32	−60.42	1.32	1.30	1.34	1.29	1.31	Segal and Barrington
St. John's	Newfoundland	Canada	47.62	−52.75	1.32	1.37	1.50	1.35	1.38	Segal and Barrington
Stephenville	Newfoundland	Canada	48.53	−58.55	1.28	1.29	1.40	1.28	1.31	Segal and Barrington
Cambridge Bay	Northwest Territories	Canada	69.10	−105.12	1.46	1.32	1.33	1.37	1.37	Kizer
Alert	Northwest Territory	Canada	82.50	−62.33	1.45	1.31	1.30	1.39	1.36	Segal and Barrington
Baker Lake	Northwest Territory	Canada	64.30	−96.00	1.44	1.31	1.32	1.35	1.35	Segal and Barrington
Clyde	Northwest Territory	Canada	70.45	−68.55	1.34	1.30	1.30	1.30	1.31	Segal and Barrington
Coppermine	Northwest Territory	Canada	67.83	−115.12	1.41	1.30	1.31	1.33	1.34	Segal and Barrington
Coral Harbour	Northwest Territory	Canada	64.20	−83.37	1.40	1.31	1.32	1.34	1.34	Segal and Barrington
Eureka	Northwest Territory	Canada	80.00	−85.93	1.50	1.31	1.30	1.45	1.38	Segal and Barrington
Fort Smith	Northwest Territory	Canada	60.02	−111.97	1.35	1.28	1.32	1.31	1.31	Segal and Barrington
Frobisher Bay	Northwest Territory	Canada	63.75	−68.55	1.39	1.31	1.29	1.31	1.32	Segal and Barrington
Hall Beach	Northwest Territory	Canada	68.78	−81.25	1.41	1.33	1.33	1.37	1.36	Segal and Barrington
Inuvik	Northwest Territory	Canada	68.30	−133.48	1.43	1.29	1.29	1.35	1.34	Segal and Barrington
Isachsen	Northwest Territory	Canada	78.78	−103.53	1.51	1.32	1.32	1.43	1.39	Segal and Barrington
Mould Bay	Northwest Territory	Canada	76.23	−119.33	1.48	1.31	1.30	1.38	1.36	Segal and Barrington
Norman Wells	Northwest Territory	Canada	65.30	−126.85	1.40	1.29	1.32	1.37	1.34	Segal and Barrington

(*continued*)

TABLE 16.A.1 *(Continued)*

City	State/Province/ Country	Country/ Region	Latitude, degrees	Longitude, degrees	Feb-Mean	May-Mean	Aug-Mean	Nov-Mean	Average	Source
Resolute	Northwest Territory	Canada	74.68	−94.92	1.45	1.32	1.31	1.39	1.37	Segal and Barrington
Sachs Harbour	Northwest Territory	Canada	71.98	−125.28	1.45	1.33	1.32	1.37	1.37	Segal and Barrington
Halifax	Nova Scotia	Canada	44.63	−63.50	1.33	1.35	1.53	1.36	1.39	Kizer
Kentville	Nova Scotia	Canada	45.07	−64.48	1.33	1.35	1.50	1.35	1.38	Kizer
Sable Island	Nova Scotia	Canada	43.93	−60.03	1.31	1.33	1.55	1.33	1.37	Segal and Barrington
Sydney	Nova Scotia	Canada	46.17	−60.05	1.33	1.35	1.50	1.35	1.38	Kizer
Central Patricia	Ontario	Canada	51.50	−90.15	1.35	1.32	1.41	1.31	1.34	Kizer
Geraldton	Ontario	Canada	49.68	−86.95	1.34	1.32	1.43	1.31	1.35	Kizer
Kingston	Ontario	Canada	44.23	−76.48	1.33	1.36	1.41	1.34	1.36	Kizer
London	Ontario	Canada	43.03	−81.15	1.33	1.38	1.44	1.33	1.37	Kizer
Moosonee	Ontario	Canada	51.27	−80.65	1.34	1.31	1.38	1.31	1.33	Segal and Barrington
North Bay	Ontario	Canada	46.37	−79.42	1.33	1.34	1.42	1.32	1.35	Kizer
Ottawa	Ontario	Canada	45.38	−75.72	1.33	1.34	1.41	1.33	1.35	Kizer
Sault Ste. Marie	Ontario	Canada	46.48	−84.50	1.33	1.34	1.46	1.32	1.36	Kizer
Sioux Lookout	Ontario	Canada	50.12	−91.90	1.34	1.32	1.41	1.30	1.34	Kizer
Toronto	Ontario	Canada	43.68	−79.63	1.33	1.37	1.43	1.33	1.36	Kizer
Trout Lake	Ontario	Canada	53.83	−89.87	1.35	1.30	1.35	1.30	1.32	Segal and Barrington
Windsor	Ontario	Canada	42.27	−82.97	1.33	1.39	1.47	1.34	1.38	Kizer
Summerside	Prince Edward Island	Canada	46.43	−63.83	1.33	1.34	1.47	1.34	1.37	Kizer

Caplan	Quebec	Canada	48.10	−65.65	1.32	1.33	1.43	1.33	1.35	Kizer
Fort Chimo	Quebec	Canada	58.10	−68.42	1.35	1.31	1.33	1.30	1.32	Segal and Barrington
Gagnon	Quebec	Canada	51.95	−68.13	1.32	1.33	1.39	1.31	1.34	Kizer
Inoucdjouac	Quebec	Canada	58.45	−78.12	1.38	1.32	1.35	1.30	1.34	Segal and Barrington
Maniwaki	Quebec	Canada	46.37	−75.98	1.31	1.32	1.43	1.30	1.34	Segal and Barrington
Montreal	Quebec	Canada	45.47	−73.75	1.33	1.34	1.42	1.33	1.36	Kizer
Nitchequon	Quebec	Canada	53.20	−70.90	1.33	1.29	1.34	1.29	1.31	Segal and Barrington
Normandin	Quebec	Canada	48.85	−72.53	1.32	1.33	1.41	1.32	1.34	Kizer
Post de La Baleine	Quebec	Canada	55.28	−77.77	1.35	1.34	1.39	1.32	1.35	Kizer
Quebec	Quebec	Canada	46.80	−71.38	1.32	1.34	1.42	1.33	1.35	Kizer
Sept Iies	Quebec	Canada	50.22	−66.27	1.31	1.31	1.38	1.30	1.32	Segal and Barrington
Val d'Or	Quebec	Canada	48.05	−77.78	1.33	1.33	1.41	1.32	1.35	Kizer
Prince Albert	Saskatchewan	Canada	53.22	−105.68	1.33	1.28	1.36	1.29	1.31	Kizer
Regina	Saskatchewan	Canada	50.93	−104.67	1.34	1.27	1.33	1.28	1.30	Kizer
Swift Current	Saskatchewan	Canada	50.27	−107.73	1.32	1.25	1.28	1.27	1.28	Kizer
Uranium City	Saskatchewan	Canada	59.57	−108.48	1.35	1.30	1.35	1.33	1.33	Kizer
Weyburn	Saskatchewan	Canada	49.67	−103.85	1.34	1.27	1.31	1.27	1.30	Kizer
Carmacks	Yukon	Canada	62.10	−136.30	1.34	1.28	1.34	1.34	1.32	Kizer
Whitehorse	Yukon	Canada	60.72	−135.07	1.29	1.26	1.23	1.27	1.26	Segal and Barrington
Guantanamo	Cuba	Caribbean	19.90	−75.15	1.47	1.47	1.47	1.47	1.47	Samson
Curacao	Netherlands Antilles	Caribbean	12.18	−68.98	1.62	1.62	1.62	1.62	1.62	Samson

(continued)

TABLE 16.A.1 (*Continued*)

City	State/Province/ Country	Country/ Region	Latitude, degrees	Longitude, degrees	Feb-Mean	May-Mean	Aug-Mean	Nov-Mean	Average	Source
San Juan	Puerto Rico	Caribbean	18.43	−66.00	1.91	2.25	2.16	2.18	2.12	Bean et al.
Swan Island		Caribbean	17.40	−83.93	1.54	1.80	1.80	1.80	1.73	Samson
Balboa	Panama	Central America	8.93	−79.57	1.62	1.62	1.62	1.62	1.62	Samson
Hong Kong	Special Administrative Region	China	22.30	114.17	1.40	1.71	1.80	1.47	1.58	Samson
Bruxelles	Belgium	Europe	50.90	4.48	1.47	1.47	1.47	1.47	1.47	Samson
Glbraltar	British Colony	Europe	36.15	−5.35	1.47	1.62	1.71	1.62	1.60	Samson
Camborne	England	Europe	50.22	−5.32	1.47	1.47	1.47	1.47	1.47	Samson
Cardington	England	Europe	52.10	−0.42	1.40	1.40	1.40	1.40	1.40	Samson
Helsinki	Finland	Europe	60.32	24.97	1.34	1.40	1.40	1.34	1.37	Samson
Bordeaux	France	Europe	44.85	−0.70	1.54	1.80	1.91	1.71	1.73	Samson
Bitburg	Germany	Europe	49.95	6.52	1.29	1.29	1.29	1.29	1.29	Samson
Gross Rohrheim	Germany	Europe	49.72	8.47	1.54	1.54	1.54	1.54	1.54	Samson
Wiesbaden	Germany	Europe	50.05	8.33	1.29	1.29	1.29	1.29	1.29	Samson
Athens	Greece	Europe	37.97	23.72	1.34	1.47	1.47	1.34	1.40	Samson
Rome	Italy	Europe	41.80	12.60	1.34	1.54	1.54	1.34	1.43	Samson
La Coruna	Spain	Europe	43.37	−8.42	1.34	1.47	1.47	1.47	1.43	Samson
Madrid	Spain	Europe	40.40	−3.68	1.29	1.29	1.29	1.29	1.29	Samson
Ostersund	Sweden	Europe	63.18	14.62	1.34	1.34	1.47	1.34	1.37	Samson
Beograd	Yugoslavia	Europe	44.78	20.53	1.34	1.47	1.54	1.34	1.42	Samson
Tananarive	Madagascar (Malagasy Republic)	Indian Ocean	−18.90	47.53	2.04	1.91	1.91	1.91	1.94	Samson
Marion Island		Indian Ocean	−46.88	37.87	1.34	1.34	1.34	1.34	1.34	Samson
Nouvelle Amsterdam Island		Indian Ocean	−37.83	77.57	1.71	1.47	1.47	1.47	1.52	Samson

Sapporo	Hokkaldo	Japan	43.05	141.33	1.29	1.29	1.40	1.29	1.31	Samson
Toyko (Tateno)	Honshu	Japan	36.05	140.13	1.47	1.47	1.47	1.47	1.47	Samson
Nicosia	Cyprus	Mediterranean	35.15	33.28	1.19	1.34	1.91	1.15	1.34	Samson
Palma	Majorca	Mediterranean	39.60	2.70	1.34	1.24	1.34	1.34	1.31	Samson
Samsun	Turkey	Asia	41.28	36.33	1.34	1.47	1.47	1.40	1.42	Samson
Mazatian	Mexico	North America	23.18	−106.43	2.04	1.80	1.62	1.80	1.80	Samson
Ship C		North Atlantic Ocean	52.75	−35.50	1.34	1.34	1.40	1.34	1.36	Samson
Ship K		North Atlantic Ocean	45.00	−16.00	1.34	1.47	1.47	1.47	1.43	Samson
Ship V		North Atlantic Ocean	34.00	164.00	1.62	1.62	1.80	1.62	1.66	Samson
Ship M		North Sea	66.00	2.00	1.34	1.34	1.40	1.34	1.36	Samson
Nandi	Fiji Islands	Pacific	−17.75	177.45	1.71	1.62	1.24	1.62	1.52	Samson
Raoul Island	Kermadec Islands	Pacific	−29.25	−177.92	1.47	1.34	1.34	1.47	1.40	Samson
Guam	Mariana Islands	Pacific	13.55	144.83	1.71	1.62	1.47	1.54	1.58	Samson
Majuro Island	Marshall Islands	Pacific	7.08	171.38	1.54	1.54	1.71	1.54	1.58	Samson
Lae	New Guinea	Pacific	−6.75	146.98	1.71	1.71	1.71	2.18	1.80	Samson
Invercargill	New Zealand	Pacific	−46.42	168.32	1.47	1.47	1.47	1.47	1.47	Samson
Canton Island	Phoenix Islands	Pacific	−2.77	−171.72	1.80	1.80	1.80	1.80	1.80	Samson
Chatham Island		Pacific	−43.97	−176.55	1.47	1.47	1.47	1.62	1.50	Samson
Koror Island		Pacific	7.33	134.48	1.19	1.19	1.29	1.24	1.22	Samson
Macquarie Island		Pacific	−54.50	158.95	1.34	1.34	1.34	1.34	1.34	Samson
Midway Island (Atoll)		Pacific	28.22	−177.37	1.47	1.47	1.62	1.47	1.50	Samson
Wake Island (Atoll)		Pacific	19.28	166.65	1.62	1.91	2.18	2.18	1.94	Samson
Clark Field	Luzon	Philippines	15.13	120.58	1.40	1.54	1.71	1.71	1.58	Samson
Okhotsk	Khabarovsk Krai	Russia	59.37	143.20	1.47	1.40	1.47	1.40	1.43	Samson
Syktyvkar	Komi Krai	Russia	61.67	50.85	1.29	1.40	1.40	1.29	1.34	Samson
Tura	Krasnoyarsk Krai	Russia	64.27	100.23	1.80	1.34	1.34	1.54	1.48	Samson

(continued)

TABLE 16.A.1 (*Continued*)

City	State/Province/ Country	Country/ Region	Latitude, degrees	Longitude, degrees	Feb-Mean	May-Mean	Aug-Mean	Nov-Mean	Average	Source
Vladivostok	Primorsky Krai	Russia	43.12	131.90	1.24	1.40	1.54	1.34	1.37	Samson
Chita	Zabaykalsky Krai	Russia	52.08	113.48	1.54	1.24	1.29	1.29	1.33	Samson
Moscow		Russia	55.82	37.62	1.29	1.29	1.40	1.29	1.31	Samson
Buenos Aires	Argentina	South America	−34.83	−58.53	1.40	1.47	1.40	1.34	1.40	Samson
Ezeiza	Argentina	South America	−34.83	−58.53	1.34	1.47	1.47	1.19	1.36	Samson
Recife	Brazil	South America	−8.12	−34.92	1.54	1.54	1.54	1.54	1.54	Samson
Antofagasta	Chile	South America	−23.47	−70.43	1.62	1.47	1.47	1.47	1.50	Samson
Puerto Montt	Chile	South America	−41.48	−72.85	1.47	1.47	1.47	1.47	1.47	Samson
Quintero	Chile	South America	−32.78	−71.53	1.40	1.47	1.47	1.47	1.45	Samson
Bogota	Columbia	South America	4.70	−74.15	1.62	1.71	1.62	1.80	1.68	Samson
Lima	Peru	South America	−12.10	−77.03	1.71	1.47	1.47	1.47	1.52	Samson
Birmingham	Alabama	United States	33.52	−86.82	1.37	1.54	1.65	1.41	1.49	Kizer
Huntsville	Alabama	United States	34.73	−86.58	1.36	1.52	1.65	1.40	1.48	Kizer
Mobile	Alabama	United States	30.73	−88.05	1.40	1.57	1.65	1.43	1.51	Kizer
Montgomery	Alabama	United States	32.38	−86.32	1.38	1.55	1.65	1.42	1.49	Kizer
Anchorage	Alaska	United States	61.17	−149.98	1.40	1.40	1.40	1.40	1.40	Samson
Annette Island	Alaska	United States	55.00	−131.58	1.31	1.31	1.37	1.30	1.32	Segal and Barrington
Barrow	Alaska	United States	71.30	−156.78	1.40	1.34	1.40	1.40	1.39	Samson
Fairbanks	Alaska	United States	64.82	−147.87	1.38	1.30	1.33	1.33	1.33	Segal and Barrington
Juneau	Alaska	United States	58.36	−134.58	1.34	1.29	1.34	1.34	1.33	Kizer
Nome	Alaska	United States	64.50	−165.43	1.54	1.47	1.47	1.47	1.48	Samson
Shemya	Alaska	United States	52.72	−174.10	1.34	1.34	1.34	1.34	1.34	Samson
Yakutat	Alaska	United States	59.50	−139.68	1.31	1.32	1.34	1.31	1.32	Segal and Barrington

Flagstaff	Arizona	United States	35.17	−111.67	1.24	1.21	1.21	1.25	1.22	Kizer
Phoenix	Arizona	United States	33.43	−112.03	1.35	1.26	1.37	1.40	1.34	Bean et al.
Prescott	Arizona	United States	34.58	−112.35	1.26	1.22	1.21	1.29	1.24	Kizer
Tucson	Arizona	United States	32.22	−110.97	1.30	1.29	1.28	1.34	1.30	Kizer
Winslow	Arizona	United States	35.03	−110.67	1.23	1.20	1.21	1.23	1.21	Kizer
Yuma	Arizona	United States	32.67	−114.67	1.36	1.37	1.47	1.44	1.41	Kizer
Fort Smith	Arkansas	United States	35.39	−94.32	1.35	1.49	1.55	1.36	1.43	Kizer
Little Rock	Arkansas	United States	34.73	−92.23	1.32	1.45	1.56	1.36	1.42	Bean et al.
Texarkana	Arkansas	United States	33.49	−93.93	1.37	1.53	1.63	1.39	1.47	Kizer
Bakersfield	California	United States	35.42	−119.00	1.35	1.38	1.52	1.46	1.42	Kizer
Eureka	California	United States	40.80	−124.18	1.36	1.34	1.43	1.43	1.39	Kizer
Fresno	California	United States	36.73	−119.78	1.34	1.36	1.44	1.44	1.39	Kizer
Los Angeles (Long Beach)	California	United States	33.82	−118.15	1.47	1.34	1.34	1.47	1.40	Samson
Red Bluff	California	United States	40.17	−122.25	1.34	1.32	1.34	1.41	1.35	Kizer
Sacramento	California	United States	38.58	−121.50	1.34	1.34	1.40	1.44	1.38	Kizer
San Diego	California	United States	32.73	−117.17	1.62	1.40	1.47	1.62	1.52	Samson
San Francisco (Oakland)	California	United States	37.73	−122.20	1.40	1.46	1.69	1.45	1.49	Bean et al.
San Jose	California	United States	37.33	−121.88	1.36	1.40	1.54	1.48	1.44	Kizer
San Luis Obispo	California	United States	35.33	−120.67	1.38	1.43	1.65	1.51	1.49	Kizer
Santa Maria	California	United States	34.90	−120.47	1.44	1.45	1.68	1.47	1.50	Bean et al.
Alamosa	Colorado	United States	37.42	−105.87	1.22	1.21	1.26	1.21	1.22	Kizer
Denver	Colorado	United States	39.77	−104.88	1.29	1.29	1.47	1.47	1.37	Samson
Grand Junction	Colorado	United States	39.10	−108.53	1.27	1.23	1.28	1.27	1.26	Bean et al.
Pueblo	Colorado	United States	38.27	−104.62	1.23	1.22	1.27	1.22	1.24	Kizer

(continued)

TABLE 16.A.1 (*Continued*)

City	State/Province/ Country	Country/ Region	Latitude, degrees	Longitude, degrees	Feb-Mean	May-Mean	Aug-Mean	Nov-Mean	Average	Source
Bridgeport	Connecticut	United States	41.22	−73.29	1.34	1.39	1.46	1.40	1.39	Kizer
Hartford	Connecticut	United States	41.77	−72.68	1.33	1.38	1.46	1.39	1.39	Kizer
New Haven	Connecticut	United States	41.34	−72.96	1.34	1.39	1.46	1.40	1.39	Kizer
Wilmington	Delaware	United States	39.75	−75.55	1.34	1.42	1.44	1.42	1.40	Kizer
Washington	District of Columbia	United States	38.83	−77.03	1.29	1.39	1.52	1.33	1.38	Bean et al.
Cocoa Beach	Florida	United States	28.23	−80.60	1.47	1.71	1.71	1.71	1.64	Samson
Jacksonville	Florida	United States	30.33	−81.67	1.40	1.56	1.62	1.46	1.50	Kizer
Key West	Florida	United States	24.55	−81.75	1.47	1.47	1.47	1.47	1.47	Samson
Melbourne	Florida	United States	28.08	−80.60	1.43	1.58	1.62	1.46	1.52	Kizer
Miami	Florida	United States	25.82	−80.28	1.44	1.52	1.57	1.48	1.50	Bean et al.
Miami Beach	Florida	United States	25.78	−80.21	1.45	1.60	1.62	1.47	1.53	Kizer
Tallahassee	Florida	United States	30.45	−84.28	1.40	1.57	1.62	1.44	1.50	Kizer
Tampa	Florida	United States	27.97	−82.53	1.45	1.55	1.61	1.48	1.52	Bean et al.
West Palm Beach	Florida	United States	26.68	−80.11	1.44	1.59	1.62	1.47	1.53	Kizer
Atlanta	Georgia	United States	33.65	−84.42	1.32	1.41	1.51	1.36	1.40	Bean et al.
Augusta	Georgia	United States	33.42	−82.05	1.37	1.53	1.62	1.44	1.48	Kizer
Columbus	Georgia	United States	32.50	−84.88	1.38	1.55	1.63	1.43	1.49	Kizer
Savannah	Georgia	United States	31.96	−81.03	1.38	1.54	1.62	1.46	1.50	Kizer
Hilo (Hawaii)	Hawaii	United States	19.73	−155.07	1.47	1.71	1.91	1.62	1.66	Samson
Honolulu/Oahu	Hawaii	United States	21.32	−157.92	1.45	1.46	1.59	1.44	1.48	Kizer
Kahului/Maui	Hawaii	United States	20.90	−156.43	1.45	1.46	1.58	1.44	1.48	Kizer
Lihue (Kauai)	Hawaii	United States	21.98	−159.35	1.47	1.62	1.62	1.47	1.54	Samson
Boise	Idaho	United States	43.57	−116.22	1.30	1.30	1.34	1.32	1.31	Bean et al.
Lewiston	Idaho	United States	46.38	−116.96	1.30	1.25	1.27	1.29	1.28	Kizer
Pocatello	Idaho	United States	42.87	−112.45	1.26	1.24	1.27	1.25	1.25	Kizer

Cairo	Illinois	United States	37.21	−89.18	1.34	1.49	1.63	1.36	1.45	Kizer
Chicago	Illinois	United States	41.87	−87.68	1.33	1.40	1.54	1.33	1.39	Kizer
Joliet	Illinois	United States	41.50	−88.17	1.32	1.36	1.49	1.34	1.37	Bean et al.
Moline	Illinois	United States	41.80	−89.98	1.33	1.40	1.53	1.32	1.39	Kizer
Peoria	Illinois	United States	40.70	−89.60	1.33	1.42	1.55	1.33	1.40	Kizer
Rantoul	Illinois	United States	40.30	−88.15	1.34	1.47	1.47	1.34	1.40	Samson
Rockford	Illinois	United States	42.28	−89.10	1.33	1.39	1.53	1.32	1.39	Kizer
Springfield	Illinois	United States	39.80	−89.65	1.33	1.44	1.57	1.34	1.41	Kizer
Evansville	Indiana	United States	38.05	−87.54	1.34	1.47	1.62	1.36	1.44	Kizer
Fort Wayne	Indiana	United States	41.07	−85.13	1.33	1.42	1.53	1.34	1.40	Kizer
Indianapolis	Indiana	United States	39.77	−86.17	1.34	1.44	1.58	1.35	1.42	Kizer
South Bend	Indiana	United States	41.68	−86.25	1.33	1.41	1.53	1.33	1.40	Kizer
Burlington	Iowa	United States	40.80	−91.22	1.33	1.41	1.53	1.32	1.39	Kizer
Davenport	Iowa	United States	41.58	−90.64	1.33	1.40	1.52	1.32	1.39	Kizer
Des Moines	Iowa	United States	41.58	−93.62	1.32	1.39	1.47	1.31	1.37	Kizer
Dubuque	Iowa	United States	42.55	−90.85	1.33	1.38	1.51	1.32	1.38	Kizer
Sioux City	Iowa	United States	42.50	−96.40	1.31	1.35	1.41	1.29	1.34	Kizer
Waterloo	Iowa	United States	42.50	−92.33	1.33	1.38	1.49	1.31	1.37	Kizer
Dodge City	Kansas	United States	37.77	−99.97	1.31	1.40	1.46	1.33	1.37	Bean et al.
Goodland	Kansas	United States	39.37	−101.67	1.26	1.29	1.32	1.25	1.28	Kizer
Topeka	Kansas	United States	39.05	−95.67	1.31	1.41	1.43	1.31	1.37	Kizer
Wichita	Kansas	United States	37.68	−97.33	1.31	1.41	1.38	1.31	1.35	Kizer
Lexington	Kentucky	United States	38.05	−84.50	1.34	1.47	1.61	1.38	1.44	Kizer
Louisville	Kentucky	United States	38.19	−85.74	1.34	1.47	1.62	1.37	1.44	Kizer
Burrwood	Louisiana	United States	28.97	−89.37	1.62	1.71	1.71	1.62	1.66	Samson
Lake Charles	Louisiana	United States	30.22	−93.15	1.41	1.60	1.67	1.46	1.53	Bean et al.

(continued)

TABLE 16.A.1 ***(Continued)***

City	State/Province/ Country	Country/ Region	Latitude, degrees	Longitude, degrees	Feb-Mean	May-Mean	Aug-Mean	Nov-Mean	Average	Source
New Orleans	Louisiana	United States	29.95	−90.07	1.41	1.59	1.68	1.43	1.52	Kizer
Shreveport	Louisiana	United States	32.47	−93.82	1.38	1.55	1.66	1.41	1.49	Kizer
Caribou	Maine	United States	46.88	−67.97	1.31	1.31	1.42	1.32	1.34	Bean et al.
Portland	Maine	United States	43.65	−70.32	1.30	1.36	1.46	1.34	1.36	Bean et al.
Baltimore	Maryland	United States	39.28	−76.62	1.34	1.43	1.44	1.42	1.41	Kizer
Frederick	Maryland	United States	39.60	−77.67	1.34	1.43	1.43	1.40	1.40	Kizer
Boston	Massachusetts	United States	42.35	−71.05	1.33	1.37	1.48	1.38	1.39	Kizer
Nantucket	Massachusetts	United States	41.25	−70.07	1.40	1.47	1.54	1.40	1.45	Samson
Pittsfield	Massachusetts	United States	42.45	−73.25	1.33	1.37	1.44	1.37	1.38	Kizer
Detroit	Michigan	United States	42.38	−83.03	1.33	1.39	1.47	1.34	1.38	Kizer
Escanaba	Michigan	United States	45.81	−86.91	1.33	1.34	1.47	1.32	1.36	Kizer
Flint	Michigan	United States	42.97	−83.73	1.29	1.31	1.39	1.30	1.32	Segal and Barrington
Grand Rapids	Michigan	United States	42.88	−85.52	1.33	1.38	1.50	1.33	1.38	Kizer
Houghton Lake	Michigan	United States	44.42	−84.83	1.33	1.36	1.48	1.32	1.37	Kizer
Lansing	Michigan	United States	42.73	−84.55	1.33	1.39	1.49	1.33	1.38	Kizer
Marquette	Michigan	United States	46.54	−87.48	1.33	1.34	1.46	1.31	1.36	Kizer
Mt. Clemens	Michigan	United States	42.60	−82.82	1.34	1.40	1.40	1.34	1.37	Samson
Muskegon	Michigan	United States	43.23	−86.27	1.33	1.38	1.50	1.32	1.38	Kizer
Sault Ste. Marie	Michigan	United States	46.47	−84.37	1.30	1.33	1.44	1.31	1.34	Bean et al.
Ypsilanti	Michigan	United States	42.25	−83.62	1.33	1.39	1.48	1.34	1.38	Kizer

Duluth	Minnesota	United States	46.81	−92.19	1.33	1.33	1.44	1.30	1.35	Kizer
International Falls	Minnesota	United States	48.60	−93.40	1.32	1.30	1.46	1.31	1.34	Bean et al.
Minneapolis	Minnesota	United States	44.98	−93.27	1.33	1.34	1.45	1.30	1.35	Kizer
Rochester	Minnesota	United States	44.02	−92.47	1.33	1.36	1.47	1.31	1.36	Kizer
St. Cloud	Minnesota	United States	45.58	−94.17	1.33	1.34	1.43	1.29	1.35	Kizer
Jackson	Mississippi	United States	32.30	−90.18	1.39	1.56	1.68	1.41	1.50	Kizer
Columbia	Missouri	United States	38.97	−92.37	1.34	1.54	1.71	1.71	1.56	Samson
Kansas City	Missouri	United States	39.08	−94.49	1.32	1.42	1.47	1.32	1.38	Kizer
Springfield	Missouri	United States	37.22	−93.30	1.34	1.47	1.54	1.35	1.42	Kizer
St. Joseph	Missouri	United States	39.75	−94.83	1.32	1.41	1.46	1.31	1.37	Kizer
St. Louis	Missouri	United States	38.70	−90.33	1.34	1.46	1.58	1.34	1.42	Kizer
Billings	Montana	United States	45.78	−108.50	1.29	1.23	1.22	1.24	1.24	Kizer
Glasgow	Montana	United States	48.18	−106.63	1.31	1.28	1.35	1.30	1.31	Bean et al.
Great Falls	Montana	United States	47.50	−111.35	1.25	1.26	1.29	1.26	1.27	Bean et al.
Havre	Montana	United States	48.55	−109.58	1.31	1.23	1.23	1.26	1.26	Kizer
Helena	Montana	United States	46.60	−112.03	1.29	1.23	1.22	1.25	1.25	Kizer
Missoula	Montana	United States	46.83	−114.17	1.29	1.23	1.23	1.27	1.25	Kizer
Grand Island	Nebraska	United States	40.97	−98.33	1.30	1.35	1.38	1.28	1.33	Kizer
Lincoln	Nebraska	United States	40.82	−96.70	1.31	1.37	1.41	1.29	1.34	Kizer
Norfolk	Nebraska	United States	42.08	−97.47	1.31	1.35	1.40	1.28	1.33	Kizer
North Platte	Nebraska	United States	41.13	−100.70	1.31	1.34	1.49	1.32	1.36	Bean et al.
Omaha	Nebraska	United States	41.25	−96.00	1.31	1.36	1.55	1.33	1.38	Bean et al.
Scottsbluff	Nebraska	United States	41.83	−103.67	1.26	1.25	1.31	1.24	1.26	Kizer
Valentine	Nebraska	United States	42.70	−100.53	1.29	1.30	1.36	1.26	1.30	Kizer
Elko	Nevada	United States	40.83	−115.83	1.27	1.24	1.25	1.28	1.26	Kizer

(continued)

TABLE 16.A.1 (*Continued*)

City	State/Province/ Country	Country/ Region	Latitude, degrees	Longitude, degrees	Feb-Mean	May-Mean	Aug-Mean	Nov-Mean	Average	Source
Ely	Nevada	United States	39.28	−114.85	1.26	1.25	1.25	1.26	1.25	Bean et al.
Las Vegas	Nevada	United States	36.07	−115.17	1.28	1.21	1.23	1.28	1.25	Bean et al.
Reno	Nevada	United States	39.52	−119.82	1.32	1.30	1.28	1.35	1.31	Kizer
Tonopah	Nevada	United States	38.08	−117.25	1.29	1.26	1.26	1.32	1.28	Kizer
Winnemucca	Nevada	United States	40.98	−117.75	1.29	1.26	1.25	1.30	1.27	Kizer
Concord	New Hampshire	United States	43.20	−71.53	1.33	1.36	1.46	1.37	1.38	Kizer
Atlantic City	New Jersey	United States	39.37	−74.58	1.34	1.42	1.46	1.43	1.41	Kizer
Newark	New Jersey	United States	40.73	−74.17	1.34	1.40	1.45	1.40	1.40	Kizer
Trenton	New Jersey	United States	40.22	−74.73	1.34	1.41	1.45	1.41	1.40	Kizer
Albuquerque	New Mexico	United States	35.05	−106.62	1.23	1.22	1.32	1.23	1.25	Bean et al.
Clayton	New Mexico	United States	36.35	−103.17	1.24	1.25	1.27	1.23	1.25	Kizer
Jornada	New Mexico	United States	32.62	−106.73	1.22	1.20	1.23	1.23	1.22	Kizer
Raton	New Mexico	United States	36.90	−104.47	1.23	1.22	1.26	1.22	1.23	Kizer
Roswell	New Mexico	United States	33.40	−104.52	1.23	1.21	1.25	1.23	1.23	Kizer
Silver City	New Mexico	United States	32.75	−108.25	1.22	1.21	1.22	1.24	1.22	Kizer
Albany	New York	United States	42.75	−73.80	1.30	1.37	1.46	1.32	1.36	Bean et al.
Binghamton	New York	United States	42.10	−75.92	1.33	1.38	1.42	1.37	1.38	Kizer
Buffalo	New York	United States	42.93	−78.72	1.30	1.33	1.43	1.31	1.34	Bean et al.
New York	New York	United States	40.77	−73.87	1.29	1.40	1.47	1.29	1.36	Samson
Rochester	New York	United States	43.17	−77.60	1.33	1.37	1.41	1.35	1.36	Kizer
Syracuse	New York	United States	43.05	−76.15	1.33	1.37	1.42	1.35	1.37	Kizer
Asheville	North Carolina	United States	35.60	−82.55	1.36	1.50	1.61	1.42	1.47	Kizer
Cape Hatteras	North Carolina	United States	35.25	−75.67	1.39	1.57	1.71	1.46	1.52	Bean et al.
Charlotte	North Carolina	United States	35.22	−80.85	1.36	1.50	1.59	1.44	1.47	Kizer
Greensboro	North Carolina	United States	36.08	−79.95	1.31	1.42	1.53	1.35	1.39	Bean et al.

Releigh	North Carolina	United States	35.78	−78.63	1.36	1.49	1.57	1.46	1.46	Kizer
Wilmington	North Carolina	United States	34.23	−77.92	1.37	1.50	1.63	1.48	1.49	Kizer
Winston Salem	North Carolina	United States	36.10	−80.25	1.35	1.49	1.56	1.44	1.46	Kizer
Bismarck	North Dakota	United States	46.77	−100.75	1.32	1.32	1.44	1.31	1.35	Bean et al.
Fargo	North Dakota	United States	46.84	−96.88	1.34	1.31	1.38	1.28	1.33	Kizer
Williston	North Dakota	United States	48.60	−103.67	1.34	1.26	1.28	1.27	1.29	Kizer
Akron	Ohio	United States	41.08	−81.52	1.33	1.41	1.45	1.35	1.39	Kizer
Cincinnati	Ohio	United States	39.18	−84.47	1.34	1.45	1.58	1.36	1.43	Kizer
Cleveland	Ohio	United States	41.44	−81.64	1.33	1.41	1.45	1.35	1.38	Kizer
Columbus	Ohio	United States	39.97	−83.00	1.34	1.44	1.51	1.36	1.41	Kizer
Dayton	Ohio	United States	39.87	−84.12	1.34	1.40	2.04	1.40	1.50	Samson
Sandusky	Ohio	United States	41.39	−82.65	1.33	1.41	1.47	1.34	1.39	Kizer
Toledo	Ohio	United States	41.57	−83.47	1.32	1.37	1.49	1.34	1.38	Bean et al.
Youngstown	Ohio	United States	41.10	−80.65	1.34	1.41	1.44	1.36	1.38	Kizer
Oklahoma City	Oklahoma	United States	35.40	−97.60	1.32	1.44	1.51	1.34	1.40	Bean et al.
Tulsa	Oklahoma	United States	36.15	−96.00	1.33	1.46	1.45	1.34	1.39	Kizer
Baker	Oregon	United States	44.87	−117.83	1.30	1.26	1.26	1.30	1.28	Kizer
Burns	Oregon	United States	43.58	−119.00	1.31	1.27	1.26	1.31	1.29	Kizer
Eugene	Oregon	United States	44.12	−123.17	1.34	1.32	1.31	1.37	1.34	Kizer
Medford	Oregon	United States	42.38	−122.87	1.34	1.32	1.27	1.36	1.32	Bean et al.
Portland	Oregon	United States	45.52	−122.68	1.34	1.32	1.32	1.36	1.33	Kizer
Roseburg	Oregon	United States	43.25	−123.33	1.35	1.32	1.32	1.38	1.34	Kizer
Harrisburg	Pennsylvania	United States	40.27	−76.88	1.34	1.41	1.43	1.40	1.39	Kizer
Philadelphia	Pennsylvania	United States	39.96	−75.22	1.34	1.41	1.44	1.41	1.40	Kizer
Pittsburgh	Pennsylvania	United States	40.35	−79.93	1.30	1.34	1.42	1.31	1.34	Bean et al.
Wilkes Barre	Pennsylvania	United States	41.25	−75.88	1.34	1.39	1.43	1.39	1.38	Kizer

(continued)

TABLE 16.A.1 (*Continued*)

City	State/Province/ Country	Country/ Region	Latitude, degrees	Longitude, degrees	Feb-Mean	May-Mean	Aug-Mean	Nov-Mean	Average	Source
San Juan	Puerto Rico	United States	18.44	−66.00	1.48	1.60	1.62	1.61	1.57	Kizer
Providence	Rhode Island	United States	41.83	−71.42	1.34	1.38	1.48	1.39	1.40	Kizer
Charleston	South Carolina	United States	32.90	−80.03	1.36	1.52	1.63	1.42	1.48	Bean et al.
Columbia	South Carolina	United States	33.95	−81.12	1.37	1.52	1.61	1.45	1.48	Kizer
Florence	South Carolina	United States	34.18	−79.80	1.37	1.51	1.61	1.46	1.48	Kizer
Greenville-Spartanburg	South Carolina	United States	34.85	−82.40	1.36	1.51	1.61	1.43	1.47	Kizer
Aberdeen	South Dakota	United States	45.50	−98.50	1.32	1.31	1.36	1.27	1.32	Kizer
Rapid City	South Dakota	United States	44.15	−103.10	1.28	1.32	1.40	1.30	1.32	Bean et al.
Sioux Falls	South Dakota	United States	43.55	−96.73	1.32	1.34	1.40	1.28	1.33	Kizer
Bristol	Tennessee	United States	36.50	−82.17	1.35	1.49	1.58	1.41	1.45	Kizer
Chattanooga	Tennessee	United States	35.05	−85.32	1.36	1.52	1.65	1.41	1.47	Kizer
Knoxville	Tennessee	United States	35.97	−83.92	1.35	1.50	1.63	1.41	1.47	Kizer
Memphis	Tennessee	United States	35.17	−89.93	1.36	1.52	1.65	1.39	1.47	Kizer
Nashville	Tennessee	United States	36.12	−86.68	1.34	1.44	1.52	1.36	1.41	Bean et al.
Abilene	Texas	United States	32.47	−99.72	1.32	1.41	1.33	1.34	1.35	Kizer
Amarillo	Texas	United States	35.23	−101.67	1.26	1.30	1.28	1.26	1.28	Kizer
Austin	Texas	United States	30.27	−97.75	1.38	1.50	1.52	1.41	1.45	Kizer
Big Spring	Texas	United States	32.23	−101.50	1.32	1.39	1.37	1.32	1.35	Bean et al.

Brownsville	Texas	United States	25.92	−97.47	1.47	1.68	1.76	1.46	1.58	Bean et al.
Corpus Christi	Texas	United States	27.80	−97.40	1.44	1.58	1.60	1.47	1.52	Kizer
Dallas	Texas	United States	32.78	−96.80	1.36	1.50	1.49	1.38	1.43	Kizer
Del Rio	Texas	United States	29.38	−100.92	1.32	1.36	1.31	1.35	1.34	Kizer
El Paso	Texas	United States	31.78	−106.50	1.26	1.21	1.31	1.25	1.26	Bean et al.
Fort Worth	Texas	United States	32.75	−97.33	1.35	1.49	1.46	1.38	1.42	Kizer
Houston	Texas	United States	29.75	−95.37	1.42	1.58	1.67	1.44	1.52	Kizer
Laredo	Texas	United States	27.49	−99.46	1.37	1.43	1.38	1.41	1.40	Kizer
Lubbock	Texas	United States	33.58	−101.85	1.27	1.30	1.28	1.28	1.28	Kizer
Midland	Texas	United States	32.00	−102.08	1.27	1.30	1.28	1.29	1.28	Kizer
Port Arthur	Texas	United States	29.90	−93.93	1.42	1.59	1.70	1.44	1.53	Kizer
San Angelo	Texas	United States	31.47	−100.43	1.31	1.38	1.32	1.34	1.34	Kizer
San Antonio	Texas	United States	29.53	−98.47	1.34	1.46	1.49	1.35	1.41	Bean et al.
Victoria	Texas	United States	28.83	−97.00	1.42	1.57	1.62	1.46	1.51	Kizer
Waco	Texas	United States	31.55	−97.13	1.37	1.51	1.51	1.40	1.45	Kizer
Wichita Falls	Texas	United States	33.92	−98.48	1.32	1.44	1.36	1.34	1.36	Kizer
Salt Lake City	Utah	United States	40.75	−111.88	1.24	1.24	1.27	1.24	1.25	Kizer
Burlington	Vermont	United States	44.48	−73.22	1.33	1.35	1.43	1.34	1.36	Kizer
Lynchburg	Virginia	United States	37.42	−79.15	1.35	1.47	1.50	1.43	1.43	Kizer
Norfolk	Virginia	United States	36.85	−76.28	1.35	1.46	1.54	1.47	1.45	Kizer
Richmond	Virginia	United States	37.53	−77.43	1.35	1.46	1.49	1.44	1.43	Kizer

(continued)

TABLE 16.A.1 (*Continued*)

City	State/Province/ Country	Country/ Region	Latitude, degrees	Longitude, degrees	Feb-Mean	May-Mean	Aug-Mean	Nov-Mean	Average	Source
Roanoke	Virginia	United States	37.27	−79.95	1.35	1.47	1.51	1.42	1.44	Kizer
Wallops Island	Virginia	United States	37.93	−75.48	1.34	1.44	1.49	1.46	1.43	Kizer
Blaine	Washington	United States	49.00	−122.75	1.34	1.34	1.48	1.34	1.37	Kizer
Port Angeles	Washington	United States	48.06	−123.32	1.34	1.35	1.49	1.35	1.38	Kizer
Quillayute	Washington	United States	48.38	−124.73	1.34	1.34	1.34	1.34	1.34	Samson
Seattle	Washington	United States	47.45	−122.30	1.40	1.40	1.40	1.54	1.43	Samson
Spokane	Washington	United States	47.62	−117.52	1.30	1.31	1.30	1.32	1.31	Bean et al.
Tacoma	Washington	United States	47.25	−122.43	1.34	1.33	1.42	1.35	1.36	Kizer
Tatoosh Island	Washington	United States	48.38	−124.73	1.36	1.43	1.49	1.37	1.41	Bean et al.
Walla Walla	Washington	United States	46.08	−118.17	1.31	1.27	1.28	1.30	1.29	Kizer
Yakima	Washington	United States	46.60	−120.52	1.33	1.30	1.32	1.33	1.32	Kizer
Beckley	West Virginia	United States	37.75	−81.17	1.34	1.47	1.52	1.40	1.43	Kizer
Charleston	West Virginia	United States	38.35	−81.63	1.34	1.46	1.51	1.39	1.42	Kizer
Elkins	West Virginia	United States	38.92	−79.83	1.34	1.45	1.46	1.40	1.41	Kizer
Huntington	West Virginia	United States	38.33	−82.38	1.34	1.46	1.54	1.39	1.43	Kizer
Parkersburg	West Virginia	United States	39.25	−81.41	1.34	1.45	1.48	1.38	1.41	Kizer
Green Bay	Wisconsin	United States	44.52	−88.02	1.33	1.36	1.49	1.32	1.37	Kizer
La Crosse	Wisconsin	United States	43.81	−91.24	1.33	1.36	1.49	1.31	1.37	Kizer
Madison	Wisconsin	United States	43.07	−89.38	1.33	1.38	1.51	1.32	1.38	Kizer
Milwaukee	Wisconsin	United States	43.03	−87.90	1.33	1.38	1.51	1.32	1.38	Kizer
Casper	Wyoming	United States	42.85	−106.32	1.26	1.23	1.26	1.23	1.24	Kizer
Cheyenne	Wyoming	United States	41.13	−104.82	1.24	1.23	1.29	1.23	1.25	Kizer
Lander	Wyoming	United States	42.80	−108.72	1.25	1.27	1.28	1.26	1.27	Bean et al.
Sheridan	Wyoming	United States	44.77	−107.00	1.28	1.23	1.23	1.24	1.24	Kizer
Yellowstone Park	Wyoming	United States	44.50	−110.50	1.27	1.23	1.23	1.24	1.24	Kizer

REFERENCES

Abramowitz, M. and Stegun, I., *Handbook of Mathematical Functions (NBS AMS 55, seventh printing with corrections)*. Washington, DC: U.S. Government Printing Office, pp. 931–933, 1968.

Achariyapaopan, T., "A Model of Geographic Variation of Multipath Fading Probability," *Bellcore National Radio Engineer's Conference Record*, pp. TA1–TA16, 1986.

Babler, G. M., "A Study of Frequency Selective Fading for a Microwave Line-of-Sight Narrowband Radio Channel," *Bell System Technical Journal*, Vol. 513, pp. 731–757, March 1972.

Barnett, W. T., "Multipath Propagation at 4, 6 and 11 GHz," *Bell System Technical Journal*, Vol. 51, pp. 321–361, February 1972.

Bean, B. R., Horn, J. D. and Ozanich, Jr., A. M., *Climatic Charts and Data of the Radio Refractive Index for the United States and the World, National Bureau of Standards Monograph 22*. Washington, DC: U.S. Government Printing Office, 1960.

Bean, B. R., Cahoon, B. A., Samson, C. A. and Thayer, G. D., *A World Atlas of Atmospheric Radio Refractivity, Environmental Science Services Administration Monograph 1*. Washington, DC: U.S. Government Printing Office, 1966.

Boithias, L., *Radio Wave Propagation*. New York: McGraw-Hill, pp. 1–49, 1987.

Cellerino, G., D'Avino, P. and Moreno, L., "Frequency Diversity Protection in Digital Radio with Hitless Switching," *IEEE International Conference on Communications (ICC) Conference Record*, Vol. 3, pp. 47.2.1–47.2.5, June 1985.

Committee TR14-11, *TIA/EIA Telecommunications Systems Bulletin 10-F, Interference Criteria for Microwave Systems*. Washington, DC: Telecommunications Industry Association, June 1994.

Dirner, P. L. and Lin, S. H., "Measured Frequency Diversity Improvement for Digital Radio," *IEEE Transactions on Communications*, Vol. 33, pp. 106–109, January 1985.

Friis, H. T., "A Note on a Simple Transmission Formula," *Proceedings of the I. R. E and Waves and Electrons*, Vol. 34, pp. 254–256, May 1946.

Friis, H. T., "Microwave Repeater Research," *Bell System Technical Journal*, Vol. 27, p. 188, April 1948.

Giger, A. J., *Low-Angle Microwave Propagation: Physics and Modeling*. Boston: Artech House, pp. 216–218, 1991.

Greenfield, P. E., "Digital Radio Performance on a Long, Highly Dispersive Fading Path," *IEEE International Conference on Communication*, pp. 1451–1454, May 1984.

Greenstein, L. J. and Shafi, M., "Outage Calculation Methods for Microwave Digital Radio," *IEEE Communications Magazine*, Vol. 25, pp. 30–39, February 1972.

Gullett, F., *Pathloss 4.0 Manual.* Coquitlam: Contract Telecommunication Engineering, p. 34 (Worksheets), 2001.

Hause, L. G. and Wortendyke, D. R., *National Telecommunications and Information Administration Report 79–18, Automated Digital System Engineering Model*. Washington, DC. United States Government Printing Office, 1979.

Hodges, D. D., Watson, R. J. and Wyman, G., "An Attenuation Time Series Model for Propagation Forecasting," *IEEE Transactions on Antennas and Propagation*, Vol. 54, pp. 1726–1733, June 2006.

International Telecommunication Union—Radiocommunication Sector (ITU-R), "Recommendation P.581-2, The Concept of 'Worst Month'," p. 1, 1990.

International Telecommunication Union—Radiocommunication Sector (ITU-R), "Recommendation P.678-1, Characterization of the Natural Variability of Propagation Phenomena," pp. 1–2, 1992.

International Telecommunication Union—Radiocommunication Sector (ITU-R), "Recommendation P.525-2, Calculation of Free-Space Attenuation," pp. 1–3, 1994.

International Telecommunication Union—Radiocommunication Sector (ITU-R), "Recommendation P.341-5, The Concept of Transmission Loss for Radio Links," pp. 1–6, 1999a.

International Telecommunication Union—Radiocommunication Sector (ITU-R), "Recommendation P.840-3, Attenuation Due to Clouds and Fog," pp. 1–7, 1999b.

International Telecommunication Union—Radiocommunication Sector (ITU-R), "Recommendation P.1510, Annual Mean Surface Temperature," pp. 1–2, 2001a.

International Telecommunication Union—Radiocommunication Sector (ITU-R), "Recommendation P.836-3, Water Vapour: Surface Density and Total Columnar Content," pp. 1–12, 2001b.

International Telecommunication Union—Radiocommunication Sector (ITU-R), "Recommendation P.841-4, Conversion of Annual Statistics to Worst-Month Statistics," pp. 1–6, 2005a.

International Telecommunication Union—Radiocommunication Sector (ITU-R), "Recommendation P.838-3, Specific Attenuation Model for Rain for Use in Prediction Methods," pp. 1–8, 2005b.

International Telecommunication Union—Radiocommunication Sector (ITU-R), "Recommendation P.526-10, Propagation by Diffraction," pp. 1–37, 2005c.

International Telecommunication Union—Radiocommunication Sector (ITU-R), "Recommendation F.752-2, Diversity Techniques for Point-to-Point Fixed Wireless Systems," pp. 1–13, 2006a.

International Telecommunication Union—Radiocommunication Sector (ITU-R), "Recommendation F.1093-2, Effects of Multipath Propagation on the Design and Operation of Line-of-Sight Digital Fixed Wireless Systems," pp. 1–7, 2006b.

International Telecommunication Union—Radiocommunication Sector (ITU-R), "Recommendation P.1410-4, Propagation Data and Prediction Methods Required for the Design of Terrestrial Broadband Radio Access systems Operating in a Frequency Range from 3 to 60 GHz," pp. 1–28, 2007a.

International Telecommunication Union—Radiocommunication Sector (ITU-R), "Recommendation P.1411-4, Propagation Data and Prediction Methods for the Planning of Short-Range Outdoor Radiocommunication Systems and Radio Local Area Networks in the Frequency Range 300 MHz to 100 GHz," pp. 1–25, 2007b.

International Telecommunication Union—Radiocommunication Sector (ITU-R), "Recommendation P.1057-2, Probability Distributions Relevant to Radio wave Propagation Modeling," pp. 1–18, 2007c.

International Telecommunication Union—Radiocommunication Sector (ITU-R), "Recommendation P.1407-3, Multipath Propagation and Parameterization of its Characteristics," pp. 1–15, 2007d.

International Telecommunication Union—Radiocommunication Sector (ITU-R), "Recommendation P.1814, Prediction Methods Required for the Design of Terrestrial Free-Space Optical Links," pp. 1–12, 2007e.

International Telecommunication Union—Radiocommunication Sector (ITU-R), "Recommendation P.1817, Propagation Data Required for the Design of Terrestrial Free-Space Optical Links," pp. 1–17, 2007f.

International Telecommunication Union—Radiocommunication Sector (ITU-R), "Recommendation P.372-9, Radio Noise," pp. 1–75, 2007g.

International Telecommunication Union—Radiocommunication Sector (ITU-R), "Recommendation P.837-5, Characteristics of Precipitation for Propagation Modelling," pp. 1–11, 2007h.

International Telecommunication Union—Radiocommunication Sector (ITU-R), "Recommendation P.833-6, Attenuation in Vegetation," pp. 1–17, 2007i.

International Telecommunication Union—Radiocommunication Sector (ITU-R), "Recommendation P.1144-4, Guide to the Application of the Propagation Methods of Radiocommunication Study Group 3," pp. 1–9, 2007j.

International Telecommunication Union—Radiocommunication Sector (ITU-R), "Recommendation P.676-7, Attenuation by Atmospheric Gases," pp. 1–23, 2007k.

International Telecommunication Union—Radiocommunication Sector (ITU-R), "Recommendation P.530-13, Propagation Data and Prediction Methods Required for the Design of Terrestrial Line-of-Sight Systems," pp. 1–48, 2009.

Kizer, G. M., *Microwave Communication*. Ames: Iowa State University Press, pp. 335–339, 367–368, 1990.

Kizer, G. M., "Microwave Radio Communication," *Handbook of Microwave Technology, Volume 2*, Ishii, T. K., Editor. San Diego: Academic Press, pp. 449–504, 1995.

Knight, W., *Microwave Path Design Reference Manual*. Plano (north Dallas): Alcatel-Lucent, A.1–C.12, June 2006.

Lane, D., *Digital Microwave Systems Applications Seminar*. Raleigh: Harris Stratex, Slide 193, May 2008.

Liebe, H. J. and Gimmestad, G. G., "Calculation of Clear Air EHF Refractivity," *Radio Science*, Vol. 13, pp. 245–251, March 1978.

Loso, F.G., Inserra, J.R., Brockel, K.H., Barnett, W.T. and Vigants, A., "U.S. Army Tactical LOS Radio Propagation Reliability," *Proceedings of the [IEEE] Tactical Communications Conference*, Vol. 1, pp. 109–117, April 1992.

Morita, K., "Prediction of Rayleigh Fading Occurrence Probability of Line-of-Sight Microwave Links," *Review of the Electrical Communication Laboratories (NTT, Japan)*, pp. 810–821, November–December 1970.

Olsen, R. L. and Segal, B., "New Techniques for Predicting the Multipath Fading Distribution on VHF/UHF/SHF terrestrial line-of-sight links in Canada," *Canadian Journal of Electrical and Computer Engineering*, Vol. 17, pp. 11–23, No. 1 1992.

Olsen, R. L. and Tjelta, T., "Worldwide Techniques for Predicting the Multipath Fading Distribution on Terrestrial LOS Links: Background and Results of Tests," *IEEE Transactions on Antennas and Propagation*, Vol. 47, pp. 157–170, January 1999.

Olsen, R. L., Tjelta, T., Martin, L. and Doble, J. E., "Towards a More Accurate Method of Predicting the Distribution of Multipath Fading on Terrestrial Microwave Links," *Electronics Letters*, Vol. 22, pp. 902–903, August 1986.

Olsen, R. L., Tjelta, T., Martin, L. and Segal, B., "Worldwide Techniques for Predicting the Multipath Fading Distribution on Terrestrial LOS Links: Comparison with Regional Techniques," *IEEE Transactions on Antennas and Propagation*, Vol. 51, pp. 23–30, January 2003.

Ranade, A. and Greenfield, P. E., "An Improved Method of Digital Radio Characterization from Field Measurements," *IEEE International Conference on Communications*, pp. C2.6.1–C2.6.5, June 1983.

Rummler, W. D., "More on the Multipath Fading Channel Model," *IEEE Transactions on Communications*, Vol. 29, pp. 346–352, March 1981.

Rummler, W. D., "A Comparison of Calculated and Observed Performance of Digital Radio in the Presence of Interference," *IEEE Transactions on Communications*, Vol. 30, pp. 1693–1700, July 1982a.

Rummler, W. D., "A Statistical Model of Mjultipath Fading on a Space Diversity Radio Channel," *IEEE International Conference on Communications (ICC) Record*, Vol. 1, pp. 3B.4.1–3B.4.6, June 1982b.

Rummler, W. D., "Characterizing the Effects of Multipath Dispersion on Digital Radios," *IEEE Global Telecommunications Conference (GLOBECOM)*, pp. 52.5.1–52.5.7., November 1988.

Rummler, W. D. and Dhafi, M., "Route Design Methods," *Terrestrial Digital Microwave Communications*. Ivanek, F, Editor. Norwood: Artech House, pp.326–329, 1989.

Samson, C. A., *Refractivity Gradients in the Northern Hemisphere, Office of Telecommunications Report 75–59*. Washington, DC: U.S. Government Printing Office, 1975.

Samson, C. A., *Refractivity and Rainfall Data for Radio Systems Engineering, Office of Telecommunications Report 76–105*. Washington, DC: U.S. Government Printing Office, 1976.

Schiavone, J. A., "Microwave Radio Meteorology: Diurnal Fading Distributions," *Radio Science*, Vol. 17, pp. 1301–1312, September–October 1982.

Segal, B. and Barrington, R. E., *The Radio Climatology of Canada Tropospheric Refractivity Atlas for Canada, Communications Research Centre Report No. 1315-E*. Ottawa: Department of Communications, December 1977.

Strohbehn, J. W., "Line-of-Sight Wave Propagation Through the Turbulent Atmosphere," *Proceedings of the IEEE*, Vol. 56, pp. 1301–1318, August 1968.

Strohbehn, J. W. and Beran, M. J., "Comments on Line-of-Sight Wave Propagation Through the Turbulent Atmosphere," *Proceedings of the IEEE*, Vol. 57, pp. 703–704, April 1969.

Tjelta, T., Olsen, R. L. and Martin, L., "Systematic Development of New Multivariable Techniques for Predicting the Distribution of Multipath Fading on Terrestrial Microwave Links," *IEEE Transactions on Antennas and Propagation*, Vol. 38, pp. 1650–1665, October 1990.

United States Air Force, *Handbook of Geophysics, Revised Edition*. New York: The Macmillan Company, pp. 4–2 and 4–3, 1960.

Vigants, A., "Space-Diversity Engineering," *Bell System Technical Journal*, Vol. 54, pp. 103–142, January 1975.

Vigants, A., "Temporal Variability of Distance Dependence of Amplitude Dispersion and Fading," *IEEE International Conference on Communications*, Vol. 3, pp. 1447–1450, May 1984.

Working Group 9, *Long Term / Short Term Objectives for Terrestrial Microwave Coordination, Recommendation WG 9.85.001*. Washington, DC: National Spectrum Managers Association, 1985.

MICROWAVE FORMULAS AND TABLES

A.1 GENERAL

TABLE A.1 General

Decibel (dB) = 10 $\log(P_O/P_I)$ = 20 $\log(E_O/E_I)$
Neper = 1/2 Ln (P_O/P_I) = Ln (E_O/E_I)
Neper = 0.1151 [dB Value]
dB = 8.686 [Neper value]

P_O = Power at the output

P_I = Power at the input

E_O = Voltage at the output

E_I = Voltage at the input

If $\log_B X = A$ then $B^A = X = \text{Antilog}_B\ A$

$\log(x)$ = common (Brigg's) logarithm = $\log_{10}(x) = \log_e(x)/\log_e(10)$

$\ln(x)$ = natural (Napierian) logarithm = $\log_e(x) = \log_{10}(x)/\log_{10}(e)$

$e \cong 2.7182818284\ldots$

dBW = 10 log (power measured in watts)
dBm = 10 log (power measured in milliwatts) = dBW + 30
dBrn = 10 log (power measured in picowatts) = dBW + 90

Adding 3dB (actually 3.01) to any dB value doubles the power (ratio).
Subtracting 3dB (actually 3.01) from any dB value halves the power (ratio).
Adding 10dB to any dB value multiplies the power (ratio) by 10.
Subtracting 10dB to any dB value divides the power (ratio) by 10.
Changing the sign of a dB value inverts the power (ratio).
Adding 1dB to any dB value multiplies the power (ratio) by 5/4 (actually 1.26).
Subtracting 1dB from any dB value multiplies the power (ratio) by 4/5 (actually 0.794).

Note: For this document the following abbreviations will be used:

Abs $(x) = |x|$ = absolute value = magnitude ignoring sign or phase

Cerfc (x) = complementary error function

Ln (x) = natural (Napierian) logarithm = $\log_e(x)$

$\log(x)$ = common (Brigg's) logarithm = $\log_{10}(x)$

sqrt $(x) = [x]^{1/2}$ = square root function

Arabic Numeral	Roman Numeral	Arabic Numeral	Roman Numeral	Arabic Numeral	Roman Numeral	Arabic Numeral	Roman Numeral
0	Nulla (N)	5	V	50	L	500	D
1	I	10	X	100	C	1000	M

Vertical lines on both sides of the numeral multiply the value by 100. A horizontal bar over the Roman numeral multiplies the value by 1000. Two horizontal bars over the numeral or one horizontal bar below the numeral multiplies the value by 1,000,000.

Digital Microwave Communication: Engineering Point-to-Point Microwave Systems, First Edition. George Kizer.

TABLE A.2 Scientific and Engineering Notation

Symbol	Prefix	Name	Multiplication Factor
		Googolplex	$10^{10^{100}}$
		(centillion)	10^{600}
		Centillion	10^{303}
		(vigintillion)	10^{120}
		(novemdecillion)	10^{114}
		(octodecillion)	10^{108}
		(septendecillion)	10^{102}
		Googol	10^{100}
		(sexdecillion)	10^{96}
		(quindecillion)	10^{90}
		(quattuordecillion)	10^{84}
		(tredecillion)	10^{78}
		(duodecillion)	10^{72}
		(undecillion)	10^{66}
		Vigintillion	10^{63}
		Novemdecillion (decillion)	10^{60}
		Octodecillion	10^{57}
		Septdecillion (nonillion)	10^{54}
		Sexdecillion	10^{51}
		Quindecillion (octillion)	10^{48}
		Quattuordecillion	10^{45}
		Tredecillion (septillion)	10^{42}
		Duodecillion	10^{39}
		Undecillion (sextillion)	10^{36}
		Decillion	10^{33}
		Nonillion (quintillion)	10^{30}
		Octillion	10^{27}
Y	Yotta	Septillion (quadrillion)	10^{24}
Z	Eta	Sextillion	10^{21}
E	Exa	Quintillion (trillion)	10^{18} = 1,000,000,000,000,000,000
P	Peta	Quadrillion (billiard)	10^{15} = 1,000,000,000,000,000
T[a]	Tera	Trillion (billion)	10^{12} = 1,000,000,000,000
G[a]	Giga	Billion (milliard)	10^{9} = 1,000,000,000
M[a]	Mega	Million	10^{6} = 1,000,000
Ma	Myria		10^{4} = 10,000
k[a]	Kilo	Thousand	10^{3} = 1000
h	Hecto	Hundred	10^{2} = 100
da	Deka (deca)	Ten	10^{1} = 10
		One (unity)	10^{0} = 1
d	Deci	Tenth	10^{-1} = 0.1
c	Centi	Hundredth	10^{-2} = 0.01
m	Milli	Thousandth	10^{-3} = 0.001
μ	Micro	Millionth	10^{-6} = 0.000,001
n	Nano	Billionith	10^{-9} = 0.000,000,001
Å		Ángstrom	10^{-10} m
p	Pico	Trillionth	10^{-12} = 0.000,000,000,001
f	Femto	Quadrillionth	10^{-15} = 0.000,000,000,000,001
a	Atto	Quintillionth	10^{-18} = 0.000,000,000,000,000,001
z	Zepto	Sextillionth	10^{-21}
y	Yocto	Septillionth	10^{-24}

[a]Note: k (kilo), in computer usage, the prefix indicates $2^{10} = 1024$; M (mega), in computer usage, the prefix indicates $2^{20} = 1{,}048{,}576$; G (giga), in computer usage, the prefix indicates $2^{30} = 1{,}073{,}741{,}824$; T (tera), in computer usage, the prefix indicates $2^{40} = 1{,}099{,}511{,}627{,}776$.

Scientific notation is a multiplicative factor of 10 to the *n*th power. Engineering notation limits the exponent *n* to multiples of 3.

In Europe, the usage of decimal points and commas are reversed relative to US usage. Common usage is to replace the US comma or European decimal point with a space.

The names in the table are usage in the United States and France. Usage in Great Britain and Germany is shown in parentheses ( ).

TABLE A.3 Emission Designator

An emission designator is a coded word defining the type of signal modulation and its bandwidth. The FCC and ITU-R format for the emission designator is **three numerals and a capital letter** to express necessary bandwidth followed by **three capital letters** describing the form of modulation. The **necessary bandwidth** (usually considered to be the channel bandwidth) uses the letter to indicate the magnitude and the decimal location.

Examples: 60 Hz = 60H0
100 kHz = 100K
70 MHz =70M0
1.99 GHz = 1G99
10.74 GHz = 10G7
10.75 GHz = 10G8

The radio signal's **form of modulation** is described by three symbols as follows:

The **first symbol** describes the manner in which the main carrier is modulated and is one of the following:

Unmodulated Carrier	N
Amplitude Modulation	
Double Sideband	A
Single Sideband, Full Carrier	H
Single Sideband, Reduced or Variable Carrier	R
Single Sideband, Suppressed Carrier	J
Independent Sideband	B
Vestigial Sideband	C
Angle Modulation	
Frequency Modulation	F
Phase Modulation	G
Combination of Amplitude and Angle Modulation	D
Pulse Modulation	
Unmodulated Pulses	P
Pulse Amplitude Modulation	K
Pulse Width Modulation	L
Pulse Position Modulation	M
Angle Modulation during pulse period	Q
Combinations of the above or other	V
Combinations of two or more modes	W
Cases not otherwise covered	X

For current generation QAM digital radios, the usual choice is D.

The **second symbol** describes the nature of the signal(s) modulating the carrier. It is one of the following:

No Modulating Signal	0
A single channel containing quantized or digital information without the use of a modulating subcarrier excluding time division multiplex	1
A single channel containing quantized or digital information with the use of a modulating subcarrier excluding time division multiplex	2
A single channel containing analog information	3
Two or more channels containing quantized or digital information	7
Two or more channels containing analog information	8
Composite system with one or more channels containing quantized or digital information together with one or more channels containing analog information	9
Cases not otherwise covered	X

For current generation digital radios, the usual choices are 1 or 7.

The **third symbol** describes the type of information being transmitted and is one of the following:

No Information Transmitted	N
Telegraphy (aural reception)	A
Telegraphy (automatic reception)	B
Facsimile	C
Data Transmission (telemetry or telecommand)	D
Telephony (including sound broadcasting)	E
Television (video)	F
Combination of the above	W
Cases not otherwise covered	X

For current generation digital radios, the usual choice is W.

TABLE A.4 Typical Commercial Parabolic Antenna Gain (dBi)

	Diameter ft (m)								
Frequency, GHz	1 (0.3)	2 (0.6)	3 (0.9)	4 (1.2)	6 (1.8)	8 (2.4)	10 (3.0)	12 (3.7)	15 (4.6)
1.9		19.3	22.5	25.3	28.9	31.1	33.0	34.6	36.6
2.1		20.0	23.8	26.3	29.8	31.9	33.8	35.4	37.4
2.4		21.1	24.4	27.3	30.7	32.9	34.5	36.9	
3.9					34.5	37.2	39.0	40.7	42.7
4.7		26.6		32.8	36.4	38.6	40.4	42.2	44.1
5.85	23.0	28.0				41.0	42.6	44.2	45.4
6.175				35.0	38.6	41.2	43.0	44.8	46.5
6.775				35.7	39.2	41.8	43.4	45.3	47.1
7.438		30.8		36.6	40.4	42.8	44.6	46.4	48.0
7.813		31.0	34.4	37.2	40.8	43.1	45.1	46.7	48.5
8.125		31.6		37.6	41.4	43.7	45.6	47.2	49.0
8.35				37.6	41.1	43.7	45.7	47.2	48.7
10.6		34.6	37.6	39.8	43.4	45.9	47.8	49.3	
11.2		34.7	38.0	40.4	43.8	46.5	48.2	49.7	
14.8	32.1	36.8		42.7	46.2	48.6	50.5		
18.7	33.6	38.7	42.1	44.6	47.9	50.5			
22.4	35.5	40.3	43.6	46.1	49.4	51.6			
28.5	36.9	41.9							
38.5	40.1	45.1							
78.5	43.8	51.0							

TABLE A.5 Typical Rectangular Waveguide

Band Designation, GHz	Nominal Frequency Range, GHz	Typical Waveguide
2	1.7–2.5	Coaxial cable
4	3.7–4.2	WR-229
5	4.4–5.0	WR-187
Lower 6	5.9–6.4	WR-137/WR-159
Upper 6	6.5–6.9	WR-137
STL	6.9–7.1	WR-137
7	7.1–7.8	WR-112
8	7.8–8.5	WR-112
10½	10.6–10.7	WR-75
11	10.7–11.7	WR-75/WR-90
13	12.7–12.7	WR-75
15	14.0–15.4	WR-62
18	17.7–18.7	WR-42
23	21.2–23.6	WR-42
31	31.0–31.3	WR-28
38	38.6–40.00	WR-28

TABLE A.6 Typical Rectangular Waveguide Data

Operating Range for TE10 Mode Frequency, GHz	EIA Designation	Attenuation Over Operating Range dB/100 m	dB/100 ft	Inside Dimensions, in.	Cutoff for for TE10 Mode Frequency, GHz
1.70–2.60	WR 430	2.59–1.69 (B)	0.788–0.516 (B)	4.300–2.150	1.375
		1.64–1.08 (A)	0.501–0.330 (A)		
2.60–3.95	WR 284	4.85–3.31 (B)	1.478–1.008 (B)	2.840–1.340	2.080
		3.08–2.10 (A)	0.940–0.641 (A)		
3.30–4.90	WR 229	6.11–4.33 (B)	1.862–1.320 (B)	2.290–1.145	2.579
		3.91–2.77 (A)	1.192–0.845 (A)		
3.95–5.85	WR 187	9.15–6.33 (B)	2.79–1.93 (B)	1.872–0.872	3.155
		5.81–4.00 (A)	1.77–1.22 (A)		
4.90–7.05	WR 159	9.48–7.35 (B)	2.89–2.24 (B)	1.590–0.795	3.714
		6.04–4.66 (A)	1.84–1.42 (A)		
5.85–8.20	WR 137	12.6–10.1 (B)	3.85–3.08 (B)	1.372–0.622	4.285
		8.04–6.36 (A)	2.45–1.94 (A)		
7.05–10.00	WR 112	18.1–14.1 (B)	5.51–4.31 (B)	1.122–0.497	5.260
		11.5–8.99 (A)	3.50–2.74 (A)		
8.20–12.40	WR 90	28.3–19.8 (B)	8.64–6.02 (B)	0.900–0.400	6.560
		18.0–12.6 (A)	5.49–3.83 (A)		
10.00–15.00	WR 75	33.0–23.1 (B)	10.07–7.03 (B)	0.750–0.375	7.873
		21.2–14.8 (A)	6.45–4.50 (A)		
12.4–18.00	WR 62	41.9–36.6 (B)	12.76–11.15 (B)	0.622–0.311	9.490
		20.1–17.6 (S)	6.14–5.36 (S)		
15.00–22.00	WR 51	56.8–41.3 (B)	17.30–12.60 (B)	0.510–0.255	11.578
		27.5–20.0 (S)	8.37–6.10 (S)		
18.00–26.50	WR 42	90.9–65.0 (B)	27.7–19.8 (B)	0.420–0.170	14.080
		43.6–31.2 (S)	13.30–9.50 (S)		
22.00–33.00	WR 34	109–75.8 (B)	33.3–23.1 (B)	0.340–0.170	17.368
		52.8–36.7 (S)	16.1–11.2 (S)		
26.50–40.00	WR 28	71.9–49.2 (S)	21.9–15.0 (S)	0.280–0.140	21.100
33.00–50.00	WR 22	102–68.6 (S)	31.0–20.9 (S)	0.224–0.112	26.350
40.00–60.00	WR 19	127–89.2 (S)	38.8–27.2 (S)	0.188–0.094	31.410
50.00–75.00	WR 15	174–128 (S)	52.9–39.1 (S)	0.148–0.074	39.900
60.00–90.00	WR 12	306–171 (S)	93.3–52.2 (S)	0.122–0.061	48.400
75.00–110.00	WR 10	328–231 (S)	100.0–70.4 (S)	0.100–0.050	59.050
90.00–140.00	WR 8	499–325 (S)	152.0–99.0 (S)	0.080–0.040	73.840

B, brass; A, aluminum; S, silver.

TABLE A.7 Typical Copper Corrugated Elliptical Waveguide Loss

Frequency, GHz	Waveguide Type CommScope/RFS	Loss dB/100 m	dB/100 ft
1.9	EW20/E20	2.0/1.3	0.61/0.39
2.1	EW20/E20	1.7/1.1	0.51/0.35
2.4	EW20/—	1.5/—	0.45/—
3.7	EW34/E38	2.2/2.3	0.68/0.70
3.9	EW34/E38	2.2/2.2	0.66/0.67
4.0	EW34/E38	2.1/2.2	0.65/0.66

Continued

TABLE A.7 (*Continued*)

Frequency, GHz	Waveguide Type CommScope/RFS	Loss	
		dB/100 m	dB/100 ft
4.7	EW43/E46	2.8/2.8	0.88/0.85
5.9	EW52/E60	4.0/4.0	1.2/1.2
6.2	EW52/E60	3.9/3.9	1.2/1.2
6.8	EW63/E65	4.4/4.4	1.4/1.3
7.4	EW64/E70	4.8/4.9	1.5/1.5
7.8	EW77/E78	5.9/5.8	1.8/1.8
8.1	EW77/E78	5.7/5.7	1.7/1.7
8.4	EW77/E78	5.6/5.6	1.7/1.7
10.6	EW90/E105	10.4/9.3	3.2/2.8
11.2	EW90/E105	10.1/9.0	3.1/2.8
12.7	EW127/E130	11.6/11.3	3.6/3.4
13.0	EW127/E130	11.5/11.2	3.5/3.4
14.8	EW132/E150	15.7/13.8	4.8/4.2
18.7	EW180/E185	19.4/19.3	5.9/5.9
22.4	EW220/E220	28.2/28.3	8.6/8.6

TABLE A.8 Typical Copper Circular Waveguide Loss

Frequency, GHz	Waveguide Type	Loss	
		dB/100 m	dB/100 ft
3.7	WC-281/−269	1.2/1.3	0.39/0.45
4.2	WC-281/−269	1.1/1.3	0.34/0.39
4.7	WC-281/−269	1.0/1.1	0.32/0.35
5.9	WC-281/−269	0.91/0.99	0.28/0.30
6.4	WC-281/−269/−205	0.91/0.98/1.6	0.28/0.30/0.50
6.8	WC-281/−269/−166	0.89/0.97/2.5	0.27/0.30/0.76
7.4	WC-281/−166	0.89/2.3	0.27/0.70
8.1	WC-281/−166	0.89/2.1	0.27/0.65
8.4	WC-281/−166	0.89/2.1	0.27/0.64
10.7	WC-281/−166/−109	0.91/1.9/4.5	0.28/0.57/1.4
11.7	WC-281/−166/−109	0.92/1.9/4.3	0.28/0.57/1.3
17.7	WC-109	3.6	1.1
20.0	WC-109	3.6	1.1

Waveguide Attenuation (Loss)

$$\text{Attn}\left(\frac{\text{dB}}{100\ \text{m}}\right) = \frac{A\left(\frac{f}{f_C}\right)^2 + B}{\sqrt{\left(\frac{f}{f_C}\right)\left[\left(\frac{f}{f_C}\right)^2 - 1\right]}} \tag{A.1}$$

$$\text{Attn}\left(\frac{\text{dB}}{100\ \text{ft}}\right) = 0.3048\text{Attn}\left(\frac{\text{dB}}{100\ \text{m}}\right) \tag{A.2}$$

f = frequency of interest (GHz);

f_C = cutoff frequency (GHz).

See Chapter 5 for general methods of determining A and B.
A and B are coefficients listed in Table A.9.

TABLE A.9 Rectangular Waveguide Attenuation Factors

Waveguide Designation	Wall Metal	Cutoff Frequency, GHz	Lowest Frequency, GHz	Highest Frequency, GHz	A	B
WR 430	Brass	1.375	1.700	2.600	0.79912	0.87227
WR 430	Aluminum	1.375	1.700	2.600	0.51666	0.53603
WR 284	Brass	2.080	2.600	3.950	1.61319	1.54625
WR 284	Aluminum	2.080	2.600	3.950	1.02226	0.98538
WR 229	Brass	2.579	3.300	4.900	2.09096	2.09398
WR 229	Aluminum	2.579	3.300	4.900	1.33704	1.34170
WR 187	Brass	3.155	3.950	5.850	3.07195	2.89720
WR 187	Aluminum	3.155	3.950	5.850	1.92858	1.87417
WR 159	Brass	3.714	4.900	7.050	3.74082	2.85964
WR 159	Aluminum	3.714	4.900	7.050	2.35605	1.86956
WR 137	Brass	4.285	5.850	8.200	5.06772	4.23786
WR 137	Aluminum	4.285	5.850	8.200	3.12730	2.90245
WR 112	Brass	5.260	7.050	10.000	7.00480	6.11684
WR 112	Aluminum	5.260	7.050	10.000	4.48911	3.81714
WR 90	Brass	6.560	8.200	12.400	9.91582	8.23680
WR 90	Aluminum	6.560	8.200	12.400	6.31386	5.22805
WR 75	Brass	7.873	10.000	15.000	11.19789	11.06060
WR 75	Aluminum	7.873	10.000	15.000	7.14940	7.17723
WR 62	Brass	9.490	12.400	18.000	21.66835	3.28603
WR 62	Silver	9.490	12.400	18.000	10.44449	1.49119
WR 51	Brass	11.578	15.000	22.000	20.04607	19.60623
WR 51	Silver	11.578	15.000	22.000	9.71048	9.48393
WR 42	Brass	14.080	18.000	26.500	31.61794	30.18296
WR 42	Silver	14.080	18.000	26.500	15.19172	14.43432
WR 34	Brass	17.368	22.000	33.000	36.60789	36.64440
WR 34	Silver	17.368	22.000	33.000	17.71324	17.78239
WR 28	Silver	21.100	26.500	40.000	23.74185	23.77586
WR 22	Silver	26.350	33.000	50.000	32.63928	34.86860
WR 19	Silver	31.410	40.000	60.000	43.23420	42.89257
WR 15	Silver	39.900	50.000	75.000	67.35574	41.32939
WR 12	Silver	48.400	60.000	90.000	60.35912	156.85742
WR 10	Silver	59.050	75.000	110.000	110.96517	110.45313
WR 8	Silver	73.840	90.000	140.000	159.76070	146.55681

TABLE A.10 CommScope Elliptical Waveguide Attenuation Factors

Waveguide Designation	Metal	Cutoff Frequency, GHz	Lowest Frequency, GHz	Highest Frequency, GHz	A	B
EW 17	Copper	1.364	1.700	2.400	0.49424	0.48626
EW 20	Copper	1.570	1.900	2.700	0.69877	0.48371
EW 28	Copper	2.200	2.600	3.400	0.77964	0.86249
EW 34	Copper	2.376	3.100	4.200	1.05075	0.76711
EW 37	Copper	2.790	3.300	4.300	1.01620	1.29877
EW 43	Copper	2.780	4.400	5.000	1.45111	0.97967
EW 52	Copper	3.650	4.600	6.425	1.81424	1.68203
EW 63	Copper	4.000	5.850	7.125	2.07262	1.94243
EW 64	Copper	4.320	5.300	7.750	2.23080	2.17665
EW 77	Copper	4.720	6.100	8.500	2.56220	2.90547
EW 85	Copper	6.460	7.700	9.800	3.98638	4.31452
EW 90	Copper	6.500	8.300	11.700	4.62183	4.78385
EW 127	Copper	7.670	10.000	13.250	5.52354	4.58571
EW 132	Copper	9.220	11.000	15.350	7.46516	5.79189
EW 180	Copper	11.150	14.000	19.700	8.52820	9.89320
EW 220	Copper	13.340	21.000	23.600	14.52826	8.35617
EW 240	Copper	15.200	22.000	26.500	14.95157	15.98567

TABLE A.11 RFS Elliptical Waveguide Attenuation Factors

Waveguide Designation	Metal	Cutoff Frequency, GHz	Lowest Frequency, GHz	Highest Frequency, GHz	A	B
E 20	Copper	1.380	1.700	2.300	0.45746	0.51945
E 30	Copper	1.800	2.500	3.100	0.51252	1.05598
E 38	Copper	2.400	3.100	4.200	0.90171	1.19017
E 46	Copper	2.880	4.400	5.000	1.37676	0.95528
ES 46	Copper	3.080	4.400	5.000	1.99642	0.42518
EP 58	Copper	3.560	4.400	6.200	1.76406	1.42354
E 60	Copper	3.650	5.600	6.425	1.77577	1.80124
E 65	Copper	4.010	5.900	7.125	1.89412	2.34046
EP 70	Copper	4.340	6.400	7.750	2.24551	2.32916
E 78	Copper	4.720	7.100	8.500	2.77861	2.24383
EP 100	Copper	6.430	8.500	10.000	3.14271	4.84624
E 105	Copper	6.490	10.300	11.700	4.55513	3.12052
E 130	Copper	7.430	10.700	13.250	5.71860	3.75977
E 150	Copper	8.640	13.400	15.350	6.93307	4.83389
E 185	Copper	11.060	17.300	19.700	9.20008	7.95227
E 220	Copper	13.360	21.200	23.600	16.17421	3.89422
E 250	Copper	15.060	24.250	26.500	16.57892	10.08753

TABLE A.12 Elliptical Waveguide Cutoff Frequencies

EW/EWP-	Range, GHz	Cutoff, GHz
CommScope		
17	1.70–2.40	1.36
20	1.90–2.70	1.57
28	2.60–3.40	2.20
34	3.10–4.20	2.38
37	3.30–4.30	2.79
43	4.40–5.00	2.78
52	4.60–6.425	3.65
63	5.85–7.125	4.00
64	5.30–7.75	4.32
77	6.10–8.50	4.72
85	7.70–9.80	6.46
90	8.30–11.70	6.50
127	10.00–13.25	7.67
132	11.00–15.35	9.22
180	14.00–19.70	11.15
220	17.00–23.60	13.34
240	18.00–26.50	15.20

Continued

TABLE A.12 (*Continued*)

E/EP-	Range, GHz	Cutoff, GHz
RFS		
20	1.70–2.30	1.38
30	2.30–3.10	1.80
38	3.00–4.20	2.40
46	3.65–5.00	2.88
S46	3.90–5.00	3.08
58	4.40–6.20	3.56
60	4.50–6.425	3.65
65	5.00–7.125	4.01
70	5.40–7.75	4.34
78	5.90–8.50	4.72
100	8.00–10.00	6.43
105	8.10–11.70	6.49
130	9.30–13.25	7.43
150	10.80–15.35	8.64
185	13.70–19.70	11.06
220	16.70–23.60	13.36
250	19.00–26.50	15.06
300	24.00–33.40	19.05
380	29.00–39.50	23.45

TABLE A.13 Circular Waveguide Cutoff Frequencies (GHz)

Waveguide	TE_{11}	TM_{01}	TE_{21}	TM_{11}/TE_{01}	TE_{31}	TM_{21}	TE_{41}
WC-281	2.460	3.213	4.083	5.121	5.612	6.863	7.105
WC-269	2.571	3.359	4.268	5.353	5.866	7.175	7.428
WC-205	3.374	4.407	5.600	7.024	7.698	9.415	9.746
WC-166	4.167	5.443	6.916	8.675	9.506	11.627	12.036
WC-109	6.346	8.289	10.532	13.211	14.477	17.706	18.330
WC-75	9.223	12.047	15.307	19.200	21.040	25.733	26.640
Waveguide	TE_{12}	TM_{02}	TM_{31}	TE_{51}	TE_{22}	TM_{12}/TE_{02}	TE_{61}
WC-281	7.123	7.376	8.524	8.570	8.962	9.374	10.021
WC-269	7.446	7.710	8.911	8.959	9.368	9.799	10.476
WC-205	9.771	10.117	11.693	11.756	12.293	12.859	13.746
WC-166	12.066	12.494	14.440	14.518	15.181	15.880	16.976
WC-109	18.376	19.028	21.991	22.110	23.119	24.183	25.853
WC-75	26.707	27.653	31.960	32.133	33.600	35.147	37.573

TABLE A.13 ***(Continued)***

Waveguide	TM_{41}	TE_{32}	TM_{22}	TE_{13}	TE_{71}	TM_{03}	TM_{51}
WC-281	10.139	10.708	11.245	11.405	11.462	11.561	11.721
WC-269	10.599	11.193	11.755	11.922	11.981	12.086	12.253
WC-205	13.907	14.688	15.424	15.644	15.722	15.859	16.078
WC-166	17.175	18.139	19.048	19.319	19.416	19.584	19.855
WC-109	26.156	27.624	29.009	29.422	29.569	29.826	30.239
WC-75	38.013	40.147	42.160	42.760	42.973	43.347	43.947
Waveguide	TE_{42}	TE_{81}	TM_{32}	TM_{61}	TE_{23}	TM_{13}/TE_{03}	TE_{52}
WC-281	12.400	12.891	13.041	13.275	13.321	13.592	14.054
WC-269	12.963	13.476	13.632	13.877	13.926	14.208	14.691
WC-205	17.010	17.683	17.888	18.210	18.273	18.644	19.278
WC-166	21.006	21.837	22.090	22.488	22.566	23.024	23.807
WC-109	31.991	33.257	33.642	34.248	34.367	35.064	36.257
WC-75	46.493	48.333	48.893	49.773	49.947	50.960	52.693

Formula

For coaxial cable velocity factor is related to group velocity using the following formulas:

$$V_G = \text{Group velocity} = V_O V_F \tag{A.3}$$

V_O = velocity of propagation in free space;

= 0.9833 ft/ns;

= 0.2998 m/ns;

V_F = velocity factor = 1/sqrt [dielectric constant (relative permittivity)]. See Table A.15.

The absolute delay D of a transmission line is given by

$$D = \frac{L}{V_G} \tag{A.4}$$

L = physical length of the transmission line.

The effective length of the cable (when compared to an RF signal traveling in free space) is given by

$$L_{EFF} = \text{Effective length} = \frac{L}{V_F} \tag{A.5}$$

The cutoff frequency for coaxial cable is given by

$$F_{CO} = \frac{7.50 \times V_F}{D(\text{in.}) + d(\text{in.})} = \frac{190 \times V_F}{D(\text{mm}) + d(\text{mm})} \tag{A.6}$$

D = inside diameter of outer conductor;

d = outside diameter of inner conductor.

Cable operation should be limited to a frequency no higher than 1/2 of the cutoff frequency.

TABLE A.14 Typical Coaxial Microwave Connectors

Connector	Description	Frequency Limit, GHz	Wrench Size, in.	Recommended Torque
BNC	Bayonet type N Connector (or Neill–Concelman)	4		
SMB	Subminiature type B	4		
N (common)	Named for Paul Neil	11	13/16 (hex nut)	14 in-lb
SMC	Subminiature type C	10	1/4	Brass: 3 in-lb
TNC (common)	Threaded type N Connector (or Neill–Concelman)	10	5/8 (hex nut)	14 in-lb
7 mm (or APC-7)	Sexless connector	18	3/4	14 in-lb
N (precision)	Named for Paul Neil	18	13/16 (hex nut)	24 in-lb
TNC (precision)	Threaded type N Connector (or Neill–Concelman)	15	5/8 (hex nut)	24 in-lb
SMA	Subminiature type A	25	5/16	Stainless steel/thick wall brass: 8 in-lb Thin wall brass: 4 in-lb
3.5 mm	Mates with SMA	27	5/16	Same as SMA
2.9 mm (or K)	Mates with SMA	40	5/16	Same as SMA
GPO	Gilbert Push On	40		
SSMA	Smaller SMA	38	1/4	Stainless steel: 8 in lb
2.4 mm	Mates with 1.85 mm	50		
1.85 mm (or V)	Mates with 2.4 mm	60		
1 mm		110		

To convert from inch-pounds (in-lbs) to Newton-meters (N-m), multiply inch-pounds by 0.113.
Caution: Both BNC and N connectors have 50- and 75-ohm versions. The 50-ohm versions have larger diameter center pins. Do not attempt to mate the 50- and 75-ohm connectors. Either a poor connection or permanently deformed connector will result.

50-Ohm Coaxial Cable Attenuation (Loss)

$$\text{Attn}\left(\frac{\text{dB}}{100\ \text{m}}\right) = A\sqrt{f} + Bf \tag{A.7}$$

$$\text{Attn}\left(\frac{\text{dB}}{100\ \text{ft}}\right) = 0.3048\text{Attn}\left(\frac{\text{dB}}{100\ \text{m}}\right) \tag{A.8}$$

f = frequency of interest (MHz).

See Chapter 5 for general methods of determining A and B.
A and B are coefficients listed in Table A.16.

TABLE A.15 Coaxial Cable Velocity Factors

Dielectric	Air	9913	Foam FEP Teflon	Foam Polyethylene	TFE Teflon	Polyethylene	Silicon	PVC	Solid Nylon
Velocity Factor	0.9997	0.84	0.80	0.78–0.80	0.69–0.71	0.66–0.67	0.58	0.55	0.45

TABLE A.16 50 Ohm Coaxial Cable Attenuation Factors

Cable Type	Diameter, in.	*A*, Conductive	*B*, Dielectric	Maximum Frequency, MHz
Flexible Foam				
1/4″	0.25″	0.565	0.00161	20,000
3/8″	0.38″	0.379	0.00177	14,000
1/2″	0.50″	0.325	0.00154	12,000
Semirigid Foam				
1/4″	0.25″	0.394	0.00120	16,000
3/8″	0.38″	0.331	0.00109	14,000
1/2″	0.50″	0.211	0.000619	9000
5/8″	0.63″	0.158	0.000597	6000
7/8″	0.88″	0.126	0.000487	5000
1–1/4″	1.25″	0.0847	0.000523	3300
1–5/8″	1.63″	0.0626	0.000453	2500
2–1/4″	2.25″	0.0517	0.000476	2200
Air Dielectric				
1/2″	0.50″	0.267	0.00241	11,000
5/8″	0.63″	0.172	0.000251	7000
1–5/8″	1.63″	0.0638	0.000282	2700
2–1/4″	2.25″	0.0528	0.000263	2300
Other				
MIL-C-17/28 (RG-58)	0.20″	1.35	0.00413	1000
MIL-C-17/60 (RG-142)	0.20″	1.21	0.00394	8000
MIL-C-17/74 (RG-213/214)	0.42″	0.531	0.00413	11,000
MIL-C-17/79 (RG-218/219)	0.87″	0.220	0.00413	1000
MIL-C-17/84 (RG-223)	0.21″	1.25	0.00413	12,400
Belden 8237	0.41″	0.555	0.00669	7000
Belden 8240	0.19″	1.12	0.0124	7000
Belden 9913	0.41″	0.395	0.00217	7000
LMR-200	0.20″	1.05	0.00108	7000
LMR-240	0.24″	0.794	0.00108	7000
LMR-300	0.30″	0.630	0.00108	7000
LMR-400	0.41″	0.401	0.000853	7000
LMR-500	0.50″	0.317	0.000854	7000
LMR-600	0.59″	0.248	0.000853	7000
LMR-900	0.87″	0.170	0.000525	7000
LMR-1200	1.20″	0.123	0.000525	3500
LMR-1700	1.67″	0.0868	0.000525	3500

LMR is the registered trade mark of Times Microwave Systems.

TABLE A.17 Frequency Bands, General Users

Band Designation	Nominal Frequency Range	ITU Designation
ULF	300 Hz–3 kHz	Hectokilometric waves
VLF	3–30 kHz	Myriametric waves
LF	30–300 kHz	Kilometric waves
MF	300 kHz–3 MHz	Hectometric waves
HF	3–30 MHz	Decametric waves
VHF	30–300 MHz	Metric waves
UHF	300 MHz–3 GHz	Decimetric waves
SHF	3–30 GHz	Centimetric waves
EHF	30–300 GHz	Millimetric waves
	300 GHz–3 THz	Decimillimetric waves
	3–30 THz	Centimillimetric waves
	30–300 THz	Micrometric waves
	300 THz–3 PHz	Decimicrometric waves

Infrared light begins about 1 THz.

Source: FCC Title 47 CFR Part 97.3, NTIA Manual for Federal Radio Frequency Management, Chapter 6.2, ITU-T B.15, ITU-R V.431-7.

TABLE A.18 Frequency Bands, Fixed Point to Point Operators

Band Designation	Nominal Frequency Range, GHz	Users
2 GHz	1.850–2.690	FCC, NTIA, Canada
2.4 GHz	2.400–2.4835	FCC (unlicensed, CFR Part 15)
4 GHz[a]	3.700–4.200	FCC
5 GHz[a]	4.400–4.940	NTIA
5.2 GHz	5.150–5.350	FCC (unlicensed, CFR Part 15)
5.8 GHz	5.725–5.850	FCC (unlicensed, CFR Part 15)
Lower 6 GHz[a]	5.925–6.425	FCC, Canada
Upper 6 GHz	6.525–6.875	FCC
STL	6.875–7.125	FCC (CFR Parts 74 and 101)
7 GHz	7.125–7.725	NTIA, Canada
8 GHz	7.725–8.500	NTIA, Canada
10½ GHz[a]	10.550–10.680	FCC
11 GHz[a]	10.700–11.700	FCC
CARS	12.700–13.250	FCC (CFR Parts 74 and 101)
15 GHz	14.500–15.350	NTIA, Canada
18 GHz	17.700–19.700	FCC, NTIA
23 GHz[a]	21.200–23.600	FCC, NTIA
LMDS[a]	27.500–31.300	FCC, NTIA
38 GHz	38.600–40.000	FCC
60 GHz	57.000–64.000	FCC (unlicensed, CFR Part 15)
70 GHz[a]	71.000–76.000	FCC, NTIA
80 GHz[a]	81.000–86.000	FCC, NTIA
90 GHz	92.000–95.000	FCC (licensed and unlicensed), NTIA

[a] Shared with satellite service

FCC, US Commercial (CFR Part 101), NTIA, US Federal Government.

TABLE A.19 Frequency Bands, Radar, Space, and Satellite Operators

Band Designation	Nominal Frequency Range
L	1–2 GHz
S	2–4 GHz
C	4–8 GHz
X	8–12 GHz
K_u	12–18 GHz
K[a]	18–27 GHz
K_a[a]	27–40 GHz
V	40–75 GHz
W	75–110 GHz
mm (millimeter)	110–300 GHz
submm (sub-millimeter)	300 GHz–3 THz

[a]For space radio communications, the K and K_a bands are often designated by the single symbol K_a.
Source: IEEE Std 521–2002, ITU-R V.431-7.

TABLE A.20 Frequency Bands, Electronic Warfare Operators

Band Designation	Nominal Frequency Range
A	0 Hz–250 MHz
B	250–500 MHz
C	500 MHz–1 GHz
D	1–2 GHz
E	2–3 GHz
F	3–4 GHz
G	4–6 GHz
H	6–8 GHz
I	8–10 GHz
J	10–20 GHz
K	20–40 GHz
L	40–60 GHz
M	60–100 GHz

EU-NATO-US ECM Band Designations.

TABLE A.21 Frequency Bands, Great Britain Operators

Band Designation	Nominal Frequency Range, GHz
L	1–2
S	2–4
C	4–8
X	8–12
K_u	12–18
K	18–26.5
K_a	26.5–40
Q	30–50
U	40–60
V	50–75
E	60–90
W	75–110
F	90–140
D	110–170

Radio Society of Great Britain (RSGB) Frequency Bands.

TABLE A.22 Signal-to-Noise Ratio for Demodulator 10^{-6} BER

Receiver Type and Number of Symbols	S/N (dB) for 10^{-6} BER	Information Density, bits/s/Hz	Transmitter Peak-to-Average Ratio, dB
CPSK (Phase Shift Keying, Coherent Demodulation)			
2	13.6	1	0.0
4	13.8	2	0.0
8	19.3	3	0.0
16	25.1	4	0.0
32	31.1	5	0.0
64	37.2	6	0.0
DPSK (Phase Shift Keying, Differential Demodulation)			
2	15.9	1	0.0
4	16.1	2	0.0
8	22.1	3	0.0
16	28.1	4	0.0
32	34.1	5	0.0
64	40.2	6	0.0
QPR (Quadrature Partial Response)			
3	17.6	1	2.0
9	17.6	2	2.0
25	22.3	3	3.3
49	24.6	4	4.6
121	28.5	5	4.3
225	30.8	6	5.7
529	34.3	7	5.2
961	37.0	8	6.3
2209	40.8	9	5.4
QAM (Quadrature Amplitude Modulation, Coherent Demodulation)			
2	13.5	1	0.0
4	13.5	2	0.0
8	17.1	3	1.3
16	20.2	4	2.6
32	23.2	5	2.3
64	26.2	6	3.7
128	29.1	7	3.2
256	32.0	8	4.3
512	34.9	9	3.4
1024	37.7	10	4.5
TCM 2D (Trellis Coded Modulation, Two Dimensions)			
32	20.9	4	2.3
128	27.2	6	3.2
256	30.3	7	4.3
512	33.3	8	3.4
TCM 4D (Trellis Coded Modulation, Four Dimensions)			
32	19.9	4.5	2.3
128	26.2	6.5	3.2
256	29.3	7.5	4.3
512	32.3	8.5	3.4

Sometimes BER performance is specified in E_b/N_o rather than S/N.

$$\frac{E_b}{N_o} = \text{energy per bit to noise power spectral density ratio (dB)}$$

$$= \text{S/N (dB)} - 10\log_{10}\left(\frac{\text{bits per symbol} \times \text{symbols per second}}{B\ (\text{Hz})}\right)$$

$$\approx \text{S/N (dB)} - 10\log_{10}[\text{spectral efficiency (bits/s/Hz)}]$$

$$\text{(assuming the modulated signal essentially fills the transmission channel)} \quad \text{(A.9)}$$

S/N (dB) = average signal-to-noise power ratio;
B = channel bandwidth.

Signal-to-noise ratio is the ratio of average powers. Most transmitters are peak power limited. Practical radio systems must consider the ratio of peak (transmit) signal power to average noise power. The above transmitter peak-to-average ratios assume square root raised cosine filtering. They represent "typical" industry values. Actual peak-to-average ratio will depend on the filtering alpha value.

The S/N conversion from 10^{-6} to 10^{-3} or to 10^{-12} is 1–3 dB depending on modulation complexity. Forward error correction coding gain (S/N improvement) is typically 2–5 dB for the 10^{-6} BER. The above equation assumes synchronous demodulation; however, asynchronous demodulation is typically used in practice. This increases the S/N by 1 or 2 dB. This is compensated for with forward error correction. Practical considerations will degrade coding gain when measured performance is compared to theoretical.

Peak-to-average ratio is for the constellation with no filtering. Nyquist filtering will increase the peak-to-average ratio.

A.2 RADIO TRANSMISSION

A.2.1 Unit Conversions

$$\text{Watts} = 0.001 \times 10^{\text{dBm}/10} \quad \text{(A.10)}$$

$$\text{dBm} = 10\ \log(1000 \times \text{Watts}) \quad \text{(A.11)}$$

$$\text{dBm} = \text{dBW} + 30 \quad \text{(A.12)}$$

$$\text{dBm} = \text{dBrn} - 90 \quad \text{(A.13)}$$

$$\text{dBm0} = \text{dBm} - \text{TLP(dB)} \quad \text{(A.14)}$$

$$\text{dBrn0} = \text{dBrn} - \text{TLP(dB)} \quad \text{(A.15)}$$

$$P(\text{W/m}^2) = 10\ P(\text{mW/cm}^2) = E(\text{V/m})H(\text{A/m})$$

$$= \frac{E(\text{V/m})^2}{(120\pi)} = (120\pi)\ H(\text{A/m})^2 \quad \text{(A.16)}$$

$$\text{V/m} = 10^{[\text{dB}(\mu\text{V/m})-120]/20} \quad \text{(A.17)}$$

$$\text{A/m} = 10^{[\text{dB}(\mu\text{A/m})-120]/20} \quad \text{(A.18)}$$

$$\text{dBm(mW)} = \text{dB}(\mu\text{V}) - 107.0 \quad \text{(A.19)}$$

$$\text{dB(mW/m}^2) = \text{dB}(\mu\text{V/m}) - 115.8 \quad \text{(A.20)}$$

$$\text{dB}(\mu\text{A/m}) = \text{dB}(\mu\text{V/m}) - 20\ \log(120\pi) \quad \text{(A.21)}$$

$$\text{dB(W/m}^2) = 10 \ \log \ [\text{(V/m)(A/m)}] \tag{A.22}$$

$$\text{dB(mW/m}^2) = \text{dB(W/m}^2) + 30.0 \tag{A.23}$$

A.2.2 Free Space Propagation Absolute Delay

1.0167 ns/ft

3.3356 ns/m

A.2.3 Waveguide Propagation Absolute Delay

$$\begin{aligned} &1.0167 \text{ ns/ft} \left(1 \Big/ \left\{ \text{sqrt} \left[1 - \left(\frac{f_C}{f} \right)^2 \right] \right\} \right) \\ &3.3356 \text{ ns/m} \left(1 \Big/ \left\{ \text{sqrt} \left[1 - \left(\frac{f_C}{f} \right)^2 \right] \right\} \right) \end{aligned} \tag{A.24}$$

f_C = cutoff frequency for mode of interest;
f = frequency of operation greater than or equal to f_C.

A.2.4 Coaxial Cable Propagation Absolute Delay

1.0167 ns/ft/Velocity factor;

3.3356 ns/m/Velocity factor;

Velocity factor (between 0 and 1) = 1/sqrt (Dielectric constant);

Dielectric constant [1 (air) or greater] = relative permittivity of dielectric.

$$\begin{aligned} \text{Polystyrene foam dielectric constant} &= 1.05 \\ \text{Teflon dielectric constant} &= 2.1 \\ \text{Polyethylene dielectric constant} &= 2.25 \end{aligned} \tag{A.25}$$

A.2.5 Free Space Propagation Wavelength

$$\begin{aligned} \lambda(\text{ft}) &= \frac{0.98357}{F(\text{GHz})} \\ \lambda(\text{m}) &= \frac{0.29980}{F(\text{GHz})} \end{aligned} \tag{A.26}$$

F = frequency of electromagnetic wave.

A.2.6 Dielectric Medium Propagation Wavelength

$$\begin{aligned} \lambda(\text{ft}) &= \frac{\left[\dfrac{0.98357}{F\ (\text{GHz})}\right]}{\text{sqrt}(\varepsilon_r)} \\ \lambda(\text{m}) &= \frac{\left[\dfrac{0.29980}{F\ (\text{GHz})}\right]}{\text{sqrt}(\varepsilon_r)} \end{aligned} \tag{A.27}$$

F = frequency of electromagnetic wave;
ε_r = medium dielectric constant (permittivity relative to free space).

A.2.7 Free Space Loss (dB)

$$96.58 + 20 \log F(\text{GHz}) + 20 \log D(\text{miles})$$

$$92.45 + 20 \log F(\text{GHz}) + 20 \log D(\text{km}) \quad (A.28)$$

F = frequency of radio wave;
D = path length.

A.2.8 Effective Radiated Power (ERP) and Effective Isotropic Radiated Power (EIRP)

$$\text{ERP(W)} = 10^{P(\text{dBW})/10} \times 10^{G(\text{dBd})/10} = 0.001[10^{P(\text{dBm})/10} \times 10^{G(\text{dBd})/10}] \quad (A.29)$$

$$\text{EIRP(W)} = 10^{P(\text{dBW})/10} \times 10^{G(\text{dBi})/10} = 0.001[10^{P(\text{dBm})/10} \times 10^{G(\text{dBi})/10}] \quad (A.30)$$

$$\text{ERP(W)} = \text{EIRP(W)} \times 10^{-2.15/10} = 0.61 \text{ EIRP(W)} \quad (A.31)$$

P = transmitter power (at the antenna input);
$G(\text{dBd})$ = antenna gain referenced to a half wave dipole;
$G(\text{dBi})$ = antenna gain referenced to an isotropic radiator.

A.2.9 Voltage Reflection Coefficient

$$\text{Voltage reflection coefficient} = R = \frac{(\text{VSWR} - 1)}{(\text{VSWR} + 1)} \quad (A.32)$$

Reflected voltage divided by incident voltage at point of reflection;
VSWR, voltage standing wave ratio.

A.2.10 Voltage Standing Wave Ratio Maximum

$$\text{VSWR maximum } (V_{max}) = \text{sqrt}\left[\frac{(1+R)}{(1-R)}\right] \quad (A.33)$$

A.2.11 Voltage Standing Wave Ratio Minimum

$$\text{VSWR minimum } (V_{min}) = \text{sqrt}\left[\frac{(1-R)}{(1+R)}\right] \quad (A.34)$$

A.2.12 Voltage Standing Wave Ratio

$$\text{VSWR} = \frac{(1+R)}{(1-R)} \quad (A.35)$$

Maximum voltage divided by minimum voltage in standing wave pattern

$$= \text{Absolute value} \left[\left(\frac{Z_L}{Z_G} \right) \text{ or } \left(\frac{Z_G}{Z_L} \right) \right], \text{ whichever is larger} \tag{A.36}$$

Z_G = originating generator internal resistance;
Z_L = end (load) terminating resistance.

Note: If the generator and load are separated by a homogeneous transmission line, at the beginning of the signal's transmission, the transmission line is the load. At the other end of the transmission line, the transmission line is the generator.

A.2.13 Power Reflection Coefficient

$$\text{Power reflection coefficient} = R^2 \leq 1 \tag{A.37}$$

Reflected power divided by incident power.

A.2.14 Reflection Loss

$$\text{Reflection loss} = -10 \log (1 - R^2) \geq 0 \tag{A.38}$$

Decibel value of incident power loss due to reflected power.

A.2.15 Return Loss

$$\text{Return loss} = -20 \log R \geq 0 \tag{A.39}$$

Decibel value of incident power divided by reflected power.

For the above formulas the following definitions apply:

$$R = \text{Absolute value} \left[\frac{(Z_L - Z_G)}{(Z_L + Z_G)} \right] \tag{A.40}$$

Z_L = load (termination) impedance (ohms);
Z_G = generator (source) impedance (ohms).

A.2.16 *Q* (Quality) Factor (Figure of Merit for Resonant Circuits or Cavities)

$$Q = \frac{f_0}{(f_U - f_L)} \tag{A.41}$$

f_0 = circuit resonant frequency;
f_U = circuit upper half power frequency*;
f_L = circuit lower half power frequency*.

*Frequency at which output power is 3 dB less than that at f_0.

A.2.17 Q (Quality) Factor (Figure of Merit for Optical Receivers)

$$Q = \text{optical signal-to-noise ratio for a given BER and transmission rate} \quad (A.42)$$

A.2.18 Typical Long-Term Interference Objectives

$$\text{Same system interference} = \text{Radio front end noise} - 6\ \text{dB} \quad (A.43)$$

$$\text{Foreign system interference} = \text{radio front end noise} - 10\ \text{dB} \quad (A.44)$$

A.2.19 Frequency Planning Carrier-to-Interference Ratio (C/I)

$$C/I(\text{dB}) = P(\text{dB}) + G(\text{dB}) + L(\text{dB}) + D(\text{dB}) \quad (A.45)$$

$$P(\text{dB}) = \text{transmitter power differential}$$

$$= P_{\text{C}}(\text{dBm}) - L_{\text{C}}(\text{dB}) - P_{\text{I}}(\text{dBm}) + L_{\text{I}}(\text{dB}) \quad (A.46)$$

$$G(\text{dB}) = \text{antenna gain differential}$$

$$= G_{\text{C}}(\text{dB}) - G_{\text{I}}(\text{dB}) \quad (A.47)$$

$$L(\text{dB}) = \text{free space loss differential}$$

$$= 20\ \log\left(\frac{d_{\text{I}}}{d_{\text{C}}}\right) \quad (A.48)$$

$$D(\text{dB}) = \text{antenna discrimination}$$

$$= D_{\text{C}}(\text{dB}) + D_{\text{I}}(\text{dB}) \quad (A.49)$$

$P_{\text{C}}(\text{dBm})$ = transmitter power of desired signal;
$P_{\text{I}}(\text{dBm})$ = transmitter power of undesired signal;
$L_{\text{C}}(\text{dB})$ = power loss of desired signal between transmitter and transmit antenna;
$L_{\text{I}}(\text{dB})$ = power loss of undesired signal between transmitter and transmit antenna;
$G_{\text{C}}(\text{dB})$ = gain of transmit antenna at site A toward site B;
$G_{\text{I}}(\text{dB})$ = gain of transmit antenna at site C toward site D;
$D_{\text{C}}(\text{dB})$ = discrimination (relative to main lob power) of receive antenna at site B toward site C;
$D_{\text{I}}(\text{dB})$ = discrimination (relative to main lobe power) of transmit antenna at site C toward site B;
d_{C} = distance from site A to site B;
d_{I} = distance from site C to site B;
Site A = transmit location of desired signal;
Site B = receive location of desired signal;
Site C = transmit location of interfering signal;
Site D = intentional receive location of interfering signal.

A.2.20 Noise Figure, Noise Factor, Noise Temperature, and Front End Noise

The minimum noise of an ideal amplifier perfectly impedance matched to its receive antenna is the noise introduced by a (hypothetical) resistor of the interface impedance (typically 50 ohms) operating at operating temperature T (usually assumed to be 290 K = 17 °C = 63 °F). In general, the noise P delivered to a matched device by the noise source resistor at temperature T may be shown to be the following:

n = noise produced by a matched resistor operating at temperature T;
= KTb (W);

K = Boltzmann's constant = 1.38×10^{-23} (joules/degree kelvin);
T = noise temperature of the resistor (degrees kelvin = degrees celsius + 273);
b = noise bandwidth of the device (Hz).

If the amplifier adds noise to the received signal, that noise is characterized by adding another noise temperature to characterize the added noise. The relationship to amplifier signal to noise ratio is the following:

$$\begin{aligned} \text{nf} &= \text{noise factor} \\ &= 1 + \left(\frac{T_e}{T_o}\right) \\ &= \frac{\text{s/n}_\text{I}}{\text{s/n}_\text{O}} \end{aligned} \tag{A.50}$$

T_o = amplifier operating ("room") temperature (nominally 290 degrees K);
T_e = amplifier additional ("excess") noise temperature (degrees K);
= device "noise temperature";
= $T_o(\text{nf} - 1)$;
s/n_I = signal-to-noise power ratio at input to amplifier;
s/n_O = signal-to-noise power ratio at output of amplifier.

$$\begin{aligned} \text{NF (dB)} &= \text{noise figure} \\ &= 10\ \log(\text{nf}) \\ &= \text{S/N}_\text{I} - \text{S/N}_\text{O} \end{aligned}$$

nf = $10^{\text{NF}/10}$;
S/N_I = signal-to-noise ratio at input to amplifier (dB);
= $10\ \log(\text{s/n}_\text{I})$;
S/N_O = signal-to-noise ratio at output of amplifier (dB);
= $10\ \log(\text{s/n}_\text{O})$.

For cascaded (series) active amplifiers:

$$\begin{aligned} \text{nf} &= \text{overall noise factor of the cascaded amplifiers} \\ &= \text{nf}_1 + \frac{(\text{nf}_2 - 1)}{g_1} + \frac{(\text{nf}_3 - 1)}{g_1\, g_2} + \cdots + \frac{(\text{nf}_n - 1)}{g_1\, g_2 \cdots g_{(n-1)}} \end{aligned} \tag{A.51}$$

nf_1 = noise factor of the first device;
nf_2 = noise factor of the second device;
nf_3 = noise factor of the third device;
g_1 = gain (power ratio) of first device;
g_2 = gain (power ratio) of second device;
g_n = gain (power ratio) of nth device.

The implied assumption is all devices are matched impedances and bandwidth shrinkage of cascaded devices is insignificant.

The noise figure of an attenuator is simply the attenuation (dB, > 0) of the attenuator.

The noise figure of a cascaded attenuator and an amplifier is the sum of the two (positive dB values).

The "front end" noise produced by an amplifier may be calculated as follows:

$$\begin{aligned} n &= \text{noise produced by a matched "internal" resistor} \\ &= K(T_o + T_e)B\ 10^6\ \text{(W)} \\ &= 1.38 \times 10^{-17} T_o \left[1 + \left(\frac{T_e}{T_o}\right)\right] B \\ &= 4.00 \times 10^{-15}\ \text{nf}\ B \end{aligned} \tag{A.52}$$

B = noise bandwidth of the device (MHz);
N = front end noise = 10 log(n).

$$\begin{aligned} N(\text{dBW}) &= 10\ \log(n) \\ &= -144 + \text{NF(dB)} + 10\ \log(B) \end{aligned} \tag{A.53}$$

$$\begin{aligned} N(\text{dBm}) &= N(\text{dBW}) + 30 \\ &= -114 + \text{NF(dB)} + 10\ \log(B) \end{aligned} \tag{A.54}$$

$$N(\text{dBW/MHz}) = -144 + \text{NF} \tag{A.55}$$

$$\text{N(dBW/4 kHz)} = -168 + \text{NF} \tag{A.56}$$

A common problem is to determine the signal associated with a known radio threshold signal-to-noise ratio. Assume that the radio receiver is limited by front end noise.

$$\begin{aligned} S(\text{dBm}) &= \text{S/N(dB)} + N(\text{dBm}) \\ &= \text{S/N(dB)} - 114 + \text{NF(dB)} + 10\ \log(B) \end{aligned} \tag{A.57}$$

S = received signal power level (dBm) at threshold;
S/N = receiver threshold signal-to-noise ratio (dB);
NF = receiver noise figure (dB);
B = receiver bandwidth (MHz).

Remember that the receiver noise figure is the noise figure of the front end amplifier plus the loss (dB) between the amplifier and the measurement location. The typical amplifier noise figure for low frequency microwave radios is about 2 dB. The typical waveguide and receiver filter loss in front of a receiver is about 2 dB. Therefore, the typical microwave radio receiver noise figure is 4 dB.

A.2.21 Shannon's Formula for Theoretical Limit to Transmission Channel Capacity

$$\begin{aligned} C &\leq W\ \log_2[1 + (\text{S/N})] \\ &\leq 3.322\ W\ \log_{10}[1 + (\text{S/N})] \\ &\lessapprox 3.322\ W\ \log_{10}(\text{S/N}) \\ &\leq 0.3322\ W[\text{S/N(dB)}] \end{aligned} \tag{A.58}$$

C = channel capacity (Mb/s);
W = channel bandwidth (MHz);
S/N = channel signal-to-noise power ratio.

In the above equations, replacing the power ratio (S/N + 1) with S/N introduces less than 1% dB error for all S/Ns ≥10 dB.

Shannon's limit may be rewritten to define the minimum S/N required to achieve a given spectral efficiency:

$$\mathrm{S/N(dB)} \geq 3\ C/W \tag{A.59}$$

$$\mathrm{S/N(dB)} \geq 3[\text{spectral efficiency (bits/s/Hz)}] \tag{A.60}$$

Assumptions are filtering is rectangular ("brick wall"), noise is Gaussian and the transmitted signal spectrum fills the transmission channel bandwidth.

A.3 ANTENNAS (FAR FIELD)

See Chapter 8 for near field considerations.

A.3.1 General Microwave Aperture Antenna (Far Field) Gain (dBi)

$$\begin{aligned} &11.1 + 20\ \log\ F(\mathrm{GHz}) + 10\ \log A(\mathrm{ft}^2) + 10\ \log\left(\frac{E}{100}\right) + 20\ \log\ \cos\left(\frac{C}{2}\right) \\ &21.5 + 20\ \log\ F(\mathrm{GHz}) + 10\ \log\ A(\mathrm{m}^2) + 10\ \log\left(\frac{E}{100}\right) + 20\ \log\ \cos\left(\frac{C}{2}\right) \end{aligned} \tag{A.61}$$

F = frequency of radio wave;
A = antenna physical area;
C = angle between incoming and outgoing radio signal paths;
E = antenna power transmission efficiency expressed as a percentage.

A.3.2 General Microwave Antenna (Far Field) Relative Gain (dBi)

$$G_{\mathrm{ref}} + 20\ \log\left(\frac{F}{F_{\mathrm{ref}}}\right) \tag{A.62}$$

G_{ref} = antenna gain (dBi) at F_{ref} (GHz);
F_{ref} = reference frequency (GHz);
F = frequency of interest (GHz).

This formula is typically used to interpolate antenna catalog data. If F_{ref} is the mid-band frequency, accuracy is typically within 0.1 dB of measured values at band edges.

A.3.3 Parabolic (Circular) Microwave Antenna (Far Field) Gain (dBi)

$$\begin{aligned} &10.1 + 20\ \log\ F + 20\ \log\ D(\mathrm{ft}) + 10\ \log\ \left(\frac{E}{100}\right) \\ &20.5 + 20\ \log\ F + 20\ \log\ D(\mathrm{m}) + 10\ \log\ \left(\frac{E}{100}\right) \end{aligned} \tag{A.63}$$

F = frequency of radio wave (GHz);
D = antenna diameter;
E = antenna power transmission efficiency expressed as a percentage, $40 \leq E \leq 60$, $E \approx 55$ typically.

A.3.4 Parabolic (Circular) Microwave Antenna Illumination Efficiency

$$10\ \log\left(\frac{E}{100}\right) = -\ 10.1 + G - 20\ \log(F) - 20\ \log[D(\text{ft})]$$
$$= -\ 20.4 + G - 20\ \log(F) - 20\ \log[D(\text{m})] \quad \text{(A.64)}$$
$$\frac{E}{100} = 10^{[10\ \log\{E/100\}]/10} \quad \text{(A.65)}$$

G = antenna far-field gain (dBi);
F = frequency of radio wave (GHz);
D = antenna diameter;
E = antenna power transmission efficiency (%).

A.3.5 Panel (Square) Microwave Antenna (Far Field) Gain (dBi)

$$11.1 + 20\ \log\ F(\text{GHz}) + 20\ \log\ S(\text{ft}) + 10\ \log\left(\frac{E}{100}\right)$$
$$21.5 + 20\ \log\ F(\text{GHz}) + 20\ \log\ S(\text{m}) + 10\ \log\left(\frac{E}{100}\right) \quad \text{(A.66)}$$

F = frequency of radio wave;
S = length of side of square;
E = antenna power transmission efficiency expressed as a percentage, typically $E \approx 100$.

A.3.6 Panel (Square) Microwave Antenna Illumination Efficiency

$$\frac{E}{100} = \frac{(N0 + N1\beta + N2\beta^2 + N3\beta^3 + N4\beta^4)}{(1 + D1\beta + D2\beta^2 + D3\beta^3 + D4\beta^4)} \quad \text{(A.67)}$$
$$\beta = \left\{\left(\frac{W}{\lambda}\right)\ \sin\left(\frac{\phi_{3\text{dB}}}{2}\right)\right\} \quad \text{(A.68)}$$
$$0.447 \leq \beta \leq 1.49$$

E = antenna power transmission efficiency (%)

$$10 \leq E \leq 100$$

W = width of the square antenna (measured along the edge)
λ = radio free space wavelength
λ (ft) = 0.98357/f (GHz)
λ (m) = 0.29980/f (GHz)
$\phi_{3\text{dB}}$ = antenna 3-dB beam width
= angle measured between the two −3 dB power values (referenced to the boresight power)
$N0$ = −0.4468979109577574
$N1$ = 2.705347403057084
$N2$ = −5.689139811168476
$N3$ = 5.017375871680245

$N4 = -1.037085334383484$
$D1 = -7.914244751535077$
$D2 = 24.1096714821637$
$D3 = -33.58979930453501$
$D4 = 18.85685129777957$

A.3.7 Angle Between Incoming and Outgoing Radio Signal Paths, C, for a Passive Reflector

Refer to Figure A.1

A = total horizontal included angle (measured in the horizontal plane) formed by the incoming and outgoing radio paths converging at the reflector or antenna;
= positive angle measured (on a map) between paths from transmitter and receiver antenna to the reflector or antenna (ignoring relative height of sites);

θ_1 = smaller vertical path angle formed by the horizontal plane and a line between the reflector or antenna and the transmitter or receiver antenna;

θ_2 = larger vertical path angle formed by the horizontal plane and a line between the reflector or antenna and the transmitter or receiver antenna;

$$B = \text{reflector bearing angle correction}$$

$$= \arctan \frac{\left[\tan\left(\frac{A}{2}\right)\right][\cos\,\theta_1 - \cos\,\theta_2]}{(\cos\,\theta_1 + \cos\,\theta_2)} \tag{A.69}$$

$$\theta_3 = \text{reflector tilt relative to perpendicular to horizontal plane} \tag{A.70}$$

$$\theta_3 = \arctan\left[\frac{\{(\cos\,B)(\sin\,\theta_1 + \sin\,\theta_2)\}}{\{\cos\,\left(\frac{A}{2}\right)(\cos\,\theta_1 + \cos\,\theta_2)\}}\right] \tag{A.71}$$

$$C = \text{angle between incoming and outgoing radio signal path}$$

$$= 2\ \arccos\left\{\frac{(\sin\,\theta_1 + \sin\,\theta_2)}{[2(\sin\,\theta_3)]}\right\} \tag{A.72}$$

$$= A \quad \text{when } \theta_1 \text{ and } \theta_2 \text{ are} < 20^\circ (< 0.1 \text{ dB gain error})$$

For the above all cosines and tangents are positive for angles between 0° and 90°. They are negative for angles between 90° and 180°. Sines are positive for paths going down from the repeater or antenna

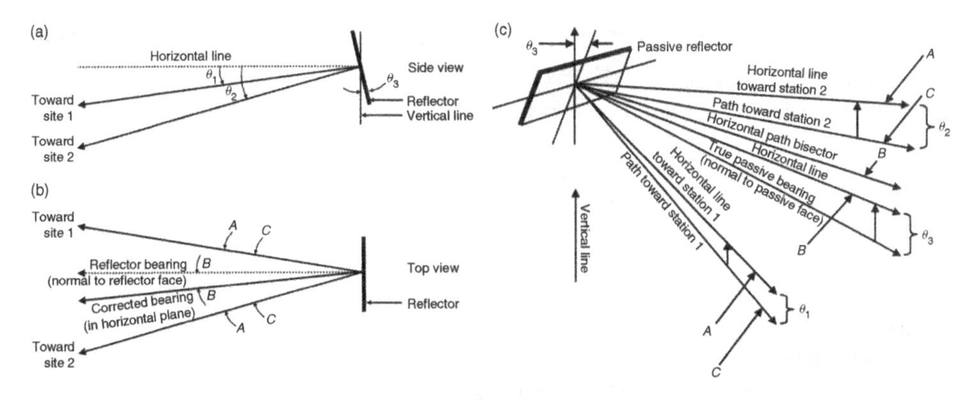

Figure A.1 Reflector geometry measured (a) using horizontal plane projections of the paths, (b) using a horizontal plane projection of reflector bearing, and (c) between actual paths towards the sites.

(toward the transmitter or receiver) or negative for path going up (toward the transmitter or receiver). B always rotates the passive bearing toward the path with the least vertical angle θ_1.

A.3.8 Signal Polarization Rotation Through a Passive Reflector, $\Delta\phi$

The definitions in the previous paragraph can be used.

$$\begin{aligned}\Delta\phi &= \text{rotation of both signal polarizations for signal passing through a} \\ &\quad \text{passive reflector relative to signal in horizontal plane} \\ &= \phi_2 + \phi_2 - \pi\,(\text{or } 180^\circ) \qquad \text{(A.73)} \\ \phi_1 &= \text{arc cos } \{[\sin\ \theta_1 - (\sin\ \theta_2\ \cos\ C)]/(\cos\ \theta_2\ \sin\ C)\}; \\ \phi_2 &= \text{arc cos}\{[\sin\ \theta_3 - (\sin\theta_1\ \cos(C/2))]/[\cos\ \theta_1\ \sin(C/2)]\}.\end{aligned}$$

$\Delta\phi$ is negative when counterclockwise as viewed from the right-hand path when facing the reflector. $\Delta\phi$ is negative when clockwise as viewed from the left-hand path when facing the reflector. These are true regardless of which paths are assigned to θ_1 and θ_2 above.

A.3.9 Signal Effects of Polarization Rotation

The definitions in the previous two paragraphs can be used.

$$\text{Received signal loss (dB)} = 10\ \log\ (\cos^2\Delta\phi) \qquad \text{(A.74)}$$

$$\text{Cross-polarization discrimination (XPD, dB)} = -10\ (\log\ \sin^2\Delta\phi) \qquad \text{(A.75)}$$

A.3.10 Passive Reflector (Far Field) Two-Way (Reception and Retransmission) Gain (dBi)

$$22.2 + 40\ \log\ F(\text{GHz}) + 20\ \log\ A(\text{ft}^2) + 20\ \log\ \cos\left(\frac{C}{2}\right)$$

$$42.9 + 40\ \log\ F(\text{GHz}) + 20\ \log\ A(\text{m}^2) + 20\ \log\ \cos\left(\frac{C}{2}\right) \qquad \text{(A.76)}$$

F = frequency of radio wave;
A = reflector area;
C = angle between incoming and outgoing radio signal paths.

A.3.11 Rectangular Passive Reflector 3-dB Beamwidth (Degrees, in Horizontal Plane)

Square projection onto both paths

One edge parallel to Earth (zero rotation)

$$\frac{49.8}{F[(\text{GHz}) \times W\ (\text{ft})\cos(C/2)]}$$

$$\frac{15.2}{F\ [(\text{GHz}) \times W\ (\text{m})\cos(C/2)]}$$

F = frequency of radio wave;
W = reflector width;
C = angle between incoming and outgoing radio signal paths.

Square projection onto both paths

Diamond shape (square with 45° rotation)

W = width of unrotated square

$$\frac{50.8}{\{F(\mathrm{GHz})\ W\ (\mathrm{ft})\cos(C/2)\}}$$

$$\frac{15.5}{\{F(\mathrm{GHz})\ W\ (\mathrm{m})\cos(C/2)\}}$$

F = frequency of radio wave;
W = reflector width;
C = angle between incoming and outgoing radio signal paths.

A.3.12 Elliptical Passive Reflector 3-dB Beamwidth (Degrees)

$$\frac{57.9}{\{F(\mathrm{GHz})D\ (\mathrm{ft})\}}$$

$$\frac{17.7}{\{F(\mathrm{GHz})D\ (\mathrm{m})\}}$$

F = frequency of radio wave;
D = smaller reflector diameter.

It is assumed that the reflector has a 45° angle to the paths of propagation so that the projection of the reflector shape onto the path is circular.

A.3.13 Circular Parabolic Antenna 3-dB Beamwidth (Degrees)

$$\frac{(57.9\ \mathrm{NBW})}{[F(\mathrm{GHz})D(\mathrm{ft})]} \approx \frac{88.0}{[F(\mathrm{GHz})D(\mathrm{ft})]}$$

$$\frac{(17.7\ \mathrm{NBW})}{[F(\mathrm{GHz})D(\mathrm{m})]} \approx \frac{26.8}{[F(\mathrm{GHz})D(\mathrm{m})]}$$

$$E(\mathrm{dB}) = -10.1 + G(\mathrm{dB}) - 20\ \log[F(\mathrm{GHz})] - 20\ \log[D(\mathrm{ft})]$$

$$= -20.4 + G(\mathrm{dB}) - 20\ \log[F(\mathrm{GHz})] - 20\ \log[D(\mathrm{m})] \quad (A.77)$$

$$X = E(\mathrm{PR}) = 10^{E(\mathrm{dB})/10} \quad (A.78)$$

$$\mathrm{NBW} = \frac{\mathrm{NBWn}}{\mathrm{NBWd}} \quad (A.79)$$

$$\mathrm{NBWn} = (C1 + C2X + C3X^2 + C4X^3 + C5X^4) \quad (A.80)$$

$$\mathrm{NBWd} = 1 + C6X + C7X^2 + C8X^3 + C9X^4 + C10X^5 \quad (A.81)$$

$C1 = 11.65806521303201$
$C2 = 96.51565982372452$
$C3 = -152.1423956242208$
$C4 = -4.217102204230871$
$C5 = 48.37581221497451$
$C6 = 49.10391941747248$
$C7 = 32.82051399322144$

$$C8 = -201.8020843245464$$
$$C9 = 130.4562878319081$$
$$C10 = -11.38859718015969$$

$E(\text{dB})$ = 10 log (antenna efficiency, power ratio);
$E(\text{PR})$ = antenna efficiency (power ratio), $0.1 \leq E \leq 1.0$;
NBW = antenna-normalized 3.01-dB bandwidth;
F = frequency of radio wave;
D = antenna diameter;
G = antenna isotropic gain (dBi).

A.3.14 Passive Reflector Far Field Radiation Pattern Envelopes

Normalized to 0 dB for $\theta = 0°$;

d = width of square or diameter of circle projected onto path;
= physical width or diameter $\times \cos(C/2)$;
C = angle between incoming and outgoing radio signal paths;
λ = free space wavelength of radio wave;
θ = azimuth of measurement point relative to path of maximum transmission (bore sight);
$X = (d/\lambda)|\sin\theta| \geq 0$;
$P(\text{dB})$ = far-field radiation pattern envelope relative power intensity.

A.3.14.1 Rectangular Reflector

Square projection onto both paths;
One edge parallel to Earth (zero rotation).

$$\text{For } X \leq 0.50 \ : \ P = 20 \log\left[\frac{\sin(\pi X)}{(\pi X)}\right] \tag{A.82}$$

$$\text{For } X > 0.50 \ : \ P = -20 \log(\pi X) \tag{A.83}$$

A.3.14.2 Diamond Reflector

Square projection onto both paths;
Diamond shape (square with 45° rotation);
d = width of unrotated square.

$$\text{For } X \leq 0.70 \ : \ P = 40 \log\left[\frac{\sin(2.221X)}{(2.221X)}\right] \tag{A.84}$$

$$\text{For } X > 0.70 \ : \ P = 6.02 - 40 \log(\pi X) \tag{A.85}$$

A.3.14.3 Elliptical Reflector

Circular projection onto both paths

$$\text{For } X \leq 0.775 \ : \ P = 20 \log\left[\frac{\sin(2.680X)}{(2.680X)}\right] \tag{A.86}$$

$$\text{For } X > 0.775 \ : \ P = 4.06 - 30 \log(\pi X) \tag{A.87}$$

A.3.15 Inner Radius for the Antenna Far-Field Region

This is sometimes called *outer radius for the antenna near-field region.*

$$d_{FF} = \text{radial distance from the antenna to the far field edge}$$

$$= \frac{2D^2}{\lambda} \quad \text{(A.88)}$$

λ = wavelength of the radio wave in the same dimensions as D;
d_{FF} (ft) = $2.033D^2$ (ft) F(GHz);
d_{FF}(m) = $6.671D^2$(m) F(GHz);
F = frequency of radio wave;
D = larger linear dimension of the antenna in a plane of projection orthogonal to the line of wave propagation;
= parabolic antenna diameter;
= passive reflector {[width $\times$ cos(α_h)] or [height $\times$ cos(α_v)], which ever is larger};
α_h = angle between path direction and perpendicular to face of reflector measured in a horizontal plan;
α_v = angle between path direction and perpendicular to face of reflector measured in a vertical plan.

As noted in Chapter 8, a more accurate version of this formula is to take into consideration the illumination efficiency of the antenna:

$$d_{FF} = \text{radial distance from the antenna to the far field edge}$$

$$= \frac{2\eta D^2}{\lambda} \quad \text{(A.89)}$$

η = illumination power efficiency (0 to 1).

The above formulas have no specified accuracy. They are rough rules of thumb. If the far-field transition is defined as the point at which the antenna boresight gain is 1 dB less than the far-field gain for that distance, the following formula may be used to estimate that limit for a circular antenna:

$$\Delta_{dB}(1 \text{ dB far-field transition distance}) = -10.46 + 8.730\eta - 4.116\eta^2 - \frac{0.4638}{\eta} \quad \text{(A.90)}$$

For a square antenna, the 1-dB transition point is given by the following:

$$\Delta_{dB}(\text{far-field transition distance}) = -8.544 + 6.188\eta - 1.954\eta^2 - \frac{0.5349}{\eta} \quad \text{(A.91)}$$

$$\Delta_{dB} = 10 \log\left[\frac{d}{\left(\frac{2D^2}{\lambda}\right)}\right]$$

d = radial distance from the antenna in the same units as D;
η = illumination efficiency (0–1).

A.4 NEAR-FIELD POWER DENSITY

A.4.1 Circular Antennas

$$S(\Delta) = \text{near-field power density(mW/cm}^2)$$
$$= 10^{(\psi/10)}$$
$$\psi = 10 \ \log_{10} [S(\Delta] = 10 \ \log_{10} \left[\frac{S(\Delta)}{S(\Delta = 1)}\right] + 10 \ \log_{10} [S(\Delta = 1)]$$

$$10 \ \log_{10} \left[\frac{S(\Delta)}{S(\Delta = 1)}\right] = -2 \ \Delta_{\text{dB}} \tag{A.92}$$

or

$$10 \ \log_{10} \left[\frac{S(\Delta)}{S(\Delta = 1)}\right] = A + B\eta + \frac{C}{\eta} + D\eta^2 + \frac{E}{\eta^2} + F\eta^3 \tag{A.93}$$

$A =$ 40.430453;
$B = -61.480406$;
$C = -0.46691971$;
$D =$ 55.376708;
$E =$ 0.04791274;
$F = -19.805638$,
whichever is smaller.

$S(\Delta = 1) = (\pi p\eta)/(16D^2)$;
$\Delta = d/(2D^2/\lambda) =$ normalized distance parameter for circular antenna
$= 0.49179d(\text{ft})/[D(\text{ft})^2 f(\text{GHz})] = 0.14990d(\text{m})/[D(\text{m})^2 f(\text{GHz})]$;
$\Delta_{\text{dB}} = 10 \ \log(\Delta)$;
$\Delta = 1(\Delta_{\text{dB}} = 0)$ normalized distance at nominal far field crossover point;
$\eta =$ antenna efficiency $(0 \le \eta \le 1) = E/100$;
$E =$ antenna efficiency (%), see earlier far-field equations
$p =$ transmitter power (mW);
$D =$ aperture diameter (cm) = 30.48 diameter (ft);
f = operating frequency;
d = perpendicular distance from the antenna aperture.

A.4.2 Square Antennas

$$S(\Delta) = \text{near-field power density(mW/cm}^2)$$
$$= 10^{(\psi 10)}$$
$$\psi = 10 \ \log_{10}[S(\Delta] = 10 \ \log_{10} \left[\frac{S(\Delta)}{S(\Delta = 1)}\right] + 10 \ \log_{10}[S(\Delta = 1)]$$

$$10 \ \log_{10} \left[\frac{S(\Delta)}{S(\Delta = 1)}\right] = -2 \ \Delta\text{dB} \tag{A.94}$$

or

$$10 \ \log_{10} \left[\frac{S(\Delta)}{S(\Delta = 1)}\right] = A + B\eta + \frac{C}{\eta} + D\eta^2 + \frac{E}{\eta^2} + F\eta^3 \tag{A.95}$$

$A = 34.223061$;
$B = -58.288613$;
$C = 0.51017224$;

$D = 64.124471$;
$E = -0.013593334$;
$F = -29.354905$,
whichever is smaller.

$S(\Delta = 1) = (p\eta)/(4W^2)$;
$\Delta = d/(2W^2/\lambda)$ = normalized distance parameter for square antenna;
$= 0.49179d(\text{ft})/[W(\text{ft})^2 f(\text{GHz})] = 0.14990d(\text{m})/[W(\text{m})^2 f(\text{GHz})]$;
$\Delta_{\text{dB}} = 10 \log(\Delta)$;
$\Delta = 1$ ($\Delta_{\text{dB}} = 0$) normalized distance at nominal far-field crossover point;
η = antenna efficiency ($0 \leq \eta \leq 1$) = $E/100$;
E = antenna efficiency (%), see earlier far-field equations;
p = transmitter power (mW);
W = antenna width (cm) = 30.48 width (ft);
f = operating frequency;
d = perpendicular distance from the antenna aperture.

A.5 ANTENNAS (CLOSE COUPLED)

A.5.1 Coupling Loss L_{NF} (dB) Between Two Antennas in the Near Field

The following formulas estimate near-field coupling loss between antennas:
$D = \Delta_{\text{dB}}$ = normalized distance between antennas;
$-10 \leq D = \Delta_{\text{dB}} \leq 0$;
$\Delta_{\text{dB}} = 10 \log(\Delta)$;
$\Delta = d/(2D^2/\lambda)$ = normalized distance parameter for the larger circular antenna;
$\Delta = d/(2W^2/\lambda)$ = normalized distance parameter for the larger square antenna;
R = ratio of smaller antenna width/larger antenna width;
$0 \leq R \leq 1$;
$N = \eta$ = illumination efficiency;
$0.25 \leq \eta = 1.0$;
L_{NF} = near-field antenna to antenna coupling loss (dB);
= value to be added to far-field free space loss.

A.5.2 Coupling Loss L_{NF}(dB) Between Identical Antennas

$$L_{NF} = C1 + C2 \times D + C3 \times N + C4 \times D^2 + C5 \times N^2 + C6 \times D \times N + C7 \times D^3 + C8 \times N^3 + C9 \times D \times N^2 + C10 \times D^2 \times N \quad (A.96)$$

Circular (Parabolic) Antennas	Square (Aligned) Antennas
$C1 = 0.5688330523739922$	$C1 = 0.246159837605786$
$C2 = -0.2843725447552475$	$C2 = -0.8316031494312162$
$C3 = -5.913339006420615$	$C3 = -6.285507130745876$
$C4 = -0.005544709076135127$	$C4 = -0.07920675867559671$
$C5 = 12.46763586356988$	$C5 = 15.99365443163881$
$C6 = 0.2328735832473644$	$C6 = 1.544481166675684$
$C7 = 0.00194088480963481$	$C7 = -0.0002143259518259517$
$C8 = -7.561944604459402$	$C8 = -10.80646533812489$
$C9 = 0.299596446278667$	$C9 = -0.1245258794169019$
$C10 = -0.0905373781148429$	$C10 = -0.02108477297350535$

A.5.3 Coupling Loss L_{NF} (dB) Between Different-Sized Circular Antennas

$$L_{NF} = C1 + C2 \times D + C3 \times R + C4 \times D^2 + C5 \times R^2 + C6 \times D \times R + C7 \times D^3 + C8 \times R^3 + C9 \times D \times R^2 + C10 \times D^2 \times R \quad (A.97)$$

$N = \eta = 1.00$	$N = \eta = 0.75$
$C1 = -0.7499780289155289$	$C1 = -0.2714292776667776$
$C2 = -0.5133628527953528$	$C2 = 0.02152531477781479$
$C3 = 3.911939824296074$	$C3 = 3.675637157518407$
$C4 = -0.07520881757131757$	$C4 = 0.0321249222999223$
$C5 = -6.863575712481961$	$C5 = -10.12064281204906$
$C6 = 0.6943731306193806$	$C6 = -0.1223080099067599$
$C7 = 0.002120276482776482$	$C7 = 0.006847157472157472$
$C8 = 3.416898148148148$	$C8 = 6.74849537037037$
$C9 = 0.08539001623376626$	$C9 = 0.334229301948052$
$C10 = -0.0209452047952048$	$C10 = -0.0458997668997669$

$N = \eta = 0.50$	$N = \eta = 0.25$
$C1 = 0.1013892010767011$	$C1 = 0.1093747294372294$
$C2 = 0.2001590773115773$	$C2 = 0.142861466033966$
$C3 = -0.6023601345413846$	$C3 = -0.5683255501443001$
$C4 = 0.05432371378621378$	$C4 = 0.03395236985236985$
$C5 = 0.4445470328282829$	$C5 = 0.6277401244588743$
$C6 = -0.4858658146020646$	$C6 = -0.3156246699134199$
$C7 = 0.005887286324786324$	$C7 = 0.002853418803418803$
$C8 = -0.01127946127946136$	$C8 = -0.192550505050505$
$C9 = 0.3789476461038961$	$C9 = 0.1949728084415584$
$C10 = -0.05209527139527139$	$C10 = -0.02924848484848485$

A.5.4 Coupling Loss L_{NF} (dB) Between Different-Sized Square Antennas (Both Antennas Aligned)

$$L_{NF} = C1 + C2 \times D + C3 \times R + C4 \times D^2 + C5 \times R^2 + C6 \times D \times R + C7 \times D^3 + C8 \times R^3 + C9 \times D \times R^2 + C10 \times D^2 \times R \quad (A.98)$$

$N = \eta = 1.00$	$N = \eta = 0.75$
$C1 = -0.9729305264180264$	$C1 = -0.3371434274059274$
$C2 = -0.8812381138306138$	$C2 = -0.1846516780441781$
$C3 = 4.742492976699226$	$C3 = 0.887764439033189$
$C4 = -0.2101820207570208$	$C4 = -0.01862155622155622$
$C5 = -8.833426902958152$	$C5 = -1.338570752164502$
$C6 = 1.227099754828505$	$C6 = 0.1831897564935065$
$C7 = -0.005192715617715618$	$C7 = 0.006001243201243201$
$C8 = 4.532586279461279$	$C8 = 0.6045138888888888$
$C9 = 0.1829265422077922$	$C9 = 0.3801099837662338$
$C10 = 0.0529536297036297$	$C10 = -0.004622077922077921$

$N = \eta = 0.50$	$N = \eta = 0.25$
$C1 = -0.0139240342990343$	$C1 = 0.1224828477078477$
$C2 = 0.1031087360787361$	$C2 = 0.169524861989862$
$C3 = -0.2748664266289267$	$C3 = -0.7053111198986198$
$C4 = 0.03496463952713952$	$C4 = 0.04124757742257742$
$C5 = 0.08828057359307374$	$C5 = 0.8015615981240979$
$C6 = -0.343841555944056$	$C6 = -0.3901483508158508$
$C7 = 0.006170593758093758$	$C7 = 0.003830542605542605$
$C8 = 0.08970959595959589$	$C8 = -0.2554187710437709$
$C9 = 0.4123693181818182$	$C9 = 0.2612280844155844$
$C10 = -0.0415508991008991$	$C10 = -0.03715820845820846$

For values of η between the values above, calculate the values for $\eta = 1.00$, 0.75, 0.50, and 0.25 and use two-dimensional cubic interpolation (A.10.1) to determine the desired value.

A.5.5 Coupling Loss L_{NF} (dB) for Antenna and Square Reflector in the Near Field

The following formulas estimate near field coupling loss between an antenna and a reflector:

$D = \Delta_{dB}$ = normalized distance between antenna and reflector;
$-8 \leq D = \Delta_{dB} \leq 0$;
$\Delta_{dB} = 10 \log(\Delta)$;
$\Delta = d/(2W^2/\lambda)$ = normalized distance parameter for the reflector width;
R = ratio of antenna width/reflector width;
$0 \leq R \leq 1$;
$N = \eta$ = illumination efficiency;
$0.25 \leq \eta \leq 1.0$;
L_{NF} = near field antenna to reflector coupling loss (dB);
= value to be added to far-field free space loss.

A.5.6 Coupling Loss L_{NF} (dB) for Circular Antenna and Square Reflector

$$L_{NF} = C1 + C2 \times D + C3 \times R + C4 \times D^2 + C5 \times R^2 + C6 \times D \times R + C7 \times D^3 + C8 \times R^3 + C9 \times D \times R^2 + C10 \times D^2 \times R \tag{A.99}$$

$N = \eta = 1.00$	$N = \eta = 0.75$
$C1 = -0.4566947811447811$	$C1 = -0.2431097426647427$
$C2 = -0.2020482792780412$	$C2 = 0.06178604780801208$
$C3 = 2.191165239698573$	$C3 = 1.058045280984448$
$C4 = -0.02878061052703909$	$C4 = 0.03989009482580911$
$C5 = -4.37696097883598$	$C5 = -2.486101521164021$
$C6 = 0.3237634547000618$	$C6 = -0.03516842648423002$
$C7 = 0.007908487654320989$	$C7 = 0.01286877104377104$
$C8 = 2.162461419753087$	$C8 = 1.312422839506173$
$C9 = 0.3109449404761905$	$C9 = 0.3230811011904762$
$C10 = -0.03379225417439703$	$C10 = -0.04815576685219542$

$N = \eta = 0.50$	$N = \eta = 0.25$
$C1 = -0.131068519320186$	$C1 = -0.05018120811287484$
$C2 = 0.1926641022755904$	$C2 = 0.2815940879028379$
$C3 = 0.4079589168136392$	$C3 = -0.09241433875794949$
$C4 = 0.07328302583659727$	$C4 = 0.09614733044733044$
$C5 = -1.326173390652558$	$C5 = -0.2993532848324522$
$C6 = -0.2101255144557823$	$C6 = -0.2734471239177489$
$C7 = 0.01526261223344557$	$C7 = 0.01694427609427609$
$C8 = 0.7796810699588479$	$C8 = 0.2686085390946506$
$C9 = 0.3159635416666667$	$C9 = 0.2477760416666666$
$C10 = -0.05155578231292517$	$C10 = -0.04260757575757575$

For values of η between the values above, calculate the values for $\eta = 1.00$, 0.75, 0.50, and 0.25 and use two-dimensional cubic interpolation (A.10.1) to determine the desired value.

A.5.7 Coupling Loss L_{NF} (dB) for Square Antenna and Square Reflector (Both Aligned)

$$L_{NF} = C1 + C2 \times D + C3 \times R + C4 \times D^2 + C5 \times R^2 + C6 \times D \times R + C7 \times D^3 + C8 \times R^3 + C9 \times D \times R^2 + C10 \times D^2 \times R \quad \text{(A.100)}$$

$N = \eta = 1.00$	$N = \eta = 0.75$
$C1 = -0.6466395109828443$	$C1 = -0.3507770017636684$
$C2 = -0.4242813704505371$	$C2 = -0.06543854400009162$
$C3 = 3.395700560098338$	$C3 = 1.723228713590936$
$C4 = -0.08242216209716208$	$C4 = 0.008616482855768571$
$C5 = -6.342335537918872$	$C5 = -3.605354938271605$
$C6 = 0.8051073917748917$	$C6 = 0.206446064471243$
$C7 = 0.004274172278338946$	$C7 = 0.01070726711560045$
$C8 = 2.923791152263375$	$C8 = 1.771116255144033$
$C9 = 0.1849970238095238$	$C9 = 0.2753497023809524$
$C10 = -0.006447186147186151$	$C10 = -0.03658038033395176$
$N = \eta = 0.5$	$N = \eta = 0.25$
$C1 = -0.1980865496232163$	$C1 = -0.07070867564534233$
$C2 = 0.1141195494056208$	$C2 = 0.2586083355092878$
$C3 = 0.8217887665009888$	$C3 = 0.03945184971407202$
$C4 = 0.05379672232529374$	$C4 = 0.09023423177351747$
$C5 = -2.054483906525573$	$C5 = -0.5848569223985893$
$C6 = -0.07965355287569573$	$C6 = -0.2614715762213976$
$C7 = 0.01389718013468013$	$C7 = 0.01650486812570146$
$C8 = 1.094843106995885$	$C8 = 0.4112782921810701$
$C9 = 0.3029285714285714$	$C9 = 0.2718556547619048$
$C10 = -0.04733208101422387$	$C10 = -0.04616696042053185$

For values of η between the values above, calculate the values for $\eta = 1.00$, 0.75, 0.50, and 0.25 and use two-dimensional cubic interpolation (A.10.1) to determine the desired value.

A.5.8 Two Back-to-Back Square Reflectors Combined Gain

The following factor Y (dB) accounts for the combined gain of two flat square reflectors relative to the gain of a single reflector the size of the smaller reflector.

The formulas are valid for $R_W \geq 1$ and for all X.

d = separation between the two reflectors
a^2 = projected area of smaller reflector
b^2 = projected area of larger reflector
projected area = physical area × [cosine (1/2 path included angle at reflector)]
a = width of the smaller reflector assuming reflector projection is square
b = width of the larger reflector assuming reflector projection is square
λ = free space wavelength of radio wave
a, b, d, and λ are in the same linear units
$X_p = d/(2a^2/\lambda)$
$R_W = b/a$
$K = (1/2)$ sqrt $(1/X_p)$, sqrt is the square root function
$p = K(R_W - 1)$
$q = K(R_W + 1)$

Y(dB) = combined gain of both reflectors relative to gain of single smaller reflector

$$= 20 \log[(U^2 + V^2)(2X_p)] \qquad \text{(A.101)}$$

$$U = q\ C(q) - p\ C(p) + (1/\pi)[\sin(\pi p^2/2) - \sin(\pi q^2/2)]$$
$$V = q\ S(q) - p\ S(p) + (1/\pi)[\cos(\pi q^2/2) - \cos(\pi p^2/2)]$$
$$C(z) = 0.5 + f(z)\sin(\pi z^2/2) - g(z)\cos(\pi z^2/2)$$
$$S(z) = 0.5 - f(z)\cos(\pi z^2/2) - g(z)\sin(\pi z^2/2)$$
$$f(z) = (1 + 0.926\,z)/(2 + 1.792\,z + 3.104\,z^2)$$
$$g(z) = 1/(2 + 4.142\,z + 3.492\,z^2 + 6.670\,z^3)$$

For $X \gg 1$, $Y(\text{dB}) \cong 6 + 40 \log R_W + 40 \log K$
For $X \ll 1$, $Y(\text{dB}) \cong 0$.

If one of the transmit or receive parabolic antennas is in the near field of the combined reflectors, the addition of the parabolic and rectangular reflector near-field correction may be needed.

A.6 PATH GEOMETRY

A.6.1 Horizons (Normal Refractivity over Spherical Earth)

d = distance to horizon from antenna;
h = antenna height above spherical Earth;
K = equivalent earth radius factor.

Geometric horizon

$$d(\text{miles}) = 1.225 \text{ sqrt}[h(\text{ft})]$$
$$d(\text{km}) = 3.570 \text{ sqrt}[h(\text{m})] \qquad \text{(A.102)}$$

Electromagnetic wave horizon

$$d(\text{miles}) = 1.225 \text{ sqrt}[Kh(\text{ft})]$$
$$d(\text{km}) = 3.570 \text{ sqrt}[Kh(\text{m})] \qquad \text{(A.103)}$$

Radio horizon (typical $K = 4/3$)

$$d(\text{miles}) = 1.414 \text{ sqrt}[h(\text{ft})]$$
$$d(\text{km}) = 4.122 \text{ sqrt}[h(\text{m})] \tag{A.104}$$

Optical horizon (typical noonday $K = 7/6$)

$$d(\text{miles}) = 1.323 \text{ sqrt}[h(\text{ft})]$$
$$d(\text{km}) = 3.856 \text{ sqrt}[h(\text{m})] \tag{A.105}$$

A.6.2 Earth Curvature (Height Adjustment Used on Path Profiles)

$$h(\text{ft}) = \frac{[d_1(\text{miles}) \times d_2(\text{miles})]}{(1.500 \times K)}$$
$$h(\text{m}) = \frac{[\text{d}_1(\text{km}) \times d_2(\text{km})]}{(12.75 \times K)} \tag{A.106}$$

d_1 = distance from one end of the path to the location of interest;
d_2 = distance from other end of the path to the location of interest;
K = equivalent earth radius factor.

A.6.3 Reflection Point

Figure A.2 is an illustration for the smooth-earth case.

h_1 = physical height of the antenna at one end of the path *above the reflection point physical height*;
h_2 = physical height of the antenna at the other end of the path *above the reflection point physical height*;
$h_1 \geq h_2$;
d_1 = distance from one end of the path to the reflection point;
d_2 = distance from the other end of the path to the reflection point;
D = distance from one end of the path to the other end;
= total path distance;
= $d_1 + d_2$;
K = equivalent earth radius factor ≥ 0;
a = physical earth radius $\cong$ 6367 km $\cong$ 3957 (statute) miles;
$C = (h_1 - h_2/h_1 + h_2) \geq 0$;
$M = D^2/[4Ka(h_1 + h_2)]$.

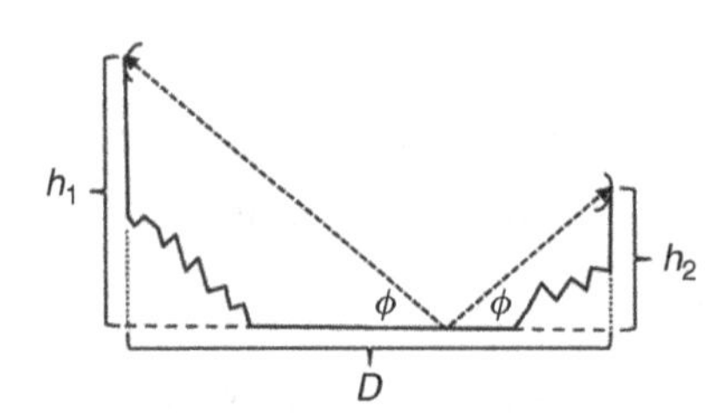

Figure A.2 Smooth-earth surface path profile reflection geometry.

$$d_1 = \left(\frac{D}{2}\right)(1 + B) \tag{A.107}$$

$$d_2 = D - d_1 \tag{A.108}$$

$$B = 2\sqrt{\frac{M+1}{3M}}\cos\left[\left(\frac{\pi}{3}\right) + \left(\frac{1}{3}\arccos\left[\frac{3C}{2}\sqrt{\frac{3M}{(M+1)^3}}\right]\right)\right]$$

$$\phi\,(\text{rad}) = \left[\frac{(h_1 + h_2)}{D}\right][1 - M(1 + B^2)]$$

All of these equations are in the same units of distance.

The following relationships also apply to the reflection point:

For d_1, d_2, and D in miles and h_1 and h_2 in feet:

$$\text{For } K = \frac{2}{3} \ : \ \left(\frac{h_1}{d_1}\right) - d_1 = \left(\frac{h_2}{d_2}\right) - d_2 \tag{A.109}$$

$$\text{For } K = \frac{4}{3} \ : \ \left(\frac{h_1}{d_1}\right) - \left(\frac{d_1}{2}\right) = \left(\frac{h_2}{d_2}\right) - \left(\frac{d_2}{2}\right) \tag{A.110}$$

$$\text{For } K = \infty \ : \ d_1 = \left[\frac{h_1}{(h_1 + h_2)}\right] D, \quad d_2 = D - d_1 \tag{A.111}$$

For d_1, d_2, and D in kilometers and h_1 and h_2 in meters:

$$\text{For } K = \frac{2}{3} \ : \ \left(\frac{h_1}{d_1}\right) - \left(\frac{d_1}{8.5}\right) = \left(\frac{h_2}{d_2}\right) - \left(\frac{d_2}{8.5}\right) \tag{A.112}$$

$$\text{For } K = \frac{4}{3} \ : \ \left(\frac{h_1}{d_1}\right) - \left(\frac{d_1}{17}\right) = \left(\frac{h_2}{d_2}\right) - \left(\frac{d_2}{17}\right) \tag{A.113}$$

$$\text{For } K = \infty \ : \ d_1 = \left[\frac{h_1}{(h_1 + h_2)}\right] D, \quad d_2 = D - d_1 \tag{A.114}$$

Figure A.3 is an illustration for the smooth inclined earth case.

All the above-mentioned definitions apply with the following exceptions:

h_1 = physical height of the antenna at one end of the path *above the extended flat terrain physical height*;

h_2 = physical height of the antenna at the other end of the path *above the extended flat terrain physical height.*

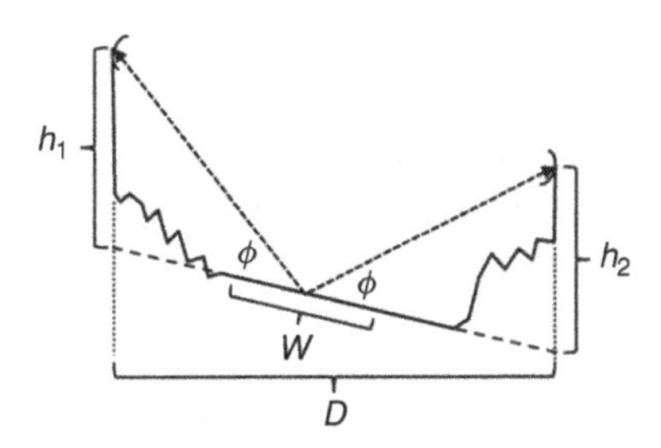

Figure A.3 Smooth inclined earth path profile reflection geometry.

The minimum width W of the flat terrain is approximately [first Fresnel zone radius (F_1)]/sin ϕ.

A.6.4 Fresnel Zone Radius (Perpendicular to the Radio Path)

$$F_n(\text{ft}) = 72.1 \text{ sqrt}\left\{\frac{[n \times d_1 \text{ (miles)} \times d_2\text{(miles)}]}{[F(\text{GHz}) \times D \text{ (miles)}]}\right\}$$

$$F_n(\text{m}) = 17.3 \text{ sqrt}\left\{\frac{[n \times d_1 \text{ (km)} \times d_2\text{(km)}]}{[F(\text{GHz}) \times D\text{(km)}]}\right\} \quad \text{(A.115)}$$

n = Fresnel zone number (an integer);
d_1 = distance from one end of the path to the reflection;
d_2 = distance from the other end of the path to the reflection;
D = total path distance = $d_1 + d_2$;
F = frequency of radio wave.

A.6.5 Fresnel Zone Projected onto the Earth's Surface

$$\text{Ln} = \frac{d\sqrt{1 + \left[\frac{(4h_1h_2)}{(n\lambda d)}\right]}}{1 + \left[\frac{(h_1 + h_2)^2}{(n\lambda d)}\right]} \quad \text{(A.116)}$$

Ln = projected length of Fresnel zone projected onto the smooth earth in the direction of the radio path (twice the projected semi-major axis).

$$\text{Wn} = \frac{\sqrt{1 + \left[\frac{(4h_1h_2)}{(n\lambda d)}\right]}}{\sqrt{1 + \left[\frac{(h_1 + h_2)^2}{(n\lambda d)}\right]}} \quad \text{(A.117)}$$

Wn = projected width of Fresnel zone projected onto the smooth earth in the direction of the radio path (twice the projected semi-minor axis);
n = Fresnel zone integer designation (1 for first Fresnel zone);
d = path length between transmitter and receiver;
h_1 = transmit antenna height above reflection point height;
h_2 = receive antenna height above reflection point height;
λ = radio wavelength;
λ (ft) = 0.98357/F (GHz);
λ (m) = 0.29980/F (GHz);
F = radio frequency;

D, h_1, h_2, and λ are all in the same distance units.

A.6.6 Reflection Path Additional Distance

$$d(\text{ft}) = \frac{\left(\frac{h}{F_1}\right)^2}{[2.033F(\text{GHz})]}$$
$$d(\text{m}) = \frac{\left(\frac{h}{F_1}\right)^2}{[6.671F(\text{GHz})]} \tag{A.118}$$
$$= \text{reflection path distance} - \text{direct path distance}$$

h = (radial)distance of reflection point below main path;
F_1 = first Fresnel zone radius;
F = Frequency of radio wave.

A.6.7 Reflection Path Additional Delay

$$T(\text{ns}) = \frac{1000}{\Delta F(\text{MHz})} \tag{A.119}$$

ΔF = frequency difference between consecutive peaks on spectrum analyzer display.

$$d(\text{ft}) = \frac{T(\text{ns})}{1.017}$$
$$d(\text{m}) = \frac{T(\text{ns})}{3.336} \tag{A.120}$$
$$= \text{reflection path distance} - \text{direct path distance}$$

A.6.8 Reflection Path Relative Amplitude

$$A_R(\text{dB}) = 20 \log \left\{ \frac{[(10^{A_{\text{PK-NL}}/20}) - 1]}{[(10^{A_{\text{PK-NL}}/20}) + 1]} \right\} \tag{A.121}$$

$A_{\text{PK-NL}}(\text{dB})$ = peak to null variation on received spectrum power (dB difference between consecutive power maximum and power minimum points on spectrum)

A.6.9 Antenna Launch Angle

$$\theta = \text{antenna launch angle}$$
$$= \arctan \phi_1 - \arcsin \phi_2 \tag{A.122}$$
$$\phi_1 = \left(\frac{d}{D}\right)$$
$$= 1.894 \times 10^{-4} \frac{d(\text{ft})}{D(\text{miles})}$$
$$= 1.000 \times 10^{-3} \frac{d(\text{m})}{D(\text{km})}$$

$$\phi_2 = \frac{[\text{sqrt}(d^2 + D^2)]}{(2Ka)}$$
$$= \frac{\{\text{sqrt}[D^2(\text{miles}) + 3.587 \times 10^{-8} d^2(\text{ft})]\}}{7913\ K}$$
$$= \frac{\{\text{sqrt}[D^2(\text{km}) + 1.000 \times 10^{-6} d^2(\text{m})]\}}{12,735\ K}$$

$d = h_F - h_N$;
h_F = height of the far-end antenna above mean sea level;
h_N = height of the near-end antenna above mean sea level;
D = distance between near-and far-end antennas;
K = equivalent earth radius factor;
a = earth radius ≅ 3957 (statute) miles ≅ 6367 km.

For d much smaller than D (nearly horizontal paths such that [d (ft)/D (miles)] <900 or [d (m)/D (km)] <170) the following has less than 1% error:

$$\theta(\text{rad}) = \frac{\{d - [D^2/(2Ka)]\}}{D}$$
$$\theta(\text{degrees}) = \frac{\{[d(\text{ft})/92.15] - [D^2(\text{miles})/(138.1K)]\}}{D(\text{miles})} \quad \text{(A.123)}$$
$$\theta(\text{degrees}) = \frac{\{[d(\text{m})/17.45] - [D^2(\text{km})/(222.3K)]\}}{D(\text{km})}$$

Angles are positive if above the horizon and negative if below.

Note that this formula can be used to determine the angle of arrival of a reflected signal after the reflection point has been determined. It can also be used to estimate beam angle movement (to estimate power fading due to antenna pattern) over an expected K factor range.

A.6.10 Antenna Height Difference

$$[h_F(\text{ft}) - h_N(\text{ft})] = 46.08\ D(\text{miles})(\theta_N - \theta_F)$$
$$[h_F(\text{m}) - h_N(\text{m})] = 8.727\ D(\text{km})(\theta_N - \theta_F) \quad \text{(A.124)}$$

h_F = height of the far-end antenna above mean sea level;
h_N = height of the near-end antenna above mean sea level;
D = distance between near-and far-end antennas;
θ_N(degrees) = launch angle of near end;
θ_F(degrees) = launch angle of far end.

Angles are positive if above the horizon and negative if below.
It is assumed launch angles are small (nearly horizontal).

A.6.11 *K* Factor (From Launch Angles)

$$K = -\frac{D(\text{miles})}{[69.12(\theta_N + \theta_F)]}$$
$$= -\frac{D(\text{km})}{[111.2(\theta_N + \theta_F)]} \quad \text{(A.125)}$$

D = distance between near-and far-end antennas;
θ_N(degrees) = launch angle of near end;
θ_F(degrees) = launch angle of far end.

Angles are positive if above the horizon and negative if below.
It is assumed launch angles are small (nearly horizontal).

A.6.12 Refractive Index and K Factor (From Atmospheric Values)

$$
\begin{aligned}
n &= \text{atmospheric index of refraction} \\
N &= 1,000,000(n-1) \\
a &= \text{earth radius} \\
K &= \text{effective earth radius factor} \\
&= \frac{1}{\left[1+a\left(\frac{dn}{dh}\right)\right]} \\
&= \frac{253}{\{253+[dN/dh(\text{N units per mile})]\}} \\
&= \frac{157}{\{157+[dN/dh(\text{N units per km})]\}}
\end{aligned}
\quad (A.126)
$$

K(light) = typically 6/5(average) to 7/5(midday);
K(radio wave) = typically 4/3(average);
N(light) = N_1;
N(radio wave) = $N_1 + N_2$;
$N_1 = (77.6\ p)/(273+T)$;
$N_2 = (373,000\ W)/(273+T)^2$;
p = atmospheric pressure (mbar);
= 33.9(atmospheric pressure in inches of mercury);
= 1.33(atmospheric pressure in millimeters of mercury);
T = atmospheric temperature in degree centigrade;
W = water vapor pressure (mbar);
= $H_R E_S$;
H_R = atmospheric relative humidity(%)/100;
E_S = atmospheric saturation vapor pressure(mbar);
= $6.108 \times 10^{[(7.500T)/(237.3+T]}$.

A.7 OBSTRUCTION LOSS

A.7.1 Knife-Edge Obstruction Loss

$$
\begin{aligned}
L_{KE} &= \text{knife-edge loss (dB)} \\
&= -10\ \log_{10}[0.25+0.5(\text{SFI}+\text{CFI})+0.5(\text{SFI}^2+\text{CFI}^2)], \quad X>0 \\
&= -10\ \log_{10}[0.25-0.5(\text{SFI}+\text{CFI})+0.5(\text{SFI}^2+\text{CFI}^2)], \quad X \le 0 \quad (A.127) \\
&= \frac{\text{received signal power without obstruction}}{\text{received signal power with obstruction}} \\
X &= h/F_1
\end{aligned}
$$

h = perpendicular distance from main beam to obstruction;
F_1 = first Fresnel zone distance;
$A = = 2^{1/2}|X|$;
$F = (1 + 0.9260A)/(2 + 1.792A + 3.104A^2)$;
$G = 1/(2 + 4.142A + 3.492A^2 + 6.670A^3)$;
$S = \sin(\pi A^2/2)$;
$C = \cos(\pi A^2/2)$.

$$\text{CFI} = 0.5 + (FS) - (GC)$$

$$\text{SFI} = 0.5 - (FC) - (GS)$$

A curved edge can be considered a knife edge if the following condition applies:

$$|\phi| \le \frac{\lambda}{4r}$$

ϕ = angle (rad) formed by a horizontal plane passing through the obstruction's edge and the ray that hits the obstruction's edge;
λ = wavelength of the radio wave;
r = radius of curvature of the obstruction's edge.

A.7.2 Rounded-Edge Obstruction Path Loss

$$L_{\text{RE}} = \text{rounded-edge loss(dB)}$$

$$= 10 \log_{10}\left[\frac{\text{received signal power without rounded-edge obstruction}}{\text{received signal power with rounded-edge obstruction}}\right] \quad (A.128)$$

$$= L_{\text{KE}} + L_{\text{RO}}$$

$$L_{\text{KE}} = \text{knife-edge loss (dB)as calculated above} \quad (A.129)$$

$$L_{\text{RO}} = \text{additional rounded obstruction loss (dB)}$$

$$\ge 0$$

$$= -6 - 20 \log_{10}(mn) + 7.2\, m^{1/2} - (2 - 17n)m + 3.6\, m^{3/2} - 0.8\, m^2 \quad \text{if } mn > 4$$

$$= +7.2\, m^{1/2} - (2 - 12.5\, n)m + 3.6\, m^{3/2} - 0.8\, m^2 \quad \text{if } mn \le 4 \quad (A.130)$$

$m = r[(d_1 + d_2)/(d_1 d_2)]/(\pi r/\lambda)^{1/3}$;
$n = h(\pi r/\lambda)^{2/3}/r$;
h = perpendicular distance from main beam to obstruction;
≥ 0 (path is obstructed or grazing);
d_1 = distance from transmit antenna to path ray intersection above obstruction;
d_2 = distance from receive antenna to path ray intersection above obstruction;
r = obstruction radius of curvature;
l = radio free space wavelength;
$= 0.98357/F(\text{GHz})(\text{ft})$;
$= 0.29980/F(\text{GHz})(\text{m})$;
F = radio operating frequency(GHz).

A.7.3 Smooth-Earth Obstruction Loss

$$X = \frac{h}{F_1} \leq 0.75$$

h = perpendicular distance from main beam to obstruction;
F_1 = first Fresnel zone distance;
P = received signal power with obstruction/received signal power without obstruction.

$$\text{For } X \leq 0, \quad P(\text{dB}) = -10 + 20X \tag{A.131}$$

$$\text{For } 0 \leq X \leq 0.75, \quad P(\text{dB}) = -10 + 20X - 6.665X^2 \tag{A.132}$$

A.7.4 Infinite Flat Reflective Plane Obstruction Loss

$$X = \frac{h}{F_1} \geq 0$$

h = perpendicular distance from main beam to obstruction;
F_1 = first Fresnel zone distance;
P = received signal power with obstruction/received signal power without obstruction.

$$P\,(\text{dB}) = 10 \log \left\{1 + C_{\text{comp}} \cos\left[\pi\left(1 + X^2\right)\right]\right\}^2 \tag{A.133}$$

C_{comp} = composite earth reflection coefficient = $10^{CS\,(\text{dB})/20}10^{CD\,(\text{dB})/20}10^{R(\text{dB})/20}$;
$C_{\text{S}}(\text{dB})$ = reflection (earth roughness scattering) coefficient;
$C_{\text{D}}(\text{dB})$ = divergence coefficient (if the earth is flat);
$R(\text{dB})$ = reflection coefficient (see Chapter 13).

A.7.5 Reflection (Earth Roughness Scattering) Coefficient

It is the relative magnitude of signal reflected from a rough earth's surface ignoring polarization, divergence and earth's dielectric constant.

$\Delta = \Delta h \sin\phi/\lambda \cong 0.01745\ \Delta h\ \phi$ (degrees)/λ for small grazing angle ϕ (since $\sin\phi \cong \phi$ (rad) within 10% for ≤ 0.785 (45°) or 1% for ≤ 0.262 (15°)).

ϕ (degrees) = grazing angle;
λ = radio wave free space wavelength.

A.7.5.1 Gaussian Model

$$C_{\text{S}}(\text{dB}) = \text{reflection coefficient (dB)} = 10\ \log[\mathrm{e}^{-16\pi^2\Delta^2}] = -685.810\Delta^2$$

$$\Delta h = \text{standard deviation of the normal distribution of reflecting surface heights} \tag{A.134}$$

A.7.5.2 Uniform Model

$$C_{\text{S}}\,(\text{dB}) = \text{reflection coefficient (dB)} = 10\ \log\left[\frac{\sin^2(2\pi\Delta)}{(2\pi\Delta)^2}\right]$$

$$\Delta h = \text{maximum difference of uniformly distributed reflecting surface heights} \tag{A.135}$$

$$\text{Reflection coefficient envelope (dB)} = 20\ \log\left[\frac{\sin(2\pi\,\Delta)}{(2\pi\,\Delta)}\right], \quad \Delta_{\text{dB}} \leq -5.8$$

$$\text{Reflection coefficient envelope (dB)} = -16.0 - 2\Delta_{\text{dB}}, \quad \Delta_{\text{dB}} > -5.8$$

$\Delta_{\text{dB}} = 10\ \log(\Delta)$.

Notice that Δh is based on difference in heights, not absolute height.

A.7.5.3 *Empirical Models*

R = reflection coefficient;
= (voltage) amplitude of the reflected signal relative to the amplitude of the incident signal;
$20\ \log_{10}(R)$ = reflected signal power/incident signal power (dB).

Exponential

$$R = \frac{1}{\left[1 + \left(\frac{P^2}{2}\right)\right]} \tag{A.136}$$

Pseudoexponential

$$R = \frac{1}{\left[1 + \left(\frac{2P^2}{3}\right)\right]^{(3/4)}} \tag{A.137}$$

Normal

$$R = \exp\left(-\frac{P^2}{2}\right) \tag{A.138}$$

$$\exp(x) = e^x$$

Longley–Rice Empirical

$$R = \exp\left(-\frac{P}{2}\right) \tag{A.139}$$

P = Rayleigh roughness parameter

= effective terrain roughness

$$= 4\pi\left(\frac{\sigma}{\lambda}\right)\ \sin\theta \tag{A.140}$$

σ = root mean square(RMS)surface height(measured from crest to trough);
λ = radio wave free space wavelength;
θ = grazing angle of incident signal relative to the mean surface plane.

A.7.6 Divergence Coefficient from Earth

It is the relative magnitude of signal reflected from a curved earth ignoring polarization, roughness and the earth's dielectric constant.

Typically, this factor is not significant except for very shallow grazing angle paths.

$$C_{\rm D} = \text{reflection coefficient}$$
$$= \text{abs}\left(\frac{\text{reflected signal magnitude}}{\text{incident signal magnitude}}\right)$$
$$= \text{sqrt}\left\{1 + \left[\frac{(2d_1 d_2)}{(Krd\ \sin(\phi))}\right]\right\} \quad \text{(A.141)}$$

ϕ(rad) = grazing angle at reflection = angle of incidence = angle of reflection;
d_1 = distance from one end of the path to the location of interest;
d_2 = distance from the other end of the path to the location of interest;
$d = d_1 + d_2$;
R_e = Earth's equatorial radius ≅ 6378 km ≅ 3963 miles;
R_p = Earth's polar radius ≅ 6357 km ≅ 3950 miles;
E^2 = eccentricity^2 = $(R_e{}^2 - R_p{}^2)/R_e{}^2$;
ψ = latitude at reflection;
r = Earth's radius at reflection = $R_p/\text{sqrt}[1 - (E^2 \cos\psi)]$;
K = equivalent earth radius factor;
C_D(dB) = 20 log(C_D);
ϕ(degrees) = $(180/\pi)\,\phi$(rad).

A.7.7 Divergence Factor for a Cylinder

$$\text{Cylinder divergence factor (dB)} = -12 - 10\ \log(\text{Rn}) \quad \text{(A.142)}$$

Rn = cylinder's radius/λ (wavelength);
= 1.017 cylinder's radius (ft) f (GHz);
= 3.336 cylinder's radius (meters) f (GHz);
f = operating frequency.

A.7.8 Divergence Factor for a Sphere

$$\text{Sphere divergence factor (dB)} = -32 - 40\ \log\ (\text{Rn}) \quad \text{(A.143)}$$

Rn = sphere's radius/λ (wavelength);
= 1.017 cylinder's radius (ft) f (GHz);
= 3.336 cylinder's radius (m) f (GHz);
f = operating frequency.

A.7.9 Signal Reflected from Flat Earth

See Chapter 13 for details.

A.7.10 Ducting

F_{min} (GHz) = lowest frequency propagated by a atmospheric duct;
dh = thickness of duct (atmospheric gradient);
dN (N units) = refractive index change across the duct;

For dh in feet:

$$F_{\text{min}}(\text{GHz}) = \frac{393}{[\text{d}h(\text{d}N - 0.0479\text{d}n)^{1/2}]} \cong \frac{2150}{\text{d}h^{3/2}} \quad \text{(A.144)}$$

For dh in meters:

$$F_{\text{min}}(\text{GHz}) = \frac{120}{[\text{d}h(\text{d}N - 0.157\text{d}n)^{1/2}]} \cong \frac{362}{\text{d}h^{3/2}}$$

A.8 MAPPING

A.8.1 Path Length and Bearing

Consider two locations, A and B. Neither location can be at an earth pole or on opposite sides of the Earth. In addition, the great circle path must not cross the ±180° longitude line (International Date Line).

θ_A = latitude of A (degrees);
θ_B = latitude of B (degrees);
Φ_A = longitude of A (degrees);
Φ_B = longitude of B (degrees);
North latitudes and east longitudes are taken as positive;
South latitudes and west longitudes are taken as negative.

The angles must be converted to decimal notation.
The angle Φ is assumed initially to be composed of Φ_d degrees, Φ_m minutes, and Φ_s seconds.

$$\Phi(\text{decimal degrees}) = \Phi_d + \left(\frac{\Phi_m}{60}\right) + \left(\frac{\Phi_s}{3600}\right) \quad \text{(A.145)}$$

$$\Phi(\text{decimal radians}) = 0.017453292 \times \Phi(\text{decimal degrees}) \quad \text{(A.146)}$$

$$\Phi(\text{decimal degrees}) = 57.295780 \times \Phi(\text{decimal radians}) \quad \text{(A.147)}$$

$$Z(\text{degrees}) = \arccos\{(\sin\theta_A \sin\theta_B) + [\cos\theta_A \cos\theta_B \cos(\phi_A - \phi_B)]\}$$

= angular difference between the two sites measured at the center of the Earth (A.148)

The above formula is derived from the spherical law of cosines and is commonly suggested for great circle path distance calculations. However, the following ("haversine") formula has significantly less round-off error:

$$Z(\text{degrees}) = 2 \times \arcsin\left\{\text{sqrt}\left\{\left(\sin\left[\frac{(\theta_A - \theta_B)}{2}\right]\right)^2 + \left[\cos\theta_A \cos\theta_B\left(\sin\left[\frac{(\phi_A - \phi_B)}{2}\right]\right)^2\right]\right\}\right\} \quad \text{(A.149)}$$

The haversine equation is preferred for all path distance calculations (except for the rare case of sites at opposite sides of the Earth).

D = great circle distance from A to B

R = radius of the Earth

$D = 0.017453292\ R\ Z$, Z in degrees, D and R both in kilometers or both in miles (A.150)

For most modern map coordinate systems (NAD83, WGS84, WGS72, WGS66, GRS80, GRS67, and IAU68), the following are the primary earth radiuses:

R_e = Earth/s equatorial radius ≅ 6378.1 km ≅ 3963.3 miles;
R_p = Earth's polar radius ≅ 6356.8 km ≅ 3950.0 miles;
E^2 = eccentricity$^2 = (R_e{}^2 - R_p{}^2)/R_e{}^2$;
R = Earth's radius on path = R_p/sqrt $[1 - (E^2 \cos\ \theta_{AVG})]$;
$\theta_{AVG} = (\theta_A + \theta_B)/2$.

If three significant figures are adequate, the following apply:
R ≅ 6367 km ≅ 3957 (statute) miles.
If site A latitude and longitude are in cells C2 and D2, respectively, and site B latitude and longitude are in cells E2 and F2, respectively, the following formula may be placed in an Excel spread sheet cell to calculate the above simplified reduced round-off error formula for path great circle distance in miles:

```
= IF(AND(ISNUMBER(C2),ISNUMBER(D2),ISNUMBER(E2),ISNUMBER(F2)),
ROUND(ABS(MAX(0.0001,(2*ASIN((((SIN(((((D2*PI()/180)–
(F2*PI()/180))/2)))^2)*COS((C2*PI()/180))*COS((E2*PI()/180))+
(SIN((((C2*PI()/180)–
(E2*PI()/180)))/2))^2)^0.5))*180/PI()*0.017453292*3957*5280)),8),CHAR(45))
```

In the above formula, it is assumed that the longitude and latitude are in decimal degrees. Longitude is positive if East (0 to 180°) and negative if West (0° to −180°). Latitude is positive if North (0 to 90°) and negative if South (0° to −90°).

Bearing is the horizontal angle (measured clockwise from true north) that points from the site of interest toward the other site. North is 0°, East is 90°, South is 180°, and West is 270°.

B_A = bearing at A toward B;
B_B = bearing at B toward A.

$$\alpha(\text{degrees}) = \text{arc}\ \cos\left\{\frac{[\sin\theta_B - (\sin\theta_A\ \cos Z)]}{(\cos\theta_A \sin Z)}\right\} \quad (A.151)$$

$$\beta(\text{degrees}) = \text{arc}\ \cos\left\{\frac{[\sin\theta_A - (\sin\theta_B\ \cos Z)]}{(\cos\theta_B\ \sin Z)}\right\} \quad (A.152)$$

$$0^\circ \le \alpha \le +180^\circ$$

$$0^\circ \le \beta \le +180^\circ$$

If $\phi_A \le \phi_B$,

$$B_A = \alpha$$

$$B_B = 360 - \beta$$

If $\phi_A > \phi_B$,

$$B_A = 360 - \alpha$$

$$B_B = \beta$$

Sometimes an angle must be converted from decimal notation to degrees, minutes, and seconds format:

Φ (decimal degrees)

$$\Phi_d(\text{degrees}) = \text{Int}\,[\Phi] \tag{A.153}$$

$$\Phi_m(\text{min}) = \text{Int}[60(\Phi - \Phi_d] \tag{A.154}$$

$$\Phi_s(\text{s}) = 3600\,[\Phi - \Phi_d - (\Phi_m\,60)] \tag{A.155}$$

Int $[x]$ rounds x down to the largest integer less than or equal to x.

$$\Phi\,(\text{degrees, min, s}) = \Phi_d, \Phi_m, \Phi_s \tag{A.156}$$

A.9 TOWERS

A.9.1 Three-Point Guyed Towers

Refer to Figure A.4.

A.9.1.1 Minimum Land Area (Tower Orientation may Limit Antenna Placement)

$$A = \{D \times [1 + \sin\,(30^\circ)]\} + E + F + \text{margin} \tag{A.157}$$

$$B = [2 \times D \times \cos(30^\circ)] + (2 \times F) + \text{margin} \tag{A.158}$$

margin = additional distance to allow for unforseen circumstances.

A.9.1.2 Any Tower Orientation

$$C = 2 \times (D + E) + \text{margin} \tag{A.159}$$

$$D = \text{tower height} \times \text{factor}$$

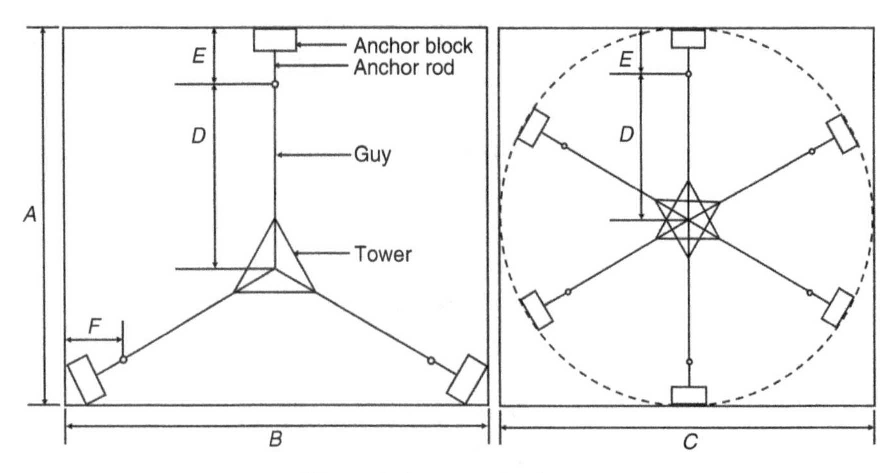

Figure A.4 Tower land area.

factor = percentage of tower height expressed as a fraction

= 0.8 typically(range is from 1.0 to 0.4 but cost increases as factor gets smaller)

$E = 15$ ft (5 m) typically

$F = 20$ ft (6 m) typically (A.160)

A.9.2 Three-Leg Self-Supporting Tower

The distances in Table A.23 are the lengths of the sides of the rectangle enclosing the tower leg pads.

A.9.3 Four-Leg Self-Supporting Tower

TABLE A.23 Three-Leg Self-Supporting Tower

Tower Height, ft	Land Width 1, ft	Land Width 2, ft
50	26.6 + margin	23.4 + margin
75	30.2 + margin	26.6 + margin
100	33.5 + margin	29.3 + margin
125	37.2 + margin	32.8 + margin
150	42.0 + margin	37.1 + margin
175	46.8 + margin	41.3 + margin
200	50.8 + margin	44.6 + margin
225	54.5 + margin	48.0 + margin
250	58.7 + margin	51.2 + margin
275	62.1 + margin	54.8 + margin
300	66.8 + margin	59.0 + margin
325	70.0 + margin	61.6 + margin
350	73.1 + margin	64.4 + margin

The distances in Table A.24 are the lengths of one side of the square enclosing the tower leg pads.

TABLE A.24 Four-Leg Self-Supporting Tower

Tower Height, ft	Land Width, ft
50	21.8 + margin
75	25.3 + margin
100	28.9 + margin
125	32.2 + margin
150	35.9 + margin
175	39.5 + margin
200	43.3 + margin
225	46.9 + margin
250	50.6 + margin
275	53.8 + margin
300	57.6 + margin
325	60.0 + margin
350	64.0 + margin

Land width includes the space necessary for the guy anchors or tower leg pads.
Margin is accommodation for ditches, fences, and easements (typically 40 ft).
Source: The land widths were adapted from White, R. F., *Engineering Considerations for Microwave Communications Systems*, San Carlos: Lenkurt Electric, p. 86, 1970.

A.10 INTERPOLATION

For all the following equations, double precision arithmetic is strongly recommended. Also the convention that multiplication occurs before addition is assumed.

A.10.1 Two-Dimensional Interpolation

Sometimes it is necessary to interpolate between tabular data points. The following methods create linear ($Y = A + BX$), quadratic ($Y = A + BX + CX^2$), or cubic ($Y = A + BX + CX^2 + DX^3$) polynomials. If the data is known to have a nonlinear relationship to X, (such as squared or logarithmic components), first transform the input X data. For example, make the substitution $Z = X^2$ or $Z = \log(X)$, respectively. Replace X below with Z. The results will be the appropriate polynomials (e.g., $Y = A + BZ + CZ^2 + DZ^3$, which are equivalent to $Y = A + B(X^2) + C(X^4) + D(X^6)$ or $Y = A + B\ \log(X) + B\ \log^2(X) + B\ \log^3(X)$).

The following discussion uses Figure A.5.

A.10.1.1 Lagrangian Interpolation This method uses a limited number of data points to produce a polynomial that exactly reproduces the input data.

A.10.1.2 Linear Interpolation Given a set of two X,Y values, (X_1, Y_1) and (X_2, Y_2) with $X_1 < X_2$, interpolate a value of Y for X such that $X_1 < X < X_2$:

$$Y = Y_1 + \left\{\left(\frac{\text{CC}}{\text{CA}}\right) \times \text{CB}\right\} \tag{A.161}$$

$\text{CA} = X_2 - X_1;$
$\text{CB} = X - X_1;$
$\text{CC} = Y_2 - Y_1.$

Special Cases: Linear interpolation is best when it interpolates a function that is nearly linear in X and Y. When graphed, this type of function is a straight line if the X and Y graph coordinates are both linear or both logarithmic.

If the graphed function is a straight line when the Y coordinates are linear but the X coordinates are logarithmic, the following modified linear interpolation formula works well:

$$Y = Y1 + \left[(\mathrm{Y}_2 - \mathrm{Y}_1)\ \frac{\log_{10}\left(\dfrac{X}{X_1}\right)}{\log_{10}\left(\dfrac{X_2}{X_1}\right)}\right] \tag{A.162}$$

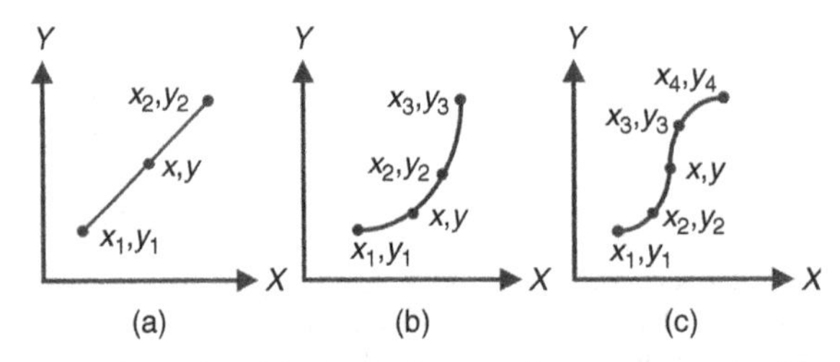

Figure A.5 Two-dimensional interpolation: (a) linear, (b) quadratic, and (c) cubic.

If the graphed function is a straight line when the Y coordinates are logarithmic but the X coordinates are linear, the following modified linear interpolation formula works well:

$$\log Y = \log_{10}(Y_1) + \left[(X - X_1) \frac{\log_{10}\left(\frac{Y_2}{Y_1}\right)}{(X_2 - X_1)} \right] \tag{A.163}$$

$$Y = 10^{\log Y}$$

When the best approach is not clear, testing all three cases with data samples is often helpful. In the above cases, the common logarithm $\log_{10}(\ )$ and base 10 were used. The formulas are applicable for any logarithm and base (e.g., ln( ) and e).

A.10.1.3 Quadratic Interpolation Given a set of three X, Y values, (X_1, Y_1), (X_2, Y_2), and (X_3, Y_3) with $X_1 < X_2 < X_3$, interpolate a value of Y for X such that $X_1 < X < X_3$ (The X values do not need to be equally spaced.):

$$Y = A + (B \times X) + (C \times X^2) \tag{A.164}$$

$C21 = X_2 - X_1;$
$C31 = X_3 - X_1;$
$C32 = X_3 - X_2;$
$K1 = C21 \times C31;$
$K2 = C21 \times C32;$
$K3 = C31 \times C32;$
$\mathrm{CA1} = (X_2 \times X_3)/K1;$
$\mathrm{CA2} = -(X_1 \times X_3)/K2;$
$\mathrm{CA3} = (X_1 \times X_2)/K3;$
$\mathrm{CB1} = -(X_2 + X_3)/K1;$
$\mathrm{CB2} = (X_1 + X_3)/K2;$
$\mathrm{CB3} = -(X_1 + X_2)/K3;$
$\mathrm{CC1} = 1/\mathrm{K1};$
$\mathrm{CC2} = -1/\mathrm{K2};$
$\mathrm{CC3} = 1/K3;$
$A = \mathrm{CA1} \times Y_1 + \mathrm{CA2} \times Y_2 + \mathrm{CA3} \times Y_3;$
$B = \mathrm{CB1} \times Y_1 + \mathrm{CB2} \times Y_2 + \mathrm{CB3} \times Y_3;$
$C = \mathrm{CC1} \times Y_1 + \mathrm{CC2} \times Y_2 + \mathrm{CC3} \times Y_3.$

A.10.1.4 Cubic Interpolation Given a set of four X, Y values, (X_1, Y_1), (X_2, Y_2), (X_3, Y_3), and (X_4, Y_4) with $X_1 < X_2 < X_3 < X_4$, interpolate a value of Y for X such that $X_1 < X < X_4$ (The X values do not need to be equally spaced.) This method will automatically perform linear or quadratic interpolation if the data is linear or quadratic:

$$Y = A + (B \times X) + (C \times X^2) + (D \times X^3) \tag{A.165}$$

$C21 = X_2 - X_1;$
$C31 = X_3 - X_1;$
$C32 = X_3 - X_2;$
$C41 = X_4 - X_1;$
$C42 = X_4 - X_2;$
$C43 = X_4 - X_3;$
$K1 = C21 \times C31 \times C41;$
$K2 = C21 \times C32 \times C42;$

$K3 = C31 \times C32 \times C43;$
$K4 = C41 \times C42 \times C43;$
$\mathrm{CA1} = (X_2 \times X_3 \times X_4)/K1;$
$\mathrm{CA2} = -(X_1 \times X_3 \times X_4)/K2;$
$\mathrm{CA3} = (X_1 \times X_2 \times X_4)/K3;$
$\mathrm{CA4} = -(X_1 \times X_2 \times X_3)/K4;$
$\mathrm{CB1} = -(X_2 \times X_3 + X_2 \times X_4 + X_3 \times X_4)/K1;$
$\mathrm{CB2} = (X_1 \times X_3 + X_1 \times X_4 + X_3 \times X_4)/K2;$
$\mathrm{CB3} = -(X_1 \times X_2 + X_1 \times X_4 + X_2 \times X_4)/K3;$
$\mathrm{CB4} = (X_1 \times X_2 + X_1 \times X_3 + X_2 \times X_3)/K4;$
$\mathrm{CC1} = (X_2 + X_3 + X_4)/K1;$
$\mathrm{CC2} = -(X_1 + X_3 + X_4)/K2;$
$\mathrm{CC3} = (X_1 + X_2 + X_4)/K3;$
$\mathrm{CC4} = -(X_1 + X_2 + X_3)/K4;$
$\mathrm{CD1} = -1/K1;$
$\mathrm{CD2} = 1/K2;$
$\mathrm{CD3} = -1/K3;$
$\mathrm{CD4} = 1/K4;$
$A = \mathrm{CA1} \times Y_1 + \mathrm{CA2} \times Y_2 + \mathrm{CA3} \times Y_3 + \mathrm{CA4} \times Y_4;$
$B = \mathrm{CB1} \times Y_1 + \mathrm{CB2} \times Y_2 + \mathrm{CB3} \times Y_3 + \mathrm{CB4} \times Y_4;$
$C = \mathrm{CC1} \times Y_1 + \mathrm{CC2} \times Y_2 + \mathrm{CC3} \times Y_3 + \mathrm{CC4} \times Y_4;$
$D = \mathrm{CD1} \times Y_1 + \mathrm{CD2} \times Y_2 + \mathrm{CD3} \times Y_3 + \mathrm{CD4} \times Y_4.$

A.10.1.5 Least Squared Error Interpolation This method uses a user-defined number of data points to produce a polynomial that approximates the input data with least squared error. If it uses the same data points as those used in LaGrangian interpolation, it achieves the same result.

Given a set of $n + 1$ (X, Y) values, $(X_1, Y_1), (X_2, Y_2) \ldots (X_{n+1}, Y_{n+1})$ with $X_1 < X_2 < \cdots < X_{n+1}$, interpolate a value of Y for X such that $X_1 < X < X_{n+1}$. (The X values do not need to be equally spaced.)

$\mathrm{Ca1} = n + 1;$
$\mathrm{Ca2} = X_1 + X_2 + X_3 + \cdots + X_{n+1};$
$\mathrm{Ca3} = X_1^2 + X_2^2 + X_3^2 + \cdots + X_{n+1}^2;$
$\mathrm{Ca4} = X_1^3 + X_2^3 + X_3^3 + \cdots + X_{n+1}^3;$
$\mathrm{Ca5} = X_1^4 + X_2^4 + X_3^4 + \cdots + X_{n+1}^4;$
$\mathrm{Ca6} = X_1^5 + X_2^5 + X_3^5 + \cdots + X_{n+1}^5;$
$\mathrm{Ca7} = X_1^6 + X_2^6 + X_3^6 + \cdots + X_{n+1}^6;$
$\mathrm{Cb1} = -(Y_1 + Y_2 + Y_3 + \cdots + Y_{n+1});$
$\mathrm{Cb2} = -(X_1 \times Y_1 + X_2 \times Y_2 + X_3 \times Y_3 + \cdots + X_{n+1} \times Y_{n+1});$
$\mathrm{Cb3} = -(X_1^2 \times Y_1 + X_2^2 \times Y_2 + X_3^2 \times Y_3 + \cdots + X_{n+1}^2 \times Y_{n+1});$
$\mathrm{Cb4} = -(X_1^3 \times Y_1 + X_2^3 \times Y_2 + X_3^3 \times Y_3 + \cdots + X_{n+1}^3 \times Y_{n+1});$
$\mathrm{Cc1} = \mathrm{Ca1} \times \mathrm{Ca3} - \mathrm{Ca2}^2;$
$\mathrm{Cc2} = \mathrm{Ca1} \times \mathrm{Ca4} - \mathrm{Ca2} \times \mathrm{Ca3};$
$\mathrm{Cc3} = \mathrm{Ca1} \times \mathrm{Ca5} - \mathrm{Ca2} \times \mathrm{Ca4};$
$\mathrm{Cc4} = \mathrm{Ca1} \times \mathrm{Ca5} - \mathrm{Ca3}^2;$
$\mathrm{Cc5} = \mathrm{Ca1} \times \mathrm{Ca6} - \mathrm{Ca3} \times \mathrm{Ca4};$
$\mathrm{Cc6} = \mathrm{Ca1} \times \mathrm{Ca7} - \mathrm{Ca4}^2;$
$\mathrm{Cc7} = \mathrm{Ca1} \times \mathrm{Cb2} - \mathrm{Ca2} \times \mathrm{Cb1};$
$\mathrm{Cc8} = \mathrm{Ca1} \times \mathrm{Cb3} - \mathrm{Ca3} \times \mathrm{Cb1};$
$\mathrm{Cc9} = \mathrm{Ca1} \times \mathrm{Cb4} - \mathrm{Ca4} \times \mathrm{Cb1};$
$\mathrm{Cd1} = \mathrm{Cc1} \times \mathrm{Cc4} - \mathrm{Cc2} \times \mathrm{Cc2};$
$\mathrm{Cd2} = \mathrm{Cc1} \times \mathrm{Cc5} - \mathrm{Cc2} \times \mathrm{Cc3};$
$\mathrm{Cd3} = \mathrm{Cc1} \times \mathrm{Cc5} - \mathrm{Cc3} \times \mathrm{Cc2};$
$\mathrm{Cd4} = \mathrm{Cc1} \times \mathrm{Cc6} - \mathrm{Cc3} \times \mathrm{Cc3};$
$\mathrm{Cd5} = \mathrm{Cc1} \times \mathrm{Cc8} - \mathrm{Cc2} \times \mathrm{Cc7};$
$\mathrm{Cd6} = \mathrm{Cc1} \times \mathrm{Cc9} - \mathrm{Cc3} \times \mathrm{Cc7};$

Cd7 = Cd1 × Cd4 − Cd3 × Cd2;
Cd8 = Cd1 × Cd6 − Cd3 × Cd5.

A.10.1.6 Linear Interpolation

$$Y = A + (B \times X) \tag{A.166}$$

$B = -\text{Cc7}/\text{Cc1}$;
$A = -(\text{Ca2} \times B + \text{Cb1})/\text{Ca1}$.

A.10.1.7 Quadratic Interpolation

$$Y = A + (B \times X) + (C \times X^2) \tag{A.167}$$

$C = -\text{Cd5}/\text{Cd1}$;
$B = -(\text{Cc2} \times C + \text{Cc7})/\text{Cc1}$;
$A = -(\text{Ca2} \times B + \text{Ca3} \times C + \text{Cb1})/\text{Ca1}$.

A.10.1.8 Cubic Interpolation

$$Y = A + (B \times X) + (C \times X^2) + (D \times X^3) \tag{A.168}$$

$D = -\text{Cd8}/\text{Cd7}$;
$C = -(\text{Cd2} \times D + \text{Cd5})/\text{Cd1}$;
$B = -(\text{Cc2} \times C + \text{Cc3} \times D + \text{Cc7})/\text{Cc1}$;
$A = -(\text{Ca2} \times B + \text{Ca3} \times C + \text{Ca4} \times D + \text{Cb1})/\text{Ca1}$.

A.10.2 Three-Dimensional Interpolation

A.10.2.1 All Data on a Rectangular Grid The following descriptions rely on Figure A.6.

A.10.2.2 Linear Interpolation Given four sets of x, y, z values located on a rectangular X,Y grid, interpolate a value of z for a set of x,y coordinates within the grid of X,Y values.

$$\begin{aligned} z(x, y) = {} & z(x_1, y_1)[(1 - \Delta X)(1 - \Delta Y)] + z(x_1, y_2)[(1 - \Delta X)\Delta Y] \\ & + z(x_2, y_1)[\Delta X(1 - \Delta Y)] + z(x_2, y_2)[\Delta X \, \Delta Y] \end{aligned} \tag{A.169}$$

$\Delta X = (x - x_1/x_2 - x_1)$;
$\Delta Y = (y - y_1/y_2 - y_1)$.

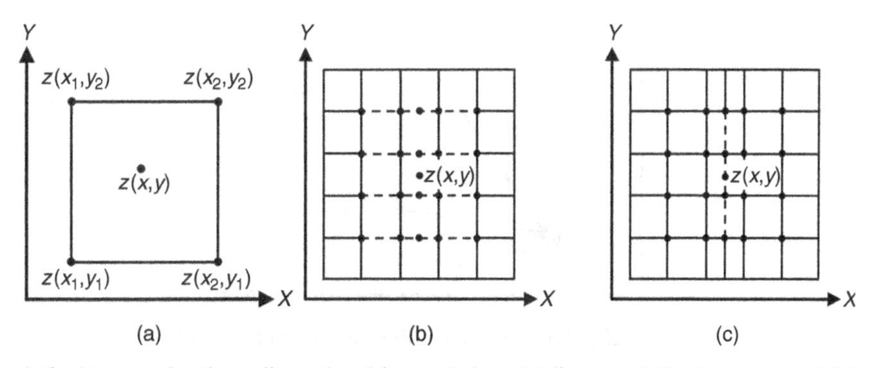

Figure A.6 Rectangular three-dimensional interpolation: (a) linear and (b) first step, and (c) second step of higher order.

A.10.2.3 Higher Order Interpolation Using two-dimensional interpolation, interpolate a set of z values for the desired x value and multiple y values (Figure A.6b). Next, again using two-dimensional interpolation, interpolate the desired z value for the desired x and y values (Figure A.6C). Linear, quadratic, or cubic interpolation may be used for either of these two-dimensional interpolations. This approach is helpful in filling out a sparse data set so linear interpolation may be used.

A.10.2.4 Data Scattered If the data is scattered over the x–y plane, the problem is much more challenging. There is no known optimal solution to this problem. Many approaches (Kriging and methods by Shepard, Lam, Watson and Philip, Cooke and Mostaghimi, Akima, Montefusco, and Casciola and Briggs) are available but all have their trade-offs. Stability and loss of data integrity are common problems (especially for sparse data). The simplest approach is to treat the data set as a group of triangles in the x–y plane and interpolate the z values based on a plane (flat surface) defined by the three-dimensional coordinates at the vertices of the triangle enclosing the x–y value of interest. The following discussion relates to Figure A.7.

Data is assumed to lie on a rectangular grid. The data points (X_1, Y_1, Z_1), (X_2, Y_2, Z_2), and (X_3, Y_3, Z_3) enclose the desired data point (X, Y) in the x–y plane. The value Z at (X, Y) is desired.

$$Z = \frac{([X \times B_1] + [Y \times B_2] + B_3)}{B_4} \tag{A.170}$$

$A_1 = X_2 - X_3;$
$A_2 = Y_2 - Y_3;$
$A_3 = Z_2 - Z_3;$
$A_4 = (X_2 \times Y_3) - (X_3 \times Y_2);$
$A_5 = (X_2 \times Z_3) - (X_3 \times Z_2);$
$A_6 = (Y_2 \times Z_3) - (Y_3 \times Z_2);$
$B_1 = -(A_3 \times Y_1) + (A_2 \times Z_1) - A_6;$
$B_2 = (A_3 \times X_1) - (A_1 \times Z_1) + A_5;$
$B_3 = (A_6 \times X_1) - (A_5 \times Y_1) + (A_4 \times Z_1);$
$B_4 = (A_2 \times X_1) - (A_1 \times Y_1) + A_4.$

The data points (X_1, Y_1, Z_1), (X_2, Y_2, Z_2), and (X_3, Y_3, Z_3) must not lie in a straight line. They must be picked in such a way that they enclose the data location of interest. One method of doing this is to pick four x, y, z points in each of the four quadrants that lie closest to X, Y.

From the database of known values, determine the following four three-dimensional points:

X_A, Y_A, Z_A such that $X_A \leq X$ and $Y_A \geq Y$ (upper left quadrant).
X_B, Y_B, Z_B such that $X_B \geq X$ and $Y_B \geq Y$ (upper right quadrant).

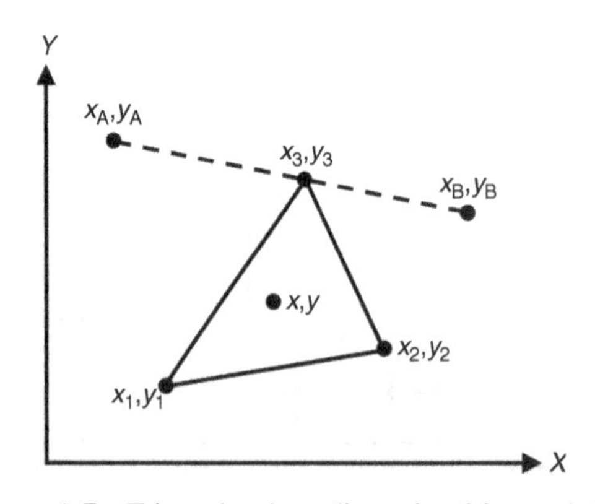

Figure A.7 Triangular three-dimensional interpolation.

X_1, Y_1, Z_1 such that $X_1 \leq X$ and $Y_1 \leq Y$ (lower left quadrant).
X_2, Y_2, Z_2 such that $X_2 \geq X$ and $Y_2 \leq Y$ (upper right quadrant).

The = portion of $\leq$ or $\geq$ is necessary to deal with the edges of a rectangular data grid. If $X_A = X_B = X_1 = X_2$ and $Y_A = Y_B = Y_1 = Y_2$ then the desired three-dimensional point is found and $Z = Z_1$. Otherwise, interpolation to determine Z is required.

A.10.2.5 Initial Linear Interpolation Perform linear interpolation to determine X_3, Y_3, and Z_3, given X_A, Y_A, Z_A and X_B, Y_B, Z_B using one of the following distinct cases:

Case L_1: If $X_A = X_B$ and $Y_A = Y_B$ then the two points are the same.

$$X_3 = X_A$$
$$Y_3 = Y_A$$
$$Z_3 = Z_A$$

Case L_2: If $X_A = X_B$ then the two points form a vertical line.

$$X_3 = X_A$$
$$Y_3 = Y_A$$
$$Z_3 = Z_A$$

Case L_3: If $Y_A = Y_B$ then the two points form a horizontal line.

$$Z_3 = Z_A + \left(\text{CC1} \times \frac{\text{CA2}}{\text{CA1}}\right)$$

$X_3 = X$;
$Y_3 = Y_A$;
$\text{CA1} = X_B - X_A$;
$\text{CA2} = X_3 - X_A$;
$\text{CC1} = Z_B - Z_A$.

Case L_4: If none of the above three cases occur, perform normal linear interpolation.

$$X_3 = X$$
$$Y_3 = Y_A + \left(\text{CB1} \times \frac{\text{CA2}}{\text{CA1}}\right)$$

$\text{CA1} = X_B - X_A$;
$\text{CB1} = Y_B - Y_A$;
$\text{CC1} = Z_B - Z_A$;
$\text{CA2} = X_3 - X_A$.

$$Z_3 = Z_A + (\text{CC1} \times \text{CC3}/\text{CC2})$$

$\text{CB2} = Y_3 - Y_A$;
$\text{CC2} = \text{sqrt}(\text{CA1} \times \text{CA1} + \text{CB1} \times \text{CB1})$;
$\text{CC3} = \text{sqrt}(\text{CA2} \times \text{CA2} + \text{CB2} \times \text{CB2})$;
$\text{sqrt}(X)$ = square root of X.

A.10.2.6 Final Triangular Interpolation After determining X_3, Y_3, and Z_3, use triangular interpolation to determine Z, given X and Y plus X_1, Y_1, Z_1 and X_2, Y_2 and Z_2 using one of the following distinct cases:

Case T_1: If $X_1 = X_2$ and $Y_1 = Y_2$ and $X_1 = X_3$ and $Y_1 = Y_3$ then all points are the same.

$$Z = Z_1 \tag{A.171}$$

Case T_2: If $X_1 = X_2$ and $Y_1 = Y_2$ then the two lower points are the same.
Subcase T_{2a}: If $X_1 = X_3$ then the two distinct points form a vertical line

$$Z = Z_1 + \left(\mathrm{CC1} \times \frac{\mathrm{CA2}}{\mathrm{CA1}}\right) \tag{A.172}$$

$\mathrm{CA1} = Y_3 - Y1;$
$\mathrm{CA2} = Y - Y_1;$
$\mathrm{CC1} = Z_3 - Z_1.$

Subcase T_{2b}: If Subcase T_{2a} does not apply, the two distinct points form a diagonal line.

$$Z = Z_1 + \left(\mathrm{CC1} \times \frac{\mathrm{CC3}}{\mathrm{CC2}}\right) \tag{A.173}$$

$\mathrm{CA1} = X_3 - X_1;$
$\mathrm{CB1} = Y_3 - Y_1;$
$\mathrm{CC1} = Z_3 - Z_1;$
$\mathrm{CA2} = X - X_1;$
$\mathrm{CB2} = Y - Y_1;$
$\mathrm{CC2} = \mathrm{sqrt}(\mathrm{CA1} \times \mathrm{CA1} + \mathrm{CB1} \times \mathrm{CB1});$
$\mathrm{CC3} = \mathrm{sqrt}(\mathrm{CA2} \times \mathrm{CA2} + \mathrm{CB2} \times \mathrm{CB2}).$

Case T_3: If none of the above cases apply, perform normal triangular interpolation.

$$Z = \frac{([X \times B_1] + [Y \times B_2] + B_3)}{B_4} \tag{A.174}$$

$A_1 = X_2 - X_3;$
$A_2 = Y_2 - Y_3;$
$A_3 = Z_2 - Z_3;$
$A_4 = (X_2 \times Y_3) - (X_3 \times Y_2);$
$A_5 = (X_2 \times Z_3) - (X_3 \times Z_2);$
$A_6 = (Y_2 \times Z_3) - (Y_3 \times Z_2);$
$B_1 = -(A_3 \times Y_1) + (A_2 \times Z_1) - A_6;$
$B_2 = (A_3 \times X_1) - (A_1 \times Z_1) + A_5;$
$B_3 = (A_6 \times X_1) - (A_5 \times Y_1) + (A_4 \times Z_1);$
$B_4 = (A_2 \times X_1) - (A_1 \times Y_1) + A_4.$

Defined X, Y, Z values at the four grid corners are necessary. In addition, data representing different domains (e.g., land or sea data that may have a different range of values) must be sufficiently "fenced" with known data points separating the domains so that the triangular interpolation does not extend between domains.

B

PERSONNEL AND EQUIPMENT SAFETY CONSIDERATIONS

B.1 GENERAL SAFETY GUIDELINES

Before starting installation or test activities, attend a "job huddle" with the supervisor or team leader to discuss the project to be executed. The huddle should including the following topics:

the work to be accomplished;

safety hazards on the job and methods used to minimize them;

compliance with local safety procedures and practices;

location of first aid supplies;

location and use of fire protection equipment;

location of the name and phone number of a doctor, the police, and the fire department and the address of the nearest hospital (this information must be prominently displayed on the work location);

telephone numbers for safety, backup, and alarm escalation;

safety and service hazards of the job and the methods to minimize them;

location of appropriate fuses, breakers, fire extinguishers, insulated gloves or poles, telephones, and exits;

shutoff operation of all appropriate electrical power;

exit or means of egress from the working area;

action necessary in case of emergency.

Create and review a formal work outline or method of procedure (MOP) for significant work activities.

Use the "two-person" approach when near live electrical circuits.

Appropriate safety checks must be made daily before work initiation.

Daily safety briefings by the job supervisor may be utilized to promote safe practices.

Electrically rated rubber matting must be used while maintenance is performed on electrical equipment or batteries.

Equipment or circuits that are de-energized must be rendered inoperative (locked) and must have tags attached at all points where such equipment or circuits can be energized. Tags must be placed to identify plainly the equipment or circuits being worked on.

Digital Microwave Communication: Engineering Point-to-Point Microwave Systems, First Edition. George Kizer.

An approved first aid kit must be located at each installation site. This requirement applies regardless of the size of the job or the number of personnel involved. Each first aid kit must be easily accessible and in a location known to all workers.

The name and phone number of a doctor, the police, and the fire department and the address of the nearest hospital must be prominently displayed on the work site

Minor injuries, cuts, and scratches must be treated promptly.

When a person may be exposed to injurious corrosive materials, suitable facilities for drenching or flushing of the eyes and body must be provided.

Good housekeeping practices must be adhered to at all times.

- All work areas must be kept neat and clean at all times.
- All work areas, storerooms, aisles, and passageways and floors must be kept neat and in a clean condition.
- All hazardous materials such as paints, aerosol cans, and flammable material must be stored in approved lockers or cabinets.

The facility operator must always be notified before the start of any job requiring dismantling or removal of equipment.

The facility operator and the work supervisor must detail the scope of work and determine what, if any, hazardous materials are present and provide disposition instructions to all concerned.

Work must comply with the appropriate Local, State, and Federal Laws' orders and regulations involving hazardous waste and/or materials.

All material designated as hazardous under the compulsory waste management systems and applicable regulations must be identified.

The work supervisor must ensure that all tools and materials utilized during the installation process are safe for normal use, are nontoxic, and must present no abnormal hazards to persons or the environment. Worn-out tools and associated materials must be disposable as normal refuse without requiring special precautions.

Periodic inspections of the building/facility and existing equipment must be made during the course of ongoing installation activities with a facility operator representative for the purpose of identifying new potentially hazardous conditions.

Unless required for possible return of defective material, all extraneous material must be promptly disposed.

All flammable materials such as wastepaper, foam plastic, cloth bags, packing boxes, packing material, and similar materials supplied during the installation must be removed from the building on a daily basis (more frequently if any accumulation of such materials creates a potential fire hazard).

The installation supervisor must ensure that any such material removed from a facility does not create a hazardous or unsightly appearance to the areas surrounding the building.

Any packing material kept must be stored in such a manner as to not create a safety or fire hazard.

Equipment removal presents certain potentially hazardous conditions (falling debris, airborne dust, etc.), which are inherent in removal operations.

The following areas must be reviewed before performing any removals:

- proper use of tools
- personnel protective equipment
- fire protective equipment
- building protection.

Epoxy resins and sealant are flammable and toxic to skin, eyes, and the respiratory tract. Skin/eye protection and good ventilation are required when using resins and sealants.

Food and beverages are not permitted in equipment areas.

Smoking and open flames are prohibited except in designated areas.

No type of portable heat source such as kerosene, gasoline, and electric must be used during installation unless expressly approved by the facility operator.

Protective ear equipment must be worn in all areas of high noise level. A good guideline to follow is to use protective ear equipment whenever normal speech and communication cannot be accomplished because of the surrounding noise level (i.e., 65 dBA).

B.2 EQUIPMENT PROTECTION

Equipment damage is defined in two categories: physical damage or breakage and damage caused by allowing tools, hardware, dirt, dust, or other foreign substances to get into the equipment.

All electrical equipment must be covered with a protective sheet, plastic shield, or mat if work is done above it.

Provide protective covering for equipment power connections when working nearby.

Use of steel tape or rulers around live equipment must not be allowed.

Skinning tools must be applied in a direction away from the body and care must be taken to avoid striking hands against apparatus or equipment.

Utilize cable stripping and sheathing bags or an approved container for all cable residue.

When connecting power cables to rack, first verify if power to cables is disconnected.

Before applying power to rack, verify proper polarity and voltage of rack strapping and applied power.

When changing polarity of rack DC power, verify that chassis ground is connected to rack rail ground (to minimize personnel hazard during lightning strikes).

DC power lug connections should be crimped to the power cable.

Bolt connection of the lug to the equipment rack should be torqued to no more than 32 in-lb.

Use insulated tools whenever working near primary power.

Remove all rings, watches, and other metal jewelry when working with electronic equipment. Short-circuiting low voltage, low impedance DC circuits can cause severe burns and/or eye and skin damage.

After receiving the equipment and after inventory and inspection, all material must be repacked, as required, to prevent deterioration or damage while in storage or during further moves.

A secure dry area must be utilized for storage that is not in close proximity to any corrosive elements.

Equipment must not be stored or placed in such a manner as to exceed the safe floor load rating of the facility as determined by the facility operator.

Do not climb or stand on cable racks or equipment frames.

Tools must not be placed or hung on cable racks or equipment frames.

All unused tools must be safely secured at all times.

Lamps or drop cords must not be hooked or suspended from forms or skinners.

When working on additions, care must be taken that existing cables are not pulled or damaged.

Cable ends must not be dropped across sensitive components.

All pins, switches, knobs, dials, and other protruding parts of electronic assemblies must be adequately protected to prevent damage during handling, movement, and storage. This includes unmated connectors that are exposed to physical or environmental damage.

Before mounting components, the bays, racks, frames, or cabinets must be checked for rust or scratches and treated for corrosion.

Care must be used in cleaning any equipment to make certain that adjustments are not disturbed and contacts are not damaged.

Installers who mount component assemblies must utilize the appropriate safeguards and equipment.

A continuous effort must be applied to locate and eliminate all possible fire hazards.

Protection from dirt, dust, and other debris is important, as damage or malfunctioning of equipment may result and not be apparent until after installation completion. Guidelines to accomplish this are the following:

Do not block or cover airflow to working equipment.

Whenever possible, equipment should be unpacked outside the equipment room to minimize the amount of dust, lint, etc. introduced into the air from the packing materials.

The exterior and interior of equipment and component parts must be free of grease, dirt, and other forms of corrosion, corrosion-inducing materials, and extraneous matter.

In an installation addition to existing equipment, cable racks, equipment frame tops, and other areas that may be disturbed should be vacuumed.

During installation operations, keep as much of the installation material covered as possible.

Remove the covering only when necessary.

Do not use bare cardboard to cover any equipment, as abrasion between projections and the cardboard produces lint and paper dust that drop into the equipment.

Cover cardboard with sheet plastic if it must be used or use some other hard, semirigid material for protection.

Keep equipment panel covers in place except when necessary to have access to the components covered.

Check panel covers periodically for proper fit.

Keep cable holes covered and airtight.

B.3 EQUIPMENT CONSIDERATIONS

Use caution near modules or cables supplying laser radiation. Exposure to laser radiation can cause eye damage.

Keep all waveguide ports terminated (into a load or shorting plate). Waveguide is an effective radiator (typically 12–16 dB return loss). Exposure to microwave energy can cause eye damage.

Ground faults between racks and/or AC-powered equipment are possible. When connecting coax cables between racks or between AC-powered test equipment and the rack in an unfamiliar facility, do not allow your body to connect the two racks or the rack and the equipment. Ground faults/loops pose a significant shock hazard. When connecting a coaxial cable to rack mounted equipment, do not place your hands on both the cable connector and the equipment connector.

SMA connectors should be finger-tightened and then tightened to 8-in-lb torque using a calibrated wrench.

Use the proper size wrench on all SMA, Type N, and TNC connectors. Never use pliers.

When externally powering DC equipment, use a current limiter on the DC supply to avoid damage to active devices.

Disconnect rack power before removing power supplies to avoid arcing or printed circuit damage.

Place all meters on their highest scale before beginning a test.

Check the input RF power specification on test equipment before connecting RF power to the instrument.

Check the power-handling capability of all components before applying RF power. For example, a 0.5-dB loss to a 100-W signal creates 11 W of heat.

Never touch devices operating at a high power level (e.g., the center conductor of a coaxial cable or connector, microstrip circuitry, finned attenuators, or terminations). They can be quite hot.

Use the right connectors when possible. Avoid adapters.

Keep all connectors and waveguide flanges clean.

Tighten all connectors and adapters to appropriate torque to reduce losses.

Check connectors for proper tightness and alignment. Never force the connection.

Be sure you have the correct type of cable for your application.

Avoid using BNC connectors above 500 MHz.

Provide appropriate mechanical support to devices (e.g., couplers, attenuators, detectors) that must be connected to other equipment.

Units with an "ESS" symbol contain electrostatic sensitive devices. These units should be stored in an antistatic container when not in use. Anyone handling these units should observe antistatic precautions. Damage to the unit may result if antistatic protection is not maintained.

Allow for at least 1 h of system warm-up time before making any final system adjustments or measurements. Failure to allow for adequate warm-up may result in improper setting and inaccurate measurements.

B.4 PERSONNEL PROTECTIVE EQUIPMENT

The installation supervisor must ensure that each installer must have in his or her possession the proper safety equipment.

The installation supervisor must act as a safety observer when hazardous work is accomplished and ensure the proper protective equipment is utilized for the task.

Workers must use appropriate eye or face protection when exposed to eye or face hazards from flying particles, molten metal, liquid chemicals, acids or caustic liquids, chemical gases or vapors, or potentially injurious light radiation (including laser radiation and fiber fragments produced from splicing activities).

Workers must wear protective footwear when working in areas where there is a danger of foot injuries due to falling or rolling objects or objects piercing the sole.

Workers must wear a protective helmet when working in areas where there is a potential for injury to the head from falling objects or low head bump hazards.

A protective helmet designed to reduce electrical shock hazard must be worn by affected personnel when near exposed electrical conductors that could contact the head.

Special application clothing for safety must be worn as applicable for each job as required (i.e., aprons for battery handling).

Loose clothing or neckties must not be worn around machines or when utilizing power tools of any type.

Leather soles and heels must not be worn on ladders, scaffolds, and stools.

Metallic objects such as rings, watchbands, key chains, neck chains, and wire-rim glasses must not be worn during work periods.

B.5 ACCIDENT PREVENTION SIGNS

The installation supervisor must ensure that accident prevention signs are used throughout the installation project. These tags should include contact information and data of posting.

Accident prevention signs can be classified into three categories: danger signs, caution signs, and safety instruction signs.

Accident prevention tags are a temporary means of warning all concerned of a hazardous condition, defective equipment, radiation hazards, etc.

Tags can be attached to the device in question by string, wire, or tape in such a way to be readily noticeable to help prevent accidental injury to personnel.

B.6 TOWER CLIMBING

All tower climbers must be trained and familiar with the US Occupational Safety and Health Administration (OSHA) and National Association of Tower Erectors (NATE) guidelines. OSHA suggests that tower climbing is the most dangerous job in the United States.

All tower climbers must have received tower climbing and first aid/CPR training. In addition, all climbers should participate in emergency drills at least once a year.

All tower climbers should have been medically qualified within the past 3 years to perform climbing activities.

The tower climbers must not climb if they are ill or adversely affected by drugs, alcohol, lack of rest, or other physically or mentally impairing condition.

Work in pairs: a climber on the tower and another person on the ground advising the climber and other personnel of current conditions.

The ground monitor is responsible for maintaining communication with the climber. He continuously monitors conditions and assesses current and potential risk. He should be tower-climbing qualified and trained in emergency procedures.

Overcommunicate. Be sure everyone in the area understands who will be on the tower and what actions will take place. The climber and a designated ground monitor must maintain constant communication.

Preplan medical treatment and evacuation. Have ready access to a first aid kit.

Confirm both the climber and the ground monitor clearly understand each other's responsibilities, the scope of the tower work, and the anticipated schedule.

Confirm that all other personnel in the area are aware of the tower-climbing activity.

Use appropriate communication equipment (e.g., walkie-talkie or cell phone with earpiece) to update the climber. Agree on hand signals for common signals (including stop, come down, and call me).

Inspect equipment, tower, and climbing structures before each climb.

Power to all RF emitters (typically antennas) must be reduced to a safe level or turned off before accessing the structure. Appropriate lockout or tagout methods must be used to secure reduced or eliminated RF power.

Verify that no anticipated activity could interact with any electrical power lines.

Verify that no anticipated activity could interact with any other devices or structures (including power and communication cables and antennas and other devices on or near the tower).

The tower must have an unobstructed climbing ladder that is secure and in good repair. The tower must also have a properly installed, maintained, and functioning safety climb system.

The tower must be in good repair with properly maintained cross-members, guys, and their deadman guy anchors, bases, and grounding systems. Check for loose members and corrosion. Review the tower with binoculars prior to climbing.

Assess other potential factors such as insects, birds, animals, power lines, other structures, and other workers and climbers.

Climbers must not be allowed on the tower during adverse weather conditions such as high winds, extreme temperatures, lightning, rain, or snow.

Tower climbers should wear a climbing hat with chin strap and safety glasses (with tint in harsh light) or mesh goggles. People who could come under the tower within the drop zone should wear hard hats and safety glasses. The drop zone is a circle around the tower base of radius one half the tower's height. Appropriate standards include ANSI Z87.1-2003, Occupational and Educational Personal Eye and Face Protection Devices, and ANSI Z89.1-2003, Industrial Head Protection.

Wear appropriate clothing: steel-shank work boots with rigid nonslip soles, heavy long pants and long-sleeved shirt, and gripping water-resistant gloves.

Use secured tool bags with closures. Attach small cords on major tools and clip them to your harness.

Gather and secure all tools before moving to another position on the tower or returning to ground level.

Use a pulley system to raise and lower large tools or devices. Let the person on the ground manage the line and do the pulling.

Wear a full-body harness with hip, seat, chest, and back D-rings. Belts and harnesses do not fully protect the climber.

Stay connected 100% of the time while climbing, descending, and working on the tower.

When ascending or descending the main portion of the tower, the climber must be attached to a safety cable or a line with a fall-arresting device (such as a rope or cable grab).

Use lanyards to attach yourself to the tower when working. Fall-arresting devices are for fall safety, not positioning. Use lanyards to secure yourself into work position. You may have to detach yourself from the attaching device when you are in work position. Do this only after you are securely attached with at least one lanyard.

B.7 HAND TOOLS

The proper tool must be used for each task. This includes the correct-size wrench, pliers, hammer, and screwdriver.

Tools with wooden handles must not be used if the handles are chipped, burred, or split.

Unsafe tools, machines, equipment, etc. must not be utilized for any task.

B.8 ELECTRICAL POWERED TOOLS

All electrical powered tools must be checked daily for frayed or worn power cords. Damaged plugs and any other item that might present a hazard must not be used.

Before use, all electrical powered tools should be checked [with a volt/ohm meter (VOM)] for shorts or grounds that would make the case or chassis hot and recorded in the site log.

All power tools must have the proper plug to give adequate grounding protection. Allowable exceptions would be only the double-insulated-type power tools (do not use if operating electronic equipment is in the area).

No "ground buster" adapter plugs are allowed.

Personnel working with or near all power tools must wear safety goggles or face shields to prevent injury from flying debris.

The following precautions must be used when utilizing electric drills:

- Ensure all bits are sharp and of the proper type for the work to be done.
- Make sure that both drill and work are properly positioned to eliminate strain and the appropriate pressure is being applied for the drill bit.
- Always check the area of work for the possibility of damage under or around the piece being drilled and act accordingly.

Power saws must be checked as follows:

- Ensure all guards and protective covers are in place.
- Position saw and material being worked to avoid strain or blade bending during the operation.
- Use the proper blade for the material to be cut and stand clear of falling ends.

All material must be secured in some manner before attempting to work on it using powered tools.

Motor-driven tools and equipment must not be plugged into existing equipment frame AC outlets.

B.9 SOLDERING IRONS

A stand must always be used to hold the soldering iron when not in use. This stand must be properly clamped or secured to prevent movement.

Route line or extension cords away from the soldering iron.

All excess solder on the iron tip must be wiped with a damp sponge or wiping pad. Do not flip off excess solder.

Allow soldering irons to cool naturally.

Eye protection is required during all soldering operations.

B.10 LADDERS

Any ladder used to support a person, material, or equipment must be adequately strong, in good condition, and properly secured.

Only approved nonconductive ladders must be used when working in or near an area of electrified equipment or where there is an electrical hazard.

Ladder rungs or steps that are not corrugated, knurled, or dimpled must have skid-resistant material applied to the rung or step.

Scaffolds can be hazardous.

Check for damaged or loose parts and slippery substances on the planks.

Do not ascend a scaffold without all horizontal and diagonal braces in position. Scaffolding must be properly assembled per the manufacturer's instructions.

B.11 HOISTING OR MOVING EQUIPMENT

When storing or moving equipment, avoid excessive shock, vibration, dust, or moisture.

When unpacking radio crates, use proper material handling equipment such as forklift, lifting sling or block, and tackle.

Use of open hooks for lifting or transporting is prohibited.

All hooks must be closed or moused.

When unbanding equipment crates, use care. Steel bands will recoil when cut.

After uncrating racks, secure racks to the superstructure to avoid damage due to seismic activity or accidental contact.

Avoid moving heavy or bulky equipment through close quarters or crowded aisles.

Each move must be planned in advance for best access and to minimize possible hazard.

When impossible to avoid close quarters, protection must be provided for equipment already in place. This includes floors, walls, and columns.

Care must be exercised to avoid possible contact and damage to existing equipment.

Schedule operations to preclude a congestion of personnel in a particular area of movement. This action will lessen the possibility of physical impact with the equipment.

Hoisting around or over existing equipment or buildings must require the use of a tag line to avoid contact with these items.

In lifting or erecting equipment by hand, enough personnel must be provided to safely share the load.

When hoisting operations are involved, ensure that the hoisting equipment is in good repair, well secured and attached to the equipment to be moved at a point designed to take the stresses imposed.

All personnel must be familiar with and use the following general rules for lifting:

- Get a good footing.
- Place feet about shoulder width apart.
- Bend at the knees to grasp weight.
- Get a firm grip/hold.
- Keep back as upright as possible.
- Lift gradually by straightening the legs.
- Wear gloves when lifting heavy objects.
- When an object is too heavy, unwieldy, or bulky to handle comfortably, get help.

When moving heavy objects/equipment, observe the following general rules:

- Have sufficient workers available to handle the weight of the equipment without undue stress.
- Use handcarts and dollies whenever appropriate.

Ensure that the equipment is positioned in such a manner that all sharp and protruding edges are guarded.

Ensure that aisles and passages are clear before starting.

Assign someone to coordinate traffic flow.

Have sufficient works available to handle the weight.

Position equipment to minimize sharp or protruding edges.

Ensure that all hoisting equipment used is of a sufficient size and strength to handle the load in a safe manner.

In hoisting, all equipment must be inspected to ensure they are in good working order.

Ropes, slings, eyebolts, chains, and cables must be checked for frayed or worn spots.

All personnel not involved in the hoisting operations should be cleared from the immediate area.

No one must be allowed under or near a suspended load.

The hoisting area must be clear of obstructions or other equipment.

B.12 BATTERIES

Current capacity of batteries is very high. Shorts can result in very hot molten metal explosions.

Remove all rings, watches, and other metal jewelry.

Use only insulated tools.

Vented cells give off highly flammable hydrogen. Avoid sparks and flames.

Insulate all battery leads in overhead ladder, racks, and shelves.

Avoid shorts during equipment reconfiguration.

Do not use CO_2 fire extinguisher on battery fire. Thermal shock may crack plastic case.

Pour electrolyte into water, not water into electrolyte.

Avoid contact with acid electrolyte. Use protective gloves, apron, and eye wear.

When working on cells, have electrolyte-neutralizing agent available:

lead acid (sulfuric acid) battery: baking soda, table soda, or bicarbonate;

nickel cadmium alkaline (potassium hydroxide) battery: boric acid or vinegar.

Do not clean cells with solvents. Use only water.

Do not mix hydrometers, filler bulbs, thermometers, or other accessories used in different types of batteries.

Fill vented cells using only distilled water.

B.13 LASER SAFETY GUIDELINES

OSHA and all safety practices must be specified in the Site Installation Specification and be followed when working with or around fiber-optic cable.

Do not look into the end of fiber-optic cable leads as unterminated optical connectors emit an invisible laser radiation that will cause serious damage to the eye.

Avoid passing beam over reflective objects.

Point fiber toward ground.

Use fiber and laser caps or covers when possible.

If fiber is removed, cover the laser or remove module power.

Consider use of protective eyewear. Be aware, this equipment is usually wavelength dependent.

Never look into a broken optical fiber cable.

Never look directly into an unterminated fiber-optic connector.

Never look at an optical fiber splice or connector unless it is absolutely known that no laser radiation is present.

Always remove electrical power from fiber-optic transmitters before disconnecting fiber-optic connectors in the path between the transmitter and the receiver.

Never connect an unterminated optical cable to a fiber-optic transmitter. Connect the cable to a test set, receiver, or termination first.

Do not touch the end of the cable connector.

Clean the end of the cable with lint-free wipes or swabs soaked in isopropyl alcohol.

Wipe cable end with dry lint-free tissue and blow dry with compressed air.

Do not use air from your mouth to dry the cable end.

Keep protective covers on fiber cable until ready to use.

Do not place protective covers in a pocket or other dirty place.

Replace protective covers on cable when cable is not in use.

Store protective covers in a clean plastic bag when not in use.

Store cables, with protective covers on, in a clean plastic bag when not in use.

Do not leave cables hanging off the edge of shelf or on the floor.

All laser sources must be disconnected when cables are being spliced.

Under no circumstances should splicing and/or termination operations be performed on light guide cable without the craft person having satisfactorily completed an approved training course.

Glass fibers are very sharp and can pierce the skin.

Do not let cut pieces of fiber stick to your clothing or drop in the work area.

Place all cut-up or broken pieces of glass on a loop of dark-colored tape.

B.14 SAFE USE OF LASERS AND LED IN OPTICAL FIBER COMMUNICATION SYSTEMS

Visible light has wavelengths between 400 and 780 nm.

Fiber optics communication system lasers and LEDs operate at wavelengths of 850, 980, 1310, 1480, and 1550 nm. As such, the lasers and LEDs produce invisible but potentially destructive light.

Besides wavelength ("color"), laser light differs from ordinary light in three ways. It is monochromatic, directional, and coherent.

Lasers pose a greater hazard than ordinary light because they can focus energy onto a small area.

Laser radiation of sufficient intensity and exposure time can cause damage to skin and eye.

Damage to the eyes is typically to the retina (partial or total loss of eye sight) or to the cornea (cataracts causing blurring or loss of sight).

Laser classifications

- Class 1—safe if not disassembled (e.g., CD player)
- Class 2/2a—eye hazard if you stare into beam (e.g., supermarket scanner)
- Class 3a—eye hazard if collected or focused into eye (e.g., laser pointer)
- Class 3b—eye hazard if direct or reflected beam is viewed (e.g., telecommunication equipment)
- Class 4—eye hazard if direct, reflected, or diffused beam is viewed. May also cause skin damage and fire hazard (e.g., laser welder)

B.15 OPTICAL FIBER COMMUNICATION SYSTEM (OFCS) SERVICE GROUPS (SGs)

SG1—no risk of damaging the eye when viewing the end of a fiber with a microscope, an eye loupe, or the unaided eye.

SG2—visible light with wavelength between 400 and 700 nm is potentially hazardous if viewed for more than 1/4 s.

SG3A—not hazardous when viewed by the unaided eye but may be hazardous when viewed with an optical instrument such as a microscope or eye loupe.

SG3B—optical power output greater than SG3A but less than +27 dBm. Damage to eyes is possible if safety precautions are not taken. Strict adherence to safety precautions recommended.

SG4—optical output is greater than +27 dBm. This equipment is quite dangerous. Strict adherence to safety precautions is required.

B.16 ELECTROSTATIC DISCHARGE (ESD)

High voltage electricity is caused when two materials are rubbed together. This can cause damage to sensitive electronic equipment.

Ground straps, wrist, or heal must be worn before and while touching or handling electrostatic sensitive (ESS) devices.

Surfaces with resistance to ground in excess of 100 MΩ (such as ordinary tile) must be covered with properly grounded static dissipating runner or waxed with a static dissipating wax.

The workbench and work surface must be covered with an antistatic or static dissipating material bonded to an earth-grounded metal bench.

All AC-powered electrical equipment must be grounded via a three-wire power cord.

Workers within 1 ft of unprotected sensitive components must be grounded.

Wrist or shoe ground straps of 250 kΩ–2 MΩ must contact skin and the grounded bench or safety ground.

Do not use ground straps near high voltages.

Test ground straps with electronic tester twice a day. Use moisturizer cream if necessary to achieve better conductivity.

Modules must be stored (even temporarily), packed, and shipped in antistatic bags or containers.

ESS devices should remain in their original containers until actually needed.

ESS device containers should be in contact with the antistatic work surface and the wrist strap before parts are removed from containers.

Common plastic, white foam, cellophane, and masking adhesive tapes must not come in contact with ESS devices or their packaging.

Remove unneeded static-producing objects from work area.

Keep other objects at safe distance from work area.

Use a topical antistat on objects necessary for work.

Minimize movement and friction.

Protect sensitive parts with protective packaging.

Place components only on a dissipative mat or dissipative work surface.

All modules should be handled as static-sensitive devices unless they are known not to contain static-sensitive parts.

Do not handle circuit boards or components unnecessarily.

For modules, handle by the plastic handle. For components, handle the body, not the leads. Do not use synthetic bristled brushes to clean modules.

Clothing must not be allowed to touch devices.

Do not use synthetic bristled brushes to clean modules.

Use only grounded antistatic (metalized) (de)soldering tools.

VOMs can damage devices if used to measure resistance.

Assure test setups have correct voltage polarity.

B.17 MAXIMUM PERMISSIBLE MICROWAVE RADIO RF EXPOSURE

OSHA limits

Maximum permissible exposure (MPE) = 10 mW/cm^2

Limit applies whether the radiation is continuous or intermittent.

Average time is 0.1 h.

Applicable frequency range is 10 MHz–100 GHz.

Reference: United States Federal Government, "OSHA Radiation Protection Guide," Code of Federal Regulations, 29 CFR Part 1910, Subpart G, Standard 1910.97, subparagraph (a) (2) (i).

Institute of Electrical and Electronics Engineers (IEEE) limits

MPE = 10 mW/cm^2 for occupational/controlled exposure*

MPE = 1 mW/cm^2 for general population/uncontrolled exposure**

*Controlled environments—a location where individuals are aware of radiation exposure.

**Uncontrolled environments—individuals have no knowledge or control over their exposure.

Average time is 30 min.

Applicable frequency range is 3–100 GHz.

Reference: ANSI/IEEE Std C95.1-2005, Standard for Safety levels with Respect to Human Exposure to Radio Frequency Electromagnetic Fields, 3 kHz to 300 GHz.

FCC limits

MPE = 5 mW/cm^2 for occupational/controlled exposure*

MPE = 1 mW/cm^2 for general population/uncontrolled exposure**

*Controlled environments—a location where individuals are aware of radiation exposure (average time is 6 min).

**Uncontrolled environments—individuals have no knowledge or control over their exposure (average time is 30 min).

Applicable frequency range is 1.5–100 GHz.

Reference: FCC Office of Engineering & Technology Bulletin 65 Edition 97-01 (August 1997), Guidelines for Human Exposure to Radiofrequency Electromagnetic Fields.

The worst-case limit is the FCC MPE limit of 1 mW/cm^2 for general population/uncontrolled exposure. Procedures described in Chapter 8 allow calculation of the maximum power that can be delivered to the listed antenna and not exceed the above FCC MPE limit of 1 mW/cm^2.

B.18 PROTECT OTHER RADIO USERS [FCC]

Operation of an intentional, unintentional, or incidental radiator is subject to the conditions that no harmful interference* is caused ... [Federal Title 47 CFR Part 15.5(b)].

The operator of an RF device shall be required to cease operating the device upon notification by a Commission representative that the device is causing harmful interference. Operation shall not resume until the condition causing the harmful interference has been corrected [Federal Title 47 CFR Part 15.5(c)].

*Harmful interference: Any emission, radiation, or induction that endangers the functioning of a radio navigation service or of other safety services or seriously degrades, obstructs, or repeatedly interrupts a radiocommunications service ... [Federal Title 47 CFR Part 15.3(m)].

B.19 PAUSE (PREVENT ALL UNPLANNED SERVICE EVENTS) AND ASK YOURSELF (VERIZON AND AT&T OPERATIONS)

Do I know why I am doing this work?

What are the possible results of my actions?

Is this activity potentially service affecting?

Can I prevent or control service interruption?

Is this the right time to do this work?

Have I identified and notified everybody—customers and internal groups—who will be directly affected by this work?

Have I coordinated with everyone who could impact the success of this effort?

Am I trained and qualified to do this work?

Do I know how to contact Tier 1 and Tier 2 support?

Should anyone be online with me?

Do I have everything I need to complete the task and to quickly restore service if something goes wrong?

Do I need a pretest and posttest procedure for this task?

Are the work orders, MOP, and supporting documentation current and error free?

Have I walked through the procedure?

Have I completed appropriate paper work, database changes, and other appropriate records?

Have appropriate labels, lockouts, and signs been correctly placed.

B.20 PROTECT YOURSELF (BELL SYSTEM OPERATIONS)

No job is so important and no service is so urgent that we cannot take time to perform our work safely.

B.21 PARTING COMMENT

Most accidents are not caused by Nature or the result of faulty equipment but are the result of human error. Lack of proper training or preparation, improper use of or lack of safety equipment, inadequate preparation, poor communication, or simple carelessness are often the cause. Take appropriate precautions. The most important aspects of the job are *common sense, preparation, and communication.* Act accordingly.

INDEX

Digital Microwave Communication: Engineering Point-to-Point Microwave Systems, First Edition. George Kizer.